Terpenoids

This unique volume covers specific aspects of the biological chemistry of terpenoids. It provides extensive information related to classification, general methods of extraction and isolation of terpenoids, synthesis and pharmacological activities of monoterpenoids, synthesis and medicinal uses of diterpenoids, biogenesis of terpenoids, synthesis and medicinal uses of sesqui terpenoids and sesterpenoids. Some terpenes are also classified as diterpene alkaloids. Most of the terpenoids with diverse molecular structures are biologically active and are used for the treatment of various diseases such as cancer, malaria, inflammation, tuberculosis and infection, and this is discussed.

Features

- Activities and biological relationships of terpenes.
- An accurate assessment of where and what terpenes can lead to.
- Discusses how microbes, in particular the actinomycetales, have well over 400 different gene clusters that produce terpenes.
- Arranged by biological activities and usage.
- Provides information on eukaryotic enzymes that have been shown to be a source of "ethnobotanical" terpenes.

Terpenoids

Chemistry, Biochemistry, Medicinal Effects, Ethno-pharmacology

Bimal Krishna Banik
Prince Mohammad Bin Fahd University, Kingdom of Saudi Arabia

Biswa Mohan Sahoo
Roland Institute of Pharmaceutical Sciences, India

Abhishek Tiwari
Pharmacy Academy, IFTM University, Moradabad, India

CRC Press
Taylor & Francis Group
Boca Raton London New York

CRC Press is an imprint of the
Taylor & Francis Group, an **informa** business

Cover image: PowerUp/Shutterstock

First edition published 2023
by CRC Press
6000 Broken Sound Parkway NW, Suite 300, Boca Raton, FL 33487-2742

and by CRC Press
4 Park Square, Milton Park, Abingdon, Oxon, OX14 4RN

© 2023 Taylor & Francis Group, LLC

CRC Press is an imprint of Taylor & Francis Group, LLC

ISBN: 978-0-367-44031-2 (HB)
ISBN: 978-1-032-41640-3 (PB)
ISBN: 978-1-003-00868-2 (EB)

DOI: 10.1201/9781003008682

Typeset in Times
by codeMantra

Contents

Preface vii

Acknowledgement ix

Authors xi

1. Classification, General Methods of Extraction and Isolation of Terpenoids 1
   Biswa Mohan Sahoo, Bimal Krishna Banik, and Abhishek Tiwari

2. Synthesis and Pharmacological Activities of Monoterpenoids 29
   Biswa Mohan Sahoo, Bimal Krishna Banik, and Abhishek Tiwari

3. Synthesis, and Medicinal Uses of Diterpenoids 87
   Biswa Mohan Sahoo, Bimal Krishna Banik, and Abhishek Tiwari

4. Name Reactions in Terpenoid Chemistry 153
   Bimal Krishna Banik, Biswa Mohan Sahoo, and Abhishek Tiwari

5. Biogenesis of Terpenoids 181
   Bimal Krishna Banik, Biswa Mohan Sahoo, and Abhishek Tiwari

6. Anticancer Natural Terpenoids 197
   Bimal Krishna Banik, Biswa Mohan Sahoo, and Abhishek Tiwari

7. Terpene Glycosides 229
   Bimal Krishna Banik, Biswa Mohan Sahoo, and Abhishek Tiwari

8. Terpenes as Starting Compounds for Other Types of Molecules Including Alkaloids 255
   Bimal Krishna Banik, Biswa Mohan Sahoo, and Abhishek Tiwari

9. Synthesis and Medicinal Uses of Triterpenoids 285
   Biswa Mohan Sahoo, Bimal Krishna Banik, and Abhishek Tiwari

10. Synthesis and Medicinal Uses of Sesquiterpenoids and Sesterpenoids 331
    Biswa Mohan Sahoo, Bimal Krishna Banik, and Abhishek Tiwari

11. Terpenes as Useful Drugs 361
    Biswa Mohan Sahoo, Bimal Krishna Banik, and Abhishek Tiwari

12. Terpenoids: A Brief Survey of Naturally Occurring Terpenes Molecules 399
    Bimal Krishna Banik, Abhishek Tiwari, and Biswa Mohan Sahoo

13. A Brief Account on Plant and Marine-Derived Terpenoids 453
    Bimal Krishna Banik, Abhishek Tiwari, and Biswa Mohan Sahoo

14. A Synthetic Overview of Terpenes and Nor Diterpenes 467
    Bimal Krishna Banik, Abhishek Tiwari, and Biswa Mohan Sahoo

15. Sesquiterpenes: A Chemical Synthesis and Biological Activity 487
    Bimal Krishna Banik, Abhishek Tiwari, and Biswa Mohan Sahoo

16. Sesquiterpenes Lactone: Biogenesis and Chemical Synthesis 517
    Bimal Krishna Banik, Abhishek Tiwari, and Biswa Mohan Sahoo

17. An Overview of Lactarane: A New Class of Bio-Active Molecules 547
Bimal Krishna Banik, Abhishek Tiwari, and Biswa Mohan Sahoo

18. Terpenes: A Potential Chiral Pool for the Synthesis of Medicinally Privileged Bioactive Molecules 557
Bimal Krishna Banik, Abhishek Tiwari, and Biswa Mohan Sahoo

19. Terpenes: A Unique Source of Oil and Fragrance in Industry 589
Bimal Krishna Banik, Abhishek Tiwari, and Biswa Mohan Sahoo

Index 611

Preface

Terpenoids are natural products and isolated from different natural sources such as plants, animals, microbes, insects and marine. Terpenoids are categorized into monoterpenes, sesquiterpenes, diterpenes, sesterpenes, and triterpenes depending on the presence of their carbon units. Most of the terpenoids with diverse molecular structures are biologically active and are used for the treatment of various diseases such as cancer, malaria, inflammation, tuberculosis and infection. This book provides extensive information related to classification, general methods of extraction and isolation of terpenoids, synthesis and pharmacological activities of monoterpenoids, synthesis and medicinal uses of diterpenoids, biogenesis of terpenoids, synthesis and medicinal uses of sesqui terpenoids and sesterpenoids. Some terpenes are also classified as diterpene alkaloids.

This book will be beneficial for undergraduate, postgraduate and research scholars in chemistry and biological science, plant physiologists, pharmacists, pharmacologists, clinicians, industrialists and biochemists. This will enable the readers to get thorough knowledge about terpenoids.

This book has 19 chapters targeting the above subjects. Banik, Sahoo and Tiwari have written these chapters.

In Chapter 1, classification, general methods of extraction and isolation of terpenoids are documented. Synthesis and pharmacological activities of monoterpenoids are described in Chapter 2. In Chapter 3, synthesis and medicinal uses of diterpenoids are focused. In Chapter 4, name reactions in terpenoid chemistry are discussed; in Chapter 5, biogenesis of terpenoids is discussed; in Chapter 6, anticancer natural terpenoids are discussed; in Chapter 7, terpene glycosides is discussed; in Chapter 8, terpenes as starting compounds for other types of molecules including alkaloids; in Chapter 9, synthesis and medicinal uses of triterpenoids are discussed; in Chapter 10, synthesis and medicinal uses of sesquiterpenoids and sesterpenoids are discussed; in Chapter 11, terpenes as useful drugs are described; in Chapter 12, a brief survey of naturally occurring terpenes molecules is presented; in Chapter 13, a brief account on plant and marine-derived terpenoids is given; in Chapter 14, a synthetic overview of terpenes and nor diterpenes; in Chapter 15, sesquiterpenes: A chemical synthesis and biological activity; in Chapter 16, sesquiterpenes lactone: biogenesis and chemical synthesis; in Chapter 17, An overview of lactarane: a new class of bio-active molecules; in Chapter 18, terpenes: A potential chiral pool for the synthesis of medicinally privileged bio-active molecules; in Chapter 19, terpenes: A unique source of oil & fragrance in industry are discussed in detail.

The authors will be pleased if this book activates the research and education lives of students, scientists and academic professionals along with creating business potential. Information provided in this book will also serve the curiosity of general people about terpenoids and natural products.

Acknowledgement

This book could not have been published without the assistance of CRC, Taylor & Francis Publisher. In particular, the authors are tremendously grateful to Ms. Hilary Lafoe and Ms. Sukirti Singh for their knowledge in science, helping attitude and encouragement.

Authors

Professor Bimal Krishna Banik

Bimal Krishna Banik conducted his Ph. D. degree work in Chemistry at the Indian Association for the Cultivation of Science, Calcutta. Then, he pursued postdoctoral research at Case Western Reserve University, Ohio and Stevens Institute of Technology, New Jersey. Dr. Banik is a FRSC and CChem of the Royal Society of Chemistry, UK. Dr. Banik was a Tenured Full Professor in Chemistry and First President's Endowed Professor in Science & Engineering at the University of Texas-Pan American (UTPA). He was also the Vice President of Research & Education Development at the Community Health Systems of Texas. At present, he is a Full Professor and Senior Researcher of Natural Sciences & Human Studies at Prince Mohammad Bin Fahd University, Kingdom of Saudi Arabia.

Professor Banik has been conducting synthetic chemistry and medicinal research on diverse cancers, antibiotics, hormones, carbohydrates, metals and salts, natural products, catalysis and environmentally benign methods. As the Principal Investigator, he has been awarded $7.25 million grants from US NIH, US NCI and US Private Foundations. At present, he has 578 publications, 513 presentation abstracts and 8,017 journal citations with 47 h-index. Due to his contributions to research, Professor Banik is designated as a "Distinguished Researcher" and a "Distinguished Scientist" by Scientific Organizations.

Dr. Biswa Mohan Sahoo

Dr. Biswa Mohan Sahoo is currently working as Professor at Roland Institute of Pharmaceutical Sciences, Berhampur, Odisha. He is also serving as Research Coordinator of Biju Patnaik University of Technology Nodal Centre of Research, Odisha. Dr. Sahoo obtained his B. Pharm. Degree from Berhampur University, Odisha, and a Master's degree from the Department of Medicinal Chemistry, L.M College of Pharmacy, Gujarat.

Dr. Sahoo received his Ph.D. degree from the School of Pharmaceutical Education and Research, Berhampur University. He has also completed his Post Graduate Diploma in Pharmaceutical Regulatory Affairs from Dept. Of Pharmaceutics, Hamdard University, New Delhi. He has published more than 60 research and review articles in peer-reviewed national and international journals of repute. His h-index is 16.

He has filed and published four patents in India and Australia and authored several book chapters in various textbooks published by known authors. He is an editorial board and advisory board member of various international journals. He is recognized as a Ph.D. Supervisor under Jawaharlal Nehru Technological University (JNTU) Kakinada, Mewar University, Rajsthan and BPUT, Odisha.

Professor Abhishek Tiwari

Dr. Abhishek Tiwari is working as Professor in the Department of Pharmaceutical Chemistry, Pharmacy Academy, IFTM University, Lodhipur-Rajput, Moradabad (U.P.), India. Dr. Tiwari obtained his B.Pharm. degree from Jiwaji University Gwalior (M.P.), M.Pharm. from M.S. Ramaiah College of Pharmacy, Bengaluru (Karnataka) and Ph.D. from Uttarakhand Technical University, Dehradun (Uttarakhand). Dr. Tiwari has received grants of more than 15 lakhs from the Department of Biotechnology, Uttarakhand; All India Council of Technical Education, India; Oil and Natural Gas Corporation and Power Finance Corporation. Dr. Tiwari has been granted 3 Indian and 7 overseas patents. In addition, he has submitted a number of patents for approval. Dr. Tiwari is an editorial member of various national and international journals. He received "Outstanding Academic and Research Award" in 2021, "Best Author Award" in 2019, "Best Teacher Award" in 2019 and "Outstanding Teacher Award" in 2014 and 2015. He has published 48 research papers in international journals. Remarkably, he has published more than 17 Indian books and a few chapters. He is a recognized Ph.D. supervisor of Uttarakhand Technical University, Dehradun, and IFTM University, Moradabad.

1

Classification, General Methods of Extraction and Isolation of Terpenoids

Biswa Mohan Sahoo*, Bimal Krishna Banik†, and Abhishek Tiwari‡

CONTENTS

1.1   Introduction 2
    1.1.1   Properties of Terpenoids...2
        1.1.1.1   Physical Properties...2
        1.1.1.2   Chemical Properties..2
        1.1.1.3   Phytochemical Properties 2
        1.1.1.4   Pharmacological Properties 2
    1.1.2   Isoprene Rule..2
    1.1.3   Violations of Isoprene Rule 3
    1.1.4   Special Isoprene Rule 3
    1.1.5   Exception to Special Isoprene Rule...4
1.2   Classification of Terpenoids 4
1.3   Nomenclature of Terpenoids 7
1.4   General Methods of Extraction of Terpenoids..9
    1.4.1   Maceration...10
        1.4.1.1   Types of Maceration 10
        1.4.1.2   Steps Involved in Maceration..10
    1.4.2   Digestion..10
    1.4.3   Infusion..11
    1.4.4   Decoction...11
    1.4.5   Percolation 11
        1.4.5.1   Types of Percolation..11
        1.4.5.2   Steps Involved in Percolation..11
    1.4.6   Soxhlet Extraction 12
    1.4.7   Automated Soxhlet Extraction 12
    1.4.8   Supercritical Fluid Extraction (SFE)..12
    1.4.9   Pressurized Liquid Extraction (PLE) 12
    1.4.10   Pressurized Hot Water Extraction (PHWE)...12
    1.4.11   Microwave-Assisted Extraction (MAE)..12
    1.4.12   Ultrasound Extraction (Sonication)...13
1.5   Solvents Used for Extraction...15
    1.5.1   Properties of Solvents Used in Extractions 16
    1.5.2   Factors to Be Considered for Selecting Solvents during Extraction 16
1.6   Isolation of Terpenoids..16
    1.6.1   Isolation of Essential Oils from Plant Parts 17
    1.6.2   Separation of Terpenoids from Essential Oils..17
        1.6.2.1   Chemical Methods..17
        1.6.2.2   Physical Methods 17

* Roland Institute of Pharmaceutical Sciences, Berhampur affiliated to Biju Patnaik University of Technology (BPUT), Rourkela, Odisha, India. drbiswamohansahoo@gmail.com.

† Department of Mathematics and Natural Sciences, College of Sciences and Human Studies, Prince Mohammad Bin Fahd University, Al Khobar, Kingdom of Saudi Arabia. bimalbanik10@gmail.com and bbanik@pmu.edu.sa

‡ Pharmacy Academy, IFTM University, Lodhipur Rajput, Moradabad, U.P. 244102.

DOI: 10.1201/9781003008682-1

1.7   General Methods of Structure Elucidation of Terpenoids.. 19
1.8   Conclusion.. 26
Acknowledgment.. 27
Abbreviations.. 27
References.. 27

1.1 Introduction

Terpenes are the natural products present in essential oils of different plant parts such as fruits, flowers, seeds, leaves, barks, roots, etc [1]. These are pleasant smelling in nature. Chemically, terpenes are the hydrocarbons, whereas terpenoids are oxygenated derivatives containing alcohols and their glycosides, ethers, aldehydes, ketones, carboxylic acids, esters, etc., hydrogenated and dehydrogenated derivatives of terpenes [2]. Terpenoids are also called as isoprenoids with general formula $(C_5H_8)_n$. Isoprene is the building block of all terpenoids and is chemically 2-methylbuta-1,3-diene (Figure 1.1). The analysis of oils of turpentine was carried out in 1818 by Houston. In 1866, Dumas proposed the term "terpene" derived from the plant resin turpentine [3]. Otto Wallach proposed that terpenoid build-up of isoprene unit in 1887 (Nobel prize in 1910). He developed the theory that the monoterpenes build-up of two, the sesquiterpenes of three and the diterpenes of four isoprene units. He also suggested that the simple terpenes possess a skeleton based on p-cymene, the sesquiterpenes are related to naphthalenes and the diterpenes are related to phenanthrenes.

Due to the development of various chromatographic and spectroscopic techniques, structures of different terpenoids are identified such as camphor, citral and pinene by Bredt (1893), Wagner (1894), Tiemann (1895) and Wackenrodder (1837), respectively [4]. In 1953, Leopold Ruzicka explained the isoprene unit based on the concept of the "isopren rule," which was later identified by Lynen and Bloch. Ruzicka explained the structure elucidation of the sesqui- and diterpenes [5]. In addition, Rowland reported the first penta terpenoid solanesol (acyclic unsaturated alcohol) in 1956.

1.1.1 Properties of Terpenoids

Terpenoids represent the largest group of secondary metabolites obtained from the natural source. These are aromatic compounds responsible for flavor and fragrance. There are a number of comprehensive studies performed on its physical, chemical, phytochemical and pharmacological properties [6,7].

1.1.1.1 *Physical Properties*

- Most of the terpenoids are colorless, pleasant smelling liquid and lighter than water. But some of them are solid (e.g., camphor).
- These are soluble in organic solvents and insoluble in water.
- Most of the terpenoids are optically active.
- They are volatile in nature.
- Boiling point of terpenoids ranges from 150°C to 180°C.

1.1.1.2 *Chemical Properties*

- Terpenoids are open chain or cyclic unsaturated compounds with one or more double bonds.
- They undergo addition reaction with olefinic reagents like hydrogen, halogen, halogen acids, NOCl (Tilden's reagent), NOBr to form addition products.
- They undergo polymerization and dehydrogenation.
- They undergo Diel's Alder reaction due to the presence of conjugated double bonds in some of the terpenoids (e.g. Carvone, Farnesene and Ocimene).
- They are easily oxidized by all of the oxidizing agents due to the presence of olefinic bonds.
- Terpenoid produces isoprene as one of the products of thermal decomposition.

1.1.1.3 *Phytochemical Properties*

- Terpenoids react with chloroform and sulfuric acid producing reddish brown coloration (Salkowski's test).
- Terpenoids treated with copper acetate solution produce emerald green color.

1.1.1.4 *Pharmacological Properties*

Terpenoids possess various pharmacological properties such as anti-viral, anti-bacterial, anti-malarial, anti-inflammatory, anti-cancer, hypoglycemic, antioxidants anthelmintic, antidiabetic, etc.

1.1.2 Isoprene Rule

Terpenoid undergoes thermal decomposition to produce isoprene as one of the productd [8]. It was proposed by Tilden in 1884. So, isoprene rule states that all terpenoid molecules are made up of two or more isoprene units (Scheme 1.1).

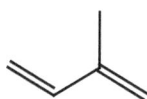

FIGURE 1.1   Structure of isoprene unit (2-methylbuta-1,3-diene).

Isoprene rule is also called as C_5 rule because all terpenoid contains carbon atom in multiple of five like hemiterpenes (C_5), monoterpenes (C_{10}), sesquiterpenes (C_{15}), diterpenes (C_{20}), sesterterpenes (C_{25}), triterpenes (C_{30}), tetraterpenes (C_{40}) and polyterpenes (C>40). Further Isoprene rule is confirmed based on the following facts and reactions (Schemes 1.2 and 1.3).

$$\text{Terpenoid} \xrightarrow{\Delta} \text{Isoprene } (C_5H_8)$$

SCHEME 1.1 Thermal decomposition of terpenoid.

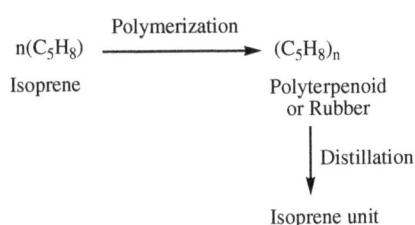

$$n(C_5H_8) \xrightarrow{\text{Polymerization}} (C_5H_8)_n$$

Isoprene   Polyterpenoid or Rubber

↓ Distillation

Isoprene unit

SCHEME 1.2 Polymerization of isoprenes.

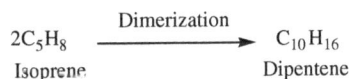

$$2C_5H_8 \xrightarrow{\text{Dimerization}} C_{10}H_{16}$$

Isoprene   Dipentene

SCHEME 1.3 Dimerization of isoprenes.

- Isoprenes undergo polymerization to produce polyterpenoiods or rubber. Similarly, distillation of rubber produces isoprene units.
- Isoprenes undergo dimerization to produce dipentene

1.1.3 Violations of Isoprene Rule

Carbon content of certain terpenoids is not multiple of five. For example, cryptone is a naturally occurring ketonic terpenoid which contains nine carbon atoms instead of ten carbon atoms (Figure 1.2). So, it cannot be divided into two isoprene units.

1.1.4 Special Isoprene Rule

Further, Ingold formulated a special isoprene rule in 1925. According to this rule, isoprene units are linked through "head to tail" fashion or 1,4-linkage in terpenoids. In isoprene, branched end is considered as head (*H*) and other part is tail (*T*) as shown in Figure 1.3.

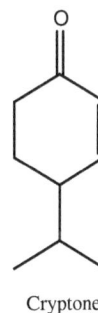

Cryptone

FIGURE 1.2 Structure of ketonic terpenoid.

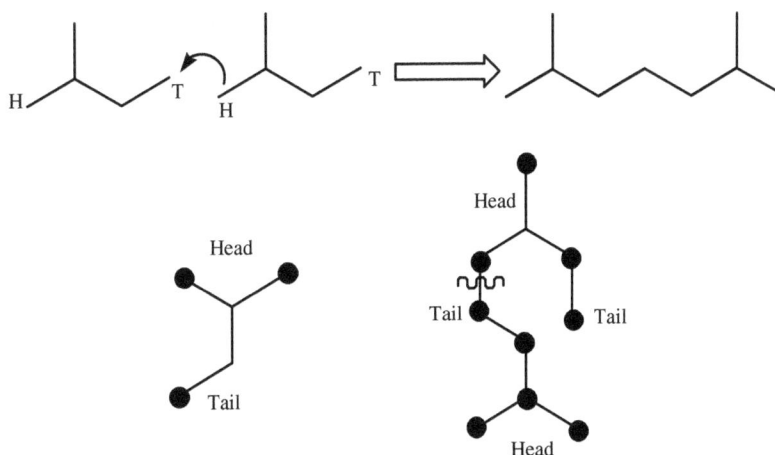

FIGURE 1.3 Head to tail fashion or 1,4-linkage in terpenoids.

Lavandulol

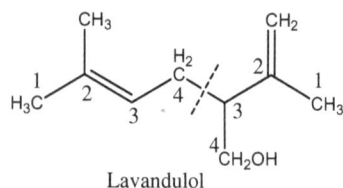

FIGURE 1.4   Structure of monoterpene alcohol.

1.1.5 Exception to Special Isoprene Rule

Lavandulol is monoterpene alcohol present in essential oils such as lavender oil (Figure 1.4). In case of Lavandulol, the two isoprene units are not joined in head to tail arrangement. Similarly, Carotenoids are joined tail to tail at their center [9].

1.2 Classification of Terpenoids

The general formula of all naturally derived terpenoid is $(C_5H_8)_n$. They are classified on the basis of a number of isoprene units present in their structure (Table 1.1). Hemiterpene is the simplest terpenoid which consists of single isoprene unit (C_5H_8) e.g. tiglic, angelic, isovaleric and senecioic acids [10–13].

Based on the presence of number of rings in their structure, terpenoids are further categorized into different types such as acyclic (presence of open chain structure), monocyclic (presence of one ring), bicyclic (presence of two rings), tricyclic (presence of three rings) and tetracyclic (presence of four rings). The general molecular formula of parent hydrocarbon of saturated terpenoid hydrocarbon is presented in Table 1.2.

Acyclic terpenoids are open-chain unsaturated compounds belonging to the general class of ethylenic hydrocarbons. Structurally, the acyclic terpenoids possess a double bond between the first and second or between the second and third of at least one of the terminal three-carbon atom groups, thus forming the end group (1) $CH_2=C(CH_3)-$ or (2) $(CH_3)_2C=$. Acyclic terpenoids are further categorized into acyclic monoterpenoid (e.g. Geraniol, Linalool, Citral, Nerol, Myrcene), sesquiterpenoids (e.g. Farnesol, β-Nerolidol), diterpenoid (e.g. Phytol), sesterterpenoids (e.g. Hippolide-E) as presented in Table 1.3.

TABLE 1.1

List of Terpenoids with Their Molecular Formula

Terpenoids	Number of Isoprene Units	Number of Carbon Atoms	Molecular Formula
Hemiterpenes	1	5	C_5H_8
Monoterpenes	2	10	$C_{10}H_{16}$
Sesquiterpenes	3	15	$C_{15}H_{24}$
Diterpenes	4	20	$C_{20}H_{32}$
Sesterpenes	5	25	$C_{25}H_{40}$
Triterpenes	6	30	$C_{30}H_{48}$
Tetraterpenes or Carotenoids	8	40	$C_{40}H_{64}$
Polyterpenes or Rubber	n	>40	$(C_5H_8)_n$

TABLE 1.2

General Molecular Formula of Parent Saturated Hydrocarbon

Type of Terpenoids	General Molecular Formula
Acyclic	C_nH_{2n+2}
Monocyclic	C_nH_{2n}
Bicyclic	C_nH_{2n-2}
Tricyclic	C_nH_{2n-4}
Tetracyclic	C_nH_{2n-6}

TABLE 1.3

List of Acyclic Terpenoids with Their Sources and Structures

Type of Terpenoids	Name of Terpenoids	Plant Sources	Structures
Acyclic monoterpenoid	Geraniol	Citronella oil (*Cymbopogon flexuosus*)	
Acyclic monoterpenoid	Linalool	Coriander (*Coriandrum sativum*)	

(Continued)

TABLE 1.3 (*Continued*)

List of Acyclic Terpenoids with Their Sources and Structures

Type of Terpenoids	Name of Terpenoids	Plant Sources	Structures
Acyclic monoterpenoid	Citral	Lemon grass oil (*Cymbopogon flexuosus*)	
Acyclic monoterpenoid	Nerol	Lemongrass (*Cymbopogon flexuosus*)	
Acyclic monoterpenoid	Myrcene	Varbena oil (*Lippia citriodora*)	
Acyclic sesquiterpenoids	Farnesol	Citronella (*Cymbopogon flexuosus*)	
Acyclic sesquiterpenoids	β-Nerolidol	Bitter orange (*Citrus aurantium*)	
Acyclic Diterpenoid	Phytol	Chlorophyll	
Acyclic sesterterpenoids	Hippolide-E	Marine sponge (*Hippospongia lachne*)	

Monocyclic terpenoids are hydrocarbons which contain a six-carbon ring and may be considered as derivatives of either cyclohexane or benzene. These terpenoids are produced readily from the acyclic terpenes by ring closure or from the bicyclic terpenes by ring fission. Based on the types of substituted six-carbon ring structures, monocyclic terpenoids are classified into the following types (Table 1.4).

Type A: Monocyclic terpenoids which contain an iso three-carbon group and a one-carbon group, usually in para position. In some of the other monocyclic terpenoids, these groups are present in ortho or meta position. So, Type A monocyclic terpenoids may be considered as derivatives of menthane i.e. isopropylmethylcyclohexane or derivatives of cymene i.e. isopropylmethylbenzene.

TABLE 1.4

List of Monocyclic Terpenoids with Their Sources and Structures

Monocyclic monoterpenoid	α-Terpineol	Petitgrain oil	
Monocyclic monoterpenoid	Carvone	Caraway (*Carum carvi*)	
Monocyclic monoterpenoid	Menthol	Peppermint (*Mentla piperita*)	
Monocyclic sesquiterpene	Cineole	Eucalyptus (*Eucalyptus globules*)	
Monocyclic sesquiterpene	Zingiberene	Ginger (*Zingiber officinale*)	
Monocyclic sesterterpenoid	Manoalide	Marine sponge (*Luffariella variabilis*)	

Type B: Monocyclic terpenoids which don't contain any iso three-carbon group. This type of monocyclic terpenoids contains two one-carbon groups in gem configuration and one or two additional isolated one-carbon or two-carbon groups, one of which is usually in the meta position. Thus, Type *B* monocyclics are considered as derivatives of dimethylcyclohexane.

Bicyclic terpenoids are categorized into four groups such as thujane group, carane group, pinane group and camphane group. These terpenoids are also classified as bicyclic monoterpenoid (e.g. *α*-pinene, Camphor), bicyclic sesquiterpenoids (e.g. *β*-Santalol), bicyclic diterpenoid (e.g. Sclareol) and bicyclic sesterterpenoid (e.g. Leucosceptrine) (Table 1.5).

Tricyclic terpenoids contain three rings in their structure. These are categorized into different types such as tricyclic sesquiterpenoids (e.g. Khushimol), tricyclic diterpenoid (e.g. Abietic acid) and tricyclic sesterterpenoid (e.g. Heliocide-H1) (Table 1.6).

Tetracyclic terpenoids contain four rings in their structure. These are categorized into different types such as tetracyclic diterpenoids (e.g. Steviol) and tetracyclic sesterterpenoid (e.g. Sesterstatin-7) as presented in Table 1.7.

1.3 Nomenclature of Terpenoids

Most of the terpene names are derived from the genus or family names of the plants or the essential oils from which the products were first isolated or in which they occur most abundantly (Table 1.8). The position of the double bond in the structure of the terpene is indicated by the Greek letter Δ (delta) to which an index number is attached referring to the first carbon atom of the doubly-linked pair. The parent compound for nomenclature is called as P-menthane. If the compound contains one double bond or two double bonds, it is named as P-menthene or P-menthadiene, respectively. Δ-1,5 indicates that the double bonds are between 1 and 2 & between 5 and 6

TABLE 1.5

List of Bicyclic Terpenoids with Their Sources and Structures

Bicyclic monoterpenoid	*α*-Pinene	Sweet orange	
Bicyclic monoterpenoid	Camphor	Cinnamomum (*Cinnamomum camphora*)	
Bicyclic sesquiterpenoids	*β*-Santalol	Sandal wood (*Santalum album*)	
Bicyclic diterpenoid	Sclareol	Clary sage (Salvia sclarea)	
Bicyclic sesterterpenoid	Leucosceptrine	*Leucosceptrum canum*	

TABLE 1.6

List of Tricyclic Terpenoids with Their Sources and Structures

Tricyclic sesquiterpenoids	Khushimol	Vetiver (Chrysopogon zizanioides)	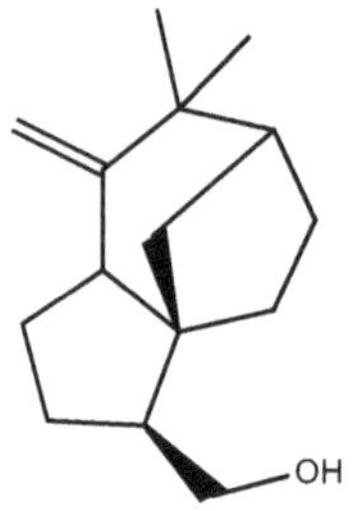
Tricyclic diterpenoid	Abietic acid	Pine tree (_Pinus sylvestris_)	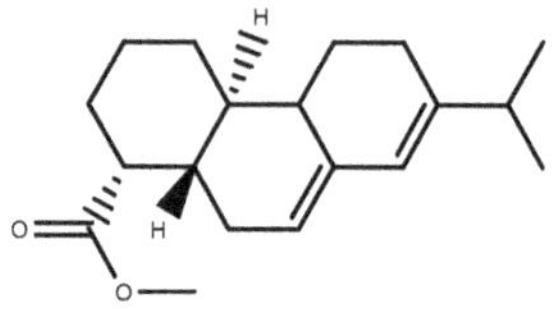
Tricyclic sesterterpenoid	Heliocide-H1	Upland cotton (_Gossypium hirsutum_)	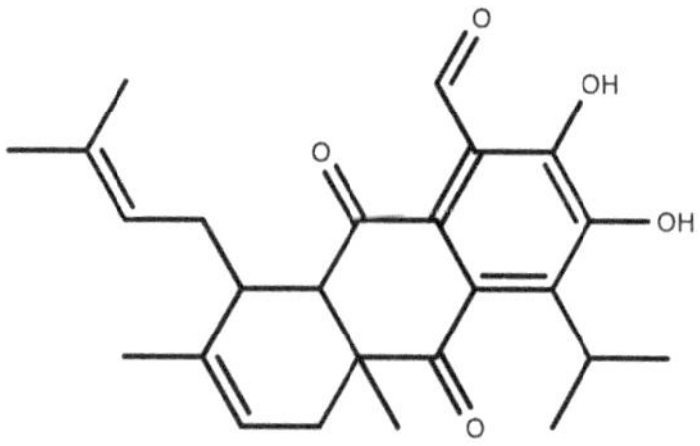

TABLE 1.7

List of Tetracyclic and Macrocyclic Terpenoids with Their Sources and Structures

Tetracyclic diterpenoids	Steviol	Stevia rebaudiana	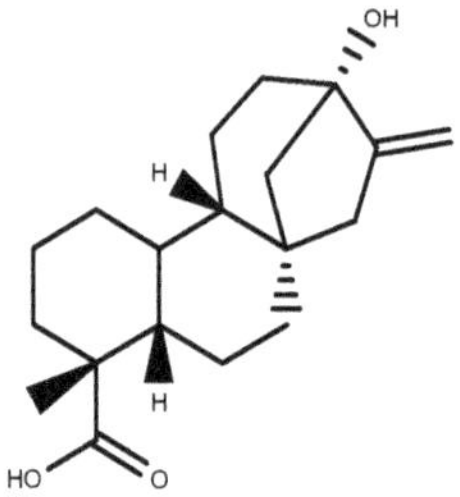
Tetracyclic sesterterpenoid	Sesterstatin-7	_Marine sponge (Hyrtios erecta)_	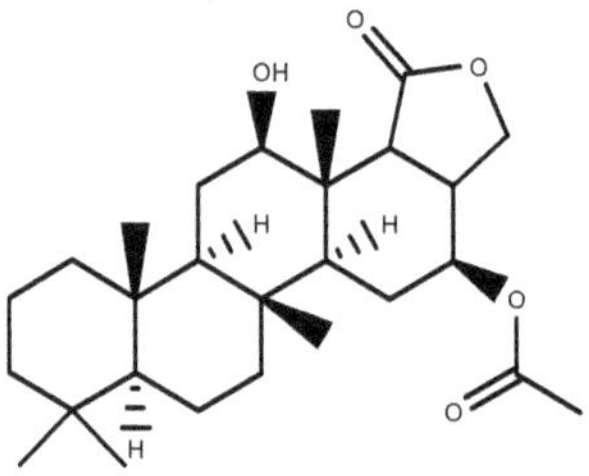
Macrocyclic Diterpenoid	Taxol	_Taxus brevifolia_	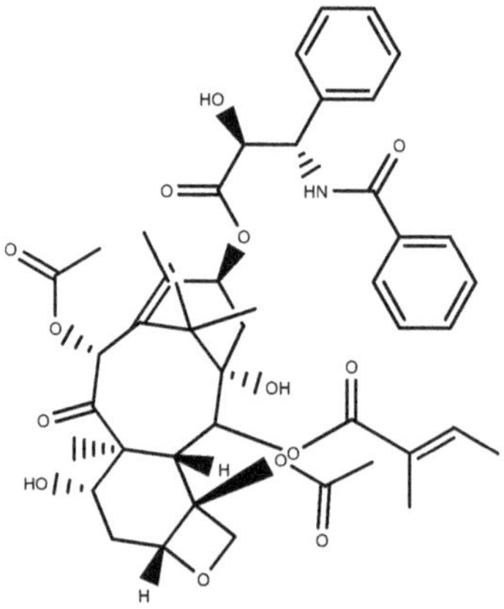

TABLE 1.8

Names of Terpenes Based on Genus/Family Name

Name of Terpenes	Name of Genus/Family
Menthol	*Mentha piperita*
Ocimene	*Ocimum sanctum*
Carvone	*Carum carvi*
Taxol	*Taxus brevifolia*
Steviol	*Stevia rebaudiana*
Camphor	*Cinnamomum camphora*
Santalol	*Santalum album*
Sclareol	*Salvia sclarea*
Leucosceptrine	*Leucosceptrum canum*
Zingiberene	*Zingiber officinale*

position. The presence of hydroxyl (–OH), aldehydes (–CHO) and ketone (C=O) group in terpenoids is represented by suffix "ol", "al" and "one" respectively. Example: geraniol, citral and carvone (Figure 1.5) [14,15].

1.4 General Methods of Extraction of Terpenoids

Extraction is the first step to separate the desired terpenoids from various natural products. Extraction process involves the separation of active constituents (terpenoids) from plant sources by using suitable solvents. The purpose of standardized extraction procedures for crude drugs is to obtain therapeutically active compounds by eliminating the inert materials on treatment with selective solvents known as menstruum [16–19].

The extraction of terpenoids from natural sources proceeds through the following stages.

1. The solvent penetrates the solid matrix or plant materials.
2. The solutes or constituents dissolve in the solvents.
3. The solutes or constituents are diffused out of the solid matrix.
4. The extracted solutes or constituents are collected.

The efficiency of extraction depends on the following factors.

a. **Properties of the extraction solvents:** polar/non-polar, solubility, boiling point, selectivity, cost and safety are considered for better extraction.

b. **Particle size of the plant materials:** Generally, small particle size increases the more surface area so that absorption of solvent is more that leads to increase in extraction.

c. **Solvent-to-solid ratio:** Greater the solvent-to-solid ratio, higher the yield of extraction.

d. **Extraction temperature:** Optimum temperature should be maintained.

e. **Duration of extraction:** Extraction efficiency increases with increase in extraction duration in a certain range of time.

The various methods used for extraction of plant materials containing terpenoids includes maceration, infusion, digestion, decoction, percolation, hot continuous extraction (soxhlet), aqueous or alcoholic extraction, counter current extraction (CCE), greener extraction methods such as super critical fluid extraction (SFE), pressurized liquid extraction (PLE) and microwave-assisted extraction (MAE) and ultrasound extraction (sonication) etc (Figure 1.6, Tables 1.9–1.11). Most of

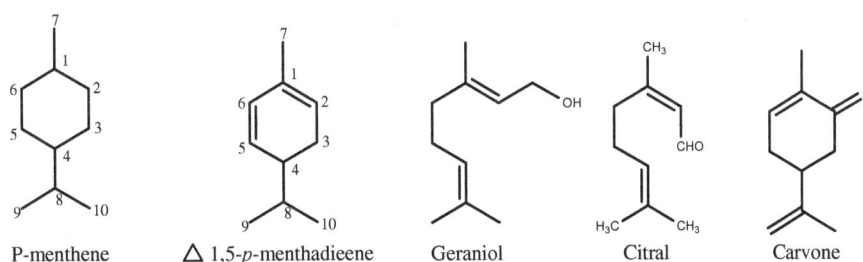

FIGURE 1.5 Nomenclature of terpenoids.

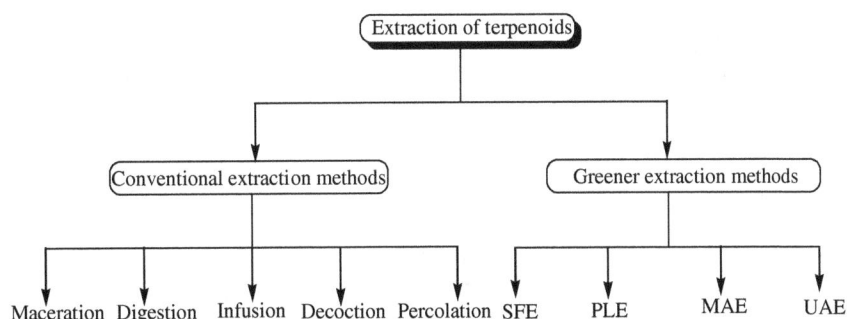

FIGURE 1.6 General methods of extraction of terpenoids.

pharmacopoeias usually refer to maceration and percolation for the extraction of active principles from crude drugs.

1.4.1 Maceration

It is an isocratic and cold extraction method. In this method, the plant material is converted into moderately coarse powder and then immersed in a particular solvent in a closed vessel. This solvent mixture is allowed to stand for 1 week and then the extracted liquid containing essential oil is strained.

Advantages

- Maceration is suitable for the extraction of thermolabile components.
- This process requires a small sample size.
- It has high swelling properties.
- It is an energy-saving process.

Disadvantages

- It is a very slow and time-consuming process.
- The consumption of solvents is more.
- Unable to extract the plant constituents exhaustively.

1.4.1.1 Types of Maceration

Maceration is categorized into following types.

a. **Simple maceration:** Simple Maceration is used for organized and unorganized crude drugs. E.g. tincture of orange and lemon.
b. **Double maceration:** In this process, the plant materials containing terpenes are macerated twice by using the menstruum which is divided into two parts in such a manner that the same volume is used for each maceration. E.g. concentrated infusion of orange.
c. **Triple maceration:** In this process, plant materials are macerated thrice by using the menstruum which is divided into three parts in such a way that the same volume for three parts is used for each maceration.

This maceration process is carried out with the help of heating or stirring. E.g. concentrated infusion of Quassia and Senna.

1.4.1.2 Steps Involved in Maceration

There are various steps involved during the maceration process to extract the active constituents from plant materials by using suitable solvents (Figure 1.7).

Step-I: Plant materials are cleaned, air dried and crushed into coarse powder. The powdered materials is then dipped into appropriate solvent in a closed container and allowed to stand for 4–6 days at room temperature with occasional agitation. Lid of the container is opened time to time to release the generated pressure and shake well till the soluble materials will be dissolved.

Step-II: The wet solid material is filtered off using a funnel with a cotton plug and then the marc is further pressed to recover as much as occluded solution as possible. Sufficient time is allowed for coagulation and settling. Finally, the settled matter is filtered using filter paper.

Step-III: The resultant extract is then concentrated under reduced pressure to attain the crude extract of the plant materials.

Step-IV: The crude extract of the plant materials is subjected for further analysis.

1.4.2 Digestion

This process is a modified form of maceration in which heat as well as pressure is applied for extraction. For this, the equipment like pressure cooker or autoclave is used as digester. The extraction of non-thermolabile materials is carried out efficiently by this method due to the high penetration power of solvents and solubility property of the crude drugs. The plant materials are treated with menstrum for a definite period of time under the specified condition of temperature and pressure.

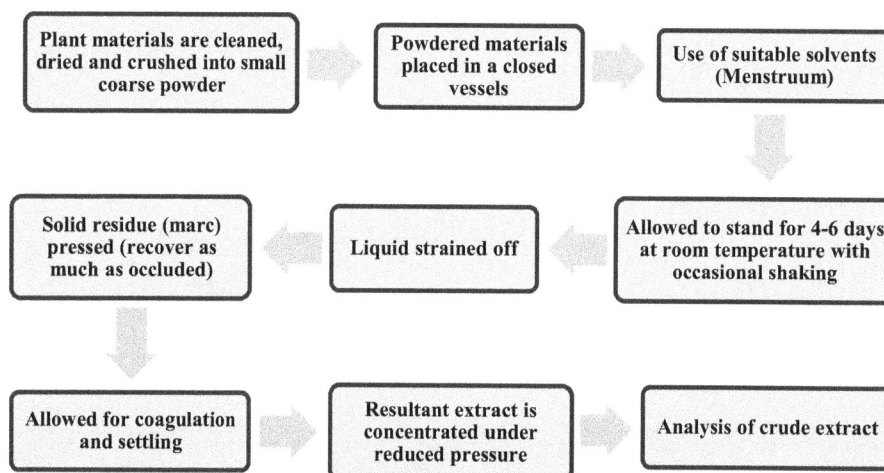

FIGURE 1.7 Flow diagram for process of maceration.

Extraction by the digestion method proceeds faster due to the generation of high temperature and pressure by digester.

1.4.3 Infusion

It is the process of extracting chemical constituents from plant materials by using solvents such as water or alcohol. In this method, the plant materials are allowed to remain suspended in the solvent over a period of time. It is used for those drugs which are soft in nature so that water may penetrate easily to the tissues and the active constituents are water-soluble. Infusion may be two types such as fresh infusion and concentrated infusion.

1.4.4 Decoction

It is the process in which water-soluble and heat stable constituents of hard and woody crude drugs can be extracted. Water is used as menstruum. It is aqueous preparations of plant materials boiled in water for 15–20 min until the water volume is reduced to half. After boiling, the liquid is cooled and filtered. It cannot be used for the extraction of thermolabile components. This process may enhance the dissolution of active constituents as compared to the maceration process.

1.4.5 Percolation

It is an extraction process in which the extraction of plant materials is carried out by downward displacement of the solvents through gravitational force. In this process, the plant parts are moistened with menstruum and packed in a percolator for 4 h. Percolator is a narrow, cone-shaped vessel open at both ends.

Merits

- It requires less time than maceration.
- Extraction of thermolabile constituents can be possible.

Demerits

- It requires more time than any other type of extraction.
- Requires more solvents and skilled persons.

1.4.5.1 Types of Percolation

There are three types of percolation such as simple percolation, modified percolation and reserved percolation.

a. **Simple percolation:** It is used for the preparation of tinctures. There are three stages involved for the preparation of tinctures by simple percolation processes such as imbibition, maceration and percolation (e.g. tincture of belladonna and cardamom).

i. **Imbibition:** The powdered plant material is moistened with a sufficient quantity of menstruum and allowed to stand for 4 h in a closed vessel. The moistened plant material is then packed into a percolator and sufficient quantity of menstruum is added to saturate the plant materials. When the liquid starts coming out from the outlet of percolator, the outlet is closed. Then a sufficient quantity of menstruum is added in order to leave a layer above the materials.

ii. **Maceration:** The moistened materials are kept in contact with menstruum for 24 h. During this period, the menstruum dissolves the active constituents of the plant material and becomes saturated with it.

iii. **Percolation:** It consists of the downward displacement of the saturated solution formed in maceration and extraction of the remaining active constituents present in the plant material by the slow passage of the menstruum through the column. After collecting 3/4th of the required volume of the finished product or when the drug is completely exhausted, the marc is pressed. Mix the expressed liquid with the percolate. Finally, a sufficient quantity of menstruum is added to produce the required volume and then filtered.

b. **Modified percolation:** It is mainly used to prepare concentrated products. It involves repeated maceration.

c. **Reserved percolation:** In this case, the extraction is carried out based on the general percolation procedure. Evaporation is performed under reduced pressure to get the consistency of a soft extract (semi-solid) such that all the water is removed. E.g. Cantharidin from cantharides.

1.4.5.2 Steps Involved in Percolation

The various steps involved in the maceration process include size reduction, imbibition, packing, maceration and percolation (Figure 1.8).

Step-I: Size reduction: The plant materials to be extracted are subjected to a suitable degree of size reduction, usually from coarse powder to fine powder.

Step-II: Imbibition: During imbibitions, the powdered plant material is moistened with a suitable amount of menstruum and allowed to stand for 4 h in a well-closed container.

Step-III: Packing: After imbibitions, the moistened plant material is packed into a percolator.

FIGURE 1.8 Steps involved in percolation.

TABLE 1.9

Comparison Study Among Conventional Extraction Methods

Extraction Method	Time for Extraction	Temperature	Characteristics of Active Constituents
Maceration	3–7 days	Room temp	• Soluble in the menstruum • Heat stable/unstable
Percolation	24 h	Room temp	• Soluble in the menstruum • Heat stable/unstable
Digestion	Few days	Moderately high	• Heat stable
Infusion	Short period	Cold or boiling water	• Readily soluble
Decoction	15 mins	Boiling water	• Water-soluble • Heat stable

Step-IV: Maceration: After packing, sufficient menstruum is added to saturate the material. The percolator is allowed to stand for 24–25 h to macerate the plant materials.

Step-V: Percolation: The lower tap is opened and the collected liquid is allowed to drip slowly at a controlled rate until 3/4th volume of the finished product is obtained.

1.4.6 Soxhlet Extraction

This method is the classic approach for extracting solid samples. It involves the consumption of less solvent as compared to maceration or percolation. This method utilizes the principle of reflux and siphoning to continuous extraction of the herbal constituents with fresh solvents. So, the sample (plant materials) is held in a porous cellulose thimble and is extracted continuously with a fresh aliquot of distilled and condensed solvents. For this purpose, the extraction is carried out at temperatures below the boiling point of solvents for a period of around 16–24 h at 4–6 cycles/h. But, there will be an increase in the chances of thermal degradation due to high temperature of heating and long extraction time (Figure 1.9 and Table 1.12). E.g. Extraction of Ursolic acid from *Cynomorii Herba*.

Merits

- Large quantity of plant materials can be extracted with much smaller quantity of solvents.
- Economic process due to less consumption of energy in short period of time.

Demerits

- It is a small-scale process
- This process depends on the nature of active constituents and solvent system.

1.4.7 Automated Soxhlet Extraction

It is an automatic continuous extraction technique with high extraction efficiency. This extraction process involves three approaches. Initially, it immerses the thimble which contains the plant materials directly into the boiling solvents. Then, the

thimble is moved above the solvent to mimic the rinse extraction step of soxhlet extraction. Finally, the concentration step involves the use of automated equipment to reduce the final volume up to 1–2 mL. This three-stage approach shortens the extraction time period to 2 h because it provides direct contact between the sample materials and solvents at the boiling point of solvents.

1.4.8 Supercritical Fluid Extraction (SFE)

It is the process of separating one component from another using supercritical fluid as the extracting solvent. Supercritical carbon dioxide is used as the extraction solvent instead of using any organic solvents because of its ideal properties such as low critical temperature (31°C), low polarity, selectivity, inertness, low cost, non-toxicity and ability to extract thermolabile substances (Figure 1.10). E.g. Extraction of essential oil from rosemary (*Rosmarinus officinalis*).

1.4.9 Pressurized Liquid Extraction (PLE)

This extraction process applies high pressure during extraction. Due to high pressure, the solvents are present in a liquid state above their boiling point which results increase in solubility and rate of diffusion of active constituents of plant materials in the solvents. PLE significantly decreases the consumption of solvents and reduces the time period of extraction as compared to other methods. It is applied for solid sample extraction.

1.4.10 Pressurized Hot Water Extraction (PHWE)

Pressurized hot water is extensively used to replace organic solvents during extraction processes. Temperatures below the critical temperature of water, but usually above 100°C are applied. Pressure should be high enough to keep the water in liquid state throughout the extraction processes.

1.4.11 Microwave-Assisted Extraction (MAE)

In this technology, microwaves generate heat by interacting with polar compounds such as water and the components present in the plant materials. This type of extraction is based on two mechanisms such as ionic conduction and dipole rotation. The transfers of heat and mass are in the same direction in MAE and it generates a synergistic effect to accelerate the extraction process and improve the yield of the active

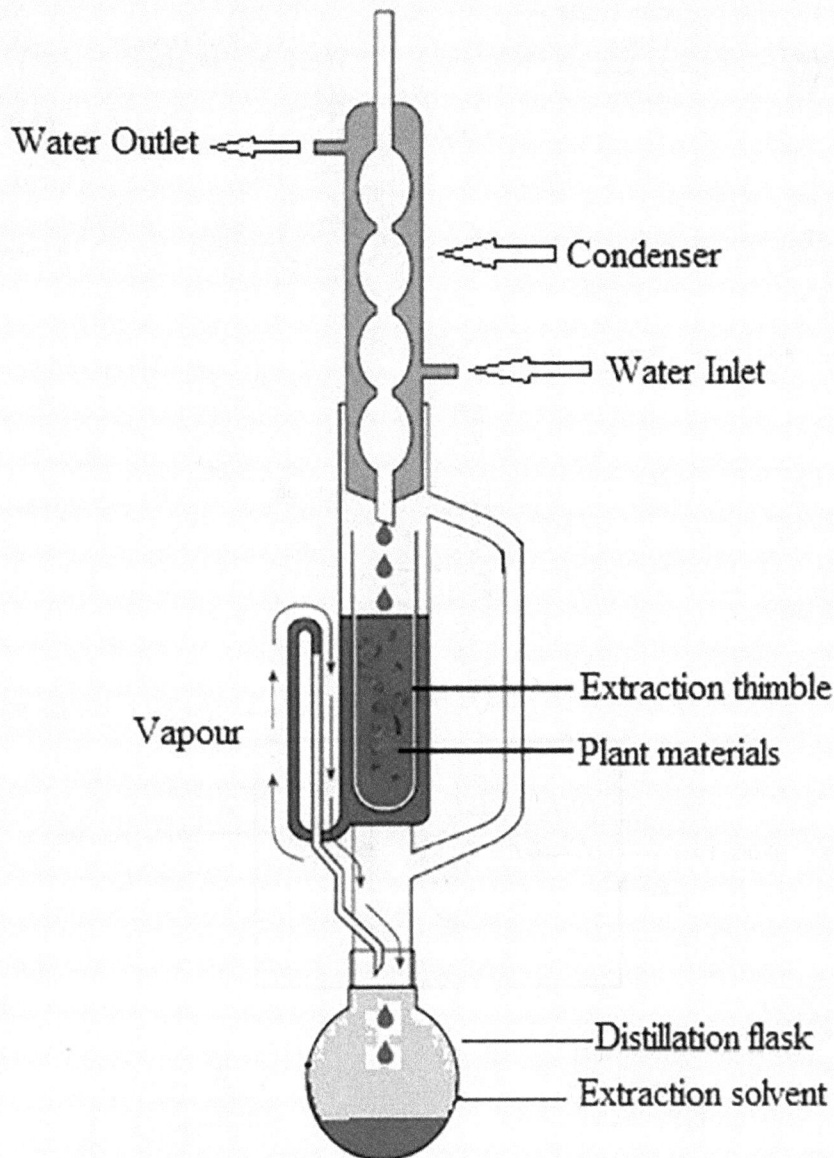

FIGURE 1.9 Extraction of terpenes by Soxhlet apparatus.

constituents. The application of MAE provides several advantages such as increase in the extract yield, decrease in thermal degradation and selective heating of plant material [20]. MAE is also regarded as a green technology because it reduces the usage of organic solvent for extraction (Figure 1.11).

There are two types of MAE methods such as solvent-free or solventless extraction (applied for volatile substances) and solvent extraction (applied for non-volatile substances).

Advantages of MAE

- It reduces solvent consumption.
- It has a shorter operational time.
- It possesses high recovery.
- It has good reproducibility.
- It involves minimal sample manipulation for the extraction process.

Disadvantages of MAE

- Additional filtration or centrifugation is required to remove the solid residue during MAE.
- The efficiency of MAE is poor when either the extracting compounds or the solvents are nonpolar, or when they are volatile.

1.4.12 Ultrasound Extraction (Sonication)

Ultrasonic-assisted extraction (UAE) is also called as ultrasonic extraction or sonication. UAE utilizes the ultrasonic wave energy of 350–450 W with frequencies >20 kHz during extraction process. Due to the presence of ultrasound in the solvent, it produces cavitation that accelerates the dissolution and diffusion of the plant constituents as well as the heat transfer. The advantage of UAE includes low solvent

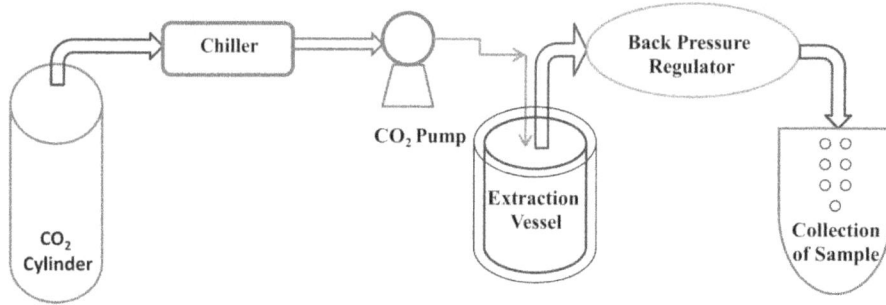

FIGURE 1.10 Supercritical fluid extraction (SFE).

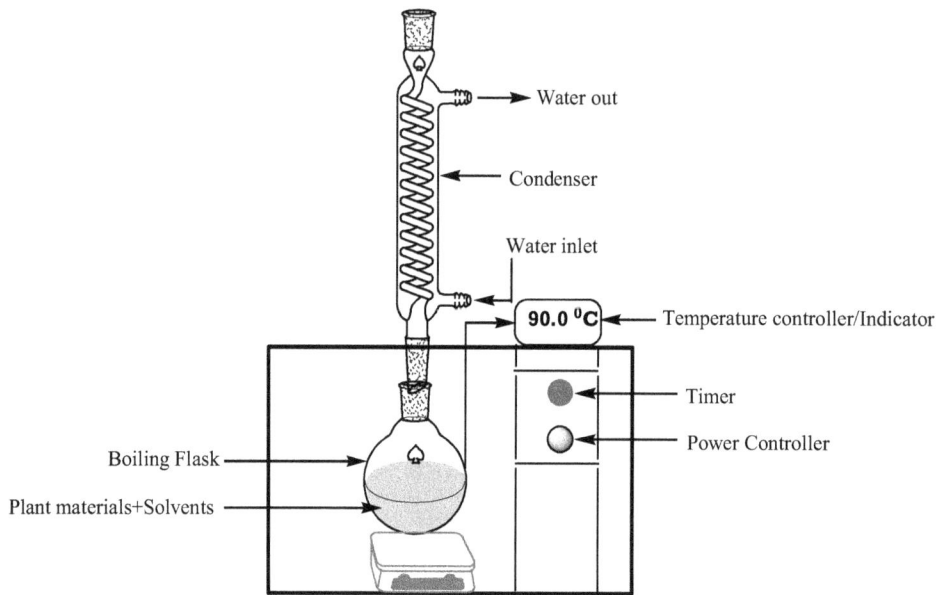

FIGURE 1.11 Microwave-assisted extraction.

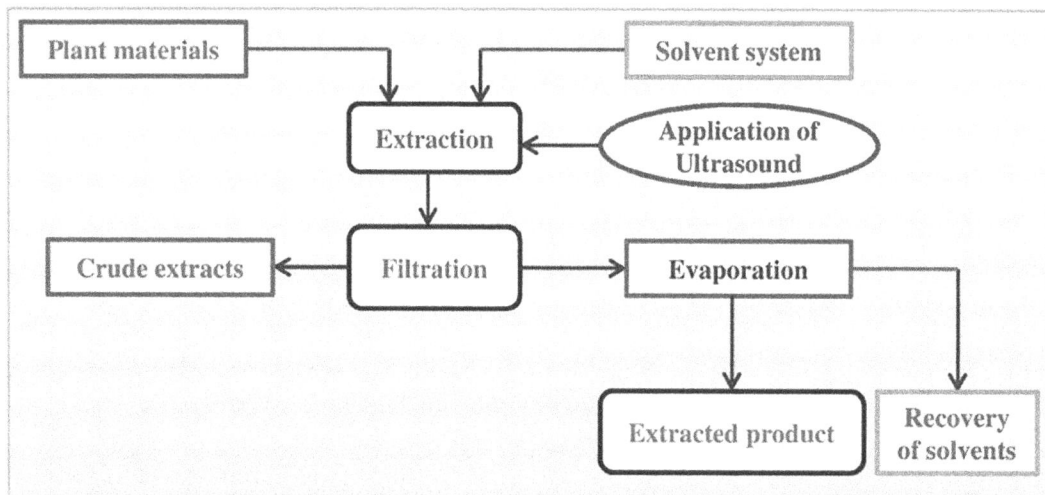

FIGURE 1.12 Ultrasound extraction (sonication).

and energy consumption, improved extraction efficiency and the reduction of extraction temperature and time [21,22]. It is applicable for the extraction of thermolabile and unstable compounds (Figure 1.12).

Advantages of UAE

- It is an inexpensive, simple and efficient method alternative to the conventional extraction technique.

- It increases the yield and makes faster kinetics of extraction.
- It is easy to operate.
- It reduces the operating temperature allowing the extraction of thermolabile compounds.
- The ultrasound apparatus is cheaper and its operation is easier as compared with other novel extraction techniques such as MAE.

Disadvantages of UAE

- The active constituents of medicinal plants through the formation of free radicals and consequently undesirable changes in the drug molecules.

Factors affecting the choice of extraction process

Various factors which affect the choice of the extraction process are represented as follows.

a. Character of active constituents present in plant materials.

b. Maceration process is applied if the chemical constituent in plant materials is soft, whereas the percolation process is followed when the constituent is hard and tough

c. Therapeutic potential or value of the active drug molecule.

d. **Cost of drug molecule:** Generally, percolation process is applied for the extraction of costly drug substances whereas cheap drug molecules are extracted by the maceration process.

e. **Stability of extracted drug molecule:** maceration or percolation should be used for extraction of thermolabile constituents

f. **Selection of solvents:** Generally, water is used as a solvent for the maceration process and non-aqueous solvents are selected for the percolation process.

g. **Concentration of product:** Dilute products such as tinctures can be made by using maceration or percolation process. But the percolation process is used for the extraction of concentrated products.

1.5 Solvents Used for Extraction

The solvents used for the extraction of medicinal plants are known as menstruum. The selection of the solvents plays a major role during the extraction process. The choice of solvent depends on the type of plant, part of the plant to be extracted, nature of the bioactive constituents and the availability of solvents. Suitable solvents for extraction are selected based on various criteria such as solubility, boiling point, selectivity, cost and safety. Generally, polar solvents such as water, methanol and ethanol are used for the extraction of polar compounds whereas nonpolar solvents such as hexane and dichloromethane are used during the extraction of nonpolar compounds. Water is used as a universal solvent for the extraction of most of terpenoids because of its high polarity and miscibility with organic solvents. Similarly, organic solvents also used for extraction are ethanol, methanol, chloroform, acetone, ether, hexane and dichloromethane etc. For liquid–liquid extraction, the convenient way is to select two miscible solvents such as water–dichloromethane, water–ether, water-hexane, etc. [23].

The compounds which are to be extracted using liquid–liquid extraction should soluble in organic solvents but not in water to separate the active constituents easily. The various solvents used for extractions are arranged according to the increasing order of polarity (Table 1.13). Among all the listed solvents, n-hexane is least polar, whereas water is highly polar. During the extraction process, the selected solvents are added according to the order of increasing polarity, starting from the least polar solvent (*n*-hexane) to the highest polarity solvent (water). It is better to select five

TABLE 1.10

Comparison between Extraction Processes

Extraction Process	Type of Solvents	Temperature	Pressure
Maceration	Aqueous (water) and non-aqueous	Room temperature	Atmospheric
Percolation	Aqueous (water) and non-aqueous	Room temperature and occasionally heating	Atmospheric
Soxhlet extraction	Organic	Heating	Atmospheric
Supercritical fluid extraction	Supercritical fluid (CO_2)	Room temperature	High

TABLE 1.11

Principles of Extraction Techniques

Extraction Techniques	Principle/Mechanism
Conventional extraction	Sample is shaken with the appropriate solvent in a container and the extracting components are separated by filtration
Soxhlet extraction	Sample is placed in a disposable, porous container (thimble) and constantly refluxing solvent flows through the thimble and leaches out analytes that are collected continuously
Microwave-assisted extraction	Rapid heating of the sample-solvent mixture in closed vessels for a more efficiency extraction.
Supercritical fluid extraction	Sample is placed in flow-through container and a supercritical fluid (e.g., CO_2) is passed through sample. After depressurization, extracted analyte is collected
Ultrasound-assisted extraction	Finely divided sample in a container is immersed in an ultrasonic bath with solvent and subjected to ultrasonic irradiation

TABLE 1.12

Advantages and Disadvantages of Extraction Techniques

Extraction Technique	Advantages	Disadvantages
Soxhlet extraction	Simple to handle and economical	Time-consuming, solvent consumption, analysis of samples is limited by the extraction step and limited extraction efficiency
Automated soxhlet extraction	Time saving, consumption of less solvents, reproducible and easy operation	Costly instrument
Supercritical fluid extraction	Environmental friendly, high speed of analysis	High analytical cost
Pressurized liquid extraction	Simple extraction protocol, short extraction time and easy operation	Costly instrument
Microwave-assisted extraction	Simple instrumentation, consumption of less solvents, short extraction period	Costly instrument
Ultrasound extraction	Simple and short extraction time	Limited extraction efficiency

TABLE 1.13

List of Solvents Used for Extraction Process

List of Solvents	Polarity
n-Hexane	0.009
Petroleum ether	0.117
Diethyl ether	0.117
Ethyl acetate	0.228
Chloroform	0.259
Dichloromethane	0.309
Acetone	0.355
n-Butanol	0.586
Ethanol	0.654
Methanol	0.762
Water	1.000

solvents during extraction such as two solvents with low polarity (*n*-hexane, chloroform), two with medium polarity (dichloromethane, *n*-butanol), and one with the highest polarity (water) [24].

1.5.1 Properties of Solvents Used in Extractions

The ideal solvents used in the extraction process possess the following properties.

- Solvents should be highly selective for the terpenoids to be extracted.
- They should possess high power for extraction.
- They should not react with the extracted components or with other compounds present in the plant material.
- Solvents must be eco-friendly during the extraction process.
- Solvents are volatile in nature.
- Solvents should be less toxic.
- Solvents should be available easily and economic.
- The generally used solvents include water, aliphatic alcohols with up to three carbon atoms, ether, hexane, chloroform and glycerine etc.

1.5.2 Factors to Be Considered for Selecting Solvents during Extraction

Various factors are considered while selecting solvents to carry out various extraction processes are given below.

a. **Selectivity:** The selected solvent should have the ability to extract the active constituent selectively and leave the inert materials.

b. **Reactivity:** Suitable solvent of extraction should not react with the active components of the extract.

c. **Viscosity:** Solvents should have low viscosity to allow ease of penetration of active phytoconstituents.

d. **Boiling temperature:** Solvent boiling temperature should be as low as possible to prevent degradation of the extracting substances by heat.

e. **Recovery:** The solvent of extraction should be recovered easily after extraction.

f. **Safety:** Ideal solvent for extraction should be nontoxic and nonflammable.

g. **Cost:** Solvents and extraction process should be economic.

1.6 Isolation of Terpenoids

There are various terpenoids such as monoterpenoid, diterpenoid, sesquiterpenoids etc present in essential oils of natural products. Hence, it is necessary to isolate the individual terpenoids for determining their physical, chemical and biological properties. The isolation of terpenoids is carried out by using the following methods [25–27].

  i. Isolation of essential oils from plant parts

    a.　Expression method

    b.　Hydro distillation (HD) method

    c.　Steam distillation (SD) method

    d.　Extraction with volatile solvents

    e.　Adsorption in purified fats (Enfluerage)

  ii. Separation of terpenoids from essential oils

    a.　Chemical methods

    b.　Physical methods (Chromatography and Fractional distillation)

1.6.1 Isolation of Essential Oils from Plant Parts

a. **Expression method:** In the expression method, plant material containing essential oils is crushed and the obtained juice is screened to remove the large particles. Then the screened juice is centrifuged in a high-speed centrifugal machine to isolate the essential oils.

b. **Hydro distillation (HD) method:** This method is commonly used for the extraction of essential oils by using water as a solvent. But, some of the compounds are decomposed during hydro distillation process due to the high temperature in the extraction medium.

c. **Steam distillation (SD) method:** It is most widely used method in which plant material is macerated and essential oils are isolated by distillation using steam instead of water to avoid decomposition during the distillation process (Figure 1.13).

   Demerits of steam distillation:
   - Essential oil may undergo decomposition during the distillation process.
   - Some of the constituents of essential oil like ester mainly responsible for the odor and fragrance of the oil and it may undergo decomposition that results in formation of inferior quality perfume.

d. **Extraction with volatile solvents:** As described above, some of the essential oils are heat sensitive and decomposed during the distillation process. To avoid this problem, essential oils are isolated by treating directly with suitable organic solvents like petroleum ether, hexane and ethanol. After isolation, solvents are removed by distillation under reduced pressure to obtain pure essential oils.

e. **Adsorption in purified fats (Enfluerage):** This method involves the following steps.
   - The fat is allowed to warm at 50°C on a glass plate.
   - Then the fat is covered with flower petals and kept for several days until it is saturated with essential oils.
   - Then the old petals are replaced by fresh petals and the process is repeated for several days.
   - Finally, the petals are removed and the fat saturated with essential oils is treated with ethanol so that all the essential oils present in fat are dissolved in ethanol.
   - Further, the fractional distillation under reduced pressure is carried out to remove the solvent (ethanol) and essential oils are separated.

1.6.2 Separation of Terpenoids from Essential Oils

1.6.2.1 Chemical Methods

Essential oils containing terpenoid hydrocarbon are treated with nitrosyl chloride in chloroform (Tilden's reagent) producing a crystalline adduct of hydrocarbons with a sharp melting point [28]. These adducts are separated and decomposed into their corresponding hydrocarbons (Scheme 1.4).

Similarly, essential oil containing alcohols are separated by their reaction with phthalic anhydride to produce diesters. The diesters are further treated with $NaHCO_3$ and KOH produces parent terpenoid alcohol. It is observed that primary alcohol reacts readily, whereas secondary alcohol reacts less readily and tertiary alcohols do not react at all (Scheme 1.5).

Terpenoid containing aldehyde and ketone are separated by reaction with common carbonyl reagents such as $NaHSO_3$, phenylhydrazine, 2,4-dinitrophenyl hydrazine, semicarbazide, hydroxyl amine etc (Scheme 1.6).

1.6.2.2 Physical Methods

Fractional distillation and chromatographic methods are most widely used physical methods for the separation of terpenoids from essential oils. During fractional distillation of essential oils, monoterpenoids are separated first followed by

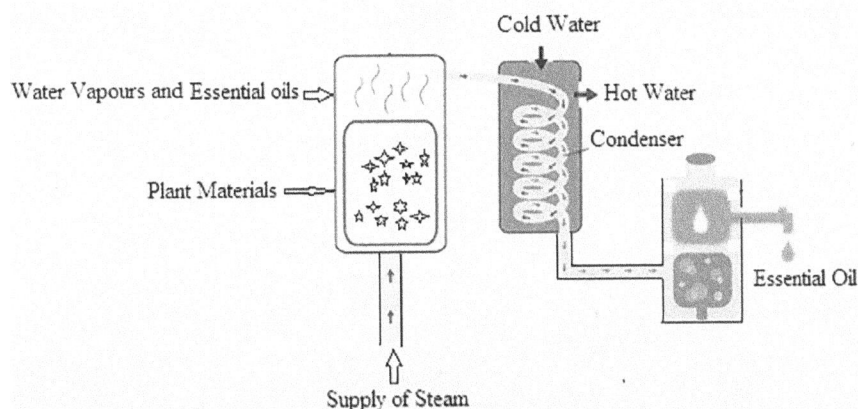

FIGURE 1.13   Isolation of essential oils by Steam distillation.

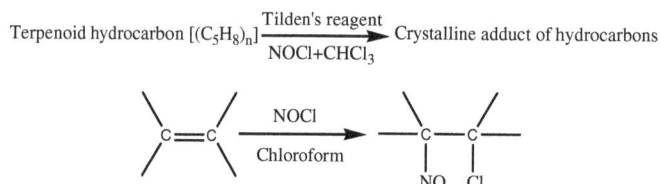

SCHEME 1.4 Treatment of terpenoid hydrocarbon with Tilden's reagent.

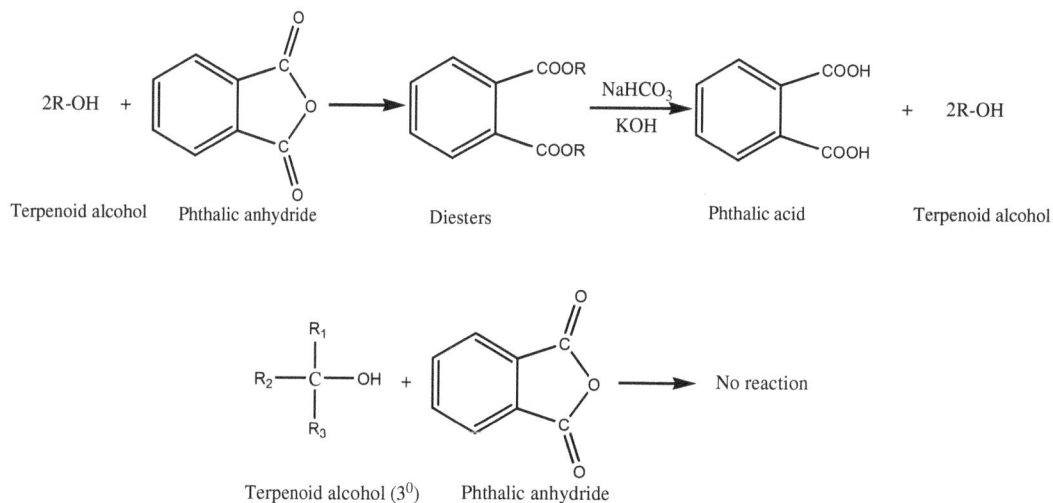

SCHEME 1.5 Separation of essential oil containing alcohols.

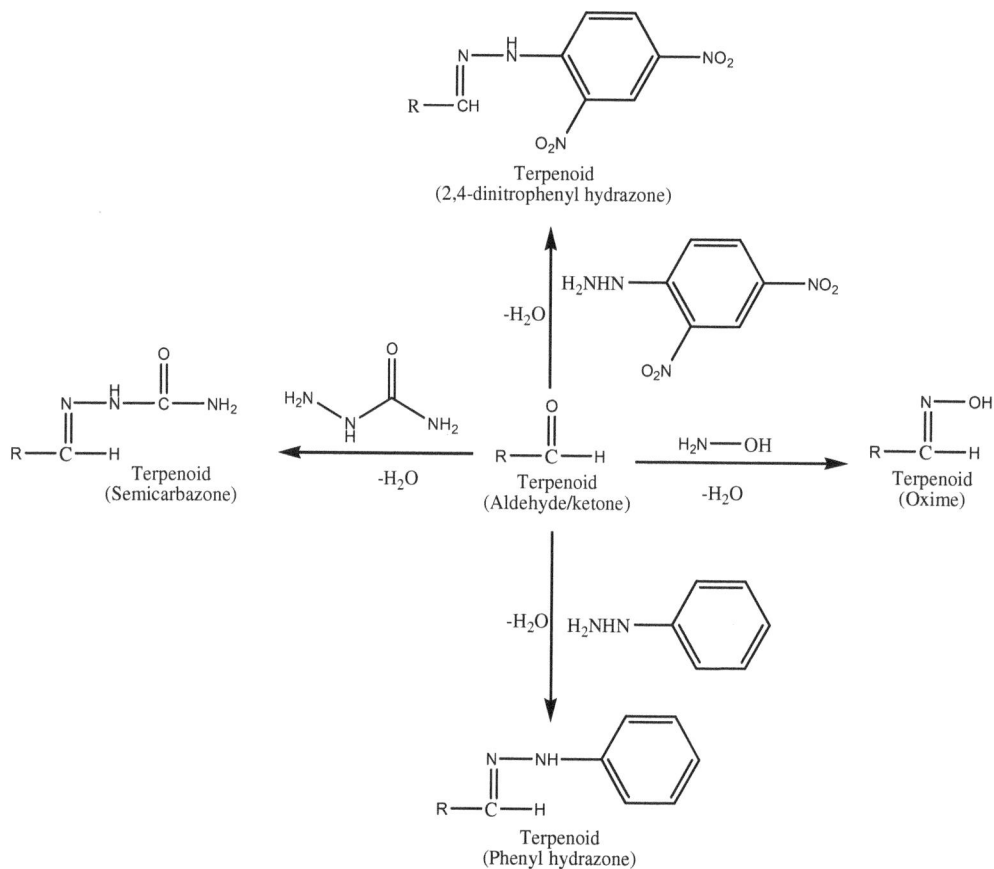

SCHEME 1.6 Separation of terpenoids containing aldehyde and ketone.

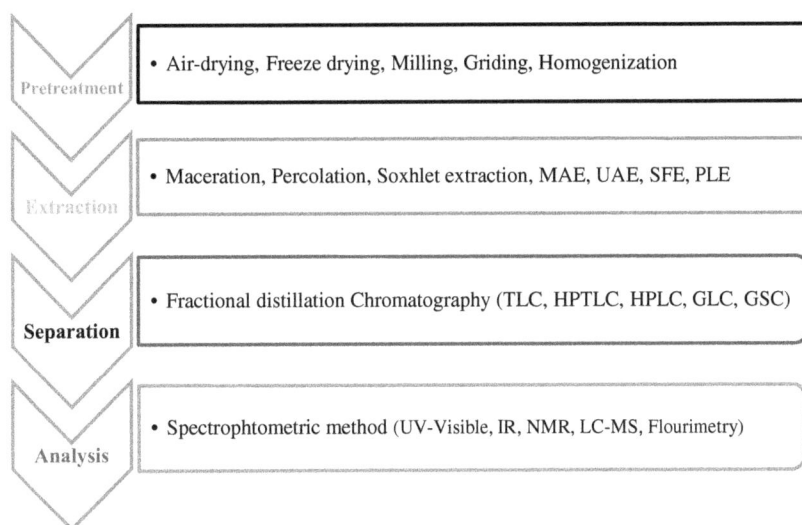

FIGURE 1.14 Steps involved in general methods of extraction and isolation of terpenoids.

the separation of their oxygenated derivatives. Similarly, the residual part containing sesquiterpenoids remains undistilled. Finally, monoterpenoids are separated by distillation under reduced pressure to avoid their decomposition.

Currently, various chromatographic methods are used efficiently for the separation of terpenoids from essential oils. Generally, two phases are involved during the separation of terpenoids in chromatographic methods which include the stationary phase and mobile phase. Based on the type of stationary phase, there are mainly two types of chromatographic method such as adsorption chromatography (stationary phase-solid) and partition chromatography (stationary phase-liquid). Similarly, based on the type of mobile phase (liquid or gas), adsorption chromatography is categorized into two types such as liquid-solid chromatography (LSC) and gas-solid chromatography (GSC) whereas partition chromatography is categorized into two types such as liquid–liquid chromatography (LLC) and gas-liquid chromatography (GLC) [29]. Depending on the nature of both stationary and mobile phase, chromatography is of two types such as normal phase chromatography (stationary phase-polar and mobile phase-nonpolar e.g. TLC, HPTLC) and reverse phase chromatography (stationary phase-nonpolar and mobile phase-polar e.g. HPLC) [30].

In the case of adsorption chromatography, the separation of terpenoids is based on their adsorption affinity toward the adsorbent or stationary phase whereas the separation of terpenoids depends on their partition coefficient in the case of partition chromatography. The components with lower adsorption affinity travel faster over the adsorbent layer under the influence of mobile phases and eluted or separated first as compared to other components (Figure 1.14) [31].

1.7 General Methods of Structure Elucidation of Terpenoids

There is development of various strategies for the structural determination of terpenoids. The methods are based on chemical degradation and analysis of spectroscopic data [32–40].

i. **Molecular formula:** Molecular formula is determined with the help of qualitative and quantitative analysis. The determination of molecular weight is carried out by means of mass spectrometry. In the case of optically active terpenoids, specific rotation can be measured.

ii. **Nature of oxygen atom:** Oxygen atom is generally present as functional groups such as alcohol, aldehyde, ketone, ether, ester or carboxylic groups in terpenoids. The presence of alcoholic group (–OH) can be determined by the formation of acetates with acetic anhydride (Scheme 1.7).

Similarly, the primary alcoholic group undergo esterification with acids more readily in comparison to secondary and tertiary alcohols.

Tertiary < secondary < Primary

Terpenoids containing carbonyl functional group (aldehyde or ketone) produce crystalline addition products like oxime, phenyl hydrazone and semicarbazone on reaction with hydroxyl amine, phenylhydrazine and semicarbazide respectively (Scheme 1.9).

The presence of –COOH group is confirmed by two facts such as terpenoid is soluble in NH_3 and it gives effervescence with $NaHCO_3$. The number of –COOH groups is estimated by titration against a standard alkali.

iii. **Presence of C-alkyl group:** The important C-alkyl group is C–CH_3 group. C–CH_3 group is detected and estimated by the oxidation of terpenoid in presence of chromic acid to produce acetic acid (Scheme 1.10) which is estimated volumetrically (Kuhn-Roth method).

iv. **Presence of unsaturation:** The presence of unsaturation or olefinic double bond and number of double bonds in terpenoids is confirmed by means of catalytic hydrogenation or addition reaction with halogens or halogen acids, per acids and nitrosyl chloride in chloroform (Tilden's reagent) as outlined

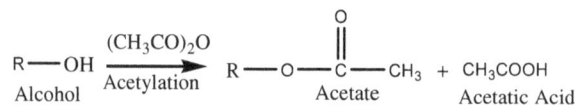

SCHEME 1.7 Acetylation of alcohol.

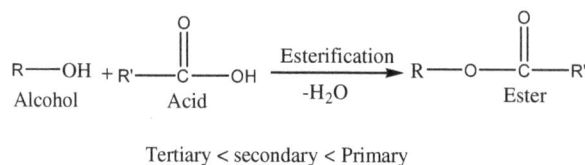

Tertiary < secondary < Primary

SCHEME 1.8 Esterification of alcohol with acids.

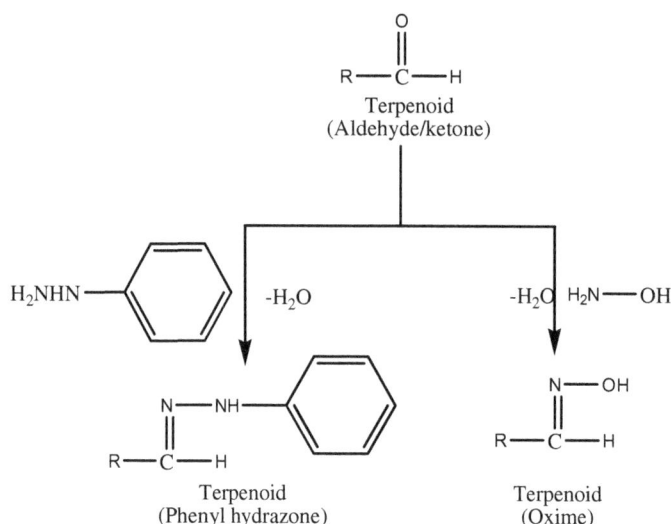

SCHEME 1.9 Reactions of aldehydes with carbonyl reagents.

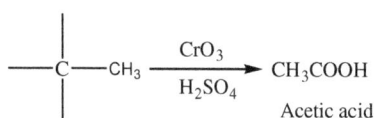

SCHEME 1.10 Oxidation of terpenoid in presence of chromic acid.

in Scheme 1.11. In 1875, Tilden introduced nitrosyl chloride as a reagent which forms crystalline derivatives of terpenes. For example, α-pinene from oil of turpentine. This reaction distinguishes one terpene from another.

v. **Presence of conjugated double bond:** The presence of conjugated double bonds in terpenoids is generally detected by Diel's Alder reaction. Example: Myrcene forms adduct with maleic anhydride (Scheme 1.12). The presence of conjugated double bonds is further confirmed by UV absorption spectra with λ_{max} 224 nm.

vi. **Presence of α, β-unsaturated carbonyl group:** The presence of α, β-unsaturated carbonyl group in terpenoids is confirmed by UV absorption spectra with λ_{max} 220–250 nm.

vii. Presence of skeletal structure and position of side chain

Terpenoids undergo dehydrogenation with sulfur, selenium, polonium or palladium to produce aromatic compounds. For example α-terpineol on dehydrogenation in presence of selenium yields p-cymene (Scheme 1.13).

Similarly, citral on heating with potassium hydrogen sulfate produces the aromatic compound *p*-cymene as presented in Scheme 1.14.

Camphor undergoes distillation in presence of $ZnCl_2$ or phosphorous pentoxide (P_2O_5) to produce p-cymene as shown in Scheme 1.15.

Thus, the carbon skeleton present in terpenoid can be presented as follows. From this, the position of side chain (methyl and isopropyl groups) in terpenoid is also confirmed (Figure 1.15).

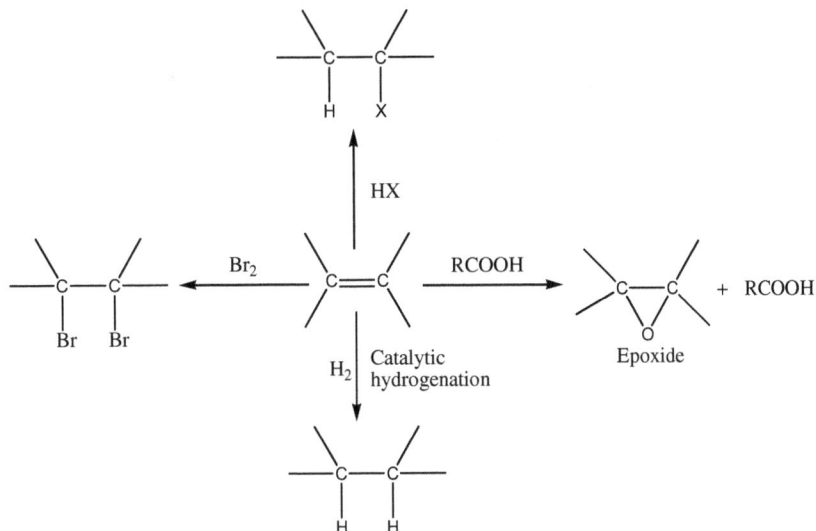

SCHEME 1.11 Addition reactions of terpenoid containing olefinic double bond.

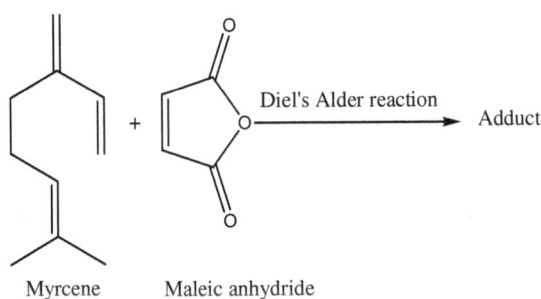

Myrcene   Maleic anhydride

SCHEME 1.12 Diel's Alder reaction of terpenoids containing conjugated double bonds.

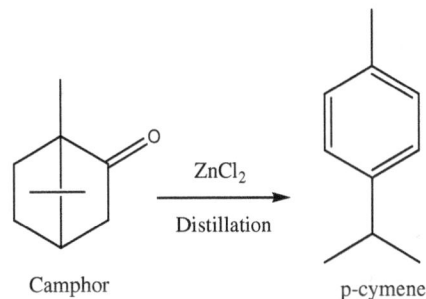

Camphor   p-cymene

SCHEME 1.15 Distillation of camphor.

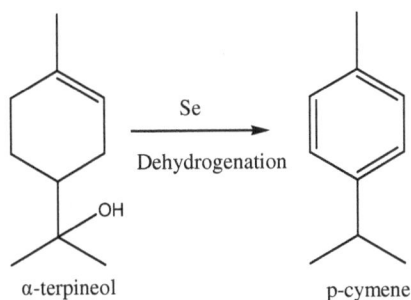

α-terpineol   p-cymene

SCHEME 1.13 Dehydrogenation of α-terpineol.

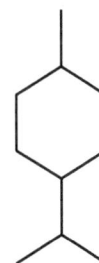

FIGURE 1.15 Carbon skeleton of terpenoid.

viii. **Oxidative degradation:**

Various reagents uses for oxidative degradation include ozone, acid, neutral or alkaline potassium permanganate, chromic acid, sodium hypobromide, osmium tetroxide, nitric acid, lead tetra acetate and peroxy acids etc. Depending on a particular group to be oxidized, the oxidizing agents are selected. Terpenoids react with ozone produces ozonide which on decomposition either by hydrolysis or catalytic reduction yield carbonyl compounds (Scheme 1.16).

Similarly, terpenoids react with nitric acid yield aromatic acid and aliphatic acid. For example, Citral undergoes oxidation in presence of potassium permanganate ($KMnO_4$) and chromic acid (CrO_3) produces laevulic acid, oxalic acid and acetone (Scheme 1.17).

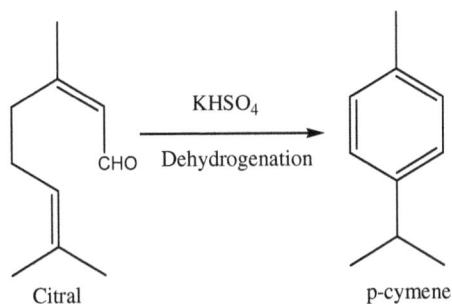

Citral   p-cymene

SCHEME 1.14 Heating of citral with potassium hydrogen sulfate.

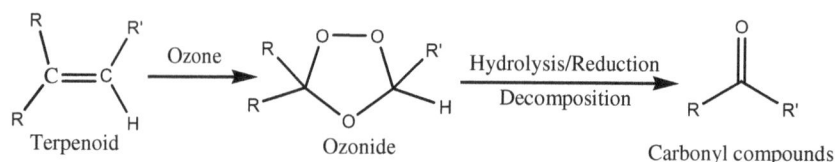

SCHEME 1.16 Oxidative degradation of terpenoids.

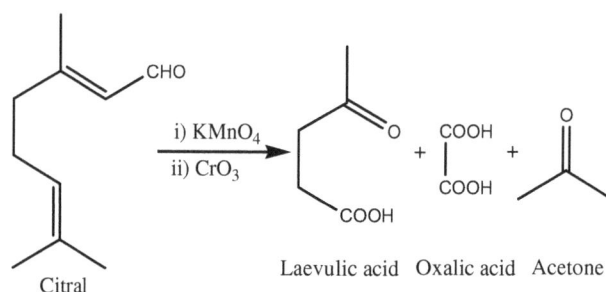

Laevulic acid   Oxalic acid   Acetone

SCHEME 1.17 Oxidation of terpenoids.

ix. Spectral study of terpenoids

Various spectroscopic methods such as ultraviolet-visible (UV-Vis), Fourier transform infrared (FT-IR), NMR (^{1}H-NMR and ^{13}C-NMR), Mass spectroscopy and elemental analysis (C, H, N) are studied extensively to determine and characterize the structure of different types of terpenoids [41–46].

a. **Ultraviolet spectrophotometry (UV):** The application of UV spectroscopy in the structure elucidation of terpenoids was extensively studied by R. B. Woodward. In case of ultra violet spectroscopy, Woodward developed a set of rules to calculate absorption maximum (λ_{max}) for the molecular structure of terpenoids. The presence of the various components of terpenoids can be determined due to their characteristic UV absorption. The UV-visible spectra are recorded over the range of 200–800 nm.

The functional groups present in terpenoids exhibit λ_{max} between 200 and 350 nm in the UV range and are termed as chromophores. These consist of olefinic bonds (double and triple bonds) and carbonyl groups including aldehydes, ketones, carboxylic acid and esters. Single chromophores such as isolated double and triple bonds do not adsorb at a wavelength higher than 200 nm and the isolated aldehydes, ketones, carboxylic acids and their esters adsorb at a higher wavelength but their intensities are too low [47].

Hence, both the above-mentioned isolated chromophores are not of much importance for the structure elucidation of terpenoids. However, UV absorption data of various terpenoids becomes valuable only when the molecule contains conjugated double bonds and α, β-unsaturated carbonyl group [48]. When more than one conjugated system is present in the molecule, the sum of the values is normally considered. The λ_{max} for acyclic dienes, homoannular and heteroannular diene is 217–228, 256–265 and 230–240 nm respectively (Figure 1.16).

Menthol, like many other terpene alcohols such as citronellol, geraniol, linalool, myrcenol, nerol and nerolidol, does not absorb UV light in the range of 290–320 nm, but absorbs well below 290 nm with the λ_{max} at 220 nm.

b. **Infrared spectroscopy (IR):** The elucidation of the structure of terpenes in the 1950s made the use of classical dehydrogenation and oxidative degradation method accompanied by interpretation of the IR spectra [49]. The correlation between ring size and the frequency of carbonyl absorption in the infrared spectrum played an important role in the identification of the ring size of ketones, lactones and cyclic anhydrides. The FT-IR spectra are recorded over the range of 400–4,000 cm^{-1}. Among the detection of functional groups, the absorption maximum (cm^{-1}) in IR spectra are 3,650–2,650 cm^{-1} (–OH group), 1,750–1,700 cm^{-1} (saturated C=O group), 1,700–1,660 cm^{-1} (α, β-unsaturated), 1,740–1,720 cm^{-1} (–CHO), 1,725–1,700 cm^{-1} (saturated acids), 1,950–1,735 cm^{-1} (esters) etc [50]. Infrared spectroscopy is used to detect the presence of various functional groups (lactone: ν_{max} 1,786 cm^{-1} and cyclohexanone: ν_{max} 1,724 cm^{-1}) present in terpenoids. In case of citral, IR spectrum exhibits C–H stretching at 3,000–2,850 cm^{-1}, CH$_3$ bending at 1,450 and 1,375 cm^{-1}, CH$_2$ bending at 1,465 cm^{-1}, C=C at 1,680–1,600 cm^{-1}) and C=O at 1,740–1,720 cm^{-1} (Figures 1.17 and 1.18 and Table 1.14).

c. **Nuclear magnetic resonance spectroscopy (NMR):** Both ^{1}H-NMR (PMR) and ^{13}C-NMR (CMR) spectra are evaluated to confirm the presence of number of hydrogens and carbons

Acyclic terpenoid:

Butadiene (λ_{max}=217nm; ε_{max}=20900)

Myrcene (λ_{max}=224nm; ε_{max}=14600)

Monocyclic terpenoid:

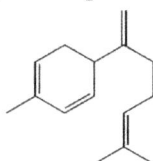

Zingiberene (λ_{max} =260nm; ε_{max} =2700)

β-Phellandrene (λ_{max}=232nm; ε_{max}=9100)

Trycyclic terpenoid:

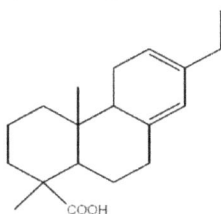

Levopimaric acid (Homoannular)
(λ_{max}=272nm; ε_{max}=7100)

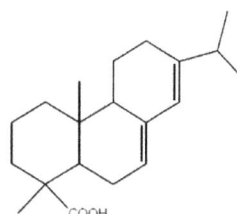

Abietic acid (Heteroannular)
(λ_{max}=237.5nm; ε_{max}=23000)

FIGURE 1.16 Ultraviolet absorption bands of some conjugated terpenoids.

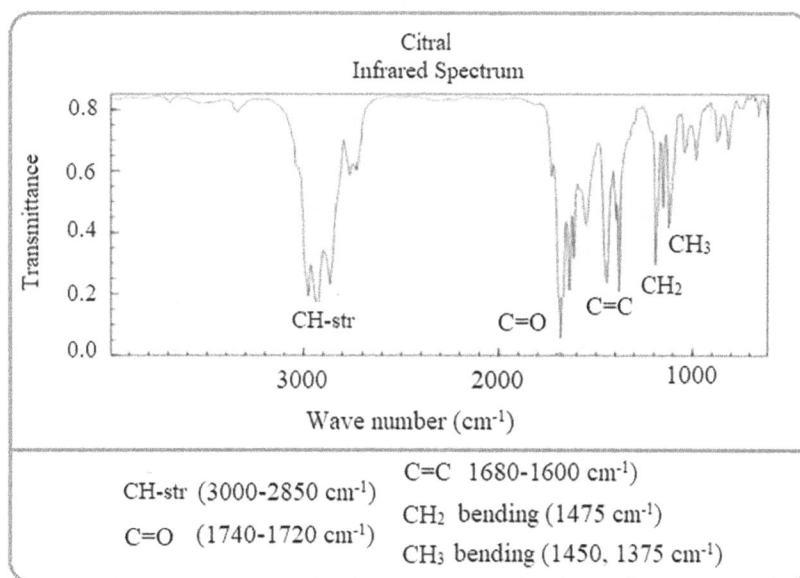

FIGURE 1.17 Infrared spectrum of terpenoid (citral).

present in terpenoid molecules. The application of Fourier transform methods to record ^{13}C-NMR spectra that useful data were collected for a range of terpenoids during the early 1970s and then applied to the structural elucidation of terpenoids. The structural studies on the taxanes illustrate the impact of NMR strategies on structure elucidation. The elucidation of the complete

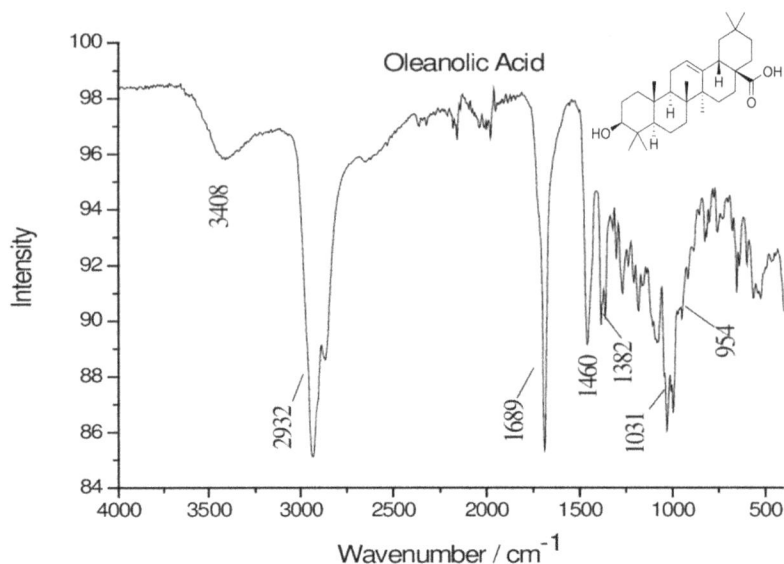

FIGURE 1.18 Infrared spectrum of oleanolic acid.

TABLE 1.14

Infrared Spectroscopy Correlation Chart

Bond	Types of Compounds	Frequency Range (cm⁻¹)	Intensity
C–H	Alkane	2,850–2,970	Strong
C–H	Alkene, aromatic	3,010–3,090	Medium
C=C	Alkene	1,610–1,680	Variable
C=C	Aromatic ring	1,500–1,600	Variable
O–H	Alcohol, phenol	3,590–3,650	Variable
C–O	Alcohol, ether, carboxylic acids	1,050–1,300	Strong
C=O	Aldehydes, ketones, carboxylic acids	1,690–1,760	Strong
C–N	Amines, amides	1,180–1,360	Strong
N–H	Amines, amides	3,300–3,500	Medium
N–O	Nitro (aliphatic)	1,540, 1,310	Variable
N–O	Nitro (aromatic)	1,520, 1,350	Variable
C≡N	Nitriles	2,210–2280	Strong
C–F	Flouroalkane	900–1,100	Variable
C–Cl	Chloroalkane	750–850	Variable
C–Br	Bromoalkane	515–685	Variable
C–I	Iodoalkane	420–510	Variable

structure of taxane carbon skeleton was reported in 1966 and involved the assignment of the ^{1}H-NMR spectrum [51].

For example ^{13}C-NMR of citral represents various chemical shift values for carbons at different positions in citral. The chemical shift of C_1 is 185–220 ppm, C_2, C_3, C_7 and C_8 is 100–150 ppm. Similarly, the chemical shift of C_4, C_9, C_{10} is present between 8 and 30 ppm and the chemical shift of C_5 and C_6 is 15–55 ppm (Figure 1.19).

^{1}H-NMR of citral represents various chemical shift values for protons at different positions in citral. Proton at C_1 has a chemical shift value of 9.4–9.8, protons at C_2 and C_3 have a chemical shift value of 5–6 ppm, protons at C_4 and C_5 have

a chemical shift value of 1.9 ppm and protons at C_6 and C_7 and C_8 have chemical shift value of 1.7 ppm (Figure 1.20).

d. **Mass spectrometry (MS):** Fast atom bombardment spectroscopy (FAB-MS) provides the information regarding exact molecular ion peak along with the fragmentation pattern of the terpenoid molecules. Mass spectra were recorded at 70 eV. It is an important tool for the structure elucidation of different types of terpenoids. FAB-MS in both positive and negative modes and a combination of GC-MS or HPLC-MS spectrometry coupled with computerized analysis and Tandem MS spectral analysis have been

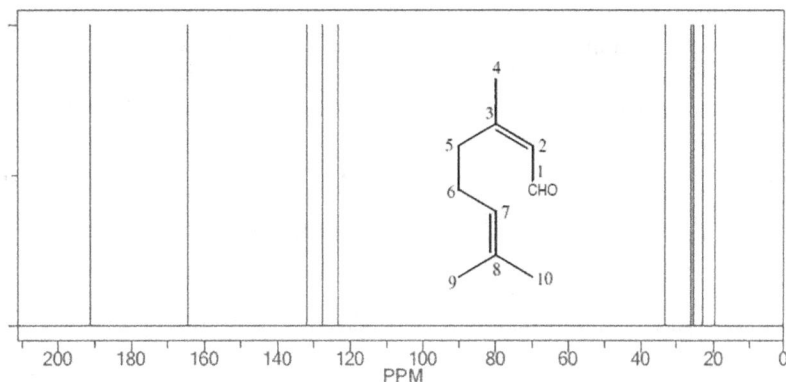

FIGURE 1.19 ^{13}C-NMR spectra of citral.

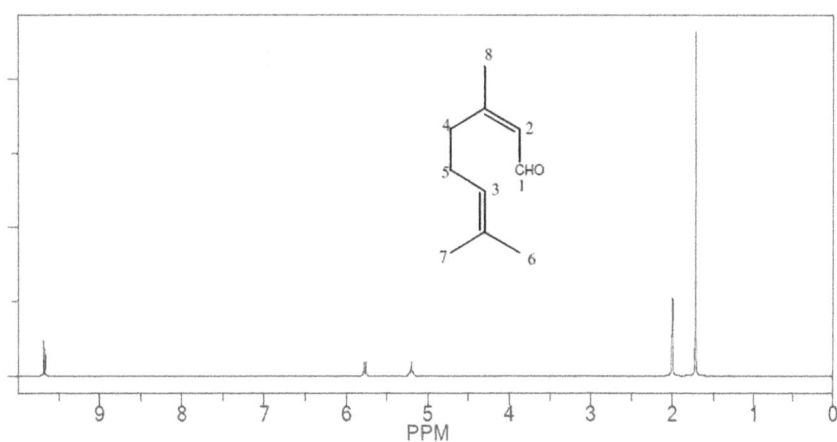

FIGURE 1.20 ^{1}H-NMR spectra of citral.

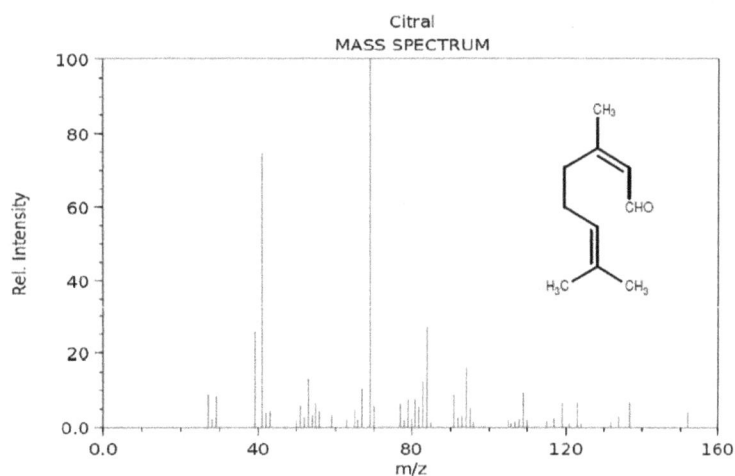

FIGURE 1.21 Mass spectrum of citral (m/z).

extensively employed for the determination of various types of terpenoid structure [52]. For example, the molecular weight of citral is 152. Its mass spectrum shows peak at m/z 29 due to haemolytic α-cleavage, peak at m/z 69 due to homolytic β-cleavage, peak at m/z 124 due to heterolytic α-cleavage (Figure 1.21).

e. **X-ray analysis:** Similarly, X-ray crystallography was the method of choice for the elucidation of terpenoid structure in 1971. It is also a very useful technique for elucidating the stereochemistry of terpenoids. The terpenoid synthase structural biology is studied with the help of X-ray analysis. It determines the crystal structure of FPP synthase, which catalyzes the 1′-4 (i.e., head-to-tail) coupling of C_5 isoprenoid units. So, X-ray crystal structure determination is required to find out the structure and function of enzymes involved in the synthesis of terpenoids [53].

f. **Elemental analysis:** It is a process by which samples of chemical compounds like terpenoids are analyzed for determining the presence of elements and also isotopic substances (Table 1.15). Elemental analysis can be performed in two ways such as qualitative and quantitative. Qualitative analysis determines the type of elements present in the sample. Whereas, quantitative analysis determines the amount of particular element present in the sample [54].

g. **Synthesis:** The structure of isolated terpenoids is finally confirmed by synthesis. In terpenoid

chemistry, many of the syntheses are unclear and in such cases, analytical data is considered in conjunction with the synthesis. For example, the structure of citral is confirmed by its synthesis as outlined in Scheme 1.18 [55–58].

1.8 Conclusion

Terpenes are a large class of natural hydrocarbon secondary metabolites built up from five-carbon isoprene units linked together most commonly in a head-to-tail arrangement. The extraction process plays a key role to separate the active constituents (terpenoids) from plant sources by using suitable solvents. The extraction of terpenoid progresses through various stages which include (1) penetration of the solvent into the solid matrix (plant materials) (2) the solute or plant constituents dissolving in the solvents (3) the solute is diffused out of the solid matrix (4) the extracted solutes or components are collected out. The various methods used for extraction of plant materials containing terpenoids include maceration, infusion, digestion, decoction, percolation, hot continuous extraction (soxhlet), aqueous or alcoholic extraction, CCE, greener extraction methods such as SFE, PLE and MAE and ultrasound extraction (sonication) etc. Chromatographic methods such as TLC, HPTLC, HPLC, GSC, GLC and column chromatography etc are applied for the separation and isolation of different types of terpenoids (acyclic, monocyclic, bicyclic) from natural sources. Various spectral methods including UV, FT-IR, NMR, LC-MS, X-ray crystallography, fluorimetry, elemental analysis (C:H:N:O) etc are used for the identification, estimation and structural elucidation of terpenoids.

TABLE 1.15

Elemental Analyses of Terpenoid

| Name of Terpenoids | Molecular Formula | Elemental Analysis Calculated (%) | | | |
		C	H	N	O	X
Geraniol	$C_{10}H_{18}O$	77.87	11.76	-	10.37	-
Linalool	$C_{10}H_{18}O$	77.87	11.76	-	10.37	-
Citral	$C_{10}H_{16}O$	78.90	10.59	-	0.51	-
Nerol	$C_{10}H_{18}O$	77.87	11.76	-	10.37	-
Myrcene	$C_{10}H_{16}$	88.16	11.84	-	-	-
Farnesol	$C_{15}H_{26}O$	81.02	11.79	-	7.20	-
Nerolidol	$C_{16}H_{28}$	87.19	12.81	-	-	-

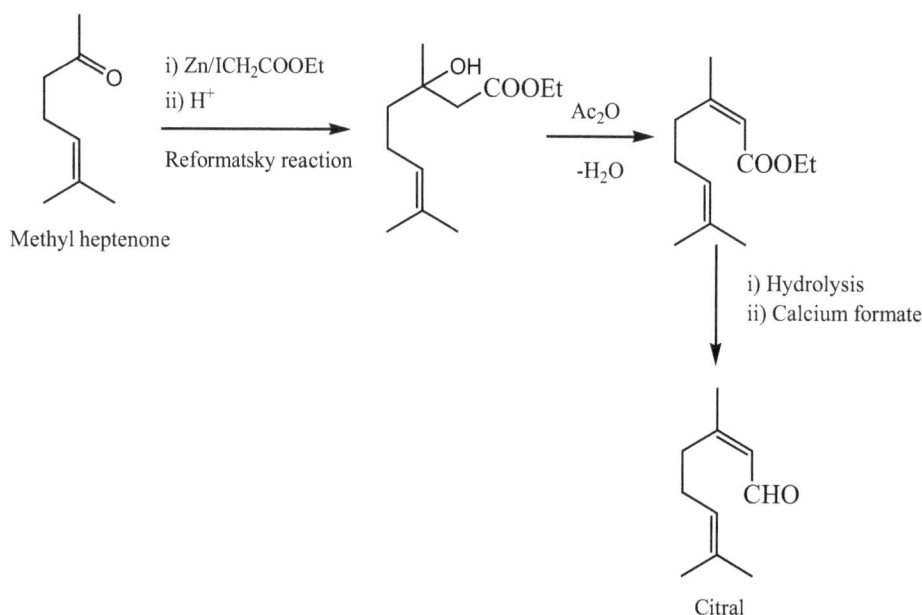

SCHEME 1.18 Synthesis of citral.

Acknowledgment

Biswa Mohan Sahoo is thankful to Roland Institute of Pharmaceutical Sciences and Bimal K. Banik is grateful to Prince Mohammad Bin Fahd University.

Abbreviations

CCE	Counter current extraction
CO_2	Carbon dioxide
CrO_3	Chromic acid
°C	Degree centigrade
GLC	Gas-liquid chromatography
GSC	Gas-solid chromatography
HD	Hydro distillation
HPLC	High-performance liquid chromatography
HPTLC	High-performance thin-layer chromatography
kHz	Kilohertz
$KMnO_4$	Potassium permanganate
KOH	Potassium hydroxide
LSC	Liquid-solid chromatography
λ_{max}	Maximum wavelength
MAE	Microwave-assisted extraction
$NaHCO_3$	Sodium bicarbonate
$NaHSO_3$	Sodium bisulfite
nm	Nanometre
NMR	Nuclear magnetic resonance
NOBr	Nitrosyl bromide
NOCl	Nitrosyl chloride
PHWE	Pressurized hot water extraction
PLE	Pressurized liquid extraction
P_2O_5	Phosphorous pentoxide
ppm	Parts per million
SD	Steam distillation
SFE	Supercritical fluid extraction
TLC	Thin-layer chromatography
UAE	Ultrasonic-assisted extraction
UV	Ultraviolet
W	Watt
$ZnCl_2$	Zinc chloride

REFERENCES

1. Chatwal, G. *Organic Chemistry of Natural Products*, volume I. **2019**, 1.1–1.155.
2. Agarwal, O. P. *Organic Chemistry of Natural Products*, volume I. **2010**, 312–433.
3. Newman, D. J.; Cragg, G. M. Natural products as sources of new drugs over the 30 years from 1981 to 2010. *Journal of Natural Products*, **2012**, 75(3), 311–35.
4. Newman, D. J.; Cragg, G. M. Natural products as sources of new drugs from 1981 to 2014. *Journal of Natural Products*, **2016**, 79(3), 629–61.
5. Ruzicka, L. The isoprene rule and the biogenesis of terpenic compounds. *Experientia*, **1953**, 9, 357–367.
6. Wallis, T. E. *Text Book of Pharmacognosy*. Delhi: CBS Publishers and Distributors, **1989**, 356–549.
7. Harrewijn, P.; van Oosten, A. M.; Piron, P. G. M. Specific properties of terpenoids. *Natural Terpenoids as Messengers*, **2000**, 109–179.
8. Cragg, G. M.; Newman, D. J.; Snader, K. M. Natural products in drug discovery and development. *Journal of Natural Products*, **1997**, 60, 52–60.
9. Grayson, D. H. Monoterpenoids. *Natural Product Reports*, **2000**, 17, 385–419.
10. Brahmkshatriya, P. P.; Brahmkshatriya, P. S. Terpenes: Chemistry, biological role, and therapeutic applications. *Natural Products*, **2013**, 2665–2691.
11. Connolly, J. D.; Hill, R. A. Triterpenoids. *Natural Product Reports*, **2010**, 27, 79–132.
12. Mahato, S. B; Sen, S. Advances in triterpenoid research, 1990–1994. *Phytochemistry*, **1997**, 44, 1185–1236.
13. Fraga, B. M. Natural sesquiterpenoids. *Natural Product Reports*, **2008**, 25, 1180–1209.
14. Seigler, D. S. *Introduction to Terpenes. Plant Secondary Metabolism*. Springer, **1998**, 312–323.
15. Grafflin, M. W. System of nomenclature for terpene hydrocarbons acyclics, monocyclics, bicyclics. *Advances in Chemistry*, **1955**, 14, 1–11.
16. Majekodunmi, S. O. Review of extraction of medicinal plants for pharmaceutical research. *MRJMMS*, **2015**, 3, 521–527.
17. Ingle, K. P.; Deshmukh, A. G.; Padole, D. A.; Dudhare, M. S.; Moharil, M. P.; Khelurkar, V. C. Phytochemicals: Extraction methods, identification, and detection of bioactive compounds from plant extracts. *Journal of Pharmacognosy and Phytochemistry*, **2017**, 6, 32–36.
18. Azwanida, N. N. A review on the extraction methods use in medicinal plants, principle, strength, and limitation. *Medicinal and Aromatic Plants*, **2015**, 4, 196.
19. Komae, H.; Hayashi, N. Separation of essential oils by liquid chromatography. *Journal of Chromatography A*, **1975**, 114, 258–260.
20. Albuquerque, B. R.; Prieto, M. A.; Barreiro, M. F.; Rodrigues, A.; Curran, T. P.; Barros, L.; Ferreira, I. C. F. R. Catechin-based extract optimization obtained from *Arbutus unedo* L. fruits using maceration/microwave/ultrasound extraction techniques. *Industrial Crops and Products*, **2017**, 95, 404–15.
21. Vinatoru, M.; Mason, T. J.; Calinescu, I. Ultrasonically assisted extraction (UAE) and microwave assisted extraction (MAE) of functional compounds from plant materials. *TrAC Trends in Analytical Chemistry*, **2017**, 97, 159–178.
22. Chemat, F.; Rombaut, N.; Sicaire, A. G.; Meullemiestre, A.; Fabiano-Tixier, A. S.; Abert-Vian, M. Ultrasound assisted extraction of food and natural products. Mechanisms, techniques, combinations, protocols and applications. A review. *Ultrasonics Sonochemistry*, **2017**, 34, 540–560.
23. Pandey, A.; Tripathi, S. Concept of standardization, extraction, and pre-phytochemical screening strategies for herbal drug. *Journal of Pharmacognosy and Phytochemistry*, **2014**, 2, 115–119.
24. Bhan, M. Ionic liquids as green solvents in herbal extraction. *International Journal of Advanced Research and Development*, **2017**, 2, 10–12.
25. Sasidharan, S.; Chen, Y.; Saravanan, D.; Sundram, K. M., Yoga Latha, L. Extraction, isolation and characterization of bioactive compounds from plants' extracts. *African Journal of Traditional, Complementary and Alternative Medicines*, **2011**, 8, 1–10.

26. Altemimi, A.; Lakhssassi, N.; Baharlouei, A.; Watson, D. G.; Lightfoot, D. A. Phytochemicals: Extraction, isolation, and identification of bioactive compounds from plant extracts. *Plants*, **2017**, 6, 42.

27. Qing-Wen, Z., Li-Gen, L., Wen-Cai, Y. Techniques for extraction and isolation of natural products: A comprehensive review. *Chinese Medical*, **2018**, 1–26.

28. Yadav, N.; Yadav, R.; Goyal, A. Chemistry of terpenoids. *International Journal of Pharmaceutical Sciences Review and Research*, **2014**, 27(2), 272–278.

29. Ozek, T.; Demirci, F. Isolation of natural products by preparative gas chromatography. *Methods in Molecular Biology*, **2012**, 864, 275–300.

30. Sciarrone, D.; Panto, S.; Donato, P.; Mondello, L. Improving the productivity of a multidimensional chromatographic preparative system by collecting pure chemicals after each of three chromatographic dimensions. *Journal of Chromatography A*, **2016**, 1475, 80–85.

31. Heftmann, F. *Chromatography: Fundamentals and Application of Chromatographic and Electrophoretic Techniques*, 5th Edition. Amsterdam: Elsevier, 1992, 281–285.

32. Banu, K. S.; Cathrine, L. General techniques involved in phytochemical analysis. *International Journal of Advanced Research in Chemical Science*, **2015**, 2, 25–32.

33. de las Heras, B.; Rodríguez, B.; Bosca, L.; Villar, A.M. Terpenoids: Sources, structure elucidation and therapeutic potential in inflammation. *Current Topics in Medicinal Chemistry*, **2003**, 3(2), 171–185.

34. Hanson, J. R. The development of strategies for terpenoid structure determination. *Natural Product Reports*, **2001**, 18, 607–617.

35. Voelter, W. Some recent aspects in the structure elucidation of natural product. *Pure & Applied Chemistry*, **1976**, 48, 105–126.

36. Manasa, G. Structural elucidation of Terpenoids by spectroscopic techniques. *Journal of Chemical and Pharmaceutical Sciences*, **2014**, 5, 36–37.

37. Breitmaier, E. *Isolation and Structure Elucidation. Terpenes*, Wiley-VCH Verlag GmbH & Co. **2006**, 160–179.

38. Abubakar, A. R.; Haque, M. Preparation of medicinal plants: Basic extraction and fractionation procedures for experimental purposes. *Journal of Pharmacy and Bioallied Sciences*, **2020**, 12, 1–10.

39. Harborne, J. B. *The Terpenoids: Phytochemical Methods*. Dordrecht: Springer, **1984**, 100–141.

40. Croteau, R.; Ronald, R. C. Terpenoids. In: *Chromatography. Fundamentals and Applications: Part E*, Amsterdam: Elsevier, **1983**, 147–190.

41. Coates, J. Interpretation of infrared spectra: A practical approach. *Encyclopedia of Analytical Chemistry*, **2000**, 10815–10837.

42. Hanson, J. R. The development of strategies for terpenoid structure determination. *Natural Product Reports*, **2001**, 18, 607–617.

43. Tilden, W. A. XLVI–On the decomposition of terpenes by heat. *Journal of the Chemical Society, Transactions*, **1884**, 45, 410.

44. Jiang, Z.; Kempinski, C.; Chappell, J. Extraction and analysis of terpenes/terpenoids. *Current Protocols in Plant Biology*, **2016**, 1, 345–358.

45. Bowman, J. M.; Samuel Braxton, M.; Churchill, M. A.; Hellie, J. D.; Starrett, S. J.; Causby, G. Y.; Ellis, D. J.; Ensley, S. D.; Maness, C. D. M.; Joshua, R. S.; Yang Hua, R. S. W.; Butcher, D. J. Extraction method for the isolation of terpenes from plant tissue and subsequent determination by gas chromatography. *Microchemical Journal*, **1997**, 56(1), 10–18.

46. Mahrath, A. J.; Hassan, G. S.; Obaid, E. K. Isolation and characterization of terpenoid derivatives from medicinal plant roots by thin layer and flash column chromatography (TLC & FCC) Techniques. *International Journal of Chemical Sciences*, **2015**, 13(2), 983–989.

47. Hanson, J. R. The development of strategies for terpenoid structure determination. *Natural Product Reports*, **2001**, 18, 607–617.

48. Singh, K. S; Majik, M. S.; Tilvi, S. Chapter-6 – Vibrational spectroscopy for structural characterization of bioactive compounds, *Comprehensive Analytical Chemistry*, **2014**, 65, 115–148.

49. Messerschmidt, R.G.; Harthcock, M.A. *Infrared Microscopy Theory and Applications*, Mercel Dekker, New York, NY, **1988**.

50. Kasai, T.; Furukawa, K.; Sakamura, S. Infrared and mass spectra of amino acid amides. *Journal of the Faculty of Agriculture of the Hokaido University*, **1979**, 59, 279.

51. Voelter, W. Some recent aspects in the structure elucidation of natural product. *Pure & Applied Chemistry*, **1976**, 48, 105–126.

52. Yadava, N.; Yadava, R.; Goyalb, A. Chemistry of terpenoids. *International Journal of Pharmaceutical Sciences Review and Research*, **2014**, 27(2), 272–278.

53. Christianson, D. W. Structural and chemical biology of terpenoid cyclases. *Chemical Reviews*, **2017**, 117, 11570–11648.

54. Jiang, Z.; Kempinski, C.; Chappell, J. Extraction and analysis of terpenes/terpenoids. *Current Protocols in Plant Biology*, **2016**, 1, 345–358.

55. Serra, S. Chapter 7: Recent developments in the synthesis of the flavors and fragrances of terpenoid origin. *Studies in Natural Products Chemistry*, **2015**, 46, 201–226.

56. Chaplin, J. A.; Gardiner, N.; Mitra, R. K.; Parkinson, C. J.; Portwig, M.; Mboniswa, B.A.; Dickson, M.D.E.; Brady, D.; Marais, S.F.; Reddy, S. Process for preparing (-)-menthol and similar compounds, WO 02/36795 A2, **2002**.

57. Yamamoto, T. Method for purifying (−)-n-isopulegol and citrus perfume compositions containing purified (−)-n-isopulegol obtained by the method, US 5,773,410, **1998**.

58. Dewick, P. M. *Medicinal Natural Products: A Biosynthetic Approach*, 2nd Edition. John Wiley and Sons, Ltd., **2002**, 507.

2

Synthesis and Pharmacological Activities of Monoterpenoids

Biswa Mohan Sahoo[*], Bimal Krishna Banik[†], and Abhishek Tiwari[‡]

CONTENTS

2.1 Introduction 30
    2.1.1 Factors Affecting Production of Monoterpenoids 31
2.2 Synthesis of Monoterpenoids 31
    2.2.1 Synthesis of Acyclic Monoterpenoids 31
        2.2.1.1 Synthesis of Myrcene 31
        2.2.1.2 Synthesis of Citral 31
        2.2.1.3 Synthesis of Geraniol 33
        2.2.1.4 Synthesis of Citronellol 33
        2.2.1.5 Synthesis of Citronellal 33
        2.2.1.6 Synthesis of Linalool 33
        2.2.1.7 Synthesis of Ocimenes 33
    2.2.2 Synthesis of Monocyclic Monoterpenoids 35
        2.2.2.1 Synthesis of Menthol 35
        2.2.2.2 Synthesis of (−)-Isopulegol 36
        2.2.2.3 Synthesis of p-Cymene 37
        2.2.2.4 Synthesis of Thymol 40
        2.2.2.5 Synthesis of Carvacrol 40
        2.2.2.6 Synthesis of Eucalyptol 41
        2.2.2.7 Synthesis of Piperitone 41
        2.2.2.8 Synthesis of Carvone 43
        2.2.2.9 Synthesis of α-Terpineol 43
        2.2.2.10 Synthesis of Pulegone and Isopulegone 44
        2.2.2.11 Synthesis of Limonene 44
        2.2.2.12 Synthesis of Perillyl Alcohol 44
        2.2.2.13 Synthesis of Phellandrene 44
        2.2.2.14 Synthesis of Carveol 45
        2.2.2.15 Synthesis of Catalpol 45
        2.2.2.16 Synthesis of Thymoquinone 46
        2.2.2.17 Synthesis of Eugenol 47
        2.2.2.18 Synthesis of Sobrerol 48
        2.2.2.19 Synthesis of Aucubin 49
    2.2.3 Synthesis of Bicyclic Monoterpenoids 49
        2.2.3.1 Synthesis of Ascaridole 49
        2.2.3.2 Synthesis of Pinene 50
        2.2.3.3 Synthesis of Camphor 50
        2.2.3.4 Synthesis of Borneol 50
        2.2.3.5 Synthesis of Sabinene Hydrate 51
        2.2.3.6 Synthesis of Fenchone 52
        2.2.3.7 Synthesis of Verbenol 52

[*] Roland Institute of Pharmaceutical Sciences, Berhampur affiliated to Biju Patnaik University of Technology (BPUT), Rourkela, Odisha, India. drbiswamohansahoo@gmail.com.

[†] Department of Mathematics and Natural Sciences, College of Sciences and Human Studies, Prince Mohammad Bin Fahd University, Al Khobar, Kingdom of Saudi Arabia. bimalbanik10@gmail.com and bbanik@pmu.edu.sa

[‡] Pharmacy Academy, IFTM University, Lodhipur Rajput, Moradabad, U.P. 244102.

DOI: 10.1201/9781003008682-2

2.3 Pharmacological Activities of Monoterpenoids.. 52
    2.3.1   Antimicrobial Activity 55
    2.3.2   Antidepressant Activity... 58
    2.3.3   Antiviral Activity 59
    2.3.4   Analgesic and Anti-Inflammatory Activity 59
    2.3.5   Antioxidant Activity.. 62
    2.3.6   Anticancer Activity 63
    2.3.7   Anti-Proliferative Activity... 65
    2.3.8   Anticonvulsant Activity... 66
    2.3.9   Sedative and Hypnotics 69
    2.3.10 Treatment of Neurodegenerative Disorders 69
    2.3.11 Cardiovascular Effects 70
    2.3.12 Herbicidal Activity 72
    2.3.13 Antidiabetic Agents... 73
    2.3.14 Miscellaneous.. 74
2.4 Conclusion... 81
Abbreviation... 81
References.. 81

2.1 Introduction

Monoterpenes are major components of essential oils which belong to the group of isoprenoids containing ten carbon atoms (two isoprene units). These are categorized into two subgroups: acyclic and cyclic (monocyclic and bicyclic). Among each group of monoterpenoids, these are structurally unsaturated hydrocarbons with various functional groups such as alcohol, aldehyde and ketone [1].

Acyclic monoterpenoids are comparatively few and possess the carbon skeleton of 2,6-dimethyloctane (Figure 2.1). The commonly available acyclic monoterpenes include myrcene, citral, ocimrne, citronellal, citronellol, nerol, geraniol, linalool, etc. Acyclic monoterpenoids are formed by the linkage of two isoprene units. Out of four double bonds present in two isoprene units, one double bond is consumed and the remaining three double bonds are present in the structure of acyclic monoterpenoids.

The acyclic monoterpenoids are easily cyclized to produce monocyclic monoterpenoids by utilizing one more double bond in the linkage. So, acyclic monoterpenoids possess three double bonds, and monocyclic monoterpenoids have two double bonds in their structure. Similarly, bicyclic and tricyclic monoterpenoids possess one and zero double bonds, respectively (Table 2.1).

Various monocyclic monoterpenoids are α-terpineol, pulegone, piperitone, limonene, thymol, menthol, carvone, ionone and eucalyptol, etc. The bicyclic monoterpenoids may be categorized into three types based on the size of the second ring. In each class of bicyclic monoterpenoids, the first ring is a six-membered while the second ring can be either a three, four, or five-membered ring. Thujane and carene are examples of the group containing (6+3)-membered rings, α- and β-pinene represent the (6+4) group where as borneol and camphor are considered under (6+5) membered rings system [2].

There is progress in the development of analytical instruments to analyze the chemical structure of various components of terpenoids. A large number of lipophilic or hydrophobic monoterpenoids are identified and isolated by solvent extraction of essential oils. In general, monoterpenoids exhibit characteristic odor and taste depending on the presence of different types of chemical constituent. Hence, these are used as cosmetic materials, food additives, insecticides, insect-repellent and medicinal agents. Various chemical reactions, synthetic procedures and microbial transformation are carried out to produce diverse functionalized compounds of monoterpenoids [3]. The successful production of monoterpenoids by microbial transformation is not only dependent on sufficient expression levels of soluble and active recombinant enzymes but also on some other factors. Various factors play a key role during the production of monoterpenoids as outlined below [4].

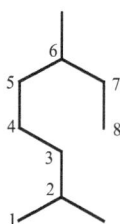

FIGURE 2.1 Skeletal structure of acyclic monoterpenoid.

TABLE 2.1

Types of Monoterpenoids with General Formula

Types of Monoterpenoids	General Formula	Number of Double Bonds
Acyclic	C_nH_{2n+2}	3
Monocyclic	C_nH_{2n}	2
Bicyclic	C_nH_{2n-2}	1
Tricyclic	C_nH_{2n-4}	0

On the other hand, larval insects and mammals are used for direct biotransformation of monoterpenoids to evaluate their fate, safety or toxicity in the organisms.

2.1.1 Factors Affecting Production of Monoterpenoids

i. Substrates of monoterpenoids
ii. Localization of the enzymes with the substrate(s) of secondary metabolite production
iii. Over expression of the recombinant enzyme
iv. Competing pathways depleting monoterpenoid precursors or catalyzing side reactions.
v. Sufficient cofactor or coenzyme for both primary and secondary metabolism.

2.2 Synthesis of Monoterpenoids

The study of biological reactions is of great importance for the development and utilization of biocatalysts as greener alternatives to standard chemical syntheses. Biocatalytic reactions replace or couple with organic syntheses to produce diverse terpenoid molecules. The extent of chemical reactivity plays a major role for the production of chemicals and pharmaceuticals with minimal impact on the environment. Synthetic biology is able to produce non-natural compounds by co-expression of enzymes.

Further, attempts were made to improve the diversity of monoterpenes via incorporation of non-natural tailoring enzymes into pathways (e.g. cytochrome P450s or glycosyltransferases). The inclusion of such "non-natural" enzyme helps to develop new chemical space successfully. For example, the incorporation of a *Mycobacterium sp.* cytochrome P450 into an engineered *E. coli* limonene producer resulted in the formation of perillyl alcohol. The ability to alter the substrate specificity of monoterpene synthases in such a way that new set of small, structurally diverse natural product libraries may be achieved. There is significant diversity of enzymes, which are responsible for the biosynthesis of terpenoids [5,6].

Hence, terpenoid synthases transform linear poly-isoprenoid pyrophosphates into thousands of structurally and stereochemically diverse products. Industrially, the process for L-menthol synthesis was developed and introduced which uses the renewable monoterpene product myrcene obtained from gum rosin (pine sap) as the starting material. The key step of this synthesis involves asymmetric isomerization of geranyldiethylamine catalyzed by the chiral catalyst complex (S)-BINAP-Rh. This breakthrough in industrial asymmetric catalytic manufacturing was the result of collaboration between the Japanese chemical company Takasago and several prominent chemists including Dr. Ryoji Noyori, who received the Nobel Prize in Chemistry for his contributions to chirally catalyzed hydrogenation reactions in 2001 [7]. It is possible to produce either enantiomer of menthol by changing the configuration of the BINAP-Rh catalyst. Microbial synthesis of myrcene using engineered *E. coli* expressing the monoterpene synthase i.e. myrcene synthase is currently under investigation as an alternative green biosynthetic approach. Limonene and perillyl alcohol have been successfully produced in *E. coli* by using a heterologous MVA pathway (Figure 2.2, Table 2.2) together with the limonene synthase and a cytochrome P450 gene, which specifically hydroxylates limonene to yield perillyl alcohol [8,9].

2.2.1 Synthesis of Acyclic Monoterpenoids

Geranyl pyrophosphate is the precursor for monoterpenes and biosynthesis is mediated by terpene synthase. The elimination of the pyrophosphate group from geranyl pyrophosphate leads to the formation of acyclic monoterpenes such as ocimene and myrcenes, whereas the hydrolysis of the phosphate groups leads to the prototypical acyclic monoterpenoid like geraniol. Similarly, the additional rearrangements and oxidation reactions generate acyclic monoterpenoids such as citral, citronellal, citronellol, linalool, etc [10].

2.2.1.1 Synthesis of Myrcene

Myrcene is an acyclic monoterpenoid with the chemical formula $C_{10}H_{16}$. It was first isolated from Bay oil and turpentine oil in 1895. Its IUPAC name is 7-Methyl-3-methylene-octa-1,6-diene. It is optically active and present in two forms such as α and β-myrcene.

Normant et al. reported the synthesis of myrcene in the year 1955. First of all, Linalool is produced by the reaction between methylheptenone and vinyl magnesium bromide (CH_2=CH–MgBr). Then, Linalool undergoes dehydration in presence of I_2 to produce myrcene as presented in Scheme 2.1.

Commercially, myrcene is produced by the pyrolysis of β-pinene at 400°C as depicted in Scheme 2.2 [11].

2.2.1.2 Synthesis of Citral

Citral is an acyclic monoterpenoid with the chemical formula $C_{10}H_{16}O$. Chemically, it is 3,7-dimethyl-2,6-octadienal. It is the major constituent of citrus fruit's peel oil. It is isolated as its crystalline bisulphate which on hydrolysis produces citral. Citral possesses two geometrical isomers such as E (trans) and Z-isomer (cis). E-isomer is known as geranial or citral-a while Z-isomer is known as neral or citral-b [12]. Both isomers differ in the arrangement of aldehyde (-CHO) group about the double bond present between C_2 and C_3 position (Figure 2.3).

Barbier et al. reported the synthesis of citral from methylheptenone in the year 1896. First of all, methylheptenone is converted into geranic ester in presence of zinc and ethyl iodoacetate via Reformatsky reaction [13]. After that, Tiemann et al. reported the conversion of geranic ester into cital by hydrolysis followed by distillation in the presence of calcium formate during the year 1898 (Scheme 2.3).

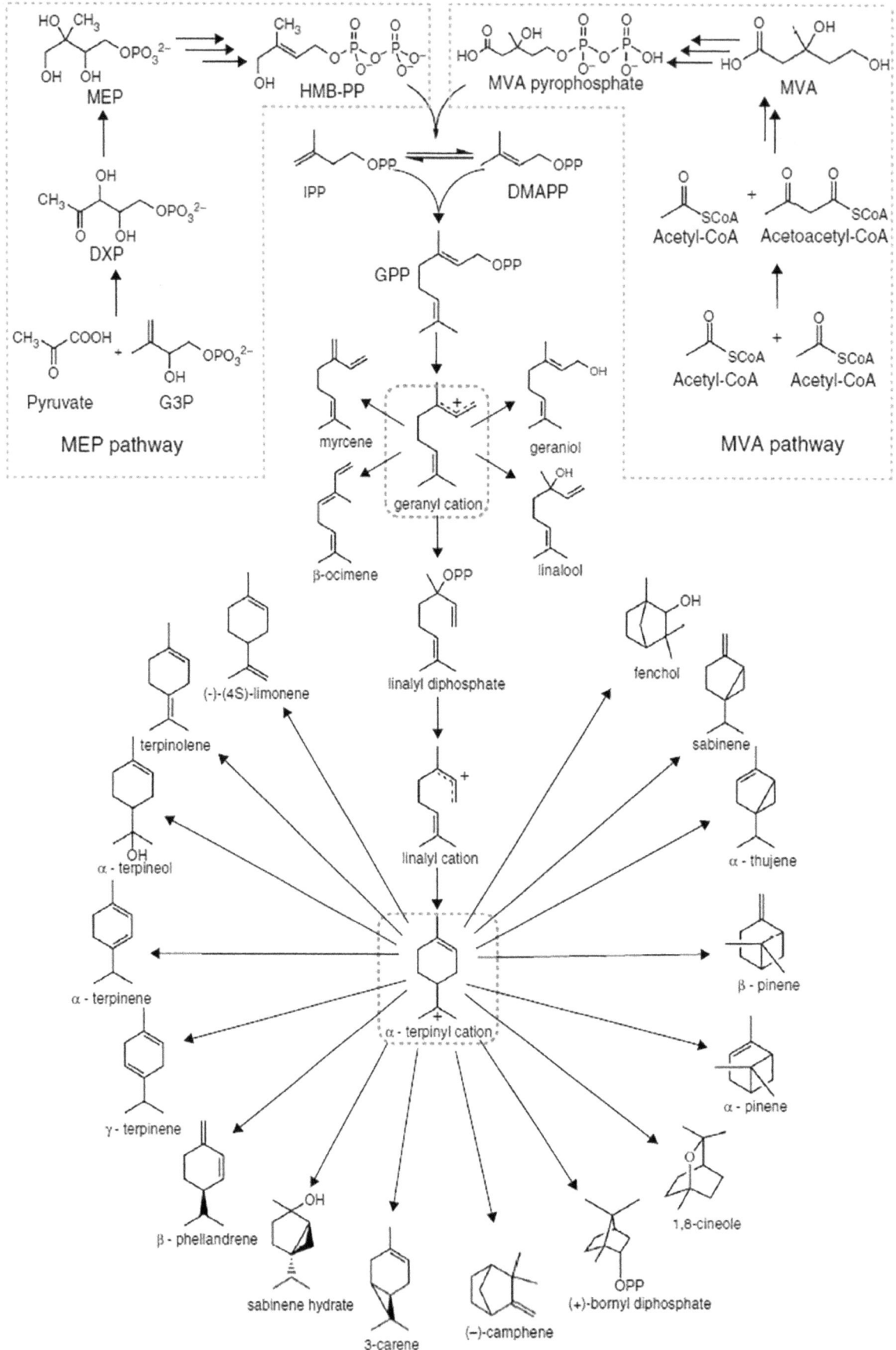

FIGURE 2.2 Production pathways of monoterpenoids.

TABLE 2.2

List of Precursor Pathways and Monoterpenoid Production Strains

Name of Monoterpenoids	Host	Precursor Pathways	mTS/C	mTS/C Source
Limonene	*E. coli*	MEP, GPPS	Limonene synthase	*Mentha spicata*
α-Pinene	*E. coli*	MVA, GPPS2	Pinene synthase	*Abes grandis*
β-Pinene	*E. coli*	MVA, GPPS2	Pinene synthase	*Abies grandis*
Myrcene	*E. coli*	MVA, GPPS	Myrcene synthase	*Quercus ilex*
Geraniol	*E. coli*	MVA, GPPS	Geraniol synthase	*Ocimum basilicum*
Linalool	*S. cerevisiae*	MVA	Linalool synthase	*Lavandula angustifolia*
Cineole	*S. cerevisiae*	MVA, ERG20	Cineole synthase	*Salvia fruticosa*
3-Carene	*E. coli*	MEP, GPPS	3-Carene cyclase	*Picea abies*

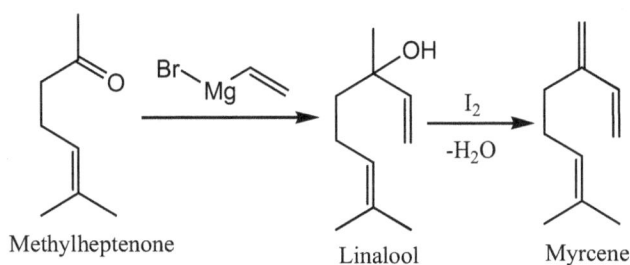

SCHEME 2.1 Synthesis of myrcene from methylheptenone.

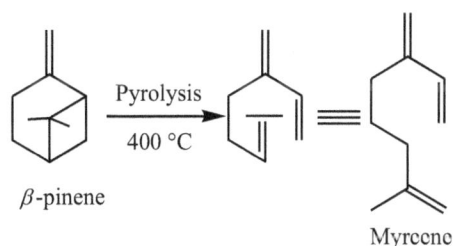

SCHEME 2.2 Synthesis of myrcene from β-pinene.

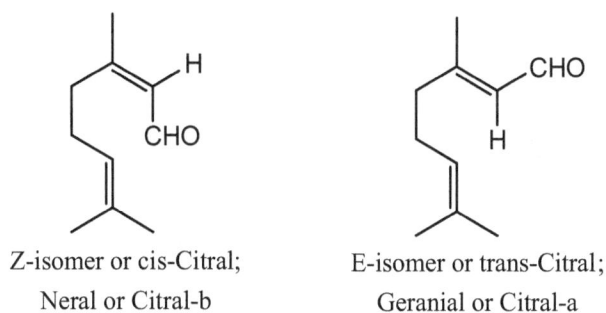

Z-isomer or cis-Citral;
Neral or Citral-b

E-isomer or trans-Citral;
Geranial or Citral-a

FIGURE 2.3 Structure of geometrical isomers of citral.

2.2.1.3 Synthesis of Geraniol

Geraniol is acyclic monoterpenoid alcohol with the molecular formula $C_{10}H_{18}O$. It is the major constituent of rose oil, palmarosa oil and citronella oil. Its IUPAC name is (2E)-3,7-dimethyl-2,6-octadien-1-ol. Brack et al. reported the synthesis of Geraniol from 4-bromo-2-methyl-but-2-ene via sigmatropic rearrangements such as Claisen rearrangement and Cope rearrangement in the year 1951 [14] (Scheme 2.4).

2.2.1.4 Synthesis of Citronellol

Citronellol is a natural acyclic monoterpenoid alcohol. It is present as two enantiomeric forms in nature such as (+)- and (−)-Citronellol. (+)-Citronellol is present in citronella oils such as *Cymbopogon nardus* (50%) whereas (−)-Citronellol is found in rose oils (18%–55%) and *Pelargonium geraniums*. Citronellol is produced by catalytic hydrogenation of Citral followed by reduction in presence of sodium-amalgm [15] (Scheme 2.5).

2.2.1.5 Synthesis of Citronellal

Citronellal is natural acyclic monoterpenoid aldehydes with the molecular formula $C_{10}H_{18}O$. It is optically active. It is isolated from the plant *Cymbopogon nardus*. Citronellal can be produced by catalytic hydrogenation of Neral or Citral-b (Scheme 2.6) [16].

2.2.1.6 Synthesis of Linalool

Linalool is an acyclic monoterpenoid with the molecular formula $C_{10}H_{18}O$. It is having octa-1,6-diene substituted by methyl groups at positions 3 and 7 and a hydroxy group at position 3 in its structure. Linalool is the isomer of geraniol and nerol but differs from them in that it is tertiary alcohol. There are two stereoisomers of *Linalool* such as (R)-(−)-linalool and (S)-(+)-linalool due to the presence of stereogenic center at C_3 of Linalool structure. (R)-(−)-linalool is known as licareol while (S)-(+)-linalool is called as coriandrol. Linalool is produced from isopentenyl pyrophosphate via the universal isoprenoid intermediate geranyl pyrophosphate, through a class of membrane-bound enzymes called monoterpene synthases. Ruchika reported the synthesis of Linalool from methylheptenone in the year 1919 as per Scheme 2.7 [17].

Similarly, Normant et al. reported the synthesis of Linalool by reaction between methylheptenone and vinyl magnesium bromide (CH₂=CH-MgBr) in the year 1955 (Scheme 2.8) [18].

2.2.1.7 Synthesis of Ocimenes

Ocimenes are the acyclic monoterpenes with the molecular formula $C_{10}H_{16}$. These are isolated from the leaves of *Ocimum basilicum* of family Labiatae. There are two isomers of ocimene such as α-ocimene and β-ocimene. β-ocimene is showing geometrical isomers including cis and trans β-ocimene. Ocimenes

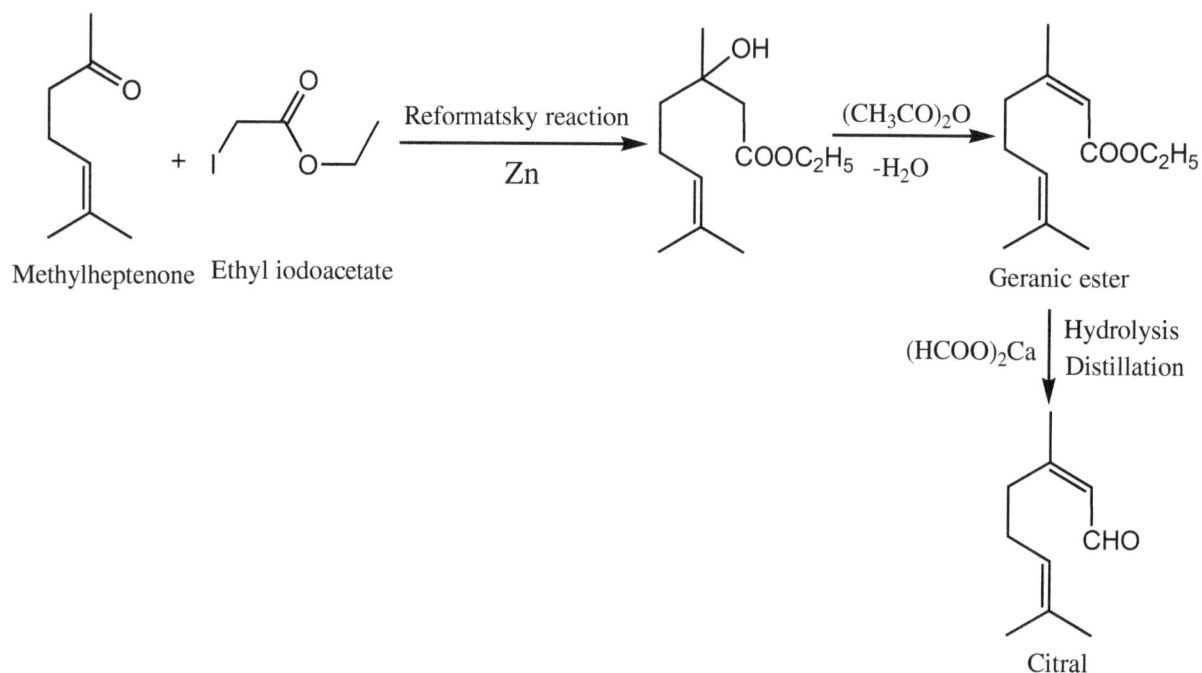

SCHEME 2.3 Synthesis of citral from methylheptenone.

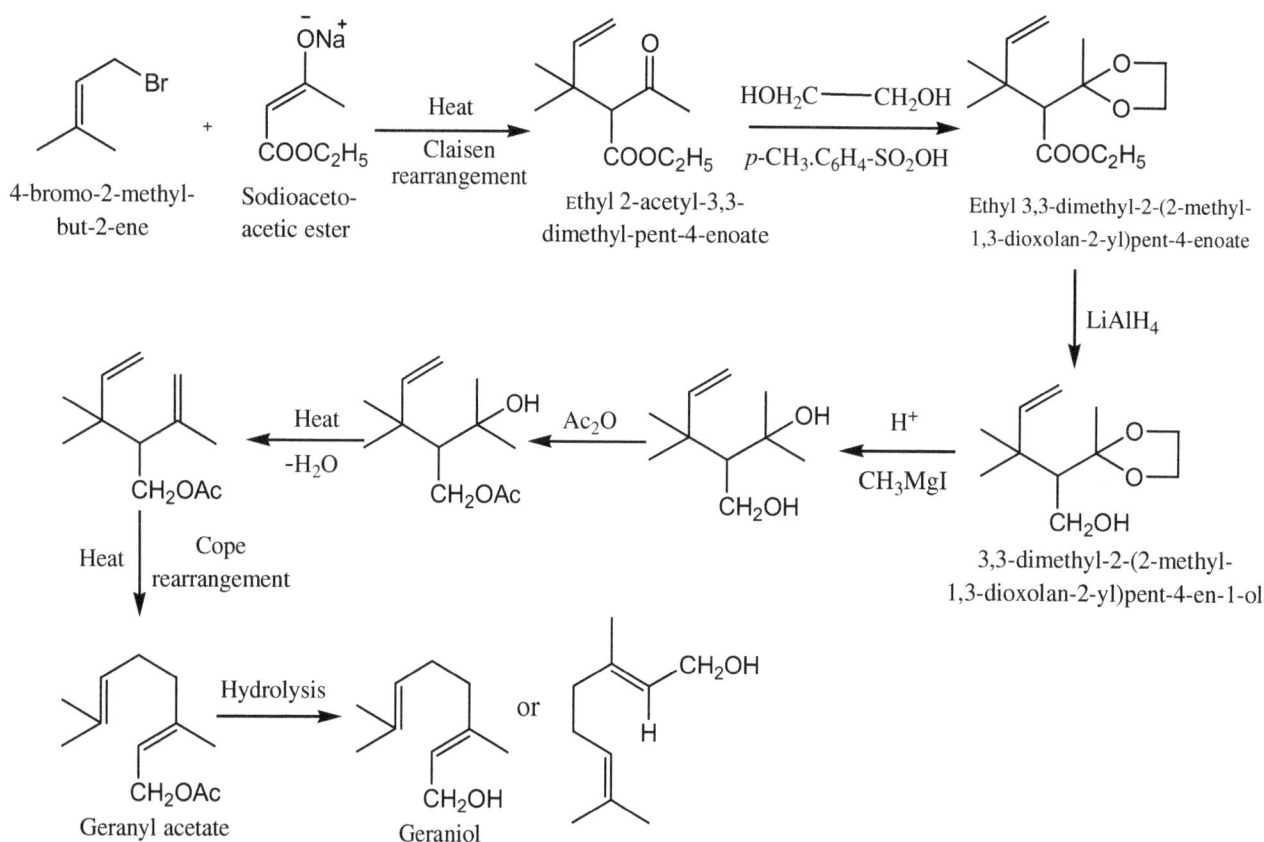

SCHEME 2.4 Synthesis of Geraniol from 4-bromo-2-methyl-but-2-ene.

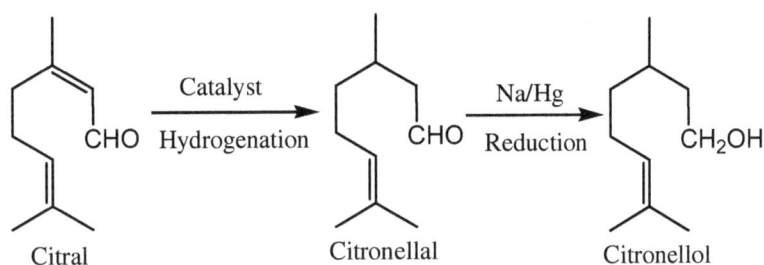

SCHEME 2.5 Synthesis of citronellol from citral.

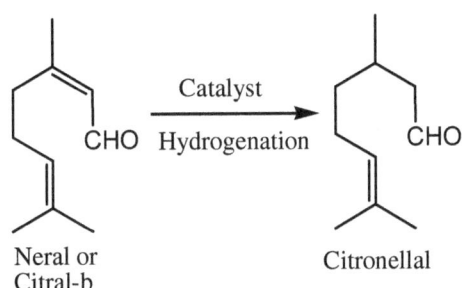

SCHEME 2.6 Synthesis of citronellol from neral or citral-b.

are isomer of myrcene and both terpenes differ only in the position of double bond. Ocimenes are produced by pyrolysis of α-pinene. Pyrolysis is carried out at elevated temperatures in absence of oxygen. Both *α*- and *β*-ocimene are selectively generated by dehydration of geraniol in a weak acidic medium (Scheme 2.9).

Both *α*- and *β*-ocimene are also selectively generated by dehydration of geraniol in a weak acidic medium (Scheme 2.10) [19].

2.2.2 Synthesis of Monocyclic Monoterpenoids

2.2.2.1 Synthesis of Menthol

Menthol is cyclic monoterpene alcohol which is present as a major constituent in the essential oils of *Mentha piperita*. Menthol is chemically known as 5-methyl-2-(1-methylethyl)cyclohexanol or 2-isopropyl-5-methylcyclohexanol or

p-methan-3-ol with the molecular formula $C_{10}H_{20}O$ and mol. wt. 156.27. Menthol exists in eight stereoisomeric forms or four enatiomeric pairs such as (±)-menthol, (±)-isomenthol, (±)-neomenthol and (±)-neoisomenthol. This is due to presence of three dissimilar chiral carbons at C_1, C_3 and C_4. Among these optical isomers, (−)-menthol or (L-menthol) is available naturally (80%) with the configuration; 1R, 3R, 4S [20]. The structure of various enantiomeric pairs of menthol may be written as follows (Figure 2.4).

Croteau et al. described the biosynthesis of menthol in plants which involve eight steps pathway from primary metabolism (Figure 2.5). They explained the anatomic structures where synthesis takes place as well as the actions of enzymes involved during biosynthesis. The universal monoterpene precursor geranyl diphosphate (GDP) is formed by the condensation of isopentenyl diphosphate (IPP) and dimethylallyl pyrophosphate (DMAPP) which in turn is converted to (−)-limonene through cyclization. (−)-Limonene is converted to (−)-trans-isopiperitenol through NADPH and oxygen-dependent hydroxylation. (−)-Isopiperitenone is formed by allylic oxidation followed by NADPII-dependent reduction to form (+)-cis-isopulegone (Figure 2.6, Table 2.3). Isomerization of (+)-cis-isopulegone leads to the formation of (+)-pulegone which is the precursor of (+)-menthofuran, (−)-menthone and (+)-isomenthone. Reduction of these ketones yield (−)-menthol, (+)-neomenthol, (+)-isomenthol and (+)-neoisomenthol [21].

Kotz and Schwarz reported the synthesis of (±)-menthol from m-cresol in the year 1907. The (±)-menthol is resolved by means of phthalic acid (Scheme 2.11) [22].

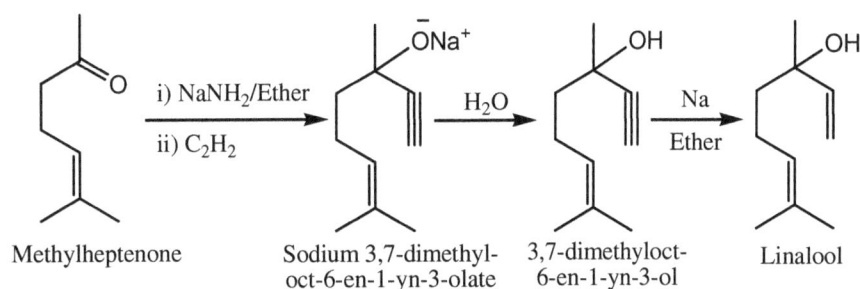

SCHEME 2.7 Synthesis of linalool from methylheptenone.

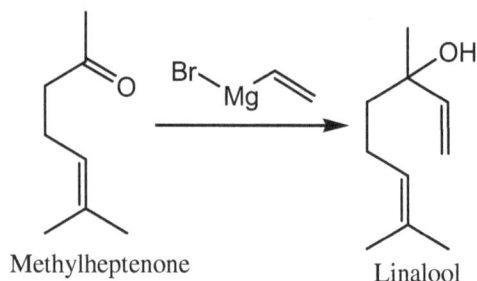

SCHEME 2.8 Synthesis of linalool.

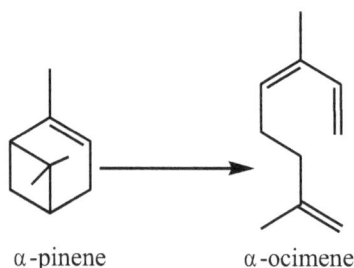

SCHEME 2.9 Synthesis of ocimene from α-pinene.

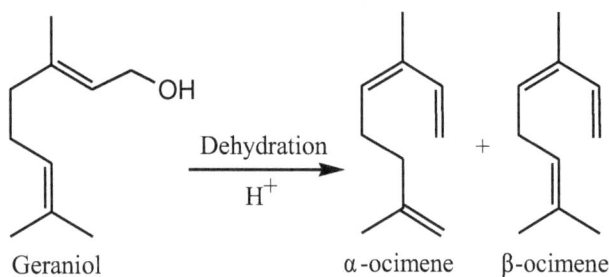

SCHEME 2.10 Synthesis of ocimene from geraniol.

Menthol is also obtained by the hydrogenation of either pulegone or thymol as presented in the following Scheme 2.12.

Coman et al. reported one-pot synthesis of menthol catalyzed by a highly diastereoselective Au/MgF$_2$ catalyst [23]. The synthesis of (±)-menthol in a single-step one-pot reaction (route B) becomes more advantageous method in terms of economic and environmental point of view. It involves the isomerization of (+)-citronellal to (−)-isopulegol in presence of ZnBr$_2$ dissolved in organic solvents (CH$_2$Cl$_2$ and benzene) or in an aqueous ZnBr$_2$ solution, followed by its hydrogenation in presence of Raney Ni to produce to menthol (Scheme 2.13, steps 1A and 2A).

Apesteguía et al. developed stable, active and highly selective bifunctional Ni/Al-MCM-41 catalysts for the one-pot synthesis of menthols from citral. Bifunctional catalysts containing Pd or Ni supported on SiO$_2$-Al$_2$O$_3$, Al-MCM-41, or zeolite HBEA were prepared and tested for conversion of citral to menthol. But, Ni (8%)/Al-MCM-41 was considered as the best catalyst which yielded more than 90% menthols at 2026.0 kPa. The one-pot synthesis of menthols from citral is successfully completed by using bifunctional metal/acid catalysts (Scheme 2.14, Table 2.4) with the ability to promote the following reaction pathway selectively [24].

a. Hydrogenation of citral to citronellal
b. Cyclization of citronellal to isopulegol
c. Hydrogenation of isopulegols to menthol

2.2.2.2 Synthesis of (−)-Isopulegol

Isopulegol is a monocyclic monoterpene with the molecular formula C$_{10}$H$_{18}$O. Its IUPAC name is 5-Methyl-2-prop-1-en-2-ylcyclohexan-1-ol. It exists in two isomeric forms such as (+)-Isopulegol and (-)-Isopulegol. Jacob et al. reported a simple green and efficient method for the synthesis of (−)-isopulegol from (+)-citronellal in the presence of a solid

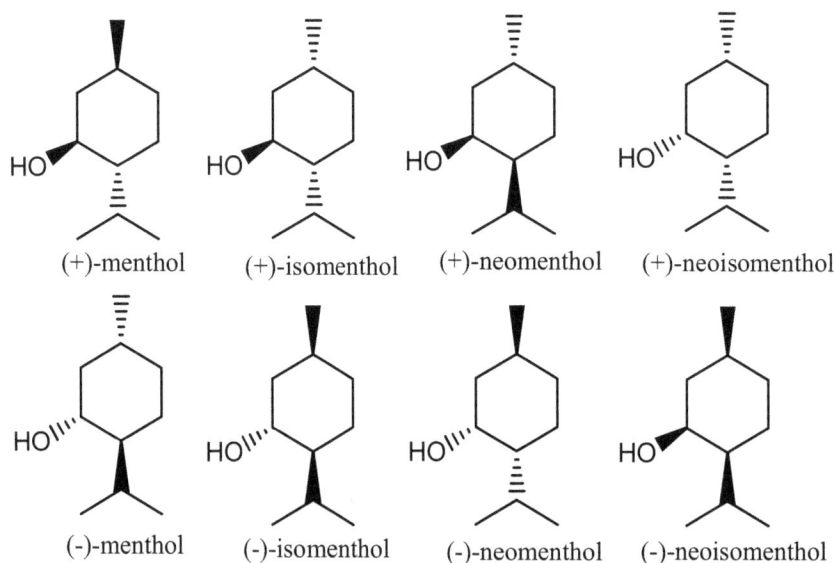

FIGURE 2.4 Structure of various stereoisomers of menthol.

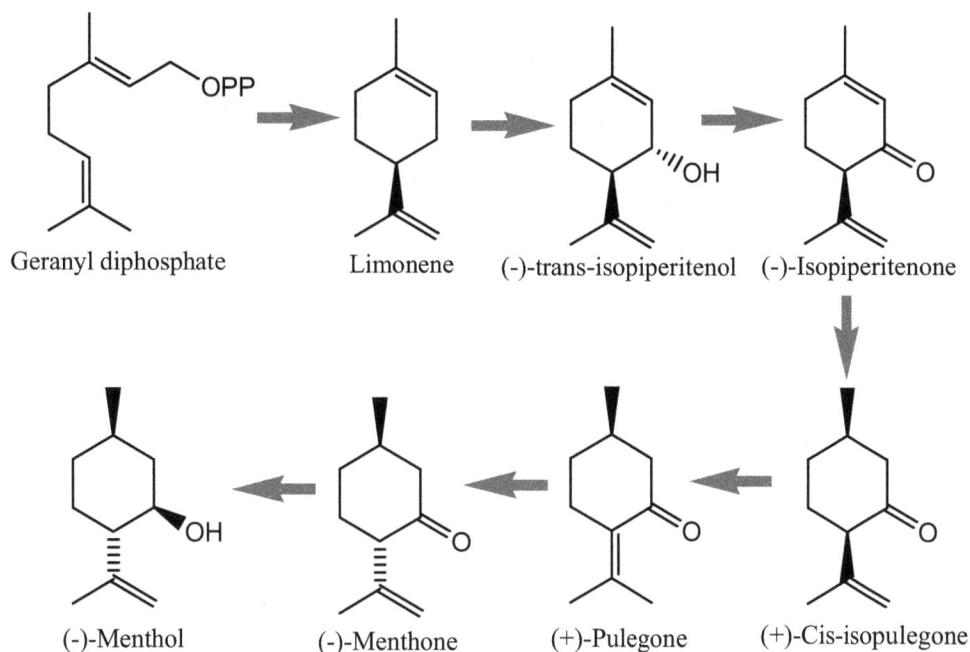

FIGURE 2.5 Synthetic pathway of (–)-menthol biosynthesis from GDP.

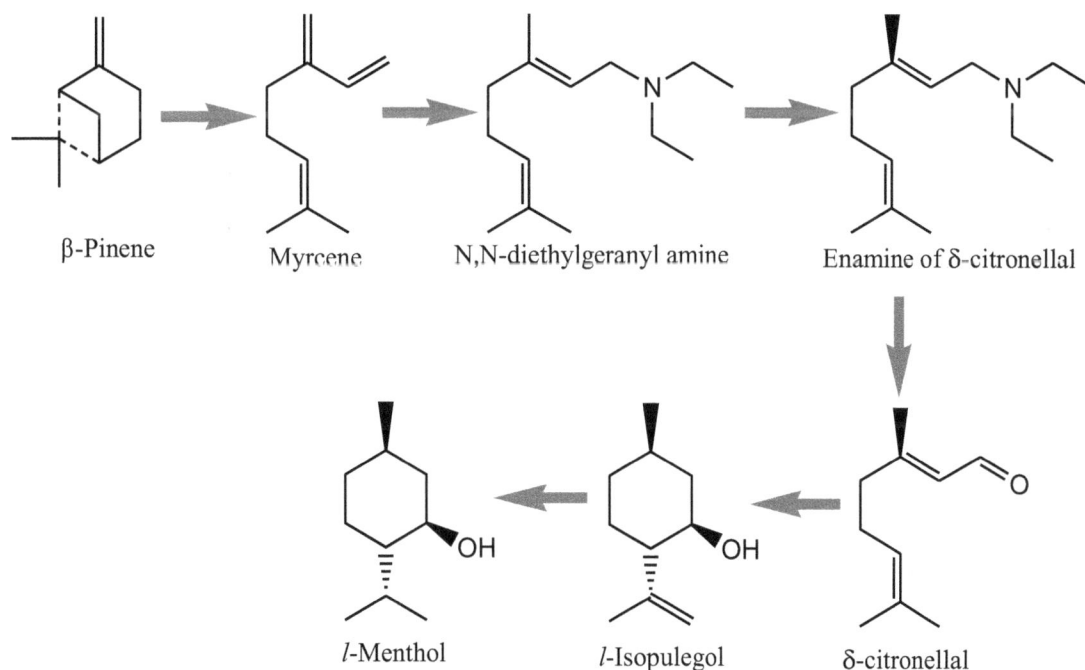

FIGURE 2.6 Synthetic pathway of *l*-menthol from *β*-pinene.

supported catalyst (SiO$_2$/ZnCl$_2$), under solvent-free conditions and MW irradiation (488 W) as depicted in Scheme 2.15 [25].

2.2.2.3 Synthesis of p-Cymene

p-Cymene is a monoterpene in which toluene is substituted by an isopropyl group at position 4. Its IUPAC name is

1-Methyl-4-(propan-2-yl)benzene. It is produced from menthol in two steps such as (1) dehydration of menthol to yield two isomeric forms of *p*-menthene (2) dehydrogenation of *p*-menthene to yield *p*-cymene (**Scheme 2.16**) [26].

Further, *α*-terpineol on heating with concentrated sulphuric acid (H$_2$SO$_4$) produces *p*-cymene. Similarly, camphor on distillation with zinc chloride or phosphrous pentoxide yield *p*-cymene (Scheme 2.17) [27].

TABLE 2.3

Comparative Study on Chemical Synthesis and Biosynthesis of L-Menthol

Chemical Synthesis	Biosynthesis

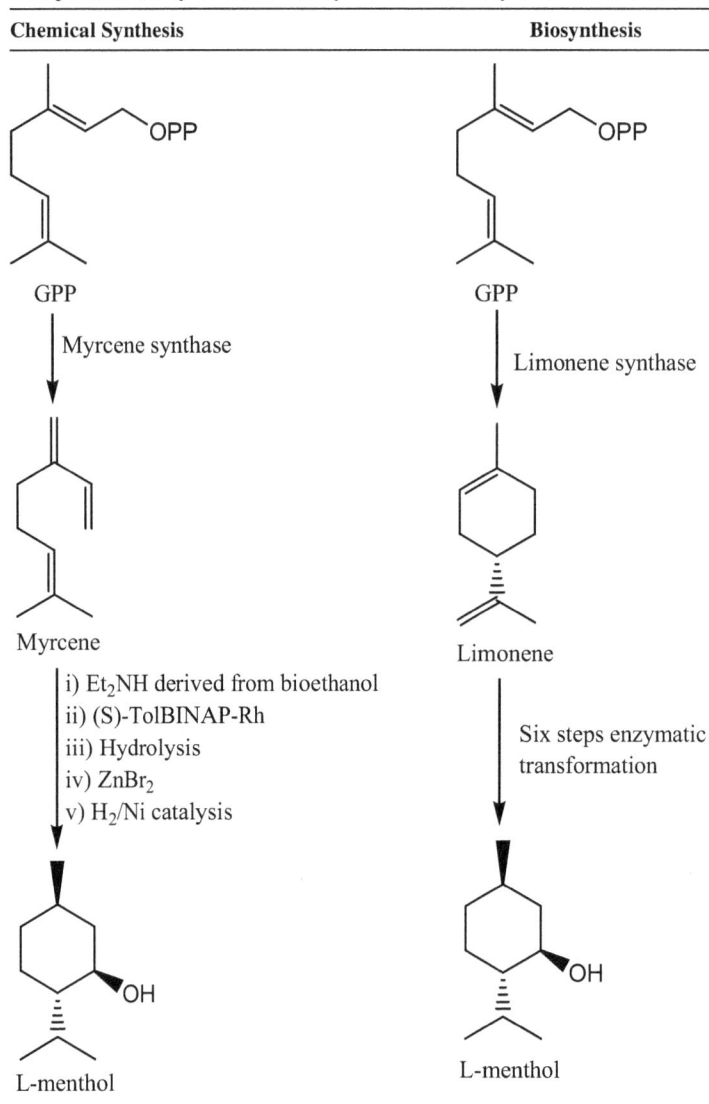

GPP

Myrcene synthase

Myrcene

i) Et$_2$NH derived from bioethanol
ii) (S)-TolBINAP-Rh
iii) Hydrolysis
iv) ZnBr$_2$
v) H$_2$/Ni catalysis

L-menthol

GPP

Limonene synthase

Limonene

Six steps enzymatic transformation

L-menthol

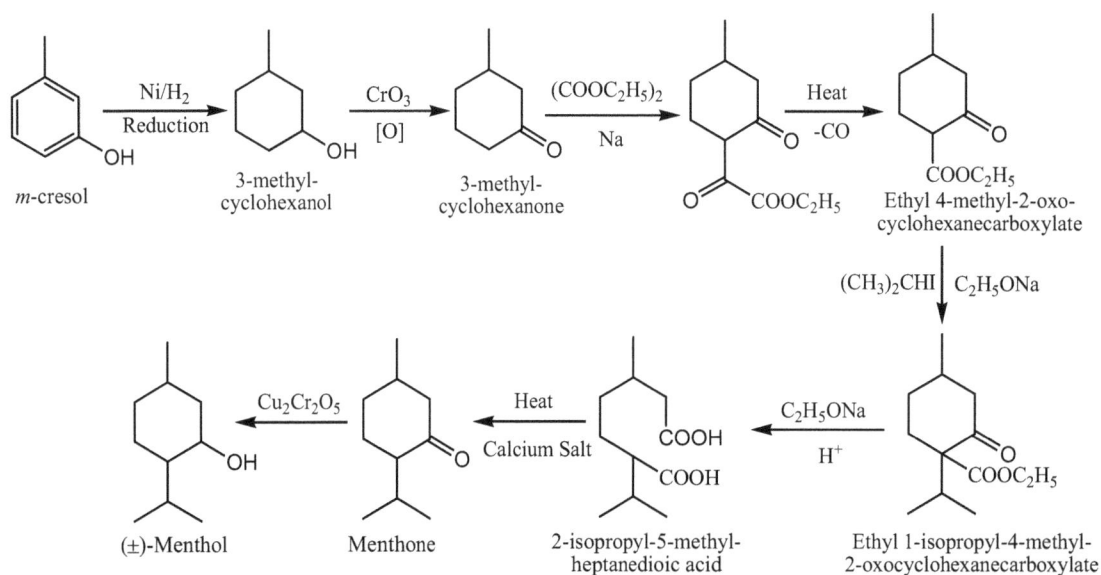

m-cresol

Ni/H$_2$
Reduction

3-methyl-cyclohexanol

CrO$_3$
[O]

3-methyl-cyclohexanone

(COOC$_2$H$_5$)$_2$
Na

Heat
-CO

COOC$_2$H$_5$
Ethyl 4-methyl-2-oxo-cyclohexanecarboxylate

(CH$_3$)$_2$CHI   C$_2$H$_5$ONa

(±)-Menthol

Cu$_2$Cr$_2$O$_5$

Menthone

Heat
Calcium Salt

2-isopropyl-5-methyl-heptanedioic acid

COOH
COOH

C$_2$H$_5$ONa
H$^+$

COOC$_2$H$_5$
Ethyl 1-isopropyl-4-methyl-2-oxocyclohexanecarboxylate

SCHEME 2.11   Synthesis of (±)-menthol from m-cresol.

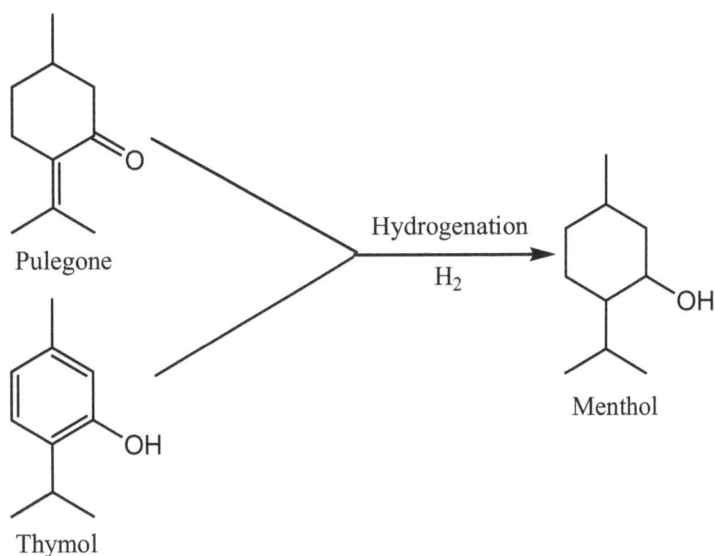

SCHEME 2.12 Synthesis of menthol from pulegone or thymol.

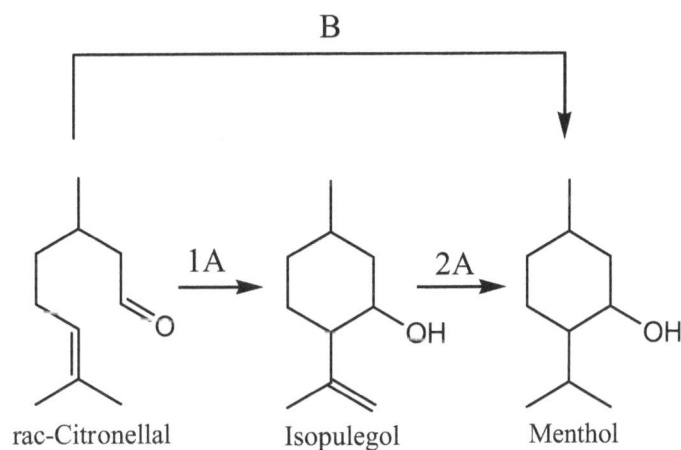

SCHEME 2.13 Synthesis of menthol from citronellal in two steps (route 1A+2A) and in one pot (route B).

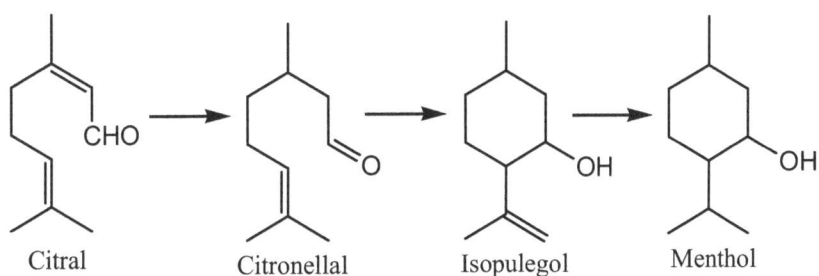

SCHEME 2.14 Reaction sequence for menthol synthesis from citral.

TABLE 2.4

Comparative Study on Reaction Rates, Product Yields and Synthesis of Menthol Using Ni/Al-MCM-41 Catalysts

| Catalyst | Pressure (kPa) | Yield (%) | Menthol Isomer Distribution (%) | | |
		Menthols	(±)-Menthol	(±)-Neomenthol	(±)-Isomenthol
Ni/Al-MCM-41	506.5	89.2	72.3	20.2	7.5
Ni/Al-MCM-41	2026.0	93.0	70.1	20.0	8.9
Ni (8%)/Al-MCM-41	2026.0	94.0	71.0	20.0	8.0

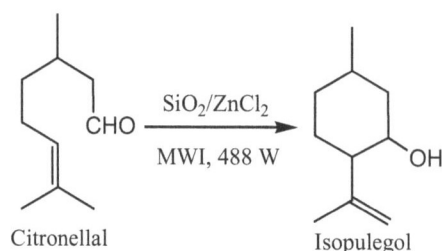

SCHEME 2.15   Synthesis of isopulegol.

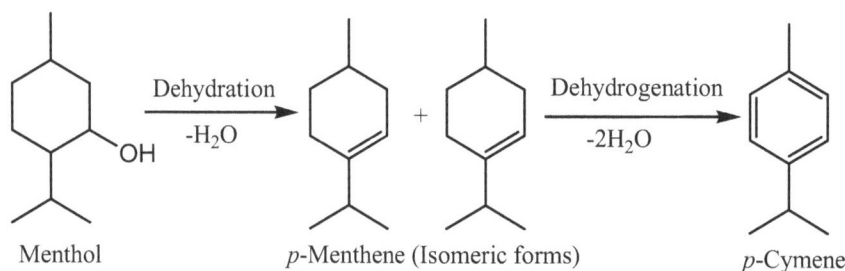

SCHEME 2.16   Synthesis of *p*-cymene from menthol.

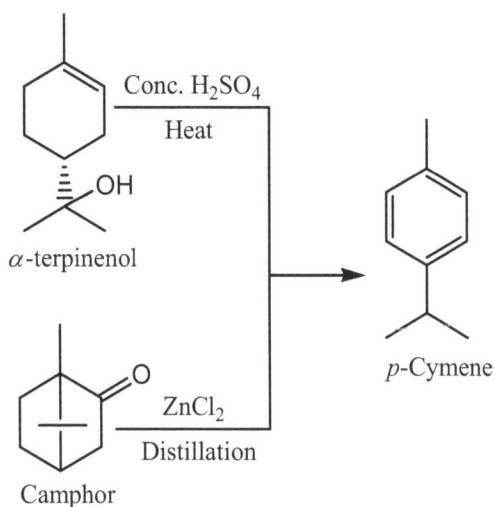

SCHEME 2.17   Synthesis of *p*-cymene from α-terpineol or camphor.

2.2.2.4 Synthesis of Thymol

Thymol is a natural monoterpenoid phenol derivative with the molecular formula $C_{10}H_{14}O$. It is isomeric with carvacrol and is having a hydroxyl group at a particular position on the phenolic ring. It is extracted from *Thymus vulgaris and Ajwain* as a white crystalline compound with a pleasant aromatic odor and strong antiseptic properties. Thymol was first isolated by Neumann from the plants of *Thymus vulgaris* in 1719.

Thymol may be prepared synthetically by various routes from different starting materials such as pulegone, menthol, m-Cresol and pipertone [28].

Route-I: It involves four steps such as (1) reduction of pulegone to produce menthol (2) Oxidation of menthol to produce a ketone (Menthone) (3) Bromination of Menthone to obtain dibromomenthone (4) heating the dibromomenthone with quinoline to yield thymol (Scheme 2.18).

Route-II: It involves two steps such as (1) Oxidation of menthol to produce a ketone (menthone) and (2) Menthone is then treated with bromine and quinoline to produce thymol (Scheme 2.19).

Route-III: It involves a reaction between *m*-cresol and isopropanol in the presence of a suitable catalyst to yield thymol (Scheme 2.20).

Route-IV: It involves the treatment of pipertone with ferric chloride produces thymol (Scheme 2.21).

Route-V: Thymol is produced from GPP in the following ways (Scheme 2.22).

2.2.2.5 Synthesis of Carvacrol

Carvacrol is a monoterpenoid phenol. Chemically, it is 2-Methyl-5-(propan-2-yl)phenol. It is present in the essential oil of *Origanum vulgare* (oregano). Carvacrol is produced from GPP in the following ways. First of all, GPP undergoes cyclization to produce γ-terpinene which on aromatization to yield *p*-cymene. Finally, *p*-cymene undergoes hydroxylation to produce Carvacrol (Scheme 2.23) [29].

Carvacrol can be produced synthetically by several methods as follows (Scheme 2.24) [30].

a. Hot treatment of carvol with phosphoric acid.
b. Heating five parts of camphor with one part of iodine.
c. Dehydrogenation of carvone in presence of Pd/C catalyst.
d. Fusion of cymol sulfonic acid with caustic potash.
e. Reaction of 1-methyl-2-amino-4-propyl benzene with nitrous acid (HNO_2).

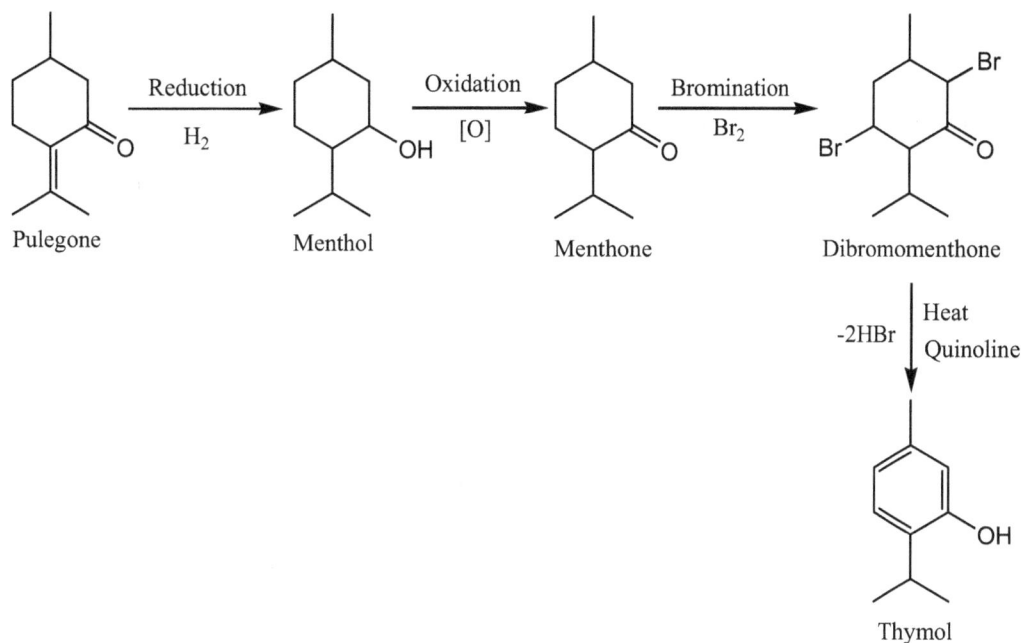

SCHEME 2.18 Synthesis of thymol from pulegone.

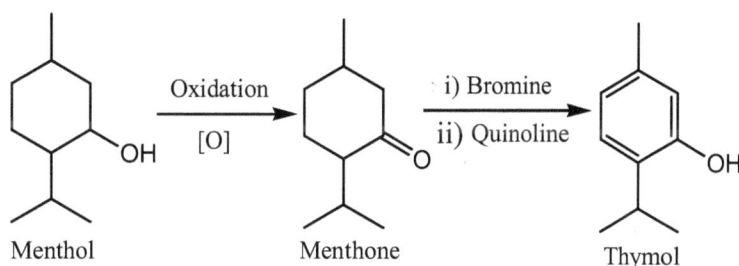

SCHEME 2.19 Synthesis of thymol from menthol.

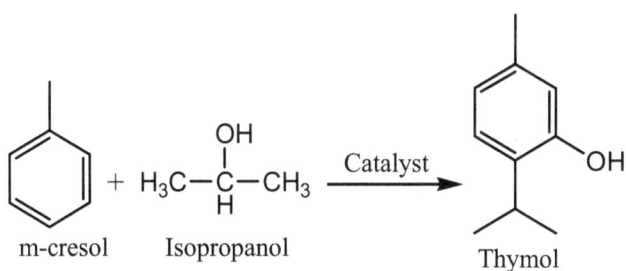

SCHEME 2.20 Synthesis of thymol from *m*-cresol.

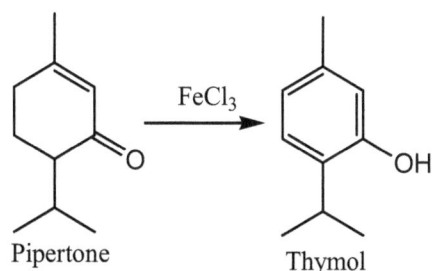

SCHEME 2.21 Synthetic route of thymol from pipertone.

2.2.2.6 Synthesis of Eucalyptol

Eucalyptol is the primary constituent of oil produced by *Eucalyptus globulus*. It is also frequently called as 1,8-cineol. It is a cyclic ether and chemically known as 1,3,3-trimethyl-2-oxabicyclo[2.2.2]octane. It is prepared synthetically by the dehydration of terpin hydrate as given below (Scheme 2.25) [31].

2.2.2.7 Synthesis of Piperitone

Piperitone is a natural monoterpene ketone which is a component of *Eucalyptus dives*. It possesses two stereoisomers such as D-form and L-form. Its IUPAC name is 6-isopropyl-3-methyl-1-cyclohex-2-enone. Synthesis of Piperitone is carried out by following two methods [32].

> **Method-I:** It involves Birch reduction of 1-isopropyl-2-methoxy-4-methyl-benzene to produce 4-isopropyl-3-methoxy-1-methyl-cyclohexa-1,4-diene which on hydrolysis under acidic condition yield piperitone. Birch reduction is an organic reaction which is used to convert arenes to cyclohexadienes (Scheme 2.26).

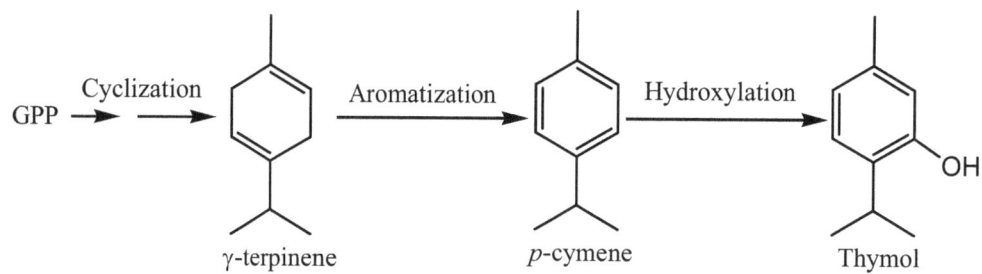

SCHEME 2.22   Synthetic route of thymol from GPP.

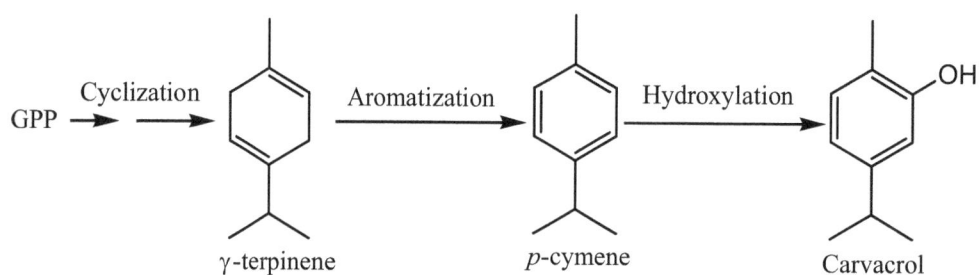

SCHEME 2.23   Synthetic route of carvacrol from GPP.

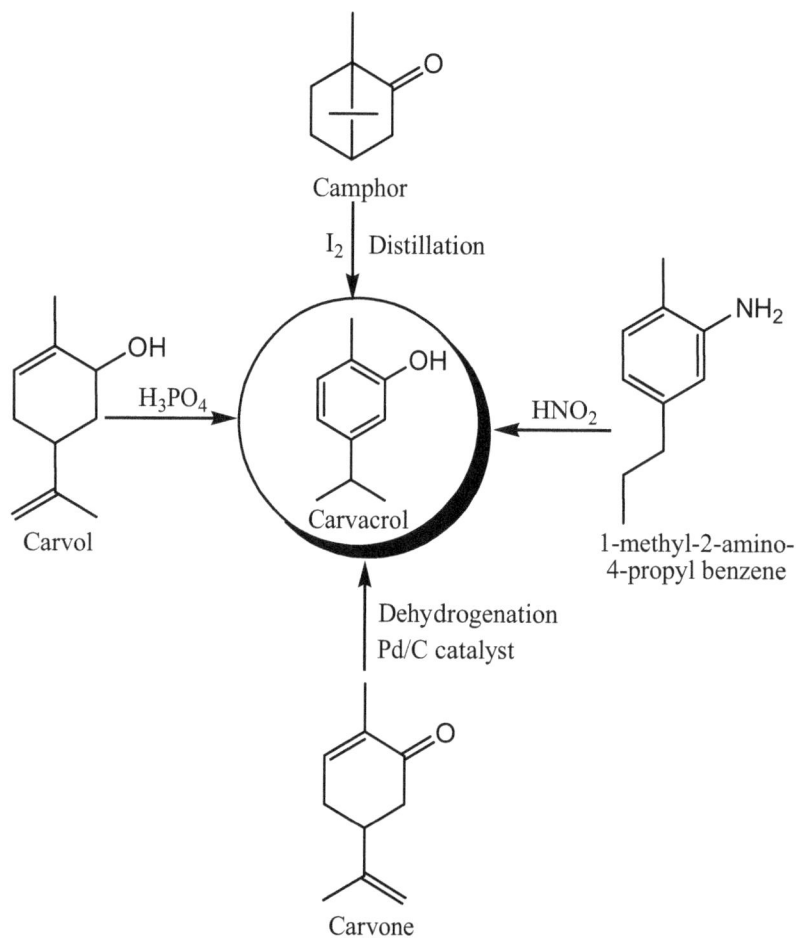

SCHEME 2.24   Various synthetic routes for production of carvacrol.

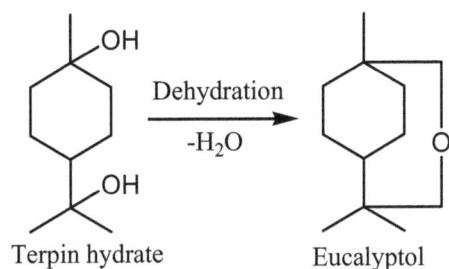

SCHEME 2.25 Synthesis of eucalyptol.

Method-II: It follows Bergamanns synthesis which involves condensation followed by cyclization of methylvinyl ketone and mesityl oxide to produce piperitone (Scheme 2.27).

2.2.2.8 Synthesis of Carvone

Carvone is present naturally in various essential oils. It was isolated from the seeds of caraway (*Carum carvi*), dill and leaves of spearmint (*Mentha spicata*). Its IUPAC name is 2-Methyl-5-(prop-1-en-2-yl)cyclohex-2-en-1-one. It possesses two mirror images or enantiomeric forms such as R-(–)-carvone, or l-carvone and S-(+)-carvone, or d-carvone. Carvone may be synthetically produced from limonene via limonene nitrosochloride which is formed by the treatment of limonene with nitrosochloride (NOCl). Then, limonene nitrosochloride is converted into carvoxime, which can be achieved by refluxing in presence of base. Finally, carvoxime undergoes acidic hydrolysis to yield carvone. The d-limonene or (+)-limonene is used as a precursor for the synthesis of S-(+)-carvone while the same procedure affords R-(–)-carvone from R-(+)-limonene (Scheme 2.28) [33].

2.2.2.9 Synthesis of α-Terpineol

Terpineol is monoterpene alcohol. It is isolated from various sources such as cajuput oil, pine oil and petitgrain oil. There are four isomers of terpineol such as α, β, γ-terpineol and terpinen-4-ol (Figure 2.7). Both β-and γ-terpineol differ by the position of the double bond. Among these isomers, α-terpineol is the major constituent [34].

Terpineol is manufactured commercially more readily from of α-pinene. So, α-pinene undergoes hydration to produce terpine hydrate which on dehydration to yield α-terpineol (Scheme 2.29).

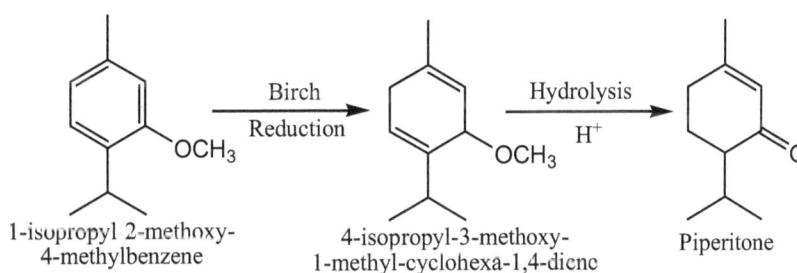

SCHEME 2.26 Synthesis of piperitone via birch reduction.

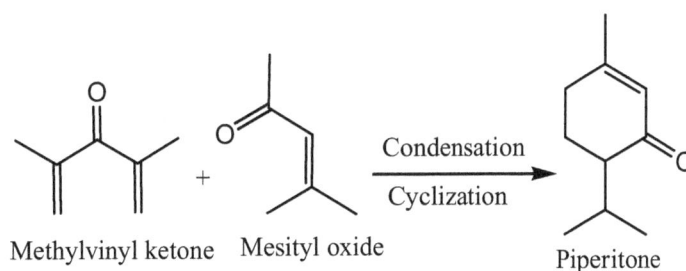

SCHEME 2.27 Bergamanns synthesis of piperitone.

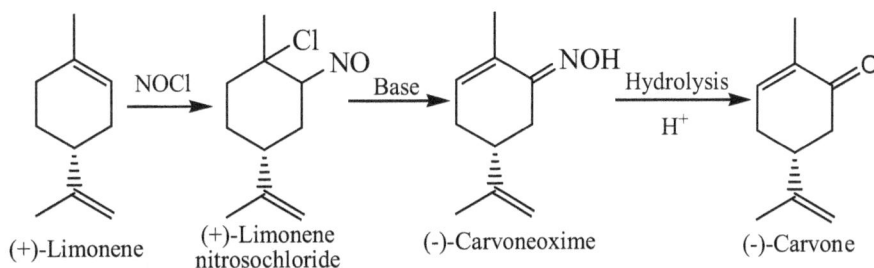

SCHEME 2.28 Synthesis of carvone from (+)-limonene.

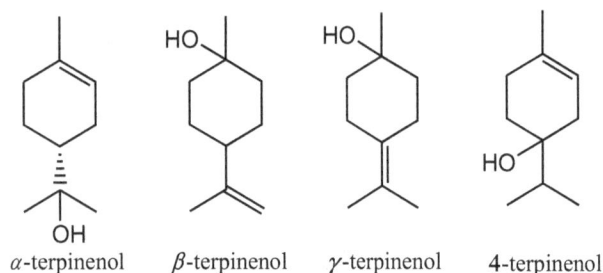

α-terpinenol   β-terpinenol   γ-terpinenol   4-terpinenol

FIGURE 2.7   Structures of isomers of terpineol.

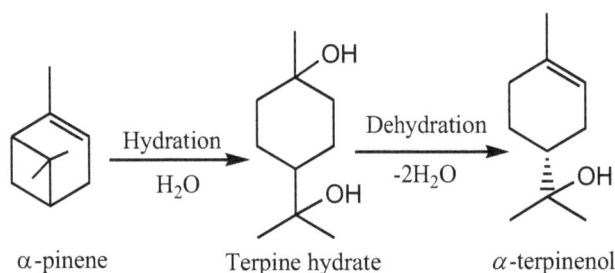

SCHEME 2.29   Synthesis of α-terpineol from α-pinene.

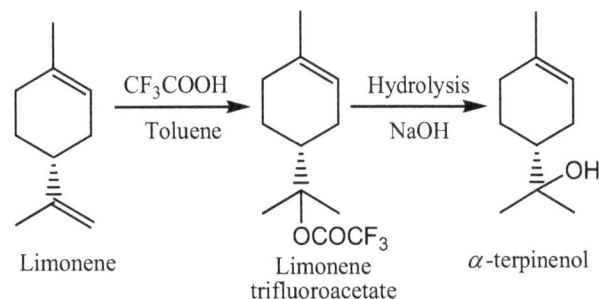

SCHEME 2.30   Synthesis of α-terpineol from limonene.

Limonene reacts with trifluoroacetic acid based on Markovnikov addition reaction to produce an intermediate trifluoroacetate, which is further hydrolyzed with sodium hydroxide at room temperature to produce α-terpineol (Scheme 2.30) [36].

2.2.2.10 Synthesis of Pulegone and Isopulegone

Pulegone and isopulegone are monocyclic monoterpene ketone. Pulegone is obtained from the essential oils of a variety of plants such as *Nepeta cataria* (catnip) and *Mentha piperita*. Its molecular formula is $C_{10}H_{16}O$ and IUPAC name is (R)-5-Methyl-2-(1-methylethylidine)cyclohexanone. Pulegone is optically active liquid with b.p. 221°C–222°C. (+)-pulegone is the main constituent of essential oil of pennyroyal. Isopulegone is isomerized into pulegone by alkaline reagents [37]. (–)-citronellol (**1**) on treatment with pyridinium chlorochromate in dry methylene chloride produces isopulegone (**4**) in one step via the intermediates citronellal (**2**) and the isopulegol (**3**). Then, treatment of (**4**) with ethanolic sodium

hydroxide yield (–)-pulegone (**5**). Pyridinium chlorochromate is acidic enough to induce cyclization of compound citronellal (**2**) to isopulegol (**3**) as presented in Scheme 2.31 [38].

2.2.2.11 Synthesis of Limonene

Limonene is a monocyclic monoterpene with the molecular formula $C_{10}H_{16}$. It is an optically active liquid with b.p 117°C. It occurs mainly as the D- or (R)-enantiomer. (+)-Limonene is present in lemon, orange and caraway oil whereas (–)-Limonene is present in peppermint oil. But its racemic form is obtained from turpentine oil. Limonene is synthesized by dehydration of terpin or α-terpineol (Scheme 2.32) [39].

2.2.2.12 Synthesis of Perillyl Alcohol

Perillyl alcohol is the naturally occurring monocyclic monoterpene. Its IUPAC name is (4-isopropenyl-1-cyclohexen-1-yl) methanol. It is found in the essential oils of various plants, such as lavender, lemongrass and peppermint. It exists in two isomeric forms such as (R)-(+)-Perillyl alcohol and (S)-(–)-Perillyl alcohol. It is a metabolite of limonene and is formed from geranyl pyrophosphate in the mevalonate pathway. The conversion of limonene to perillyl alcohol is achieved via hydroxylation by enzymes, cytochrome P450. Commercially, Perillyl alcohol is prepared from (+)-limonene. First of all, (+)-limonene undergoes acetylation to produce limonene acetate. This compound is then treated with Pd/K_2CO_3 yield perillyl alcohol (Scheme 2.33) [40,41].

2.2.2.13 Synthesis of Phellandrene

Phellandrene is the monocyclic monoterpene with $C_{10}H_{16}$. These are isolated from *Eucalyptus phellandra* are two types such as α- and β-phellandrene. In the case of α-phellandrene, both double bonds are endocyclic while in the case of β-phellandrene, one of the double bond is exocyclic. The IUPAC names for α- and β-phellandrene is 2-methyl-5-(1-methylethyl)-1,3-cyclohexadiene, 3-methylene-6-(1-methylethyl)cyclohexene respectively.

The biosynthesis of phellandrene starts with dimethylallyl pyrophosphate and isopentenyl pyrophosphate condensing in an SN1 reaction to form geranyl pyrophosphate. The resultant monoterpene undergoes cyclization to form a menthyl cationic species. Then, hydride shift forms an allylic carbocation. Finally, an elimination reaction occurs at one of two positions to yield either α-phellandrene or β-phellandrene (Scheme 2.34) [42].

Sen et al. reported the synthesis of enantiomerically pure (S)(+)-α-Phellandrene from (R)(–)-Carvone. The synthesis of α-Phellandrene from Carvone involves two transformation steps which include (1) selective hydrogenation of the double bond at C_8 of the isopropylidene group of Carvone by using triphenyl phosphine and rhodium chloride trihydrate as the soluble transition metal complex for the catalytic transfer of hydrogen to the terminal olefin (2) reductive introduction of the double bond at C_5 by preparation of the tosyl hydrazone derivative (Scheme 2.35) [43].

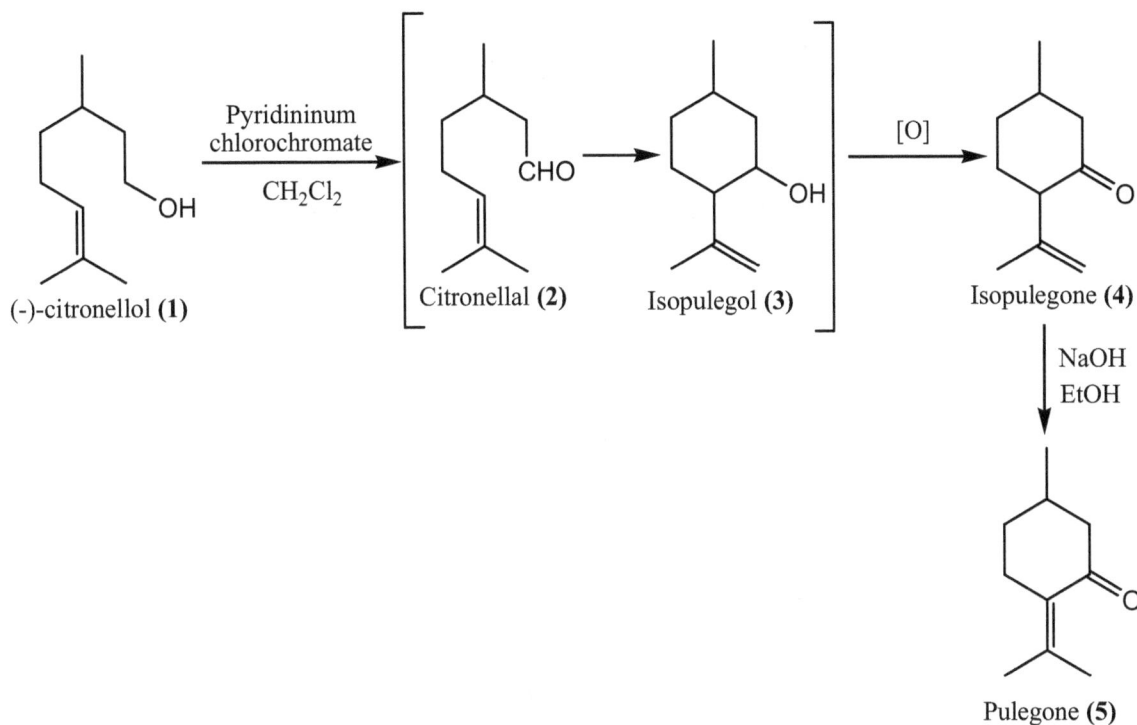

SCHEME 2.31 Synthesis of pulegone and isopulegone.

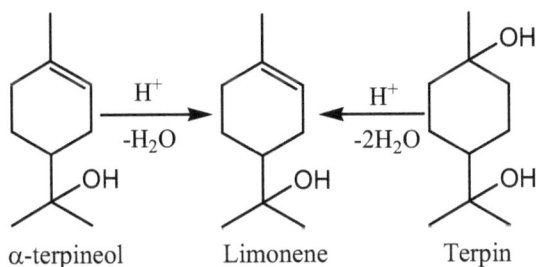

SCHEME 2.32 Synthesis of limonene from α-terpineol or terpin.

2.2.2.14 Synthesis of Carveol

Carveol is a natural unsaturated, monocyclic monoterpenoid alcohol. Its IUPAC name is 2-methyl-5-(prop-1-en-2-yl)cyclohex-2-en-1-ol. There are two forms of Carveol such as (1R, 5R) or (–)-cis form and (1S, 5R) or (–)-trans-carveol. Among these isomers, cis-(–)-carveol is the main constituent of spearmint essential oil (*Mentha spicata*).

Song et al. reported the stereodivergent synthesis of carveol and dihydrocarveol through ketoreductases/ene-reductases catalyzed asymmetric reduction (Scheme 2.36). Stereodivergent asymmetric synthesis is an efficient approach to produce various types of stereoisomers of a desired product. Due to the presence of multiple stereogenic centers in the structure of carvone, stereodivergent strategy is applied for the synthesis of all possible stereoisomers of carveol and dihydrocarveol by using ketoreductases and ene-reductases. The asymmetric reduction of (R)-carvone and (S)-carvone catalyzed by Prelog and anti-Prelog stereo-preferred ketoreductases first produced all four possible stereoisomers of carveol with medium to high diastereomeric excesses up to >99% [44].

2.2.2.15 Synthesis of Catalpol

Catalpol is an iridoid glucoside. It is a monoterpene attached with a glucose molecule. First, catalpol was isolated from genus *Catalpa* in 1962. Later, it was found to be present in

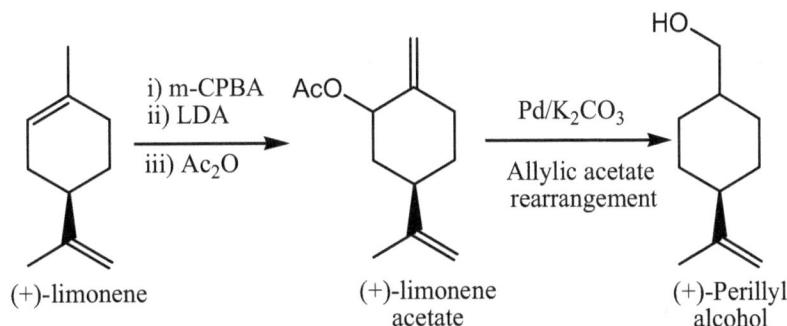

SCHEME 2.33 Synthesis of perillyl alcohol from (+)-limonene.

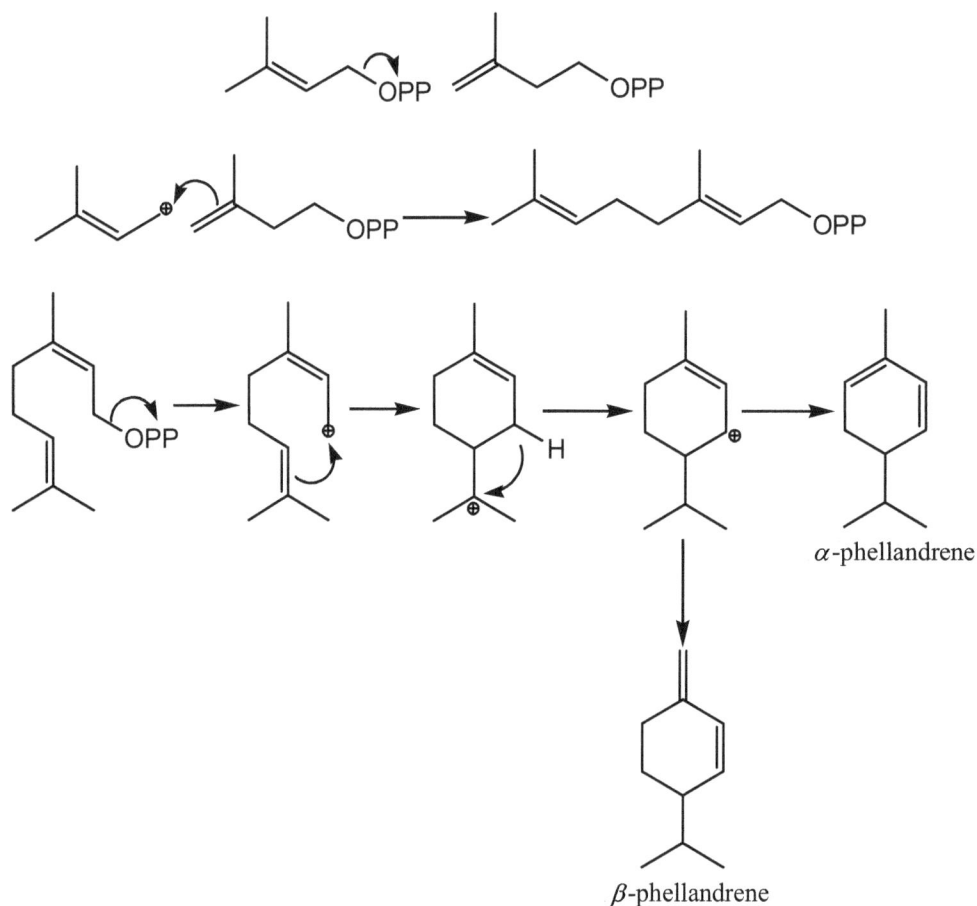

SCHEME 2.34 Synthesis of phellandrene.

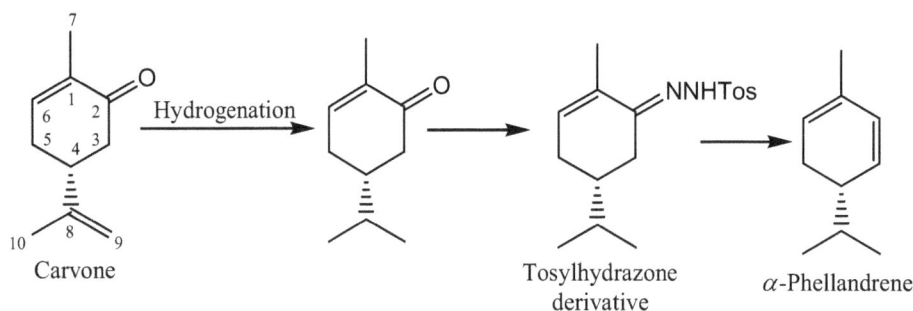

SCHEME 2.35 Synthesis of (S)(+)-α-phellandrene from (R)(–)-carvone.

larger quantities in several plants in genus *Rehmannia* of family Orobanchaceae in 1969. Its IUPAC name is (1aS, 1bS, 2S, 5aR, 6S, 6aS)-6-Hydroxy-1a-(hydroxymethyl)-1a, 1b, 2,5a, 6,6a-hexahydrooxireno[4,5]cyclopenta[1,2-c]pyran-2-yl-β-D-glucopyranoside. S.R. Jensen has described the possible biosynthetic pathway for catalpol. First of all, epi-iridodial is derived from the precursor geraniol (Scheme 2.37). Then, a glucose molecule is added at C_1 of the iridoid skeleton and the oxidation of the aldehyde at C_4 of epi-iridotrial yields 8-epiloganic acid. A subsequent hydrolysis at C_8 produces mussaenosidic acid, followed by dehydration to yield deoxyngeniposidic acid. Then, geniposidic acid is produced via hydrolysis at C_{10} followed by decarboxylation to remove the carboxylic acid at

C_4 to get bartsioside. The very widely accepted precursor to catalpol, aucubin, is then furnished via hydroxylation at C_6. Finally, the epoxidation with the alcohol at C_{10} produces catalpol [45].

2.2.2.16 Synthesis of Thymoquinone

Thymoquinone (TQ) is an unsaturated monocyclic monoterpene with the molecular formula $C_{10}H_{12}O_2$. It is the main ingredient of the essential oil present in *Nigella sativa*. Its IUPAC name is 2-isopropyl-5-methylbenzo-1,4-quinone. The synthesis of thymoquinone involves various steps as presented in Scheme 2.38 [46].

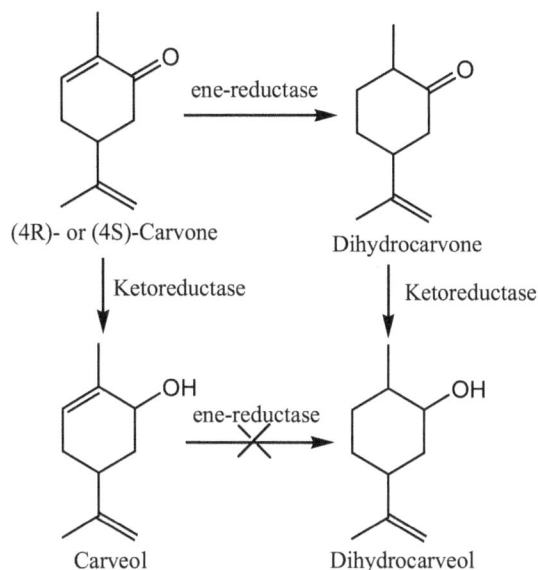

SCHEME 2.36 Synthesis of carveol.

i. Cyclization of Geranyl diphosphate to form γ-terpinene.

ii. Aromatization of γ-terpinene into p-cymene.

iii. Hydroxylation p-cymene to yield carvacrol.

iv. Further hydroxylation of carvacrol leads to the formation of thymohydroquinone.

v. Oxidation of thymohydroquinone to produce thymoquinone.

2.2.2.17 Synthesis of Eugenol

Eugenol is a monocyclic monoterpenoid with the molecular formula $C_{10}H_{12}O_2$. Its IUPAC name is 2-methoxy-4-(prop-2-en-1-yl)phenol. It is a colorless to pale yellow, aromatic oily liquid extracted from essential oils of clove, nutmeg, cinnamon and basil. The biosynthesis of eugenol starts with the amino acid (tyrosine). L-tyrosine is first converted to 4-coumaric acid by the enzyme tyrosine ammonia lyase (TAL). 4-coumaric acid is then converted to caffeic acid by p-coumarate 3-hydroxylase

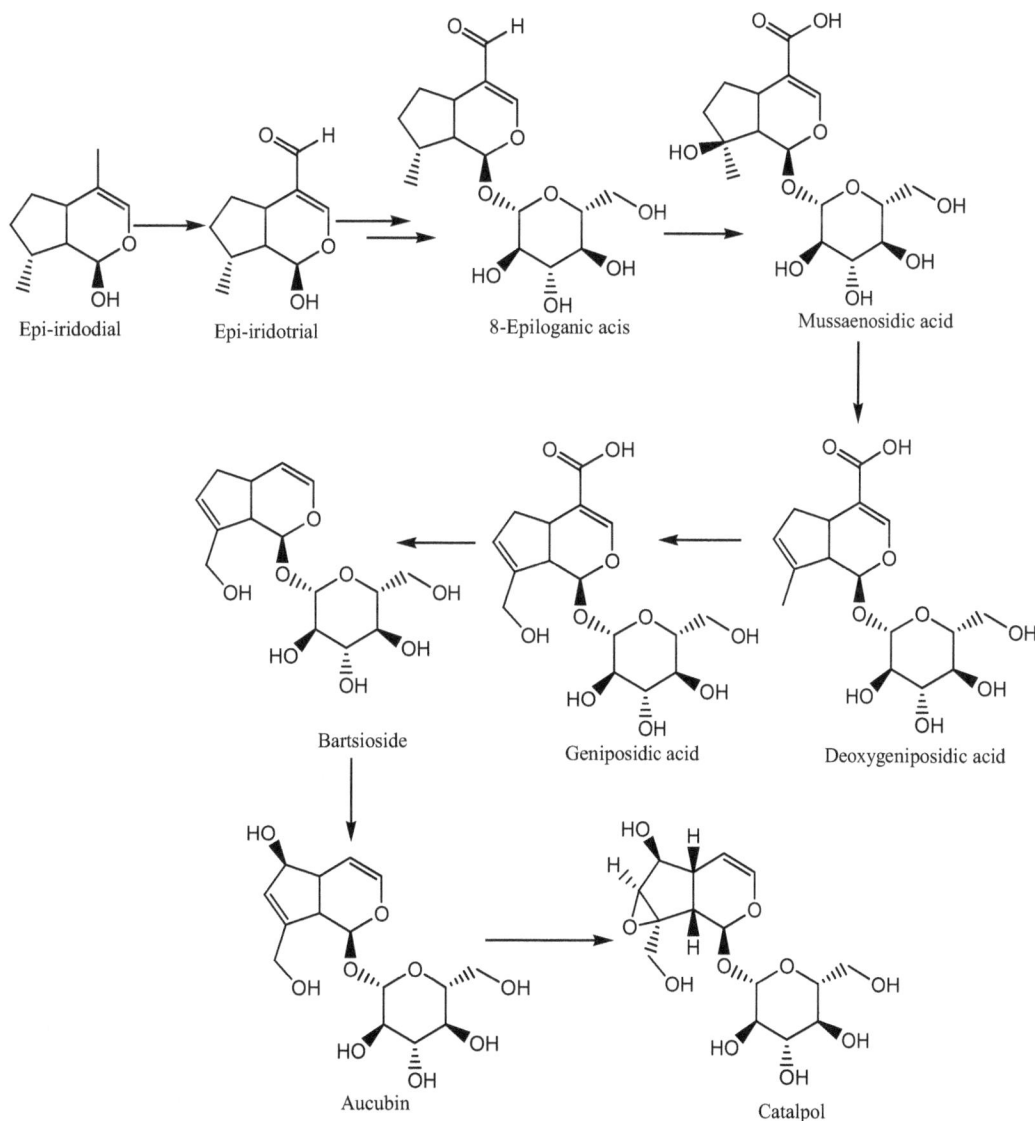

SCHEME 2.37 Biosynthesis of catalpol as proposed by Jensen.

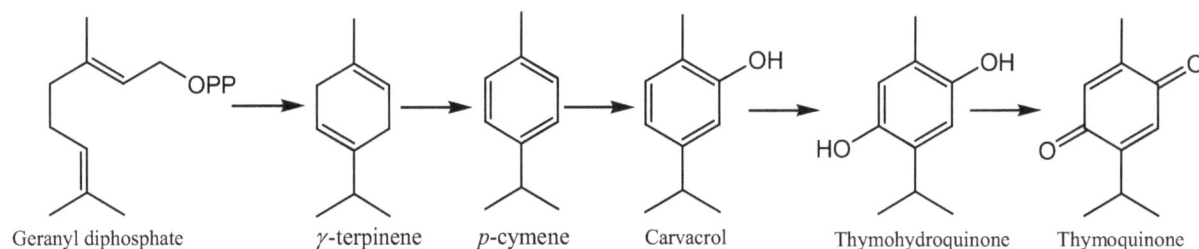

SCHEME 2.38 Biosynthetic pathway of thymoquinone.

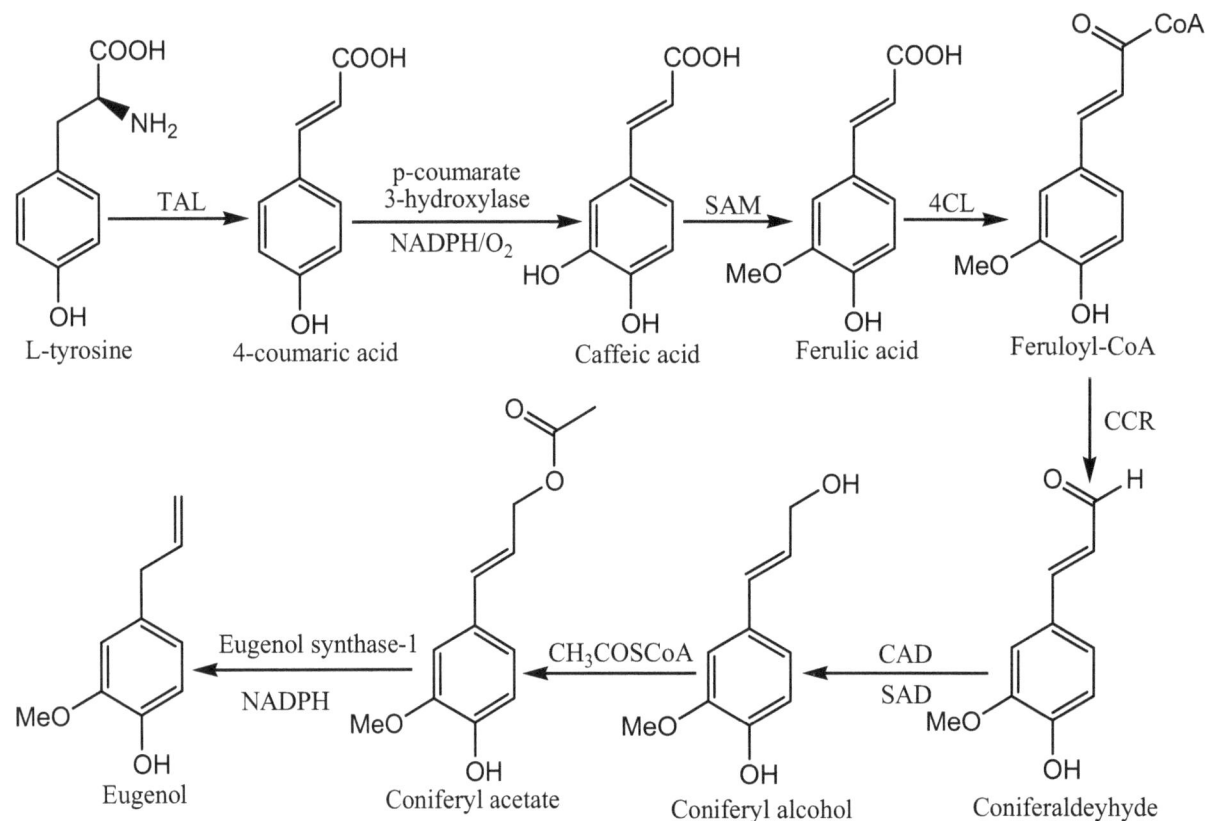

SCHEME 2.39 Synthesis of eugenol from L-tyrosine.

in presence of NADPH and O_2. Then, S-Adenosyl methionine (SAM) is used to methylate the caffeic acid to form ferulic acid, which is in turn converted to feruloyl-CoA by the action of enzyme 4-hydroxycinnamoyl-CoA ligase (4CL). Next, feruloyl-CoA is reduced to coniferaldehyde in presence of cinnamoyl-CoA reductase (CCR). Coniferaldeyhyde is then further reduced to coniferyl alcohol in the presence of cinnamyl-alcohol dehydrogenase (CAD) or sinapyl-alcohol dehydrogenase (SAD). Coniferyl alcohol is then converted to an ester in the presence of the substrate CH3COSCoA to produce coniferyl acetate. Finally, coniferyl acetate is converted to eugenol via the enzyme eugenol synthase-1 and NADPH (Scheme 2.39) [47].

2.2.2.18 Synthesis of Sobrerol

Sobrerol is the hydration product of epoxide of α-pinene. It is also a product of the oxidation of terebenthene (oil of turpentine) in sunlight, discovered by Ascanio Sobrero. Chemically, it is 1-methyl-4-(2-hydroxy-isopropyl)-cyclohex-1,2-en-6-ol. Chen et al reported the convenient and highly efficient synthesis of optically pure trans-(+)-sobrerol (**1**), starting from methyl 3,5-dihydroxy-4-methyl benzoate (**2**) in various steps with overall yield of 26%. The key intermediate (**4**) is prepared with high yield by selective esterification of compound (**3**) using lipase as catalyst. A critical step to stereoselectively inverse the configuration of compound (**7**) is obtained under Mistunobu conditions (Scheme 2.40) [48].

Note: Reagents and conditions: (1) H_2, 5%Rh/Al_2O_3, ethanol, 105°C for 24h, 65%. (2) Lipase (Ps. fluorescens), r.t. for 72h, 93%. (3) (1) t-BuPh2SiCl, imidazole, DMAP, DMF, r.t. for 20h. (2) K_2CO_3, MeOH, r.t. for 7h, 84%. (4) Ph3P, DEAD, THF, r.t. for 15h, 92%. (5) TBAF, THF, r.t. for 20h, 90%. (6) p-Nitrobenzoic acid, Ph3P, DEAD, toluene, r.t. for 9h, 78%. (7) K_2CO_3, MeOH, r.t. for 8h, 93%. (8) MeMgBr, THF, 6h, 85%.

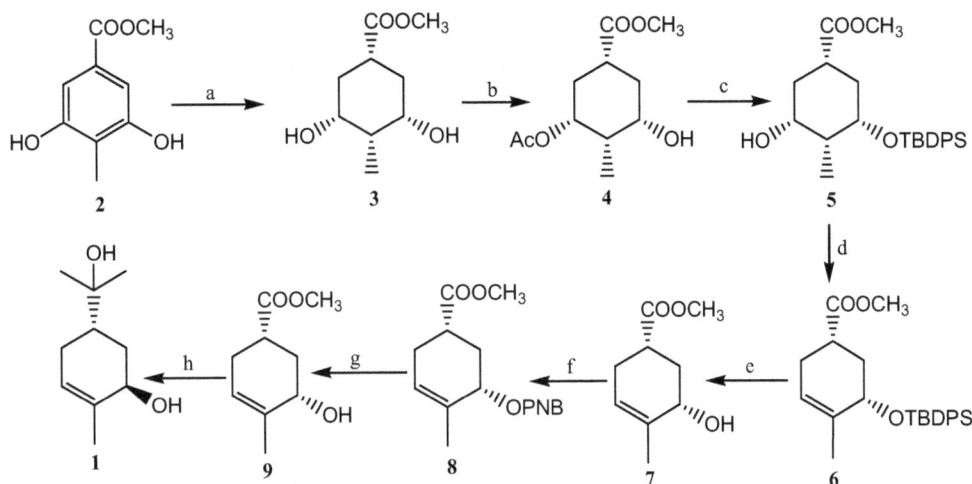

Note: Reagents and conditions: a) H_2, 5%Rh/Al$_2$O$_3$, ethanol, 105°C for 24 h, 65%. b) lipase (Ps. fluorescens), r.t. for 72 h, 93%. c) (1) t-BuPh$_2$SiCl, imidazole, DMAP, DMF, r.t. for 20 h. (2) K$_2$CO$_3$, MeOH, r.t. for 7 h, 84%. d) Ph$_3$P, DEAD, THF, r.t. for 15 h, 92%. e) TBAF, THF, r.t. for 20 h, 90%. f) p-Nitrobenzoic acid, Ph$_3$P, DEAD, toluene, r.t. for 9 h, 78%. g) K$_2$CO$_3$, MeOH, r.t. for 8h, 93%. h) MeMgBr, THF, 6 h, 85%.

SCHEME 2.40 Synthetic route to sobrerol under Mistunobu conditions.

2.2.2.19 Synthesis of Aucubin

Aucubin is a monoterpenoid-based compound of iridoid glycoside class. It possesses a cyclopentan-[C]-pyran skeleton. It is found in *Aucuba Japonica* (Garryaceae) and *Eucommia ulmoides* (Eucommiaceae).

Geranyl pyrophosphate (GPP) is the precursor for generating different types of iridoids. GPP is generated through the mevalonate pathway or the methylerythritol phosphate pathway. The initial steps of the pathway involve the fusion of three molecules of acetyl-CoA to produce the C_6 compound i.e 3-hydroxy-3-methylglutaryl-CoA (HMG-CoA). HMG-CoA is then reduced in two steps by the enzyme HMG-CoA reductase. The resulting mevalonate is then sequentially phosphorylated by two separate kinases such as mevalonate kinase and phosphomevalonate kinase, to produce 5-pyrophosphomevalonate. Phosphomevalonate decarboxylase through a concerted decarboxylation reaction affords isopentenyl pyrophosphate (IPP). IPP is the basic C_5 building block which is added to prenyl phosphate cosubstrates to form longer chains. IPP is isomerized to the allylic ester dimethylallyl pyrophosphate (DMAPP) by IPP isomerase. Through a multi-step process, including the dephosphorylation DMAPP, IPP and DMAPP are combined to form the C_{10} compound called geranyl pyrophosphate (GPP).

The biosynthesis studies suggest that the synthetic sequence from 10-hydroxygerinol to 8-epi-iriotrial involves the dephosphorylation of GPP that leads to a geranyl cation which is then hydroxylated to form 10-hydroxygeraniol. Then, 10-hydroxylgeraniol is isomerized to 10-hydroxynerol which is oxidized

to produce trialdehyde in presence of NAD. Finally, the trialdehyde undergoes double Michael addition to yield 8-epi-iridotrial.

The cyclizaton reaction to form the iridoid pyran ring may be proceeded through two routes.

Route-1: Hydride nucleophilic attack on C_1 leads to 1-O-carbonyl atom attack on C_3 to yield lactone ring.

Route-2: Loss of proton from C_4 leads to the formation of a double bond between C_3 and C_4 and consequently the 3-O-carbonyl atom will attach to C_1.

The cyclized lactone intermediate is glycosylated to form boschnaloside which is then hydroxylated on C_{10}. Boschnaloside is then oxidized to geniposidic acid followed by decarboxylation to produce bartisioside. Bartisioside is then hydroxylated to form aucubin (Scheme 2.41) [49,50].

2.2.3 Synthesis of Bicyclic Monoterpenoids

2.2.3.1 Synthesis of Ascaridole

Ascaridole is a bicyclic monoterpenoids with molecular formula $C_{10}H_{16}O_2$. It has unusual bridging peroxide functional group in its structure. It is the chief constituent of the oil of Mexican tea, *Dysphania ambrosioides*. Zeigler et al. reported the synthesis of Ascaridole by irradiation of α-terpinene in presence of UV light and chlorophyll (Scheme 2.42) [51].

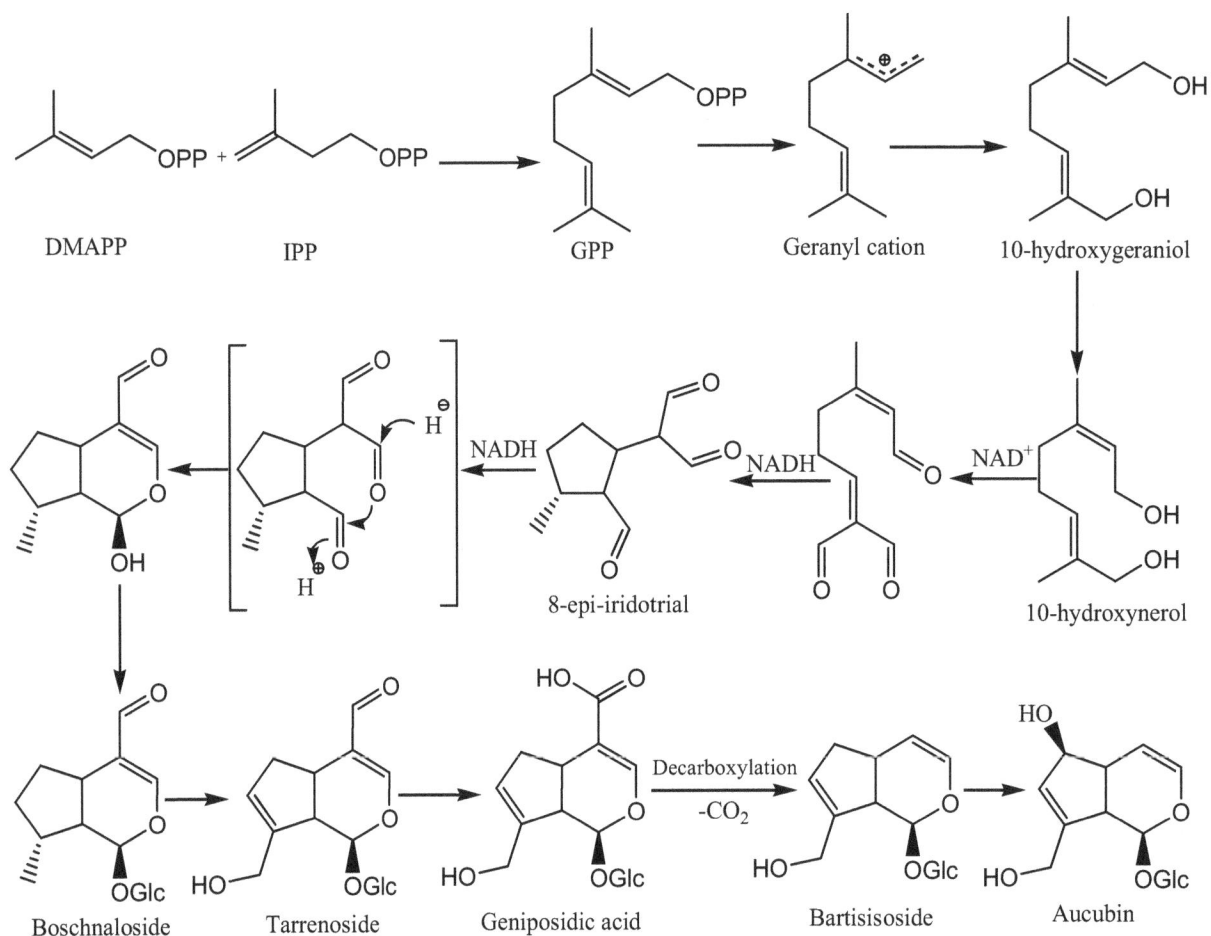

SCHEME 2.41   Synthetic route to Aucubin via GPP.

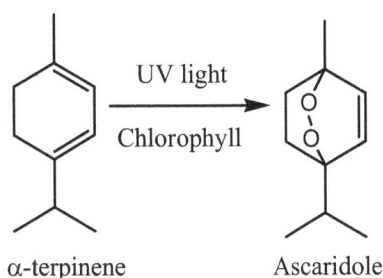

SCHEME 2.42   Synthesis of ascaridole.

2.2.3.2 Synthesis of Pinene

Pinene is a bicyclic monoterpene with the molecular formula $C_{10}H_{16}$. There are two structural isomers of pinene found in nature such as α-pinene and β-pinene. The racemic mixture of pinene is found in Eucalyptus oil [52].

Both α- and β-pinene are produced from geranyl pyrophosphate, via cyclization of linaloyl pyrophosphate followed by loss of proton from carbocation (Scheme 2.43) [53].

2.2.3.3 Synthesis of Camphor

Camphor is a bicyclic monoterpene with the molecular formula $C_{10}H_{16}O$. The main source of Camphor is *Cinnamomum*

camphora. There are two possible enantiomers of camphor such as (*R*)- and (*S*)-form.

It is produced biosynthetically from geranyl pyrophosphate, via cyclization of linaloyl pyrophosphate to yield bornyl pyrophosphate. Then bornyl pyrophosphate undergoes hydrolysis to produce borneol which is on oxidation to obtain camphor (Scheme 2.44) [54].

Camphor is produced commercially from α-pinene present in the turpentine oil via various steps sequentially as follows (Scheme 2.45) [55].

a. Treatment of α-pinene with HCl
b. Isomerization of pinene hydrochloride
c. Treatment of bornyl chloride with aquous KOH
d. Oxidation of borneol with HNO_3

2.2.3.4 Synthesis of Borneol

Borneol is a bicyclic monoterpene with molecular formula $C_{10}H_{18}O$. Its hydroxyl group is placed in an endo position in the structure. There are two enantiomers of borneol such as d-(+)- and l-(−)-borneol. Both isomers are found in nature. Borneol is present in various plant species such as *Heterotheca, Artemisia, Callicarpa, Dipterocarpaceae, Blumea balsamifera*

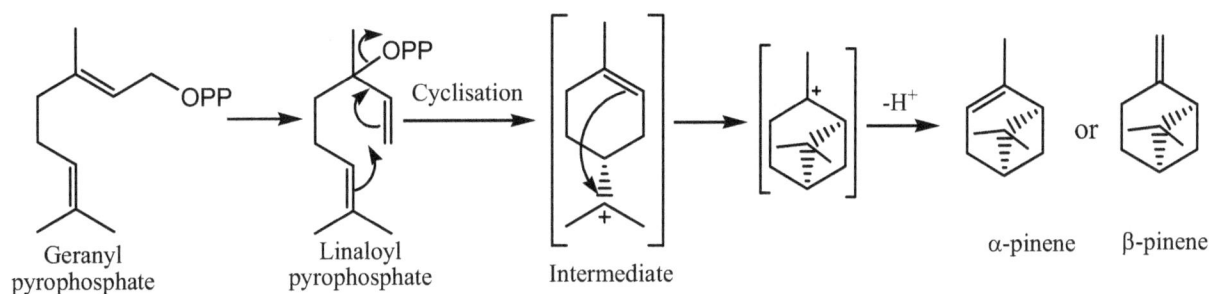

SCHEME 2.43   Synthesis of pinene from GPP.

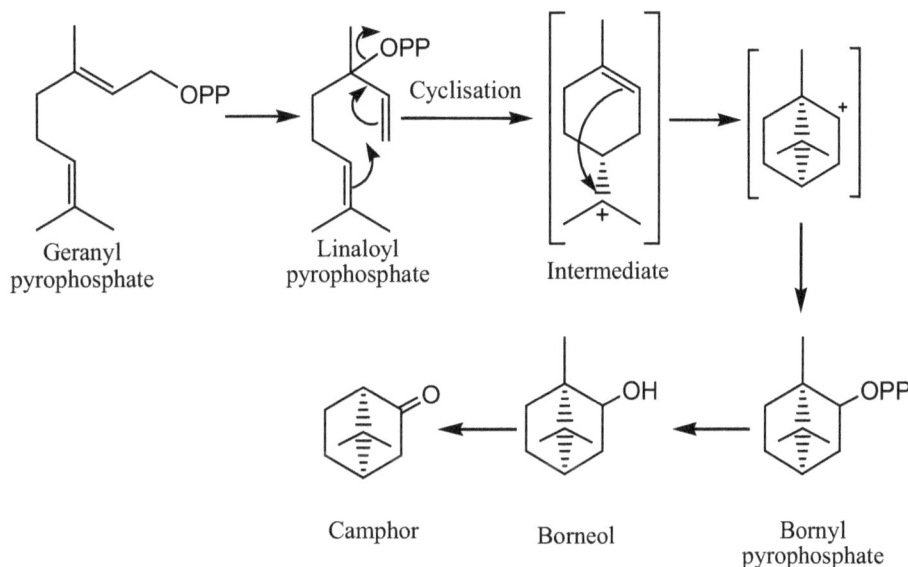

SCHEME 2.44   Synthesis of camphor from GPP.

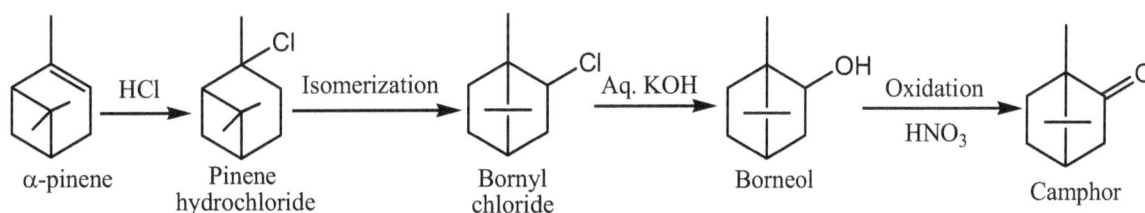

SCHEME 2.45   Synthesis of camphor from α-pinene.

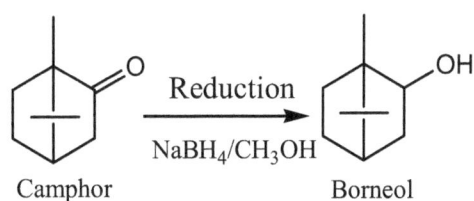

SCHEME 2.46   Synthesis of borneol from camphor.

and *Kaempferia galangal*. Camphor on reduction in presence of NaBH$_4$ produces Borneol. It is a fast and irreversible process (Scheme 2.46). In other hand, Borneol can be synthesized by reduction of camphor based on a reversible process called Meerwein–Ponndorf–Verley reduction [56].

2.2.3.5 Synthesis of Sabinene Hydrate

Sabinene is a natural bicyclic monoterpene of thujene type with the molecular formula C$_{10}$H$_{16}$. Its IUPAC name is 4-methylene-1-(1-methylethyl)bicyclo[3.1.0]hexane. It is isolated from the essential oils of several of plant sources including holm oak and Norway spruce. It possesses strained ring system with a cyclopentane ring fused to a cyclopropane ring.

Galopin et al. reported the short and efficient synthesis of (±)-trans-sabinene hydrate. As sabinene is having 5+3 bicyclic skeleton in its structure, 3-isopropyl-2-cyclopentenone is used as an intermediate. The double bond of intermediate can be functionalized into a cyclopropane ring and the ketonic

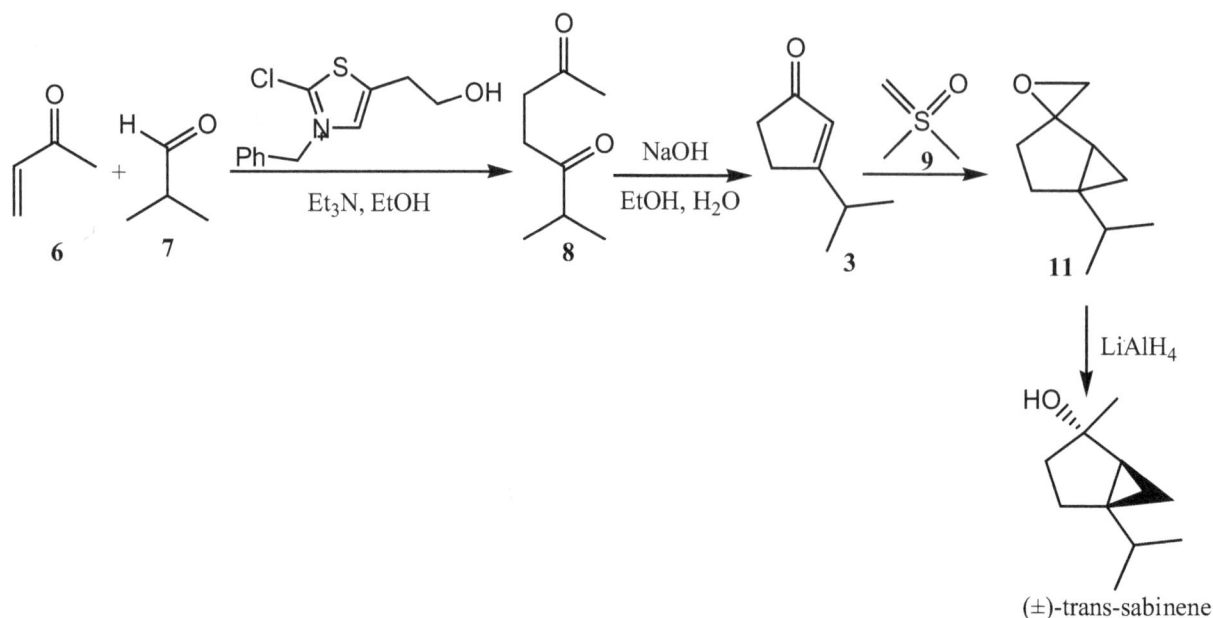

SCHEME 2.47 Synthesis of sabinene hydrate.

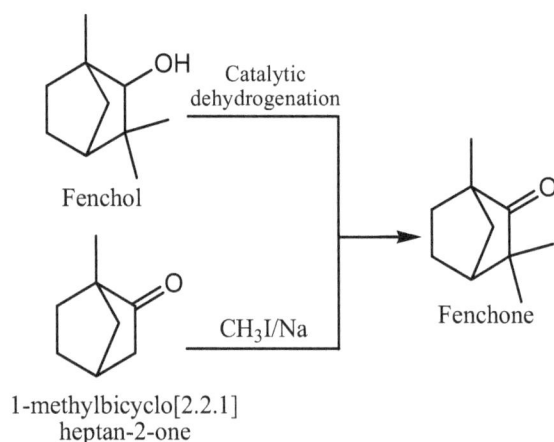

SCHEME 2.48 Synthetic route of fenchone.

group can be transformed into the tertiary alcohol of sabinene hydrate (Scheme 2.47) [57].

2.2.3.6 Synthesis of Fenchone

Fenchone is a bicyclic monoterpene ketone with the molecular formula $C_{10}H_{16}O$. It is a colorless oily liquid with b.p 192°C–193°C. Its IUPAC name is 1,3,3-trimethyl-bicyclo[2.2.1] heptan-2-one. It is a constituent of absinthe and the essential oil of fennel. It is a mixture of the enantiomers such as *d*-fenchone and *l*-fenchone. It is optically active. Fenchone can be prepared through catalytic dehydrogenation of fenchol. Similarly, 1-methylbicyclo[2.2.1]heptan-2-one undergoes methylation in presence of CH_3I/Na to yield fenchone (Scheme 2.48) [58].

2.2.3.7 Synthesis of Verbenol

Verbenol is a group of stereoisomeric bicyclic monoterpene alcohols with the molecular formula $C_{10}H_{16}O$. Its IUPAC name

is 4,6,6-trimethylbicyclo[3.1.1]hept-3-en-2-ol. There are four stereoisomers of verbenol such as (+)-cis, (−)-cis, (+)-trans and (−)-trans verbenol. The cis-verbenol is present in the aggregation pheromones of bark beetles *Ips* and *Dendroctonus*. The synthesis of (−)-cis-verbenol from (−)-α-pinene, involves the following steps (Scheme 2.49) [59].

  i. Acetoxylation of (−)-*α*-pinene (**1**) in presence of esorated tetra-acetate to yield compound (**2**).
  ii. Hydroxylation of compound (**2**) to produce (−)-trans verbenol (**3**).
  iii. Oxidation of (−)-trans verbenol (**3**) in presence of pyridinium chlorochromate (PCC) to obtain (−)-verbenone (**4**).
  iv. Reduction of (−)-verbenone (**4**) in presence of sodium borohydride ($NaBH_4$) to produce (−)-cis-verbenol (**5**).

Similarly, the reduction of (1R, 5R)- or (1S, 5S)-verbenones with $LiAlH_4$ or $NaBH_4$ yielded the mixture of cis- and trans-verbenols. Another procedure was reported for stereospecific preparation of 1S, 4S, 5S-cis-verbenol by reducing 1S, 5S-verbenone with (i-Bu)$_2$AlH in CH_2Cl_2 at −70°C (Scheme 2.50) [60].

2.3 Pharmacological Activities of Monoterpenoids

The pharmacological benefits of medicinal plants are mainly due to the presence of bioactive phytoconstituents in the plant tissues as primary and secondary metabolites. These constituents are identified as alkaloids, essential oil (terpenoids), glycosides, flavonoids, saponins and steroids etc. Essential oil is a complex mixture of volatile plant constituents which is characterized by low molecular weight components such as terpenes,

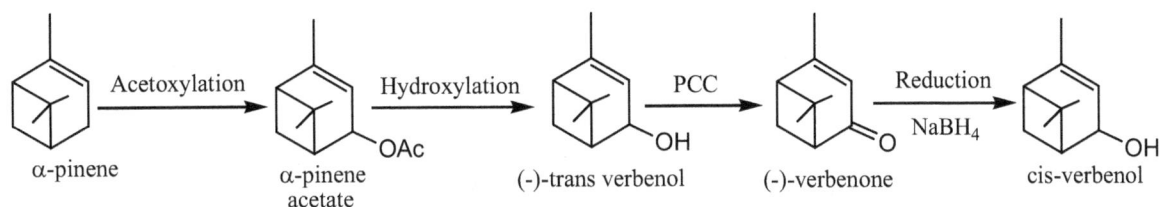

SCHEME 2.49 Synthesis of verbenol from α-pinene.

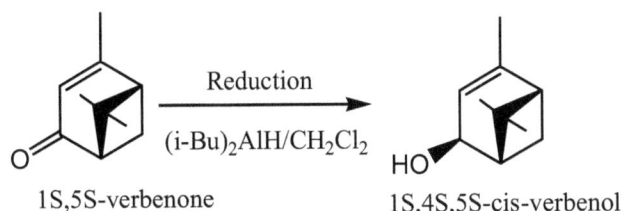

SCHEME 2.50 Synthesis of 1S,4S,5S-cis-verbenol.

terpenoids and other aromatic compounds. Terpenoids possess diverse chemical structures with various pharmacological

effects and different mechanisms of action as presented in Tables 2.5 and 2.6 [61,62].

The lipophilic nature of essential terpenoids enables the constituents to cross the cell membrane and then enter the cytoplasm. Then these constituents interact with cell organelles and cytoplasmic materials to impart their biological response or activities. Most of the medicinal plants contain essential oils and are widely used medicinally for the treatment of various disorders such as respiratory tract infections, cancer, malaria, inflammation, microbial infection, tuberculosis, Parkinsonism, rheumatoid arthritis, Alzheimer's disease, cardiovascular diseases, hyperglycemia, hyperlipidemia and HIV etc (Figures 2.8–2.10) [63,64].

TABLE 2.5

Biological Source, Pharmacological Properties and MIC Values for Antimicrobial Activity of Monocyclic Monoterpenes

Monocyclic Monoterpene	Biological Source	Pharmacological Activities	MIC Value (mg/mL)
Carvone	Leaves of *Perovskia angustifolia*	Antibacterial Anticancer	0.8 (MSSA) 0.008–0.064 (*Klebsiella pneumoniae*)
Limonene	Citrus plants Aerial parts of Hyssopus cuspidatus	Antibacterial Antifungal Anti-inflammatory Antioxidant	0.03(MIC$_{90}$:*Mycoplasma pneumoniae*) 0.008-0.064 (*Klebsiella pneumoniae*)
Thymol	Thymus pubescens	Antibacterial, Anticancer	0.8 (MSSA) 3.17 (MRSA) 0.008–0.064 (*Klebsiella pneumoniae*)
α-Terpineol	Leaves of *Perovskia angustifolia*, Flowers of *Salvia lavandulifolia*, Aerial parts of *Salvia officinalis*	Antibacterial	0.1–0.4 (*Porphylomonas* sp.) 0.4–0.8 (*Prevotella*) 0.1–0.4 (*Fusobacterium nucleatum*)
1,8-Cineole	Aerial parts of *Thymus pubescens, Rosmarinus officinalis*, Aerial parts of *Coridothymus capitatus*, Aerial parts of *Hyssopus cuspidatus*, *Salvia officinalis*	Antibacterial Anticancer, Anti-inflammatory Antioxidant	>2.0 (*Enterococcus faecalis*) 0.4 (*Streptococcus salivarius*) 0.4 (*Streptococcus sanguinis*) 1.25 (*B. subtilis, S. aureus, E. coli*) 2.5 (*Pseudomonas aeruginosa*) 5.0 (*Staphylococcus epidermidis*)
p-Cymene	Aerial parts of *Dicyclophora persica* Aerial parts of *Coridothymus capitatus, Lavandula latifolia*	Antibacterial Antifungal	1.8 (*Bacillus subtilis*); 7.2 (*Enterococcus faecalis, S. aureus; E. coli*) 3.6 (*Staphylococcus epidermidis*); 0.6 (*Aspergillus niger*)
Carvacrol	*Thymus pubescens*	Antibacterial, Antifungal, Antiviral, Cytotoxic, Antitumor	0.4 (MSSA) 1.6 (MRSA) 0.008–0.064 (*K. pneumonia*)
Eugenol	*Syzygium aromaticum*	Antibacterial, Antiviral, Cytotoxic,	0.01 (*Escherichia coli*) 0.002 (*Helicobacrter pylori*) 0.008–0.064 (*K. pneumonia*)
Menthol	*Mentha piperita*	Anesthetic, Antibacterial Antiviral Cytotoxic Antioxidant Antiseptic	0.008–0.064 (*Klebsiella pneumonia*)

TABLE 2.6

Biological Source, Pharmacological Properties and MIC Values for Antimicrobial Activity of Bicyclic Monoterpenes

Bicyclic Monoterpene	Biological Source	Pharmacological Activities	MIC Value (mg/mL)
α-Pinene	Aerial parts of *Acinos rotundifolius*, Aerial parts of *Hyssopus cuspidatus*	Antibacterial Antiviral Antifungal Hypotensive	1.2 (VRSA) 2.5 (*Pseudomonas aeruginosa*) 2.5 (*Escherichia coli*) >2.0 (*Enterococcus faecalis*) 0.4 (*Streptococcus salivarius*) 0.4 (*Streptococcus sanguinis*) 7.5 (*Bacillus subtilis*) >15.0 (*Staphylococcus aureus*) 15.0 (*Escherichia coli, S. epidermidis*) 0.0625 (B. subtilis, E. coli) 0.0313 (*S. aureus, S. epidermidis*) 0.117 (*Cryptococcus neoformans*) 0.004 (MRSA) 0.128 (*Candida albicans*) 0.064 (*Aspergillus niger*) 0.008–0.064 (*K.pneumoniae strains*)
β-Pinene	Aerial parts of *Acinos rotundifolius*, Aerial parts of *Hyssopus cuspidatus*	Antibacterial Antiviral Antifungal Hypotensive	0.0625 (*B. subtilis, E. coli*) 0.0313 (*Staphylococcus aureus*) 0.187 (*Cryptococcus neoformans*) 0.006 (MRSA) 0.008–0.064 (*Klebsiella pneumonia*)
Carene	*Pinus sylvestris, Asarum heterotropoides* (root), *Juniperus communis* (leaves)	Antibacterial Antifungal Local anesthetic	0.18–0.7 (*S. aureus, E. coli*) 0.001–0.005 (*Candida* sp.) 0.001–0.01 (*Aspergillus niger*)
Isoborneol	*Thymus pubescens*	Antibacterial Antiviral	0.008–0.064 (*Klebsiella pneumonia*)
Verbenol	*Ferula hermonis* (rhizome and roots)	Antifungal Neuroprotective	0.128 (*Candida albicans*) 0.032 (*Tricophyton mentagrophytes*) 0.128 (*Aspergillus niger*)

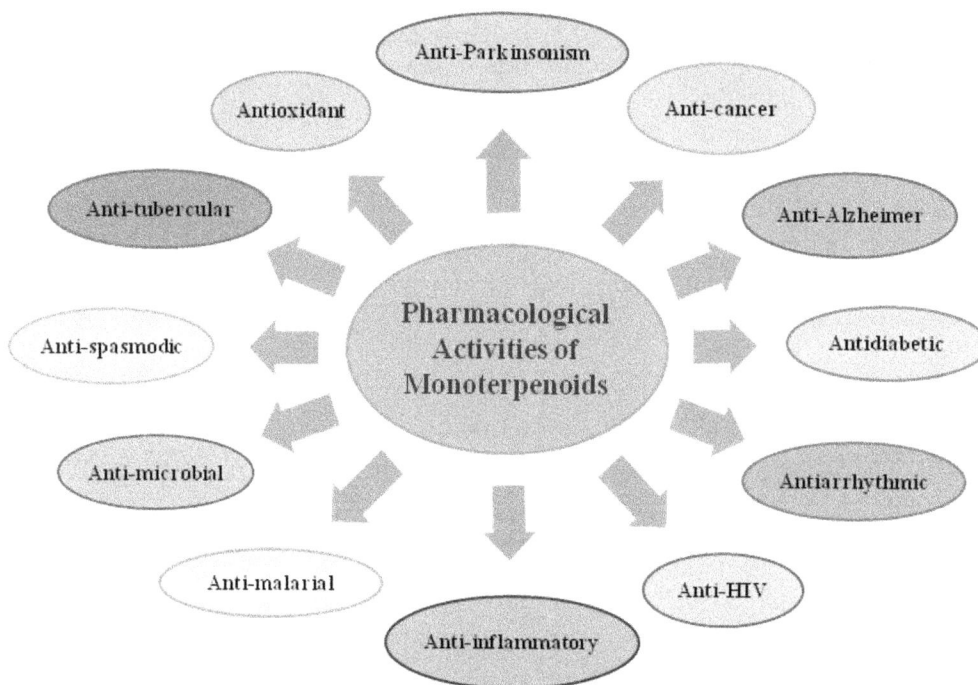

FIGURE 2.8 Pharmacological activities of monoterpenoids.

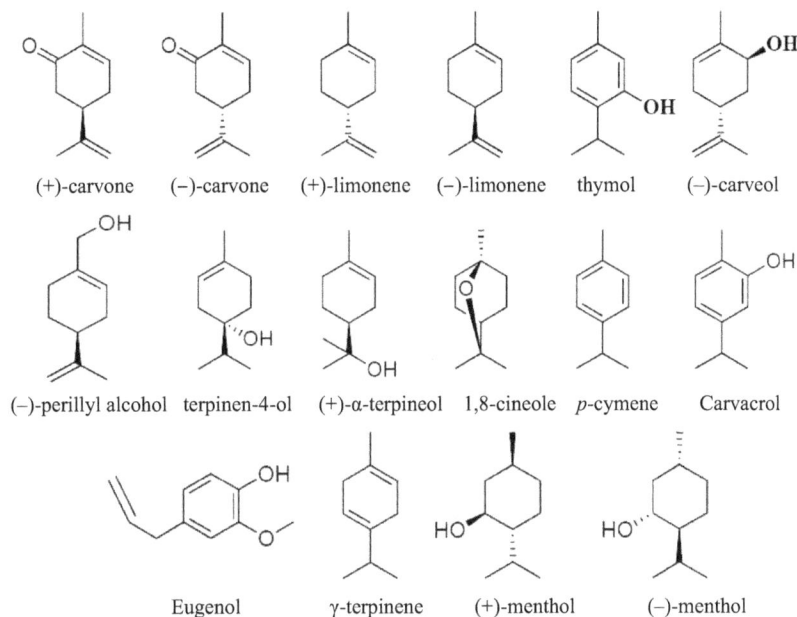

FIGURE 2.9 Structures of monocyclic monoterpenes.

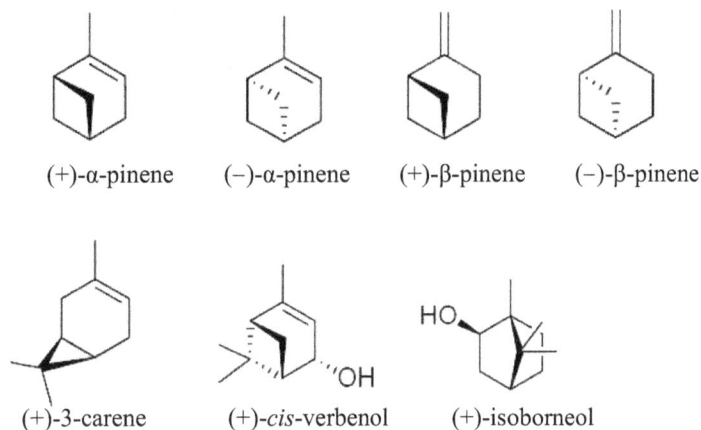

FIGURE 2.10 Structures of bicyclic monoterpenes.

2.3.1 Antimicrobial Activity

(−)-Linalool, (−)-citronellol, (+)-citronellal, (+)-citronel-lyl acetate, linalyl acetate, geraniol and citral are monoter-penes which exhibit antimicrobial activity (Figure 2.11). The antimicrobial activity of these compounds increases due to the presence of an oxygen-containing functional group. (−)-Linolool showed significant antibacterial properties against *Pseudomonas aeruginosa, Escherichia coli, Mycoplasma and Staphylococcus aureus.* (−)-Linalool and linalyl acetate exhibit significant anti-inflammatory activity by decreasing

edema in the carrageenan-induced paw edema model in rats (Table 2.7). Linalyl acetate is considered as a pro-drug of (−)-linalool [65,66].

Lai et al. reported terpene derivatives as a potential agent against antimicrobial resistance (AMR) pathogens. They have been reported to have a potent antimicrobial activity which exhibits both bacteriostatic and bactericidal effects against tested pathogens. Both *in-vivo* and *in-vitro* experiments were conducted to evaluate the antimicrobial activity of either syn-thetic or naturally derived terpenoids (Table 2.8). Combination therapy with terpenes has been widely observed in the case

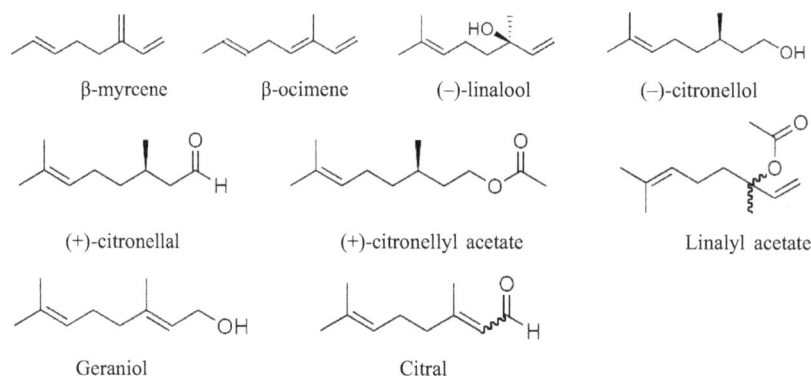

FIGURE 2.11 Structures of acyclic monoterpenes.

TABLE 2.7

Biological Source, Pharmacological Properties and MIC Values for Antimicrobial Activity of Acyclic Monoterpenes

Acyclic Monoterpene	Biological Source	Pharmacological Activities	MIC Value (mg/mL)
β-Myrcene	Aerial parts of *Thymus pubescens, Rosmarinus officinalis, Ocimum basilicum*	Antibacterial Cardiotonic Diuretic	>2.0 (*Enterococcus faecalis*) 0.4 (*Streptococcus salivarius*) 1.5 (*Streptococcus sanguinis*)
β-Ocimene	Aerial parts of *Dicyclophora persica, Ocimum basilicum, Pimenta acris*	Antibacterial Antifungal	1.8 (*Bacillus subtilis*) 7.2 (*E. faecalis, S.aureus, E. coli*) 3.6 (*S. epidermidis*)
Linalool	Bergamot fruit (*Citrus bergamia*), Aerial parts of *Coridothymus capitatus*	Antibacterial Antimycoplasmal Antiviral Anti-inflammatory Antiedematous	0.1–0.8 (*Porphylomonas* sp.) 0.2–1.6 (*Prevotella* sp.) 0.1–0.2 (*F. nucleatum*) >5.0 (*Pseudomonas aeruginosa*) 2.5 (*Escherichia coli*) 0.015 (MIC$_{90}$: *M. pneumoniae*) 0.008–0.064 (*K. pneumoniae*)
Linalyl acetate	*Citrus bergamia*, Flowers of *Matricaria recutita, Boswellia carteri, Lavandula officinalis,* Leaves of *Melaleuca alternifolia*	Antibacterial Anti-inflammatory Antiedematous	0.015 (MIC$_{90}$: *M. pneumoniae*)
Citronellol	*Thymus pubescens*, aerial parts of *Dracocephalum Moldavica*	Antibacterial Antioxidant	0.008–0.064 (*K. pneumoniae*)
Citronellal	*Thymus pubescens*	Antifungal	25.0–50.0 (*Pyricularia grisea*)
Geraniol	*Thymus pubescens*, aerial parts of *Dracocephalum moldavica*	Antibacterial Antiviral Cytotoxic Antitumor	2.5 (*Pseudomonas aeruginosa*) 1.2 (*Escherichia coli*) 0.008–0.064 (*K. pneumonia*)
Citral	Aerial parts of *Dracocephalum moldavica*	Antibacterial Antiviral Cytotoxic Anti-inflammatory	0.008–0.064 (*K. pneumoniae*)

of current clinical practice, particularly in antifungal drugs. It was observed that the antifungal activity of fluconazole is potentiated by combining monoterpenes like thymol and carvacrol when subjected to fluconazole-sensitive *C. albicans, C. tropicalis* and *C. glabrata* and *fluconazole resistant C. albicans, C. krusei, C. glabrata, C. tropicalis* and *C. parapsilopsis*. The combination analysis showed that the tested strains exhibit a Fractional Inhibitory Concentration (FIC) index of less than 0.5 FIC is a term used to express the degree

of synergy interaction between antibacterial drugs. FIC<0.5 denotes positive synergism while FIC>0.5 implies negative synergism [67,68].

Startseva et al. reported the antifungal activity of monoterpenoids from the menthane series and their thio derivatives (Figure 2.12, Table 2.9). These are obtained by the reactions of corresponding monoterpenoids with thiols in the presence of Lewis acids. Based on (+)-limonene-1,2-oxide and methyl ether of mercaptoacetic acid in the presence of sodium methylate, the active anti-micotic agent with a wide spectrum of action was synthesized [69].

Antifungal activity of menthol was tested against *Fusarium verticillioides* by using the semisolid agar susceptibility technique. The growth of *F. verticillioides* was reduced by 75% when 200 ppm menthol was applied [70]. It was observed that menthol was lethal to the spores of *Rhizopus stolonifer* after 48 h treatment by using shake culture method with MIC value of 400 μg/mL. Similarly, different essential oil components are tested against five fungi such as *Aspergillus niger, F. oxysporum, Penicillium digitatum, R. stolonifer* and *Mucor* spp. [71] menthol exhibited potential activity against *R. stolonifer* and *Mucor* spp. with MIC value of 200 μg/mL, where as other components such as citral, cinnamic aldehydes citronellal and geraniol were most active against *A. niger, F. oxysporum* and *P. digitatum* with MIC value of 100 μg/mL [72].

TABLE 2.8

Summary of Antimicrobial Activity of Some Terpenoid Class

Terpenoids	Tested Microorganism	Mode of Action
Carvacrol Thymol Geraniol	Resistant *Enterobacter aerogenes*	Efflux pump inhibition
Linalyl acetate (+)- Menthol Thymol	*S. aureus* *E. coli*	Growth inhibition
Carvacrol (+)-Carvone	*E. coli* *S. typhimurium*	Growth inhibition
α-Terpinoel (±)-Linalool α-Phellandrene	*T. mentagrophytes* *Trichophyton violaceum* *Microsporium gypseum* *Histoplasma capsulatum*	Growth inhibition

Duan et al. reported the synthesis and antifungal activity of novel Myrtenal-based 4-methyl-1,2,4-triazole-thioethers. Myrtenal is a natural bicyclic monoterpene. It contains an aldehyde group, a carbon-carbon double bond and a four-membered ring. It was present in essential oils of various medicinal plants such as *Artemisia douglasiana, Ferula hermonis, Lepidiummeyenii, Otacanthus* and *Mentha haplocalyx*. The myrtenal derivatives bearing 1,2,4-triazole moiety were designed and synthesized by multi-step reactions. First of all, myrtenal (2) was prepared regioselectively by oxidation of α-pinene (1) in presence of selenium dioxide (SeO$_2$) with anhydrous ethanol as oxidant and 1,4-dioxane as solvent. Myrtenic acid (3) was prepared by further oxidation of myrtenal (2) by using the combination of NaClO$_2$-H$_2$O$_2$ as oxidant in aqueous solution. Myrtenic acid (3) undergoes chlorination in presence of chlorinating reagent, thionyl chloride (SOCl$_2$) to produce myrtenyl chloride (4). Myrtenal-based 4-methylthiosemicarbazide was (5) then prepared by the N-acylation reaction of myrtenyl chloride (4) with 4-methylthiosemicarbazide. Finally, Myrtenal-based 4-methyl-1,2,4-triazole-thioethers (6a-n) were synthesized with 72%–95% yields by the one-pot sequential processes which involve the cyclization reaction of the intermediate (5) followed by the nucleophilic substitution reaction with different alkyl halides under microwave irradiation (Scheme 2.51) [73].

The antifungal activity of the target compounds was evaluated by using *in-vitro* method against *Fusarium oxysporum* f. sp. *cucumerinum, Physalospora piricola, Alternaria solani, Cercospora arachidicola* and *Gibberella zeae* at 50 μg/mL. Compounds 6c (R=*i*-Pr), 6l (R=*o*-NO$_2$ Bn) and 6a (R=Et) exhibited significant antifungal activity against

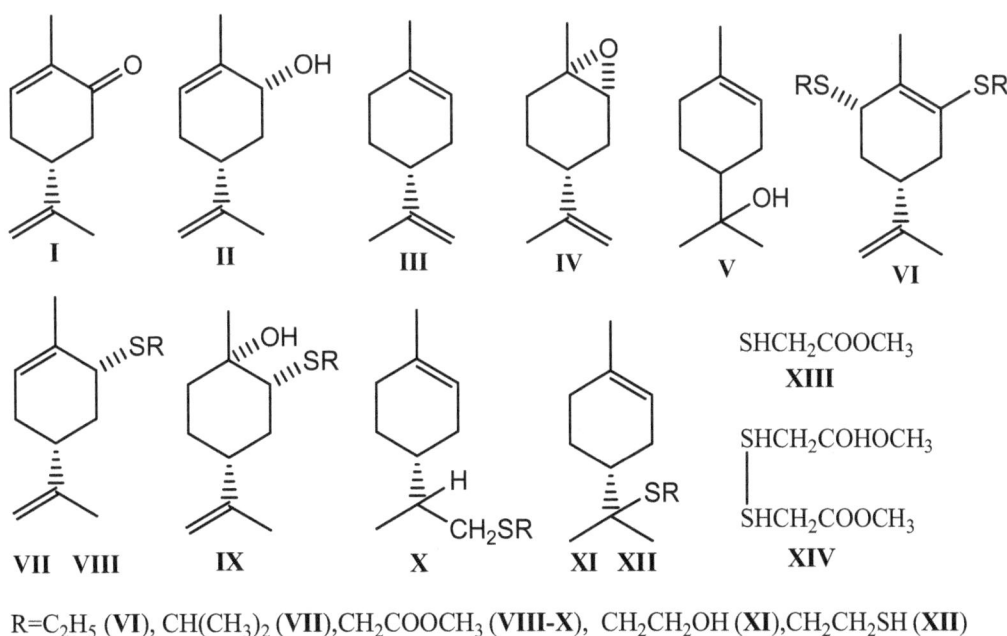

R=C$_2$H$_5$ (**VI**), CH(CH$_3$)$_2$ (**VII**),CH$_2$COOCH$_3$ (**VIII-X**), CH$_2$CH$_2$OH (**XI**),CH$_2$CH$_2$SH (**XII**)

FIGURE 2.12 Structures of menthane series and their thio-derivatives.

TABLE 2.9

Antifungal Activity Level for Monoterpenoids (I–V) and Thio-Derivatives (VI–XIV)

Compounds	Candida albicans Non-Pathogenic	Candida albicans Pathogenic	Candida parapsilosis	Rhodotorula rubra	Aspergillus niger	Penicillium tardum	Candida kruzei
I	+/–	+	+/–	+	–	+/–	2+
II	2+	2+	+/–	+/–	+/–	2+	+
III	2+	+	2+	2+	2+	3+	+
IV	+	+/–	+/–	+	–	+/–	–
V	2+	+	2+	3+	+/–	+	+
VI	–	No data	No data	–	–	No data	–
VII	–	–	+/–	+/–	+	–	+/–
VIII	+/–	–	2+	+	–	+/–	+
IX	No data	3+	No data	3+	3+	No data	No data
X	+/–	No data	No data	+/–	No data	No data	–
XI	–	–	–	+/–	+/–	+/–	–
XII	+/–	–	+	–	–	+/–	–
XIII	2+	2+	2+	2+	2+	2+	2+
XIV	2+	2+	2+	2+	3+	2+	+/–

Note: Sign (+) denotes the zone of 1–3 mm growth inhibition (low activity level); sign (2+) denotes the zone of sign, (3+) denotes the zone of growth inhibition ≥5 mm, which corresponds to a high activity level; (–) denotes growth inhibition less than 1 mm (very low activity level).

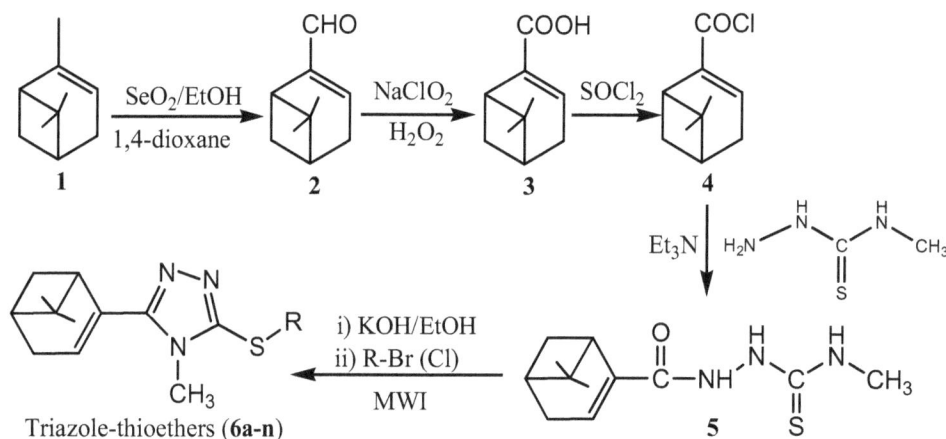

SCHEME 2.51 Synthesis of Myrtenal-based 4-methyl-1,2,4-triazole-thioethers (6a-n).

P. piricola with inhibition rates of 98.2%, 96.4% and 90.7%, respectively. This result represented that the myrtenal derivatives exhibit better or comparable antifungal activity as compared to fungicide azoxystrobin with 96.0% inhibition rate (Table 2.10). Azoxystrobin is used as a positive control.

2.3.2 Antidepressant Activity

Depressive disorder is a common mental disorder associated with a significant negative impact on quality of life and cognitive function. It is associated with an increased risk of disability, suicide and heart failure. According to

information from the World Health Organization (WHO), depression is estimated to be the second known cause of world disability by the year 2020 and it will probably make the greatest contribution to the burden of disease through the year 2030 [74].

Monoterpene like β-pinene and linalool were tested at dose of 100 mg/kg for determining their antidepressant effects. Thymol (15–30 mg/kg) demonstrated the antidepressant-like effect on a chronic unpredictable mild stress model of depression in mice. The most promising monoterpene derivative for treating depression is (S)-mecamylamine (Figure 2.13). It is an antagonist of nicotinic acetylcholine receptors [75].

TABLE 2.10

Antifungal Activity of the Target Compounds (**6a–n**) at 50 µg/mL

| Comp. Code | R | Percentage (%) Inhibition Rate Against Fungi | | | | |
		F. cucumerinum	*C. arachidicola*	*P. piricola*	*A. solani*	*G. zeae*
6a	Et	30.9	36.7	90.7	28.9	27.6
6b	*n*-Pr	35.6	46.7	54.6	47.8	30.6
6c	*i*-Pr	37.9	36.7	98.2	28.9	40.9
6d	*n*-Bu	47.2	46.7	36.8	47.8	36.5
6e	Bn	20.5	33.3	43.3	38.6	45.6
6f	*o*-Me-Bn	20.5	33.3	70.5	32.9	60.2
6g	*m*-Me-Bn	23.2	36.7	48.1	47.1	49.0
6h	*p*-Me-Bn	30.0	25.4	20.0	18.3	33.3
6i	*o*-Cl-Bn	40.0	33.1	30.0	26.7	50.0
6j	*m*-Cl-Bn	30.0	40.8	30.0	18.3	56.7
6k	*p*-Cl-Bn	20.0	25.4	20.0	18.3	36.7
6l	*o*-NO$_2$-Bn	37.9	46.7	96.4	45.1	42.4
6m	*m*-NO$_2$-Bn	58.8	36.7	51.1	50.5	37.9
6n	*p*-NO$_2$-Bn	30.9	26.7	36.8	42.4	30.6
Myrtenal	-	25.0	28.2	37.8	41.3	31.7
Azoxystrobin	-	87.5	92.3	96.0	91.3	90.9

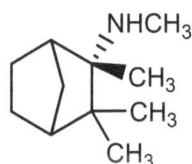

FIGURE 2.13 Chemical structure of (S)-mecamylamine.

2.3.3 Antiviral Activity

Herpes simplex virus (HSV) is differentiated into two antigenic types such as type-1 (HSV-1) and type-2 (HSV-2). HSV infects and replicates cells at the site of entry and the mucocutaneous surface. HSV-1 infection is very common and mostly affects adult people, whereas HSV-2 is a sexually transmitted infection which causes genital herpes. The infection with HSV-2 increases the risk of acquiring and transmitting HIV infection [76,77].

Paul et al. reported the antiviral activity of monoterpenes β-pinene and limonene against herpes simplex virus *in-vitro*. All antiviral assays were performed by using RC-37 cells (Table 2.11). Cytotoxicity was determined in a neutral red assay and antiviral assays were performed with HSV-1 strain KOS.

TABLE 2.11

Cytotoxicity of *β*-Pinene and Limonene on RC-37 Cells

Concentration (µg/mL)	*β*-Pinene	Limonene
10	100.0±2.7	100.0±3.5
25	95.5±8.4	98.4±6.2
50	99.8±1.0	81.8±6.6
75	61.0±9.9	18.1±15.3
100	5.4±0.7z	15.1±6.5

Note: Viability of the drug-treated cells was determined in a standard neutral red assay. Results are given in % compared to untreated controls and represent the mean of three independent experiments.

TABLE 2.12

Antiviral Effect of Serial Dilutions of Monoterpenes against HSV-1

Concentration (µg/mL)	*β*-Pinene	Limonene
1	82.3±18.0	94.7±8.5
2.5	67.5±26.9	99.0±9.6
5	23.1±6.2	64.3±16.1
7.5	6.1±12.3	27.8±19.8
10	0.0±0.0	8.2±13.8
25	0.0±0.0	0.0±0.0
50	0.0±0.0	0.0±0.0

Note: Results are given in % compared to untreated virus controls and represent the mean of three independent experiments.

The mode of antiviral action was evaluated at different periods during the viral replication cycle. Acyclovir was considered as positive control. Both tested terpenoids interacted with HSV-1 in a dose-dependent manner thereby inactivating viral infection (Table 2.12). Neutral red dye uptake was determined by measuring the optical density (OD) of the eluted neutral red at 540nm in a spectrophotometer. The mean OD of the cell-control wells was assigned a value of 100%. The cytotoxic concentration of the drug which reduced viable cell number by 50% (TC$_{50}$) was determined from dose-response curves (Table 2.13). The inhibition of HSV replication was measured by plaque reduction assay [78].

2.3.4 Analgesic and Anti-Inflammatory Activity

Hyperalgesia refers to the enhancement in sensitivity to pain caused by nociceptors or peripheral nerves. Analgesics are the agents which relieve the pain by acting selectively on the CNS and peripheral pain mediators without changing consciousness. Analgesics may be narcotic or non-narcotic. Inflammation is a complex biological response which occurs when the body is exposed to infective agents or to physical or chemical injury [79].

TABLE 2.13

Selectivity Indices (SI) of Monoterpenes for HSV-1

Monoterpene	Max. Noncytotoxic concentration (µg/mL) ± SD	TC_{50} (µg/mL) ± SD	IC_{50} (µg/mL) ± SD	Selectivity Index (SI)
β-Pinene	60±6.9	85±5.3	3.5±4.0	24.3
Limonene	40±6.4	60±11.0	5.9±5.3	10.2

Note: Experiments were repeated independently two times and data presented are the mean of three experiments.

The analgesic effect of menthol was evaluated by monitoring 13 quasistatic tester (QST) parameters in twelve healthy male volunteers. Menthol was administered to the right hand at a concentration of 400 mg and the obtained results were compared to baseline measurements. Both cold and heat-mediated hyperalgesia was recorded over 180 min and a decrease of cold pain (6.93°C±6.5°C/20.23°C±6.8°C) and heat-pain (45.5°C±2.1°C/44.1°C±1.5°C) was noted. At 225 min, however, cold pain was not statistically significant (9.5°C±7.9°C), whereas heat induced pain was decreased significantly [80,81].

Quintans et al. reported the analgesic activity of monoterpenes and also provided an overview of their mechanisms of action. It can be concluded that the acyclic monoterpenes mainly exhibit analgesic activity by modulating the opioid system, as observed in the case of myrcene, linalool, citronellol and citronellal (Table 2.14). In addition, they are also able to decrease the release of NO, which was also demonstrated by citral (Table 2.15). While in the case of the monocyclic monoterpenes, the majority of compounds demonstrated analgesic effects without the participation of the opioid system, such as with (+)-pulegone, (+)-limonene, (−)-carvone, carvacrol and hydroxydihydrocarvone (Tables 2.16 and 2.17). The exceptions to this rule were (−)-menthol, α-phellandrene and thymoquinone, which were effective as kappa opioid receptor agonists [82,83].

α-Pinene, found in oils of coniferous trees and rosemary, showed anti-inflammatory activity by decreasing the activity

TABLE 2.14

Chemical Structures of Monoterpenes

Acyclic Monoterpenes	Monocyclic Monoterpenes	Bicyclic Monoterpene

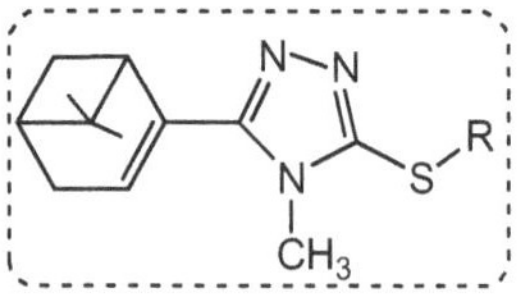

Neral (*cis*-citral)

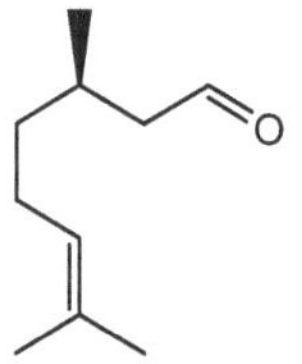

Neral (*cis*-citral)

(α)-Phellandrene

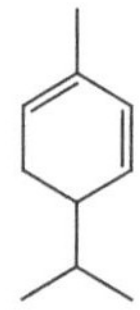

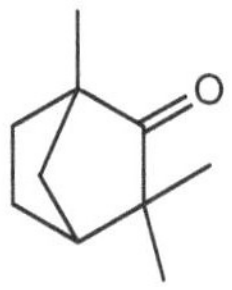

(-)-Fenchone

Geranial (*trans*-citral)

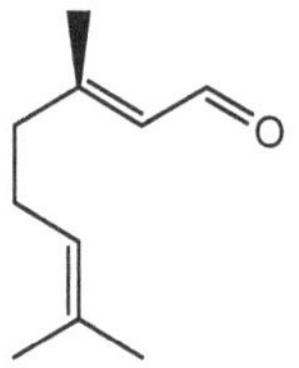

Geranial (*trans*-citral)

Thymoquinone

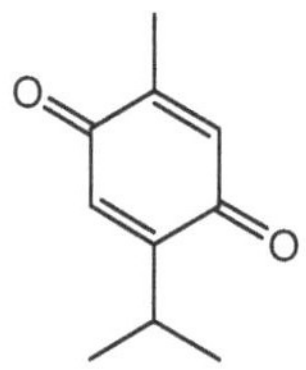

(α)-Phellandrene

(−)-Fenchone

Thymoquinone

1,8-Cineol

TABLE 2.15

Chemical and Pharmacological Aspects of Acyclic Monoterpene

Acyclic Monoterpene	Dose/Concentration (mg/kg)	Route of Administration	Animals Used for Study	Mechanism of Action
Citral	50–200	i.p.	Male Swiss mice	Peripheral anti-nociceptive property
Citronellal	50–200	i.p.	Male Swiss mice	Opioid system
Citronellol	25–100	i.p.	Male Swiss mice	Peripheral (inhibition of TNF-α and NO synthesis) and central systems (Opioid system)
(−)-Linalool	25–100	s.c.	Male CD1 mice	Opioidergic and cholinergic systems
Myrcene	5–450	p.o.	Swiss mice	Analgesic peripheral

TABLE 2.16

Chemical and Pharmacological Aspects of Monocyclic Monoterpene

Monocyclic Monoterpene	Dose/Concentration (mg/kg)	Route of Administration	Animals Used for Study	Mechanism of Action
Carvacrol	25–100	i.p.	Male Swiss mice	Inhibition of peripheral mediators and central mechanisms non-opioid
(−)-Carvone	50–200	i.p.	Male Swiss mice	Decreased peripheral nerve excitability
p-Cymene	25–100	i.p.	Male Swiss mice	Opioid system
(+)-limonene	25–100	i.p.	Male Swiss mice	Inhibition of the synthesis of inflammatory mediators
(+)-Menthol	1–10	p.o.	Male Swiss mice	Without anti-nociceptive activity
(−)-Menthol	10–50	p.o.	Male Swiss mice	Selective activation of k-opioid receptors
α-Terpineol	25–100	i.p.	Male Swiss mice	Peripheral and central action

TABLE 2.17

Chemical and Pharmacological Aspects of Bicyclic Monoterpene

Bicyclic Monoterpene	Dose/Concentration (mg/kg)	Route of Administration	Animals Used for Study	Mechanism of Action
1,8-Cineol	0.2–0.5	i.p.	Male Swiss mice	Spinal and supraspinal action
β-Pinene	0.05–0.2	i.p.	Male Swiss mice	Partial agonist of μ-opioid receptors
Rotundifolone	125–250	i.p.	Male Swiss mice	Opioid system

of mitogen-activated protein kinases (MAPKs), expression of nuclear factor kappa B (NF-κB) and production of interleukin-6 (IL-6), tumor necrosis factor-α (TNF-α) and nitric oxide (NO) in lipopolysaccharide (LPS)-induced macrophages. In human chondrocytes, α-pinene inhibited IL-1β-induced inflammation pathway by suppressing NF-κB, c-Jun N-terminal kinase (JNK) activation and expression of iNOS and matrix metalloproteinases (MMP)-1 and -13, suggesting its role as an anti-osteoarthritic agent [84,85].

The monoterpene γ-terpinene, present in the essential oil of many plants including Eucalyptus, reduces the acute inflammatory response. It reduced carrageenan induced paw edema, migration of neutrophil into lung tissue, and IL-1β and TNF-α production and inhibited fluid extravasation [86].

Monoterpene *p*-cymene treatment reduced elastase-induced lung emphysema and inflammation in mice. It reduced the alveolar enlargement, number of macrophages, and levels of proinflammatory cytokines such as IL-1β, IL-6, IL-8 and IL-17 in bronchoalveolar lavage fluid (BALF). Mechanistically, p-cymene blocks NF-κB and MAPK signaling pathways [87].

Monoterpene *d*-limonene was reported to reduce allergic lung inflammation in mice probably via its antioxidant properties. It also reduced carrageenan-induced inflammation by reducing cell migration, cytokine production and protein extravasation. *d*-Limonene exerted an anti-osteoarthritic

effect by inhibiting IL-1β-induced NO production in human chondrocytes. *d*-Limonene treatment reduced doxorubicin-induced production of two proinflammatory cytokines, TNF-α and prostaglandin E₂ (PGE₂) [88,89].

Borneol, a bicyclic monoterpene present in Artemisia, Blumea and Kaempferia, has been used in traditional medicine. Borneol alleviated acute lung inflammation by reducing inflammatory infiltration, histopathological changes and cytokine production in LPS-stimulated mice. It suppressed phosphorylation of NF-κB, IκBα, p38, JNK and ERK. Borneol inhibited migration of leukocytes into the peritoneal cavity in carrageenan-stimulated mice, suggesting its anti-inflammatory function [90,91].

Silva et al. reported the anti-inflammatory and anti-ulcer activities of carvacrol, a phenolic monoterpene constituent of essential oils produced by oregano and other aromatic plants and spices. The pharmacological evaluation was carried out in experimental models of edema induced by different phlogistic agents and gastric lesions induced by acetic acid. In models of paw edema induced by dextran or histamine, carvacrol was effective at 50 mg/kg. Carvacrol at doses of 25, 50 and 100 mg/kg showed a healing capacity on gastric lesions induced by acetic acid (60%, 91% and 81%, respectively) after 14 days of treatment (Table 2.18). These results suggest that carvacrol acts on different pharmacological targets, probably interfering

TABLE 2.18

Effects of Daily Oral Treatment with Carvacrol or Cimetidine for 14 days on Gastric Lesions Induced by 80% Acetic Acid in Rats

Treatments	Dose (mg/kg)	Length(mm)	Ulcerated Area (mm²)	Ulcer Volume (mm³)	Healing Rate (%)
Vehicle	-	12.09±0.61	12.35±1.70	32.49±3.58	-
Carvacrol	25	2.40±0.30[a]	5.88±0.38[a]	13.04±1.03[a]	60
	50	2.08±0.64[a]	2.41±0.73[a]	2.70±0.76[a]	91
	100	2.28±0.20[a]	3.14±0.36[a]	6.32±2.23[a]	81
Cimetidine	100	1.18±0.21[a]	0.84±0.40[a]	0.97±0.52[a]	97

Note: Data are mean±SEM values. Statistical calculations were by one-way analysis of variance followed by Tukey's test.

[a] p<0.001 compared with vehicle.

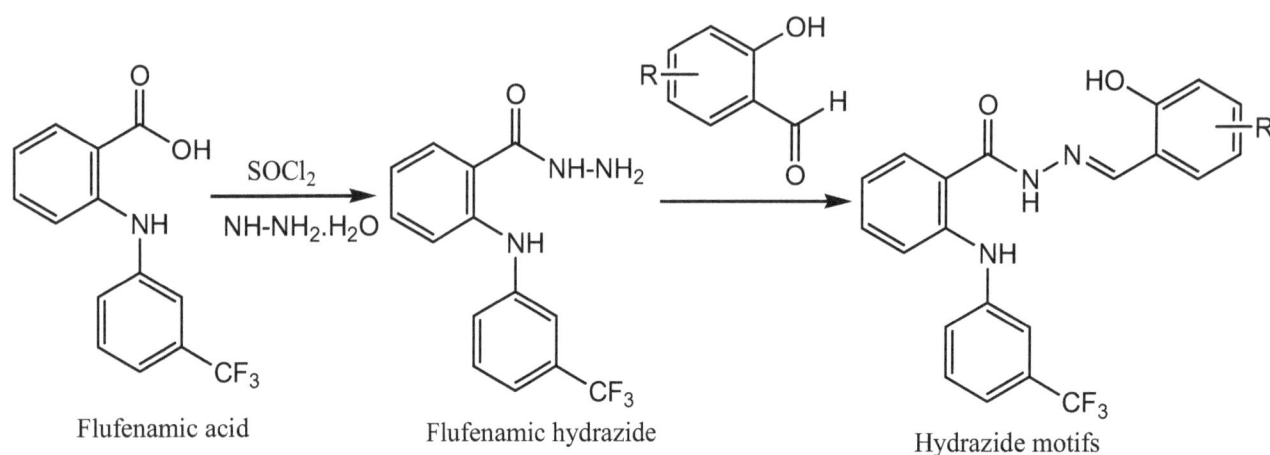

SCHEME 2.52 Synthesis of (E)-N'-(2-hydroxy-3-isopropyl-6-methylbenzylidene-2-((3-(trifluromethyl) phenyl) amino)benzohydrazide.

in the release and/or synthesis of inflammatory mediators, such as the prostanoids, and thus favoring the healing process for gastric ulcers [92].

2.3.5 Antioxidant Activity

Bendre et al. reported three novel compounds derived from flufenamic acid. Synthesis of hydrazide motifs was achieved by reaction between 2-((3-(trifluromethyl)phenyl)amino) benzo-hdrazide (Flufenamic hydrazide) and naturally occurring phenolic monoterpenoids based aldehydes (Scheme 2.52). The developed methodology offers mild and faster reaction conditions with excellent product yield. The synthesized compounds have been screened for their antioxidant activity using DPPH radical scavenger assay and ABTS assay (Table 2.19). All the compounds exhibited good antioxidant activity as compared with standards that is, ascorbic acid and α-Tocopherol [93].

Ratnamala et al. reported the synthesis, characterization and antioxidant activity of Carvacrol Containing Novel thiadiazole and oxadiazole moieties. The synthesis of these compounds was carried out according to the procedure outlined in Scheme 2.53. Firstly, carvacrol was converted to its 4-nitroso derivatives by treating ethanolic solution of phenol with concentrated hydrochloric acid and sodium nitrite at 0°C. The nitroso compound was reduced in an ammonical solution by passing H_2S gas. The 4-aminophenol was further reacted with acetic anhydride to form acetamide derivative. This Acetamide derivative was then treated with ethyl-bromoacetate to obtain ester-type compound and further converted to their hydrazide compound by treatment with hydrazine hydrate. The hydrazide compound reacts with CS_2 in the presence of ethanolic KOH to produce potassium-2-(2-(4-acetamido-5-isopropyl-2-methylphenoxy) acetyl) hydrazine carbodithiaoate. This hydrazide salt on treatment with cold conc. H_2SO_4 at 5°C and NaOH solution separately produces N-(2-isopropyl-4-((5- mercapto-1,3,4-thiadiazol-2-yl)-5-methoxy)-5-methylphenyl)acetamide and N-(2-isopropyl-4-((5-mercapto-1,3,4-oxadiazol-2-yl)-5-methoxy)5-methylphenyl)acetamide. Finally, the synthesized derivatives were evaluated for their *in-vitro* antioxidant

activity by using radical scavenger DPPH assay method. All the compounds exhibited remarkable antioxidant activity. Out of which, some of the compounds showed better or similar antioxidant activity as compared to the standard compound ascorbic acid [94].

TABLE 2.19

Antioxidant Activity of Hydrazide Motifs

Structure	Antioxidant Activity (% RSA)	
	DPPH Radical Scavenger Assay	ABTS Assay
1,8-Cineol	82.53	73.95
	78.64	86.30
	99.71	80.32

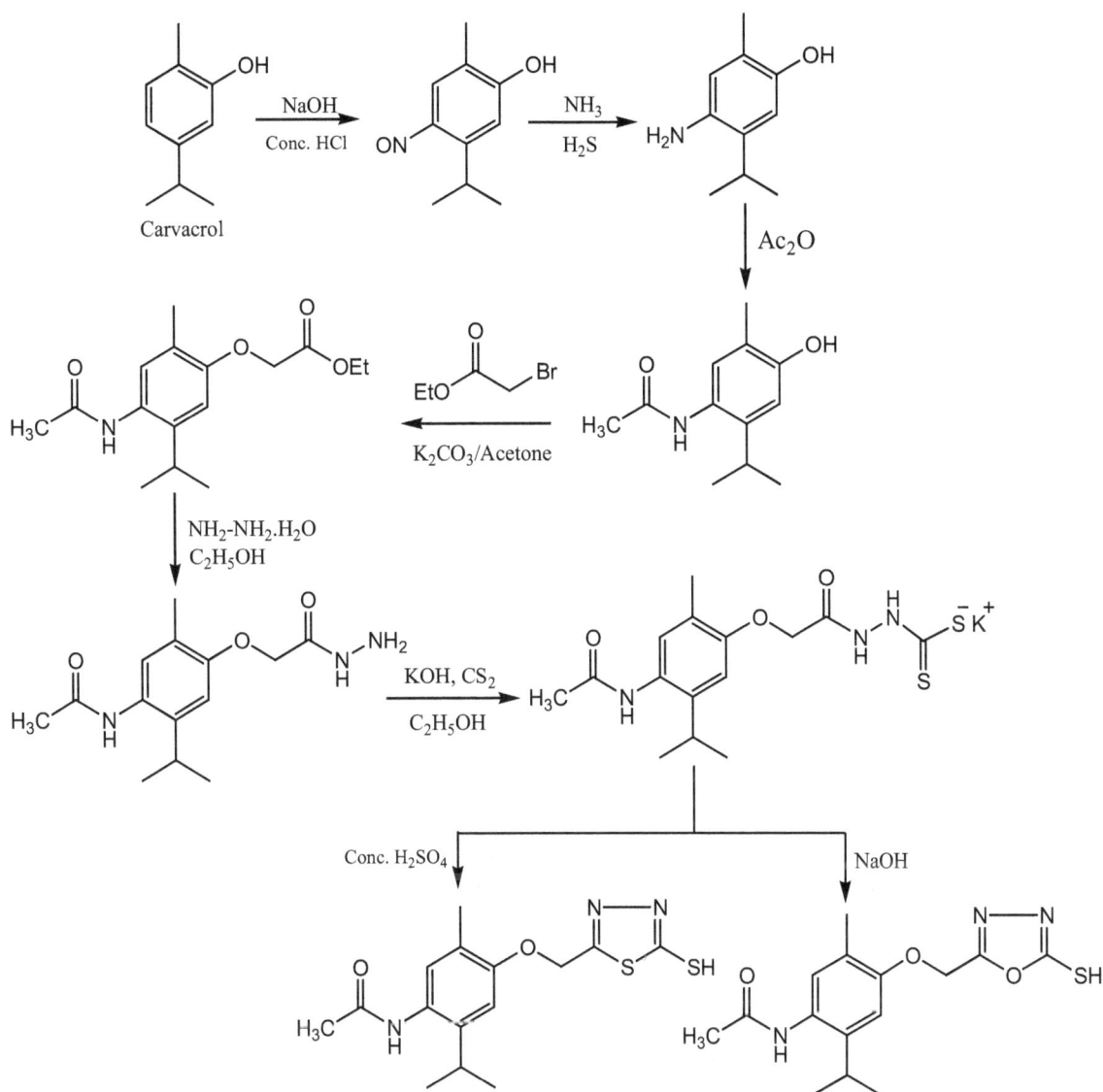

SCHEME 2.53 Carvacrol Containing Novel thiadiazole and oxadiazole moieties.

2.3.6 Anticancer Activity

Cancer represents malignant growth and tumor results due to uncontrolled cell division. It is established that thymoquinone is significantly effective against various cancers such as lung cancer, colorectal cancer, prostate cancer, breast cancer, glioblastoma and other types of cancer. It is reported that thymoquinone causes p53-independent apoptotic cell deaths by the stimulation of caspase-8 and caspase-9, followed by the down-regulation of caspase-3 in the caspase ladder. It also controls the ratio of Bax/Bcl2 by up-regulating the apoptosis favoring the Bax and down-regulating the apoptotic opposing Bcl2 proteins in p53-null HL-60 cell lines simultaneously [95,96].

The antitumor activity of TQ in breast cancer was found to be mediated through different modes of mechanisms, which includes induction of apoptosis, stimulation of the immune system and inhibition of VEGF expression. The anticancer activity of TQ is investigated in prostate cancer both *in-vivo* and *in-vitro*. It was studied that the TQ-induced growth

inhibition in PC-3 and C4-2B prostate cancer cells is due to the generation of ROS. The results suggested that the up-regulation of GADD45a and AIF and down-regulation of Bcl-2-related proteins might have a role in cell death by TQ in PC-3 cells. TQ down-regulated the expression of proliferation marker PCNA and cyclin D1 in A549 lung cancer cell lines *in-vitro*. It is found that TQ significantly decreased the proliferation of human LoVo colon cancer cell and also decreased the levels of p-PI3K, p-Akt, p-GSK3b and β-catenin and thereby down-regulated the expression of COX-2 that resulting in the reduction of PGE2 levels (Figure 2.14) [97–100].

Liu et al. reported the monoterpenoids, citral and geraniol as moderate inhibitors of CYP2B6 hydroxylase activity. Among the monoterpenoids screened, borneol, camphor, cineole, isoborneol, menthol and perillaldehyde showed slight inhibition of CYP2B6 catalyzed bupropion hydroxylation, displaying greater than 50% inhibition at 50 μM (Figure 2.15). Citral and geraniol strongly inhibited CYP2B6 hydroxylase activity in a competitive manner, with Ki values

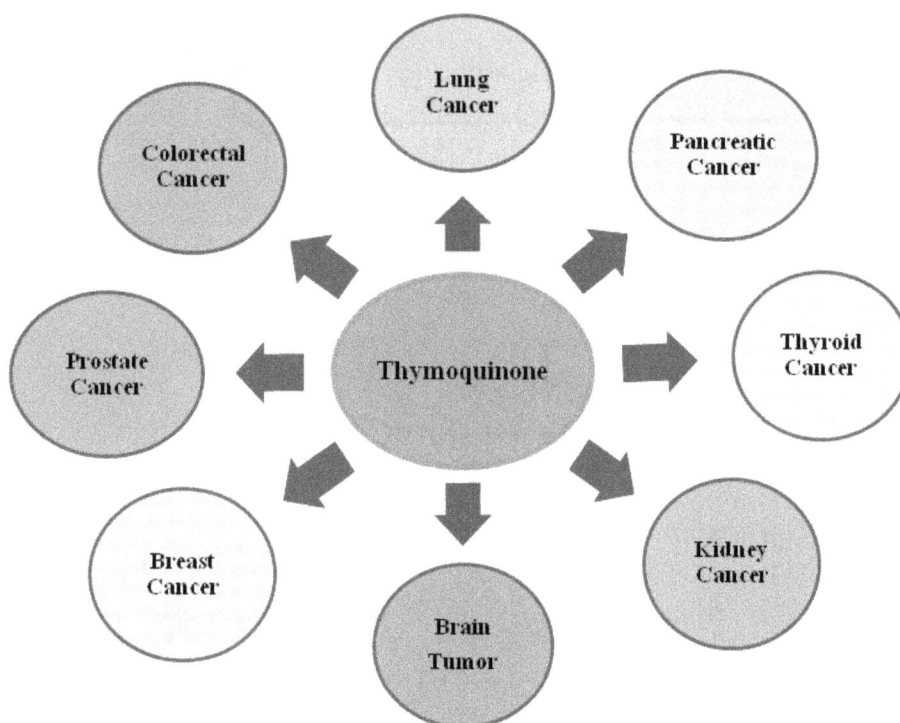

FIGURE 2.14 Therapeutic potentials of thymoquinone for prevention and treatment of different types of cancer.

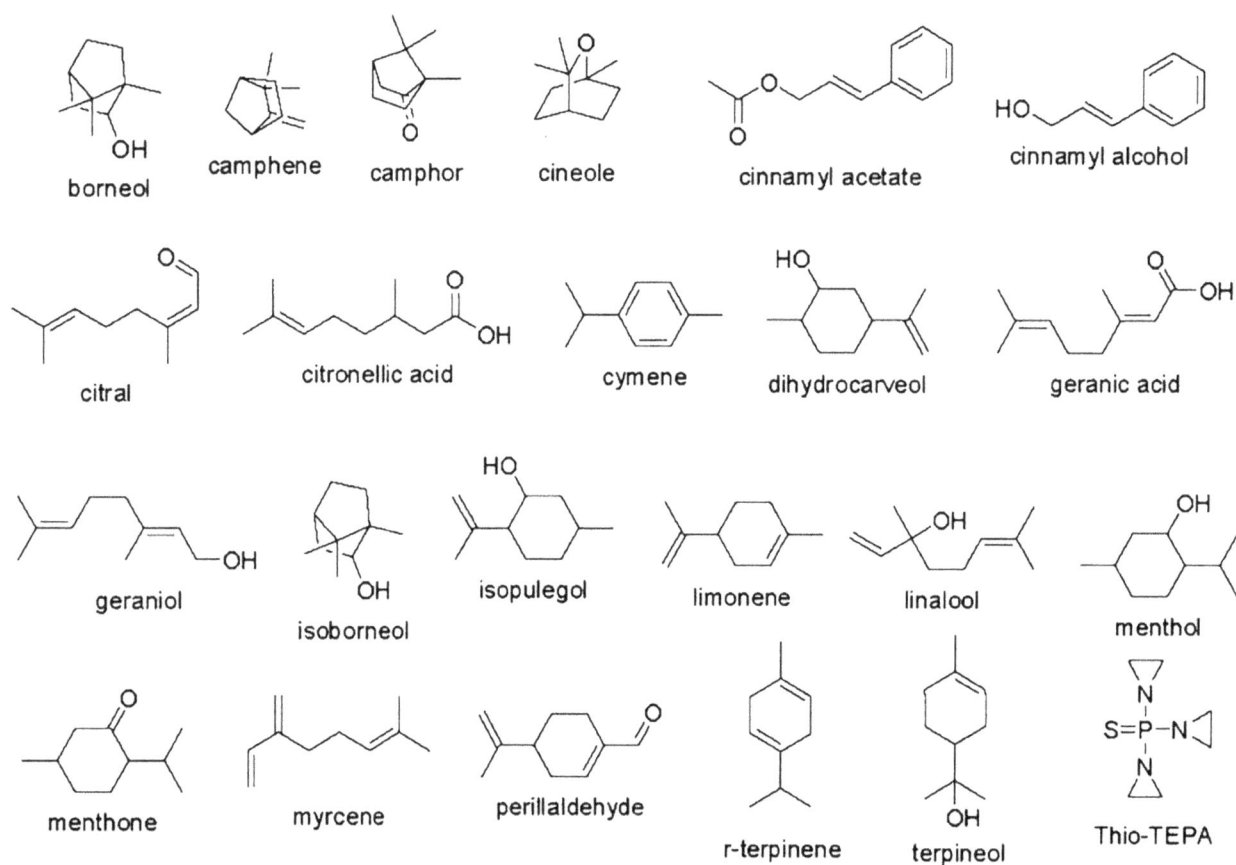

FIGURE 2.15 Chemical structures of the 22 monoterpenoids investigated and of thio-TEPA.

TABLE 2.20

Inhibition of CYP2B6-Catalyzed Bupropion Hydroxylation
by Selected Monoterpenoids

Microsomal Protein (mg/mL)	Ki (μM)		Type of Inhibition
	Citral	Geraniol	
0.25	9.1	6.6	Competitive
0.5	11.4	6.1	Competitive

of 6.8 and 10.3 μM respectively which are higher than the Ki (1.8 μM) of the well-known CYP2B6-selective inhibitor thio-TEPA (Table 2.20) [101].

Perillyl alcohol induces apoptosis without affecting the rate of DNA synthesis in both liver and pancreatic tumor cells. It also significantly reduces the appearance of secondary tumors which suggests that it may have chemopreventive activity against mammary cancer. It reduces the growth rate of transplanted prostatic carcinomas in nude mice. The anti-tumor activity of Perillyl alcohol is due to modulation of cell proliferation, cell death, or cell differentiation. The treatment of perillyl alcohol (800 μM) on cultured human MIA PaCa2 pancreatic carcinoma cells, significantly increase the percentage of cells undergoing apoptosis (Table 2.21) [102].

Apoptosis is referred as programmed cell death. Oxidative stress is the major factor to promote apoptosis. The Bcl-2 proteins are the regulators for cell apoptosis under both normal and oxidative conditions. Ling-ai et al. reported the effect of catalpol on H_2O_2-induced apoptosis in the rat embryonic ventricular myocardial cell line (H9c2). It was examined that the potential mechanisms of catalpol-associated protection are due to alteration in the mitochondrial-dependent caspase pathway and changes in Bcl-2 and Bax expression. Mitochondria are the common integrators and transducers of various pro-apoptotic signals and its disruption and the subsequent release of pro-apoptotic proteins like cytochrome-c is the critical component of the apoptosis process. The experimental study showed that preincubation of H9c2 with catalpol (0.1–10 μg/mL) for 24h significantly reduced the decrease in viability associated with exposure to H_2O_2 (100 μM) for 24h as determined by MTT and LDH assay. MDA levels were decreased significantly and SOD activities were significantly increased by catalpol pretreatment in a dose-dependent manner. Catalpol decreased the H_2O_2-induced apoptosis in H9c2 cells in a concentration-dependent manner as indicated by Hoechst 33258 and flow-cytometric analysis [103].

TABLE 2.21

Effects of Perillyl Alcohol on Tumor Cell Proliferation and Tumor Volume

Treatment	% BrDU Positive Tumor Cells	Tumor Volume (cm³)
Perillyl alcohol (5%)	29.7±6.0	3.02±0.62
Control (no drug)	35.1±11.7	1.43±0.63[a]

Note: The data is represented as mean±SEM, *n*=5.

[a] $p < 0.05$ vs control by student's t-test.

2.3.7 Anti-Proliferative Activity

Szakonyi et al. reported the stereoselective synthesis of monoterpene-based 1,2,4- and 1,3,4-oxadiazole derivatives. It was accomplished starting from α, β-unsaturated carboxylic acids, prepared previously through the oxidation of (–)-2-carene-3-aldehyde [derived in a two-step reaction from (–)-(S)-perillaldehyde] using commercially available (–)-myrtenal [104].

There are several methods for the preparation of 1,2,4-oxazoles starting from compounds bearing carboxyl function. Carboxylic acids **1** and **2** were coupled with amidoximes in the presence of CDI in DCM with excellent yields, followed by TBAF-catalyzed ring closure of O-acylated amidoximes **3–5** at room temperature. The obtained α, β-unsaturated 3-aryl-1,2,4-oxazoles **6–8** were stereoselectively dihydroxylated with the OsO4/NMO system resulting in **9–11** as single diastereoisomers (Scheme 2.54) [105].

The synthesis of the isomeric 1,3,4-oxadiazole ring system was also started from monoterpenic acids **1** and **2**. In this case, however, we have found different reactivity between the pinane and carane ring system under the ring-closure process. Myrtenic acid **1** was coupled with benzhydrazide in the presence of CDI. Cyclodehydration of N, N′-diacylhydrazine **12** with $POCl_3$ at 80°C furnished compound **13** with moderate yield. Dihydroxylation of **13** with OsO_4/NMO afforded **14** in a highly diastereoselective reaction (Scheme 2.55) [106].

Starting from (–)-2-carene-3-carboxylic acid **2**, subsequent ring closure of N, N′-diacylhydrazine **15** resulted in 1,3,4-oxazole **16** with ring rearrangement and loss of chirality of the carane ring system (Scheme 2.56). Dihydroxylation of **16** under mild conditions provided exo-dihydroxylated racemic oxadiazole **17** with moderate yield (Table 2.22).

Sun et al. reported the synthesis and in vitro anti-proliferative activity of β-elemene monosubstituted derivatives in HeLa cells mediated through arrest of cell cycle at the G_1 phase. Chemically, it is 5S, 7R, 10S-(–)-(1-methyl-1-vinyl-2,4-diisopropenyl-cyclohexane. It is a natural sesquiterpene extracted from the medicinal herb *Curcuma wenyujin*. The monosubstituted β-elemene ether and amine derivatives were synthesized according to the procedure outlined in Scheme 2.57. The synthesis began with chlorination of β-elemene with NaClO, the resulting product chlorinated β-elemene (**2**) reacted further with alcohols or amine to yield the desired β-elemene monosubstituted ether or amine derivatives 3, 4 (a–h). Their anti-proliferative activities were evaluated by WST-1 method on HeLa cell lines. Results are shown in Table 2.23. In general, the proliferative activity for amine derivatives is much higher than that of the ether derivatives, especially for isopropyl amine, cyclopentyl amine or cyclohexyl amine [107].

Myrcene, the acyclic monoterpene, also exhibits significant anti-proliferative and cytotoxic effects in various tumor cell lines such as MCF-7 (breast carcinoma), HeLa (human cervical carcinoma), A549 (human lung carcinoma), HT-29 (human colon adenocarcinoma), P388 (leukemia) and Vero (monkey kidney).

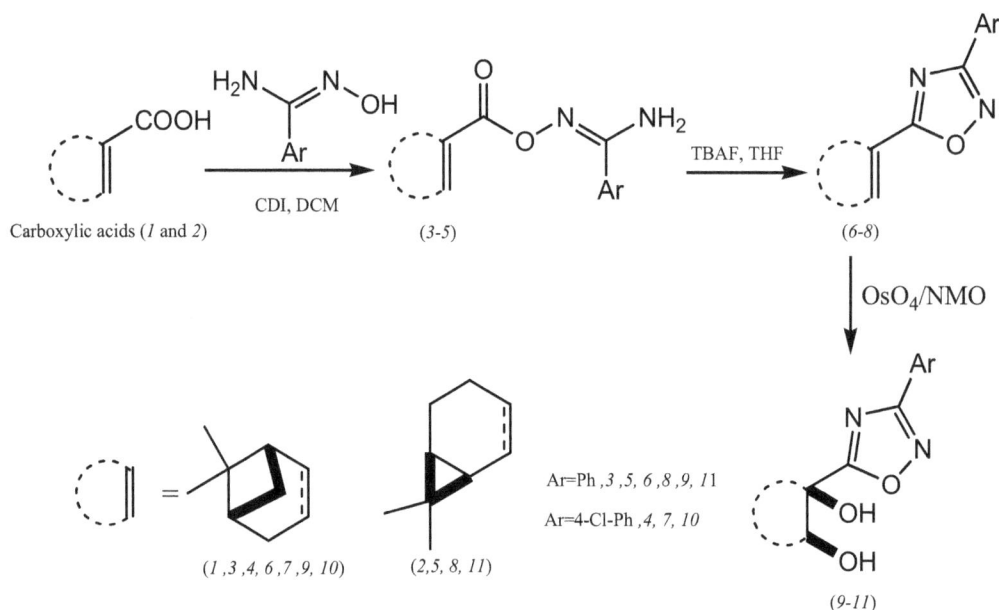

SCHEME 2.54 Stereoselective synthesis of monoterpene-based 1,2,4-oxadiazoles.

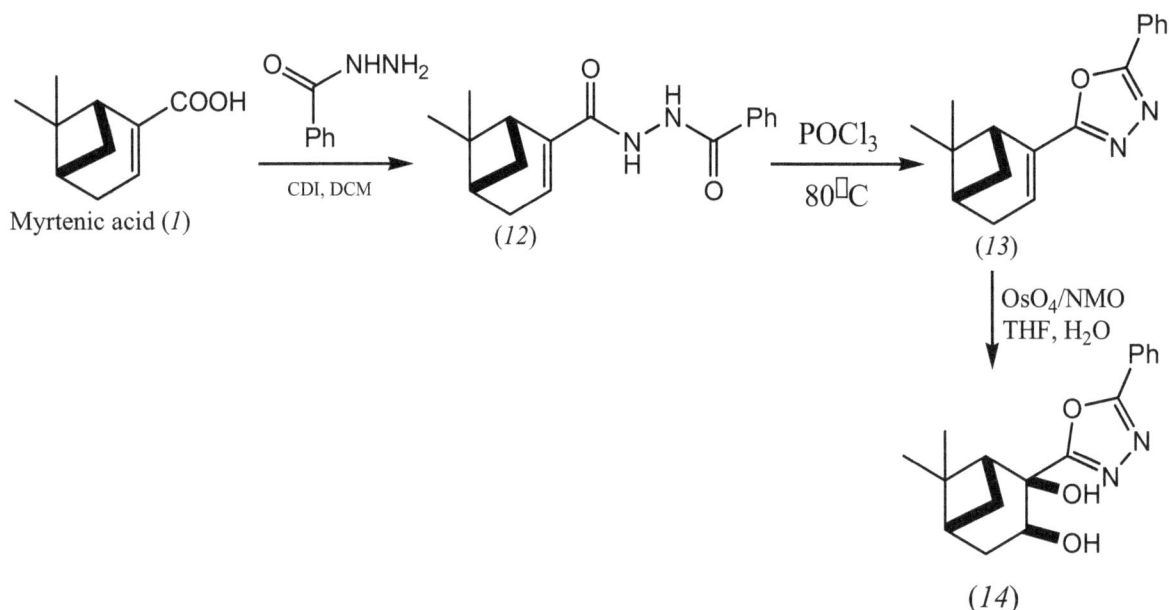

SCHEME 2.55 Stereoselective synthesis of pinane-based 1,3,4-oxadiazole.

2.3.8 Anticonvulsant Activity

Epilepsy is one of the most common neurological conditions that show a prevalence rate of 1%–2% of the world population. Epilepsy is characterized by manifestation of seizure. Seizures result from the abnormal discharge of groups of neurons within the brain, usually within a focal point. The current developments of new anticonvulsant drugs require the appropriate choice of animal models of epilepsy for the determination of pharmacological and toxicological activities as well as novel mechanisms of action of naturally derived and synthetically produced terpenoids [108–112].

Citronellol is an acyclic monoterpenoid present in the essential oils of various aromatic plant species such as *Cymbopogon winterianus*. It occurs naturally as two isomeric optical forms such as *R*-(+)-isomer and *S*-(–)-isomer. The experiments were performed to evaluate the profile of citronellol anticonvulsant activity in three seizure models such as pentylenetetrazol (PTZ) and picrotoxin-induced convulsions and maximal electroshock-induced seizures (MES) in mice. In addition, the effects of citronellol on the nerve compound action potential (CAP) were also examined. The CAP was recorded with the single-sucrose gap method. Administration of citronellol (400 mg/kg) significantly reduced the number of animals of

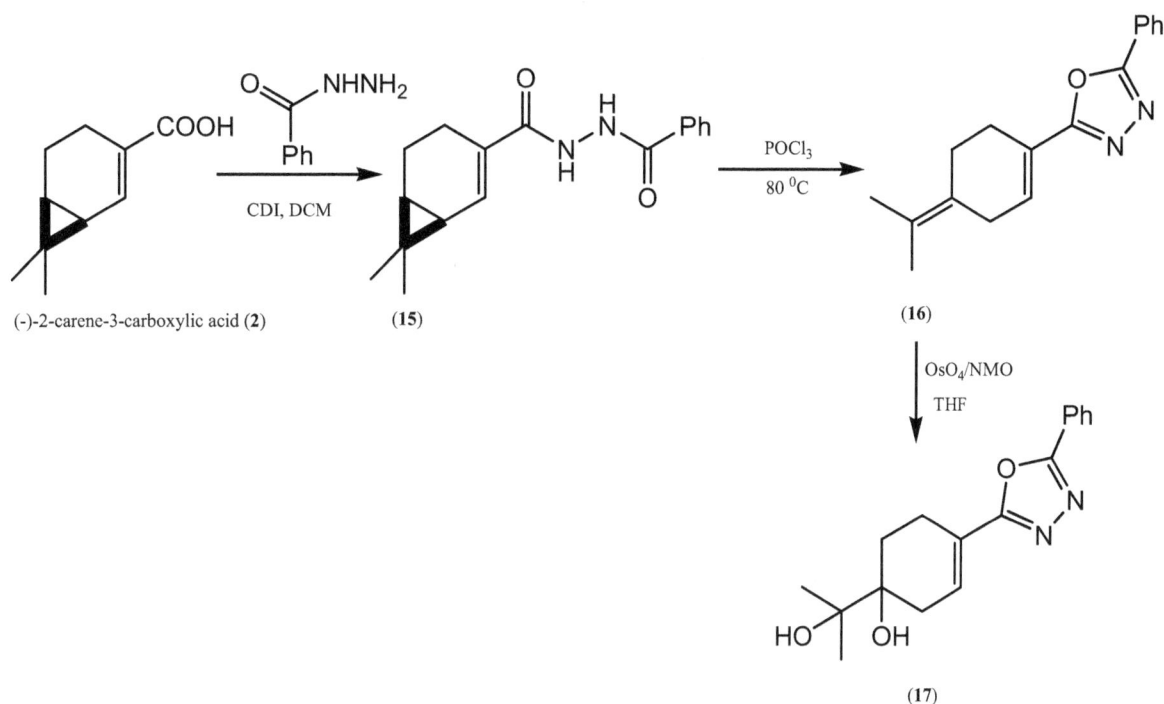

SCHEME 2.56 Rearrangement of the carane ring toward cycloxene-based 1,3,4-oxadiazoles

TABLE 2.22

Anti-Proliferative Activities of the Tested Monoterpene Analogs

Analog	Conc. (μM)	Inhibition (%)±SEM (Calculated IC$_{50}$ Value (μM))			
		HeLa	A 2780	MCF7	MDA-MB-231
3	10	29.83±1.65	96.66±0.23	35.07±2.64	19.06±2.98
	30	98.46±0.08	96.60±0.37	93.00±0.68	84.58±1.87
4	10	37.96±2.20	96.28±0.32	49.99±0.80	33.33±1.94
	30	98.17±0.19	96.82±0.18	96.08±0.52	89.95±0.90
5	10	21.94±2.94	91.95±0.26	28.62±1.98	20.31±1.45
	30	96.31±0.33	93.77±0.27	85.27±1.84	58.06±1.75
6	10	-^{a}	56.80±3.18	17.30±1.74	19.61±2.23
	30	47.39±2.99	93.34±0.69	23.87±2.37	29.68±2.54
7	10	-	15.35±1.79	19.80±1.99	-
	30	24.18±2.42	36.92±1.39	23.44±1.86	24.66±1.38
8	10	-	12.18±1.70	14.62±2.49	11.71±0.74
	30	12.31±2.25	36.58±1.38	35.75±2.79	20.27±2.62
9	10	-	-	-	11.45±2.75
	30	10.83±2.40	34.56±1.83	-	21.57±1.69
10	10	-	17.93±1.06	-	11.43±1.68
	30	26.01±0.74	48.27±0.64	44.00±1.34	24.67±2.13
11	10	12.36±1.23	-	-	-
	30	13.41±1.87	28.44±0.92	25.32±1.10	10.36±2.14
12	10	-	-	-	15.51±2.77
	30	-	31.13±2.48	-	15.92±2.69
13	10	-	11.76±0.76	-	-
	30	-	48.66±1.97	-	18.81±2.59
14	10	-	-	-	-
	30	-	29.12±2.32	14.46±2.31	16.28±1.85
Cisplatin	10	42.61±2.33	83.57±1.21	53.03±2.29	67.51±1.01
	30	99.93±0.26	95.02±0.28	86.90±1.24	87.75±1.10

^{a} Growth inhibition values less than 10% are considered negligible and not given numerically.

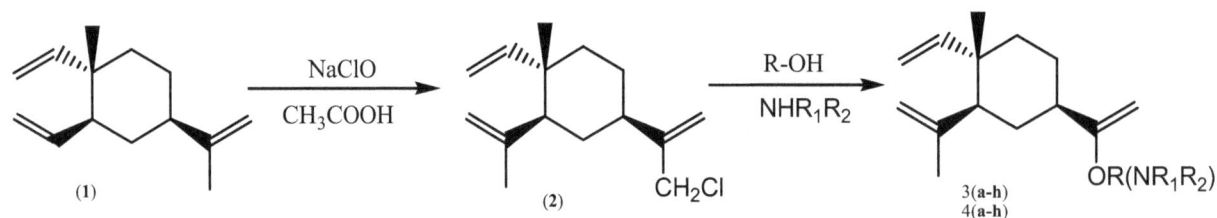

Comp. Code	R	Comp. Code	R_1R_2
3a	Me	4a	BuH
3b	Et	4b	t-BuH
3c	Bu	4c	H, Cyclohexyl
3d	t-Bu	4d	H, Cyclopentyl
3e	$(CH_2)_{10}CH_3$	4e	H, Cyclohexyl
3f	Cyclohexyl	4f	Imidazole
3g	Ph-CH$_2$	4g	Et
3h	$CH_2CH_2OCH_3$	4h	i-Prh

SCHEME 2.57   Synthetic procedure of β-elemene monosubstituted derivatives.

TABLE 2.23

Anti-Proliferative Activity of β-Elemene Monosubstituted
Derivatives in HeLa Cell Lines

Compounds	IC_{50} (μM)
1	236.2 ± 3.2
3a	211.9 ± 2.8
3b	105.9 ± 24.4[b]
3c	265.9 ± 17.9
3h	237.6 ± 40.3
4a	10.81 ± 0.21[b]
4b	25.37 ± 0.45[b]
4c	3.39 ± 1.40[b]
4d	0.52 ± 0.20[b]
4e	5.90 ± 2.42[b]
4f	15.67 ± 1.80[b]
4g	14.35 ± 1.05[b]
4h	0.04 ± 0.01[b]

Note: The IC_{50} were calculated according to the absorbance. Data are presented as mean$\pm$SD, $n=3$. Differences were considered significant when $p<0.05$ (a) and $p<0.01$ (b).

convulsion induced by PTZ and eliminated the extensor reflex of MES test in about 80% of the experimental animals. In addition, administration of citronellol exhibited protection in the pentylenetetrazol and picrotoxin tests by increasing the latency of clonic seizures. Adult male albino Swiss mice weighing 24–30 g, with 3 months of age were used throughout this study. The animals were randomly housed in appropriate cages at 24°C$\pm$2°C on a 12 h light cycle with free access to food and tap water. They were used in groups of eight animals each. All drugs were injected intraperitoneally (i.p.). Diazepam at 3 mg/kg (i.p.) was used as positive control. The results demonstrate that citronellol possesses significant anticonvulsant activity probably due to the reduction of neuronal excitability mainly through the voltage-dependent Na$^+$ channels (Table 2.24) [113].

Sousa et al reported the anticonvulsant activity of the racemate and enantiomers of linalool. The pretreatment of the mice with (S)-(+)-, (R)-(−)- and rac-linalool increased the latency of convulsions significantly in the PTZ model. Only rac-linalool had an effect at the dose of 200 mg/kg. The enantiomers and their racemic mixture were effective in inhibiting the convulsant effect of PTZ at the dose of 300 mg/kg. The linalools

TABLE 2.24

Time-Course of Citronellol on CAP Amplitude Blockade

Citronellol (mM)	Control (mV)	Amplitude after 10 minutes (mV)	Amplitude after 20 minutes (mV)	Amplitude after 30 minutes (mV)
0.03	22.1 ± 1.6	22.1 ± 1.2	21.3 ± 1.1	20.7 ± 1.5
0.32	23.6 ± 2.3	22.6 ± 1.9	22.3 ± 2.2	21.2 ± 2.5
3.2	22.1 ± 0.8	16.1 ± 2.2[a]	14.4 ± 2.5[a]	13.3 ± 3.0[a]
4.5	21.9 ± 2.2	17.4 ± 2.0	6.3 ± 1.3[a]	3.1 ± 1.5[a]
6.4	20.4 ± 1.7	6.4 ± 0.7[a]	2.2 ± 0.7[a]	1.4 ± 0.4[a]

Note: Each value represents the mean$\pm$S.E.M of at least four nerve bundles.

[a] $p<0.05$ (statistical significance).

displayed pharmacological activity close to that of diazepam. In the PIC seizure model, (R)-(–)-linalool and rac-linalool exhibited activity at the dose of 200 mg/kg, but the rac-linalool was more potent than (R)-(–)-linalool whereas (S)-(+)-linalool had no effect at this dose. On the other hand, this enantiomer was effective at the dose of 300 mg/kg, but less potent than (R)-(–)-linalool and rac-linalool. In the MES model, linalools decreased the convulsion time of the mice at the doses of 200 and 300 mg/kg. But, rac-Linalool presented maximum effect at the dose of 300 mg/kg. Using the parameter of tonic hind convulsions, only (R)-(–)-linalool showed protection from tonic extension at the dose of 200 mg/kg. When the (+)- and (–)-enantiomers and rac-linalool were administered at the dose of 300 mg/kg, they were also effective in preventing tonic convulsions induced by transcorneal electroshock in the animals. From the results, it was observed that, the (+)- and (–)-forms of Linalool were equipotent and the rac-linalool was more effective than phenytoin [114].

α-Terpineol is a monoterpenoid alcohol and investigated for its anticonvulsant activity. This compound increased the latency to convulsions induced by pentylenetetrazole at doses of 100 and 200 mg/kg and decreased the incidence of hind limb extension produced by MES in a dose-related manner at doses of 200 and 400 mg/kg. The reference drug phenobarbital also produced a significant protection at a dose of 25 mg/kg [115].

2.3.9 Sedative and Hypnotics

Various essential oils and monoterpenes possess therapeutic properties such as central nervous system (CNS) relaxant and tranquilizer. Different structurally related monoterpene alcohols were evaluated in mice to investigate their pharmacological potential in CNS. Isopulegol, neoisopulegol, (+/–)-isopinocampheol, (–)-myrtenol, (–)-cis-myrtanol, (+)-p-menth-1-en-9-ol and (+/–)-neomenthol exhibited the depressant effect in the pentobarbital-induced sleep test and indicate sedative property. (–)-menthol, (+)-dihydrocarveol and (+/–)-isoborneol were found to be ineffective in this test. The results indicate that these monoterpenes have the profile of sedative property and this pharmacological effect is influenced by the structural characteristics of the molecules [116].

Linalool is a monoterpene and is used in traditional medical systems as hypno-sedatives. The psychopharmacological evaluations of linalool administered through i.p. and i.c.v. route reported the sedative and anticonvulsant central effects in various mouse models. For this study, mice were placed in an inhalation chamber for 60 min, in an atmosphere saturated with

1% or 3% linalool. Immediately after inhalation, animals were evaluated for determining locomotion, barbiturate-induced sleeping time, body temperature and motor coordination (rota-rod test). The 1% and 3% linalool increased (p < 0.01) pento-barbital sleeping time and reduced (p < 0.01) body temperature. The 3% linalool decreased (p < 0.01) locomotion. Motor coordination was not affected. Therefore, linalool inhaled for 1 h appears to induce sedation without significant impairment in motor abilities [117].

2.3.10 Treatment of Neurodegenerative Disorders

Neurodegenerative disorders (NDDs) are characterized by progressive dysfunctions of the nervous system which results, due to the loss of neural structure and function, finally lead to neuronal death. Various NDDs like Alzheimer's disease (AD), Parkinson's disease (PD), Huntington's disease (HD) are a heterogeneous group of disorders with the progressive and severe loss of neurons. Hence, the investigation of monoterpenoids which can modulate the targets of NDDs is highly significant. Further developments including detections of monoterpenoids and their derivatives with high neuroprotective or neurotrophic activity as well as the results of qualified clinical trials are needed to draw conclusions regarding the safety and efficacy of these agents [118].

Iridoids class of monoterpenoid compounds constructed from 10-carbon skeleton of isoprene building units (Figure 2.16). The structural marker of these compounds is the cis fused cyclopental[c]pyran system, which is present in nature in the form of glycosides, aglycones, as secoiridoids or bisiridoid forms. For secoiridoid C7–C8 bond of the iridoid skeleton is cleaved following a series of oxidation steps to give rise to compounds like secologanin that also serve as a precursor for the synthesis of alkaloids [119]. The biosynthetic pathways of terpenoids involve some pivotal intermediates like mevalonic acid. Despite the initiating primary metabolite is a 2 carbon [C] metabolite, acetyl Co-A, the basic skeleton of all terpenoids by definition is 5-C isoprene units in the form of IPP and dimethyl allyl pyrophosphate (DMAPP). The precursor of all terpenoids in the further steps of reaction is the geranyl pyrophosphate (GPP), which is made from two isoprene units. The sesquiterpenes (15C), Diterpenes (20C) and triterpenes (30C) are typical examples of terpenoids which arise by condensation of these isoprene units via a range of both cyclic and acyclic monoterpenes of biological significance, while the 8-hydroxy geraniol unique pathways ⇒ iridoids and their derivatives (Scheme 2.58). Iridoids and structurally related monoterpenes display a wide range of pharmacological effects

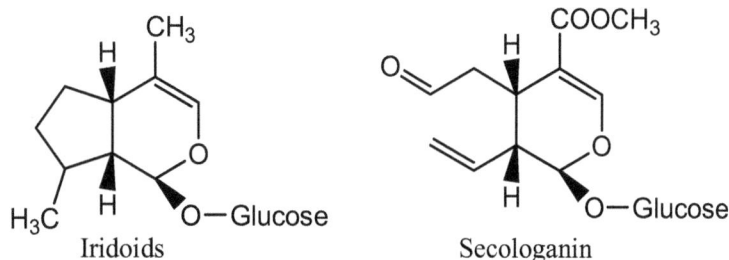

FIGURE 2.16 Structure of iridoids class of monoterpenoid.

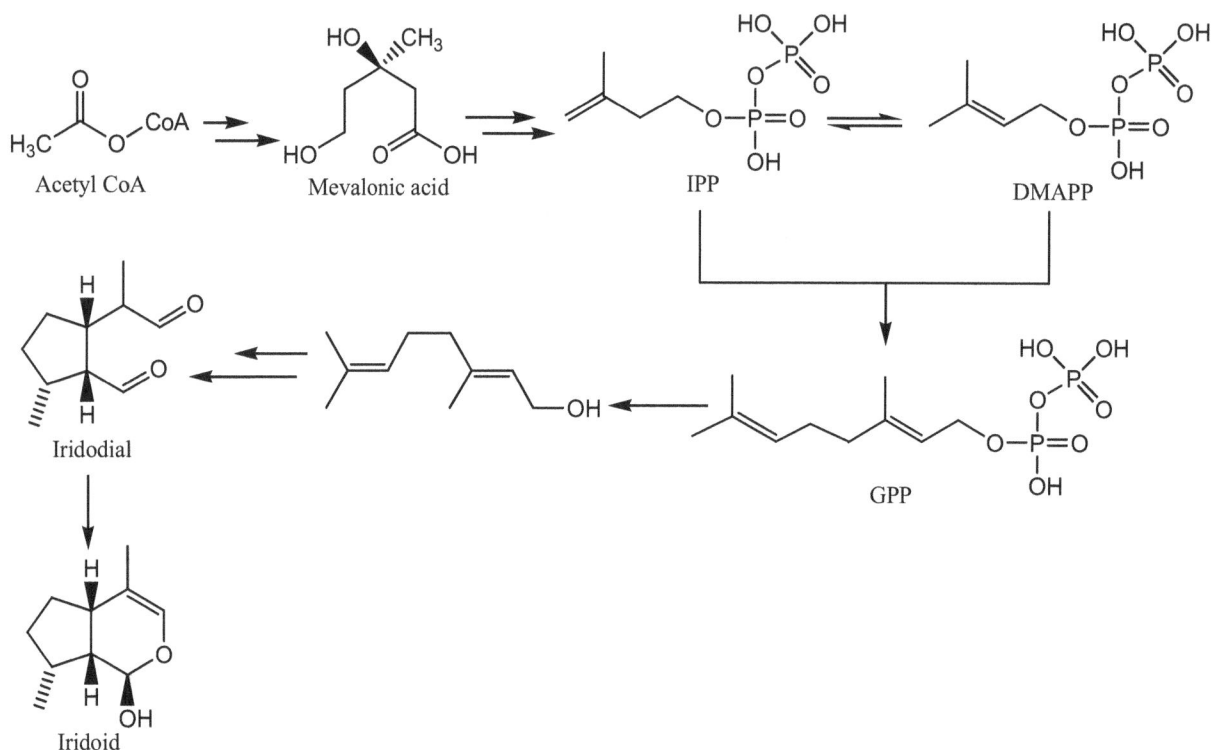

SCHEME 2.58 Synthetic route to generation of iridoids.

which make them potential modulators of AD. A mechanistic approach of scrutiny addressing their effects in Alzheimer's brain including increase in protein phosphorylation signaling, amyloid beta formation [120].

Alzheimer's disease (AD) is a progressive, age-related and neurodegenerative disease which is characterized by progressive loss of memory and other cognitive-related impairments. Jiaming Li et al reported the design, synthesis and evaluation of genipin derivatives for the treatment of Alzheimer's disease. These are evaluated for their inhibitory activity against acetylcholinesterase (AChE). Compound **13a** (methyl-1-methoxy-7-((4-((3,5,6-trimethylpyrazin-2-yl)methyl)piperazin-1-yl) methyl)-1,4a, 5,7a-tetrahydrocyclopenta[c]pyran-4-carboxylate) (Figure 2.17) displayed the most potent AChE inhibitory activity in this series with IC$_{50}$ value of 218 nM. It was investigated to determine the neuroprotective action of these compounds against PC12 cells injured by amyloid β-protein 1–42 (Aβ1–42). Among the tested compounds, **8a** (ethyl 7-((4-benzylpiperazin-1-yl) methyl)-1-methoxy-1,4a, 5,7a-tetrahydrocyclopenta[c]pyran-

4-carboxylate dihydrochloride) exhibited higher inhibition rate of % Inhibition equal to 22.29 as compared to the positive reference, Donepezil with % inhibition of 17.65 (Tables 2.25 and 2.26) [121].

2.3.11 Cardiovascular Effects

The cardiovascular effects of carvacrol were studied both *in-vivo* and *in-vitro*. In normotensive rats, carvacrol reduced the blood pressure and heart rate and inhibited hypertension induced by L-NAME at a dose of 100 mg/kg (i.p.). Neves et al. demonstrated that carvacrol induced an endothelium-independent relaxation, possibly involving inhibition of Ca^{2+} influx through the membrane. Carvacrol causes the influx of Ca^{2+} in endothelial cells by increasing intracellular Ca^{2+} and leading to activation of K$^+$ channels sensitive to Ca^{2+} medium (IKCa) and low (SKCa) conductance [122,123].

Bastos et al. performed experiments *in-vitro* using preparations of isolated rings of superior mesenteric artery of rats and

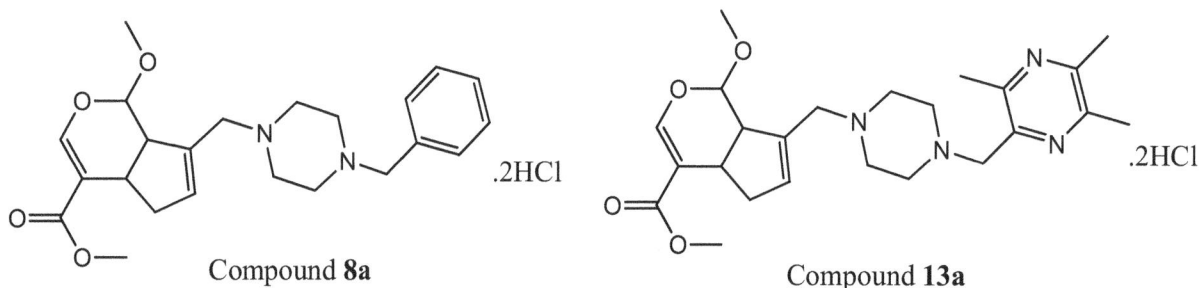

Compound **8a**

Compound **13a**

FIGURE 2.17 Structure of genipin derivatives (8a, 13a).

TABLE 2.25

In-Vitro Inhibitory Activity of Genipin Derivatives against AChE

Compounds	IC_{50} (nM, 3 minutes)	IC_{50} (nM, 5 minutes)	IC_{50} (nM, 7 minutes)
4a	1,919	2,496	2,561
5a	1,592	1,921	1,824
6a	>10,000	>10,000	>10,000
7a	1,587	1,924	2,210
8a	>10,000	>10,000	>10,000
9a	7,881	7,630	6,762
10a	2,852	2,519	2,556
11a	>10,000	>10,000	>10,000
12a	401	802	1,283
13a	121	253	281
14a	610	769	832
15a	741	984	1,535
Genipin	>10,000	>10,000	>10,000
Donepezil	157	281	293

TABLE 2.26

Effect of Genipin Derivatives on Protecting PC12 Cells Against Aβ1-42

Group	Conc. (μM)	MTT(570 nm OD)	Inhibition Rate of Cell Damage (%)
8a	1	0.446±0.02	7.43
	3.2	0.469±0.01[b]	14.55
	10	0.484±0.02[c]	19.20
	32	0.494±0.02[c]	22.29
8b	1	0.457±0.01[b]	10.84
	3.2	0.460±0.01[b]	11.76
	10	0.465±0.02[b]	13.31
	32	0.436±0.01	4.33
13a	1	0.439±0.02	5.26
	3.2	0.442±0.01	6.19
	10	0.447±0.02	7.74
	32	0.450±0.02	8.67
13b	1	0.430±0.01	2.48
	3.2	0.433±0.02	3.41
	10	0.437±0.02	4.64
	32	0.441±0.01	5.88
Genipin	32	0.436±0.02	4.33
Donepezil	32	0.479±0.01[c]	17.65
Aβ1-42	5.5	0.422±0.01[a]	-
Control	-	0.745±0.03	-

[a] p<0.001 vs control group.

[b] p<0.05.

[c] p<0.01 vs Aβ1-42 group.

Note: $n=4$, $x±s$.

presented that citronellol was able to induce vasorelaxation. The relaxation of the vessel was due to the inhibition of Ca^{2+} influx through the membrane and the release of Ca^{2+} from intracellular stores [124].

Touvay et al. evaluated the effects of limonene and sobrerol on pulmonary hypertension and right ventricular hypertrophy induced by monocrotaline (MCT) in rats. After daily oral administrations of both limonene and sobrerol were able to significantly decrease the changes induced by MCT at a dose of 400 mg/kg of rat [125].

Linalool possesses optical isomers such as (+) and (−)-linalool. Both were evaluated independently for their effects on blood pressure and heart rate, administered by inhalation. The results demonstrated that the optical isomers exhibited opposite effects. (+)-Linalool showed a stimulating effect whereas (−)-linalool produced a depressing effect on the cardiovascular system.

Menezes et al. evaluated the hypotensive activity of (±)-linalool in non-anesthetized normotensive rats and reported that this monoterpene was able to induce hypotension associated with tachycardia [126].

Johnson et al. studied the effect of menthol on human forearm cutaneous vessels and rat arteries *in-vitro*. The results showed that vasodilatation induced by menthol appears to be independent of cholinergic/nitric oxide pathways and mediated by activation of transient receptor potential melastatin-8 (TRPM8) [127]. Baylie et al. reported that menthol was able to inhibit the currents for L-type Ca^{2+} in right ventricular myocytes isolated from rabbits.

El Tahir et al. observed that α-pinene and p-cymene showed hypotension and bradycardia after intravenous administration in urethane anesthetized rats. This effect may be due to inhibition of vasomotor center with the consequent decrease in the sympathetic outflow reaching the peripheral blood vessels and the heart which results in decreases in both arterial blood pressure and the heart rate [128,129]. Menezes et al. demonstrated that the (+)-α-pinene and (−)-β-pinene exhibited hypotensive effect associated with tachycardia after intravenous administration in non-anesthetized normotensive rats.

The cardiovascular effects of α-terpineol were first reported by Saito et al. They demonstrated that α-terpineol exhibited hypotensive effect in rats at a dose of 5 mg/kg administered intravenously. Magalhaes et al. reported that α-terpineol induced vasorelaxation in perfused rat mesenteric vascular bed which was abolished in the presence of L-NAME. This result suggested the involvement of NO in this vasorelaxation.

Saito et al. observed that perillyl alcohol possesses hypotensive effects at doses of 1 and 5 mg/kg, when administered intravenously in rats. Rotundifolone is the main constituent of the essential oil of *Mentha x villosa* [130]. Guedes et al. reported that the intravenous administration of rotundifolone reduced blood pressure and heart rate significantly in rats. They performed the experiments *in-vitro* using isolated preparations of atrial and aortic rings from rats to investigate the mechanisms involved in these responses. These experiments demonstrated that rotundifolone (Figure 2.18) was able to

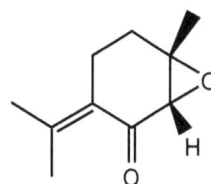

FIGURE 2.18 Structure of rotundifolone.

TABLE 2.27

Cardiovascular Effects of Monoterpenes

Monoterpenes	Tissue and/or Species	Study Type	Effects
Carvacrol	Isolated canine and human ventricular cardiomyocytes	In-vitro	Cardiac arrhythmias
	Anesthetized rat	In-vivo	Decreased heart rate
Citronellol	Normotensive rat	In-vivo	Hypotension
	Rat mesenteric artery	In-vitro	Vasorelaxation
1,8-Cineole	Normotensive anesthetized and conscious rats	In-vivo	Hypotension
Limonene	Rat	In-vivo	Reduction and prevention of cardiovascular injuries caused by pulmonary hypertension
Sobrerol	Rat	In-vivo	Regulate the development of pulmonary hypertension
(+)-Linalool	Human (Inhalation)	In-vivo	Cardiovascular system stimulant
Menthol	Human forearm cutaneous vessels and rat arteries	In-vitro	Vasorelaxation
p-Cymene	Urethane anesthetized rat	In-vivo	Hypotension and bradycardia
Thymol	Isolated canine and human ventricular cardiomyocytes	In-vitro	Cardiac arrhythmias
(+)-α-Pinene	Urethane anesthetized rat	In-vivo	Hypotension and bradycardia
(−)-β-Pinene	Non-anesthetized normotensive rat	In-vivo	Hypotension and tachycardia
Terpineol	Rat isolated mesenteric vascular bed	In-vitro	Vasorelaxation
Eucalyptol	Left ventricular papillary muscle of rat	In-vitro	Negative inotropic effect
Perillyl alcohol	Rat	In-vivo	Hypotension
Rotundifolone	Normotensive rats	In-vivo	Hypotension and bradycardia

induce negative inotropic and chronotropic effects in atrium and in aorta vasorelaxation. The vasorelaxation was due to the inhibition of Ca^{2+} influx through the membrane and the release of Ca^{2+} from intracellular stores (Table 2.27) [131].

2.3.12 Herbicidal Activity

Vaughn et al. performed the synthesis and herbicidal activity of modified monoterpenes structurally similar to Cinmethylin.

It is a benzylether derivative of the monoterpene 1,4-cineole (Figure 2.19). Benzyl ethers of monoterpene alcohols are prepared by treating monoterpene alcohol (50 g) and benzyl chloride (200 mL). To this, powdered KOH (100 g) was added. The reaction mixture was heated at 70°C with stirring for 3 h. Then the distilled water (500 mL) was added and the organic phase of the resulting biphasic mixture was extracted three times by using ethyl acetate (200 mL aliquots) as solvent with the help of separating funnel. The ethyl acetate fraction was

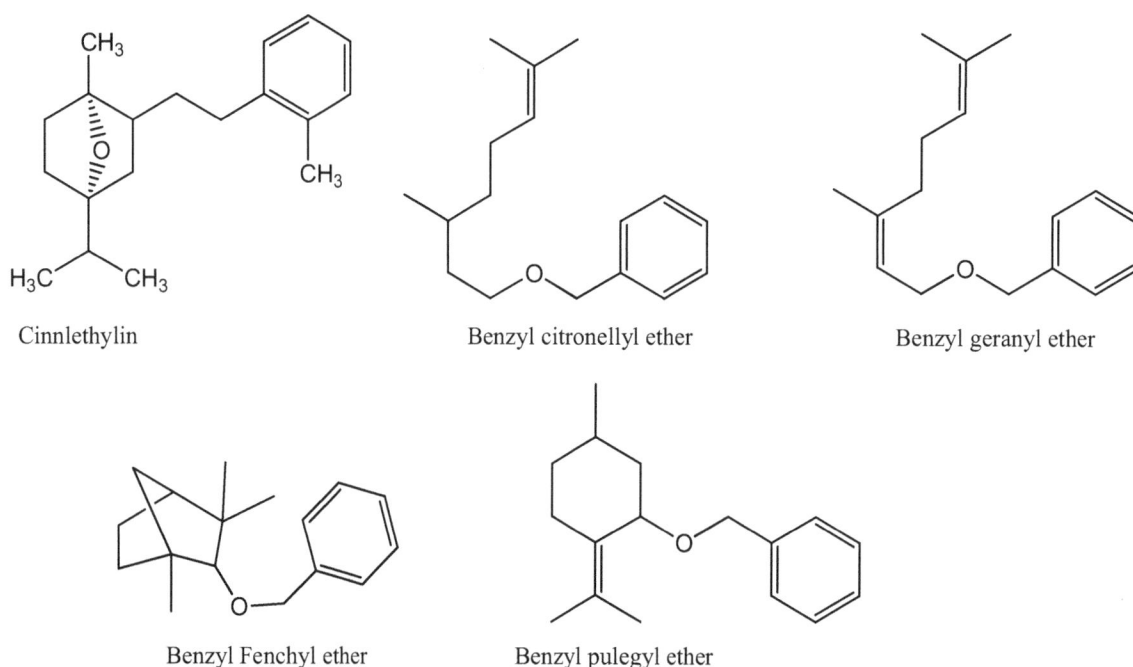

Cinnlethylin   Benzyl citronellyl ether   Benzyl geranyl ether

Benzyl Fenchyl ether   Benzyl pulegyl ether

FIGURE 2.19 Chemical structure of cinmethylin and benzyl monoterpene ethers.

separated from the aqueous phase and concentrated by using roto-evaporation and the resultant ethers (>80% of the total fraction) were purified (>99%) by column chromatography on Sephadex LH-20 eluted with methanol and then on silica gel (70–200 mesh) eluted with hexane:ethyl acetate (25:1). The mechanism of action of Cinmethylin appears to result from the inhibition of mitosis in meristematic regions of susceptible plants. Monoterpenes containing different functional groups have been reported to inhibit the growth of the green alga (*Chlorella*) [132].

Zhao et al. carried out the design, synthesis and evaluation of novel cis-*p*-menthane type Schiff base compounds as effective herbicides. A series of novel p-menthane type Schiff base derivatives were designed and synthesized (Schemes 2.59 and 2.60). First of all, Cis- and trans-N, N'-diacetyl-p-menthane-1,8-diamines have been synthesized with a yield of 64% by Kovals'skaya and coworkers through the interaction of terpin hydrate with acetonitriles under the conditions of the Ritter reaction. In place of terpin hydrate, turpentine or pinene is also used as starting material. Hydrolysis of N, N'-diacetyl-p-menthane-1,8-diamines in NaOH solution of ethylene glycol at the reaction temperature higher than170°C led to the convenient formation of cis-p-menthane-1,8-diamines with a yield of 84%. Finally, p-menthane-1,8-diamine react smoothly with various benzaldehyde derivatives. After stirring in ethanol for about 24h at room temperature, over 95% of p-menthane-1,8-diamine could be transformed to the corresponding Schiff base. Benzaldehyde

derivatives substituted with electron withdrawing groups could reduce the electron density of peripheral carbonyl groups and evidently accelerated this nucleophilic addition reaction. The pre-emergence herbicidal activities against annual ryegrass were investigated. The results indicated that some of p-menthane type Schiff's base derivatives showed promising herbicidal activity against ryegrass as presented in Table 2.28 [133].

2.3.13 Antidiabetic Agents

Diabetes is a disease state which impairs the body's ability to maintain blood glucose or blood sugar level. Diabetes or hyperglycemia is also considered as metabolic disorder in which glucose level in the blood is increased. There are three types of diabetes such as type-1, type-2 and gestational (GDM). Type-1 diabetes can develop at any age but frequently occurs in children and adolescents. In case of type-1 diabetes, pancreatic β-cells are destroyed through an autoimmune mediated reaction so that pancreas is unable to produce sufficient insulin to control blood glucose level. So, type-1 diabetes is called as insulin-dependent diabetes mellitus (IDDM) or juvenile onset diabetes. Whereas, type-2 diabetes (T2D) occurs more commonly in adults. It is characterized by insulin resistance and called as non-insulin-dependent diabetes mellitus (NIDDM) or adult-onset diabetes. Gestational diabetes involves high blood glucose level during pregnancy and the complication is associated with both mother and child [134,135].

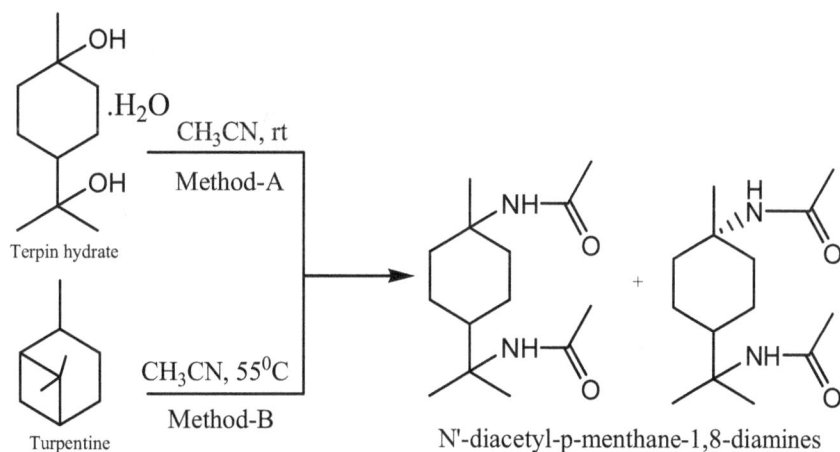

SCHEME 2.59 Synthesis of N, N'-diacetyl-p-menthane-1,8-diamines from terpinhydrate (a) or turpentine (b).

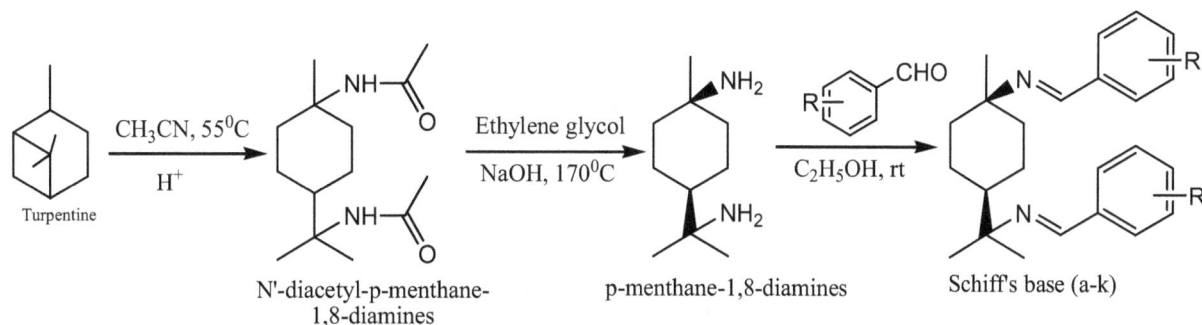

SCHEME 2.60 Synthesis of Schiff base derivatives of cis-p-menthane-1,8-diamines.

TABLE 2.28

Herbicidal Activity of p-Menthane Type Schiff's Base Derivatives against Annual Ryegrass

Compd.	Inhibition Rate of Root Growth at Different Concentrations					
	10[a]	5	2.5	1.25	0.625	0.3125
a	100.0[b]	100.0	83.3	62.2	41.4	14.3
b	94.9[b]	85.5	78.0	58.0	35.0	14.0
c	100.0[b]	92.8	67.0	34.2	26.3	20.3
d	100.0[b]	100.0	88.3	54.3	15.5	12.5
e	100.0[b]	100.0	100.0	99.3	98.0	94.1
f	100.0[b]	100.0	100.0	100.0	88.0	71.0
g	100.0[b]	100.0	100.0	100.0	100.0	100.0
h	100.0[b]	100.0	100.0	100.0	100.0	100.0
i	93.8[b]	81.7	87.5	65.8	45.5	35.4
j	100.0[b]	100.0	100.0	99.7	86.2	79.2
k	67.8[b]	50.3	54.8	35.4	46.4	37.7
Glyphosate	100.0[b]	100.0	100.0	100.0	95.7	80.1

[a] All data in this raw are the concentration of different p-menthane type Schiff base derivatives solutions (mmol/L).

[b] All data in raw 11a to Glyhosate are the inhibition rate (%) of root growth or shoot growth at different concentration of 11a to Glyhosate.

Both *in-vitro* and *in-vivo* studies are performed to evaluate the antidiabetic activity of monoterpenes [136]. The experiments are carried out to determine the antidiabetic effects of monoterpenes (*in-vitro*) include cell culture studies using pancreatic cells, muscles, adipocytes and liver cells as well as enzyme/protein-based assays (Table 2.29). Genipin and geniposide as well as iridoids and their glycosides have displayed potential antidiabetic effects *in-vitro* (Figure 2.20). The *in-vivo* experiments on the antidiabetic effects of monoterpenes include streptozotocin (STZ)-induced diabetic, high fat diet (HFD) fed and spontaneously obese mouse models (Table 2.30).

2.3.14 Miscellaneous

Both natural monoterpenes and their synthetic derivatives exhibited various pharmacological activities such as antifungal, antibacterial, antioxidant, anticancer, antiarrhythmic, local anesthetic, anti-nociceptive, anti-inflammatory, antihistaminic, antidiabetics and anti-spasmodic etc [13]. Acyclic monoterpenes such as β-myrcene and its configurational isomers of β-ocimene are found in the oils of basil (*Ocimum basilicum*, Labiatae), bay (*Pimenta acris*, Myrtaceae) and hops (strobiles of *Humulus lupulus*, Cannabaceae) as well as several other essential oils. β-Myrcene is an acyclic unsubstituted monoterpene and is reported to exhibit estrogenic activity. This compound also exhibits cardiotonic and diuretic properties and can be administered orally, parenterally or rectally [137–140].

Naigre et al. reported the comparison of antimicrobial properties of monoterpenes and their carbonylated products. Monoterpenes and their carbonylated products were evaluated for their antibacterial and antifungal properties. These compounds exhibit both bacteriostatic and fungistatic activities. Among the tested compounds, (1R, 2S, 5R)-isopulegol (Figure 2.21), its carbonylated product, and (R)-carvone showed significant bactericidal activities against *Enterococcus faecium* and *Escherichia coil* above a concentration of μL/mL. It seems that the presence of an oxygenated function in the framework increases the antimicrobial properties. MBCs (Minimal Bactericidal Concentrations) and MFCs (Minimal Fungicidal Concentrations) of (1R, 4R)-isolimonene, (1R, 2S, 5R)-isopulegol, (R)-carvone and their carbonylated products (final concentrations of 200 μL/mL, 20, 10 and 2 μL/mL) were determined using a membrane filtration method. Monoterpenes had a poor activity against *P.aeruginosa* (MICs ≥ 6.25 μL/mL) and *E. faecium* (MICs ≥ 12.5 μL/mL). Activities against *S. aureus* and *E.coli* were variable with MICs ranging from 0.78 to values above 200 μL/mL. It was observed that carbonylated

TABLE 2.29

Antidiabetic Effect of Monoterpenes (In-Vitro)

Monoterpenes	Models	Outcome
Cymene	Advanced glycation end products (AGEs)	100 μM-Inhibit AGE formation; inhibit glycation specific decline in BSA α-helix content and β-sheet.
Genipin	C2C12 myotubes	10 μM-Stimulate glucose uptake; promote GLUT4 translocation; increase insulin receptor IRS-1, AKT and GSK3 β-phosphorylation.
Geniposide	Rat INS-1 pancreatic β-cells	Prevent cell damage induced by high (25 mM) glucose through the AMPK pathway.
Gentiopicroside	HL1C hepatoma cells	50 and 100 μM-Suppress Pck1 expression; induce phosphorylation of components in the insulin signaling cascade (Akt and Erk1/2 phosphorylation).
Paeoniflorin	3T3-L1 adipocytes treated with tumor necrosis factor (TNF)-α	50 μg/mL-Increase insulin-stimulated glucose, promote serine phosphorylation of IRS-1 and insulin-stimulated phosphorylation of AKT.
(R)-(+)-Limonene	3T3-L1 cell culture; α-amylase and α-glucosidase enzymes	Increase GLUT1 expression at mRNA level; Weak enzyme inhibition (mM range).
Sweroside	HL1C hepatoma cells	Suppress Pck1 expression and induce phosphorylation of components in the insulin signaling cascade (Akt and Erk1/2 phosphorylation).
Swertiamarin	Steatosis in HepG2 cells induced by 1 mM oleic acid	25 μg/mL-Maintain membrane integrity; prevent apoptosis; increase the expressions of major insulin signaling proteins (insulin receptor, PI3K and pAkt) with concomitant reduction in p307 IRS-1.
Thujone	Palmitate-induced insulin resistance in skeletal muscle (Soleus muscles)	Ameliorate palmitate oxidation and enhance insulin-stimulated glucose transport; restore (partially) GLUT4 translocation and AS160 phosphorylation; increase AMPK phosphorylation.

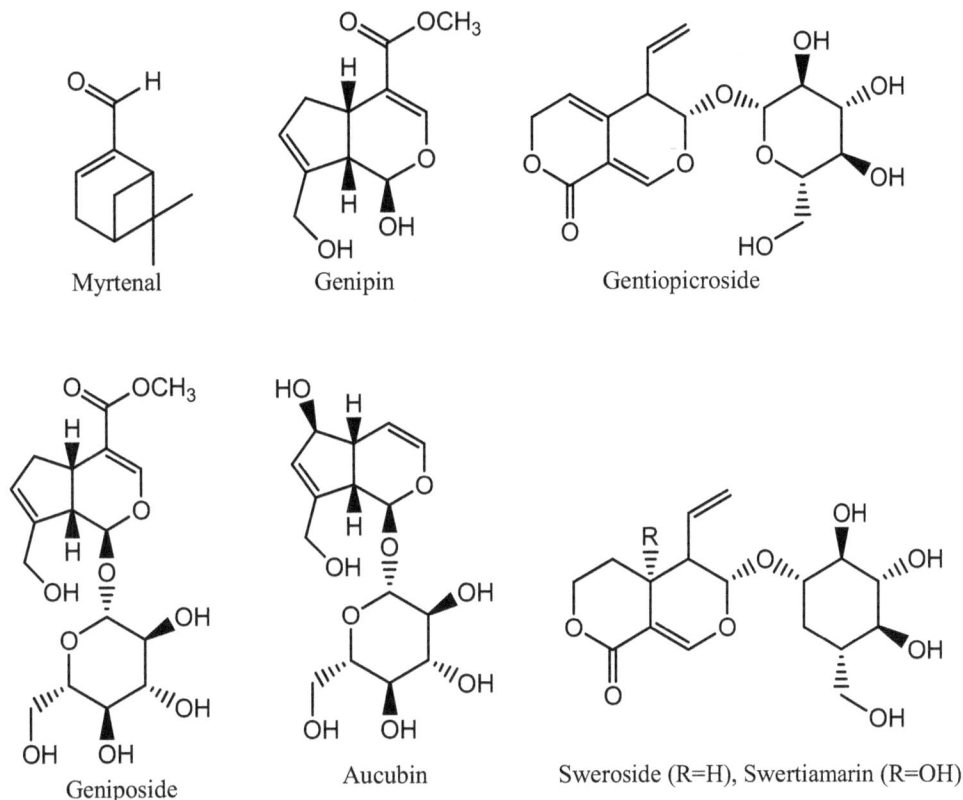

FIGURE 2.20 Structures of monoterpenes as antidiabetic agents.

TABLE 2.30

Antidiabetic Effects of Monoterpenes (In-Vivo)

Monoterpenes	Models	Outcome
Aucubin	STZ-induced diabetic rats-5 mg/kg, i.p. twice daily for the first 5 days followed by single injections daily for 10 days	Lower blood glucose; reverse lipid peroxidation and the decreased in activities of antioxidant enzymes in liver and kidneys; increase immunoreactive β-cells.
Borneol	STZ-induced diabetic rats-25 or 50 mg/kg, p.o. for 30 days.	Lower blood glucose and HbA1c; increase blood insulin; restore body weight loss; increase liver glycogen level
Carvacrol	STZ-induced diabetes in rats-25 and 50 mg/kg, p.o. for 7 days.	Suppress serum glucose, total cholesterol, ALA, AST and lactate dehydrogenase
Carvone	STZ-induced diabetic rats-50 mg/kg, p.o. for 30 days.	Reduce plasma glucose, HbA1c; improve the levels of hemoglobin and insulin. Reverse activities of carbohydrate metabolic enzymes.
Catalpol	STZ-diabetic rats-0.1 mg/kg, i.p.	Enhance glucose uptake in the isolated soleus muscle of diabetic rats; increase glycogen synthesis.
Citronellol	STZ-induced diabetic rats-25, 50 and 100 mg/kg, p.o. for 30 days.	Improve the levels of insulin, hemoglobin and hepatic glycogen with significant decrease in glucose and HbA1c levels.
Cymene	STZ-induced diabetic rats-20 mg/kg, p.o. for 60 days	Improve HbA1c and nephropathic parameters (albumin excretion rate, serum creatinine and creatinine clearance rate).
Genipin	Aging rats-25 mg/kg, i.p. for 12 days	Ameliorate systemic and hepatic insulin resistance.
Geniposide	STZ-induced diabetic rats-injection (50 μM, 10 μL) to the lateral ventricle	Prevent spatial learning deficit; reduce tau phosphorylation.
Geraniol	STZ-induced diabetic rats-100, 200 and 400 mg/kg, p.o. for 45 days	Improve the levels of insulin, hemoglobin and decrease plasma glucose.
D-Limonene	STZ-induced diabetic rats-50, 100 and 200 mg/kg, p.o. for 45 days.	Decreased activity of glycolytic enzyme, glucokinase and liver glycogen.
Menthol	STZ-nicotinamide induced diabetes in rats-25, 50 and 100 mg/kg, p.o. for 45 days	Reduce blood glucose and glycosylated hemoglobin levels; increase the total hemoglobin, plasma insulin and liver glycogen levels; protect hepatic and pancreatic islets.
Myrtenal	STZ-induced diabetic rats-80 mg/kg, p.o. for 28 days	Decrease plasma glucose; increase plasma insulin levels; up-regulate IRS2, Akt and GLUT2 in liver; increase IRS2, Akt and GLUT4 protein expression in skeletal muscle.
Thymol	HFD-fed rats-14 mg/kg, p.o. for 4 weeks.	Decrease body weight gain, visceral fat-pad weights, lipids, ALT, AST, LDH, blood urea nitrogen, glucose, insulin and leptin levels; decrease serum lipid peroxidation and increase antioxidant levels.

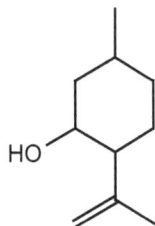

FIGURE 2.21 Structures of isopulegol.

products exhibit improved activity against *S. aureus*, *E. fae-cium*, *E. coli* and especially toward *P. aeruginosa* (12.5 μL/mL ≤ MICs ≤ 50 pL/mL) [141].

Geraniol is an acyclic isoprenoid monoterpene isolated from the essential oils of aromatic plants such as *Cinnamomum tenuipilum* and *Valeriana officinalis*. The potential application of geraniol as a drug is based on its pharmacological properties, including antitumor, anti-inflammatory, antioxidative, and antimicrobial activities, and hepatoprotective, cardioprotective and neuroprotective effects, etc. Oral administration of Geraniol (GE) with a dose of 25, 50 and 75 mmol/kg to mice, inhibits lung cancer cell growth *in-vivo* via an increase of cell apoptosis (Table 2.31). Geraniol has an inhibitory effect on the pamidronate-induced inflammatory response by stimulating production of IL-10. Geraniol also decreases MetS-induced inflammation and free radical injury by suppressing visceral adiposity and decreasing fasting blood glucose and the glycemic excursion. In an STZ-induced diabetic rat model, GE could be used to treat diabetes, as GE administration for 45 days decreased plasma glucose and hemoglobin HbA1c and could restore the insulin response. Oral administration of GE ameliorated the neuromuscular disorder via the increase of tyrosine hydroxylase immunoreactive expression and the decrease of α-synuclein expression. GE administration at a dose of 100 or 200 mg/kg alleviated allergic asthma in an ovalbumin-sensitized rat model. In experimental mice models, GE showed an anti-nociceptive activity following pain caused by inflammation. The anti-nociceptive activity is due to a decrease in peripheral nerve excitability. GE treatment could relieve pain by suppressing neurotransmitters and mediators of inflammation production. Also, GE showed gastroprotective and gastric healing effects on acute ethanol-induced ulcers and chronic acetic acid-induced ulcers [142].

Carvacrol or cymophenol is a phenolic monoterpenoid present in many plant essential oils such as Satureja, Lippia,

Oreganum and Thymus. Chemically, it is 5-isopropyl-2-meth-ylphenol (Figure 2.22). Considerable and significant research has been established to determine the biological and pharmacological properties of carvacrol for its potential use in clinical applications. Diverse and various activities of carvacrol are observed such as antioxidant, antimicrobial, antitumor, anti-mutagenic, anti-genotoxic, analgesic, antispasmodic, anti-inflammatory, angiogenic, anti-parasitic, anti-platelet, AChE inhibitory, anti-elastase, insecticidal, anti-hepatotoxic or hepatoprotective [143].

Linalool is unsaturated monoterpene alcohol. Linalool exists in two enantiomeric forms, R-(−) and S-(+)-linalool (Figure 2.23). Linalool can be produced semi-synthetically from natural pinene or through total chemical synthesis on an industrial scale. The partial synthesis of linalool is based either on α- or β-pinene. α-Pinene is hydrogenated to cis-pinane and subsequently oxidized to a cis/trans mixture of pinane hydroperoxide, which is in turn reduced to the corresponding pinanols and the latter finally pyrolyzed to the respective d- or l-linalools. Isolation of linalool as a single compound from an essential oil mixture can be achieved by using various chromatographic techniques such as column chromatography or through fractional distillation. It is the chemical intermediate during biosynthesis of vitamin E. Linalool exhibits several biological activities such as antibacterial, anti-inflammatory, anti-nociceptive, anticancer and anti-plasmodic effects. The antimicrobial activity of linalool against various pathogens using the microtitre plate assay (minimum inhibitory concentration) method indicates that linalool exhibits some degree of activity against *Escherichia coli* (MIC value: 51.9 μM) and *C. albicans* (MIC value: 38.9 μM). The chemopreventive activity of linalool was also studied using the 7,12-dimethylbenz[a] anthracene (DMBA) induced rat mammary carcinogenesis model [144–147].

(−)-Citronellol demonstrated antioxidant properties in 2,2-diphenym-1-picrylhydrazyl (DPPH) assay. Citral is an acyclic monoterpene that has been examined for its antiviral activity against Herpes simplex virus type 1 (HSV-1) *in-vitro*. Citral is found to exhibit significant anti-HSV-1 activity by direct inactivation of free virus particles. Citral is also exhibiting strong fungicidal activities. This compound is able to form a charge-transfer complex with an electron donor of fungal cells that results in fungal death. Both citral and geraniol exhibit antifungal activity against *Cryptococcus neoformans* [148,149].

The essential oil of *Chenopodium ambrosioides* (EOCA) consists of monoterpenes such as ascaridole, isoascaridole,

TABLE 2.31

Cytotoxic and Antitumor Activities of Geraniol

Cancer Type	Suppressive Effect	Cellular Processes	Function Study
Colon cancer	Anti-proliferation	Apoptosis, DNA damage and cell cycle arrest	In-vitro
Endometrial carcinoma	Anti-migration	Antiangiogenesis	In-vitro & in-vivo
Liver cancer	Anti-proliferation	Apoptosis	In-vitro & in-vivo
Lung cancer	Anti-proliferation	Apoptosis	In-vitro
Pancreatic cancer	Anti-proliferation	Apoptosis	In-vitro
Prostate cancer	Anti-proliferation	Cell cycle arrest	In-vitro
Skin cancer	Anti-proliferation	Apoptosis	In-vitro

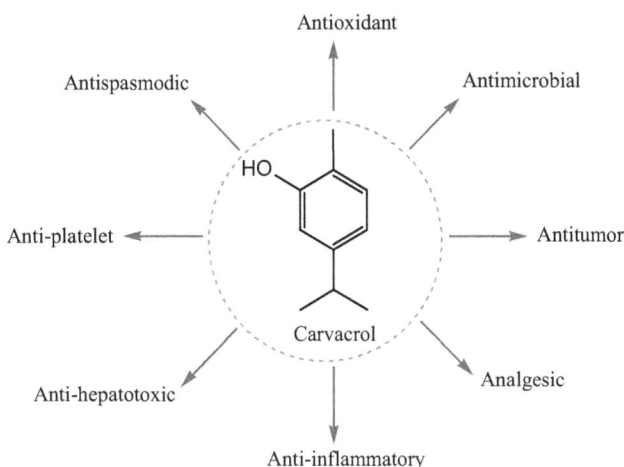

FIGURE 2.22 Pharmacological activities of carvacrol.

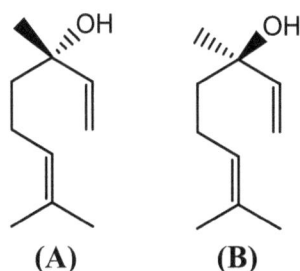

FIGURE 2.23 Structure of linalool [A: (S)-(+)-linalool and B: (R)-(–)-linalool)].

α-terpinene and p-cymene. EOCA shows a potent radical scavenging activity at concentrations ranging from 20 to 5 μg/mL. It exhibits antiradical activity with an IC_{50} of 9.7 μg/mL, as compared to ascorbic acid (IC_{50} 12.7 μg/mL). It was observed that the essential oils which contain oxygenated monoterpenoids (ascaridole and isoascaridole) have greater antioxidant properties than non-oxygenated ones. The potential anticancer activity of EOCA was assessed against HT29 (human colon adeno-carcinoma) cells. At concentrations of 50 and 25 μg/mL, EOCA showed good cytotoxic activity, with growth inhibition of 100% and 56% ± 3.45 respectively [150].

α, β-epoxy-carvone (EC) (Figure 2.24) is a monoterpene and is present in the essential oil of *Carum carvi*, *Kaempferia galangal* etc. EC caused a significant decrease in the motor activity of animals when compared with the control group, up

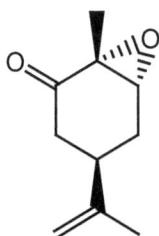

FIGURE 2.24 Structure of α,β-epoxy-carvone.

to 120 min after the administration at doses of 200, 300 or 400 mg/kg injected by i.p. route in mice. It was observed that there is significant increase in the sleeping time of animals with doses of 300 or 400 mg/kg. The EC reduced the remaining time of the animals on the rotating rod (Rota-rod test) in the dose of 400 mg/kg [151].

Choi et al. reported the anti-ischemic and anti-inflammatory activity of (S)-cis-verbenol. It decreased the expression levels of mRNA and proteins of proinflammatory cytokines such as IL-1β, IL-6 and TNF-α in ischemic brain and immune-stimulated glial cells. In addition to this, (S)-cis-verbenol significantly prevented neuronal cell death caused by oxygen-glucose deprivation (OGD, 1 h) and subsequent re-oxygenation (5 h). In DPPH and DHR123 fluorescence assays, (S)-cis-verbenol did not exhibit direct ROS scavenging activity [152].

Thymoquinone (TQ) is one of the active constituent of *Nigella sativa*. It exhibits several biological effects such as its antidiabetic, antioxidant, anticancer, hypolipidemic and anti-inflammatory activities etc. It was observed that thymoquinone is effective therapeutically for controlling diabetes and hyperlipidemia by decreasing oxidative stress and inflammatory responses. Abdelmeguid et al. indicated that TQ decreases the glucose level and also increased the release of pancreatic insulin in diabetic rats. Sankaranarayanan et al. reported that TQ (80 mg/kg, gastric gavage) improved the glycemic effects of STZ plus nicotinamide in rats. TQ (50 mg/kg for 30 d, gastric gavage) decreased blood glucose levels in the gestational diabetic hamster by inhibiting the synthesis of gluconeogenic enzymes. TQ is used efficiently for the treatment of cardiovascular disorders through an enhancement in the activity of aryl-esterase accompanied by decreasing the activity of HMG-CoA reductase. The anti-inflammatory effect of TQ is associated with its inhibitory effects on cyclooxygenase and 5-lipoxygenase. Similarly, antioxidant activity is associated with the scavenging activity against reactive oxygen species (ROS) (Table 2.32) [153–157].

Various eugenol derivatives are synthesized by structural modifications of eugenol. These derivatives are produced by esterification reactions of the hydroxyl group (–OH) of eugenol with various carboxylic acids (Scheme 2.61) and also involve the addition reactions in the double bond of the allyl group of eugenol (Scheme 2.62) [158].

The evaluation of the antibacterial activity of eugenol derivatives using the zone of inhibition (ZOI) technique, measured in millimeter (mm). This technique demonstrates the potential for inhibition of microbial growth by given drug substances. The eugenol derivatives exhibited promising antibacterial activity with lower minimum inhibitory concentration of 500 μg/mL as compared to eugenol (1,000 μg/mL). These derivatives were active against various bacterial strains such as *Escherichia coli* and *Staphylococcus aureus* (Table 2.33).

The antioxidant activity of eugenol derivatives was evaluated with DPPH (2,2-diphenyl-1-picrylhydrazyl). Radical scavenging activity is one of the most commonly used methods for evaluating the antioxidant activity of drug substances. The ability to capture free radicals by the eugenol derivatives (**1–19**) against DPPH was expressed as an inhibitory concentration of 50% (IC_{50}), which represents the concentration

TABLE 2.32

A Summary of Cardioprotective Effects of Thymoquinone

Dose	Rout of Administration	Experimental Model	Pharmacological Effects
20 mg/kg	Gastric gavage	Hyper-cholesterolemic rabbit	Prevented the progression of atherosclerosis by lowering the levels of LDL-C, TC/HDL-C and TG, MDA and increasing the level of HDL-C.
3.5 mg/kg	p.o.	Hyper-cholesterolemic rabbit	Prevented the progression of atherosclerosis by lowering the levels of LDL-C, TC/HDL-C, and TG, MDA and increasing the level of HDL-C.
10 mg/kg	p.o.	Hyper-cholesterolemic rabbit	Prevented the progression of atherosclerosis by decreasing the levels of LDL-C, TC/HDL-C and TG, MDA and PC and also increasing the level of HDL-C.
5 mg/kg	i.p.	STZ-diabetic rat	Prevented the progression of diabetes by decreasing the levels of expression of COX-2 enzyme MDA and increasing the level of SOD in the pancreatic tissue.
20 mg/kg	p.o.	Diabetic mice during gestation and lactation	Prevented the progression diabetes in their offspring by decreasing the levels of blood glucose, free radicals, plasma proinflammatory cytokines (IL-1β, IL-6 and TNF-α) and lipids.
50 mg/kg	Gastric gavage	Gestational diabetic hamster	Prevented the progression of diabetes by inhibiting the synthesis of gluconeogenic enzymes.
80 mg/kg	Gastric gavage	STZ-diabetic rat	Prevented the progression of diabetes by decreasing the activities of the glucose-6-phosphtse, fructose-1,6-bisphosphatase and MDA.
3 mg/mL	i.p.	STZ-diabetic rat	Prevented the progression of diabetes and also improved the toxic activities of STZ, such as heterochromatin aggregates, DNA damage, segregated nucleoli and fragmentation and vacuolization of mitochondria by decreasing the level of MDA and increasing the level of SOD.

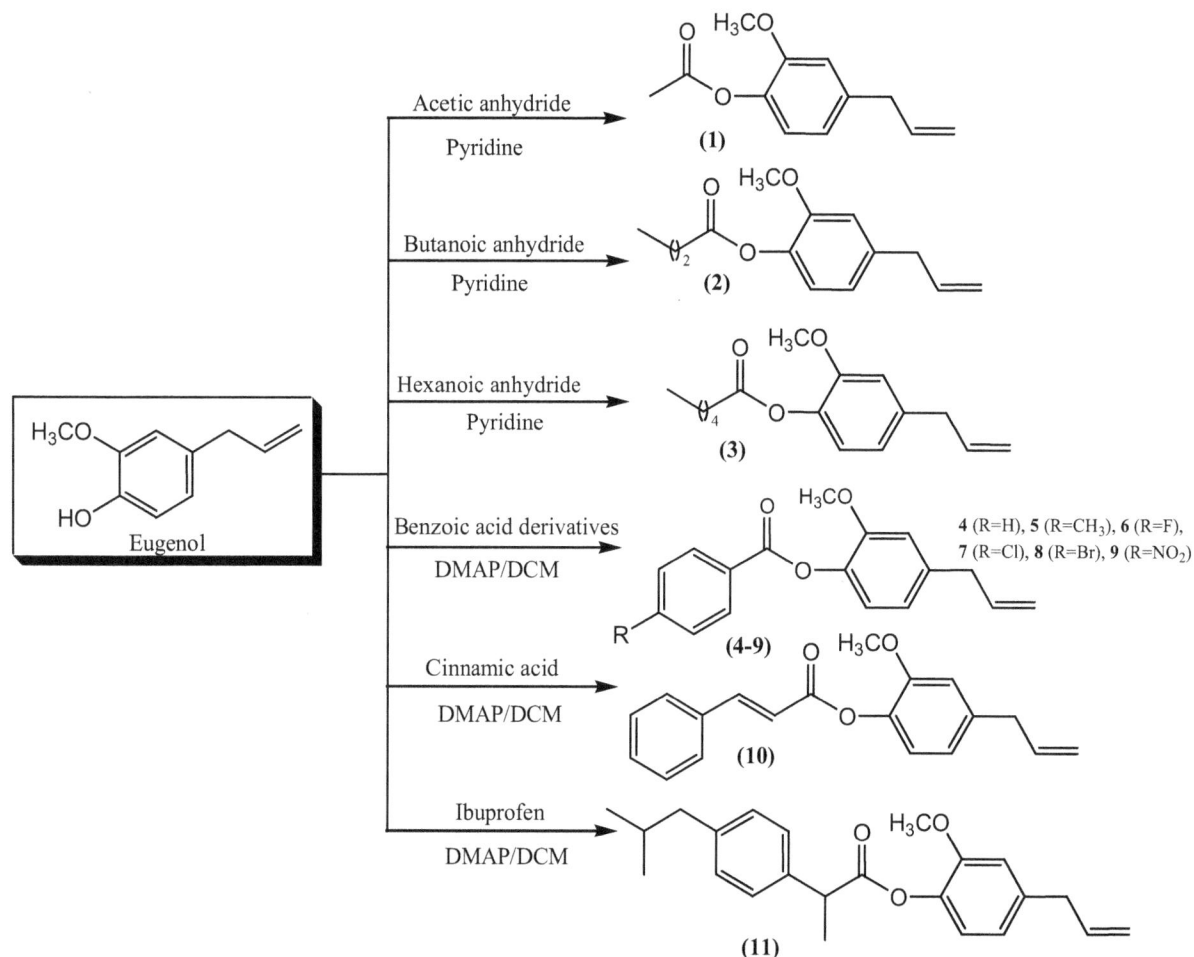

SCHEME 2.61   Eugenol derivatives (1–11) through reactions in the hydroxyl groups.

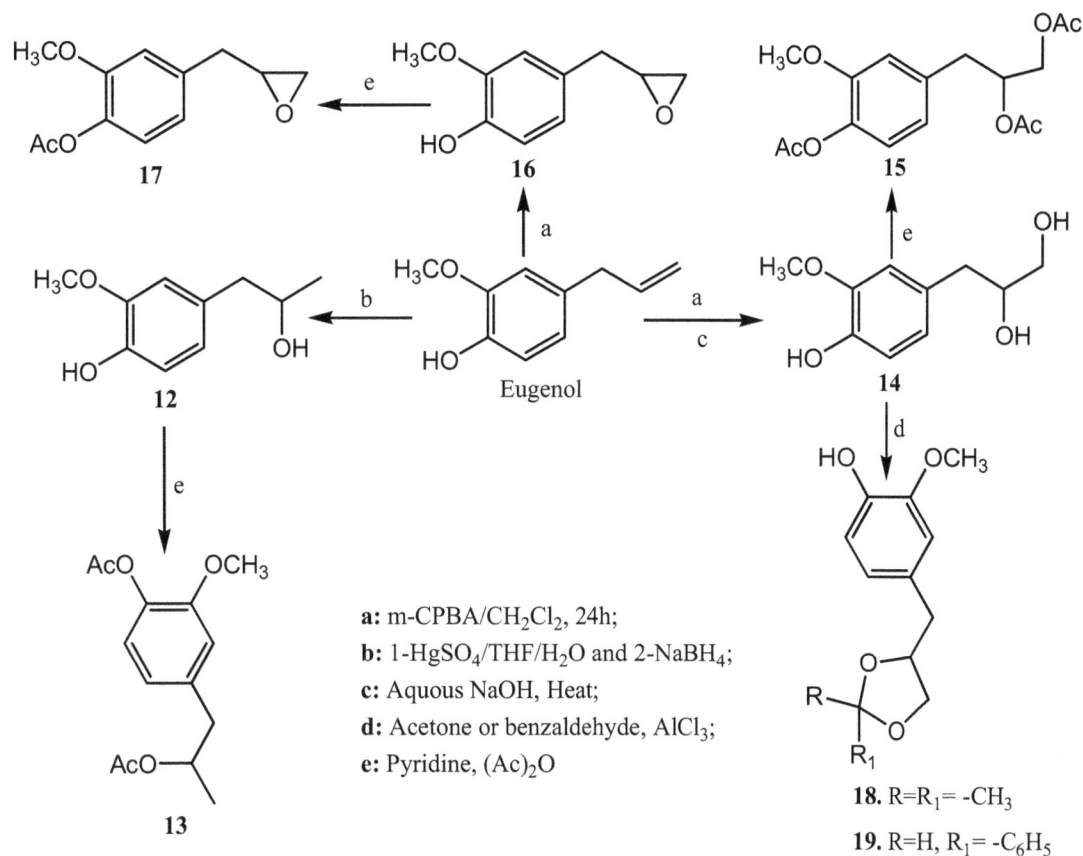

SCHEME 2.62 Eugenol derivatives (12–19) through reactions in the double bond.

a: m-CPBA/CH$_2$Cl$_2$, 24h;
b: 1-HgSO$_4$/THF/H$_2$O and 2-NaBH$_4$;
c: Aquous NaOH, Heat;
d: Acetone or benzaldehyde, AlCl$_3$;
e: Pyridine, (Ac)$_2$O

18. R=R$_1$= -CH$_3$
19. R=H, R$_1$= -C$_6$H$_5$

TABLE 2.33

Effect of Eugenol Derivatives against Six Bacterial Strains

Compounds	Zone of Inhibition (mm)					
	P. aeruginosa	*E. coli*	*S. aureus*	*Streptococcus*	*K. pneumoniae*	*B. cereus*
1	0	0	0	0	0	0
2	0	0	0	0	0	0
3	0	0	0	0	0	0
4	0	0	0	0	0	12
5	0	0	0	6	0	12
6	0	0	0	0	0	0
7	0	0	0	0	0	0
8	0	12	6	0	0	0
9	0	0	0	11	0	0
10	0	0	0	12	6	0
11	0	0	0	0	0	0
12	0	0	0	10	0	0
13	0	0	0	12	0	8
14	0	0	0	9	0	0
15	0	12	0	10	6	0
16	0	0	10	10	15	20
17	6	0	0	10	6	12
18	0	0	0	9	0	0
19	0	0	0	0	0	0
Methyl eugenol	0	0	0	0	0	0
Eugenol	12	0	0	6	11	12
Tetracycline	0	10	20	10	9	0

TABLE 2.34

Minimum Inhibitory Concentration (MIC) Presented by Eugenol Derivatives (1–19) against Different Bacterial Strains

	Minimum Inhibitory Concentration MIC (µg/mL)					
Compounds	***P. aeruginosa***	***E. coli***	***S. aureus***	***Streptococcus***	***K. pneumoniae***	***B. cereus***
4	NA	NA	NA	NA	NA	NA
5	NA	NA	NA	100	NA	NA
8	NA	500	1,000	NA	NA	NA
9	NA	NA	NA	100	NA	NA
10	NA	NA	NA	100	NA	NA
12	NA	NA	NA	100	NA	NA
13	NA	NA	NA	100	NA	NA
14	NA	NA	NA	NA	NA	NA
15	NA	1000	NA	NA	NA	NA
16	NA	NA	NA	NA	500	500
17	NA	NA	NA	NA	NA	1000
18	NA	NA	NA	NA	NA	NA
19	NA	NA	NA	NA	NA	NA
Eugenol	1,000	0	1,000	1,000	1,000	1,000
Penicillin	125	250	250	250	250	62.5

NA, No activity at the concentrations analyzed.

required to capture 50% of the radicals in the medium (Table 2.34). Trolox and gallic acid were used as positive controls. Eugenol and its derivatives have the capacity to transfer electrons or hydrogen atoms by neutralizing free radicals and thereby block the oxidative process. In the case of antioxidant activity, it was observed that the eugenol derivatives exhibited significant antioxidant activity with $IC_{50} > 100 \mu g/mL$ when compared with the eugenol ($IC_{50} = 4.38 \mu g/mL$). This activity is due to the esterification of the hydroxyl group (–OH) of the eugenol molecule. On the other hand, the chemical modification in the double bond of the eugenol molecule affects the capacity of capturing radicals more than the starting molecule (eugenol). The eugenol derivatives exhibit the proximal antioxidant capacity with $IC_{50} = 19.30 \mu g/mL$ as compared to commercial standards such as Trolox ($IC_{50} = 16.00 \mu g/mL$) (Table 2.35) [159,160].

Allegra et al. reported the action of sobrerol (Figure 2.25) on mucocilliary transport. Sobrerol is a mucolytic agent [161]. Its pharmacological action was evaluated in frog palate preparation

by measuring the mucociliary transport velocity. The mucus samples are collected from patients suffering from a chronic bronchial inflammatory disease characterized by dense bronchial secretion. The patients were examined before and after 4 days of treatment with sobrerol (600 mg/day). From the results (Table 2.36), it was reported that sobrerol facilitates the mucociliary transport of bronchial secretion of patients [162,163].

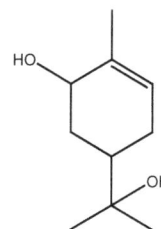

FIGURE 2.25 Structure of sobrerol.

TABLE 2.35

IC$_{50}$ Presented by Eugenol Derivatives (1–19)

Compounds	IC$_{50}$ (µg/mL)	Compounds	IC$_{50}$ (µg/mL)
1	>100	13	>100
2	>100	14	20
3	>100	15	>100
4	>100	16	19.3
5	>100	17	>100
6	>100	18	32
7	>100	19	30.37
8	>100	Methyl eugenol	>100
9	>100	Eugenol	4.38
10	>100	Gallic acid	0.64
11	>100	Trolox	16
12	51.12	Isoeugenol	50.7

TABLE 2.36

Relative Rate of Mucocilliary Transport of Mucus Samples of Patients before and after 4 days of Treatment with Sobrerol

	Relative Rate of Mucocilliary Transport	
Case No.	**Before Treatment**	**After Treatment**
1	0.82	1.18
2	0.45	0.78
3	0.61	0.80
4	0.65	0.75
5	0.62	0.73
6	0.53	0.70
7	0.62	0.81
8	0.60	0.87
Mean ± SEM	0.612 + 0.03	0.827 + 0.05
T		3.68
$t_{0.01}$		2.97 p < 0.01

2.4 Conclusion

The structural modifications of natural products manufacture novel biologically active compounds with diverse structures like monoterpenoids either synthetically or via biotransformation. In general, monoterpenes belong to the terpenoid class of natural products and are bio-synthesized through the mevalonic acid pathway. Their pharmacological activities are determined by using both *in-vitro* and *in-vivo* methods. In general, the extraction procedure influences the chemical composition of these terpenoids and the active chemical constituents present in it also contribute to their different levels of pharmacological activities including anticancer, anti-malaria, anti-microbial, antiviral, anti-tubercular, anti-Parkinsonism, anti-Alzheimer's disease, antioxidant, analgesic, anti-inflammatory, etc. The structures of the most biologically active monoterpenoids can be utilized as templates for further investigations to find out their mechanism of actions in relation to their pharmacological activities.

Abbreviation

AChE	Acetylcholine
AD	Alzheimer's disease
AGEs	Advanced glycation end products
ALT	Alanine amino transferase
AMR	Antimicrobial resistance
AST	Aspartate amino transferase
CAP	Compound action potential
CNS	Central nervous system
DCM	Dichloromethane
DMAPP	Dimethyl allyl pyrophosphate
DMBA	Dimethylbenz[a]anthracene
DNA	Deoxyribonucleic acid
DPPH	2,2-diphenym-1-picrylhydrazyl
EC	α, β-Epoxy-carvone
EOCA	Essential oil of *Chenopodium ambrosioides*
FIC	Fractional inhibitory concentration
GDM	Gestational diabetes mellitus
GDP	Geranyl diphosphate
GE	Geraniol
GPP	Geranyl pyrophosphate
HD	Huntington's disease
HFD	High fat diet
HIV	Human immunodeficiency virus
HNO$_2$	Nitrous acid
HSV-1	Herpes simplex virus type-1
i.p	Intraperitoneally
kg	kilogram
IDDM	Insulin-dependent diabetes mellitus
IL	Interleukin
IPP	Isopentenyl pyrophosphate
LDH	Lactic dehydrogenase
LPS	Lipopolysaccharide
MES	Maximal electroshock-induced seizures
mg	Milligram
MICs	Minimum (or minimal) inhibitory concentration
MAPKs	Mitogen-activated protein kinases
μg	Microgram
μL	Microliter
MCT	Monocrotaline
mL	Milli liter
MMP	Matrix metallo proteinases
MRSA	Methicillin-resistant Staphylococcus aureus
MSSA	Methicillin-susceptible Staphylococcus aureus
NADPH	Nicotinamide adenine dinucleotide phosphate (Reduced)
NDDs	Neurodegenerative disorders
OD	Optical density
PD	Parkinson's disease
%	Percent
POCl$_3$	Phosphoryl chloride
PTZ	Pentylenetetrazol
QST	Quasistatic tester
SI	Selectivity indices
STZ	Streptozotocin
TC50	Median toxic dose
T2D	Type-2 diabetes
TNF	Tumor necrosis factor
TRPM8	Transient receptor potential melastatin-8
μg	Microgram
μm	Micrometer
WHO	World Health Organization

REFERENCES

[1] Grayson, D. H. Monoterpenoids. *Natural Product Reports*, **2000**, 17, 385–419.

[2] Zebec, Z.; Wilkes, J.; Jervis, A. J.; Scrutton, N. S.; Takano, E.; Breitling, R. Towards synthesis of monoterpenes and derivatives using synthetic biology. *Current Opinion in Chemical Biology*, **2016**, 34, 37–43.

[3] Cragg, G. M.; Newman, D. J. Natural products: A continuing source of novel drug leads. *Biochimica et Biophysica Acta – General Subjects*. **2013**, 1830(6), 3670–95.

[4] Baser, K.H.C.; Buchbauer, G. *Handbook of Essential Oils: Science, Technology, and Applications*, 2nd Edition. Taylor and Francis Group, New York, **2010**.

[5] Zebec, Z.; Wilkes, J.; Jervis, A. J.; Scrutton, N. S.; Takano, E.; Breitling, R. Towards synthesis of monoterpenes and derivatives using synthetic biology. *Current Opinion in Chemical Biology*, **2016**, 34, 37–43.

[6] Charlwood, B.V.; Banthorpe, D.V. The biosynthesis of monoterpenes. *Progress in Phytochemistry*, **1978**, 5, 65–125.

[7] Masumoto, N.; Korin, M; Ito, M. Geraniol and linalool synthases from wild species of perilla. *Phytochemistry*, **2010**, 71, 1068–1075.

[8] Yamada, Y.; Kuzuyama, T.; Komatsu, M.; Shin-Ya, K.; Omura, S.; Cane, D. E.; Ikeda, H. Terpene synthases are widely distributed in bacteria. *Proceedings of the National Academy of Sciences of the United States of America*, **2015**, 112, 857–862.

[9] Himmelberger, J. A.; Cole, K. E.; Dowling, D. P. *Chapter 3.14-Biocatalysis: Nature's Chemical Toolbox, Green Chemistry: An Inclusive Approach*, **2018**, 471–512.

[10] Banthorpe, D. V.; Charlwood, B. V.; Francis, M. J. O. Biosynthesis of monoterpenes. *Chemical Reviews*, **1972**, 72, 2, 115–155.

[11] Kolicheski, M. B.; Cocco, L. C.; Mitchell, D.; Kaminski, M. Synthesis of myrcene by pyrolysis of β-pinene: Analysis of decomposition reactions. *Journal of Analytical and Applied Pyrolysis*, **2007**, 80(1), 92–100.

[12] Finar, I. L. *Organic Chemistry*, Volume 2: Stereochemistry and the Chemistry Natural Product, 5th Edition. Pearson, **2002**, 375–377.

[13] Takabe, K.; Yamada, T.; Katagiri, T. A facile synthesis of citral. *Synthetic Communications*, **1983**, 13(4), 297–301.

[14] Francis, M. J.; Banthorpe, D. V.; Le Patourel, G. N. Biosynthesis of monoterpenes in rose flowers. *Nature*, **1970**, 228(5275), 1005–1006.

[15] Yagi, H.; Shirado, M.; Hidai, M.; Uchida, Y. Synthesis of citronellol from isoprene. *Journal of Japan Oil Chemists' Society*, **1977**, 26(4), 232–235.

[16] Gaich, T.; Mulzer, J. 2.7 Chiral pool synthesis: Starting from terpenes. *Comprehensive Chirality*, **2012**, 2, 163–206.

[17] Woronuk, G.; Demissie, Z. A.; Rheault, M. R.; Mahmoud, S.S. Biosynthesis and therapeutic properties of Lavandula essential oil constituents. *Planta Medica*, **2011**, 77(1), 7–15.

[18] Cseke, L.; Dudareva, N.; Pichersky, E. Structure and evolution of linalool synthase. *Molecular Biology and Evolution*, **1998**, 15(11), 1491–1498.

[19] Eisenacher, M.; Beschnitt, M.; Holderich, W. Novel route to a fruitful mixture of terpene fragrances in particular phellandrene starting from natural feedstock geraniol using weak acidic boron based catalyst. *Catalysis Communications*, **2012**, 26(5), 214–217.

[20] Farco, J. A.; Grundmann, O. Menthol-pharmacology of an important naturally medicinal "cool". *Mini-Reviews in Medicinal Chemistry*, **2013**, 13, 124–131.

[21] Croteau, R. B.; Davis, E. M.; Ringer, K. L.; Wildung, M.R. (−)-Menthol biosynthesis and molecular genetics. *Naturwissenschaften*, **2005**, 92, 562–577.

[22] Dewick, P. M. *Medicinal Natural Products: A Biosynthetic Approach*, 3rd Edition. John Wiley & Sons, **2009**.

[23] Negoi, A.; Wuttke, S.; Kemnitz, E.; Macovei, D.; Parvulescu, V. I.; Teodorescu, C. M.; Coman, S. M. One-pot synthesis of menthol catalyzed by a highly diastereoselective Au/MgF2 catalyst. *Angewandte Chemie International Edition*, **2010**, 49, 8134–8138.

[24] Trasarti, A. F.; Marchi, A. J.; Apesteguía, C. R. Design of catalyst systems for the one-pot synthesis of menthols from citral. *Journal of Catalysis*, **2007**, 247, 155–165.

[25] Jacob, R. G.; Perin, G.; Loi, L. N.; Lenardao, E. J. Green synthesis of (−)-isopulegol from (+)-citronellal: Application to essential oil of citronella. *Tetrahedron Letters*, **2003**, 44(18), 3605–3608.

[26] Martín-Luengo, M. A.; Yates, M.; Martínez Domingo, M. J.; Casal, B.; Iglesias, M.; Esteban, M.; Ruiz-Hitzky, E. Synthesis of p-cymene from limonene, A renewable feedstock. *Applied Catalysis B: Environmental*, **2008**, 81(3–4), 218–224.

[27] Zografos, A. L. *From Biosynthesis to Total Synthesis: Strategies and Tactics for Natural Products*. John Wiley & Sons, **2016**, 201–210.

[28] Shanmugapriya, K.; Palanichamy, M.; Arabindoo, B.; Murugesan, V. A novel route to produce thymol by vapor phase reaction of m-cresol with isopropyl acetate over Al-MCM-41 molecular sieve. *Journal of Catalysis*, **2004**, 2(10), 347–357.

[29] Mazza, G.; Kiehn, F. A.; Marshall, H. H. Monarda: A source of geraniol, linalool, thymol and carvacrol-rich essential oils. In Janick, J.; Simon, J. E. (eds.). *New Crops*. New York: Wiley, 1993, 628–631.

[30] Husnu Can Baser, K.; Buchbauer, G. *Handbook of Essential Oils: Science, Technology, and Applications*, CRC Press, 2009, 212.

[31] Gilles, M.; Zhao, J.; An, M.; Agboola, S. Chemical composition and antimicrobial properties of essential oils of three Australian Eucalyptus Species. *Food Chemistry*, **2010**, 119(2), 731–737.

[32] Agarwal, O. P. *Chemistry of Organic Natural Products, Terpenoids*, 4th Edition, Volume 1. Krishna Prakashan Media (P), Ltd., 2010, 361–362.

[33] Demidova, Y. S.; Suslov, E. V.; Simakova, O. A.; Volcho, K. P.; Salakhutdinov, N. F.; Simakova, I. L.; Murzin, D. Y. Selective one-pot carvone oxime hydrogenation over titania supported gold catalyst as a novel approach for dihydrocarvone synthesis. *Journal of Molecular Catalysis A: Chemical*, **2016**, 420, 142–148.

[34] Smith, M. B. *Chapter 8 Synthetic Strategies, Organic Synthesis*, 4th Edition. 2017, 419–482.

[35] Prakoso, T.; Hanley, J.; Soebianta, M. N. Synthesis of terpineol from α-pinene using low-price acid catalyst. *Catalysis Letters*, **2018**, 148, 725–731.

[36] Yuasa, Y.; Yuasa, Y. A practical synthesis ofd-α-terpineol via markovnikov addition of d-limonene using trifluoroacetic acid. *Organic Process Research & Development*, **2006**, 10(6), 1231–1232.

[37] Black, C.; Buchanan, G. L.; Jarvie, A. W. A synthesis of (±)-pulegone. *Journal of the Chemical Society*, **1956**, 2971–2973.

[38] Corey, E. J.; Ensley, H. E.; Suggs, J. W. Convenient synthesis of (S)-(-)-pulegone from (-)-citronellol. *The Journal of Organic Chemistry*, **1976**, 41(2), 380–381.

[39] Kousar, S.; Nadeem, F.; Khan, O.; Shahzadi, A. Chemical synthesis of various limonene derivatives-a comprehensive review. *International Journal of Chemical and Biochemical Sciences*, **2017**, 11, 102–112.

[40] Noma, Y.; Asakawa, Y. Biotransformation of monoterpenoids, comprehensive natural products II. *Chemistry and Biology*, **2010**, 3, 669–801.

[41] Geoghegan, K.; Evans, P. Synthesis of (+)-perillyl alcohol from (+)-limonene. *Tetrahedron Letters*, **2014**, 55(8), 1431–1433.

[42] Sen, A.; Grosch, W. The synthesis of enantiomerically pure (S)(+)-α-phellandrene from (R)(−)-carvone. *Flavour and Fragrance Journal*, **1990**, 5(4), 233–234.

[43] LaFever, R. E.; Croteau, R. Hydride Shifts in the biosynthesis of the p-menthane monoterpenes α-terpinene, γ-terpinene and β-phellandrene. *Archives of Biochemistry and Biophysics*, **1993**, 301, 361–366.

[44] Guo, J.; Zhang, R.; Ouyang, J.; Zhang, F.; Qin, F.; Liu, G.; Zhang, W.; Li, H.; Ji, X.; Jia, X.; Qin, B.; You, S. Stereodivergent synthesis of carveol and dihydrocarveol through Ketoreductases/Ene-reductases catalysed asymmetric reduction. *ChemCatChem*, **2018**, 10(23), 5496–5504.

[45] Damtoft, S. Biosynthesis of catalpol. *Phytochemistry*, **1994**, 35(5), 1187–1189.

[46] Ahmad, A.; Mishra, R. K.; Vyawahare, A.; Kumar, A.; Rehman, M. U.; Qamar, W.; Khan, A. Q.; Khan, R.

Thymoquinone (2-Isopropyl-5-methyl-1, 4-benzoquinone) as a chemopreventive/anticancer agent: Chemistry and biological effects. *Saudi Pharmaceutical Journal*, **2019**, 27, 1113–1126.

[47] Harakava, R. Genes encoding enzymes of the lignin biosynthesis pathway in Eucalyptus. *Genetics and Molecular Biology*, **2005**, 28(3 Suppl), 601–607.

[48] Wang, Q.; Li, Y.; Chen, Q. A convenient, large scale synthesis of trans-(+)-sobrerol, *Synthetic Communications: An International Journal for Rapid Communication of Synthetic Organic Chemistry*, **2003**, 33(12), 2125–2134.

[49] Nangia, A.; Prasuna, G.; Rao, P. Synthesis of cyclopenta[c] pyran skeleton of iridoid lactones. *Tetrahedron*, 53(43), **1997**, 14507–14545.

[50] Damtoft, S.; Jensen, S.; Jessen, C.; Knudsen, T. Late stages in the biosynthesis of aucubin in Scrophularia. *Phytochemistry*, **1993**, 35(5), 1089–1093.

[51] Johnson, M. A.; Croteau, R. Biosynthesis of ascaridole: Iodide peroxidase-catalyzed synthesis of a monoterpene endoperoxide in soluble extracts of *Chenopodium ambrosioides* fruit. *Archives of Biochemistry and Biophysics*, **1984**, 1(15), 254–266.

[52] Noma, Y.; Asakawa, Y. Biotransformation of monoterpenoids by microorganisms, insects, and mammals. In: Baser, KHC, Buchbauer, G (eds). *Handbook of Essential Oils: Science, Technology, and Applications*. CRC Press: Boca Raton, FL, 2010, 585–736.

[53] Sarria, S.; Wong, B.; Martín, H. G.; Keasling, J. D.; Peralta-Yahya, P. Microbial synthesis of pinene. *ACS Synthetic Biology*, **2014**, 3(7), 466–475.

[54] Tholl, D. Biosynthesis and biological functions of terpenoids in plants. *Advances in Biochemical Engineering/ Biotechnology*, **2015**, 148, 63.

[55] Fairlie, J. C.; Hodgson, G. L.; Money, T. Synthesis of (±)-camphor. *Journal of the Chemical Society, Perkin Transactions*, **1973**, 119, 2109–2112.

[56] Grimshaw, J. Reduction of carbonyl compounds, carboxylic acids and their derivatives. *Electrochemical Reactions and Mechanisms in Organic Chemistry*, **2000**, 330–370.

[57] Galopin, C. C. A short and efficient synthesis of (±)-trans-sabinene hydrate. *Tetrahedron Letters*, **2001**, 42(33), 5589–5591.

[58] Boyle, P. H.; Cocker, W.; Grayson, D. H.; Shannon, P. V. R. Total synthesis of (±)-fenchone, a synthesis of (+)-fenchone and of (+)-cis-2,2,5-trimethyl-3-vinylcyclo-pentanone, a photoisomer of (–)-trans-caran-4-one. *Journal of the Chemical Society D*, **1971**, 8, 395–396.

[59] Mori, K.; Mizumachi, N.; Matsui, M. Synthesis of optically pure (1S, 4S, 5S)-2-pinen-4-ol (cis-verbenol) and its antipode, the pheromone of ipsbark beetles. *Agricultural and Biological Chemistry*, **1976**, 40(8), 1611–1615.

[60] Ishmuratov, G. Y.; Tukhvatshin, V. S.; Yakovleva, M. P. Stereospecific synthesis of cis-verbenol. *Russian Journal of Organic Chemistry*, **2016**, 52, 755–756.

[61] Yang, W.; Chen, X.; Li, Y.; Guo, S.; Wang, Z.; Yu, X. Advances in pharmacological activities of terpenoids. *Natural Product Communications*, **2020**, 15(3), 1–13.

[62] Buchbauer, G.; Jäger, W.; Jirovetz, L.; Ilmberger, J.; Dietrich, H. Therapeutic properties of essential oils and fragrances. *American Chemical Society Symposium*, **1993**, 525, 159–165.

[63] Polya, G. *Biochemical Targets of Plant Bioactive Compounds: A Pharmacological Reference Guide to Sites of Action and Biological Effects*. CRC Press, 2013, 860.

[64] *Goodman and Gilman, Pharmacological Basis of Therapeutics*, Macmillan, **1965**, 982–983.

[65] Viollon, C.; Chaumont, J. P. Antifungal properties of essential oils and their main components upon *Cryptococcus neoformans*. *Mycopathology*, **1994**, 3, 151–153.

[66] Furneri, P. M.; Mondello, L.; Mandalari, G.; Paolino, D.; Dugo, P.; Garozzo, A.; Bisignano, G. In vitro antimycoplasmal activity of Citrus bergamia essential oil and its major components. *European Journal of Medicinal Chemistry*, **2012**, 52, 66–69.

[67] Mahizan, N. A.; Yang, S.-K.; Moo, C. L.; Ai-Lian Song, A.; Chong, C. M.; Chong, C. W.; Abushelaibi, A.; Erin Lim, S. H.; Lai, K. S. Terpene derivatives as a potential agent against antimicrobial resistance (AMR) pathogens. *Molecules*, **2019**, 24, 2631–2622.

[68] Chou, T. C. Theoretical basis, experimental design, and computerized simulation of synergism and antagonism in drug combination studies. *Pharmacological Reviews*, **2006**, 58, 621–681.

[69] Startseva, V. A.; Nikitina, L. E.; Sirazieva, E.V.; Dorofeeva, L. Y.; Lisovskaya, S. A.; Glushko, N. P.; Garaev, R. A.; Akulova, I. V. Synthesis and biological activity of monoterpenoids belonging to menthane series. *Chemistry for Sustainable Development*, **2009**, 17, 533–539.

[70] Dambolena, J. S.; Lopez, A. G.; Cánepa, M. C.; Theumer, M. G.; Zygadlo, J. A.; Rubinstein, H. R. Inhibitory effect of cyclic terpenes (limonene, menthol, menthone and thymol) on *Fusarium verticillioides* MRC 826 growth and fumonisin-B1 biosynthesis. *Toxicon*, **2008**, 51, 37–44.

[71] Moleyar, V.; Narasimham, P. Antifungal activity of some essential oil components. *Food Microbiology*, **1986**, 3, 331–336.

[72] Edris, A. E.; Farrag, E. S. Antifungal activity of peppermint and sweet basil essential oils and their major aroma constituents on some plant pathogenic fungi from the vapor phase. *Nahrung*, **2003**, 47, 117–121.

[73] Lin, G. S.; Duan, W. G.; Yang, L. X.; Huang, M.; Lei, F. H. Synthesis and antifungal activity of novel Myrtenal-based 4-methyl-1,2,4-triazole-thioethers. *Molecules*, **2017**, 22(193), 2–10.

[74] Ohayon, M. M. Specific characteristics of the pain/depression association in the general population. *The Journal of Clinical Psychiatry*, **2004**, 65, 5–9.

[75] Salakhutdinov, N. F.; Volcho, K. P.; Yarovaya, O. I. Monoterpenes as a renewable source of biologically active compounds. *Pure and Applied Chemistry*, **2017**, 89(8), 2–13.

[76] Whitley, R. J.; Roizman, B. Herpes simplex virus infections. *Lancet*, **2001**, 357, 1513–1518.

[77] De Clercq, E. Antiviral drugs in current clinical use. *Journal of Clinical Virology*, **2004**, 30, 115–133.

[78] Astani, A.; Schnitzler, P. Antiviral activity of monoterpenes beta-pinene and limonene against herpes simplex virus *in-vitro*. *Iranian Journal of Microbiology*, **2014**, 6(3), 149–155.

[79] de Cássia da Silveira e Sá, R.; Andrade, L. N.; de Sousa, A review on anti-inflammatory activity of monoterpenes. *Molecules*, **2013**, 18(1), 1227–1254.

84

Terpenoids

[80] Stengel, M.; Binder, A.; Klebe, O.; Wasner, G.; Schattschneider, J.; Baron, R. Topical menthol: Stability of a sensory profile in a human surrogate model. *European Journal of Pain*, **2007**, 11(S1), S59–S207.

[81] Binder, A.; Stengel, M.; Klebe, O.; Wasner, G.; Baron, R. Topical high concentration (40%) menthol-Somato sensory profile of a human surrogate pain model. *Journal of Pain,* **2011**, 12, 764–773.

[82] Guimaraes, A. G.; Quintans, J. S. S.; Quintans-Junior, L. J. Monoterpenes with analgesic activity-a systematic review. *Phytotherapy Research*, **2013**, 27(1), 1–15.

[83] Peana, A. T.; D'Aquila, P. S.; Panin, F.; Serra, G.; Pippia, P.; Moretti, M. D. Anti-inflammatory activity of linalool and linalyl acetate constituents of essential oils. *Phytomedicine*, **2002**, 9, 721–726.

[84] Kim, D. S.; Lee, H. J.; Jeon, Y. D.; Han, Y. H.; Kee, J. Y.; Kim, H. J.; Shin, H. J.; Kang, J.; Lee, B. S., Kim, S. H.; Kim, S. J.; Park, S. H.; Choi, B. M.; Park, S. J.; Um, J. Y.; Hong, S. H. Alpha-pinene exhibits anti-inflammatory activity through the suppression of MAPKs and the NF-κB pathway in mouse peritoneal macrophages. *The American Journal of Chinese Medicine*, **2015**, 43, 731–742.

[85] Rufino, A. T.; Ribeiro, M.; Judas, F.; Salgueiro, L.; Lopes, M. C.; Cavaleiro, C.; Mendes, A. F. Anti-inflammatory and chondroprotective activity of (+)-α-pinene: Structural and enantiomeric selectivity. *Journal of Natural Products*, **2014**, 77, 264–269.

[86] De Oliveira Ramalho, T. R.; de Oliveira, M. T.; de Araujo Lima, A. L.; Bezerra-Santos, C. R.; Piuvezam, M. R. Gamma-terpinene modulates acute inflammatory response in mice. *Planta Medica*, **2015**, 81, 1248–1254.

[87] Games, E.; Guerreiro, M.; Santana, F. R.; Pinheiro, N. M.; de Oliveira, E. A.; Lopes, F. D.; Olivo, C. R.; Tibério, I. F.; Martins, M. A.; Lago, J. H.; Prado, C. M. Structurally related monoterpenes *p*-Cymene, carvacrol and thymol isolated from essential oil from leaves of lippia sidoides cham. (Verbenaceae) protect mice against elastase-induced emphysema. *Molecules*, **2016**, 21, E1390.

[88] Rufino, A. T.; Ribeiro, M.; Sousa, C.; Judas, F.; Salgueiro, L.; Cavaleiro, C.; Mendes, A. F. Evaluation of the antiinflammatory, anti-catabolic and pro-anabolic effects of Ecaryophyllene, myrcene and limonene in a cell model of osteoarthritis. *European Journal of Pharmacology*, **2015**, 750, 141–150.

[89] Rehman, M. U.; Tahir, M.; Khan, A. Q.; Khan, R.; Oday-OHamiza; Lateef, A.; Hassan, S. K.; Rashid, S.; Ali, N.; Zeeshan, M.; Sultana, S. *d*-Limonene suppresses doxorubicin-induced oxidative stress and inflammation via repression of COX-2, iNOS and NFκB in kidneys of Wistar rats. *Experimental Biology and Medicine*, **2014**, 239, 465–476.

[90] Jiang, J.; Shen, Y. Y.; Li, J.; Lin, Y. H.; Luo, C. X.; Zhu, D. Y. (+)-Borneol alleviates mechanical hyperalgesia in models of chronic inflammatory and neuropathic pain in mice. *European Journal of Pharmacology*, **2015**, 757, 53–58.

[91] Almeida, J. R.; Souza, G. R.; Silva, J. C.; Saraiva, S. R.; Júnior, R. G.; Quintans de, S.; Barreto de, S.; Bonjardim, L. R.; Cavalcanti, S. C.; Quintans, L. J., Jr. Borneol: A bicyclic monoterpene alcohol, reduces nociceptive behavior and inflammatory response in mice. *Scientific World Journal*, **2013**, 18, 808460.

[92] Silva, F. V.; Guimaraes, A. G.; Elayne, R. S.; Sousa-Neto, B. P.; Machado, F. D. F.; Quintans-Junior, L. J.; Arcanjo, D. R.; Oliveira, F. A.; Oliveira, R. C. M. Anti-inflammatory and anti-ulcer activities of carvacrol, a monoterpene present in the essential oil of Oregano. *Journal of Medicinal Food*, **2012**, 15(11), 984–991.

[93] Patil, M.; Bendre, R. Synthesis, characterization and antioxidant potency of naturally occurring phenolic mono-terpenoids based hydrazide motifs. *Medicinal Chemistry*, **2018**, 8, 177–180.

[94] Suresh, D. B.; Jamatsing, D. R.; Pravin, S. K.; Ratnamala, S. B. Synthesis, characterization and antioxidant activity of carvacrol containing novel thiadiazole and oxadiazole moieties. *Modern Chemistry and Applications*, **2016**, 4(4), 1–4.

[95] Imran, M.; Rauf, A.; Khan, I. A.; Shahbaz, M.; Qaisrani, T. B.; Fatmawati, S.; Abu-Izneid, T.; Imran, A.; Rahman, K. U.; Gondal, T. A. Thymoquinone: A novel strategy to combat cancer: A review. *Biomedicine & Pharmacotherapy*, **2018**, 106, 390–402.

[96] Goyal, S. N.; Prajapati, C. P.; Gore, P. R.; Patil, C. R.; Mahajan, U. B.; Sharma, C.; Talla, S. P.; Ojha, S. K. Therapeutic potential and pharmaceutical development of thymoquinone: A multitargeted molecule of natural origin. *Frontiers in Pharmacology*, **2017**, 8(656), 1–19.

[97] Gali-Muhtasib, H.; Diab-Assaf, M.; Boltze, C.; Al-Hmaira, J.; Hartig, R.; Roessner, A.; Schneider-Stock, R. Thymoquinone extracted from black seed triggers apoptotic cell death in human colorectal cancer cells via a p53-dependent mechanism. *International Journal of Oncology*, **2004**, 25(4), 857–866.

[98] Subramanian, J.; Govindan, R. Lung cancer in never smokers: A review. *Journal of Clinical Oncology*, **2007**, 25(5), 561–570.

[99] Yang, J.; Kuang, X.; Lv, P.; Yan, X. Thymoquinone inhibits proliferation and invasion of human nonsmall-cell lung cancer cells via ERK pathway. *Tumor Biology*, **2015**, 36(1), 259–269.

[100] Wirries, A.; Breyer, S.; Quint, K.; Schobert, R.; Ocker, M. Thymoquinone hydrazone derivatives cause cell cycle arrest in p53-competent colorectal cancer cells. *Experimental and Therapeutic Medicine*, **2010**, 1(2), 369–375.

[101] Seo, K.-A.; Kim, H.; Ku, H.-Y.; Ahn, K.-J.; Park, S. J.; Bae, S. K.; Shin, S.-J.; Liu, K. H. The monoterpenoids citral and geraniol are moderate inhibitors of CYP2B6 hydroxylase activity. *Chemico-Biological Interactions,* **2008**, 174, 141–146.

[102] Crowell, P. L.; Ayoubi, A. S.; Burke, Y. D. Antitumorigenic effects of limonene and perillyl alcohol against pancreatic and breast cancer. *Advances in Experimental Medicine and Biology*, **1996**, 401, 131–136.

[103] Hu, L.-a.; Sun, Y.-k.; Zhang, H.-s.; Zhang, J.-s.; Hu, J. Catalpol inhibits apoptosis in hydrogen peroxide-induced cardiac myocytes through a mitochondrial-dependent caspase pathway. *Bioscience Reports*, **2016**, 36, 1–9.

[104] Gonda, T.; Berdi, P.; Zupko, I.; Fulop, F.; Szakonyi, Z. Stereoselective Synthesis, Synthetic and pharmacological application of monoterpene based 1,2,4- and 1,3,4-oxadiazoles. *International Journal of Molecular Sciences*, **2018**, 19(81), 1–11.

[105] Saleh, M.; Hashem, F.; Glombitza, K. Cytotoxicity and in vitro effects on human cancer cell lines of volatiles of

Apium graveolens var filicinum. *Pharmaceutical and Pharmacological Letters*, **1998**, 8, 97–99.

[106] Silva, S. L. d.; Figueiredo, P. M.; Yano, T. Cytotoxic evaluation of essential oil from *Zanthoxylum rhoifolium* Lam. leaves. *Acta Amazonica*, **2007**, 37, 281–286.

[107] Sun, Y.; Liu, G.; Zhang, Y.; Zhu, H.; Ren, Y.; Shen, Y. M. Synthesis and in-vitro anti-proliferative activity of Beta-elemene mono-substituted derivatives in HeLa cells mediated through arrest of cell cycle at theG1phase. *Bioorganic and Medicinal Chemistry*, **2009**, 17, 1118–1124.

[108] Noebels, J. L. The biology of epilepsy genes. *Annual Review of Neuroscience*, **2003**, 26, 599–625.

[109] Rogawski, M. A.; Porter, R. J. Antiepileptic drugs: Pharmacological mechanisms and clinical efficacy with consideration of promising developmental stage compounds. *Pharmacological Reviews*, **1990**, 42, 223–286.

[110] Fisher, R. S. Animal models of the epilepsies. *Brain Research Reviews*, **1989**, 14, 245–278.

[111] Kupferberg, H. J. Strategies for identifying and developing new anticonvulsant drugs. *Pharmaceutisch Weekblad*, **1992**, 14, 132–138.

[112] Gale, K. GABA and epilepsy: Basic concepts from preclinical research, *Epilepsia*, **1992**, 33, S3–S12.

[113] de Sousa, D. P.; Ramos Goncalves, J. C.; Quintans-Junior, L.; Santos Cruz, J.; Machado Araujo, D. A.; de Almeida, R. N. Study of anticonvulsant effect of citronellol, a monoterpene alcohol, in rodents. *Neuroscience Letters*, **2006**, 401, 231–235.

[114] de Sousa, D. P.; Nóbrega, F. F. F.; Santos, C. C. M. P.; de Almeida, R. N. Anticonvulsant activity of the linalool enantiomers and racemate: Investigation of chiral influence. *Natural Product Communications*, **2010**, 5(12), 1847–1851.

[115] de Sousa, D. P.; Quintans, L.; de Almeida, R. N. Evolution of the anticonvulsant activity of α-terpineol. *Pharmaceutical Biology*, **2007**, 45(1), 69–70.

[116] de Sousa, D. P.; Raphael, E.; Brocksom, U.; Brocksom, T. J. Sedative effect of monoterpene alcohols in mice: A preliminary screening. *A Journal of Biosciences: Zeitschrift für Naturforschung*, **2007**, 62(7–8), 563–566.

[117] Linck, V. M.; da Silva, A. L; Figueiró, M.; Piato, A. L.; Herrmann, A. P.; Dupont Birck F.; Caramão, E. B.; Nunes, D. S.; Moreno, P. R.; Elisabetsky, E. Inhaled linalool-induced sedation in mice. *Phytomedicine*, **2009**, 16(4), 303–307.

[118] Monczor, M. Diagnosis and treatment of Alzheimer's disease. *Current Medicinal Chemistry – Central Nervous System Agents*, **2005**, 5, 5–13.

[119] Tholl, D. Biosynthesis and biological functions of terpenoids in plants. *Advances in Biochemical Engineering/Biotechnology*, **2015**, 148, 63–106.

[120] Kochar Kaur, K.; Allahbadia, G. N.; Singh, M. Monoterpenes-A class of terpenoid Group of Natural Products-as a source of natural anti-diabetic agents in the future-A Review. *CPQ Nutrition*, **2019**, 3(4), 1–21.

[121] Huang, W.; Wang, Y.; Li, J.; Zhang, Y.; Ma, X.; Zhu, P.; Zhang, Y. Design, synthesis and evaluation of genipin derivatives for the treatment of Alzheimer's disease. *Chemical Biology & Drug Design*, **2019**, 93(2), 110–122.

[122] Viana Santos, M. R.; Moreira, F. V.; Fraga, B. P.; Quintans-Junior, L. J. Cardiovascular effects of monoterpenes: A review. *Revista Brasileira de Farmacognosia*, **2011**, 21(4), 764–771.

[123] Peixoto-Neves, D.; Silva-Alves, K. S.; Gomes, M. D.; Lima, F. C.; Lahlou, S.; Magalhães, P. J.; Ceccatto, V. M.; Coelho-De-Souza, A. N.; Leal-Cardoso, J. H. Vasorelaxant effects of the monoterpenic phenol isomers, carvacrol and thymol, on rat isolated aorta. *Fundamental & Clinical Pharmacology*, **2010**, 24, 341–350.

[124] Bastos, J. F.; Moreira, I. J.; Ribeiro, T. P.; Medeiros, I. A.; Antoniolli, A. R.; De Sousa, D. P.; Santos, M. R. Hypotensive and vasorelaxant effects of citronellol, a monoterpene alcohol, in rats. *Basic & Clinical Pharmacology & Toxicology*, **2010**, 106, 331–337.

[125] Touvay, C.; Vilain, B.; Carré, C.; Mencia-Huerta, J. M.; Braquet, P. Effect of limonene and sobrerol on monocrotaline induced lung alterations and pulmonary hypertension. *International Archives of Allergy and Immunology*, **1995**, 107, 272–274.

[126] Menezes, I. A. C.; Barreto, C. M. N.; Antoniolli, A. R.; Santos, M. R. V.; De Sousa, D. P. Hypotensive activity of terpenes found in essential oils. *Journal of Bioscience*, **2010**, 65, 562–566.

[127] Johnson, C. D.; Melanaphy, D.; Purse, A.; Stokesberry, S. A.; Dickson, P.; Zholos, A. V. Transient receptor potential melastatin 8 channel involvement in the regulation of vascular tone. *The American Journal of Physiology-Heart and Circulatory Physiology*, **2009**, 296, 1868–1877.

[128] El Tahir, K. E. H.; Ageel, A. M. Effect of the volatile oil of Nigella sativa on the arterial blood pressure and heart rate of the guinea-pig. *Saudi Pharmaceutical Journal*, **1994**, 2, 163–168.

[129] El Tahir, K. E. H.; Al-Ajmi, M. F.; Al-Bekairi, A. M. Some cardiovascular effects of the dethymoquinonated Nigella sativa volatile oil and its major components α-pinene and p-cymene in rats. *Saudi Pharmaceutical Journal*, **2003**, 113, 104–110.

[130] Saito, K.; Okabe, T.; Inamori, Y.; Tsujibo, H.; Miyake, Y.; Hiraoka, K.; Ishida, N. The biological properties of monoterpenes: Hypotensive effects on rats and antifungal activities on plant pathogenic fungi of monoterpenes. *Mokuzai Gakkaishi*, **1996**, 42, 677–680.

[131] Guedes, D. N.; Silva, D. F.; Barbosa-Filho, J. M.; Medeiros, I. A. Calcium antagonism and the vaso-relaxation of the rat aorta induced by rotundifolone. *Brazilian Journal of Medical and Biological Research*, **2004**, 37, 1881–1887.

[132] Vaughn, S. F.; Spencer, G. E. Synthesis and herbicidal activity of modified monoterpenes structurally similar to cinnlethylin. *Weed Science*, **1996**, 44, 7–11.

[133] Xua, S.-c.; Zhua, S.-J.; Wang, J.; Bi, L.-W.; Chen, Y.-X.; Lua, Y.-J.; Gu, Y.; Zhao, Z.-D. Design, synthesis and evaluation of novel cis-p-menthane type Schiff base compounds as effective herbicides. *Chinese Chemical Letters*, **2017**, 28, 1509–1513.

[134] McCall, A. L.; Farhy, L. S. Treating type-1 diabetes: From strategies for insulin delivery to dual hormonal control. *Minerva Endocrinology*, **2013**, 38, 145–163.

[135] Upadhyay, J.; Polyzos, S. A.; Perakakis, N.; Thakkar, B.; Paschou, S. A.; Katsiki, N.; Underwood, P.; Park, K. H.; Seufert, J.; Kang, E. S. Pharmacotherapy of type-2 diabetes: An update. *Metabolism*, **2018**, 78, 13–42.

[136] Habtemariam, S. Antidiabetic potential of monoterpenes: A case of small molecules punching above their weight. *International Journal of Molecular Sciences*, **2018**, 19(4), 1–23.

[137] Koziol, A.; Stryjewska, A.; Librowski, T.; Salat, K.; Gawel, M.; Moniczewski, A.; Lochynski, S. Pharmacological properties and potential applications of natural monoterpenes. *Mini-Reviews in Medicinal Chemistry*, **2014**, 14(14), 1156–1168.

[138] Park, S. N.; Lim, Y. K.; Freire, M. O.; Cho, E.; Jin, D.; Kook, J. K. Antimicrobial effect of linalool and α-terpineol against periodontopathic and cariogenic bacteria. *Anaerobe*, **2012**, 18, 369–372.

[139] Knobloch, K.; Pauli, A.; Iberl, B.; Weigand, H.; Weis, N. Antibacterial and antifungal properties of essential oil components. *Journal of Essential Oil Research*, **1989**, 1, 119–128.

[140] Cristiane, B. S.; Silvia, S.; Guterres, V. W.; Elfrides, E. S. Antifungal activity of the lemongrass oil citral against *Candida* spp. *The Brazilian Journal of Infectious Diseases*, **2008**, 27, 137–49.

[141] Naigre, R.; Kalck, P.; Roques, C.; Roux, I.; Michel, G. Comparison of antimicrobial properties of monoterpenes and their carbonylated products. *Planta Medica*, **1996**, 62, 275–277.

[142] Lei, Y.; Fu, P.; Jun, X.; Cheng, P. Pharmacological properties of geraniol–A review. *Planta Medica*, **2019**, 85, 48–55.

[143] Mohammedi, Z. Carvacrol: An update of biological activities and mechanism of action. *Open Access Journal of Chemistry*, **2017**, 1(1), 53–62.

[144] Guy, P. P. K; Viljoen, A. M. Linalool – A review of a biologically active compound of commercial importance. *Natural Product Communications*, **2008**, 3(7), 1184–1191.

[145] Van Zyl, R. L.; Seatlholo, S. T.; van Vuuren, S. F.; Viljoen, A. M. The biological activities of 20 nature identical essential oil. *Journal of Essential Oil Research*, **2006**, 18, 129–133.

[146] Peana, A. T.; D'Aquila, P. S.; Panin, F.; Serra, G.; Pippia, P.; Moretti, M. D. L. Anti-inflammatory activity of linalool and linalyl acetate constituents of essential oils. *Phytomedicine*, **2002**, 9, 721–726.

[147] Semikolenov, V. A.; Ilyna, I. I., Simakova, I. L. Linalool synthesis from α-pinene: Kinetic peculiarities of catalytic steps. *Applied Catalysis A: General*, **2001**, 211, 91–107.

[148] Orhan, I. E.; Ozcelik, B.; Kartal, M.; Kan, Y. Antimicrobial and antiviral effects of essential oils from selected Umbelliferae and Labiatae plants and individual essential oil components. *Turkish Journal of Biology*, **2012**, 36, 239–246.

[149] Librowski, T.; Moniczewski, A. Strong antioxidant activity of carane derivatives. *Pharmacological Reports*, **2010**, 62, 178–184.

[150] Al-kaf, A. G.; Crouch, R. A.; Denkert, A.; Porzel, A.; Al-Hawshabi, O. S. S.; Awadh Ali, N. A.; Setzer, W. N.; Wessjohann, L. Chemical composition and biological activity of essential oil of *Chenopodium ambrosioides* from Yemen. *American Journal of Essential Oils and Natural Products*, **2016**, 4(1), 20–22.

[151] de Sousa, D. P.; Nóbrega, F. F. F; Claudino, F. S.; de Almeida, R. N.; Leite, J. R.; Mattei, R. Pharmacological effects of the monoterpene α, β-epoxy-carvone in mice. *Revista Brasileira de Farmacognosia*, **2007**, 17(2), 170–175.

[152] Choi, I.-Y.; Lim, J. H.; Hwang, S.; Kim, W.-K. Anti-ischemic and anti-infl ammatory activity of (S)-cis-verbenol. *Free Radical Research*, **2010**, 44(5), 541–551.

[153] Farkhondeh, T.; Samarghandian, S.; Borji, A. An overview on cardioprotective and anti-diabetic effects of thymoquinone. *Asian Pacific Journal of Tropical Medicine*, **2017**, 10(9), 849–854.

[154] Abdelmeguid, N. E.; Fakhomy, R.; Kamal, S. M.; Al Wafai, R. J. Effects of *Nigella sativa* and thymoquinone on biochemical and subcellular changes in pancreatic β-cells of streptozotocin-induced diabetic rats. *Journal of Diabetes*, **2010**, 2(4), 256–266.

[155] Sankaranarayanan, C.; Pari, L. Thymoquinone ameliorates chemical induced oxidative stress and b-cell damage in experimental hyperglycemic rats. *Chemico-Biological Interactions*, **2011**, 190(2–3), 148–154.

[156] Fararh, K. M.; Shimizu, Y.; Shiina, T.; Nikami, H.; Ghanem, M. M.; Takewaki, T. Thymoquinone reduces hepatic glucose production in diabetic hamsters. *Research in Veterinary Science*, **2005**, 79(3), 219–223.

[157] Ahmad, S.; Beg, Z. H. Hypolipidemic and antioxidant activities of thymoquinone and limonene in atherogenic suspension fed rats. *Food Chemistry*, **2013**, 138(2–3), 1116–1124.

[158] Kaufman, T. S. The multiple faces of Eugenol. A versatile starting material and building block for organic and bioorganic synthesis and a convenient precursor toward biobased fine chemicals. *Journal of the Brazilian Chemical Society*, **2015**, 26, 1055–1085.

[159] da Silva, F. F. M.; Queiroz Monte, F. J.; de Lemos, T. L. G.; do Nascimento, P. G. G.; de Medeiros Costa, A. K.; de Paiva, L. M. M. Eugenol derivatives: Synthesis, characterization, and evaluation of antibacterial and antioxidant activities. *Chemistry Central Journal*, **2018**, 12(34), 1–9.

[160] Brand-Williams, W.; Cuvelier, M. E.; Berset, C. Use of a free radical method to evaluate antioxidant activity. *LWT - Food Science and Technology*, **1995**, 28, 25–30.

[161] Allegra, L.; Bossi, R.; Braga, P. C. Action of sobrerol on mucociliary transport. *Respiration*, **1981**, 42(2), 105–109.

[162] Henderson, G. G.; Eastburn, W. J. S. The conversion of pinene into sobrerol. *Journal of the Chemical Society, Transactions*, **1909**, 95, 1465–1466.

[163] Puchelle, E.; Tournier, J. M.; Petit, A.; Zahm, J. M.; Lauque, D.; Vidailhet, M.; Sadoul, P. The frog palate for studying mucus transport velocity and mucociliary frequency. *European Journal of Respiratory Diseases*, **1983**, 128(Pt 1), 293–303.

3

Synthesis, and Medicinal Uses of Diterpenoids

Biswa Mohan Sahoo*, Bimal Krishna Banik†, and Abhishek Tiwari‡

CONTENTS

3.1 Introduction 87
3.2 Synthesis of Diterpenoids 88
3.3 Medicinal Uses of Diterpenoids...96
    3.3.1 Treatment of Cardiovascular Diseases 96
    3.3.2 Treatment of Cancer..99
    3.3.3 Treatment of Inflammation...112
    3.3.4 Management of Alzheimer's Disease..115
    3.3.5 Treatment of Diabetes and Hyperlipidemia 117
    3.3.6 Treatment of Microbial Infection 121
    3.3.7 Treatment of Tuberculosis 124
    3.3.8 Treatment of Pain and Smooth Muscle Spasms 128
    3.3.9 Treatment of Leishmaniasis 129
3.4 Conclusion..143
References...145

3.1 Introduction

Diterpenes are hydrocarbons with molecular formula $C_{20}H_{32}$. Diterpenoids represent a large and structurally diverse class of compounds derived from four isoprene units linked in a head-tail fashion. According to the number of rings present in diterpenes, these are broadly categorized into phytane, cembrene, labdane, abietane, pimarane, clerodane, and stemarene diterpenoids (Table 3.1). Diterpenes are also classified into different types such as linear or acyclic (Phytol), bicyclic (Sclareol, *Salvia sclarea*), tricyclic (Abietic acid, *Pinus sylvestris*), tetracyclic (tigliane), pentacyclic (Cafestol), and macrocyclic (Taxol, *Taxus brevifolia*) depending on the presence of their skeletal core. These compounds are mainly present in polyoxygenated form with hydroxyl (–OH) and keto (C=O) groups in nature [1].

The molecular structure of diterpenes contains skeletons with 20 carbon atoms (C_{20}). It may be categorized into acyclic: phytane, bicyclic: labdane and clerodane, tricyclic: abietane, pimarane and cassane, tetracyclic: gibberellane, kaurane and vouacapane, macrocyclic: lathyrane and taxane (Figure 3.1).

Several diterpenes are produced in plants, animals, cyanobacteria, and fungi via the HMG-CoA (3-hydroxy-3-methylglutaryl-CoA) reductase pathway, mevalonate, or deoxyxylulose phosphate pathways in which geranylgeranyl pyrophosphate act as a primary intermediate. So, diterpenes are derived by the addition of one IPP unit with FPP to produce geranylgeranyl-pyrophosphate (GGPP). From GGPP, structurally diverse terpenoids are generated primarily by two classes of enzymes such as diterpene synthases and cytocromes P450 (CYPs). GGPP is the precursor for the synthesis of the phytane by the action of the enzyme geranylgeranyl reductase [2].

The evaluation of biological activities by *in-vitro* cell-based assays and *in-vivo* animal studies indicates the medicinal effects of diterpenes against a variety of diseases. Diterpenes possess various pharmacological activities including anti-inflammatory, antidiabetic, antihyperlipidemic, anticancer,

TABLE 3.1

Classification of Diterpenes

Types of Diterpenes	Number of Rings	Examples of Diterpenes
Acyclic	0	Phytane (Phytol)
Monocyclic	1	Cembrene-A
Bicyclic	2	Sclarene, Labdane
Tricyclic	3	Abietane (abietic acid), Taxadiene (Taxol)
Tetracyclic	4	Stemarene, Stemodene

* Roland Institute of Pharmaceutical Sciences, Berhampur affiliated to Biju Patnaik University of Technology (BPUT), Rourkela, Odisha, India. drbiswamohansahoo@gmail.com.
† Department of Mathematics and Natural Sciences, College of Sciences and Human Studies, Prince Mohammad Bin Fahd University, Al Khobar, Kingdom of Saudi Arabia. bimalbanik10@gmail.com and bbanik@pmu.edu.sa
‡ Pharmacy Academy, IFTM University, Lodhipur Rajput, Moradabad, U.P. 244102.

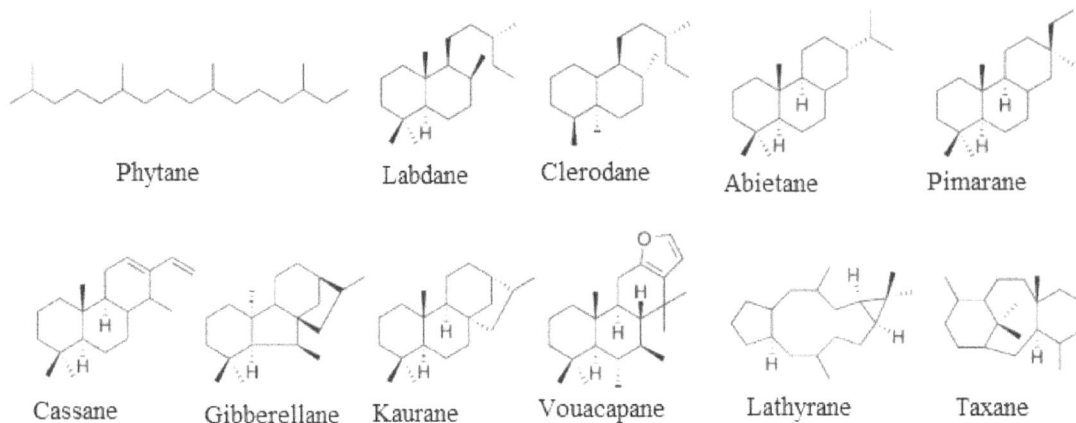

FIGURE 3.1 Molecular structure of diterpenes.

antiviral, anti-leishmanial, antimicrobial, anti-Alzheimer's, antispasmodic activities, etc [3,4].

3.2 Synthesis of Diterpenoids

Most of the naturally occurring terpenoids are synthesized from isopentenyl diphosphate as the starting material in the MEP pathway. This pathway also generates dimethylallyl diphosphate, from pyruvate and D-glyceraldehyde 3-phosphate (GAP). When coupled together, they form one molecule of geranylgeranyl diphosphate (GGDP) in presence of geranylgeranyl diphosphate synthase (GGDPS).

Terpenoids are biosynthesized from a universal 5-carbon building block, Isopentenyl diphosphate (IPP). Further, IPP can be derived from two pathways such as (1) Classical cytosolic mevalonic acid (MVA) pathway (2) Plastidial methylerythritol 4-phosphate (MEP) pathway.

The MVA pathway in the cytosol, starting from 3-acetyl-CoA to yield IPP finally, is responsible for synthesizing sesquiterpenoids and sterols. Whereas, the MEP pathway producing IPP and dimethylallyl diphosphate (DMAPP) from pyruvate and GAP are mainly responsible for generating monoterpenoids and diterpenoid constituents.

Ginkgolides are biologically active terpenic lactones present in the leaves of *Ginkgo biloba*. These are diterpenoids with 20 carbon skeletons. The biosynthetic pathway ginkgolides can be described in three major steps such as (1) Plastid MEP pathway to provide universal isoprenoid precursors; (2) Condensing steps to produce direct precursors, and (3) Modifying procedures to yield ginkgolides.

There are seven enzymes involved in the MEP pathway to biosynthesize the GAP into IPP and DMAPP. Among the seven enzymes in the ginkgo MEP pathway, the condensation of pyruvate and GAP into 1-deoxy-D-xylulose 5-phosphate (DXP) is catalyzed by 1-deoxy-D-xylulose 5-phosphate synthase (GbDXS), which initiates the biosynthesis of ginkgolides. 1-deoxy-D-xylulose-5-phosphate reducto-isomerase (DXR) is the second enzyme of the MEP pathway for ginkgolide biosynthesis, which catalyzes the conversion of DXP into MEP. 2-C-methyl-D-erythritol 4-phosphate cytidyltransferase (MECT), the third enzyme, catalyzes

the formation of 4-(cytidine 5′-diphospho)-2-C-methyl-D-erythritol from MEP. The following two steps are then catalyzed by 4-(cytidine 5′-diphospho)-2-C-methyl-D-erythritol kinase (CMEK) and 2-C-methyl-D-erythritol-2,4-cyclodiphosphate synthase (MECS), respectively. Then, 2-C-methyl-D-erythritol-2,4-cyclodiphosphate is catalyzed by 1-hydroxy-2-methyl-2-(E)-butenyl-4-diphosphate synthase (HDS), forming 1-hydroxy-2-methyl-2-(E)-butenyl-4-diphosphate (HMBPP). In the final step of the MEP pathway, HMBPP is reduced by HMBPP reductase (HDR) to produce both IPP and its isomer DMAPP.

In plastids, the key universal diterpene precursor geranylgeranyl diphosphate (GGDP) is biosynthesized from IPP and DMAPP by catalysis of geranylgeranyl diphosphate synthase (GGPPS). Levopimaradiene is biosynthesized from GGDP via a series of reactions, and levopimaradiene synthase (LPS) performs the first committed step of cyclization of GGPP in ginkgolide biosynthesis. C ring is further oxidized and rearranged to produce the ginkgolides [5]. The plastid MEP biosynthetic pathway of ginkgolide is shown in Figure 3.2.

The basic cassane skeleton (**6**) may be derived from pimarane through methyl migration from C-13 to C-14 (14′) in the biosynthetic pathway. Pimaranes are formed through the cyclization of the labdane pyrophosphate (**12**). Biosynthesis of Labdane-type of diterpene involves an initial cyclization of GGPP promoted by a class-II diTPS (diterpene synthase) to produce a cyclic diphosphate intermediate followed by conversion of this intermediate (**13**) into the final diterpene skeleton by a class-I diTPS [6] (Figure 3.3).

Phytane is the isoprenoid alkane. It's IUPAC name is 2,6,10,14-tetramethyl hexadecane. It is a head-to-tail linked regular isoprenoid with molecular formula $C_{20}H_{42}$. It is produced when phytol (a constituent of chlorophyll) loses its hydroxyl group, whereas phytol loses one-carbon atom to produce pristane. So, chlorophyll is considered the major source of phytane and pristane. Chlorophyll is the photosynthetic pigment in plants, algae, and cyanobacteria. Both Pristane and phytane are formed by diagenesis of phytol under oxic and anoxic conditions, respectively [7] (Figure 3.4).

Edward et al. reported the total synthesis of the Cyathane diterpenoid, (±)-Sarcodonin-G, isolated from the fungus *Sarcodon scabrosus*. The cyathane family of diterpenoids

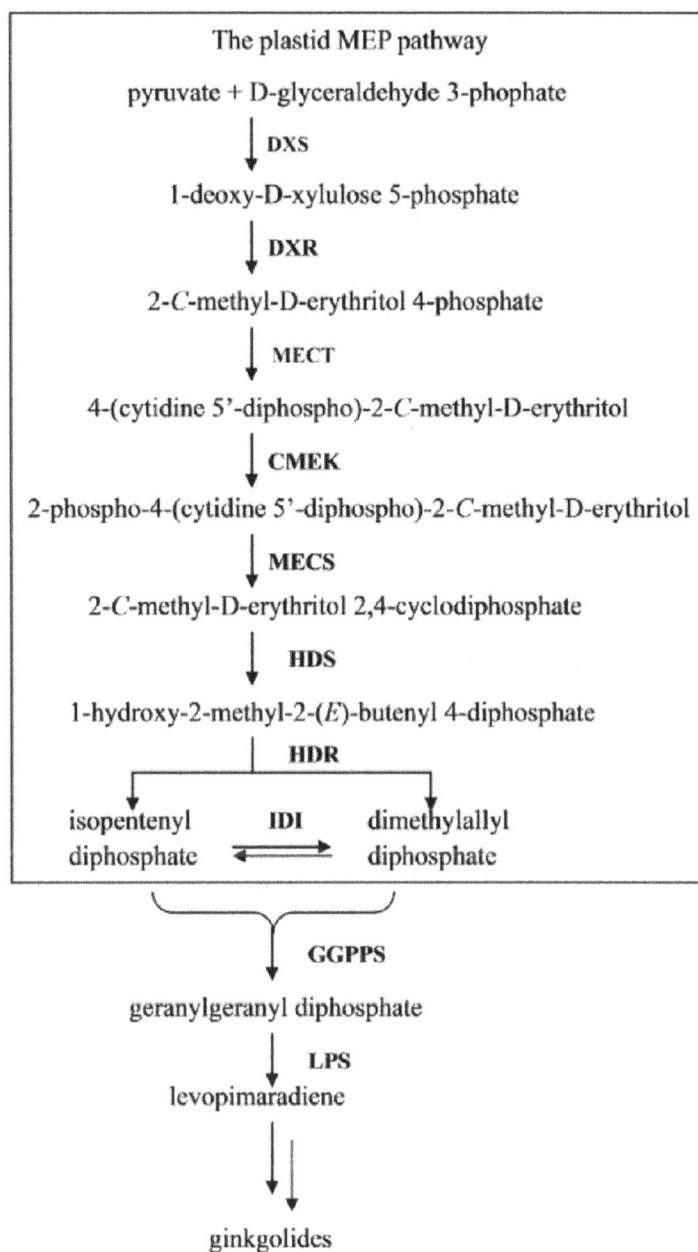

FIGURE 3.2 Methyl-erythritol-4-phosphate pathway in the isoprenoid biosynthesis of *Ginkgo biloba*. DXS, 1-deoxy-D-xylulose-5-phosphate synthase; DXR, 1-deoxy-D-xylulose-5-phosphate reductoisomerase; MECT, 2-C-methyl-D-erythritol 4-phosphate cytidyltransferase; CMEK, 4-(cytidine-5′-diphospho)-2-C-methyl-D-erythritol kinase; MECS, 2-C-methyl-D-erythritol-2,4-cyclodiphosphate synthase; HDS, 1-hydroxy-2-methyl-2-(E)-butenyl-4-diphosphate synthase; HDR, 1-hydroxy-2-methyl-2-(E)-butenyl-4-diphosphate reductase; IDI, Isopentenyl diphosphate isomerase; GGPPS, Geranylgeranyl diphosphate synthase; and LPS, Levopimaradiene synthase.

shares the tricyclic carbon skeleton (Figure 3.5). The first group of these structurally unique natural products was structurally characterized by Ayer and Taube in 1972 [8].

The asymmetric synthesis of the diterpenoid marine toxin, (+)-Acetoxycrenulide is carried out in various steps. First of all, the enantioselective synthesis of the (+)-Acetoxycrenulide (XIII) is performed by the conversion of (R)-citronellol (I) into the butenolide (VI) whose stereogenic center is provided by the terpenic alcohol. Three contiguous chiral C-atoms are subsequently set by conjugate addition of the enantiopure

allylphosphonamide reagent (VII). The resulting product (VIII) is transformed into the selenium derivative (X), whose oxide gives by thermal activation in dimethylacetamide at 220°C via 1,2-elimination and Claisen rearrangement of the cyclooctenone core product (XI). Finally, Simmons-Smith cyclopropanation and further transformations produce the titled compound (XIII) [9] (Figures 3.5 and 3.6).

Cembrene-A is a natural monocyclic diterpene. It's IUPAC name is (1E, 5E, 9E, 12R)-1,5,9-trimethyl-12-(prop-1-en-2-yl) cyclotetradeca-1,5,9-triene. It is isolated from corals of the

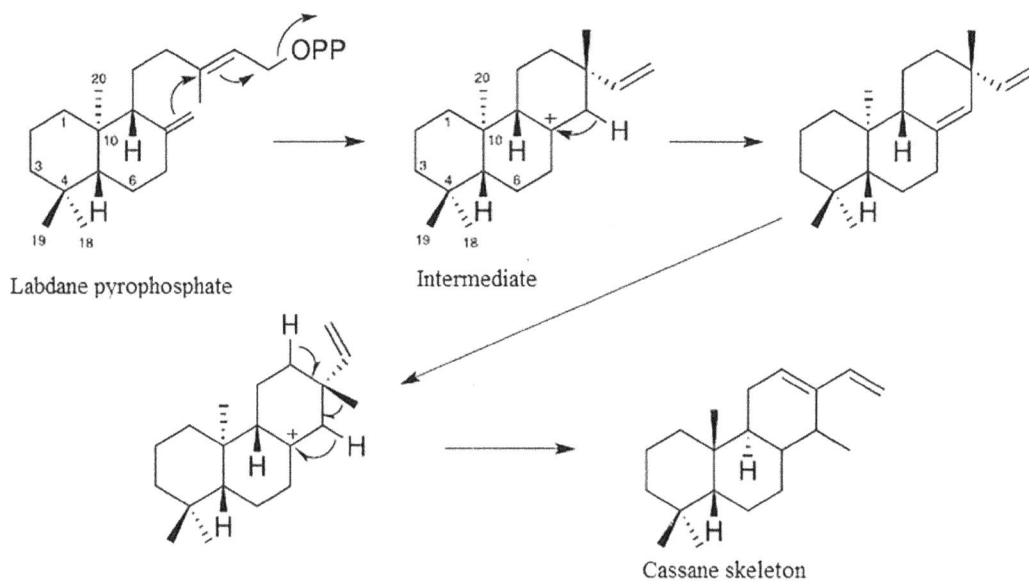

FIGURE 3.3   Biosynthetic pathway of the basic cassane skeleton.

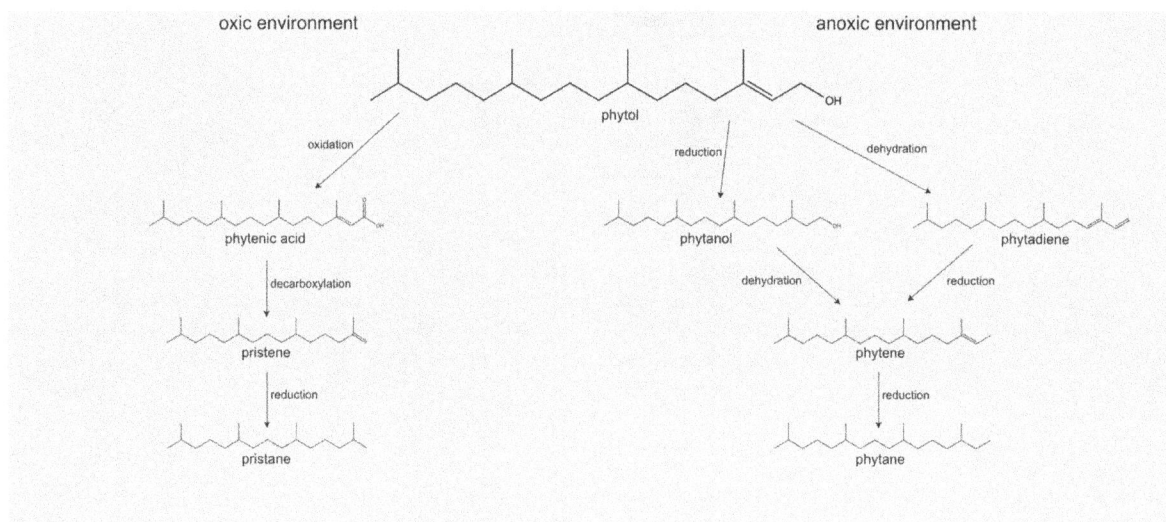

FIGURE 3.4   Formation of Pristane and phytane from phytol.

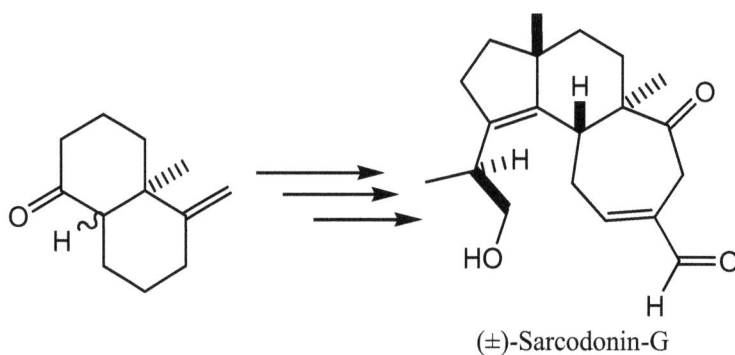

FIGURE 3.5   Synthesis of (±)-Sarcodonin-G.

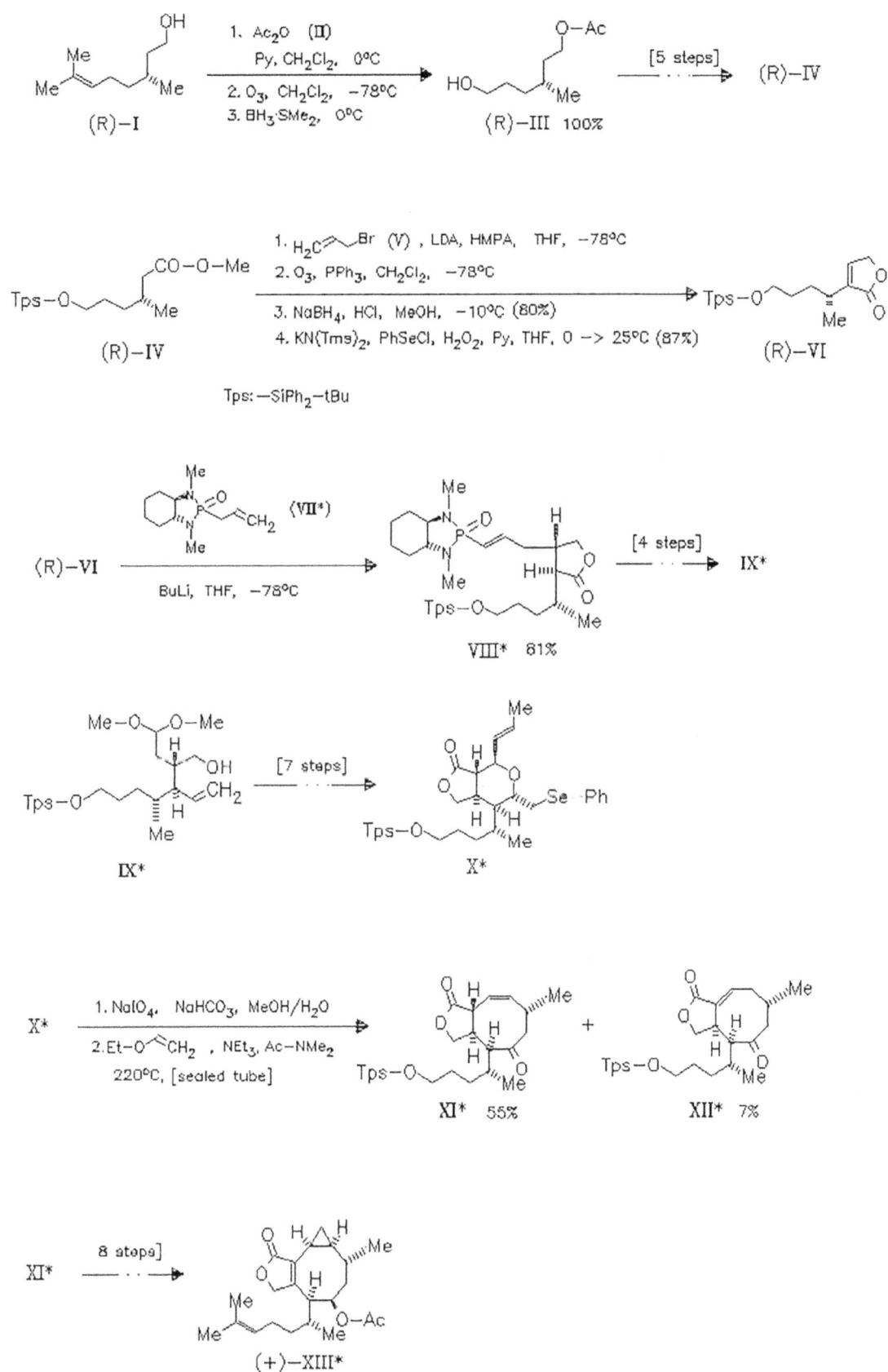

FIGURE 3.6 Synthesis of titled compound (XIII).

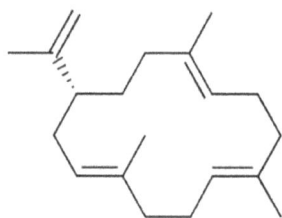

FIGURE 3.7 Structure of cembrene-A.

genus *Nephthea*. It is colorless oil with a faint wax-like odor. Cembrenes are biosynthesized by macrocyclization of geranylgeranyl pyrophosphate [10] (Figure 3.7).

Labdane is a natural bicyclic diterpene. The labdanes were so-called because the first members of this class were originally obtained from labdanum (resin obtained from the gum rockrose). It's IUPAC name is (4aR, 5S, 6S, 8aS)-1,1,4a, 6-tetramethyl-5-[(3R)-3-methylpentyl]decalin. Labdane exhibits various biological activities including antibacterial, antifungal, anti-protozoal, anti-inflammatory activities, etc. Sclarene is a diterpene present in the foliage of *Podocarpus hallii*. It consists of a labdane skeleton with double bonds at C-8(17), C-13(16), and C-14. It's IUPAC name is (4aS, 8S, 8aS)-4,4,8a-trimethyl-7-methylidene-8-(3-methylidenepent-4-enyl)-2,3,4a, 5,6,8-hexahydro-1H-naphthalene. It is derived from a hydride of labdane [11].

Labdane Sclarene

Abietane is a diterpene and chemically, it is 13α-isopropylpodocarpane. It forms the structural basis for a variety of natural compounds like abietic acid, carnosic acid, and ferruginol. These compounds are collectively called abietanes or abietane diterpenes [12] (Figures 3.8 and 3.9).

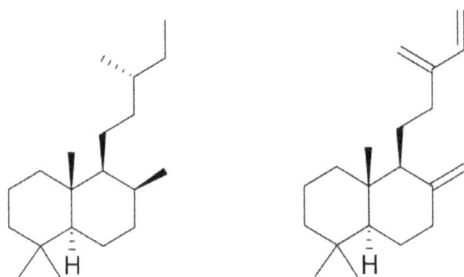

Labdane   Sclarene

FIGURE 3.8 Structure of labdane and sclarene.

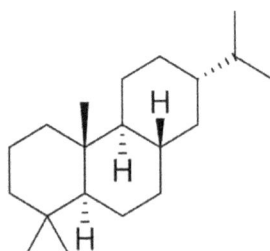

FIGURE 3.9 Structure of abietane.

Taxanes

Taxanes are the diterpenoids containing a taxadiene core. Taxadiene is produced from geranylgeranyl pyrophosphate by taxadiene synthase. Taxadiene is considered the intermediate for the synthesis of taxol. These compounds are produced by plants of the genus Taxus (yew trees) and are widely used as chemotherapeutic agents [13] (Figure 3.10).

Stemarene is a diterpene hydrocarbon. It's IUPAC name is (1R, 2S, 7S, 10S, 13R)-2,6,6,12-tetramethyl-tetracyclo[11.2.1.01,10.02,7]hexadec-11-ene. It can be produced biosynthetically through enzyme extracts from rice [14] (Figure 3.11).

Stemodene is a labdane-related diterpene. It's IUPAC name is (4aS, 6aS, 8S, 11aR, 11bS)-4,4,11b-trimethyl-9-methylenetetradecahydro-8,11a-methanocyclohepta[a]naphthalene. Morrone et al. reported that diterpene synthase (OsKSL11) from rice produces stemod-13(17)-ene from syn-copalyl diphosphate. OsKSL11 represents the first identified stemodene synthase, which catalyzes the various steps involved in the biosynthesis of the stemodane family of diterpenoid [15] (Figure 3.12).

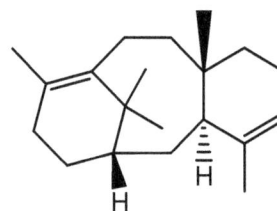

FIGURE 3.10 Structure of taxadiene.

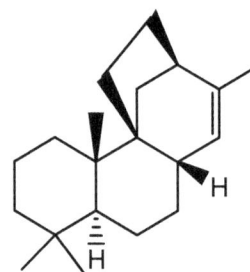

FIGURE 3.11 Structure of stemarene.

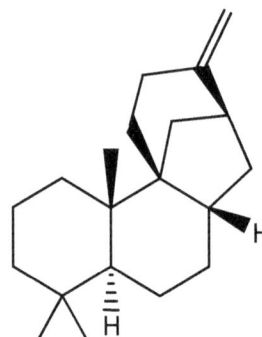

FIGURE 3.12 Structure of stemodene.

Steviol is a diterpene with the molecular formula $C_{20}H_{30}O_3$. It was first isolated from the plant *Stevia rebaudiana* in 1931. Stevioside and its hydrolysis products steviol and isosteviol are used as leads in the field of medicinal chemistry to provide diverse pharmacological activities such as antitumor, antibacterial, antihyperglycemic, anti-inflammatory, etc. Hydrolysis of stevioside under alkaline or acidic conditions, generates *ent*-kaurane diterpenoid steviol or *ent*-beyerane diterpenoid isosteviol, respectively [16,17] (Figure 3.13).

Halimane is a diterpenoid and its skeleton is formed by cyclization of geranylgeranyl diphosphate (GGPP) which is catalyzed by class-II diterpene cyclases (DTCs). These enzymes are characterized by having an aspartate-rich DXDD motif and are differentiated from class-I terpene synthases (TPS) enzymes by not having the characteristic aspartate-rich DDXXD motif that binds divalent metal ions required for catalysis of diphosphate ionization. The DTCs catalyze the GGPP by cyclization based on the acid-base mechanism. The process starts with 14,15 double bond protonation of E, E, E-GGPP followed by carbon-carbon double bond anti-addition (C_{10} on C_{15}, then C_6 on C_{11}) to provide four possible bicyclic products.

Chernov et al. reported the synthesis of diterpene β-Carbolines. Its synthesis is based on the Pictet-Spengler reaction of tryptophan or tryptamine. It is methyl lambertianate (**I**) of labdane-type terpenoid. The derivatives of this compound exert neurotropic activity. The first method of the synthesis of β-carbolines involves a two-step transformation of methyl 16-formyllambertianate (**II**) subjected to reductive amination with tryptamine to produce terpenotryptamine

(**III**). Intramolecular aminomethylation with formaldehyde yields the required N-labdanocarboline (**IV**). The second method of the synthesis of β-Carboline involves a direct reaction of aldehyde (**II**) with tryptamine and L-tryptophan methyl ester. In both cases, the reaction proceeds in boiling benzene and is catalyzed by p-toluenesulfonic acid. The experiments with tryptamine resulted in a mixture (1:1) of labdanocarbolines of (R)- and (S)-series in a 78% yield isolated as N-benzoyl derivatives (**V, VI**). The reaction of aldehyde **II** with L-tryptophan methyl ester under the described conditions led to a mixture of stereoisomeric 1,3-disubstituted labdanocarbolines resolved through N-benzoyl derivatives (**VII, VIII**). The reaction of compound acetoxyaldehyde (**IX**) with tryptamine or tryptophan methyl ester is not stereoselective and results in both cases in an equimolar mixture of (1R) and (1S) isomers, which were isolated as N-benzoyl derivatives (**X, XI**) and (**XII, XIII**) [18] (Figure 3.14).

Mulinane

The mulinane class of diterpenoids is a set of tricyclic (5–6–7), biologically active natural products whose members exhibit a variety of oxidation states. The synthesis of four mulinanes employing a divergent approach that relies on a diastereoselective anionic oxy-Cope rearrangement to set the relative configuration of the C8 stereocenter and an unprecedented vinylogous Saegusa dehydrogenation reaction to address C-ring functionality [19] (Figure 3.15).

Pallavicinin and neopallavicinin are two modified labdane-type diterpenoids. Neopallavicinin is the diastereomer of pallavicinin. These were first isolated by Wu et al. from

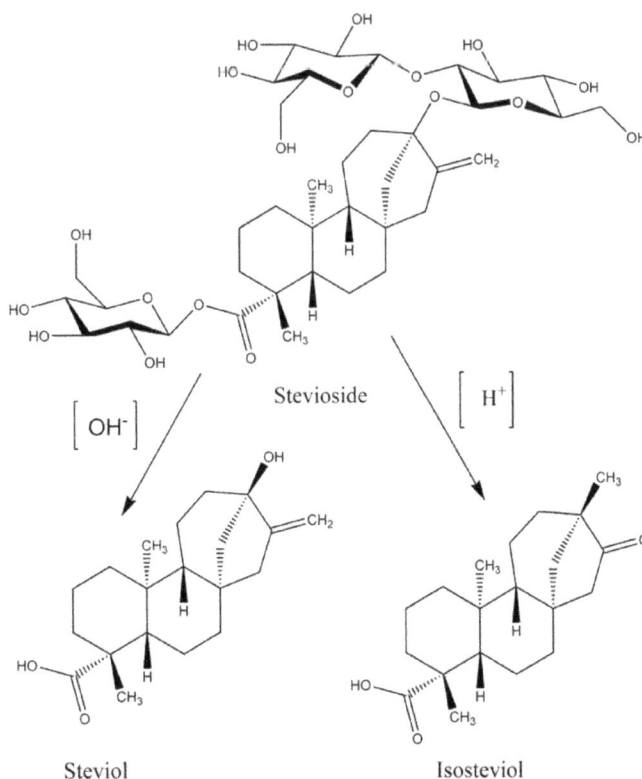

FIGURE 3.13 Formation of steviol and isosteviol from stevioside.

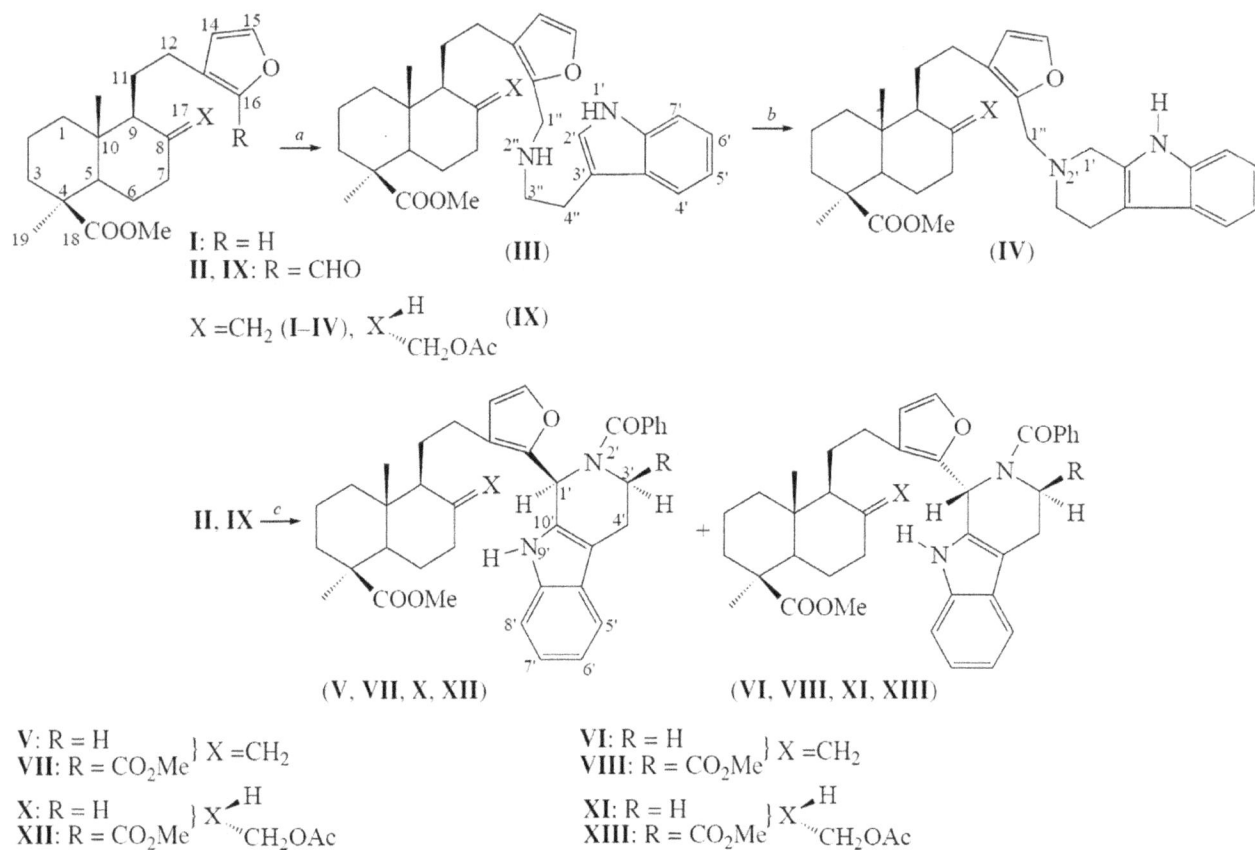

FIGURE 3.14 Synthesis of β-carbolines. Reagents and conditions: (a) tryptamine, benzene, 20°C, 18h, NaBH₄, MeOH, 30min; (b) CH₂O, TsOH, benzene, 80°C, 1h; (c) tryptamine or tryptophan methyl ester, TsOH, benzene, 80°C, 2h, then C₆H₅COCl, 20°C, 6h.

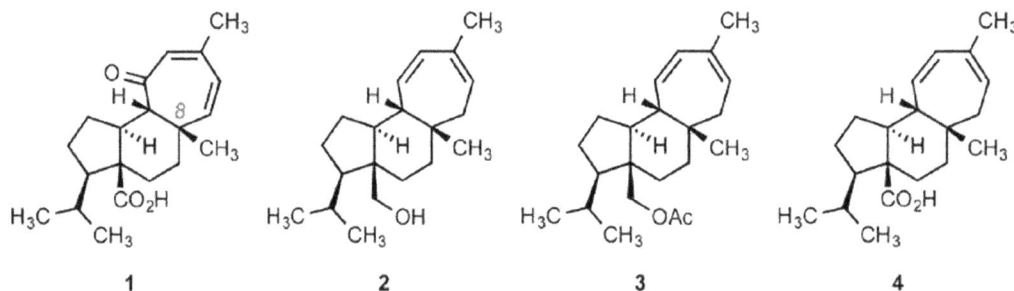

FIGURE 3.15 Structure of mulinane class.

liverwort *Pallavicinia subciliata* in 1994. Wong reported the first total synthesis of pallavicinin (**5**) and neopallavicinin (**6**) in 2006. Through a biomimetic pathway, (±)-pallavicinin and (±)-neopallavicinin were synthesized from (±)-Wieland–Miescher ketone via Grob fragmentation, intramolecular aldol cyclization, and intramolecular Michael reaction. The synthesis was stereocontrolled efficiently and followed the stereochemistry of the Wieland–Miescher ketone [20,21] (Figure 3.16).

The first asymmetric total synthesis of (−)-pallavicinin and (+)-neopallavicinin was proposed by Jia in 2015. The first enantioselective synthesis of these diterpenoids has been achieved

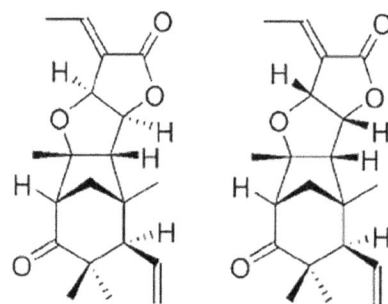

FIGURE 3.16 Structure of pallavicinin (5) and neopallavicinin (6).

in 15 steps. The synthesis include a palladium-catalyzed enantioselective decarboxylative allylation to form the chiral all-carbon quaternary stereocenter, a palladium-catalyzed oxidative cyclization to assemble the [3.2.1]-bicyclic moiety, and an unprecedented LiBHEt₃-induced fragmentation/protonation of an α-hydroxy epoxide to form the α-furan ketone with the desired configuration [22,23].

Harringtonolide

Zhai et al. reported the total synthesis of the diterpenoid (+)-Harringtonolide. It was first isolated from the seeds of *Cephalataxus harringtonia* by Buta and co-workers in 1978 and subsequently from the bark of the Chinese species *Cephalotaxus hainanensis* by Sun et al. in 1979 [24]. The key transformations include an intramolecular Diels–Alder reaction and a rhodium-complex-catalyzed intramolecular [3+2] cycloaddition to install the tetracyclic core as well as a highly efficient tropone formation. It was found to have potent cytotoxic activities with an IC₅₀ of 43 nm for KB tumor cells and cause necrosis [25,26] (Figure 3.17).

Peucelinendiol

Peucelinendiol is a diterpenoid, isolated from the roots of *Peucedanum oreoselinum*. The synthesis of peucelinendiol is carried out by using geraniol as starting material [27].

The reaction proceeds through the key intermediates i.e anion of phenylthiogeraniol and (–)-geraniol epoxide obtained by the Sharpless procedure. Its structure was established as (6S, 7R)-7-hydroxymethyl-2,6,10,14-tetramethylpentadeca-2,9,13-trien-6-ol. Similarly, Isopeucelinendiol, epipeucelinendiol, and epiisopeucelinendiol were also synthesized [28] (Figure 3.18).

Dawei et al. described a concise approach to synthesize the tetracyclic framework of the marine diterpenoid plumisclerin-A [29]. Starting from commercially available iridoid genipin, a highly selective intermolecular Diels–Alder reaction was utilized to construct the A/B/C tricycle. Later, a four-step sequence involving stereoselective epoxidation and Dauben oxidative rearrangement was developed to introduce the requisite enone moiety. The unique tricyclo[4,3,1,01,5]decane motif was successfully generated via a late-stage SmI2-mediated reductive coupling [30] (Figure 3.19).

Neodolastane

The neodolastane diterpenoids comprise a group of 44 compounds that include guanacastepenes, heptemerones, plicatilisins, radianspenes, 2,15-epoxy-5,13-dihydroxyneodolast-3-en-14-one, and sphaerostanol. These diterpenoids are characterized by a tricyclic neodolastane skeleton which consists of fused five, seven, and six-membered rings [31] (Figure 3.20).

The neodolastane skeleton was first proposed by Vidari as an intermediate in the biosynthesis of trichoaurantianolide-A in 1995. The structures of guanacastepenes A-O (**7–21**) were determined by X-ray crystallography. This family is characterized by the typical tricyclic C₅–C₇–C₆ neodolastane skeleton [32] (Figure 3.21).

The neodolastane skeleton is related to the dolastane skeleton of marine origin by transannular cyclization or to the neodolabellanes by a Wagner–Meerwein migration of a methyl group from the dolabellane skeleton. Clardy and Hiersemann proposed the biogenesis of the neodolastane skeleton as a series of enzyme-catalyzed ring closures and Wagner–Meerwein migrations from geranylgeranyl pyrophosphate (Figures 3.22 and 23).

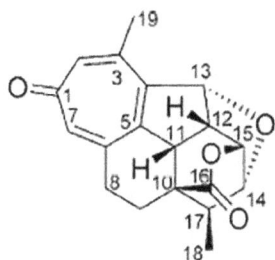

FIGURE 3.17 Structure of (+)-Harringtonolide.

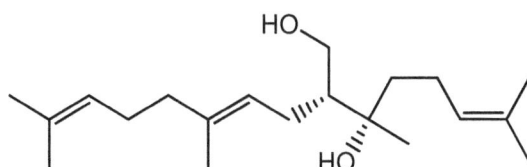

FIGURE 3.18 Structure of peucelinendiol.

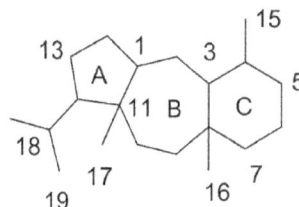

FIGURE 3.20 Structure of neodolastane.

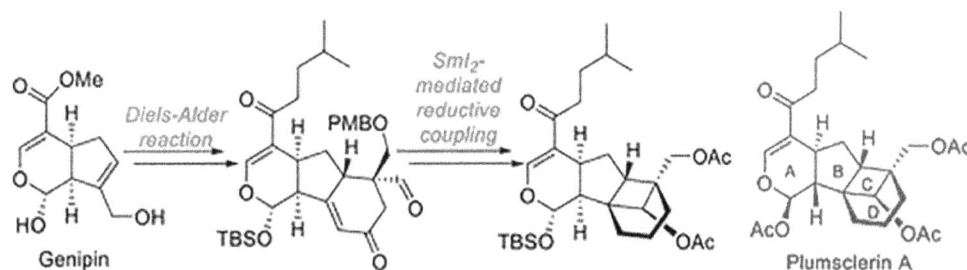

FIGURE 3.19 Synthesis of plumisclerin-A.

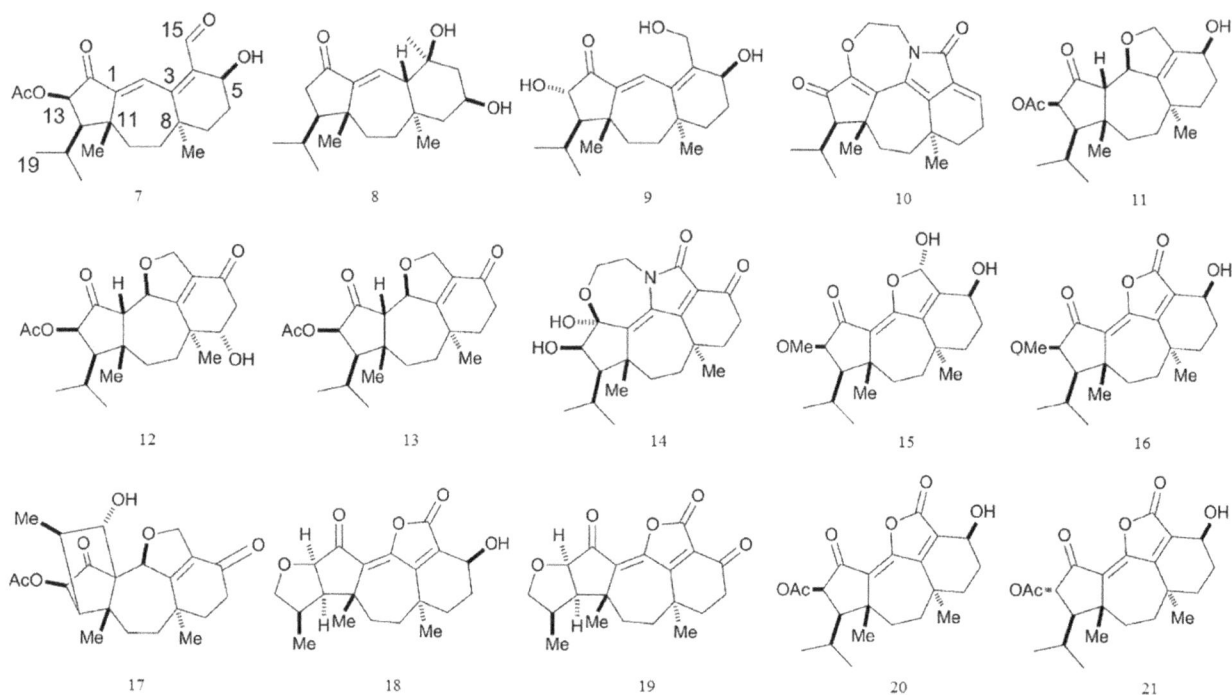

FIGURE 3.21 Structures of guanacastepenes.

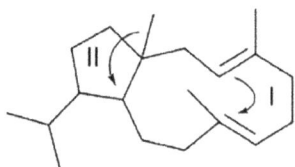

FIGURE 3.22 Structures of dolabellane skeleton.

The striatals and striatins are the group of diterpenoids isolated from cultures of the bird's nest fungus *Cyathus striatus*. The striatals possess a cyathane skeleton. The biosynthesis of striatal-A was studied by adding [1-^{13}C]- and [2-^{13}C]-D-glucose to resting cells of *C. striatus*. The optimal glucose concentration in the incubation solution (0.5%) was determined in preliminary experiments. Under the conditions described in the experimental section, approximately 7 mg of striatin-A were produced within 16 h [33] (Figure 3.24).

3.3 Medicinal Uses of Diterpenoids

There is a continuous approach in research for the development of therapeutic agents from natural products of plant sources in recent years. Naturally occurring diterpenes exhibit various medicinal activities such as anti-inflammatory, antimicrobial, antidiabetic, antihyperlipidemic, anticancer, diuretic, antiviral, antispasmodic, and neuroprotective activities, etc [34–36] (Figure 3.25).

3.3.1 Treatment of Cardiovascular Diseases

Treatment of cardiovascular diseases can be performed in different ways including the administration of diuretics, β-adrenergic blockers, angiotensin-converting enzyme (ACE) inhibitors, angiotensin-II receptor subtype-1 antagonists, and Ca^{2+} channel blockers. The naturally occurring antispasmodic phytochemicals are used as a potential therapy for cardiovascular diseases. Various diterpenes are reported to have pronounced cardiovascular effects. For example, grayanotoxin-I produces positive inotropic responses and forskolin is a well-known activator of adenylate cyclase. Similarly, eleganolone and 14-deoxyandrographolide exhibit vasorelaxant properties, and marrubenol inhibit smooth muscle contraction by blocking L-type calcium channels. These diterpenoids exhibit vasorelaxant activity and inhibit vascular contraction by blocking the influx of extracellular Ca^{2+}. However, Kaurane and pimarane-type diterpenes decreased the arterial blood pressure in normotensive rats [37] (Figure 3.26).

Arterial hypertension is a common and progressive disorder that causes a major risk for cardiovascular diseases. The hypotensive action demonstrated by the diterpene involves the blockade of autonomic ganglia and the renin-angiotensin system as well as the inhibition of β-adreno receptors. The subtype β$_1$-adrenoreceptor is present in cardiac muscle and its activation by circulating catecholamines induces positive inotropic and chronotropic effects [38].

Tirapelli et al. reported the hypotensive action of naturally occurring diterpenes. Naturally occurring diterpenes such as forskolin and stevioside, exhibit vaso-relaxant activity and inhibit vascular contraction by different mechanisms of action. It was reported that the administration of stevioside of 750 mg/

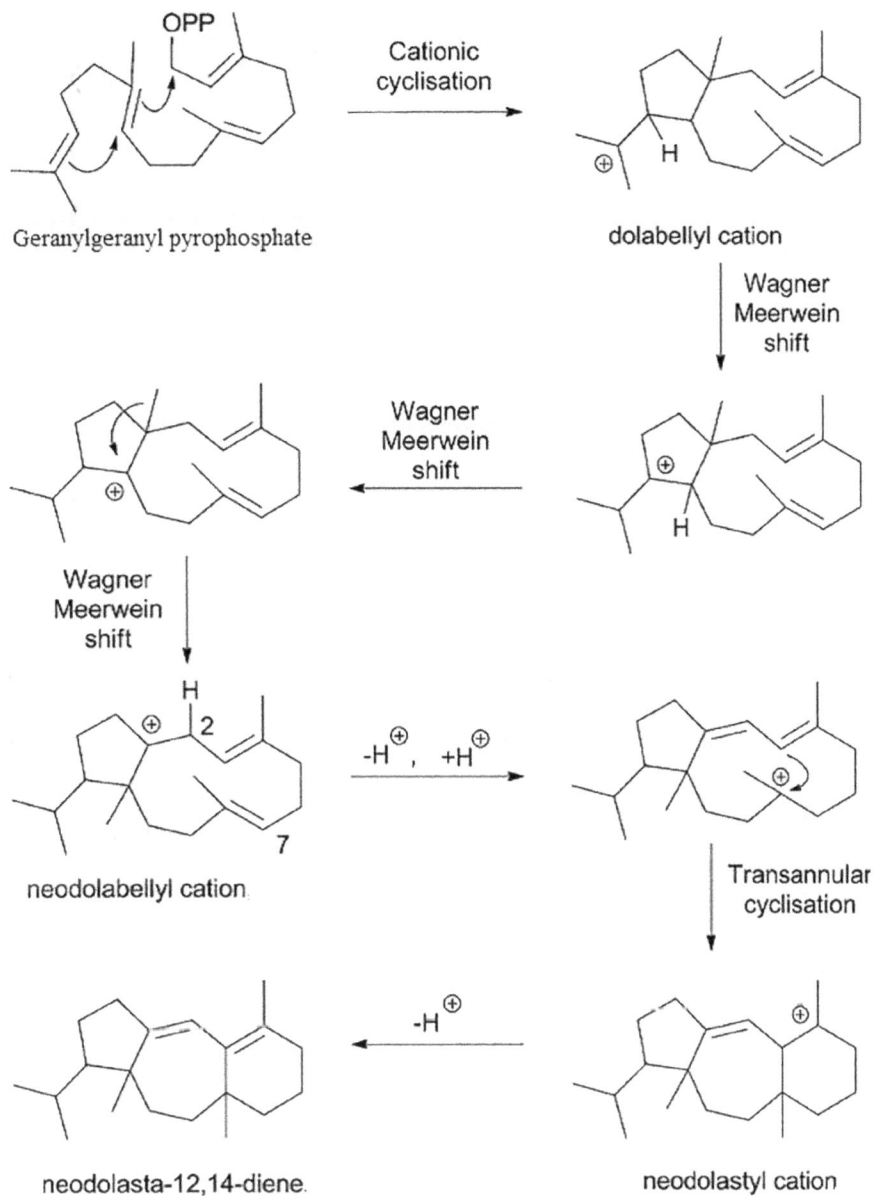

FIGURE 3.23 Proposed biogenesis of the neodolastane skeleton.

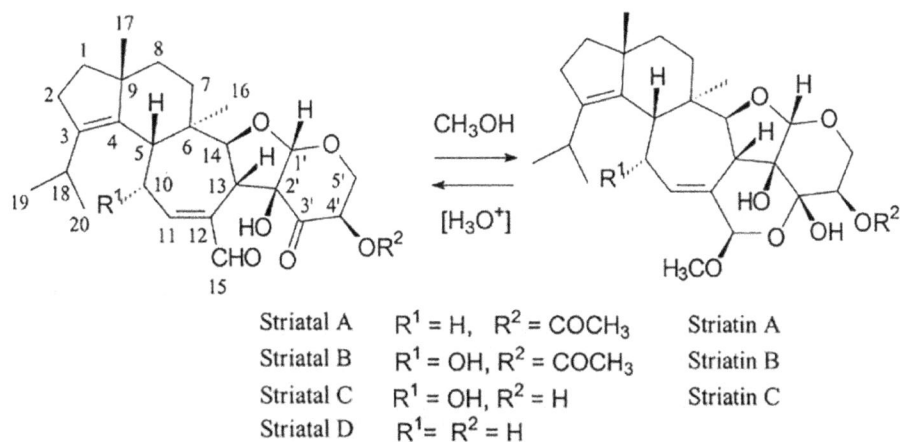

Striatal A	R¹ = H, R² = COCH₃	Striatin A
Striatal B	R¹ = OH, R² = COCH₃	Striatin B
Striatal C	R¹ = OH, R² = H	Striatin C
Striatal D	R¹ = R² = H	

FIGURE 3.24 Structures of striatals and striatins.

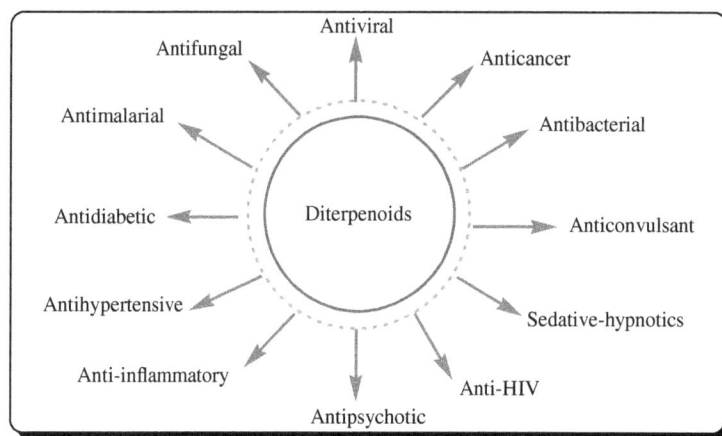

FIGURE 3.25 Medicinal uses of diterpenoids.

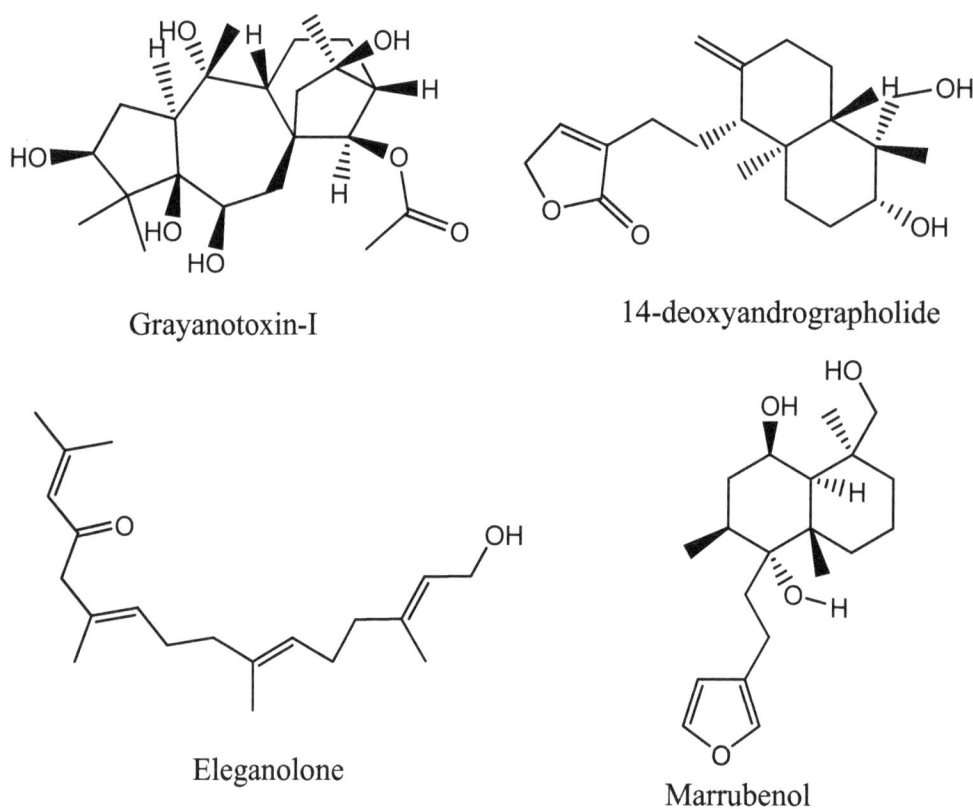

Grayanotoxin-I

14-deoxyandrographolide

Eleganolone

Marrubenol

FIGURE 3.26 Chemical structures of diterpenoids showing activity against cardiovascular diseases.

day for 3 days failed to alter blood pressure significantly in normotensive rats [39].

Forskolin is the main active constituent of the Ayurvedic herb *Coleus forskohlii*. It is the labdane-type diterpene and is chemically known as 7 beta-acetoxy-8,13-epoxy-1-alpha, 6 beta, 9 alpha-trihydroxy-labd-14-ene-11-one. The chief mechanism of action of forskolin is to increase cyclic adenosine monophosphate (cAMP) and cAMP-mediated functions, via activation of the enzyme adenylate cyclase. It was demonstrated that forskolin lowers blood pressure via the relaxation of vascular smooth muscle. Intravenous administration of forskolin (4 μg/kg/min) decreased vascular resistance with improved

left ventricular contractility and induced a 20% reduction in arterial blood pressure. In addition to this, intravenous administration of forskolin (10 μg/kg/min) increased the blood flow to the brain in rabbits. Forskolin lowers the intracellular pressure (IOP) by reducing the aqueous inflow. This is due to increases in intracellular cAMP by stimulating the enzyme adenylate cyclase by topical ocular application of forskolin [40].

Similarly, 14-deoxy-11,12-didehydroandrographolide antagonizes the positive chronotropic effect elicited by the β-adrenoreceptor agonist isoproterenol in isolated rat right atria. This result suggests that the bradycardic effect is elicited by the diterpene *in-vivo* due to the direct blocking action of β_1-adrenoreceptor [41].

The intravenous administration of diterpene 14-deoxy-11,12-didehydroandrographolide (1.7–6.7 mmol/kg) in normotensive rats caused a significant decrease in both blood pressure and heart rate in a dose-dependent manner. The cardiovascular effects of trans-dehydrocrotonin involve its direct action in the heart. The intravenous injection of trans-dehydrocrotonin at 10 and 15 mg/kg induced remarkable hypotension and bradycardia in normotensive rats. Two kaurane-type diterpenes such as ent-kaur-16-en-19-oic acid and ent-kaur-16-en-15-one-19-oic acid displayed cardiovascular effects. The vascular relaxation effects of ent-kaur-16-en-19-oic acid demonstrated that it blocks the extracellular Ca^{2+} influx by interacting with both voltage and receptor-operated channels. The diterpene ent-pimara-8(14), 15-dien-19-oic acid (1–15 mg/kg) produced a dose-dependent decrease in mean arterial pressure from conscious normotensive rats [42].

The cardiovascular effects of the diterpene 8(17), 12E, 14-labdatrien-18-oic acid were demonstrated using a combined approach of *in-vivo* and *in-vitro* methods. The intravenous administration of this diterpene at 5, 10, 20, and 30 mg/kg in normotensive rats induced short-term hypotension. Intravenous administration of the diterpene, Labd-8 (17)-en-15-oic acid with a dose of 1–10 mg/kg induces immediate and dose-dependent decreases in mean arterial blood pressure in normotensive rats. This effect occurs through the activation of β_2-adrenergic vascular receptors [43,44].

Rebaudiosides (A, B, C, D, E) are isolated from *S. rebaudiana* Bertoni leaves. All of these isolated diterpenoid glycosides possess the same chemical skeletal structure (steviol) but differ in the residues of carbohydrates at positions C_{13} and C_{19}. Rebaudioside-A is a steviol glycoside that is 200 times sweeter than sugar [45] (Tables 3.2 and 3.3).

Odek et al. reported the antihypertensive effect of the clerodane diterpene ajugarin-I on experimentally hypertensive rats. Ajugarin-I was isolated from *Ajuga remota* of the family Labitae. The administration of ajugarin-I in the hypertensive rats, lowered the blood pressure from 165.00 to ± 1.57 mmHg to a normotensive level of about 116.33 ± 1.67 mmHg at 10 mg/kg in 2 weeks [46] (Figure 3.27).

3.3.2 Treatment of Cancer

Cancer is a leading cause of death worldwide. It is defined as the uncontrolled growth of abnormal cells anyplace in the body. These abnormal cells are called cancer cells, malignant cells, or tumor cells. There are different types of cancer such as breast cancer, bladder cancer, endometrial cancer, melanoma (skin cancer), lung cancer, liver cancer, thyroid cancer, pancreatic cancer, prostate cancer, kidney cancer, lymphoma (cancer of the immune system), leukemia (cancer of bone

TABLE 3.2

Hypotensive Action of Diterpenes in Rats

Diterpenes	Dose/Administration	Proposed Mechanism(s) of Action
Stevioside	8 and 16 mg/kg/i.v (infusion) 25 mg/kg/i.p 100 and 200 mg/kg i.v (bolus)	Reduces vascular resistance via inhibition of extracellular Ca^{2+} influx and the release of a vasodilator prostaglandin.
14-Deoxy-11, 12-didehydro andrographolide	1.7–6.7 mmol/kg/i.v (bolus)	Activation of the autonomic ganglia, renin-angiotensin system, and cardiac β-adrenoreceptor.
Trans-dehydro crotonin	10 and 15 mg/kg/i.v (bolus)	Direct negative chronotropic and inotropic effects in the heart by a non-adrenergic non-cholinergic mechanism.
Ent-kaur-16-en-19-oic acid	10 mg/kg/i.v (bolus)	Reduces the vascular resistance via inhibition of extracellular influx of Ca^{2+}.
Ent-kaur-16-en-15-one-19-oic acid	10 mg/kg/i.v (bolus)	Reduces the vascular resistance via inhibition of extracellular influx of Ca^{2+}.
Kolavenic acid	30 mg/kg/i.v (bolus)	Activation of muscarinic receptor.
Ent-pimara-8(14),15-dien-19-oic acid	1–15 mg/kg/i.v (bolus)	Reduces the vascular resistance via inhibition of Ca^{2+} influx extracellularly and the release of NO and a vasodilator prostaglandin.
8(17),12E,14-labdatrien-18-oic acid	5–30 mg/kg/i.v (bolus)	Reduces vascular resistance via inhibition of influx of Ca^{2+} extracellularly and the release of NO.
Labd-8 (17)-en-15-oic acid	1–10 mg/kg/i.v (bolus)	Reduces vascular resistance mainly via inhibition of influx of Ca^{2+} extracellularly.

TABLE 3.3

Cardiovascular Effects of Diterpenes in Humans

Diterpenes	Dose (Oral)	Period of Treatment	Cardiovascular Action
Stevioside	750 mg daily	12 months	Reduction in systolic and diastolic blood pressure.
	1,500 mg daily	24 months	Reduction in systolic and diastolic blood pressure.
	3.75, 7.5 or 15 mg/kg/day	24 weeks	No effect
	750 mg daily	3 days	No effect
	250 mg daily	3 months	No effect
Rebaudioside-A	1,000 mg/daily	16 weeks	No effect
	1,000 mg/daily	4 weeks	No effect

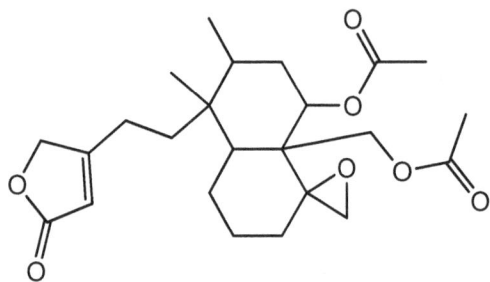

FIGURE 3.27 Structure of ajugarin-I.

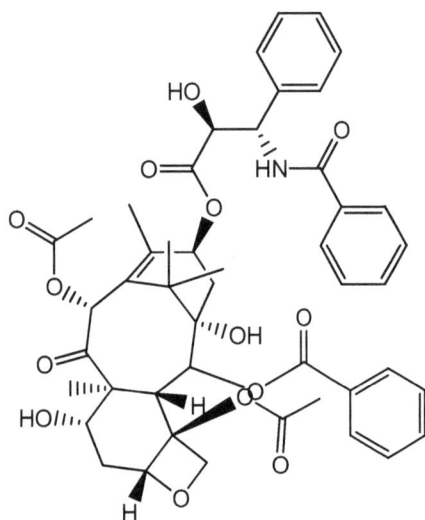

FIGURE 3.28 Structure of diterpene paclitaxel.

marrow), brain tumor, colon, and rectal cancer, etc. The signs and symptoms of cancer depend on the specific type and grade of cancer. Treatment of cancer may include chemotherapy, radiation therapy, or surgery [47].

The diterpene paclitaxel (Taxol) is a well-known anti-proliferative agent. The activity of paclitaxel was first reported in the 1960s when the crude extract of bark from the Pacific yew (*Taxus brevifolia*) was evaluated for cytotoxic activity as part of the natural products screening program of the U.S. National Cancer Institute (NCI). Paclitaxel displays significant activity against several human tumor xenografts from human leukemias and solid tumors such as cancer of the breast, ovary, pancreas, lung, endometrium, and brain. It is a mitotic inhibitor [48]. It binds to tubulin and inhibits the disassembly of microtubules resulting in the inhibition of cell division. It also induces apoptosis by blocking the function of the apoptosis inhibitor protein Bcl-2 (B-cell Leukemia 2) [49] (Figure 3.28).

There are 14 diterpenoids isolated from *Euphorbia fischeriana*. Depending on the presence of carbon skeleton and substituents, they can be classified into eight subtypes such as ent-abietane (22–28), tigliane (29–31), daphnane, ingenane (32–35), ent-atisane (36–39), ent-rosane (40–42), ent-kaurane (43), and lathyrane (44). These compounds have been found to inhibit the proliferation of several cancer cells with promising IC_{50} [50] (Figures 3.29–3.35, Tables 3.4 and 3.5).

Sclareol is a fragrant chemical compound found in *Salvia sclarea*, from which it derives its name. It's IUPAC name is (1R, 2R, 4aS, 8aS)-1-[(3R)-3-hydroxy-3-methylpent-4-enyl]-2,5,5,8a-tetramethyl-3,4,4a, 6,7,8-hexahydro-1H-naphthalene-2-ol. It is classified as a bicyclic diterpene alcohol. It is an amber-colored solid with a sweet, balsamic scent. Sclareol is also able to kill human leukemic cells and colon cancer cells

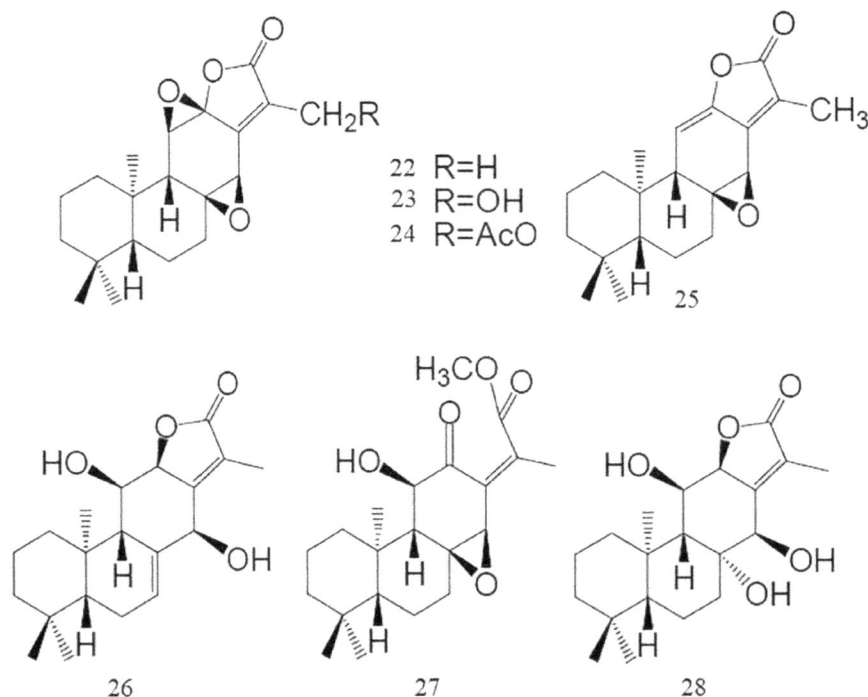

22 R=H
23 R=OH
24 R=AcO

25

26   27   28

FIGURE 3.29 Structure of ent-Abietane-type diterpenoids.

R₁

29 Ac

30 (C₁₅H₃₁)CO

31

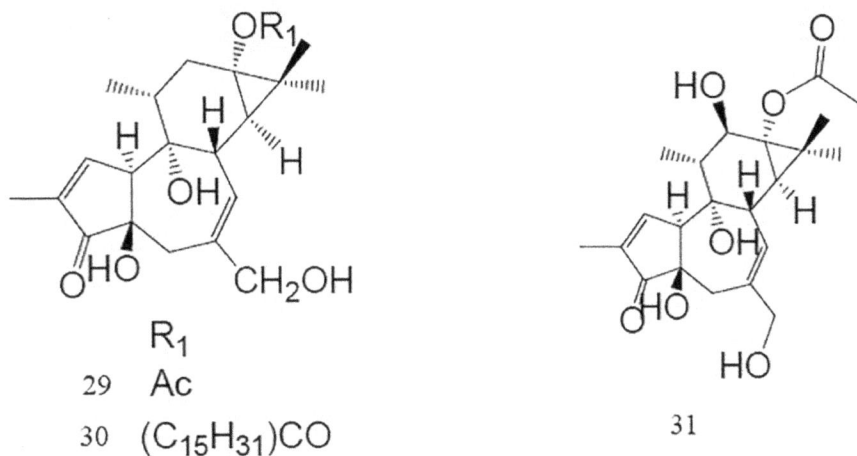

FIGURE 3.30 Structure of tigliane.

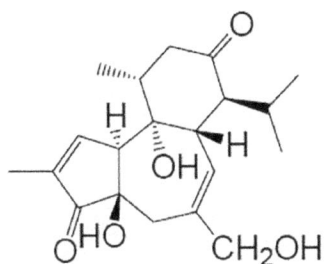

FIGURE 3.31 Structure of daphnane.

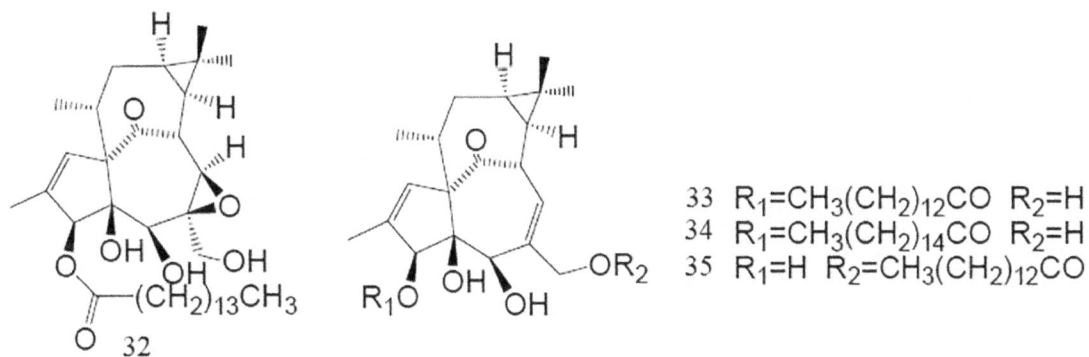

33 R₁=CH₃(CH₂)₁₂CO R₂=H
34 R₁=CH₃(CH₂)₁₄CO R₂=H
35 R₁=H R₂=CH₃(CH₂)₁₂CO

(CH₂)₁₃CH₃

32

FIGURE 3.32 Structure of ingenane.

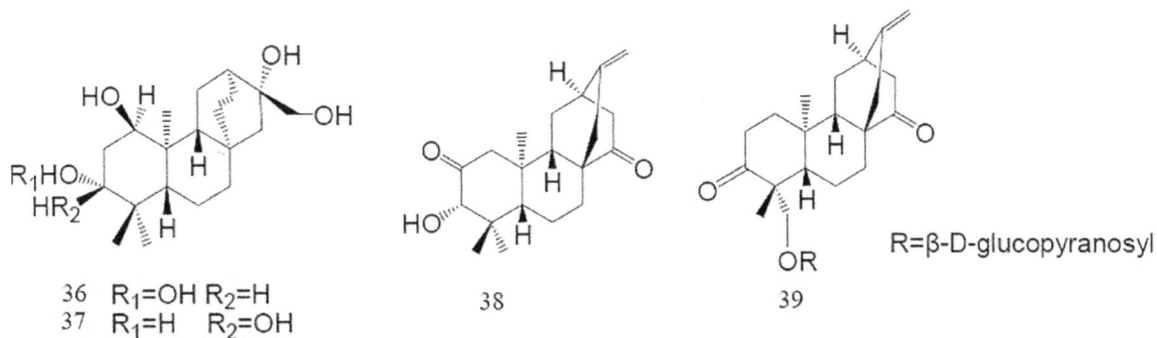

36 R₁=OH R₂=H
37 R₁=H R₂=OH

38

39

R=β-D-glucopyranosyl

FIGURE 3.33 Structure of ent-atisane.

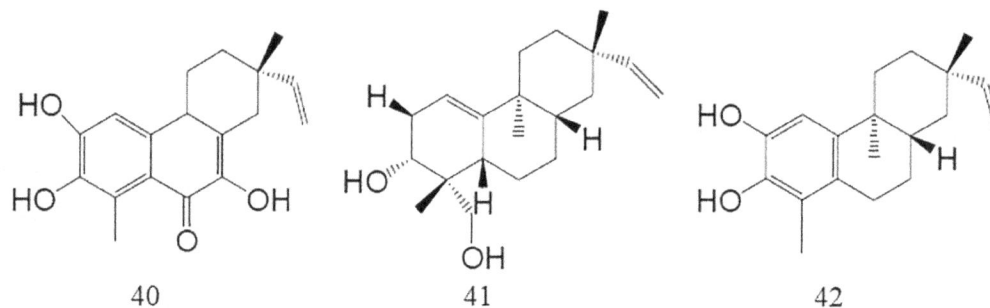

FIGURE 3.34 Structure of ent-rosane.

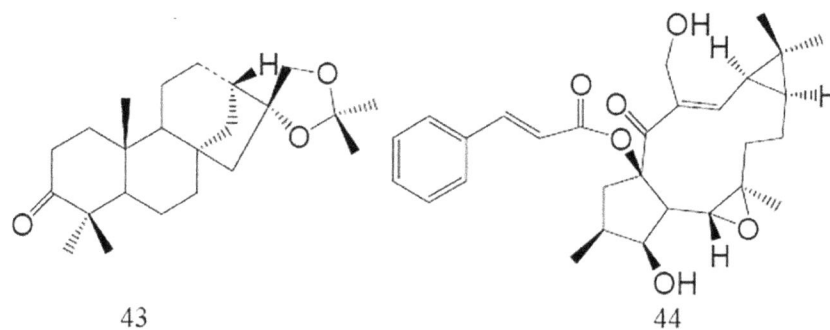

FIGURE 3.35 Structure of ent-kaurane and lathyrane.

TABLE 3.4

In Vitro Cytotoxic Activity Study of Diterpenoids

Compounds	Subtype	Type of Cancer	Cell Lines (IC_{50})
Jolkinolide-A	Ent-abietane	Liver	HepG-2 (80.12 μM/48 h)
		Breast	MCF-7 (56.34 μM/48 h)
		Cervical	Hela (>100 μM/48 h)
Jolkinolide-B	Ent-abietane	Human leukemic	K562 (12.1 μg/mL/24 h)
		Lung	A549 (28.24 μM)
		prostate	PC3 (244 μM/48 h)
Prostratin	Tigliane	Liver	HepG-2 (11.77 μM/48 h)
		Breast	MCF-7 (17.4 μM/48 h)
		Gastric	SGC-7901 (25.4 μM)
Langduin-A	Daphnane	Liver	HepG-2 (35 μM/48 h)
		Breast	MCF-7 (19.4 μM/48 h)
		Gastric	SGC-7901 (21.3 μM/48 h)
Ingenol 3-palmitate	Ingenane	Lung	A549 (2.88 μg/mL/72 h)
		Liver	BEL7402 (25.87 μg/mL/72 h)
		Colon	HCT116 (14.38 μg/mL/72 h)
		Breast	MDA-MB-231 (22 μg/mL/72 h)
Ent-1β,3 β,16β, 17-tetra- hydroxy-atisane	Ent-atisane	Breast	MCF-7 (23.21 μM)
Ent-kaurane-3-oxo-16 β, 17-acetonide	Ent-kaurane	Liver	Hep-3B (8.15 μM)

TABLE 3.5

Diterpenoids from *E. fischeriana* Inhibit Mammosphere Formation in MCF-7 Cells

Bioactive Ingredient	Subtype
Jolkinolide-b	Ent-abietane
Ingenol-3-myristinate	Ingenane
Ingenol-3-palmitate	Ingenane
Euphorin-E	Ent-abietane
Euphorin-H	Ent-abietane
Yuexiandajisu-E	Ent-abietane
Ingenol-20-myristinate	Ingenane
Ent-3-hydroxyatis-16-ene-2,14-dione	Ent-atisane
19-O-β-D-glucopyranosyl-ent-atis-16-ene-3,14-dione	Ent-atisane
Euphorin C	Ent-rosane
Ebractenoid C	Ent-rosane
Ebractenoid F	Ent-rosane
Jolkinol-A	Lathyrane

by apoptosis. Sclareol has shown significant cytotoxic activity against both human leukemic and breast cell lines and also enhances the effect of anticancer drugs (doxorubicin, etoposide, and cisplatinum) against MDD2 breast cancer cell lines [51,52] (Figure 3.36).

Chen et al. reported anti-proliferative and apoptosis-inducible activity of labdane and abietane diterpenoids in HeLa cells. Two abietane diterpenoids such as 18-hydroxyferruginol (8,11,13-abietatriene-12,18-diol) and hinokiol (8,11,13-abietatriene-3,12-diol) and a labdane diterpenoid, kayadiol [8(17), 13-labdiene-15,18-diol] were isolated from the pulp of *Torreya nucifera*. The cytotoxic activity of three labdane and abietane diterpenoid compounds (Figure 3.37) was determined by

using the MTT assay after culturing of HeLa cells in the presence of the chemical compounds up to 100 μM for 2 days. The treatment of Hinokiol against HeLa cells resulted in a slight decrease in cell viability at 100 μM. Whereas, the inhibitory effect of 18-hydroxyferruginol treatment on cell viability was greater than that of hinokiol. The labdane compound kayadiol was found to have the most effective inhibitory effect against a wide variety of human cancer cells. Kayadiol was found to have IC$_{50}$ of 30 μM in HeLa cells, and also to exhibit antiproliferative effects toward six other human cancer cell lines, with IC$_{50}$ values of 30–50 μM by using the MTT assay method. Kayadiol showed activation of caspases-3 and -9, as well as an increase in the depolarization of mitochondrial membrane potential and the Bax/Bcl-2 ratio toward HeLa cells [53] (Table 3.6).

Sarcodonin-G (SG) is one of the cyathane diterpenoids and is isolated from the mushroom *Sarcodon scabrosus*. It exhibited anti-proliferative activity against human cancer cells (HeLa). SG exhibited IC$_{50}$ of 20 μM, estimated by MTT assay 2 days after culture of cells with the chemical. After the culture, the activity of mitochondrial succinic dehydrogenase was measured by further incubation of the cells with 0.5 mg/mL 3-(4,5-dimethylthiazol-2-yl)-2,5-diphenyl tetrazolium bromide (MTT) for 4 h, followed by measurement of absorbance at 570 nm with a reference wavelength at 655 nm. Cell survival was calculated from absorbance and presented as a percentage of the surviving cells. SG treatment of HeLa cells resulted in a dose-dependent generation of apoptotic events such as DNA-laddering (≤100 μM). SG-treated HeLa cells demonstrated the activation of caspase-3 and caspase-9 and also an increase in

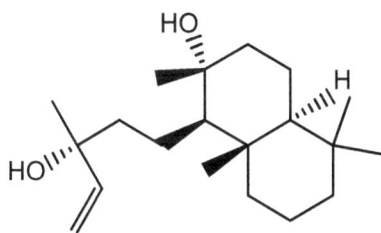

FIGURE 3.36 Structure of sclareol.

TABLE 3.6

Anti-Proliferative Activity of Kayadiol in Various Cancer Cell Lines

Cell Lines	IC50 Values (μM)
HeLa	30.0
U251SP	32.7
HAC-2	34.4
T-Tn	38.2
HEC-1	42.2
T-98	49.4
HLE	50.3

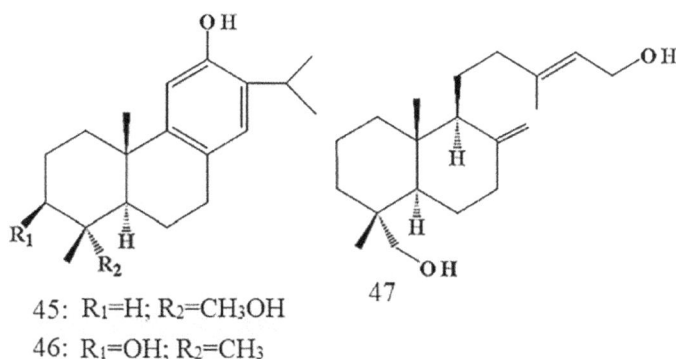

45: R₁=H; R₂=CH₃OH
46: R₁=OH; R₂=CH₃

47

FIGURE 3.37 Chemical structures of (45) 18-hydroxyferruginol, (46) hinokiol, and (47) kayadiol

the ratio of Bax/Bcl-2. This was analyzed by Western blot analysis [54] (Figure 3.38).

The anticancer potential of diterpenoids was demonstrated for Caseanigrescens-A (IC$_{50}$ of 1.4 μM), B (IC$_{50}$ of 0.83 μM), C (IC$_{50}$ of 1.0 μM), and D (IC$_{50}$ of 1.0 μM) against the A2780 human ovarian cancer cell line. Labdane diterpenes such as Aulacocarpinolide (IC$_{50}$ of 12.5 μg/mL) and aulacocarpin B (IC$_{50}$ of 25 μg/mL) were found to be cytotoxic toward murine leukemia L1210 cells [55] (Figure 3.39).

Diterpenes such as dictyotalide-A, dictyotalide-B, nordictyotalide and 4-acetoxydictyolactone, were isolated from *D. dichotoma*. These compounds exhibited significant cytotoxic activity against mouse melanoma cells (B16) with IC$_{50}$ values of 2.57, 0.58, 1.58, and 1.57 μg/mL, respectively [56] (Figure 3.40).

Similarly, Dictyolactone was also isolated from *D. dichotoma* and Neodictyolactone was isolated from *D. linearis*. Among these diterpenes, Dictyolactone displayed significant cytotoxic activity against P-388 cells, P-388/DOX cells, and KB cells with EC$_{50}$ values of 2.8, 2.4, and 4.9 μg/mL, while Neodictyolactone was less active against those cells with EC$_{50}$ values of 3.4, 3.9, and 6.2 μg/mL, respectively [57] (Figure 3.41).

Two new diterpenes such as Dictyone and dictyone acetate were isolated from *D. dichotoma*. These diterpenes exhibited moderate cytotoxic activity against three proliferating mouse cell lines, a normal fibroblast line NIH3T3, and two virally transformed forms SSVNIH3T3 and KA3IT with IC$_{50}$ values ranging from 5 to 35 μg/mL [58] (Figure 3.42).

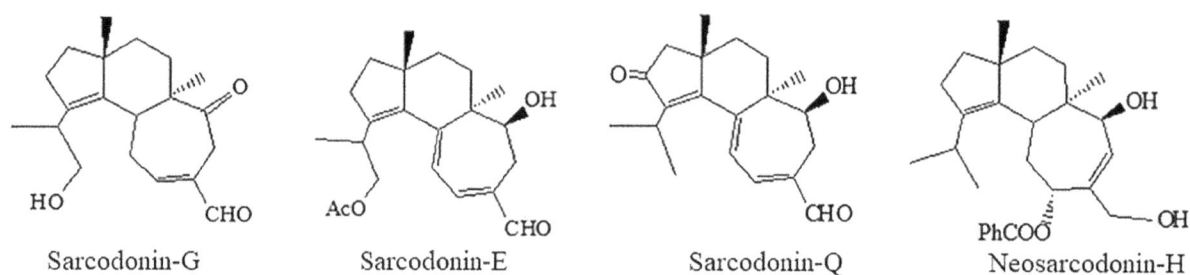

Sarcodonin-G   Sarcodonin-E   Sarcodonin-Q   Neosarcodonin-H

FIGURE 3.38 Chemical structures of cyathane diterpenes.

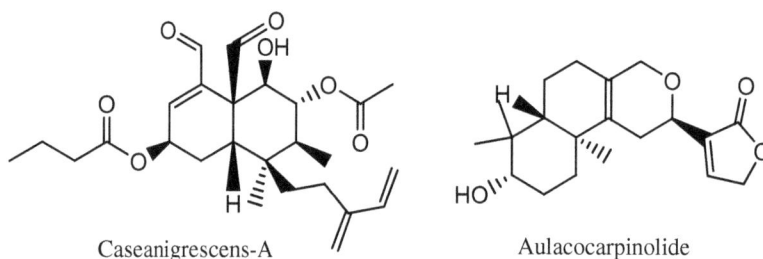

Caseanigrescens-A   Aulacocarpinolide

FIGURE 3.39 Structures of caseanigrescens-A and aulacocarpinolide.

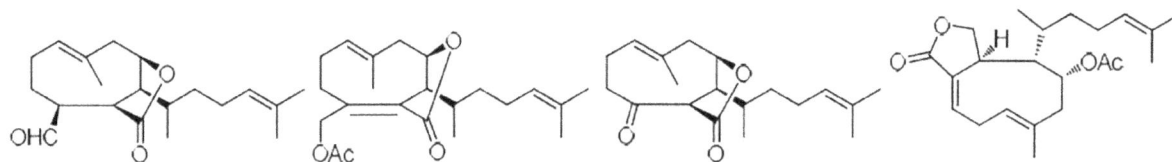

FIGURE 3.40 Structures of diterpenes (dictyotalide-A, dictyotalide-B, nordictyotalide, and 4-acetoxydictyolactone).

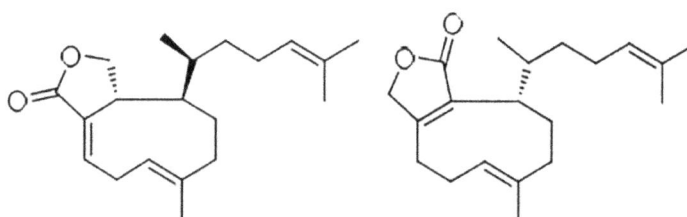

FIGURE 3.41 Structures of dictyolactone and neodictyolactone.

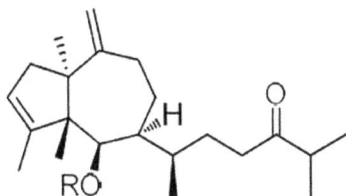

FIGURE 3.42 Structures of dictyone (R=H) and dictyone acetate (R=Ac).

Fourteen new cassane diterpenoids, caesalminaxins A-L (48–61) were isolated from the seeds of *Caesalpinia minax*. Among the new diterpenoids, compounds **50** and **51** possess a rare spiro C/D ring system whereas compound **52** contains a spiro A/B ring system. The isolated compounds and Bonducellpin-D were tested for cytotoxicity against the human cancer cell lines HepG-2, K562, HeLa, and DU145 by using the MTT method. These compounds exhibited moderate activity against four tested human cancer cell lines. Camptothecin is used as a positive control [59] (Figure 3.43, Table 3.7).

Three new ent-abietane diterpenoids, called fischerianoids A-C (62–64), were isolated and identified from the ethyl acetate extracts of roots of the medicinally used plant *Euphorbia fischeriana*. These compounds demonstrated selective inhibitory potency against certain human tumor cell lines including HL-60 (acute leukemia), MM-231 (breast cancer), SMMC-7721 (hepatic cancer), A-549 (lung cancer), HEP3B (hepatic cancer), SW-480 (colon cancer), and one normal colonic epithelial cell NCM460 with IC_{50} values ranging from 8.50 ± 0.13 to $35.52\pm0.08\,\mu M$. Cisplatin was used as the positive control [60] (Figure 3.44, Table 3.8).

Sharma et al. reported the anticancer activity of andrographolide (65) and its derivatives like neoandrographolide (66), 14-deoxyandrographolide (67), 14-deoxy-11,12-didehydroandrographolide (68), 14-deoxy-14,15-didehydroandrographolide (69), andrograpanin (70), isoandrographolide (71), 14-acetylandrographolide (72) and 19-o-acetylanhydroandrographolide (73). Andrographolide is a labdane type of diterpenoid lactone. Chemically, andrographolide is designated as (3-[2-[decahydro-6-hydroxy-5-(hydroxymethyl)-5,8a-dimethyl-2-methylene-1-napthalenyl] ethylidene]dihydro-4-hydroxy-2(3H)-furanone) [61] (Figure 3.45, Table 3.9).

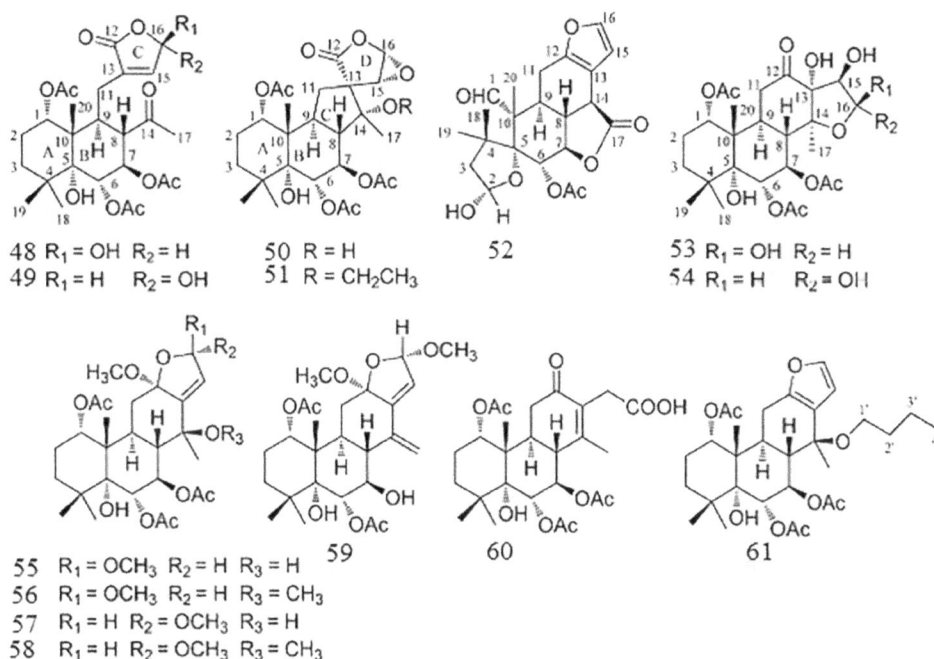

48 R_1 = OH R_2 = H
49 R_1 = H R_2 = OH

50 R = H
51 R = CH₂CH₃

52

53 R_1 = OH R_2 = H
54 R_1 = H R_2 = OH

55 R_1 = OCH₃ R_2 = H R_3 = H
56 R_1 = OCH₃ R_2 = H R_3 = CH₃
57 R_1 = H R_2 = OCH₃ R_3 = H
58 R_1 = H R_2 = OCH₃ R_3 = CH₃

59

60

61

FIGURE 3.43 Structures of cassane diterpenoids.

TABLE 3.7

Cytotoxic Activities of the Isolated Compounds against Four Cancer Lines[a,b]

Compounds	HepG-2	K562	HeLa	DU145
51	>50	9.9+1.7	12.1+0.8	17.1+1.4
55	11.5+2.9	9.2+0.9	>50	>50
Bonducellpin-D	8.0+1.3	>50	>50	12.0+0.8
Camptothecin	0.6+0.1	0.4+0.1	0.5+0.1	0.8+0.1

[a] Result is expressed as IC_{50} values in μM.
[b] Results are represented as mean±SD based on three independent experiments.

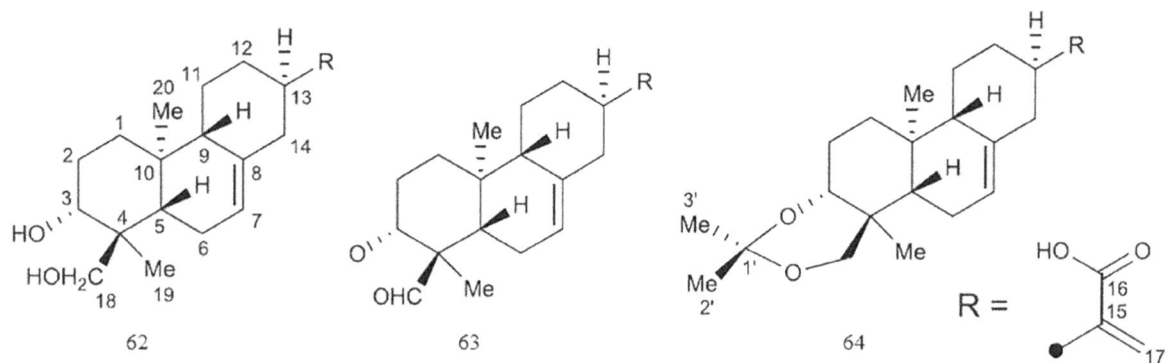

FIGURE 3.44 Structures of ent-abietane diterpenoids.

TABLE 3.8

Cytotoxic Activity of Ent-abietane Diterpenoids against Several Human Tumor Cell Lines

Entry	IC_{50} (µM)						
	HL-60	**MM-231**	**SMMC-7721**	**A-549**	**HEP3B**	**SW-480**	**NCM460**
62	>40	12.10±0.21	32.48±0.13	>40	15.95± 0.15	>40	>40
63	28.78±0.17	9.12±0.21	>40	>40	8.50±0.13	35.52±0.08	>40
64	>40	25.45±0.12	>40	>40	27.34±0.05	>40	>40
Cisplatin	1.60±0.10	3.82±0.23	2.78±0.15	2.81±0.35	2.97±0.21	1.45±0.12	0.87±0.10

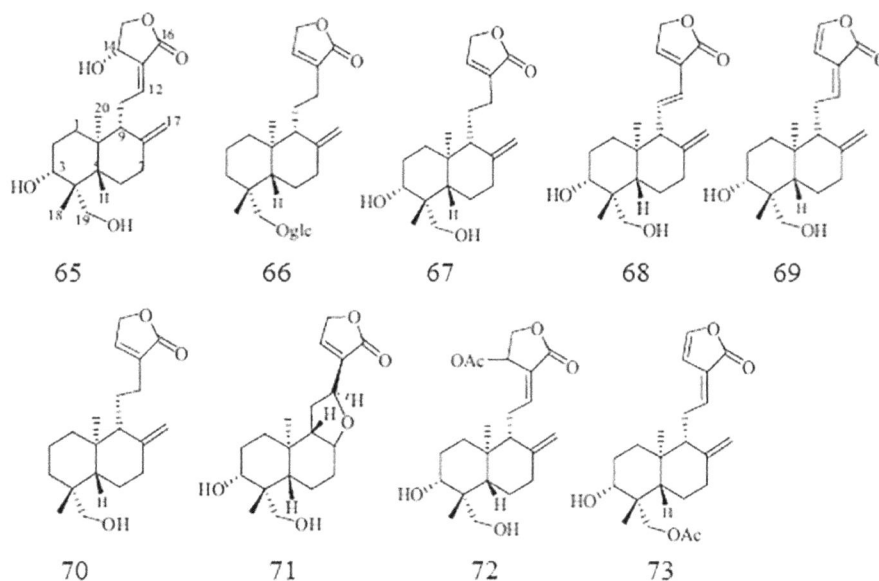

FIGURE 3.45 Structure of bioactive chemical constituents of *Andrographis paniculata*.

TABLE 3.9

Anticancer Potential of Andrographolide and Its Derivatives against Different Cell Lines

Compounds	Activity against Cancer Cell Lines	GI_{50} (µM)
Andrographolide	Breast, CNS, colon, lung, ovarian, prostate, and renal	3-30
Andrographolic acid	Ovarian (SK OV3)	4.0

Heliosterpenoids-A and B are two novel jatrophane-derived diterpenoids with a 5/6/4/6-fused tetracyclic ring skeleton system. These diterpenoids were isolated from the whole plants of *Euphorbia helioscopia*. These were found to be potent inhibitors of P-glycoprotein (ABCB1) and also exhibited cytotoxicity against MDA-MB-231 cell lines. The P-glycoprotein (P-gp) inhibitory effects of compounds **74** and **75** were tested by using an Adriamycin (ADM)-resistant human breast adenocarcinoma cell line (MCF-7/ADR). Cyclosporin A (CsA) was used as a positive control. Both compounds exhibited inhibitory activity with IC_{50} values of 1.28 and 1.02 μM respectively as compared to CsA (IC_{50} = 0.49 μM). Moreover, the cytotoxicity of Heliosterpenoids-A and B were also evaluated against five human cancer lines (MDA-MB-231, A549, Hela, U118MFG, and RKO) and with Adriamycin as the positive control (IC_{50} = 0.31 μM) by MTT assay. Heliosterpenoids-A exhibited modest cytotoxicity against MDA-MB-231 cell lines with an IC_{50} value of 24.7 μM [62–64] (Figure 3.46).

Patrícia et al. reported the cytotoxic activity of Royleanone diterpenes from *Plectranthus madagascariensis* Benth. Several extracts were prepared from *Plectranthus madagascariensis* by using various solvents like acetone, methanol, and supercritical CO_2. For this extraction, different techniques such as maceration, ultrasound-assisted, and supercritical fluid extraction were used. High-performance liquid chromatography with a diode array detector was used to determine the chemical composition of the extracts. From the extracts, the major compounds identified such as rosmarinic acid (**76**) and abietane diterpenes 7α, 6β-dihydroxyroyleanone (**77**), 7α-formyloxy-6β-hydroxyroyleanone (**78**), 7α-acetoxy-6β-hydroxyroyleanone (**79**), and coleon U (**80**). The cytotoxic activity of these compounds was evaluated by using human cancer cell lines, including breast (MDA-MB-231, MCF-7), colon (HCT116), and lung (NCI-H460, NCI-H460/R) cancer, and also in healthy lung (MCR-5) cells. Doxorubicin (DOX) was used as positive control in MDA-MB-231, MCF-7, and HCT116 cells whereas Paclitaxel (PCX) was used as a positive control in NCI-H460, NCI-H460/R, and MCR-5 cells [65] (Figure 3.47, Table 3.10).

Yoshida et al. reported the antitumor activity of daphnane-type diterpene Gnidimacrin isolated from *Stellera chamaejasme* L. It was found to inhibit cell growth of human leukemias, stomach cancers, and non-small cell lung cancers *in-vitro* at concentrations of 10^{-9} to 10^{-10} M. It was determined by an MTT-based chemosensitivity assay [66] (Figure 3.48, Table 3.11).

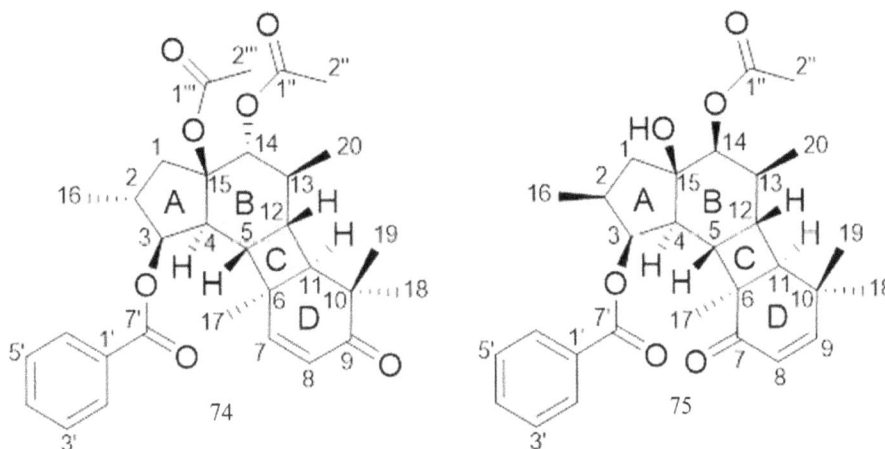

FIGURE 3.46 Structure of compounds 74 and 75.

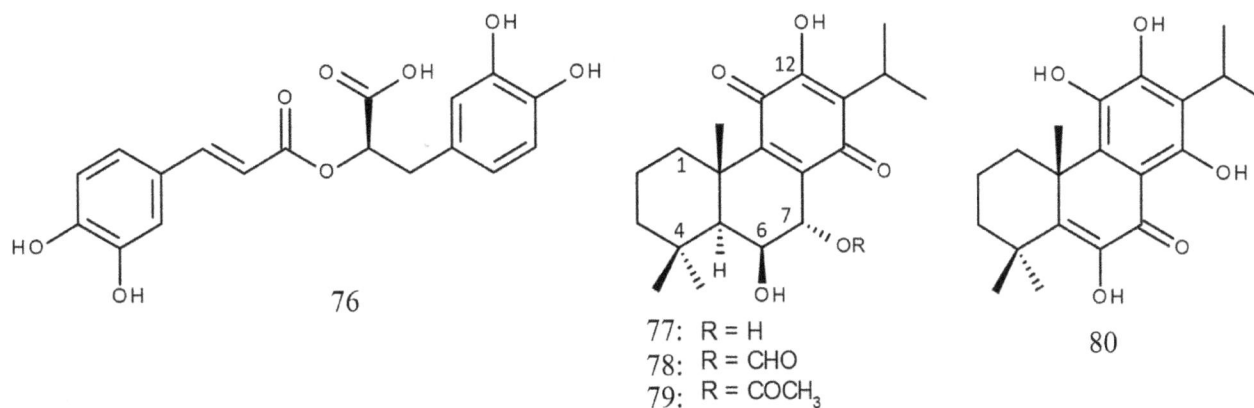

FIGURE 3.47 Chemical structure of the major components in *P. madagascariensis* extracts.

TABLE 3.10

Growth Inhibition (GI$_{50}$/µM) of Human Cell Lines by *P. madagascariensis* Extracts

Tested Compounds	Sulforhodamine-B Assay Cell Lines			MTT Assay Cell Lines		
	MDA-MB-231	MCF-7	HCT116	NCI-H460	NCI-H460/R	MCR-5
76	>100	nt	nt	>100	>100	>100
77	>100	26.0±0.6	≥50	[b]25 ±2	25 ±2	91±13
78	>100	[a]7.9±0.8	7.9±1.2	[b]14.9±2.9	nt	nt
79	>100	[a]6.4±0.4	-	[b]2.7±0.4	3.1±0.4	[b]8.6±0.4
80	46.9	[a]5.5±0.8	-	[b]3.0±0.2	nt	nt
Positive control	0.072±0.0021 (DOX)	0.16±0.0018 (DOX)	0.125± 0.0013 (DOX)	0.0006±0.0001 (PCX)	0.117±0.013 (PCX)	0.523±0.001 (PCX)

Note: nt, not tested.

Significant selectivity toward cancer cells: [a] $p<0.05$, [b] $p<0.001$ (in NCI-H460, NCIH460/R, and MCR-5 cells).

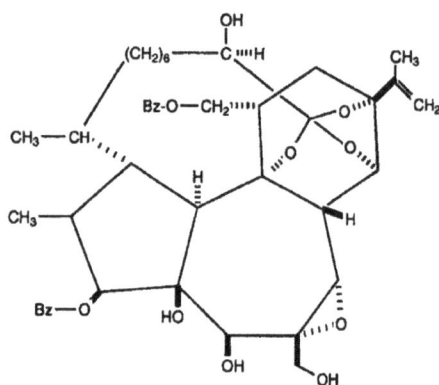

FIGURE 3.48 Structure of gnidimacrin isolated from *S. chamaejasme* L. (Bz=benzoyl).

TABLE 3.11

Antitumor Activity of Gnidimpcrin against Human Tumor Cell Lines

Cell Lines	Characteristics	IC$_{50}$ (µg/mL)	M
K562	CML	0.00019±0.00003	2.5×10^{-10}
HL60	PML	0.00022±0.00005	2.8×10^{-10}
MKN-28	Stomach cancer	0.00280±0.00350	3.6×10^{-10}
MKN-45	Stomach cancer	0.00027±0.00004	3.5×10^{-10}
MKN-74	Stomach cancer	>1	$>1.3\times10^{-10}$
PC-7	NSCLC	0.00046±0.00029	5.9×10^{-10}
PC-9	NSCLC	>1	$>1.3\times10^{-10}$
PC-14	NSCLC	0.00067±0.00023	8.6×10^{-10}
N231	SCLC	>1	$>1.3\times10^{-10}$
H69	SCLC	>1	$>1.3\times10^{-10}$
HLE	Hepatoma	>1	$>1.3\times10^{-10}$

Cells were cultured in the presence or absence of gnidimacrin at various concentrations for 4 days. Proliferation of cells was measured by MTT assay. IC$_{50}$: The concentration of Gnidimacrin required to inhibit cell growth to 50% of the control.

Herical exhibits cytotoxic and hemolytic properties. The pronounced effects on the different cells could be observed after 24 h of incubation. Balb/3T3 cells showed the highest

TABLE 3.12

Cytotoxic Activity of Herical

Cell Lines	Concentration Inducing Lysis of 90% of the Cells	
	[µg/mL]	[nmol/mL]
Ehrlich ascites tumor cells (mouse)	20–50	40–100
BALB/3T3 (mouse embryonic)	0.5–1	1–2
L-1210 (lymphocytic leukemia, mouse)	5–10	10–20
HeLa-S3 (epitheloid carcinoma, cervix, human)	2–5	4–10
KB cells (epidermoid carcinoma, oral, human)	1–2	2–4

sensitivity against Herical, whereas cells of the ascitic form of Ehrlich carcinoma were not or only weakly affected up to concentrations of 20 µg/mL. The cytotoxic activity of herical is shown in Table 3.12. Herical exhibits hemolytic properties on porcine erythrocytes. At 50 µg/mL more than 50% of the porcine erythrocytes were hemolyzed [67].

Antimalarial Activity

Parasitic diseases affect about 30% of the world's population. Among parasitic diseases, malaria is considered one of the most devastating infectious diseases that claim more lives than any other parasitic infection. Malaria is endemic to more than 100 nations and remains one of the main leading causes of death in children less than 5 years of age worldwide. Natural products play a major role in the treatment of malaria. Artemisinin is one of the most clinically used antimalarial drugs [68].

Similarly, the abietane diterpenoids are widely produced by conifers belonging to the families Araucariaceae, Cupressaceae, Pinaceae, and Podocarpaceae. The abietane-type diterpenoid (+)-ferruginol is a bioactive compound isolated from the tree *Podocarpus ferruginea*. It was demonstrated that ferruginol and some phthalimide-containing analogs have potential antimalarial activity (Table 3.13). These compounds were evaluated against malaria strains 3D7 and K1, and the cytotoxicity was measured against a mammalian cell line. Compound 3 exerted potent activity with an EC$_{50}$=86 nM (3D7 strain), 201 nM (K1 strain) and low cytotoxicity in mammalian cells (SI>290) [69] (Figure 3.49).

TABLE 3.13

Antiplasmodial Activity and Cytotoxicity (EC50-lM) of Ferruginol and Analogs on *P. falciparum* 3D7 Strain, K1 Strain, and HepG2, RAJI, BJ, and HEK293 Cell Lines

| Compounds | *P. falciparum*[a] | | | | Cell Lines[b] | | | |
| | 3D7 | | K1 | | HepG2 | RAJI | BJ | HEK293 |
	EC_{50}[c]	SI[d]	EC_{50}[c]	SI[d]	(EC_{50}[c])	(EC_{50}[c])	(EC_{50}[c])	(EC_{50}[c])
DHAA	2.47	4.6	1.33	8.6	>25	11.49	19.57	>25
81	1.36	>18.4	15–50	n.c.	>25	>25	>25	>25
82	0.086	>290	0.201	>124	>25	>25	>25	>25
83	2.09	>12.0	15–50	n.c.	>25	>25	>25	>25
84	15–50	n.c.	4.0	>6.3	>25	>25	>25	>25
85	6.67	2.6	15–50	n.c.	>25	>25	17.43	>25
86	15–50	n.c.	15–50	n.c.	>25	>25	>25	>25
87	15–50	n.c.	15–50	n.c.	>25	>25	>25	>25
88	15–50	n.c.	15–50	n.c.	>25	>25	>25	>25
89	15–50	n.c.	15–50	n.c.	>25	>25	>25	>25
90	15–50	n.c.	15–50	n.c.	>25	>25	>25	>25
91	15–50	n.c.	15–50	n.c.	>25	>25	>25	>25
Chloroquine	0.023	n.c.	0.69	n.c.	>25	>25	>25	>25
Gambogic acid	-	-	-	-	<2.0	0.24	0.21	0.21
Staurosporine	-	-	-	-	<2.0	<2.0	<2.0	<2.0

[a] 3D7 (chloroquine-sensitive) and K1 (chloroquine-resistant).

[b] BJ, normal human foreskin fibroblast; HEPG2, human liver carcinoma cell line; HEK293, human embryonic kidney cell line; RAJI, Burkitt lymphoma cell line.

[c] EC_{50} (μM): concentration corresponding to 50% growth inhibition of the parasite or cells.

[d] Selectivity index (SI): EC_{50} lower values of cytotoxic activity/EC_{50} values of antimalarial activity.

NC, not calculated.

The aqueous ethanolic extracts of powdered twigs of *Euphorbia esula* afforded 16 new diterpenoids, named euphorbesulins A-P. These euphorbesulins include presegetane (93–95), jatrophane (96–109), paraliane (110), and isopimarane (111) diterpenoids. Compounds 93–95 represent a rare type of presegetane diterpenoid. Diterpenoid **99** displayed low nanomolar antimalarial activity, whereas the remaining compounds exhibited only moderate or no antimalarial activity (Table 3.14). Artemisinin is used as positive control [70] (Figure 3.50).

Andrographolide (AND) is the diterpene lactone compound, isolated from the methanolic fraction of the plant *Andrographis paniculata*. This compound was found to have potent antiplasmodial activity when tested in isolation and combination with curcumin and artesunate against the erythrocytic stages of *Plasmodium falciparum in-vitro* and *Plasmodium berghei* ANKA *in-vivo*. IC_{50}s for artesunate (AS), andrographolide (AND), and curcumin (CUR) were found to be 0.05, 9.1, and 17.4 μM, respectively. AND was found to have synergistic action with curcumin (CUR) and addictively interactive with artesunate (AS). Andrographolide-curcumin exhibited better antimalarial activity *in-vivo*, not only by reducing parasitemia (29%) as compared to the control group (81%), but also by extending the life span by 2–3 folds [71].

Endale et al. reported the *in-vivo* antimalarial activity of a labdane diterpenoid from the leaves of *Otostegia integrifolia* Benth. The methanolic (80%) leaf extracts *O. integrifolia* was tested for antimalarial activity *in vivo against Plasmodium berghei*. This extract displayed potent antiplasmodial activity that results in the isolation of a labdane diterpenoid and

was identified as otostegindiol. Otostegindiol exhibited significant ($p < 0.001$) antimalarial activity at doses of 25, 50 and 100 mg/kg with chemosuppression values of 50.13%, 65.58% and 73.16%, respectively (Table 3.15). The acute toxicity study revealed that the crude extract possesses no toxicity in mice up to a maximum dose of 5,000 mg/kg [72] (Figure 3.51).

Yamada et al. reported the antimalarial activity of Kalihinol-A and new relative diterpenoids. New isocyano diterpenoid such as Δ^9-kalihinol-Y and two new isothiocyano diterpenoids including 10-epikalihinol-I and 5,10-bisisothiocy anatokalihinol-G were isolated along with kalihinol-A, kalihinene and 6-hydroxykalihinene (Table 3.16). These diterpenoids were isolated from the Okinawan Sponge, *Acanthella* sp. Among these tested compounds, Kalihinol-A was found to have particularly potent ($EC_{50} = 1.2 \times 10^{-9}$ M) and selective (selective index, SI=317) *in-vitro* antimalarial activity [73] (Figure 3.52).

Antifilarial Activity

Lymphatic filariasis is a debilitating disease with an adverse social and economic impact. It is caused by *Wuchereria bancrofti*, *Brugia malayi* and *B. timori*. The infection remains unabated in spite of treatment with existing antifilarial drugs diethylcarbamazine (DEC) and ivermectin which are chiefly microfilaricides [74].

Singh et al. reported the antifilarial activity of diterpenoids, isolated from *Taxodium distichum*.

Four molecules such as 3-acetoxylabda-8(20), 13-diene-15-oic acid (K001), Beta-sitosterol (K002), labda-8(20),

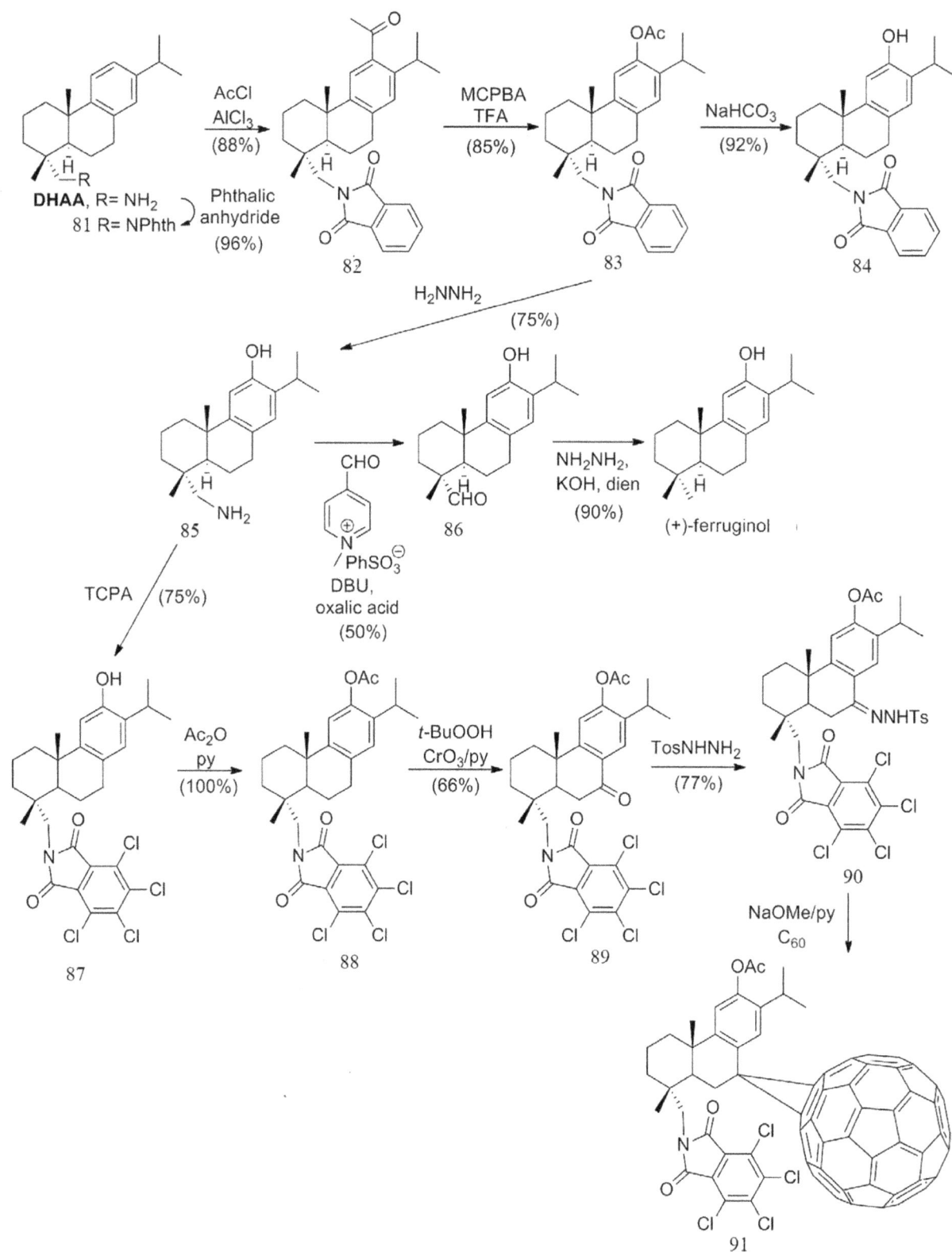

FIGURE 3.49 Scheme for synthesis of ferruginol analogs.

TABLE 3.14

Antimalarial Activity of Euphorbesulins

Tested Compounds	IC$_{50}$ (μM)	Tested Compunds	IC$_{50}$ (μM)
93	2.41±0.44	100	>5
94	>5	101	>5
95	>5	102	>10
96	>5	103	>5
97	>5	104	>5
98	0.12±0.04	105	N.A
99	>5	Artemisinin	7±0.1 nM (>10 μM)

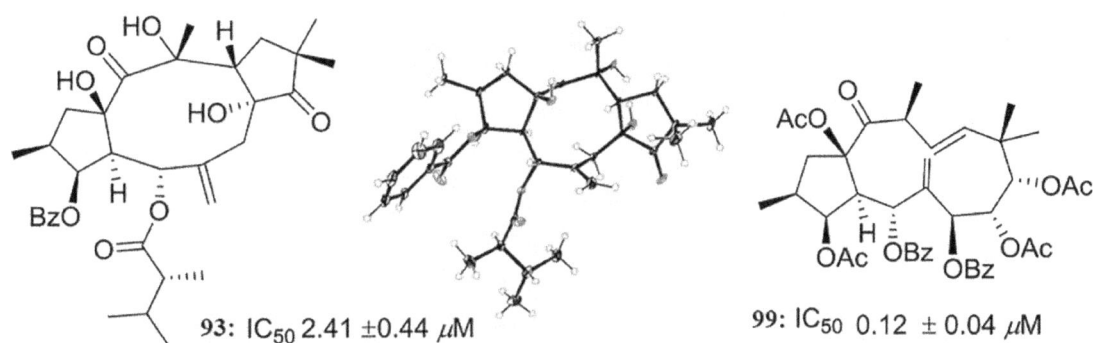

FIGURE 3.50 Structure of euphorbesulins.

TABLE 3.15

Percentage Suppression of the 80% Methanol Leaf Extract of *Otostegia integrifolia* against *Plasmodium berghei* in Mice

Test Sample	Dose (mg/kg/day)	Percent Parasitaemia±SEM	Percent Suppression
Extract	200	14.04±2.36[a]	42.37
Extract	400	6.00±0.53[a]	75.28
Extract	600	4.34±0.77[a]	80.52
Chloroquine	25	0.00[a]	100
Vehicle	0.2 mL	24.36±1.51	-

[a] The mean value is significant ($p<0.001$) when compared with vehicle-treated group; data are expressed as means±SEM for five mice per a group.

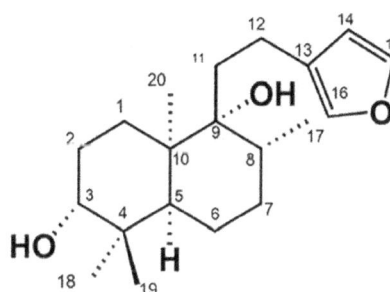

FIGURE 3.51 Structure of otostegindiol.

TABLE 3.16

Cytotoxicity of Compounds (2–6) to FM3A Cells and *Plasmodium falciparum* (Antirnalarial Activity)

Compounds	*Plasmodium falciparum* EC$_{50}$ (M)	FM3A Cells EC$_{50}$ (M)	Selective Index (SI)[b]
10-Epikalihinol-I	>1.8×10⁻⁶ (62%)[a]	-	-
5,10-bis-isothiocyanato-Kalihinol-G	2.6×10⁻⁶	7.0×10⁻⁷	-
Kalihinol-A	1.2×10⁻⁹	3.8×10⁻⁷	317
Kalihinene	1.0×10⁻⁸	3.7×10⁻⁸	4
6-hydroxykalihinene	8.0×10⁻⁸	1.2×10⁻⁶	15
Mefloquine[c]	3.2×10⁻⁸	2.9×10⁻⁹⁶	90

[a] Growth percentage at concentration indicated.

[b] Selectivity index (SI) defined as ratio of FM3A cells cytotoxicity to *P. falciparum*.

[c] Antimalarial standard.

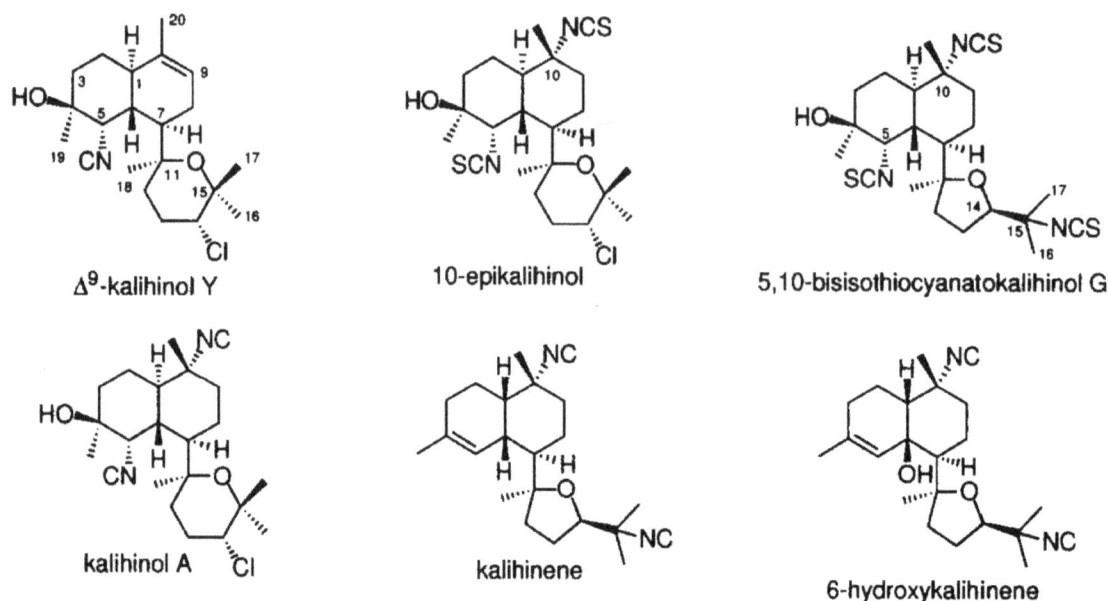

Δ⁹-kalihinol Y   10-epikalihinol   5,10-bisisothiocyanatokalihinol G

kalihinol A   kalihinene   6-hydroxykalihinene

FIGURE 3.52 Structure of kalihinol and analogs.

13-diene-15-oic acid (K003) and Metasequoic acid-A (K004) were isolated from the ethanolic extract of aerial parts of *T. distichum.* These compounds were evaluated for antifilarial activity against *B. malayi* by 3-(4,5-dimethylthiazol-2-yl)-2,5-diphenyltetrazolium bromide (MTT) reduction and motility assays *in-vitro* and in two animal models such as *Meriones unguiculatus* and *Mastomys coucha,* harboring *B. malayi* infection. A001 (crude extract) was effective in killing microfilariae and adult worms *in vitro.* The diterpenoid K003 produced 100% reduction in motility of both microfilariae and adult worms (Table 3.17). It also exhibited 80% inhibition in MTT reduction potential of adult female worms [75] (Figure 3.53).

3.3.3 Treatment of Inflammation

Inflammation is the part of the body's defense mechanism that plays a key role in the healing process. Inflammation may be three types such as acute (inflammation lasts for few days), subacute (inflammation lasts for 2–6 weeks) and chronic inflammation (inflammation lasts for months or years). The inflammatory mediators are the messenger that acts on blood vessels or cells to promote inflammatory responses. Various inflammatory mediators include prostaglandins (PGs), inflammatory cytokines such as IL-1β, TNF-α, IL-6 and IL-15 and chemokines such as IL-8 and GRO-alpha. The treatment of inflammation depends on the cause and severity [76].

TABLE 3.17

In-Vitro Activity of *Taxodium distichum* and the Reference Drugs Ivermectin and Diethylcarbamazine (DEC) on Adult Worms (AW) and Microfilariae (Mf) of *Brugia malayi* Assessed by Motility Assay (MA) and MTT Reduction Assay

Antifilarial Agent	LC_{100} (µg/mL)[a] for AW in MA	IC_{50} (µg/mL)[b] for AW in MA	Mean % Inhibition in MTT Reduction by AW	LC_{100} (µg/mL) for Mf in MA	IC_{50} (µg/mL) for Mf in MA	CC_{50} (µg/mL)[c]	SI for AW in MA	SI for Mf in MA
A001	15.63	10.00	47.00	3.91	1.95	250	25	128.29
K001	62.50	41.23	76.89	250	88.38	697	16.91	7.88
K002	>250	>250	38.05	>250	>250	-	-	-
K003	125	74.33	82.05	31.25	18.58	180	2.42	9.69
K004	125	52.56	85.02	31.25	11.05	190	3.61	17.19
Ivermectin (µM)	5	3.05	5.80	2.5	1.57	250	81.96	159.23
DEC-C (µM)	1000	314.98	62.54	500	297.30	9000	28.57	30.27

Motility score of parasite in Control wells (parasite+DMSO only)=4 (i.e. parasites were motile and highly active); MTT absorbance values (OD_{510nm}) of control wells (parasites+DMSO only)=0.64.

Abbreviations: SI Selectivity Index (CC_{50}/IC_{50}), DEC-C Diethylcarbamazine-citrate.

[a] LC_{100}=100 % reduction in motility indicates death of parasite.

[b] IC_{50}=concentration of the agent at which 50% inhibition in motility of the parasites is achieved.

[c] CC_{50}=concentration at which 50% of cells are killed.

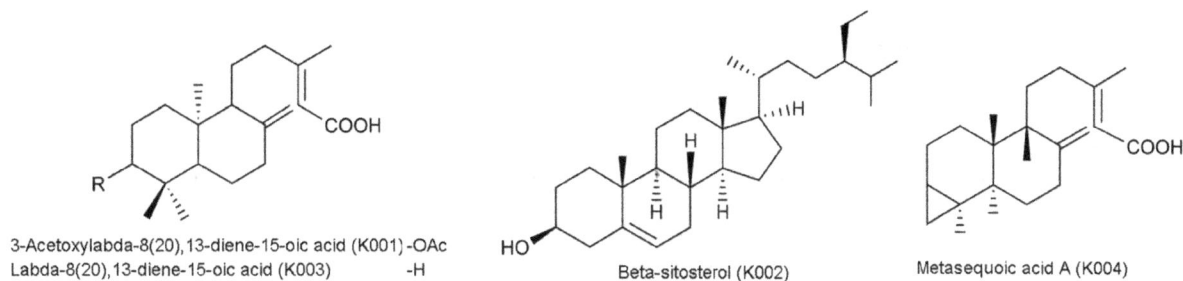

3-Acetoxylabda-8(20),13-diene-15-oic acid (K001) -OAc
Labda-8(20),13-diene-15-oic acid (K003)   -H

Beta-sitosterol (K002)

Metasequoic acid A (K004)

FIGURE 3.53 Structure of the compounds isolated from *Taxodium distichum*.

Chua et al. reported that the extracts of *Andrographis paniculata* contain labdane diterpenes that reduced liver inflammation. The biologically active diterpene constituents present in the leaves of *Andrographis paniculata* include andrographolide, dehydroandrographolide, and neoandrographolide. These compounds interfere with COX activity and inflammatory cytokine release. Andrographolide suppresses the receptor activator of NF-κB ligand-induced osteoclastogenesis via attenuation of NF-κB and extracellular signal-regulated kinase/MAPK signaling pathways *in-vitro*. It also inhibits the release of TNF-α and the production of both IL-6 and IL-17 [77]. Parichatikanond et al. demonstrated the anti-inflammatory effects of andrographolide may be due to down-expression of genes in the inflammatory cascade [78] (Figure 3.54).

Yang et al. reported that the diterpene triptolide from *Tripterygium wilfordii* inhibits the production and gene expression of a range of cytokines and chemokines, including IL-1β, IL-6, TNF-α, and IFN-γ, in vitro, while it also decreases PGE2 production via COX-2 gene suppression. Further, it was demonstrated that triptolide suppressed production and mRNA levels of promatrix metalloproteinase (MMP)-1 and -3 on human synovial fibroblasts induced by IL-1α, and cytokine-induced MMP-3, MMP-13 and aggrecanase-1 gene expression in chondrocytes and synovial fibroblasts. In animal models of arthritis, triptolide decreased the arthritic score and incidence along with delaying injury onset. It also increased the production of the anti-inflammatory cytokine transforming growth factor (TGF)-β [79] (Figure 3.55).

Abietic acid is an abietane diterpenoid with molecular formula $C_{20}H_{30}O_2$ [80]. It is abieta-7,13-diene substituted by a carboxyl (–COOH) group at position 18 [81]. Both abietic and dehydroabietic acids occur in the oleoresin of the *Abies*

grandis and *Pinus contorta*. Abietic acid possesses anti-inflammatory property in macrophages via activation of PPAR-γ. Macrophages play major role in modulating the initiation and perpetuation of the inflammatory response. The activation of macrophages promotes the synthesis and release of eicosanoids such as prostaglandins (PGs), leukotrienes (LTs), nitric oxide (NO), and mediators involved in the onset of inflammation [82]. Activated macrophages also secrete proinflammatory cytokines such as tumor necrosis factor α (TNF-α) and interleukin-1β (IL-1β). Proinflammatory cytokines induce the expression of enzymes such as COX-2 and iNOS in macrophages and other cells that results release of inflammatory mediators [83] (Tables 3.18 and 3.19). Abietic acid was found to reduce the release of TNF-α and IL-1β at highest dose (100 μM). This effect is also similar to that of COX and 5-LOX inhibitors [84,85] (Figure 3.56).

Heras et al. reported a novel diterpenoid labdane which inhibits the eicosanoid generation from stimulated macrophages but enhances the release of arachidonic acid. The diterpenoid *ent*-8α-hydroxy-labda-13(16), 14-diene (labdane F2) was obtained from *Sideritis javalambrensis*. Labdane-F2 inhibited the generation of prostaglandin-E_2 in cultured mouse peritoneal macrophages, treated with zymosan, ionophore A23187, or arachidonic acid and in J774 macrophage-like cells activated by bacterial lipopolysaccharide (LPS). Labdane also prevents the induction of both iNOS and COX-2 by LPS in J774 cells without cell toxicity [86] (Figure 3.57).

Various types of marine diterpenoids are reported as potential anti-inflammatory agents. Eunicellane-based diterpenes include krempfielins, hirsutalins, klymollins, klysimplexin, klysimplexin sulfoxide, simplexin, and cladieunicellin [87]. These compounds are isolated and identified from soft corals belonging to the genera *Cladiella* or *Klyxum*. Some of these

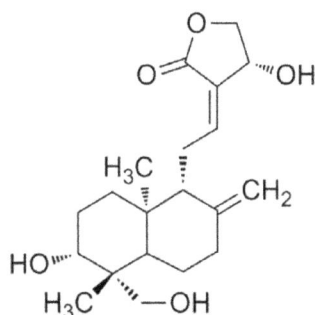

FIGURE 3.54 Structure of andrographolide.

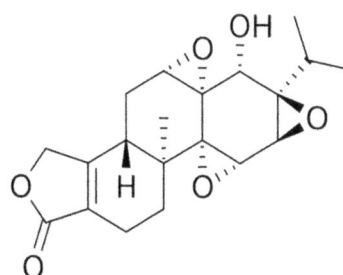

FIGURE 3.55 Structure of triptolide.

TABLE 3.18

Effect of Abietic Acid on Nitrite, PGE$_2$, TNF-α and IL-1β Production by Lipopolysaccharide (LPS)-Stimulated Peritoneal Macrophages

Treatment	Nitrite (μm)	PGE$_2$ (ng/mL)	TNF-α (ng/mL)	IL-1β (pg/mL)
Cells alone	4.6±0.5	0.07±0.1	3.4±0.3	22.0±0.001
LPS	43.6±2.6	3.74±0.3	16.2±1.4	200.0±2.3
Abietic acid (1 μM)	41.4±2.5	2.45±0.9	19.3±1.8	180.1±2.1
Abietic acid (10 μM)	37.2±1.9	1.65±0.7[a]	15.9±1.7	170.0±2.2
Abietic acid (100 μM)	20.1±0.7[b]	1.25±0.3[b]	8.9±1.1[b]	110.0±3.3[a]
Dexamethasone (1 μM)	19.0±0.5[b]	0.8±0.1[b]	2.0±0.4[c]	30.0±3.1[c]

Results are expressed as mean±SEM, $n=8$, from two separate experiments. Abietic acid was added to cells 30 min before LPS and incubation continued for 24 h.

[a] $p < 0.05$.

[b] $p < 0.001$ vs LPS (analysis of variance followed Dunnet's test).

[c] $p < 0.001$

TABLE 3.19

Effects of Abietic Acid and Reference Drugs on PGE$_2$ and LTC$_4$ Release (ng/mL) from Calcium Ionophore A23187-Stimulated Peritoneal Macrophages

Treatment	PGE$_2$	LTC$_4$
Control	3.2±0.8	3.0±0.3
Cells alone	0.3±0.01	0.2±0.1
Abietic acid (1 μM)	2.5±0.3	3.1±0.2
Abietic acid (10 μM)	2.0±0.1[a]	3.2±0.1
Abietic acid (100 μM)	0.9±002[b]	2.9±0.3
Indometacin (10 μM)	0.4±0.01[c]	ND

Results are expressed as mean±SEM, $n=8$, from two separate experiments.

ND, not determined.

[a] $p < 0.05$.

[b] $p < 0.01$.

[c] $p < 0.001$ vs A23187 (analysis of variance followed Dunnet's test).

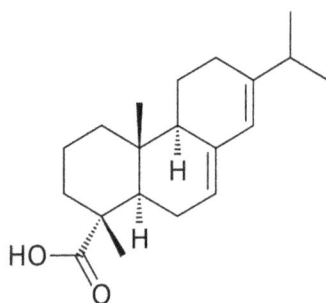

FIGURE 3.56 Structure of of abietic acid.

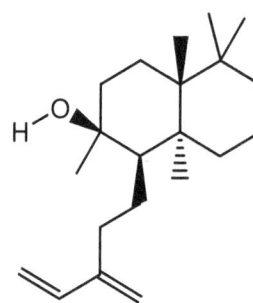

FIGURE 3.57 Structure of labdane with inflammatory activity.

compounds have been shown to inhibit the up-regulation of inducible nitric oxide synthases (iNOS), COX-2, or IL-6 proteins in RAW 246.7 macrophages stimulated with lipopolysaccharide (LPS) [88]. Excavatolide-B is a briarane diterpenoid, isolated from the coral *Briareum excavatum*, demonstrates *in-vitro* and *in-vivo* anti-inflammatory activity [89]. The presence of 8,17-epoxide and 12-hydroxyl groups in Excavatolide-B is responsible for the anti-inflammatory effect [90]. Cembrane diterpenoids are isolated from corals of the genera *Sinularia, Lobophytum, Eunicea, and Sarcophyton*. Different groups of cembrane diterpenoids include gibberosenes, grandilobatin, querciformolides, sarcocrassocolides, crassumolides, crassarines, sinularolides, durumolides, and columnariols [91].

These compounds exhibit anti-inflammatory activity by inhibiting the expression of iNOS and/or COX-2 by LPS-stimulated RAW264.7 cells [92–94] (Figure 3.58).

Twelve new diterpenoids based on two rare skeletal types namely, paralianones A-D (**112–115**) and pepluanols A-H (**116–118**), along with five known compounds were isolated from the acetone extract of *Euphorbia peplus*. These diterpenoids were evaluated for their anti-inflammatory activity by using lipopolysaccharide (LPS) stimulated mouse macrophage cellular model. Among all the tested compounds, Compounds **114, 115, 127, 119**, and **122** exhibited moderate inhibitory effects on NO inhibition, with IC$_{50}$ values ranging from 29.9 to 38.3 μM (Table 3.20). MG-132, an inhibitor of proteasome, was used as positive control [95] (Figure 3.59).

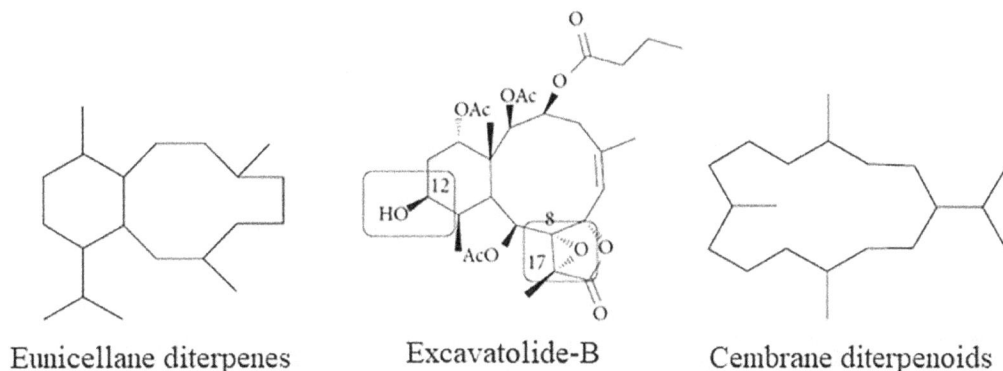

Eunicellane diterpenes   Excavatolide-B   Cembrane diterpenoids

FIGURE 3.58 Structure of marine diterpenoids.

TABLE 3.20

Inhibitory Effects on LPS-Stimulated No Production in RAW264.7 Cells

Tested Compounds	IC_{50} (µM)	Tested Compounds	IC_{50} (µM)
111	43.2	10	>50
112	>50	11	36.6
113	33.7	12	>50
114	38.3	13	29.9
115	>50	14	>50
116	>50	15	>50
117	>50	16	37.1
118	>50	17	47.5
9	>50	MG-132	0.18

3.3.4 Management of Alzheimer's Disease

Alzheimer's disease (AD) is one of the major well-known neurodegenerative disorder and characterize 50%–60% of dementia in patients. The prevalence rate of AD is positively correlated with age which affects more than 40% of people over 65 years old. It presents behavioral and cognitive deficits, caused by the accumulation of amyloid beta (Aβ) aggregates in the brain. It is also accompanied by an innate immune response mediated by microglia and astroglia, known as neuroinflammation. Neuroinflammation is induced by beta-amyloid (Aβ) which plays a major role in the pathogenesis of AD. So, the inhibition of Aβ-induced neuroinflammation serves as a potential strategy for the treatment of AD [96].

Oridonin is an organic heteropentacyclic compound and ent-kaurane diterpenoid with molecular formula $C_{20}H_{28}O_6$ [97]. It is isolated from the leaves of the medicinal herb *Rabdosia rubescens*. Wang et al. demonstrated that Oridonin inhibits the glial activation and decreases the release of inflammatory cytokines in the hippocampus of $Aβ_{1-42}$-induced AD mice [98]. In addition, Oridonin inhibits the NF-κB pathway and $Aβ_{1-42}$-induced apoptosis. Furthermore, Oridonin attenuates memory deficits in $Aβ_{1-42}$-induced AD mice [99]. From this study, it was concluded that Oridonin inhibits neuroinflammation and attenuate memory deficits induced by $Aβ_{1-42}$. It suggests that Oridonin might be a promising candidate for the treatment of AD [100] (Figure 3.60).

Ginkgolides are cyclic diterpenes with 20-carbon skeletons of labdane type [101]. These are commonly isolated from root bark and leaves of *Ginkgo biloba*. The class of ginkgolides was first isolated from the tree *Ginkgo biloba* in the year 1932 [102]. The structure elucidation of ginkgolides was accomplished by Maruyama et al in the year 1967 [103]. Ginkgolides are biosynthesized from geranylgeranyl pyrophosphate. These are categorized into Ginkgolide A, B, C, J and M. Ginkgolide-B is a diterpenoid trilactone with six five-membered rings [104–106].

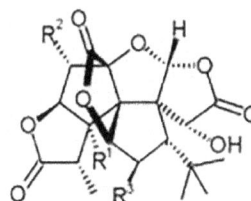

Types of Ginkgolides	R_1	R_2	R_3
Ginkgolide A	OH	H	H
Ginkgolide-B	OH	OH	H
Ginkgolide-C	OH	OH	OH
Ginkgolide-J	OH	H	OH
Ginkgolide-M	H	OH	OH

Ginkgolide-B protects hippocampal neurons from Aβ-induced apoptosis by increasing the production of brain-derived neurotrophic factor and reducing apoptotic death of neuronal cells in hemorrhagic rat brain. Vitolo et al. reported that ginkgolide-J is the most potent inhibitor of Aβ-induced hippocampal neuronal cell death among the ginkgolides in EGb761 [107].

Cryptotanshinone is a labdane-type diterpene and is derived from the roots of *Salvia miltiorrhiza*. It can easily cross the blood-brain barrier (BBB) and affects the cognitive function in mice. It was found to reduce the Aβ production by up-regulating α-secretase, which cleaves APPs in the middle of the Aβ sequence. In addition to this, cryptotanshinone protects the neuronal cell damage by inhibiting Aβ aggregation [108–110] (Figure 3.61).

Owona et al. reported the protective effects of Forskolin on behavioral deficits and neuropathological changes in a mouse model of cerebral amyloidosis. Forskolin is a diterpene of the labdane family and is mainly found in the root of a plant called *Coleus forskohlii*. It was reported that forskolin suppresses the production of tumor necrosis factor (TNF) in microglia by

	R	R'	R"
112	OH	OAc	H
113	OAc	OAc	OAc
114	H	OAc	H
115	H	=O	H
119	OAc	OAc	H

	R	R'	R"	R'''
116	OBz	OAc	OH	OH
117	OBz	OH	OAc	OH
118	OH	OAc	OAc	OH
120	OBz	OAc	OAc	OH
121	OBz	OAc	OH	OAc
122	OBz	OAc	OAc	OAc

123

124

125: R= OH
126: R= OAc

127: R= H
128: R= OAc

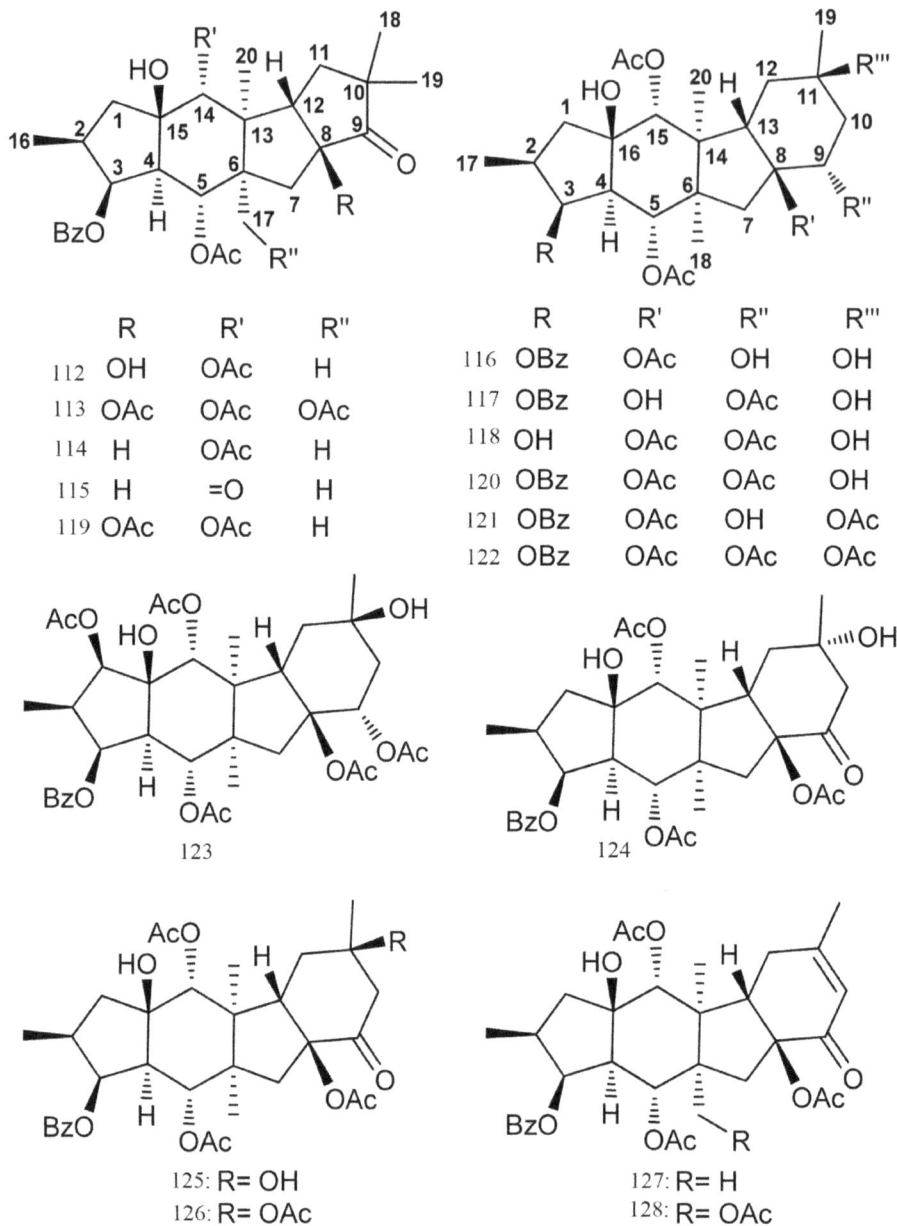

FIGURE 3.59 Structure of paralianones A–D (112–115) and pepluanols A–H (116–118).

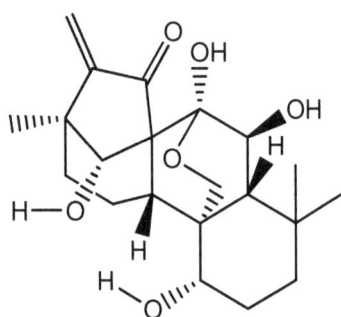

FIGURE 3.60 Structure of oridonin.

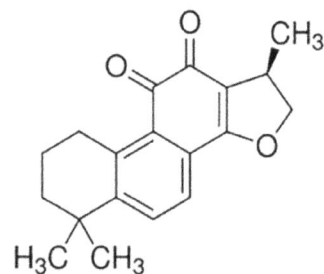

FIGURE 3.61 Structure of cryptotanshinone.

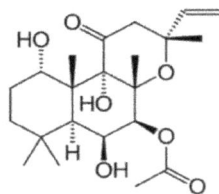

FIGURE 3.62 Molecular structure of forskolin.

reducing the nuclear translocation and DNA binding activity of NF-κB. Because forskolin is a potent inhibitor of NF-κB, it could be of therapeutic utility for the treatment of AD by directly inhibiting the production and deposition of Aβ peptides in the brain. Forskolin reduced the plaque numbers in cortex versus CMC control (CMC, 181.9 ± 11.44; forskolin, 128.00 ± 10.89, $p < 0.01$, $n = 6$/group) [111] (Figure 3.62).

3.3.5 Treatment of Diabetes and Hyperlipidemia

Diabetes mellitus is a group of metabolic disorders. It is characterized by hyperglycemia, and occurs due to the defects in the secretion and action of insulin. Venkatachalapathi et al. performed the evaluation of a labdane diterpene forskolin isolated from *Solena amplexicaulis* (Lam.) Gandhi (Cucurbitaceae) revealed the promising antidiabetic and antihyperlipidemic pharmacological properties. The *in-vivo* evaluation of antidiabetic and antihyperlipidemic properties of methanolic extract of *S. amplexicaulis* (MeOHSa) and Forskolin was carried out by using streptozotocin (STZ) induced diabetic rats. The hypoglycemic effect of Forskolin may be due to the enhanced glucose-mediated stimulus to release insulin. This effect is produced by the elevation of CAMP which activates the signaling pathways in β-cells viz., Protein Kinase (PKA) pathway and guanine nucleotide pathway regulated by CAMP [112] (Table 3.21).

Rebaudioside-A (Reb-A) is a major constituent of *Stevia rebaudiana*. It was recently proposed as an insulinotropic agent. It was investigated to evaluate the antihyperglycemic effect of Reb-A on the activities of hepatic enzymes of carbohydrate metabolism in streptozotocin (STZ)-induced diabetic rats. Diabetic rats showed significant ($p < 0.05$) increase in the levels of plasma glucose and glycosylated hemoglobin and significant ($p < 0.05$) decrease in the levels of plasma insulin and hemoglobin. The activities of gluconeogenic enzymes such as glucose-6-phosphatase and fructose-1,6-bisphosphatase were significantly ($p < 0.05$) increased while hexokinase and glucose-6-phosphate dehydrogenase were significantly ($p < 0.05$) decreased in the liver along with glycogen. The oral treatment with Reb-A to diabetic rats significantly ($p < 0.05$) decreased blood glucose and reversed these hepatic carbohydrate metabolizing enzymes in a significant manner (Table 3.22). Histopathology changes of pancreas confirmed the protective effects of Reb-A in diabetic rats So, these results reveal that Reb-A possesses an antihyperglycemic activity and also provide the evidence for its traditional usage to control diabetes [113] (Figure 3.63).

Gao et al. reported the antihyperglycemic activity of Clerodane diterpenoids. These are isolated from the vines of *Tinospora crispa* which include clerodane diterpenoids such as tinosporols A-C (**130–132**) and tinosporoside A (**133**), together with Compound **129** (Borapetoside-E). Borapetoside-E significantly reduced the serum glucose levels at dose-dependent manners in alloxan-induced hyperglycemic mice and db/db type-2 diabetic mellitus (T2DM). T2DM is a chronic metabolic disorder characterized by deregulation of glucose and lipid metabolism. The hyperglycemic activity was evaluated *in-vivo* in mice. Metformin served as a positive control. The alloxan injection markedly increases the serum glucose level from 7.0 ± 0.5 to 24.3 ± 3.7 mM/L. But, the intravenously administrated compound **129** (20 mg/kg) in alloxan-induced

TABLE 3.21

Effect of MeOHSa and Forskolin on Various Biochemical Markers in STZ Induced Diabetic Rats

Parameter	Control	Diabetic Control	Diabetic+MeOHSa (600 mg/kg b.w.)	Diabetic+Forskolin (10 mg/kgb.w.)	Diabetic+Glibenclamide (600 mg/kg b.w.)
			Treatment Group		
Hemoglobin (%)	14.5 ± 0.4	6.9 ± 0.2^a	$12.5 \pm 0.5^{b,d}$	13.3 ± 0.4^d	13.9 ± 0.3^d
SGOT (IU/L)	41.6 ± 2.8	112.3 ± 3.6^a	49.4 ± 1.5^d	44.2 ± 1.7^d	42.6 ± 4.7^d
SGPT (IU/L)	28.2 ± 6.4	102.6 ± 1.4^a	57.1 ± 5.2^{bd}	30.6 ± 2.5^d	32.9 ± 7.5^d
HDL (mg/dL)	52.8 ± 8.7	11.4 ± 2.0^a	43.5 ± 7.6^e	51.8 ± 5.0^d	49.1 ± 0.1^d
LDL (mg/dL)	82.8 ± 5.6	231.9 ± 2.5^a	95.2 ± 4.3^d	84.5 ± 1.3^d	88.2 ± 8.7^d
TC (mg/dL)	156.3 ± 4.6	280.4 ± 1.4^a	161.6 ± 5.2^d	157.8 ± 2.7^d	159.3 ± 2.5^d
TG (mg/dL)	103.3 ± 1.1	185.7 ± 1.6^a	114.7 ± 8.1^d	107.4 ± 8.1^d	109.8 ± 5.3^d
Glycogen (mg/g liver tissue)	19.0 ± 1.9	5.2 ± 1.4^a	14.4 ± 0.5^f	18.8 ± 3.7^d	18.6 ± 0.7^d

Note: Values are expressed as mean $\pm$ SEM ($n = 6$).

Abbreviations: SGOT-serum glutamate oxaloacetate transaminase; SGPT, serum glutamate pyruvate transaminase; HDL, high-density lipoproteins; LDL, low density lipoproteins; TC, total cholesterol; TG, triglycerides.

[a] $p < 0.001$, significantly different from the control group.
[b] $p < 0.01$, significantly different from the control group.
[c] $p < 0.05$, significantly different from the control group.
[d] $p < 0.001$, significantly different from the diabetic control group.
[e] $p < 0.01$, significantly different from the diabetic control group.
[f] $p < 0.05$, significantly different from the diabetic control group.

TABLE 3.22

Effect of Reb-A on the Activities of Hexokinase, Glucose-6-Phosphatase, and Fructose-1,6-Bisphosphatase in Normal and Experimental Rat Liver

Groups	Hexokinase (U[a])	Glucose-6-Phosphatase (U[b])	Fructose-1,6-Bisphosphatase (U[c])
Normal	0.52 ± 0.03^a	0.14 ± 0.01^a	0.47 ± 0.01^a
Normal + Reb-A (200 mg/kg BW)	0.53 ± 0.04^a	0.13 ± 0.01^a	0.45 ± 0.02^a
Diabetic rats	0.21 ± 0.01^b	0.27 ± 0.02^b	0.65 ± 0.03^b
Diabetic + Reb-A (200 mg/kg BW)	0.42 ± 0.02^c	0.19 ± 0.01^c	0.41 ± 0.02^c
Diabetic + glibenclamide (600 µg/kg BW)	0.43 ± 0.02^c	0.20 ± 0.01^c	0.42 ± 0.01^c

Note: Values are means ± SD for six rats. Values not sharing a common marking (a, b, c) differ significantly at $p < 0.05$ (DMRT).

[a] µmoles of glucose phosphorylated/h/mg protein.
[b] µmoles of inorganic phosphorous liberated/min/mg protein.
[c] µmoles of inorganic phosphorous liberated/h/mg protein.

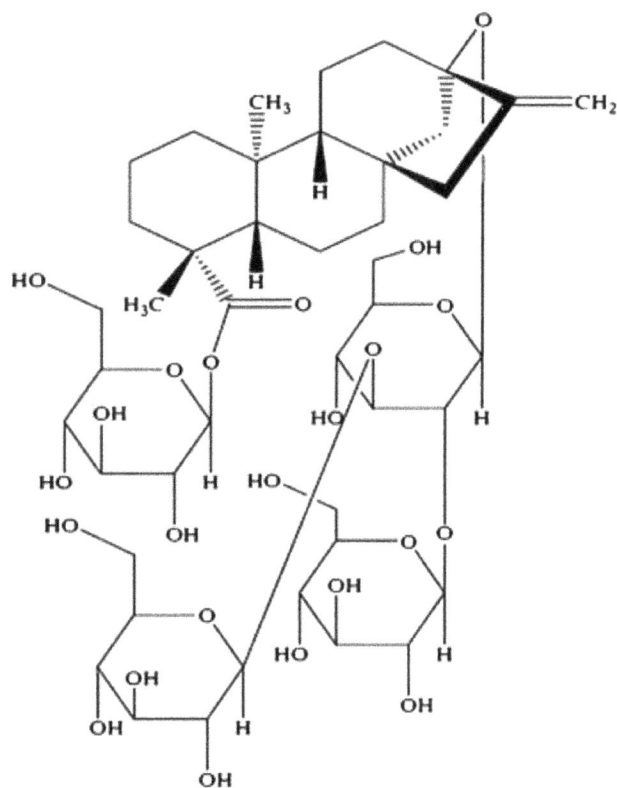

FIGURE 3.63 Structure of rebaudioside-A.

hyperglycemic mice, induce a significant decrease of blood glucose level to 21.7 ± 6.7 mM/L. It was observed that the increase in dose of of the compound **129** (40 mg/kg) caused a further decrease in glucose level to 15.9 ± 9.2 mM/L. This result indicates that the compound exerts its action in dose-dependent manner [114] (Figure 3.64).

Lee et al. reported the Abietane diterpenoids of *Rosmarinus officinalis* and their diacylglycerol acyltransferase-inhibitory activity. The inhibition of acyl-CoA:diacylglycerol acyltransferase (DGAT) has been proposed as one of the drug targets for treating obesity and type 2 diabetes. The MeOH-soluble extract of *Rosmarinus officinalis* yielded two new diacylglycerol acyltransferase (DGAT) inhibitory-abietane diterpenoids such as 7β-hydroxy-20-deoxo-rosmaquinone (**134**) and 7β-methoxy-20-deoxo-rosmanol (**135**), along with six known components, like carnosol (**136**), 7α-methoxyrosmanol (**137**), 7β-methoxyrosmanol (**138**), 12-methoxy-canosic acid (**139**), rosmanol (**140**), and rosmadial (**141**). These compounds inhibited DGAT1 activity with the IC$_{50}$ values ranging from 39.5 ± 0.6 to 144.2 ± 3.1 µM (Table 3.23). In addition to this, carnosol (**136**) exhibits inhibition of *de novo* intracellular triacylglycerol synthesis in human hepatocyte HepG2 cells. Kuraridine is used as positive control [115] (Figure 3.65).

Sen et al. reported two natural diterpenoids such as Alysine-A and Alysine-B, isolated from *Teucrium alyssifolium*. These compounds exert antidiabetic effects via

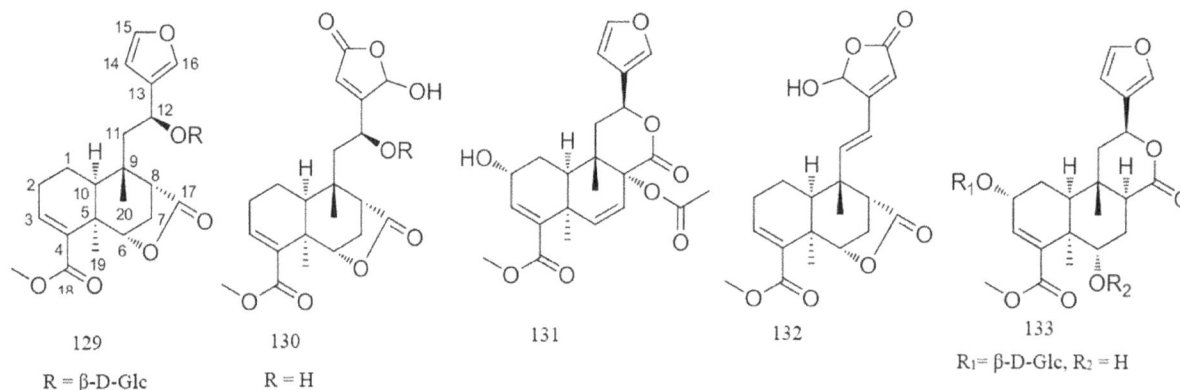

129
R = β-D-Glc

130
R = H

131

132

133
R₁ = β-D-Glc, R₂ = H

FIGURE 3.64 Structure of clerodane diterpenoids.

enhanced glucose uptake and suppressed glucose absorption. The key enzymes and proteins within the regulation of glucose homeostasis were investigated to explore the antihyperglycemic impact and the mechanism of the test compounds. Both compounds increased the expression of the genes that were activated by the insulin in the tested cell lines. The insulin receptor (INSR), IRS2, protein kinase (PK), GLUT4, and phosphoinositide 3-kinase (PI3K) expression levels were significantly increased in the 3T3-L1, C2C12, and HepG2 cells. These are significant genes involved in the insulin signaling pathway which favor the antidiabetic activity. Both Alysine-A and Alysine-B were tested for their impact on insulin secretion under normal or hyperglycemic conditions using the BTC6 cells. Glucose transport test was performed with the Caco-2 cell line. Glibenclamide was used as the positive control [116] (Figure 3.66, Tables 3.24–3.29).

Carney et al. reported a new Daphnane diterpenoid (Maprouneacin) with potent antihyperglycemic activity. Maprouneacin was isolated from ethanolic extract of *Maprounea Africana*. It was evaluated by using the *in-vivo* noninsulin-dependent diabetes mellitus db/db mouse model (Table 3.30). Maprouneacin displayed potent glucose-lowering properties when administered by the oral route [117] (Figure 3.67).

TABLE 3.23

Inhibitory Activity of Compounds (**134-141**) against DGAT1

Compounds Code	Compounds Name	DGAT1 Inhibitory Activity IC_{50} (μM)[a]
134	7β-Hydroxy-20-deoxo-rosmaquinone	39.5±0.6
135	7β-Methoxy-20-deoxo-rosmanol	97.2±3.4
136	Carnosol	62.5±2.1
137	7α-Methoxy-rosmanol	63.0±1.4
138	7β-Methoxy-rosmanol	92.7±3.4
139	12-Methoxy-canosic acid	85.2±2.5
140	Rosmanol	142.6±1.9
141	Rosmadial	144.2±3.1
Positive control	Kuraridine	9.8±0.3

[a] The IC_{50} values were determined via regression analyses and expressed as the mean±SD of the three replicates.

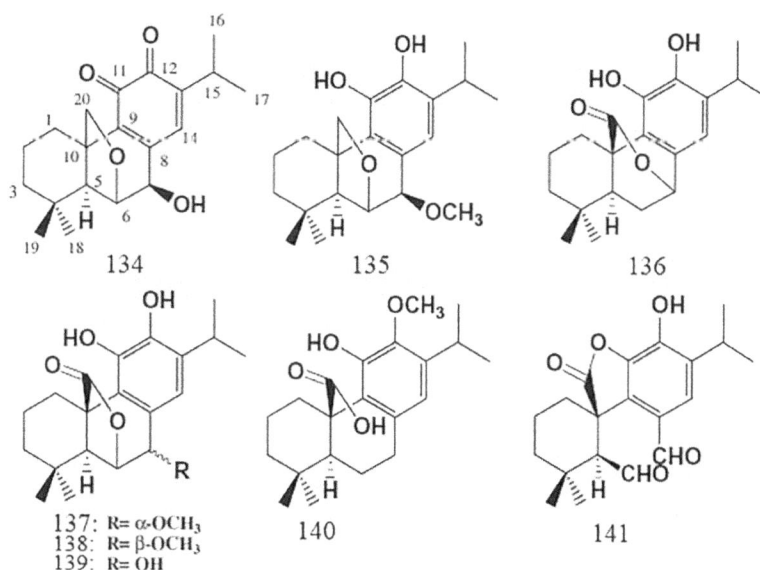

FIGURE 3.65 Chemical structures of compounds (134–141) isolated from *R. officinalis*.

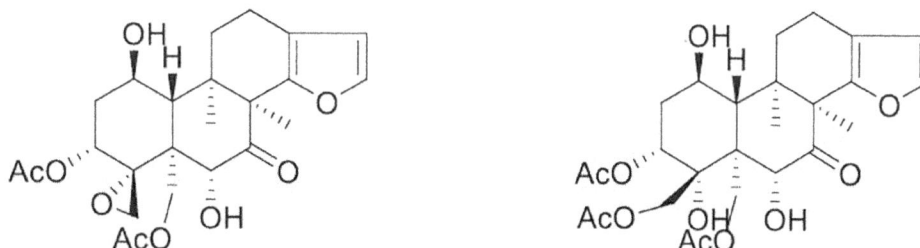

FIGURE 3.66 Structures of the test compounds (Alysine-A and Alysine-B).

TABLE 3.24

Effects of the Test Compounds on LDH Activity in the Extracellular Medium of the Cell Lines

Treatments	LDH Activity in the Extracellular Medium (mU/mL)				
	BTC6	C2C12	3T3-L1	HepG2	Caco-2
Alysine-A	57.4±3.62	54.4±0.72	58.8±3.73	81.5±1.80	47.8±2.67
Alysine-B	66.9±3.68	49.6±2.71	56.1±5.58	62.7±1.78	38.2±3.39
Control	61.2±7.67	52.0±4.3	50.6±6.48	78.2±3.21	41.6±5.43

TABLE 3.25

Remaining Glucose Concentrations in the Extracellular Medium of the C2C12 Muscle, HepG2 Liver, and 3T3-L1 Adipocyte Cells after 2-h Incubation with the Test Compounds

Treatments	Remaining Glucose Concentration in the Extracellular Medium (µg/mL)		
	C2C12	HepG2	3T3-L1[d]
Alysine-A	2.02±0.19[a]	2.48±0.13[a]	10.38±1.22[b]
Alysine-B	3.36±0.40[a]	3.39±0.48[a]	11.94±1.40[c]
Control	12.4±2.1	10.8±2.8	13.4±1.1
Insulin	2.25±0.10[a]	2.27±0.07[a]	8.23±0.62[a]
Metformin	1.84±0.06[a]	2.05±0.19[a]	7.30±0.98[a]

[a] $p < 0.0001$.

[b] $p < 0.001$.

[c] $p < 0.05$.

[d] Adipocyte differentiation of the preadipocyte cells were confirmed by observing 8.4-fold induction of the G3PDH activity (preadipocytes: 4.94±0.72 and differentiated adipocytes: 41.6±2.68).

TABLE 3.26

Quantities of Glycogen Stored by the Positive Control and Test Compounds in the C2C12 Muscle and HepG2 Liver Cell Lines

Treatments	Intracellular Glycogen Concentration per 1.0×10^6 Cells (µg/mL)	
	C2C12	HepG2
Alysine-A	0.717±0.080[a]	0.591±0.087[a]
Alysine-B	0.507±0.071[a]	0.432±0.051[a]
Insulin	0.534±0.031[a]	0.305±0.018[a]
Control	0.232±0.011	0.185±0.015[a]

[a] $p < 0.0001$.

TABLE 3.27

Temporal Glucose Concentrations on the Apical and Basolateral Surfaces of the Caco-2 Cell Monolayer Treated either with the Positive Control or Test

Treatments	Glucose Concentration (µg/mL) Apical Side			Glucose Concentration (µg/mL) Basolateral Side		
	0 min	30 min	60 min	0 min	30 min	60 min
Alysine-A	1.70±0.16	1.88±0.02	1.65±0.06	0.30±0.00	3.80±0.32[a]	7.29±0.51[a]
Alysine-B	1.73±0.03	1.58±0.09	1.69±0.10	0.11±0.01	3.53±0.24[a]	14.06±0.91[a]
Insulin	1.63±0.10	1.61±0.03	1.61±0.06	0.09±0.00	4.55±0.37[a]	6.79±0.42[a]
Control	1.68±0.08	1.70±0.11	1.61±0.04	0.13±0.05	0.27±0.17	0.50±0.32

[a] $p < 0.0001$.

TABLE 3.28

Glucose Concentrations Released by α-Glycosidase at
the Apical Surface of the Caco-2 Cells in the Presence
of the Different Test Compounds

Treatments	Glucose Concentration (µg/mL)
Alysine-A	0.080±0.03
Alysine-B	0.081±0.08
Control	0.084±0.07

TABLE 3.29

Effect of Alysine-A and Alysine-B on the Insulin Secretion in
the BTC6 Cell

Treatments	Insulin Concentration (ng/mL)	
	Glucose (8 mM)	Glucose (33 mM)
Alysine-A	ND	ND
Alysine-B	ND	ND
Glibenclamide	4.1±0.33	6.8±0.71
Control	ND	ND

ND, not detectable.

3.3.6 Treatment of Microbial Infection

There are various types of microbes including bacteria, fungi, virus and prozoa that cause different types of diseases such as tuberculosis, ringworm, chickenpox and malaria respectively. Treatment of microbial infection depend on the cause of the infection [118].

The diterpenoids aframolins-A and B, labdan-8(17), 12(E)-diene-15,16-dial, and aframodial isolated from the seeds of *Aframomum longifolius*. These compounds were tested for their antimicrobial activities. Aframodial is chemically known as (2E)-2-[2-(5,5,8a-trimethylspiro[3,4,4a, 6,7,8-hexahydro-1H-naphthalene-2,2′-oxirane]-1-yl)ethylidene]butanedial. But only aframodial was active against *Cryptococcus neoformans*, *Staphylococccus aureus*, and methycillin-resistant *Staphylococcus aureus* with similar MIC value of 20 µg/mL [119] (Figure 3.68).

Coll et al. reported the antimicrobial activity of the diterpenes flexibilide and sinulariolide derived from *Sinularia flexibilis* Quoy and Gaimard 1833 (Coelenterata: Alcyonacea, Octocorallia). Antimicrobial activity of the diterpenes was determined by using the disc assay method with antibiotics as controls and the minimum inhibitory concentrations

TABLE 3.30

Changes in Glucose Levels, Body Weight, and Food Consumption for Vehicle, Positive Control, and Maprouneacin

Treatment	Change in Plasma Glucose Levels, Predose vs Post-Dose (mg/dL)		Body Weight, Average (g/mouse)		Food Intake (g/mouse)
	3 h	27 h	0 h	24 h	Average
Vehicle	−28	−27	40.2±0.5	40.1±0.4	6.7
Metformin[a]	−132	−160	41.0 ±0.8	40.9±0.8	7.3
Maprouneacin[b]	−254	−246	37.7±0.7	37.8±0.8	0.1

[a] Single dose, 250 mg/kg.

[b] Single dose, 0.5 mg/kg.

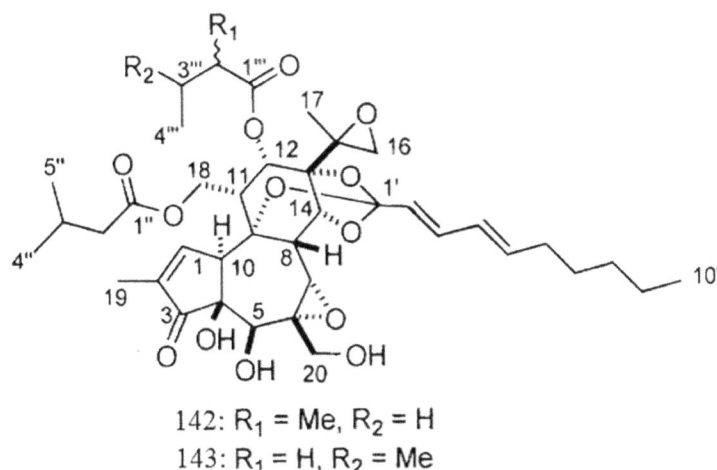

142: R₁ = Me, R₂ = H
143: R₁ = H, R₂ = Me

FIGURE 3.67 Structure of daphnane diterpenoid.

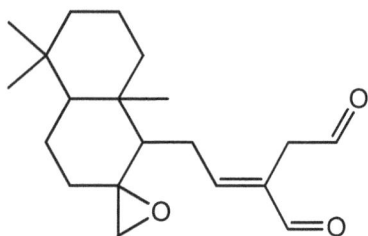

FIGURE 3.68 Structure of aframodial.

(MIC) of the diterpenes were determined using the tube dilution technique. Diterpenes such as sinulariolide and flexibilide exhibited significant antimicrobial activity and inhibited the growth of Gram-positive bacteria (Table 3.31). Flexibilide was effective even at concentrations as low as 5 ppm, whereas sinulariolide was effective at concentrations of 10 ppm [120] (Figure 3.69).

Horminone is an abietane diterpenoid. It is abieta-8,12-diene substituted by hydroxy groups at positions C_7, C_{12} and oxo groups at positions C_{11}, C_{14}. Its IUPAC name is (4bS,

8aS, 10R)-1,10-dihydroxy-4b, 8,8-trimethyl-2-propan-2-yl-5,6,7,8a, 9,10-hexahydrophenanthrene-3,4-dione. It was isolated from the roots of *Hminium pyrmzicum*. The MIC value of horminone is determined against various bacterial strains (*Siaphylococrcrs aum*, *Vibno cbokrae*, *Candida albicans*, and *Psedmonas mginosa*) by the broth micro-dilution method. It exhibited higher antibacterial activity against *Staphylococcus aureus* and *Vibrio cholerae* with MIC values of 7.8 µg/mL. The MIC values of Horminone was also determined against *Candida albicans* (250 µg/mL), *Pseudomonas aeruginosa* (250 µg/mL), and *Escherichia coli*, *Shigella dysenteriae*, *Salmonella typhimurium*, *Streptococcus faealis* (≥ 500 µg/mL) [121] (Figure 3.70).

Mulinolic acid is a diterpenoid, and isolated from the aerial parts of *Mulinum crassifolium*. Diterpenoid acids such as mulin-12,14-dien-11-on-20-oic acid (**144**) and mulin-12-ene-11,14-dion-20-oic acid (**145**), are closely related to mulinolic acid (**146**). These are isolated from *Azorella compacta*. It was found that these compounds exhibited antimicrobial activity. *In-vitro* antimicrobial activities were determined against methicillin-sensitive and methicillin-resistant *S. aureus*,

TABLE 3.31

Antimicrobial Activity of 5 ppm *S. flexibilis* Diterpenes

Treatments	Zone of Inhibition (mm)					
	Bacillus sp.	*S. aureus*	*P. aeruginosa*	*P. vulgaris*	*E. coli*	*C. albicans*
Flexibilide	25	25	-	-	-	-
Dihydroflexibilide	-	-	-	-	-	-
Sinulariolide	13	12	-	-	-	-
Epi-sinulariolide	6	6	-	-	-	-
Epi-sinulariolide acetate	5	-	-	-	-	-
Distilled water	-	-	-	-	-	-
Ethanol	-	-	-	-	-	-

Note: Values represent the diameter (mm) of the diterpene inhibition zone, while a negative (-) sign means absence of antimicrobial activity, after 48 h incubation.

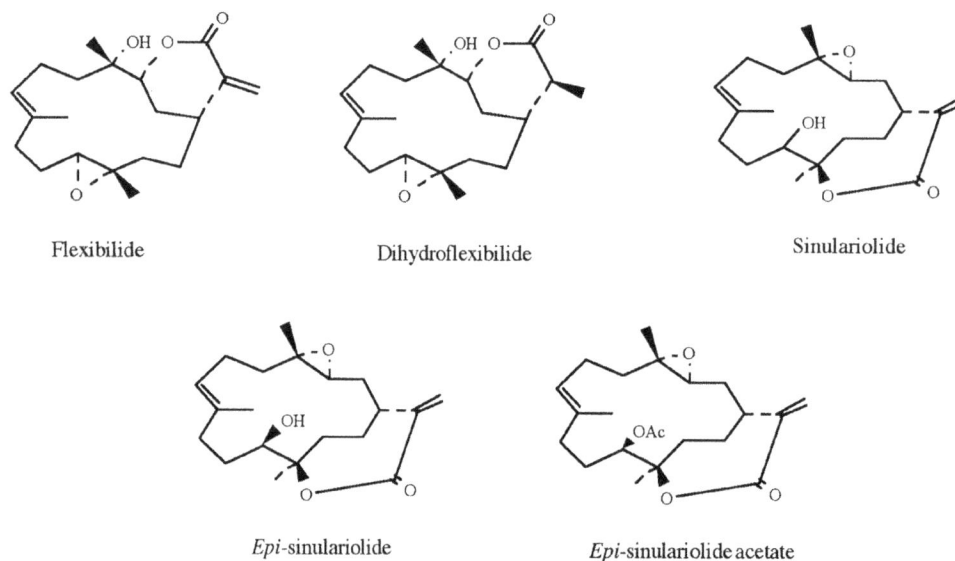

Flexibilide

Dihydroflexibilide

Sinulariolide

Epi-sinulariolide

Epi-sinulariolide acetate

FIGURE 3.69 Structures of flexibilide, dihydroflexibilide, sinulariolide, epi-sinulariolide and epi-sinulariolide acetate.

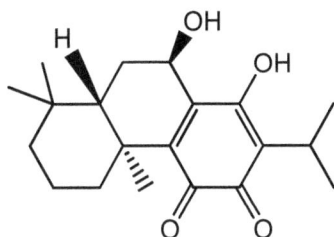

FIGURE 3.70 Structures of horminone.

Vancomycin-resistant *E. faecium, E.coli* and *C. albicans* by using the agar diffusion method. The zones of growth inhibition were measured in millimeters using a hand-held Fowler Ultra-cal II digital caliper. DMSO (10 μL) was used as solvent control which did not exhibit any growth inhibition. In an agar diffusion test, the spots of 200 μg of compounds (**144**) and (**145**) in 10 μL of DMSO produced mostly hazy zones of inhibition of 8–12 mm in diameter against methicillin-susceptible and methicillin-resistant strains of *S. aureus*, a vancomycin-resistant strain of *E. faecium*, and a strain of *E. coli* with increased membrane permeability for compounds with high molecular weight. But, compound (**146**) is completely inactive against all of these bacteria at the same concentration [122].

Under the same experimental conditions, clear zones of inhibition of 16–18 mm in diameter were observed for antibiotic disks containing 30 μg of Vancomycin against methicillin-susceptible and methicillin-resistant *S. aureus*, for 10 μg of Bacitracin against vancomycin-resistant *E. faecium*, for 30 μg of Cefoxitin against *E. coli* with increased membrane permeability, and for 100 μg of Nystatin against *C. albicans* [123] (Figure 3.71).

Teixeira et al. reported the antimicrobial and antibiofilm action of Casbane diterpene from *Croton nepetaefolius* against oral bacteria. Oral infections are caused by a wide range of microorganisms that colonize the tooth surface at or below the gingival margin. This colonization can lead to pathological states which include dental caries and periodontal diseases with loss of tooth. The antibacterial activity of Casbane diterpene (CD) was evaluated *in-vitro* against

Streptococcus oralis, S. mutans, S. salivarius, S. sobrinus, S. mitis and *S. sanguinis.* The CD isolated from Croton nepetaefolius exhibited antimicrobial effect on planktonic forms and biofilms of oral pathogens, with MIC values of 62.5 mg/mL against *Streptococcus oralis* and the values between 125 and 500 mg/mL against *S. mutans, S. salivarius, S. sobrinus, S.mitis* and *S. sanguinis* (Table 3.32). It exhibited an inhibitory effect on *S. mutans* biofilm formation at 250 mg/mL, and a decrease on viable cell of 94.28% as compared to the normal biofilm growth [124] (Figure 3.72).

Guanacastepene

The fermentation extracts of culture CR115, an unknown plant endophyte originally isolated from Costa Rica, were found to be active against antibiotic-resistant bacteria. The metabolite responsible for activity was identified as a novel diterpenoid antibiotic guanacastepene with molecular weight 374.47 and molecular formula $C_{22}H_{30}O_5$. Mechanistic study is carried out in an *E. coli imp* strain that suggested the membrane damage as the primary mode of bactericidal action (Table 3.33). This compound also lysed human RBCs and caused leakage of intracellular potassium from *E. coli imp* [125] (Figure 3.73).

Sa: *S. aureus*, Ef: *E. faecium*, Ec: *E. coli*, Ca: *C. albicans*, BIA (Biochemical Induction Assay): detects DNAdamaging activity.

Sample solubilized and serially diluted in DMSO at 10–0.078 mg/mL. Five microliters (5 μL) of each concentration was spotted on the agar surface.

Plate was incubated for 18 h at 37°C and zone of growth inhibition recorded in millimeter.

H, hazy; T, toxic; 3+, strong DNA damage. MIC (determined by Microbroth Dilution Method) against *E. coli imp* was 62.5 μg/mL. Guanacastepene was inactive in the BIA (Table 3.34).

Data presented represent % of untreated control after 10 min drug pretreatment and 5 min pulse labeling. Incorporation = precursor incorporated into TCA-precipitable material. H-Tdr, H-Udr and H-AA are tritiated thymidine, uridine and amino acids respectively (Table 3.35).

Herical is a striatal type of diterpenoid. It was produced by fermentation of *Hericium ramosum* [126] (Figure 3.74). The antifungal and antibacterial activities of herical in the disc/

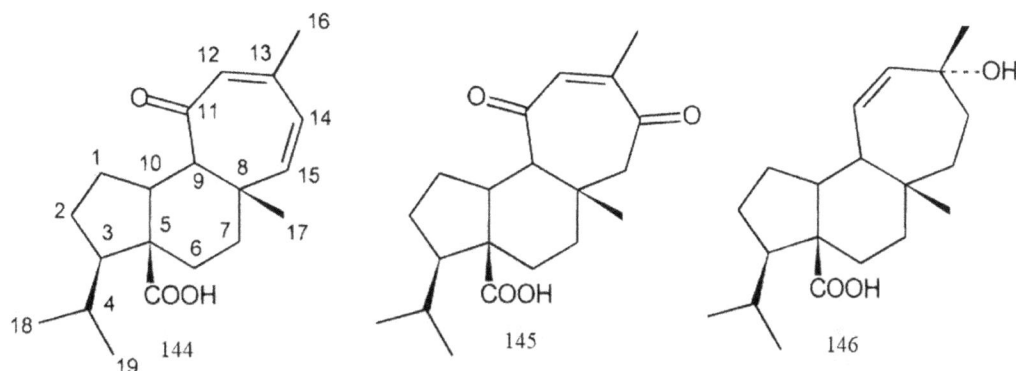

FIGURE 3.71 Structures of mulinolic acid and analogs.

TABLE 3.32

Minimal Inhibitory Concentration (MIC) of Casbane Diterpene (CD) at Different Times of Growth (12, 18 and 24 h) and Minimal Bactericidal Concentration (MBC) against Oral Pathogens

Bacteria	MIC (mg/mL)/time (h)			MBC (mg/mL)
	12 h	18 h	24 h	
S. mutans	125	250	250	500
S. mitis	125	125	125	500
S. sobrinus	250	500	500	500
S. oralis	62.5	62.5	62.5	125
S. salivarius	250	250	250	500
S. sanguinis	125	250	250	500

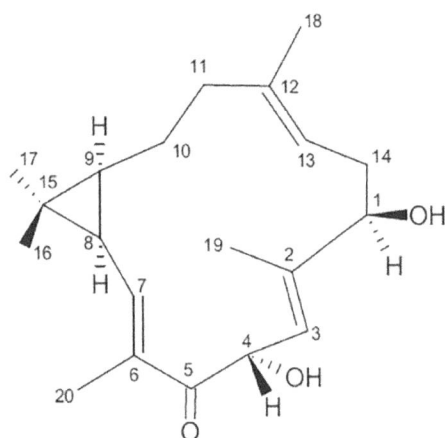

FIGURE 3.72 Structure of casbane diterpene extracted from the stalks of *Croton nepetaefolius*.

TABLE 3.33

Antimicrobial Activity of Guanacastepene

(Conc. μg/spot)	Sa375	Sa310	Ef379	Ec389	Ec442	Ca54	BIA
50	7:10H	6:9H	8H	8H	5H	5:9H	9T
25	3:8H	7H	6H	7H	5H	8H	8T
12.5	7H	7H	6H	7H	5H	7H	7T
6.25	6H	6H	-	7H	5H	6H	7T
3.1	6H	6H	-	6H	-	6H	7H
1.6	5H	5H	-	6H	-	-	7H
0.8	-	-	-	6H	-	-	7H
0.4	-	-	-	6H	-	-	7H
Methicillin	15:22H	-	-	-	-	-	-
Vancomycin	15	17	-	-	-	-	-

agar diffusion and the serial dilution assay were measured and presented in Table 3.36.

3.3.7 Treatment of Tuberculosis

Tuberculosis is primarily caused by *Mycobacterium tuberculosis*. It is a major source of morbidity and mortality worldwide with almost 2 million deaths annually. So, new drugs for the treatment of tuberculosis are necessary due to increase in the multidrug resistant (MDR) and extensively drug-resistant ((XDR) strains [127].

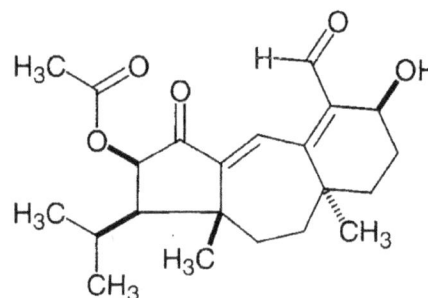

FIGURE 3.73 Structure of guanacastepene.

TABLE 3.34

Effects of Guanacastepene and Known Antimicrobials on the Incorporation of Radiolabeled Precursors into Macromolecules in *E. coli imp*

Compounds	Conc. (μg/mL)	H-Tdr	H-Udr	H-AA
Guanacastepene	128	20	37	27
	64	21	44	20
	32	93	76	92
	16	93	90	100
Ciprofloxacin	0.25	3	90	95
Rifampin	0.25	93	2	12
Chloramphenicol	8	98	95	16
Polymyxin-B	8	1	2	3

TABLE 3.35

Membrane Damaging Effects on Human RBC and *E. coli imp*

Compounds	μg/mL	RBC Lysis (%)[a]	K-Leakage (%)[a]
Guanacastepene	128	5	NT
	64	29	75
	32	83	74
	16	92	60
	8	90	NT
	4	3	NT
Polymyxin-B	32	4	60
Amphotericin-B	4	60	NT
	2	10	NT
	1	1	NT

NT, not tested.

[a] Drug treatment periods for RBC lysis and potassium leakage from *E. coli* were 2 and 1 hour respectively.

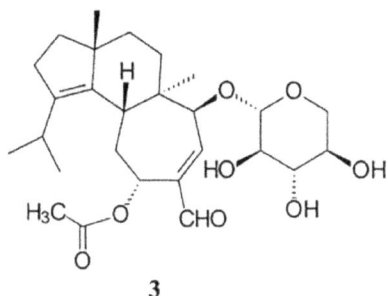

FIGURE 3.74 Structure of herical.

TABLE 3.36

Antifungal Spectrum of Herical (3) in the Plate Diffusions Assay

Test Organism	Diameter Inhibition Zone [mm] 100 µg/disc
Absidia glauca (+)	12
Absidia glauca (D)	20
Alternaria porri	28
Aspergillus ochraceus	15
Epicoccum purpurascens	10
Fusarium fujikoroi	12
Mucor miehei	31
Nematospora coryli	10
Neurospora crassa	17
Paecilomyces varioti	13
Penicillium notatum	15
Pythium debaryanum	13
Saccharomyces cerevisiae	12
Ustilago nuda	13

To reduce the side effects from the use of chemically synthesized or clinically used anti-tubercular drugs, various investigations were carried out to focus on natural products as a source of potential anti-tuberculosis drugs. Among various phytoconstituents, several classes of diterpenoids were found to exert significant anti-tubercular activity [128].

Silva et al. reported that the labdane diterpenes such as copalic acid (**147**), 3β-acetoxy-copalic acid (**148**), 3β-hydroxy-copalic acid (**149**) and ent-agathic acid (**150**) were isolated from *Copaifera langsdorffii* oleoresin. These four compounds were subjected for structural modifications by reduction with hydrogen/palladium, esterification with diazomethane,

esterification with methanol/sulfuric acid and conversion into sodium salt. All these compounds were assayed *in-vitro* against *Mycobacterium tuberculosis* (H37Rv, ATCC 27294). The four compounds exhibited minimum inhibitory concentration (MIC) value of 125 µg/mL. A methylated derivative of compound 3β-hydroxy-copalic acid, and a sodium salt of copalic acid demonstrated MIC values of 25 µg/mL (71.7 µM) and 6.25 µg/mL (19.2 µM) respectively. It was observed that the sodium salt of copalic acid displayed similar activity in comparison with streptomycin (MIC 6.25 µg/mL) and a better activity as compared to pyrazinamide (MIC 3.12 µg/mL; 25.34 µM) [129] (Figures 3.75 and 3.76, Tables 3.37 and 3.38).

Xiaochi et al. reported the anti-tuberculosis effects of diterpenoids isolated from the roots of *Euphorbia ebracteolata*. By using various chromatographic techniques, 15 diterpenoids were isolated from the roots of E. ebracteolata which include rosane derivatives (**160–171**), isopimarane (**172**), abietane (**173**), and lathyrane (**174**). The inhibitory effects of isolated diterpenoids against *Mycobacterium tuberculosis* were evaluated by using an Alamar blue cell viability assay method. Among the tested compounds, two rosane-type diterpenoids **162** and **167** displayed moderate inhibitory effects with the MIC values of 18 and 25 µg/mL respectively. For the potential inhibitor **162**, the inhibitory effect against the target enzyme GlmU was evaluated, which displayed a moderate inhibitory effect with the IC$_{50}$ of 12.5 µg/mL (Table 3.39). Kanamycin is used as positive control [130] (Figure 3.77).

Rodriguez et al. reported the antimycobacterial activity of diterpene (Serrulatane), isolated from the West Indian Sea Whip *Pseudopterogorgia elisabethae*. Two serrulatane-based diterpenes such as erogorgiaene (**175**) and 7-hydroxyerogorgiaene (**176**), were obtained from the hexane extract of the coral. Erogorgiaene (**175**) induced 96% growth inhibition for *Mycobacterium tuberculosis* H37Rv at a concentration of 12.5 µg/mL. On the other hand, 7-hydroxyerogorgiaene (**176**) inhibited 77% of mycobacterial growth at a concentration of 6.25 µg/mL [131] (Figure 3.78).

Ricardo et al. reported the diterpenes synthesized from the natural Serrulatane Leubethanol and their *in-vitro* activities against *Mycobacterium tuberculosis*. Leubethanol (Leub) is a natural serrulatane diterpenoid, isolated from the methanolic extracts of *Leucophyllum frutescens*. Seventeen new derivatives of the natural diterpene leubethanol, including some potential pro-drugs, with changes in the functionality of the aliphatic chain or modifications of the aromatic ring and

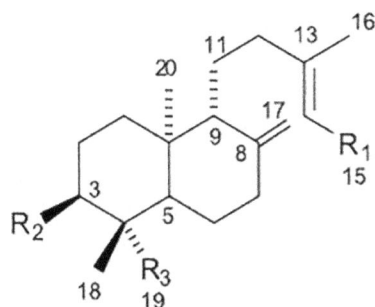

	R$_1$	R$_2$	R$_3$
147.	COOH;	H ;	CH$_3$
148.	COOH;	AcO;	CH$_3$
149.	COOH;	OH ;	CH$_3$
150.	COOH;	H ;	COOH

FIGURE 3.75 Structures of the natural diterpenes.

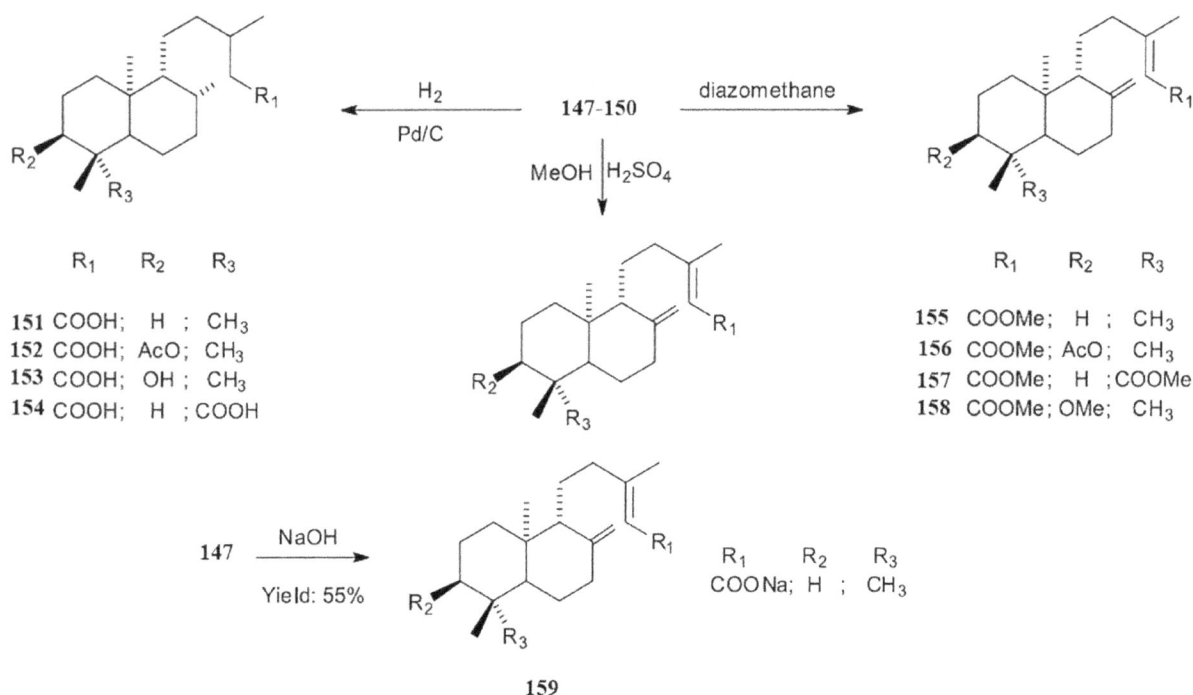

FIGURE 3.76 Scheme of the structural modifications of labdane diterpenes.

TABLE 3.37

Antibacterial Spectrum of Herical (3) in the Serial Dilution Assay

| Test Organisms | Minimal Inhibitory Concentration (MIC) | |
	[µg/mL]	[nmol/mL]
Escherichia coli	>100	>200
Arthrobacter citreus	10–20	20–40
Bacillus brevis	2–5	4–10
Bacillus subtilis	5–10	10–20
Corynebacterium insidiosum	10–20	20–40
Sarcina lutea	10–20	20–40
Streptomyces spec.	20–50	40–100

TABLE 3.39

MIC Values of Diterpenoids (**1–15**) against *Mycobacterium tuberculosis*

Compounds	MIC (µg/mL)	Compounds	MIC (µg/mL)
160	125±1.3	168	150±1.2
161	150±2.2	169	100±1.1
162	15±0.35	170	125±0.9
163	200±2.4	171	150±1.3
164	100±1.5	172	200±1.6
165	125±1.5	173	150±1.4
166	200±1.4	174	300±2.1
167	25±0.8	Kanamycin	10±0.6

Note: Data were presented as means±SD of three parallel measurements.

TABLE 3.38

Anti-tubercular Activity Results for All Compounds and First Line Drugs

Compounds	MIC (µg/mL)	Compounds	MIC (µg/mL)
147	125	10	125
148	125	11	125
149	125	12	125
150	125	13	125
151	62.5	14	25
152	125	15	6.25
153	125	Ethambutol	1.64
154	250	Pyrazinamide	3.12
155	125	Isoniazid	0.06

the phenolic group were synthesized. These compounds were tested for their anti-tubercular activity against the H37Rv strain of Mycobacterium tuberculosis *in-vitro* by using MABA technique. Ethambutol (EMB) was used as positive control [132] (Figures 3.79 and 3.80).

The phytochemical investigation of the acetone extract of the aerial parts of *Leucas stelligera* afforded four new compounds (**179–182**) belonging to the labdane diterpene series. These compounds exhibited selective antimycobacterial activity against *Mycobacterium tuberculosis* (H37Ra) with IC_{50} values in the range of 5.02–9.80µg/mL and IC_{90} values ranging from 10.85 to 46.52µg/mL as compared to isoniazid with

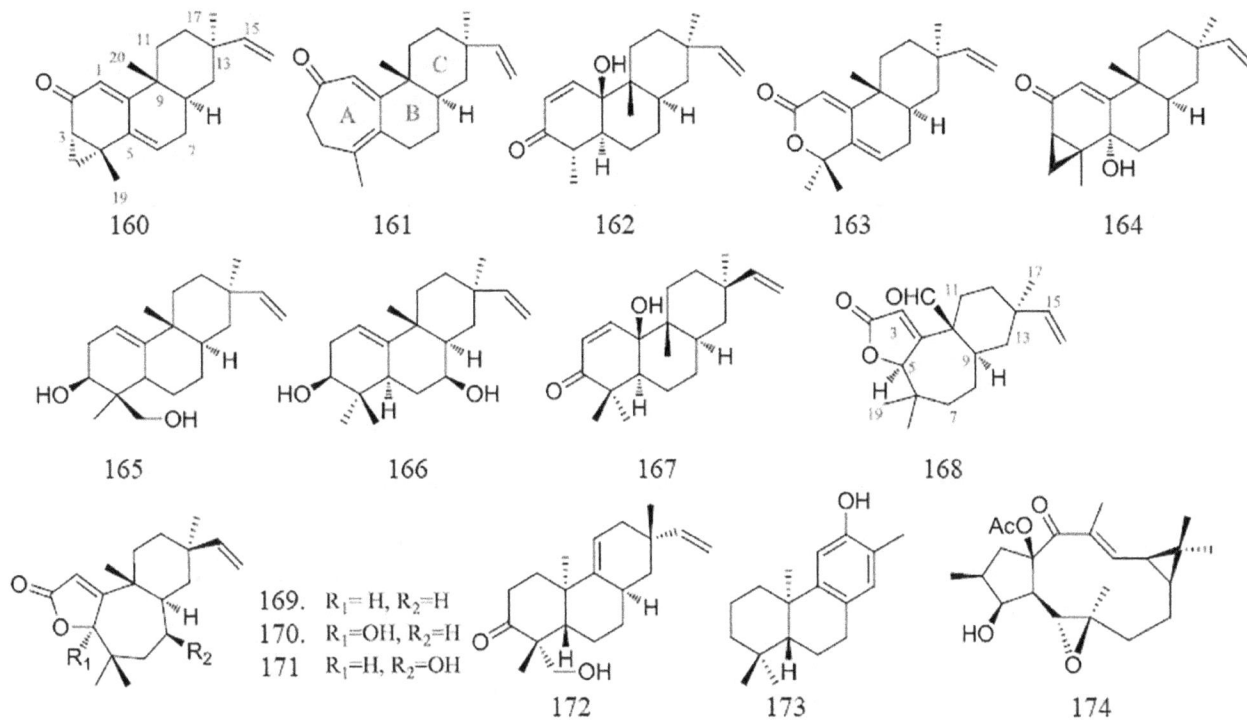

FIGURE 3.77 Structure of diterpenoids isolated from the roots of *Euphorbia ebracteolata*.

169. $R_1 = H, R_2 = H$
170. $R_1 = OH, R_2 = H$
171. $R_1 = H, R_2 = OH$

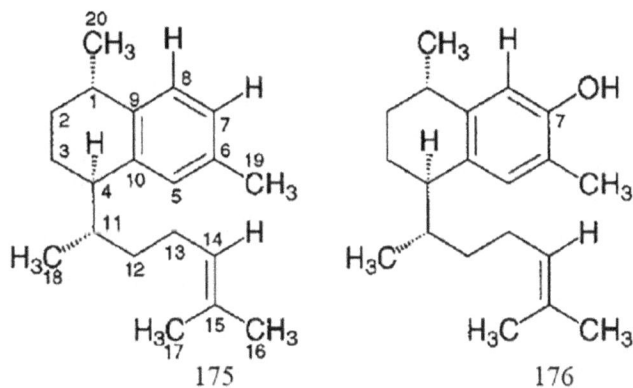

FIGURE 3.78 Structure of erogorgiaene (175) and 7-hydroxyerogorgiaene (176).

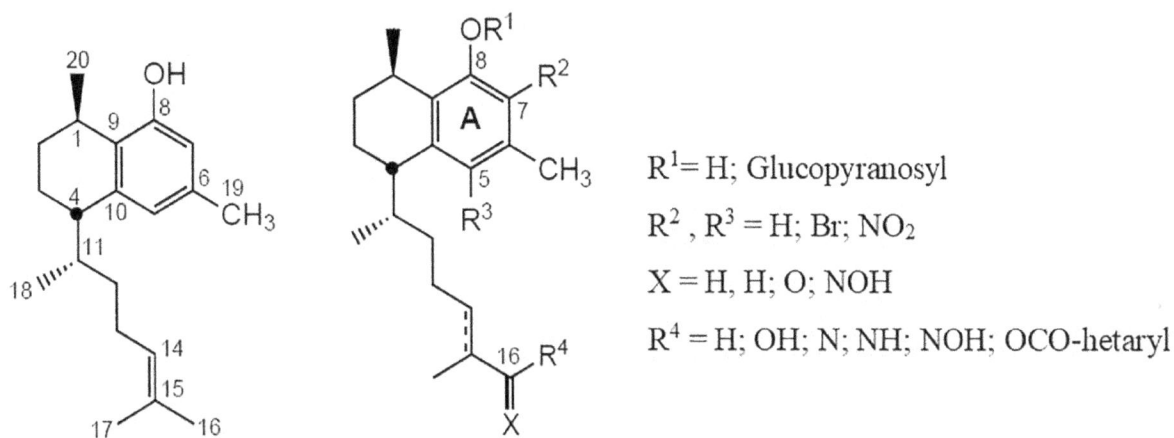

$R^1 = H$; Glucopyranosyl

$R^2, R^3 = H$; Br; NO_2

$X = H, H$; O; NOH

$R^4 = H$; OH; N; NH; NOH; OCO-hetaryl

FIGURE 3.79 Leubethanol modifications on ring A and the alkyl side chain.

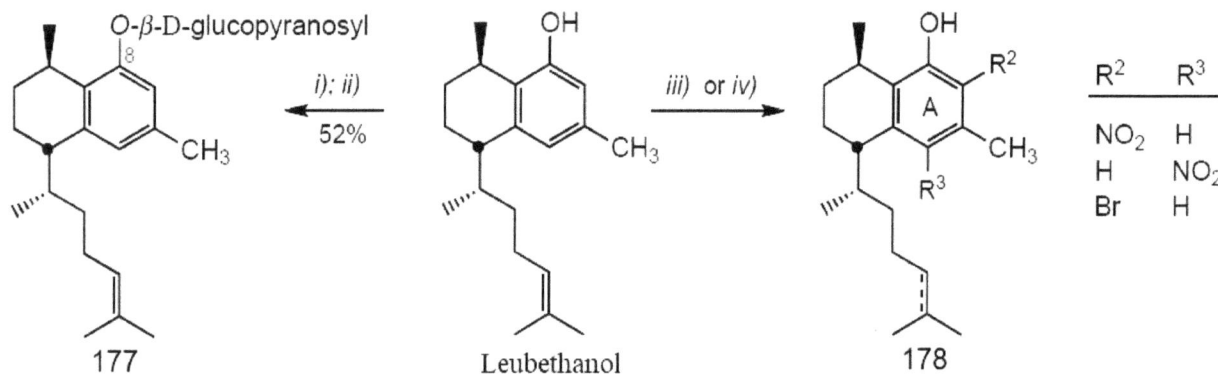

i) 2,3,4,6-Tetra-O-acetyl-α-D-glucopyranosyl trichloroacetimidate, $BF_3.OEt$ / CH_2Cl_2; ii) EtONa / EtOH; iii) 60% HNO_3; iv) NBS / CCl_4.

FIGURE 3.80 Modifications of leubethanol on the phenolic hydroxyl and aromatic ring A.

TABLE 3.40

In Vitro Antimycobacterial Activity of Compounds against *M. tuberculosis* H37Ra

Tested Samples	IC_{50} (μg/mL)	IC_{90} (μg/mL)
179	5.02±1.07	19.67±3.27
180	5.55±0.92	14.88±2.93
181	5.95±1.09	10.85±2.47
182	9.8±1.94	46.52±7.91

IC_{90} value of 0.055±0.003 μg/mL (Table 3.40). Isoniazid was used as positive control [133] (Figure 3.81).

3.3.8 Treatment of Pain and Smooth Muscle Spasms

Pain is defined as an unpleasant sensory and emotional experience associated with potential tissue damage. Pain has sensory, affective and a cognitive component associated with the anticipation of future harm. The various neurotransmitters and receptor play a major role in the modulation of pain processes in the central nervous system (CNS) and peripheral nervous system which include vanilloid, opioid, non-opioids (i.e. β-adrenergic, dopaminergic), glutamate, protein kinase-C, potassium ion (K^+) channels, and nitric oxide/cyclic guanosine monophosphate pathways etc [134].

Vouacapane diterpenes are commonly isolated from the fruits of different *Pterodon* species. The antinociceptive

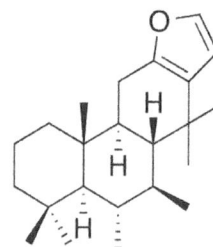

FIGURE 3.82 Structure of vouacapane.

activity of the vouacapans was evaluated in chemical and thermal models of nociception in mice that include acetic acid-induced abdominal constriction, paw compression, formalin and hot-plate test [135]. The antinociceptive effect of vouacapans may involve multiple mechanisms of action such as opioid, catecholaminergic, vanilloid, glutamate, serotonergic systems etc. This compound exhibited the antinociceptive effect (500 μmol/kg, i.p.) when compared with the control group. Results suggested that vouacapane's antinociceptive response may be associated with dopamine [136] (Figure 3.82).

Borrelli et al. reported the antispasmodic effects and structure-activity relationships (SAR) study of labdane diterpenoids, isolated from *Marrubium globosum* ssp. Libanoticum. The chloroform extract obtained from *M. globosum* aerial parts reduced the acetylcholine-induced contractions in the isolated mouse ileum. The extract contains four new compounds such as 13-epicyllenin-A, 13,15-diepicyllenin-A, marrulibacetal,

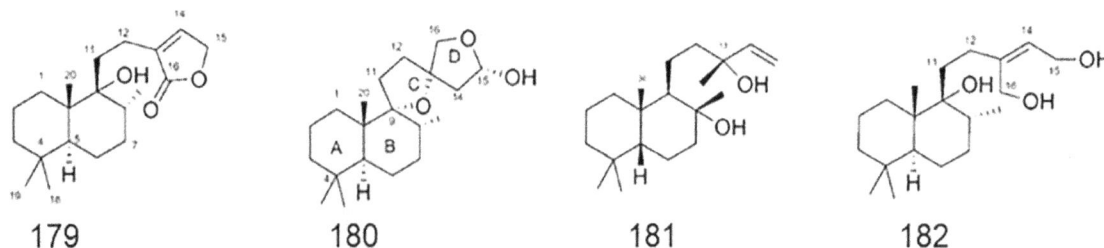

FIGURE 3.81 Phtoconstituents of *L. stelligera*.

and marrulactone. Marrulibacetal reduced significantly the acetylcholine-induced contraction at a concentration of 0.001–10 µg/mL in a concentration-dependent manner. Compounds such as 13-epicyllenin A, 13,15-diepicyllenin-A completely inhibited the acetylcholine-induced contraction at a concentration of 100 µg/mL, while marrulactone was found to be less active [137] (Figures 3.83 and 84).

Salvinorin-A is the *trans*-neoclerodane diterpene with chemical formula $C_{23}H_{28}O_8$. It is isolated from the leaves of *Salvia divinorum*. It is a potent psychotomimetic agent and selective agonist of kappa opioid receptors in the brain. It causes dysphoria by inhibting the release of dopamine in the striatum. It exhibits CNS activity at doses of 200–500 µg [138,139] (Figure 3.85).

3.3.9 Treatment of Leishmaniasis

Leishmania is a genus of protozoan parasites which are transmitted by the sandfly bite and cause the diseases collectively known as leishmaniasis. This disease is characterized by clinical manifestations which may occur from self-healing

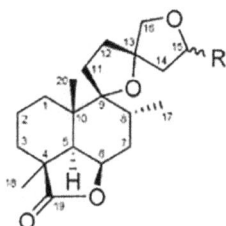

FIGURE 3.83 Structure of 13-epicyllenin-A (R=α-OH), 13,15-diepicyllenin-A (R=β-OH).

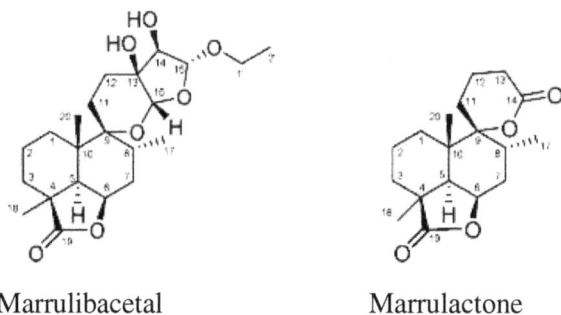

Marrulibacetal   Marrulactone

FIGURE 3.84 Structure of marrulibacetal and marrulactone.

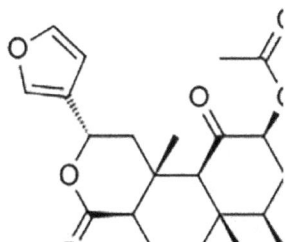

FIGURE 3.85 Structure of salvinorin-A.

cutaneous and mucocutaneous skin ulcers to a fatal visceral form of the disease called visceral leishmaniasis or kala-azar. According to the World Health Organization (WHO), nearly 1.5–2 million cases of cutaneous leishmaniasis occur every year worldwide [140,141].

Adriana et al. reported the anti-leishmanial activity of diterpene acids, isolated from copaiba oil (*Copaifera reticulate*). Various diterpene acids, isolated from *Copaifera* include agathic acid, hydroxycopalic acid, kaurenoic acid, methyl copalate, pinifolic acid and polyaltic acid. Significant anti-leishmanial activity of diterpene acids was evaluated against axenic amastigote, promastigote and intracellular amastigote forms of the parasite (*Leishmania amazonensis*). Hydroxycopalic acid and methyl copalate demonstrated the most activity, with IC_{50} values of 2.5 and 6.0 µg/mL, respectively, against promastigote forms of *L. amazonensis* (Table 3.41). Similarly, the most active diterpene acids which inhibit the proliferation of axenic amastigotes were pinifolic acid and kaurenoic acid, with IC_{50} values of 4.0 and 3.5 µg/mL, respectively. Amphotericin-B displayed IC_{50} of 0.06 µg/mL against promastigotes after 72 h of treatment. But, Amphotericin-B showed IC_{50} of 0.23 µg/mL against amastigote forms. The ANOVA ($p < 0.05$) indicated the significant differences between the isolated compounds and the control group [142,143] (Figure 3.86).

Antiviral Activity

Dichotomane diterpenes such as Da-1 and AcDa-1 were isolated from *D. menstrualis*. Both compounds exhibited significant anti-retroviral activity against HIV-1 replication with EC_{50} values of 40 and 70 µM respectively. Da-1 was also isolated from *D. pfaffii* and exhibited inhibitory activity against HSV-1 replication with an EC_{50} value of 5.10 µM [144,145] (Figure 3.87).

Chikungunya virus (CHIKV) is an arthropod-borne virus causing an infectious disease which is characterized by fever and arthralgia.

Litaudon et al. reported the antiviral activity of Flexibilane and Tigliane Diterpenoids, isolated from *Stillingia lineate*. These compounds were investigated for selective antiviral activity against CHIKV, Semliki Forest virus, Sindbis virus, HIV-1 and HIV-2 viruses. Compounds **194–197** were found to be the most potent and are selective inhibitors of CHIKV, HIV-1, and HIV-2 replication. Compound **196** inhibited CHIKV replication with an EC50 value of 1.2 µM on CHIKV and a selectivity index of >240, while compound **197** inhibited HIV-1 and HIV-2 with EC_{50} values of 0.043 and 0.018 µM, respectively [146] (Figure 3.88).

Tigliane diterpenoids are the largest group of phorboids. These terpenoids possess a 5/7/6/3-tetracyclic ring system, in which rings A and B, and B and C, are trans-fused where as rings C and D are cis-fused. These compounds were evaluated in a virus-cell-based assay against CHIKV and exhibited significant anti-CHIKV activities with EC_{50} values between 20 nM and 5 µM [147] (Figure 3.89).

Prostratin [13-O-acetyl-12-deoxyphorbol] is a phorbol ester that was first isolated from Strathmore weed Pimelea prostrate. Prostratin contains four rings such as A, B, C and D. Ring A is trans linked to the 7-membered ring B while Ring C is a 6 membered and is cis linked to the cyclopentane ring

TABLE 3.41

Anti-leishmanial and Cytotoxic Activity of Diterpenes from Copaiba Oil

Compounds	µg/mL			
	IC$_{50}$		CC$_{50}$	IS
	Promastigote	Amastigote	Erythrocytes	CC$_{50}$/IC$_{50}$
Agathic acid	28.0±1.5b	17.0±2.0b	350.0±3.4	20.6
Hydroxycopalic acid	2.5±0.4a	18.0±1.5b	40.0±2.4	2.2
Kaurenoic acid	28.0±0.7b	3.5±0.5a	140.0±17.0	40.0
Methyl copalate	6.0±0.9a	14.0±1.0b	500.0±3.0	35.7
Pinifolic acid	70.0±8.0b	4.0±0.4a	>500.0	>125.0
Polyaltic acid	35.0±2.0b	15.0±1.0b	>500.0	>33.3
Amphotericin-B	0.06±0.0	0.23±0.0	ND	ND

CC$_{50}$, cytotoxic concentration of 50%; IC$_{50}$, inhibitory concentration of 50%; ND, not determined; SI, selectivity index between erythrocytes and amastigote forms.

a   Significant difference when compared to others diterpenes acids (ANOVA test; $p < 0.05$).

b   No significant difference when compared to hidroxycopalic acid and methyl copalate (promastigote test), kaurenoic acid and pinifolic acid (amastigote test) (ANOVA test; $p < 0.05$). Significant differences between compound's activity and control cell growth, ANOVA ($p < 0.05$). Values represent the mean±standard deviation of at least three experiments performed in duplicate.

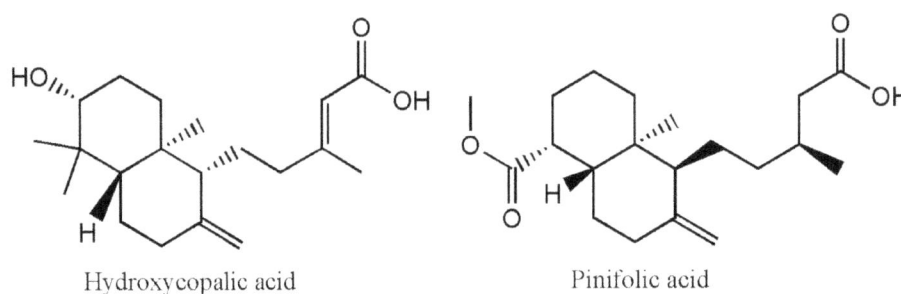

FIGURE 3.86   Molecular structures of diterpene acids isolated from copaiba oil.

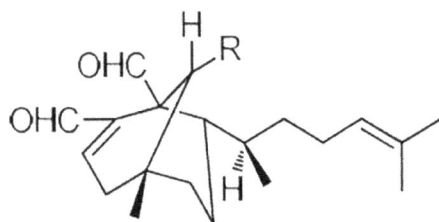

FIGURE   3.87   Chemical   structures   of   Da-1   (R=OH)   and AcDa-1((R=OAc).

D. Prostratin inhibits HIV-1 infections by down regulating HIV-1 cellular receptors through the activation of protein kinase-C (PKC) pathway and reduces the HIV-1 latency [148] (Figure 3.90).

Zou et al. reported five new clerodane diterpenoid dichrocephnoids (A–E) together with three known analogs. These compounds were isolated from the whole herb of *Dichrocephala benthamii*. Dichrocephnoid-A (**198**) possesses a very rare pentacyclic ring system with highly oxygenated functionalities. All of the isolated compounds exhibited inhibitory activity against HIV-1 integrase (Table 3.42). But, Dichrocephnoid-C (**200**) exhibited the most promising inhibition of the enzyme with IC$_{50}$ value of 12.35±1.27 mM.

Baicalein is used as Positive control and 1% DMSO is used as negative control [149] (Figure 3.91).

Spirocaesalmin is a novel rearranged vouacapane diterpenoid which exhibits significant activity against respiratory syncytial virus (RSV). This compouns possesses a new carbon skeleton with a spiro-CD ring system. RSV is a major pathogen of some respiratory infections. The selectivity against RSV virus was evaluated by the selectivity index (SI)=TC$_{50}$/IC$_{50}$. Both IC$_{50}$ and TC$_{50}$ were expressed in µg/mL. Normally, the selectivity index SI>4 is considered significant. Ribavirin was used as a positive control. Spirocaesalmin has been found to exhibit significant activity against respiratory syncytial virus (IC$_{50}$=19.5±1.5 µg/mL, TC$_{50}$=126.9±2.0 µg/mL and SI=6.5) in cell culture whereas the corresponding values for the positive control (ribavirin) are 3.6±0.2, 62.5±1.9 µg/mL and 17.4, respectively [150] (Figure 3.92).

Huang et al. reported a series of diterpenoid derivatives based on podocarpic acid and evaluated as anti-influenza-A virus agents. Influenza viral infection causes a contagious respiratory illness called as flu. Influenza type-A virus (IAV) infections have caused major pandemics in the past with serious impacts on global health condition. Two diterpenoids such as (+)-podocarpic acid (PA) and (+)-totarol, are abietane-type tricyclic phenolic diterpenoids. These derivatives exhibited nanomolar activities against an H1N1 influenza-A virus (A/

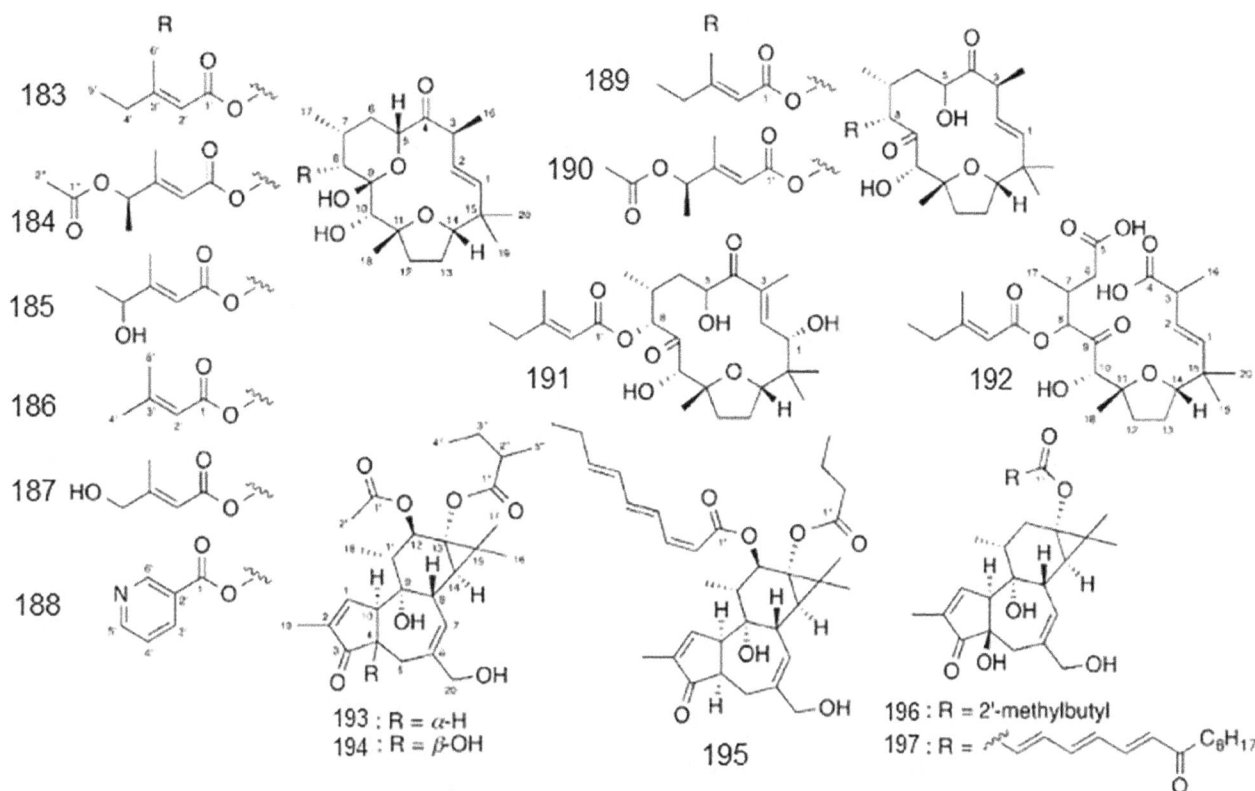

FIGURE 3.88 Structures of diterpenoids with antiviral activity against chikungunya virus.

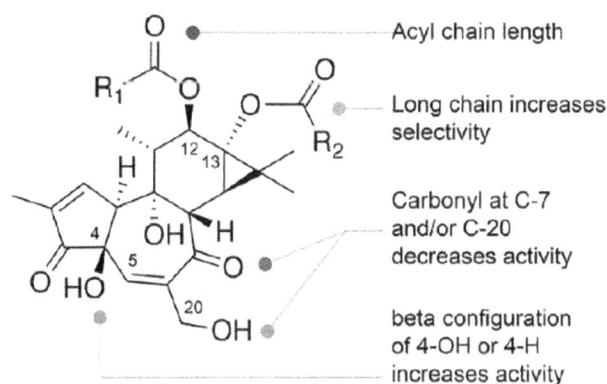

FIGURE 3.89 Structure activity relationships (SAR) of tiglianes.

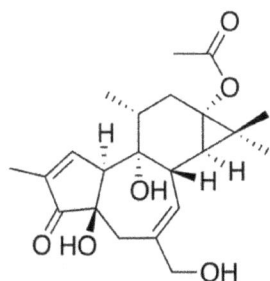

FIGURE 3.90 Structure of prostratin.

TABLE 3.42

Inhibition of Compounds (1–8) against HIV-1 Integrase ($n=3$)

Compound Code	Comp. Name	(IC_{50} in mM)
198	Dichrocephnoid-A	36.82±2.43
199	Dichrocephnoid-B	34.66±1.69
200	Dichrocephnoid-C	12.35±1.27
201	Dichrocephnoid-D	>50
202	Dichrocephnoid-E	>50
-	Baicalein	1.36±0.42

Note: Value present mean±SD of triplicate experiments.

Puerto Rico/8/34) that was resistant to two anti-influenza drugs such as oseltamivir and amantadine. These compounds inhibit the influenza virus by targeting the viral hemagglutinin-mediated membrane fusion. These results indicated that podocarpic acid derivatives may serve as potential drug candidates to fight against drug-resistant influenza-A virus infections [151] (Figure 3.93).

Reagents and conditions: (1) NaOH, MeOH, Me$_2$SO$_4$, rt, 30 min, 70°C, 15 min; (2) BnBr, K$_2$CO$_3$, Ac$_2$O, 90°C, MW, 30 min; (3) acrylic acid, DCC, DMAP, DCM, 90°C, 15 min; (4) NaH, DMF, succinic or glutaric anhydride, 0°C, 30 min; (5) NaH, DMF, BrCH$_2$CH$_2$COOH, rt, overnight; (6) RCOOH, DCC, DMAP, DCM, 110°C, MW, 45 min or anhydride, pyridine, rt, overnight; (7) NaH, Me$_2$SO$_4$, 0°C, 50 min, rt, overnight; (8) BOP, MeNH$_2$·HCl, TEA, THF, rt, overnight; (9) CrO$_3$, Ac$_2$O, AcOH, 0°C, 1 h, rt, overnight (Figure 3.94, Table 3.43).

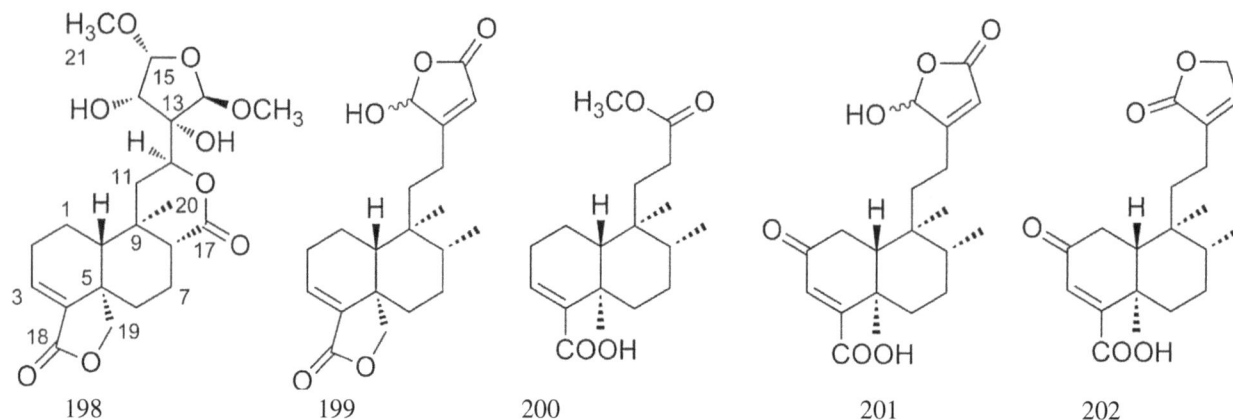

FIGURE 3.91 Structure of clerodane diterpenoid dichrocephnoids (A–E).

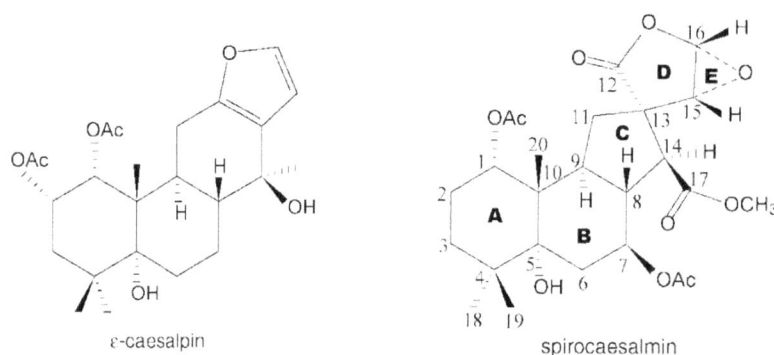

FIGURE 3.92 Chemical formulae of ε-caesalpin and spirocaesalmin.

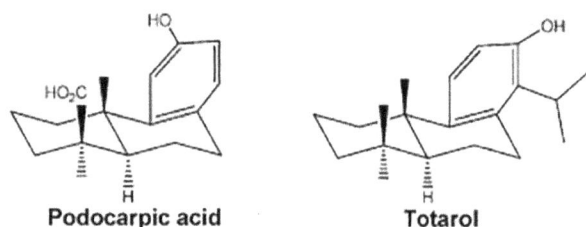

FIGURE 3.93 Structure of diterpenoids with anti-influenza activity.

Taxoquinone is an abietane type of diterpenoid, isolated from *Metasequoia glyptostroboides*. It was evaluated for its therapeutic efficacy as a potential inhibitor of influenza-A (H_1N_1) virus and 20S proteasome by using cytopathic reduction assay and proteasome inhibition assay respectively. Taxoquinone (500 μg/mL) exhibited a significant antiviral effect against the H_1N_1 influenza virus in cytopathic reduction assay on MDCK cell line. Similarly, it displayed its anticancer effect in terms its ability to inhibit 20S human proteasome with an IC_{50} value of 8.2 ± 2.4 μg/μL. Epoxomicin was used as the positive control [152] (Figure 3.95).

Anthelmintic Activity

seven-keto-sempervirol is a diterpenoid, isolated from *Lycium chinense*. This compound contains skeletons of 20 carbon atoms with four isoprene units. The scaffold consists of a single phenyl group, one hydroxyl group and one carbonyl group. As per Lipinski's rule of 5, this compound possesses only one violation to the rule (log $p > 5$). However, three out of the four core rules are satisfied. It exerted anthelmintic activity against both *Schistosoma mansoni* and *Fasciola hepatica* trematodes. By using a microtiter plate-based helminth fluorescent bioassay (HFB), this activity is specific (Therapeutic index = 4.2, when compared to HepG2 cell lines) and moderately potent ($LD_{50} = 19.1$ μM) against *S. mansoni* schistosomula cultured *in vitro* [153] (Figure 3.96).

Neuroprotective Activity

The ethanolic extract of the whole plant of *Euphorbia helioscopia* afforded 17 new jatrophane diterpenoid esters, helioscopianoids A–Q (**226–242**), along with eight known compounds (**243–250**). Their structures were elucidated by extensive spectroscopic methods and $Mo_2(OAc)_4$-induced electronic circular dichroism (ECD) analysis, and the structures of compounds **226, 227**, were confirmed by X-ray crystallography. Compounds **226–242** were evaluated for inhibitory effects on P-glycoprotein (P-gp) in an adriamycin (ADM)-resistant human breast adenocarcinoma cell line (MCF-7/ADR) and neuro-protective effects against serum deprivation-induced androtenone-induced PC12 cell damage. Compounds **233** and **241** increased the accumulation of ADM in MCF-7/ADR cells by approximately 3-fold at a concentration of 20 μmol/L. Compound **8** could attenuate rotenone-induced PC12 cell

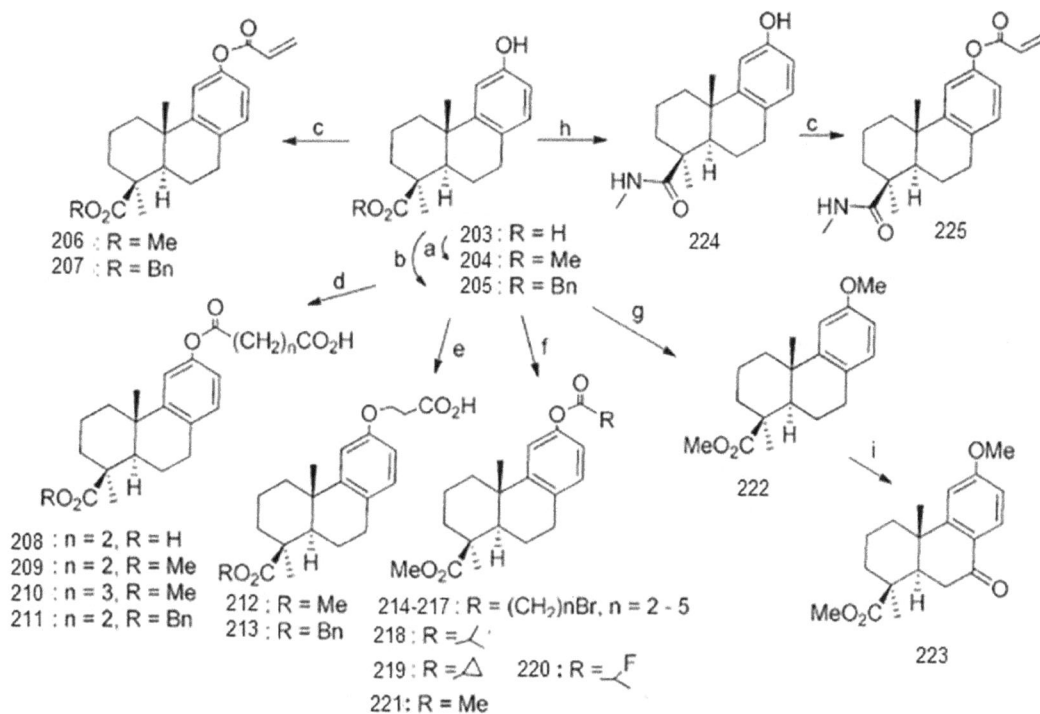

FIGURE 3.94 Scheme for synthesis of Podocarpic acid derivatives.

damage, and compounds **227**, **233**, and **237** exhibited neuroprotective activities against serum deprivation-induced PC12 cell damage. Nerve Growth Fator (NGF, 0.05 mg/mL) was used as a positive control [154] (Figure 3.97, Table 3.44).

Immunomodulatory Effects

Fuchs et al. reported the immunomodulatory effects of diterpene quinine derivatives, isolated from the roots of *Horminum pyrenaicum* in human PBMC (peripheral blood mononuclear cell). Four abietane diterpene quinone derivatives (horminone, 7-O-acetylhorminone, inuroyleanol and its 15,16-dehydro-derivative), two nor-abietane diterpene quinones (agastaquinone and 3-deoxyagastaquinone) and two abeo 18(4→3)-abietane diterpene quinones (agastol and its 15,16-dehydro-derivative) could be identified. For screening the immunomodulatory effect of these compounds, the interferon gamma-(IFN-γ)-dependent immunometabolic pathways of tryptophan breakdown via indoleamine 2,3-dioxygenase-1 (IDO-1) and neopterin formation by GTP-cyclohydrolase-1 (GTP-CH-I) represent prominent targets. Because IFN-γ-related signaling is strongly sensitive to oxidative triggers. These diterpenoids were able to suppress the above-mentioned pathways with different potency in a dose-dependent manner [154] (Figure 3.98, Tables 3.45 and 3.46).

Treatment of Asthma

Fengsen et al.carried out investigation of the anti-asthmatic activity of Oridonin on a mouse model of asthma. Asthma is a chronic airway inflammatory disease which is associated with various cellular components. It is characterized by inflammation, eosinophilia and mucus hypersecretion in the airways and lungs. For this study, BALB/c mice were sensitized by using ovalbumin (OVA) and then the sensitized mice were treated with Oridonin prior to OVA challenge. This *in-vivo* study indicated that Oridonin decreased the OVA-induced airway hyper-responsiveness significantly ($p < 0.05$) [154].

Rossi et al. reported that Salvinorin-A inhibits airway hyperreactivity induced by ovalbumin sensitization. It has been demonstrated that Salvinorin-A inhibits the acute inflammatory response by affecting the production of leukotrienes (LTs). LTs play a major role in the pathogenesis of asthma. The administration of Salvinorin-A (10 mg/kg) before each allergen exposure significantly reduces the OVA-induced LT increase in the air pouch. So, Salvinorin-A represents as a promising drug candidate for the treatment of allergic condition like asthma [155].

Dong et al. reported the suppressive effect of Carnosol on Ovalbumin-induced allergic asthma. Carnosol is a diterpenoid, found in rosemary extracts (*Rosmarinus officialis*). The effect of carnosol on a murine allergic asthma model was investigated. It inhibited the degranulation of RBL-2H3 mast cells. The treatment of Carnosol inhibited the increase in the number of eosinophils in the bronchoalveolar lavage fluids (BALF) of mice treated with ovalbumin [156].

Miscellaneous

There are different types of crenulidane diterpenes (crenulacetal-A, B and C), isolated from the marine algae of the genus *Dictyota*. Crenulacetal-A and B were isolated from *D. spinulosa*. Whereas, crenulacetal-C is isolated from the brown alga *Dictyota dichotoma*. Among these three diterpenes, crenulacetal-C exhibits significant pesticidal activity against *Polydora websterii* (harmful lugworm) at a concentration of 1.5 ppm [157] (Figure 3.99).

TABLE 3.43

Anti-IAV Activities of Diterpenoids

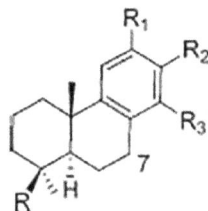

Compd	R	R₁ (R₂, R₃)	EC₅₀[b,c]	CC₅₀[d]
206	CO₂H	OH	18.2	91.2
207	CO₂Me	OH	0.16±0.17	>40
208	CO₂Bn	OH	1.59±2.17	>40
209	CO₂Me	(structure)	0.64±0.44	>40
210	CO₂Bn	(structure)	0.48±0.55	>40
211	CO₂H	(structure)	18.7	>40
212	CO₂Me	(structure)	0.18±0.21	>40
213	CO₂Me	(structure)	0.37±0.25	>40
214	CO₂Bn	(structure)	>6.5	>40
215	CO₂Me	(structure)	>25	>40
216	CO₂Bn	(structure)	NA[f]	>40
217	CO₂Me	(structure)	0.21±0.17	>40
218	CO₂Me	(structure)	0.57±0.27	>40
219	CO₂Me	(structure)	0.42±0.13	>40
220	CO₂Me	(structure)	0.19±0.15	>40
221	CO₂Me	(structure)	0.32±0.35	>40
222	CO₂Me	(structure)	0.15±0.13	>40
223	MeNHCO	OH	>34.8	>40
224	MeNHCO	(structure)	>29.3	>40
225	Me	H (R₂ = OH) (R₃ = CMe₂)	3.5	21.0
Oseltamivir			>20	
Amantadine			>20	

[a] An influenza virus strain A/Puerto Rico/8/1934, PR8, was used in the assays.

[b] Concentration (μM) required to protect cells from the cytocidal effect of PR8 by 50%.

[c] The EC50 (μM) is presented as mean±SD from at least three independent tests.

[d] Concentration (μM) that reduced the viability of MDCK cells by 50%.

[e] Known compounds.

[f] NA, not active.

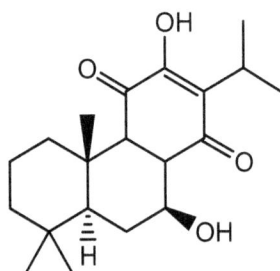

FIGURE 3.95 Chemical structure of taxoquinone.

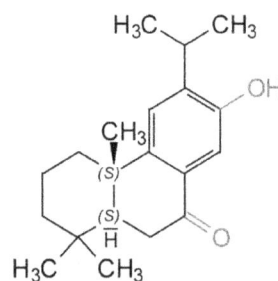

FIGURE 3.96 Structire of 7-keto-sempervirol.

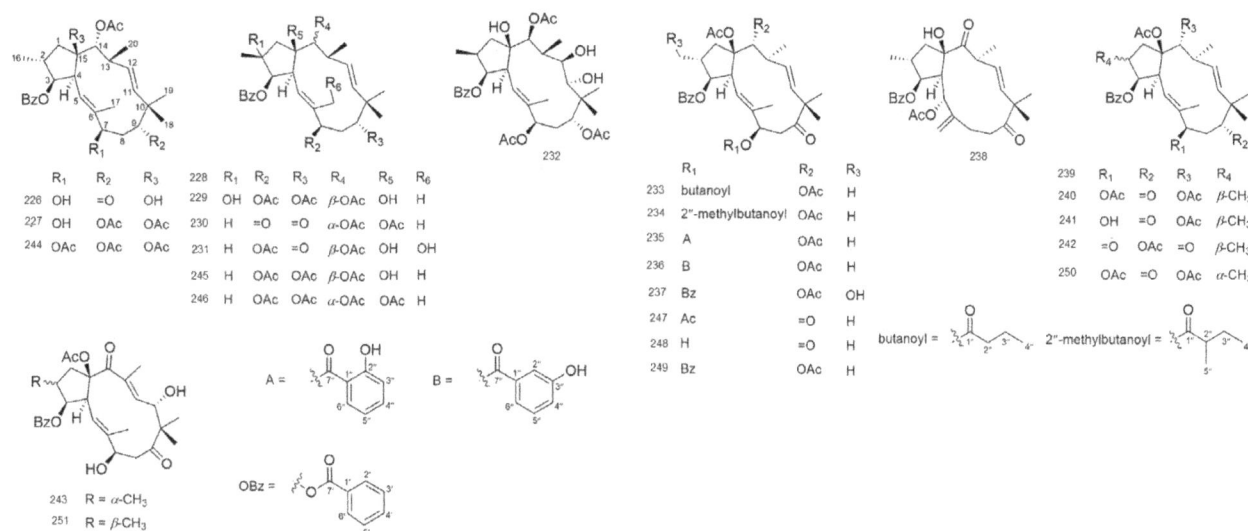

FIGURE 3.97 Structures of diterpeniods extracted from *Euphorbia helioscopia.*

TABLE 3.44

Neuroprotective Effects of Compounds **2**, **8**, and **12** on the Survival Rate of PC12 Cells Injured by Serum Deprivation (10 μmol/L mean±SD, $n=9$)

Group	Survival Rate (% of Control)
Control	100.0±4.4
Model	56.5±7.0
NGF	76.2±5.2[b]
227	75.4±4.6[b]
246	74.2±5.8[b]
238	72.3±3.9[b]

[a] $p < 0.001$ vs. control.
[b] $p < 0.001$ vs. model.
[c] $p < 0.01$ vs. model.
[d] $p < 0.1$ vs. model.

Similarly, piscicidal diterpene such as acetoxycrenulide was isolated from *D. dichotoma* and Mediterranean *Dictyota* species. This compound exhibited strong fish antifeedant activity due to its piscicidal activity. In addition to this, acetoxycrenulide displayed anti-microfouling activity against three marine bacterial strains D41, 4M6, and TC5 with EC_{50} values of 82±28, 69±17, and 154±20 μM, respectively [158] (Figure 3.100).

Zhang et al. reported the antimicrobial and anti-parasitic activities of abietane diterpenoids, isolated from *Cupressus sempervirens*. The isolated four abietane diterpenoids include 6-deoxytaxodione (1), taxodione (2), ferruginol (3), and sugiol (4). The *in-vitro* anti-leishmanial assay was performed on a culture of *Leishmania donovani* promastigotes by using Alamar Blue assay method to measure quantitatively the proliferation of various human and animal cell lines, bacteria and fungi (Figure 3.101). Compounds 1 and 2 exhibited potent anti-leishmanial activity with IC_{50} values of 0.077 and 0.025 μg/mL, respectively, against *L. donovani* promastigotes. Compound 1 was found to be >21- and

>1.4-fold potent than standard anti-leishmanial drugs such as pentamidine (IC_{50} 1.62 μg/mL) and amphotericin-B (IC_{50} 0.11 μg/mL) respectively. Whereas, compound 2 was found to be >64- and >4.4-fold more potent than those observed for the standard anti-leishmanial drugs. In addition to this, both compound 1 and 2 displayed potent antibacterial activities with IC_{50} value of 0.80 and 0.85 μg/mL respectively against methicillin-resistant *Staphylococcus aureus*. Compounds 1–3 were also evaluated against *L. donovani* amastigotes in THP1 macrophage cultures. Among the tested compounds, only compound 3 showed cytotoxicity toward transformed THP1 cells (Table 3.47), which was found to be equipotent to pentamidine. However, compound 1 demonstrated the best activity without any cytotoxicity against THP1 cells. The antimalarial activities of compounds 1 and 2 against *P. falciparum* D6 and W2 strains were attributed by low IC_{50} values of 1.9–3.0 μg/mL, as compared with standard antimalarial drugs such as chloroquine and artemisinin (Table 3.48). Among all the compounds, the antimicrobial activity was strongly displayed by compound 2 against *C. albicans, C. glabrata, C. krusei, A. fumigatus, C. neoformans, S. aureus*, MRSA, and Mycobacterium intracellulare. Compound 2 exhibited potent inhibition against C. glabrata and C. neoformans with IC_{50} values of 0.8 μg/mL compared to the standard drug amphotericin-B with IC_{50} values of 0.24 and 0.36 μg/mL respectively. 6-Deoxytaxodione was found to be weakly active with IC_{50} value of 0.85 μg/mL as compared to compound taxodione against most of these organisms (Table 3.49) [159].

Tanshinones constituents are naturally occurring diterpene and are isolated from the roots of *Salvia miltiorrhiza*. Various Tanshinones include dihydrotanshinone, tanshinone-I, or tanshinone-IIA. These compounds are found to inhibit the production of prostaglandin and leukotriene. Dihydrotanshinone-I has been reported to have cytotoxic activity toward various tumor cells. Tanshinone-I is anti-inflammatory. It modulates or prevents breast cancer metastasis by regulating the adhesion molecules. Similarly, Tanshinone-IIA exhibits anti-inflammatory

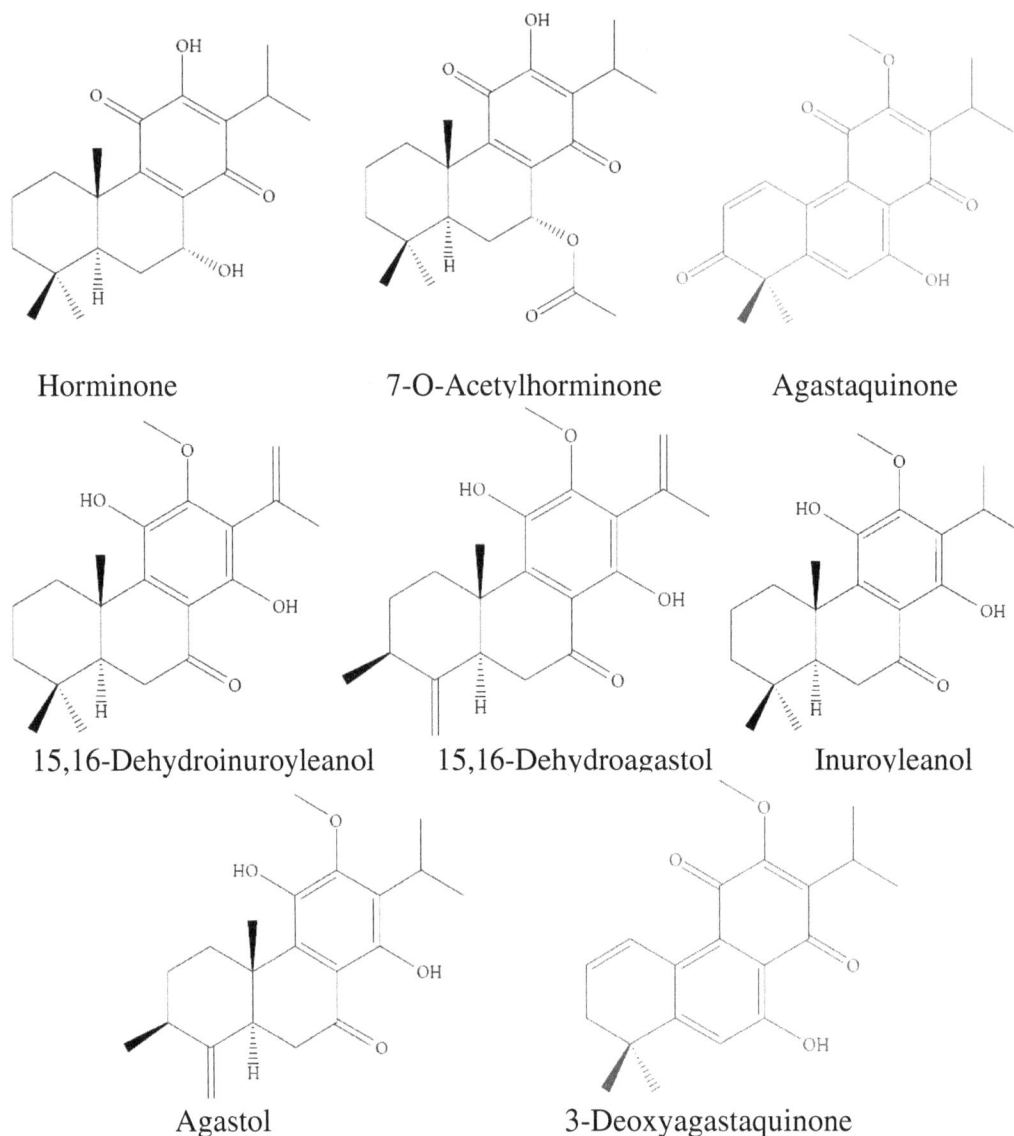

FIGURE 3.98 Structures of diterpenes isolated from the roots of *Horminum pyrenaicum.*

TABLE 3.45

IC_{50} of *Horminum pyrenaicum* Extract Sub-Fractions (HOR-a3 to HORa6) that Had the Largest Effect on Neopterin Formation and IDO-1 Activity, as Indicated by the Kynurenine to Tryptophan Ratio in Phytohemagglutinin Stimulated PBMC

	IC_{50} (µg/mL)	
Extract Fractions	**Neopterin Formation**	**IDO-1 Activity**
HOR-a3	63.8	7.02
HOR-a4	121	18.4
HOR-a6	159	34.0

HOR, horminum root extract fractions.

TABLE 3.46

IC_{50} of the Isolated Compounds on IDO-1 Activity, as Indicated by the Kynurenine to Tryptophan Ratio in Phytohemagglutinin Stimulated PBMC

Treatments	**IC_{50} (µM)**
Horminone	346
7-O-Acetylhorminone	71.7
Agastaqiunone	46.3
Inuroyleanol	n.d.
Agastol	165
3-Deoxyagastaquinone	79.5
1-Methyl-D-tryptophan (positive control)	9.3

Note: n.d.: concentration-dependent activity could not be determined in the tested concentration range.

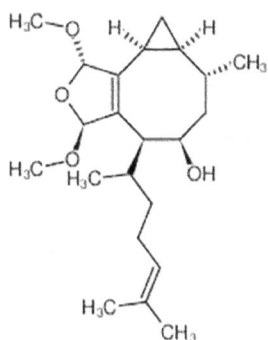

FIGURE 3.99 Structure of crenulacetal B.

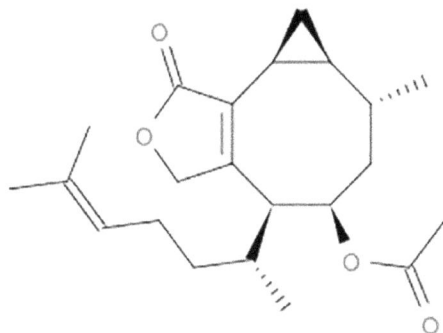

FIGURE 3.100 Structure of piscicidal diterpene.

activity by inhibiting the NF-κB signaling via inhibition of IκBα phosphorylation [160–162] (Figure 3.102).

Halimane is a diterpenoid, found in marine organisms such as Porifera, sponges of the genus *Agelas* and *Raspailia*. It is used as tuberculosis biomarker. It also exhibits antibacterial, antifungal, antimalarial, cytotoxic activities, inhibition of adenosine transfer rabbit erythrocytes, Ca^{2+} channel antagonistic action and α_1 adrenergic blockade etc. This compound produces antitumor activity against several cell lines such as pancreatic adenocarcinoma, human colon carcinoma, bladder, lung and cervix cancer etc [163] (Figure 3.103).

The clerodane diterpenes such as (5R, 8R, 9S, 10R)-12-oxo-ent-3,13(16)-clerodien-15-oic acid (**251**), 16-hydroxy-clerod-3,13(14)-diene-15,16-olide (**252**) demonstrated potent concentration-dependant suppressive effect on the viability of axenically cultured amastigotes of *L. aethiopica*. These compounds were most active with LD_{50} values of 1.08 and 4.12μg/mL respectively. Compounds 1 and 2 found to be 20 and 10 times more selective respectively to leishmania amastigotes than the permissive host cell line, THP-1 cells or the promastigotes stage of the parasites. The standard anti-leishmanial drug, amphotericin-B (AMB), displayed some toxicity to THP-1 cells at its higher concentrations [164] (Figure 3.104, Table 3.50).

Ghosh et al. reported the anti-leishmanial activity of Labdane diterpenes isolated from *Alpinia nigra* seeds. These diterpenes were tested against *Leishmania donovoni* promastigotes *in-vitro*. These compounds were found to be more active against

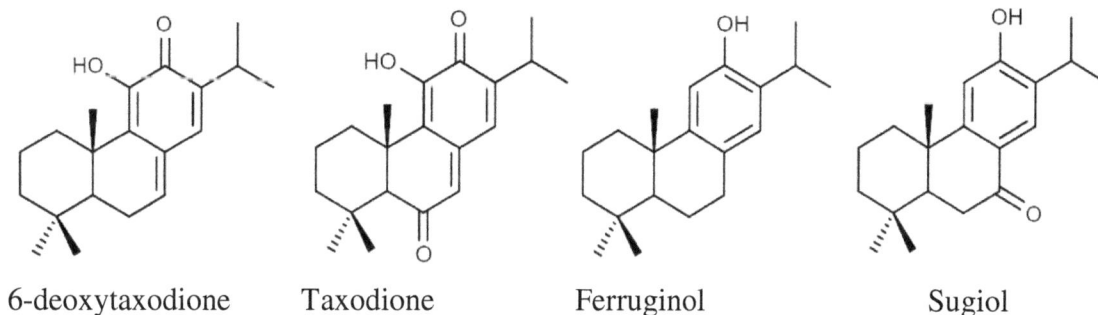

6-deoxytaxodione   Taxodione   Ferruginol   Sugiol

FIGURE 3.101 Structure of abietane diterpenoids isolated from *Cupressus sempervirens.*

TABLE 3.47

Anti-Leishmanial Activity (IC_{50}/IC_{90} in μg/mL) of Extract, and Abietane Diterpenes Isolated from *C. sempervirens*

| Extract/Compound | Leishmania donovani | | | | Cytotoxicity (Transformed THP1 Cells) | |
| | Promastigotes | | Macrophage Amastigotes | | | |
	IC50	IC90	IC50	IC90	IC50	IC90
C. sempervirens EtOH extract	0.65	1.13	NT	NT	NT	NT
6-deoxytaxodione	0.077	0.196	2.72	3.14	>10	>10
Taxodione	0.025	0.056	5.40	>10	>10	>10
Ferruginol	NT	NT	2.36	3.15	5.79	7.39
Sugiol	NT	NT	>10	>10	>10	>10
Pentamidine	1.62	2.10	2.39–2.74	4.18–5.41	>10	>10
Amphotericin-B	0.11	0.19	0.15–1.09	0.31–1.96	>10	>10

IC_{50}, half-maximal inhibitory concentration; IC_{90}, inhibitory concentration of 90%; NT, not tested.

TABLE 3.48

Antimalarial Activities of *C. sempervirens* extract

| Extract/ Compound | *P. falciparum* (µg/mL) | | | | Vero (µg/mL) |
| | D6[a] | | W2[b] | | |
	IC_{50}	SI_c	IC_{50}	SI_c	TC_{50}
C. sempervirens EtOH extract	2.9	10.3	3.2	9.4	30
6-deoxytaxodione	3.0	>1.8	2.8	>1.7	NC
Taxodione	2.6	>1.8	1.9	>2.5	NC
Chloroquine	0.02	-	0.15	-	NC
Artemisinin	0.005	-	0.002	-	NC

NC, not cytotoxic (up to the maximum dose tested; 4.7 µg/mL for pure compounds and 47.6 µg/mL for crude extract and column fractions); SI, selectivity index; TC_{50}, median toxic concentration.

[a] Chloroquine-sensitive clone.

[b] Chloroquine-resistant clone.

[c] Selectivity index = IC_{50} VERO cells/IC_{50} *P. falciparum*.

TABLE 3.49

Antimicrobial Activity of *C. sempervirens* Extract

| Organisms | IC_{50} (µg/mL) of extract/compounds | | | | |
	C. sempervirens EtOH Extract	6-Deoxytaxodione	Taxodione	Amphotericin-B	Ciprofloxacin
C. albicans	-	-	4.47	0.17	NT
C. glabrata	-	2.21	<0.8	0.24	NT
C. krusei	-	7.66	3.10	0.68	NT
A. fumigatus	-	16.18	3.94	0.73	NT
C. neoformans	3.71	1.85	<0.8	0.36	NT
S. aureus	12.48	1.34	0.80	NT	0.11
MRSA	11.45	0.85	0.80	NT	0.10
M. intracellulare	-	-	20.00	NT	0.30

Note: not active; MRSA, methicillin-resistant Staphylococcus aureus; NT, not tested.

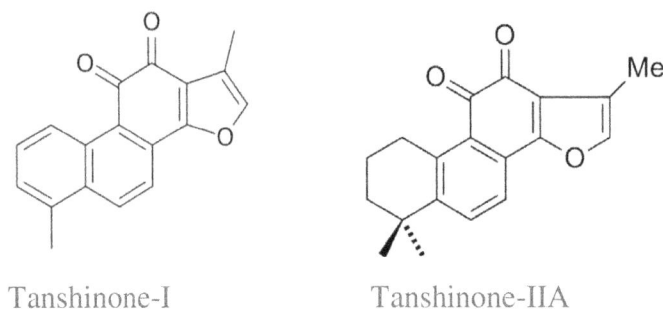

FIGURE 3.102 Structure of diterpenes isolated from roots of *Salvia miltiorrhiza*.

the leishmania promastigotes. MTT assay displayed the significant leishmanicidal efficacy of compound-A (IC_{50} = 63.0 nM) and compound-B (IC_{50} = 90.5 nM) against the promastigotes. Human red blood cells hemolysis experiment suggests both the compounds are fairly safe [165] (Figure 3.105).

Dictyol-E was isolated from *D. dichotoma* and displayed weak antibacterial activity against the marine bacterial strains *Pseudoalteromonas* sp., *Paracoccus* sp. and *Polaribacter* sp. with EC_{50} values of 100, 133, and 92 µM, respectively. It also exhibited a significant inhibitory effect on rat liver microsomal diacylglycerol acyltransferase with an IC_{50} value of 46.0 µM.

Dictyol-H was isolated from *D. divaricata* and displayed moderate antitumor activity against the KB9 cell line with an IC_{50} value of 22 µg/mL [166] (Figure 3.106).

Hoffmann et al. reported a library of 14 minimally cytotoxic diterpenoid-like compounds with $CC_{50} > 70$ mM on HepG2 human liver cells. These compounds were screened against *Mycobacterium smegmatis*, *Staphylococcus aureus*, and *Escherichia coli* to determine antimicrobial activity. Some compounds with a phenethyl alcohol core substituted with a *β*-cyclocitral derivative demonstrated antimycobacterial activity, with the most active being compound **1** (MIC = 23.4

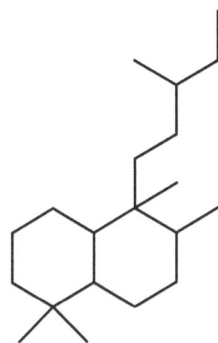

FIGURE 3.103 Structure of Halimane.

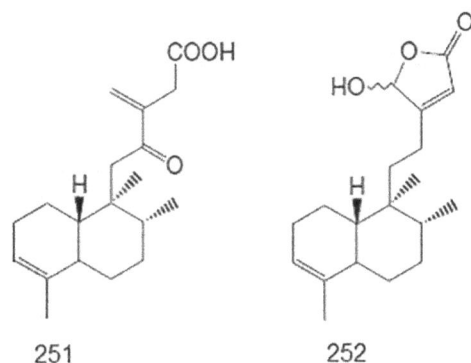

FIGURE 3.104 Structure of the clerodane diterpenes (251 and 252).

mg/L, $IC_{50=}0.6$ mg/L). Lower activity was exhibited against *S. aureus*, while no activity was displayed against *E. coli*. Low cytotoxicity was re-confirmed on HepG2 cells and additionally on RAW 264.7 murine macrophages (SI for both cell lines > 38). The sub-lethal (IC_{50} at 6 h) effect of compound **253** on *M. smegmatis* was examined through untargeted metabolomics and compared to untreated bacteria and bacteria treated with sub-lethal (IC_{50} at 6 h) concentrations of the anti-tuberculosis drugs ethambutol, isoniazid, kanamycin, and streptomycin. The study revealed that compound 1 acts differently from the reference antibiotics and that it significantly affects amino acid, nitrogen, nucleotides and folate-dependent one-carbon metabolism of *M. smegmatis*, giving some insights into the mode of action of this molecule [167] (Tables 3.51 and 3.52).

Harziane diterpenoids, typically with a 4/7/5/6-fused tetracyclic carbon skeleton, are isolated from the fungal genus *Trichoderma*. Harziandione possess 4/7/5/6 tetracyclic scaffold. It was the first harziane diterpenoid isolated from the liquid culture of *T. barzianum* in 1992. Ten harziane diterpenoids were reported from different *Trichoderma* species. Harziane diterpenoids exhibit antifungal activity against

Sclerotium rolfsii, antibacterial activity against *Escherichia coli* and *Staphylococcus aureus*, cytotoxic activity against HeLa and MCF-7 cell lines and inhibitory effect of the growth against some marine phytoplankton species. Harzianol-I exhibited significant antibacterial effect against the growth of *Staphylococcus aureus* (EC_{50} = 7.7 ± 0.8 μg/mL), *Bacillus subtilis* (EC_{50} = 7.7 ± 1.0 μg/mL), and *Micrococcus luteus* (EC_{50} = 9.9 ± 1.5 μg/mL) [168,169] (Tables 3.53 and 3.54).

A new diterpenoid furnolactone with molecular formula $C_{20}H_{22}O_6$ has been isolated from the stems of *Tinospora cordifolia*. Its spectral characteristics are very similar to the clerodane diterpenoids. Columbin is a furanolactone diterpene isolated from the roots of *Jateorhiza* and *Tinospora* species. Columbin is used for medicila purpose due to its structural similarity with the known kappa opioid receptor (KOR) agonist salvinorin-A (trans-neoclerodane diterpene). It is the main component of *Salvia divinorum* and is a highly selective κ-opioid receptor (KOR) agonist. KOR is considered as major target for several diseases such as anxiety, depression, and drug addiction. It was demonstrated that Columbin possesses cytotoxic and anti-inflammatory activities [170,171] (Figure 3.107).

Cassanes containing a furan ring is called furanocassanes or vouacapanes. These compounds represent the important group of tetracycliccassanes and their structure is characterized by a skeleton constructed from the fusion of three cyclohexane rings and one furan ring. Vouacapane diterpenes are commonly isolated from the fruits of different *Pterodon* species by the Soxhlet method. For this extraction process, various solvents such as petroleum ether, hexane and ethanol are used [172].

Duarte et al. (1996) assessed the possible involvement of biogenic amines in the antinociceptive effect of compound 6α, 7β-Dihydroxyvouacapan-17β-oic acid (274) by using acetic acid-induced abdominal contortion on mice. This compound exhibited an antinociceptive effect (500 μmol/kg, i.p.) when compared with the control group. These results suggested that vouacapane's antinociceptive response may be associated with dopamine [173] (Figure 3.108).

Galceran et al. assessed the anti-inflammatory and analgesic potential of 6α, 7β-dihydroxyvouacapan-17β-oic acid (**275**). The oral administration of 50 mg/kg (equivalent to 143.5 μmol/kg) of compound **275** inhibited the inflammatory mechanisms triggered by carrageenan and prostaglandin E2. But it could not significantly inhibit the edema produced by dextran. The acetic acid-induced contortion test showed dose-dependent inhibition in mice treated orally with this compound (50, 200 or 400 mg/kg p.o.). In the formalin test, compound 21 showed antinociceptive activity on neurogenic and inflammatory pain models in mice given oral treatment (50 and 100 mg/kgp.o.). However, in the 100 mg/kg (or 287 μmol/kg) dose (p.o.), this

TABLE 3.50

Anti-leishmanial Activity of Diterpenes and AMB with LD_{50} Values in μg/mL

Drugs	*L. aethiopica* Promastigotes	Amastigotes	THP-1
AMB	0.189 ± 0.01 (*n*=3)	0.183 ± 0.04 (*n*=6)	>10
251	>100	4.12 ± 0.28 (*n*=5)	>100
252	11.67 ± 0.88 (*n*=3)	1.08 ± 0.25 (*n*=5)	12.13 ± 1.42 (*n*=4)

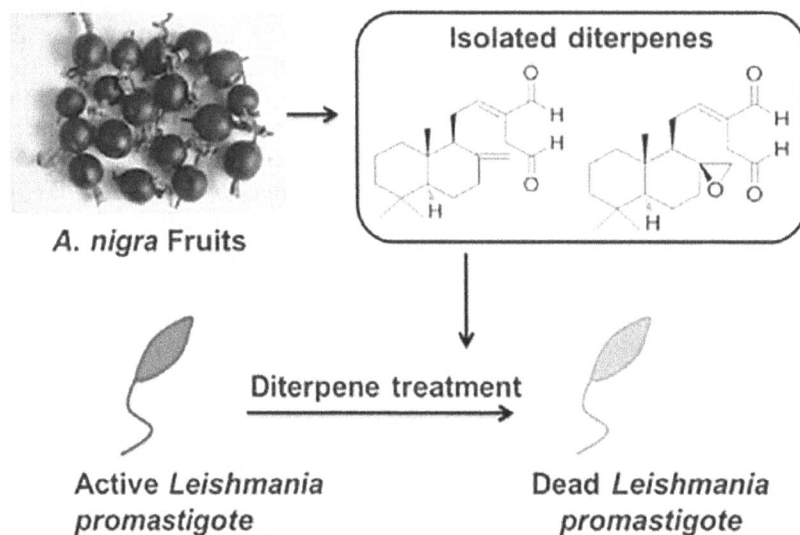

FIGURE 3.105 Labdane diterpenes isolated from *Alpinia nigra* seeds.

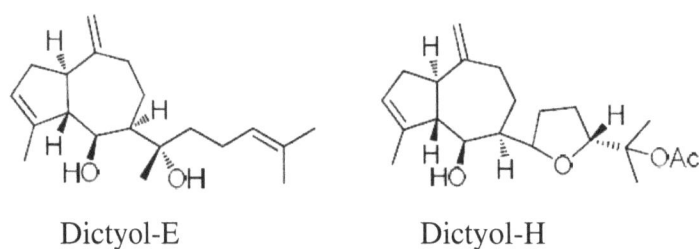

Dictyol-E   Dictyol-H

FIGURE 3.106 Structure of Dictyol-E and Dictyol-H.

TABLE 3.51

Physicochemical Properties and Antibacterial Activity of Compound Library

Compd.	Structures	RB	HBA	HBD	cLogP	MW	LRV	*S. aureus*	*E. coli*	*M. smegmatis*
253		4	2	1	3.90	274.40	0	62.5	>250	23.4
254		4	2	0	4.01	272.38	0	>250	>250	62.5
255		4	2	1	3.88	274.40	0	250	>250	31.25
256		4	2	0	4.01	272.38	0	250	>250	125
257		4	1	0	4.98	284.44	0	>250	>250	250

TABLE 3.51 (*Continued*)

Physicochemical Properties and Antibacterial Activity of Compound Library

Compd.	Structures	RB	HBA	HBD	cLogP	MW	LRV	*S. aureus*	*E. coli*	*M. smegmatis*
258		4	1	1	5.12	300.48	1	250	>250	250
259		4	1	0	5.24	298.46	1	62.5	>250	250
260		6	4	1	3.76	334.45	0	>250	>250	>250
261		6	4	0	3.95	332.43	0	>250	>250	>250
262		0	2	0	3.42	256.34	0	>250	>250	>250
263		1	3	0	3.42	286.37	0	>250	>250	>250
264		1	2	0	4.06	272.38	0	>250	>250	125
265		1	1	0	4.69	258.40	0	250	>250	62.5
266		1	1	0	4.70	258.40	0	>250	>250	>250
PE		2	1	1	1.64	122.16	0	NA	NA	250

Chemical structures and physicochemical properties of compound library calculated by Swiss ADME. RB, number of rotatable bonds; HBA, number of H-bond acceptors; HBD, number of H-bond donors; cLogP, consensus Log P; MW, molecular weight; and LRV, Lipinski's rule violations. For predicted good oral bioavailability, values should be <10, <10, <5, ≤5, <500, and ≤1 respectively. Minimum inhibitory concentration (MIC; mg/L) of compound library against *Staphylococcus aureus*, *Escherichia coli* and *Mycobacterium smegmatis*. Values of >250 are considered to indicate no significant antimicrobial activity.

TABLE 3.52

Antimycobacterial and Cytotoxic Activity of Compound **253**

Compound	MIC[a]	Viability of HepG2 Cells at MIC[b] (%)	Viability of RAW Cells at MIC[b] (%)	IC$_{50}$[c]	CC$_{50}$ on HepG2 Cells[a]	CC$_{50}$ on RAW Cells[a]	Selectivity Index[d]
253	23.4 [85.3]	104±0.14	88±0.29	0.6(0.2−1.3) [2.2(0.8−4.8)]	>27.4 [>100]	>27.4 [>100]	>38.5

[a] $p<0.001$
[b] $p<0.01$
[c] $p<0.05$
[d] $p<0.001$

TABLE 3.53

Cytotoxicity of Title Compounds

Compds	Structures	IC$_{50}$ (Mean±SD, µM)		
		NCI–H1975	HepG2	MCF-7
267	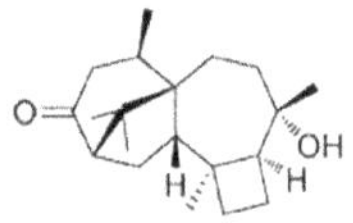	>80	>80	>80
268	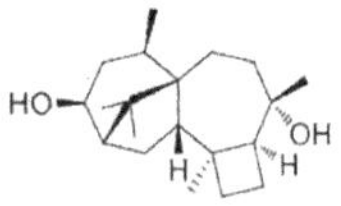	>80	>80	>80
269	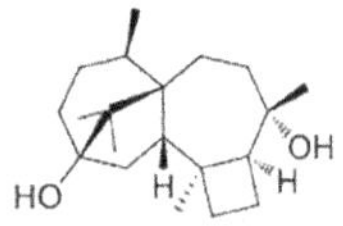	>80	>80	>80
270	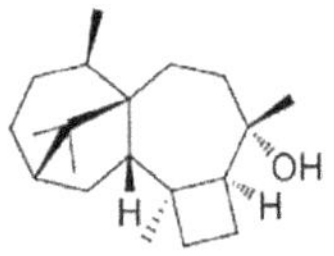	58.72±0.51	60.88±1.70	53.92±0.68
271	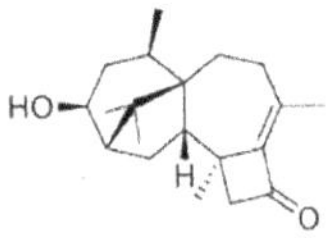	>80	>80	>80
272	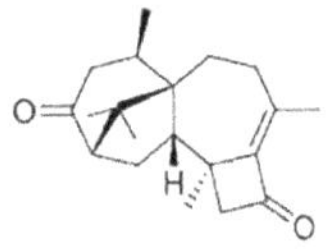	>80	>80	>80
273	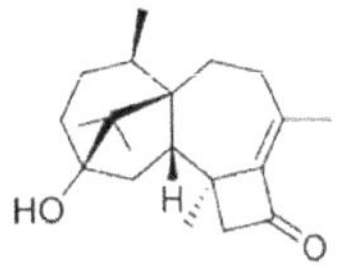	>80	>80	>80
Taxol	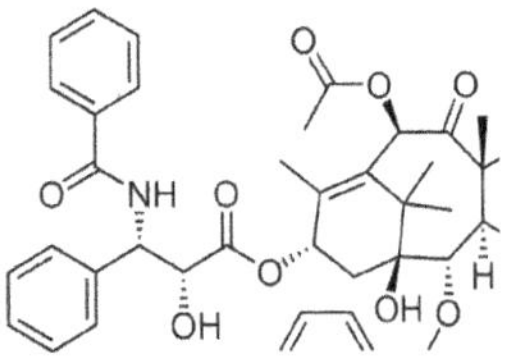	(4.79±0.18)×10^{-3}	(25.60±1.70)×10^{-3}	(41.15±6.73)×10^{-3}

compound was not able to increase the latency time during the hot-plate test. These results suggested that compound 21 has peripheral anti-inflammatory and analgesic effects [174] (Figure 3.109).

Omena et al. assessed the larvicidal activity of three vouacapane diterpenes against stage 4 *Aedes aegypti* larvae. Compounds **275**, **276**, and 6α-hydroxyvouacapan-7β, 17β-lactone (**277**) presented LC$_{50}$ of 14.69 µg/mL (42.2 nmol/mL), 21.76 µg/mL (60 nmol/mL), and 50.08 µg/mL (151.6 nmol/mL), respectively. The substances with LC$_{50}$ values lower than 100 µg/mL are considered as active compounds against *Aedes aegypti* [175] (Figure 3.110).

TABLE 3.54

Antibacterial Activity of Title Compounds

	EC$_{50}$ (Mean±SD, µg/mL)		
Compds	*S. aureus*	*B. subtilis*	*M. luteus*
267	>200	>200	>200
269	>200	>200	>200
270	7.7±0.8	7.7±1.0	9.9±1.5
273	>200	>200	>200
Ampicillin	<0.5	<0.5	<0.5

Pimenta et al. carried out the larvicidal activity of 6α-acetoxyvouacapane (**278**) against stage 3 *Aedes aegypti* larvae, in concentrations ranging from 12.5 to 500 µg/mL. Compound 30 showed median lethal concentration (LC$_{50}$) of 186.21 µg/mL (540.5 nmol/mL) [176] (Figure 3.111).

Vieira et al. demonstrated the anti-proliferative activity of Vouacapane-6α, 7β, 14β, 19-tetraol via MTT (3-(4,5-dimethylthiazol-2-yl)2,5-diphenyl tetrazoliumbromide) colorimetric assay by using SK-MEL-37 human melanoma cells. This compound displayed IC$_{50}$ of 32 µmol/L in concentrations ranging from 1.4 to 92 µmol/L. Doxorubicin was the positive control that exhibited IC$_{50}$ value of 35 µmol/L [177] (Figure 3.112).

Reis et al. assessed the blood vessel relaxation and the possible action mechanisms of methyl-6α-acetoxy-7β-hydroxyvouacapan-17β-oate in isolated mouse aorta preparations. The results suggested that this compound induces endothelium-independent vascular relaxation by blocking the L-type Ca^{2+}channel (Cav1.2). So, it suggested that title compound possesses cardiovascular effect [178] (Figure 3.113).

Oliveira et al. performed the leishmanicidal activity of compound methyl-6α-acetoxy-7β-hydroxyvouacapan-17β-oate against promastigotes of *Leishmania amazonensis* and *L.braziliensis*, in concentrations ranging from 8.0 to 128.0 µg/mL. This compound showed parasite growth inhibitory concentration 50% (IC$_{50}$) less than 30µg/ml (equivalent to 74.16 nmol/mL). Whereas, the positive control, Amphotericin-B exhibited an IC$_{50}$ value of 5.41 nmol/L [179].

Grayanane diterpenoids possess a unique 5/7/6/5 tetracyclic system and were isolated from the flowers, fruits or roots of *P. formosa* of family Ericaceae. According to the junction between ring A and ring B, the grayananes can be divided into two types such as A/B-trans grayananes and A/B-cis grayananes. But, a large number of A/B-trans-grayananes

were identified in nature. These trans-grayananes are further subdivided into 10-oxygen-grayananes, 10-hydro-grayananes and 10-dehydro-grayananes based on the oxygenation of the C-10 position. These compounds exhibit diverse biological activities such as agalgesic, antinociceptive, anticancer, antiviral, antifeedant, insecticidal, etc [180] (Figure 3.114, Table 3.55).

Rhodomollins-A and B were screened *in-vitro* in A/95e359 influenza virus-infected MDCK cells, and rhodomollin B demonstrated the modest activity with an IC$_{50}$ value of 19.24mM. In an acetic acid-induced writhing test, mollolide-A exhibited a significant effect with a 73% reduction at a dose of 20 mg/kg. Pierisformosoid-D exhibited 38.3% inhibitory effect against KCNQ2 potassium channel at a concentration of 10mM. In the year 2013, the cAMP regulation activities of pierisformotoxins A-D have been evaluated, and the results showed that peierisformotoxins-B and C considerably decreased the cAMP level in N1E-115 neuroblastoma cells at the concentrations of 50 and 10mM, respectively. In 2014, principinols A-F were tested *in-vitro* for the inhibition on PTP1B, and principinols-D and E exhibited significant inhibitory activity against PTP1B with IC$_{50}$ value of 24.46±6.14 and 3.14±0.12mM respectively [181,182].

3.4 Conclusion

Diterpenes are hydrocarbons with diverse molecular structures. These compounds are derived from four isoprene units linked in a head-tail fashion. Diterpenes are categorized into acyclic, monocyclic, bicyclic, tricyclic and tetracyclic. The diterpenes contain skeletons with 20 carbon atoms in their structures. Naturally, these compounds are present in polyoxygenated form with hydroxyl (–OH) and keto (C=O) groups. These are produced from plants, animals, cyanobacteria, and fungi via the reductase pathway, mevalonate, or deoxyxylulose phosphate pathways in which geranylgeranyl pyrophosphate act as a primary intermediate. Diterpenes possess several pharmacological activities such as anti-inflammatory, antidiabetic, antihyperlipidemic, anticancer, antiviral, anti-leishmanial, antimicrobial, anti-Alzheimer's, antispasmodic activities, etc. The biological activities of the diterpenes are reported based on the evaluation by *in-vitro* cell-based assays and *in-vivo* animal studies against a variety of disease conditions.

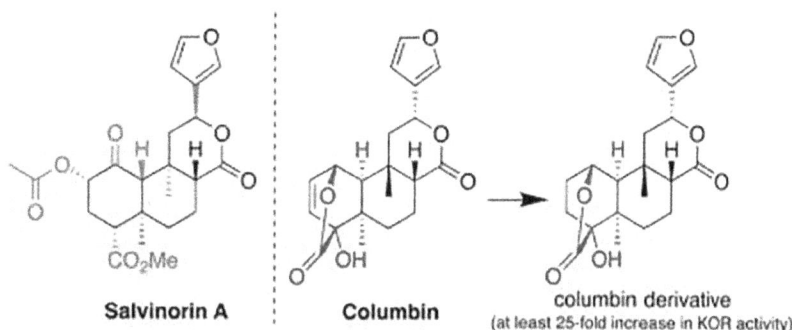

Salvinorin A   **Columbin**   columbin derivative (at least 25-fold increase in KOR activity)

FIGURE 3.107 Structure of diterpenoid isolated from the stems of *Tinospora cordifolia*.

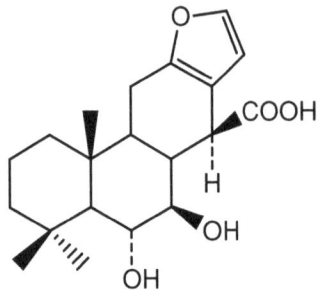

FIGURE 3.108   Structure of 6α,7β-Dihydroxyvouacapan-17β-oic acid (274).

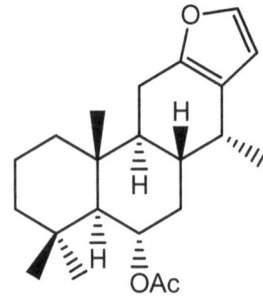

FIGURE 3.111   Structure of 6α-acetoxyvouacapane.

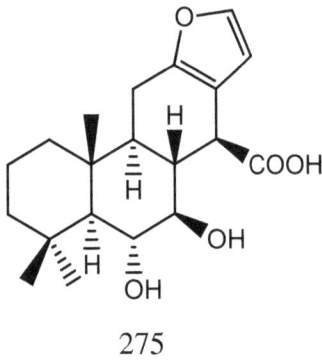

275

FIGURE 3.109   Structure of dihydroxyvouacapan-17β-oic acid (275).

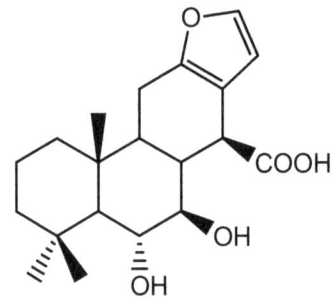

FIGURE 3.112   Structure of vouacapane-6α,7β,14β,19-tetraol.

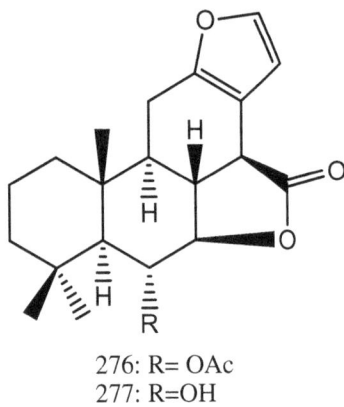

276: R= OAc
277: R=OH

FIGURE 3.110   Structure of vouacapane diterpenes with larvicidal activity.

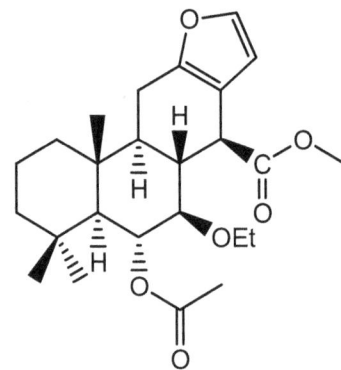

FIGURE 3.113   Structure of methyl-6α-acetoxy-7β-hydroxyvouacapan-17β-oate.

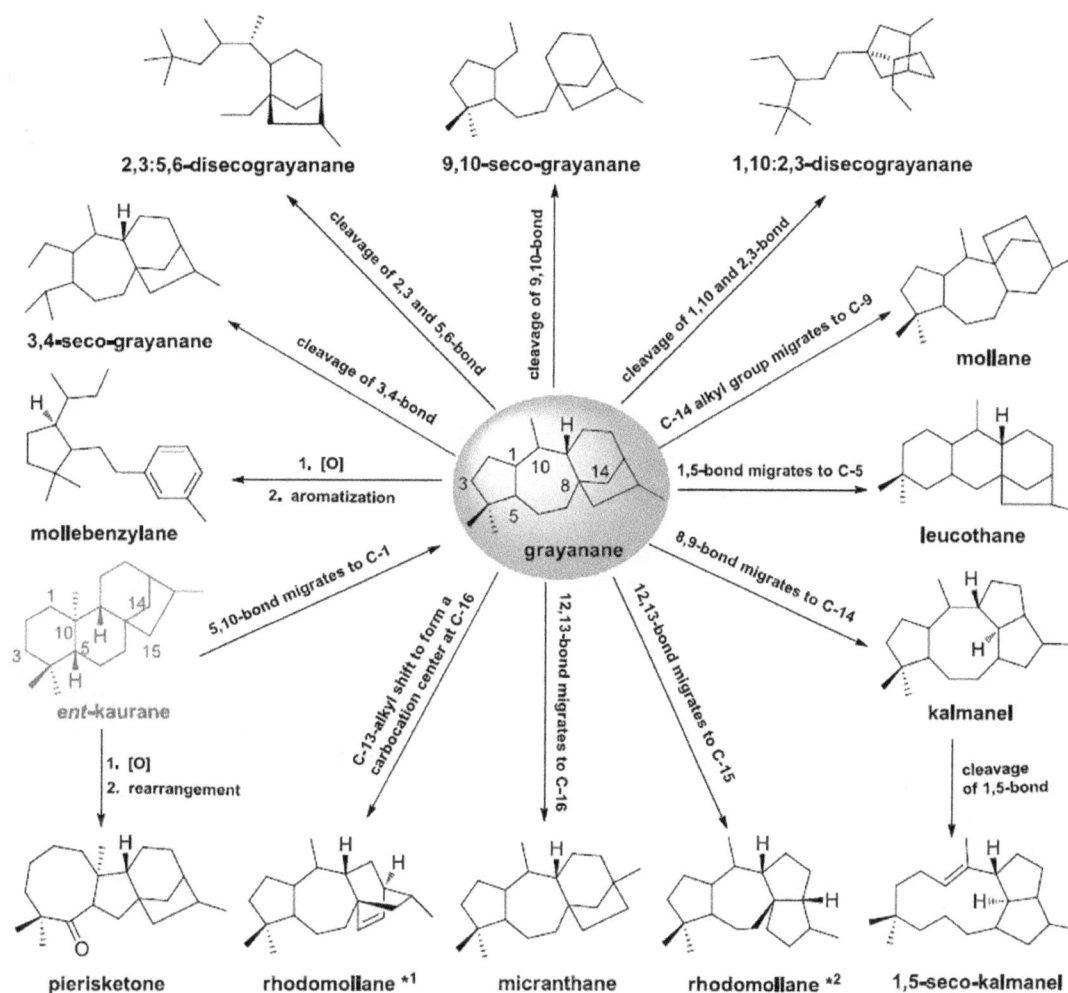

FIGURE 3.114 Grayanoids skeletons from Ericaceae (in the last 7 years) and their proposed biosynthetic relationships.

TABLE 3.55

Name, Skeleton Type and Origin of Grayanane Diterpenoids

Type of Skeleton	Name of Grayanane Diterpenoids
A/B-trans-grayanane	Pierisoid-C
	Pierisoid-D
	Pierisformosoid-D
	Pierisformosoid-F
	Pierisformosoid-G
	Pierisformosin-E
	Pierisformosin-G
	Mollfoliagein-A
A/B-cis-grayanane	Rhodomoside-E
	Rhododecorumin-X
	Principinol-E
	Principinol-F

REFERENCES

[1] Davis, E. M.; Croteau, R. Cyclization enzymes in the biosynthesis of monoterpenes, sesquiterpenes, and diterpenes. *Topics in Current Chemistry*, **2000**, 209, 53–95.

[2] Pattanaik, B.; Lindberg, P. Terpenoids and their biosynthesis in cyanobacteria. *Life*, **2015**, 5(1), 269–293.

[3] Vuorela, P.; Leinonen, M.; Saikku, P. Natural products in the process of finding new drug candidates. *Current Medicinal Chemistry*, **2004**, 11(11), 1375–1389.

[4] Lanzotti, V. Diterpenes for therapeutic use. *Natural Products*, **2013**, 3173–3191.

[5] Zeng, Z.; Zhu, J.; Chen, L.; Wen, W.; Yu, R. Biosynthesis pathways of ginkgolides. *Pharmacognosy Reviews*, **2013**, 7(13), 47–52.

[6] Oliveira, L. A. R.; Oliveira, G. A. R.; Borgesd, L. L.; Bara, M. T. F.; Silveira, D.. Vouacapane diterpenoids isolated from Pterodon and their biological activities. *Revista Brasileira de Farmacognosia*, **2017**, 27, 663–672.

[7] Rontani, J. F.; Bonin, P. Production of pristane and phytane in the marine environment: Role of prokaryotes. *Research in Microbiology*, **2011**, 162(9), 923–933. doi: 10.1016/j.resmic.2011.01.012.

[8] Piers, E.; Gilbert, M.; Cook, K. L. Total synthesis of the cyathane diterpenoid (±)-sarcodonin-G. *Organic Letters*, **2000**, 2(10), 1407–1410.

[9] Wang, T. Z.; Emmanuel Pinard, A.; Paquette, L. A. Asymmetric synthesis of the diterpenoid marine toxin (+)-acetoxycrenulide. *Journal of the American Chemical Society*, **1996**, 118, 1309–1318.

[10] Vanderah, D. J.; Rutledge, N.; Schmitz, F. J.; Ciereszko, L. S. Marine natural products: Cembrene-A and cembrene-C from a soft coral, Nephthea species. *Journal of Organic Chemistry*, **1978**, 43(8), 1614–1616.

[11] Cocker, J. D.; Halsall, T. G. The chemistry of gum labdanum. II. The structure of labdanolic acid. *Journal of the Chemical Society*, **1956**, 4262–4271.

[12] San Feliciano, A.; Gordaliza, M.; Salinero, M. A.; Miguel del Corral, J. M. Abietane acids: Sources, biological activities, and therapeutic uses. *Planta Medica*, **1993**, 59(6), 485–490.

[13] Lin, X.; Hezari, M.; Koepp, A. E.; Floss, H. G.; Croteau, R. Mechanism of taxadiene synthase, a diterpene cyclase that catalyzes the first step of taxol biosynthesis in Pacific Yew. *Biochemistry*, **1996**, 35(9), 2968–2977.

[14] Mohan, R. S.; Yee, N. K. N.; Coates, R. M.; Ren, Y.-Y.; Stamenkovic, P.; Mendez, I.; West, C. A. Biosynthesis of cyclic diterpene hydrocarbons in rice cell suspensions: Conversion of 9,10-syn-labda-8(17), 13-dienyl Diphosphate to 9β-Pimara-7,15-diene and Stemar-13-ene. *Archives of Biochemistry and Biophysics*, **1996**, 330(1), 33–47.

[15] Morrone, D.; Jin, Y.; Xu, M.; Choi, S.-Y.; Coates, R. M.; Peters, R. J. An unexpected diterpene cyclase from rice: Functional identification of a stemodene synthase. *Archives of Biochemistry and Biophysics*, **2006**, 448(1–2), 133–140.

[16] Wanga, M.; Li, H.; Xu, F.; Gao, X.; Li, J.; Xu, S.; Zhan, D.; Wu, X.; Xu, J.; Hua, H.; Li, D. Diterpenoid lead stevioside and its hydrolysis products steviol and isosteviol: Biological activity and structural modification. *European Journal of Medicinal Chemistry*, **2018**, 156, 885–906.

[17] Ullah, A.; Munir, S.; Mabkhot, Y.; Badshah, S. L. Bioactivity profile of the diterpene isosteviol and its derivatives. *Molecules*, **2019**, 24, 678–702.

[18] Chernov, S. V.; Shul'ts, E. E.; Shakirov, M. M.; et al. Synthesis of diterpene β-carbolines. *Doklady Chemistry*, **2001**, 381, 362–365. doi: 10.1023/A:1013329127494.

[19] Reber, K. P.; Xu, J.; Guerrero, C. A. Synthesis of mulinane diterpenoids. *The Journal of Organic Chemistry*, **2015**, 80(4), 2397–2406. doi: 10.1021/jo502602u.

[20] Huei-Ju, L.; Chia-Li, W. Note: Neopallavicinin from the Taiwanese liverwort *Pallavicinia subciliata*. *Journal of Asian Natural Products Research*, **1998**, 1(3), 177–182.

[21] Peng, X. -S.; Wong, H. N. C. Total synthesis of (±)-pallavicinin and (±)-neopallavicinin. *Chemistry-An Asian Journal*, **2006**, 1(1–2), 111–120. doi: 10.1002/asia.200600061.

[22] Bian, Y. -C.; Peng, X. -S.; Wong, H. N. C. Total synthesis of Pallavicinia diterpenoids: An overview. *Tetrahedron Letters*, **2016**, 57(50), 5560–5569. doi: 10.1016/j.tetlet.2016.11.025.

[23] Huang, B.; Guo, L.; Jia, Y. Protecting-group-free enantioselective synthesis of (–)-pallavicinin and (+)-neopallavicinin. *Angewandte Chemie International Edition*, **2015**, 54(46), 13599–13603. doi: 10.1002/anie.201506575.

[24] Buta, J. G.; Flippen, J. L.; Lusby, W. R. Isolation of (+)-Harringtonolide from the seeds of *Cephalataxus harringtonia*. *The Journal of Organic Chemistry*, **1978**, 43, 1002–1003.

[25] Sun, N. J.; Xue, Z.; Liang, X. T.; Huang, L. Isolation of (+)-Harringtonolide from the bark of the Chinese species *Cephalotaxus hainanensis*. *Acta Pharmaceutica Sinica B*, **1979**, 14, 39–44.

[26] Zhang, H. -J.; Hu, L.; Ma, Z.; Li, R.; Zhang, Z.; Tao, Z. H. Total synthesis of the diterpenoid (+)-harringtonolide. *Angewandte Chemie International Edition*, **2016**, 55(38), 11638–11641. doi: 10.1002/anie.201605879.

[27] Lemmich, E. Peucelinendiol, a new acyclic diterpenoid from *Peucedanum oreoselinum*. *Phytochemistry*, **1979**, 18(7), 1195–1197.

[28] Moran Joaquín, R.; Victoria, A.; Manuel, G. Synthesis and structure of the diterpenoid peucelinendiol. *Bulletin of the Chemical Society of Japan*, **1988**, 61(12), 4435–4437.

[29] Martín, M. J. Plumisclerin-A, a diterpene with a new skeleton from the soft coral *Plumigorgia terminosclera*. *Organic Letters*, **2010**, 12, 912–914.

[30] Gao, Y.; Wei, Y.; Ma, D. Synthetic studies toward plumisclerin-A. *Organic Letters*, **2019**, 21(5), 1384–1387.

[31] Williams, D. R.; Pinchman, J. R. Studies of neodolastanes – Synthesis of the tricyclic core of the trichoaurantianolides. *Canadian Journal of Chemistry*, **2013**, 91(1), 21–37. doi: 10.1139/v2012-088.

[32] Markovic, D.; Kolympadi, M.; Deguin, B.; Francois-Hugues, P.; Turks, M. The isolation and synthesis of neodolastane diterpenoids. *Natural Product Reports*, **2014**, 32(2), 230–255. doi: 10.1039/c4np00077c.

[33] Anke, T.; Rabe, U.; Schu, P.; Eizenhöfer, T.; Schrage, M.; Steglich, W. Studies on the biosynthesis of striatal-type diterpenoids and the biological activity of herical. *Zeitschrift Für Naturforschung C*, **2002**, 57(3–4), 263–271. doi: 10.1515/znc-2002-3-411.

[34] Balunas, M. J.; Kinghorn, A. D. Drug discovery from medicinal plants. *Life Sciences*, **2005**, 78(5), 431–441.

[35] Fabricant, D. S.; Farnsworth, N. R. The value of plants used in traditional medicine for drug discovery. *Environmental Health Perspectives*, **2001**, 109(1), 69–75.

[36] Zhang, L.; Lu, S. Overview of medicinally important diterpenoids derived from plastids. *Mini-Reviews in Medicinal Chemistry*, **2017**, 17(12), 988–1001.

[37] Tirapelli, C. R.; Ambrosio, S. R.; da Costa, F. B.; de Oliveira, A. M. Diterpenes: A therapeutic promise for cardiovascular diseases. *Recent Patents on Cardiovascular Drug Discovery*, **2008**, 3(1), 1–8.

[38] Tirapelli, C. R.; Ambrosio, S. R.; de Oliveira, A. M.; Tostes, R. C. Hypotensive action of naturally occurring diterpenes: A therapeutic promise for the treatment of hypertension. *Fitoterapia*, **2010**, 81, 690–702.

[39] Hanson, R.; De Oliveira, B. H. Stevioside and related sweet diterpenoid glycosides. *Natural Product Reports*, **1993**, 10(3), 301–309.

[40] Caprioli, J.; Sears, M. Forskolin lowers intraocular pressure in rabbits, monkeys, and man. *Lancet*, **1983**, 1(8331), 958–960.

[41] Zhang, C. Y.; Tan, B. K. H. Hypotensive activity of aqueous extract of *Andrographis paniculata* in rats. *Clinical and Experimental Pharmacology and Physiology*, **1996**, 23, 675–678.

[42] Zhang, C.; Kuroyangi, M.; Tan, B. K. Cardiovascular activity of 14-deoxy-11,12- didehydroandrographolide in the anaesthetized rat and isolated right atria. *Pharmacological Research*, **1998**, 38(6), 413–417.

[43] Silva, R. M.; Oliveira, F. A., Cunha, K. M.; Maia, J. L.; Maciel, M. A.; Pinto, A. C. Cardiovascular effects of trans-dehydrocrotonin, a diterpene from *Croton cajucara* in rats. *Vascular Pharmacology*, **2005**, 43(1), 11–18.

[44] Carakostas, M. C.; Curry, L. L.; Boileau, A. C.; Brusick, D. J. Overview: The history, technical function and safety of rebaudioside A, a naturally occurring steviol glycoside, for use in food and beverages. *Food and Chemical Toxicology*, **2008**, 46(7), 40–46.

[45] Maki, K. C.; Curry, L. L; Carakostas, M. C.; Tarka, S. M.; Reeves, M. S.; Farmer, M. V. The hemodynamic effects of rebaudioside A in healthy adults with normal and low-normal blood pressure. *Food and Chemical Toxicology*, **2008**, 46(7), 40–46.

[46] Odek-Ogunde, M.; Rajab, M. S. Antihypertensive effect of the clerodane diterpene ajugarin-I on experimentally hypertensive rats. *East African Medical Journal*, **1994**, 71(9), 587–590.

[47] Arruebo, M.; Vilaboa, B. S. -G.; Lambea, J.; Tres, A.; Valladares, M.; González-Fernández, Á. Assessment of the evolution of cancer treatment therapies. *Cancers*, **2011**, 3, 3279–3330.

[48] De Furia, M. D. Paclitaxel (Taxol): A new natural product with major anticancer activity. *Phytomedicine*, **1997**, 4(3), 273–282.

[49] Horwitz, S. B. Taxol (paclitaxel): Mechanisms of action. *Annals of Oncology*, **1994**, 5(Suppl 6), S3–S6.

[50] Jian, B.; Zhang, H.; Han, C.; Liu, J. Anti-cancer activities of diterpenoids derived from *Euphorbia fischeriana* Steud. *Molecules*, **2018**, 23, 387. doi: 10.3390/molecules23020387.

[51] Dimas, K.; Papadaki, M.; Tsimplouli, C.; Hatziantoniou, S.; Alevizopoulos, K., et al. Labd-14-ene-8,13-diol (sclareol) induces cell cycle arrest and apoptosis in human breast cancer cells and enhances the activity of anticancer drugs. *Biomedical and Pharmacology Journal*, **2006**, 60(3), 127–133.

[52] Islam, M. T. Diterpenes and their derivatives as potential anticancer agents. *Phytotherapy Research*, **2017**, 31(5), 691–712.

[53] Chen, S.-p.; Dong, M.; Kita, K.; Shi, Q.-W.; Cong, B.; Guo, W.-z.; Sugaya, S.; Sugita, K. u.; Suzuki, N. Anti-proliferative and apoptosis-inducible activity of labdane and abietane diterpenoids from the pulp of *Torreya nucifera* in HeLa cells. *Molecular Medicine Reports*, **2010**, 3, 673–678.

[54] Dong, M.; Chen, S.-P.; Kita, K.; Ichimura, Y.; Guo, W.-Z.; Lu, S.; Sugaya, S.; Hiwasa, T.; Takiguchi, M.; Mori, N.; Kashima, A.; Morimura, K.; Hirota, M.; Suzuki, N. Anti-proliferative and apoptosis-inducible activity of Sarcodonin G from *Sarcodon scabrosus* in HeLa cells. *International Journal of Oncology*, **2009**, 34, 201–207.

[55] Ayafor, J. F.; Tchuendem, M. H.; Nyasse, B.; Tillequin, F.; Anke, H. Novel bioactive diterpenoids from *Aframomum aulacocarpos*. *Journal of Natural Products*, **1994**, 57(7), 917–923.

[56] Chen, J.; Li, H.; Zhao, Z.; Xia, X.; Li, B.; Zhang, J.; Yan, X. Diterpenes from the Marine Algae of the Genus Dictyota. *Marine Drugs*, **2018**, 16(5), 159. doi: 10.3390/md16050159.

[57] Finer, J.; Clardy, J.; Fenical, W.; Minale, L.; Riccio, R.; Battaile, J.; Kirkup, M.; Moore, R. E. Structures of dictyodial and dictyolactone, unusual marine diterpenoids. *The Journal of Organic Chemistry*, **1979**, 44, 2044–2047.

[58] Gedara, S. R.; Abdelhalim, O. B.; Elsharkawy, S. H.; Salama, O. M.; Shier, T. W.; Halim, A. F. Cytotoxic hydroazulene diterpenes from the brown alga Dictyota dichotoma. *Zeitschrift für Naturforschung C: A Journal of Biosciences*, **2003**, 58, 17–22.

[59] Zheng, Y.; Zhang, S. -W.; Cong, H. -J.; Huang, Y. -J.; Xuan, L. -J. Caesalminaxins A–L, cassane diterpenoids from the seeds of Caesalpinia minax. *Journal of Natural Products*, **2013**, 76(12), 2210–2218. doi: 10.1021/np400545v.

[60] Li, M.; He, F.; Zhou, Y.; Wang, M.; Tao, P.; Tu, Q.; Lv, G.; Chen, X. Three new ent-abietane diterpenoids from the roots of Euphorbia fischeriana and their cytotoxicity in human tumor cell lines. *Archives of Pharmacal Research*, **2019**, 42, 512–518.

[61] Sharma, V.; Sharma, T.; Kaul, S.; Kapoor, K. K.; Dhar, M. K. Anticancer potential of labdane diterpenoid lactone "andrographolide" and its derivatives: a semi-synthetic approach. *Phytochemistry Reviews*, **2017**, 16(3), 513–526. doi: 10.1007/s11101-016-9478-9.

[62] Shi, Q. W.; Su, X. H.; Kiyota, H. Chemical and pharmacological research of the plants in genus Euphorbia. *Chemical Reviews*, **2008**, 108, 4295–4327.

[63] Lu, Z. Q., et al. Cytotoxic diterpenoids from *Euphorbia helioscopia*. *Journal of Natural Products*, **2008**, 71, 873–876.

[64] Hegazy, M. E. F., et al. Bioactive jatrophane diterpenes from *Euphorbia guyoniana*. *Phytochemistry*, **2010**, 71, 249–253.

[65] Matias, D.; Nicolai, M.; Saraiva, L.; Pinheiro, R.; Faustino, C.; Diaz Lanza, A.; Pinto Reis, C.; Stankovic, T.; Dinic, J.; Pesic, M.; Rijo, P. Cytotoxic activity of royleanone diterpenes from *Plectranthus madagascariensis* Benth. *ACS Omega*, **2019**, 4, 8094–8103.

[66] Yoshida, M.; Feng, W.; Saijo, N.; Ikekawa, T. Antitumor activity of daphnane-type diterpene gnidimacrin isolated from *Stellera chamaejasme* L. *International Journal of Cancer*, **1996**, 66(2), 268–273.

[67] Anke, T.; Rabe, U.; Schu, P.; Eizenhöfer, T.; Schrage, M.; Steglich, W. Studies on the biosynthesis of striatal-type diterpenoids and the biological activity of herical. *Zeitschrift Für Naturforschung C*, **2002**, 57(3–4), 263–271. doi: 10.1515/znc-2002-3-411.

[68] Pan, W. H.; Xu, X. Y.; Shi, N.; Tsang, S. W.; Zhang, H. J. Antimalarial activity of plant metabolites. *International Journal of Molecular Sciences*, **2018**, 19(5), 1382. doi: 10.3390/ijms19051382.

[69] González, M. A.; Clark, J.; Connelly, M.; Rivas, F. Antimalarial activity of abietane ferruginol analogues possessing a phthalimide group. *Bioorganic & Medicinal Chemistry Letters* **2014**, 24, 5234–5237.

[70] Zhou, B., et al. Euphorbesulins A-P, structurally diverse diterpenoids from *Euphorbia esula*. *Journal of Natural Products*, **2016**, 79, 1952–1961.

[71] Mishra, K.; Dash, A. P.; Dey, N. Andrographolide: A novel antimalarial diterpene lactone compound fromandrographis paniculata and its interaction with curcumin and artesunate, *Journal of Tropical Medicine*, **2011**, Article ID 579518, 1–6.

[72] Endale, A.; Bisrat, D.; Animut, A.; Bucar, F.; Asres, K. In-vivo antimalarial activity of a labdane diterpenoid from the leaves of *Otostegia integrifolia* Benth. *Phytotherapy Research*, **2013**, 27(12), 1805–1809. doi: 10.1002/ptr.4948.

[73] Miyaoka, H.; Shimomura, M.; Kimura, H.; Yamada, Y.; Kim, H.-S.; Yusuke, W. Antimalarial activity of kalihinol A and new relative diterpenoids from the Okinawan sponge, *Acanthella* sp. *Tetrahedron*, **1998**, 54(44), 13467–13474. doi: 10.1016/s0040-4020(98)00818-7.

[74] Sahare, K. N.; Singh, V. Anti-filarial activity of ethyl acetate extract of *Vitex negundo* leaves in vitro. *Asian Pacific Journal of Tropical Medicine*, **2013**, 6(9), 689–692.

[75] Kushwaha, V.; Saxena, K.; Verma, R., et al. Antifilarial activity of diterpenoids from *Taxodium distichum*. *Parasites Vectors*, **2016**, 9, 312. doi: 10.1186/s13071-016-1592-4.

[76] Kumar, V.; Abbas, A. K.; Aster, J. C.; Robbins, S. L. *Inflammation and Repair. Robbins Basic Pathology*. Saunders, Philadelphia, London. **2012**, 29–74.

[77] Chua, L.S. Review on liver inflammation and anti-inflammatory activity of Andrographis paniculata for hepatoprotection. *Phytotherapy Research*, **2014**, 28, 1589–1598.

[78] Parichatikanond, W.; Suthisisang, C.; Dhepakson, P.; Herunsalee, A. Study of anti-inflammatory activities of the pure compounds from *Andrographis paniculata* (Burm.f.) Nees and their effects on gene expression. *International Immunopharmacology*, **2010**, 10, 1361–1373.

[79] Yang, C. L. H.; Or, T. C. T.; Ho, M. H. K.; Lau, A. S. Y. Scientific basis of botanical medicine as alternative remedies for rheumatoid arthritis. *Clinical Reviews in Allergy & Immunology*, **2013**, 44, 284300.

[80] DeWitt, D. L. Prostaglandin endoperoxide synthase: Regulation of enzyme expression. *Biochimica et Biophysica Acta*, **1991**, 1083, 121–134.

[81] MacMicking, Q. W.; Xie, N.; Nathan, C. Nitric oxide and macrophage function. *Annual Review of Immunology*, **1997**, 15, 323–350.

[82] Moilanen, E.; Vapaatalo, H. Nitric oxide in inflammation and immune response. *Annals of Medicine*, **1995**, 27, 359–367.

[83] Moncada, S.; Palmer, R. M.; Higgs, E. A. Nitric oxide: Physiology, pathophysiology, and pharmacology. *Pharmacological Reviews*, **1991**, 43, 109–142.

[84] Vane, J. R.; Bakle, Y. S.; Botting, R. M. Cycloxygenases 1 and 2. *Annual Review of Pharmacology and Toxicology*, **1998**, 38, 97–120.

[85] Winter, C. A.; Risley, E. A.; Nuss, G. W. Carrageenan-induced oedema in hind paw of the rats as an assay for anti-inflammatory drugs. *Proceedings of the Society for Experimental Biology and Medicine*, **1962**, 111, 544–547.

[86] Pang, L.; de las Heras, B.; Hoult, J. R. S. A novel diterpenoid labdane from *Sideritis javalambrensis* inhibits eicosanoid generation from stimulated macrophages but enhances arachidonate release. *Biochemical Pharmacology*, **1996**, 51, 863–868.

[87] Wu, S. -L.; Su, J. -H.; Wen, Z. -H., et al. Simplexins A-I, eunicellin based diterpenoids fromthe soft coral *Klyxum simplex*. *Journal of Natural Products*, **2009**, 72(6), 994–1000.

[88] Chen, B.-W.; Chang, S.-M.; Huang, C.-Y., et al. Hirsutalins AH, eunicellin-based diterpenoids from the soft coral *Cladiella hirsute*. *Journal of Natural Products*, **2010**, 73(11), 1785–1791.

[89] Chen, B.-W.; Chao, C.-H.; Su, J.-H.; Wen, Z.-H.; Sung, P.-J.; Sheu, J.-H. Anti-inflammatory eunicellin-based diterpenoids from the cultured soft coral Klyxum simplex. *Organic & Biomolecular Chemistry*, **2010**, 8(10), 2363–2366.

[90] Tai, C.-J.; Su, J.-H.; Huang, M.-S.; Wen, Z.-H.; Dai, C.-F.; Sheu, J.-H. Bioactive eunicellin-based diterpenoids from the soft coral Cladiella krempfi. *Marine Drugs*, **2011**, 9(10), 2036–2045.

[91] Hsu, F.-J.; Chen, B.-W.; Wen, Z.-H., et al. Klymollins A-H, bioactive eunicellin-based diterpenoids from the formosan soft coral Klyxum molle. *Journal of Natural Products*, **2011**, 74(11), 2467–2471.

[92] Chen, Y.-H.; Tai, C.-Y.; Kuo, Y.-H., et al. Cladieunicellins A-E, new eunicellins from an Indonesian soft coral *Cladiella* sp. *Chemical & Pharmaceutical Bulletin*, **2011**, 59(3), 353–358.

[93] Liaw, C.-C.; Shen, Y.-C.; Lin, Y.-S.; Hwang, T.-L.; Kuo, Y.-H.; Khalil, A. T. Frajunolides E-K, briarane diterpenes from Junceella fragilis. *Journal of Natural Products*, **2008**, 71(9), 1551–1556.

[94] Yang, B.; Zhou, X.-F.; Lin, X.-P., et al. Cembrane diterpenes chemistry and biological properties. *Current Organic Chemistry*, **2012**, 16(12), 1512–1539.

[95] Wan, L.-S.; Chu, R.; Peng, X.-R.; Zhu, G.-L.; Yu, M.-Y.; Li, M.-H. Pepluane and paraliane diterpenoids from *Euphorbia peplus* with potential anti-inflammatory activity. *Journal of Natural Products*, **2016**, 79(6), 1628–1634. doi: 10.1021/acs.jnatprod.6b00206.

[96] Richards, S. S.; Hendrie, H. C. Diagnosis, management, and treatment of Alzheimer disease: A guide for the internist. *Archives of Internal Medicine*, **1999**, 159(8), 789–798. doi: 10.1001/archinte.159.8.789.

[97] Wang, S.; Yang, H.; Yu, L.; Jin, J.; Qian, L.; Zhao, H., et al. Oridonin attenuates Aβ1–42-induced neuroinflammation and inhibits NF-κB pathway. *PLoS One*, **2014**, 9(8), e104745. doi: 10.1371/journal.pone.0104745.

[98] Hebert, L. E.; Scherr, P. A.; Bienias, J. L.; Bennett, D. A.; Evans, D. A. Alzheimer disease in the US population: Prevalence estimates using the 2000 census. *Archives of Neurology*, **2003**, 60, 1119–1122.

[99] Selkoe, D. J. The cell biology of beta-amyloid precursor protein and presenilin in Alzheimer's disease. *Trends in Cell Biology*, **1998**, 8, 447–453.

[100] Selkoe, D. J. Alzheimer's disease: Genes, proteins, and therapy. *Physiological Reviews*, **2001**, 81, 741–766.

[101] Tanzi, R. E.; Bertram, L. Twenty years of the Alzheimer's disease amyloid hypothesis: A genetic perspective. *Cell*, **2005**, 120, 545–555.

[102] Stromgaard, K.; Nakanishi, K. Chemistry and biology of terpene trilactones from ginkgo biloba. *Angewandte Chemie*, **2004**, 43(13), 1640–1658.

[103] Maruyama, M.; Terahara, A.; Itagaki, Y.; Nakanishi, K. The ginkgolides. I. Isolation and characterization of the various groups. *Tetrahedron Letters*, **1967**, 8(4), 299–302.

[104] Zen, Z.; Zhu, J.; Chen, L.; Wen, W.; Yu, R. Biosynthesis pathways of ginkgolides. *Pharmacognosy Reviews*, **2013**, 7(13), 47–52.

[105] Bate, C.; Tayebi, M.; Williams, A. Ginkgolides protect against amyloid-beta1–42-mediated synapse damage in vitro. *Molecular Neurodegeneration*, **2008**, 3, 1.

[106] Howes, M. J. R.; Perry, N. S. L.; Houghton, P. J. Plants with traditional uses and activities, relevant to the management of Alzheimer's disease and other cognitive disorders. *Phytotherapy Research*, **2003**, 17, 1–18.

[107] Vitolo, O.; Gong, B.; Cao, Z.; Ishii, H.; Jaracz, S.; Nakanishi, K.; Arancio, O.; Dzyuba, S. V.; Lefort, R.; Shelanski, M. Protection against beta-amyloid induced abnormal synaptic function and cell death by Ginkgolide J. *Neurobiology of Aging*, **2009**, 30, 257–265.

[108] Mei, Z.; Situ, B.; Tan, X.; Zheng, S.; Zhang, F.; Yan, P.; Liu, P. Cryptotanshinione upregulates alpha-secretase by activation PI3K pathway in cortical neurons. *Brain Research*, **2010**, 1348, 165–173.

[109] Mei, Z.; Yan, P.; Situ, B.; Mou, Y.; Liu, P. Cryptotanshinione inhibits beta-amyloid aggregation and protects damage from beta-amyloid in SH-SY5Y cells. *Neurochemical Research*, **2012**, 37, 622–628.

[110] Yoo, K.-Y.; Park, S.-Y. Terpenoids as potential anti-Alzheimer's disease therapeutics. *Molecules*, **2012**, 17, 3524–3538.

[111] Owona, B. A.; Zug, C.; Schluesener, H. J.; Zhang, Z.-Y. Protective effects of forskolin on behavioral deficits and neuropathological changes in a mouse model of cerebral amyloidosis. *Journal of Neuropathology & Experimental Neurology*, **2016**, 75(7), 618–627. doi: 10.1093/jnen/nlw043.

[112] Venkatachalapathi, A.; Thenmozhi, K.; Karthika, K.; Ali, M. A.; Paulsamy, S.; AlHemaid, F.; Elshikh, M. S. Evaluation of a labdane diterpene forskolin isolated from *Solena amplexicaulis* (Lam.) Gandhi (Cucurbitaceae) revealed promising antidiabetic and antihyperlipidemic pharmacological properties. *Saudi Journal of Biological Sciences*, **2019**, 26, 1710–1715.

[113] Saravanan, R.; Vengatash babu, K.; Ramachandran, V. Effect of Rebaudioside A, a diterpenoid on glucose homeostasis in STZ-induced diabetic rats. *Journal of Physiology and Biochemistry*, **2012**, 68(3), 421–431. doi: 10.1007/s13105-012-0156-0.

[114] Gao, Y.; Niu, Y. F.; Wang, F.; Hai, P.; Wang, F.; Fang, Y. D.; Xiong, W. Y.; Liu, J. K. Clerodane diterpenoids with antihyperglycemic activity from *Tinospora crispa*. *Natural Products and Bioprospecting*, **2016**, 6(5), 247–255. doi: 10.1007/s13659-016-0109-3.

[115] Cui, L.; Kim, M. O.; Seo, J. H.; Kim, I. S.; Kim, N. Y.; Lee, S. H.; Lee, H. S. Abietane diterpenoids of *Rosmarinus officinalis* and their diacylglycerol acyltransferase-inhibitory activity. *Food Chemistry*, **2012**, 132(4), 1775–1780. doi: 10.1016/j.foodchem.2011.11.138.

[116] Sen, A.; Ayar, B.; Yilmaz, A.; Acar, O. O.; Turgut, G. C.; Topcu, G. Natural diterpenoid alysine A isolated fromTeucrium alyssifoliumexertsantidiabetic effect via enhanced glucose uptake and suppressed glucose absorption. *Turkish Journal of Chemistry*, **2019**, 43, 1350–1364.

[117] Carney, J. R.; Krenisky, J. M.; Williamson, R. T.; Luo, J.; Carlson, T. J.; Hsu, V. L.; Moswa, J. L. Maprouneacin: A new daphnane diterpenoid with potent anti-hyperglycemic Activity from *Maprounea africana*. *Journal of Natural Products*, **1999**, 62(2), 345–347. doi: 10.1021/np980356c.

[118] Sandjo, L. P.; Kuete, V. In: *Diterpenoids from the Medicinal Plants of Africa. Medicinal Plant Research in Africa*, 1st Edition. Oxford: Elsevier, **2013**, 105–133.

[119] Grennan, D.; Varughese, C.; Moore, N. M. Medications for treating infection. *JAMA*. **2020**, 323(1), 100. doi: 10.1001/jama.2019.17387.

[120] Aceret, T. L.; Coll, J. C.; Uchio, Y.; Sammarco, P. W. Antimicrobial activity of the diterpenes flexibilide and sinulariolide derived from *Sinularia flexibilis* Quoy and Gaimard 1833 (Coelenterata: Alcyonacea, Octocorallia). *Comparative Biochemistry and Physiology Part-C*, **1998**, 120, 121–126.

[121] Batista, O.; Duarte, A.; Nascimento, J.; Simões, M. F.; de la Torre, M. C.; Rodríguez, B. Structure and antimicrobial activity of diterpenes from the roots of *Plectranthus hereroensis*. *Journal of Natural Products*, **1994**, 57(6), 858–861.

[122] Loyola, L. A.; Borquez, J.; Morales, G.; San-Martín, A. Mulinolic acid, a diterpenoid from *Mulinum crassifolium*. *Phytochemistry*, **1996**, 43(1), 165–168.

[123] Wachter, G. A.; Matooq, G.; Hoffmann, J. J.; Maiese, W. M.; Singh, M. P.; M. Gloria; Timmermann, B. N. Antibacterial diterpenoid acids from *Azorella compacta*. *Journal of Natural Products*, **1999**, 62, 1319–1321.

[124] Sa, N. C.; Arruda Cavalcante, T. T.; Ximenes Araujo, A.; dos Santos, H. S.; Jane Ribeiro Albuquerque, M. R.; Nogueira Bandeira, P.; Silva da Cunha, R. M.; Sousa Cavada, B.; Holanda Teixeira, E. Antimicrobial and antibiofilm action of Casbane Diterpene from *Croton nepetaefolius* against oral bacteria. *Archieves of Oral Biology*, **2012**, 57, 550–555.

[125] Singh, M. P.; Janso, J. E.; Luckman, S. W.; Brady, S. F.; Clardy, J.; Greenstein, M.; Maiese, W. M. Biological activity of guanacastepene, a novel diterpenoid antibiotic produced by an unidentified fungus CR115. *The Journal of Antibiotics*, **2000**, 53(3), 256–261.

[126] Anke, T.; Oberwinkler, F.; Steglich, W.; Hofle, G. Antibiotics from basidiomycetes, 1. The striatins-new antibiotics from the basidiomycete *Cyathus striatus* 3338Đ3343. (Huds. ex Pers.) Willd. *The Journal of Antibiotics*, **1977**, 30, 221–225.

[127] Bass, J. B. Jr; Farer, L. S.; Hopewell, P. C.; O'Brien, R.; Jacobs, R. F.; Ruben, F.; Snider, D. E. Jr, Thornton, G. Treatment of tuberculosis and tuberculosis infection in adults and children. American Thoracic Society and The Centers for Disease Control and Prevention. *American Journal of Respiratory and Critical Care Medicine*, **1994**, 149(5), 1359–1374. doi: 10.1164/ajrccm.149.5.8173779.

[128] Sansinenea, E.; Ortiz, A. Antitubercular natural terpenoids: Recent developments and syntheses. *Current Organic Synthesis*, **2014**, 11(4), 545.

[129] Silva, A. N.; Soares, A. C. F.; Cabral, M. M. W. Antitubercular activity increase in labdane diterpenes from *Copaifera oleoresin* through structural modification. *Journal of the Brazilian Chemical Society*, **2017**, 28(6), 1106–1112.

[130] Yu, Z.; Wei, Y.; Tian, X.; Yan, Q.; Yan, Q.; Huo, X.; Ma, X. Diterpenoids from the roots of *Euphorbia ebracteolata* and their anti-tuberculosis effects. *Bioorganic Chemistry*, **2018**, 77, 471–477.

[131] Rodríguez, A. D.; Ramírez, C. Serrulatane diterpenes with antimycobacterial activity isolated from the west indian sea whip *Pseudopterogorgia elisabethae*. *Journal of Natural Products*, **2001**, 64(1), 100–102. doi: 10.1021/np000196g.

[132] Escarcena, R.; Perez-Meseguer, J.; del Olmo, E.; Alanis-Garza, B.; Garza-González, E.; Salazar-Aranda, R.; de Torres, N. W. Diterpenes synthesized from the natural serrulatane leubethanol and their in-vitro activities against *Mycobacterium tuberculosis*. *Molecules*, **2015**, 20, 7245–7262. doi: 10.3390/molecules20047245.

[133] Kulkarni, R. R.; Shurpali, K.; Puranik, V. G.; Sarkar, D.; Joshi, S. P. Antimycobacterial labdane diterpenes from *Leucas stelligera*. *Journal of Natural Products*, **2013**, 76(10), 1836–1841. doi: 10.1021/np400002p.

[134] Garland, E. L. Pain processing in the human nervous system: A selective review of nociceptive and biobehavioral pathways. *Primary Care: Clinics in Office Practice*, **2012**, 39, 561–571.

[135] Duarte, I. D. G.; Alves, D. L. F.; Veloso, D. P.; Craig, M. N. Evidence of the involve-ment of biogenic amines in the antinociceptive effect of a vouacapan extracted from *Pterodon polygalaflorus* Benth. *Journal of Ethnopharmacology*, **1996**, 55, 13–18.

[136] Oliveira, L. A. R.; Oliveira, G. A. R.; Borges, L. L.; Bara, M. T. F.; Silveira, D. Vouacapane diterpenoids isolated from Pterodon and their biologicalactivities. *Revista Brasileira de Farmacognosia*, **2017**, 27, 663–672.

[137] Rigano, D.; Aviello, G.; Bruno, M.; Formisano, C.; S. Rosselli; Capasso, R.; Senatore, F.; Izzo, A. A.; Borrelli, F. Antispasmodic effects and structure-activity relationships of labdane diterpenoids from *Marrubium globosum* ssp. Libanoticum. *Journal of Natural Products*, **2009**, 72, 1477–1481.

[138] Siebert, D. J. Salvia divinorum and salvinorin A: New pharmacologic findings. *Journal of Ethnopharmacology*, **1994**, 43, 53–56.

[139] Castro, A.; Coll, J.; Pant, A. K.; Pakrash, O. Neo-Clerodane diterpenoids from *Ajuga macrosperma* var. breviflora. *Natural Product Communications*, **2015**, 10(6), 857–860. PubMed PMID: 26197499.

[140] Murray, H. W.; Berman, J. D.; Davies, C. R.; Saravia, N. G. Advances in leishmaniasis. *Lancet*, **2005**, 366, 1561–1577.

[141] Seifert, K. Structures, targets and recent approaches in anti-Leishmania drug discovery and development. *The Open Medicinal Chemistry Journal*, **2011**, J 5, 31–39.

[142] Santos, A. O.; Ueda-Nakamura, T.; Dias-Filho, B. P.; Veiga-Jr, V. F.; Nakamura, C. V. Copaiba Oil: An alternative to development of newdrugs against leishmaniasis. *Evidence-Based Complementary and Alternative Medicine*, **2012**, 1–7. doi: 10.1155/2012/898419.

[143] Tiuman, T. S.; Santos, A. O.; Ueda-Nakamura, T; Dias-Filho, B. P.; Nakamura, C. V. Recent advances in leishmaniasis treatment. *International Journal of Infectious Diseases*, **2011**, 15, 525–532.

[144] Pereira, H. S.; Leao-Ferreira, L. R.; Moussatche, N.; Teixeira, V. L.; Cavalcanti, D. N.; Costa, L. J.; Diaz, R.; Frugulhetti, I. C. Antiviral activity of diterpenes isolated from the Brazilian marine alga *Dictyota menstrualis* against human immunodeficiency virus type 1 (HIV-1). *Antiviral Research*, **2004**, 64, 69–76.

[145] Abrantes, J. L.; Barbosa, J.; Cavalcanti, D.; Pereira, R. C.; Frederico Fontes, C. L.; Teixeira, V. L.; Moreno Souza, T. L.; Paixao, I. C. The effects of the diterpenes isolated from the Brazilian brown algae *Dictyota pfaffii* and *Dictyota menstrualis* against the herpes simplex type-1 replicative cycle. *Planta Medica*, **2010**, 76, 339–344.

[146] Olivon, F.; Palenzuela, H.; Girard-Valenciennes, E.; Neyts, J.; Pannecouque, C.; Roussi, F.; Grondin, I.; Leyssen, P.; Litaudon, M. Antiviral activity of flexibilane and tigliane diterpenoids from *Stillingia lineate*. *Journal of Natural Products*, **2015**, 78(5), 1119–1128. doi: 10.1021/acs.jnatprod.5b00116.

[147] Remy, S.; Litaudon, M. Macrocyclic diterpenoids from euphorbiaceae as a source of potent and selective inhibitors of chikungunya virus replication. *Molecules*, **2019**, 24, 2336–2351.

[148] Abbas Miana, G.; Riaz, M.; Shahzad-ul-Hussan, S.; Zafar Paracha, R.; Zafar Paracha, U. Prostratin: An overview. *Mini Reviews in Medicinal Chemistry*, **2015**, 15(13), 1122–1130. doi: 10.2174/1389557515666150511154108.

[149] Song, B.; Ding, Xin-hui Tian, G.; Li, L.; Zhou, C.; Zhang, Q.-b.; Wang, M.-h.; Zhang, T.; Z.-m. Zou. Anti-HIV-1 integrase diterpenoids from *Dichrocephala benthamii*. *Phytochemistry Letters*, **2015**, 14, 249–253.

[150] Jiang, R.-W.; Ma, S.-C.; But, P. P. H.; Mak, T. C. W. Isolation and characterization of spirocaesalmin, a novel rearranged vouacapane diterpenoid from *Caesalpinia minax* Hance. *Journal of the Chemical Society, Perkin Transactions 1*, **2001**, 2920–2923.

[151] Dang, Z.; Jung, K.; Zhu, L.; Xie, H.; Lee, K. H.; Chen, C. H.; Huang, L. Phenolic diterpenoid derivatives as anti-influenza a virus agents. *ACS Medicinal Chemistry Letters*, **2015**, 6(3), 355–358. doi: 10.1021/ml500533x.

[152] Bajpai, V. K.; Ahmad Rather, I.; Chul Kang, S.; Park, Y.-H. A diterpenoid taxoquinone from *Metasequoia glyptostroboides* with pharmacological potential. *Indian Journal of Pharmaceutical Education and Research*, **2016**, 50(3), 458–464.

[153] Edwards, J.; Brown, M.; Peak, E.; Bartholomew, B.; Nash, R. J.; Hoffmann, K. F. The diterpenoid 7- keto-sempervirol, derived from lycium chinense, displays anthelmintic activity against both *Schistosoma mansoni* and *Fasciola hepatica*. *PLOS Neglected Tropical Diseases*, **2015**, 9(3), 1–19. doi: 10.1371/journal.pntd.0003604.

[154] Mai, Z.; Ni, G.; Liu, Y.; Zhang, Z.; Li, L.; Chen, N.; Yu, D. Helioscopianoids A-Q, bioactive jatrophane diterpenoid esters from *Euphorbia helioscopia*. *Acta Pharmaceutica Sinica B*, **2018**, 8(5), 805–817.

[155] Rossi, A.; Caiazzo, E.; Bilancia, R.; Riemma, M. A.; Pagano, E.; Cicala, C.; Roviezzo, F. Salvinorin-A inhibits airway hyperreactivity induced by ovalbumin sensitization. *Frontiers in Pharmacology*, **2017**, 7. doi: 10.3389/fphar.2016.00525.

[156] Lee, J.-E.; Im, D.-S. Suppressive effect of carnosol on ovalbumin-induced allergic asthma. *ACS Medicinal Chemistry Letters*, **2020**, 1–6.

[157] Takikawa, M.; Uno, K.; Ooi, T.; Kusumi, T.; Akera, S.; Muramatsu, M.; Mega, H.; Horita, C. Crenulacetal-C, a marine diterpene, and its synthetic mimics inhibiting *Polydora websterii*, a harmful lugworm damaging pearl cultivation. *Chemical and Pharmaceutical Bulletin*, **1998**, 29, 462–466.

[158] Kusumi, T.; Muanza-Nkongolo, D.; Goya, M.; Ishitsuka, M.; Iwashita, T.; Kakisawa, H. Structures of crenulacetals A, B, C, and D. The new diterpenoids from the brown algae of *Dictyotaceae*. *The Journal of Organic Chemistry*, **1986**, 17, 384–387.

[159] Zhang, J.; Rahman, A. A.; Jain, S.; Jacob, M. R.; Khan, S. I.; Tekwani, B. L.; Ilias, M. Antimicrobial and antiparasitic abietane diterpenoids from Cupressus sempervirens. *Research and Reports in Medicinal Chemistry*, **2012**, 2, 1–6.

[160] Lee, W. Y.; Cheung, C. C.; Liu, K. W. Cytotoxic effects of tanshinones from *Salvia miltiorrhiza* on doxorubicin-resistant human liver cancer cells. *Journal of Natural Products*, **2010**, 73(5), 854–859.

[161] Yoon, Y.; Kim, Y. O.; Jeon, W. K.; Park, H. J.; Sung, H. J. Tanshinone-IIA isolated from *Salvia miltiorrhiza* BUNGE induced apoptosis in HL60 human premyelocytic leukemia cell line. *Journal of Ethnopharmacology*, **1999**, 68(1–3), 121–127.

[162] Bian, W.; Chen, F.; Bai, L.; Zhang, P.; Qin, W. Dihydrotanshinone-I inhibits angiogenesis both in-vitro and in-vivo. *Acta Biochimica et Biophysica Sinica*, **2008**, 40(1), 1–6.

[163] Roncero, A. M.; Tobal, I. E.; Moro, R. F.; Díez, D.; Marcos, I. S. Halimane diterpenoids: Sources, structures, nomenclature and biological activities. *Natural Product Reports*, **2018**, 9, 1–37.

[164] Habtemariam, S. In vitro antileishmanial effects of antibacterial diterpenes from two Ethiopian Premna species: *P. schimperi* and *P. oligotricha*. *BMC Pharmacology*, **2003**, 3, 1–6.

[165] Ghosh, S.; Singh, R. K.; Dubey, V. K.; Rangan, L. Antileishmanial activity of labdane diterpenes isolated from *Alpinia nigra* seeds. *Letters in Drug Design & Discovery*, **2017**, 14(1), 119–124.

[166] Alarado, A. B.; Gerwick, W. H. Dictyol-H: A new tricyclic diterpenoid from the brown seaweed *Dictyota dentata*. *Journal of Natural Products*, **2004**, 48, 132–134.

[167] Crusco, A.; Baptista, R.; Bhowmick, S.; Beckmann, M.; Mur, L. A. J.; Westwell, A. D.; Hoffmann, K. F. The antimycobacterial activity of a diterpenoid-like molecule operates through nitrogen and amino acid starvation. *Frontiers in Microbiology*, **2019**, 10, 1444. doi: 10.3389/fmicb.2019.01444.

[168] Li, W.-Y.; Liu, Y.; Lin, Y.-T.; Liu, Y.-C.; Guo, K.; Li, S.-H. Antibacterial harziane diterpenoids from a fungal symbiont Trichoderma atroviride isolated from *Colquhounia coccinea* var. mollis. *Phytochemistry*, **2020**, 170, 112198. doi: 10.1016/j.phytochem.2019.112198.

[169] Ghisalberti, E. L.; Hockless, D. C. R.; Rowland, C.; White, A. H. Harziandione, a new class of diterpene from *Trichoderma harzianum*. *Journal of Natural Products*, **1992**, 55(11), 1690–1694.

[170] Hanuman, J. B.; Bhatt, R. K.; Sabata, B. K. A diterpenoid furanolactone from *Tinospora cordifolia*. *Phytochemistry*, **1986**, 25(7), 1677–1680. doi: 10.1016/s0031-9422(00)81234-0.

[171] Yilmaz, A.; Saylor Crowley, R.; Sherwood, A. M.; Prisinzano, T. E. Semisynthesis and kappa opioid receptor activity of derivatives of columbin, a furanolactone diterpene. *Journal of Natural Products*, **2017**, 80(7), 2094–2100. doi: 10.1021/acs.jnatprod.7b00327.

[172] Duarte, I. D. G.; Alves, D. L. F.; Craig, M. N. Possible participation of endogenousopioid peptides on the mechanism involved in analgesia induced by vouacapano. *Life Sciences*, **1992**, 50, 891–897.

[173] Duarte, I. D. G.; Alves, D. L. F.; Veloso, D. P.; Craig, M. N. Evidence of the involve-ment of biogenic amines in the antinociceptive effect of a vouacapan extracted from *Pterodon polygalaflorus* Benth. *Journal of Ethnopharmacology*, **1996**, 55, 13–18.

[174] Galceran, C. B.; Sertie, J. A. A.; Lima, C. S.; Carvalho, J. C. T. Anti-inflammatory and analgesic effects of 6α, 7β-dihydroxyvouacapan-17β-oic acid isolated from *Pterodon emarginatus* Vog. fruits. *Inflammopharmacology*, **2011**, 19, 139–143.

[175] Omena, M. C.; Bento, E. S.; Paula, J. E.; Santana, A. E. G. Larvicidal diterpenes from *Pterodon polygalaeflorus*. *Vector-Borne and Zoonotic Diseases*, **2006**, 6, 216–222.

[176] Pimenta, A. T. A.; Santiago, G. M. P.; Arriaga, A. M. C.; Menezes, G. H. A.; Bezerra, S. B. Estudo fitoquímico e avaliaç̧ ão da atividade larvicida de *Pterodon polygalaeflorus* Benth. (Leguminosae) sobre *Aedes aegypti*. *Revista Brasileira de Farmacognosia*, **2006**, 16, 501–505.

[177] Vieira, C. R.; Marques, M. F.; Soares, P. R.; Matuda, L.; Oliveira, C. M. A.; Kato, L.; Silva, C. C.; Guillo, L. A. Antiproliferative activity of *Pterodon pubescens* Benth. seed oil and its active principle on human melanoma cells. *Phytomedicine*, **2008**, 15, 528–532.

[178] Reis, F. C.; Andrade, D. M. L.; Neves, B. J.; Oliveira, L. A. R.; Pinho, J. F.; Silva, L. P.; Cruz, J. S.; Bara, M. T. F.; Andrade, C. H.; Rocha, M. L. Blocking the L-type Ca²⁺channel (Cav1.2) is the key mechanism for the vascular relaxing effect of Pterodonspp. and its isolated diterpene methyl-6α-acetoxy-7β-hydroxyvouacapan-17β-oate. *Pharmacological Research*, **2015**, 100, 242–249.

[179] Oliveira, L. A. R.; Oliveira, G. A. R.; Lemes, G. F.; Romão, W.; Vaz, B. G.; Albuquerque, S.; Gonç̧ alez, C.; Lião, L. M.; Bara, M. T. F. Isolation and structural characterization of two new furanoditerpenes from *Pterodon emarginatus* (Fabaceae). *Journal of the Brazilian Chemical Society*, **2017**, doi: 10.21577/0103-5053.20170118.

[180] Chun-Huan, L.; Jia-Yao, Z.; Xiu-Yun, Z.; Sheng-Hong, L.; Jin-Ming, G. An overview of grayanane diterpenoids and their biological activities from the Ericaceae family in the last seven years. *European Journal of Medicinal Chemistry*, **2019**, 166, 400–416. doi: 10.1016/j.ejmech.2019.01.079.

[181] Niu, C. S.; Li, Y.; Liu, Y. B.; Ma, S. G.; Liu, F.; Cui, L.; Yu, H. B.; Wang, X. J.; Qu, J.; Yu, S. S. Grayanane diterpenoids with diverse bioactivities from the roots of *Pieris formosa*. *Tetrahedron*, **2018**, 74, 375–382.

[182] Li, C. -H.; Zhang, J. -Y.; Zhang, X. -Y.; Li, S. -H.; Gao, J. -M. An overview of grayanane diterpenoids and their biological activities from the Ericaceae family in the last seven years. *European Journal of Medicinal Chemistry*, **2019**, 166, 400–416. doi: 10.1016/j.ejmech.2019.01.079.

4 Name Reactions in Terpenoid Chemistry

Bimal Krishna Banik[*], Biswa Mohan Sahoo[†], and Abhishek Tiwari[‡]

CONTENTS

4.1 Introduction 154
    4.1.1 Click Reactions..154
    4.1.2 Molecular Rearrangements 155
    4.1.3 Wagner–Meerwein Rearrangement..157
    4.1.4 Wittig Reaction...158
    4.1.5 Mannich Reaction...158
    4.1.6 Oppenauer Oxidation 159
    4.1.7 Grignard's Reaction..160
    4.1.8 Reformatsky Reaction 160
    4.1.9 Pauson–Khand Reaction 160
    4.1.10 Diels–Alder Reaction 161
    4.1.11 Baeyer–Villiger Oxidation..162
    4.1.12 Claisen–Schmidt Condensation 162
    4.1.13 Claisen Rearrangement 163
    4.1.14 Dieckmann Condensation 164
    4.1.15 Beckmann Rearrangement 164
    4.1.16 Pinacol–Pinacolone Rearrangement 164
    4.1.17 Birch Reduction..165
    4.1.18 Wurtz Reaction...167
    4.1.19 Wolff–Kishner Reduction...168
    4.1.20 Wolff Rearrangement 169
    4.1.21 Meerwein–Ponndorf–Verley (MPV) Reduction..169
    4.1.22 Curtius Rearrangement 170
    4.1.23 Luche Reduction...170
    4.1.24 Favorskii Rearrangement 170
    4.1.25 Knoevenagel Reaction 170
    4.1.26 Perkin Reaction 171
    4.1.27 Lossen Rearrangement 172
    4.1.28 Cope Rearrangement...172
    4.1.29 Arndt–Eistert Reaction 174
    4.1.30 Miscellaneous...175
4.2 Conclusion..177
References...177

[*] Department of Mathematics and Natural Sciences, College of Sciences and Human Studies, Prince Mohammad Bin Fahd University, Al Khobar, Kingdom of Saudi Arabia. Corresponding author e-mail address: bimalbanik10@gmail.com.

[†] Roland Institute of Pharmaceutical Sciences, Berhampur affiliated to Biju Patnaik University of Technology (BPUT), Rourkela, Odisha, India. drbiswamohansahoo@gmail.com.

[‡] Faculty of Pharmacy, Pharmacy Academy, IFTM University, Lodhipur Rajput, Moradabad-244102, Uttar Pradesh, India.

DOI: 10.1201/9781003008682-4

4.1 Introduction

The name reactions are chemical reactions named after their discoverers or developers. There are systematic approaches for naming reactions based on their reaction mechanisms. Grignard reaction, Wittig reaction, Claisen condensation, Claisen–Schmidt condensation, Friedel–Crafts acylation, Friedel–Crafts alkylation, Wolff–Kishner reduction, Baeyer–Villiger oxidation, Beckmann rearrangement, Hofmann reaction, Knoevenagel reaction, Mannich reaction, Oppenauer oxidation, Perkin reaction, pinacol–pinacolone rearrangement, Reformatsky reaction, Dieckmann *condensation*, Wurtz reaction, Meerwein–Pondorf reduction, Birch reduction, Biginelle reaction, Curtius rearrangement, Favorskii rearrangement, Noyori reduction, Luche reduction and Diels–Alder reaction are the well-known examples of name reactions (Figure 4.1) [1–4].

Terpenes are mainly present as constituents of essential oils and occur as secondary metabolites. These are mainly produced naturally from plants and other sources such as insects, fungi, marine microorganisms, etc [5]. Most of the naturally occurring terpenes are unsaturated hydrocarbons with five-unit carbon (C_5) structure called isoprene [6]. The isoprene units are well arranged in a uniform pattern, i.e. head-to-tail fashion. Depending on the presence of isoprene units (C_5H_8), terpenes are categorized into monoterpene ($C_{10}H_{16}$), sesquiterpene ($C_{15}H_{24}$), diterpene ($C_{20}H_{32}$), sesterpene ($C_{25}H_{40}$), triterpene ($C_{30}H_{48}$), tetraterpenes ($C_{40}H_{64}$), etc [7]. The transformations and chemical modification of terpenes can be achieved by using rearrangement reactions of carbon chain that convert these substances into new organic compounds called terpenoids [8]. The chemical modifications of terpenes may include various chemical reactions such as hydrogenation, cyclization, isomerization, polymerization, rearrangement, condensation, hydration, oxidation, hydroformylation, ring contraction, etc [9,10].

4.1.1 Click Reactions

Thiol-ene reactions are known to proceed via free-radical addition reaction, and these reactions predominately generate an anti-Markovnikov product. The reaction mechanism initially proceeds through the attack of a sulfenyl radical on the olefin to afford a new carbon radical. The as-formed radical subsequently reacts with another thiol substrate to afford the thioether product and a new sulfenyl radical. It allows the propagation step to continue the radical cycle. As there is a breaking of an S-H bond during the reaction, the lifetime of the intermediate carbon radical and the structure of the thiol affect the result of the overall reaction rate [11]. In addition to this, the reaction depends on the structure of the olefin because the formation of the C-S bond proceeds through a reversible process (Figure 4.2).

Firdaus et al. reported the thiol-ene (Click) reactions as efficient tools for terpene modification. They focused on the application of thiol-ene (click) chemistry for the production of a wide range of molecules that are derived from terpenes. The thiol-ene reaction is considered as an efficient approach and

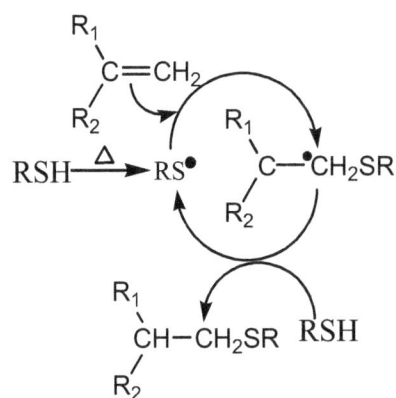

FIGURE 4.2 General mechanism of the thiol-ene addition reaction.

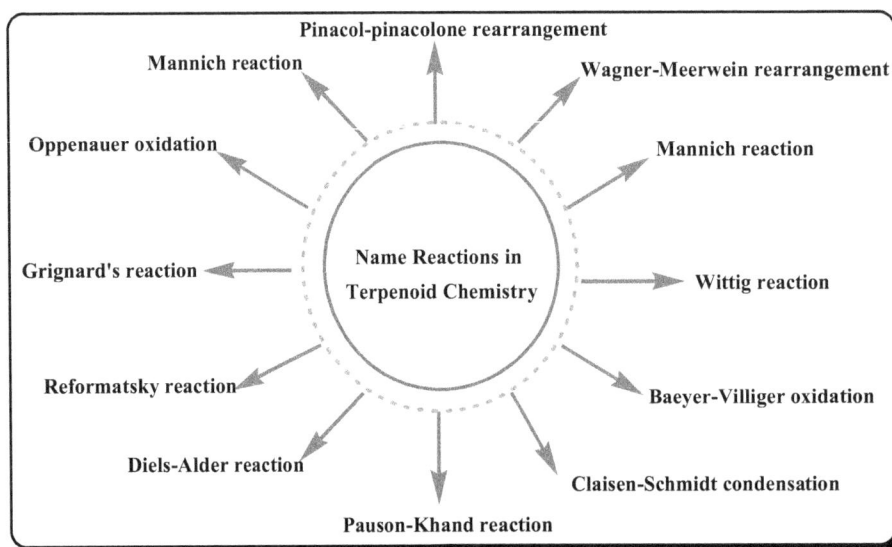

FIGURE 4.1 Name reactions involved in terpenoid chemistry.

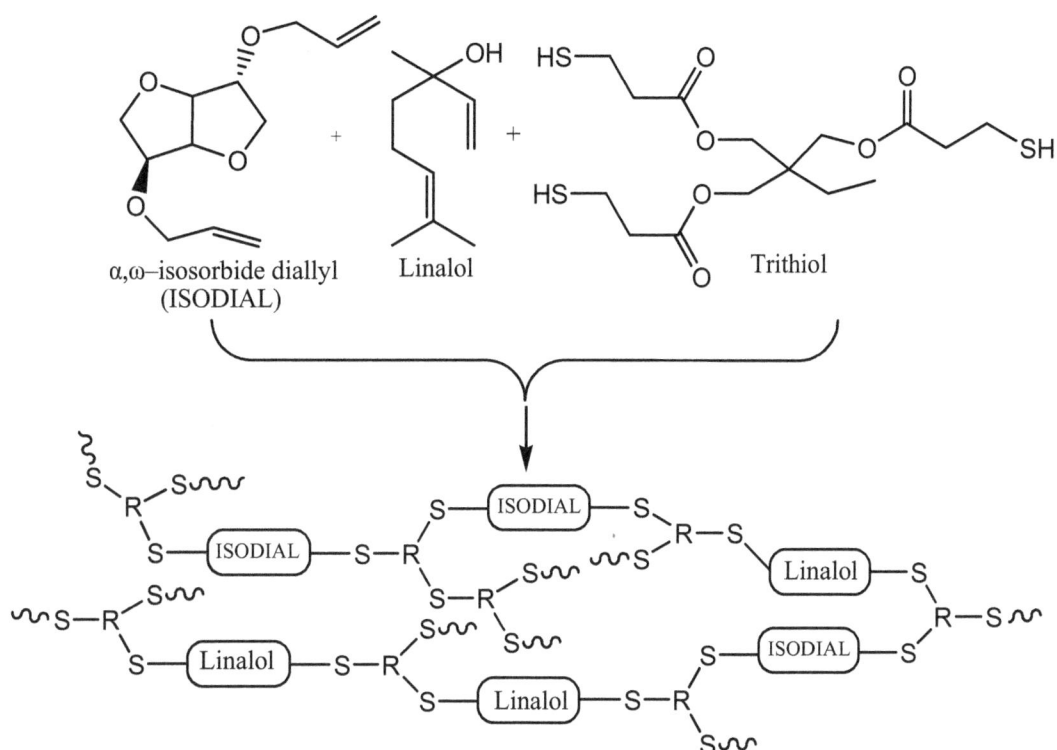

FIGURE 4.3 Synthesis of an ISODIAL–linalool network via thiol-ene reactions.

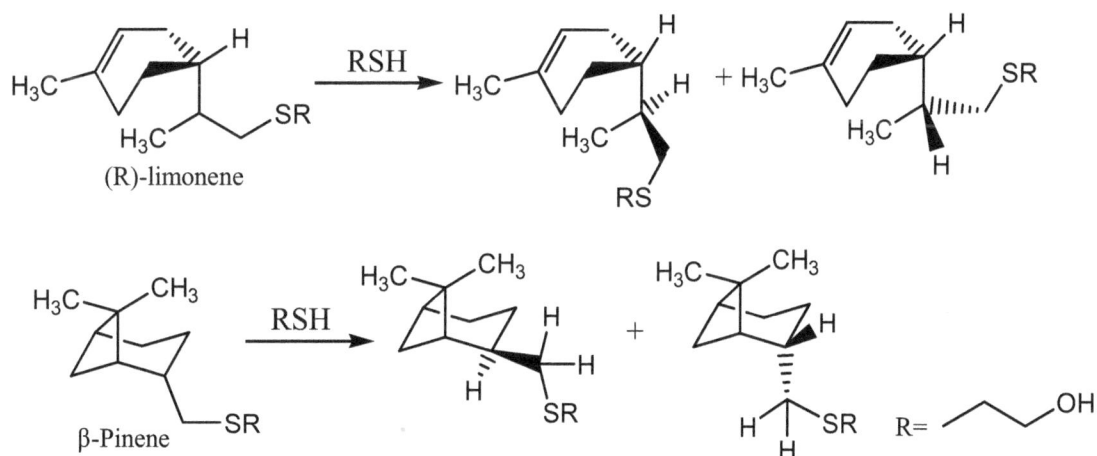

FIGURE 4.4 Hydrogen-abstraction step in the addition of 2-mercaptoethanol to (R)-limonene (top) and β-pinene (bottom).

versatile tool for noncatalytic carbon-heteroatom bond formation [12]. The utilization of thiol-ene chemistry in a polymerization reaction provides several advantages including faster reaction rate, decreased oxygen inhibition, the formation of homogenous polymer structures and low polymer shrinkage (Figures 4.3 and 4.4).

Firdaus and Meier developed new classes of monomers and polymers from limonene (Figure 4.5). They described the addition of cysteamine hydrochloride to limonenes as an efficient approach for the synthesis of new renewable monomers that contain amino groups for the preparation of polyamides (Figure 4.6) [13].

4.1.2 Molecular Rearrangements

The rearrangements of different organic compounds like terpenes are performed by using acids, iodine, boron trifluoride etherate, thionyl chloride-pyridine, manganese (III) acetate and silicon dioxide, etc. This type of rearrangement reactions is observed during the synthesis of natural products like diterpenes and triterpenes. The mechanism of molecular rearrangements is helpful to suggest the structures of generated products [14].

During the synthesis of sesquiterpenes, the rearrangement of cyclohexadienyl alcohols is carried out in the presence of thionyl

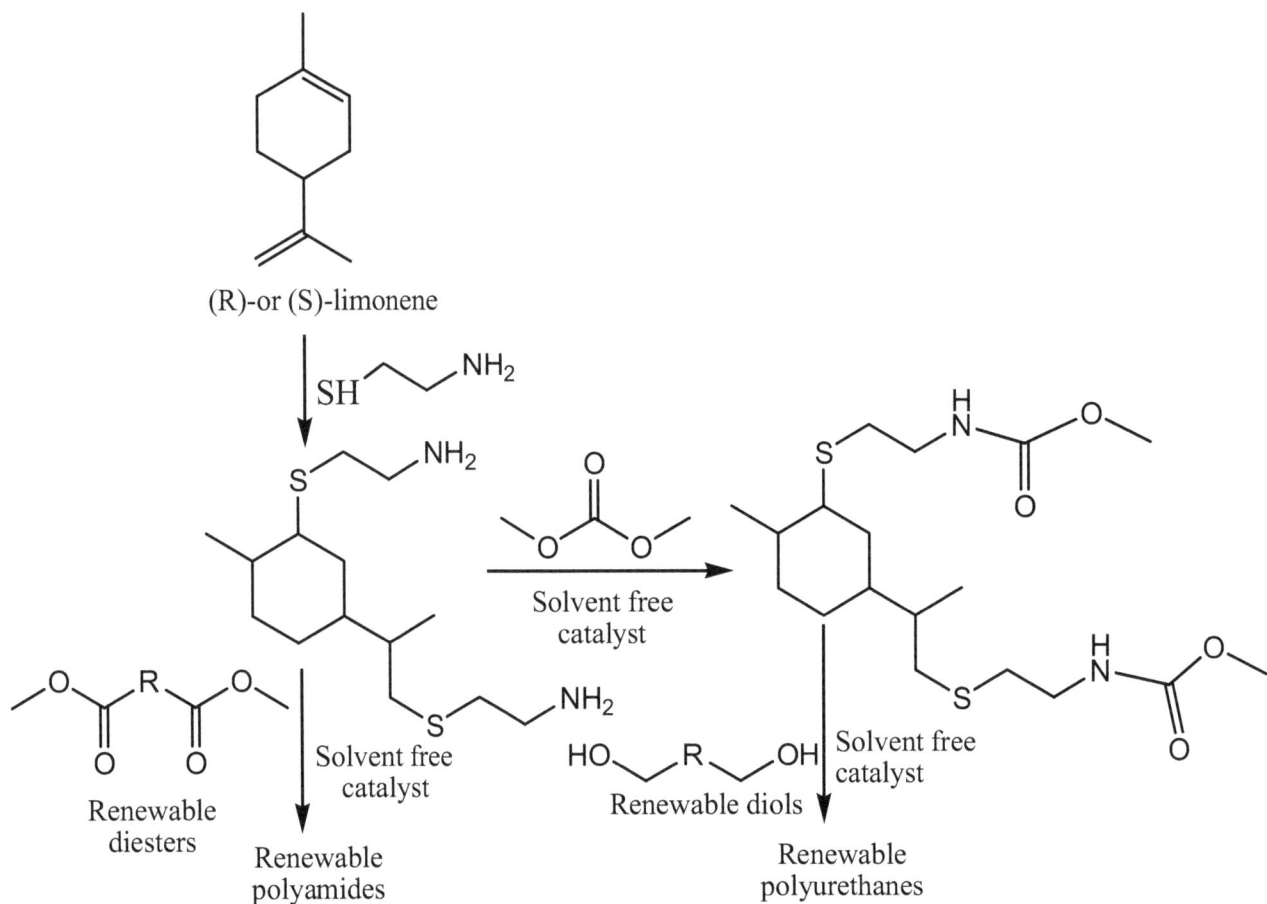

FIGURE 4.5 Synthetic route to limonene-based polyamides and polyurethanes *via* thiol-ene reactions.

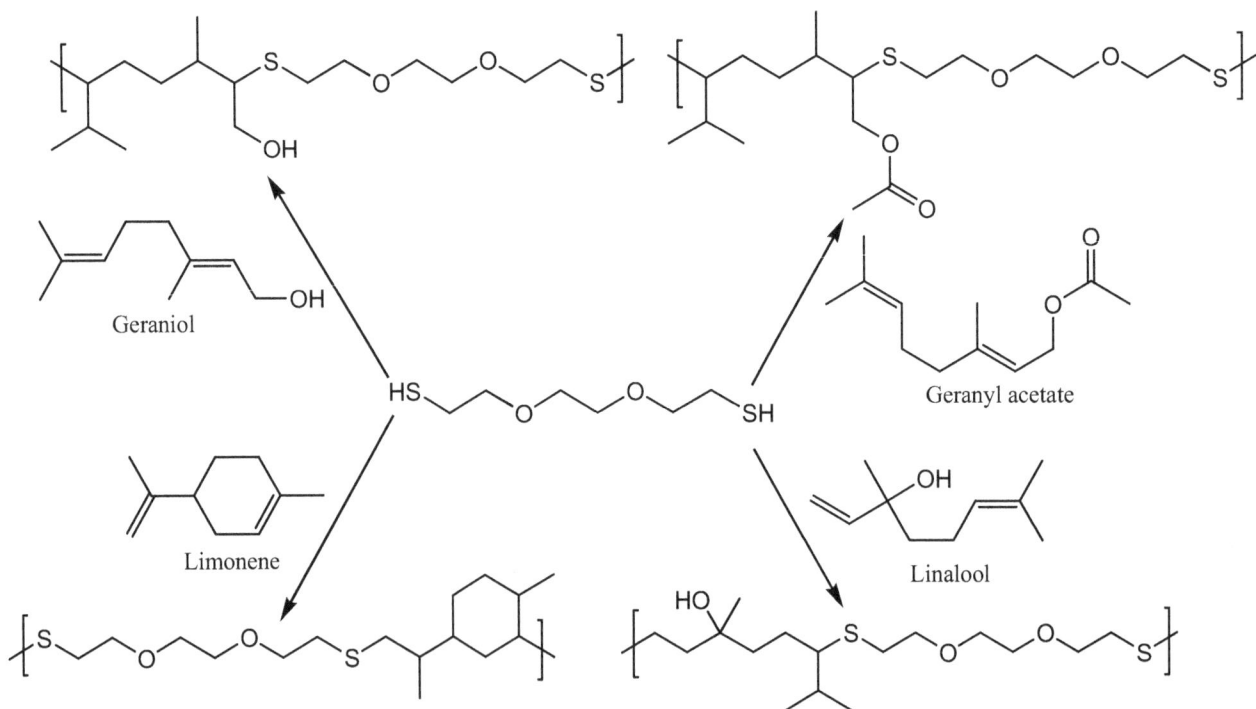

FIGURE 4.6 Synthesis of terpene-based polythioethers through thiol-ene photopolymerization.

chloride and pyridine. The alcohol produced from α-santonine undergoes aromatization with thionyl chloride and pyridine that produces hyposantonin via the intermediates (Figure 4.7).

Treatment of sesquiterpene cartol with thionyl chloride and pyridine leads to ring contraction that produces acoradienes (Figure 4.8) [15].

4.1.3 Wagner–Meerwein Rearrangement

This rearrangement reaction was first reported by Wagner in 1899, followed by Meerwein in 1914. It is a class of 1,2-rearrangement of carbocation intermediates, in which the migratory group (phenyl, vinyl, alkyl, H, etc.) rearranges from an adjacent carbon atom to the carbocation center to form a more stable carbocation or to alleviate the ring strain.

The Wagner–Meerwein rearrangement is commonly observed in fused cyclic compounds, like bicyclic terpene systems. In addition to this, rearrangement also occurs for monocyclic compounds (e.g., transformation of α-campholic acid to lauric acid), as depicted in Figure 4.9 [16].

In the case of terpene molecules, the carbon skeleton can be changed or the functional groups in the molecule can be transferred in the molecule sometimes through the effect of chemical reagents like acid. During the elimination reaction, addition reaction or nucleophilic substitution reaction of bicyclic terpenoids, there is the involvement of Wagner–Meerweein rearrangement reaction. The assembly of terpenes is catalyzed by terpene synthases that are well established to operate via carbocation intermediates. The key features of the catalysis of terpene synthases are the 1,2-rearrangement of carbocation intermediates from which the migratory group, such as alkyl, aryl, or vinyl group, rearranges from a neighboring carbon to the carbocation center to form a more stable carbocation or to alleviate the ring strain. Such 1,2-migration reactions are also known as Wagner–Meerwein rearrangements (Figure 4.10). This type of rearrangement is involved in the biosynthesis of monoterpene camphene and sesquiterpene vetispiradiene (Figure 4.11) [17].

Talatisamine is a C19-diterpenoid alkaloid, isolated from Aconitum species (Figure 4.12). It exhibits K+ channel inhibitory and antiarrhythmic activities. The synthesis of talatisamine is accomplished by utilizing Wagner–Meerwein rearrangement reaction [18].

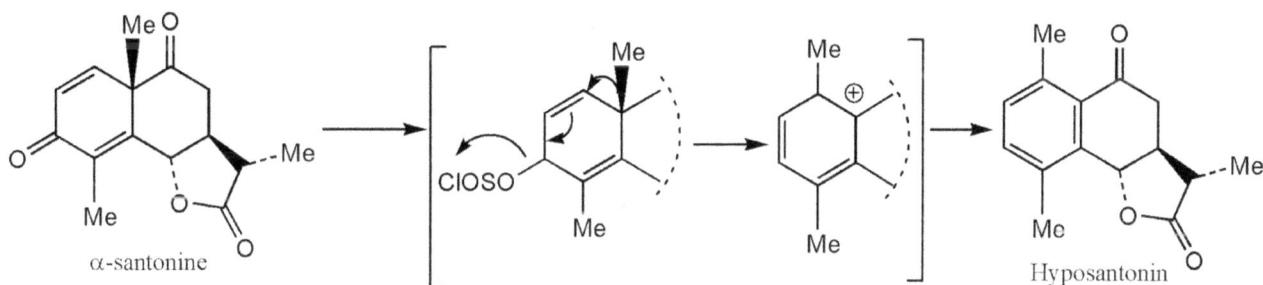

FIGURE 4.7 Synthesis of hyposantonin.

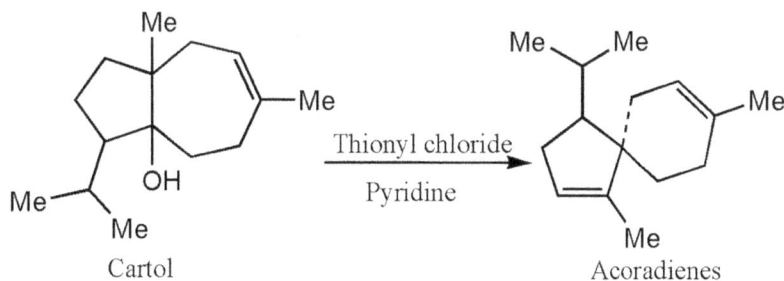

FIGURE 4.8 Synthesis of acoradiene.

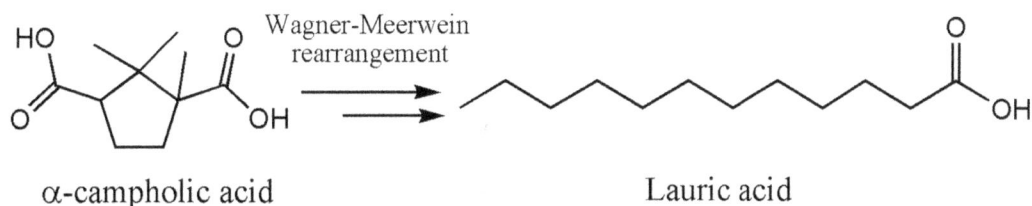

FIGURE 4.9 Transformation of α-campholic acid to lauric acid.

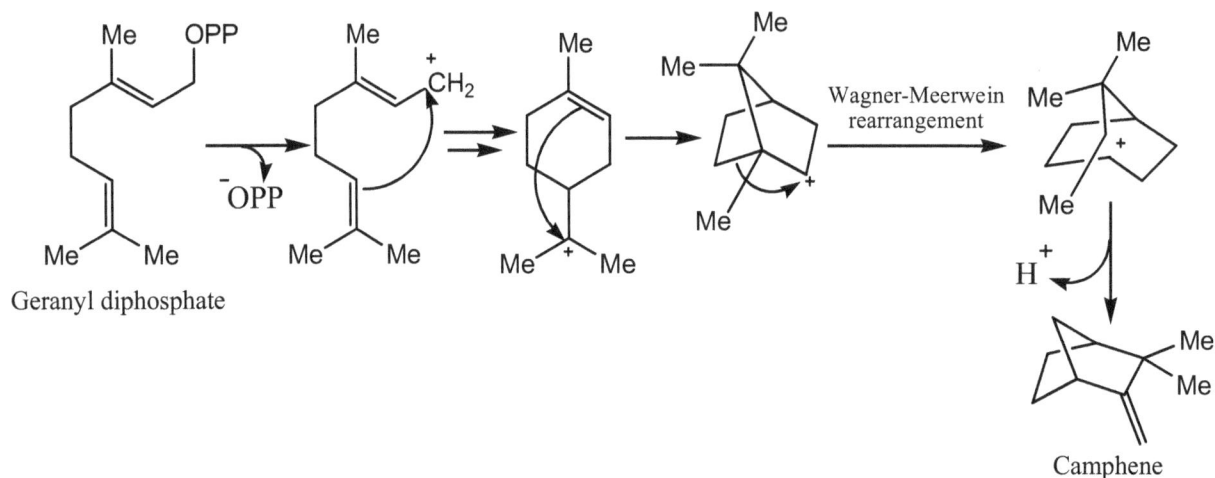

FIGURE 4.10 Synthesis of camphene from geranyl diphosphate via W-M rearrangement.

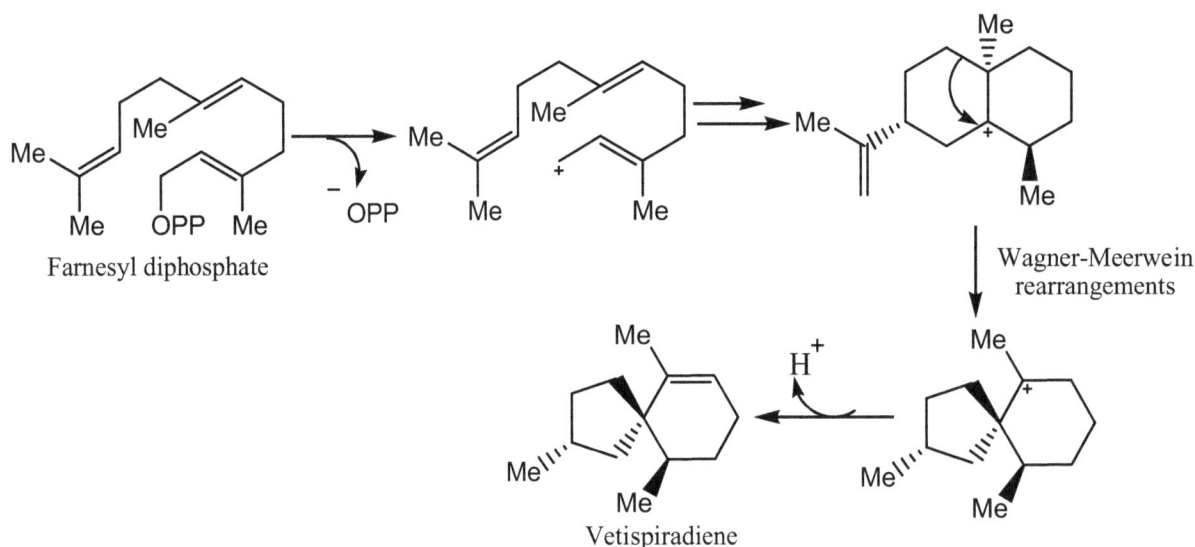

FIGURE 4.11 Synthesis of vetispiradiene from farnesyl diphosphate via W-M rearrangement.

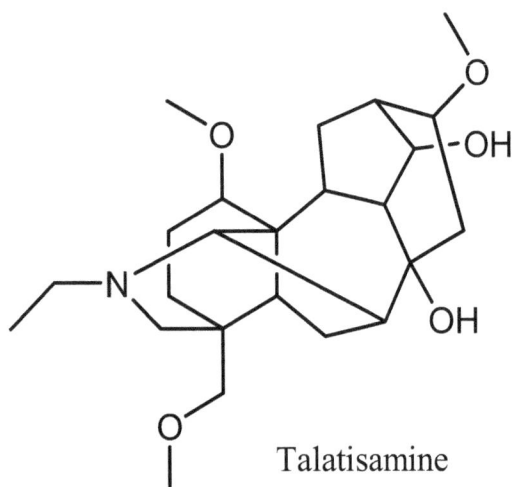

FIGURE 4.12 Structure of talatisamine.

4.1.4 Wittig Reaction

G. Wittig discovered Wittig reaction for converting carbonyl compounds into olefins. Pommer demonstrated the utilization of the Wittig reaction for linking terpenoid building blocks to give vitamin-A and carotenoids on an industrial scale [19]. Building block with a C_{15} ylide reacts with β-formylcrotyl acetate to produce vitamin A acetate through Wittig reaction (Figure 4.13).

4.1.5 Mannich Reaction

Mannich reaction is a carbon–carbon bond-forming reaction that involves the addition of resonance stabilized carbon nucleophiles to iminium salts and imines.

Citral is an acyclic monoterpene aldehyde and is chemically called 3,7-dimethyl-2,6-octadienal. Amine peroxides are synthesized by subjecting citral to Mannich-type reaction with

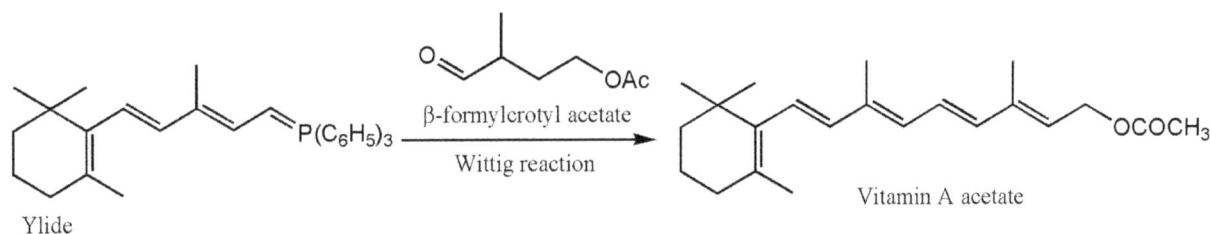

FIGURE 4.13 Synthesis of vitamin A acetate through Wittig reaction.

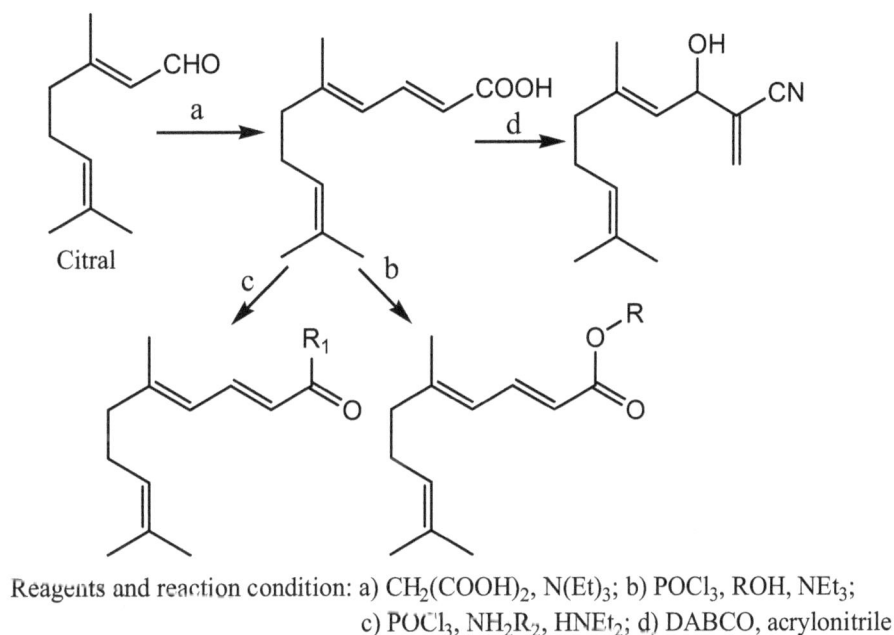

Reagents and reaction condition: a) CH$_2$(COOH)$_2$, N(Et)$_3$; b) POCl$_3$, ROH, NEt$_3$;
c) POCl$_3$, NH$_2$R$_2$, HNEt$_2$; d) DABCO, acrylonitrile

FIGURE 4.14 Scheme Synthesis of conjugated acid derivatives from citral.

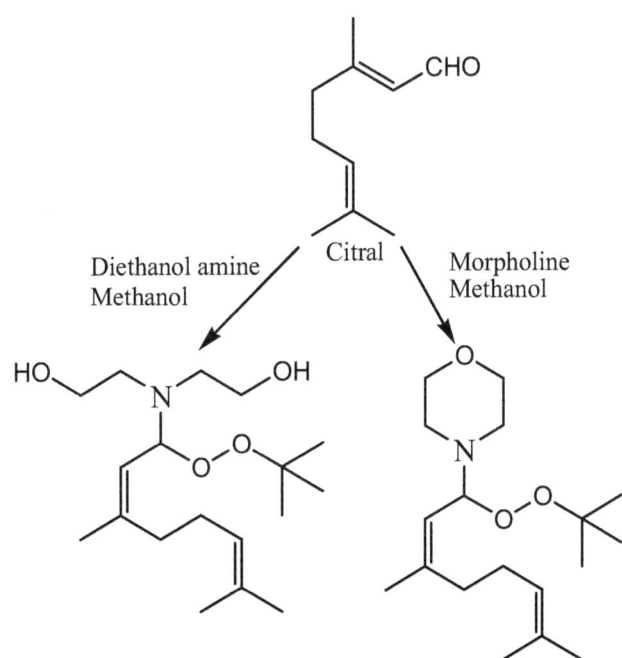

FIGURE 4.15 Synthesis of amine peroxides from citral.

amines and *t*-butyl hydroperoxide. The peroxy-amines are obtained by the Mannich reaction of citral imine (Figure 4.14). Similarly, amine peroxides are synthesized by subjecting citral to Mannich-type reaction with amines and *t*-butyl hydroperoxide (Figure 4.15) [20].

4.1.6 Oppenauer Oxidation

The Oppenauer oxidation is commonly used in various industrial processes for the synthesis of terpenes. The formation of unsaturated carbonyl compounds during the rearrangement of epoxides results from the Oppenauer oxidation of the allylic alcohol. This reaction is possible because (1) some epoxide rearrangement catalysts are also capable of catalyzing the Oppenauer oxidation and (2) by-products of the rearrangement-dihydrocarvone and aldehyde (Figure 4.16).

The Oppenauer oxidation of alcohols is a very useful method and is widely used for the synthesis of terpenoids. In addition to this, the oxidation of various saturated and unsaturated alcohol, such as geraniol, is also obtained under mild condition by using alkylaluminum R$_3$Al (R=Me, Et, *i*-Bu and 2-methylbutane)/t-butyl hydroperoxide system (Figure 4.17) [21].

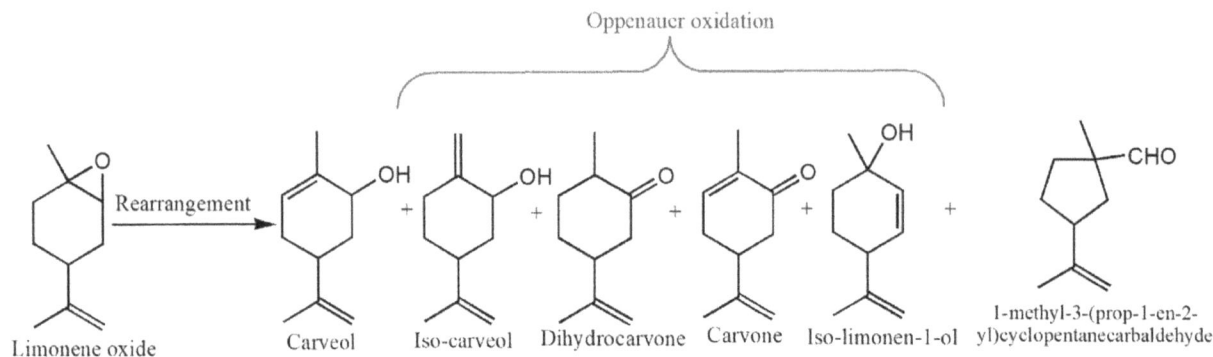

FIGURE 4.16 Oppenauer oxidation of limonene epoxide.

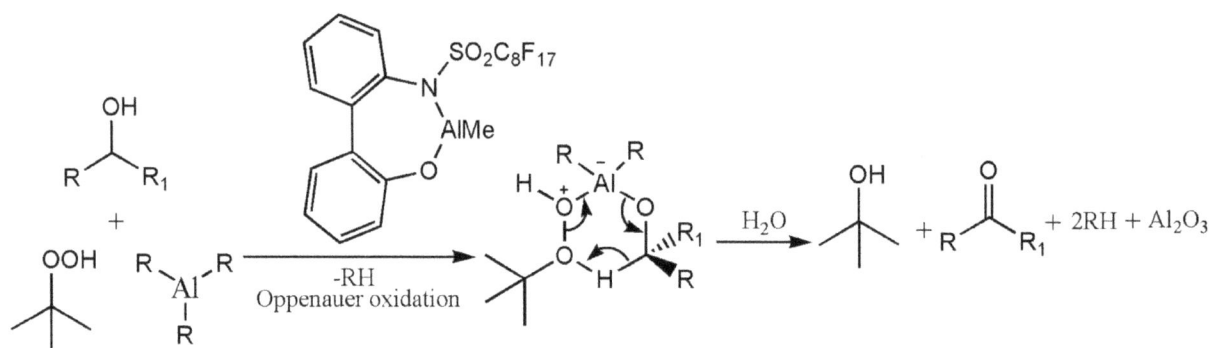

FIGURE 4.17 Oppenauer oxidation of alcohol (geraniol).

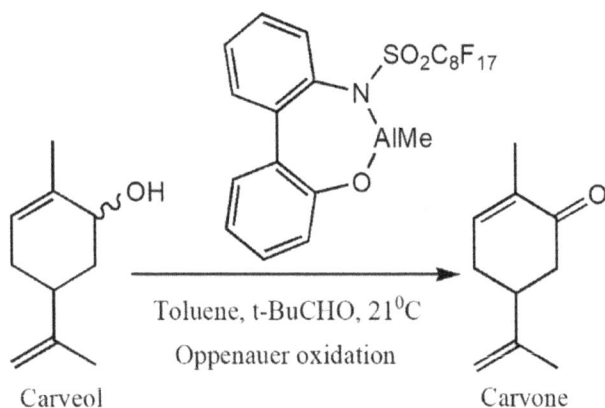

FIGURE 4.18 Oppenauer oxidation of carveol.

Modified aluminum catalyst is found to be highly effective for Oppenauer oxidation of alcohols. For example, the oxidation of carveol in toluene with *t*-BuCHO (1.2 equiv) as a hydride acceptor in the presence of modified aluminum catalyst (1 mol%) proceed smoothly at 21°C to produce carvone with 94% yield (Figure 4.18) [22].

4.1.7 Grignard's Reaction

This type of reaction is widely applied in the chemistry of terpenoids. The reaction was successfully employed by Perkin

et al. to synthesize a large number of compounds which are related to terpenoids. By Grignard's reagent, methyl or isopropyl groups can be introduced into a compound with a carbonyl group. The naturally occurring terpenoid, α-terpineol is tertiary alcohol that can be prepared by applying Grignard's reaction (Figure 4.19) [23].

4.1.8 Reformatsky Reaction

Reformatsky reaction is the reaction of a carbonyl compound (aldehyde or ketone), with α-halogeno ester in the presence of zinc to produce β-hydroxy ester. Reformatsky reaction is applied for synthesizing several terpenoids.

Halogen substituted ester is treated with a carbonyl compound (aldehyde, ketone or ester) in the presence of zinc to produce a hydroxy ester. The latter compound when treated with dilute acid yields hydroxy acid which may be further converted into an unsaturated acid or a hydrocarbon (Figure 4.20) [24].

4.1.9 Pauson–Khand Reaction

Pauson–Khand reaction (PKR) is a versatile reaction used for the synthesis of cyclopentenone and its derivatives. This reaction was discovered in 1973. In this reaction, alkyne, alkene and carbon monoxide are subjected to [2+2+1] cyclo-addition reaction mediated by a hexacarbonyldicobalt alkyne complex to produce α, β-cyclopentenones in a single step. PKR is also applied for the total synthesis of terpenoids [25].

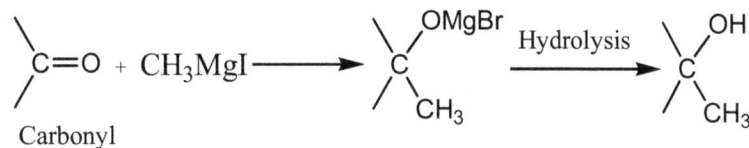

FIGURE 4.19 Synthesis of terpenoids via Grignard's reaction.

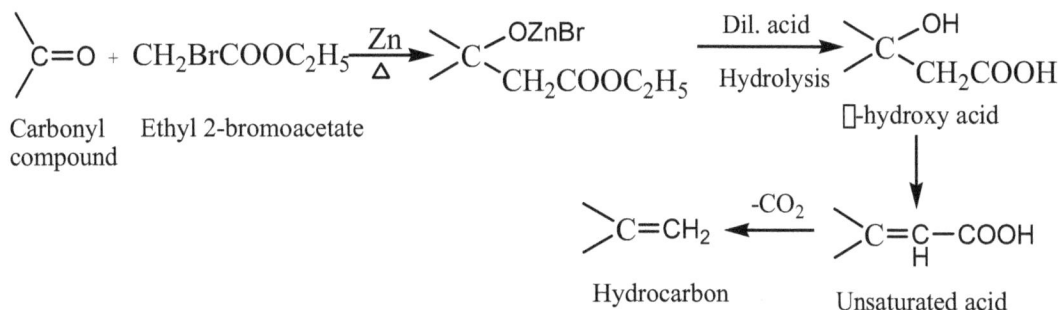

FIGURE 4.20 Reformatsky reaction for synthesizing several terpenoids.

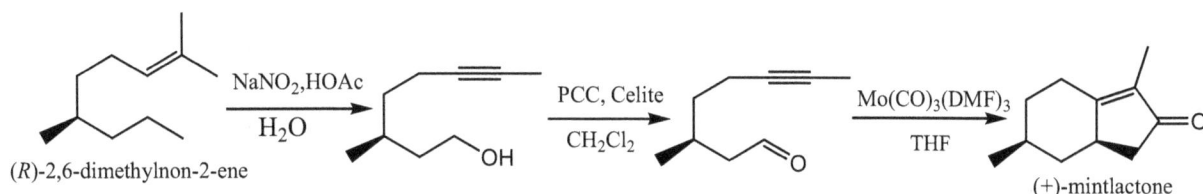

FIGURE 4.21 Total synthesis of (+)-mintlactone.

In 2009, Zhai et al. reported the total synthetic pathway of (+)-mintlactone starting from (−)-citronellol by using molybdenum-mediated intramolecular hetero-Pauson–Khand reaction as a key step. This pathway started with (−)-citronellol which upon treatment with nitrous acid following Abidi's protocol to produce alkynol directly in three steps. The latter compound undergoes oxidation upon treatment with PCC in CH_2Cl_2 at ambient temperature to yield ynal as suitable precursor for PKR. The ynal is then treated with freshly prepared organomolybdenum $Mo(CO)_3(DMF)_3$ as catalyst, in THF at ambient temperature to produce the natural product (+)-mintlactone (Figure 4.21) [26].

Epoxydictymene is a diterpene that occurs naturally in the brown alga Dictyota dichotoma. Schreiber et al. reported the total synthesis of epoxydictymene in 1994 by employing Pauson–Khand reaction. The synthesis starts from the commercially available (+)-puelgone.

4.1.10 Diels–Alder Reaction

Diels–Alder reaction is used extensively in organic reactions. It is a chemical reaction between a conjugated diene and a substituted alkene (dienophile) to afford substituted cyclohexene derivative (Figure 4.22) [27]. Fringuelli et al. reported the use of $AlCl_3$ as a catalyst for the Diels–Alder reaction of R-(-)-carvone with isoprene at 25°C. The use of $AlCl_3$ decreases the surface area for catalysis. The major problem of using $AlCl_3$

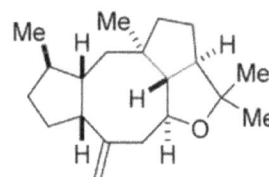

FIGURE 4.22 Structure of epoxydictymene.

as the catalyst at 25°C is that the polymerization of isoprene and carvone occurs at a similar rate to that of the Diels–Alder reaction. So, $EtAlCl_2$ and Et_2AlCl are selected to replace $AlCl_3$ (Figure 4.23).

Further, Diels–Alder reaction of R-(−)-carvone with isoprene unit is performed by using different Lewis acids as catalysts. The reaction takes place predominantly in an *anti*-orientation with respect to the isopropylene group of R-(−)-carvone (Figure 4.24).

S-(+)-carvone undergoes Diels–Alder reaction in the presence of diethylaluminum chloride proceeded by the approach of the diene anti to the isopropenyl group to give the trimethylsilyl enol ether with high *cis*-diastereoslectivity, which provided enone by Saegusa oxidation. Prior to transformation of the carbonyl group at C-4 into a methyl group, the carbonyl group at C-8 was masked selectively as dithioacetal (Figure 4.25) [28].

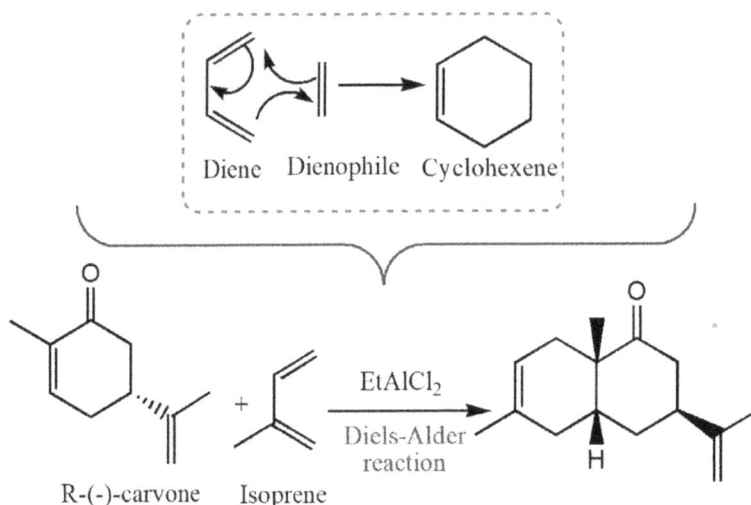

FIGURE 4.23 Diels–Alder reaction of *R*-(–)-carvone using AlCl₃ as catalyst.

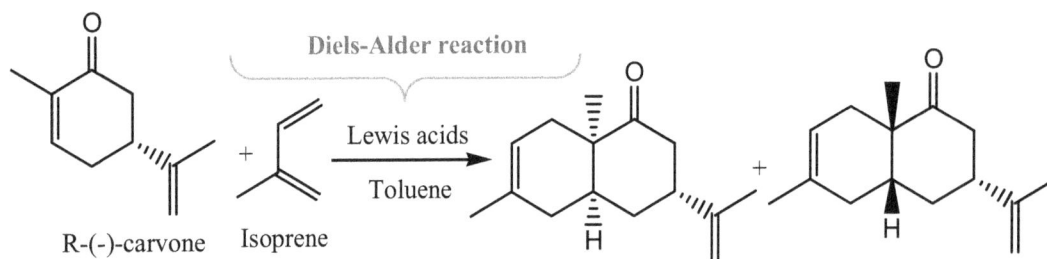

FIGURE 4.24 Diels–Alder reaction of *R*-(–)-carvone using Lewis acids as catalysts.

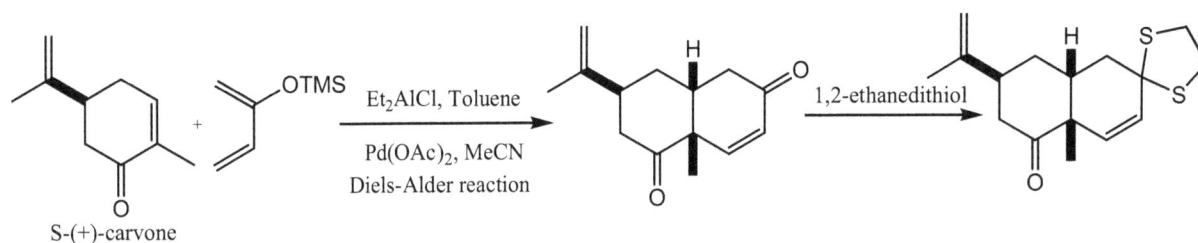

FIGURE 4.25 Diels–Alder reaction of *S*-(+)-carvone.

4.1.11 Baeyer–Villiger Oxidation

The Baeyer–Villiger Oxidation is the oxidative cleavage of a carbon–carbon bond adjacent to a carbonyl, which converts ketones to esters and cyclic ketones to lactones. The Baeyer–Villiger can be carried out with peracids, such as MCBPA, or with hydrogen peroxide and a Lewis acid.

Regioisomeric lactones are produced by microbiological Baeyer–Villiger oxidation of enantiomers of menthone and dihydrocarvone. The dihydrocarvone enantiomers are substrates for the Baeyer–Villiger monooxygenase, and the (1R, 4S)-iso-dihydrocarvone is formed from (4S)-carvone. First, it is isomerized into (1S, 4S)-dihydrocarvone before the ring-opening reactions take place. Baeyer–Villiger oxidation occurs due to the insertion of the oxygen atom near

a carbonyl group and the most substituted adjacent carbon atom. The same Baeyer–Villiger monooxygenase catalyze the oxidation of (1R, 4R)-dihydrocarvone and (1S, 4R)-iso-dihydrocarvone but with a different regiospecificity that results the formation of (3S, 6R)-6-isopropenyl-3-methyl-2-oxo-oxepanone. This Baeyer–Villiger type of monooxygenase is NADPH-dependent and exhibit optimal activity at pH 8 (Figure 4.26) [29].

4.1.12 Claisen–Schmidt Condensation

Claisen–Schmidt condensation is the reaction between an aldehyde or ketone with α-hydrogen and an aromatic carbonyl compound lacking α-hydrogen in the presence of base (Figure 4.27).

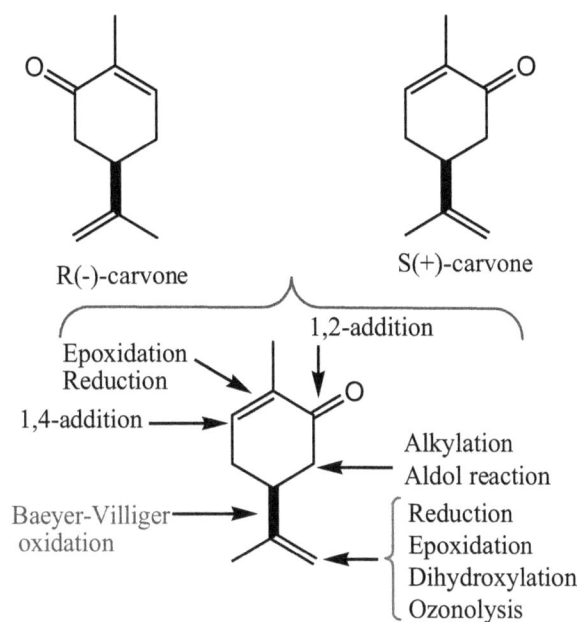

FIGURE 4.26   Versatile transformation of carvone.

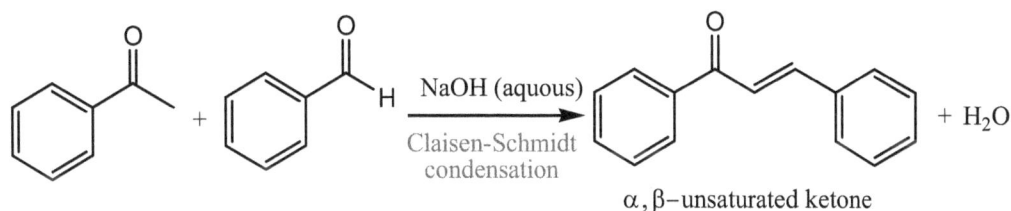

FIGURE 4.27   Claisen–Schmidt condensation.

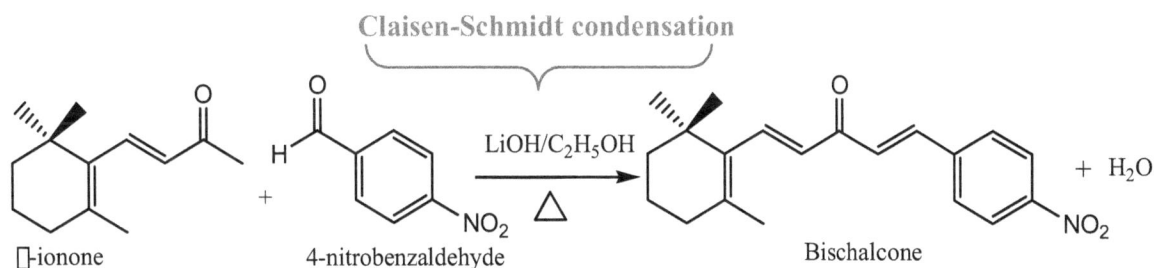

FIGURE 4.28   Synthetic scheme for Bischalcone by Claisen–Schimidt condensation reaction.

A new terpenoid-like bischalcone (1*E*, 4*E*)-1-(4-nitrophenyl)-5-(2,6,6-trimethylcyclohex-1-enyl)-penta-1,4-dien-3-one was synthesized from a β-ionone and 4-nitrobenzaldehyde by Claisen–Schmidt condensation reaction [30] (Figure 4.28).

4.1.13 Claisen Rearrangement

Claisen rearrangement was proposed in the name of German chemist Rainer Ludwig Claisen in 1912. The key intermediate required for the synthesis of monoterpenoid (e.g. citral, linalool, geraniol, etc.) is 2-methyl-2-hepten-6-one. It can be

attained via a synthesis based on the Claisen rearrangement (Figure 4.29).

Similarly, pseudoionone (**4**) is the key intermediate for the synthesis of farnesol and nerolidol (Figure 4.30).

(+)-Furodysin and (+)-furodysinin (Figure 4.31) are the sesquiterpenes isolated from marine sponges of the genus *Dysidea*. Whereas (-)-furodysin and (-)-furodysinin are present in *D. herbacea*. Furodysinin is synthesized from (1*R*, 2*R*)-(+)-limonene oxide via key reactions including a tandem Claisen rearrangement-ene reaction and trapping of a conjugate adduct of an enone to provide a bicyclic precursor of the sesquiterpene.

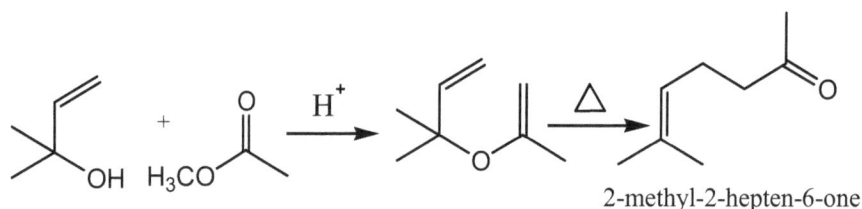

FIGURE 4.29 Claisen rearrangement.

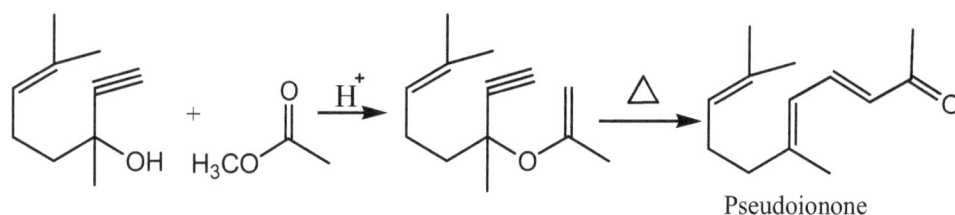

FIGURE 4.30 Synthesis of pseudoionone via Claisen rearrangement.

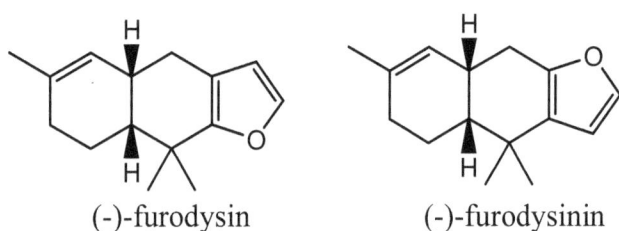

FIGURE 4.31 Structure of (+)-Furodysin and (+)-furodysinin.

Terpenes are biosynthesized through mevalonate pathway. The pathway starts with Claisen condensation between two molecules of aetyl CoA to afford compound-I followed by aldol condensation with the third molecule of aetyl CoA to yield compound-II which on hydrolysis produces isoprene (Figure 4.32) [31].

4.1.14 Dieckmann Condensation

Dieckmann condensation reaction takes place in ring formation during the synthesis of different terpenoids. This condensation involves the intramolecular reaction of diesters with a base to afford β-keto esters. Atropurpuran is a non-alkaloidal diterpene that possesses pentacyclic skeleton. It is isolated from the roots of Aconitum hemsleyanum whereas arcutines is a C20-diterpenoid alkaloid. $AlCl_3$-mediated Dieckmann cyclization strategy is applied for retrosynthetic analysis of these compounds (Figure 4.33).

Iridoids are monoterpenoids and used *for the* treatment of liver disorder, upper respiratory tract infection and dyspepsia. In the synthetic procedure of Iridoids, the intramolecular cyclization of dicarbonyl substrates affords the iridoid scaffold that utilizes both Diels–Alder and Michael addition reaction mechanisms. Further, the iridoid synthase catalyzes the cyclization of the iridoids via a Michael addition rather than a Diels–Alder reaction (Figure 4.34) [32].

4.1.15 Beckmann Rearrangement

Beckmann rearrangement reaction involves the rearrangement of aldoximes and ketoximes in the presence of acids (Lewis acids) to afford corresponding amides. Camphor and menthone are naturally occurring terpenoids that possess ketone moiety. These compounds are converted to their corresponding lactams by Beckmann rearrangement (Figure 4.35). The selectivity of the rearrangement reaction depends on the configuration of the oxime [33,34].

4.1.16 Pinacol–Pinacolone Rearrangement

Pinacol–pinacolone rearrangement is an acid-catalyzed conversion of a 1,2-diol to a carbonyl compound (aldehyde or ketone) with the elimination of water (Figure 4.36).

Liphagal is a tetracyclic meroterpenoid. It is isolated from marine sponges. It can be synthesized from the readily available sclareolide by a pinacol–pinacolone rearrangement (Figure 4.37) [35].

Scabronine G is diterpenoid and is isolated from the mushroom Sarcodon. This compound increases nerve growth factor synthesis in vitro. The pinacol–pinacolone type of rearrangement with a sulfur atom instead of a second oxygen atom has been explored for the synthesis of scabronine using Hg (II) catalysis in a regioselective fashion (Figure 4.38).

Onitin is a phenolic sesquiterpene found in Onychium siliculosum. It is chemically known as 4-hydroxy-6-(2-hydroxyethyl)-2,2,5,7-tetramethyl-3H-inden-1-one. The synthesis of onitin is accomplished by using pinacol–pinacolone rearrangement. The total synthesis of onitin involves the use of dimethylated indanone as the key intermediate. Further, the success of this synthetic route is the Suzuki–Miyaura cross-coupling reaction of aryl bromide to introduce the vinyl side chain followed by formyl selective Wacker oxidation to generate the aldehyde. The target aldehyde is obtained via an acid-catalyzed pinacol–pinacolone type rearrangement of the epoxide. The epoxide is generated from the oxidation of a

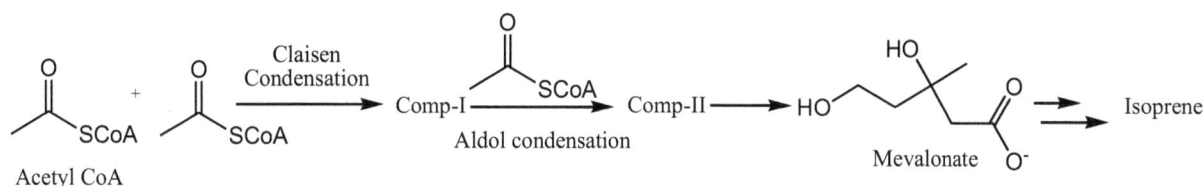

FIGURE 4.32 Production of isoprene from aetyl CoA.

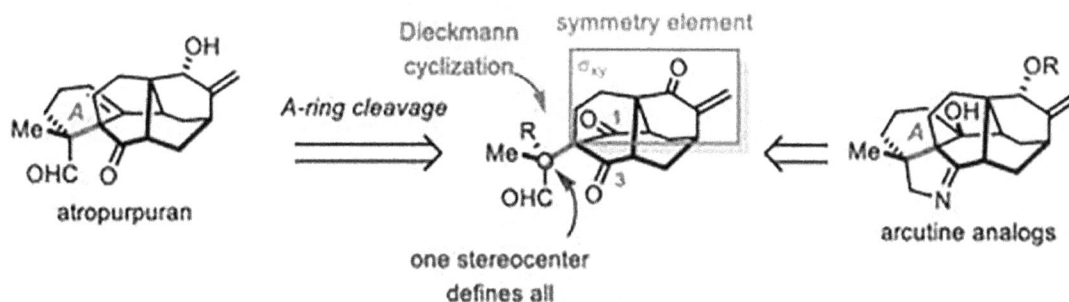

FIGURE 4.33 Retrosynthetic pathway for atropurpuran and arcutines via Dieckmann cyclization.

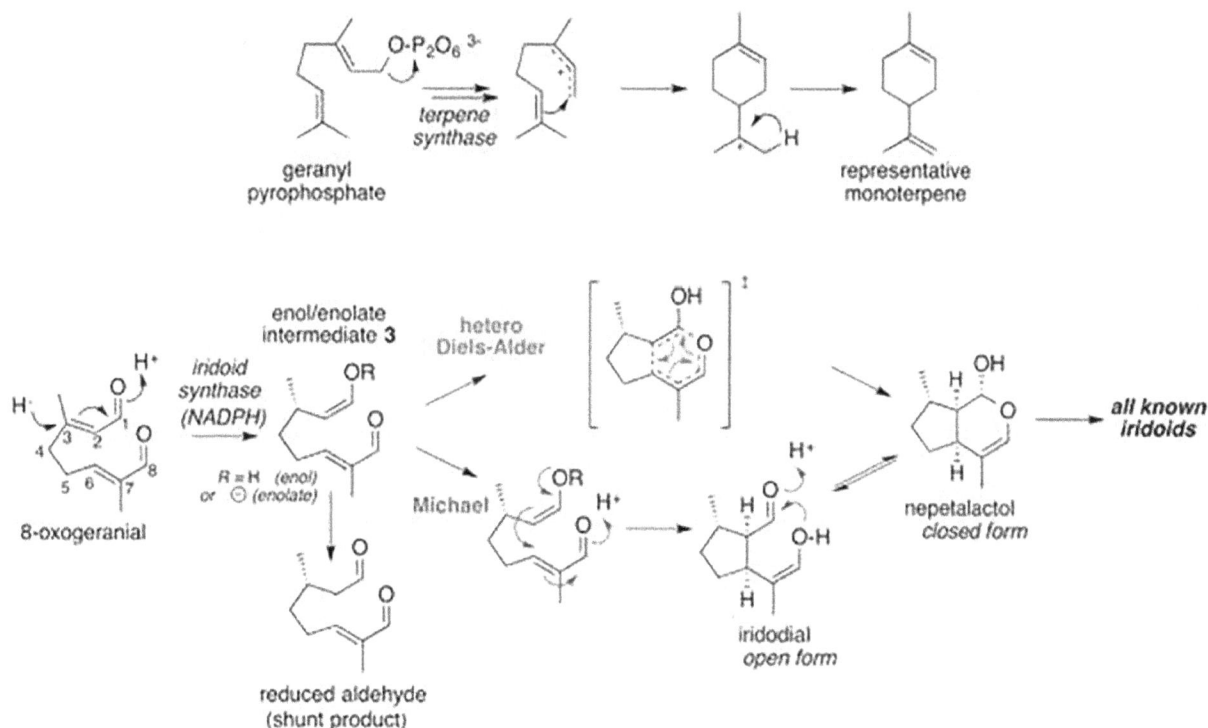

FIGURE 4.34 Generation of iridoid scaffold utilizing both Diels–Alder and Michael addition reactions.

styrene derivative by oxone. Finally the demethylation of the aldehyde followed by chemoselective reduction afford onitin (Figure 4.39) [36].

4.1.17 Birch Reduction

Birch reaction was first reported by Arthur Birch in 1944. It involves 1,4-reduction of aromatic compounds to their 1,4-hexadienes by alkali metals in ammonia in the presence of alcohol. The alkali metal serves as a source of electrons and the alcohol as a source of protons. Birch reduction is generally involved in biosynthesis of terpenes. Terpenes are derived from the sequential coupling of isopentylpyrophosphate (IPP) and its configurational isomer dimethylallylpyrophosphate (DMAPP). The mevalonate (MVA) pathway is considered for the biosynthesis of IPP and DMAPP (Figure 4.40) [37].

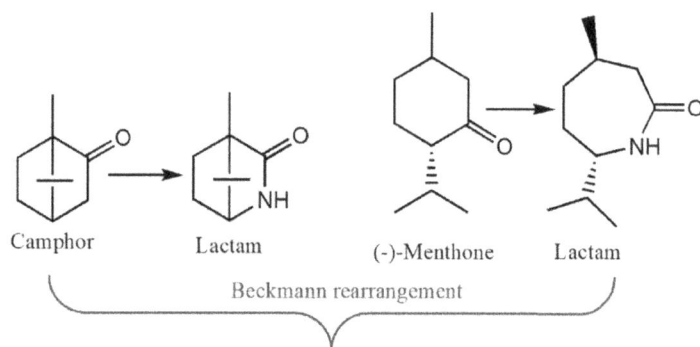

FIGURE 4.35 Beckmann rearrangement reaction of Camphor and menthone.

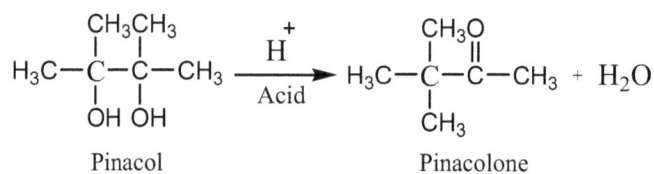

FIGURE 4.36 Pinacol–pinacolone rearrangement.

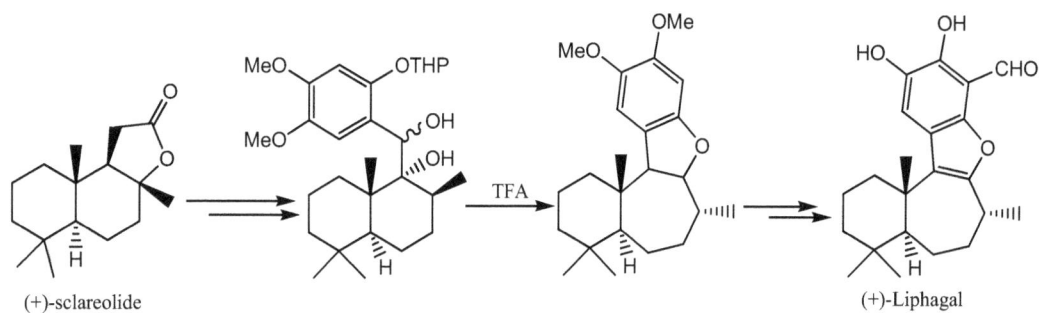

FIGURE 4.37 Synthesis of liphagal from sclareolide via pinacol–pinacolone rearrangement.

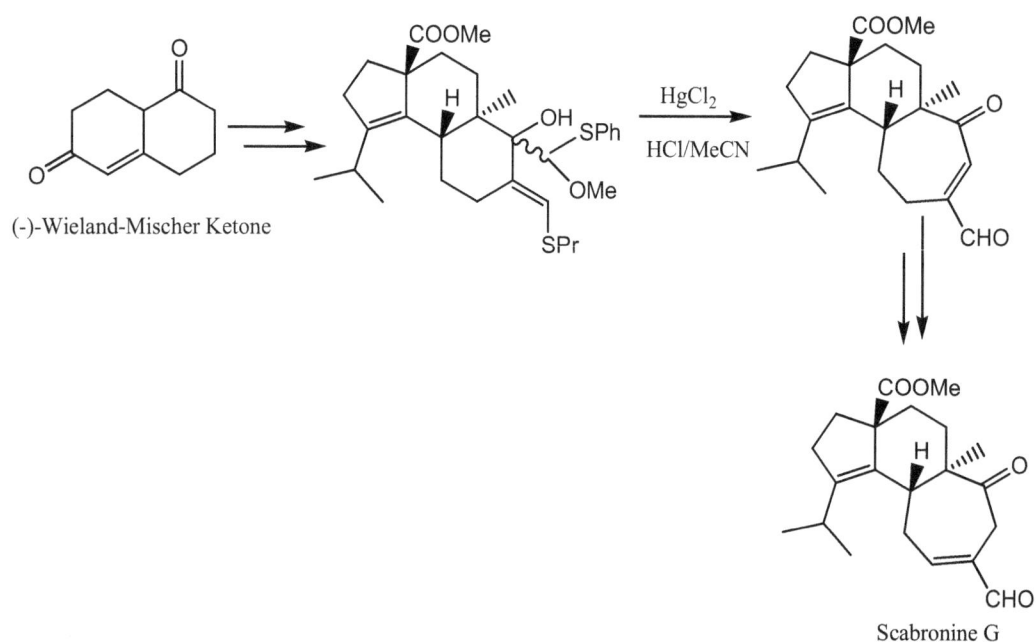

FIGURE 4.38 Synthesis of scabronine via pinacol–pinacolone rearrangement.

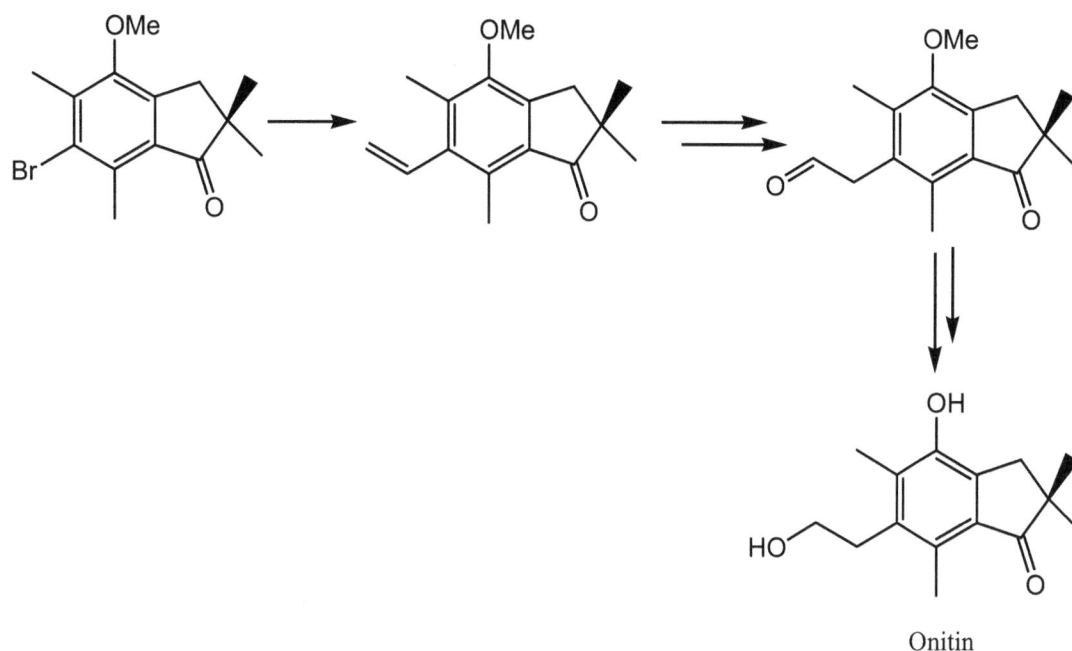

FIGURE 4.39 Synthesis of onitin via pinacol–pinacolone rearrangement.

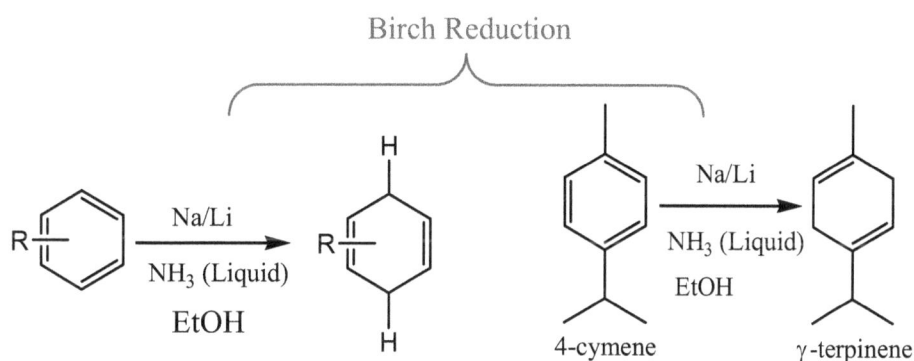

FIGURE 4.40 Birch reduction of terpenes.

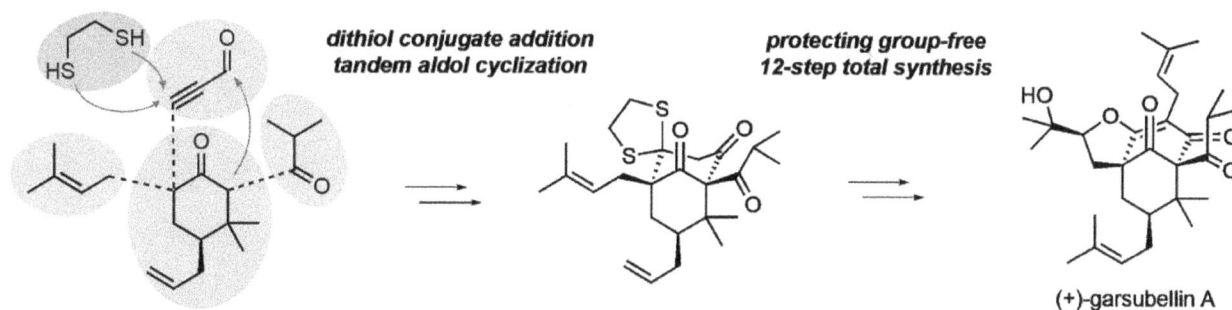

FIGURE 4.41 Synthesis of garsubellin A via Wurtz reaction.

4.1.18 Wurtz Reaction

Garsubellin A is a meroterpene and it possesses potent neurotrophic activity by inducing choline acetyltransferase, a key
enzyme for the synthesis of acetylcholine. It is found to be
a potential therapeutic agent for the treatment of Alzheimer's

disease. The synthesis of garsubellin A is performed in 18-step
sequence starting from 3,5-dimethoxyphenol. The transformations of garsubellin A include dearomative allylation, diastereoselective vinylogous lactonization, iodocarbocyclization,
transannular Wurtz and bridgehead functionalization reactions (Figure 4.41) [38].

4.1.19 Wolff–Kishner Reduction

Wolff–Kishner reduction is carried out to convert carbonyl functionalities into methylene group. Citronellal is a monoterpenoid aldehyde and is isolated from the plant Cymbopogon. It is converted to the hydrazone and semicarbazone, either of which on Wolff–Kishner reduction afford 2,6-dimethyloct-2-ene (Figure 4.42) [39].

Carvone is a monoterpene and is present in spearmint (Mentha spicata) and caraway (Carum carvi. Carvone undergoes reduction due to the presence of three double bonds. The product of reduction mainly depends on the reagents and conditions used. Catalytic hydrogenation of carvone can produce either carvomenthol or carvomenthone. Carvone undergoes a reduction in the presence of zinc and acetic acid to produce dihydrocarvone. Meerwein–Ponndorf–Verley (MPV) reduction using propan-2-ol and aluminum isopropoxide effects reduction of the carbonyl group only to afford carveol. In the presence of hydrazine and potassium hydroxide, Carvone produces limonene via a Wolff–Kishner reduction (Figure 4.43).

Camphor is a cyclic monoterpene ketone with the molecular formula $C_{10}H_{16}O$. It can be converted into bornane via Wolff–Kishner reduction. Bornane is solid with a melting point of 156°C. It is optically inactive (Figure 4.44) [40].

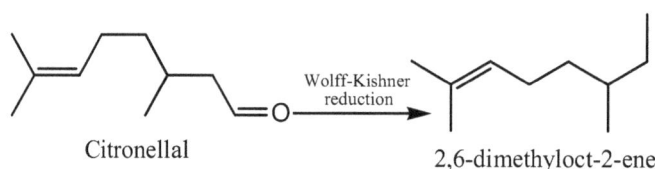

FIGURE 4.42 Wolff–Kishner reduction of Citronellal.

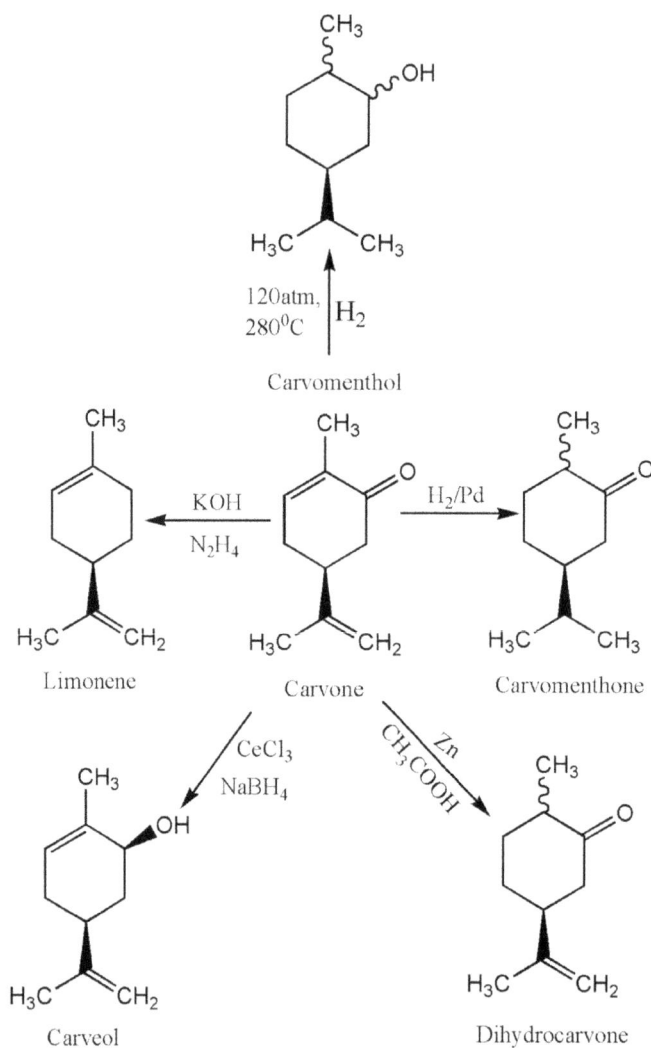

FIGURE 4.43 Production of limonene from Carvone via a Wolff–Kishner reduction.

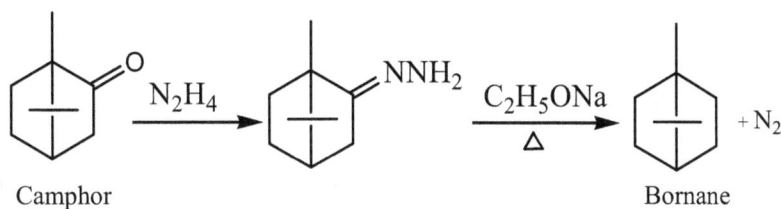

FIGURE 4.44   Production of bornane from Camphor via Wolff–Kishner reduction.

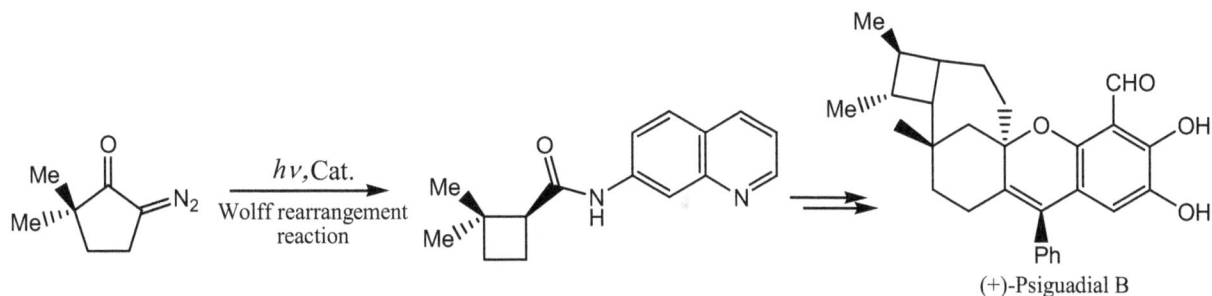

FIGURE 4.45   Synthesis of (+)-Psiguadial B via Wolff rearrangement.

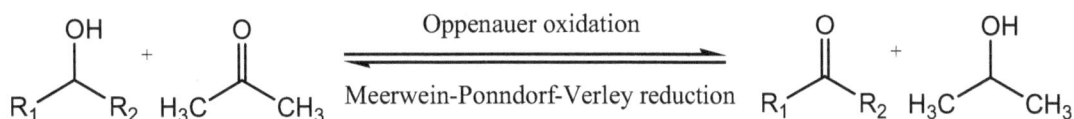

FIGURE 4.46   Meerwein–Ponndorf–Verley reduction using aluminum alkoxide as catalyst.

4.1.20 Wolff Rearrangement

Wolff rearrangement is a conversion of an α-diazoketone to a ketene with the loss of molecular nitrogen accompanying 1,2-rearrangement using a silver oxide catalyst or thermal or photochemical conditions. German chemist Ludwig Wolff discovered this reaction in 1902. (+)-Psiguadial B is a diformyl phloroglucinol meroterpenoid. It exhibits antiproliferative activity against the HepG2 human hepatoma cancer cell line. It is synthesized by using Wolff rearrangement reaction. Similarly, the synthesis of the Phenalenone diterpene salvilenone is carried out by using Wolff rearrangement reaction (Figure 4.45) [41].

4.1.21 Meerwein–Ponndorf–Verley (MPV) Reduction

Meerwein–Ponndorf–Verley reduction involves the reduction of ketones and aldehydes to their corresponding alcohols using aluminum alkoxide as a catalyst in the presence of alcohol (Figure 4.46) [42].

Meerwein–Ponndorf–Verley-Oppenauer is applied as a non-catalytic reaction of cyclic monoterpenic ketones and alcohols (menthone, camphor, menthol and borneol) under supercritical conditions. Borneol is a bicyclic monoterpenoid, present in the extract of Valeriana officinalis. Borneol exists as two enantiomers such as (+)-borneol and (–)-borneol. But the naturally occurring d-(+)-borneol is optically active. It can be synthesized by reduction of camphor via Meerwein–Ponndorf–Verley reduction in the presence of supercritical isopropanol (Figure 4.47).

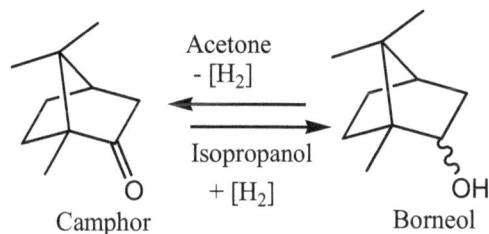

FIGURE 4.47   Meerwein–Ponndorf–Verley reduction of camphor.

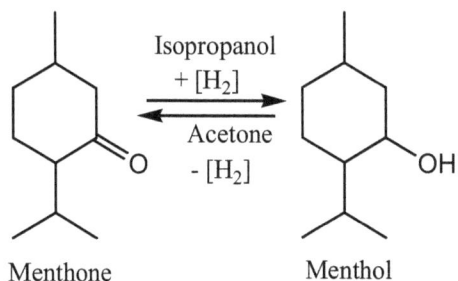

FIGURE 4.48   Meerwein–Ponndorf–Verley reduction of menthone.

Similarly, menthol is obtained from menthone via Meerwein–Ponndorf–Verley reduction in the presence of supercritical isopropanol with high selectivity (89%–98%) (Figure 4.48) [43].

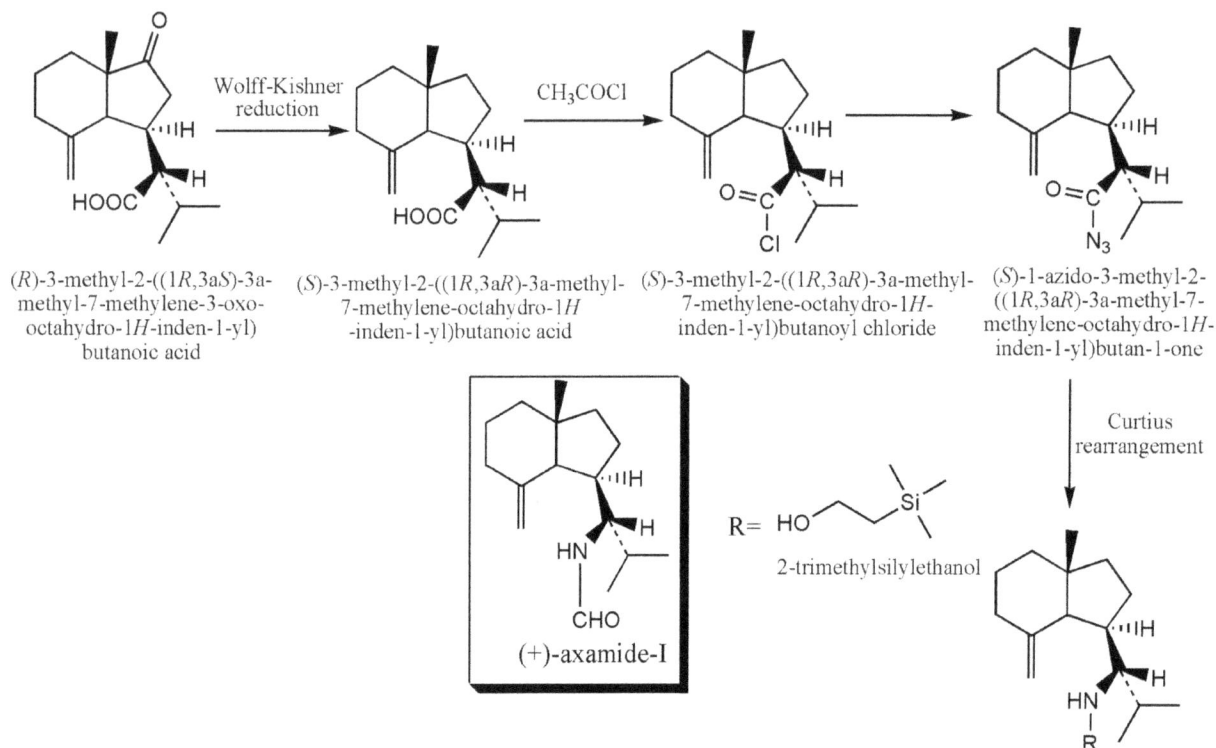

FIGURE 4.49 Production of (+)-axamide-I via Curtius rearrangement.

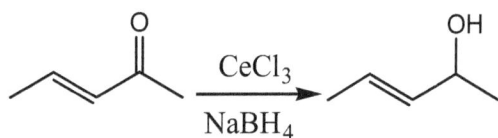

FIGURE 4.50 Luche reduction.

4.1.22 Curtius Rearrangement

Curtius rearrangement was first defined by Theodor Curtius in 1885. It involves the thermal decomposition of an acyl azide to an isocyanate with the loss of nitrogen gas. This type of *rearrangement is applied to transformation of* terpenoids. (+)-axamide-I, (+)-axisonitrile-1 and (+)-axisothiocyanate-1 are the sesquiterpenoids. These compounds are isolated from the marine sponge *Axinella canrzabina*. Removal of the keto function from (R)-3-methyl-2-((1R, 3aS)-3a-methyl-7-methylene-3-oxo-octahydro-1H-inden-1-yl)butanoic acid by Wolff–Kishner reduction afford (S)-3-methyl-2-((1R, 3aR)-3a-methyl-7-methylene-octahydro-1H-inden-1-yl)butanoic acid (90%, mp120–121°C), which is converted via the acid chloride into the acyl azide. Curtius rearrangement of the latter substance, followed by treatment of the resultant isocyanate with 2-trimethylsilylethanol afforded the crystalline carbamate (mp 79.5–80°C, 89%) (Figure 4.49) [44].

4.1.23 Luche Reduction

Luche reduction is the selective organic reduction of α, β-unsaturated ketones to allylic alcohols with sodium borohydride ($NaBH_4$) and cerium chloride ($CeCl_3$) (Figure 4.50) [45].

Kaurane-type diterpenes are found to be inhibitors of leishmania pteridine reductase I. Diterpenic acid A is esterified under mild conditions via Steglich esterification to produce 2-oxo-menth-6-en-5β-yl *ent*-kaurenoate. This ester adduct is converted into 2β-hydroxy-menth-6-en-5β-yl *ent*-kaurenoate through the selective 1,2 reduction of α, β-unsaturated ketones using Luche conditions (Figure 4.51) [46,47].

4.1.24 Favorskii Rearrangement

Favorskii rearrangement refers to the rearrangement of cyclopropanones and α-halo ketones that leads to the formation of carboxylic acid derivatives. This rearrangement reaction is named after the Russian chemist Alexei Yevgrafovich Favorskii. The construction of the first fragment started from the monoterpenoid (+)-pulegone and comprised an initial Favorskii rearrangement/ozonolysis protocol (Figure 4.52).

Transformation of (R)-pulegone to a mixture of pulegonic acids *via* Favorskii rearrangement (Figure 4.53) [48].

4.1.25 Knoevenagel Reaction

Knoevenagel reaction refers to a nucleophilic addition reaction of an active hydrogen compound with carbonyl group (aldehyde or ketone) in the presence of a base followed by dehydration in which a water molecule is eliminated. This reaction was proposed by German chemist Emil Knoevenagel in 1890. It is also considered as modified Aldol condensation reaction (Figure 4.54) [49].

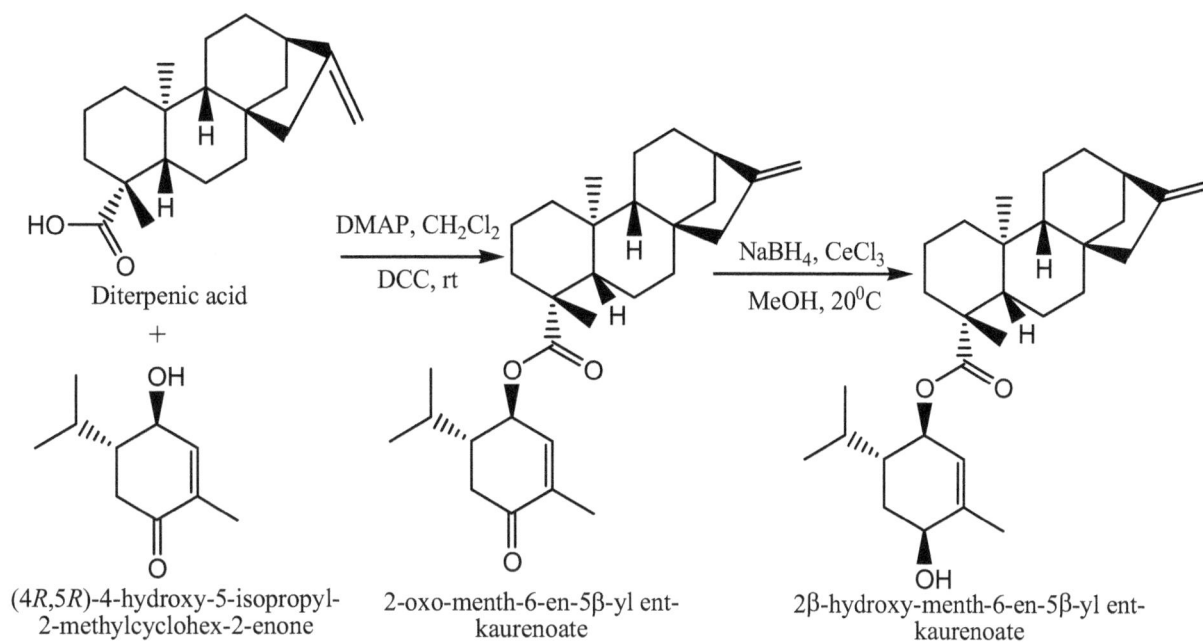

FIGURE 4.51 Production of Kaurane-type diterpenes using Luche conditions.

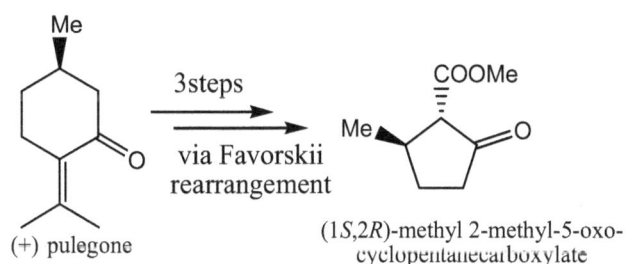

FIGURE 4.52 Favorskii rearrangement of (+)-pulegone.

Coprinol is a cuparane-type terpenoid, isolated from the cultures of a *Coprinus* sp. The synthesis of coprinol is started with the transformation of 2-methoxy-3,5-dimethylbenzaldehyde to cinnamic acid via the Knoevenagel condensation reaction (Figure 4.55) [50].

4.1.26 Perkin Reaction

Perkin Reaction involves the aldol condensation of an aromatic aldehyde and an acid anhydride in the presence of an alkali salt of the acid to produce α, β-unsaturated aromatic

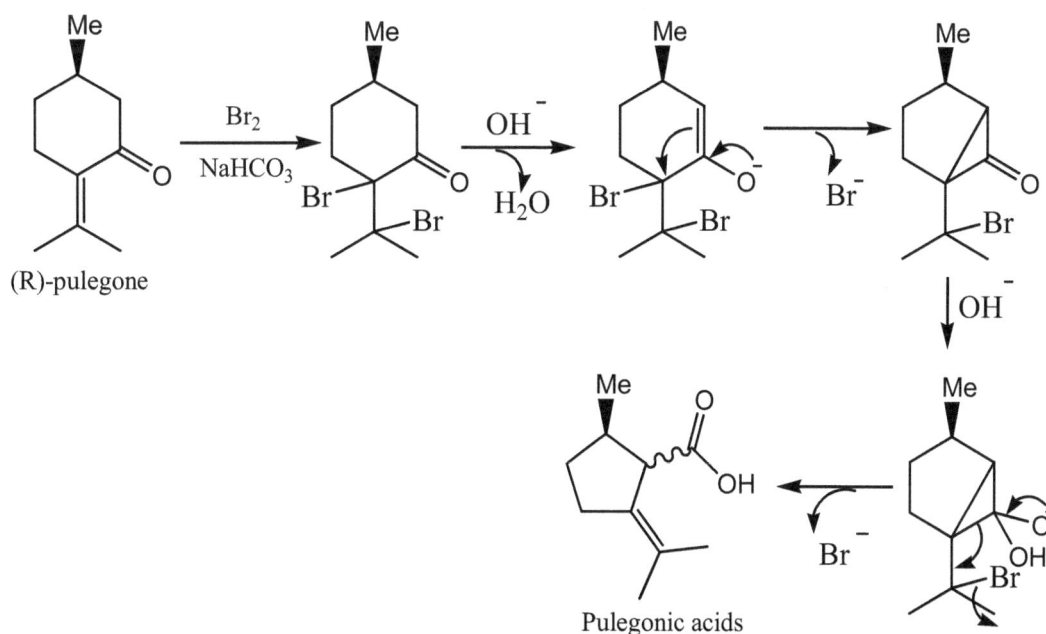

FIGURE 4.53 Transformation of (R)-pulegone *via* Favorskii rearrangement.

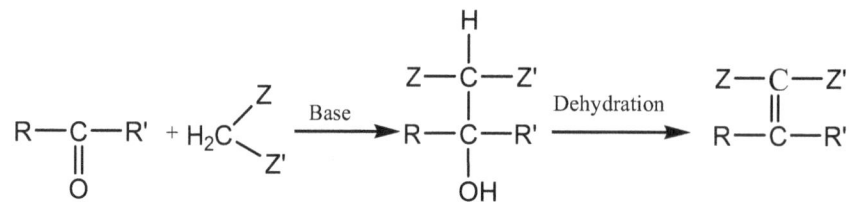

Z, Z' (electron withdrawing groups) = CO₂R, COR, CHO, CN, NO₂, etc.

FIGURE 4.54 Knoevenagel reaction.

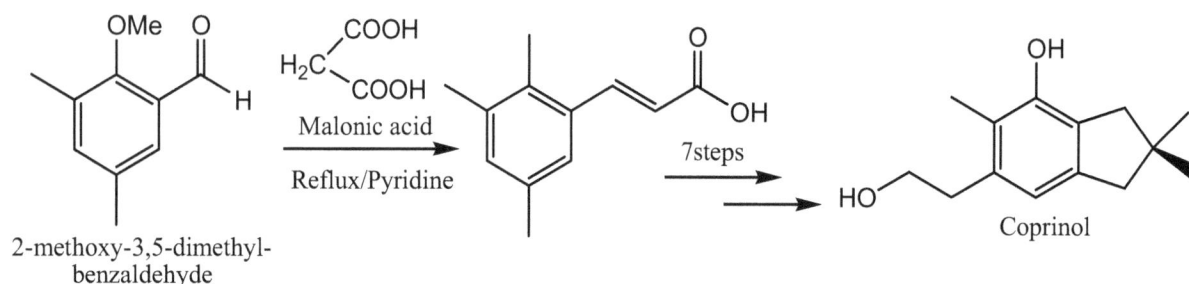

FIGURE 4.55 Production of coprinol via Knoevenagel condensation.

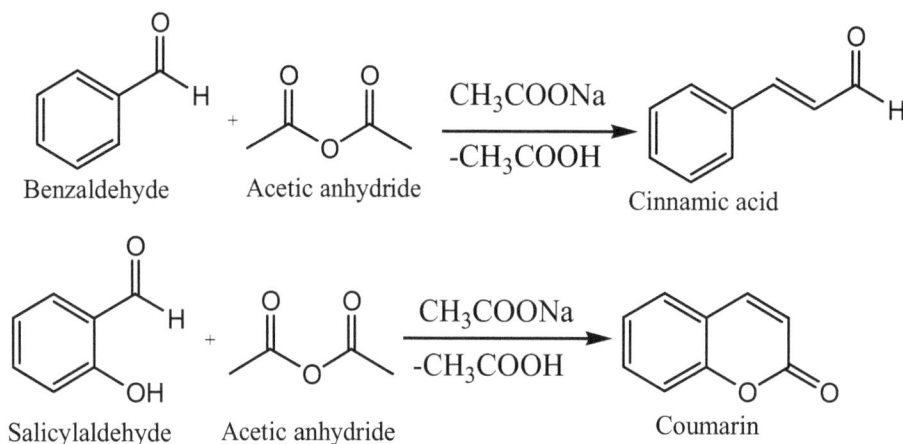

FIGURE 4.56 Production of Cinnamic acid and Coumarin via Perkin reaction.

acid (Figure 4.56). Perkin reaction was developed by English chemist William Henry Perkin in the year 1868 [51].

It is observed that Perkin's reaction can be applied to β-cyclocitral due to the presence of π-bonds in conjugation with the aldehyde group so that the electronegativity of the aldehyde group that favors Perkin's reaction in the presence of base. The presence of methyl group in β-cyclocitral further increases electronegativity of the aldehyde group [52].

4.1.27 Lossen Rearrangement

Lossen rearrangement is the intramolecular conversion of hydroxamic acids or their O-acetyl, O-aroyl and O-sulfonyl derivatives into isocyanates under thermal condition or in the presence of acid or base catalysts. Isocyanate is further

converted to the corresponding primary amine with water. The synthesis of the illudalane sesquiterpene illudinine is successfully completed by utilizing this rearrangement reaction (Figure 4.57) [53].

4.1.28 Cope Rearrangement

Cope rearrangement is a [3,3]-sigmatropic rearrangement of 1,5-dienes under thermal conditions to produce regioisomeric 1,5-dienes. This reaction proceeds through an intramolecular pathway. The synthesis of amphilectane, serrulatane diterpenoids is accomplished by using this rearrangement reaction (Figure 4.58). These compounds are isolated from Pseudopterogorgia elisabethae [54].

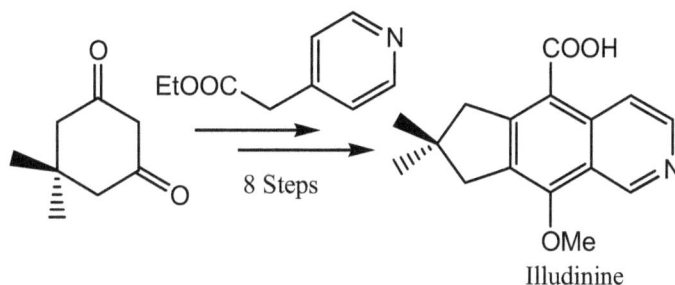

FIGURE 4.57 Synthesis of illudinine via Lossen rearrangement.

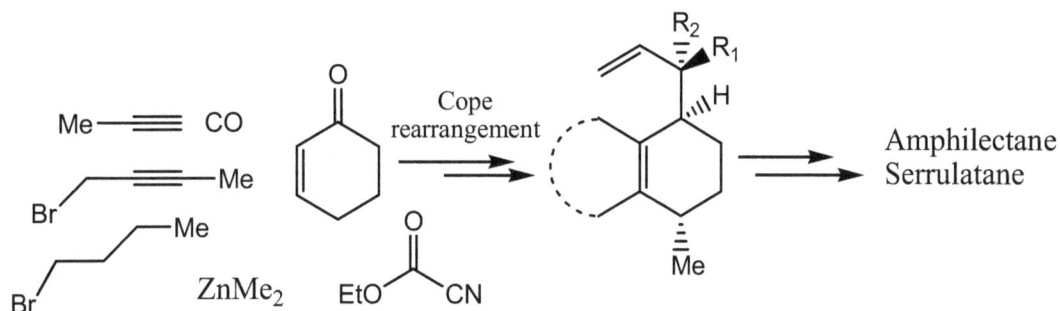

FIGURE 4.58 Synthesis of amphilectane, serrulatane via Cope rearrangement.

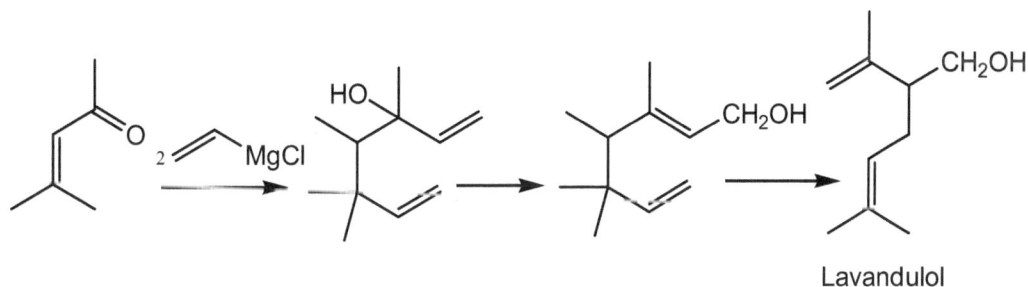

FIGURE 4.59 Synthesis of Lavandulol via Cope rearrangement.

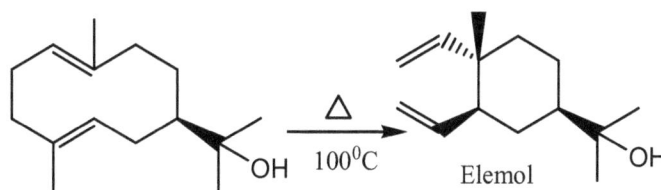

FIGURE 4.60 Synthesis of Elemol via Cope rearrangement.

Lavandulol is a monoterpene alcohol present in essential oils (lavender oil). Mesityl oxide reacts with excess vinyl-magnesium chloride to furnish alcohol which undergoes the Cope rearrangement to produce lavandulol (Figure 4.59) [55].

Elemol is a sesquiterpenoid and is the active constituent of the extracts of Chamaecyparis obtusa leaves. Chemically, it is 2-[(1R, 3S, 4S)-4-ethenyl-4-methyl-3-prop-1-en-2-ylcyclohexyl]propan-2-ol. Cope rearrangement is also utilized for the synthesis of *Elemol* (Figure 4.60) [56].

β-*sinensal* is a sesquiterpenoid, active constituent of the Citrus trifoliata. Chemically, it is (2E, 6E)-2,6-dimethyl-10-methylidenedodeca-2,6,11-trienal. Claisen–Cope rearrangement reactions is applied for the synthesis of β-sinesal (Figure 4.61) [57].

Citral is an acyclic monoterpene aldehyde. Chemically, it is (2E)-3,7-dimethylocta-2,6-dienal. It is also produced via Claisen–Cope rearrangement reactions (Figure 4.62).

The key intermediate for the synthesis of terpenoids (citral, linalool, geraniol) is 2-methyl-2-hepten-6-one. This compound is obtained via a synthesis based on the Claisen rearrangement (Figure 4.63) [58].

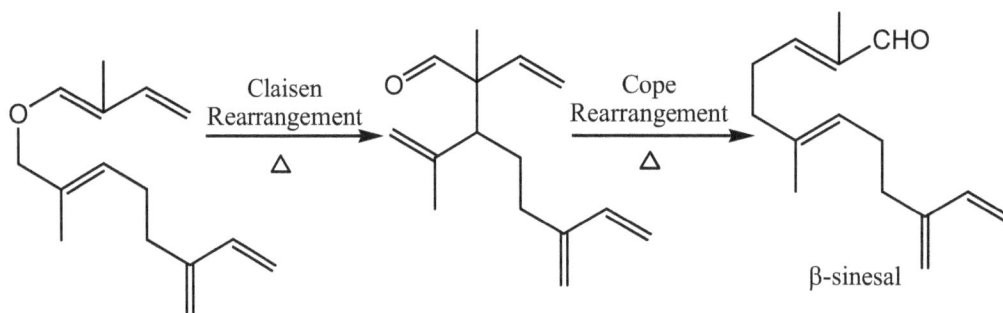

FIGURE 4.61 Synthesis of β-sinensal via Claisen–Cope rearrangement.

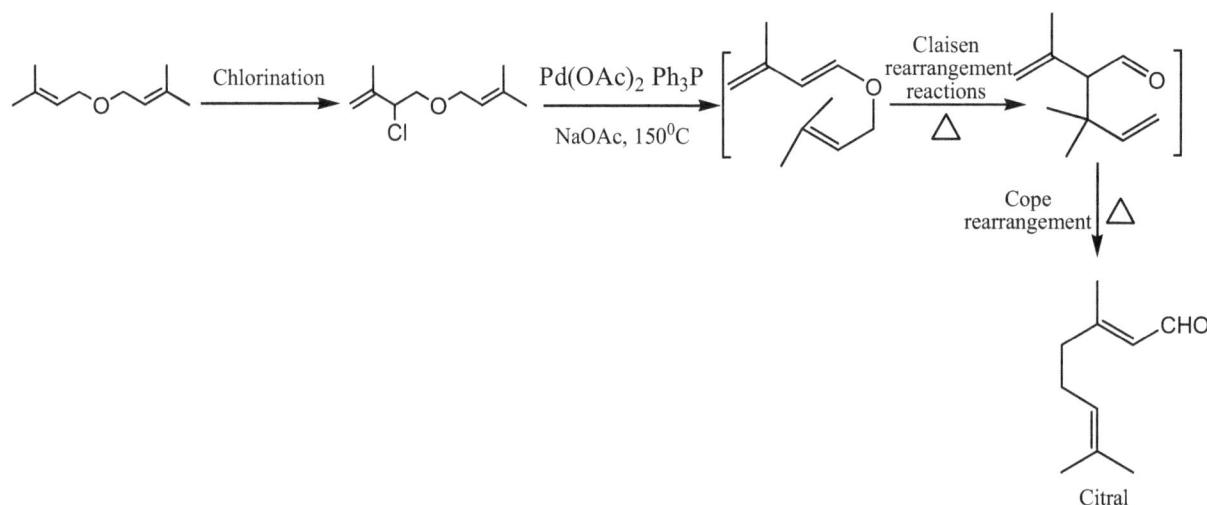

FIGURE 4.62 Synthesis of Citral via Claisen–Cope rearrangement.

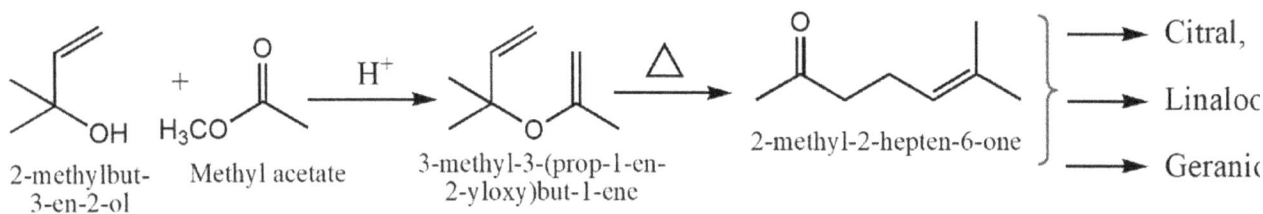

FIGURE 4.63 Synthesis of terpenoids based on the Claisen rearrangement.

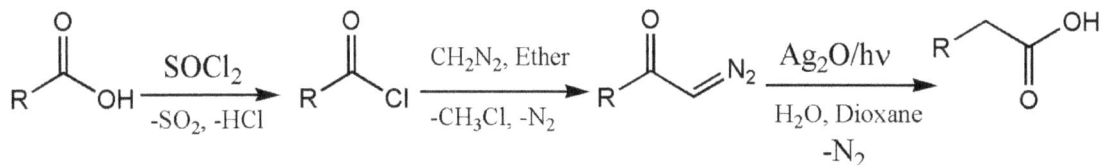

FIGURE 4.64 Arndt–Eistert reaction.

4.1.29 Arndt–Eistert Reaction

Arndt–Eistert reaction involves the conversion of carboxylic acid to one-carbon homologated carboxylic acid via α-diazoketone and subsequently Wolf rearrangement of the

intermediate in the presence of water and silver oxide (Ag$_2$O) catalyst or thermal or photochemical conditions. The reaction is named after the German chemists Fritz Arndt and Bernd Eistert. This reaction is applied for the preparation of β-amino acids from α-amino acids (Figure 4.64) [59].

The total synthesis of natural products like phenalenone diterpene salvilenone and β-ionone have been accomplished by utilizing this type reaction. β-ionone is (C$_{13}$) sesquiterpene present in plant essential oils. It is used as precursor for the synthesis of vitamin A. The starting material for synthesis of (Z)-4-oxo-β-ionone is 3-methyl-3-(5-methylfuran-2-yl)buta-noic acid (Figure 4.65) [60].

4.1.30 Miscellaneous

Conjugated acid derivatives containing nitrogen have been synthesized from the simple acyclic monoterpene citral by different reactions such as Wittig reaction, Baylis Hillman, amide and ester condensations. The chiral tricyclic terpene containing a 6,6,6-tricyclic moiety and 3,3-dimethyl-7-oxooc-tylidenyl side chain undergoes a double ring-closing reaction to produce two chiral pentacyclic terpenes in a ratio of 4:3 via an intramolecular Michael addition followed by aldol condensation under basic conditions. Three new stereogenic

centers are introduced in the initial Michael annulation reaction (Figure 4.66).

Pentacyclic triterpenes are extensively present in *Diospyros* plants. These terpenes are biosynthesized through the cycliza-tion of squalene. Michael–Aldol annulation reaction is involved for the synthesis of chiral pentacyclic terpenes (δ-Oleanolic acid, Erythrodiol, (+)-Myriceric acid A) (Figure 4.67) [60].

Iridoids include a large group of monoterpenoids. These are characterized by skeletons in which a six-membered ring, containing an oxygen atom, is fused to a cyclopentane ring called iridane skeleton. Iridoid synthase utilizes the linear monoterpene 10-oxogeranial as substrate and couples an ini-tial NAD(P)H-dependent reduction step with a subsequent cyclization step via a Diels–Alder cycloaddition or a Michael addition (Figure 4.68) [61].

Nimbolide is one of the active constituent of the leaf extract of *Azadirachta indica*. It is a terpenoid lactone. It exhibits potential anticancer activity against several types of cancer. The synthesis of nimbolide is carried out sequentially which

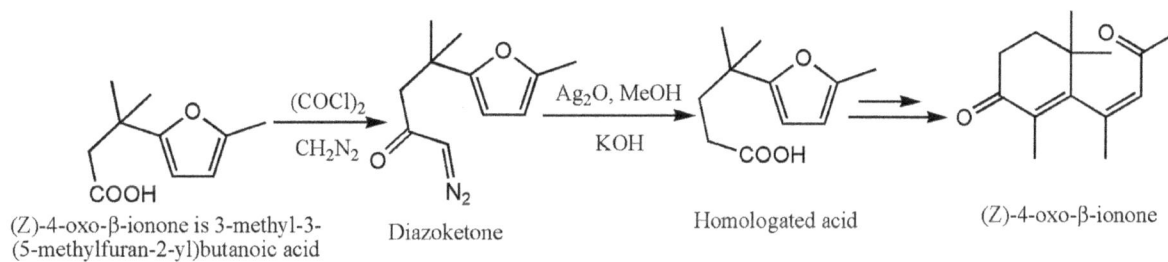

FIGURE 4.65 Synthesis of (Z)-4-oxo-β-ionone via Arndt–Eistert reaction.

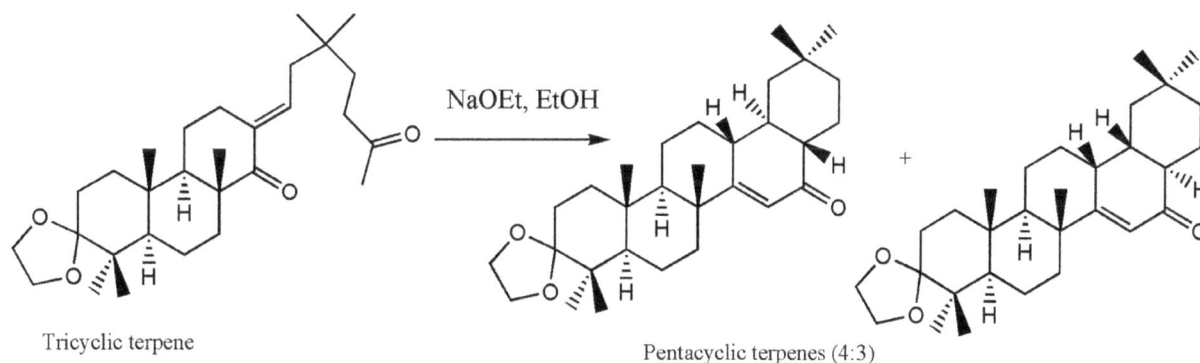

FIGURE 4.66 Generation of pentacyclic terpenes via an intramolecular Michael addition.

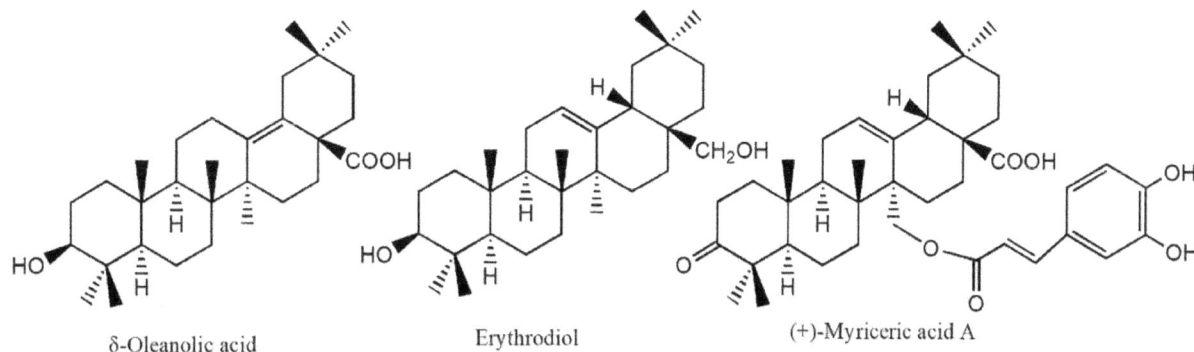

FIGURE 4.67 Structures of bioactive pentacyclic terpenes.

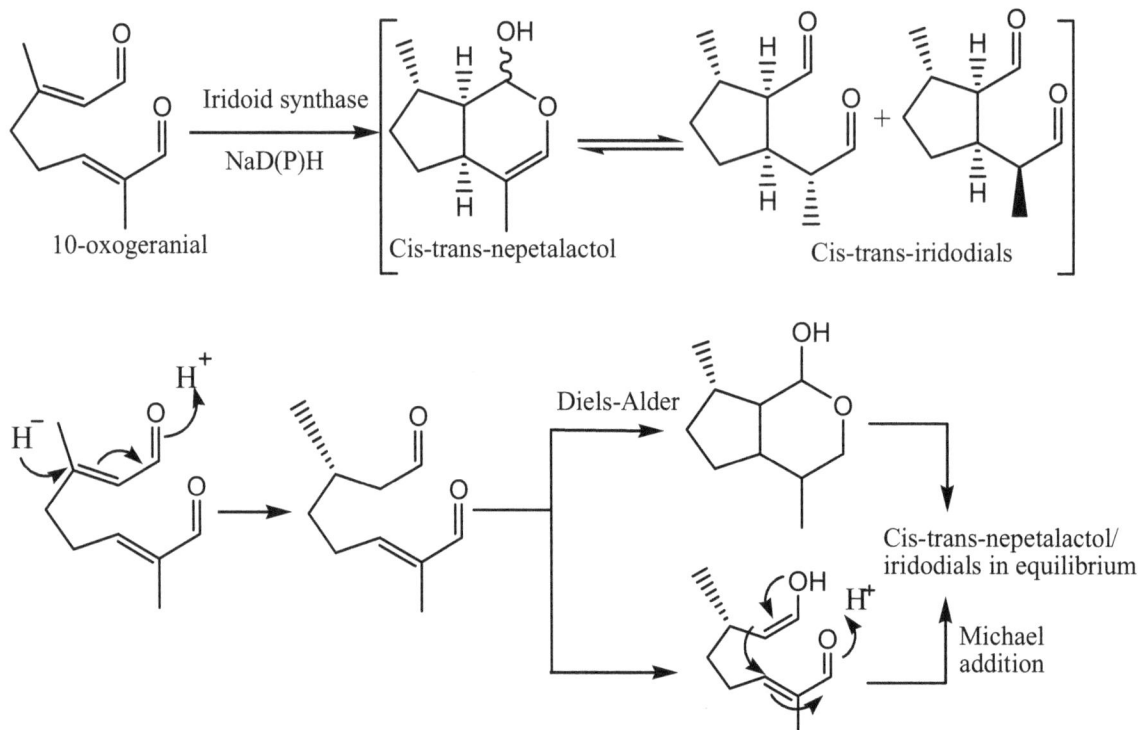

FIGURE 4.68 Generation of Iridoids via a Diels–Alder cycloaddition or a Michael addition.

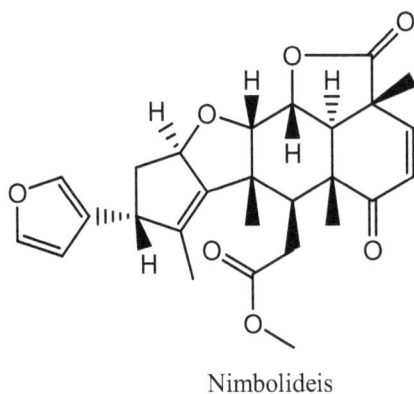

Nimbolideis

FIGURE 4.69 Synthesis of Nimbolide *via* Michael addition-aldol condensation reaction.

involves Michael addition reaction followed by aldol condensation (Figure 4.69) [62].

Secologanin is chemically a glycosylated monoterpenoid. The key transformation in the synthesis of Secologanin involves the sequential anti-selective organocatalytic Michael reaction/Fukuyama reduction followed by cyclization to produce an optically active dihydropyran ring [63]. The monoterpene indole alkaloids are biosynthetically generated via Pictet–Spengler reaction of Secologanin and tryptamine to afford strictosidine catalyzed by the enzyme strictosidine synthase (Figure 4.70).

(+)-propindilactone G is a nortriterpenoid. It is isolated from the stems of *Schisandra propinqua*.

The key steps involved in the synthesis of (+)-propindilactone G include asymmetric Diels–Alder reaction, a Pauson–Khand reaction, Pd-catalyzed reductive hydrogenolysis reaction and an oxidative heterocoupling reaction. These reactions enabled the synthesis of (+)-propindilactone G in 20 steps (Figure 4.71) [64].

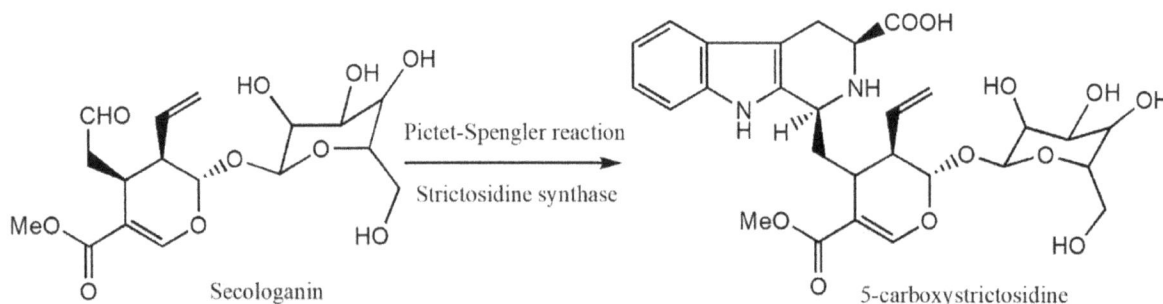

Secologanin   5-carboxystrictosidine

FIGURE 4.70 Transformation of Secologanin.

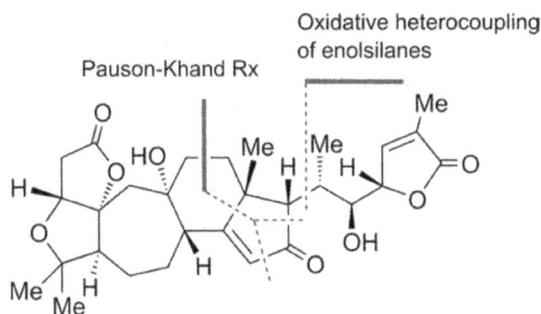

FIGURE 4.71 Structure of (+)-propindilactone G.

4.2 Conclusion

Several rearrangements and name reactions (Wittig reaction, Claisen condensation, Wolff–Kishner reduction, Baeyer–Villiger Oxidation, Beckmann rearrangement, Knoevenagel reaction, Mannich reaction, Oppenauer oxidation, Perkin Reaction, pinacol–pinacolone rearrangement, Reformatsky reaction, Dieckmann *condensation*, Wurtz reaction, Meerwein–Pondorf reduction, Birch Reduction, Curtius rearrangement, Favorskii rearrangement, Luche reduction, etc) are applied for the synthesis of terpenoids in the laboratory, pilot plants or on industrial scale. 2-Methyl-2-hepten-6-one is found to be one of the intermediate of terpenoids (citral, menthone, carvone, linalool, myrcene, menthol, geraniol etc) and their synthesis is carried out via these rearrangement reactions. As there is difficult to extract, separate and isolate the large quantities of essential oil containing terpenoids from natural sources, it is advantageous to utilize the above-mentioned reaction mechanisms to obtain terpenes with diverse molecular structures.

REFERENCES

1. Grayson, D. H. Monoterpenoids. *Natural Product Reports*, **2000**, 17, 385–419.
2. Connolly, J. D.; Hill, R. A. Triterpenoids. *Natural Product Reports*, **2010**, 27, 79–132.
3. Fraga, B. M. Natural sesquiterpenoids. *Natural Product Reports*, **2008**, 25, 1180–1209.
4. Paul, M. D. *Medicinal Natural Products: A Biosynthetic Approach*, 2nd Edition. John Wiley & Sons, Ltd, Chichester, **2001**.
5. Volcho, K. P.; Tatarova, L. E.; Suslov, E. V.; Korchagina, D. V.; Salakhutdinov, N. F.; Barkhash, V. A. Reactions of some terpenes and their derivatives with acylating agents in the presence of aluminosilicate catalysts. *Russian Journal of Organic Chemistry*, **2001**, 37(10), 1418–1429. doi: 10.1023/a:1013475222294.
6. Erman, W. F. *Chemistry of the Monoterpens*, part B. New York, **1985**, 958.
7. McCulley, C. H.; Tantillo, D. J. Predicting rearrangement-competent terpenoid oxidation levels. *Journal of the American Chemical Society*, **2020**, 142(13), 6060–6065.
8. Christianson, D. W. Structural and chemical biology of terpenoid cyclases. *Chemical Reviews*, **2017**, 117(17), 11570–11648.
9. Tetsuo, M.; Hazime, T.; Mitsuhiro, H. Synthesis of terpenoids such as geraniol and farnesol by cross coupling reactions. *Journal of Japan Oil Chemists' Society*, **1990**, 39(2), 83–89. doi: 10.5650/jos1956.39.2_83.
10. Tholl, D. Biosynthesis and biological functions of terpenoids in plants. *Advances in Biochemical Engineering/Biotechnology*, **2015**, 148, 63–106. doi: 10.1007/10_2014_295. PMID: 25583224.
11. Kolb, H. C.; Finn, M. G.; Sharpless, K. B. Click chemistry: Diverse chemical function from a few good reactions. *Angewandte Chemie International Edition*, **2001**, 40, 2004–2021.
12. Firdaus, M. Thiol–Ene (click) reactions as efficient tools for terpene modification. *Asian Journal of Organic Chemistry*, **2017**, 6, 1702–1714.
13. Firdaus, M.; Meier, M. A. R.; Biermann, U.; Metzger, J. O. Renewable co-polymers derived from castor oil and limonene. *European Journal of Lipid Science and Technology*, **2014**, 116(1), 31–36. doi: 10.1002/ejlt.201300206.
14. Banerjee, A. K.; Vera, W. Molecular rearrangements during terpene syntheses. *Synthetic Communications*, **1996**, 26, 4641–4646.
15. Lardan, A.; Reichstern, T. *Helvetica Chimica Acta*, **1962**, 45, 943–962.
16. Wagner-Meerwein Rearrangement. *Comprehensive Organic Name Reactions and Reagents*. **2010**, doi: 10.1002/9780470638859.conrr655.
17. Lin, C. I.; McCarty, R. M., Liu, H. W. The enzymology of organic transformations: A survey of name reactions in biological systems. *Angewandte Chemie International Edition*, **2017**, 56(13), 3446–3489. doi: 10.1002/anie.201603291.
18. Kamakura, D.; Todoroki, H.; Urabe, D.; Hagiwara, K.; Inoue, M. Total synthesis of talatisamine. *Angewandte Chemie International Edition*, **2020**, 59(1), 479–486. doi: 10.1002/anie.201912737.
19. Pommer, H. The Wittig reaction in industrial practice. *Angewandte Chemie International Edition in English*, **1977**, 16(7), 423–429. doi: 10.1002/anie.197704233.
20. Singh, S.; Khandare, R. P.; Sharma, M.; Bhasin, V. K.; Bhat, S. V. Monoterpene citral derivatives as potential antimalarials. *Natural Product Communications*, **2014**, 9(3), 299–302. PMID: 24689199.
21. Butala, R. R. Reference module in chemistry, molecular sciences and chemical engineering. *Aluminum Organometallics*, **2013**, 13–20, doi: 10.1016/b978-0-12-409547-2.03861-0.
22. Ooi, T.; Otsuka, H.; Miura, T.; Ichikawa, H.; Maruoka, K. Practical Oppenauer (OPP) oxidation of alcohols with a modified aluminum catalyst. *Organic Letters*, **2002**, 4(16), 2669–2672.
23. Imamoto, T.; Takiyama, N.; Nakamura; Hatajima, T.; Kamiya, Y. Reactions of carbonyl compounds with Grignard reagents in the presence of cerium chloride. *Journal of the American Chemical Society*, **1989**, 111(12), 4392–4398.
24. Ishihara, J.; Hatakeyama, S. Recent developments in the Reformatsky-Claisen rearrangement. *Molecules*, **2012**, 17(12), 14249–14259. doi: 10.3390/molecules171214249.
25. Muraki, S. *Abstracts of Papers: 41st Symposium on the Chemistry of Terpenes, Essential Oils, and Aromatics*, **1968**, 13.
26. Abidi, S. J. *Tetrahedron Letters*, **1986**, 27, 267–270.

27. Briou, B.; Améduri, B.; Boutevin, B. Trends in the Diels–Alder reaction in polymer chemistry. *Chemical Society Reviews*, **2021**, 50, 11055–11097.

28. Shing, T. K. M.; Lo, H. Y.; Mak, T. C. W. Diels-Alder reaction of R-(–)-carvone with isoprene. *Tetrahedron*, **1999**, 55(15), 4643–4648. doi: 10.1016/s0040-4020(99)00145-3.

29. Alphand, V.; Furstoss, R. Microbiological transformations. A surprising regioselectivity of microbiological Baeyer-Villiger oxidations of menthone and dihydrocarvone. *Tetrahedron: Asymmetry*, **1992**, 3(3), 379–382. doi: 10.1016/s0957-4166(00)80281-5.

30. Fernandes, W. B.; Malaspina, L. A.; Martins, F. T.; Liao, L. M.; Camargo, A. J.; Lariucci, C.; Noda-Perez, C.; Napolitano, H. B. Conformational variability in a new terpenoid-like bischalcone: Structure and theoretical studies. *Journal of Structural Chemistry*, **2013**, 54(6), 1112–1121. doi: 10.1134/S0022476613060164.

31. Nowicki, J. C. Cope and related rearrangements in the synthesis of flavour and fragrance compounds. *Molecules*, **2000**, 5, 1033–1050. doi: 10.3390/50801033.

32. Lindner, S.; Geu-Flores, F.; Bräse, S.; Sherden, N. H.; O'Connor, S. E. Conversion of substrate analogs suggests a Michael cyclization in iridoid biosynthesis. *Chemistry & Biology*, **2014**, 21(11), 1452–1456. doi: 10.1016/j.chembiol.2014.09.010.

33. Winnacker, M.; Rieger, B. Recent progress in sustainable polymers obtained from cyclic terpenes: Synthesis, properties, and application potential. *ChemSusChem*, **2015**, 8, 2455–2471. doi: 10.1002/cssc.201500421.

34. Winnacker, M.; Vagin, S.; Auer, V.; Rieger, B. Synthesis of novel sustainable oligoamides via ring-opening polymerization of lactams based on (–)-menthone. *Macromolecular Chemistry and Physics*, **2014**, 215(17), 1654–1660. doi: 10.1002/macp.201400324.

35. de Oliveira, K. T. *Studies in Natural Products Chemistry.* Volume 42: *The Synthesis of Seven-Membered Rings in Natural Products*, **2014**, 421–463. doi: 10.1016/B978-0-444-63281-4.00014-8.

36. Waters, S. P.; Tian, Y.; Li, Y.-M.; Danishefsky, S. J. Total synthesis of (–)-scabronine G, an inducer of neurotrophic factor production. *Journal of the American Chemical Society*, **2005**, 127(39), 13514–13515.

37. Birch, A. J. The birch reduction. *Journal of the Chemical Society*, **1944**, 430–436.

38. Siegel, D. R.; Danishefsky, S. J. Total synthesis of garsubellin A. *Journal of the American Chemical Society*, **2006**, 128(4), 1048–1049. doi: 10.1021/ja057418n. PMID: 16433500.

39. O'Donnell, G. W.; Sutherland, M. D. Terpenoid chemistry. XIII. Derivatives of optically pure (+)-citronellal. *Australian Journal of Chemistry*, **1966**, 19(3), 525–528. doi: 10.1071/CH9660525.

40. Simonsen, J. L. *The Terpenes*, 2nd Edition. Cambridge University Press. Vol. 1, **1953**, 394.

41. Chapman, L. M.; Beck, J. C.; Wu, L.; Reisman, S. E. Enantioselective total synthesis of (+)-Psiguadial B. *Journal of the American Chemical Society*, **2016**, 138(31), 9803–9806.

42. de Graauw, C. F.; Peters, J. A.; Van Bekkum, H.; Huskens, J. Meerwein-Ponndorf-Verley reductions and oppenauer oxidations: an integrated approach. *ChemInform*, **2010**, 26(9), 1995.

43. Philippov, A. A., Chibiryaev, A. M. Meerwein–Ponndorf–Verley–Oppenauer non-catalytic reaction of monoterpenoids in supercritical fluids. *Russian Journal of Physical Chemistry B*, **2019**, 13, 1117–1120. https://doi.org/10.1134/S1990793119070054.

44. Ghosh, A. K.; Brindisi, M.; Sarkar, A. The Curtius rearrangement: Applications in modern drug discovery and medicinal chemistry. *ChemMedChem*, **2018**, 13(22), 2351–2373. doi: 10.1002/cmdc.201800518.

45. Fruhmann, P.; Hametner, C.; Mikula, H.; Adam, G.; Krska, R.; Fröhlich, J. Stereoselective luche reduction of deoxynivalenol and three of its acetylated derivatives at C8. *Toxins*, **2014**, 6(1), 325–336. doi: 10.3390/toxins6010325.

46. Neises, B.; Steglich, W. Simple method for the esterification of carboxylic acids. *Angewandte Chemie International Edition*, **1978**, 17, 522–524.

47. Stastna, E.; Cerny, I.; Pouzar, V.; Chodounská, H. Stereoselectivity of sodium borohydride reduction of saturated steroidal ketones utilizing conditions of Luche reduction. *Steroids*, **2010**, 75, 721–725.

48. Favorskii, E. Transformation of (R)-pulegone *via* Favorskii rearrangement. *Russian Journal of Physical Chemistry A*, **1894**, 26, 590.

49. Rupainwar, R.; Pandey, J.; Smrirti, S.; Ruchi R. The importance and applications of knoevenagel reaction (brief review). *Oriental Journal of Chemistry*, **2019**, 35(1). doi: 10.13005/ojc/350154.

50. Suresh, M.; Kumar, N.; Veeraraghavaiah, G.; Hazra, S.; Singh, R. B. Total synthesis of coprinol. *Journal of Natural Products*, **2016**, 79(10), 2740–2743. doi: 10.1021/acs.jnatprod.6b00277.

51. Li, J. J. Perkin reaction (cinnamic acid synthesis). *Name Reactions*, **2002**, 276–277. doi: 10.1007/978-3-662-04835-1_217.

52. Limaye, P. A.; Huddar, P. H.; Ghate, S. M. Application of Perkin's reaction to terpene aldehyde β-cyclocitral. *Asian Journal of Chemistry*, **1993**, 5(1), 230–231.

53. Morrison, A. E.; Hoang, T. T.; Birepinte, M.; Dudley, G. B. Synthesis of Illudinine from Dimedone. *Organic Letters*, **2017**, 19(4), 858–861. doi: 10.1021/acs.orglett.6b03887.

54. Yu, X.; Su, F.; Liu, C.; Yuan, H.; Zhao, S.; Zhou, Z.; Quan, T.; Luo, T. Enantioselective total syntheses of various amphilectane and serrulatane diterpenoids via cope rearrangements. *Journal of the American Chemical Society*, **2016**, 138(19), 6261–6270. doi: 10.1021/jacs.6b02624.

55. Tomiczek, B. M.; Grenning, A. J. Aromatic cope rearrangements. *Organic & Biomolecular Chemistry*, **2021**, 19(11), 2385–2398. doi: 10.1039/d1ob00094b.

56. Nowicki, J. Claisen, Cope and related rearrangements in the synthesis of flavour and fragrance compounds. *Molecules*, **2000**, 5(8), 1033–1050. doi: 10.3390/50801033.

57. Buchi, G.; Wuest, H. Stereoselective synthesis of alpha sinensal. *Journal of the American Chemical Society*, **1974**, 96(24), 7573–7574. doi: 10.1021/ja00831a042.

58. Takabe, K.; Yamada, T.; Katagiri, T. A facile synthesis of citral. *Synthetic Communications*, **1983**, 13(4), 297–301. doi: 10.1080/00397918308066979.

59. Candeias, N. R. *Comprehensive Organic Synthesis II: 3.19 The Wolff Rearrangement.* **2014**, 944–991. doi: 10.1016/b978-0-08-097742-3.00325-6.

60. Lu, J.; Koldas, S.; Fan, H.; Desper, J.; Day, V. W.; Hua, D. H. A one-pot intramolecular tandem Michael–Aldol annulation reaction for the synthesis of chiral pentacyclic terpenes. *Synthesis,* **2019**. doi: 10.1055/s-0039-1690521.

61. El-Naggar, L. J.; Beal, J. L. Iridoids. A review. *Journal of Natural Products,* **1980**, 43(6), 649–707. doi: 10.1021/np50012a001.

62. Bacigaluppo, J. A.; Colombo, M. I.; Preite, M. D.; Zinczuk, J.; Rúveda, E. A. The Michael-Aldol condensation approach to the construction of key intermediates in the synthesis of terpenoid natural products. *Pure and Applied Chemistry,* **1996**, 68(3), 683–686. doi: 10.1351/pac199668030683.

63. Rakumitsu, K.; Sakamoto, J.; Ishikawa, H. Total syntheses of (–)-secologanin, (–)-5-carboxystrictosidine, and (–)-rubenine. *Chemistry-A European Journal,* **2019**, 201902073. doi: 10.1002/chem.201902073.

64. You, L.; Liang, X.-T.; Xu, L.; Wang, Y.-F.; Zhang, J.-J.; Su, Q.; Li, Y.; Zhang, B.; Yang, S.-L.; Chen, J.-H.; Yang, Z. Asymmetric total synthesis of propindilactone G. *Journal of the American Chemical Society,* **2015**, 150716092227002. doi: 10.1021/jacs.5b06480.

5

Biogenesis of Terpenoids

Bimal Krishna Banik[*], Biswa Mohan Sahoo[†], and Abhishek Tiwari[‡]

CONTENTS

5.1 Introduction 181
5.2 Biogenesis Pathway 182
    5.2.1 Biosynthesis of Carvone 185
    5.2.2 Biosynthesis of Limonene 185
    5.2.3 Biosynthesis of Geraniol 188
    5.2.4 Biosynthesis of Iridoids 188
    5.2.5 Biogenesis of Pimarane 188
    5.2.6 Biogenesis of Lycopene 188
    5.2.7 Biosynthesis of Terpene Indole Alkaloids 190
    5.2.8 Biosynthesis of Monoterpenoid by Engineered Microbes 190
    5.2.9 Strategies for Improving Biosynthesis of Terpenoids 192
        5.2.9.1 Protein Engineering Strategies for Biosynthesis of Terpenoids 192
        5.2.9.2 Biosynthesis of Terpenoids via Genome Engineering 192
        5.2.9.3 Biosynthesis of Monoterpenes via Recombinant Technology 192
    5.2.10 Regulation of Biosynthesis of Terpenoids 194
5.3 Conclusion 194
Abbreviation 195
References 195

5.1 Introduction

Terpenoids are a large group of natural products produced by several plants, algae, fungi, sponges, bacteria, cyanobacteria and animals. Terpenoids are the major constituents of essential oils with diverse chemical structures and biological activities including antimicrobial, antifungal, antiviral, anti-parasitic, anti-hyperglycemic, antiallergenic, anti-inflammatory, antispasmodic, antimalarial, anti-cancer, etc [1]. Several environmental conditions such as temperature, sunlight, habitat and harvest time affect the composition and quantity of the active constituents present in essential oils of plant origin [2]. Plant sources such as eucalyptus, caraway, clove, citrus, lemon grass, peppermint mainly contain terpenes that are extracted,

isolated from different parts of the plant (foliage, leaves, fruits and seeds) by steam distillation and subsequent fractional distillation. Various chromatographic methods (TLC, HPTLC, column and HPLC) are utilized for identifying and separating individual terpenoids (monoterpenes, sesquiterpenes, diterpenes, sesterpenes, triterpenes, etc) [3]. In 1887, O. Wallach proposed that terpenoids are the polymerized products of 2-methyl-1,3-butadiene or isoprene with molecular formula of C_5H_8. Further, L. Ruzicka explained the isoprene rule and the biogenesis of terpenoids in 1953 [4].

Terpenoids occur with several chemical structures in a sequential arrangement of linear hydrocarbons or chiral carbocyclic skeletons with different chemical modifications such as hydroxyl, ketone, aldehyde and peroxide groups [5]. The efficiency of terpenoid biosynthesis is affected by high

[*] Department of Mathematics and Natural Sciences, College of Sciences and Human Studies, Prince Mohammad Bin Fahd University, Al Khobar, Kingdom of Saudi Arabia. Corresponding author e-mail address: bimalbanik10@gmail.com.

[†] Roland Institute of Pharmaceutical Sciences, Berhampur affiliated to Biju Patnaik University of Technology (BPUT), Rourkela, Odisha, India. Corresponding author e-mail address: drbiswamohansahoo@gmail.com.

[‡] Faculty of Pharmacy, Pharmacy Academy, IFTM University, Lodhipur Rajput, Moradabad-244102, Uttar Pradesh, India

TABLE 5.1

List of Enzymes Involved in Production of Monoterpenoids

Monoterpenoid	Host Organisms	Precursor Pathways	Monoterpene Synthases/ Cyclases (mTS/C)	Biological Sources
Limonene	*E. coli*	MEP, GPPS	Limonene synthase	Mentha spicata
α-Pinene	*E. coli*	MVA, GPPS2	Pinene synthase	Abes grandis
β-Pinene	*E. coli*	MVA, GPPS2	Pinene synthase	Abes grandis
Myrcene	*E. coli*	MVA, GPPS	Myrcene synthase	Quercus ilex
Sabinene	*E. coli*	MVA, GPPS2	Sabinene synthase	Salvia pomifera
Geraniol	*E. coli*	MVA, GPPS	Geraniol synthase	Ocimum basilicum
Linalool	*S. cerevisiae*	MVA	Linalool synthase	Lavandula angustifolia
Cineole	*S. cerevisiae*	MVA, ERG20, IDI	Cineole synthase	Salvia fruticose
3-Carene	*E. coli*	MEP, GPPS	3-Carene cyclase	Picea abies

volatility of the terpenoids, cytotoxicity, lack of enzymatic activity and selectivity.

The complexity and diversity of terpenoid structures depend on the action of the terpene synthases and cyclases responsible for their synthesis. The biosynthetic mevalonate and non-mevalonate pathways are involved in the production of terpenoid precursors [6]. These enzymes are categorized into different groups and gaining importance due to their catalytic mechanism in designing synthetic approaches to improve the production of terpenoids. These biocatalysts transform the acyclic substrates into cyclic and acyclic terpenoids (Table 5.1) [7].

All the terpenoids are produced from repeating units of five carbon building blocks called isoprenes. The biosynthetic pathway of terpenoids comprises mevalonic acid as an intermediate and is hence called as mevalonate pathway [8]. Acetyl-CoA is considered as the biogenetic precursor for the synthesis of terpenoids that proceed through a series of reactions to condense three molecules of acetyl-CoA to generate mevalonic acid. GDP (geranyl diphosphate), FDP (farnesyl diphosphate) and GGDP (geranylgeranyl diphosphate) are the key factors for the successive synthesis of terpenoid end products. Isopentenyl diphosphate (IDP) and dimethylallyl diphosphate (DMADP) are major substrates for producing most of the terpenoids (Figure 5.1) [9].

Biosynthesis of terpenoids in the plants may be divided successively into the following four stages [10].

a. Origin of the isoprene units such as IPP and DMAPP (mevalonoid and non-mevalonoid pathways).

b. Stepwise linear polymerization of the isoprene units to form the acyclic polyprenyl precursors like geranyl PP, farnesyl PP, geranylgeranyl PP and farnesylgeranyl PP. Whereas the squalene and C40 terpenoids are formed by the dimerization of C15 (FPP) and C20 (GGPP) linear terpenoids, respectively.

c. Proper folding, cyclization, rearrangements of these polyprenyl precursors, and also sometimes shedding from and addition to the carbon skeletons to form an array of innumerable terpenoids of diverse skeleton patterns.

d. In most cases, functionalization (e.g., various regiospecific and/or stereospecific hydroxylations, oxidations,

reductions, etc.) on the skeletal frames in the presence of the specific enzymes leading to different types of terpenoids.

5.2 Biogenesis Pathway

There are two biosynthetic pathways to produce terpenoids such as (1) Mevalonate pathways (Mevalonic acid formed from three molecules of acetyl-CoA) and (2) Mevalonate-independent pathways via 1-deoxy–D-xylulose-5-phosphate arises from a combination of two glycolytic pathway intermediate called pyruvic acid and glyceraldehydes-3-phosphate [11]. The isoprenoid biosynthetic pathway can be grouped into the central carbon pathway, upstream isoprenoid pathway and downstream isoprenoid pathway. To enhance the biosynthesis of isoprene, the isoprene synthetic pathway is divided into two modules such as the upstream endogenous MEP pathway and the downstream isoprene pathway made up of the isoprene synthase (Figure 5.2) [12].

In plants, terpenoids are synthesized by two pathways, i.e., mevalonic acid (MVA) pathway and methylerythritol 4-phosphate (MEP) pathway. In the mevalonate pathway, it has been reported that isopentenyl diphosphate (IPP) is synthesized from acetyl-CoA and then isomerized to dimethylallyl diphosphate (DMAPP). MVA pathway mostly provides the precursors for the biosynthesis of sesquiterpenoids, polyprenols, phytosterols, brassinosteroids and triterpenoids in cytosol, whereas, the MEP pathway synthesizes IPP and DMAPP from pyruvate and D-glyceraldehyde 3-phosphate [13]. MEP pathway are rather applied for the biosynthesis of hemiterpenoids, monoterpenoids, diterpenoids, carotenoids, etc.

Isoprenoids are natural products with diverse structures and functions. These compounds are derived from two C_5 units called isopentenyl diphosphate (IPP) and its allylic isomer dimethylallyl diphosphate (DMAPP). In plants, the precursors for the biosynthesis of the isoprenes are produced by cytosolic mevalonate and plastidial pathways. Plastidial geranyl diphosphate synthase (GDS) condenses two C_5 intermediate subunit terpenoids derived MEP pathway to produce acyclic precursor GPP. The synthesis of IPP is mediated via the methyl erythritol phosphate (MEP) pathway in the plastid (Figure 5.3) [14].

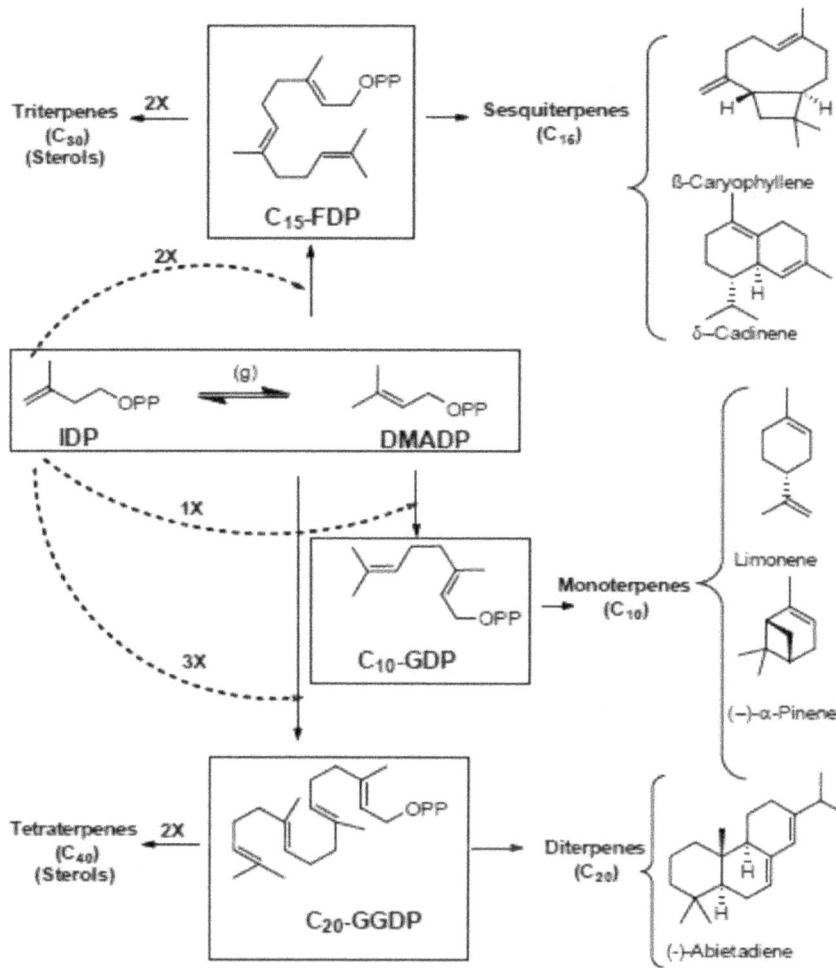

FIGURE 5.1 General scheme for biosynthesis of terpenoids.

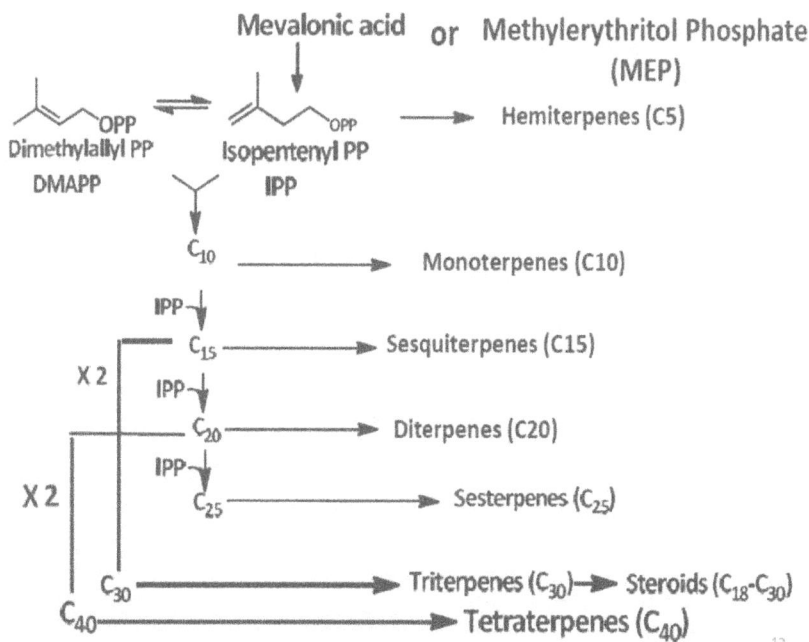

FIGURE 5.2 Biosynthesis of terpenes.

The sequential elongation reactions with the addition of one, two or three IPP units lead to the biosynthesis of geranyl diphosphate (GPP) (C_{10}), farnesyl diphosphate (FPP) (C_{15}) and geranyl geranyl diphosphate (GGPP) (C_{20}), which are the starting points of biosynthetic pathways for the generation of monoterpenes, sesquiterpenes and diterpenes respectively [15]. The key steps involved in the biosynthesis of monoterpenoids include several enzyme systems such as geranyl diphosphate synthase (GDS), (–)-limonene synthase (LS), (–)-cytochrome P_{450} (–)-limonene-3-hydroxylase (L_3OH), (–)-trans-isopiperitenol dehydrogenase (IPD), isopiperitenone reductase (IPR), (+)-cis-isopulegone isomerase (IPI), (+)-menthofuran synthase (MFS), (+)-pulegone reductase (PR), (–)-menthone:(–)-(3R)-menthol reductase (MMR) and (–)-menthone:(+)-(3S)-neomenthol reductase (MNMR) (Figure 5.4) [16,17].

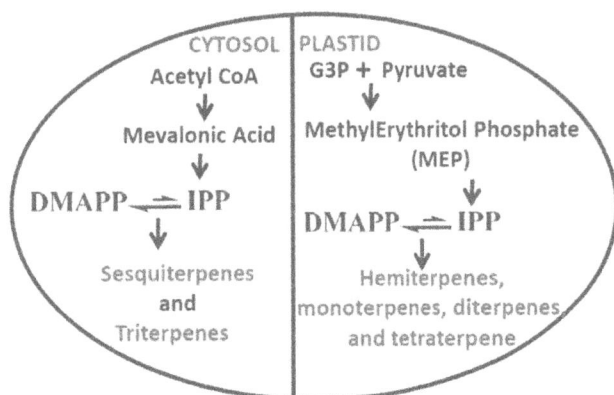

FIGURE 5.3 Biosynthesis of terpenes are via cytosolic and plastidial pathways.

The biosynthesis of different monoterpenes (limonene, terpineol and cineole) starts from geranyl diphosphate (GDP). The α-terpinyl cation is deprotonated to produce limonene. Limonene synthase converts GDP to (–)-limonene, α- and β-pinene. α-terpinyl cation affords a-terpineol, which undergoes heterocyclization to yield1,8-cineole (Figure 5.5) [18].

The biosynthesis of several acyclic monoterpenes (geraniol, linalool, myrcene and ocimene) begins from geranyl diphosphate (GDP). The geranyl and linalyl cation produce corresponding terpenes (Figure 5.6) [18].

The biosynthesis of several monocyclic (terpinene, terpineol, phellandrene and limonene) and bicyclic monoterpenes (sabinene, pinene and fenchol) starts from geranyl diphosphate. The geranyl, terpinyl, pinyl, fenchyl, bornyl and linalyl cation produce corresponding terpenes (Figure 5.7) [19].

The biosynthesis of several diterpenes (abietadiene, casbene and kaurene) involves the ionization and cyclization of GGDP directly, and preliminary cyclization to CDP by protonation of the terminal double bond. Various enzymes are involved during biosynthesis of these diterpenes include abietadiene synthase, casbene synthase and kaurene synthase B). Abietadiene synthase catalyzes the conversion of GGDP to (+)-CDP and the cyclization of this intermediate afford (+)-Abietadiene (Figure 5.8) [20].

The biosynthesis of different sesquiterpenes (bisabolene, aristolochene, germacrene, humulene) starts from farnesyl diphosphate (FDP). Various enzymes are involved during the biosynthesis of these sesquiterpenes including (E)-α-bisabolene synthase, epi-aristolochene synthase, germacrene C synthase, γ-humulene synthase. The acyclic sesquiterpene such as β-farnesene is also derived from FDP in the presence of (E)-β-farnesene synthase (Figure 5.9) [21].

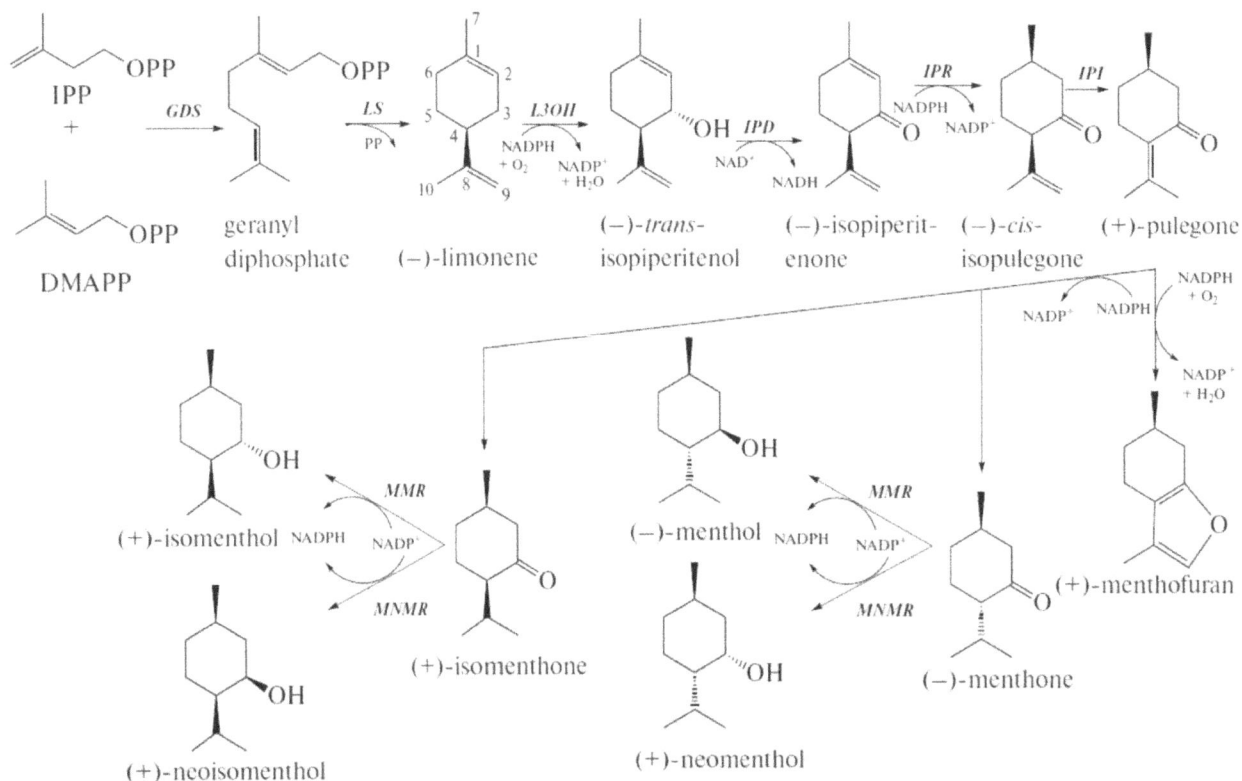

FIGURE 5.4 Biosynthetic pathways for monoterpenoids in plant.

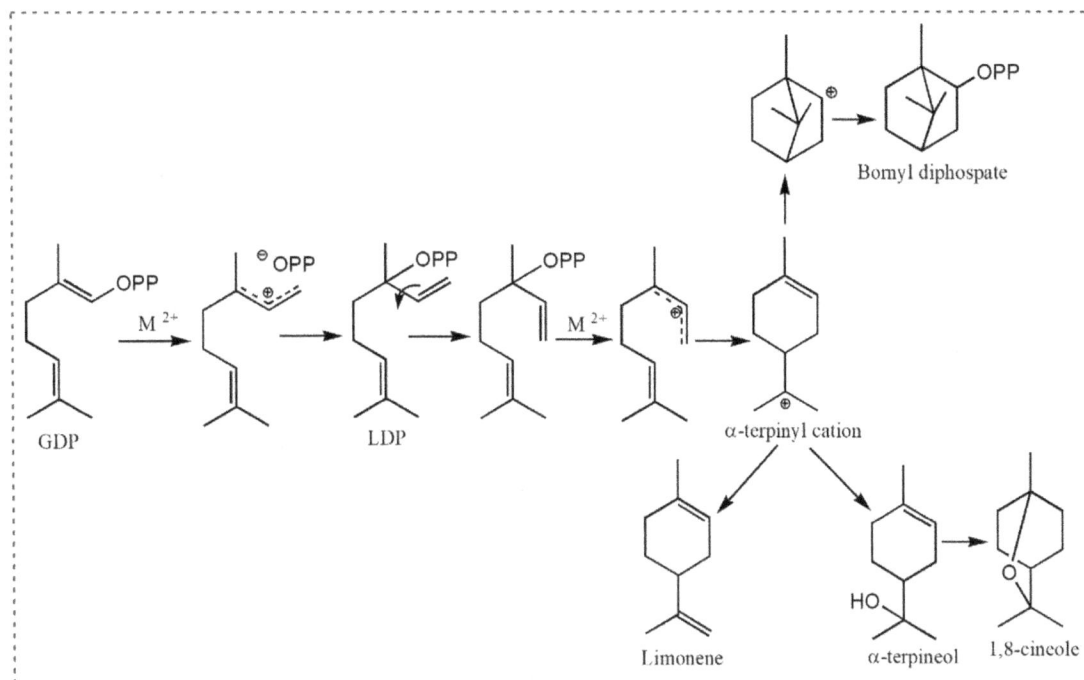

FIGURE 5.5 Biosynthesis of representative monoterpenes from GDP.

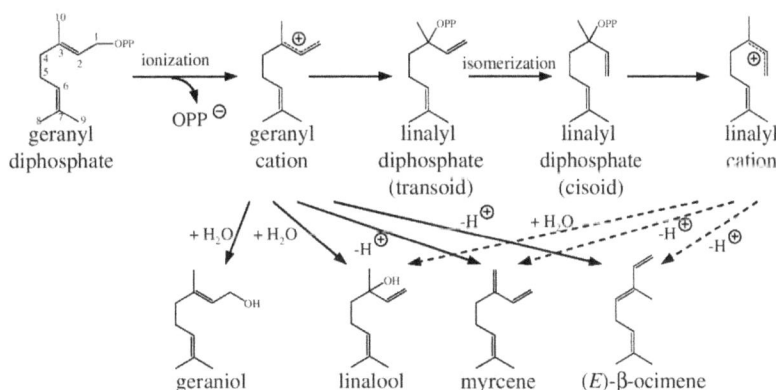

FIGURE 5.6 Biosynthesis of acyclic monoterpenes from GDP.

5.2.1 Biosynthesis of Carvone

Carvone is the major constituent of spearmint essential oil that is synthesized mainly in the epidermal oil glands of Mentha spicata. The biosynthetic pathway for (−)-carvone starts from geranyl pyrophosphate in spearmint. The biosynthesis of (−)-carvone involves the cyclization of C_{10} isoprenoid intermediate geranyl pyrophosphate to produce olefin (−)-limonene followed by hydroxylation to (−)-trans-carveol and subsequent dehydrogenation to yield (−)-carvone. The enzymes involved during synthesis of carvone include (−)-limonene cyclase, (−)-limonene hydroxylase and (−)-trans-carveol dehydrogenase (Figure 5.10) [22].

The biosynthesis of (−)-carvone involves four steps pathway starting from the C_5 isoprenoid presursors isopentenyl diphosphate (IPP) and dimethylallyl diphosphate (DMAPP). In biosynthetic pathway of (−)-carvone, four enzymes are involved including geranyl diphosphate synthase, limonene synthase, cytochrome P_{450} limonene hydroxylase and carveol dehydrogenase. The most common reactions involved during terpenoid biosynthesis are prenyltransfer, cyclization, oxygenation and redox transformation (Figure 5.11) [23].

5.2.2 Biosynthesis of Limonene

The biogenesis of cyclic terpenoids is commonly carried out to furnish intermediate carbenium ions. But in some cases

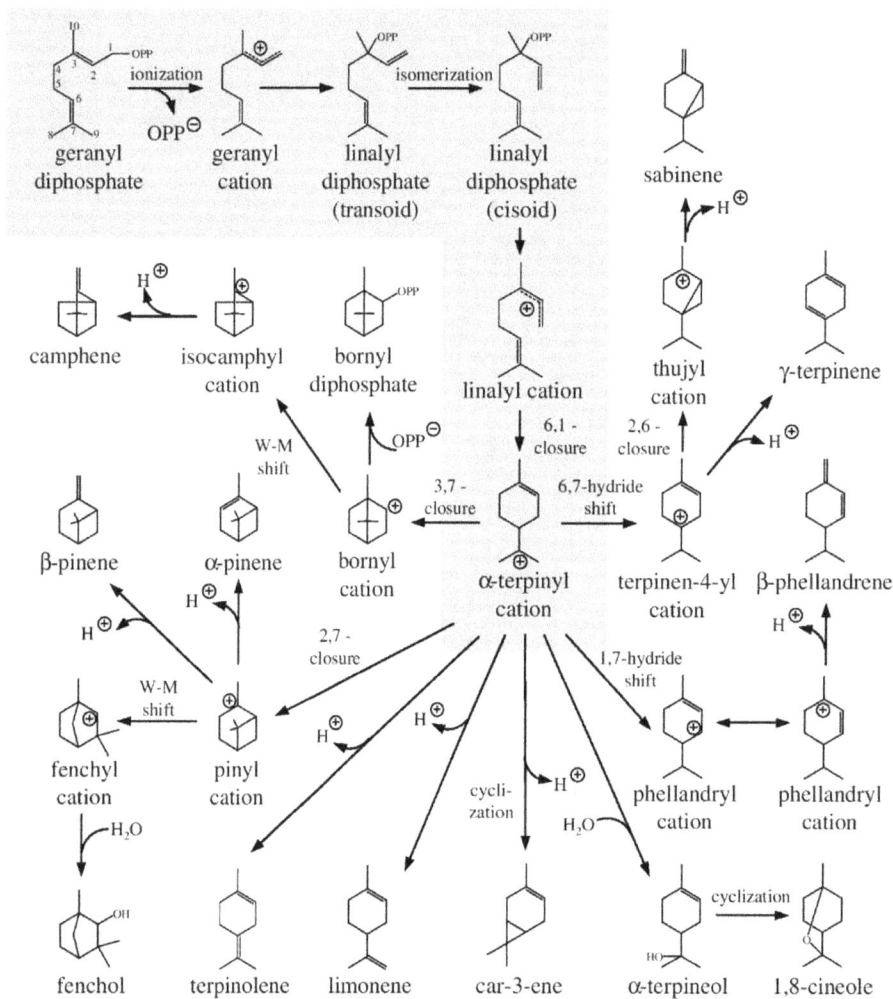

FIGURE 5.7   Biosynthesis of monocyclic and bicyclic monoterpenes from GDP.

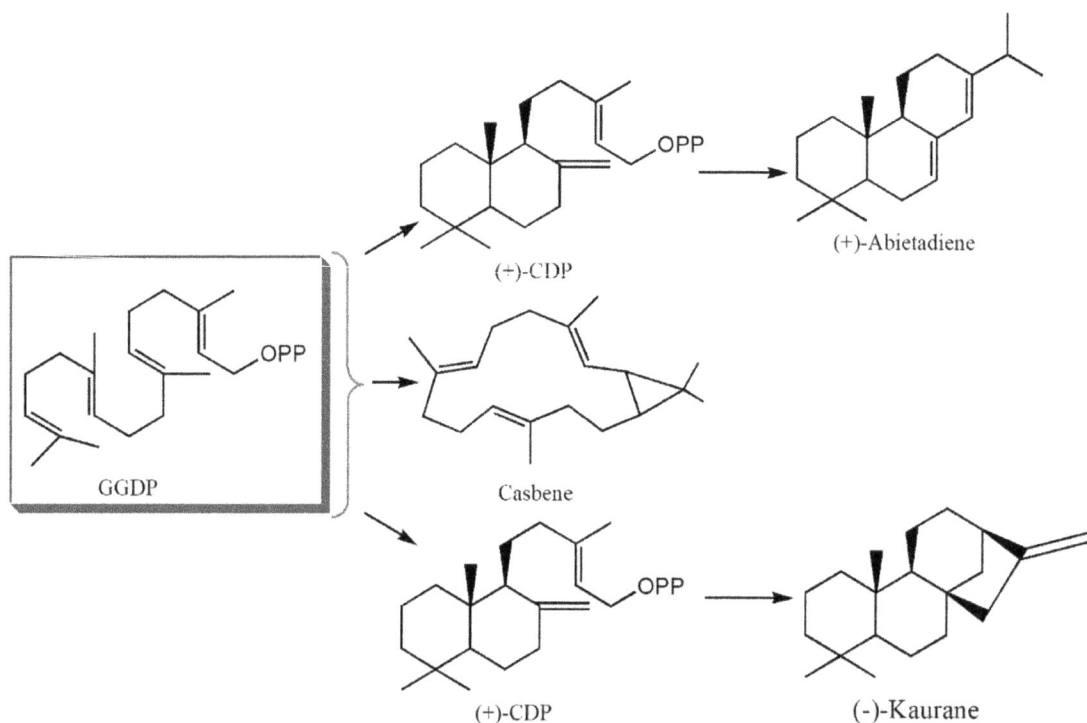

FIGURE 5.8   Biosynthesis of representative diterpenes from GGDP.

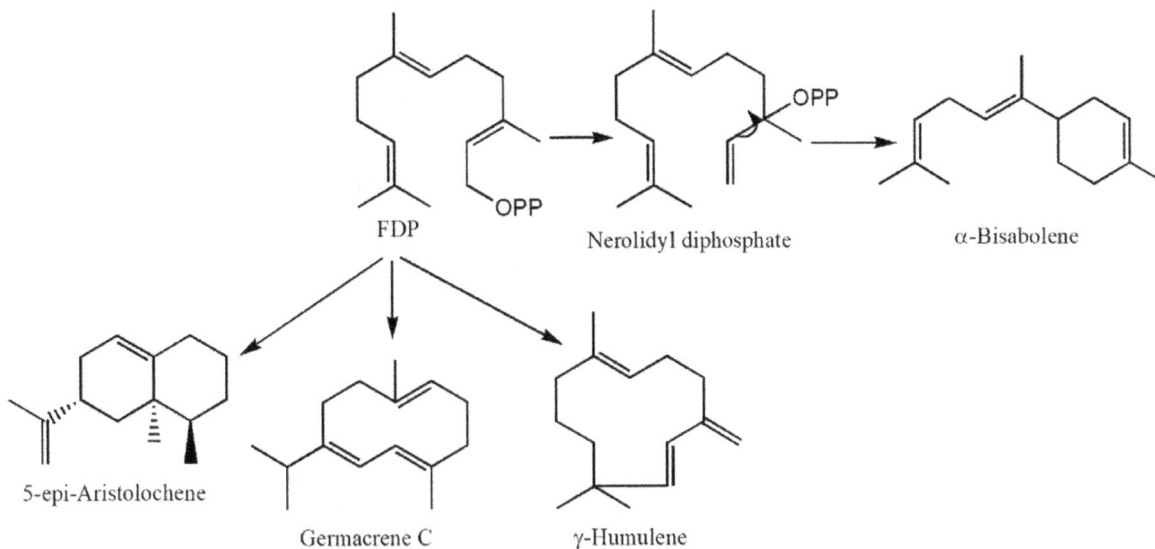

FIGURE 5.9 Biosynthesis of sesquiterpenes from FDP.

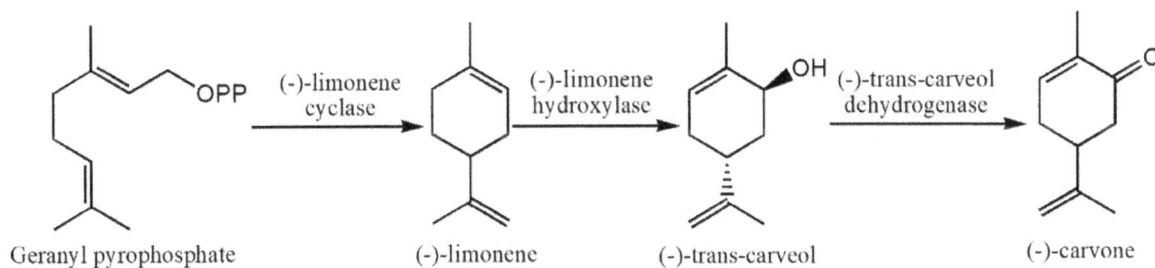

FIGURE 5.10 Biosynthetic route to carvone.

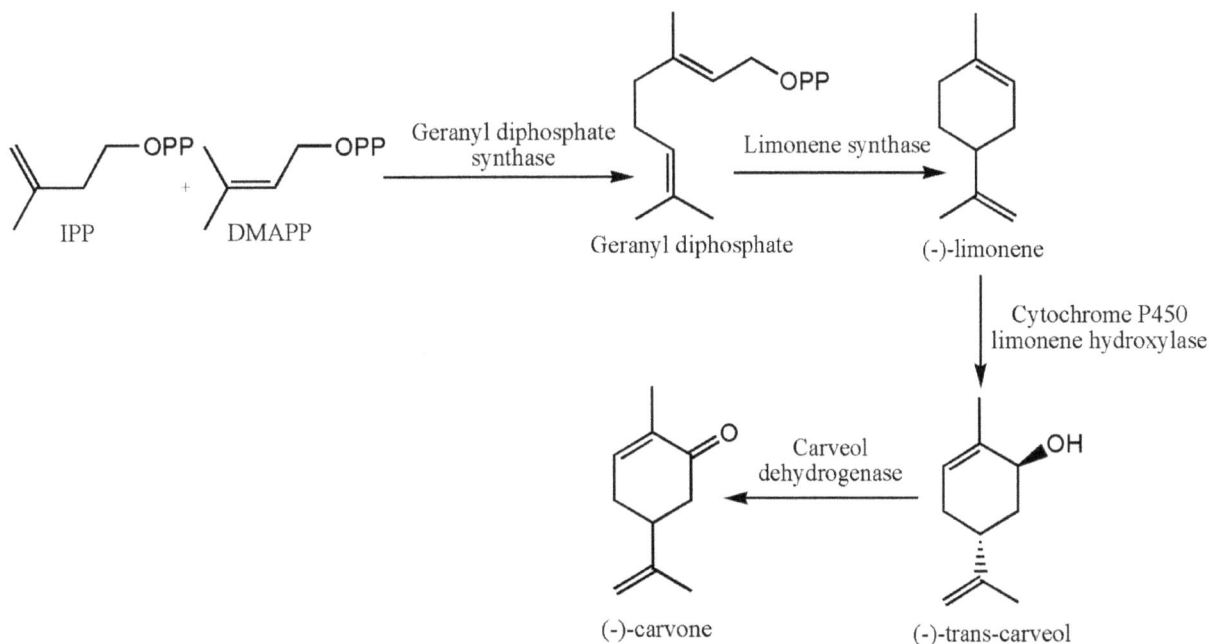

FIGURE 5.11 Biosynthetic pathway of (−)-carvone.

like monocyclic mono-terpenoids (limonene), the allylic cation remains after the separation of the pyrophosphate anion cyclizes to a cyclohexyl cation which is deprotonated to produce (*R*)- or (*S*)-limonene (Figure 5.12) [24,25].

5.2.3 Biosynthesis of Geraniol

NADP[+] dependent geraniol dehydrogenase is an enzyme which catalyzes the oxidation of geraniol to afford geranial. Geraniol is biosynthesized from geranyl pyrophosphate in the presence of geraniol synthase (Figure 5.13) [26].

5.2.4 Biosynthesis of Iridoids

Iridoids are the group of monoterpenoids. Secologanin is a seco-iridoid monoterpene synthesized from geranyl pyrophosphate in the mevalonate pathway. It is formed through several steps starting from the condensation of IPP and DMAPP (Figure 5.14) [27].

5.2.5 Biogenesis of Pimarane

The biogenesis of polycyclic diterpenes like pimarane is found to be produced from *iso*-geranyl-geranyl pyrophosphate. After dissociation

of the pyrophosphate anion, the remaining acyclic allylic cation undergoes a 1,3-sigmatropic hydrogen shift and thereby cyclizes to a monocyclic carbenium ion which, itself, isomerizes to the ionic precursor of the pimarane skeleton (Figure 5.15) [28].

5.2.6 Biogenesis of Lycopene

Lycopene is a bright red carotenoid hydrocarbon present in tomatoes, carrots, watermelons, grapefruits and papayas. Lycopene is used as a precursor for the generation of geranial and neral in red lycopene-containing tomato and watermelon fruits. Geranial (trans-citral) and Lycopene are produced from geranyl diphosphate (GDP). Lycopene is further degraded in vivo into geranial in these fruits. Bold lines on double bonds represent the possible cleavage or breaking of bonds (Figure 5.16) [29].

Farnesyl acetone, geranyl acetone, dihydro-apo-farnesal and 6-methyl hept-5-en-3-one are derived from phytoene or phytofluene. β-Ionone, β-cyclocitral and dihydroactinodiolide are the oxidative breakdown products of β-carotene. δ-Carotene appears to give rise to α-ionone, probably by the cleavage of the ε-ionone ring of δ-carotene. Tetraterpenes with 40 carbons are derived from phytoene formed by two C_{20} GGDP in a head-to-head condensation reaction (Figure 5.17)

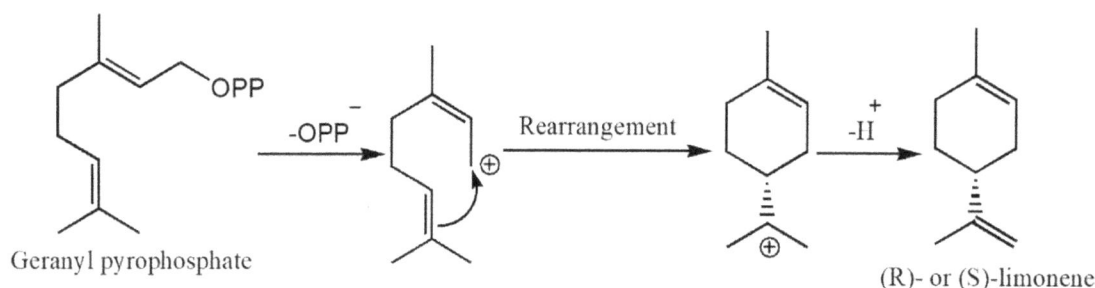

FIGURE 5.12   Biosynthetic route to limonene.

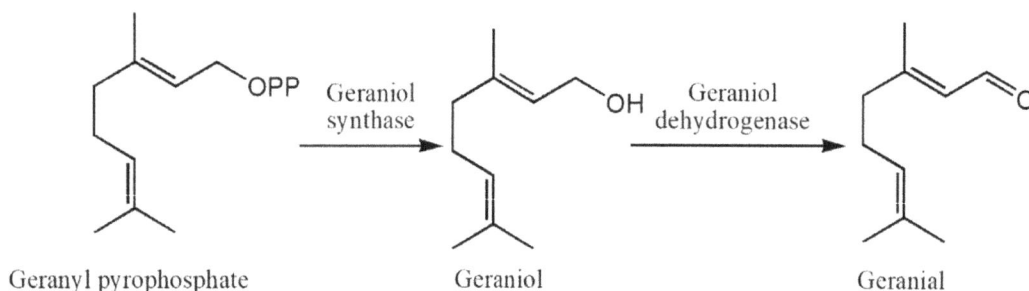

FIGURE 5.13   Biosynthesis of geraniol.

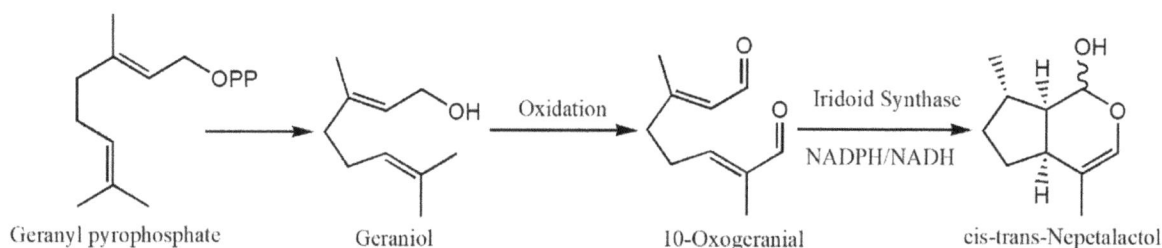

FIGURE 5.14   Biosynthesis of iridoids.

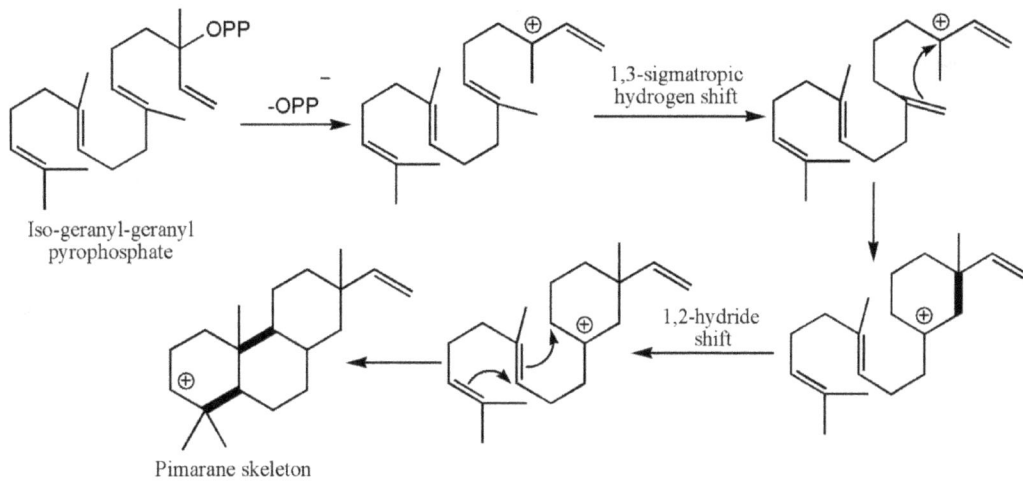

FIGURE 5.15 Biosynthetic route to pimarane.

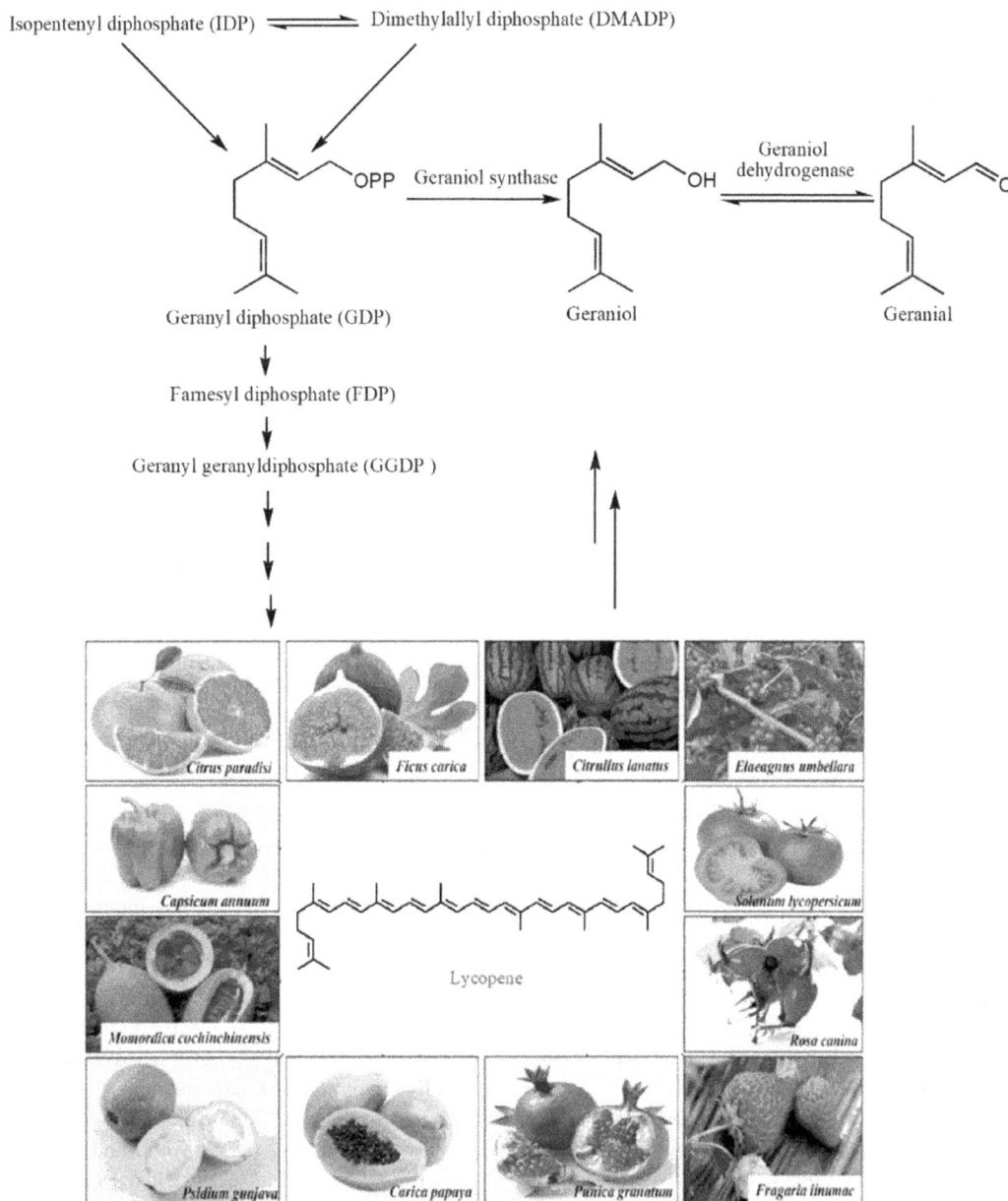

FIGURE 5.16 Biosynthetic route to lycopene.

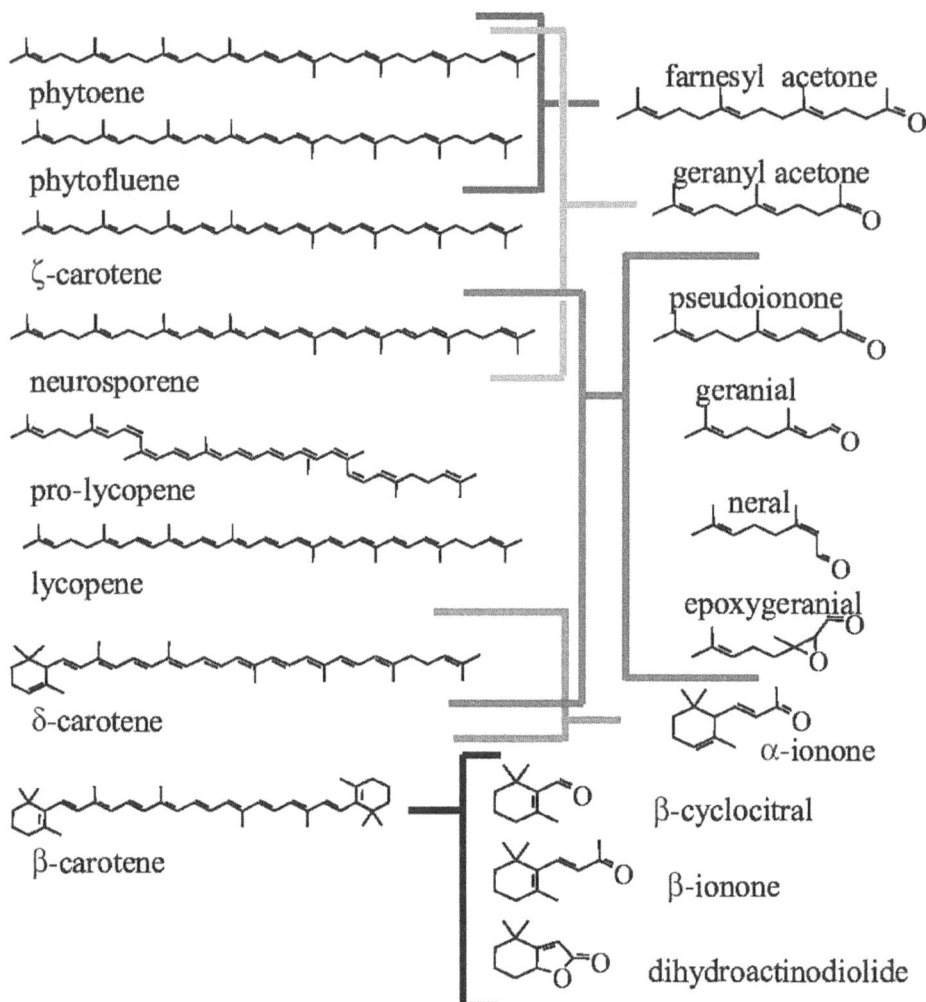

FIGURE 5.17 Biogenesis of terpenoid hydrocarbons.

[30]. Triterpenes are the 30-carbon hydrocarbons. These compounds are biosynthesized via the head-to-head condensation of two molecules of FDP to form squalene. Squalene serves as the precursor for formation of triterpenoids (e.g. hopanoids). Hopanoids are pentacyclic triterpenoid present in bacteria. Naturally cyanobacteria produces several diterpenes (e.g., tolypodiol). Tolypodiol is a diterpenoid produced from *Tolypothrix nodosa* and was found to have anti-inflammatory properties. GGDP is used as substrate for formation of phytyl by the action of the enzyme geranylgeranyl reductase.

5.2.7 Biosynthesis of Terpene Indole Alkaloids

Catharanthus roseus is a rich source of terpene indole alkaloids. The terpene indole alkaloids are derived from tryptophan and the iridoid terpene secologanin. Tryptophan decarboxylase is the pyridoxal-dependent enzyme that converts tryptophan to tryptamine. The utility of an iridoid monoterpene in these indole alkaloid pathways was first proposed after the structures of iridoid terpenes were elucidated. Secologanin was identified as the specific precursor of various iridoids. Various enzymes involved in the biosynthesis of secologanin

terpene include DXP synthase, DXP reductoisomerase, MEP synthase, geraniol-10-hydroxylase (G10H) and secologanin synthase (SLS), etc. These enzymes are observed in both vascular cells and epidermal cells of the plant (Figure 5.18) [31].

5.2.8 Biosynthesis of Monoterpenoid by Engineered Microbes

Borneol is a bicyclic monoterpenoid alcohol. It can be synthesized from bornyl diphosphate (BPP). BPP is further derived from neryl diphosphate (NPP), which is catalyzed by bornyl diphosphate synthase (BPPS). The endogenous phosphatases in *E. coli* were hypothesized to dephosphorylate BPP to produce borneol. In the case of *S. cerevisiae*, the biosynthesis of monoterpenoids is mainly carried out in both cytoplasm and organelles. GPP is the precursor of linalool and is synthesized from IPP and DMAPP. Biosynthesis of linalool is generally catalyzed by linalool synthase (LIS) from GPP, the linear C10 polyprenyl precursor [32]. Linalool is an acyclic monoterpenoid with a pair of enantiomers such as (S)-linalool and (R)-linalool. Li et al. reported the production of acyclic monoterpenoid (citronellol) in *S. cerevisiae*. Citronellol can be

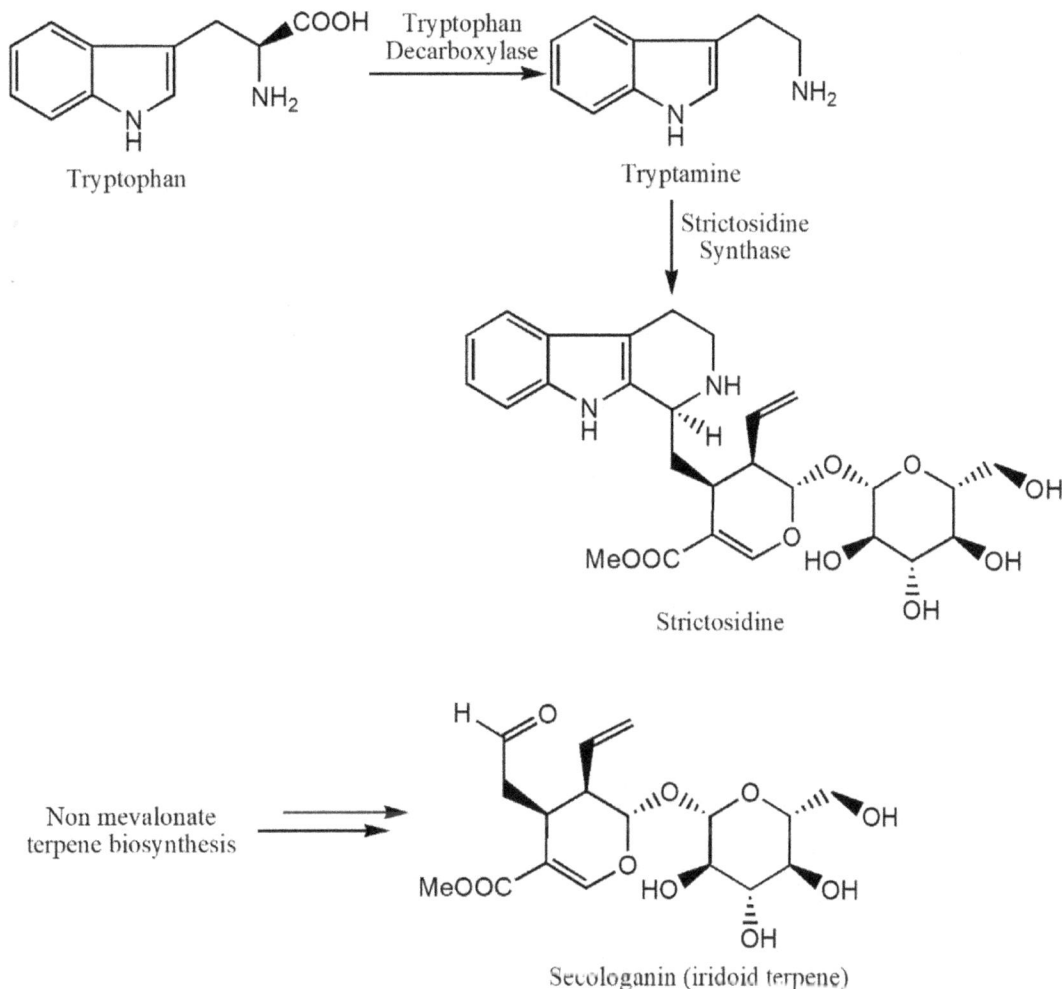

FIGURE 5.18 Steps involved in biosynthesis of terpene indole alkaloid.

synthesized from geraniol. Similarly, *S. cerevisiae* and *E. coli* are the most commonly used microbial hosts for the biosynthesis of acyclic monoterpenoid (geraniol) [33]. The compartmentalization strategy is effective to improve the biosynthesis of geraniol in *S. cerevisiae*. Pinene is a bicyclic monoterpenoid with two structural isomers such as α-pinene and β-pinene. This compound is biosynthesized from GPP. Nerol is an acyclic monoterpenoid alcohol. It can be synthesized from GPP or NPP catalyzed by nerol synthase [34]. To improve the production level of nerol, both metabolic and protein engineering strategies are coupled to improve *de novo* biosynthesis of nerol in *E. coli* from glucose. It is observed that acyclic monoterpenes are derived from GPP and NPP while the cyclic terpenoids are produced from NPP (Table 5.2) [35].

The production of terpenoids has been reported to occur naturally in some marine cyanobacteria. In the case of *Prochlorococcus* and *Synechococcus,* the production of terpenoids is found to be influenced by cell volume, carbon content of the cells, temperature of the environment and intensity of light [36]. The enzyme systems that are essential for the production of different monoterpenes in plants are absent in cyanobacteria [37]. Limonene is an example of a plant monoterpene that serves as a precursor for natural products. But

TABLE 5.2

Production of Monoterpenoids in Microorganisms

Terpenoids	Types	Hosts	Substrates
Linalool	Acyclic monoterpenoid	*Escherichia coli*	Glycerol
Geraniol	Acyclic monoterpenoid	*E. coli*	Glycerol
Limonene	Cyclic monoterpenoid	*S. cerevisiae*	Alcoholic beverage
Pinene	Bicyclic monoterpenoid	*E. coli*	Sucrose
α-Terpineol	Cyclic monoterpenoid	*S. cerevisiae*	Glucose
Citronellol	Acyclic monoterpenoid	*S. cerevisiae*	Glucose
Borneol	Bicyclic monoterpenoid	*E. coli*	Glucose
Sabinene	Bicyclic monoterpene	*S. cerevisiae*	Ethanol

limonene is not produced natively in cyanobacteria. Recently, it is reported that heterologous expression of limonene synthase (LimS), the enzyme that catalyzes the transformation of GDP to limonene was demonstrated in cyanobacteria and also the production of limonene was observed [38]. There are now numerous studies on the production of limonene in cyanobacteria by utilizing *limS*-genes from different plant sources such as *Schizonepeta tenuifolia*, *Sitka spruce*, *Mentha spicata* that have been codon-optimized and heterologously expressed in

TABLE 5.3

Heterologous production of terpenoids in cyanobacteria

Host Strain	Terpenoids	Gene and Source
Synechocystis	Isoprene	*Synechocystis* codon optimized isoprene synthase (*IspS*) from *Pueraria montana* (kudzu)
Synechocystis	Isoprene	*Synechocystis* codon optimized isoprene synthase (*IspS*) from *Pueraria montana* (kudzu)
Synechocystis	Isoprene	MVA pathway genes encoding the enzymes, Hmg-CoA synthase (HmgS) and Hmg-CoA reductase (HmgR) from *Enterococcus faecalis*
Synechocystis	Limonene	*Synechocystis* codon optimized limonene synthase (*LimS*) from *Schizonepeta tenuifolia*, and overexpression with native genes *dxs*, *crtE* and *idi*
Synechocystis	β-phellandrene	Synechocystis codon-optimized β-phellandrene synthase (β-PHLS) gene from Lavandula angustifolia (lavender)
Synechocystis	β-caryophyllene	β-caryophyllene synthase gene (QHS1) from Artemisia annua
Anabaena sp. PCC 7120	Farnesene	Anabaena codon-optimized farnesene synthesize (FaS) gene from Picea abies (Norway Spruce)
Synechococcus sp. PCC 7002	Bisabolene	*Synechococcus* codon-optimized (*E*)-a-bisabolene synthase from *Abies grandis* (Grand Fir)

the cyanobacteria *Synechocystis*, *Anabaena* sp. PCC 7120 and *Synechococcus* sp. PCC 7002 respectively (Table 5.3).

All these engineered cyanobacteria are observed to produce limonene successfully. β-phellandrene is the plant monoterpene that is present in the essential oils of lavender [39]. It is a product of the plant chloroplast localized MEP pathway and is directly synthesized from GDP by β-phellandrene synthase (β-PHLS). Whereas Cyanobacteria lack β-PHLS and cannot synthesize β-phellandrene naturally. But this monoterpene is produced from cyanobacteria through a codon-optimized β-PHLS gene from *Lavandula angustifolia* (lavender) that was heterologously expressed in *Synechocystis*. Similarly, Linalool is produced naturally through the MEP pathway of plant source and is synthesized from GDP in the presence of linalool synthase (LinS) [40]. Whereas Cyanobacteria lacks the gene encoding LinS. So as to produce linalool from cyanobacteria, a linalool synthase gene from Norway Spruce was introduced to *Anabaena* sp. PCC 7120. The overexpression and secretion of linalool using CO_2 as the carbon source was reported from the engineered cyanobacteria. The existence of a terpenoid biosynthetic pathway is also confirmed in *Synechocystis* sp. PCC 6714 through ^{13}C-labeling studies [41].

5.2.9 Strategies for Improving Biosynthesis of Terpenoids

The utility of terpenoids is rapidly increasing due to their therapeutic potentials. But the limitation is low product yields and high costs of extraction from plants and other natural sources. To overcome these problems, several strategies are involved to extract terpenoids such as (1) DNA-based engineering (promoter engineering) (2) RNA engineering (synthetic RNA switches) (3) protein and cofactor engineering (4) metabolic pathway engineering (modular pathway engineering and compartmentalization engineering) (5) genome engineering (6) cellular engineering (transporter engineering) [42].

5.2.9.1 Protein Engineering Strategies for Biosynthesis of Terpenoids

The advances in enzyme structure and modification via protein engineering accelerate the biosynthesis of terpenoids.

The intra-module protein engineering strategy is used to improve the rate-limiting *dxs/dxr/idi*, while inter-module engineering involves promoter replacement and inducer adjustment to enhance the synthesis of isoprene (Table 5.4, Figure 5.19) [43].

5.2.9.2 Biosynthesis of Terpenoids via Genome Engineering

Plasmids are used as an expression system for both exogenous and endogenous pathway genes as it is easy to use or manipulate and incorporate into the host strain. The production of terpenoids in desired hosts is carried out by expressing and tuning the exogenous pathway genes. Gene expression is augmented by transcription initiation, post-translational protein processing through the interaction between transcriptional and translational factors, repressors, activators, or enhancers. The expression of exogenous pathway genes using chromosomal integration is highly stable as compared to the plasmid-mediated expression system. The post-transcriptional regulation affects the level of translated proteins as the level of mRNA transcribed does not always correspond to translated proteins. The overexpression of genes associated with efflux pumps provides positive results in the production of monoterpenes, sesquiterpenes and diterpenes. Niu et al. reported that the replacement of the native promoter of *acrAB* in *E. coli* with the P37 strong promoter ensures the improved tolerance to pinene. The segregation of the multi-gene pathway is an efficient technique to optimize the expression of a biosynthetic pathway. The construction of a pyruvate dehydrogenase coupled with gene deletions for enhanced acetyl-CoA flux through the MVA pathway resulted in the production of limonene in *S. cerevisiae* (Table 5.5) [44].

5.2.9.3 Biosynthesis of Monoterpenes via Recombinant Technology

The monoterpenes are biosynthesized from geranyl pyrophosphate (GPP) and neryl pyrophosphate (NPP) catalyzed by recombinant Li3CARS. The biosynthesis of isoprenoids starts with the synthesis of the common C5 isoprene units-isopentenyl pyrophosphate (IPP) and dimethylallyl pyrophosphate (DMAPP) through the plastidial 2-C-methyl-D-erythritol-4- phosphate

TABLE 5.4

Biosynthesis of Terpenoids via Protein Engineering Strategies

Strain	Terpenoids	Description
E. coli	Lycopene	Enzyme engineering of mevalonate 3-kinase via single amino acid mutations into 5-phosphomevalonate 3-kinase
E. coli	(−)-Carvone	Optimization of pathway enzymes with quantification concatemer method
E. coli	Pinene	Computer-aided directed evolution of geranyl diphosphate and pinene synthase
E. coli	Viridiflorol	Random mutation by error-prone PCR and enzyme truncation on viridiflorol synthase
E. coli	Levopimaradiene	Combinatorial evolution in geranylgeranyl diphosphate synthase and levopimaradiene synthase
E. coli	Albaflavenol	Fusion of terpene synthases and cytochrome P450
E. coli	Isopentenol	Improved activity of phosphomevalonate decarboxylase in an isopentenyl diphosphate
E. coli	Astaxanthin	Fusion of *crtZ* and *crtW*
E. coli	Steviol	Fusion of *CYP714A2* and P450 reductase 2 (*AtCPR2*) from *A. thaliana*
E. coli	Geraniol	Fusion tagging of geraniol synthase with a laboratory evolved N-terminal of a chloramphenicol acetyltransferase leader sequence
E. coli	Nerol and borneol	Site-directed mutagenesis of truncated bornyl diphosphate synthase from *Lippia dulcis*
S. cerevisiae	Geranylgeraniol	Overexpression of fused geranylgeranyl pyrophosphate synthase from *Pantoea agglomerans* and farnesyl diphosphate synthase (*PaGGPPS-Erg20*) and fused geranylgeranyl pyrophosphate synthase from *Pantoea agglomerans* and diacylglycerol diphosphate phosphate (*PaGGPPSDPP1*)
S. cerevisiae	Miltiradiene	Engineering a chimeric protein through the fusion of terpene synthase 1 from *Coleus forskohlii* and kaurene synthases-like from *Salvia miltiorrhia*
S. cerevisiae	Isoprene	Directed evolution of isoprene synthase

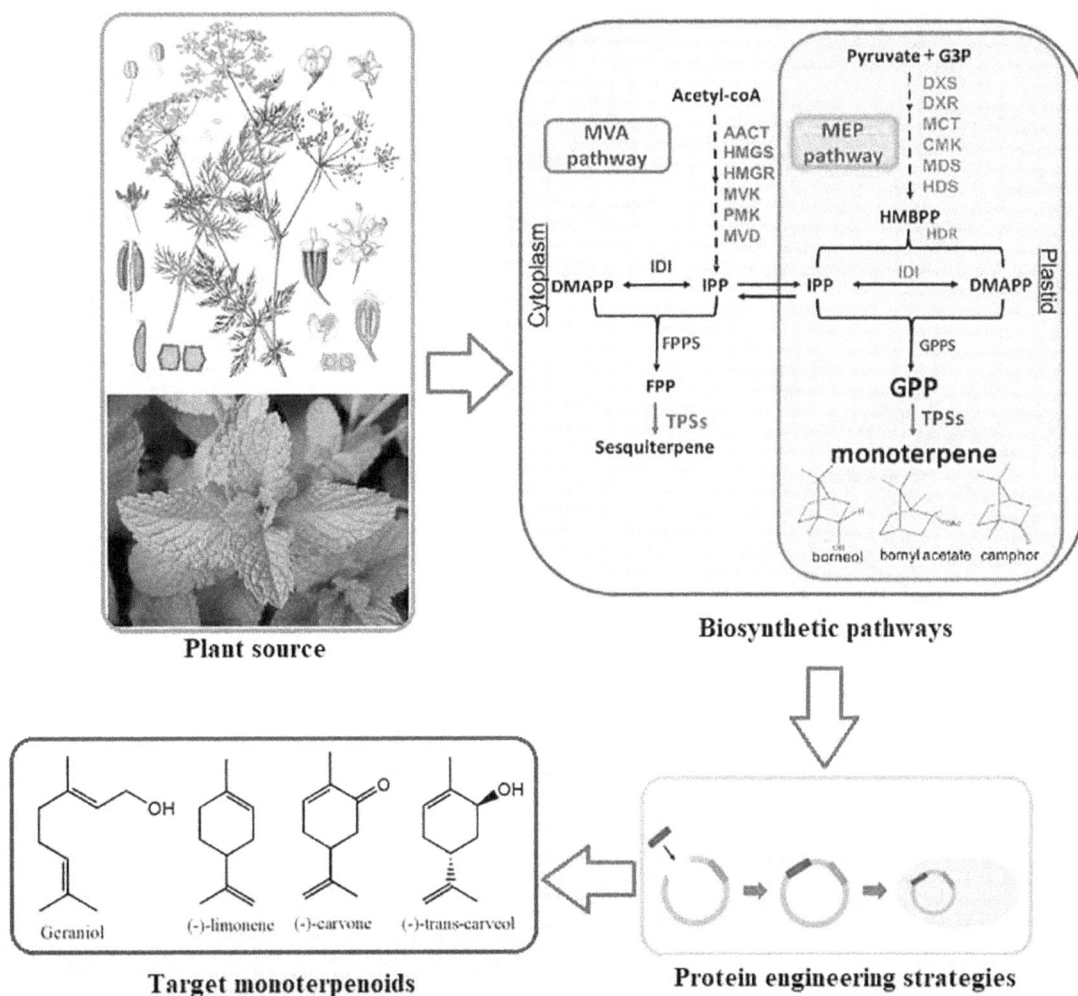

FIGURE 5.19 Production of monoterpenoids via protein engineering strategies.

(MEP) and cytosolic mevalonate (MVA) pathways. Burke et al. reported that the condensation of IPP and DMAPP in head-to-tail fashion by pernyltransferases to produce linear precursors of most isoprenoids, such as C10 geranyl pyrophosphate (GPP), C10 neryl pyrophosphate (NPP) and C15 farnesyl pyrophosphate (FPP) (Figure 5.20) [45].

TABLE 5.5

Biosynthesis of Terpenoids via Genome Engineering

Strains	Terpenoids	Description
S. cerevisiae	Crocetin	Development of a temperature-responsive strain coupled with chromosomal integration of pathway genes
E. coli	β-carotene	Integration of pathway genes
S. cerevisiae	Zerumbone	Integration of pathway genes
S. cerevisiae	β-carotene	Chromosomal integration of β-carotene biosynthetic pathway genes from *Xanthophylomyces dendrorhous*
E. coli	Pinene	Modular co-culture of MVA pathway and a TIGR mediated gene cluster

5.2.10 Regulation of Biosynthesis of Terpenoids

The exposure to sunlight leads to the down-regulation of MVA pathway genes. To support the production of terpenoids for protection against temperature stress, carbon flux through the MEP pathway increases under elevated temperatures. The compartmentalization of MEP and MVA pathways and associated downstream pathways allows for the subcellular regulation and coordination of photosynthesis-dependent and independent terpenoid biosynthetic routes. The expression of the MEP and MVA pathway genes is developmentally regulated by external stimuli [46,47].

5.3 Conclusion

Terpenoids are the hydrocarbons present in plants, algae, fungi, sponges and animals. These compounds possess diverse chemical structures and biological activities. Biogenesis of terpenoids is carried out through different pathways such as mevalonic acid (MVA) pathway and methylerythritol

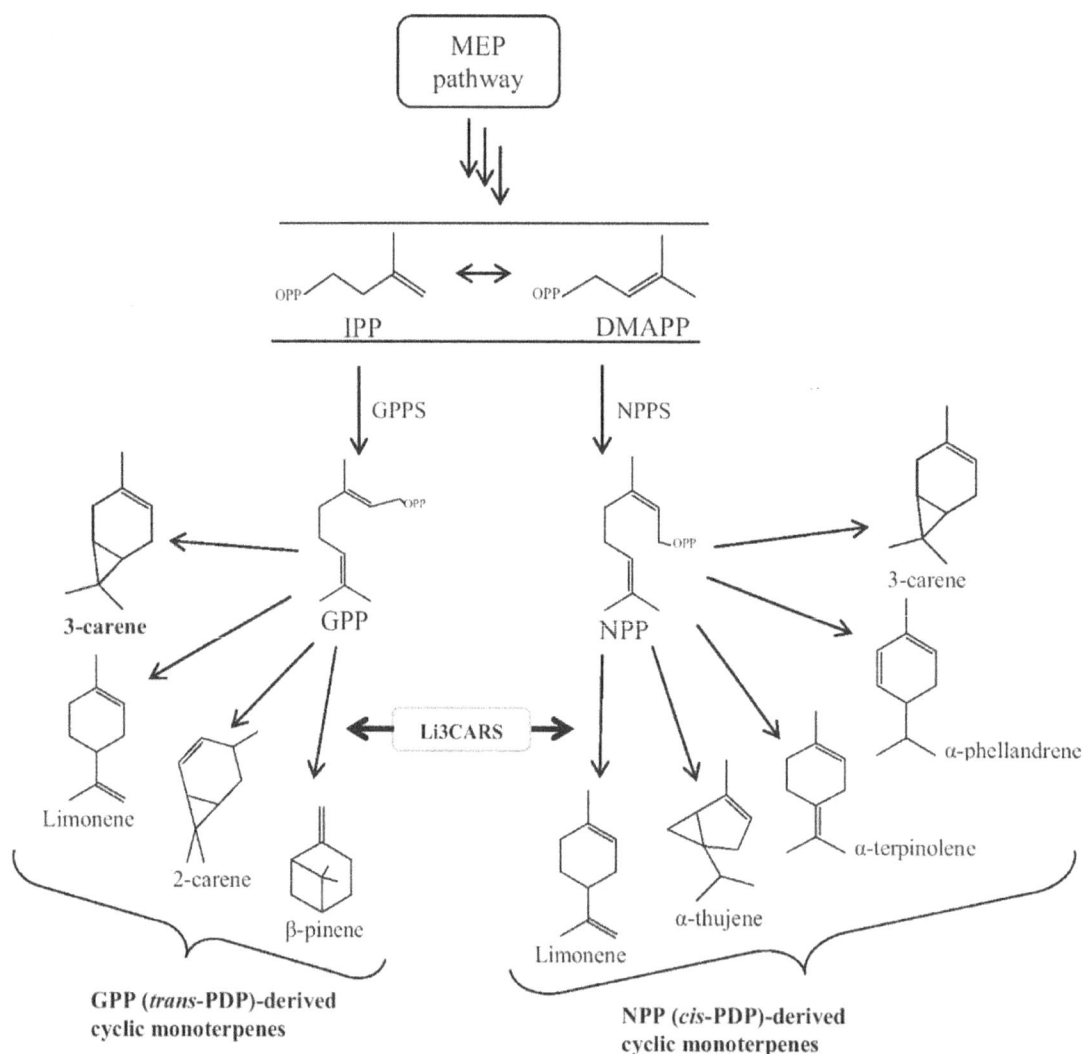

FIGURE 5.20 Biosynthesis of monoterpenes catalyzed by recombinant Li3CARS.

4-phosphate (MEP) pathway. MVA pathway generally provides the precursors for the biosynthesis of sesquiterpenoids, polyprenols, brassinosteroids and triterpenoids in cytosol. Whereas, the MEP pathway synthesizes IPP and DMAPP from pyruvate and D-glyceraldehyde 3-phosphate. MEP pathways are mainly applied for the biosynthesis of hemiterpenoids, monoterpenoids, diterpenoids, carotenoids, etc. The biosynthesis of several monoterpenes (e.g. limonene, terpineol, cineole) begins from geranyl diphosphate (GDP). The biosynthesis of several diterpenes (e.g. abietadiene, casbene and kaurene) involves the ionization and cyclization of GGDP. The biosynthesis of different sesquiterpenes (e.g. bisabolene, aristolochene, germacrene, humulene) starts from farnesyl diphosphate (FDP). Biogenesis of terpenoids requires terpene synthases (TPSs) using DMADP, GDP, FDP, or GGDP as the substrate. TPSs catalyze the generation of carbocation intermediates with multiple rearrangements of the carbon backbone that leads to the formation of terpenoids from a single substrate. The modifications in terpenoid biosynthesis are accomplished by a class of heme-containing enzymes called cytochrome P450 monooxygenases. To enhance biogenesis, metabolic flux needs to be redirected for the generation of specific terpenoids. For easy recovery of the product from the culture, it is essential to focus on products that are either volatile or hydrophobic compounds that can be excreted from the cells and are easily separated from the growth medium. Further investigation is required to improve terpenoid biosynthesis by applying genetic engineering, recombinant technology, etc.

Abbreviation

β-PHLS	β-phellandrene synthase
BPP	Bornyl diphosphate
BPPS	Bornyl diphosphate synthase
DMADP	Dimethylallyl diphosphate
DMAPP	Dimethylallyl pyrophosphate
DNA	Deoxyribonucleic acid
E. coli	*Escherichia coli*
FDP	Farnesyl diphosphate
FPP	Farnesyl diphosphate
GDP	Geranyl diphosphate
GDS	Geranyl diphosphate synthase
GGDP	Geranylgeranyl diphosphate
GPP	Geranyl diphosphate
HPLC	High-performance liquid chromatography
HPTLC	High performance thin layer chromatography
IDP	Isopentenyl diphosphate
IPD	Isopiperitenol dehydrogenase
IPP	Isopentenyl diphosphate
LinS	Linalool synthase
MEP	2-C-methyl-D-erythritol-4- phosphate
MFS	Menthofuran synthase
MVA	Mevalonic acid
NPP	Neryl diphosphate
PR	Pulegone reductase
RNA	Ribonucleic acid
TLC	Thin Layer Chromatography

REFERENCES

[1] Kirby, J.; Keasling, J. D. Biosynthesis of plant isoprenoids: Perspectives for microbial engineering. *Annual Review of Plant Biology*, **2009**, 60, 335–355.

[2] Tholl, D. Terpene synthases and the regulation, diversity and biological roles of terpene metabolism. *Current Opinion in Plant Biology*, **2006**, 9, 297–304.

[3] McConkey, M. E.; Gershenzon, J.; Croteau, R. B. Developmental regulation of monoterpene biosynthesis in the glandular trichomes of peppermint. *Plant Physiology*, **2000**, 122, 215–223.

[4] Rohmer, M.; Knani, M.; Simonin, P.; Sutter, B.; Sahm, H. Isoprenoid biosynthesis in bacteria: A novel pathway for the early steps leading to isopentenyl diphosphate. *Biochemical Journal*, **1993**, 295, 517–524.

[5] Rohmer, M. The discovery of a mevalonate-independent pathway for isoprenoid biosynthesis in bacteria, algae and higher plants. *Natural Product Reports*, **1999**, 16, 565–574.

[6] Ramak, P.; Osaloo, S. K.; Sharifi, M.; Ebrahimzadeh, H.; Behmanesh, M. Biosynthesis, regulation and properties of plant monoterpenoids. *Journal of Medicinal Plants Research*, **2014**, 8, 983–991.

[7] Zhao, Y.; Yang, J.; Qin, B.; Li, Y.; Sun, Y.; Su, S.; Xian, M. Biosynthesis of isoprene in *Escherichia coli* via methylerythritol phosphate (MEP) pathway. *Applied Microbiology and Biotechnology*, **2011**, 90, 1915–1922.

[8] Martin, V. J. J.; Pitera, D. J.; Withers, S. T.; Newman, J. D.; Keasling, J. D. Engineering a mevalonate pathway in *Escherichia coli* for production of terpenoids. *Nature Biotechnology*, **2003**, 21, 796–802.

[9] Bohlmann, J.; Meyer-Gauen, G.; Croteau, R. Plant terpenoid synthases: Molecular biology and phylogenetic analysis. *Proceedings of the National Academy of Sciences of the United States of America*, **1998**, 95, 4126–4133.

[10] Bouvier, F.; Rahier, A.; Camara, B. Biogenesis, molecular regulation and function of plant isoprenoids. *Progress in Lipid Research*, **2005**, 44, 357–429.

[11] Berthelot, K.; Estevez, Y.; Deffieux, A.; Peruch, F. Isopentenyl diphosphate isomerase: A checkpoint to isoprenoid biosynthesis. *Biochimie*, **2012**, 94, 1621–1634.

[12] Aji kumar, P. K.; Tyo, K.; Carlsen, S.; Mucha, O.; Phon, T. H.; Stephanopoulos, G. Terpenoids: Opportunities for biosynthesis of natural product drugs using engineered microorganisms. *Molecular Pharmaceutics*, **2008**, 5, 167–190.

[13] Ramawat, K. G.; Mérillon, J.-M. *Natural Products: Biosynthesis of Terpenoids*, (Chapter 121). **2013**, 2693–2732. doi: 10.1007/978-3-642-22144-6_121.

[14] Wise, M. L.; Croteau, R. Monoterpene biosynthesis. In *Comprehensive Natural Products Chemistry: Isoprenoids Including Carotenoids and Steroids*, Cane, D. E., ed., Oxford: Elsevier, **1999**, Vol. 2, 97–153.

[15] Mahmoud, S. S.; Croteau, R. B. Menthofuran regulates essential oil biosynthesis in peppermint by controlling a downstream monoterpene reductase. *Proceedings of the National Academy of Sciences of the United States of America*, **2001**, 98, 8915–8920.

[16] Estevez, J. M.; Cantero, A.; Reindl, A.; Reichler, S.; Leon, P. 1-Deoxy-D-xylulose 5-phosphate synthase, a limiting enzyme for plastidic isoprenoid biosynthesis in plants. *Journal of Biological Chemistry*, **2001**, 276(25), 22901–22909.

[17] Lange, B. M.; Ahkami, A. Metabolic engineering of plant monoterpenes, sesquiterpenes and diterpenes-current status and future opportunities. *Plant Biotechnology*, **2013**, 1, 169–196.

[18] Tholl, D. Biosynthesis and biological functions of terpenoids in plants. *Advances in Biochemical Engineering/Biotechnology*, **2015**, 148, 63–106.

[19] Haagen-Smit, A. J. The biogenesis of terpenes. *Annual Review of Plant Physiology*, 4(1), 305–324. doi: 10.1146/annurev.pp.04.060153.001513.

[20] Ambid, C.; Moisseeff, M.; Fallot, J. *Biogenesis of Monoterpenes*, 1(3), 91–93. doi: 10.1007/bf00272360.

[21] Ruzicka, L. The isoprene rule and the biogenesis of terpenic compounds. *Experientia*, **1953**, 9, 357–367.

[22] Winter, J. M.; Tang, Y. Synthetic biological approaches to natural product biosynthesis. *Current Opinion in Biotechnology*, **2012**, 23, 736–743.

[23] Talapatra, S. K.; Talapatra, B. *Chemistry of Plant Natural Products ‖ Biosynthesis of Terpenoids: The Oldest Natural Products.* (Chapter 5). **2015**, 317–344. doi: 10.1007/978-3-642-45410-3_5.

[24] Duetz, W. A.; Bouwmeester, H.; van Beilen, J. B.; Witholt, B. Biotransformation of limonene by bacteria, fungi, yeasts, and plants. *Applied Microbiology and Biotechnology*, **2003**, 61, 269–277.

[25] Rasoul-Amini, S.; Fotooh-Abadi, E.; Ghasemi, Y. Biotransformation of monoterpenes by immobilized microalgae. *Journal of Applied Phycology*, **2011**, 23, 975–981.

[26] Hassan, M.; Maarof, N. D.; Ali, Z. M.; Noor, N. M.; Othman, R.; Mori, N. Monoterpene alcohol metabolism: Identification, purification, and characterization of two geraniol dehydrogenase isoenzymes from polygonum minus leaves. *Bioscience, Biotechnology and Biochemistry*, **2012**, 76(8), 1463–1470. doi: 10.1271/bbb.120137.

[27] Kouda, R.; Yakushiji, F. Recent advances in iridoid chemistry: Biosynthesis and chemical synthesis. *Chemistry – An Asian Journal*, **2020**, 15(22), 3771–3783.

[28] Robinson, D. R.; West, C. A. Biosynthesis of cyclic diterpenes in extracts from seedlings of *Ricinus communis*. I. Identification of diterpene hydrocarbons formed from mevalonate. *Biochemistry*, **1970**, 9(1), 70–79. doi: 10.1021/bi00803a010.

[29] Srivastava, S.; Srivastava, A. K. Lycopene: Chemistry, biosynthesis, metabolism and degradation under various abiotic parameters. *Journal of Food Science and Technology*, **2015**, 52, 41–53. doi: 10.1007/s13197-012-0918-2.

[30] Giglio, S.; Chou, W. K. W.; Ikeda, H.; Cane, D. E.; Monis, P. T. Biosynthesis of 2-methylisoborneol in cyanobacteria. *Environmental Science & Technology*, **2011**, 45, 992–998.

[31] O'Connor, S. E.; Maresh, J. J. Chemistry and biology of monoterpene indole alkaloid biosynthesis. *Natural Product Reports*, **2006**, 23, 532–547.

[32] Carter, O. A.; Peters, R. J.; Croteau, R. Monoterpene biosynthesis pathway construction in *Escherichia coli*. *Phytochemistry*, **2003**, 64(2), 425–433. doi: 10.1016/s0031-9422(03)00204-8.

[33] Martin, V. J.; Pitera, D. J.; Withers, S. T.; Newman, J. D.; Keasling, J. D. Engineering a mevalonate pathway in Escherichia coli for production of terpenoids. *Nature Biotechnology*, **2003**, 21, 796–802.

[34] Zhang, Y.; Wang, J.; Cao, X.; Liu, W.; Yu, H.; Ye, L. Highlevel production of linalool by engineered *Saccharomyces cerevisiae* harboring dual mevalonate pathways in mitochondria and cytoplasm. *Enzyme and Microbial Technology*, **2020**, 134, 109462. doi: 10.1016/j.enzmictec.2019.109462.

[35] Kong, S.; Fu, X.; Li, X.; Pan, H.; Guo, D. De novo biosynthesis of linalool from glucose in engineered *Escherichia coli*. *Enzyme and Microbial Technology*, 140, 109614. doi: 10.1016/j.enzmictec.2020.109614.

[36] Duetz, W. A.; Bouwmeester, H.; van Beilen, J. B.; Witholt, B. Biotransformation of limonene by bacteria, fungi, yeasts, and plants. *Applied Microbiology and Biotechnology*, **2003**, 61, 269–277.

[37] Kiyota, H.; Okuda, Y.; Ito, M.; Hirai, M. Y.; Ikeuchi, M. Engineering of cyanobacteria for the photosynthetic production of limonene from CO_2. *Journal of Biotechnology*, **2014**, 185, 1–7.

[38] Halfmann, C.; Gu, L.; Zhou, R. Engineering cyanobacteria for the production of a cyclic hydrocarbon fuel from CO_2 and H_2O. *Green Chemistry*, **2014**, 16, 3175–3185.

[39] Davies, F. K.; Work, V. H.; Beliaev, A. S.; Posewitz, M. C. Engineering limonene and bisabolene production in wild type and a glycogen-deficient mutant of *Synechococcus* sp. PCC 7002. *Frontiers in Bioengineering and Biotechnology*, **2014**, 2, doi: 10.3389/fbioe.2014.00021.

[40] Bentley, F. K.; García-Cerdán, J. G.; Chen, H.-C.; Melis, A. Paradigm of monoterpene (β-phellandrene) hydrocarbons production via photosynthesis in cyanobacteria. *BioEnergy Research*, **2013**, 6, 917–929.

[41] Formighieri, C.; Melis, A. Regulation of β-phellandrene synthase gene expression, recombinant protein accumulation, and monoterpene hydrocarbons production in Synechocystis transformants. *Planta*, **2014**, 240, 309–324.

[42] Gao, Q.; Wang, L.; Zhang, M.; Wei, Y.; Lin, W. Recent advances on feasible strategies for monoterpenoid production in *Saccharomyces cerevisiae*. *Frontiers in Bioengineering and Biotechnology*, **2020**, 8, 609800. doi: 10.3389/fbioe.2020.609800.

[43] Daletos, G.; Stephanopoulos, G. Protein engineering strategies for microbial production of isoprenoids. *Metabolic Engineering Communications*, **2020**, 11, e001292.

[44] McConkey, M. E.; Gershenzon, J.; Croteau, R. B. Developmental regulation of monoterpene biosynthesis in the glandular trichomes of peppermint. *Plant Physiology*, **2000**, 122(1), 215–224. doi: 10.1104/pp.122.1.215.

[45] Adal, A. M.; Sarker, L. S.; Lemke, A. D.; Mahmoud, S. S. Isolation and functional characterization of a methyl jasmonate-responsive 3-carene synthase from Lavandula x intermedia. *Plant Molecular Biology*, 93(6), 641–657. doi: 10.1007/s11103-017-0588-6.

[46] Hemmerlin, A. Post-translational events and modifications regulating plant enzymes involved in isoprenoid precursor biosynthesis. *Plant Science*, **2013**, 203, 41–54.

[47] Alex, D.; Bach, T. J.; Chye, M. L. Expression of *Brassica juncea* 3-hydroxy-3-methylglutaryl CoA synthase is developmentally regulated and stress-responsive. *The Plant Journal*, **2000**, 22, 415–426.

6

Anticancer Natural Terpenoids

Bimal Krishna Banik*, Biswa Mohan Sahoo†, and Abhishek Tiwari‡

CONTENTS

6.1 Introduction 197
6.2 Monoterpenes as Anticancer Agents 199
6.3 Sesquiterpenes as Anticancer Agents 206
6.4 Diterpenes as Anticancer Agents 212
6.5 Triterpenes as Anticancer Agents 217
6.6 Tetraterpenes as Anticancer Agents 221
6.7 Conclusion 223
Abbreviation 223
References 223

6.1 Introduction

Cancer is considered as a principal public health problem in India and in other parts of the world. It is one of the major causes of mortality and morbidity around the globe, and the number of cases is increasing continuously which is estimated to be 21 million by 2030. In 2017, it was estimated that the US has approximately 1,688,780 new cancer diagnoses cases and 600,920 cancer deaths [1,2]. As per World Health Organization (WHO) data, above 14.1 million new cancer cases and 8.2 million deaths were reported globally in the year 2012 and over 70% of new cancer cases have been estimated during the next 20 years [3]. Cancer is characterized by uncontrolled growth of abnormal cells and that finally leads to metastasis. The uncontrolled proliferation of a normal cell leads to genetic instabilities and alterations within the cells and tissues that transform normal cells into malignant cells [4].

There are three stages involved in the development of cancer including initiation, promotion/progression and metastasis. In cancer development, the first step is initiation in which there is a change in the genetic materials of the cells called mutation that makes the normal cells into cancerous cells. The change in the genetic materials of cells may arise spontaneously or induced by exposure to carcinogens (cancer-causing agents). The genetic alterations can result in dysregulation of biochemical signaling pathways associated with cellular proliferation, survival and differentiation that can be influenced by several factors such as rate and type of carcinogenic metabolism and the response of the DNA repair function. Whereas the promotion step is a lengthy and reversible process in which the actively proliferating preneoplastic cells accumulate. Similarly, progression is the phase between a preneoplastic lesion and the development of invasive cancer. It is the stage of neoplastic transformation in which the genetic and phenotypic changes and cell proliferation occur. It involves the increase in tumor size where the cells may undergo mutations with invasive and metastatic potential. Finally, metastasis involves the spread of cancer cells from the primary site to other parts of the body through the blood stream or the lymphatic system (Figure 6.1) [5].

Several factors involved to cause cancer include external factors and internal factors. Various external factors are radiation, smoking, chewing tobacco, pollutants in drinking water, food, air, chemicals, certain metals and infectious agents. Similarly, internal factors include genetic mutations, body immune system and hormonal disorders [6].

Hallmarks of cancer consist of an organizing principle for rationalizing the complexities of this disease. Hallmarks include sustaining proliferative signaling, evading growth suppressors, resisting cell death, enabling replicative immortality, inducing angiogenesis and activating invasion and metastasis. Hanahan and Weinberg reported that activating invasion and metastasis is the hallmark of cancer (Figure 6.2) [7].

There are several types of cancer observed in human beings such as lung cancer, liver cancer, ovary cancer, colon and rectum cancer leukemia, prostate cancer, breast cancer, thyroid cancer, pancreas cancer, colon and rectum cancer. Among these, lung cancer is reported as top list in male and breast cancer in the case of females [8]. Approximately, 80% of the world's

---

* Department of Mathematics and Natural Sciences, College of Sciences and Human Studies, Prince Mohammad Bin Fahd University, Al Khobar, Kingdom of Saudi Arabia. Corresponding author e-mail address: bimalbanik10@gmail.com.

† Roland Institute of Pharmaceutical Sciences, Berhampur affiliated to Biju Patnaik University of Technology (BPUT), Rourkela, Odisha, India. Corresponding author e-mail address: drbiswamohansahoo@gmail.com.

‡ Faculty of Pharmacy, Pharmacy Academy, IFTM University, Lodhipur Rajput, Moradabad-244102, Uttar Pradesh, India.

DOI: 10.1201/9781003008682-6

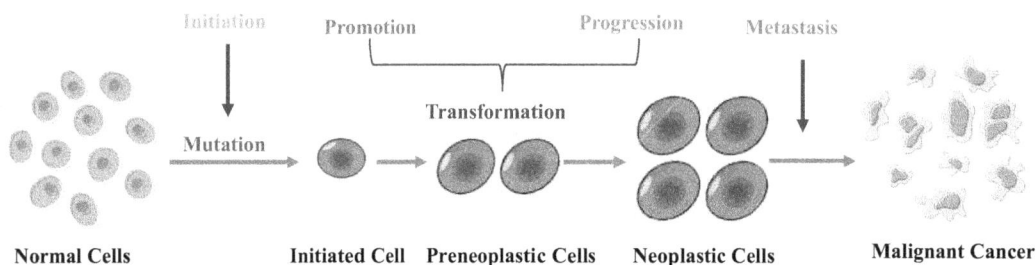

FIGURE 6.1 Development of cancer cells.

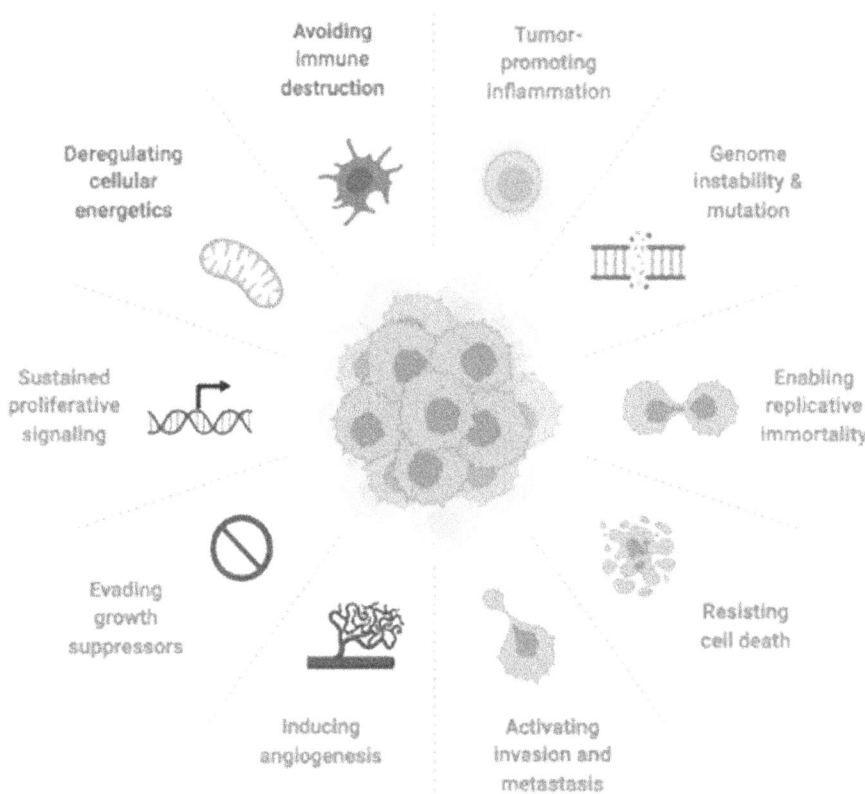

FIGURE 6.2 Hallmarks of cancer.

population utilize traditional medicines and more than 60% of clinically approved anticancer drugs are obtained from natural sources. Natural products are the major source of medicinal agents with diverse therapeutic potentials. Medicinal plants serve as nature's gift to help human beings [9]. Phytochemicals and their analogues are present in different parts of the plant sources including flower, pericarp, fruits, seeds, roots, rhizomes, stem, leaf, embryo, bark etc (Table 6.1). Natural products are a rich source of compounds like terpenoids which are applied clinically as anticancer therapy. Structurally, terpenoids are categorized into different subclasses such as monoterpenoids, sesquiterpenoids, diterpenoids, triterpenoids and tetraterpenoids. Both in-vivo and in-vitro anticancer potentials are reported for several terpenoids including Lupeol Andrographolide, Lycopene, β-Caryophyllene, Betulinic acid, Ursolic acid, Gossypol, Artemisinin, β-elemene, Cucurbitane and Deoxyelephantopin etc (Figure 6.3, Table 6.1) [10–12].

The anticancer activity of terpenoids appears promising and provides more opportunities for the potential treatment

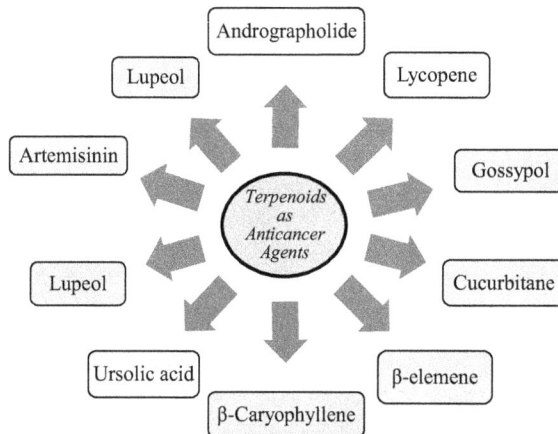

FIGURE 6.3 Terpenoids with anticancer activities.

TABLE 6.1

List of Medicinal Plants and Their Phytochemicals for Treatment of Cancer

Name of Plants	Parts of Plant	Phytochemicals	Type of Terpenoid	Treatment of Cancer
Aegle marmelos	Bark, root	Lupeol	Pentacyclic triterpenoid	Lymphoma, leukemia, melanoma and breast cancer
Andrographis paniculata	Whole plant	Andrographolide	Diterpene lactone	Colon cancer
Solanum lycopersicum	Fruit	Lycopene	Tetraterpene	Prostate and colon cancer
Syzygium aromaticum	Aerial parts	β-Caryophyllene	Bicyclic sesquiterpene	Breast cancer
Ziziphus rugosa	Pericarp and seed	Betulinic acid	Pentacyclic lupane-type triterpenoid	Cytotoxicity against human melanoma cells
Vaccinium macrocarpon	Fruit	Ursolic acid	Pentacyclic triterpenoid	Cervical, prostate cancer
Gossypium hirsutum	Whole plant	Gossypol	Terpenoid aldehyde	Breast, stomach, liver, prostate and bladder cancer
Artemisia annua	Whole plant	Artemisinin	Sesquiterpene lactone	Liver, breast and pancreatic cancer
Curcuma wenyujin	Leaves	β-Elemene	Sesquiterpene	Melanoma and breast cancer
Momordica charantia	Leaves, roots	Cucurbitane	Triterpene	Colon cancer and breast cancer
Elephantopus carolinianus	Whole plant	Deoxyelephantopin	Sesquiterpene lactone	Melanoma

of different types of cancers such as breast cancer, liver and, lymphoma, leukemia, melanoma, prostate cancer, colon cancer, liver and pancreatic cancer. Development of naturally derived anticancer drugs plays a major role in cancer research. The anticancer activities of terpenoids have been evaluated experimentally [13,14]. They exhibit promising in vitro as well as in-vivo anticancer activity against different types of cancer cells (e.g. leukemia, breast cancer, lung cancer, colorectal carcinoma, liver cancer and bladder carcinoma) (Figure 6.4) [15].

Most of the anticancer drugs exhibit their action on cancer cells by regulating cell signaling, cell cycle, metastasis, angiogenesis, apoptosis, etc [16]. Apoptosis is called programmed cell death whose activation is regulated by a series of genes. Whereas the terpenoids are inhibiting the proliferation of tumor cell and induce tumor cell death

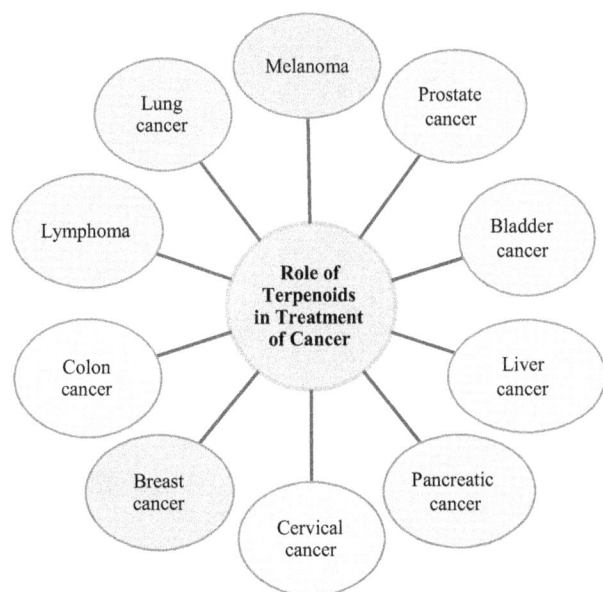

FIGURE 6.4 Role of terpenoids in treatment of different types of cancer.

by preventing the multiple cancer-specific targets such as proteasome, NF-κB, antiapoptotic protein Bcl-2, etc. (Figure 6.5) [17–19].

6.2 Monoterpenes as Anticancer Agents

Monoterpenes mainly consist of two isoprene units with molecular formula $C_{10}H_{16}$. Monoterpenes possess several structures such as linear or acyclic (e.g. myrcene and ocimene) and rings (monocyclic e.g. limonene and bicyclic e.g. thujane, carane, pinane) [20]. Whereas the modified terpenes with oxygen functionality or missing a methyl group, are called monoterpenoids (e.g. linalool, carvone, carvacrol, geraniol, geranial, perillyl alcohol, perillaldehyde and menthol) [21]. The anticancer properties of the monoterpenoids have been demonstrated experimentally. These compounds exhibit promising in vitro anticancer activity against different cancer cells such as leukemia, colorectal carcinoma, hepatoma, bladder and breast cancer etc [22]. Monoterpenes exhibit cytotoxic activity against tumor cell lines which appear to exert this effect by inducing apoptosis caused by oxidative stress (Figure 6.6, Table 6.2) [23,24].

Terpenoids display a broad spectrum of inhibitory activities against various cancerous cells. Terpenoids are permeable through the cell wall and cell membrane due to lipophilic properties. The interaction of terpenoids with polysaccharides, fatty acids and phospholipids causes loss of membrane integrity and leakage of the cellular contents that leads to cell death. Other important mechanisms of action include the denaturation of cellular proteins [25].

Linalool is a naturally occurring terpene alcohol. Chemically, it is 3,7-dimethylocta-1,6-dien-3-ol. Naturally, Linalool exists in two enantiomeric forms such as (S)-(+)-linalool and (R)-(–)-linalool. Linalool and linalyl acetate are the major constituents of lavender and rosewood essential oil. They are cytotoxic against 153BR, HNDF and HMEC-1 cancer cell lines. Yuan et al. reported that Linalool inhibits the angiogenic activity of endothelial cells by down-regulating intracellular ATP levels

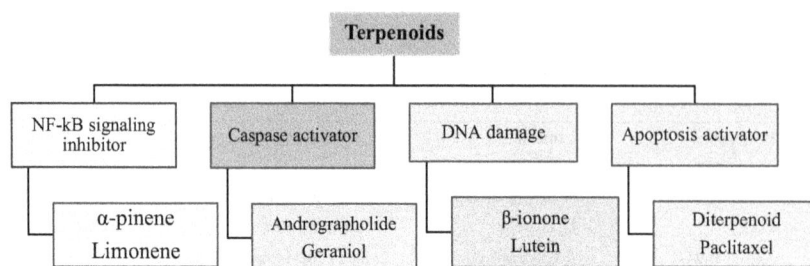

FIGURE 6.5 Molecular targets of terpenoids in cancer.

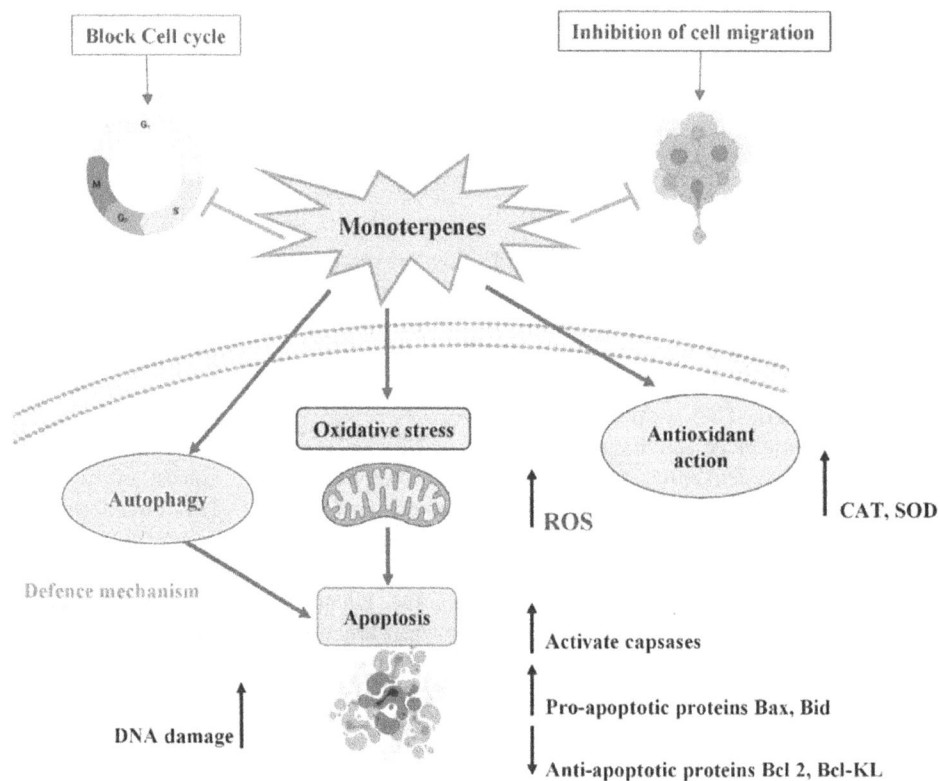

FIGURE 6.6 Mechanism action of monoterpenes.

TABLE 6.2

Cytotoxicity of Monoterpenes against Several Cell Lines

Terpenoids	Chemical Structure	Tumor Cell Lines
Terpineol		Human laryngeal epidermoid carcinoma (HepG2)
Thymol		Human promyelocytic leukemia (HL-60)

(Continued)

TABLE 6.2 (*Continued*)

Cytotoxicity of Monoterpenes against Several Cell Lines

Terpenoids	Chemical Structure	Tumor Cell Lines
Thujone	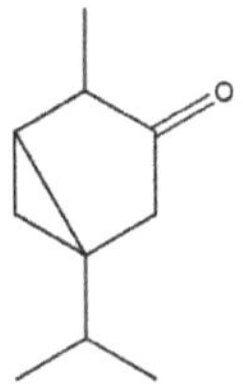	Human glioblastoma multiforme (T98G)
Geraniol	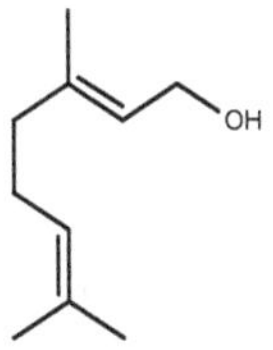	Human breast cancer (MCF-7)
Bornyl acetate	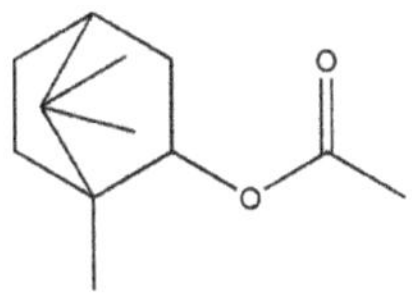	Human gastric cancer (SGC-7901)
Camphene	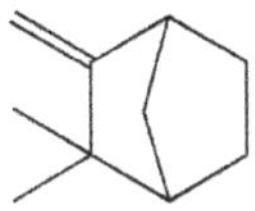	Human melanoma (A2058) Human breast cancer (SK-BR-3) Human cervical cancer (HeLa)
Carvacrol	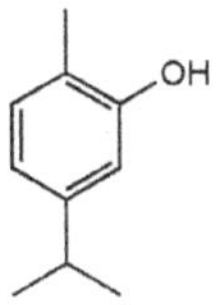	Human breast cancer (MDA-MB-231)
Citral Globularifolin		Human squamous carcinoma (A431) Human glioblastoma (U-87 MG)
Hinokitiol	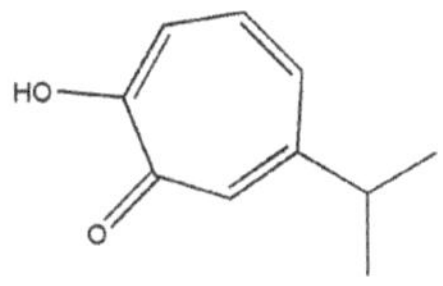	Human lung adenocarcinoma (A549)
Limonene	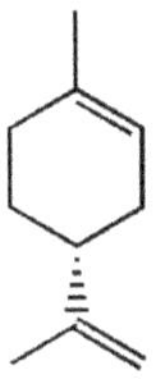	Human colon adenocarcinoma (Caco2) Human malignant melanoma (A375)
Linalool	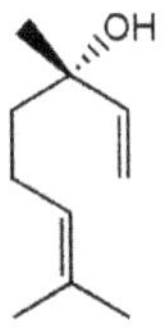	Human ovarian carcinoma (HeyA8)

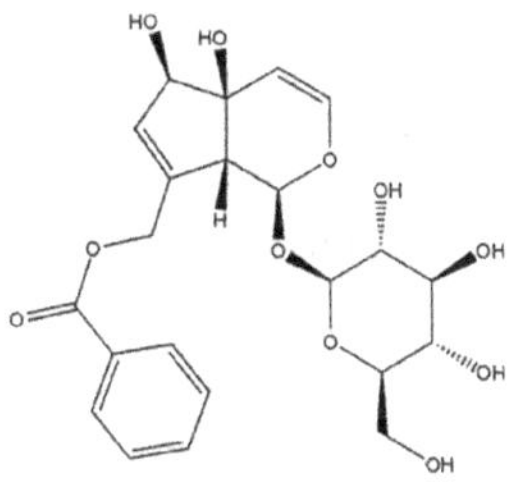

(Continued)

TABLE 6.2 (*Continued*)

Cytotoxicity of Monoterpenes against Several Cell Lines

Terpenoids	Chemical Structure	Tumor Cell Lines
Myrtenal		Human melanoma (SkMel-5)
Perillyl alcohol		Human hepatoma (SK-Hep-1)
α-Pinene		Murine melanoma (B16F10-Nex2)

and activating TRPM8. The linalool-induced ERK phosphory-lation inhibits the expression of the bone morphogenetic pro-tein (BMP)-2 and linalool-induced TRPM8 activation caused the inhibition of β1 integrin and focal adhesion kinase (FAK) signaling. ATP supplementation completely reversed linalool-induced ERK phosphorylation (Figure 6.7) [26,27].

Carvacrol is the major active ingredient of oregano oil. It arrests the G2/M phase of the cell cycle and decreases cyclin B1 expression. It was reported that carvacrol may induce apoptosis in colon cancer cells through the mitochondrial apoptotic pathway and the MAPK and PI3K/Akt signaling pathways [28]. The anti-proliferative activity of carvacrol is confirmed by nuclear condensation, FITC-Annexin V assay. The assay results reported that there is modulation in expres-sion of Bax, Bcl-2 and caspase activation. The mechanism involved in carvacrol-induced apoptosis leads to an increase in the level of reactive oxygen species. Carvacrol inhibits the

Notch signaling in PC-3 cells via down-regulation of Notch-1, and Jagged-1. Cell cycle analysis revealed that carvacrol arrest cell cycle in G0/G1 which is associated with decline in expres-sion of cyclin D1 and Cyclin-Dependent Kinase 4 (CDK4) and augment the expression of CDK inhibitor p21 [29]. Horvathova et al. demonstrated that carvacrol exerted cytotoxic effects in K562, HepG2 and colonic Caco-2 cells and significantly reduced the level of DNA damage induced in these cells by the strong oxidant H_2O_2 (Figure 6.8) [30,31].

Limonene is a monocyclic monoterpene and is found in the essential oils of citrus fruits. Limonene occurs in two opti-cally active forms such as D-limonene and L-limonene. Both are mirror images of one another chemically. D-Limonene is the major constituent of several citrus fruits such as lemon, orange, mandarin, lime and grapefruit) and has been focused for anticancer research. D-Limonene has been shown to inhibit 3-hydroxy-3-methylglutanyl coenzyme A (HMGCoA)

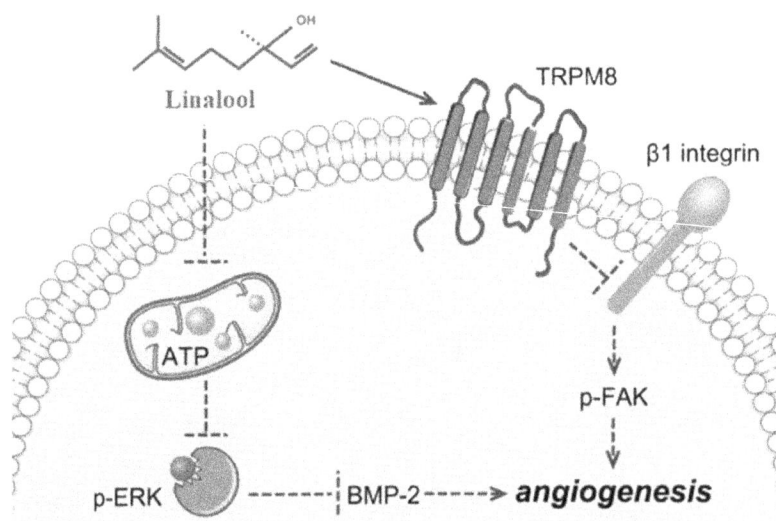

FIGURE 6.7 Inhibitory effect of linalool on angiogenesis.

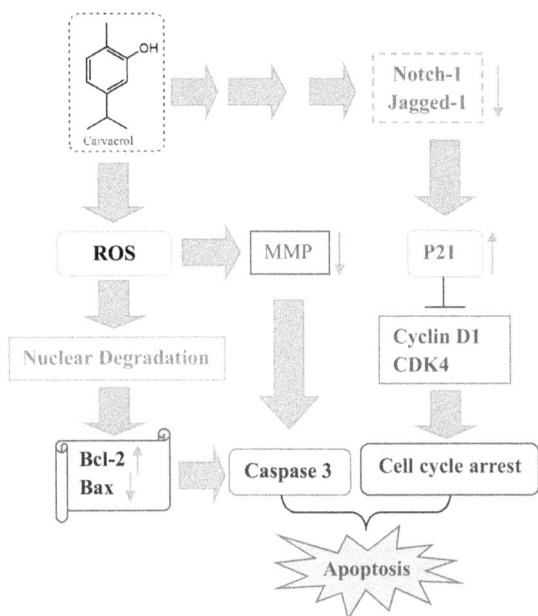

FIGURE 6.8 Anticancer effect of carvacrol.

reductase, which leads to the inhibition of protein isoprenylation of small G proteins including p21 and its membrane localization. This mechanism is believed to contribute to the efficacy of D-limonene in chemoprevention and cancer therapy. It prevents liver cancer by increasing the levels of hepatic enzymes that can detoxify carcinogens. D-limonene inhibits tumor growth and metastasis, probably via its anti-angiogenic, proapoptotic and anti-oxidant effects. In particular,

the combination of D-limonene and cytotoxic agents, such as fluorouracil (5-FU) and docetaxel, appeared to be more effective than either single treatment via a mechanism involving reactive oxygen species (ROS) generation [32].

Pattanayak et al. also reported that limonene inhibited the activity of HMG-CoA reductase due to greater binding affinity with the receptor and thus reduced the possibility of cancer growth. HMG-CoA reductase catalyzes the synthesis of mevalonate, which is the key step in the synthesis of prenyl groups and is the rate-limiting step in biosynthesis cholesterol. The expression levels of HMG-CoA reductase are mainly upregulated in various types of cancer in human beings (Figure 6.9) [33].

Marianna et al. described the antitumor activity of monoterpenes found in essential oils. Linalyl acetate, α-terpineol and camphor caused inhibition of the growth of the human colon cancer cell lines (HCT-116). α-Terpineol is a monoterpene alcohol that exhibits significant cytotoxicity against hepatocellular carcinomic human cell line (Hep G2), epithelioid carcinomic cell line (HeLa), human lymphoblastic leukemia T-cell line (MOLT-4), human chronicmyelogenous leukemia cell line (K-562). Hassan et al. suggested that α-terpineol inhibits the growth of tumor cells through a mechanism that involves inhibition of the NF-κB pathway (Figure 6.10) [34].

Menthol is a naturally occurring cyclic monoterpene alcohol which is present as an active constituent in the essential oils of Mentha species. Chemically, it is 5-methyl-2-propan-2-ylcyclohexan-1-ol [35,36]. Wang and collaborators showed that menthol inhibited the proliferation and motility of prostate cancer DU145 cells. Li and collaborators demonstrated that menthol induced cell death via the TRPM8 channel in a human bladder cancer cell line. Okamoto and collaborators

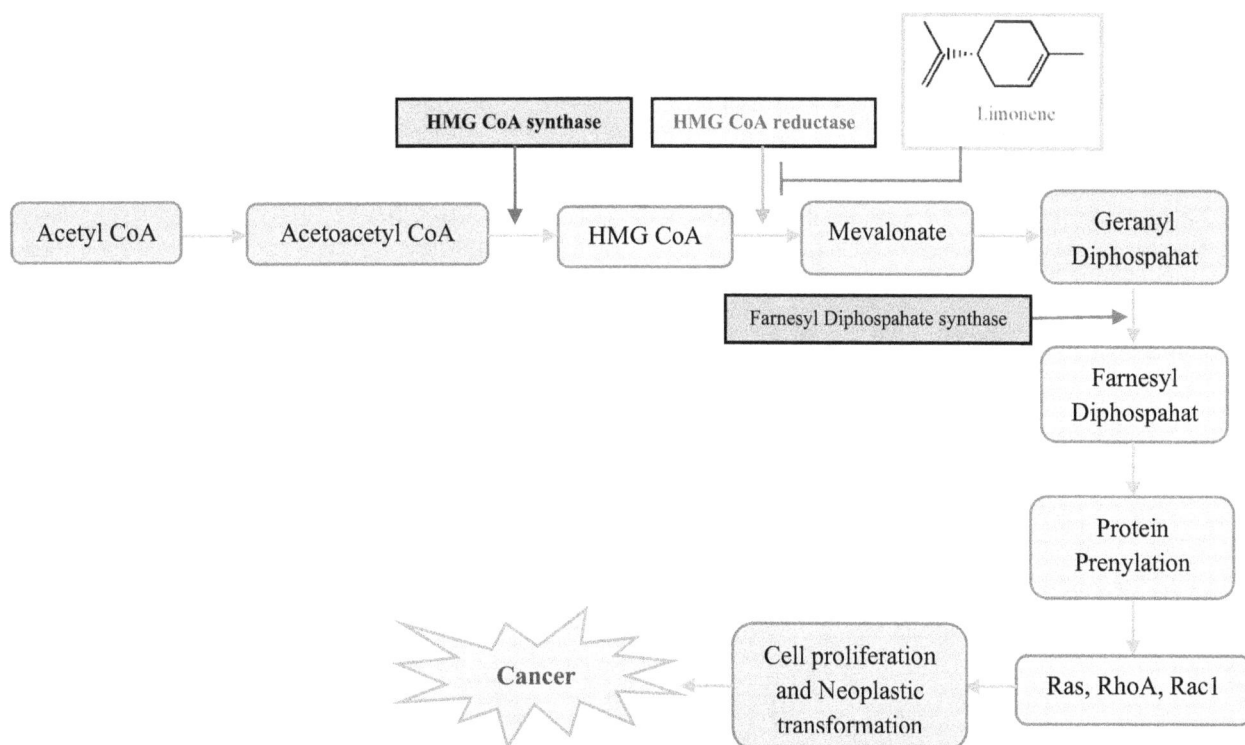

FIGURE 6.9 Mode of action of limonene via inhibition of HMG-CoA reductase.

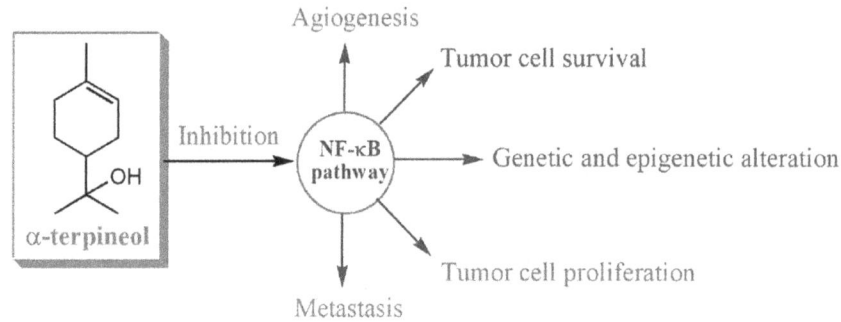

FIGURE 6.10   Antitumor activity of α-terpineol via inhibition of the NF-κB pathway.

also studied the role of menthol in the blockade of TRPM8 activity and found that it reduced the invasion potential of oral squamous carcinoma cell lines [37].

Yin et al. demonstrated the chemopreventive effect of menthol on carcinogen-induced cutaneous carcinoma through inhibition of inflammation and oxidative stress in mice. Menthol inhibits TPA-induced skin hyperplasia and inflammation and significantly suppresses the expression of cyclooxygenase-2. Furthermore, the pretreatment with menthol inhibits the formation of reactive oxygen species (ROS) and affects the activities of anti-oxidant enzymes in the skin. The expressions of NF-κB, Erk and p38 are down-regulated by the administration of menthol. So, the inflammation and oxidative stress together play a vital role in the chemopreventive effect of menthol on murine skin tumorigenesis (Figure 6.11) [38].

Borneol is a bornane type bicyclic monoterpenoid. It is chemically called 1,7,7-trimethylbicyclo[2.2.1]heptane substituted by a hydroxy group at position 2. Borneol exists in two enantiomers such as (+)-borneol or d-borneol and (−)-borneol or l-borneol. Su and collaborators demonstrated that borneol potentiates selenocystine-induced apoptosis in human hepatocellular carcinoma cells by enhancement of cellular uptake and activation of ROS-mediated DNA damage (Figure 6.12) [39].

The essential oil of Chenopodium ambrosioides comprises a large amount of Ascaridole. It is a natural bicyclic monoterpenoid that has a bridging peroxide functional group. Chemically, it is 1-methyl-4-propan-2-yl-2,3-dioxabicyclo[2.2.2]oct-5-ene. It was proposed that the anticancer mechanism of ascaridole

includes oxidative DNA damage and arrest G2/M phase of cell cycle. Further, ascaridole mediated the generation of ROS and interacted with intracellular targets that enhance apoptosis and cell death. Bezerra et al. investigated the cytotoxicity and antitumor activity of ascaridole in HL-60 and HCT-8 cell lines. It was observed that the IC_{50} values of 6.3 and 18.4 µg/mL are obtained in HL-60 and HCT-8 cell lines respectively. Results from in vivo studies using sarcoma 180 as a tumor model demonstrated inhibition rates of 33.9% at 10 mg/kg and 33.3% at 20 mg/kg (Figure 6.13) [40].

Thymol is a natural monoterpenoid with a molecular formula of $C_{10}H_{14}O$. It is the phenol derivative of p-Cymene. Chemically it is known as 2-isopropyl-5-methylphenol. Thymol exhibits the anticancer effect in acute promyelocytic leukemia cells (HL-60) by arresting the cell cycle at G0/G1 phase and mitochondrial depolarization. The cytotoxic effect of thymol is initiated via the production of ROS followed by deterioration of mitochondrial membrane. Thymol can also increase the expression of Bax protein and reduce the Bcl-2 protein that leads to activation of caspases and apoptosis. The chemopreventive effect of thymol on murine B16-F10 melanoma cells was evaluated by Paramasivam and collaborators. Thymol exhibited cytotoxicity with an IC_{50} value of 88.5 µg/mL (Figure 6.14) [41,42].

Myrcene is an acyclic monoterpene with a molecular formula of $C_{10}H_{16}$. Myrcene exists in two isomeric forms such as β-myrcene (7-methyl-3-methylene-1,6-octadiene) and α-myrcene (2-methyl-6-methylene-1,7-octadiene). Silva et al. investigated

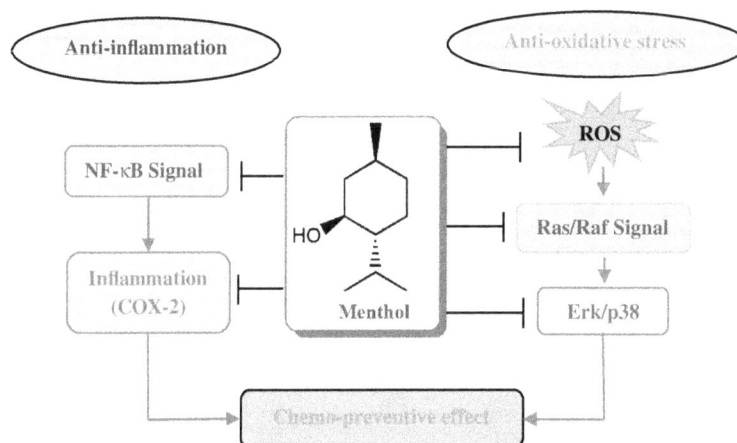

FIGURE 6.11   Chemopreventive effect of menthol.

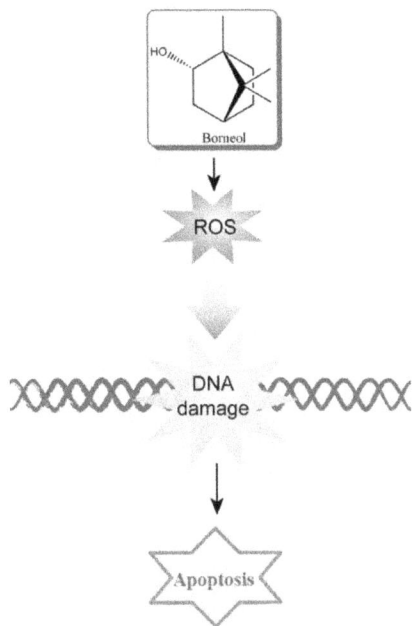

FIGURE 6.12 Activation of ROS-mediated DNA damage by borneol.

the cytotoxicity of myrcene against HeLa (human cervical carcinoma), A-549 (human lung carcinoma), HT-29 (human colon adenocarcinoma) and Vero (monkey kidney) cell lines as well as mouse macrophages. From several studies, it was demonstrated that β-myrcene inhibited the growth of breast cancer cells in vitro with IC$_{50}$ of 291 µM in MCF-7 cells (Figure 6.15) [43,44].

Geraniol is acyclic monoterpene alcohol with a molecular formula of C$_{10}$H$_{18}$O. It is the active constituent of essential oils of aromatic plants such as Cymbopogon and rose. Chemically, it is (2E)-3,7-dimethylocta-2,6-dien-1-ol. It was first isolated by O. Jacobsen in 1871. Carnesecchi et al. demonstrated that geraniol modulated DNA synthesis and potentiated 5-fluorouracil effects on human colon tumor xenografts. Zheng and collaborators suggested that geraniol showed promise as a chemopreventive agent because it showed strong GST-inducing activity in the mucosa of the small intestine and the large intestine. Geraniol also suppressed the growth of MCF-7 breast cancer cells and PC-3 prostate cancer cells. It was reported to interfere with membrane functions, ion homeostasis, inhibition of DNA synthesis and reduction in the size of colon tumors. Further, it is demonstrated that Geraniol regulates the multiple signaling molecules and pathways that are involved in the cell cycle, cell survival, proliferation, apoptosis and autophagy.

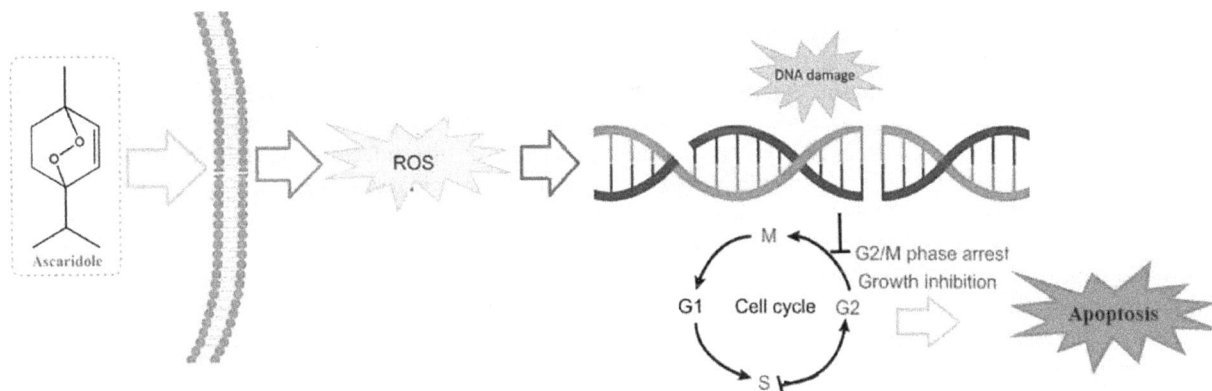

FIGURE 6.13 Mode of action of Ascaridole.

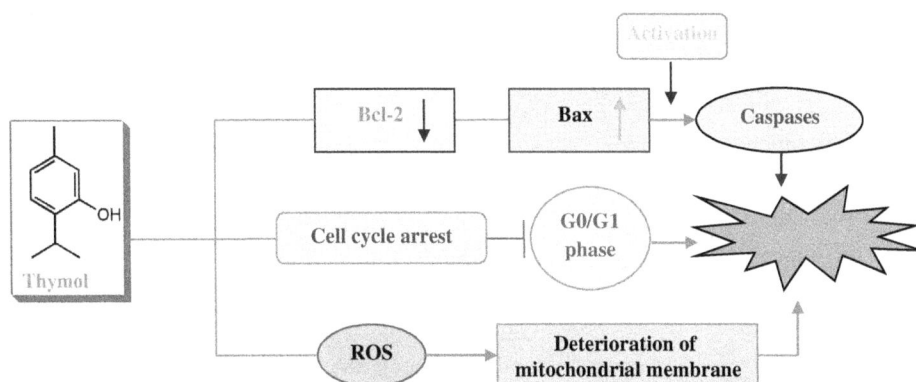

FIGURE 6.14 Mode of action of thymol.

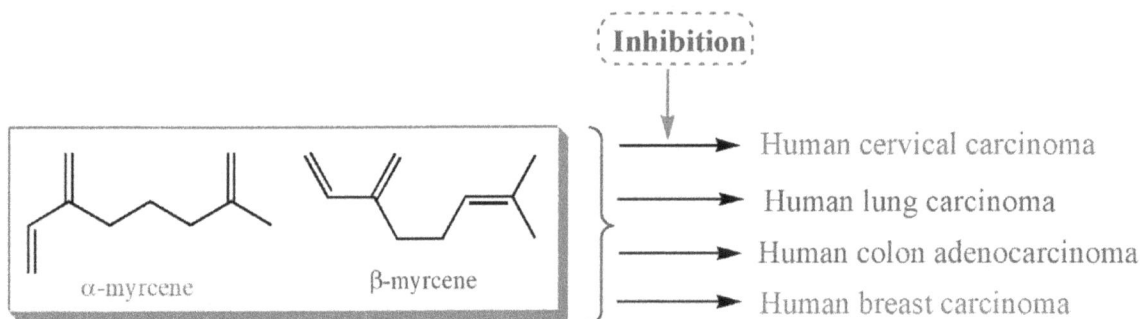

FIGURE 6.15 Mode of action of myrcene.

The molecular mechanisms by which geraniol exerts its anticancer effects may include decrease in the expression level of Bcl-2, VEGF, c-fos, COX-2, NF-κB, cyclin D1 and elevation of Bax, caspase-3 and caspase-9 activity. It has been reported that geraniol inhibits the HMG-CoA reductase in different types of cancer cells which reduces the prenylation of small G-proteins such as Ras or RhoA (Figure 6.16, Tables 6.3 and 6.4) [45–48].

Perillyl alcohol (POH) is a monocyclic terpenoid with the molecular formula $C_{10}H_{16}O$. Chemically, it is (4-prop-1-en-2-ylcyclohexen-1-yl)methanol. It is obtained through the mevalonate pathway in plants. It is isolated from leaves, stems and fruits of *Perilla frutescens*. It consists of a cyclohexene ring substituted by a hydroxymethyl and a prop-1-en-2-yl group at positions 1 and 4 respectively. Yeruva et al. demonstrated that perillyl alcohol presented dose-dependent cytotoxicity with cell cycle arrest and apoptosis. Perillic acid is a metabolite of perillyl alcohol. Perillic acid also induces cell cycle arrest and apoptosis via increase in p21, Bax and caspase-3 in non-small cell lung cancer cells (Figure 6.17, Table 6.5) [49–51].

Cantharidin (CTD) is a monoterpenoid with an epoxy-bridged cyclic dicarboxylic anhydride structure. It is a toxin extracted from coleoptera beetles and is commonly known as "Spanish fly". The anticancer property of CTD has been demonstrated experimentally. It exhibited potential in vitro anticancer activity against a broad spectrum of cancer cells such as breast cancer, colorectal carcinoma, hepatoma, bladder

carcinoma and leukemia. It has been demonstrated that the mitogen-activated protein kinases (MAPKs)/ERK/JNK/p38 signaling pathways were found to be involved in cantharidin-triggered apoptosis in different cancer cells (Figure 6.18 and Table 6.6) [59,60].

6.3 Sesquiterpenes as Anticancer Agents

Sesquiterpenes are the class of terpenes which consist of three isoprene units with the molecular formula $C_{15}H_{24}$. It has been demonstrated that sesquiterpenes and their derivatives exhibit potential activity against several cancers such as blood, breast, lung, colon, bladder, pancreatic, prostate, cervical, brain, liver, oral etc [61]. The mechanistic approaches associated with sesquiterpenes for their potential anticancer activities include modulation of nuclear factor kappa (NF-kB) activity, inhibitory action against lipid peroxidation, retardation of the production of reactive oxygen (ROS) and reactive nitrogen species (RNS) (Figure 6.19) [62].

Artemisinin is an active terpenoid isolated from the medicinal herb *Artemisia annua*. Chemically, artemisinin is a sesquiterpene trioxane lactone containing a peroxide bridge, which is essential for its anticancer activity. Artemisinin and its derivatives (ARTs) such as artesunate, artemether, dihydroartemisinin (Figure 6.20) inhibit the proliferation of various

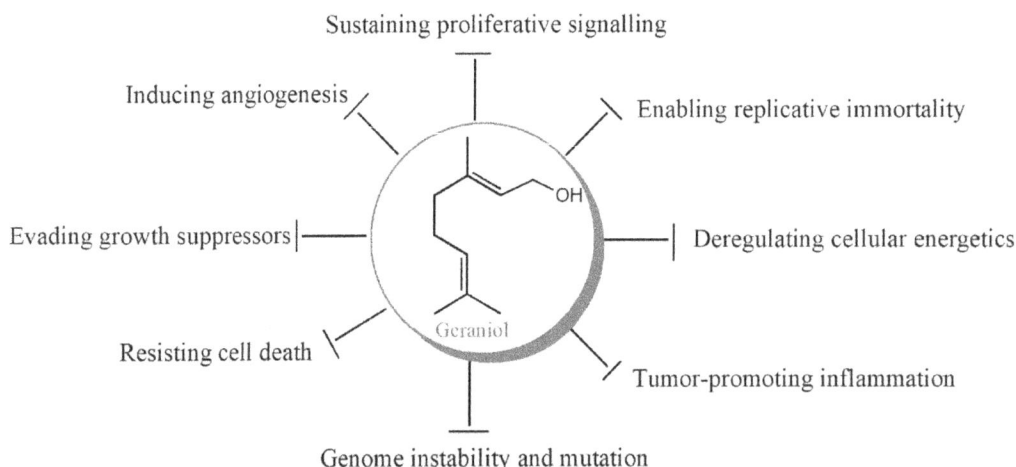

FIGURE 6.16 Hallmarks of cancer.

TABLE 6.3

Effects of Geraniol on Cancer Hallmarks

Hallmarks of Cancer	Affected Molecules or Pathways
Sustaining proliferative signaling	RhoA, Ras
Evading growth suppressors	pAKT, pmTOR
Tumor-promoting inflammation	COX-2, TNF-α, IL-1β, NF-κB, SOD
Inducing angiogenesis	VEGF, VEGFR-2
Genome instability and mutation	DNA damage
Resisting cell death	Caspase-3, Bcl-2, Bax
Deregulating cellular energetics	HMG-CoA reductase
Enabling replicative immortality	Cyclin-A, cyclin-B, CDK

TABLE 6.4

Antitumor Activity of Geraniol against Several Cancers

Types of Cancer	Hallmarks of Cancer				
	Sustaining Proliferative Signaling	Resisting Cell Death	Deregulating Cellular Energetics	Inducing Angiogenesis	Enabling Replicative Immortality
Lung cancer	Ras	Caspase-3	HMG-CoA reductase	N.D.	N.D.
Colon cancer	PKC	Bcl-2 Caspase-3	N.D.	VEGFR-2	TS ODC
Skin cancer	Ras	Bax Bcl-2	G6PD	N.D.	N.D.
Liver cancer	RhoA	N.D.	HMG-CoA reductase	N.D.	N.D.
Oral cancer	c-fos	Bax Bcl-2 Caspase-3	N.D.	VEGF	cyclin-D1

N.D., Not determined.

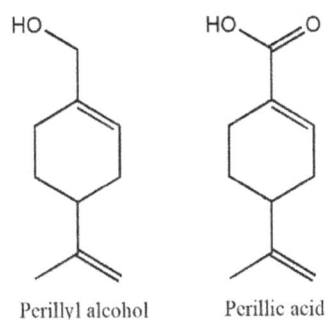

FIGURE 6.17   Structure of perillyl alcohol and perillic acid.

types of cancer cells such as leukemia, breast cancer, ovarian cancer, prostate cancer, colon cancer, hepatoma, gastric cancer, melanoma, lung cancer, etc. (Figure 6.21). The anticancer mechanism of ARTs is considered to be related to the cleavage of iron or heme-mediated peroxide bridge. Results indicate that they are toxic to cancer cells with IC$_{50}$ in the range of 10–20 μM [63,64].

Currently, Artesunate is under Phase-II clinical trial and evaluated against different types of cancers such as colorectal cancer, uveal melanoma, cervical intraepithelial neoplasia and laryngeal cancers. This compound catalyzes the production of free radicals to target the iron group contents [65].

Vernolide-A and vernodaline are the sesquiterpene lactone, present in the plant *Vernonia cinerea* L. of family Asteraceae [66]. The biological evaluation reported that vernolide-A exhibit potent cytotoxicity against human KB, DLD-1, NCI-661 and Hela tumor cell lines. Pratheeshkumar et al. investigated the effect of vernolide-A on the induction of apoptosis as well as its regulatory effect on the activation of transcription factors in B16F-10 melanoma cells. The induction of apoptosis is due to the enhancement of caspase 3 and caspase 9 while the inhibition of Bcl-2 and Bcl-xL results in the release of cytochrome c into the cytosol [67]. Further, the anti-proliferative and antimetastatic activities are mainly based on targeting the extracellular signal-regulated kinase 1 (ERK-1), extracellular signal-regulated kinase 2 (ERK-2), nuclear factor-κB (NF-κB), signal transducer and activator of transcription 3 (STAT3), matrix metalloproteinase 2 (MMP-2) and matrix metalloproteinase 9 (MMP9). The activity of vernolide-A and vernodaline is due to thiol reactivity through the α-methylene-γ-lactone group of sesquiterpene lactones (Figure 6.22) [68].

Galbanic acid is the active constituent of Ferula assafoetida. It is considered under monosubstituted coumarins class but it presents more complex structure due to the presence of sesquiterpene moiety [69]. Chemically, it is 3-[(1S, 2S, 3S)-2,3-dimethyl-2-[(2-oxochromen-7-yl)oxymethyl]-6-propan-2-ylidenecyclohexyl]propanoic acid. It was first isolated in 1962 from the species *Ferula gummosa*. The antitumor effect of Galbanic acid was observed in prostate, ovary, breast and lung cancers [70]. Recently, it was demonstrated that galbanic acid is able to inhibit H460 cells through a mechanism involving the caspase activation and Mcl-1 inhibition and become a potential anticancer agent for the treatment of lung cancer. Kim et al. reported that Galbanic acid potentiates TRAIL-induced

TABLE 6.5

List of Monoterpenes with Antitumor Activity [52–57]

Compounds	Mechanism of Action	Cell Line Tested
Linalyl acetate	Cell cycle arrest and induction of apoptosis	Human colon cancer cell lines HCT-116 (p53+/+)
		Human colon cancer cell lines HCT-116 (p53–/–)
Camphor	Cell cycle arrest and induction of apoptosis	Human colon cancer cell lines HCT-116 (p53+/+)
		Human colon cancer cell lines HCT-116 (p53–/–)
α-Terpineol	Inhibition of NF-κB pathway	Human chronic myelogenous leukemia cell line
Menthol	Inhibition gene expression of topoisomerases-I	Murine leukemia WEHI-3 cells (*in vivo*)
Borneol	Potentiates selenocystine-induced apoptosis and activation of ROS-mediated DNA damage	Caco-2 (Colon malignant cell line)
		HepG2 (Hepatocellular carcinomic human cell line)
Carvacrol	Cell cycle arrest and induction of apoptosis	CEM (Acute T lymphoblastic leukemia)
Thymol	Cell cycle arrest and induction of apoptosis	CEM (Acute T lymphoblastic leukemia)
Thymoquinone	Induction of apoptosis via p53-dependent pathway	HeLa (Epithelioid carcinomic cell line)
Myrcene	Cell cycle arrest and induction of apoptosis	HepG2 (Hepatocellular carcinomic human cell line)
Limonene	Induction of apoptosis and antiangiogenic effect	SW480 (Human colorectal adenocarcinoma)
Perillic acid	Cell cycle arrest and induction of apoptosis	H520 (Squamous lung cell carcinoma)
Perillyl alcohol	Antimetastatic activity	C6 (Glial cell line)
1,8-Cineole/Eucalyptol	Induction of apoptosis	KB (Human papilloma cell line)
Terpinen-4-ol	Cell cycle arrest and induction of apoptosis	B16 (Melanoma cells)
Citral	Induction of apoptosis	NB4 (Acute promyelocytic leukemia cell line)
Carvone	Induction of apoptosis	Hep-2 (Larynx epidermoid carcinoma)
α-Pinene	Induction of apoptosis	U937 (Histiocytic lymphoma cells)
β-Pinene	Immuno-modulatory effect	BW5147 (Murine T-cell lymphoma)
Geraniol	Cell cycle arrest	Caco-2 (Human colon cancer cell line)
Linalool	Induction of apoptosis by activation of p53 and CDKIs	NB4 (Acute myeloid leukemia)

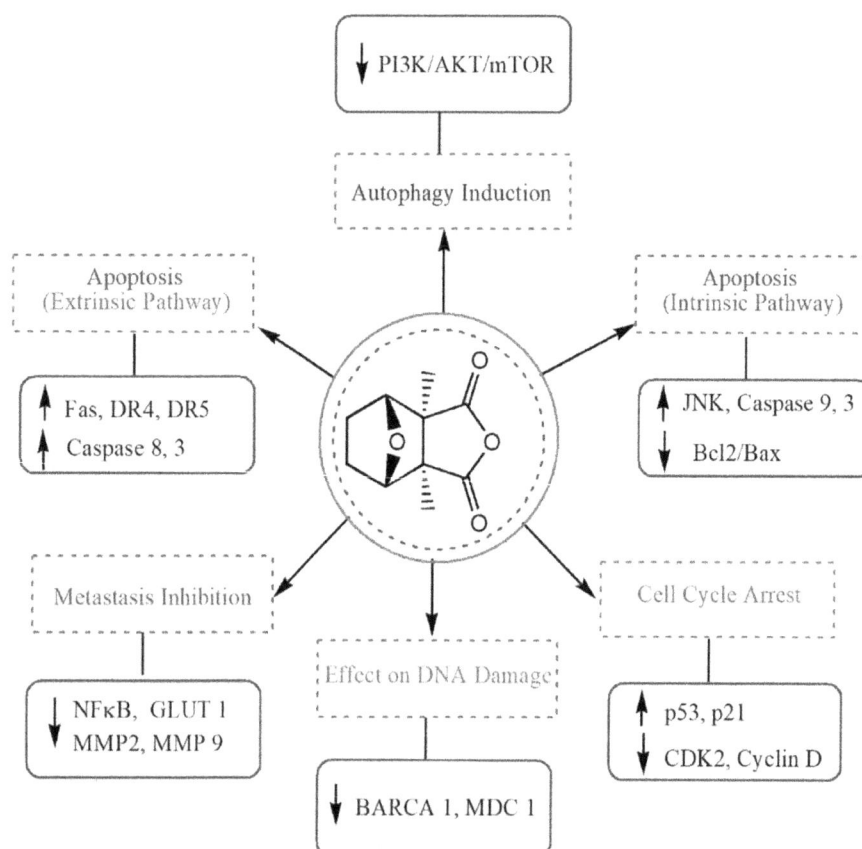

FIGURE 6.18 Molecular mechanism of cantharidin.

TABLE 6.6

Anticancer Effect and Molecular Mechanism of Cantharidin

Types of Cancer	Cell Line	Anticancer Effects	Molecular Mechanism	Evaluation Model
Melanoma	A431	• Apoptosis • Cell cycle arrest • DNA damage	• Reduction of cyclin D, cyclin E and CDK6 expression level • Arresting G0/G1 phase of cell cycle via elevation of p21 • Increased expression level of DR4, DR5	In vitro, In vivo
Lung	H460	• Apoptosis, • Cell cycle arrest • DNA damage	• Upregulation of DNA damaging genes DNIT3 and GADD45A	In vitro
Melanoma	A375.S2	• Growth inhibition • Invasion and migration • Inhibition	• Inhibition of migration & invasion via MAPK signaling pathway through NF-kB and AKT down-regulation resulting in reduction of MMP-2/-9 enzymatic activity and expression level	In vitro
Breast	MCF-7	• Apoptosis • Adhesion inhibition	• Adhesion inhibition by $\alpha2$ integrin down-regulation through PKC dependent pathway	In vitro
Bladder	TSGH 8301	• Apoptosis • Cell cycle arrest • DNA damage	• Activation of caspase-8, -9, -3, Increased generation of ROS • Down-regulation of Bcl-2 • Arrest of G0/G1 phase of cell cycle in association with decreased cyclin E	In vitro
Gastric cancer	BGC823	• Apoptosis	Suppression of growth and migration by suppressing PI3k/Akt signaling pathway	In vitro

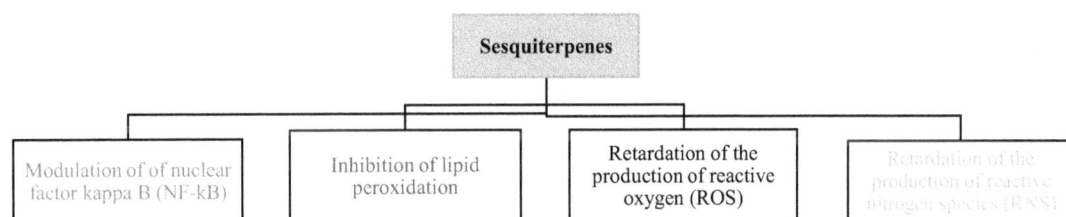

FIGURE 6.19 Mechanistic approaches of sesquiterpenes for anticancer activities.

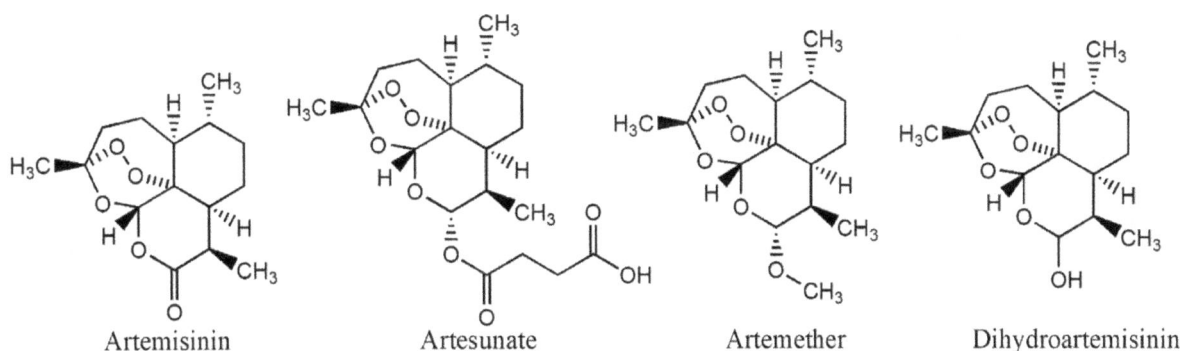

FIGURE 6.20 Molecular structures of artemisinin and its derivatives.

apoptosis in resistant non-small cell lung cancer cells via inhibition of MDR1 and activation of caspases and DR5 (Figure 6.23) [71].

Hamdi et al. isolated sesquiterpenes from the Rhizomes of *Curcuma zedoaria* and evaluated their Cytotoxicity against mouse myelomonocytic leukemia cells (WEHI-3), promyelocytic human leukemia cells (HL-60) and normal human umbilical vein endothelial cells (HUVEC) cell lines by MTT assay. They identified three guaiane type sesquiterpenes such as curcumenol, isoprocurcumenol and procurcumenol, one caraborane type sesquiterpene called curcumenone,

and one germacrane type sesquiterpene named as zederone based on UV, IR, NMR and GC-MS spectroscopic analysis (Figure 6.24). Procurcumenol is a sesquiterpenoid and chemically known as (3S, 3aS, 8aR)-3-hydroxy-3,8-dimethyl-5-propan-2-ylidene-2,3a, 4,8a-tetrahydro-1H-azulen-6-one. Whereas Curcumenone is known as 1-methyl-7-(3-oxobutyl)-4-propan-2-ylidenebicyclo[4.1.0]heptan-3-one and Zederone is (8E)-5,9,14-trimethyl-4,12-dioxatricyclo[9.3.0.0^{3,5}]tetradeca-1(11), 8,13-trien-2-one. Procurcumenol demonstrated the strongest inhibition of WEHI-3 and HL-60 cells with IC$_{50}$ values of 25.6 and 106.8 μM respectively. However, all these

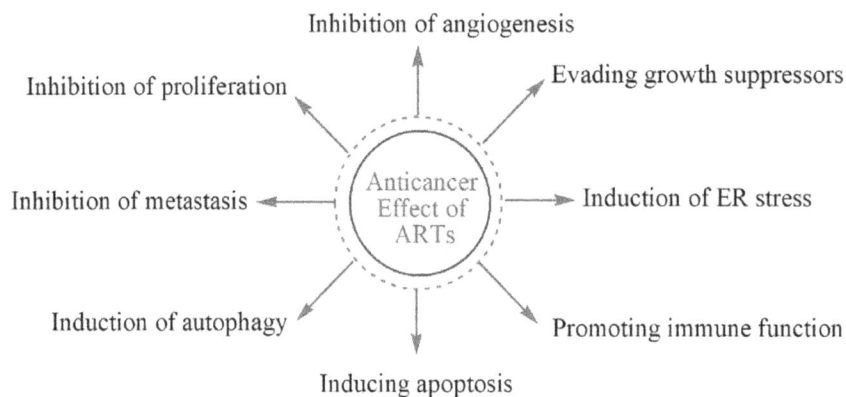

FIGURE 6.21 Anticancer effect of artemisinin and its derivatives (ARTs).

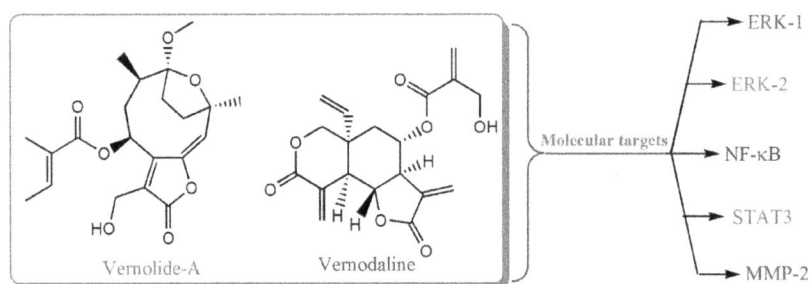

FIGURE 6.22 Molecular targets of vernolide-A and vernodaline.

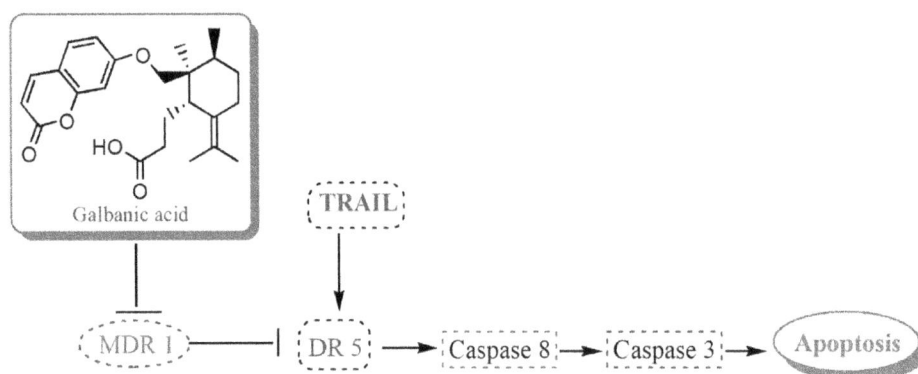

FIGURE 6.23 Antitumor effect of galbanic acid via apoptosis.

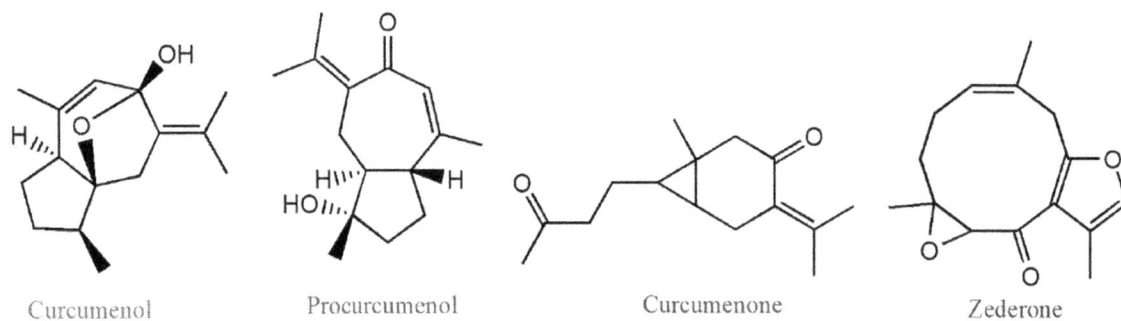

FIGURE 6.24 Structures of the isolated compounds.

compounds exhibited cytotoxicity toward HUVEC cells with IC_{50} values in the range of 16.3–50.0 μM (Table 6.7) [72].

Elemene is a sesquiterpenoid which comprises the mixture of β, δ and γ components. Among these, β-elemene exhibits anticancer activity against various types of cancer cells including leukemia, brain, breast, prostate, ovarian, cervical, colon, laryngeal and lung carcinoma cells. The in-vitro anticancer effects of β-elemene are dose-dependent with IC_{50} values of hundreds of micromoles, depending on the cancer cell types. β-Elemene has similar inhibitory effects on cell proliferation in both cisplatin-resistant (A2780/CP) and cisplatin-sensitive (A2780) ovarian carcinoma cell lines and partially reverses drug resistance to adriamycinin breast cancer MCF-7/ADM cells. The inhibition of β-elemene-induced cancer cell proliferation is mainly due to the apoptotic cell death and cell cycle arrest (Figure 6.25) [73,74].

Codonolactone (CLT) is a sesquiterpene lactone, isolated from *Chloranthus henryi* Hemsl. It has been demonstrated to inhibit the invasion, migration and metastasis of breast cancer cell by down-regulating the transcriptional activity of Runx2. Runx2 is a transcriptional factor and is considered as a potential target for antimetastatic agents to control breast cancer. Runx2 is also involved in the regulation of MMP in case of metastatic breast cancer cells. It was also reported that CLT inhibited the ability of invasion and migration in metastatic breast cancer cells. The results obtained from western blotting, Runx2 transcription factor assay and chromatin immunoprecipitation assay reported that the binding ability of Runx2 to sequences of the MMP-13 promoter was inhibited by CLT. These reports suggested that the antimetastatic effect of CLT in breast cancer was due to the inhibition of MMPs which is associated with a down-regulation of Runx2 transcriptional activity (Figure 6.26) [75].

TABLE 6.7

Anti-Proliferative Activity of Isolated Compounds

Compounds	WEHI-3 in μg/mL (μM)	HL60 in μg/mL (μM)	HUVEC in μg/mL (μM)
Procurcumenol	6 ± (25.6)	25 ± (106.8)	16.3 ± 1.0 (69.6 ± 4.2)
Curcumenol	66 ± (282.0)	39 ± (166.6)	25.9 ± 1.4 (110.6 ± 5.9)
Isoprocurcumenol	81 ± (246.1)	>100	45.1 ± 3.0 (192.7 ± 12.8)
Curcumenone	35 ± (149.5)	99 ± (423.0)	50.0 ± 8.6 (213.6 ± 35.4)
Zederone	>100	>100	42.1 ± 2.7 (171.1 ± 10.9)

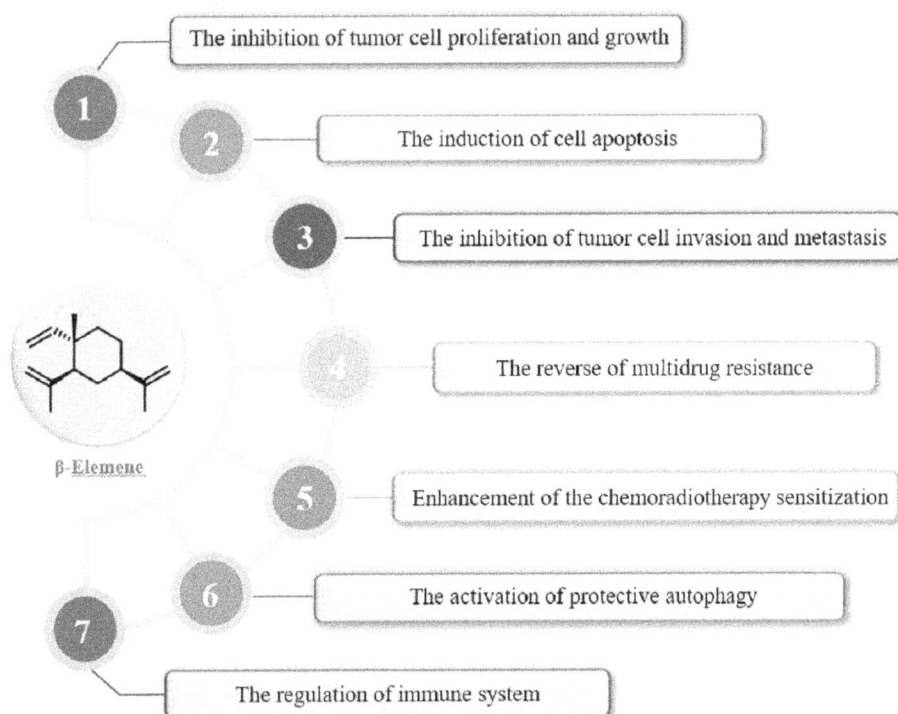

1. The inhibition of tumor cell proliferation and growth
2. The induction of cell apoptosis
3. The inhibition of tumor cell invasion and metastasis
4. The reverse of multidrug resistance
5. Enhancement of the chemoradiotherapy sensitization
6. The activation of protective autophagy
7. The regulation of immune system

β-Elemene

FIGURE 6.25 Anticancer effects of β-elemene.

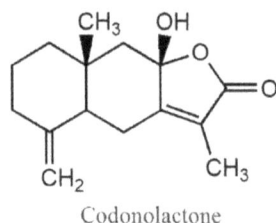

FIGURE 6.26 Chemical structure of codonolactone.

Parthenolide (PTL) is a sesquiterpene lactone of the germacranolide class which occurs naturally in the plant feverfew (*Tanacetum parthenium*). Chemically, it is (1*S*, 2*S*, 4*R*, 11*S*)-4,8-dimethyl-12-methylidene-3,14-dioxatricyclo[9.3.0.0^{2,4}]tetradec-7-en-13-one. It was reported that PTL induces cytotoxicity in various cancer cells including prostate, pancreatic, breast and colorectal cancers, multiple myeloma and leukemia. The anticancer effect of PTL involves the inhibition of proliferation, induction of apoptosis, cell cycle arrest and inhibition of metastasis [76].

The molecular mechanisms of PTL are associated with DNA-binding inhibition of two transcription factors such as the nuclear factor κB (NF-κB) and the signal transducer and activator of transcription 3 (STAT3) as well as the proapoptotic activation of p53 along with depletion of reduced glutathione (GSH), generation of reactive oxygen species (ROS) and activation of c-Jun N-terminal kinase (JNK) (Figure 6.27, Tables 6.8 and 6.9) [77].

6.4 Diterpenes as Anticancer Agents

Diterpenes are natural products that comprise four isoprene units with the molecular formula $C_{20}H_{32}$. Based on the presence of carbon skeleton and substituents, diterpenes are grouped into eight subclasses such as *ent*-abietane, daphnane, tigliane, ingenane, *ent*-atisane, *ent*-rosane, *ent*-kaurane and lathyrane [81]. Diterpenes are investigated for their cytotoxic effect against several cancer cells such as liver, lung, prostate, breast, ovary, cervical, colon and blood.

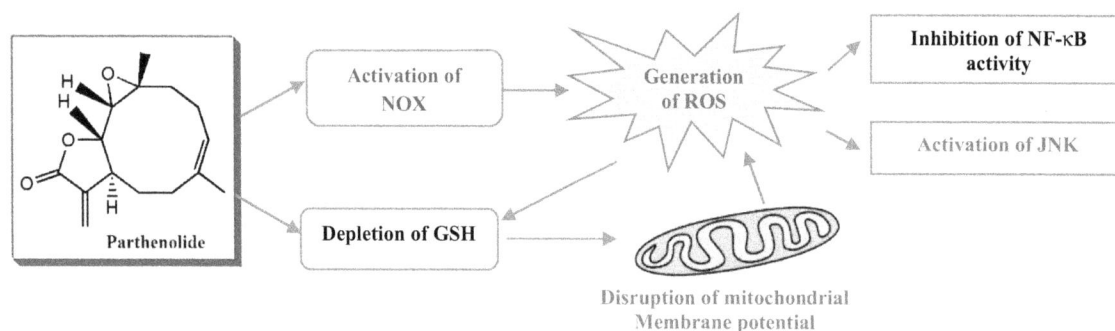

FIGURE 6.27 Molecular mechanism of PTL in cancer cells.

TABLE 6.8

List of Sesquiterpene Lactones with Anticancer Potential [78,79]

Sesquiterpene Lactones	Class	Ring Size	Biological Source	Molecular Structure
Parthenolide	Germacranolides	10-Membered	*Tanacetum parthenium*	
Alantolactone	Eudesmanolides	6/6-Fused bicyclic	*Inula helenium*	
Arglabin	Guaianolides	5/7-Fused bicyclic	*Artemisia myriantha*	

(Continued)

TABLE 6.8 (*Continued*)

List of Sesquiterpene Lactones with Anticancer Potential [78,79]

Sesquiterpene Lactones	Class	Ring Size	Biological Source	Molecular Structure
Helenalin	Pseudoguaianolides	5/7-Fused bicyclic	*Arnica montana*	
Xanthiatin	Xanthanolides	7-Membered	*Xanthium spinosum*	
Carabrol	Carabranolide	6/3-Tricyclic	*Carpesium faberi*	

TABLE 6.9

Mechanistic Approaches of Sesquiterpene Lactones with Anticancer Activity [80]

Sesquiterpene Lactone	Mechanism of Action	Cell Lines
Parthenolide	Induction of apoptosis via generation of ROS	Primary human AML cells, blast crisis CML (bcCML)
	Induction of apoptosis via generation of ROS, release of mitochondrial cytochrome c and activation of caspase	Chronic lymphocytic leukemia (CLL)
	Induction of autophagy via generation of ROS, depletion of GSH, activation of JNK and inhibition of NF-kB activity	Human breast cancer cell line MDA-MB231
	Generation of ROS with subsequent activation of JNK	Non-small lung cancer cell lines (A549 and H522)
	Induction of apoptosis via generation of ROS	Multiple myeloma cell lines
Helenalin	Induction of apoptosis via generation of ROS and disruption of mitochondrial membrane potential	Activated CD4+ T cells
Alantolactone	Induction of apoptosis via depletion of GSH, generation of ROS and mitochondrial dysfunction	Human glioblastoma cell lines U87
	Induction of apoptosis via generation of ROS and disruption of mitochondrial membrane potential	RKO human colon cancer
Costunolide	Induction of apoptosis via generation of ROS, induction of mitochondrial permeability transition and release of cytochrome c	Human promyelocytic leukemia cell line HL-60
Eupalinin A	Induction of autophagic cell death via generation of ROS and reduction of mitochondrial membrane potential	Human promyelocytic leukemia HL-60
Telekin	Induction of apoptosis via generation of ROS and loss of mitochondrial membrane potential	Human hepatocellular carcinoma cells HepG2
Salograviolide A	Induction of apoptosis via generation of ROS and disruption of mitochondrial membrane potential	Human colon cancer cell lines HT-29

It has been observed that diterpenes inhibit the proliferation of several cancer cells with promising IC_{50} values (Table 6.10) [82].

Jolkinolide B is a diterpene lactone and is isolated from the roots of *Euphorbia fischeriana*. It was demonstrated that Jolkinolide B inhibits the JAK2/STAT3 pathway in human Leukemic HL-60 and THP-1 cells at the molecular level (Figure 6.28). In addition to this, Jolkinolide B induces apoptosis in mouse melanoma B16F10 cells by altering glycolysis. Similarly, 17-hydroxy-Jolkinolide B inhibits the signal transducers and activators of transcription three signaling by covalently cross-linking janus kinases and induces apoptosis of human cancer cells. Whereas 17-acetoxyjolkinolide B irreversibly Ikappab kinase (IKK) and induces apoptosis of tumor cells (Figure 6.29, Table 6.11) [86].

Deregulation of the cell cycle is one of the essential features of cancer. The abnormal expression and activity of the cyclins, cyclin-dependent kinases (CDK) and tumor suppressor proteins may directly affect the progression of cell cycle and tumorigenesis. It has been demonstrated that diterpenoids are found to induce cell cycle arrest at G1 and G2-M phase (Figure 6.30, Table 6.12) [89].

TABLE 6.10

In Vitro Study on Cytotoxic Effect of Diterpenoids in Cancer Cell Lines [83–85]

Diterpenes	Class	Type of Cancer	Cell Lines ((IC$_{50}$)
Jolkinolide A	Ent-abietane	Liver	HepG-2 (80.12 μM)
		Breast	MCF-7 (56.34 μM)
		Cervical	Hela (>100 μM)
Jolkinolide B	Ent-abietane	Liver	HepG-2 (24.43 μM)
		Breast	MCF-7 (22.76 μM)
		Cervical	Hela (23.12 μM)
17-Hydroxyjolkinolide B	Ent-abietane	Liver	HepG-2 (42.13 μM)
		Breast	MCF-7 (25.33 μM)
		cervical	Hela (35.11 μM)
17-Acetoxyjolkinolide B	Ent-abietane	Blood	U937 (0.74 μM)
		colon	Colo205 (2.34 μM)
		gastric	HGC (3.64 μM)
		breast	MCF-7 (8.74 μM)
Prostratin	Tigliane	Liver	HepG-2 (11.77 μM)
		Breast	MCF-7 (17.4 μM)
		Gastric	SGC-7901 (25.4 μM)
13-O-Acetylphorbol	Tigliane	Liver	HepG-2 (32.3 μM)
		Breast	MCF-7 (18.1 μM)
		Gastric	SGC-7901 (24.91 μM)
12-Deoxyphorbol 13-Palmitate	Tigliane	Liver	Hep-3B (12.01 μM)
		Lung	A549 (9.38 μM)
Langduin A	Daphnane	Liver	HepG-2 (35 μM)
		Breast	MCF-7 (19.4 μM)
		Gastric	SGC-7901 (21.3 μM)
Ingenol-6,7-epoxy-3-tetradecanoate	Ingenane	Lung	A549 (3.35 μg/mL)
		Liver	BEL7402 (13.05 μg/mL)
		Colon	HCT116 (14.62 μg/mL)
		Breast	MDA-MB-231 (14.42 μg/mL)
Ingenol-3-myristinate	Ingenane	Lung	A549 (2.85 μg/mL)
		Liver	BEL7402 (15.72 μg/mL)
		Colon	HCT116 (16.05 μg/mL)
		Breast	MDA-MB-231 (18.91 μg/mL)
Ingenol 3-palmitate	Ingenane	Lung	A549 (2.88 μg/mL)
		Liver	BEL7402 (25.87 μg/mL)
		Colon	HCT116 (14.38 μg/mL)
		Breast	MDA-MB-231 (22 μg/mL)
Ent-1β,3β,16 β,17-tetrahydroxyatisane	Ent-atisane	Breast	MCF-7 (23.21 μM)
Ent-1β,3α,16β,17-tetrahydroxyatisane	Ent-atisane	Breast	MCF-7 (15.42 μM)
Ent-kaurane-3-oxo-16β,17-acetonide	Ent-kaurane	Liver	Hep-3B (8.15 μM)

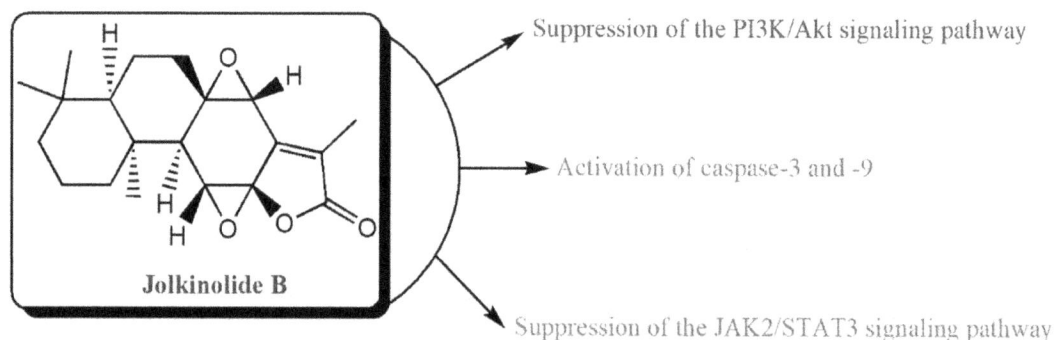

FIGURE 6.28 Mechanistic approach of diterpenoids for anticancer effect.

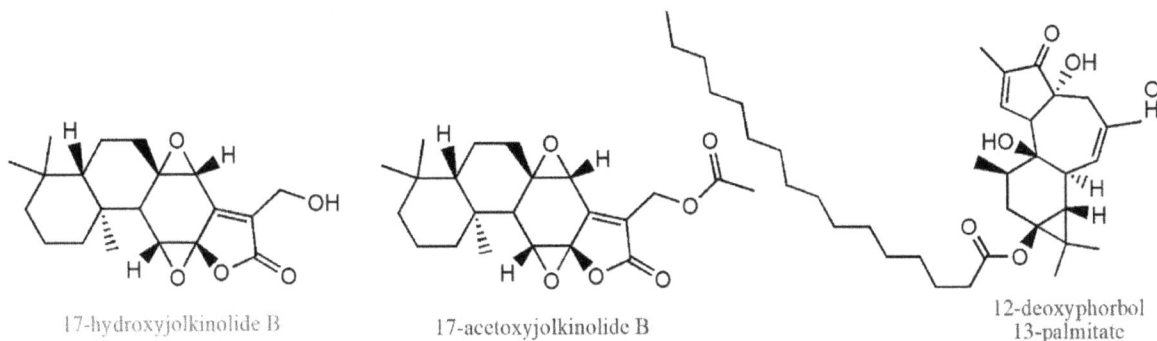

17-hydroxyjolkinolide B 17-acetoxyjolkinolide B 12-deoxyphorbol 13-palmitate

FIGURE 6.29 Chemical structure of diterpenoids.

TABLE 6.11

Mode of Action Displayed by Diterpenoids in Inducing Apoptosis [87,88]

Diterpenes	Type of Cancer	Cell Lines	Mode of Action
Jolkinolide B	Breast	MDA-MB-231	• Suppression of the PI3K/Akt signaling pathway
	Human leukemic	U937	• Activation of caspase-3 and -9 • Suppression of PI3K/Akt and XIAP pathways
		HL-60	• Suppression of the JAK2/STAT3 signaling pathway • Triggering of caspase-3, -8 and -9 activation
17-Hydroxy-jolkinolide B	Liver	HepG2	• Inhibit STAT3 activation by direct inhibition of JAK
17-Acetoxy-jolkinolide B	Lung	A549	• Inhibit tumor NF-KB activation
12-Deoxy-phorbol13-palmitate	Gastric	BGC823	• Activation of caspase-3 and -9

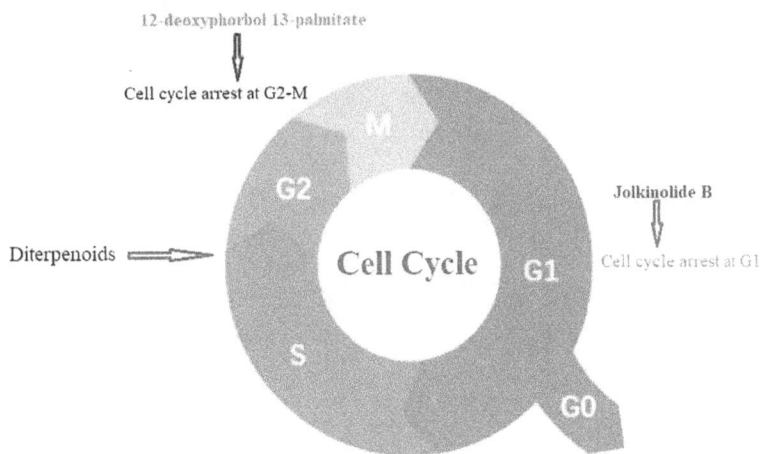

FIGURE 6.30 Deregulation of the cell cycle by diterpenoids.

TABLE 6.12

Effects of Diterpenoids on Cell Cycles

Diterpenes	Type of Cancers	Cell Lines	Effects
Jolkinolide B	Human leukemic	K562	Cell cycle arrest at G1
	Prostate	LNCap	Cell cycle arrest at G1
12-Deoxyphorbol13-palmitate	Gastric	BGC823	Cell cycle arrest at G2-M checkpoint

Oridonin is an ent-kaurane class of diterpenoid with molecular formula of $C_{20}H_{28}O_6$ (Figure 6.31). It is isolated from the herb *Rabdosia rubescens*. Oridonin has been recently demonstrated to have a therapeutic effect on various solid tumors, including liver cancer, skin carcinoma, osteoma and colorectal cancers. It inhibits the growth of primary adult T-cell leukemia, acute lymphoblastic leukemia, chronic lymphocytic leukemia, non-Hodgkin's lymphoma and multiple myeloma [90].

Andrographolide is a labdane class of bicyclic diterpenoid lactone (Figure 6.32) [91]. It is present as a major bioactive chemical constituent of the plant *Andrographis paniculata*. Rajagopal et al. reported that andrographolide inhibits the growth of several cancer cells such as breast cancer, colon cancer, lung cancer, melanoma, leukemia, prostate cancer and ovarian cancer [92]. Peng et al. demonstrated that andrographolide inhibits the breast cancer via suppression of COX-2 expression and angiogenesis through inactivation of p300 signaling and VEGF pathway. Treatment of andrographolide significantly enhanced NK cell activity in normal and tumor-bearing animals. Its administration significantly enhanced the mitogen-induced proliferation of splenocyte, thymocyte and bone marrow cells (Figure 6.33) [93].

Triptolide is a diterpenoid epoxide with a molecular formula of $C_{20}H_{24}O_6$ (Figure 6.34). It is isolated from *Tripterygium wilfordii*. The *in-vivo* anticancer activity of triptolide has been confirmed in xenograft animal models in preclinical studies. Triptolide interferes with a number of transcription factors such as NF-kB, p53, NF-AT and HSF-1. In addition to this, triptolide has been observed to induce DNA damage in cancer cells [94,95].

Tanshinones are the diterpenoids and major components of *Salvia miltiorrhiza* (Figure 6.35). Both in vitro and in vivo anticancer effects of tanshinone IIA has been studied in various human carcinoma cells such as leukemia, breast cancer, colon cancer and hepatocellular carcinoma. From the evaluation study, it was observed that tanshinone IIA also exhibits synergetic effects when administered with other anticancer agents such as doxorubicin and cisplatin. It inhibits the invasion and metastasis of cancer cells by reducing the levels of MMP2, MMP9 and NF-kB, and increasing the levels of tissue inhibitor of matrix metalloproteinase protein (TIMP) 1 and TIMP 2 [96].

Pseudolaric acid B (PAB) is a diterpene with a molecular formula of $C_{23}H_{28}O_8$ (Figure 6.36). It is the major constituent present in the root bark of *Pseudolarix kaempferi*. PAB exhibits anticancer effects by inhibiting angiogenesis and reducing hypoxia-inducible factor 1α by promoting proteasome-mediated degradation. The cytotoxicity of PAB is investigated

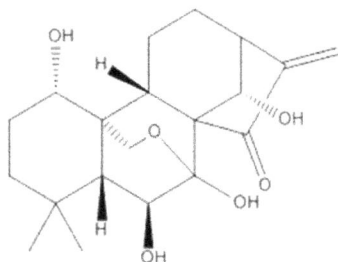

FIGURE 6.31 Molecular structure of oridonin.

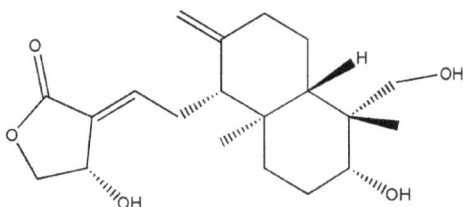

FIGURE 6.32 Molecular structure of andrographolide.

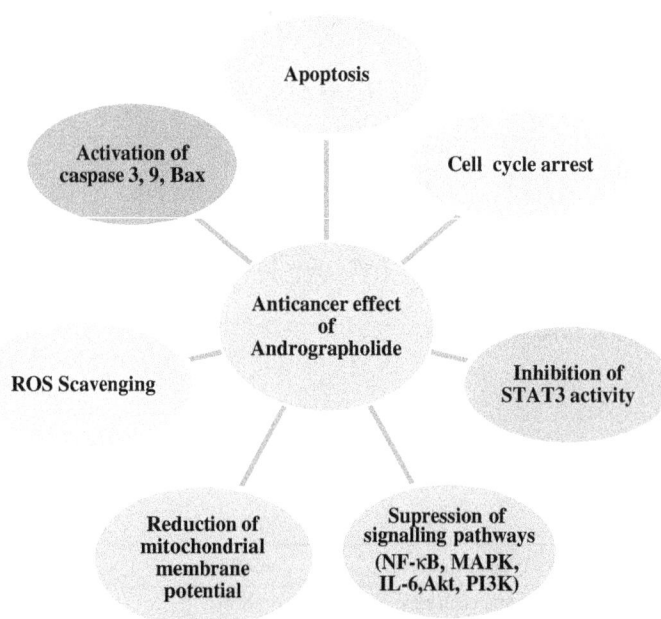

FIGURE 6.33 Anticancer mechanisms of andrographolide.

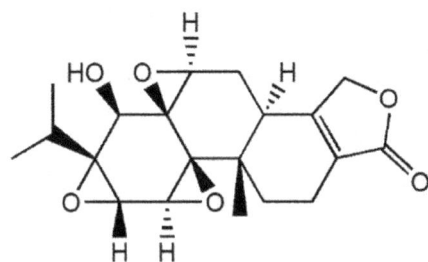

FIGURE 6.34 Structure of triptolide.

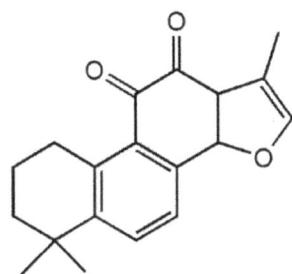

FIGURE 6.35 Molecular structure of tanshinone IIA.

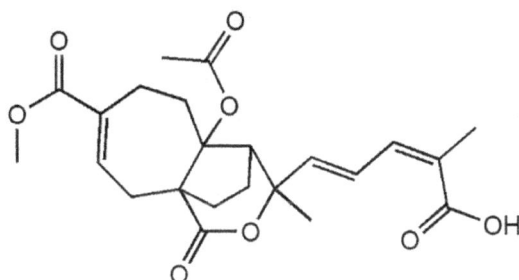

FIGURE 6.36 Molecular structure of pseudolaric acid B.

toward broad-spectrum cancer cell lines (IC$_{50}$ ~ 1 μM) of lung, colon, breast, brain and renal origins [97–99].

Excisanin A is a diterpenoid with a molecular formula of C$_{20}$H$_{30}$O$_5$ (Figure 6.37). It is obtained from *Isodon macrocalyxin*. It was tested on human Hep3B liver cancer cells and the results indicate the inhibition of growth mediated by apoptosis through inhibition of the AKT signaling pathway [100]. Deng et al. demonstrated that excisanin A inhibits the invasive behavior of breast cancer cells by modulating the integrin β1/FAK/PI3K/AKT/β-catenin signaling pathway. Further, excisanin A suppresses the proliferation of cancer cells by arresting the G1 phase of cell cycle [101].

6.5 Triterpenes as Anticancer Agents

Triterpenes are natural products composed of three terpene units or six isoprene units with the molecular formula C$_{30}$H$_{48}$. Triterpenoids can be divided into two classes such as tetracyclic

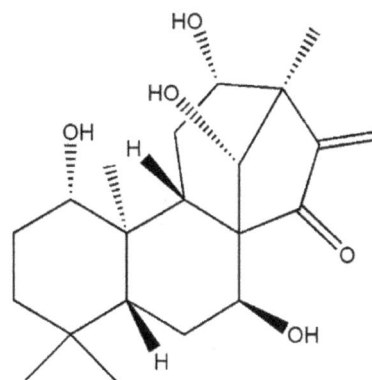

Excisanin A

FIGURE 6.37 Structure of excisanin A.

and the pentacyclic [102]. Tetracyclic triterpenoids are further grouped into lanostane, euphane, protostane, dammarane and tirucallane. Similarly, the pentacyclic triterpenes are classified into lupane, oleanane and ursane. The mechanisms by which triterpenoids regulate several transcription and growth factors, inflammatory cytokines and intracellular signaling pathways involved in proliferation, apoptosis and angiogenesis of cancer cells. The chemopreventive and anticancer effect of triterpenoids are mainly based on preclinical study on animal models of colon, breast, prostate and melanoma cancers [103].

Lupeol is a pentacyclic triterpenoid of lupane class with the molecular formula of C$_{30}$H$_{50}$O (Figure 6.38). It is isolated from the leaves of *Elephantopus scaber* L. The anticancer effect of lupeol is due to down-regulation of the expression of anti-apoptotic proteins (Bcl-2, Bcl-xL). Lupeol exhibits anticancer activity not only in EGFR wild-type non-small cell lung cancer (NSCLC) cells but also in erlotinib-resistant H1975 cells by the suppression of EGFR/STAT3 signaling pathway. Evaluation reports suggest that the inhibition of STAT3 activity by lupeol is responsible for the induction of the apoptosis of NSCLC cells [104,105].

Pristimerin is a quininemethide triterpenoid with a molecular formula of C$_{30}$H$_{40}$O$_4$. It is isolated from the species of Celastraceae and Hippocrateaceae families [106]. Yousef et al. demonstrated the anticancer potential of *Pristimerin* in colorectal cancer cells by inducing the arrest of G1 phase of cell cycle

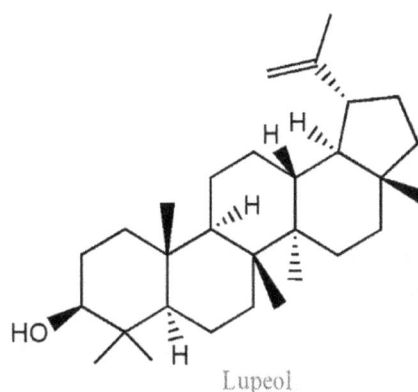

Lupeol

FIGURE 6.38 Structure of lupeol.

and apoptosis [107]. *Pristimerin also exhibits anticancer effect through* various molecular targets or signaling pathways such as cyclins, reactive oxygen species (ROS), nuclear factor kappa B (NF-κB), mitogen-activated protein kinase (MAPK) and PI3K/AKT/mammalian target of rapamycin (mTOR) pathways. Pristimerin-induced apoptotic effects are mainly due to mitochondrial dysfunction, activation of both extrinsic and intrinsic caspases, and cleavage of poly ADP-ribose polymerase (PARP). It has been demonstrated that pristimerin can induce caspase-dependent apoptosis in human glioma cancer cells, pancreatic cancer cells and hepatoma cancer cells. The apoptotic effect of pristimerin is due to down-regulation of Bcl-2 through reactive oxygen species (ROS) in human prostate cancer LNCaP and PC-3 cells [108]. Guo et al. reported that ROS-induced apoptosis by pritimerin was also reported in hepatocellular carcinoma

HepG2 cells that involve EGFR and Akt proteins. Mori et al. reported that pristimerin displayed potent anti-proliferative activity in human osteosarcoma cells by inducing apoptosis via PI3K/AKT/mTOR pathway (Figure 6.39, Table 6.13) [109].

Celastrol is a pentacyclic triterpenoid with a chemical formula of $C_{29}H_{38}O_4$. It belongs to the family of quinone methides. It is isolated from the root extracts of *Tripterygium wilfordii* and *Tripterygium regelii*. Wei et al. reported that the antitumor activity of celastrol is mediated through inhibition of proliferation, invasion and migration in cholangiocarcinoma via PTEN/PI3K/Akt pathway. It is demonstrated that Celastrol displays its efficacy against various cancers in preclinical studies. It stabilizes the endogenous inhibitor of NF-κB by suppressing either upstream IKK activity or downstream proteasome degradation pathway (Figure 6.40) [110].

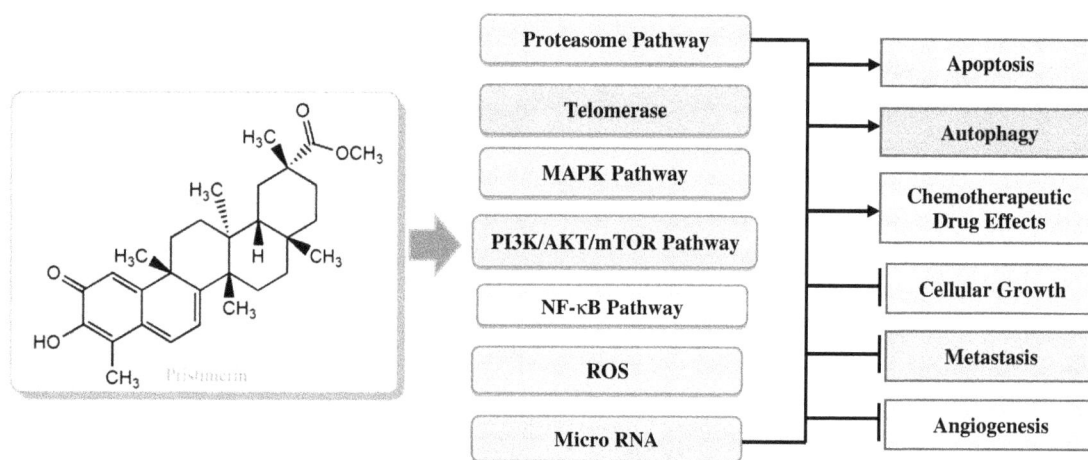

FIGURE 6.39 Mode of action of *Pristimerin.*

TABLE 6.13

Anticancer Mechanisms of Pristimerin in Different Cell Lines

Type of Cancer	Cell Lines	Mechanism of Action
Cervical cancer	HeLa	Activated ROS-dependent JNK, Bax and PARP-1
Leukemia	HL-60	Interfere with DNA synthesis
Ovarian carcinoma	OVCAR-5 MDAH-2774	Inhibit prosurvival signaling proteins Akt, mTOR and NF-kB
Osteosarcoma	MNNG	Decreased expression of Akt, mTOR and NF-κB
Oral cancer	CAL-27 SCC-25	G1 phase arrest and MAPK/Erk1/2 and Akt signaling inhibition
Osophageal squamous cell carcinoma	EC9706 EC109	Inhibit NF-κB pathway, synergistic effect with 5-FU
Prostate cance	LNCaP PC-3	Down-regulate Bcl-2 through an ROS-dependent ubiquitin-proteasomal degradation pathway
Breast cancer	SKBR3	Down-regulate HER2
Colorectal cancer	HCT-116	Inhibit the AKT/FOXO3a pathway via decreasing cyclinD1 and Bcl-X
Hepatocellular carcinoma	HepG2	Generate ROS, induce the release of cytochrome c and down-regulate the EGFR protein
Pancreatic cancer	BxPC-3 PANC-1 AsPC-1	Inhibit the translocation and DNA-binding activity of NF-κB
Myeloma	H929 U266	Inhibit IKK phosphorylation of IκB and proteosome
Glioma	U87	Activation of JNK through overproduction of ROS

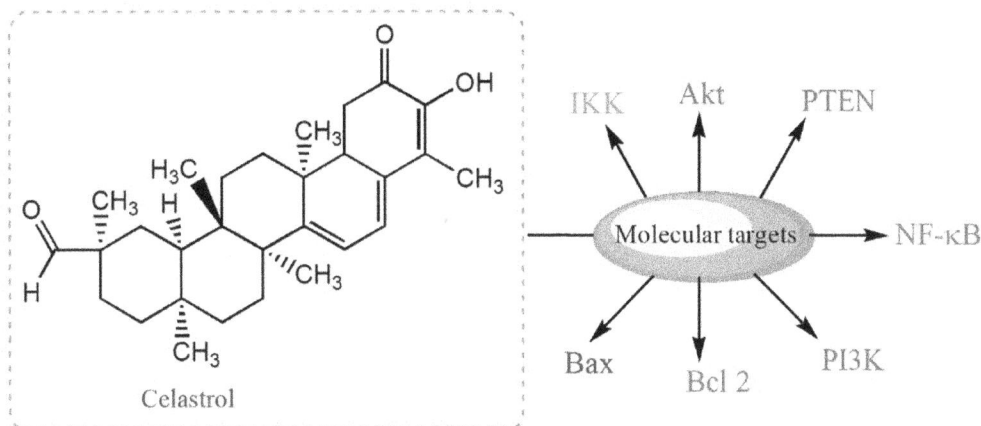

FIGURE 6.40 Molecular targets of celastrol.

Betulinic acid (BA) is a naturally occurring lupane class of pentacyclic triterpenoid with a molecular formula of $C_{30}H_{48}O_3$ [111]. Commercially it is obtained from bark of *Betula alba*. Abdel-Aziz et al. reported that BA exerts the cytotoxic effect against multidrug-resistant tumor cells via targeting autocrine motility factor receptor (AMFR) [112]. BA has found to exhibit synergistic effects with taxol to induce breast cancer cells G2/M checkpoint arrest and induction of apoptosis [113]. Wang et al. demonstrated that the administration of BA significantly elevates glucose-regulated protein 78 (GRP78) mediated endoplasmic reticulum (ER) stress that resulting in the activation of protein kinase R-like ER kinase (PERK) in apoptotic pathway (Figure 6.41) [114].

Oleanolic acid (OA) is a natural pentacyclic triterpenoid. It is olean-12-en-28-oic acid substituted by a β-hydroxy group at position 3 [115]. Several studies have demonstrated that OA exhibits its action at various stages of tumor development to inhibit the initiation and promotion of tumor as well as to induce the differentiation and apoptosis of tumor cells. From the preclinical studies, it was observed that the mechanism of action of OA and its derivatives may include anticancer cell proliferation, inducing apoptosis tumor cells, inducing autophagy, regulating cell cycle regulatory proteins, inhibiting vascular endothelial growth, anti-angiogenesis, inhibition of tumor cell migration and invasion (Figure 6.42) [116].

Ursolic acid (UA) is a pentacyclic triterpenoid with a molecular formula of $C_{30}H_{48}O_3$ (Figure 6.43). It is obtained from berries, leaves, flowers and fruits of medicinal plants, such as *Boerhaavia diffusa* L. (Nyctaginaceae), *Rosemarinus*

officinalis L. (Lamiaceae), *Eriobotrya japonica* Lindl. (Rosaceae), *Calluna vulgaris* L. (Ericaceae), *Ocimum sanctum* L. (Lamiaceae) and *Syzygium cumini* L. (Myrtaceae). UA was found to activate the immune system by activating the cell-mediated immune responses in B16F-10 melanoma-bearing mice [117]. The intraperitoneal administration of UA with a dose of 50 µmol/kg body weight for five consecutive days was found to produce increased NK cell activity in metastatic tumor-bearing animals [118].

Nomilin is a triterpenoid with a molecular formula of $C_{28}H_{34}O_9$ (Figure 6.44). It belongs to the class of compounds called limonoids. It is obtained from the extracts of citrus fruits. It exhibits promising anticancer properties. It is found to be active against several human cancer cell lines such as leukemia (HL-60), ovary (SKOV-3), cervix (HeLa), stomach (NCI-SNU-1), liver (Hep G2) and breast (MCF-7). The growth-inhibitory effect of nomilin against MCF-7 cells was significant, and the anti-proliferative activity was also dose- and time-dependent. With the use of flow cytometry, it was found that nomilin could induce apoptosis in MCF-7 cells [119].

Avicin (D and G) is the triterpenoid saponins isolated from *Acacia victoriae* (Bentham). Avicin D has been approved by the United States Food and Drug Administration for phase I clinical trial in human cancer patients. Avicin D induces autophagy by activation of AMPK. Similarly, avicin G is found to be potent inhibitors of TNF-induced NF-kB. NF-kB regulates the expression of antiapoptotic members of the Bcl-2 family. The other signaling pathways such as PI3K/Akt, and STAT3 and their downstream targets like c-myc, survivin, Bcl2 and Mcl 1 are the regulators of cell survival and apoptosis (Figure 6.45) [120,121].

Huang et al. performed the isolation of terpenoids from *Alisma orientalis* and evaluated their anti-proliferative activities. Fourteen protostane-type triterpenoids were isolated from the rhizome of *A. orientalis*. Among these triterpenoids, alisol A, alisol A 24-acetate, alisol B, alisol B 23-acetate and alisol G exhibited inhibitory effects on cancer cell lines (Figure 6.46, Table 6.14). Compounds alisol B and alisol B 23-acetate displayed the highest potential with IC_{50} values of 16.28, 14.47 and 6.66 µM for alisol B and 18.01, 15.97 and 13.56 µM for alisol B 23-acetate respectively in HepG2, MDA-MB-231

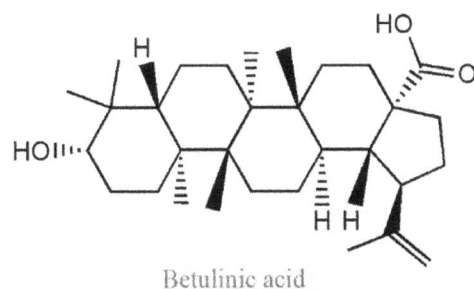

FIGURE 6.41 Structure of betulinic acid.

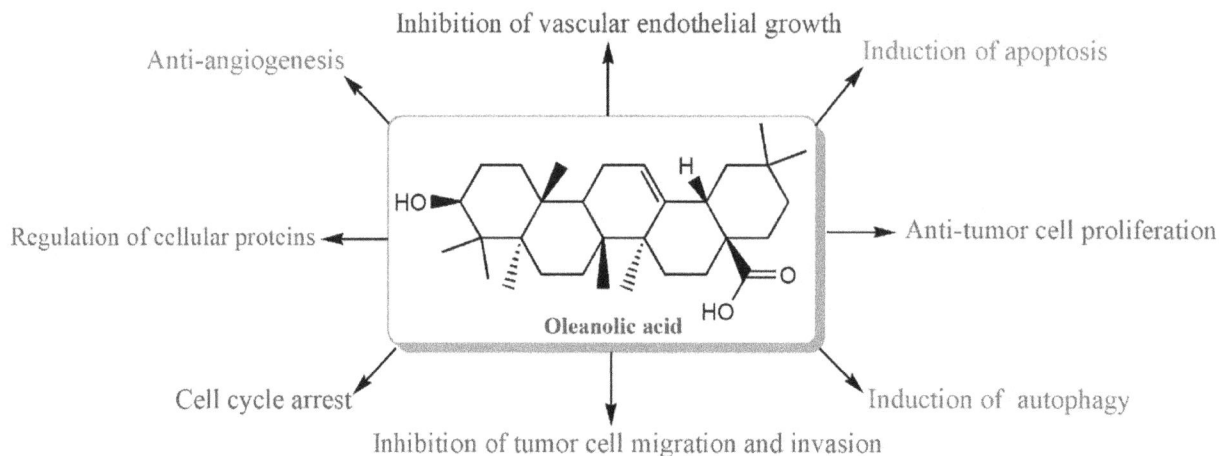

FIGURE 6.42 Molecular mechanism of action of oleanolic acid.

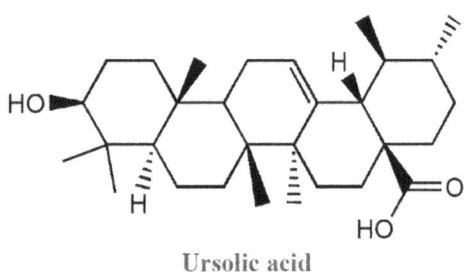

FIGURE 6.43 Structure of ursolic acid.

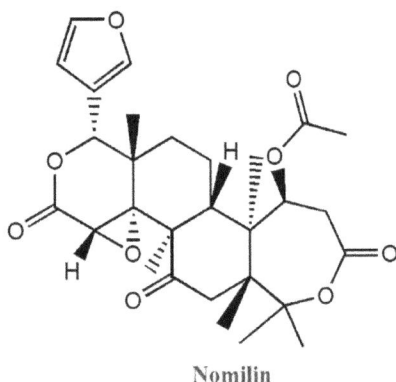

FIGURE 6.44 Structure of nomilin.

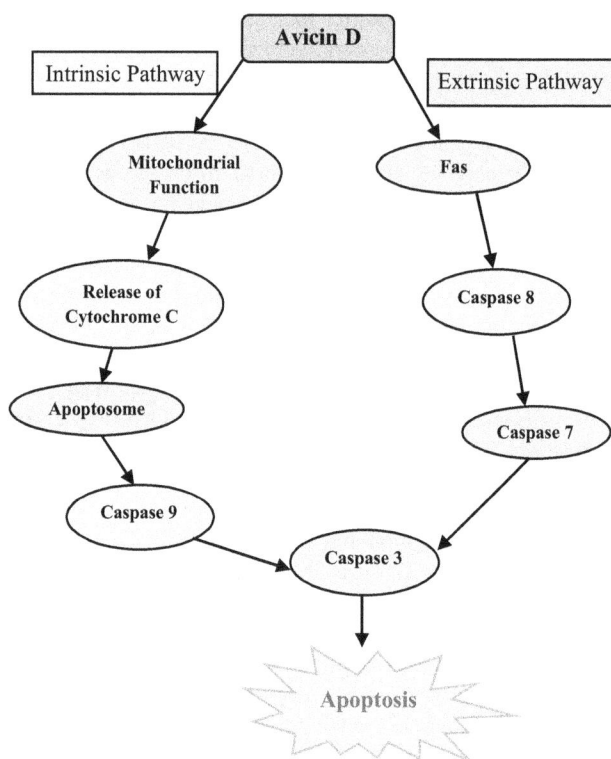

FIGURE 6.45 Avicin D-induced tumor cell death.

and MCF-7 cells. Further, the evaluation study reported that these compounds induced apoptosis effectively. Although compounds alisol A and alisol A 24-acetate induced minimal apoptosis and delayed the G2/M phase in HepG2 cells. Further study demonstrated that alisol and its derivatives also enhanced LC3II expression which indicates the occurrence of autophagy [122].

Cucurbitacins are tetracyclic terpenes with steroidal structures (Figure 6.47) [123]. It is the active constituent present in *Cucurbita andreana*. Cucurbitacins induce cell cycle arrest mainly at G2/M and S phase. It was demonstrated that Cucurbitacins exhibit synergistic effects when combined with

anticancer drugs like gemcitabine, cisplatin, doxorubicin, 5-flouroracil, paclitaxel and docetaxel in both in vitro and in vivo models [124,125].

Pachymic acid is a lanostane-type of triterpenoid with a molecular formula of $C_{33}H_{52}O_5$ which is derived from *Poria cocos* (Figure 6.48). From the evaluation results, it was revealed that Pachymic acid exhibits cytotoxicity against human lung cancer A549 cells, human prostate cancer DU145 cells and colon carcinoma HT29 cells and induces apoptosis in DU145, LNCaP prostate cancer cells and A549 cells. The anticancer effects of Pachymic acid is due to activation of PARP, caspases-9 and caspases-3. It also displays inhibitory activities on both DNA topoisomerase I, II, MMP9 and NF-κB [126,127].

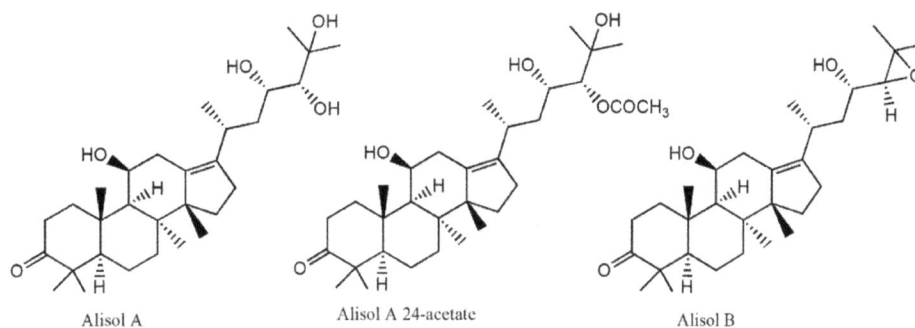

FIGURE 6.46 Structure of terpenoids isolated from *Alisma orientalis*.

TABLE 6.14

IC$_{50}$ of Alisol and Its Derivatives in Cancer Cell Lines

Compounds	IC$_{50}$ (mean ± SD, μM)		
	HepG2	MDA-MB-231	MCF-7
Alisol A	21.68 ± 2.24	18.34 ± 2.98	21.91 ± 8.71
Alisol A 24-acetate	18.94 ± 2.59	22.48 ± 9.14	20.69 ± 1.97
Alisol B	16.28 ± 1.46	14.47 ± 2.26	6.66 ± 2.98
Alisol B 23-acetate	18.01 ± 0.94	15.97 ± 1.31	13.56 ± 3.09
Alisol G	28.86 ± 0.32	33.22 ± 12.09	31.14 ± 8.71

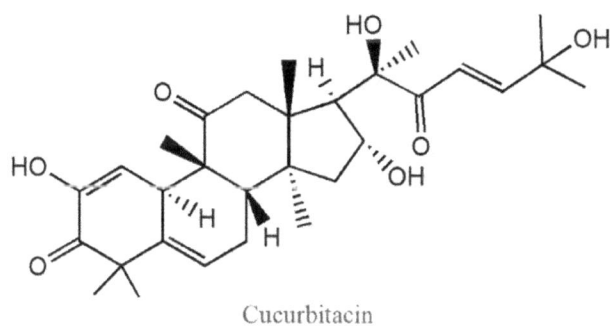

FIGURE 6.47 Structure of cucurbitacins.

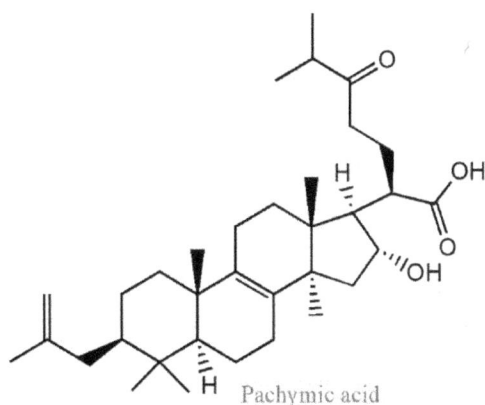

FIGURE 6.48 Structure of pachymic acid.

6.6 Tetraterpenes as Anticancer Agents

The most common tetraterpenoids are the carotenoids. Carotenoids are naturally occurring fat-soluble color pigments. From the preclinical studies, it was revealed that several carotenoids exhibit anti-carcinogenic activity including α-carotene, β-carotene, lycopene, lutein, zeaxanthin, β-cryptoxanthin, fucoxanthin, canthaxanthin and astaxanthin, etc.

Lycopene is an open-chain hydrocarbon which contains 11 conjugated and 2 non-conjugated double bonds arranged in a linear fashion (Figure 6.49). It is derived primarily from tomatoes, papayas, watermelons and grapefruits. The anticancer effect of lycopene is due to its anti-oxidant effects via scavenging ROS which allows lycopene to prevent lipid peroxidation and DNA damage. Lycopene exhibits anti-angiogenic effects due to the attenuated activities of MMP2 and u-PA via enhancing protein expression of tissue inhibitors of MMP2 and plasminogen activator inhibitor-1 [128].

Astaxanthin is a keto-carotenoid with the molecular formula $C_{40}H_{52}O_4$ (Figure 6.50). It is a blood-red pigment and is produced naturally in freshwater microalgae. It induces apoptosis through down-regulation of antiapoptotic protein (Bcl-2, p-Bad and survivin) expression and upregulation of proapoptotic ones (Bax/Bad and PARP). From evaluation results, it was reported to have anticancer effects on colorectal cancer, melanoma, or gastric carcinoma cell lines [129].

Lutein is a carotenoid with a molecular formula $C_{40}H_{56}O_2$ (Figure 6.51). The rich source of lutein are egg yolks, spinach, corn, kiwi fruit and grapes. Evaluation results reported that lutein promotes growth inhibition of breast cancer cells through increased cell type-specific ROS generation and alternation of several signaling pathways [130].

β-Carotene is red-orange colored pigment, present abundantly in fungi, plants and fruits (Figure 6.52, Table 6.15). Biochemically, it is synthesized from eight isoprene

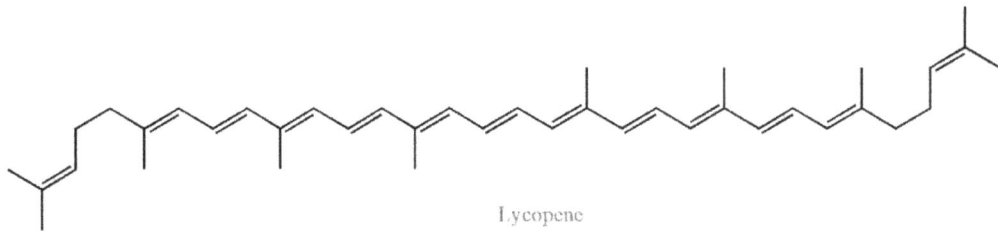

FIGURE 6.49 Structure of lycopene.

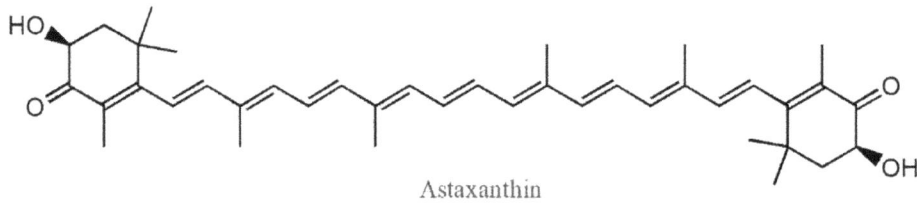

FIGURE 6.50 Structure of astaxanthin.

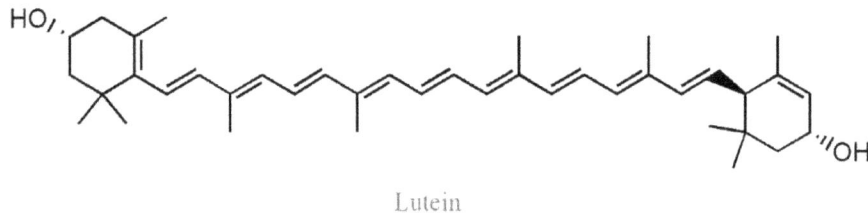

FIGURE 6.51 Structure of lutein.

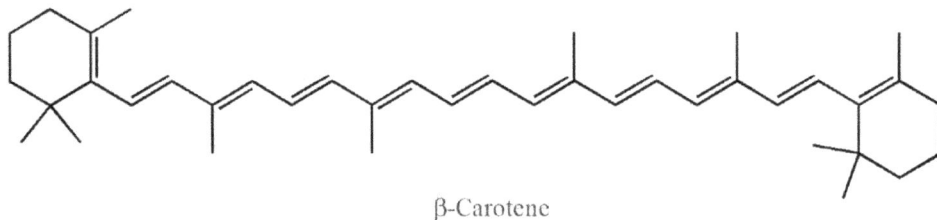

FIGURE 6.52 Structure of β-carotene.

TABLE 6.15

Molecular Mechanisms of Carotenoids for Anticancer Effects

Carotenoids	Cell Lines	Mechanisms
β-Carotene	HL-60	Activation of intrinsic pathway and NF-kB
	MCF-7	Activation of PPAR-γ and production of ROS
Lycopene	LNCaP, DU145	Reduction of mitochondrial membrane permeability, increase in release of cytochrome c; enhanced expression of Bax and down-regulation of Bcl-2 expression
Neoxanthin	PC-3	Activation of caspase-3 and PARP cleavage
Zeaxanthin	C918	Activation of intrinsic pathway
Violaxanthin	MCF-7	Induction of early apoptosis
Astaxanthin	HCT-116	Modified ratio of Bax/Bcl-2 and Bcl-xL level and increased phosphorylation of p38, JNK and ERK1/2
Fucoxanthin	HL-60	Generation of ROS, activation of caspase-9 and 3
Siphonaxanthin	HL-60	Down-regulated expression of Bcl-2 protein and increased activation of caspase-3
Halocynthiaxanthin	HL-60, MCF-7	DNA fragmentation and decreased expression of Bcl-2

units with a molecular formula $C_{40}H_{56}$. It acts as a precursor of vitamin A [131]. The evaluation study revealed that β-Carotene arrest G1 phase of cell cycle [132]. Upadhyaya et al. reported that β-carotene was found to exert potential anticancer effect against human promyelocytic leukemia (HL-60) cells [133].

6.7 Conclusion

Cancer is considered as a life-threatening disease and is reported to be the leading cause of death worldwide. So, the plant sources produce bioactive secondary metabolites like terpenoids which are extensively studied for their anticancer activities. Terpenoids mainly display their anticancer effects by targeting various pathways. From the preclinical and clinical studies, these compounds are found to be safe and effective for the treatment of cancer. The anticancer properties exhibited by terpenoids are observed promising that will potentially offer more opportunities for the treatment of cancer. Further, it is required to correlate the mechanism of action and structure-activity relationship (SAR) study of terpenoids as anticancer agents. In future, efforts should be made to identify the suitable targets of terpenoids to fight against cancer cells.

Abbreviation

AMFR	Autocrine motility factor receptor
AMPK	AMP-activated protein kinase
ATP	Adenosine triphosphate
CDK4	Cyclin-dependent kinase 4
CLL	Chronic lymphocytic leukemia
COX-2	Cyclooxygenase-2
DNA	Deoxyribonucleic acid
EGFR	Epidermal growth factor receptor
ER	Endoplasmic reticulum
ERK	Extracellular signal-regulated kinase
FAK	FAK focal adhesion kinase
5-FU	Fluorouracil
G6PD	Glucose-6-phosphate dehydrogenase
HMG-CoA	3-Hydroxy-3-methylglutaryl-CoA
IC50	Half maximal inhibitory concentration
IL	Interleukin
JNK	c-Jun N-terminal kinase
MAPK	Mitogen-activated protein kinase
MMP	Matrix metalloproteinases
NF-κB	Nuclear factor-kappa B
NK cell	Natural killer cells
PKC	Protein kinase C
PARP	Poly ADP-ribose polymerase
RNS	Reactive nitrogen species
ROS	Reactive oxygen species
SAR	Structure activity relationship
SOD	Superoxide dismutase
STAT3	Signal transducer and activator of transcription 3
TIMP	Tissue inhibitor of matrix metalloproteinase protein
VEGF	Vascular endothelia growth factor
WHO	World Health Organization

REFERENCES

1. Hanahan, D.; Weinberg, R. A. Hallmarks of cancer: The next generation. *Cell*, **2011**, 144, 646–674.
2. Sung, H.; Ferlay, J.; Siegel, R. L.; Laversanne, M.; Soerjomataram, I.; Jemal, A.; Bray, F. Global cancer statistics 2020: GLOBOCAN estimates of incidence and mortality worldwide for 36 cancers in 185 countries. *CA: A Cancer Journal for Clinicians*, **2021**, 71, 209–249.
3. Schirrmacher, V. From chemotherapy to biological therapy: A review of novel concepts to reduce the side effects of systemic cancer treatment (review). *International Journal of Oncology*, **2019**, 54(2), 407–419. doi: 10.3892/ijo.2018.4661.
4. Pucci, C.; Martinelli, C.; Ciofani, G. Innovative approaches for cancer treatment: Current perspectives and new challenges. *Ecancer Medical Science*, **2019**, 13, 961. doi: 10.3332/ecancer.2019.961.
5. Kuete, V.; Omosa, L. K.; Midiwo, J. O.; Karaosmanoglu, O.; Sivas, H. Cytotoxicity of naturally occurring phenolics and terpenoids from *Kenyan flora* towards human carcinoma cells. *Journal of Ayurveda and Integrative Medicine*, **2019**, 10, 178–184.
6. Fares, J.; Fares, M. Y.; Khachfe, H. H., et al. Molecular principles of metastasis: A hallmark of cancer revisited. *Signal Transduction and Targeted Therapy*, **2020**, 5, 28. doi: 10.1038/s41392-020-0134-x.
7. Hanahan, D.; Weinberg, R. A. Hallmarks of cancer: The next generation. *Cell*, **2011**, 144, 646–674.
8. Singh, S.; Sharma, B.; Kanwar, S. S., et al. Lead phytochemicals for anticancer drug development. *Frontiers in Plant Science*, **2016**, 7, 1667. doi: 10.3389/fpls.2016.01667.
9. Huang, M.; Lu, J. J.; Huang, M. Q.; Bao, J. L.; Chen, X. P.; Wang, Y. T. Terpenoids: Natural products for cancer therapy. *Expert Opinion on Investigational Drugs*, **2012**, 21(12), 1801–1818. doi: 10.1517/13543784.2012.727395.
10. Chopra, B.; Dhingra, A. K.; Dhar, K. L.; Nepali, K. Emerging role of terpenoids for the treatment of cancer: A review. *Mini-Reviews in Medicinal Chemistry*, **2021**, 21(16), 2300–2336. doi: 10.2174/1389557521666210112143024.
11. Nishino, H. Phytochemicals in hepatocellular cancer prevention. *Nutrition and Cancer*, **2009**, 61, 789–791.
12. Wagner, K. H.; Elmadfa, I. Biological relevance of terpenoids. Overview focusing on mono-, di- and tetraterpenes. *Annals of Nutrition and Metabolism*, **2003**, 47, 95–106.
13. Rabi, T.; Bishayee, A. Terpenoids and breast cancer chemoprevention. *Breast Cancer Res Treat*, **2009**, 115, 223–239.
14. Gould, M. N. Cancer chemoprevention and therapy by monoterpenes. *Environmental Health Perspectives*, **1997**, 105(Suppl 4), 977–979.
15. Crowell, P. L. Prevention and therapy of cancer by dietary monoterpenes. *Journal of Nutrition*, **1999**, 129, 775S–778S.
16. Laszczyk, M. N. Pentacyclic triterpenes of the lupane, oleanane and ursane group as tools in cancer therapy. *Planta Medica*, **2009**, 75, 1549–1560.
17. Yang, H.; Dou, Q. P. Targeting apoptosis pathway with natural terpenoids: Implications for treatment of breast and prostate cancer. *Current Drug Targets*, **2010**, 11(6), 733–744. doi: 10.2174/138945010791170842.
18. Orlowski, R. Z.; Baldwin, A. S., Jr. NF-kappa B as a therapeutic target in cancer. *Trends in Molecular Medicine*, **2002**, 8, 385–389.

19. D'Aguanno, S.; Del Bufalo, D. Inhibition of anti-apoptotic Bcl-2 proteins in preclinical and clinical studies: Current overview in cancer. *Cells*, **2020**, 9(5), 1287. doi: 10.3390/cells9051287.

20. Rama Rao, M. Terpenoids as potential targeted therapeutics of pancreatic cancer: Current advances and future directions. *Breaking Tolerance to Pancreatic Cancer Unresponsiveness to Chemotherapy*, **2019**, 111–116. doi: 10.1016/B978-0-12-817661-0.00007-X.

21. Silva, B. I. M.; Nascimento, E. A.; Silva, C. J.; Silva, T. G.; Aguiar, J. S. Anticancer activity of monoterpenes: A systematic review. *Molecular Biology Reports*, **2021**, 48(7), 5775–5785. doi: 10.1007/s11033-021-06578-5.

22. Zielińska-Błajet, M.; Feder-Kubis, J. Monoterpenes and their derivatives—Recent development in biological and medical applications. *International Journal of Molecular Sciences*, **2020**, 21, 7078.

23. Sobral, M. V.; Xavier, A. L.; Lima, T. C.; De Sousa, D. P. Antitumor activity of monoterpenes found in essential oils. *The Scientific World Journal*, **2014**, 2014, 953451.

24. Jaafari, A.; Tilaoui, M.; Mouse, H. A.; M'Bark, L. A.; Aboufatima, R.; Chait, A.; Lepoivre, M.; Zyad, A. Comparative study of the antitumor effect of natural monoterpenes: Relationship to cell cycle analysis. *Brazilian Journal of Pharmacognosy*, **2012**, 22, 534–540.

25. Huang, M.; Lu, J. J.; Huang, M. Q.; Bao, J. L.; Chen, X. P.; Wang, Y. T. Terpenoids: Natural products for cancer therapy. *Expert Opinion on Investigational Drugs*, **2012**, 21(12), 1801–1818.

26. Prashar, A.; Locke, I. C.; Evans, C. S. Cytotoxicity of lavender oil and its major components to human skin cells. *Cell Proliferation*, **2004**, 37, 221–229.

27. Becker, V.; Hui, X.; Nalbach, L. Linalool inhibits the angiogenic activity of endothelial cells by down regulating intracellular ATP levels and activating TRPM8. *Angiogenesis*, **2021**, 24, 613–630. doi: 10.1007/s10456-021-09772-y.

28. Fan, K.; Li, X.; Cao, Y.; Qi, H.; Li, L.; Zhang, Q.; Sun, H. Carvacrol inhibits proliferation and induces apoptosis in human colon cancer cells. *Anticancer Drugs*, **2015**, 26(8), 813–823.

29. Fahad, K.; Vipendra, S. K.; Mohd, S.; Mohd, K. A.; Irfan, A. A. Carvacrol induced program cell death and cell cycle arrest in androgen-independent human prostate cancer cells via inhibition of notch signaling. *Anti-Cancer Agents in Medicinal Chemistry*, **2019**, 19(13), 1588–1608.

30. Horvathova, E.; Sramkova, M.; Labaj, J.; Slamenovam, D. Study of cytotoxic, genotoxic and DNA-protective effects of selected plant essential oils on human cells cultured in vitro. *Neuroendocrinology Letters*, **2006**, 27, 44–47.

31. Horvathova, E.; Turcaniova, V.; Slamenova, D. Comparative study of DNA-damaging and DNA-protective effects of selected components of essential plant oils in human leukemic cells K562. *Neoplasma*, **2007**, 54, 478–483.

32. Sun, J. D-Limonene: safety and clinical applications. *Alternative Medicine Review*, **2007**, 12, 259–264.

33. Pattanayak, M.; Seth, P. K.; Smita, S.; Gupta, S. K. Geraniol and limonene interaction with 3-hydroxy-3-methylglutaryl-CoA (HMG-CoA) reductase for their role as cancer chemopreventive agents. *Journal of Proteomics & Bioinformatics*, **2009**, 2, 466–474.

34. Vieira Sobral, M.; Lira Xavier, A.; Cardoso Lima, T.; de Sousa, D. P. Antitumor activity of monoterpenes found in essential oils. *Scientific World Journal*, **2014**, 1–35, Article ID 953451.

35. Lu, H. F.; Liu, J. Y.; Hsueh, S. C., et al. (–)-Menthol inhibits WEHI-3 leukemia cells in vitro and in vivo. *In Vivo*, **2007**, 21(2), 285–289.

36. Li, Q.; Wang, X.; Yang, Z.; Wang, B.; Li, S. Menthol induces cell death via the TRPM8 channel in the human bladder cancer cell line T24. *Oncology*, **2009**, 77(6), 335–341.

37. Edris, A. E. Pharmaceutical and therapeutic potentials of essential oils and their individual volatile constituents: A review. *Phytotherapy Research*, **2007**, 21, 308–323.

38. Liu, Z.; Shen, C.; Tao, Y.; Wang, S.; Wei, Z.; Cao, Y.; Wu, H.; Fan, F.; Lin, C.; Shan, Y.; Zhu, P.; Sun, L.; Chen, C.; Wang, A.; Zheng, S.; Lu, Y. Chemopreventive efficacy of menthol on carcinogen-induced cutaneous carcinoma through inhibition of inflammation and oxidative stress in mice. *Food and Chemical Toxicology*, **2015**, 82, 12–18. doi: 10.1016/j.fct.2015.04.025.

39. Su, J.; Lai, H.; Chen, J. Natural borneol, a monoterpenoid compound, potentiates selenocystine-induced apoptosis in human hepatocellular carcinoma cells by enhancement of cellular uptake and activationof ROS-mediated DNA damage. *PLoS One*, **2013**, 8, Article ID e63502.

40. Abbasi, R.; Efferth, T.; Kuhmann, C.; Opatz, T.; Hao, X.; Popanda, O.; Schmezer, P. The endoperoxide ascaridol shows strong differential cytotoxicity in nucleotide excision repair-deficient cells. *Toxicology and Applied Pharmacology*, **2012**, 259, 302–310.

41. Kang, S. H.; Kim, Y. S.; Kim, E. K.; Hwang, J. W.; Jeong, J. H.; Dong, X.; Lee, J. W.; Moon, S. H.; Jeon, B. T.; Park, P. J. Anticancer effect of thymol on AGS human gastric carcinoma cells. *Journal of Microbiology and Biotechnology*, **2016**, 26(1), 28–37. doi: 10.4014/jmb.1506.06073.

42. Zeng, Q.; Che, Y.; Zhang, Y.; Chen, M.; Guo, Q.; Zhang, W. Thymol isolated from *Thymus vulgaris* L. inhibits colorectal cancer cell growth and metastasis by suppressing the WNT/β-catenin pathway. *Drug Design, Development and Therapy*, **2020**, 14, 2535–2547. doi: 10.2147/DDDT.S254218.

43. Saleh, M. M. H.; Hashem, F. A.; Glombitza, K. W. Cytotoxicity and in vitro effects on human cancer cell lines of volatiles of *Apium graveolens* var. filicum. *Pharmaceutical and Pharmacological Letters*, **1998**, 8, 97–99.

44. Bai, X.; Tang, J. Myrcene exhibits antitumor activity against lung cancer cells by inducing oxidative stress and apoptosis mechanisms. *Natural Product Communications*, **2020**, 15, 1–7. doi: 10.1177/1934578X20961189.

45. Duncan, R. E.; Lau, D.; El-Sohemy, A.; Archer, M. C. Geraniol and beta-ionone inhibit proliferation, cell cycle progression and cyclin-dependent kinase 2 activity in MCF-7 breast cancer cells independent of effects on HMG-CoA reductase activity. *Biochemical Pharmacology*, **2004**, 68, 1739–1747.

46. Cho, M.; So, I.; Chun, J. N.; Jeon, J. H. The antitumor effects of geraniol: Modulation of cancer hallmark pathways (review). *International Journal of Oncology*, **2016**, 48(5), 1772–1782. doi: 10.3892/ijo.2016.3427.

47. Kim, S. J.; So, I.; Kim, T. W. Geraniol inhibits prostate cancer growth by targeting cell cycle and apoptosis pathways. *Biochemical and Biophysical Research Communications*, **2011**, 407, 129–134.

48. Carnesecchi, S.; Bras-Goncalves, R.; Bradaia, A. Geraniol, a component of plant essential oils, modulates DNA synthesis and potentiates 5-fluorouracil efficacy on human colon tumor xenografts. *Cancer Letters*, **2004**, 215(1), 53–59.

49. Alonso-Gutierrez, J.; Chan, R.; Batth, T. S.; Adams, P.; Keasling, J.; Petzold, C.; Lee, T. S. Metabolic engineering of Escherichia coli for limonene and perillyl alcohol production. *Metabolic Engineering*, **2013**, 19, 33–41.

50. Satomi, Y.; Miyamoto, S.; Gould, M. N. Induction of AP-1 activity by perillyl alcohol in breast cancer cells. *Carcinogenesis*, **1999**, 20, 1957–1961.

51. Yeruva, L.; Pierre, K. J.; Elegbede, A.; Wang, R. C.; Carper, S. W. Perillyl alcohol and perillic acid induced cell cycle arrest and apoptosis in non-small cell lung cancer cells. *Cancer Letters*, **2007**, 257, 216–226.

52. Itani, W. S.; El-Banna, S. H.; Hassan, S. B.; Larsson, R. L.; Gali-Muhtasib, A.; U, H. Anti-colon cancer components from Lebanese sage (*Salvia libanotica*) essential oil: Mechanistic basis. *Cancer Biology & Therapy*, **2008**, 7, 1765–1773.

53. Okamoto, Y.; Ohkubo, T.; Ikebe, T.; Yamazaki, J. Blockade of TRPM8 activity reduces the invasion potential of oral squamous carcinoma cell lines. *International Journal of Oncology*, **2012**, 40, 1431–1440.

54. Bezerra, D. P.; Marinho Filho, J. D. B.; Alves, A. P. N. N., et al. Antitumor activity of the essential oil from the leaves of *Croton regelianus* and its component ascaridole. *Chemistry & Biodiversity*, **2009**, 6, 1224–1231.

55. Paramasivam, A.; Sambantham, S.; Shabnam, J., et al. Anticancer effects of thymoquinone in mouse neuroblastoma (Neuro-2a) cells through caspase-3 activation with down regulation of XIAP. *Toxicology Letters*, **2012**, 213, 151–159.

56. Silva, S. L.; Figueiredo, P. M.; Yano, T. Cytotoxic evaluation of essential oil from *Zanthoxylum rhoifolium* Lam. Leaves. *Acta Amazonica*, **2007**, 37, 281–286.

57. Zheng, G. Q.; Kenney, P. M.; Lam, L. K. T. Potential anticarcinogenic natural products isolated from lemongrass oil and galanga root oil. *Journal of Agricultural and Food Chemistry*, **1993**, 41, 153–156.

58. Huang, M.; Lu, J. J.; Huang, M. Q.; Bao, J. L.; Chen, X. P.; Wang, Y. T. Terpenoids: Natural products for cancer therapy. *Expert Opinion on Investigational Drugs*, **2012**, 21(12), 1801–1818.

59. Petronelli, A.; Pannitteri, G.; Testa, U. Triterpenoids as new promising anticancer drugs. *Anticancer Drugs*, **2009**, 20, 880–892.

60. Liu, D.; Chen, Z. The effects of cantharidin and cantharidin derivates on tumour cells. *Anti-Cancer Agents in Medicinal Chemistry*, **2009**, 9, 392–396.

61. Abu-Izneid, T.; Rauf, A.; Mohammad Ali, S.; Anees Ahmed, K.; Muhammad, I.; Maksim, R.; Md. Sahab, U.; Mohamad Fawzi, M.; Kannan, R. R. Sesquiterpenes and their derivatives-natural anticancer compounds: An update. *Pharmacological Research*, **2020**, 161, 105165–. doi: 10.1016/j.phrs.2020.105165.

62. Ren, Y.; Yu, J.; Kinghorn, A. D. Development of anticancer agents from plant derived sesquiterpene lactones. *Current Medicinal Chemistry*, **2016**, 23, 2397–2420.

63. Lai, H. C.; Singh, N. P.; Sasaki, T. Development of artemisinin compounds for cancer treatment. *Investigational New Drugs*, **2013**, 31(1), 230–246.

64. Nakase, I.; Lai, H.; Singh, N. P.; Sasaki, T. Anticancer properties of artemisinin derivatives and their targeted delivery by transferrin conjugation. *International Journal of Pharmaceutics*, **2008**, 354(1–2), 28–33.

65. Krishna, S.; Ganapathi, S.; Ster, I. C.; Saeed, M. E. M.; Cowan, M.; Finlayson, C.; Kovacsevics, H.; Jansen, H.; Kremsner, P. G.; Efferth, T., et al. A randomised, double blind, placebo-controlled pilot study of oral artesunate therapy for colorectal cancer. EBioMedicine, **2015**, 2, 82–90.

66. Kuo, Y. H.; Kuo, Y. J.; Yu, A. S.; Wu, M. D.; Ong, C. W.; Yang Kuo, L. M.; Huang, J. T.; Chen, C. F.; Li, S. Y. Two novel sesquiterpene lactones, cytotoxic vernolide-A and -B, from *Vernonia cinerea*. *Chemical and Pharmaceutical Bulletin*, **2003**, 51, 425–426.

67. Nguyen, N. H.; Nguyen, M. T.; Little, P. J.; Do, A. T.; Tran, P. T.; Vo, X. N.; Do, B. H. Vernolide-A and vernodaline: Sesquiterpene lactones with cytotoxicity against cancer. *Journal of Environmental Pathology, Toxicology and Oncology*, **2020**, 39(4), 299–308.

68. Pratheeshkumar, P.; Kuttan, G. Vernolide-A, a sesquiterpene lactone from *Vernonia cinerea*, induces apoptosis in B16F-10 melanoma cells by modulating p53 and caspase-3 gene expressions and regulating NF-κB-mediated bcl-2 activation. *Drug and Chemical Toxicology*, **2011**, 34(3), 261–270. doi: 10.3109/01480545.2010.520017.

69. Iranshahi, M.; Kalategi, F.; Rezaee, R.; Shahverdi, A. R.; Ito, C.; Furukawa, H.; Tokuda, H.; Itoigawa, M. Cancer chemopreventive activity of terpenoid coumarins from *Ferula* species. *Planta Medica*, **2008**, 74, 147–150.

70. Oh, B. S.; Shim, E. A.; Jung, J. H.; Jung, D. B.; Kim, B.; Shim, B. S.; Yazdi, M. C.; Iranshahi, M.; Kim, S. H. Apoptotic effect of galbanic acid via activation of caspases and inhibition of Mcl-1 in H460 non-small lung carcinoma cells. *Phytotherapy Research*, **2015**, 29, 844–849.

71. Yoon Hyeon, K.; Eun Ah, S.; Ji Hoon, J.; Ji Eon, P.; Jin Suk, K.; Ja Il, K.; Bum Sang, S.; Sung-Hoon, K. Galbanic acid potentiates TRAIL induced apoptosis in resistant non-small cell lung cancer cells via inhibition of MDR1 and activation of caspases and DR5. *European Journal of Pharmacology*, **2019**, 847, 91–96. doi: 10.1016/j.ejphar.2019.01.028.

72. Zhang, S.; Won, Y.-K.; Ong, C.-N.; Shen, H.-M. Anticancer potential of sesquiterpene lactones: Bioactivity and molecular mechanisms. *Current Medicinal Chemistry – Anti-Cancer Agents*, **2005**, 5, 239–249.

73. Lu, J. J.; Dang, Y. Y.; Huang, M.; Xu, W. S.; Chen, X. P.; Wang, Y. T. Anti-cancer properties of terpenoids isolated from *Rhizoma curcumae* – A review. *Journal of Ethnopharmacology*, **2012**, 143(2), 406–411.

74. Liu, J.; Zhang, Y.; Qu, J.; Xu, L.; Hou, K.; Zhang, J.; Qu, X.; Liu, Y. Beta-elemene induced autophagy protects human gastric cancer cells from undergoing apoptosis. *BMC Cancer*, **2011**, 11, 183.

75. Wang, W.; Chen, B.; Zou, R.; Tu, X.; Tan, S.; Lu, H.; Liu, Z.; Fu, J. Codonolactone, a sesquiterpene lactone isolated from *Chloranthus henryi* Hemsl, inhibits breast cancer cell

invasion, migration and metastasis by downregulating the transcriptional activity of Runx2. *International Journal of Oncology*, **2014**, 45, 1891–1900. doi: 10.3892/ijo.2014.2643.

76. Fang, J.; Seki, T.; Maeda, H. Therapeutic strategies by modulating oxygen stress in cancer and inflammation. *Advanced Drug Delivery Reviews*, **2009**, 61, 290–302.

77. Fonrose, X.; Ausseil, F.; Soleilhac, E.; Masson, V.; David, B.; Pouny, I.; Cintrat, J. C.; Rousseau, B.; Barette, C.; Massiot, G.; Lafanechère, L. Parthenolide inhibits tubulin carboxypeptidase activity. *Cancer Research*, **2007**, 67, 3371–3378.

78. Guzman, M. L.; Rossi, R. M.; Karnischky, L.; Li, X.; Peterson, D. R.; Howard, D. S.; Jordan, C. T. The sesquiterpene lactone parthenolide induces apoptosis of human acute myelogenous leukemia stem and progenitor cells. *Blood*, **2005**, 105, 4163–4169.

79. Nasim, S.; Crooks, P. A. Antileukemic activity of aminoparthenolide analogs. *Bioorganic & Medicinal Chemistry Letters*, **2008**, 18, 3870–3873.

80. Cai, J. W.; Xin, M. X.; Wu, Q. S.; Ma, J.; Yang, P.; Xie, H. Y.; Huang, D. S. Parthenolide induces proliferation inhibition and apoptosis of pancreatic cancer cells in vitro. *Journal of Experimental & Clinical Cancer Research*, **2010**, 29, 108.

81. Jian, B.; Zhang, H.; Han, C.; Liu, J. Anti-cancer activities of diterpenoids derived from *Euphorbia fischeriana* Steud. *Molecules*, **2018**, 23(2), 387. doi: 10.3390/molecules23020387.

82. Wang, M.; Li, J.; Qing, W.; Wang, Q. H.; Yang, B. Y.; Kuang, H. X. Antitumor activity of abietane type diterpene from *Euphorbia fischeriana* Steud. *Chinese Pharmaceutical Journal*, **2015**, 8, 1261–1262.

83. Saraste, A.; Pulkki, K. Morphologic and biochemical hallmarks of apoptosis. *Cardiovascular Research*, **2000**, 45, 528–537.

84. Wang, M.; Wei, Q.; Wang, Q. H.; Xia, Y. G.; Yang, B. Y.; Kuang, H. X. Antitumor activity of tigliane type diterpene from *Euphorbia fischeriana* Steud. *Acta Chinese Medicine and Pharmacology*, **2013**, 4, 11–14.

85. Yan, S. S.; Li, Y.; Wang, Y.; Shen, S. S.; Gu, Y.; Wang, H. B.; Qin, G. W.; Yu, Q. 17-Acetoxyjolkinolide B irreversibly inhibits IkappaB kinase and induces apoptosis of tumor cells. *Molecular Cancer Therapeutics*, **2008**, 7, 1523–1532.

86. Xu, H. Y.; Chen, Z. W.; Hou, J. C.; Du, F. X.; Liu, J. C. Jolkinolide B induces apoptosis in MCF-7 Cells through inhibition of the PI3k/Akt/Mtor signaling pathway. *Oncology Reports*, **2013**, 29, 212–218.

87. Luo, H.; Wang, A. Induction of apoptosis in K562 cells by jolkinolide B. *Canadian Journal of Physiology and Pharmacology*, **2006**, 84, 959–965.

88. Gao, C.; Yan, X.; Wang, B.; Yu, L.; Han, J.; Li, D.; Zheng, Q. Jolkinolide B induces apoptosis and inhibits tumor growth in mouse melanoma B16f10 cells by altering glycolysis. *Scientific Reports*, **2016**, 6, 36114.

89. Giacinti, C.; Giordano, A. Rb and cell cycle progression. *Oncogene*, **2006**, 25, 5220–5227.

90. Ikezoe, T.; Yang, Y.; Bandobashi, K., et al. Oridonin, a diterpenoid purified from *Rabdosia rubescens*, inhibits the proliferation of cells from lymphoid malignancies in association with blockade of the NF-kappa B signal pathways. *Molecular Cancer Therapeutics*, **2005**, 4, 578–586.

91. Lu, X. L.; Zhang, S. L.; Wang, Z. S. Analysis of andrographolide compounds. I. Ion pair high performance liquid chromatographic analysis of andrographolide derivatives. *Acta Pharmaceutica Sinica*, **1981**, 16, 182–189.

92. Rajagopal, S.; Kumar, R. A.; Deevi, D. S.; Satyanarayana, C.; Rajagopalan, R. Andrographolide, a potential cancer therapeutic agent isolated from *Andrographis paniculata*. *Journal of Experimental Therapeutics and Oncology*, **2003**, 3, 147–158. doi: 10.1046/j.1359-4117.2003.01090.x.

93. Peng, Y.; Wang, Y.; Tang, N.; Sun, D; Lan, Y.; Yu, Z.; Zhao, X.; Feng, L.; Zhang, B.; Jin, L.; Yu, F.; Ma, X.; Lv, C. Andrographolide inhibits breast cancer through suppressing COX-2 expression and angiogenesis via inactivation of p300 signaling and VEGF pathway. *Journal of Experimental & Clinical Cancer Research*, **2018**, 37(1), 248. doi: 10.1186/s13046-018-0926-9.

94. Liu, Q. Triptolide and its expanding multiple pharmacological functions. *International Immunopharmacology*, **2011**, 11, 377–383.

95. Chang, W. T.; Kang, J. J.; Lee, K. Y. Triptolide and chemotherapy cooperate in tumor cell apoptosis. A role for the p53 pathway. *Journal of Biological Chemistry*, **2001**, 276, 2221–2227.

96. Liu, J. J.; Lin, D. J.; Liu, P. Q. Induction of apoptosis and inhibition of cell adhesive and invasive effects by tanshinone IIA in acute promyelocytic leukemia cells in vitro. *Journal of Biomedical Science*, **2006**, 13, 813–823.

97. Chiu, P.; Leung, L. T.; Ko, B. C. Pseudolaric acids: Isolation, bioactivity and synthetic studies. *Natural Product Reports*, **2010**, 27, 1066–1083.

98. Yu, J.; Li, X.; Tashiro, S., et al. Bcl-2 family proteins were involved in pseudolaric acid B-induced autophagy in murine fibrosarcoma L929 cells. *Journal of Pharmacology Science*, **2008**, 107, 295–302.

99. Li, M. H.; Miao, Z. H.; Tan, W. F. Pseudolaric acid B inhibits angiogenesis and reduces hypoxia-inducible factor 1alpha by promoting proteasome-mediated degradation. *Clinical Cancer Research*, **2004**, 10, 8266–8274.

100. Qin, J.; Tang, J.; Jiao, L.; Ji, J.; Chen, W. D.; Feng, G. K.; Gao, Y. H.; Zhu, X. F.; Deng, R. A diterpenoid compound, excisanin A, inhibits the invasive behavior of breast cancer cells by modulating the integrin β1/FAK/PI3K/AKT/β-catenin signaling. *Life Science*, **2013**, 4(93), 655–663. doi: 10.1016/j.lfs.2013.09.002.

101. Muhammad Torequl, I. Diterpenes and their derivatives as potential anticancer agents. *Phytotherapy Research*, **2017**, 31(5), 691–712. doi: 10.1002/ptr.5800.

102. Petronelli, A.; Pannitteri, G.; Testa, U. Triterpenoids as new promising anticancer drugs. *Anti-Cancer Drugs*, **2009**, 20(10), 880–892. doi: 10.1097/cad.0b013e328330fd90.

103. Patlolla Jagan, M. R.; Rao Chinthalapally, V. Triterpenoids for cancer prevention and treatment: Current status and future prospects. *Current Pharmaceutical Biotechnology*, **2012**, 13(1), 147–155. doi: 10.2174/138920112798868719.

104. Pitchai, D.; Roy, A.; Ignatius, C. In vitro evaluation of anticancer potentials of lupeol isolated from *Elephantopus scaber* L. on MCF-7 cell line. *Journal of Advanced Pharmaceutical Technology & Research*, **2014**, 5(4), 179–184. doi: 10.4103/2231-4040.143037.

105. Shigematsu, H.; Gazdar, A. F. Somatic mutations of epidermal growth factor receptor signaling pathway in lung cancers. *International Journal of Cancer*, **2006**, 118, 257–262.

106. Jia-jun, L.; Yan-yan, Y.; Hong-mei, S.; Yun, L.; Chao-yue, S.; Hu-biao, C.; Jian-ye, Z. Anti-cancer effects of pristimerin and the mechanisms: A critical review. *Frontiers in Pharmacology*, **2019**, 10, 1–10. doi: 10.3389/fphar.2019.00746.

107. Yousef, B.; Guerram, M.; Hassan, H.; Hamdi, A.; Zhang, L.-Y.; Jiang, Z.-Z. Pristimerin demonstrates anticancer potential in colorectal cancer cells by inducing G1 phase arrest and apoptosis and suppressing various pro-survival signaling proteins. *Oncology Reports*, **2016**, 35, 1091–1100. doi: 10.3892/or.2015.4457.

108. Guo, Y.; Zhang, W.; Yan, Y. Y.; Ma, C. G.; Wang, X.; Wang, C.; et al. Triterpenoid pristimerin induced HepG2 cells apoptosis through ROS-mediated mitochondrial dysfunction. *JBUON*, **2013**, 18, 477–485.

109. Mori, Y.; Shirai, T.; Terauchi, R.; Tsuchida, S.; Mizoshiri, N.; Hayashi, D. Antitumor effects of pristimerin on human osteosarcoma cells in vitro and in vivo. *OncoTargets and Therapy*, **2017**, 10, 5703–5710. doi: 10.2147/OTT.S150071.

110. Zhu, B.; Wei, Y. Antitumor activity of celastrol by inhibition of proliferation, invasion, and migration in cholangiocarcinoma via PTEN/PI3K/Akt pathway. *Cancer Medicine*, **2019**, 1–14. doi: 10.1002/cam4.2719.

111. Saeed, M. E. M.; Mahmoud, N.; Sugimoto, Y.; Efferth, T.; Abdel-Aziz, H. Betulinic acid exerts cytotoxic activity against multidrug-resistant tumor cells via targeting autocrine motility factor receptor (AMFR). *Frontiers in Pharmacology*, **2018**, 9, 1–13. doi: 10.3389/fphar.2018.00481.

112. Zhang, X.; Hu, J. Y.; Chen, Y. Betulinic acid and the pharmacological effects of tumor suppression. *Molecular Medicine Reports*, **2016**, 14, 4489–4495.

113. Cai, Y.; Zheng, Y.; Gu, J.; et al. Betulinic acid chemosensitizes breast cancer by triggering ER stress-mediated apoptosis by directly targeting GRP78. *Cell Death & Disease*, **2018**, 9, 636. doi: 10.1038/s41419-018-0669-8.

114. Fulda, S.; Debatin, K. M. Sensitization for anticancer drug-induced apoptosis by betulinic acid. *Neoplasia*, **2005**, 7, 162–170.

115. Liu, J. Pharmacology of oleanolic acid and ursolic acid. *Journal of Ethnopharmacology*, **1995**, 49, 57–68.

116. Tang, Z. Y.; Li, Y.; Tang, Y. T.; Ma, X. D.; Tang, Z. Y. Anticancer activity of oleanolic acid and its derivatives: Recent advances in evidence, target profiling and mechanisms of action. *Biomedicine & Pharmacotherapy*, **2022**, 145, 112397. doi: 10.1016/j.biopha.2021.112397.

117. Shih, Y. H.; Chein, Y. C.; Wang, J. Y.; Fu, Y. S. Ursolic acid protects hippocampal neurons against kainate-induced excitotoxicity in rats. *Neuroscience Letters*, **2004**, 362, 136–140.

118. Kuttan, G.; Pratheeshkumar, P.; Manu, K. A.; et al. Inhibition of tumor progression by naturally occurring terpenoids. *Pharmaceutical Biology*, **2011**, 49, 995–1007.

119. Tian, Q; Miller, E. G.; Ahmad, H.; Tang, L.; Patil, B. S. Differential inhibition of human cancer cell proliferation by citrus limonoids. *Nutrition and Cancer*, **2001**, 40, 180–184.

120. Wang, H.; Haridas, V.; Gutterman, J. U.; Xu, Z. X. Natural triterpenoid avicins selectively induce tumor cell death. *Communicative & Integrative Biology*, **2010**, 3(3), 205–208. doi: 10.4161/cib.3.3.11492.

121. Haridas, V.; Arntzen, C. J.; Gutterman, J. U. Avicins, a family of triterpenoid saponins from *Acacia victoriae* (Bentham), inhibit activation of nuclear factor- B by inhibiting both its nuclear localization and ability to bind DNA. *Proceedings of the National Academy of Sciences*, **2001**, 98(20), 11557–11562. doi: 10.1073/pnas.191363498.

122. Xu, W.; Li, T.; Qiu, J.-F.; Wu, S.-S.; Huang, M.-Q.; Lin, L.-G.; Zhang, Q.-W.; Chen, X.-P.; Lu, J.-J. Anti-proliferative activities of terpenoids isolated from *Alisma orientalis* and their structure-activity relationships. *Anti-Cancer Agents in Medicinal Chemistry*, **2015**, 15, 228–235.

123. Miró, M. Cucurbitacins and their pharmacological effects. *Phytotherapy Research*, **1995**, 9, 159–168.

124. Jayaprakasam, B.; Seeram, N. P.; Nair, M. G. Anticancer and anti-inflammatory activities of cucurbitacins from *Cucurbita andreana*. *Cancer Letter*, **2003**, 189, 11–16.

125. Chen, X.; Bao, J.; Guo, J.; Ding, Q.; Lu, J.; Huang, M.; Wang, Y. Biological activities and potential molecular targets of cucurbitacins: A focus on cancer. *Anticancer Drugs*, **2012**, 23(8), 777–787.

126. Gapter, L.; Wang, Z.; Glinski, J. Induction of apoptosis in prostate cancer cells by pachymic acid from *Poria cocos*. *Biochemical and Biophysical Research Communications*, **2005**, 332, 1153–1161.

127. Ling, H.; Jia, X.; Zhang, Y., et al. Pachymic acid inhibits cell growth and modulates arachidonic acid metabolism in nonsmall cell lung cancer A549 cells. *Molecular Carcinogenesis*, **2010**, 49, 271–282.

128. Thoppil, R. J.; Bishayee, A. Terpenoids as potential chemopreventive and therapeutic agents in liver cancer. *World Journal of Hepatology*, **2011**, 3(9), 228–249. doi: 10.4254/wjh.v3.i9.228.

129. Faraone, I.; Sinisgalli, C.; Ostuni, A.; Armentano, M. F.; Carmosino, M.; Milella, L.; Russo, D.; Labanca, F.; Khan, H. Astaxanthin anticancer effects are mediated through multiple molecular mechanisms: A systematic review. *Pharmacological Research*, **2020**, 155, 104689. doi: 10.1016/j.phrs.2020.104689.

130. Gong, X.; Smith, J. R.; Swanson, H. M.; Rubin, L. P. Carotenoid lutein selectively inhibits breast cancer cell growth and potentiates the effect of chemotherapeutic agents through ROS-mediated mechanisms. *Molecules*, **2018**, 23(4), 905. doi: 10.3390/molecules23040905.

131. Toma, S.; Losardo, P. L.; Vincent, M.; Palumbo, R. Effectiveness of beta-carotene in cancer chemoprevention. *European Journal of Cancer Prevention*, **1995**, 4(3), 213–224. doi: 10.1097/00008469-199506000-00002.

132. Niranjana, R.; Gayathri, R.; Nimish Mol, S.; Sugawara, T.; Hirata, T.; Miyashita, K.; Ganesan, P. Carotenoids modulate the hallmarks of cancer cells. *Journal of Functional Foods*, **2015**, 18, 968–985. doi: 10.1016/j.jff.2014.10.017.

133. Upadhyaya, K.; Radha, K.; Madhyastha, H. Cell cycle regulation and induction of apoptosis by β-carotene in U937 and HL-60 leukemia cells. *Journal of Biochemistry and Molecular Biology*, **2007**, 40, 1009.

7

Terpene Glycosides

Bimal Krishna Banik*, Biswa Mohan Sahoo†, and Abhishek Tiwari‡

CONTENTS

7.1   Introduction 229
7.2   Synthesis of Glycosides 229
7.3   Monoterpenoid Glycosides 231
7.4   Sesquiterpenoid Glycosides 232
7.5   Diterpenoid Glycosides 234
7.6   Sesterpene Glycosides 236
7.7   Triterpenoid Glycosides 237
7.8   Tetraterpenoid Glycosides 244
7.9   Miscellaneous 244
7.10 Hydrolysis of Terpene Glycosides 246
7.11 Applications of Terpene Glycosides 249
7.12 Conclusion 249
Abbreviations 250
References 251

7.1 Introduction

Glycosides are defined as compounds in which one or more sugars (glycine) are combined with nonsugar molecules (aglycone) through glycosidic linkage [1]. Depending on whether the glycosidic bond lies above or below the plane of the cyclic molecule of sugar, glycosides are classified as β-glycosides or α-glycosides. If the glycone group of any glycoside is glucose then the molecule is called as glucoside. Glycosides can also be classified according to the chemical nature of the aglycone part. For example Steviol glycosides or Stevioside in which steviol is the aglycone part. The process by which glycosides are formed is called glycosylation. Terpene glycosides are terpenoids in which one or more hydroxy functions are glycosylated [2]. Terpene glycosides are present in a variety of plant sources. Several classes of terpenes are glycosylated by which their physicochemical properties and biological activities are affected (Figure 7.1) [3]. As terpene glycosides exhibit promising physiological effects, these are isolated in large numbers and their structures are elucidated by analytical data of FT-IR,^{1}H- NMR,^{13}C-NMR, 2D-NMR, LC-MS spectra, circular dichroism (CD) spectra and X-ray analysis. Glycosylated terpenoids are broadly distributed in plant sources and comprise diverse structures due to the combination of aglycones and glycones (carbohydrates) [4,5].

These are naturally occurring organic compounds with water solubility property. These are non-toxic, odorless and used as complexing agent, solubilizing agent or excipients in food and pharmaceutical industry [6]. Monoterpene glycosides have attracted much attention because these are water soluble, stable at storage condition and odorless pro-aroma molecules which can break down into the desired aroma compounds under controlled conditions [7]. Similarly, steviol glycosides are the diterpene glycosides isolated from the leaves of the stevia plant (*Stevia rebaudiana* Bertoni). These glycosides are the sweeteners and their sweetness range from 40 to 450 times sweeter than sucrose whereas the sweetness intensity and quality mainly depend on the glycosylation pattern of the diterpene aglycone [8]. Whereas the triterpene glycosides such as ginsenosides and oleanolic acid have attracted much attention because they exhibit several biological activities such as anti-viral, anti-inflammatory and neuroprotective effects [9].

7.2 Synthesis of Glycosides

Glycosides are synthesized chemically by different methods that are based on the activation of a leaving group present in the anomeric carbon center. Among these procedures, the classical Koenigs–Knorr reaction has some limitations such as

* Department of Mathematics and Natural Sciences, College of Sciences and Human Studies, Prince Mohammad Bin Fahd University, Al Khobar, Kingdom of Saudi Arabia. Corresponding author e-mail address: bimalbanik10@gmail.com.

† Roland Institute of Pharmaceutical Sciences, Berhampur affiliated to Biju Patnaik University of Technology (BPUT), Rourkela, Odisha, India. Corresponding author e-mail address: drbiswamohansahoo@gmail.com.

‡ Faculty of Pharmacy, Pharmacy Academy, IFTM University, Lodhipur Rajput, Moradabad-244102, Uttar Pradesh, India.

DOI: 10.1201/9781003008682-7

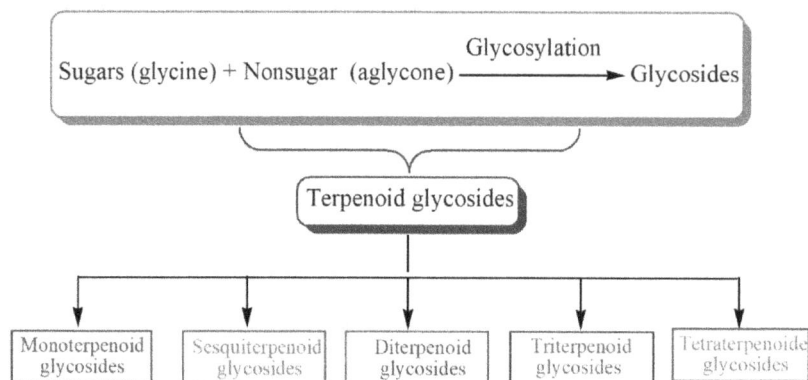

FIGURE 7.1   Types of terpene glycosides.

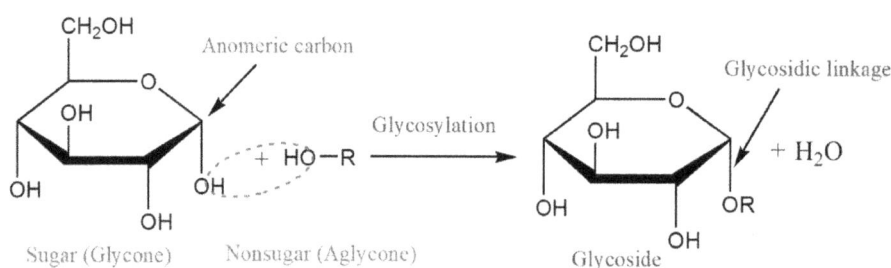

FIGURE 7.2   Synthesis of glycosides via glycosylation.

low yields, poor selectivity, necessity of protecting groups and laborious purification procedures [10]. The selection of chemical glycosylation or enzymatic glycosylation is based on several factors such as (1) conditions of the chemical reaction that may decompose the labile aglycone or unwanted anomers can be produced (2) the usage of heavy metal catalysts is undesirable because of their high price or toxicity (3) glycosylation by enzymes can help to overcome the sterical hindrance because unprotected, less bulky glycosyl residues are employed (4) the enzymatic system can glycosylate highly water soluble substrates (5) expensive aglycones can be quantitatively recovered if an enzymatic reaction is used (Figure 7.2) [11].

The enzymatic glycosylation of terpenoids is more useful due to the regio- and stereoselectivity of the biocatalyst as well as the mildness of the reaction conditions as compared to the chemical methods. Several types of biocatalysts utilized for the enzymatic glycosylation of terpenoids include glycosyltransferases (GTs), glycosidases, transglycosidases and glycoside phosphorylases [12]. The glycosyltransferases are used to differentiate enantiomers as a means of chiral separation. Naturally occurring glycosides are produced by the action of GTs. Kinetic resolution method is used for the separation of a racemic mixture of monoterpenols. Transglycosidases and glycoside phosphorylases have limited acceptor substrate spectrum and display rather low regiospecificity if the acceptor molecule carries several hydroxyl groups but can use low-cost donor sugar substrates that achieve moderate to good yields [13]. In a biphasic aqueous-organic system, the transdiglycosidase α-rhamnosyl-β-glucosidase from *Acremonium* sp. DSM 24697 effectively transglycosylates the rutinose moiety from

the flavonoid hesperidin to 2-phenylethanol, or the terpenoids geraniol and nerol. Whereas Glycosidases tolerate a very wide range of acceptor substrates but their use is impeded by poor yields and poor regioselectivity. Though glycosyltransferases are able to glycosylate a multitude of acceptor molecules with high regio- and enantioselectivity but the high expense of the glycosyl donors like UDP-glucose limits their application in the food and pharmaceutical industry [14].

Glycosyltransferases are the universal group of transferases that catalyze the relocation of a sugar moiety from an activated nucleotide diphosphate sugar donor, usually UDP-glucose to acceptor molecules. The first monoterpene-specific GTs are reported from grapevine (*Vitis vinifera*) and kiwi (*Actinidia delicosa*). GTs can be categorized into two types such as retaining GTs that transfer sugar residues with retention of the anomeric configuration at C1 and the inverting GTs that relocate sugar moieties with inversion of the anomeric configuration. It has been reported that 27 GTs from Arabidopsis glycosylate a diversity of monoterpenes (geraniol, linalool, terpineol and citronellol), sesquiterpenes and diterpenes etc [15]. The *in vivo* glycosylation of terpene alcohols is catalyzed by UDP-glucose-dependent GTs. It is a straightforward approach for the production of monoterpene glycosides (Figure 7.3) [16,17].

Recently, research groups are using recombinant GT proteins encoded by genes from *Eucalyptus perriniana*, *Rauvolfia serpentine*, *Sorghum bicolor* and *M. truncatula* that are multisubstrate enzymes to glucosylate monoterpenols. Similarly, Iridoids are cyclic monoterpenes derived from 10-oxogeranial. UGT85A24 from *Gardenia jasminoides* and UGT8 from

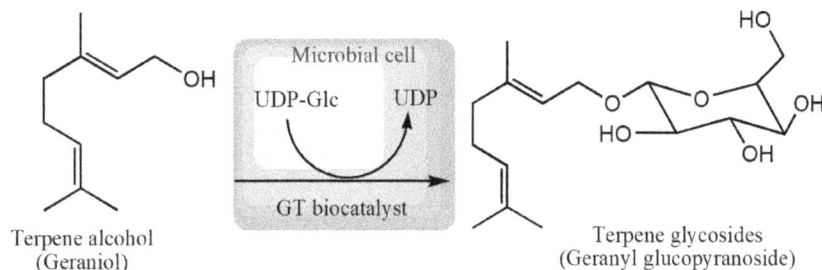

FIGURE 7.3 Biotransformation of monoterpene glycosides via GT biocatalyst.

TABLE 7.1

List of Enzymes Used for Production of Terpene Glycoside

Name of Enzymes	Terpene Glycosides
Almond β-glucosidase	Geranyl β-glucoside
Pichia etchellsii β-glucosidase	Geranyl, neryl, citronellyl β-glucoside
Trichoderma citrinoviride β-glucosidase	Geranyl β-glucoside
Hydrolase of *Gibberella fujikuroi*	Rebaudioside A
Bovine colostrum β-galactosyltransferase	Asiaticoside β-galactosides
Bacillus firmus β-cyclodextrin glucanotransferase	Stevioside α-glucosides
Bovine colostrum β-galactosyltransferase	Stevioside β-galactosides

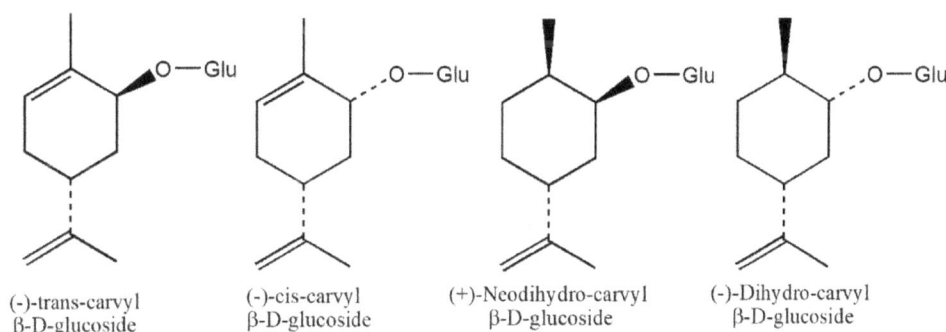

FIGURE 7.4 Structures of monoterpeglycosides.

Catharanthus roseus are found to catalyze the glucosylation of the iridoids 7-deoxyloganin and 7-deoxyloganetic acid respectively. Further, the glycosylation of the apocarotenoid crocetin in fully developed stigma of *Crocus sativus* is catalyzed by UGTCs2 (Table 7.1) [18–21].

7.3 Monoterpenoid Glycosides

Monoterpenoids such as linalool, citronellol and geraniol are generally identified as powerful aromatic compounds present in flowers, leaves and fruits of the plant sources. During the 1980s, researchers demonstrated that most of the monoterpenes present in the plants are chemically bound to carbohydrates in the form of *O*-β-D-glucopyranosides, 6-*O*-α-L-rhamnopyranosyl-β-D-glucopyranosides, 6-*O*-α-L-arabinofuranosyl-β-D-glucopyranosides and 6-*O*-β-D-apiofuranosyl-β-Dglucopyranosides [21]. Dihydro-carvyltetraacetyl-β-D-glucoside is present in Spearmint

(*Mentha species*). Other monoterpene β-D-glucosides include (+)-neodihydrocarvyl β-D glucoside, (−) dihydrocarvyl β-D glucoside, (−)-trans-carvyl β-D-glucoside and (−)cis-carvyl β-D-glucoside. Various monoterpene alcohols that serve as aglycones of Mentha glucosides include neodihydro-carveol, dihydro-carveol, trans-carveol, cis-carveol (Figure 7.4) [22].

Sakata et al. reported the synthesis and properties of menthyl glycosides. The glycoside is produced via a condensation reaction of menthol with acetobromoglucose based on the Koenig-Knorr method. l-Menthol undergoes condensation with acetobromo glucose via Koenig-Knorr reaction to afford l-menthyl-2,3,4,6-tetra-O-acetyl-β-D-glucopyranoside which on deacetylation produces 1-menthyl-β-D-glucopyranoside with 46% yield. The water solubility of menthyl monosaccharide is poor whereas the menthyl disaccharide is found to have good water solubility. So it is useful in the preparation of food additives and medicines like uroterpenoids. Further, these glycosides are preferably utilized for optical resolution of dl-menthol because of their high crystallinity (Figure 7.5) [23].

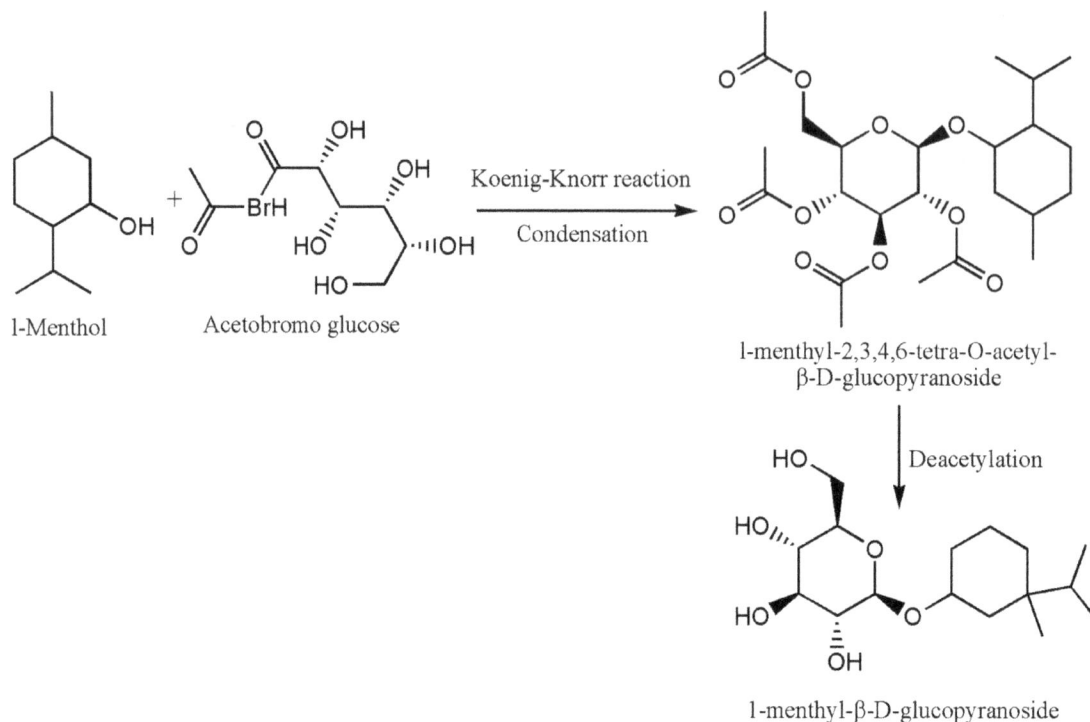

FIGURE 7.5 Synthesis of menthyl glycosides.

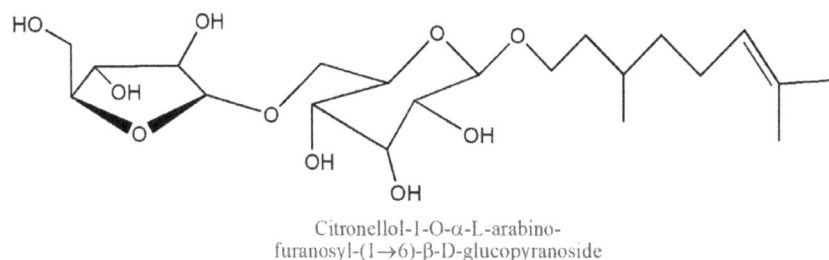

Citronellol-1-O-α-L-arabino-
furanosyl-(1→6)-β-D-glucopyranoside

FIGURE 7.6 Structure of monoterpene glycosides from the roots of *Sanguisorba officinalis* L.

Wang et al. demonstrated the isolation of monoterpene glycosides from the roots of *Sanguisorba officinalis* L. and evaluated their hemostatic activities. Column chromatography and preparative HPLC were used for the isolation of terpene glycosides. Based on chemical and spectral studies, the structures of isolated compounds were identified as citronellol-1-*O*-α-L-arabinofuranosyl-(1→6)-β-D-glucopyranoside, geraniol-1-*O*-α-L-arabinofuranosyl-(1→6)-β-D-glucopyranoside, geraniol-1-*O*-α-Larabinopyranosyl-(1→6)-β-D-glucopyranoside, 3β-[(α-L-arabinopyranosyl)oxy]-19α-hydroxyolean-12-en-28-oic acid 28-β-D-glucopyranoside. The hemostasis assay was carried out by using a Goat Anti-Human α2-plasmin inhibitor kit. It utilizes a solid phase sandwich enzyme-linked quantitative immunoabsorbent assay (ELISA) with a purified antibody specific for α2-plasmin inhibitors (Figure 7.6) [24].

The most common type of glycosides present in grapes is produced by the combination of a terpene molecule with a glucose molecule or disaccharide. The glycosylation of free forms of terpenes afford compounds that are highly soluble in an aqueous medium. The most abundant molecules in glycosidic combinations include rutinose (α-1-rhamnopyranosyl-(1→6)-d-glucopyranose), 6-*O*-α-1-arabinofuranosyl-β-d-glucopyranose, 6-*O*-α-d-apiofuranosyl-β-d-glucopyranose. These disaccharides produce rutinosides, arabinosyl glucopyranosides and apiosyl glucopyranosides, respectively, on combining with monoterpene aglycones such as linalool, geraniol, farnesol and nerol (Figure 7.7) [25].

7.4 Sesquiterpenoid Glycosides

Brasilane is an oxygenated sesquiterpenoid glycoside with the molecular formula $C_{23}H_{39}NO6$ (Figure 7.8). Chemically, it is *N*-[(2*S*, 3*R*, 4*R*, 5*S*, 6*R*)-4,5-dihydroxy-6-(hydroxymethyl)-2-[(2*E*)-2-(1,6,6-trimethyl-2,3,3*a*,   5,7,7*a*-hexahydro-1*H*-inden-4-ylidene)propoxy]oxan-3-yl]acetamide. It has been isolated from several sources such as basidiomycetes, ascomycetes, algae and sea hare [26].

α-L-Rhamnopyranosyl-(1→6)-β-D-glucopyranoside

α-D-Arabinofuranosyl-(1→6)-β-D-glucopyranoside

R groups

Geranyl　　　　Neryl　　　　Linalyl　　α-Terpineol

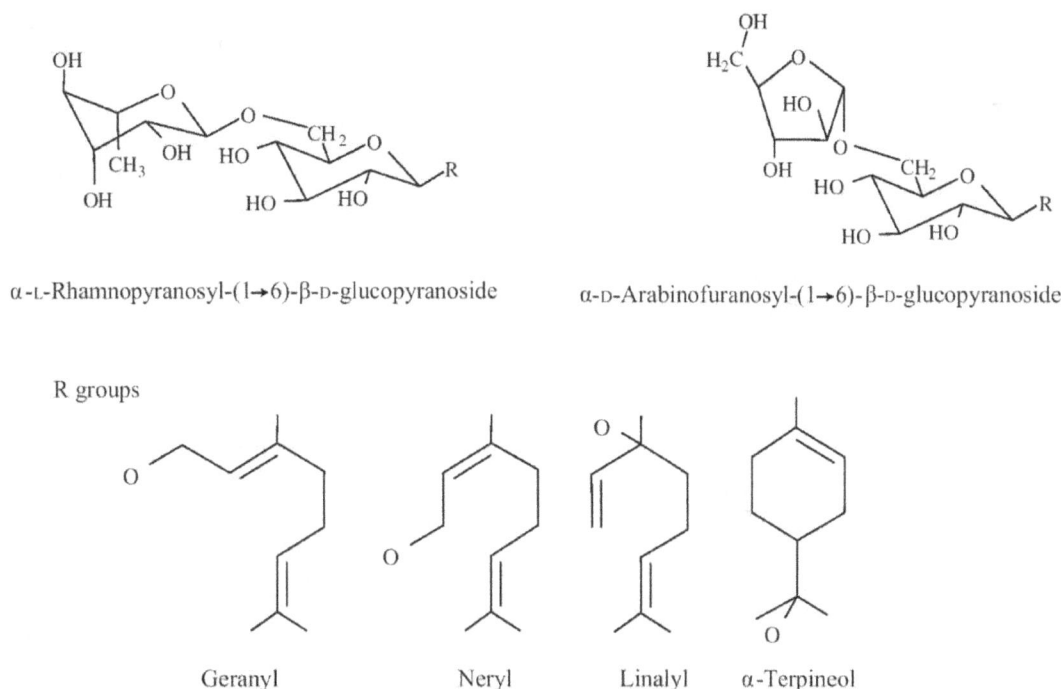

FIGURE 7.7 Structure of terpene glycosides in grapes.

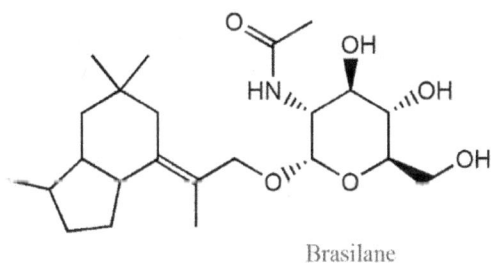

Brasilane

FIGURE 7.8 Structure of brasilane.

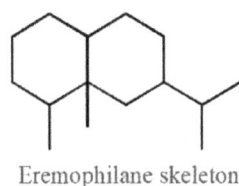

Eremophilane skeleton

FIGURE 7.9 Structure of eremophilane-type glycoside.

Eremophilane-type glycosides belong to sesquiterpenoids with a C15 skeleton system. 8-[(β-Dglucopyranosyl)oxy]eremophila-1(10), 8,11-trien-2-one is identified as first eremophilane-type glycoside (Figure 7.9). It is isolated from the extract of the whole plant of *Ligularia virgaurea*. Other eremophilane-type glycosides include fukinoside A, culcitiolide J, petasitosides A-C, Alticolosides A-I. The structures of eremophilane-type glycoside are elucidated by extensive analysis of their ^{1}H-NMR, ^{13}C-NMR, 2D NMR and HR ESI mass spectra. Further, the application of CD spectra and X-ray analysis are very useful to investigate the stereostructures of these natural products [27].

The isolated eremophilane-type glycosides demonstrate their significant application as chemotaxonomic markers. It is reported that these compounds exhibit moderate cytotoxicity against C4-2B and A375 human cancer cell lines. They also exhibit other activities like anti-inflammatory and anti-allergic (Table 7.2) [28].

Ptaquiloside is a norsesquiterpene glycoside isolated from spores, roots and rhizomes of different sources of ferns such as Bracken ferns (*Pteridium* sp.), Lip ferns (*Cheilanthes* sp.) and Brake ferns (*Pteris* sp.). Among these ferns, Bracken ferns are used as traditional medicine. It is found to have carcinogenic and clastogenic properties (Figure 7.10) [29].

Shi et al. isolated three new sesquiterpene glycosides (codonopsesquilosides A-C) from the aqueous extract of the dried roots of *Codonopsis pilosula* (Figure 7.11). These glycosides are categorized into three types such as C15 carotenoid, gymnomitrane and eudesmane. The structure and absolute configurations of these glycosides are confirmed by spectral studies using X-ray crystallographic analysis, FT-IR, ^{1}H-NMR, ^{13}C-NMR, 2D-NMR, LC-MS and CD spectra respectively [30].

Zhang et al. demonstrated the isolation of six new sesquiterpene glycosides from the whole plant of *Sarcandra glabra*. In these glycosides, sesquiterpene aglycone (eudesmanolide, elemanolide, lindenane and germacranolide) are present along with one known compound (chloranoside A) (Figure 7.12). The structures of these glycosides are elucidated by chemical and spectroscopic methods including COSY, HMQC, HMBC, NMR, etc. The new sesquiterpene glycoside has been represented as 1β, 5α, 8βH-eudesman-4(15), 7(11)-dien-8α, 12-olide-1-O-β-D-glucopyranoside and

TABLE 7.2

List of Eremophilane-Type Glycosides

Eremophilane-Type Glycosides	Biological Source	Structures
Fukinoside A	*Petasites japonicus*	R= 6'-OSO₃Glc
Culcitiolide J	*Senecio culcitioides*	
Petasitosides A	*Petasites japonicas*	
Alticolosides A	*L. alticola*	

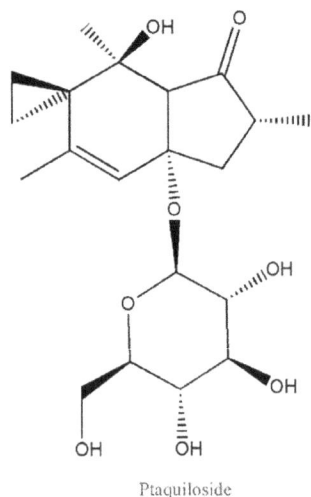

Ptaquiloside

FIGURE 7.10 Structure of ptaquiloside.

its trivial name is sarcaglaboside A. These compounds are evaluated for their hepatoprotective activity against D-galactosamine-induced toxicity in WB-F344 rat hepatic epithelial stem-like cells [31].

Picrotoxane sesquiterpene lactone glycosides (nepalactones A and B) are isolated from the root barks of the plant Coriaria nepalensis (Figure 7.13). Their structures are elucidated based

on HRESIMS, 1D, 2D NMR and ¹³C-NMR spectral data. These glycosides are subjected for evaluation to determine their activity to induce neurite outgrowth in rat pheochromocytoma (PC12) cells. From the evaluation result, it is reported that nepalactone A significantly enhance the nerve growth factor (NGF)-mediated neurite outgrowth in PC12 cells. This result suggested that the attachment of the sugar residue with the picrotoxane skeleton might reduce the neurotrophic activity [32].

7.5 Diterpenoid Glycosides

Fuscosides A–D are the diterpenoid glycosides of new structural classes from the Caribbean gorgonian *Eunicea fusca*. The first glycosylation of fuscol and eunicol has been achieved by using a modified Koenigs–Knorr glycosylation to synthesize new fuscosides and eunicosides. *E. fusca* is the biological source of aglycone precursor fuscol and the biosynthetically related dilophol diterpene eunicol. Chemically fuscol is (3*E*, 5*E*)-6-[(1*S*, 3*R*, 4*R*)-4-ethenyl-4-methyl-3-prop-1-en-2-ylcyclohexyl]-2-methylhepta-3,5-dien-2-ol. Whereas eunicol is (3*E*, 5*E*)-6-[(1*S*, 3*E*, 7*E*)-4,8-dimethylcyclodeca-3,7-dien-1-yl]-2-methylhepta-3,5-dien-2-ol (Figure 7.14). Fuscosides A and B are found to be effective anti-inflammatory agents by reducing phorbol myristate acetate (PMA)-induced edema in a mouse-ear model with potencies equivalent to indomethacin [33,34].

Several diterpene glycosides are isolated from the extract of the leaves of *Stevia rebaudiana*. These glycosides are known as steviol glycosides including stevioside, rebaudiosides A-F, rubusoside and dulcoside A (Figure 7.15). The structures of these glycosides are identified based on spectroscopic (¹H-NMR, ¹³C-NMR and MS) and chemical studies [35].

Nohara et al. reported the new acyclic diterpene glycosides (capsianosides) isolated from the methanolic extracts of aerial parts and fruits of paprika (*Capsicum annuum* L.) and pimiento (*Capsicum annuum* L). Both plants are used as vegetables and spices. The structures of the isolated compounds are elucidated by 2D- NMR, ¹H-NMR and ¹³C-NMR spectral methods. The position of olefinic methylene is determined by HMBC. The structure of capsianoside XIV is identified as 17-*O*-α-L-rhamnopyranosyl-(1→4)-[α-L-rhamnopyranosyl-(1→6)]-β-D-glucopyranoside-6*E*, 10*E*, 14*Z*-(3*S*)-19-hydroxygeranyl linalool as presented in Figure 7.16 [36].

Masateru et al. demonstrated the isolation of new labdane-type diterpene glycoside (viteoside A) from the methanolic extract of fruit of *Vitex rotundifolia* (Figure 7.17). It is used as a folk medicine for the treatment of headache. The structure was characterized by FT-IR, ¹H-NMR, ¹³C-NMR, LC-MS spectra as (8R, 10S)-6β-D-acetoxy-3β, 9α-dihydroxy-labda-13Z-en-16,15-olide-3-O-β-D glucopyranoside [37].

Chen et al. demonstrated the isolation of diterpene glycosides (Brevistylumsides A and B) from ethanolic extract of *Illicium brevistylum* of family Illiciaceae (Figure 7.18). The structures of the newly isolated compounds were confirmed on the basis of spectroscopic analysis using IR, HR-ESI-MS, ¹H-NMR, ¹³C-NMR, 2D-NMR, DEPT, COSY, HMBC and HSQC data. Traditionally these compounds are used for the treatment of rheumatoid arthritis and traumatic injuries [38].

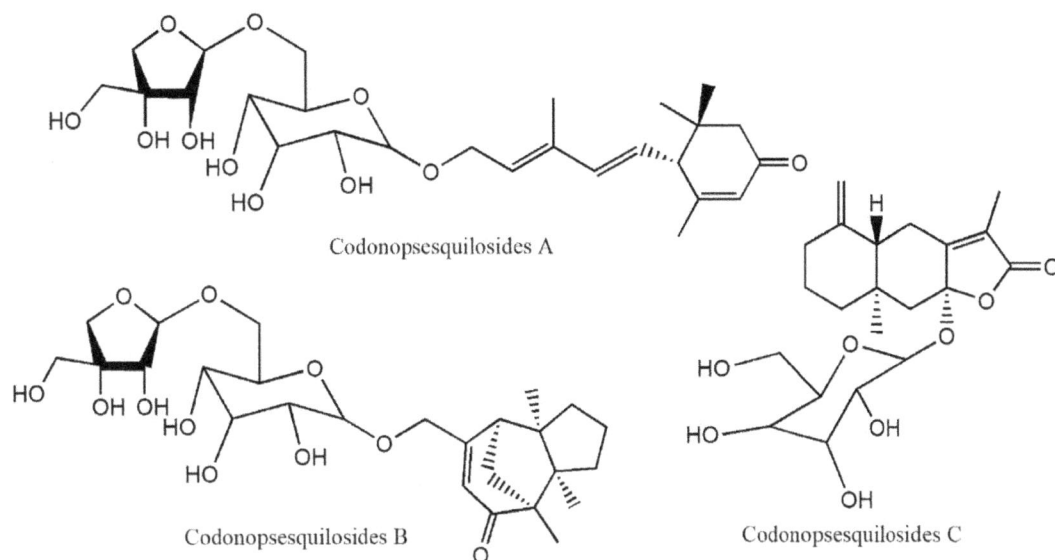

FIGURE 7.11 Structures of codonopsesquilosides A–C.

Codonopsesquilosides A

Codonopsesquilosides B

Codonopsesquilosides C

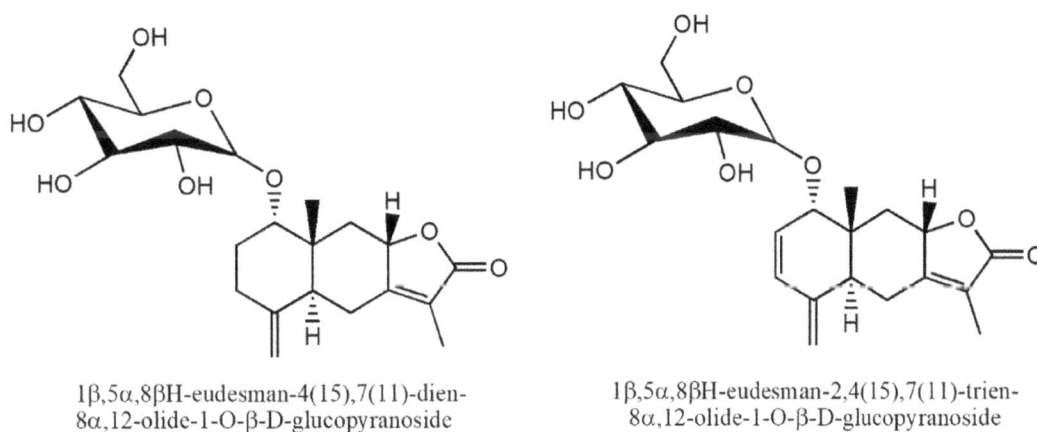

1β,5α,8βH-eudesman-4(15),7(11)-dien-
8α,12-olide-1-O-β-D-glucopyranoside

1β,5α,8βH-eudesman-2,4(15),7(11)-trien-
8α,12-olide-1-O-β-D-glucopyranoside

FIGURE 7.12 Structures of sesquiterpene glycosides.

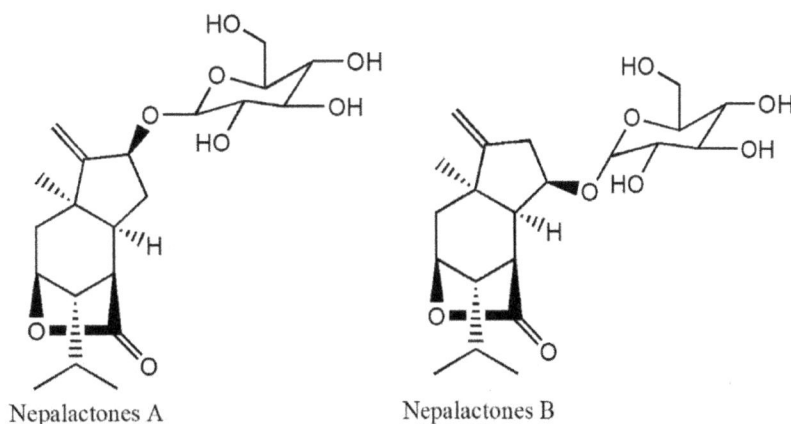

Nepalactones A

Nepalactones B

FIGURE 7.13 Chemical structures of nepalactones A and B.

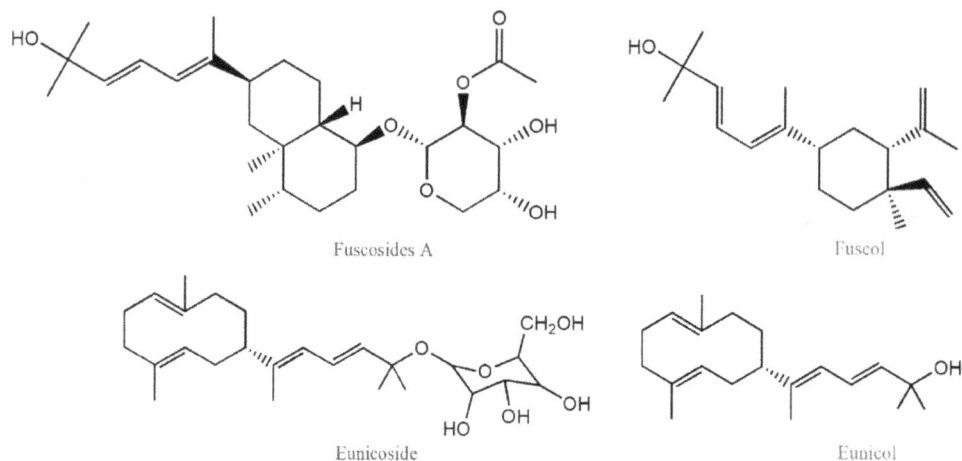

FIGURE 7.14 Chemical structures of Fuscosides A and eunicoside.

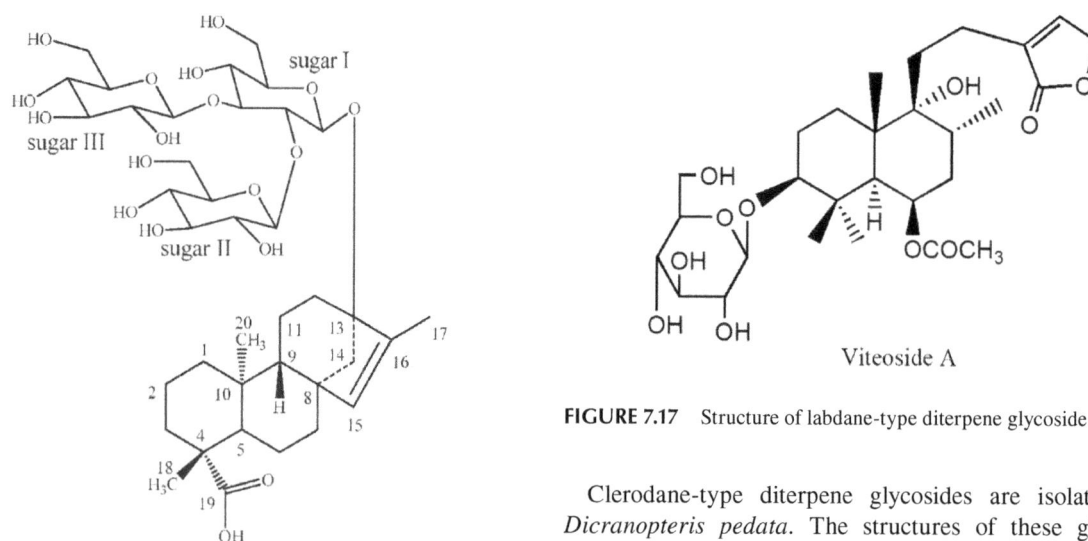

FIGURE 7.15 Structure of 13-[(2-*O*-β-D-glucopyranosyl-3-*O*-β-D-glucopyranosyl-β-D-glucopyranosyl)oxy]*ent*-kaur-15-en-19-oic acid.

FIGURE 7.16 Structure of capsianoside XIV.

FIGURE 7.17 Structure of labdane-type diterpene glycoside.

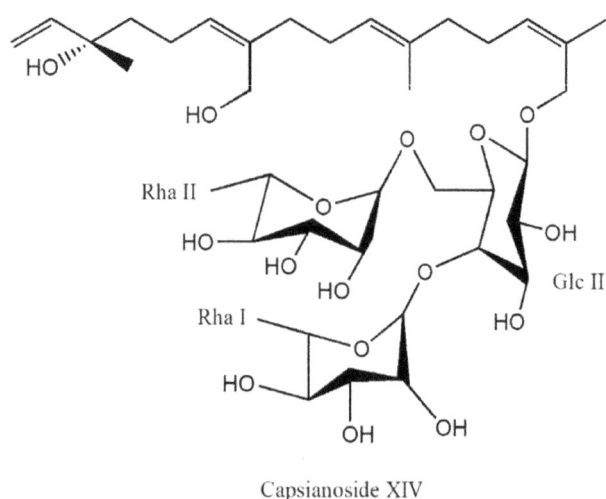

Clerodane-type diterpene glycosides are isolated from *Dicranopteris pedata*. The structures of these glycosides are elucidated by spectroscopic studies and the absolute confguration is determined by ECD calculations. From this analytical reports, clerodane-type diterpene glycosides are identified as (5*R*, 6*S*, 8*R*, 9*S*, 10*R*)-6-*O*-[β-D-glucopyranosyl-(1→4)-α-l-rhamnopyranosyl]cleroda-3,13(16), 14-diene, (5*R*, 6*S*, 8*R*, 9*S*, 10*R*, 13*S*)-6-*O*-[β-D-glucopyranosyl-(1→4)-α-l-rhamnopyranosyl]-2-oxoneocleroda-3,13-dien-15-ol, (5*R*, 6*S*, 8*R*, 9*S*, 10*R*)-6-*O*-[β-D-glucopyranosyl-(1→4)-α-l-rhamnopyranosyl]-(13*E*)-2-oxoneocleroda-3,14-dien-13-ol (Figure 7.19). These glycosides exhibit inhibitory activities against liver cancer (SMMC-7721), breast cancer (MCF-7) and colon cancer (SW480) cell lines [39].

7.6 Sesterpene Glycosides

Manoalide is a sesterterpenoid glycoside with the molecular formula $C_{25}H_{36}O_5$ (Figure 7.20). It resembles with that of the eremophilane type of sesquiterpenoids. It was first isolated from a marine sponge (*Luffariella variabilis*) in 1980. Chemically, it is (2*R*)-2-hydroxy-3-[(2*R*, 6*R*)-6-hydroxy-5-[(*E*)-4-methyl-6-(2,6,6-trimethylcyclohexen-1-yl)hex-3-enyl]-3,6-dihydro-2*H*-pyran-2-yl]-2*H*-furan-5-one. It is a potent

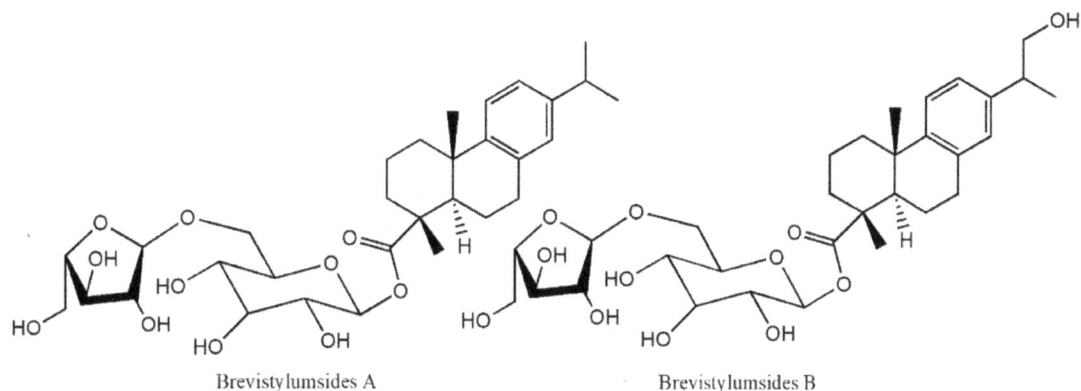

FIGURE 7.18 Structure of brevistylumsides A and B.

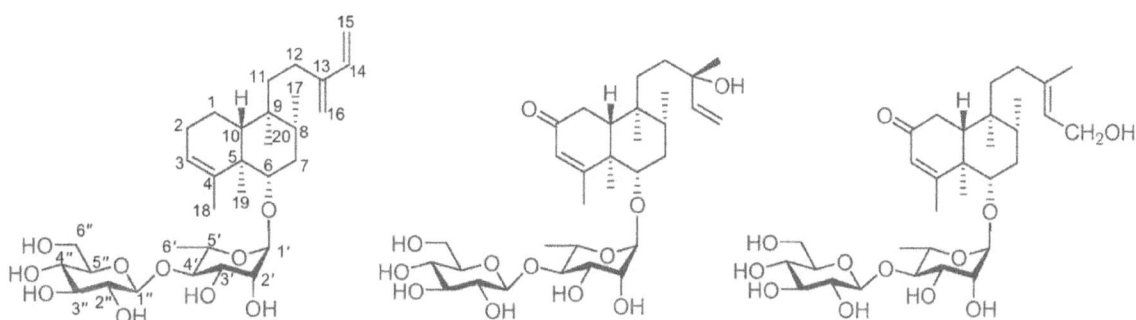

FIGURE 7.19 Structures of clerodane-type diterpene glycosides.

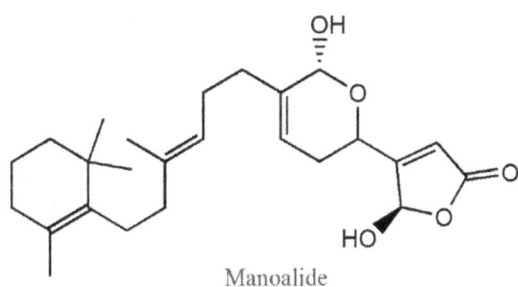

Manoalide

FIGURE 7.20 Structure of manoalide.

analgesic and anti-inflammatory agent. The anti-inflammatory activity of manoalide is due to the inhibition of Phospholipase A2 (PLA2) via irreversible binding to several lysine residues. Further, Manoalide induces the activation of caspase 3 which is a hallmark of apoptosis in oral cancer cells [40].

7.7 Triterpenoid Glycosides

Triterpene glycosides are isolated from Black cohosh (*Actaea racemose* L.). Several triterpene glycosides include actein, 23-epi-26 deoxyactein, cimicifugoside, cimigoside, cimifugine, racmoside, cimiracemoside A etc. These are used as medicinal agents to relieve the symptoms of gynecologic

problems. Triterpene glycosides from black cohosh have been shown to inhibit growth of MCF7 and MDA-MB-453 breast cancer and induce cell cycle arrest at the G1 phase [41]. Actein is a triterpenoid glycoside found in *Actaea elata* and *Actaea cimicifuga* (Figure 7.21). It has been observed to inhibit the proliferation of human breast cancer cells. It induces the increase in caspase 3 activity [42].

Cimicifugoside is a triterpenoid glycoside obtained from the rhizomes of Cimicifuga simplex with molecular formula $C_{37}H_{54}O_{11}$ (Figure 7.22). It is a nucleoside transport inhibitor. It also exhibits immunosuppressive activity [43].

Cimigoside is the active ingredient of ***Cimicifuga racemosa* of family** Ranunculaceae with molecular formula $C_{37}H_{56}O_{11}$ (Figure 7.23). It has anti-cholesterol activity, hypotensive and causes peripheral vasodilation [44].

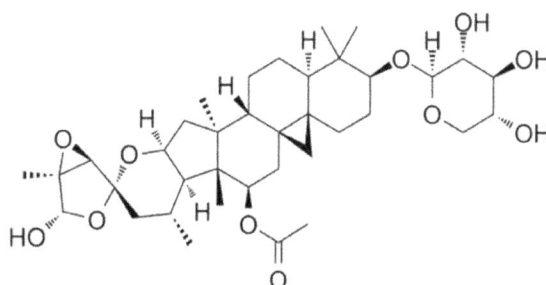

FIGURE 7.21 Structure of triterpene glycoside (actein).

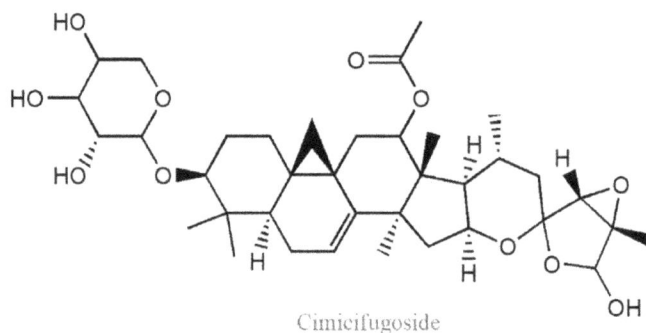

FIGURE 7.22   Structure of cimicifugoside.

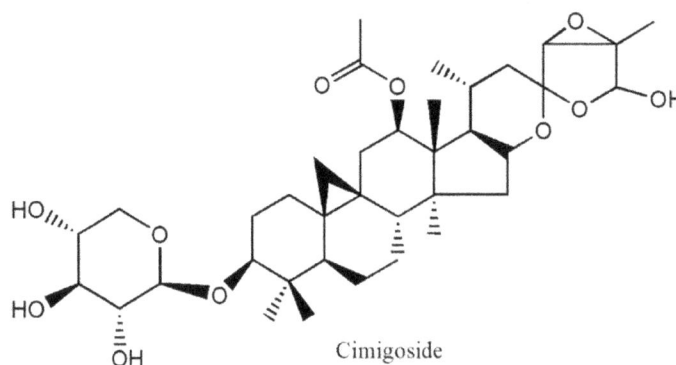

FIGURE 7.23   Structure of cimigoside.

Cimifugine is a natural product present in Eranthis cilicica, Ostericum grosseserratum with molecular formula $C_{16}H_{18}O6$ (Figure 7.24). Chemically, it is (2S)-7-(hydroxymethyl)-2-(2-hydroxypropan-2-yl)-4-methoxy-2,3-dihydrofuro[3,2-g]chromen-5-one. It is indicated for the treatment of pains associated with rheumatism, uterine cramps and migraine [45].

Cimiracemoside A is a natural product present in Actaea racemosa with molecular formula $C_{37}H_{56}O_{11}$ (Figure 7.25). Similarly, Cimiracemoside C is an active component of *Cimicifuga racemosa* with molecular formula $C_{35}H_{56}O_9$. It activates the AMPK pathway and also exhibits potential activity against diabetes [46].

Kurihara et al. reported the isolation of five new oleanane-type triterpene glycosides (strogins I–V) from the leaves of *Staurogyne merguensis* Kuntze. Strogins have sweet taste as presented in Figure 7.26 [47,48].

Ginsenoside is a triterpene glycoside with the molecular formula $C_{30}H_{52}O_2$ (Figure 7.27). It is the major active component

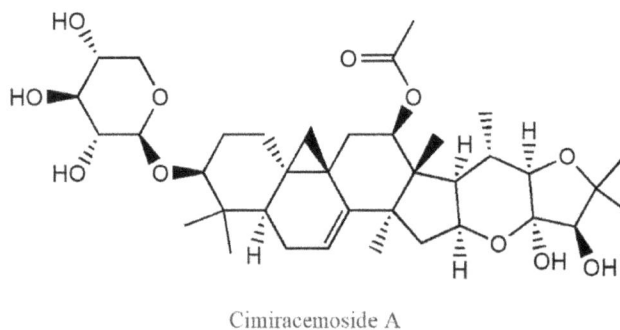

FIGURE 7.24   Structure of cimifugine.

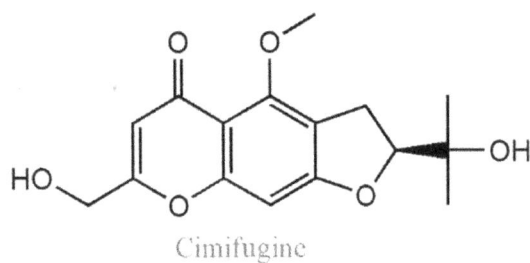

FIGURE 7.25   Structure of cimiracemoside A.

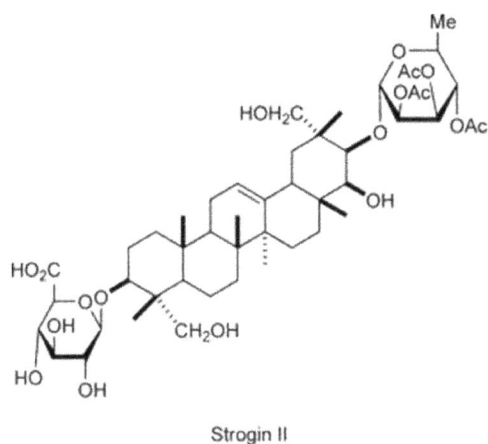

FIGURE 7.26 Structure of oleanane-type triterpene glycosides (strogins II).

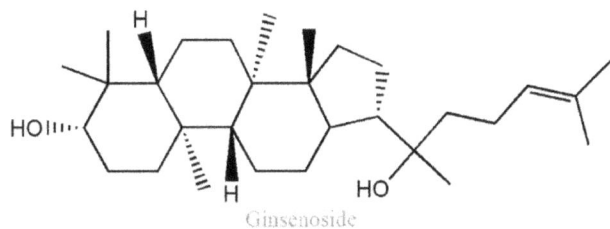

FIGURE 7.27 Structure of ginsenoside.

isolated from the root of *Panax ginseng*. It has therapeutic potential for protecting the cardiovascular system, central nervous system, endocrine and immune systems as presented in Figure 7.28 [49,50].

Khan et al. reported the isolation of ammarane-type triterpene glycoside (polysciasoside A) from the leaves of *Polyscias fulva* of family Araliaceae (Figure 7.29). The structure of the isolated compound was identified as 12-oxo-3β, 16β, 20(S)-trihydroxydammar-24-ene-3-O-α-L-rhamnopyranosyl-(1→2)-β-D-glucopyranoside as per spectral studies (FT-IR,^{1}H-NMR,^{13}C-NMR, HRESIMS, COSY, TOCSY, HMBC and HMQC). Traditionally, this glycoside is used in medicine for its wound-healing properties [51].

Orazio et al. reported the new ursane-type triterpene glycoside (Isomadecassoside) with molecular formula $C_{48}H_{78}O_{20}$ (Figure 7.30). It is isolated from the leaves of *Centella asiatica* of family Apiaceae. The structure of isomadecassoside is confirmed by analytical studies using HR-ESIMS, 1D- and 2D NMR spectra. Evaluation results revealed that isomadecassoside inhibit the nitrite production in LPS-stimulated macrophages at non-cytotoxic concentrations that indicate the anti-inflammatory effect of isomadecassoside [52].

Yoshikawa et al reported the isolation of five triterpene oligoglycosides (escins-Ia, Ib, IIa, IIb and IIIa) from the seeds of *Aesculus hippocastanum* L (Figure 7.31). The chemical structures of these compounds are determined on the basis of their chemical and physicochemical profile. It was evaluated to

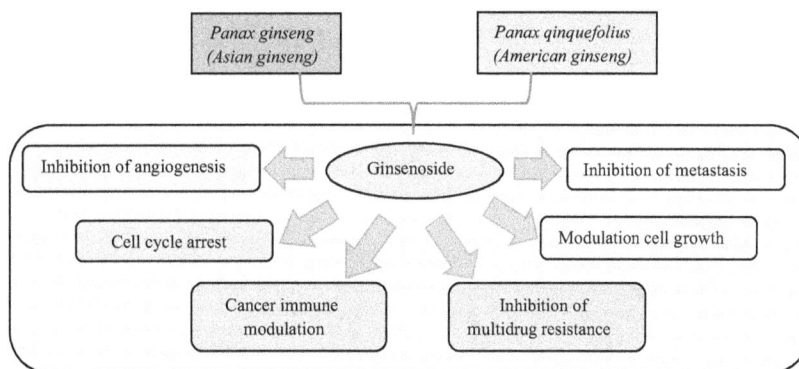

FIGURE 7.28 Mode of action of ginsenoside.

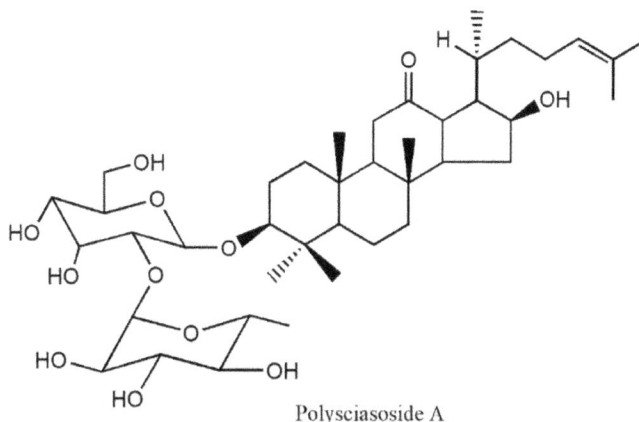

FIGURE 7.29 Structure of ammarane-type triterpene glycoside (polysciasoside A).

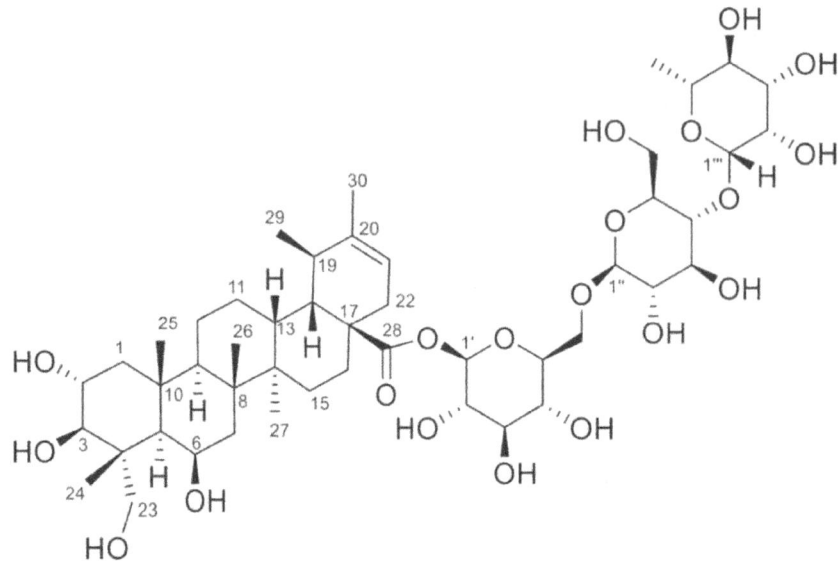

FIGURE 7.30 Structure of ursane-type triterpene glycoside (isomadecassoside).

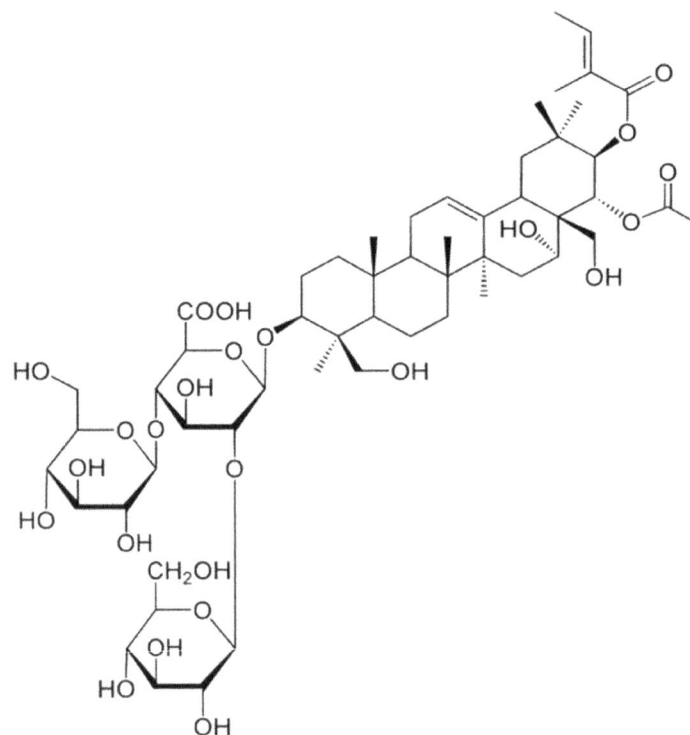

FIGURE 7.31 Structure of triterpene oligoglycosides.

determine the anti-inflammatory effect of escins on inflammation induced by carrageenan in rats. The oral glucose tolerance test was also performed to monitor the inhibitory effects of escins on the elevation of blood glucose levels in mice. Among tested compounds, escins-IIa and IIb displayed better activities for both bioassays [53].

Echinoside A is a triterpene glycoside isolated from sea cucumbers (*Actinopyga echinites*) with molecular formula

$C_{54}H_{88}NaO_{26}S$ (Figure 7.32). Chemically it is a sulfonylated holostane tetrasaccharide. It exhibits potential anticancer and antifungal activities. Evaluation studies suggested that it might inhibit the binding of topoisomerase 2α to DNA so as to induce apoptosis of cancer cells [54].

Cyclosophoside A is a cycloartane triterpene glycoside. It is isolated from the seeds of *Cassia sophera* L (Figure 7.33). It responds positively to the Liebermann–Burchard and Molish

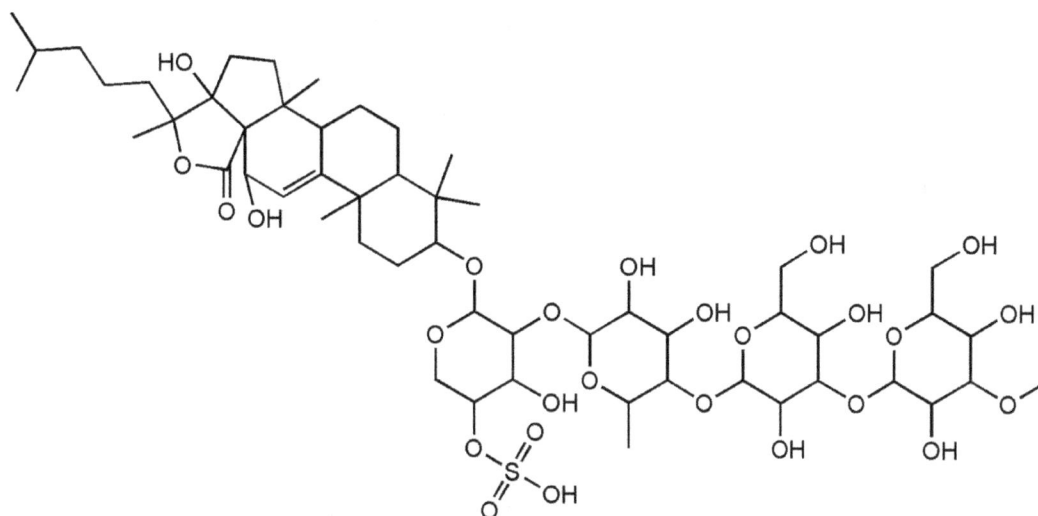

FIGURE 7.32 Structure of echinoside A.

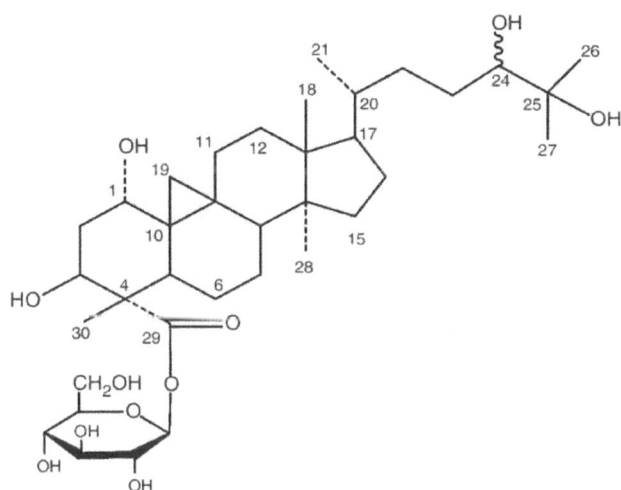

FIGURE 7.33 Structure of cyclosophoside A.

reactions. The structure of this glycoside is determined as 29-*O*-*β*-D-glucopyranoside of 1*α*, 3*β*, 24*ξ*, 25-tetrahydroxy-9, 19-cycloartan-29-oic acid based on physicochemical properties and spectroscopic methods (^{1}H-NMR,^{13}C-NMR, 1D, 2D NMR, LC-MS, DEPT, HMBC). It was demonstrated that these glycosides of C. sophera L. display therapeutic effectiveness for the treatment of diabetes [55].

Patagonicoside A is a disulfated triterpene glycoside isolated from the sea cucumber *Psolus* patagonicus (Figure 7.34). The chemical structure and purity of this isolated compound is confirmed by RP-HPLC, ^{1}H- and ^{13}-C-NMR. It is subjected to evaluation for antiproliferative, cytotoxic and hemolytic activities. It is able to suppress the growth of cancer cell lines such as Hep3B, MDA-MB231, A549 and also induce the activation of NF-κB. Further, Patagonicoside A exhibit hemolytic activity with half maximal inhibitory concentration (IC$_{50}$) value of 80 μM [56].

Holothurins are the non-sulfated triterpenoid glycosides isolated from the sea cucumber *Actinopyga agassizii* (Figure 7.35). It consists of a linear 3-O-Me-Glc-(1→3)-Glc-(1→4)-Xyl chain and branching of a linear trisaccharide at C$_2$ of the

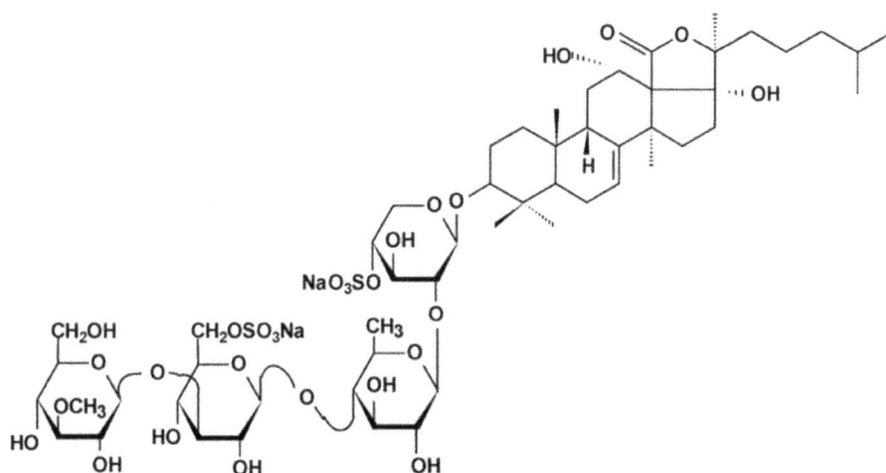

FIGURE 7.34 Structure of patagonicoside A.

xylose unit. The triterpenoid glycosides isolated from holothurians represent a sugar chain of two to six monosaccharide units linked to the C-3 of the aglycone, which is usually based on a "holostanol" skeleton [3β,

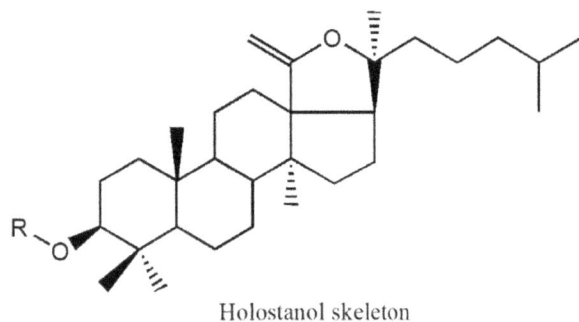

FIGURE 7.35   Structure of "holostanol" skeleton.

20S-dihydroxy-5α-lanostano-18,20-lactone]. The structure of these glycosides is determined by ^{1}H-NMR and ^{13}C-NMR and FAB-MS spectral studies [57].

Ilexdunnoside A and B are new triterpene glycosides isolated from the roots of *Ilex dunniana* Levl (Figure 7.36). The structures of these glycosides are established by ^{1}H-NMR, HRESIMS, HMBC spectral analysis. It is reported that Ilexdunnoside A and B exhibit an inhibitory effect on lipopolysaccharide (LPS) induced NO production in BV2 microglial cells in vitro with IC$_{50}$ value of 11.60 and 12.30 μM respectively [58].

Bedir et al. reported cycloartane-type triterpene glycosides that are isolated from *Astragalus aureus* (Figure 7.37). Their structures are elucidated based on 1D, 2D-NMR, ESIMS, HSQC, HMBC, COSY and HRMS analytical studies. These compounds are tested for their cytotoxic activities against several human cancer cell lines such as breast cancer (MCF7), human colon carcinoma (HT-29), human lung (A549), leukemia (HL-60), human prostate cancer (PC3) [59].

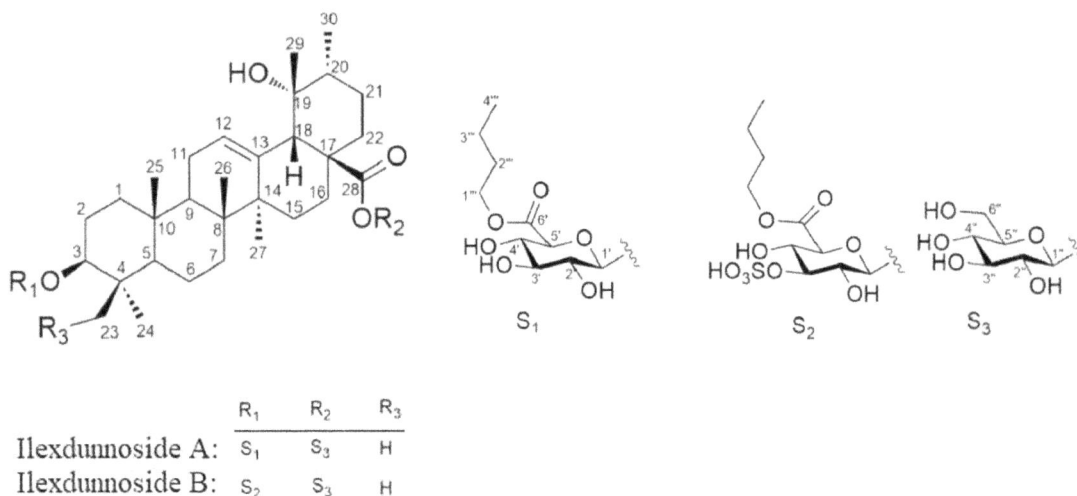

FIGURE 7.36   Structures of ilexdunnoside A and B.

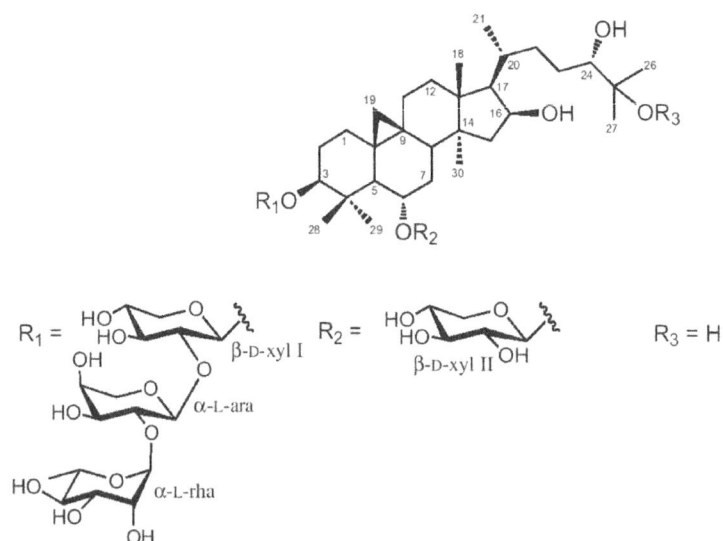

FIGURE 7.37   Structures of cycloartane-type triterpene glycosides.

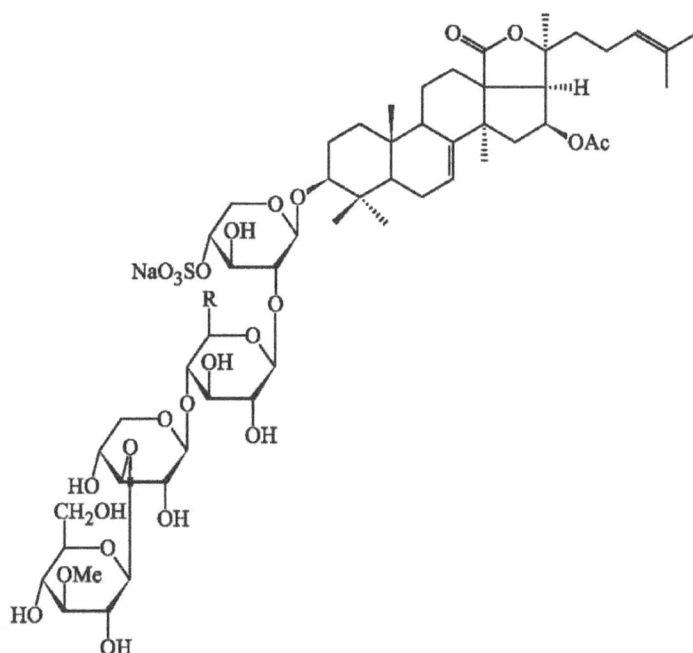

FIGURE 7.38 Structures of violaceuside A (R=CH₃) and B (R=CH₂OH).

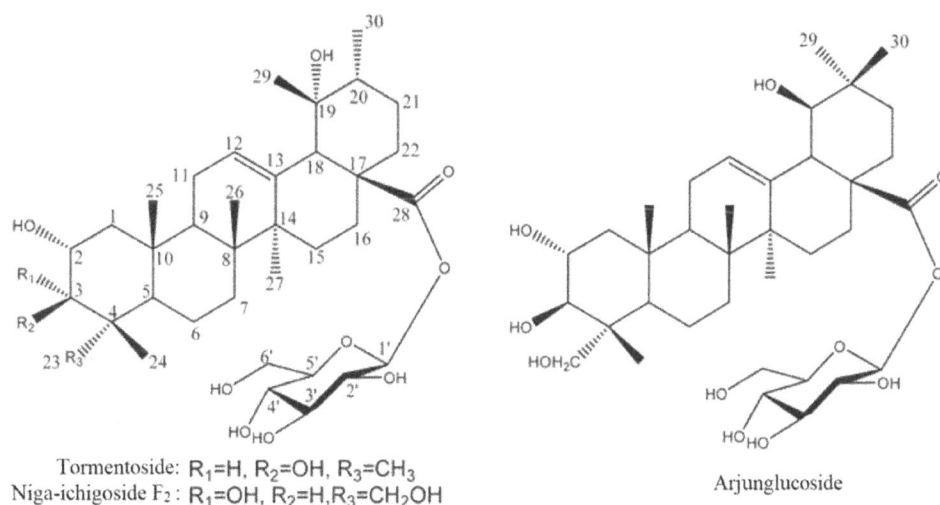

Tormentoside: R₁=H, R₂=OH, R₃=CH₃
Niga-ichigoside F₂ : R₁=OH, R₂=H,R₃=CH₂OH

Arjunglucoside

FIGURE 7.39 Chemical structure of triterpene glycosides.

Zhang et al. performed the isolation of two new triterpene glycosides (violaceusides A and B) from the sea cucumber *Pseudocolochirus violaceus* (Figure 7.38). Based on 2D NMR,^{13}C-NMR,^{1}H-NMR, ESI-MS, TOCSY, HMBC, GC-MS techniques and chemical analysis, the structures of the two new glycosides are established as 16β-acetoxy-3-O-[3-O-methyl-β-D-glucopyranosyl-(1→3)-β-D-xylopyranosyl-(1→4)-β-D-quinovopyranosyl-(1→2)-4-O-sodiumsulphate-β-D-xylopyranosyl]-holosta-7,24-diene-3β-ol and 16β-acetoxy-3-O-[3-O-methyl-β-D-glucopyranosyl-(1→3)-β-D-xylopyranosyl-(1→4)-β-D-glucopyranosyl-(1→2)-4-O-sodiumsulphate-β-D-xylopyranosyl]-holosta-7,24-diene-3β-ol

respectively. Violaceusides A and B exhibited significant cytotoxic activity against two cancer cell lines such as HL-60 (leukemia) and BEL-7402 (hepatoma) with IC$_{100}$ of 100 and 10 µM respectively. When tested by using the MTT colorimetic assay, both glycosides exhibited 84.7% and 87.9% cell division inhibitions of the HL-60 cells at a concentration of 0.01 µM respectively [60].

Triterpene glycosides are isolated from the EtOAc fraction of *Cornus kousa* fruit (Figure 7.39). Based on IR, NMR and FAB-MS spectroscopic data, the chemical structures of the compounds are established as tormentoside, niga-ichigoside F₂ and arjunglucoside. These compounds are used as a

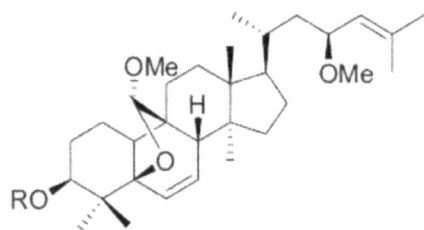

FIGURE 7.40 Chemical structures of Cucurbitane type of triterpene glycosides (R=Glc).

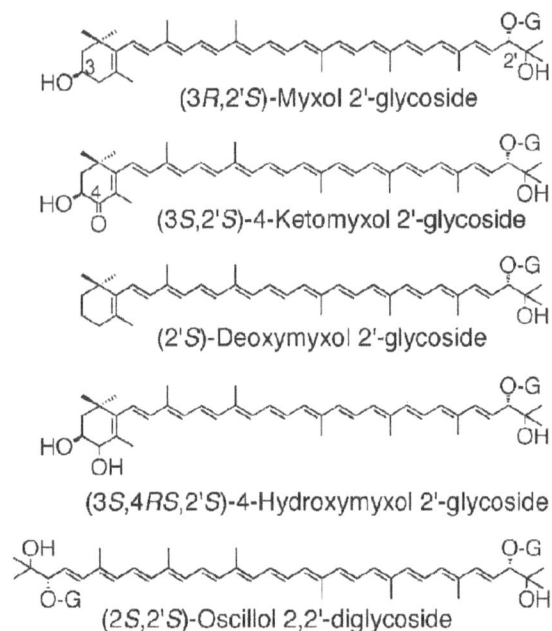

FIGURE 7.41 Structure of myxol glycosides and oscillol diglycoside.

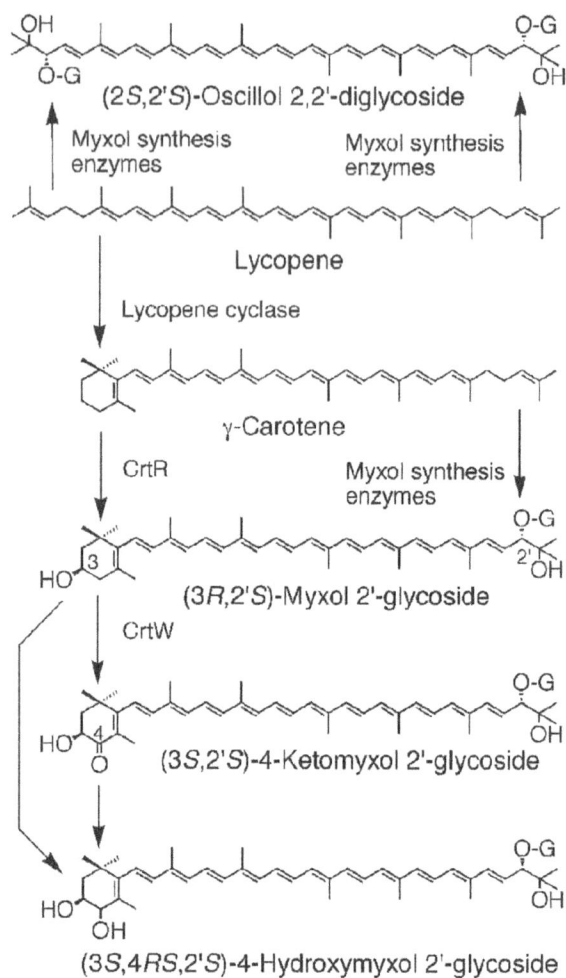

FIGURE 7.42 Synthesis of myxol glycosides and oscillol diglycoside from lycopene.

hemostatic agent and also used for the treatment of diarrhea. It has been reported that the fruit extract exhibits immuno-regulatory property [61].

Cucurbitane type of triterpene glycosides (momorchara-coside A, goyaglycoside, charantosides H, J, K) are isolated from EtOAc extract of Momordica charantia fruits (Figure 7.40). The chemical structures of these compounds are identi-fied by 1D, 2D NMR and HRESIMS spectroscopic analysis. However, the configurations of these compounds are deter-mined by ROESY correlations. The isolated compounds are evaluated for their inhibition against α-glucosidase. It was revealed that these compounds exhibit moderate inhibitory activities with IC_{50} values ranging from 28.40 to 63.26 μM as compared to the standard drug acarbose with IC_{50} value of 87.65±6.51 μM [62].

7.8 Tetraterpenoid Glycosides

The carotenoid glycosides (myxol glycosides and oscil-lol diglycosides) are present in cyanobacteria. Myxol glyco-sides and oscillol diglycoside are synthesized from lycopene

(Figure 7.41). Lycopene is cyclized to γ-carotene. Myxol glyco-sides are located in the cytoplasmic and the outer membranes. In case of oscillol 2,2′-diglycoside, myxol synthesis enzymes catalyze both end groups of lycopene (Figure 7.42). The left half (β-end group) of the γ-carotene is hydroxylated by CrtR (β-carotene hydroxylase) [63].

7.9 Miscellaneous

Kim et al. reported the column chromatographic isolation of the methanolic extract of the seeds of *Amomum xanthi-oides* (Zingiberaceae) that afford new glycosides such as (1) two new monoterpene glycosides such as (1S, 4S, 5S)-5-exo-hydroxycamphor 5-O-β-D-glucopyranoside and (1R, 4R, 5S)-5-endo-hydroxycamphor 5-O-β-D-glucopyranoside (2) diterpene glycoside e.g. amoxanthoside A (3) hedychiol A (4) pygmol (5) (1S, 4R, 6R)-(+)-6-endo-hydroxycamphor and (6) dihydroyashabushiketol (Figure 7.43). These isolated com-pounds were tested for their in vitro cytotoxicity against four human cancer cell lines such as A549 (non-small-cell lung adenocarcinoma), SK-OV-3 (ovarian cancer cells), SK-MEL-2

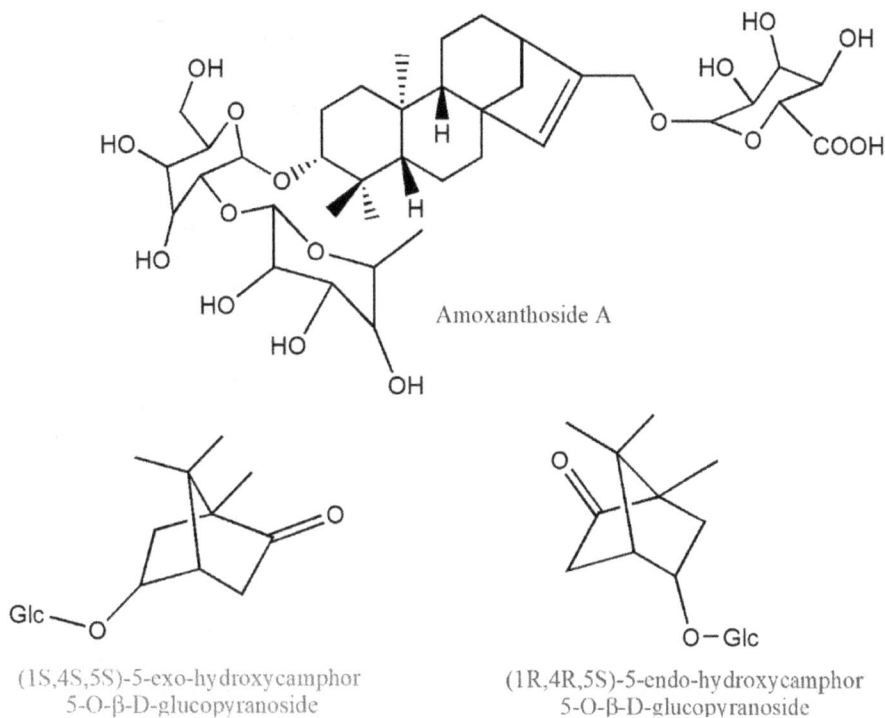

Amoxanthoside A

(1S,4S,5S)-5-exo-hydroxycamphor
5-O-β-D-glucopyranoside

(1R,4R,5S)-5-endo-hydroxycamphor
5-O-β-D-glucopyranoside

FIGURE 7.43 Structure of isolated compounds of *Amomum xanthioides*.

TABLE 7.3

In Vitro Cytotoxicity of Terpenoid Glycosides against Human Cancer Cell Lines

	IC$_{50}$ (μM)			
	A549	**SK-OV-3**	**SK-MEL-2**	**HCT-15**
Amoxanthoside A	>100.0	>100.0	>100.0	>100.0
(1S,4S,5S)-5-Exo-hydroxy-camphor5-O-β-D-gluco-pyranoside	>100.0	>100.0	>100.0	>100.0
(1R,4R,5S)-5-Endo-hydroxy camphor 5-O-β-D-gluco-pyranoside	>100.0	>100.0	>100.0	>100.0
Hedychiol A	24.29	26.72	11.08	29.20
Pygmol	22.64	27.19	24.11	16.42
(1S,4R,6R)-(+)-6-Endo-hydroxycamphor	62.64	57.19	70.11	46.42
Dihydroyashabushiketol	12.41	17.62	11.73	14.29
Doxorubicin	0.16	0.38	0.04	0.82

(skin melanoma), HCT-15 (colon cancer cells) by using a sulforhodamine B bioassay. Doxorubicin was used as a positive control (Table 7.3) [64,65].

Deng et al. reported the terpene glycosides isolated from *Sanguisorba officinalis* of Rosaceae family and evaluated their anti-inflammatory effects (Figure 7.44). The structures of isolated terpene glycosides were elucidated by NMR and HRESIMS spectral data. The isolated compounds exhibited anti-inflammatory properties in vitro by reducing the production of inflammatory mediators, such as nitric oxide (NO) as well as tumor necrosis factor-α (TNF-α) and interleukin-6 (IL-6). The in vivo anti-inflammatory assay is based on the zebrafish experimental platform. The isolated terpene glycoside exhibits significant anti-inflammatory activity by inhibiting the migration of macrophages around the neuromast at the dose of 15 μg/mL [66].

Pizza et al reported the isolation of sesquiterpene glycosides and polyhydroxylated triterpenoids from methanolic extracts of *Eriobo tryajaponica*. The hypoglycemic effects of these compounds were studied by using genetically induced diabetic mice (C57BL/KS-db/db/Ola) and normoglycemic rats. From evaluation results, it was revealed that the tested compounds exhibited marked inhibition of glycosuria [67].

Zhang et al. reported the new terpenoid glycosides obtained from the aerial parts of *Rosmarinus officinalis* L of family Lamiaceae. This plant extract exhibits several biological activities such as hepatoprotective, antibacterial, diuretic, anti-diabetic, anti-nociceptive, anti-inflammatory, antitumor, antioxidant, etc. The structures of the isolated compounds are elucidated by chemical analysis and spectral methods (UV, FT-IR, HRESI-TOF-MS, 1D, 2D NMR, ^{1}H and ^{13}C NMR). Among these, officinoterpenosides A1 is diterpenoid glycoside

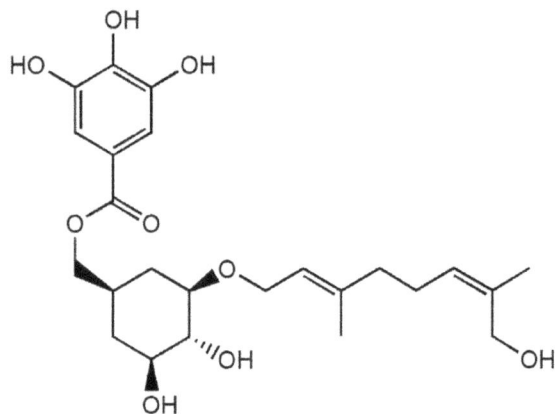

FIGURE 7.44 Structure of terpene glycosides isolated from *Sanguisorba officinalis*.

and officinoterpenosides A_2, B, C are triterpenoid glycosides as presented in Figure 7.45 [68].

7.10 Hydrolysis of Terpene Glycosides

Glycosides containing monoterpenes play a major biochemical role in plants. These serve as intermediaries in the biosynthesis of terpenes. Monoterpene glycosides are more soluble and less lipophilic than the free forms. Therefore these compounds can cross the cell barriers more easily and move toward the areas of accumulation, usage, or transformation. Glycolated monoterpenes might be involved in the formation of the cell wall, in a similar manner to mannosyl-1-phosphoryl-polyisoprenol in the synthesis of mannans by bacteria. GC analysis of hydrolyzed products and comparison with standards are used

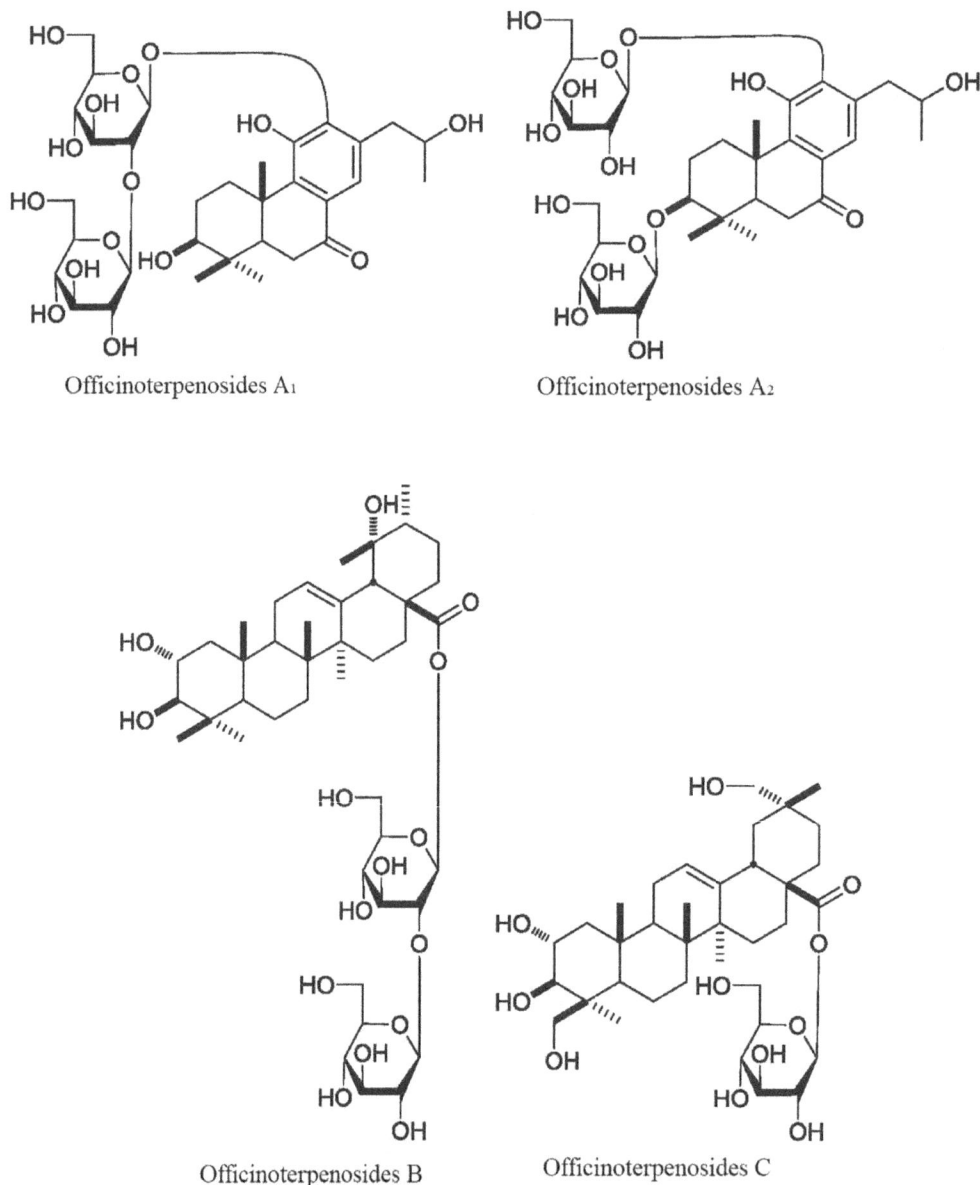

Officinoterpenosides A_1

Officinoterpenosides A_2

Officinoterpenosides B

Officinoterpenosides C

FIGURE 7.45 Structures of terpenoid glycosides isolated from aerial parts of *R. officinalis*.

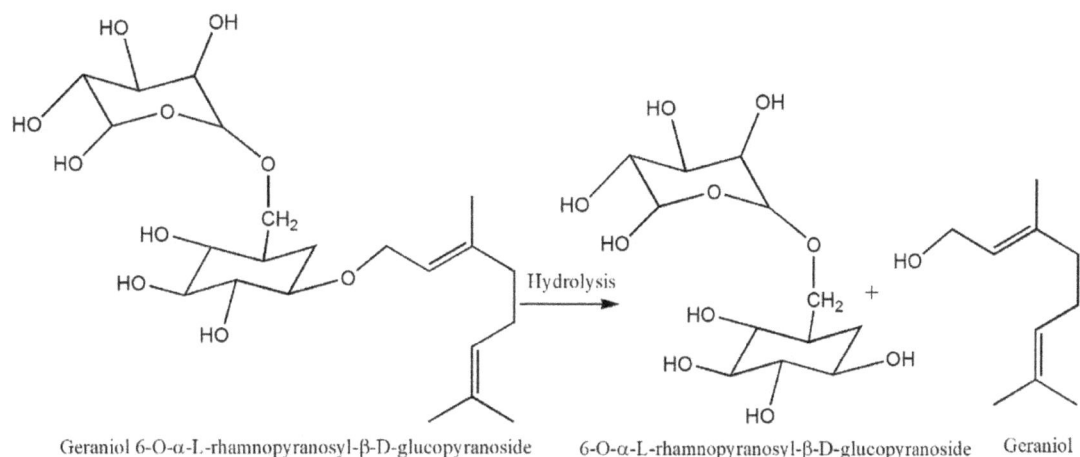

Geraniol 6-O-α-L-rhamnopyranosyl-β-D-glucopyranoside   6-O-α-L-rhamnopyranosyl-β-D-glucopyranoside   Geraniol

FIGURE 7.46 Hydrolysis of glycosides of geraniol.

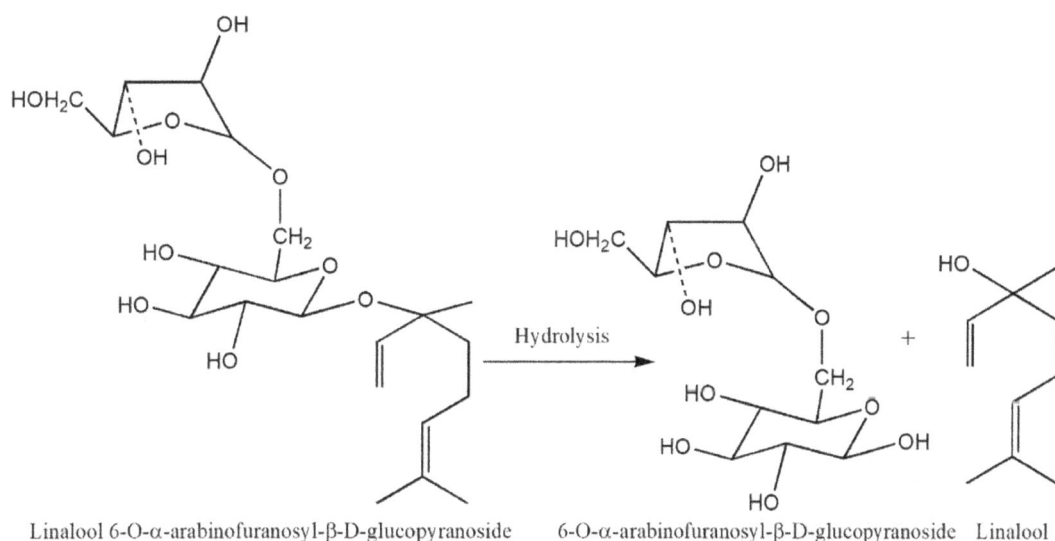

Linalool 6-O-α-arabinofuranosyl-β-D-glucopyranoside   6-O-α-arabinofuranosyl-β-D-glucopyranoside   Linalool

FIGURE 7.47 Hydrolysis of linalool glycosides.

for the identification of sugar residues. The treatment of glycosides with cellulase and the analysis of the obtained sugar solution by HPLC using an optical rotation detector (ORD) is used for the determination of the absolute configuration of carbohydrate residue [69].

The monoterpene alcohols (geraniol, citronellol, linalool and menthol) are volatile and provide fragrance with flavor. So, the increase in stability and water solubility of these volatile compounds can be achieved by glycosylation to produce non-odorous and non-volatile sugar conjugates (Figure 7.46). But the terpene alcohols are readily released from their glycosidic precursors by acidic and enzymatic hydrolysis (Figure 7.47). Due to this type of controlled and delayed release mechanism, the aroma and fragrance property of terpenes play a vital role in the cosmetic industry (Figure 7.48) [70].

The hydrolysis of monosaccharide glycosides requires the action of β-glucosidase, whereas the hydrolysis of disaccharide

glycosides requires the sequential action of suitable exo-glycosidase to remove the outermost sugar moiety followed by the action of β-glucosidase to remove the remaining glucose units (Figure 7.49). It is also observed that the endo-glycosidase is capable of hydrolyzing this linkage thereby liberating the disaccharide and aglycon (Figure 7.50) [71].

It is observed that grapes have remarkable β-glucosidase activity but low activities of the α-rhamnosidase, α-arabinosidase and β-xylosidase. Several commercially available fungal β-glucosidases are reported to perform hydrolysis of the glycosylated compounds. At first, disaccharide glycosides are cleaved by the action of α-L-rhamnosidase, α-L-arabinosidase and β-D-apiosidase to liberate the corresponding terpene monoglucosides by cleavage of the (1→6) glycoside bond. In the second step, the aglycone-carbohydrate bond is cleaved by β-D-glucosidase to release monoterpene alcohols (e.g. linalool) as presented in Figure 7.51 [72,73].

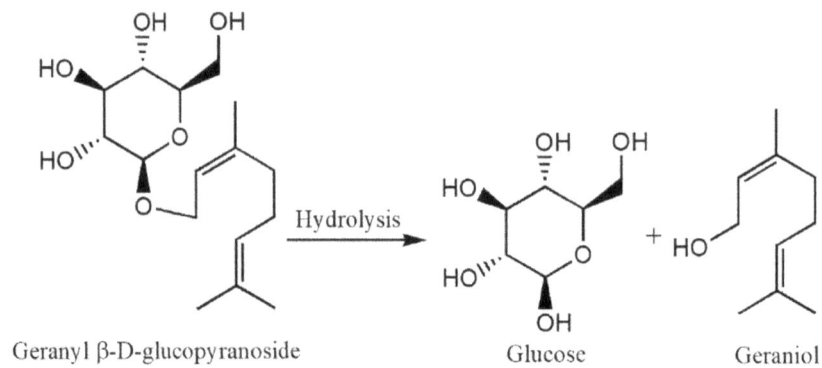

FIGURE 7.48 Hydrolysis of the geranyl-*β*-D-glucopyranoside.

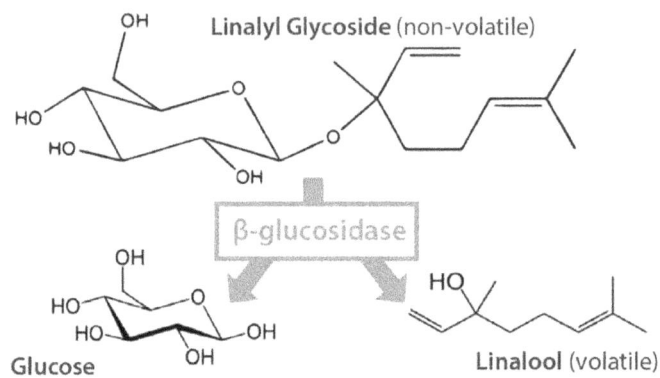

FIGURE 7.49 Hydrolysis of linalool glycosides via *β*-glucosidase.

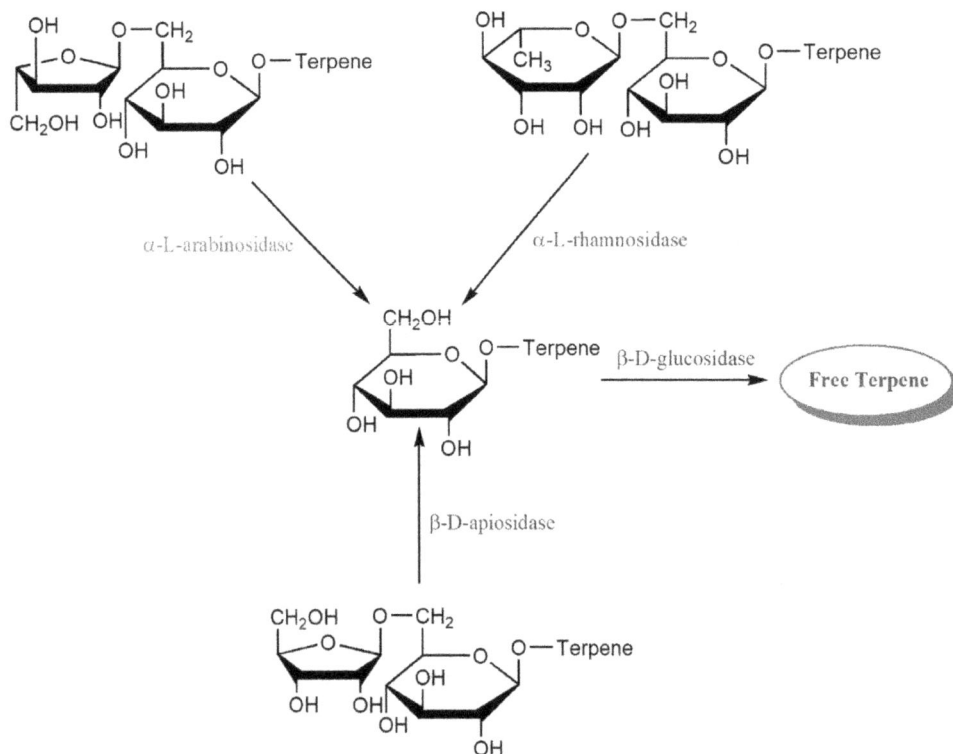

FIGURE 7.50 Enzymatic hydrolysis of disaccharide glycosides.

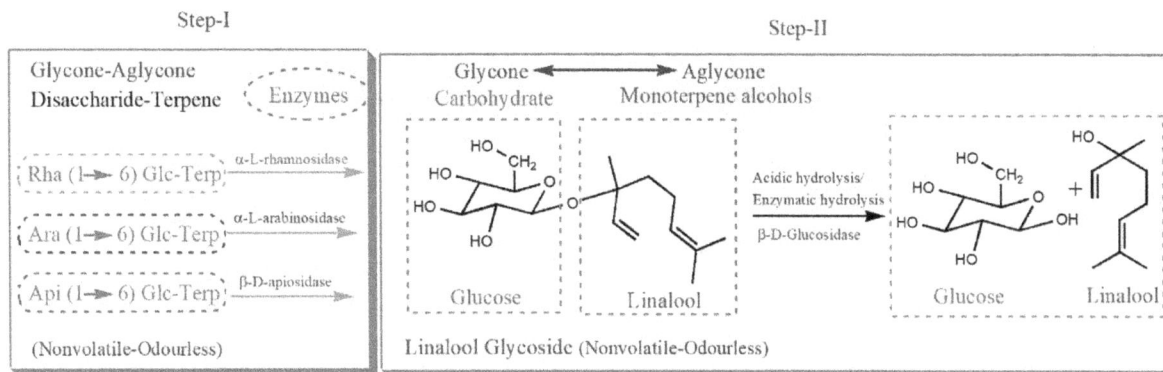

FIGURE 7.51 Steps involved in hydrolysis of glycosides.

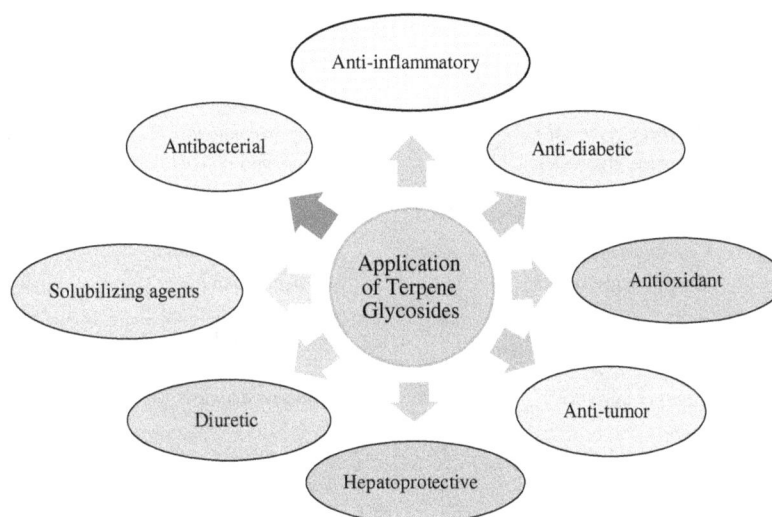

FIGURE 7.52 Application of terpene glycosides.

7.11 Applications of Terpene Glycosides

Terpene glycosides used as solubilizing agents include mogroside V, paenoiflorin, geniposide, rubusoside, rebaudioside A and stevioside [74]. These are used to enhance the solubility of medicinal agents such as paclitaxel, camptothecin, curcumin, capsaicin, cyclosporine, erythromycin, nystatin, itraconazole, celecoxib, clofazimine, digoxin, oleandrin, nifedipine and amiodarone etc. The test results reported that the use of monoterpene glycoside (paenoiflorin) and diterpene glycoside (rubusoside) increases the solubility [75]. Gas chromatography and HPLC are used to identify the presence of sugar residues and also determine the absolute configuration of carbohydrate residues respectively. The most reported eremophilane glycosides is β-D-glucopyranose unit. The poisonous nature of foxglove is due to the presence of terpenoid glycosides that exhibit strong stimulant action on heart muscle (Figure 7.52) [76]. Evaluation study reports suggest that several terpene glycosides exhibit their diverse biological activities such as anticancer, antioxidant, antibacterial, antifungal, anti-diabetic, antipyretic, antinociceptive, anti-inflammatory etc as presented in Table 7.4 [77].

7.12 Conclusion

Terpene glycosides are widely distributed in natural sources that comprise several compounds with diverse chemical structures due to the presence of aglycones (terpene alcohols) and glycone parts (carbohydrates). The extraction and isolation of terpene glycosides are carried out by different chromatographic methods (TLC, HPTLC, HPLC, Gas chromatography, column chromatography). Further, the structures of the terpene glycosides are established by various analytical methods such as FT-IR, 1D NMR, 2D NMR, ^{1}H NMR, ^{13}C NMR, LC-MS, GC-MS, DEPT, COSY, HMBC and HSQC etc. Terpene glycosides are non-toxic, odorless and used as complexing agent, solubilizing agent or excipients in the food and pharmaceutical industry. These glycosides also exhibit several biological activities such as anti-viral, anti-inflammatory, antioxidant, antibacterial, antipyretic, antinociceptive, anti-inflammatory and neuroprotective effects. It is expected that further investigations of these compounds will lead to develop new chemical entities with high therapeutic potentials.

TABLE 7.4

Functions of Terpene Glycosides

Name of Terpenoid Glycoside	Type of Terpenoid Glycoside	Biological Source	Functions
Menthyl glycosides	Monoterpene glycosides	*Mentha piperita*	Anti-microbial
Amoxanthoside A	Monoterpene glycosides	*Amomum xanthioides*	Anticancer
Carvyl-β-D-glucosides	Monoterpene glycosides	*Mentha* species	Anticancer Antioxidant
Fukinoside A	Eremophilane-type sesquiterpenoid glycosides	*Petasites japonicus*	Treatment of rheumatism
Eudesmanolide	Sesquiterpene glycosides	*Sarcandra glabra*	Hepatoprotective
Nepalactones A, B	Sesquiterpene glycosides	*Coriaria nepalensis*	Neurotrophic activity
Codonopsesquilosides A-C	Sesquiterpene glycosides	*Codonopsis pilosula*	Boost immune system
Fuscosides A–D	Diterpene glycosides	*Eunicea fusca*	Anti-inflammatory
Brevistylumsides A and B	Diterpene glycosides	*Illicium brevistylum*	Treatment of rheumatoid arthritis
Clerodane-type glycosides	Diterpene glycosides	*Dicranopteris pedata*	Antioxidant Antibacterial Antipyretic Antinociceptive Anti-inflammatory
Viteoside A	Diterpene glycosides	*Vitex rotundifolia*	Antioxidant
Stevioside	Diterpene glycosides	*Stevia rebaudiana*	Non-caloric sweetener
Officinoterpenosides A_1	Diterpene glycosides	*Rosmarinus officinalis*	Antioxidant Anti-inflammatory
Capsianoside	Diterpene glycosides	*Capsicum annuum*	Anticancer
Charantosides H	Cucurbitane-type triterpene glycosides	*Momordica charantia*	α-Glucosidase inhibitory activity
Cycloartane glycosides	Triterpene glycosides	*Astragalus aureus*	Cytotoxic activity against human breast cancer
Echinoside A	Triterpene glycosides	*Actinopyga echinites*	Antitumor Antifungal
Patagonicoside A	Triterpene glycosides	*Psolus patagonicus*	Antiproliferative Cytotoxic Hemolytic activity
Holothurins	Triterpenoid glycosides	*Sea cucumbers*	Antifungal
Ilexdunnoside A, B	Triterpene glycoside	*Ilex dunniana* Levl.	Anti-inflammatory
Polysciasoside A	Triterpene glycoside	*Polyscias fulva*	Wound-healing property
Violaceuside A	Triterpene glycoside	*Pseudocolochirus violaceus*	Anticancer
Isomadecassoside	Ursane-type triterpene glycoside	*Centella asiatica*	Anti-inflammatory
Cyclosophoside A	Cycloartane triterpene glycoside	*Cassia sophera*	Treatment of diabetes
Myxol glycosides	Carotenoid glycosides	Cyanobacteria	Photosynthesis
Oscillol diglycosides	Carotenoid glycosides	Cyanobacteria	Photosynthesis

---

Abbreviations

AMPK	AMP-activated protein kinase
CD	Circular dichroism
COSY	Correlation spectroscopy
DEPT	Distortionless enhancement by polarization transfer
DNA	Deoxyribonucleic acid
ECD	Electronic circular dichroism
FAB-MS	Fast atom bombardment-mass spectrometry
FT-IR	Fourier transform infrared spectroscopy
GC-MS	Gas chromatography-mass spectrometry
GTs	Glycosyltransferases
HMBC	Heteronuclear multiple bond correlation
HPLC	High-performance liquid chromatography
HR ESI	High resolution electrospray ionization
HR-ESI-MS	High resolution electrospray ionization mass spectrometry
HSQC	Heteronuclear single quantum coherence
IL-6	Interleukin-6
LPS	lipopolysaccharide
MS	Mass spectrometry
NGF	Nerve growth factor
NMR	Nuclear magnetic resonance
NFκB	Nuclear factor kappa B
NO	Nitric oxide
ORD	Optical rotation detector
ROESY	Rotating-frame overhauser enhancement spectroscopy
TNF-α	Tumor necrosis factor-α
TOCSY	Total correlation spectroscopy
TOF-MS	Time-of-flight mass spectrometry
μM	micrometer

REFERENCES

1. Kytidou, K.; Artola, M.; Overkleeft, H. S.; Aerts, J. M. F. G. Plant glycosides and glycosidases – A treasure-trove for therapeutics. *Frontiers in Plant Science*, **2020**, 11, 1–28. doi: 10.3389/fpls.2020.00357.

2. Pfander, H.; Stoll, H. Terpenoid glycosides. *Natural Product Reports*, **1991**, 8, 69–95.

3. Anisimov, M. M. Triterpene glycosides and structure-functional properties of biomembranes. *Nauchnye doklady vysshei shkoly. Biologicheskie Nauki*, **1987**, 10, 49–63.

4. Karliner, J.; Djerassi, C. Terpenoids. LVII. Mass spectral and nuclear magnetic resonance studies of pentacyclic triterpene hydrocarbons. *The Journal of Organic Chemistry*, **1966**, 31(6), 1945–1956. doi: 10.1021/jo01344a063.

5. Patil, S. P. Proton NMR and HR-LC/MS based phytochemical analysis of methanolic fraction of *Alectra parasitica* A. rich rhizomes. *Heliyon*, **2020**, 6(1), 1–7. doi: 10.1016/j.heliyon.2020.e03171.

6. Bouvier, F.; Rahier, A.; Camara, B., Biogenesis, molecular regulation and function of plant isoprenoids. *Progress in Lipid Research*, **2005**, 44, 357–429.

7. Moon, H. I.; Lee, J. H. Neuroprotective effects of triterpene glycosides from *Glycine max* against gutamate induced toxicity in primary cultured rat cortical cells. *International Journal of Molecular Sciences*, **2012**, 13, 9642–9648.

8. Xu, K.; Chu, F.; Li, G.; Xu, X.; Wang, P.; Song, J.; Zhou, S.; Lei, H. Oleanolic acid synthetic oligoglycosides – A review on recent progress in biological activities. *Pharmazie*, **2014**, 69, 483–495.

9. Genus, J. M. C. Stevioside. *Phytochemistry*, **2003**, 64, 913–921.

10. Demchenko, A. V. General aspects of the glycosidic bond formation. In Demchenko, A. V. (Ed.), *Handbook of Chemical Glycosylation – Advances in Stereoselectivity and Therapeutic Relevance*, Wiley-VCH Verlag, Weinheim, **2008**, 1–27.

11. Křen, V., Thiem, J. Glycosylation employing bio-systems – from enzymes to whole cells. *Chemical Society Reviews*, **1997**, 26, 463–473.

12. Rivas, F.; Parra, A.; Martinez, A.; Garcia-Granados, A. Enzymatic glycosylation of terpenoids. *Phytochemistry Reviews*, **2013**, 12, 327–339.

13. Lim, E.-K., Plant glycosyltransferases – Their potential as novel biocatalysts. *Chemistry – A European Journal*, **2005**, 11, 5486–5494.

14. Caputi, L.; Lim, E. K.; Bowles, D. J. Discovery of new biocatalysts for the glycosylation of terpenoid scaffolds. *Chemistry – A European Journal*, **2008**, 14, 6656–6662.

15. Lairson, L. L.; Henrissat, B.; Davies, G. J.; Withers, S. G. Glycosyltransferases – Structures, functions, and mechanisms. *Annual Review of Biochemistry*, **2008**, 77, 521–555.

16. Singh, S.; Phillips, G. N. Jr.; Thorson, J. S. The structural biology of enzymes involved in natural product glycosylation. *Natural Product Reports*, **2012**, 29, 1201–1237.

17. Rivas, F.; Parra, A.; Martinez, A.; Garcia-Granados, A. Enzymatic glycosylation of terpenoids. *Phytochemistry Reviews*, **2013**, 12, 327–339.

18. Maicas, S.; Mateo, J. J. Hydrolysis of terpenyl glycosides in grape juice and other fruit juices – A review. *Applied Microbiology and Biotechnology*, **2005**, 67(3), 322–335. doi: 10.1007/s00253-004-1806-0.

19. Fernández-González, M.; Di Stefano, R.; Briones, A. Hydrolysis and transformation of terpene glycosides from Muscat must by different yeast species. *Food Microbiology*, **2003**, 20(1), 35–41. doi: 10.1016/s0740-0020(02)00105-3.

20. Gunata, Y. Z.; Bitteur, S.; Brillouet, J. M.; Bayonove, C. L.; Cordonnier, R. Sequential enzymatic hydrolysis of potentially aromatic glycosides from grape. *Carbohydrate Research*, **1988**, 184, 139–149.

21. Turner, G. W.; Croteau, R. Organization of monoterpene biosynthesis in Mentha. Immunocytochemical localizations of geranyl diphosphate synthase, limonene-6-hydroxylase, isopiperitenol dehydrogenase, and pulegone reductase. *Plant Physiology*, **2004**, 136, 4215–4227.

22. Shimizu, S.; Shibata, H.; Karasawa, D.; Kozaki, T. Carvyl- and dihydrocarvyl-β-D-glucosides in spearmint (studies on terpene glycosides in mentha plants, Part II). *Journal of Essential Oil Research*, **1990**, 2, 81–86.

23. Sakata, I.; Iwamura, H. Studies on terpene glycosides. I. Synthesis and properties of menthyl glycosides. *Agricultural and Biological Chemistry*, **1979**, 43(2), 307–312. doi: 10.1271/bbb1961.43.307.

24. Sun, W.; Zhang, Z.-L.; Liu, X.; Zhang, S.; He, L.; Wang, Z.; Wang, G.-S. Terpene glycosides from the roots of *Sanguisorba officinalis* L. and their hemostatic activities. *Molecules*, **2012**, 17(7), 7629–7636. doi: 10.3390/molecules17077629.

25. Williams, P. J.; Strauss, C. R.; Wilson, B.; Massy-Westropp, R. A. Studies on the hydrolysis of *Vitis vinifera* monoterpene precursor compounds and model monoterpene ß-D-glucosides rationalizing the monoterpene composition of grapes. *Journal of Agricultural and Food Chemistry*, **1982**, 30, 1219–1223.

26. Feng, J.; Surup, F.; Hauser, M.; Miller, A.; Wennrich, J.-P.; Stadler, M.; Cox, R. J.; Kuhnert, E. Biosynthesis of oxygenated brasilane terpene glycosides involves a promiscuous-acetylglucosamine transferase. *Chemical Communications*, **2020**. doi: 10.1039/D0CC03950K.

27. Xia, G.; Huang, Y.; Xia, M.; Wang, L.; Kang, N.; Ding, L.; Chen, L.; Qiu, F. A new eremophilane glycoside from the fruits of *Physalis pubescens* and its cytotoxic activity. *Natural Product Research*, **2017**, 31, 2737–2744.

28. Hou, C.; Kulka, M.; Zhang, J.; Li, Y.; Guo, F. Occurrence and biological activities of eremophilane-type sesquiterpenes. *Mini-Review in Medicinal Chemistry*, **2014**, 14, 664–677.

29. Yamada, K.; Ojika, M.; Kigoshi, H. Ptaquiloside, the major toxin of bracken, and related terpene glycosides – Chemistry, biology and ecology. *Natural Product Report*, **2007**, 24, 798–813.

30. Jiang, Y.; Liu, Y.; Guo, Q.; Xu, C.; Zhu, C.; Shi, J. Sesquiterpene glycosides from the roots of *Codonopsis pilosula*. *Acta Pharmaceutica Sinica B*, **2016**, 6(1), 46–54. doi: 10.1016/j.apsb.2015.09.007.

31. Li, Y.; Zhang, D.-M.; Li, J.-B.; Yu, S.-S.; Li, Y.; Luo, Y.-M. Hepatoprotective sesquiterpene glycosides from *Sarcandra glabra*. *Journal of Natural Products*, **2006**, 69(4), 616–620. doi: 10.1021/np050480d.

32. Wang, Y. Y.; Tian, J. M.; Zhang, C. C.; Luo, B.; Gao, J. M. Picrotoxane sesquiterpene glycosides and a coumarin derivative from *Coriaria nepalensis* and their neurotrophic activity. *Molecules*, **2016**, 21(10), 1344. doi: 10.3390/molecules21101344.

33. Shin, J.; Fenical, W. Fuscosides A-D – Anti-inflammatory diterpenoid glycosides of new structural classes from the *Caribbean gorgonian Eunicea fusca. The Journal of Organic Chemistry*, **1991**, 56, 3153–3158.

34. Marchbank, D. H.; Kerr, R. G. Semisynthesis of fuscoside B analogues and eunicosides, and analysis of anti-inflammatory activity. *Tetrahedron*, **2011**, 67(17), 3053–3061. doi: 10.1016/j.tet.2011.03.006.

35. Chaturvedula, V. S.; Upreti, M.; Prakash, I. Diterpene glycosides from *Stevia rebaudiana. Molecules*, **2011**, 16(5), 3552–3562. doi: 10.3390/molecules16053552.

36. Jong-Hyun, L.; Kiyota, N.; Ikeda, T.; Nohara, T. Three new acyclic diterpene glycosides from the aerial parts of paprika and pimiento. *Chemical and Pharmaceutical Bulletin*, **2008**, 56(4), 582–584.

37. Ono, M.; Ito, Y.; Nohara, T. A labdane diterpene glycoside from fruit of *Vitex rotundifolia. Phytochemistry*, **1998**, 48(1), 207–209. doi: 10.1016/s0031-9422(97)00863-7.

38. Chen, Y.; Dong, Z.; Huang, D.; Chen, W.; Yao, F.; Xue, D.; Sun, L. Brevistylumsides A and B, diterpene glycosides from *Illicium brevistylum. Phytochemistry Letters*, **2015**, 14, 27–30. doi: 10.1016/j.phytol.2015.08.014.

39. Gao, B. B.; Ou, Y. F.; Zhu, Q. F. Clerodane-type diterpene glycosides from *Dicranopteris pedata. Natural Products and Bioprospecting*, **2021**, 11, 557–564. doi: 10.1007/s13659-021-00315-y.

40. Sipkema, D.; Franssen, M. C. R.; Osinga, R., et al. Marine sponges as pharmacy. *Marine Biotechnology*, **2005**, 7, 142–162. doi: 10.1007/s10126-004-0405-5.

41. Qiu, S. X.; Dan, C.; Ding, L. S.; Peng, S.; Chen, S. N.; Farnsworth, N. R.; Nolta, J.; Gross, M. L.; Zhou, P. A triterpene glycoside from black cohosh that inhibits osteoclastogenesis by modulating RANKL and TNF alpha signaling pathways. *Chemistry & Biology*, **2007**, 14(7), 860–869. doi: 10.1016/j.chembiol.2007.06.010.

42. Yue, G. G.; Xie, S.; Lee, J. K.; Kwok, H. F.; Gao, S.; Nian, Y.; Wu, X. X.; Wong, C. K.; Qiu, M. H.; Lau, C. B. New potential beneficial effects of actein, a triterpene glycoside isolated from *Cimicifuga* species, in breast cancer treatment. *Science Report*, **2016**, 6, 35263. doi: 10.1038/srep35263.

43. Lai, G. F.; Wang, Y. F.; Fan, L. M.; Cao, J. X.; Luo, S. D. Triterpenoid glycoside from *Cimicifuga racemosa. Journal of Asian Natural Products Research*, **2005**, 7(5), 695–699. doi: 10.1080/1028602042000324817.

44. Khare, C. *Cimicifuga racemosa* (Linn.) Nutt. In: Khare, C. (Eds) *Indian Medicinal Plants*. Springer, New York, NY. **2007**. doi: 10.1007/978-0-387-70638-2_349.

45. Da, Y. M.; Niu, K. Y.; Liu, S. Y.; Wang, K.; Wang, W. J.; Jia, J.; Qin, L. H.; Bai, W. P. Does *Cimicifuga racemosa* have the effects like estrogen on the sublingual gland in ovariectomized rats?. *Biological Research*, **2017**, 50(1), 11. doi: 10.1186/s40659-017-0115-x.

46. Bedir, E.; Khan, I. A. Cimiracemoside A: A new cyclolanostanol xyloside from the rhizome of *Cimicifuga racemosa. Chemical & Pharmaceutical Bulletin*, **2000**, 48(3), 425–427. doi: 10.1248/cpb.48.425.

47. Izawa, K. *Comprehensive Natural Products II || Human–Environment Interactions-Taste.* **2010**, 631–671. doi: 10.1016/b978-008045382-8.00108-8.

48. Sugita, D.; Inoue, R.; Kurihara, Y. Sweet and sweetness-inducing activities of new triterpene glycosides, strogins. *Chemical Senses*, **1998**, 23(1), 93–97. doi: 10.1093/chemse/23.1.93.

49. Shin, B. K.; Kwon, S. W.; Park, J. H. Chemical diversity of ginseng saponins from *Panax ginseng. Journal of Ginseng Research*, **2015**, 39(4), 287–298. doi: 10.1016/j.jgr.2014.12.005.

50. Elyakov, G. B.; Strigina, L. I.; Uvarova, N. I.; Vaskovsky, V. E.; Dzizenko, A. K.; Kochetkov, N. K. Glycosides from ginseng roots. *Tetrahedron Letters*, **1964**, 5, 3591–3597.

51. Bedir, E.; Toyang, N. J.; Khan, I. A.; Walker, L. A.; Clark, A. M. A new dammarane-type triterpene glycoside from *Polyscias fulva. Journal of Natural Products*, **2001**, 64(1), 95–97. doi: 10.1021/np0003589.

52. Chianese, G.; Masi, F.; Cicia, D.; Ciceri, D.; Arpini, S.; Falzoni, M.; Pagano, E.; Taglialatela-Scafati, O. Isomadecassoside, a new ursane-type triterpene glycoside from *Centella asiatica* leaves, reduces nitrite levels in LPS-stimulated macrophages. *Biomolecules*, **2021**, 11(4), 494. doi: 10.3390/biom11040494.

53. Yoshikawa, M.; Harada, E.; Murakami, T.; Matsuda, H.; Wariishi, N.; Yamahara, J.; Murakami, N.; Kitagawa, I. Escins-Ia, Ib, IIa, IIb, and IIIa, bioactive triterpene oligoglycosides from the seeds of *Aesculus hippocastanum* L.: Their inhibitory effects on ethanol absorption and hypoglycemic activity on glucose tolerance test. *Chemical and Pharmaceutical Bulletin*, **1994**, 42(6), 1357–1359. doi: 10.1248/cpb.42.1357.

54. Yu, B.; Chen, X.; Shao, X.; Li, W.; Zhang, X. Total synthesis of echinoside A, a representative triterpene glycoside of sea cucumbers. *Angewandte Chemie International Edition*, **2017**. doi: 10.1002/anie.201703610.

55. Zhao, Y.; Liu, J.-P.; Lu, D.; Li, P.-Y. A novel cycloartane triterpene glycoside from the seeds of *Cassia sophera* L. *Natural Product Research*, **2007**, 21(6), 494–499. doi: 10.1080/14786410601130042.

56. Careaga, V. P.; Bueno, C.; Muniain, C.; Laura, A.; Maier, M. S. Antiproliferative, cytotoxic and hemolytic activities of a triterpene glycoside from *Psolus patagonicus* and its desulfated analog. *Chemotherapy*, **2009**, 55(1), 60–68. doi: 10.1159/000180340.

57. Chludil, H. D. Biologically active triterpene glycosides from sea cucumbers (holothuroidea, echinodermata). *Studies in Natural Products Chemistry. Bioactive Natural Products (Part I)*, **2003**, 28, 587–615. doi: 10.1016/s1572-5995(03)80150-3.

58. Shi, Y.-S.; Zhang, Y.; Hu, W.-Z.; Chen, X.; Fu, X.; Lv, X.; Zhang, L.-H.; Zhang, N.; Li, G. Anti-inflammatory triterpene glycosides from the roots of *Ilex dunniana* Levl. *Molecules*, **2017**, 22(7), 1206–1214. doi: 10.3390/molecules22071206.

59. Gülcemal, D.; Alankuş-Çalışkan, O.; Perrone, A.; Ozgökçe, F.; Piacente, S.; Bedir, E. Cycloartane glycosides from *Astragalus aureus. Phytochemistry*, **2011**, 72(8), 761–768. doi: 10.1016/j.phytochem.2011.02.006.

60. Zhang, S.-Y.; Yi, Y.-H.; Tang, H.-F.; Li, L.; Sun, P.; Wu, J. Two new bioactive triterpene glycosides from the sea cucumber *Pseudocolochirus violaceus. Journal of Asian Natural Products Research*, **2006**, 8(1–2), 1–8. doi: 10.1080/10286020500034972.

61. Lakoon-Jung, L. D. Y.; Cho, J. G. Isolation of triterpene glycosides from the fruit of *Cornus kousa*. *Journal of the Korean Society for Applied Biological Chemistry*, **2009**, 52, 45–49. doi: 10.3839/jksabc.2009.008.

62. Gao, Y.; Chen, J.-C.; Peng, X.-R.; Li, Z.-R.; Su, H.-G.; Qiu, M.-H. Cucurbitane-type triterpene glycosides from *Momordica charantia* and their glucosidase inhibitory activities. *Natural Products and Bioprospecting*, **2020**, 1–9. doi: 10.1007/s13659-020-00241-5.

63. Takaichi, S.; Mochimaru, M. Carotenoids and carotenogenesis in cyanobacteria: Unique ketocarotenoids and carotenoid glycosides. *Cellular and Molecular Life Sciences*, **2007**, 64, 2607–2619.

64. Choi, J. W.; Kim, K. H.; Lee, I. K.; Choi, S. U.; Lee, K. R. Phytochemical constituents of *Amomum xanthioides*. *Natural Product Sciences*, **2009**, 15, 44–49.

65. Kim, K. H.; Choi, J. W.; Choi, S.; Lee, K. R. Terpene glycosides and cytotoxic constituents from the seeds of *Amomum xanthioides*. *Planta Medica*, **2010**, 76, 464–466.

66. Guo, D. L.; Chen, J. F.; Tan, L.; Jin, M. Y.; Ju, F.; Cao, Z. X.; Deng, F.; Wang, L. N.; Gu, Y. C.; Deng, Y. Terpene glycosides from *Sanguisorba officinalis* and their anti-inflammatory effects. *Molecules*, **2019**, 24(16), 2906. doi: 10.3390/molecules24162906.

67. De Tommasi, N.; De Simone, F.; Cirino, G.; Cicala, C.; Pizza, C. Hypoglycemic effects of sesquiterpene glycosides and polyhydroxylated triterpenoids of *Eriobotrya japonica*. *Planta Medica*, **1991**, 57(5), 414–416. doi: 10.1055/s-2006-960137.

68. Zhang, Y.; Adelakun, T. A.; Qu, L.; Li, X.; Li, J.; Han, L.; Wang, T. New terpenoid glycosides obtained from *Rosmarinus officinalis* L. aerial parts. *Fitoterapia*, **2014**, 99, 78–85. doi: 10.1016/j.fitote.2014.09.004.

69. Gunata, Y. Z.; Bayonove, C. L.; Cordonnier, R. E.; Arnaud, A.; Galzy, P. Hydrolysis of grape monoterpenyl glycosides by *Candida molischiana* and *Candida wickerhamii* β-glucosidases. *Journal of the Science of Food and Agriculture*, **1990**, 50(4), 499–506. doi: 10.1002/jsfa.2740500408.

70. Maicas, S.; Mateo, J. J. Hydrolysis of terpenyl glycosides in grape juice and other fruit juices: A review. *Applied Microbiology and Biotechnology*, **2005**, 67, 322–335.

71. Gunata, Y. Z.; Bayonove, C. L.; Tapiero, C.; Cordonnier, R. E. Hydrolysis of grape monoterpenyl β-D-glucosides by various β-glucosidases. *Journal of Agricultural and Food Chemistry*, **1990**, 38, 1232–1236.

72. Duerksen, J. D.; Halvorson, H. Purification and properties of an inducible β-glucosidase of yeast. *Journal of Biological Chemistry*, **1958**, 233, 1113–1120.

73. Schwab, W.; Fischer, T.; Wüst, M. Terpene glucoside production: Improved biocatalytic processes using glycosyltransferases. *Engineering in Life Sciences*, **2015**, 15(4), 376–386. doi: 10.1002/elsc.201400156.

74. Wang, X.; Song, H.; Yang, Y. Chemical profile of terpene glycosides from Meili grape detected by GC–MS and UPLC–Q-TOF-MS. *European Food Research and Technology*, **2020**, 246, 2323–2333. doi: 10.1007/s00217-020-03576-y.

75. Kim, K.; Choi, J. W.; Choi, S.; Lee, K. R. Terpene glycosides and cytotoxic constituents from the seeds of *Amomum xanthioides*. *Planta Medica*, **2010**, 76(5), 461–464.

76. Shimizu, S.; Shibata, H.; Karasawa, D.; Kozaki, T. Carvyl- and dihydrocarvyl-β-D-glucosides in spearmint (studies on terpene glycosides in mentha plants, Part II). *Journal of Essential Oil Research*, **1990**, 2(2), 81–86. doi: 10.1080/10412905.1990.9697828.

77. Cox-Georgian, D.; Ramadoss, N.; Dona, C.; Basu, C. Therapeutic and medicinal uses of terpenes. *Medicinal Plants*, **2019**, 333–359. doi: 10.1007/978-3-030 31269-5_15.

8

Terpenes as Starting Compounds for Other Types of Molecules Including Alkaloids

Bimal Krishna Banik*, **Biswa Mohan Sahoo**†, **and Abhishek Tiwari**‡

CONTENTS

8.1 Introduction 255
8.2 Monoterpenes as Starting Compounds 255
    8.2.1 Polymerization of Terpenes 261
8.3 Sesquiterpenoids as Starting Compounds 263
8.4 Miscellaneous 264
8.5 Terpenoid Alkaloids 268
    8.5.1 Monoterpenoid Alkaloids 269
    8.5.2 Diterpenoids Alkaloids 275
    8.5.3 Triterpenoid Alkaloids 278
    8.5.4 Biological Activities Terpenoid Alkaloids 278
8.6 Conclusion 278
Abbreviation 280
References 280

8.1 Introduction

Terpenoids are the naturally occurring largest and diverse group of organic compounds that are utilized as starting materials for the production of other types of molecules including alkaloids [1]. Terpenoids are mostly obtained from plant sources and few of them have been produced from other sources like fungus, marine organisms, insects and sponges [2]. Terpenoids are isolated from all parts of the plant such as flowers, leaves, twigs, stems, seeds, roots, wood, fruits or bark etc [3]. These compounds are accumulated in secretory cells, holes, trenches, glandular trichomes, or epidermic cells. Monoterpenes and sesquiterpenes are mainly present in essential oils of the medicinal plant and the larger molecular weight terpenes such as triterpene, are present in balsam and resin. The monocyclic monoterpenes are commonly used as building blocks for the synthesis of polycyclic terpene and are also utilized for the transformation of other compounds [4]. Monoterpenes are also applied for the synthesis of chiral selenium and tellurium compounds. Terpenoids are utilized as suitable starting materials for the manufacture of natural aroma chemicals in the food and pharmaceutical industries [5]. The study of these secondary metabolites provides better opportunities for researchers to design and develop new drug candidates from natural sources against several disease conditions such as cancer, Alzheimer's, malaria, inflammation, depression, tuberculosis, parkinsonian, fungal and viral infections etc [6]. Due to structural diversity, these compounds are of great interest to study their phytochemistry, synthesis and pharmacological activities [7].

8.2 Monoterpenes as Starting Compounds

Monoterpene and its derivatives represent a large group of naturally occurring organic compounds which consist of two isoprene units linked in a head-to-tail fashion in their structures [8]. Monoterpene derivatives with heteroatom like oxygen atoms are known as monoterpenoids. These compounds are used as a renewable source of starting materials for producing biologically active compounds (antibacterial, antiviral, anticonvulsant, antidepressant, anti-Alzheimer, anti-Parkinsonian, analgesic and anti-inflammatory) [9].

Patrusheva et al demonstrated the synthesis of chiral oxygen-containing heterocycles with diverse structures by reactions of monoterpenoids with carbonyl compounds (R-CHO).

---

* Department of Mathematics and Natural Sciences, College of Sciences and Human Studies, Prince Mohammad Bin Fahd University, Al Khobar, Kingdom of Saudi Arabia. Corresponding author e-mail address: bimalbanik10@gmail.com.

† Roland Institute of Pharmaceutical Sciences, Berhampur affiliated to Biju Patnaik University of Technology (BPUT), Rourkela, Odisha, India. Corresponding author e-mail address: drbiswamohansahoo@gmail.com.

‡ Faculty of Pharmacy, Pharmacy Academy, IFTM University, Lodhipur Rajput, Moradabad-244102, Uttar Pradesh, India.

DOI: 10.1201/9781003008682-8

Based on the type of catalyst used in the reaction and also the structure and type of monoterpenes and carbonyl components, these reactions lead to the formation of mono-, bi-, tri- or tetracyclic compounds as presented in Figure 8.1 [10].

The natural bicyclic monoterpenoids containing bicycle[2.2.1] moiety in their structure are considered as starting molecules for the synthesis of antiviral agents. For example, the substituted 1-norbornylamines (Figure 8.2) are synthesized from (±)-camphor that exhibits potential activity against

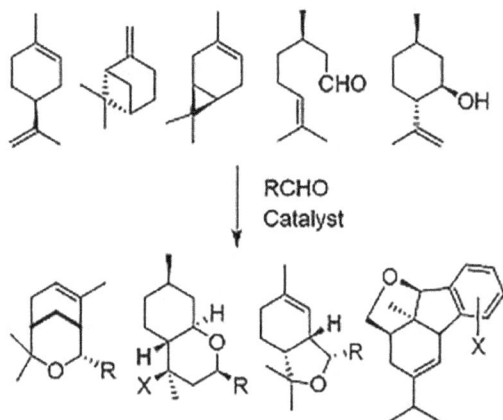

FIGURE 8.1 Synthesis of chiral oxygen-containing heterocycles.

FIGURE 8.2 Synthesis of substituted 1-norbornylamines from (±)-camphor.

influenza virus A and moderate activity against the African swine fever virus [11].

It was demonstrated that the methyl-substituted camphor ethylamine (Figure 8.3) is found to be new series of anti-influenza virus drugs that can be developed as backup scaffolds for the generation of M2 ion channel inhibitors of influenza A virus. Similarly, a new class of anti-HIV agents are developed by substituting coumarins with camphanic acid [12].

Further, the imines based on natural (+)-camphor are demonstrated as a promising source of antiviral agents. So, a series of aliphatic imines (Figure 8.4), iminoalcohols and compounds containing aromatic and heteroaromatic moieties are synthesized and evaluated for their activity against influenza A viruses [13,14].

Isopulegol is a p-menthane type of monoterpenoid with the molecular formula $C_{10}H_{18}O$. Chemically, it is 5-methyl-2-prop-1-en-2-ylcyclohexan-1-ol. (–)-isopulegol is used as starting material for synthesis of (2R, 4R, 4aR, 7R, 8aR)-4,7-dimethyl-2-(thiophen-2-yl)-octahydro-2*H*-chromen-4-ol (Figure 8.5). The newly synthesized compound exhibits significant biological activity (1 mg/kg oral) with low acute toxicity ($LD_{50} > 4,500$ mg/kg) and prolonged effect for at least 24 h [15].

Verbenol (2-pine-4-ol) is a group of stereoisomeric bicyclic monoterpene alcohol. It is the active constituent found in *Thymus cilicicus, Hyssopus officinalis*. Several chiral heterocyclic compounds are synthesized via the interaction of monoterpenoids verbenol epoxide and diol with aromatic aldehydes (Figure 8.6). These transformations are catalyzed by using montmorillonite clays. These compounds are tested to detect their analgesic activity [16,17].

Myrtenal is a natural product found in *Cyperus articulatus, Forsythia viridissima*. It belongs to the bicyclic monoterpenoid aldehyde. Chemically, it is (1R, 5S)-6,6-dimethylbicyclo[3.1.1] hept-2-ene-2-carbaldehyde. (–)-myrtenal substituted compound is synthesized via the interaction between (–)-myrtenal

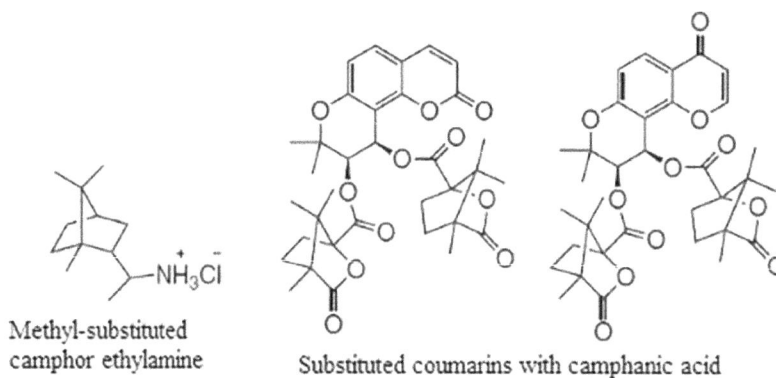

Methyl-substituted camphor ethylamine   Substituted coumarins with camphanic acid

FIGURE 8.3 Structure of methyl-substituted camphor ethylamine.

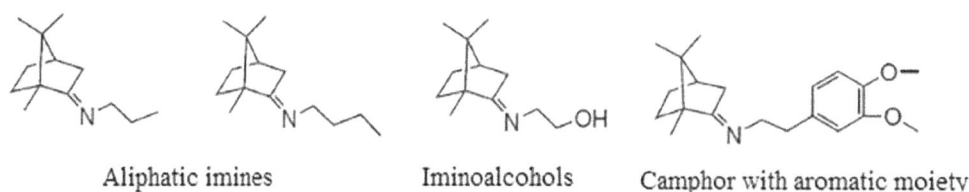

Aliphatic imines   Iminoalcohols   Camphor with aromatic moiety

FIGURE 8.4 Structures of aliphatic imines, iminoalcohols.

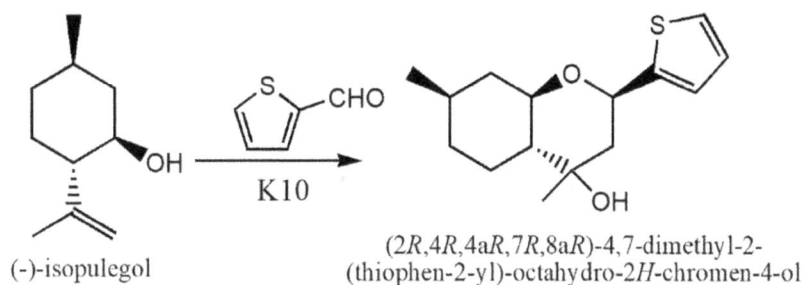

FIGURE 8.5 Structures of chromen-4-ol from (–)-isopulegol.

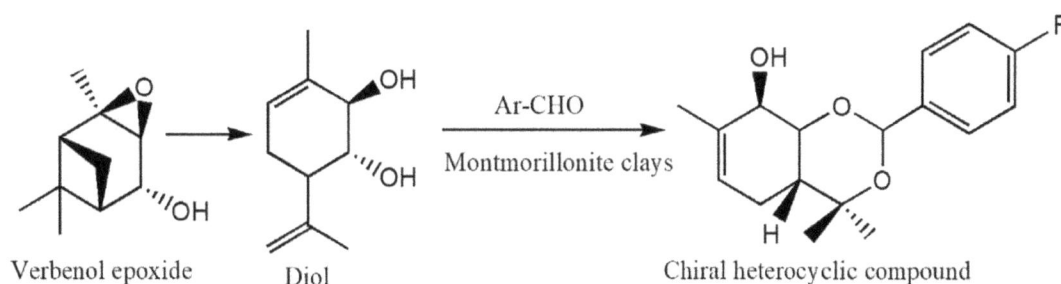

FIGURE 8.6 Synthesis of chiral heterocyclic compounds from verbenol.

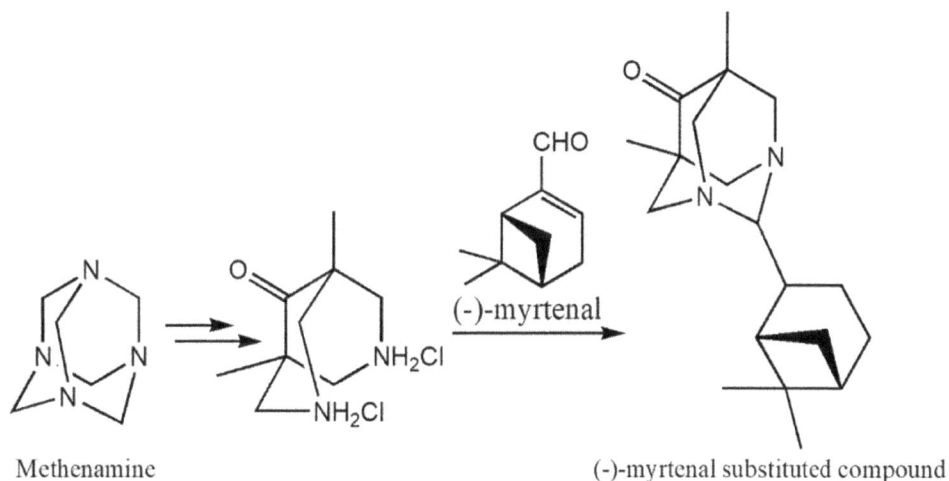

FIGURE 8.7 Synthesis of (–)-myrtenal substituted compound.

and dimethylbispidinone derived from methenamine (Figure 8.7). The newly obtained compounds exhibit significant analgesic activity [18].

Myrtenalyl adamantane compound is synthesized by the interaction of (–)-myrtenal and 2-aminoadamatane with subsequent reduction in presence of NaBH₄ (Figure 8.8). The anxiolytic activity is observed for this compound at a dose of 1 mg/kg in mice by the elevated plus maze test. The anxiolytic effect on mice is found to depend on the absolute configuration of myrtenal used for the synthesis and sex of the animals [19].

Carene is a bicyclic monoterpene that consists of fused cyclohexene and cyclopropane rings. It is a natural product present in Nepeta nepetella, Xylopia aromatica. Chemically, it is 3,7,7-trimethylbicyclo[4.1.0]hept-3-ene. Carene-based

compound is synthesized via the interaction of (+)-2-carene with vanillin in the presence of clay (Figure 8.9). It is observed that this compound reduces the symptoms of Parkinson's disease (PD) at a dose of 30 mg/kg in the neuroprotective model of PD caused by the administration of neurotoxin (MPTP) [20].

Carvone is a monoterpene which is present in spearmint (*Mentha spicata*) and caraway (*Carum carvi*). Chemically, it is 2-methyl-5-prop-1-en-2-ylcyclohex-2-en-1-one. Sarpong et al reported the enantiospecific total synthesis of the *ent*-3,4-*seco*-atisane diterpenoid (–)-crotogoudin in several steps from (*S*)-carvone as presented in Figure 8.10 [21].

Halcomb et al reported the enantiospecific synthesis of phomactin A in 27 steps from (*R*)-(+)-pulegone (Figure 8.11).

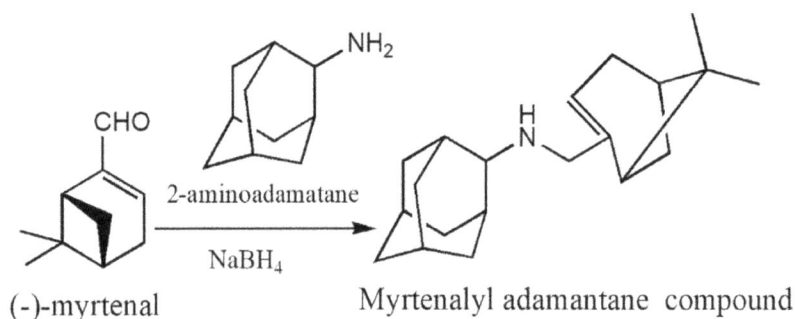

FIGURE 8.8 Synthesis of (–)-myrtenal based adamantane compound.

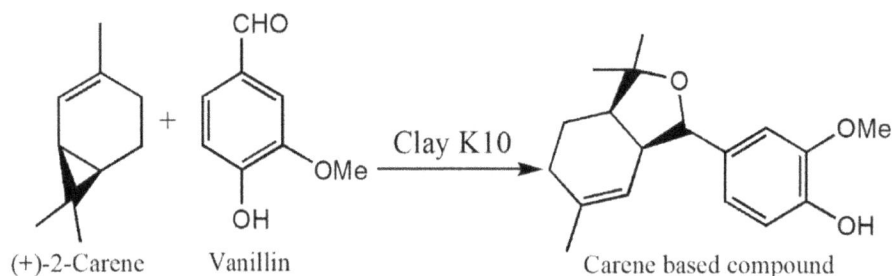

FIGURE 8.9 Synthesis of Carene-based compound.

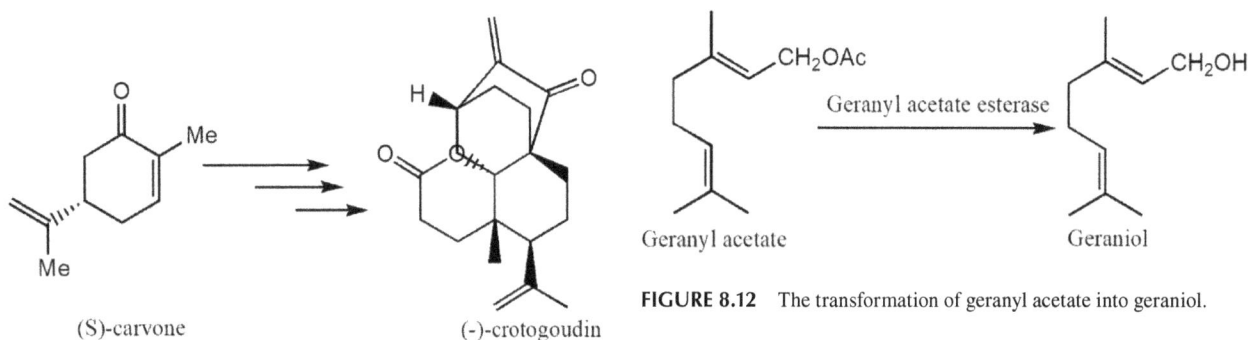

FIGURE 8.10 Synthesis of *ent*-3,4-*seco*-atisane diterpenoid (–)-crotogoudin.

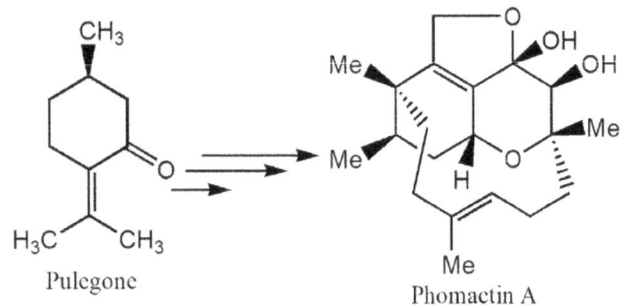

FIGURE 8.12 The transformation of geranyl acetate into geraniol.

(–)-Citronellol is used as an acyclic, chiral pool terpene building block. It is easily transformed into both citronellal and citronellic acid via the oxidation process. Similarly, linalool is most readily available as the (–) enantiomer that can be easily prepared in either enantiomeric form through asymmetric epoxidation of geraniol, mesylation and reductive ring-opening reaction. Geraniol is acyclic monoterpene alcohol with the molecular formula $C_{10}H_{18}O$. Chemically, it is 3,7-dimethylocta-trans-2,6-dien-1-ol. Dubey et al. studied the transformation of geranyl acetate into geraniol in presence of geranyl acetate esterase as presented in Figure 8.12 [23].

Gramatica et al. reported that geraniol is considered as starting material to obtain enantiomerically pure form of (R)-(+)-citronellol by microbiological (*Saccharomyces cerevisiae*) reduction process [24]. Luan et al. also demonstrated that geraniol can be stereoselectively reduced to (S)-citronellol, E/Z-isomerization to nerol and oxidation to neral/geranial [25].

Chiral pool terpene syntheses are influenced by several factors such as the current availability of the starting terpene building blocks, current state of the art in synthetic methodology and working methods designed by the researcher. Treatment of (–)-limonene with thiophenol and O_2 led to a

FIGURE 8.11 Synthesis of phomactin A from (*R*)-(+)-pulegone.

Phomactins are the diterpenoids first isolated from cultures of the fungus (*Phoma* sp.) in 1991, whereas Pulegone is a naturally occurring compound of monoterpene class. It is the active constituent present in the Mentha species (*Mentha pulegium* L) [22].

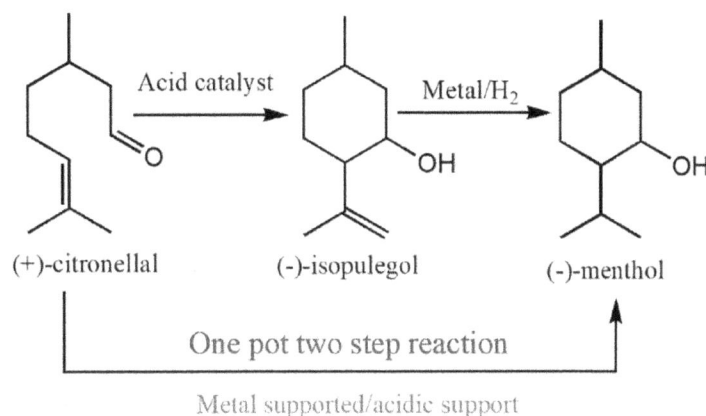

FIGURE 8.13 Production of menthol from citronellal.

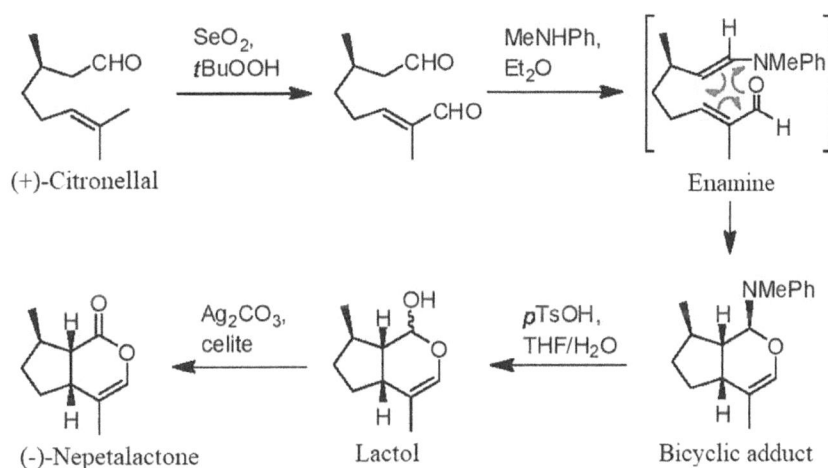

FIGURE 8.14 Synthesis of (–)-Nepetalactone.

cascade peroxidation forming the bicyclic hydroperoxide. The bicyclic monoterpene (+)-fenchone is converted into its corresponding oxime and then subjected to Beckmann fragmentation to afford nitrile [26,27]. (R)-(+)-limonene can be obtained from orange peel oil. Biotransformation of this monoterpene by the basidiomycete *Pleurotus sapidus* led to the formation of cis/trans-carveol and carvone [28].

Luque et al. reported the one-step microwave-assisted asymmetric cyclization/hydrogenation of citronellal to menthols using supported nanoparticles on mesoporous materials. Pt/Ga-MCM-41 is found to be the prime catalyst for this reaction (Figure 8.13). It involves the cyclization and isomerization of (+)-citronellal to (–)-isopulegol by using ZnBr$_2$ as a catalyst followed by the hydrogenation of the *exo*-double bond in the isopulegol to produce (–)-menthol. The reaction reactions are performed under microwave radiation (CEM DISCOVER model). Microwave synthesizer is connected with PC and monitored by sampling aliquots of the reaction mixture [29].

The monoterpene (–)-nepetalactone is a natural product present in *Nepeta nuda*, *Nepeta sibirica*. Chemically, it is 4,7-dimethyl-5,6,7,7a-tetrahydro-4aH-cyclopenta[c]pyran-1-one. (+)-citronellal is the starting material for the synthesis of (–)-nepetalactone (Figure 8.14) [30].

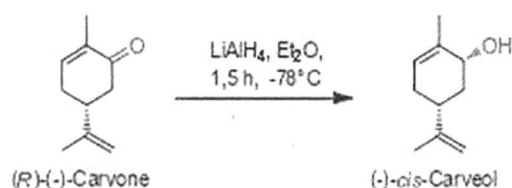

FIGURE 8.15 Transformation of (–)-cis-carveol from R-(–)-carvone.

(–)-cis-carveol is the monocyclic terpene with the molecular formula C$_{10}$H$_{16}$O. (–)-cis-carveol is the major constituent of *Xylopia aromatica*, *Origanum syriacum*. Chemically, it is (1R, 5R)-2-methyl-5-prop-1-en-2-ylcyclohex-2-en-1-ol. It is the enantiomer of a (+)-cis-carveol. (–)-cis-carveol is produced by chemo- and diastereoselective reduction of R-(–)-carvone. The transformation of the carbonyl group into alcohol can be performed using lithium aluminum hydride, Luche-reagent (sodium borohydride-cerium chloride) or with DIBAH (diisobutyl aluminum hydride). The screening results reported that (–)-cis-carveol has exhibited positive results in the prevention of breast cancer (Figure 8.15) [31].

The transformation of (R)-(–)-carvone into (R)-(+)-3-methyl-6-isopropenyl-cyclohept-3-enone-1 is based on

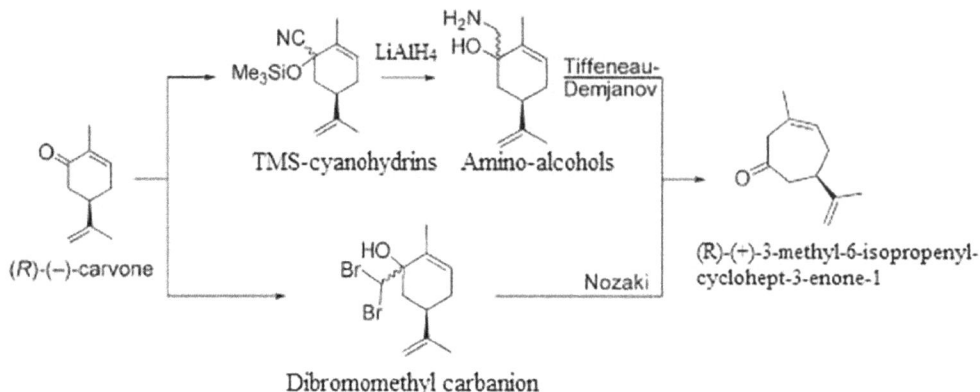

FIGURE 8.16   Ring expansions of carvone by the Tiffeneau–Demjanov rearrangement and Nozaki reaction.

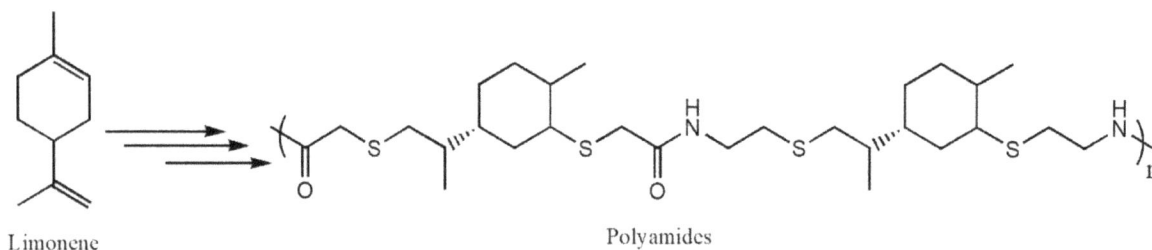

FIGURE 8.17   Polyamide synthesized from limonene.

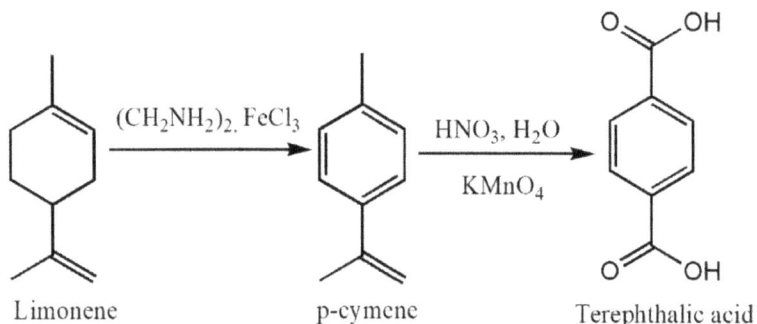

FIGURE 8.18   Synthesis of terephthalic acid from limonene.

chemo- and regioselective Tiffeneau–Demjanov ring expansion reaction. It involves the addition of TMS-cyanide with (R)-(–)-carvone to provide TMS-cyanohydrins which on reduction with c furnishes the target amino-alcohols that leads to the non-conjugated cycloheptenone [(R)-(+)-3-methyl-6-isopropenyl-cyclohept-3-enone-1] *via* Tiffeneau–Demjanov rearrangement. Similarly, the addition of the dibromomethyl carbanion to carvone *via* Nozaki ring expansion and subsequent rearrangement reaction produces (R)-(+)-3-methyl-6-isopropenyl-cyclohept-3-enone-1 (Figure 8.16) [32].

Limonene is an aliphatic hydrocarbon of cyclic monoterpene class with the molecular formula $C_{10}H_{16}$. It is the major constituent present in the oil of citrus fruits. Chemically, it is 1-methyl-4-prop-1-en-2-ylcyclohexene. The synthesis of renewable polyamides with molecular weights up to

$Mn = 12 kDa$ can be obtained *via* selective combination of thiol-modified limonene (Figure 8.17) [33].

Terephthalic acid is produced by selective oxidation of *p*-cymene (yield: 51%) in presence of Mn-Fe mixed-oxide heterogeneous catalyst with O_2 as the oxidant. p-Cymene is a monoterpene with the molecular formula $C_{10}H_{14}$. It consists of a benzene ring substituted with a methyl group and an isopropyl group at a para position in its structure. p-Cymene is synthesized from biodegradable terpenes (limonene, or eucalyptol) (Figure 8.18) [34].

Limonene undergoes isomerization to produce terpinolene and also hydrogenates to furnish 1-menthene, and then each compound undergoes transformation to yield p-cymene mutually. During the reaction, the formation of α-terpinolene demonstrates that isomerization reactions occurs at the Pt/Al_2O_3 catalyzed system in SC-ethanol (Figure 8.19) [35].

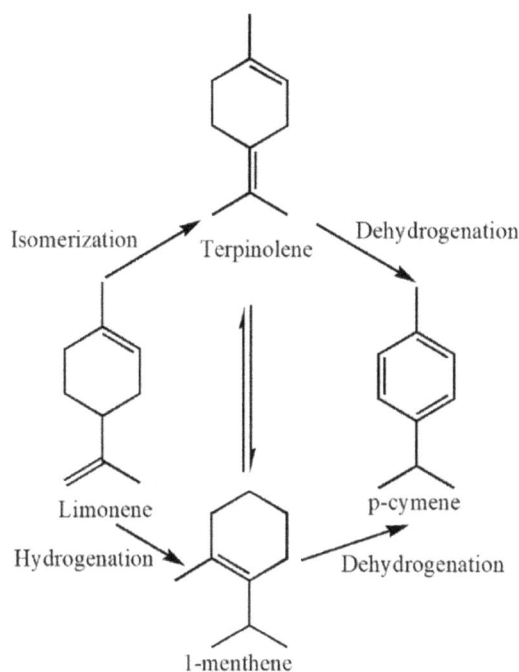

FIGURE 8.19 Possible reaction pathways over Pt/Al$_2$O$_3$ in SC-ethanol.

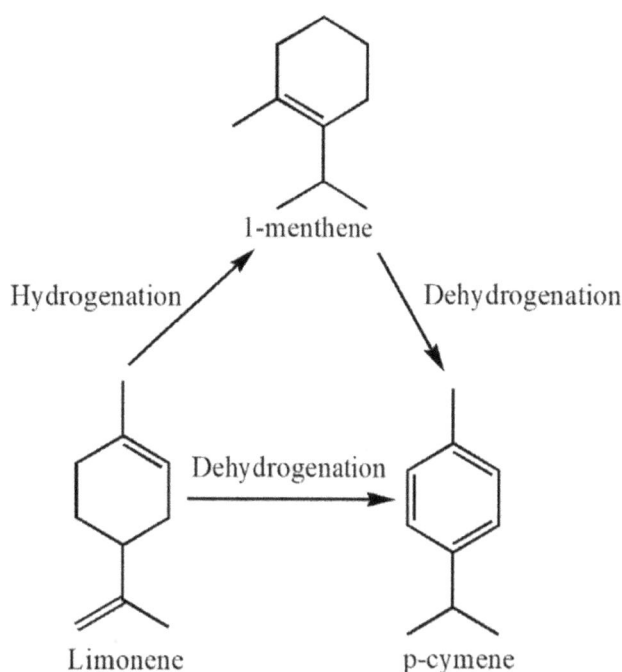

FIGURE 8.20 Possible reaction pathways over Pd/Al$_2$O$_3$ in SC-ethanol.

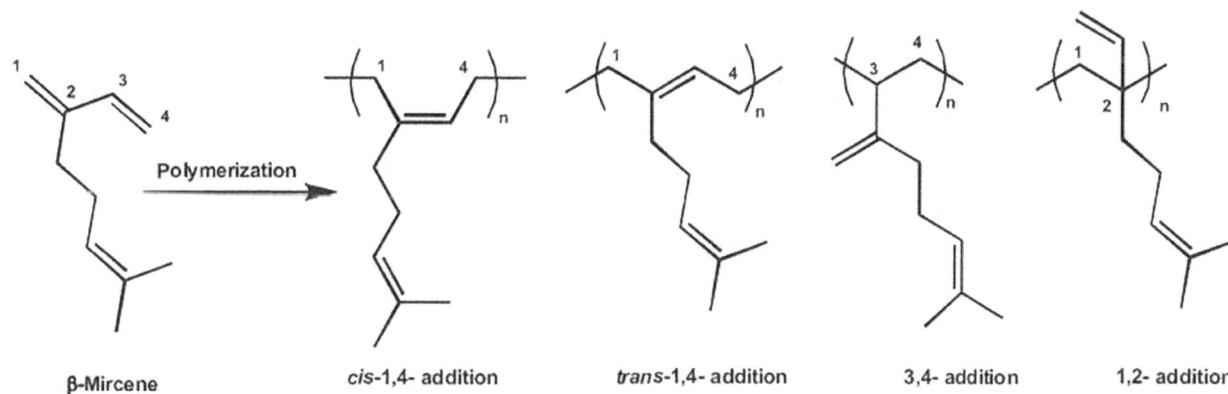

FIGURE 8.21 Polymerization reaction of myrcene.

In the case of SC-ethanol medium, Pd/Al$_2$O$_3$ catalyst promotes the production of p-cymene. Isomerization reactions do not take place in presence of this catalyst. On the other hand, the selectivity of p-cymene is found to be low over the Pt/Al$_2$O$_3$ due to the formation of several undesired products for all reaction conditions as compared to Pd/Al$_2$O$_3$ (Figure 8.20) [36].

8.2.1 Polymerization of Terpenes

Terpenes have emerged as suitable starting compounds to serve as building blocks for the synthesis of polymers. The polymers obtained from terpenes are mostly elastomers. Depending on the type of terpenes (acyclic terpenes: myrcene, ocimene and cyclic terpenes: α-pinene, β-pinene, limonene and phellandrene), polymerization can be classified into two main groups

such as cyclic and acyclic type. In the polymerization of terpenes, the mechanisms of reaction involve the polymerization of monomers via free radical, cationic, anionic, coordination and coordinative chain-growth mechanisms [37].

The polymerization reaction of myrcene takes place via a radical mechanism. Marvel et al reported the anionic polymerization using nBuLi to produce polymers [31]. The final structure of polymyrcene is influenced by the solvent and temperature (Figure 8.21). When the polymerization is carried out in nonpolar solvents like hexane, polymers with about 85% of 1,4-cis units and 15% of 3,4 defects are produced. If a polar solvent (THF) is used, the presence of 3,4-units increased up to 39%–44%. The anionic copolymerization with styrene, isoprene, or farnesene has been described to furnish block-like copolymers that are used as precursors for bio-based thermoplastic elastomers [38].

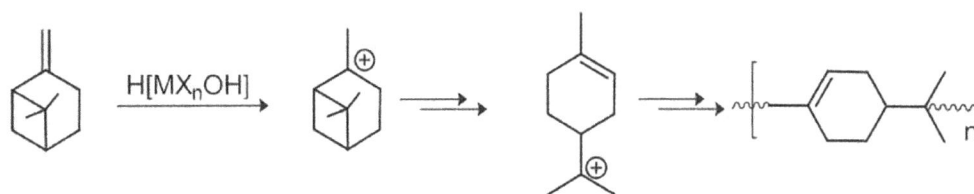

FIGURE 8.22 Mechanism for the cationic polymerization of β-pinene.

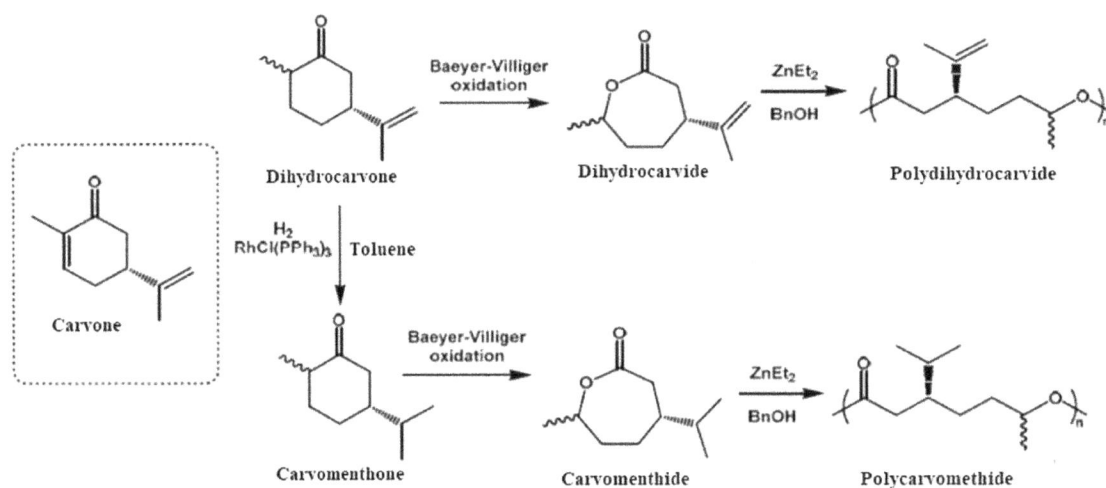

FIGURE 8.23 Production of terpene-based polymers from carvone.

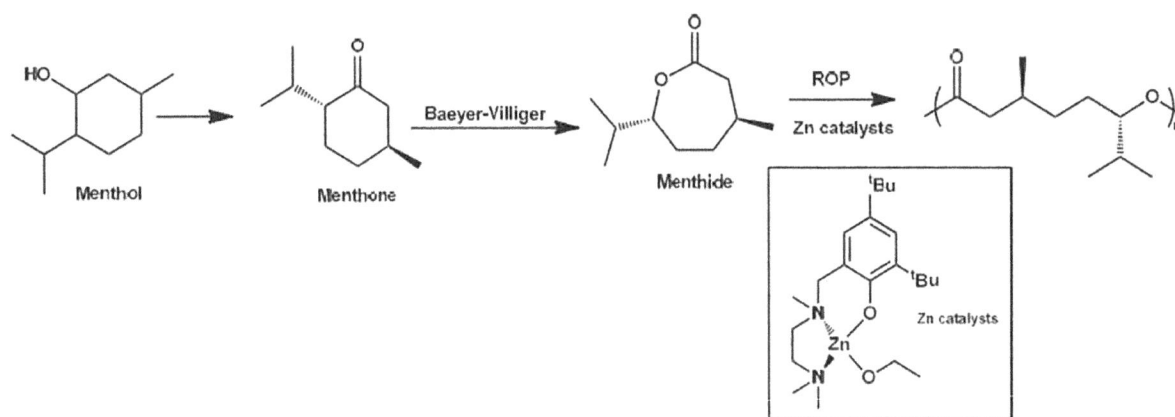

FIGURE 8.24 Production of terpene-based polymers from menthol.

In the case of pinene, both α- and β-isomers can be polymerized. But β-pinene reacts more readily due to facile ring opening and the relief of ring strain. The polymerization of Pinene generates terpene resins that can be used as thermoplastics and coatings pressure-sensitive adhesives. Lewis acids such as metal chlorides are mainly used as initiators. The activity of the catalysts depends on the strength of the Lewis acidity and the effectiveness decreases in the order of $AlBr_3 > AlCl_3 > Zr$ $Cl_4 > AlCl_3 \cdot Et_3O > BF_3 > BF_3 \cdot Et_2O > SnCl_4 > SbCl_3 > BiCl_3 > Zn$ Cl. The presence of water favors the reaction condition due to the formation of the intermediate species (HMXnOH). This

species is obtained from the reaction between the Lewis acid and water. It is the first step of the reaction polymerization of β-pinene (Figure 8.22). The transfer of the proton to the monomer to produce a cyclic carbocation, which then isomerizes. It is a key step in polymerization [39].

Carvone is a sustainable and commonly available starting material for the synthesis of organic compounds (Figure 8.23). The transformation of carvone and menthol into lactones is carried out via the Baeyer Villager oxidation process (Figure 8.24). Polymers are produced via ring-opening polymerization (ROP) reaction [40].

(a)

(b)

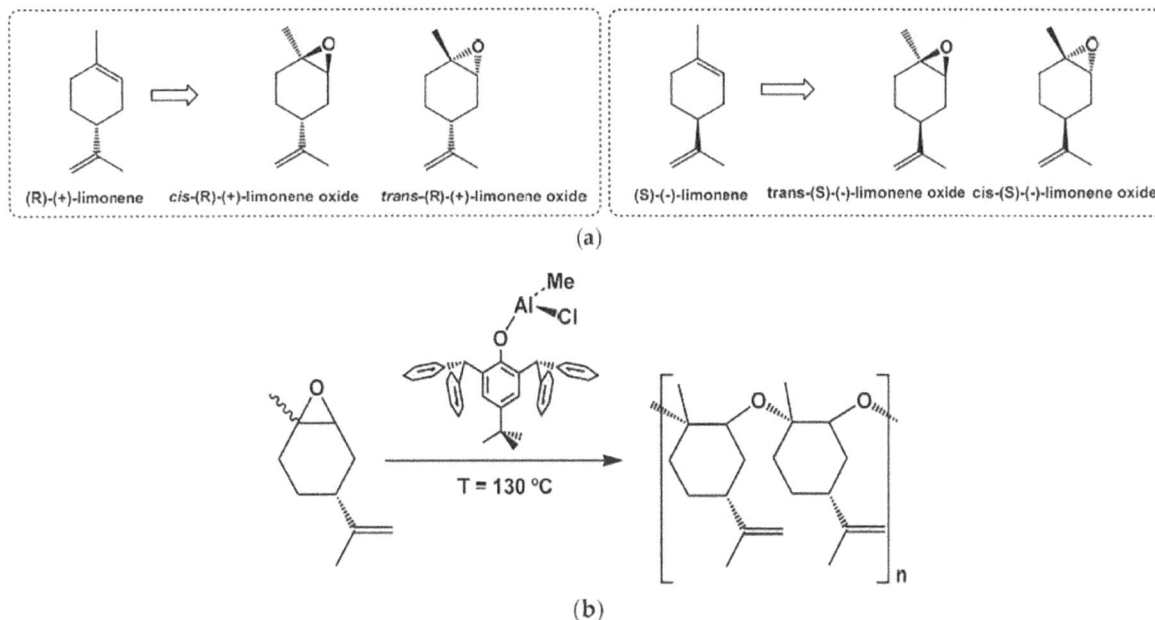

FIGURE 8.25 Polymerization of limonene via ROP using [AlMeCl(OR)] as catalyst.

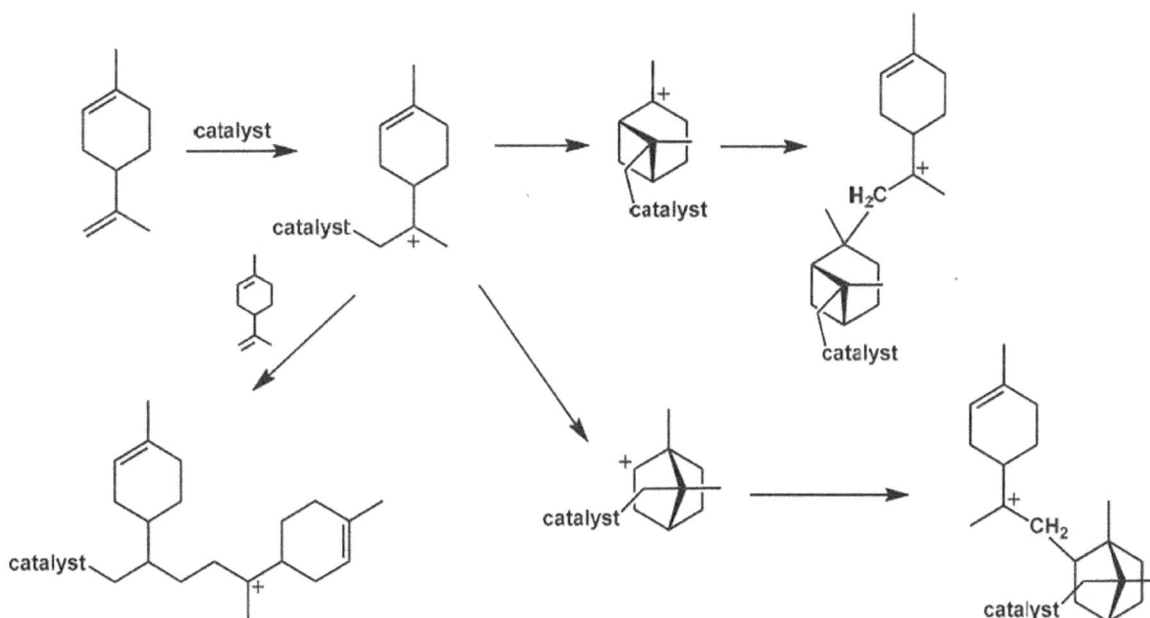

FIGURE 8.26 Cationic mechanism of limonene polymerization.

Terpene epoxides are also used as monomers. So, limonene 1,2-oxides are used to produce polymers [41]. This epoxide can be obtained via oxidation of the limonene. Limonene oxide can be polymerized due to the presence of the suitable functional group. If the polymerization takes place via the oxirane group, a polyether is generated via the ROP process. In this process, the active aluminum catalyst for ROP is [AlMeCl(OR)] (OR=2,6-(CHPh$_2$)$_2$-4-^{t}Bu-C$_6$H$_2$O). This catalyst is able to polymerize (+)-limonene oxide efficiently in 30 min with a preference for cis-isomer polymerization as presented in Figures 8.25 and 8.26 [42].

8.3 Sesquiterpenoids as Starting Compounds

Newhouse et al. reported the synthesis of a degraded limonoid (melazolide B) from the natural terpene precursor. Limonoids are the oxygenated modified triterpenes that are biosynthesized from the acetate-mevalonate pathway in citrus fruits. In this synthetic pathway, the chemoselective carbonyl α, β-dehydrogenation and a Wharton reduction are key strategic steps. α-Ionone is a natural product of sesquiterpenoids class and is present in Chromolaena odorata and Swertia japonica (Figure 8.27) [43].

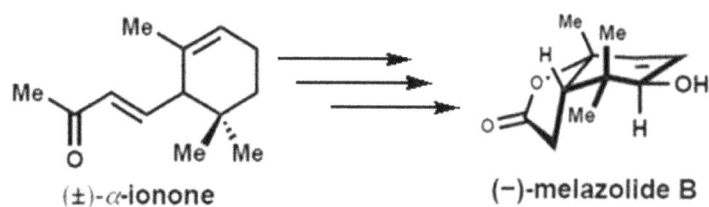

FIGURE 8.27 Synthesis of melazolide B from α-ionone.

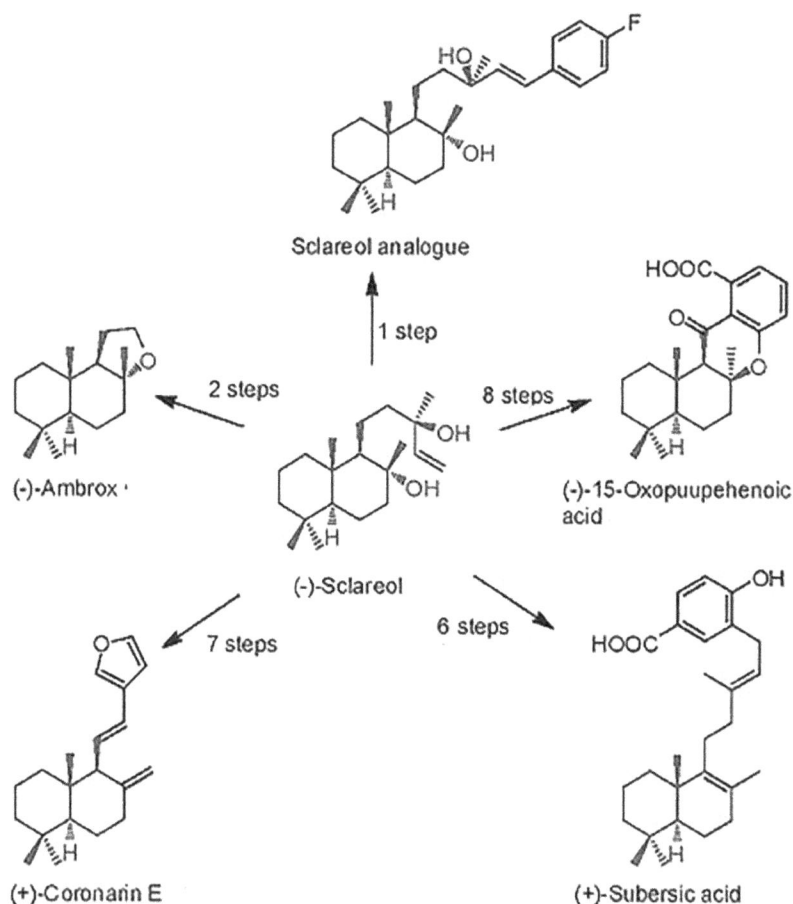

FIGURE 8.28 (−)-Sclareol-derived natural products and derivatives.

Diterpenoids as Starting Compounds

Sclareol is a labdane type bicyclic diterpenoid alcohol with molecular formula $C_{20}H_{36}O_2$ (Figure 8.28). The structure of sclareol comprises labd-14-ene substituted by hydroxy groups at positions 8 and 13. It is isolated from *Salvia sclarea*. It possesses anticancer activity. (−)-sclareol is considered as starting material for generating several compounds. (−)-Ambrox is produced on large scale by partial synthesis from (−)-sclareol. Similarly, (−)-15-oxopuupehenoic acid, (+)-coronarin E, (+)-subersic acid and (+)-scalarolide are prepared from (−)-sclareol [44].

8.4 Miscellaneous

Terpene indole alkaloids are a class of natural products with diverse chemical structures. These compounds exhibit a wide range of biological activities such as anticancer, antimalarial and anti-arrhythmic agents. Vincristine and vinblastine are used clinically to treat cancers such as Hodgkin's disease and Kaposi's sarcoma. Ajmalicine is a monoterpenoid indole alkaloid with the molecular formula $C_{21}H_{24}N_2O_3$. It is isolated from Rauvolfia and Catharanthus species. It is a selective α1-adrenoceptor antagonist used for the treatment of high blood pressure (Figure 8.29). Yohimbine is an indole alkaloid with alpha2-adrenoceptor antagonist activity. It is produced by *Corynanthe johimbe* and *Rauwolfia serpentina*. Similarly, Strychnine is a monoterpenoid indole alkaloid with molecular formula $C_{21}H_{22}N_2O_2$. It is used as rodenticide. *Catharanthus roseus* produces several terpenoid indole alkaloids (vincristine and vinblastine). These are used as chemotherapy drugs in combination with other drugs to treat lymphoma and leukemia. In the case of *Catharanthus roseus*, terpenoid indole

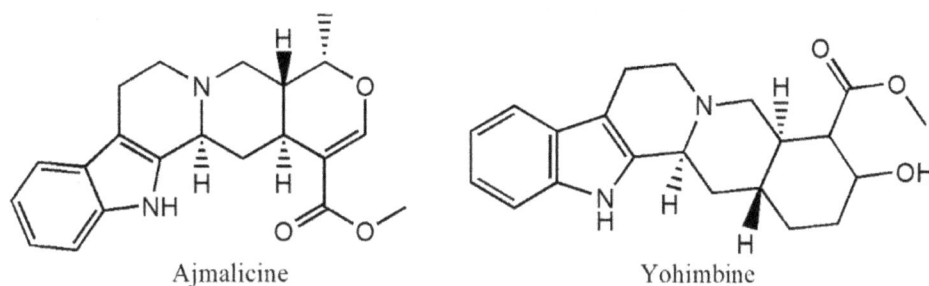

FIGURE 8.29 Structures of terpene indole alkaloids.

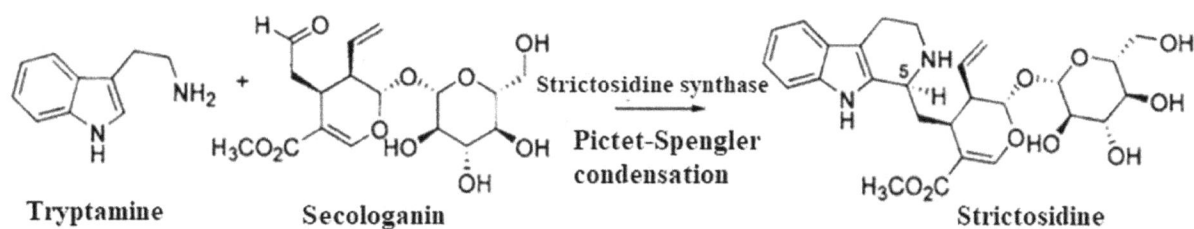

FIGURE 8.30 Biosynthesis of terpene indole alkaloid.

alkaloids are derived from the precursor strictosidine, which is a fusion product of the shikimate pathway derived tryptamine moiety and the plastidic nonmevalonate pathway derived secologanin moiety [45].

Terpene indole alkaloids are derived from tryptophan. Tryptophan decarboxylase is a pyridoxal-dependent enzyme that converts tryptophan to tryptamine. Strictosidine synthase catalyzes the stereoselective Pictet-Spengler condensation of tryptamine and the aldehyde secologanin to produce strictosidine. The Pictet-Spengler cyclization is essential for the biosynthesis of alkaloids as presented in Figures 8.30 and 8.31 [46–48].

Epoxidation of the terminal isoprene group by monooxygenases is utilized for the biosynthesis of steroids. This biocatalytic process also reduces the activation barrier for subsequent cyclization steps, as protonation of an oxirane is facilitated compared with that of a terminal isoprene group (Figure 8.32) [49].

Norcamphor is a bicyclic terpene ketone ($C_7H_{10}O$). It is used as a building block in organic synthesis. It is the precursor to produce norborneols [50]. Further, it is oxidized into the corresponding lactone via chemical and enzymatic methods. The chemically synthesized lactone is further polymerized via cationic ring-opening polymerization (CROP). The chemical conversion of norcamphor into its lactone derivatives via Baeyer–Villiger (BV) oxidation reaction, and the feasibility to polymerize these lactones through ROP using methanesulfonic acid (MSA) as a catalyst [51]. BV oxidation is reported as an efficient reaction to convert cyclohexanones to lactones [52]. The oxidation of norcamphor by BV using *m*-CPBA as an oxidizing agent can result in the formation of lactones as presented in Figure 8.33 [53].

Scianowski et al. reported terpenes as green starting materials for generating new, chiral organoselenium and organotellurium compounds (Figure 8.34). Organoselenium compounds

are mainly used as reagents and catalysts in several organic transformations. Synthetic protocols are followed to utilize camphor in designing new terpene chalcogenides. The selenium moiety can be introduced into the camphene skeleton in various positions such as C_2, C_3, C_9 and C_{10}. These compounds are classified based on the location of the selenium atom in the structure of the terpene scaffold [54].

Martinez et al demonstrated the synthesis of C10-Se-substituted camphor from (1*R*)-3,3-dimethyl-2-methylenenorbornan-1-ol by using electrophilic selenium reagents (Figure 8.35). Further, the reaction proceeds by adding phenyl selenenyl chloride as electrophile to the carbon-carbon double bond, followed by Wagner-Meerwein rearrangement that enables to get only (1*S*)-10-(phenylselanyl)camphor enantioselectively under mild reaction conditions in high yield [55].

Koizumi et al reported the synthesis of C_{10} camphor derivatives. 2-*exo*-hydroxy-10-bornyl group is used as a chiral ligand for the synthesis of optically pure haloselenuranes. Selenide is produced by the reaction of (*S*)-10-bromo-2-*exo*-borneol with an appropriate selenolate. Here, selenolate is produced by the reaction of diselenide with sodium borohydride [48]. Finally, Selenide undergoes oxidation with *t*-BuOCl enabled to obtain selenurane (Figure 8.36) [56,57].

Sharpless et al. presented the [2,3]-Sigmatropic rearrangement of Se-terpenyl compounds (Figure 8.37). The oxidation of geranyl selenol, selenide and diselenide with H_2O_2 led to the formation of linalool. Diselenide is oxidized to produce seleninic acid which undergoes rearrangement to produce alcohol [58].

Monoterpenes are also used as substrates to synthesize various tellurium-containing compounds. Clive et al. demonstrated the transformation of carvone 7,8-epoxide to telluride (Figure 8.38) in presence of sodium benzenetellurolate (PhTeNa) which is then selectively reduced to alcohol with triphenyl tin hydride (Ph₃SnH) [59].

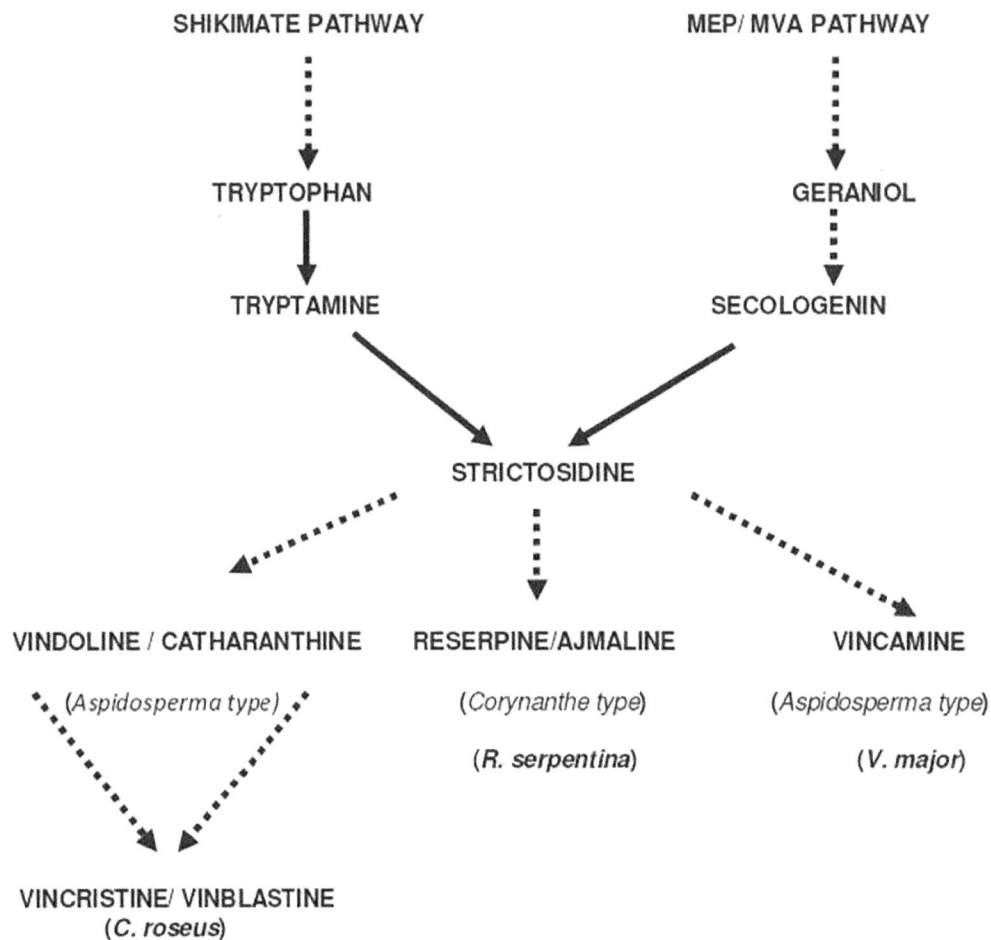

FIGURE 8.31 Strictosidine as universal precursor for terpene indole alkaloids.

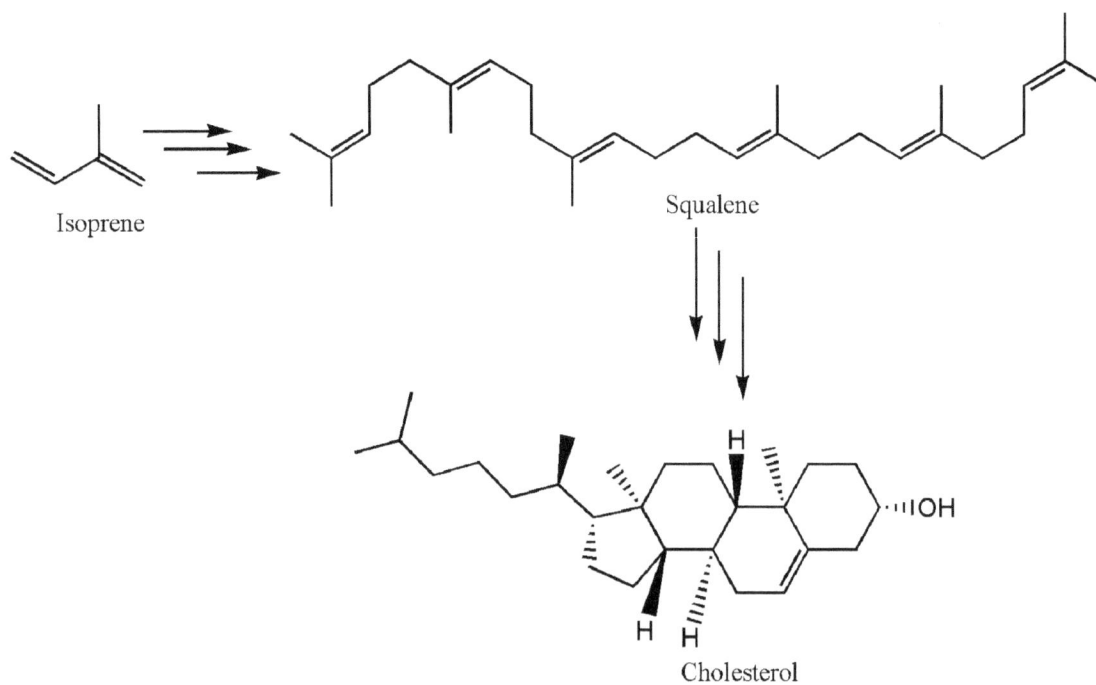

FIGURE 8.32 Biosynthesis of cholesterol from squalene.

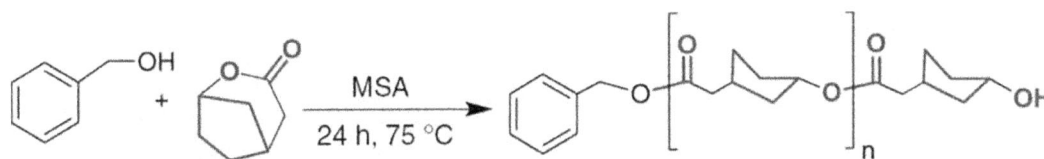

FIGURE 8.33 Ring-opening polymerization of lactone using MSA as a catalyst.

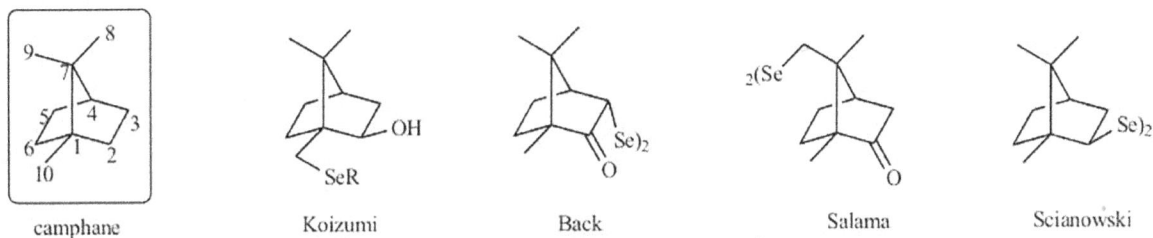

FIGURE 8.34 Possible Se-functionalization of the camphane skeleton.

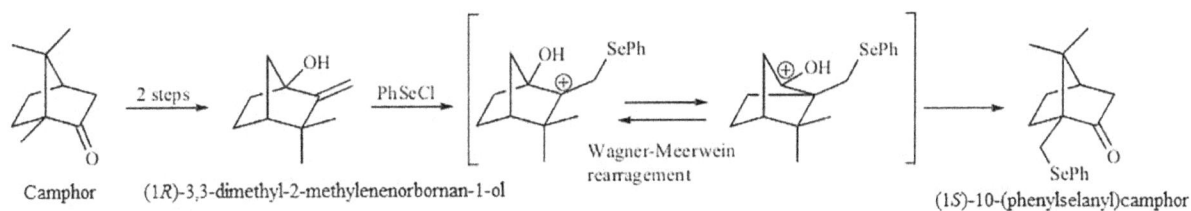

FIGURE 8.35 Synthesis of (1*S*)-10-(phenylselanyl)camphor.

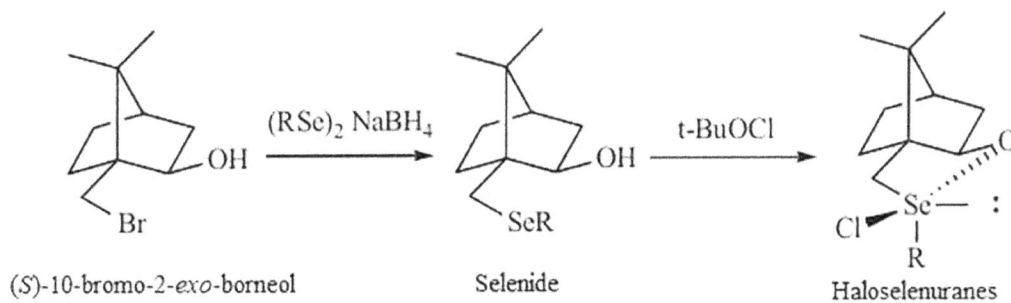

FIGURE 8.36 Scheme for the synthesis of haloselenuranes.

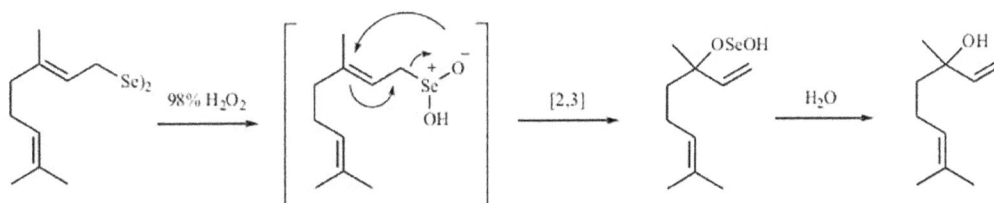

FIGURE 8.37 Conversion of geranyl diselenide to linalool.

8.5 Terpenoid Alkaloids

Terpenoid alkaloids are plant natural products with diverse

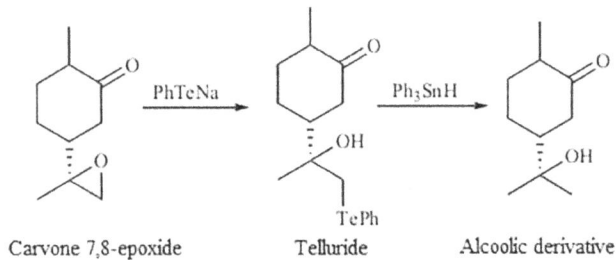

Carvone 7,8-epoxide   Telluride   Alcoolic derivative

FIGURE 8.38 Transformation of carvone 7,8-epoxide.

chemical structures and promising biological activities [60]. Depending on biogenesis, alkaloids are mainly classified into two types such as true alkaloids and pseudoalkaloids. Majority of the alkaloids are true alkaloids that are derived from α-amino acid precursors [61]. Other alkaloids such as terpenes and steroids are called as pseudoalkaloids because a relatively late amination reaction occurs in a transamination process by donating a nitrogen atom of the amino acid source. A large number of alkaloids based on mono- and diterpenoid skeleton are produced in the amination reaction. So terpenoid alkaloids are classified into hemiterpenoid alkaloids (alkaloids with a C_5 unit), monoterpenoid alkaloids (alkaloids with a C_{10} unit, e.g. actinidine), sesquiterpenoid alkaloids (alkaloids with a C_{15} unit, e.g. nupharidine), diterpenoid alkaloids (alkaloids with a C_{20} unit, e.g. aconitine), sesterterpenoid alkaloids (alkaloids with a C_{25} unit) and triterpenoid alkaloids (alkaloids

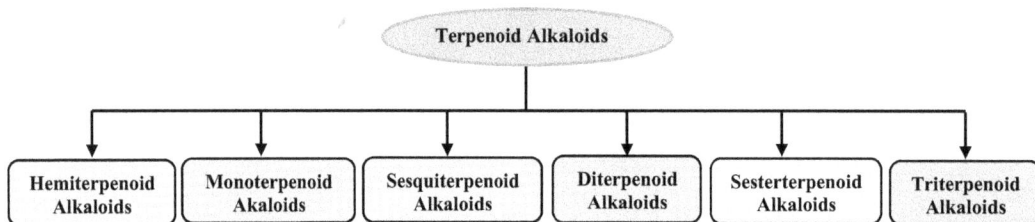

FIGURE 8.39 Classification of Terpenoid alkaloids.

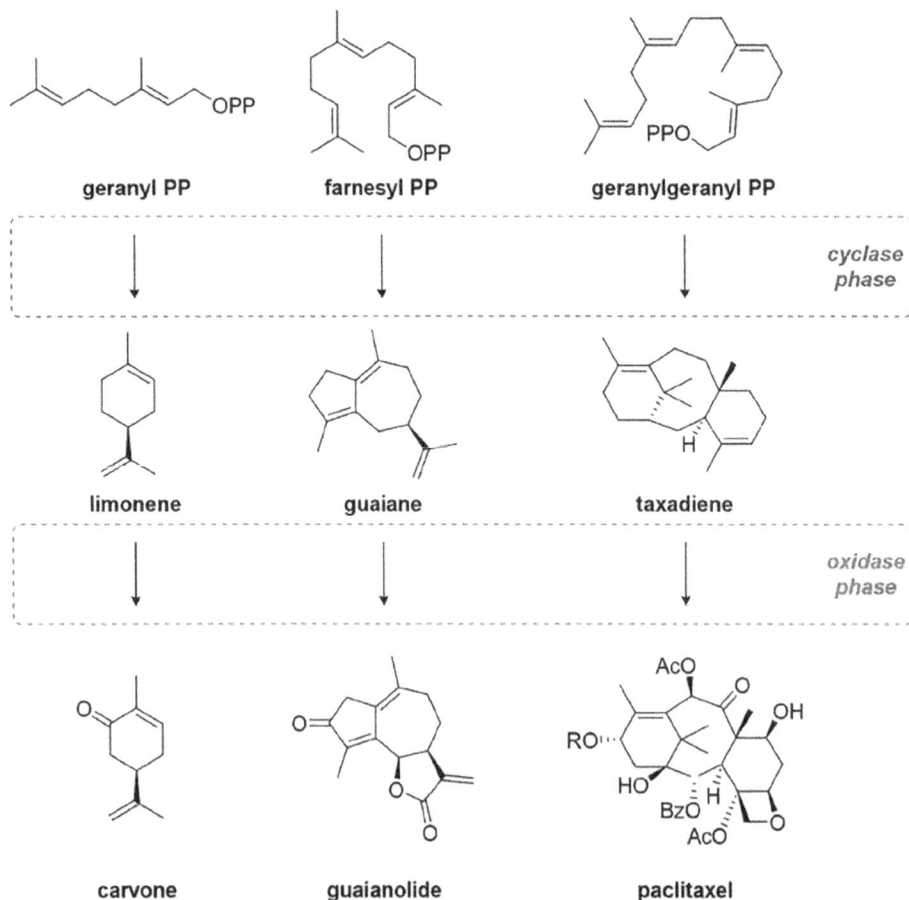

FIGURE 8.40 Biosynthesis of terpenoids in cyclase and oxidase phases.

with a C_{30} unit). Monoterpenoid alkaloids are commonly present in the plant source of family Gentianaceae whereas the diterpenoid alkaloids are mainly found in the plant source of family Ranunculaceae (Figures 8.39 and 8.40) [62].

8.5.1 Monoterpenoid Alkaloids

Monoterpenoid alkaloids are mainly obtained from several monoterpenes. These alkaloids are produced by addition of nitrogen atoms to an existing monoterpene of secologanin type. Various monoterpenoid alkaloids are isolated and characterized based on their structural properties. The more common and biologically active monoterpenoid alkaloids include actinidine, skytanthine, gentianine, etc. Most of the monoterpenoid alkaloids are isolated from the genus *Incarvillea* of family Gentianaceae. Actinidine is one of the rare types of monoterpenoid alkaloid [63]. It is a steam-volatile methylcyclopentane pyridine derivative present mainly in the essential oils of the species of *Actinidia* and *Incarvillea*. Delavayine B

and delavayine C are isolated from the aerial part of *Incarvillea delavayi*. Two monoterpenoid alkaloid dimers (argutane A and argutane B) are isolated from the roots of *I. arguta*. Similarly, other two monoterpenoid alkaloids such as argutine A and argutine B are isolated from *I. arguta*, Argutine A and argutine B have α, β-unsaturated carbonyl group and an amide moiety respectively in their structures as presented in Table 8.1 and Figure 8.41 [64].

4-nor-7,8-dehydro-10-hydroxy-skytanthine is isolated from *Argylia radiate* [65]. Delavayine A is a skytanthine-type alkaloid isolated from *I. delavayi*. Delavayine A comprises benzoate ester in its structure [66]. Similarly, incarvillateine E is another skytanthine-type alkaloid isolated from the aerial parts of *I. sinensis*. It possesses three incarvilline moieties in its structure. Incarvine E, incarvine F are isolated from *I. sinensis*, whereas mairines A–C and isoincarvilline are isolated from *I. mairei*. Mairines A–C contain δ-skytanthine N-oxide in their structures as depicted in Table 8.2 and Figure 8.42 [67].

TABLE 8.1

List of Actinidine-Type Monoterpenoid Alkaloids

Monoterpenoid Alkaloids	Molecular Formula	Plant Source
Delavayine B	$C_{21}H_{25}NO_4$	*Incarvillea delavayi*
Delavayine C	$C_{16}H_{24}NO_4$	*Incarvillea delavayi*
Incargutosine C	$C_{13}H_{17}NO_2$	*Incarvillea arguta*
Incargutosine D	$C_{13}H_{17}NO_2$	*Incarvillea arguta*
10-Acetoxy-actinidine	$C_{12}H_{15}NO_2$	*Argylia radiata*
Argutane A	$C_{20}H_{22}N_2O_2$	*Incarvillea arguta*
Argutane B	$C_{20}H_{22}N_2O_2$	*Incarvillea arguta*
Argutine A	$C_{13}H_{15}NO$	*Incarvillea arguta*
Argutine B	$C_{10}H_{12}N_2O$	*Incarvillea arguta*

TABLE 8.2

List of Skytanthine-Type Monoterpenoid Alkaloids

Monoterpenoid Alkaloids	Molecular Formula	Plant Source
4-Nor-7,8-dehydro-10-hydroxy-skytanthine	$C_{10}H_{17}NO$	*Argylia radiata*
Delavayine A	$C_{19}H_{28}NO_2$	*Incarvillea delavayi*
Incarvillateine E	$C_{53}H_{81}N_3O_8$	*Incarvillea sinensis*
Incarvine E	$C_{21}H_{35}NO_3$	*Incarvillea sinensis*
Incarvine F	$C_{29}H_{47}NO_6$	*Incarvillea sinensis*
Isoincarvilline	$C_{11}H_{21}NO$	*Incarvillea mairei*
Mairine A	$C_{11}H_{21}NO$	*Incarvillea mairei*
Mairine B	$C_{11}H_{21}NO$	*Incarvillea mairei*
Mairine C	$C_{11}H_{19}NO_2$	*I. mairei*

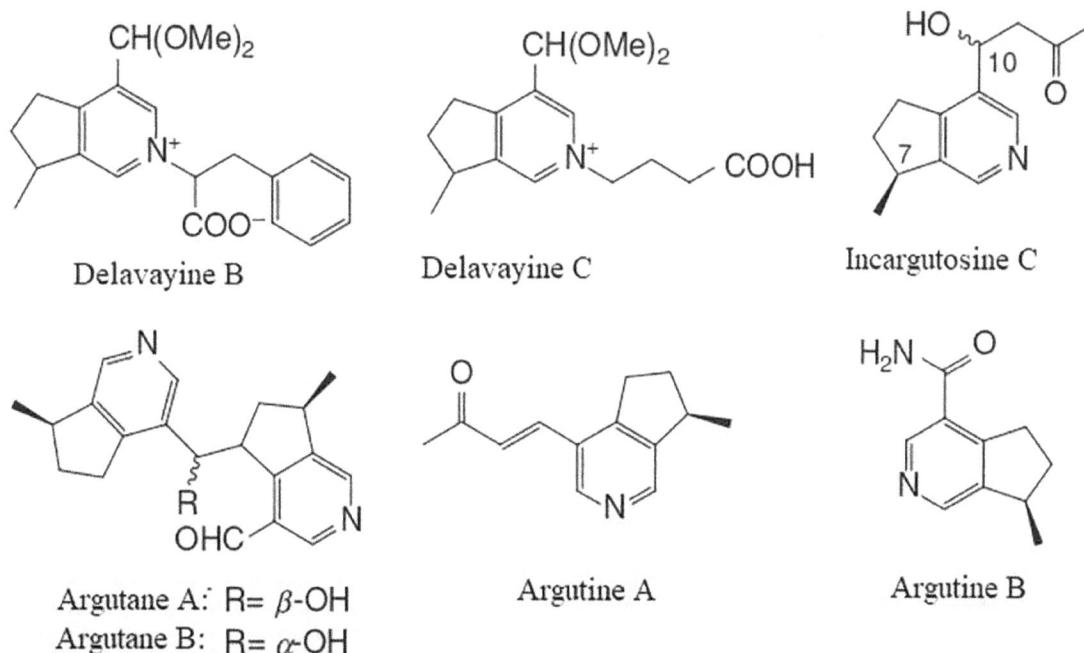

FIGURE 8.41 Structures of actinidine-type monoterpenoid alkaloids.

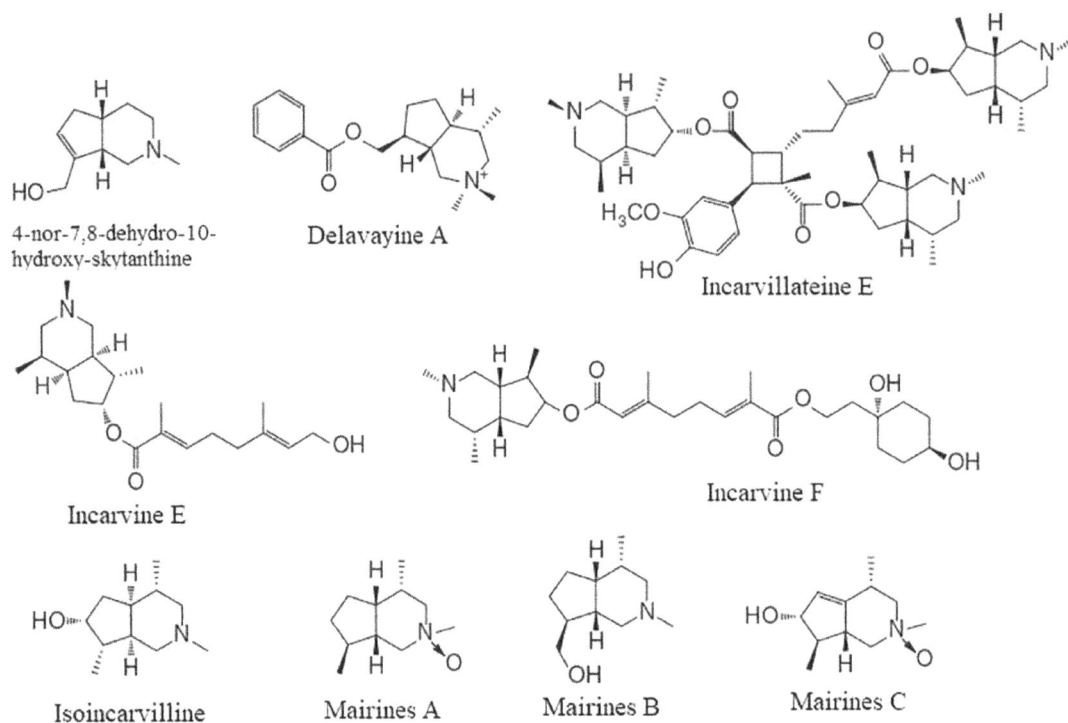

FIGURE 8.42 Structures of Skytanthine-type monoterpenoid alkaloids.

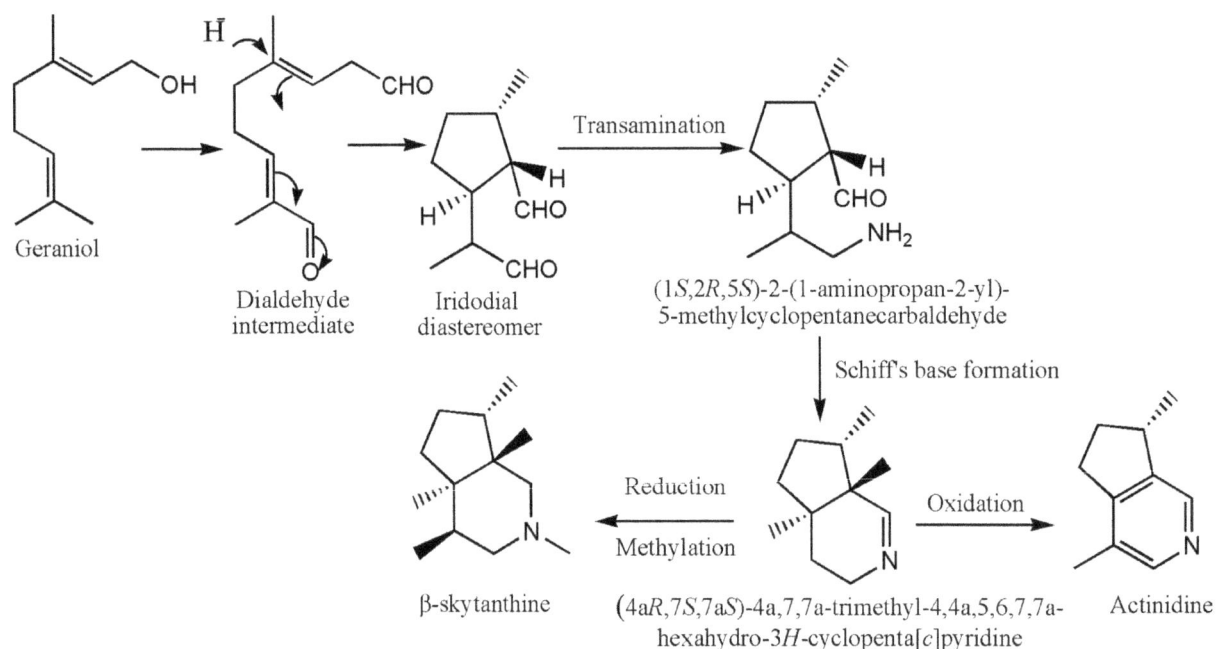

FIGURE 8.43 Biosynthesis of β-skytanthine and actinidine alkaloids from geraniol.

Monoterpenoid alkaloids are structurally related to iridoid compounds in which the oxygen heterocycle is replaced by a nitrogen-containing ring. Geraniol is proposed as starting compound for the biosynthesis of actinidine and β-skytanthine. The amination via an amino acid of an iridodial diastereomer which is produced from geraniol after forming a dialdehyde intermediate yield (1S, 2R, 5S)-2-(1-aminopropan-2-yl)-5-

methylcyclopentanecarbaldehyde and its cyclization product (4aR, 7S, 7aS)-4a, 7,7a-trimethyl-4,4a, 5,6,7,7a-hexahydro-3H-cyclopenta[c]pyridine. Later compound upon reduction and methylation produce β-skytanthine, whereas its oxidation produces actinidine as presented in Figure 8.43 [68].

Valerian is the active constituent of *Valeriana officinalis* of family Caprifoliaceae. It is used for the treatment of insomnia,

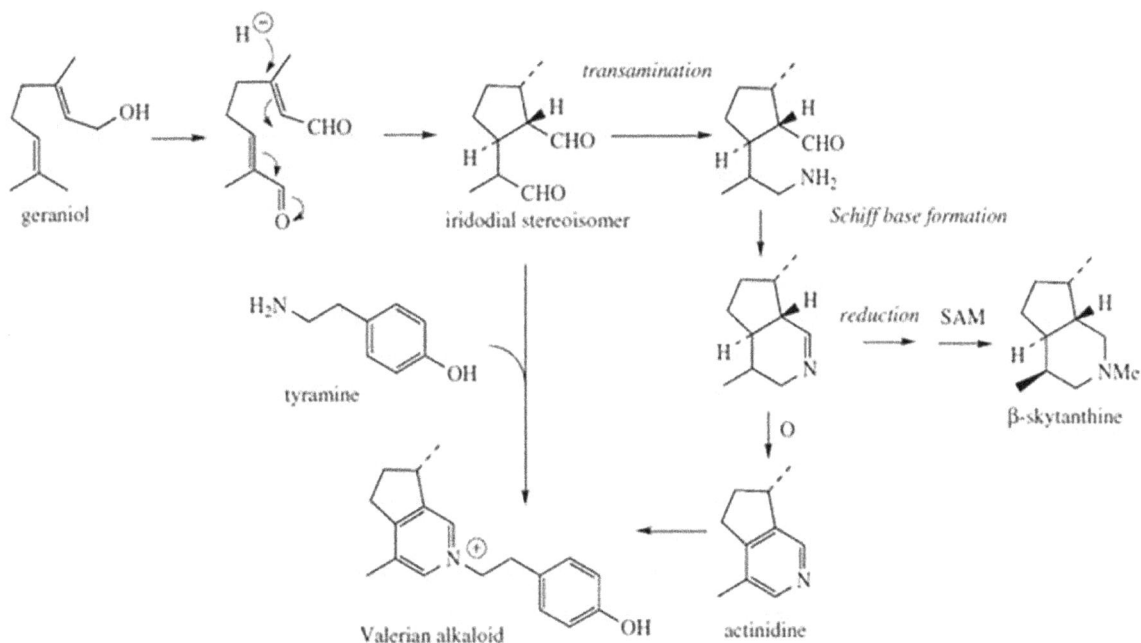

FIGURE 8.44 Valerian alkaloid from geraniol.

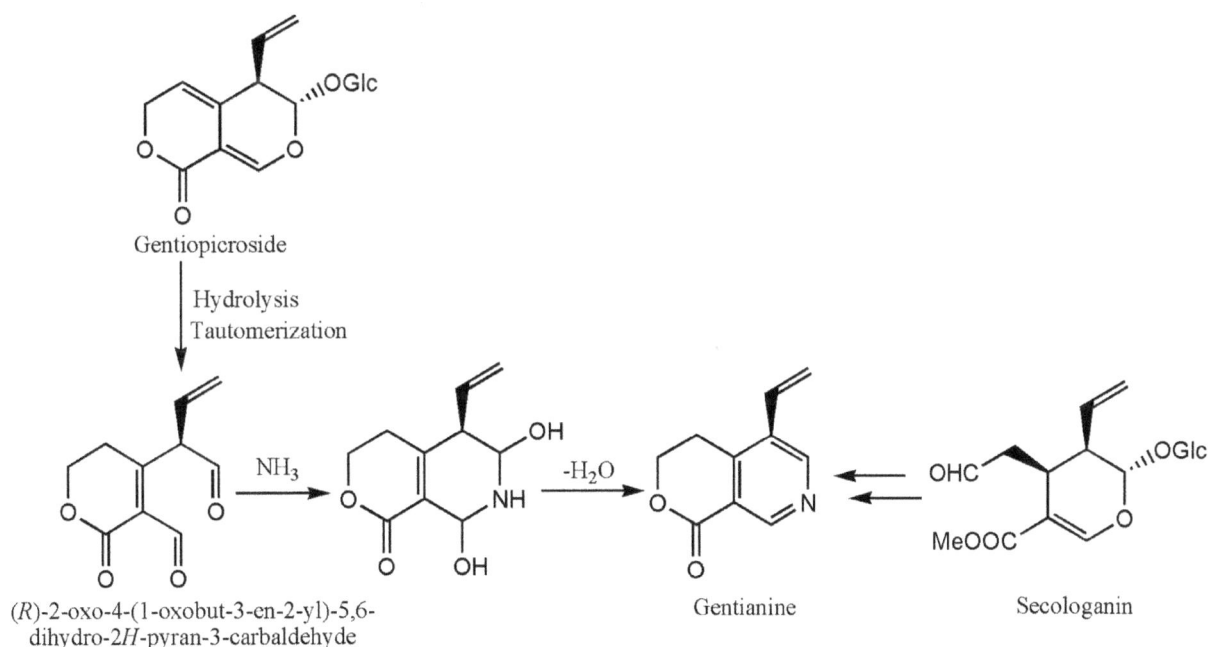

FIGURE 8.45 Biosynthetic pathways of gentianine from gentiopicroside and secologanin.

migraine, fatigue and abdominal cramps. Valerian produces sedation and hypnosis in dose-dependent manner (Figure 8.44) [69].

Gentianine-type monoterpenoid alkaloids are mainly found in plant genus including Anthocleista, Gentiana, Fagraea and Swertia of the family Gentianaceae [70]. Gentianine is a bitter, crystalline monoterpenoid alkaloid that has a wide range of pharmacological activities such as analgesic, central muscle-weakening action, inhibition of provoked aggression. Gentianine is produced from iridoid compounds such

as gentiopicroside or secologanin. Gentiopicroside undergoes enzymatic hydrolysis of the β-glucoside followed by tautomerization to produce (R)-2-oxo-4-(1-oxobut-3-en-2-yl)-5,6-dihydro-2H-pyran-3-carbaldehyde, which upon hemiaminal formation and subsequent dehydration to produce gentianine. Similarly, compound secologanin reacts with ammonia followed by cyclization to give gentianine as depicted in Figure 8.45 [71].

The biosynthetic pathway from *l*-citronellal to an enol half acetal is presented in Figure 8.46. Actinidine is formed from

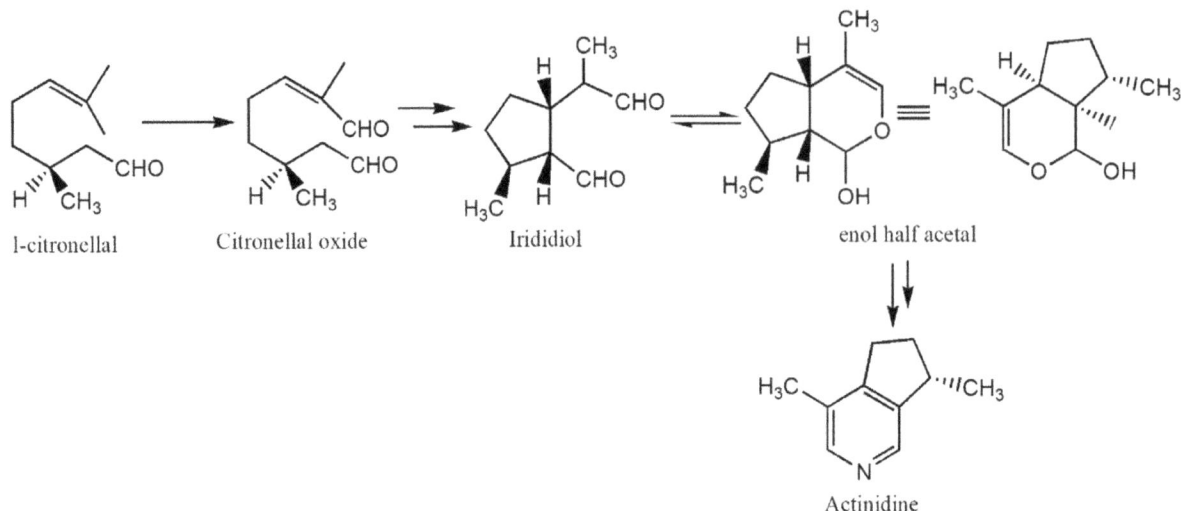

FIGURE 8.46 Biosynthetic pathway of actinidine.

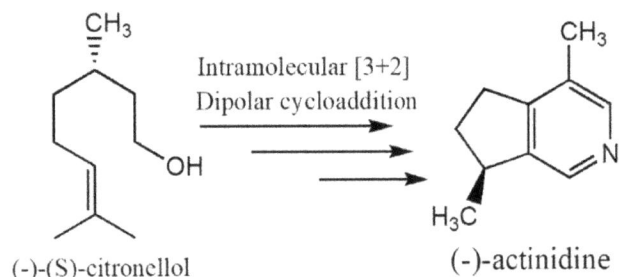

FIGURE 8.47 Synthesis of (−)-actinidine from citronellol.

this enol half acetal type by the incorporation of nitrogen. Actinidine is an iridoid produced naturally in plants and animals. It was the first cyclopentanoid monoterpene alkaloid to be discovered [72].

(−)-Actinidine is the alkaloid of iridane series and is synthesized by using intramolecular [3+2] dipolar cycloaddition of silyl nitronates, generated from (3*R/S*, 6*S*)-2,6-dimethyl-3-nitro-8-phenylthioocta-1,7*7E*- and-1,7*Z*-dienes. The key nitro compounds are generated from (−)-(*S*)-citronellol (Figure 8.47).

Further, the iridoid natural products such as iridomyrmecin, isoiridomyrmecin, teucriumlactone and dolicholactone are prepared from citronellol using a divergent diastereoselective approach. Several key steps involved in this reaction include a highly diastereoselective enamine/enal cycloaddition and the selective reduction of masked aldehyde functionalities by ionic hydrogenation (Figure 8.48) [73].

Similarly, the ant-associated iridoids such as nepetalactol, actinidine, dolichodial, isoiridomyrmecin and dihydronepetalactone are prepared from citronellal using a divergent approach (Figure 8.49) [74].

Sesquiterpenoid Alkaloids

Majority of the sesquiterpene alkaloids belong to two groups such as dendrobines and guaipyridines. The biogenetic pathways for these compounds are shown in Figure 8.50.

Dendrobine is the major component of the *Dendrobium nobile*. It was first isolated in 1932 and its structure was identified in the 1960s. Dendrobine is used as tonic for improving the vocal cords and also exhibits antipyretic, hypotensive activities [75].

Several sesquiterpenoid alkaloids of dendrobine class include dendramine, dendrine, nobilonine, dendroxine, hydroxyl-dendroxine, etc. All of these dendrobine-type alkaloids contain basic skeletons comprising one picrotoxane-type sesquiterpenoid combined with five-membered C_2–C_9-lined N-heterocycle and C_3–C_5-linked lactonic ring system. The pharmacological effects of *Dendrobium* alkaloids include regulation of hepatic lipid and gluconeogenesis, neuroprotection, antitumor, anti-inflammatory, anti-diabetic, antiviral, etc. (Figure 8.51) [76].

Guaipyridine is a sesquiterpenoid alkaloid with a bicyclic molecular structure in which a seven-membered carbocyclic ring is fused with a pyridine ring. There is the presence of two stereogenic centers in its structure. The biogenesis is presented in Figure 8.52. The pathway starts from farnesyl pyrophosphate to generate the basic guaiane skeleton. Further, the cleavage of a C=C double bond at the bicycle junction produces the xanthane sesquiterpene skeleton followed by inclusion of a nitrogen atom between the two carbonyl groups that affords the guaipyridine framework [77].

Chiral enantiopure cycloheptenone intermediate is produced from the naturally occurring (*R*)(−)carvone. Then, cycloheptenone undergoes alkylation to produce xanthane sesquiterpene carbon skeleton which on further treatment with NH_3 cyclizes to afford guaipyridine (Figure 8.53) [78].

Van der Gen et al. have demonstrated two synthetic routes for the synthesis of guaipyridines. The reaction starts from different guaiane sesquiterpenes basically by cleavage of double bond to produce xanthane carbon skeletons followed by the inclusion of nitrogen atom and cyclization. Guaiol is a guaiane sesquiterpenoid with molecular formula $C_{15}H_{26}O$. It is isolated from *Philotheca fitzgeraldii*, *Aristolochia asclepiadifolia*. Chemically it is 2-[(3*S*, 5*R*, 8*S*)-3,8-dimethyl-1,2,3,4,5,6,7,8-octahydroazulen-5-yl]propan-2-ol (Figure 8.54) [79].

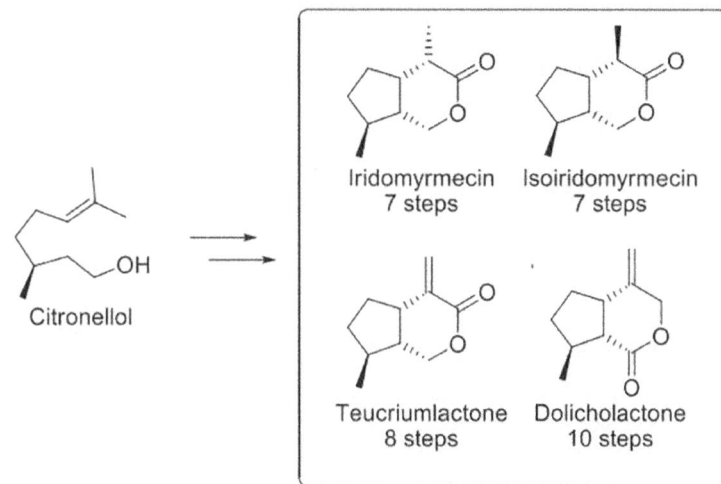

FIGURE 8.48 Preparation of iridoids from citronellol.

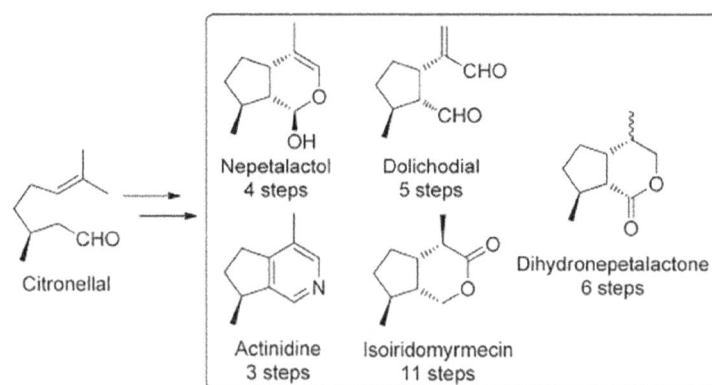

FIGURE 8.49 Preparation of ant-associated iridoids from citronellal.

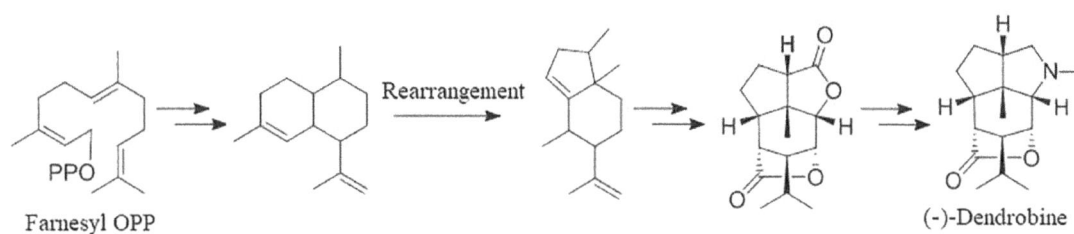

FIGURE 8.50 Synthesis of dendrobines.

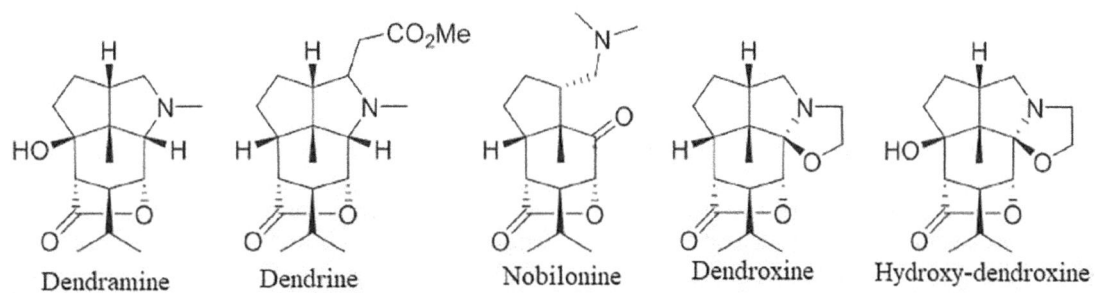

FIGURE 8.51 Structures of sesquiterpenoid alkaloid of dendrobine class.

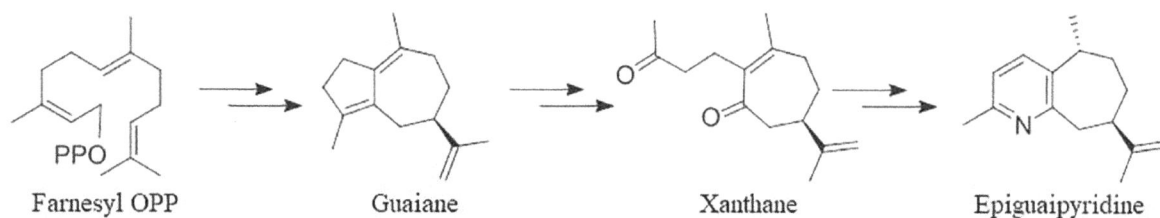

FIGURE 8.52 Biogenesis of guaipyridine.

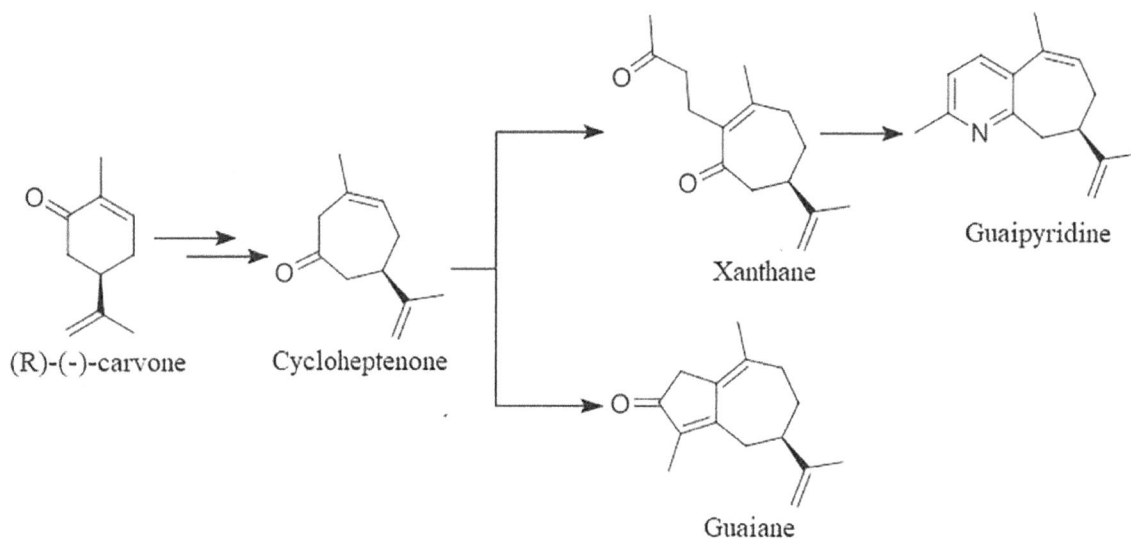

FIGURE 8.53 Synthetic approach for production of xanthane, guaiane and guaipyridines.

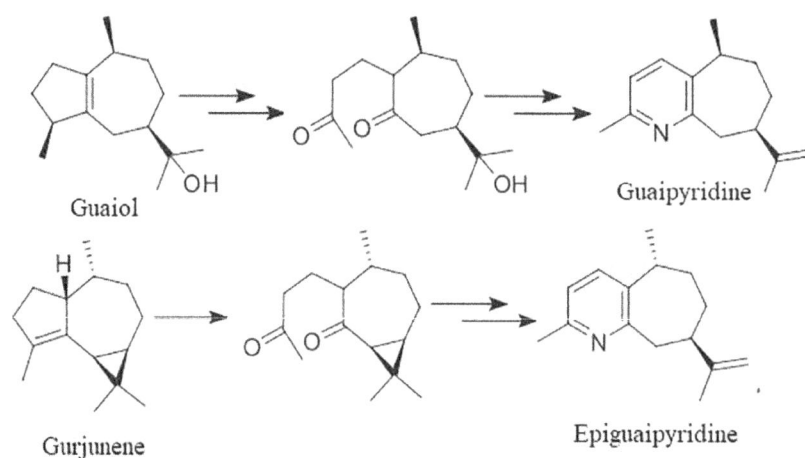

FIGURE 8.54 Van der Gen syntheses of guaipyridines.

Guaipyridine sesquiterpene alkaloids contain fused pyridine ring and seven-membered carbocycle ring system. Guaipyridine sesquiterpene alkaloids are isolated from the leaves of *Artemisia rupestris*. These alkaloids include patchoulipyridine, guaipyridine, epiguaipyridine, Cananodine, rupestines A, F. The structures of these alkaloids are elucidated on the basis of spectral data (FT-IR, NMR, LC-MS) (Figure 8.55) [80].

Craig and Henry utilized the enantiopure R-(−)-citronellene as the starting material. After cleavage of the propylidene group and re-construction with a chiral auxiliary, it allows the ring-closing metathesis (RCM) reaction to generate cycloheptene ring [81]. Finally, cananodine is synthesized by sequential of reactions. Cananodine is a guaipyridine alkaloid with the molecular formula $C_{15}H_{23}NO$. Chemically, it is 2-[(5R,

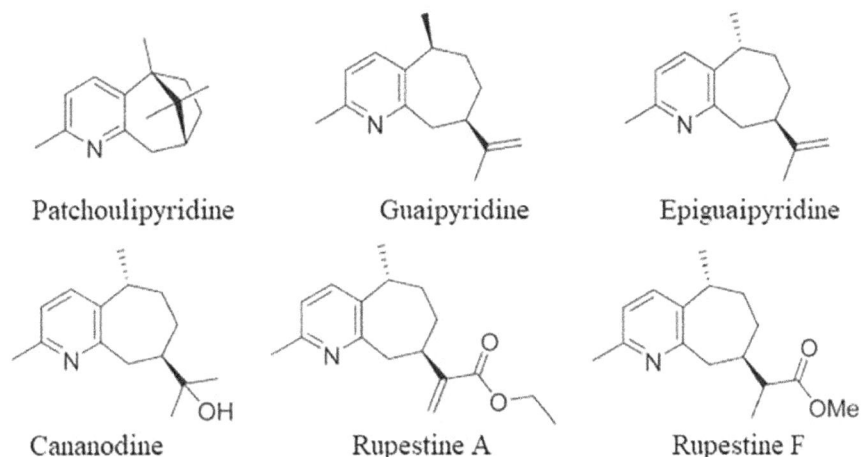

Patchoulipyridine   Guaipyridine   Epiguaipyridine

Cananodine   Rupestine A   Rupestine F

FIGURE 8.55 Structures of Guaipyridine sesquiterpene alkaloids.

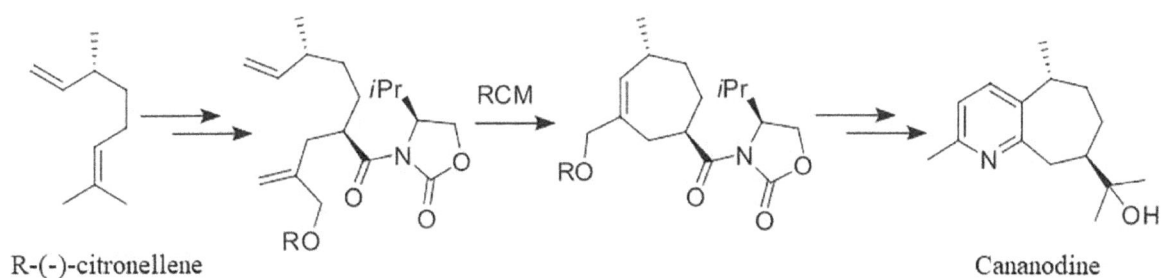

R-(-)-citronellene   Cananodine

FIGURE 8.56 Craig-Henry synthesis of cananodine.

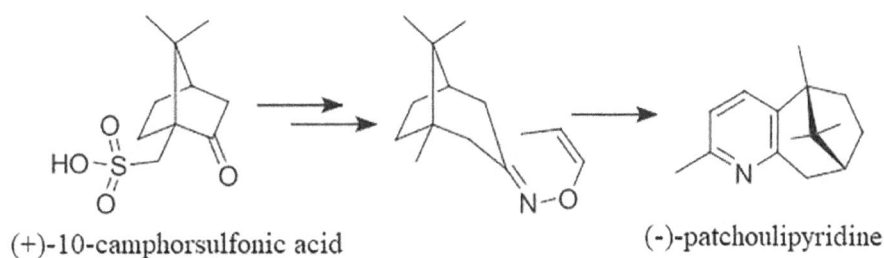

(+)-10-camphorsulfonic acid   (-)-patchoulipyridine

FIGURE 8.57 Synthesis of patchoulipyridine.

8R)-2,5-dimethyl-6,7,8,9-tetrahydro-5H-cyclohepta[b]pyridin-8-yl]propan-2-ol. It is the active constituent of *Cananga odorata* and found to be active against liver cancer (Hep G2 cell line) (Figure 8.56) [82].

Koyama et al. demonstrated the synthesis of (−)-patchoulipyridine from (+)-10-camphorsulfonic acid as starting material. (−)-Patchoulipyridine is sesquiterpene alkaloid with the molecular formula $C_{15}H_{21}N$. Chemically, it is (1R, 9R)-1,5,12,12-tetramethyl-6-azatricyclo[7.2.1.0^{2,7}]dodeca-2(7), 3,5-triene. It is isolated from the essential oils of *Pogostemon patchouli* (Figure 8.57) [83].

Three novel rotundine sesquiterpenoid alkaloids are isolated from the rhizomes of *Cyperus rotundus* including rotundine A, B and C (Figure 8.58). The structures of these alkaloids are

elucidated by spectral (FT-IR, NMR, LC-MS) and chemical analysis [84].

8.5.2 Diterpenoids Alkaloids

The majority of the diterpenoid alkaloids are isolated from the species of Aconitum and Delphinium of Ranunculaceae family. Anthriscifolcine F and G are the ranaconitine-type C_{18}-diterpenoid alkaloids. This type of diterpenoid alkaloid possesses an oxygen functionality at C-7 and hydroxyl group at C_{16} position as presented in Figure 8.59 [85].

Anthriscifolmine I and J are hetidine-type C_{20}-diterpenoid alkaloid (Figure 8.60). It is isolated from *Delphinium anthriscifolium*. It is more complex than the atisine type alkaloid. The

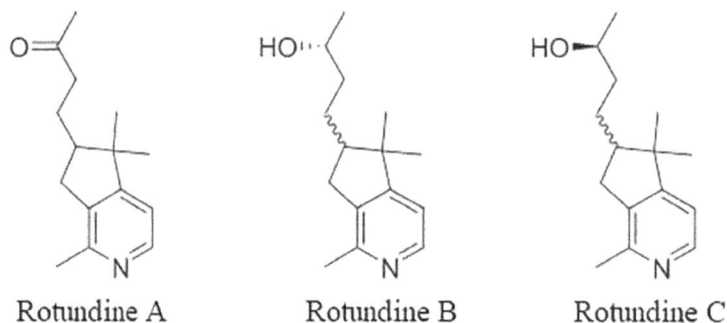

FIGURE 8.58 Structures of rotundine sesquiterpene alkaloids.

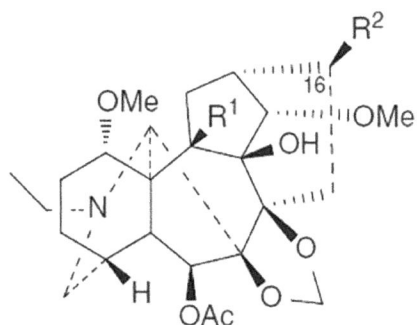

FIGURE 8.59 Structure of Anthriscifolcine F $(R_1=R_2=OH)$ and G $(R_1=H,$ $R_2=OH)$.

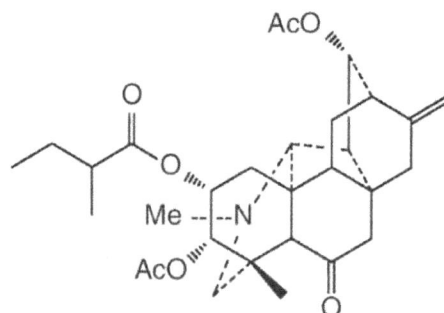

FIGURE 8.60 Structure of Anthriscifolmine I.

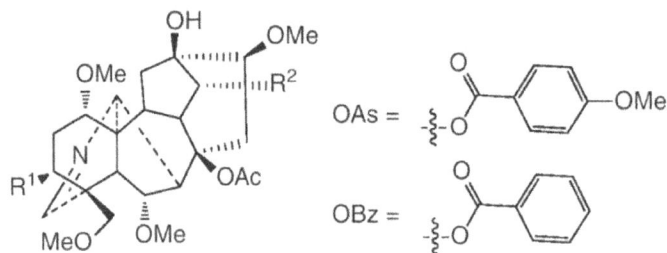

FIGURE 8.61 Structures of macrorhynine $(R_1=OH, R_2=OBz)$, ouvrardianine A $(R_1=OH, R_2=OBz)$.

structures of the two new alkaloids are elucidated on the basis of spectral data (NMR, HRESIMS) [86,87].

Ouvrardianine B is an aconitine-type C19-diterpenoid alkaloid. This type of diterpenoid alkaloid doesn't have an oxygen functionality at C-7. It contains formyl moiety that is attached to the nitrogen [87]. The aconitine-type C19-diterpenoid alkaloids such as macrorhynine and ouvrardianine A (Figure 8.61) with a rare C-19 imino unit, are isolated from *Aconitum macrorhynchum* and *Aconitum ouvrardianum* respectively [88].

Liangshantine is a diterpenoid alkaloid isolated from *Aconitum liangshanicum*. It possesses α, β-unsaturated carbonyl group at ring A. The aconitine-type C19-diterpenoid alkaloid (aconitorientaline) is isolated from the roots of Aconitum orientale. Aconitorientaline contains an epoxy group between C_1 and C_{19} (Table 8.3, Figure 8.62) [89,90].

TABLE 8.3

List of Diterpenoid Alkaloids

Diterpenoid Alkaloids	Molecular Formula	Plant Source
Anthriscifolcine F	$C_{25}H_{37}NO_8$	*Delphinium anthriscifolium*
Anthriscifolcine G	$C_{25}H_{37}NO_7$	*Delphinium anthriscifolium*
Ouvrardianine A	$C_{32}H_{41}NO_9$	*Aconitum ouvrardianum*
Ouvrardianine B	$C_{33}H_{43}NO_{10}$	*Aconitum ouvrardianum*
Macrorhynine C	$C_{33}H_{43}NO_{11}$	*Aconitum macrorhynchum*
Aconitramine D	$C_{32}H_{43}NO_6$	*Aconitum transsectum*
Aconitramine E	$C_{33}H_{47}NO_8$	*Aconitum transsectum*
Liangshantine	$C_{26}H_{37}NO_7$	*Aconitum liangshanicum*
Aconitorientaline	$C_{25}H_{39}NO_7$	*Aconitum orientale*
Umbrosumine A	$C_{38}H_{53}N_3O_{11}$	*Delphinium umbrosum*
Yunnanenseine A	$C_{45}H_{47}NO_{15}$	*Delphinium yunnanense*

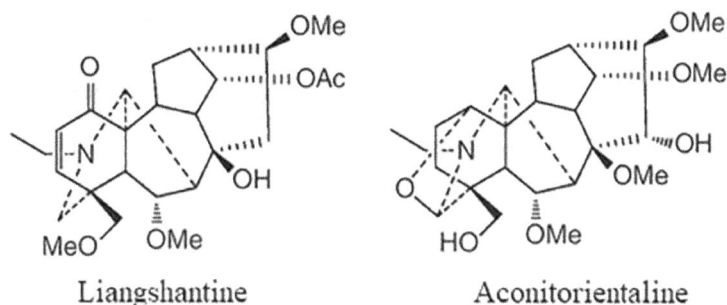

FIGURE 8.62 Structures of liangshantine and aconitorientaline.

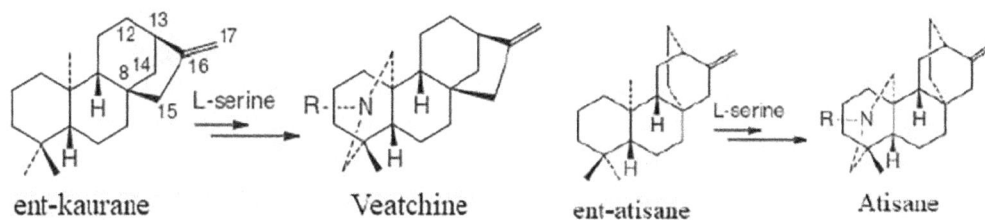

FIGURE 8.63 Biosynthetic pathway of veatchine and atisane-type diterpenoid alkaloids.

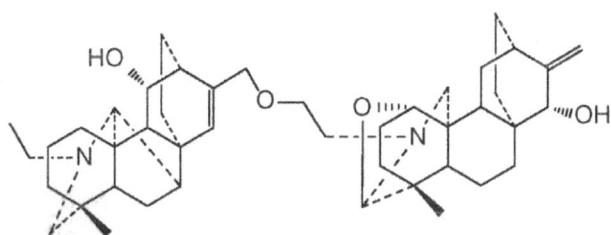

FIGURE 8.64 Molecular structure of piepunin.

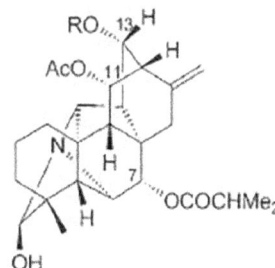

FIGURE 8.66 Molecular structure of yunnanenseines B (R=Ac) and C (R=H).

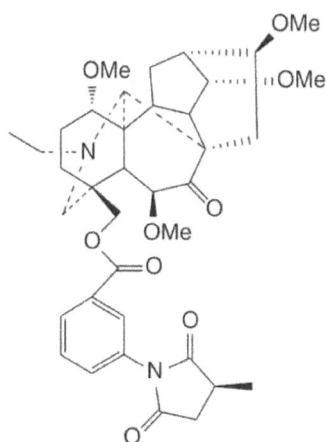

FIGURE 8.65 Molecular structure of yunnanenseine A.

The veatchine-type and atisane-type C_{20} diterpenoid alkaloids are derived from the amination of the ent-kaurane-type and ent-atisane-type diterpenoids, respectively (Figure 8.63) [91,92].

Piepunin is a novel C_{20}-diterpenoid alkaloid. It is isolated from the roots of *Aconitum piepunense*. The structure of Piepunin is established by interpretation of spectral data of

NMR, ESI-MS, FT-IR, etc. It contains both atisine and denudatine skeletons in its structure (Figure 8.64) [93].

Yunnanenseine A is C_{19} diterpenoid alkaloids, isolated from *Delphinium yunnanense*. The structure of yunnanensine A is determined as (1α, 6β, 14α, 16β)-20-ethyl-1,6,14, 16-tetramethoxy-7-oxo-8,17-cyclo-7,17-secoaconitan-18-yl-2-(2β-2-methylsuccinimide)benzoate (Figure 8.65). Whereas yunnanenseines B and C are hetisine type C_{20} diterpenoid alkaloids. These compounds are identified based on a comparison of the respective spectroscopic data (FT-IR, NMR, LC-MS). Yunnanensine B is isolated as amorphous powder, and the molecular formula is determined as $C_{28}H_{37}NO_7$ based on the HR-ESI-MS data (Figure 8.66) [94].

Aconicarmisulfonine A is a novel sulfonated C_{20}-diterpenoid alkaloid. It is isolated from the aqueous extract of the roots of *Aconitum carmichaelii*. The structure of Aconicarmisulfonine A is elucidated by extensive analysis of spectroscopic data of NMR combined with ECD calculation and single crystal X-ray diffraction study. Songorine is proposed to be the biosynthetic precursor of Aconicarmisulfonine A. Songorine is

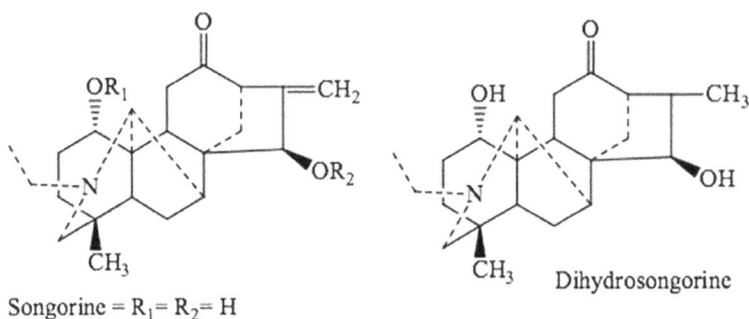

FIGURE 8.67   Structures of Songorine and its derivatives.

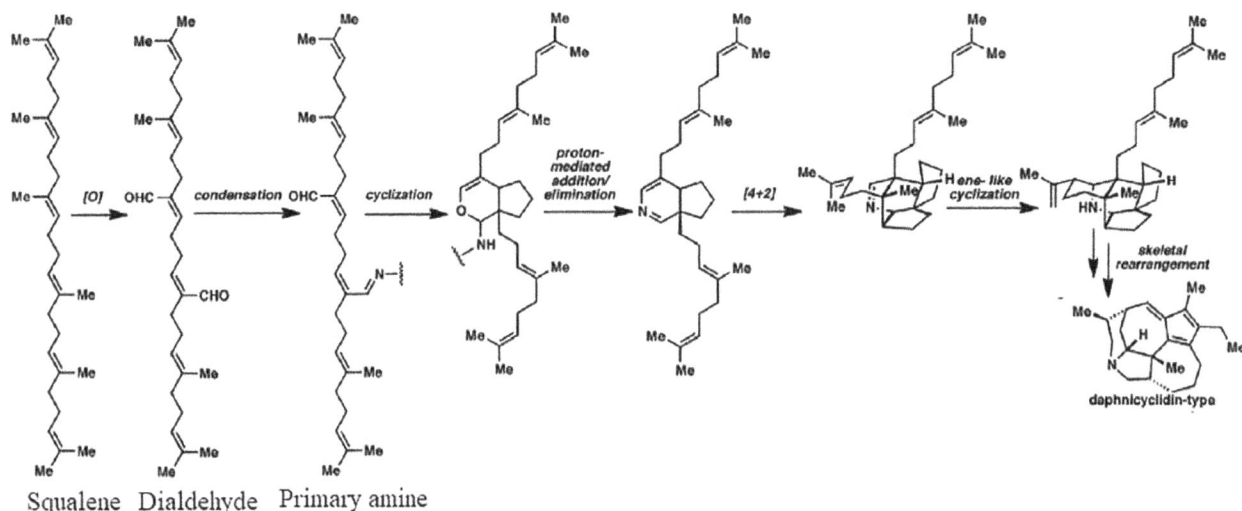

FIGURE 8.68   Biosynthetic pathway for the production of *Daphniphyllum alkaloids.*

a C_{20} diterpenoid alkaloid with molecular formula $C_{22}H_{31}NO_3$ (Figure 8.67). It is isolated from *Aconitum soongaricum*. Songorine and its derivatives exhibit several pharmacological activities such as anti-arrhythmic, anti-cardiac-fibrillation, anti-nociceptive, anti-inflammatory and anxiolytic effects [95].

8.5.3 Triterpenoid Alkaloids

Daphniphyllum alkaloids are derived from triterpenes. In 1909, Yagi et al. reported the first C_{30} type *Daphniphyllum* alkaloid (Daphnimacrine). The biosynthesis of these alkaloids starts with squalene. Squalene is a triterpene with the molecular formula $(C_5H_8)_6$. First of all, squalene is oxidized chemoselectively into the dialdehyde. Condensation of this dialdehyde with a primary amine followed by cyclization and addition/elimination leads to the formation of 6,5-fused ring system containing a dihydropyridine ring. Further, Diels–Alder cycloaddition reaction followed by an ene-type cyclization completes the "aza" cyclase phase that produces the carbon skeleton of the *Daphniphyllum* alkaloids (Figure 8.68) [96].

Squalene undergoes epoxidation to produce Squalene epoxide in presence of Squalene epoxidase which on cationic π-cyclization produces lanosterol. Further, the oxidation of lanosterol causes skeletal rearrangements to generate cyclopamine skeleton (Figure 8.69) [97].

8.5.4 Biological Activities Terpenoid Alkaloids

Monoterpenoid alkaloids such as Gentianine and Actinidine exhibit Analgesic, central muscle-weakening action and gastric protein digestion enhancing activities etc. Similarly, Dendrobine is a sesquiterpenoid alkaloid which is used as a tonic for improving the vocal cords. Songorine is a diterpenoid alkaloid that exhibits anti-inflammatory, anticancer and antiparasitic activities etc (Table 8.4, Figure 8.70) [98,99].

8.6 Conclusion

Terpenoids are the natural present abundantly in various medicinal plants. These are mainly present in the form of volatile oils in higher medicinal plants and mainly exist in the medicinal plant groups of family compositae, ranunculaceae, araliaceae, oleaceae, magnoliaceae, lauraceae, labiatae, pinaceae, umbelliferae, celastraceae, acanthaceae, etc. Terpenoids are not only extracted and isolated from the plant sources but also generated based on metabolic engineering, synthetic biology and biotransformation that act as new sources for the synthesis of terpenoid and their derivatives. Due to the presence of diverse chemical structures in terpenoids, these are used as templates for further investigations to develop new

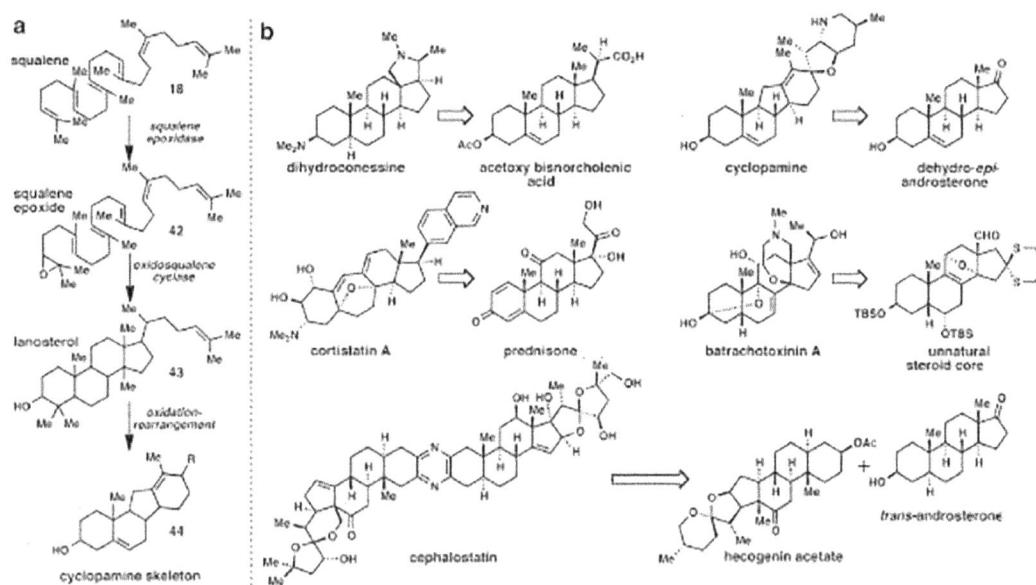

FIGURE 8.69 Generation of cyclopamine skeleton from Squalene.

TABLE 8.4

Biological Activities Terpenoid Alkaloids

Name	Source	Type	Uses
Gentianine	Strychnos angolensis	Monoterpenoid alkaloid	Analgesic
Actinidine	*Actinidia polygama*	Monoterpenoid alkaloid	Enhances gastric protein digestion
Subditine	*Nauclea subdita*	Monoterpenoid Indole alkaloid	Induce apoptosis
Valerian	*Valeriana officinalis*	Monoterpenoid alkaloid	Treatment of insomnia, migraine, fatigue
Dendrobine	*Dendrobium nobile*	Sesquiterpenoid alkaloid	Tonic for improving vocal chords
Cananodine	*Cananga odorata*	Sesquiterpenoid alkaloid	Treatment of liver cancer
Songorine	*Aconitum soongaricum*	Diterpenoid alkaloid	Anti-arrhythmic Anti-cardiac-fibrillation Anti-nociceptive Anti-inflammatory Anxiolytic effects

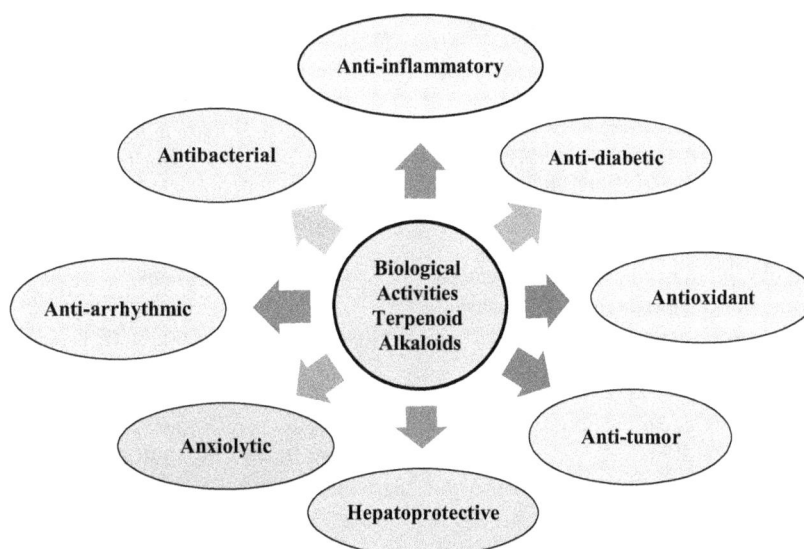

FIGURE 8.70 Biological activities terpenoid alkaloids.

chemical entity and improvements in existing treatment strategies. Evaluation study reports suggest that terpenoid and their derivatives exhibit significant biological activities such as anti-tumor, anti-inflammatory, antibacterial, antiviral, antioxidant, antiaging and antimalarial effects etc. Terpenoids and their transformed compounds are also widely used in the pharmaceuticals, food and cosmetics industries.

Abbreviation

BV	Baeyer–Villiger
CROP	Cationic ring-opening polymerization
ECD	Equivalent circulating density
ESI-MS	Electrospray ionization mass spectrometry
FT-IR	Fourier-transform infrared spectroscopy
H_2O_2	Hydrogen peroxide
HIV	Human immunodeficiency virus
HRESIMS	High resolution electrospray ionization mass spectrometry
LD50	Median lethal dose
MPTP	1-methyl-4-phenyl-1,2,3,6-tetrahydropyridine
MSA	Methanesulfonic acid
NaBH4	Sodium borohydride
NMR	Nuclear magnetic resonance
PD	Parkinson's disease
Ph3SnH	Triphenyl tin hydride
RCM	Ring-closing metathesis
THF	Tetrahydrofuran
t-BuOCl	Tert-butyl hypochlorite

REFERENCES

[1] Yang, W.; Chen, X.; Li, Y.; Guo, S.; Wang, Z.; Yu, X. Advances in pharmacological activities of terpenoids. *Natural Product Communications*, **2020**, 15, 15.

[2] Bouvier, F.; Rahier, A.; Camara, B. Biogenesis, molecular regulation and function of plant isoprenoids. *Progress in Lipid Research*, **2005**, 44, 357–429.

[3] Maimone, J. T.; Baran, P. S. Modern synthetic efforts toward biologically active terpenes. *Nature Chemical Biology*, **2007**, 3, 396–407.

[4] Dogenhardt, J.; Kollner, T. G.; Gershenzon, J. Monoterpene and sesquiterpene synthases and the origin of terpene skeletal diversity in plants. *Phytochemistry*, **2009**, 70, 1621–1637.

[5] Ruzicka, L. Isoprene rule and the biogenesis of terpenic compounds. *Experientia*, **1953**, 9, 357–367.

[6] Simeó, Y.; Sinisterra, J. V. Biotransformation of terpenoids: A green alternative for producing molecules with pharmacological activity. *Mini-Reviews in Organic Chemistry*, **2009**, 6(2), 128–134.

[7] Bergman, M. E.; Davis, B.; Phillips, M. A. Medically useful plant terpenoids: Biosynthesis, occurrence, and mechanism of action. *Molecules*, **2019**, 24, 3961.

[8] Christianson, D. W. Structural and chemical biology of terpenoid cyclases. *Chemical Reviews*, **2017**, 117, 11570–11648.

[9] Cox-Georgian, D.; Ramadoss, N.; Dona, C.; Basu, C. Therapeutic and medicinal uses of terpenes. *Medicinal Plants*, **2019**, 333–359. doi: 10.1007/978-3-030-31269-5_15.

[10] Patrusheva, O. S.; Volcho, K. P.; Salakhutdinov, N. F. Synthesis of oxygen-containing heterocyclic compounds based on monoterpenoids. *Russian Chemical Reviews*, **2018**, 87, 771.

[11] García Martínez, A.; Teso Vilar, E.; García Fraile, A.; de la Moya Cerero, S.; Rodríguez Herrero, M. E.; Martínez Ruiz, P.; Subramanian, L. R.; García Gancedo, A. Synthesis of substituted 1-norbornylamines with antiviral activity. *Journal of Medicinal Chemistry*, **1995**, 38(22), 4474–4477. doi: 10.1021/jm00022a012.

[12] Zhao, X.; Zhang, Z.-W.; Cui, W.; Chen, S.; Zhou, Y.; Dong, J.; Jie, Y.; Wan, J.; Xu, Y.; Hu, W. Identification of camphor derivatives as novel M2 ion channel inhibitors of influenza A virus. *MedChemComm*, **2015**, 6(4), 727–731. doi: 10.1039/C4MD00515E.

[13] Sokolova, A. S.; Yarovaya, O. I.; Baev, D. S.; Shernyukov, A. V.; Shtro, A. A.; Zarubaev, V. V.; Salakhutdinov, N. F. Synthesis and in vitro study of novel borneol derivatives as potent inhibitors of the influenza A virus. *European Journal of Medicinal Chemistry*, **2017**, 127, 661.

[14] Sokolova, A. S.; Yarovaya, O. I.; Shernyukov, A. V.; Gatilov, Y. V.; Razumova, Y. V.; Zarubaev, V. V.; Tretiak, T. S.; Pokrovsky, A. G.; Kiselev, O. I.; Salakhutdinov, N. F. Discovery of a new class of antiviral compounds: camphor imine derivatives. *European Journal of Medicinal Chemistry*, **2015**, 105, 263.

[15] Nazimova, E.; Pavlova, A.; Mikhalchenko, O.; Ilina, I.; Korchagina, D.; Tolstikova, T.; Volcho, K.; Salakhutdinov, N. Discovery of highly potent analgesic activity of isopulegol-derived (2R,4aR,7R,8aR)-4,7-dimethyl-2-(thiophen-2-yl)octahydro-2H-chromen-4-ol. *Medicinal Chemistry Research*, **2016**, 25, 1369.

[16] Volcho, K. P.; Salakhutdinov, N. F. Transformations of terpenoids on acidic clays. *Mini-Reviews in Organic Chemistry*, **2008**, 5, 345.

[17] Stekrova, M.; Maki-Arvela, P.; Kumar, N.; Behravesh, E.; Aho, A.; Balme, Q.; Volcho, K. P.; Salakhutdinov, N. F.; Murzin, D. Y. Prins cyclization: Synthesis of compounds with tetrahydropyran moiety over heterogeneous catalysts. *Journal of Molecular Catalysis A: Chemical*, **2015**, 410, 260.

[18] Ponomarev, K.; Pavlova, A.; Suslov, E.; Ardashov, O.; Korchagina, D; Nefedov, A.; Tolstikova, T.; Volcho, K.; Salakhutdinov, N. Synthesis and analgesic activity of new compounds combining azaadamantane and monoterpene moieties. *Medicinal Chemistry Research*, **2015**, 24(12), 4146–4156. doi: 10.1007/s00044-015-1464-z.

[19] Kapitsa, I. G.; Suslov, E. V.; Teplov, G. V.; Korchagina, D. V.; Komarova, N. I.; Volcho, K. P.; Voronina, T. A.; Shevela, A. I.; Salakhutdinov, N. F. *Pharmaceutical Chemistry Journal*, **2012**, 46, 263.

[20] Ilina, I. V.; Volcho, K. P.; Korchagina, D. V.; Salnikov, G. E.; Genaev, A. M.; Karpova, E. V.; Salakhutdinov, N. F. Unusual reactions of (+)-car-2-ene and (+)-car-3-ene with aldehydes on K10 clay. *Helvetica*, **2010**, 93(11), 2135–2150. doi: 10.1002/hlca.201000145.

[21] Finkbeiner, P.; Murai, K.; Röpke, M.; Sarpong, R. Total synthesis of terpenoids employing a 'Benzannulation of Carvone' strategy: Synthesis of (–)-crotogoudin. *Journal of the American Chemical Society*, **2017**, doi: 10.1021/jacs.7b06823.

[22] Tang, Y.; Cole, K. P.; Buchanan, G. S.; Li, G.; Hsung, R. P. Total synthesis of phomactin A. *Organic Letters*, **2009**, 11, 1591–1594.

[23] Dubey, V. S.; Luthra, R. Biotransformation of geranyl acetate to geraniol during palmarosa (Cymbopogon martinii, Roxb. wats. var. motia) inflorescence development. *Phytochemistry*, **2001**, 57, 675–680.

[24] Gramatica, P.; Manitto, P.; Maria Ranzi, B.; Delbianco, A.; Francavilla, M. Stereospecific reduction of geraniol to (R)-(+)-citronellol by *Saccharomyces cerevisiae*. *Cellular and Molecular Life Sciences*, **1982**, 38, 775–776.

[25] Luan, F.; Mosandla, A.; Müncha, A.; Wüst, M. Metabolism of geraniol in grape berry mesocarp of *Vitis vinifera* L. cv. Scheurebe: Demonstration of stereoselective reduction, E/Z-isomerization, oxidation and glycosylation. *Phytochemistry*, **2005**, 66, 295–303.

[26] Ho, T. L. *Enantioselective Synthesis Natural Products from Chiral Terpenes*. Wiley, New York, **1992**.

[27] Money, T.; Wong, M. K. C. The use of cyclic monoterpenoids as enantiopure starting materials in natural product synthesis. *Studies in Natural Products Chemistry*, **1995**, 16, 123–288.

[28] Simeó, Y.; Sinisterra, J. V. Biotransformation of terpenoids: A green alternative for producing molecules with pharmacological activity. *Mini-Reviews in Organic Chemistry*, **2009**, 6(2), 128–134.

[29] Alina Mariana, B.; Juan Manuel, C.; Rafael, L.; Antonio Angel, R. One-step microwave-assisted asymmetric cyclisation/hydrogenation of citronellal to menthols using supported nanoparticles on mesoporous materials. *Organic & Biomolecular Chemistry*, **2010**, 8(12), 2845–2849. doi: 10.1039/c003600e.

[30] Caniard, A.; Zerbe, P.; Legrand, S.; Cohade, A.; Valot, N.; Magnard, J. L.; Bohlmann, J.; Legendre, L. Discovery and functional characterization of two diterpene synthases for sclareol biosynthesis in *Salvia sclarea* (L) and their relevance for perfume manufacture. *BMC Plant Biology*, **2012**, 12(119), 1–13.

[31] Chen, J.; Lu, M.; Jing, Y.; Dong, J. The synthesis of L-carvone and limonene derivatives with increased antiproliferative effect and activation of ERK pathway in prostate cancer cells. *Bioorganic & Medicinal Chemistry*, **2006**, 14, 6539–6547.

[32] Alves, L. d. C.; Desiderá, A. L.; de Oliveira, K. T.; Newton, S.; Ley, S. V.; Brocksom, T. J. A practical decagram scale ring expansion of (R)-(-)-carvone to (R)-(+)-3-methyl-6-isopropenyl-cyclohept-3-enone-1. *Organic and Biomolecular Chemistry*, **2015**, 13(28), 7633–7642. doi: 10.1039/C5OB00525F.

[33] Firdaus, M.; Meier, M. A. R. Renewable polyamides and polyurethanes derived from limonene. *Green Chemistry*, **2012**, 15(2), 1–11. doi: 10.1039/c2gc36557j.

[34] Neaţu, F.; Culica, G.; Florea, M.; Parvulescu, V. I.; Cavani, F. Synthesis of terephthalic acid by p-cymene oxidation using oxygen: Toward a more sustainable production of biopolyethylene terephthalate. *ChemSusChem*, **2016**, 9(21), 3102–3112.

[35] Bogel-Łukasik, E.; Bogel-Łukasik, R.; da Ponte, M. N. Pt- and Pd-catalysed limonene hydrogenation in high-density carbon dioxide. *Monatshefte Fur Chemie*, **2009**, 140, 1361–1369. doi: 10.1007/s00706-009-0196-5.

[36] Emre, Y.; Mesut, A. P-Cymene production from orange peel oil using some metal catalyst in supercritical alcohols. *The Journal of Supercritical Fluids*, **2017**, S0896844617303261. doi: 10.1016/j.supflu.2017.08.015.

[37] Marvel, C. S.; Hwa, C. C. Polymyrcene. *Journal of Polymer Science*, **1960**, 45, 25–34.

[38] Cawse, J. L.; Stanford, J. L.; Still, R. H. Polymers from renewable sources. 4. Polyurethane elastomers based on myrcene polyols. *Journal of Applied Polymer Science*, **1986**, 31, 1549–1565.

[39] Cawse, J. L.; Stanford, J. L.; Still, R. H. Polymers from renewable sources. 5. Myrcene-based polyols as rubber-toughening agents in glassy polyurethanes. *Polymer*, **1987**, 28, 368–374.

[40] Palenzuela, M.; Sánchez-Roa, D.; Damián, J.; Sessini, V.; Mosquera, M. E. Polymerization of terpenes and terpenoids using metal catalysts. *Advances in Organometallic Chemistry*, **2021**, 75, 55–93.

[41] Lowe, J. R.; Martello, M. T.; Tolman, W. B.; Hillmyer, M. A. Functional biorenewable polyesters from carvone-derived lactones. *Polymer Chemistry*, **2011**, 2, 702–708.

[42] Parrino, F.; Fidalgo, A.; Palmisano, L.; Ilharco, L. M.; Pagliaro, M.; Ciriminna, R. Polymers of limonene oxide and carbon dioxide: Polycarbonates of the solar economy. *ACS Omega*, **2018**, 3, 4884–4890.

[43] Liu, Y.; Alexander, W. S.; Zhao, Y.; Lee, J.; Newhouse, T. R. Synthesis of (–)-melazolide B, a degraded limonoid, from a natural terpene precursor. *Tetrahedron*, **2022**, 1, doi: 10.1016/j.tchem.2022.100011.

[44] Wang, P.; Li, J.; Yu, C.-L.; Xiao, X.; Wu, P.-Y.; Zeng, B.-B. Concise synthesis of (+)-subersic acid from (–)-Sclareol. *Tetrahedron*, **2015**, 71(28), 4647–4650. doi: 10.1016/j.tet.2015.04.100.

[45] Zenk, M. H. Enzymatic synthesis of ajmalicine and related indole alkaloids. *Journal of Natural Products*, **1980**, 43(4), 438–451.

[46] Zhu, J.; Wang, M.; Wen, W.; Yu, R. Biosynthesis and regulation of terpenoid indole alkaloids in *Catharanthus roseus*. *Pharmacognosy Reviews*, **2015**, 9(17), 24–28. doi: 10.4103/0973-7847.156323.

[47] Dewick, P.M. *Medicinal Natural Products: A Biosynthetic Approach*, 2nd Edition. Wiley, **2002**, 356.

[48] Leonard, J. Recent progress in the chemistry of monoterpenoid indole alkaloids derived from secologanin. *Natural Product Reports*, **1999**, 16, 319–338.

[49] Kleinfelter, D. C. Schleyer PvR. 2-norbornanone. *Organic Syntheses*, **1962**, 42, 79–82.

[50] Messiha, H. L.; Ahmed, S. T.; Karuppiah, V.; Suardiaz, R.; Ascue Avalos, G. A.; Fey, N., et al. Biocatalytic routes to lactone monomers for polymer production. *Biochemistry*, **2018**, 57, 1997–2008.

[51] Dove, A. P. Organic catalysis for ring-opening polymerization. *ACS Macro Letters*, **2012**, 1, 1409–1412.

[52] Maresh, J. J.; Giddings, L. A.; Friedrich, A., et al. Strictosidine synthase: Mechanism of a Pictet-Spenger catalyzing enzyme. *Journal of the American Chemical Society*, **2008**, 130(2), 710–723. doi: 10.1021/ja077190z.

[53] Thomsett, M. R.; Storr, T. E.; Monaghan, O. R.; Stockman, R. A.; Howdle, S. M. Progress in the synthesis of sustainable polymers from terpenes and terpenoids. *Green Materials*, **2016**, 4(3), 1–70. doi: 10.1680/jgrma.16.00009.

[54] Scianowski, J.; Rafinski, Z.; Wojtczak, A. Syntheses and reactions of new optically active terpene diselenides. *European Journal of Organic Chemistry*, **2006**, 3216–3225.

[55] Martinez, A. G.; Vilar, E. T.; Fraile, A. G.; de la Moya Cerero, S.; Maroto, L. B. A new convenient procedure for the preparation of enantiopure C10-S- and C10-Se-substituted camphor-derived sulphides and selenides. *Tetrahedron Letters*, **2001**, 42, 5017–5019.

[56] Takahashi, T.; Kurose, N.; Kawanami, S.; Arai, Y.; Koizumi, T. Optically pure haloselenouranes. First synthesis and nucleophilic substitutions. *The Journal of Organic Chemistry*, **1994**, 59, 3262–3264.

[57] Saito, S.; Zhang, J.; Koizumi, T. Synthesis and structure of novel haloselenurane-Lewis acid complexes. *The Journal of Organic Chemistry*, **1998**, 63, 6029–6030.

[58] Sharpless, K. B.; Lauer, R. F. Facile thermal rearrangements of allyl selenides and diselenides. [1,3] and [2,3] shifts. *The Journal of Organic Chemistry*, **1972**, 37(24), 3973–3974.

[59] Clive, D. L. J.; Chittattu, G. J.; Farina, V.; Kiel, W. A.; Menchen, S. M.; Russel, C. G.; Singh, A.; Wong, C. K.; Curtis, N. J. Organic tellurium and selenium chemistry. Reduction of tellurides, selenides and selenoacetals with triphenyltin hydride. *Journal of the American Chemical Society*, **1980**, 102(13), 4438–4447.

[60] Nakamura, M.; Kido, K.; Kinjo, J.; Nohara, T. Two novel actinidine-type monoterpene alkaloids from *Incarvillea delavayi*. *Chemical and Pharmaceutical Bulletin*, **2000**, 48, 1826–1827.

[61] Fu, J. J.; Qin, J. J.; Zeng, Q.; Huang, Y.; Zhang, W. D.; Jin, H. Z. Two new monoterpene alkaloid derivatives from the roots of *Incarvillea arguta*. *Archives of Pharmacal Research*, **2011**, 34, 199–202.

[62] Bianco, A.; Bonadies, F.; Ciancialo, V., et al. Monoterpene alkaloids from *Argylia radiata*. *Natural Product Letters*, **2002**, 16, 77–80.

[63] Fu, J. J.; Jin, H. Z.; Shen, Y. H.; Zhang, W. D.; Xu, W. Z.; Zeng, Q.; Yan, S. K. Two novel monoterpene alkaloid dimmers from *Incarvillea arguta*. *Helvetica Chimica Acta*, **2007**, 90, 2151–2155.

[64] Devkota, K. P.; Sewald, N. Terpenoid alkaloids derived by amination reaction. In: Ramawat, K.; Mérillon, J. M. (Eds.) *Natural Products*. Springer, Berlin, Heidelberg. **2013**. doi: 10.1007/978-3-642-22144-6_30.

[65] Chi, Y. M.; Nakamura, M.; Zhao, X. Y., et al. A monoterpene alkaloid from *Incarvillea sinensis*. *Chemical and Pharmaceutical Bulletin*, **2005**, 53, 1178–1179.

[66] Huang, D. S.; Zhang, W. D.; Pei, Y.; Peng, X. Y.; Huang, Z. S.; Li, H. L.; Shen, Y. H. Two new alkaloids from *Incarvillea sinensis*. *Helvetica Chimica Acta*, **2009**, 92, 1558–1561.

[67] Su, Y. Q.; Shen, Y. H.; Lin, S.; Tang, J.; Tian, J. M.; Liu, X. H.; Zhang, W. D. Two new alkaloids from *Incarvillea mairei* var. grandiflora. *Helvetica Chimica Acta*, **2009**, 92, 165–169.

[68] Xing, A. T.; Tian, J. M.; Liu, C. M.; Li, H. L.; Zhang, W. D.; Shan, L. Three new monoterpene alkaloids and a new caffeic acid ester from *Incarvillea mairei* var. multifoliolata. *Helvetica Chimica Acta*, **2010**, 93, 718–723.

[69] Patocka, J.; Jakl, J. Biomedically relevant chemical constituents of *Valeriana officinalis*. *Journal of Applied Biomedicine*, **2010**, 8(1), 11–18. doi: 10.2478/ v10136-009-0002-z.

[70] Koch, M. Gentianine and swertiamarin from *Anthocleista procera*. *Trav Lab*, **1965**, 50–94.

[71] Guo, X. F.; Chen, C. P. Isolation and identification of gentianine from *Swertia davidi* Franch. *Chemical Abstracts*, **1980**, 94, 1293.

[72] Stepanov, A. V.; Lozanova, A. V.; Veselovsky, V. V. Stereocontrolled synthesis of the alkaloid (–)-actinidine. *Russian Chemical Bulletin*, **1998**, 47, 2286–2291. doi: 10.1007/ BF02494297.

[73] Fischman, C. J.; Adler, S.; Hofferberth, J. E. Divergent diastereoselective synthesis of iridomyrmecin, isoiridomyrmecin, teucrimulactone, and dolicholactone from citronellol. *The Journal of Organic Chemistry*, **2013**, 78(14), 7318–7323. doi: 10.1021/jo400884g.

[74] Beckett, J. S.; Beckett, J. D.; Hofferberth, J. E. A divergent approach to the diastereoselective synthesis of several ant-associated iridoids. *Organic Letters*, **2010**, 12(7), 1408–1411. doi: 10.1021/ol100077z.

[75] Brocksom, T.; de Oliveira, K.; Desiderá, A. The chemistry of the sesquiterpene alkaloids. *Journal of the Brazilian Chemical Society*, **2017**, 28(6), 933–942. doi: 10.21577/0103-5053.20170049.

[76] Mou, Z.; Zhao, Y.; Ye, F.; Shi, Y.; Kennelly, E. J.; Chen, S.; Zhao, D. Identification, biological activities and biosynthetic pathway of dendrobium alkaloids. *Frontiers in Pharmacology*, **2021**, 12, 605994. doi: 10.3389/ fphar.2021.605994.

[77] van Der Gen, A.; van Der Linde, L. M.; Witteveen, J. G. Synthesis of guaipyridine and some related sesquiterpene alkaloids. *Recueil des Travaux Chimiques des Pays-Bas*, **1972**, 91(12), 1433–1440. doi: 10.1002/recl.19720911207.

[78] Vasas, A.; Hohmann, J. Xanthane sesquiterpenoids: Structure, synthesis and biological activity. *Natural Product Reports*, **2011**, 28(4), 824–842. doi: 10.1039/c0np00011f.

[79] Van der Gen, A.; Van der Linde, L. M.; Witteveen, J. G. *Recueil des Travaux Chimiques des Pays-Bas*, **1972**, 91, 1433.

[80] Su, A.; Wu, H.-K.; He, H.; Slukhan, U.; Aisa, H. A. New guaipyridine alkaloids from *Artemesia rupestris* L. *Helvetica Chimica Acta*, **2010**, 93, 33–38.

[81] Shelton, P.; Ligon, T. J.; Dell Née Meyer, J. M.; Yarbrough, L.; Vyvyan, J. R. Synthesis of cananodine by intramolecular epoxide opening. *Tetrahedron Letters*, **2017**, 58(35), 3478–3481. doi: 10.1016/j.tetlet.2017.07.080.

[82] Craig, D.; Henry, G. D. Total synthesis of the cytotoxic guaipyridine sesquiterpene alkaloid (+)-cananodine. *European Journal of Organic Chemistry*, **2006**, 16, 3558.

[83] Koyama, J.; Okatani, T.; Ogura, T.; Tagahara, K.; Irie, H. *Chemical and Pharmaceutical Bulletin*, **1991**, 39, 481.

[84] Jeong, S. J.; Miyamoto, T.; Inagaki, M.; Kim, Y. C.; Higuchi, R. Rotundines A-C, three novel sesquiterpene alkaloids from *Cyperus rotundus*. *Journal of Natural Products*, **2000**, 63(5), 673–675. doi: 10.1021/np990588r.

[85] Stepanov, A. V.; Lozanova, A. V.; Veselovsky, V. V. Stereocontrolled synthesis of the alkaloid (–)-actinidine. *Russian Chemical Reviews*, **1998**, 47, 2286–2291. doi: 10.1007/BF02494297.

[86] Song, L.; Liu, X. Y.; Chen, Q. H.; Wang, F. P. New C19- and C18-diterpenoid alkaloids from *Delphinium anthriscifolium* var. savatieri. *Chemical and Pharmaceutical Bulletin*, **2009**, 57, 158–161.

[87] Liu, X. Y.; Song, L.; Chen, Q. H.; Wang, F. P. Two new C20-diterpenoid alkaloids from *Delphinium anthriscifolium* var. savatieri. *Natural Product Communications*, **2010**, 5, 1005–1008.

[88] Yang, L. J.; Luo, Y. P.; Chen, W.; Zhao, J. F.; Yang, X. D.; Li, L. A novel diterpenoid alkaloid from *Aconitum macrorhynchum*. *Journal of Asian Natural Products Research*, **2009**, 11, 727–730.

[89] Jing, H.; Yang, X. D.; Zhao, J. F.; Yang, S.; Zhang, H. B.; Li, L. Ouvrardianines A and B, two new norditerpenoid alkaloids from *Aconitum ouvrardianum*. *Helvetica Chimica Acta*, **2009**, 92, 514–519.

[90] Zhang, Z. T.; Liu, X. Y.; Chen, D. L.; Wang, F. P. New diterpenoid alkaloids from *Aconitum liangshanicum*. *Helvetica Chimica Acta*, **2010**, 93, 811–817.

[91] Mericli, A. H.; Cagal-Yurdusever, N.; Ozcelik, H.; Zapp, J.; Kiemer, A. K. A new diterpenoids alkaloid from the roots of a white-flowering *Aconitum orientale* sample. *Helvetica Chimica Acta*, **2012**, 95, 314–319.

[92] Zhao, P. Z.; Gao, S.; Fan, L. M. Approach to the biosynthesis of atisine-type diterpenoids alkaloids. *Journal of Natural Products*, **2009**, 72, 645–649.

[93] Cai, L.; Song, L.; Chen, Q. H.; Liu, X. Y.; Wang, F. P. Piepunine, a novel bis-diterpenoid alkaloid from the roots of *Aconitum piepunense*. *Helvetica Chimica Acta*, **2010**, 93, 2251–2255.

[94] Chen, F.-Z.; Chen, Q.-H.; Wang, F.-P. Diterpenoid alkaloids from *Delphinium yunnanense*. *Helvetica Chimica Acta*, **2011**, 94(2), 254–260. doi: 10.1002/hlca.201000171.

[95] Khan, H.; Nabavi, S. M.; Sureda, A.; Mehterov, N.; Gulei, D.; Berindan-Neagoe, I.; Taniguchi, H.; Atanasov, A. G. Therapeutic potential of songorine, a diterpenoid alkaloid of the genus *Aconitum*. *European Journal of Medicinal Chemistry*, **2017**, S022352341730867X. doi: 10.1016/j.ejmech.2017.10.065.

[96] Zhong, J.; Wang, H.; Zhang, Q.; Gao, S. The chemistry of *Daphniphyllum* alkaloids. *The Alkaloids: Chemistry and Biology*, **2021**, 113–176. doi: 10.1016/bs.alkal.2021.01.001.

[97] Cherney, E. C.; Baran, P. S. Terpenoid-alkaloids: Their biosynthetic twist of fate and total synthesis. *Israel Journal of Chemistry*, **2011**, 51(3–4), 391–405. doi: 10.1002/ijch.201100005.

[98] Lam, Y.; Ng, T. B.; Yao, R. M.; Shi, J.; Xu, K.; Sze, S. C. Evaluation of chemical constituents and important mechanism of pharmacological biology in *Dendrobium* plants. *Evidence-Based Complementary and Alternative Medicine*, **2015**, 841752. doi: 10.1155/2015/841752.

[99] Yang, W.; Chen, X.; Li, Y.; Guo, S.; Wang, Z.; Yu, X. Advances in pharmacological activities of terpenoids. *Natural Product Communications*, **2020**, 15(3), 1934578X2090355. doi: 10.1177/1934578X20903555.

9

Synthesis and Medicinal Uses of Triterpenoids

Biswa Mohan Sahoo*, Bimal Krishna Banik†, and Abhishek Tiwari‡

CONTENTS

9.1 Introduction 286
9.2 Synthesis of Triterpenoids 287
    9.2.1 Biosynthesis of Tetracyclic Triterpenoid 288
        9.2.1.1 Synthesis of Shionone 288
        9.2.1.2 Synthesis of Lanosterol 288
        9.2.1.3 Synthesis of Mogrosides 288
        9.2.1.4 Synthesis of Dammarane-Type Triterpenoids 289
        9.2.1.5 Biosynthesis of Cycloartane 289
    9.2.2 Biosynthesis of Pentacyclic Triterpenoid 290
        9.2.2.1 Biosynthesis of Hopanoids 291
        9.2.2.2 Biosynthesis of Oleanane and Ursane Triterpenoids 291
        9.2.2.3 Biosynthesis of Taraxerol 294
        9.2.2.4 Biosynthesis of Lupeol 294
        9.2.2.5 Biosynthesis of Betulinic Acid 294
        9.2.2.6 Biosynthesis of Friedelin 294
        9.2.2.7 Biosynthesis of Celastrol 294
        9.2.2.8 Biosynthesis of Oleanolic Acid 296
        9.2.2.9 Biosynthesis of Glycyrrhizin 296
9.3 Medicinal Uses of Triterpenoids 297
    9.3.1 Triterpenoids as Anticancer Agents 297
        9.3.1.1 Lupeol as Anticancer Agent 298
        9.3.1.2 Limonoids as Anticancer Agent 299
        9.3.1.3 Tirucallane as Anticancer Agent 299
        9.3.1.4 Cycloartane Triterpene as Anticancer Agent 299
        9.3.1.5 Protostane Triterpene as Anticancer Agent 299
        9.3.1.6 Ursolic Acid as Anticancer Agent 299
        9.3.1.7 Betulinic Acid as Anticancer Agent 301
        9.3.1.8 Ganoderic Acid D as Anticancer Agent 301
        9.3.1.9 Dehydrotrametenolic Acid as Anticancer Agent 301
        9.3.1.10 Impatienside A and Bivittoside D as Anticancer Agents 302
        9.3.1.11 Ananosic Acid B as Anticancer Agents 303
        9.3.1.12 Daedaleasides as Anticancer Agents 304
        9.3.1.13 Inonotsuoxides as Anticancer Agent 304
        9.3.1.14 CDDO-Me as Anticancer Agent 305
        9.3.1.15 β-Escin as Anticancer Agent 305
        9.3.1.16 Cimicifoetisides as Anticancer Agents 305
        9.3.1.17 Nimbolide as Anticancer Agent 306
        9.3.1.18 3-O-Acetyl-11-Keto-β-Boswellic Acid as Anticancer Agent 306
        9.3.1.19 Asiatic Acid as Anticancer Agent 306

* Roland Institute of Pharmaceutical Sciences, Berhampur affiliated to Biju Patnaik University of Technology (BPUT), Rourkela, Odisha, India. Corresponding author e-mail address: drbiswamohansahoo@gmail.com.

† Department of Mathematics and Natural Sciences, College of Sciences and Human Studies, Prince Mohammad Bin Fahd University, Al Khobar, Kingdom of Saudi Arabia. Corresponding author e-mail address: bimalbanik10@gmail.com.

‡ Faculty of Pharmacy, Pharmacy Academy, IFTM University, Lodhipur Rajput, Moradabad-244102, Uttar Pradesh, India.

DOI: 10.1201/9781003008682-9

         9.3.1.20 Celastrol as Anticancer Agent 307

         9.3.1.21 Glycyrrhizin as Anticancer Agent 307

    9.3.2   Triterpenoids as Anti-Diabetic Agents 307

         9.3.2.1 Lupeol as Anti-Diabetic Agent 308

         9.3.2.2 Oleanolic Acid as Anti-Diabetic Agent 308

         9.3.2.3 Viburodorol A as Anti-Diabetic Agent 310

         9.3.2.4 Ursolic Acid A as Anti-Diabetic Agent 310

    9.3.3   Triterpenoids as Antimicrobial Agents 311

         9.3.3.1 *β*-Amyrin as Antimicrobial Agents 311

         9.3.3.2 Betulinic Acid as Antimicrobial Agents 312

    9.3.4   Triterpenoids as Anti-Inflammatory Agents 312

         9.3.4.1 Avicins as Anti-Inflammatory Agents 312

         9.3.4.2 Ginsenosides as Anti-Inflammatory Agents 312

         9.3.4.3 Escin as Anti-Inflammatory Agent 312

         9.3.4.4 Glycyrrhizin as Anti-Inflammatory Agent 313

         9.3.4.5 Lupeol as Anti-Inflammatory Agent 314

         9.3.4.6 Oleanolic Acid as Anti-Inflammatory Agent 315

         9.3.4.7 Ursolic Acid as Anti-Inflammatory Agent 315

         9.3.4.8 Platycodin D as Anti-Inflammatory Agent 315

         9.3.4.9 Saikosaponins as Anti-Inflammatory Agents 315

    9.3.5   Triterpenoids as Antioxidant 316

         9.3.5.1 Fulgic Acids as Anti-Oxidants 317

         9.3.5.2 Multiflorane Triterpenoid as Anti-Oxidants 317

         9.3.5.3 Madecassic Acid as Anti-Oxidants 318

    9.3.6   Triterpenoids as Hepatoprotective 318

         9.3.6.1 Triterpenoids Saponis as Hepatoprotective 318

         9.3.6.2 Ceanothic Acid and Zizybrenalic Acid as Hepatoprotective 318

    9.3.7   Triterpenoids as Cardioprotective Agent 319

         9.3.7.1 Lupeol as Cardioprotective Agent 319

         9.3.7.2 Betulinic Acid as Cardioprotective Agent 319

    9.3.8   Triterpenoids as Antiviral Agents 319

    9.3.9   Miscellaneous 320

9.4   Conclusion 323

Abbreviations 323

References 324

9.1 Introduction

Triterpenoids are natural compounds with diverse chemical structures and biological activities. Triterpenoids are commonly produced as secondary metabolites in different plant sources. These compounds are generally produced in specific plant tissues or during several developmental stages of the plants such as germination of seeds, vegetative growth and differentiation, fruiting, nodulation, etc [1]. Various abiotic and biotic stresses also stimulate the production of triterpenoids. Naturally, triterpenoids are biosynthesized via the mevalonate pathway (MVA) or 2-*C*-methyl-*d*-erythritol 4-phosphate (MEP) pathway. MVA pathway is primarily found in the cytoplasm and peroxisome of eukaryotes, whereas the MEP pathway is present in most of the bacteria and green algae [2]. Triterpenes are primarily composed of three terpene units with the molecular formula $C_{30}H_{48}$. Most of the triterpenes are pentacyclic (oleanane, ursane, taraxerane, taraxastane and lupine) and some are monocyclic (achilleol A, marneral and camelliol C), bicyclic (lansic acid, myrrhanone A and lamesticumin A), tricyclic (achilleol B, thalianol and kadcotrione A), tetracyclic (dammarane, cucurbitane and euphane).

These are widely distributed in seeds, roots, bark and leaves of plants [3]. Pentacyclic triterpenes are mainly based on C_{30} skeleton that contain either five six-membered rings (6-6-6-6-6) (ursanes and lanostanes) or four six-membered rings and one five-membered ring (6-6-6-6-5) (lupanes and hopanes), whereas tetracyclic triterpenoids contain three six-membered and one five-membered ring (6-6-6-5) system as presented in Figure 9.1 [4]. Biosynthetically, triterpenoids are produced from six isoprene units (C_5H_8) that are shared by the common C_{30} acyclic precursor squalene. Most of the triterpenoids are alcohol that can combine with sugars to furnish glycosides called terpenoid glycosides [5].

Several spraying reagents are used in thin-layer chromatography (TLC) to detect the triterpenes include thionyl chloride, phosphomolibdic acid, silicotungstic acid, stannous chloride, arsenic chloride, antimonnium chloride, rhodamine and vanillin phosphoric acid etc (Table 9.1). TLC is performed on silica gel using most commonly used solvents (chloroform, methanol and water) with different proportions [6,7].

Further, different separation methods are used to isolate individual triterpenes that depend on the type of compounds to be separated. Free aglycones are routinely separated by various chromatographic techniques such as preparative thin layer

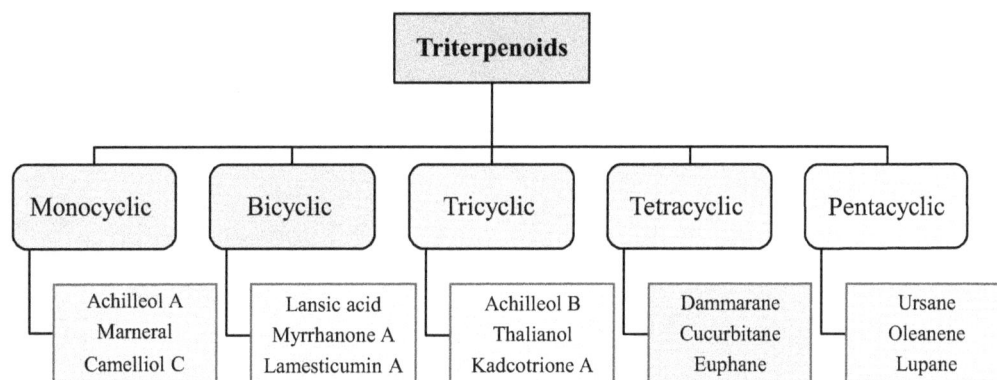

FIGURE 9.1 Classification of triterpenoids.

TABLE 9.1

Spraying Reagents Used for Detection of Triterpene in TLC

Liebermann-Burchard	Blue, green, pink, brown and yellow in visible light
Vanillin-sulfuric acid	Blue, blue-violet and yellowish
Vaniline-phosphoric acid	Red-violet in visible light and reddish or blue in UV-365
Anisaldehyde-sulfuric acid	Blue red-violet in visible light and reddish or blue in UV-365
Komarowsky reagent	Blue, yellow and red
Antimony (III) chloride (Carr-Price reagent)	Pink, purple in visible light, red-violet, green and blue in UV-365
Ehrlich reagent	Red coloration of furostanol derivatives

chromatography (PTLC), high-performance thin layer chromatography (HPTLC), column chromatography (CC), flash chromatography, medium pressure liquid chromatography (MPLC), high-performance liquid chromatography (HPLC) and vacuum liquid chromatography (VLC) [8]. In addition to this, advance techniques are used such as gas chromatography-mass spectrometry (GC-MS), droplet counter-current chromatography (DCCC) or capillary column chromatography. The structures of the newly isolated or synthesized compounds are established by various spectroscopic methods such as FT-IR, 2D-COSY NMR, ^{1}H-NMR, ^{13}C-NMR, HREIMS and LC-MS as depicted in Figure 9.2 [9].

Further, triterpenes are evaluated extensively to identify their biological properties. The potential biological activities of triterpenoids include anticancer, anti-diabetic, hepatoprotective, immunomodulatory, antimicrobial, antimycotic, cardiotonic, analgesic and anti-inflammatory as shown in Figure 9.3 [10–13].

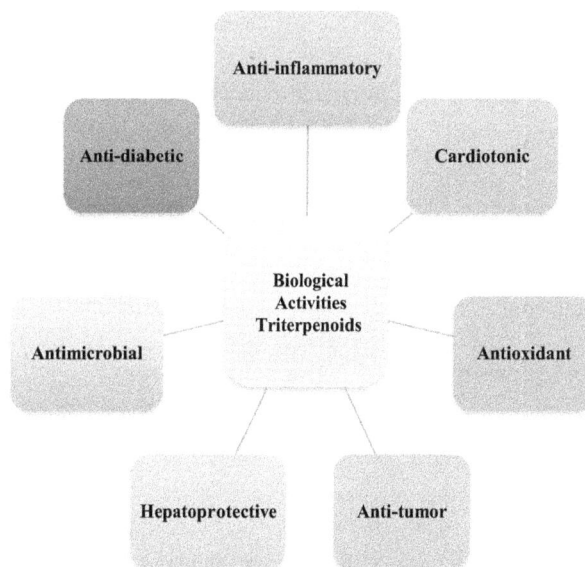

FIGURE 9.3 Biological activities of triterpenoids.

9.2 Synthesis of Triterpenoids

Triterpenoids are synthesized from isopentenyl pyrophosphate (IPP) through the intermediate called squalene (30-carbon). IPP can be synthesized through the mevalonic acid (MVA) pathway or the methylerythritol phosphate (MEP) pathway in plants (Figure 9.4) [14]. The former reaction takes place in the cytosol, whereas the latter reaction

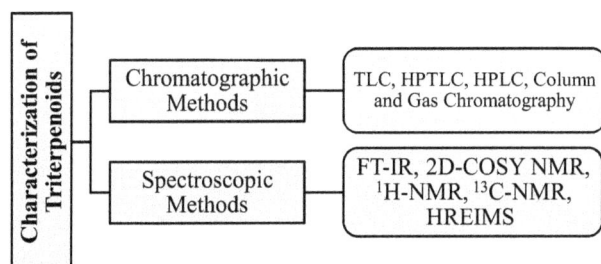

FIGURE 9.2 Characterization of triterpenoids.

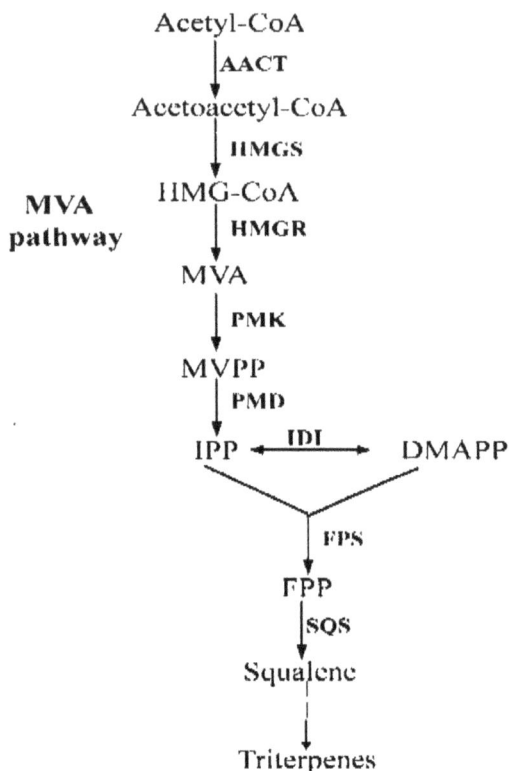

FIGURE 9.4 Mevalonic acid pathway for synthesis of triterpenes via squalene.

takes place in the plastids of plants (Figure 9.5). From this intermediate, several triterpenoids are synthesized that include protostanes, lanostanes, holostanes, cycloartanes, dammaranes, euphanes, tirucallanes, tetranortriterpenoids, quassinoids, lupanes, oleananes, friedelanes, ursanes, hopanes, isomalabaricanes and saponins [15].

Triterpenoid carbon frameworks are cyclized by members of the oxidosqualene cyclase family. Oxidosqualene cyclases convert oxidosqualene to one or more cyclic triterpene alcohols with up to six carbocyclic rings (Figure 9.6) [16,17].

9.2.1 Biosynthesis of Tetracyclic Triterpenoid

9.2.1.1 Synthesis of Shionone

Shionone is a tetracyclic triterpenoid with the molecular formula $C_{30}H_{50}O$. It is isolated from *Aster tataricus*. It constitutes six-membered tetracyclic skeleton and 3-oxo-4-monomethyl structure. For the biosynthesis of the Shionone-type triterpenoids, 2,3-oxidosqualene is converted into dammarenyl cation via an enzymatic reaction and then converted into the baccharenyl cation. Shionone (enol form) is produced when enzymes act on the baccharenyl cation (Figure 9.7). It is observed that Shionone enhances the secretion of sputum and reduces the xylene-induced ear edema [18].

9.2.1.2 Synthesis of Lanosterol

Lanosterol is a tetracyclic triterpenoid with the chemical formula $C_{30}H_{50}O$. It is found in *Eleutherococcus sessiliflorus* and *Euphorbia mellifera*. It is lanosta-8,24-diene substituted by

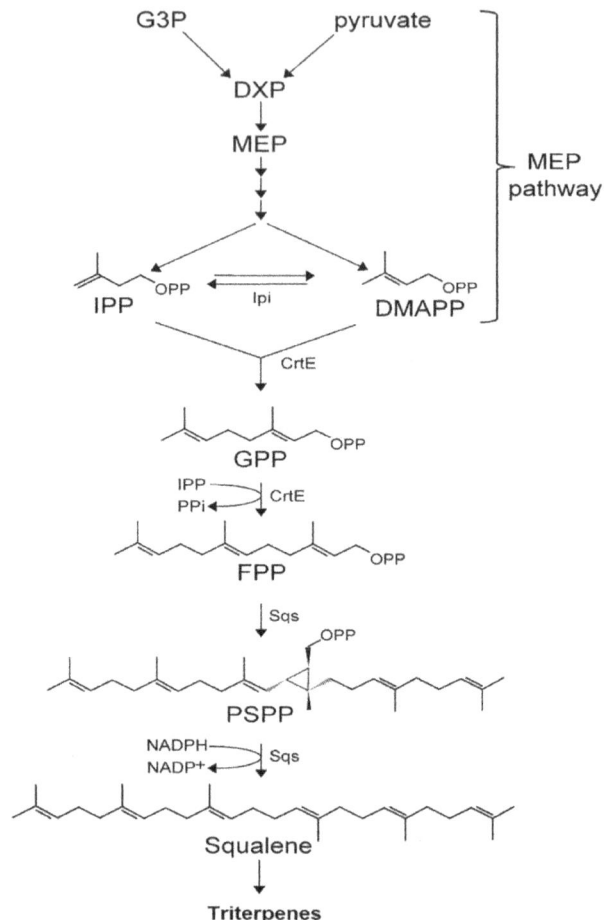

FIGURE 9.5 MEP pathway for synthesis of triterpenes via squalene.

a β-hydroxy group at 3β-position. It is derived from the chair-boat-chair conformation of 2,3-oxidosqualene. So, Squalene is first oxidized to 2,3-oxidosqualene in the presence of enzyme squalene mono-oxidase or squalene epoxidase [19]. Further, 2,3-oxidosqualene is converted to lanosterol in the presence of enzyme oxidosqualene cyclase or lanosterol synthase. The energy required for this reaction is obtained from nicotinamide adenine dinucleotide phosphate (NADPH). Finally, lanosterol is converted into lanostane-type triterpenoids in the presence of cytochrome p450 enzyme (Figure 9.8) [20].

Antcamphorol A is a lanostane-type triterpenoid with molecular formula $C_{31}H_{50}O_3$. It is found in *Taiwanofungus camphoratus*. The molecular scaffold of antcamphorol can be synthesized from its precursor lanosterol. In the later stage, it involves oxidation and cyclization of the D-ring side chain of the intermediate to produce tetrahydrofuran moiety of antcamphorol (Figure 9.9) [21].

9.2.1.3 Synthesis of Mogrosides

Mogrosides are the Cucurbitane-type tetracyclic triterpenoid saponins, isolated from plant *Siraitia grosvenorii, Siraitia siamensis* of family Cucurbitaceae. Squalene is first oxidized to 2,3-oxidosqualene in the presence of squalene epoxidase and the mogrosides are synthesized from 2,3-oxidosqualene via a series of steps catalyzed by several enzyme

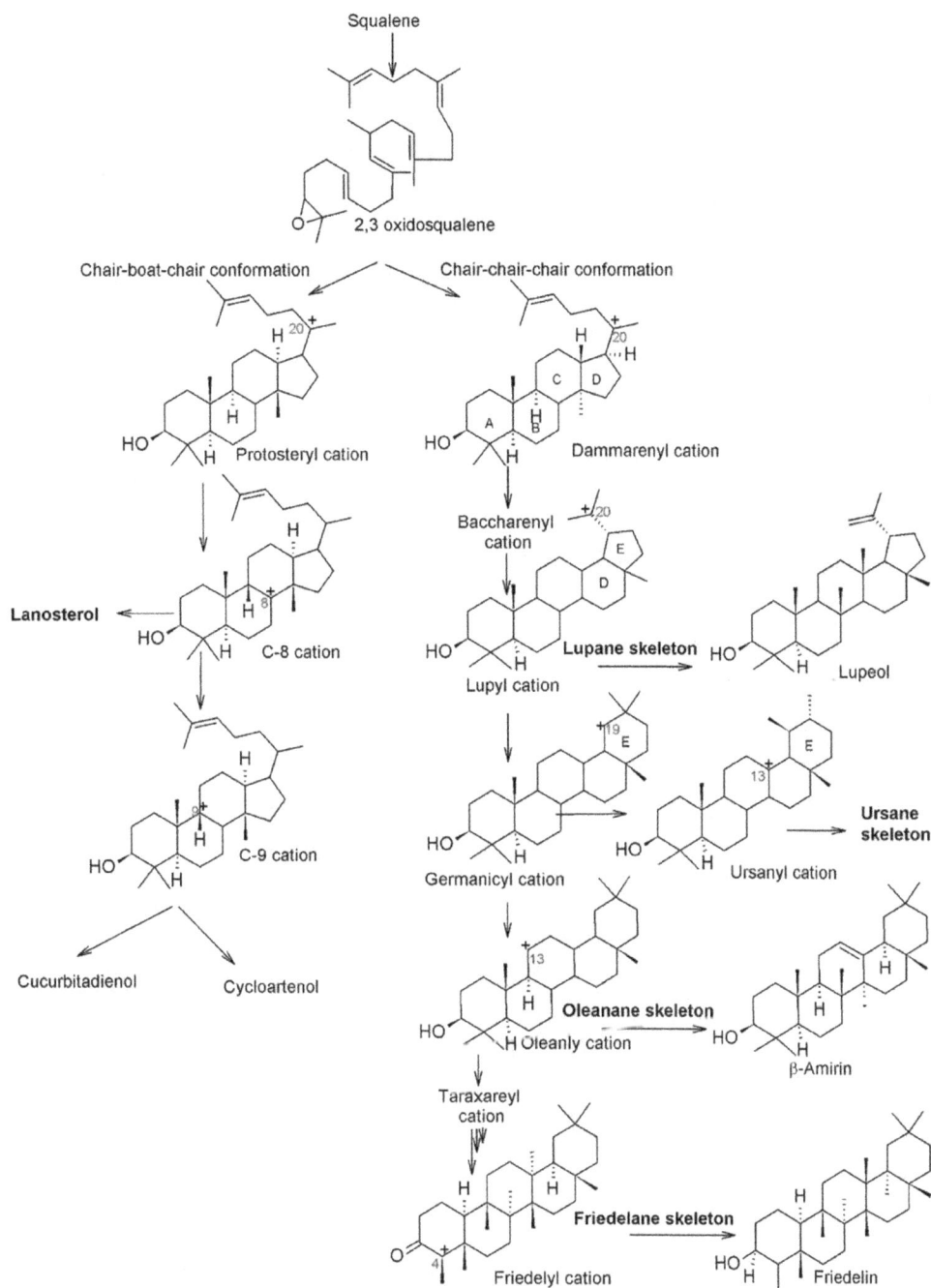

FIGURE 9.6 Biosynthesis of triterpenoids from squalene.

systems including cucurbitadienol synthase, Cytochrome P450s and UDP glycosyltransferases (UGTs) as presented in Figure 9.10 [22].

9.2.1.4 Synthesis of Dammarane-Type Triterpenoids

Dammarane is the dammarane-type of tetracyclic triterpene. It is isolated from the tropical plants of family Dipterocarp. For the biosynthesis of dammarane type of tetracyclic triterpene from the squalene precursor, squalene is first converted into 2,3-oxidosqualene and the dammarane diol synthase converts it into dammarendiol, which is transformed into protopanaxadiol

and then changed into dammarane-type triterpenoids via enzymatic reaction (Figure 9.11). Several disease conditions that are treated and prevented by dammarane-type triterpenoids include hyperlipidemia, cardiovascular disorder, bone diseases, diabetes mellitus, metabolic syndrome and cancer [23].

9.2.1.5 Biosynthesis of Cycloartane

Cycloartane is a triterpene with the molecular formula $C_{30}H_{52}$. It is also known as 4,4,14-trimethyl-9,19-cyclo-5α, 9β-cholestane. Its derivative cycloartenol is used as a precursor for the synthesis of most of the plant steroids as presented in Figure 9.12 [24].

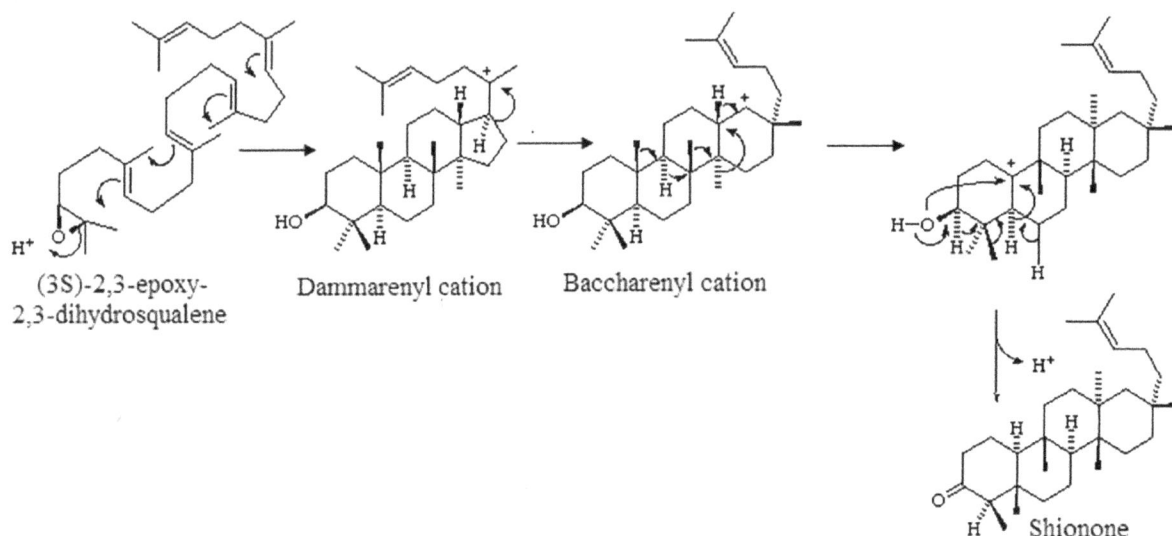

FIGURE 9.7 Synthesis of shionone from 2,3-oxidosqualene.

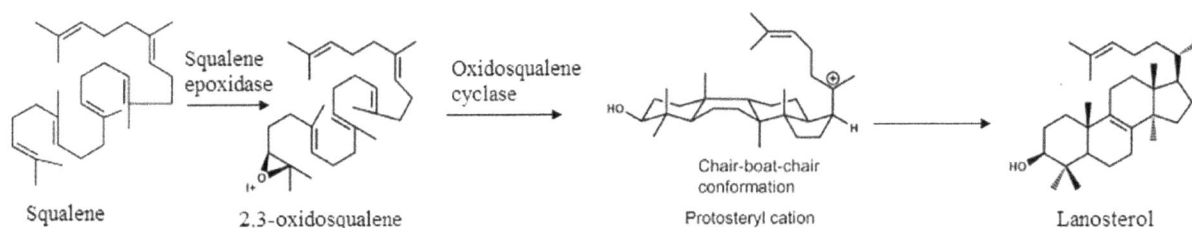

FIGURE 9.8 Synthesis of lanosterol from squalene.

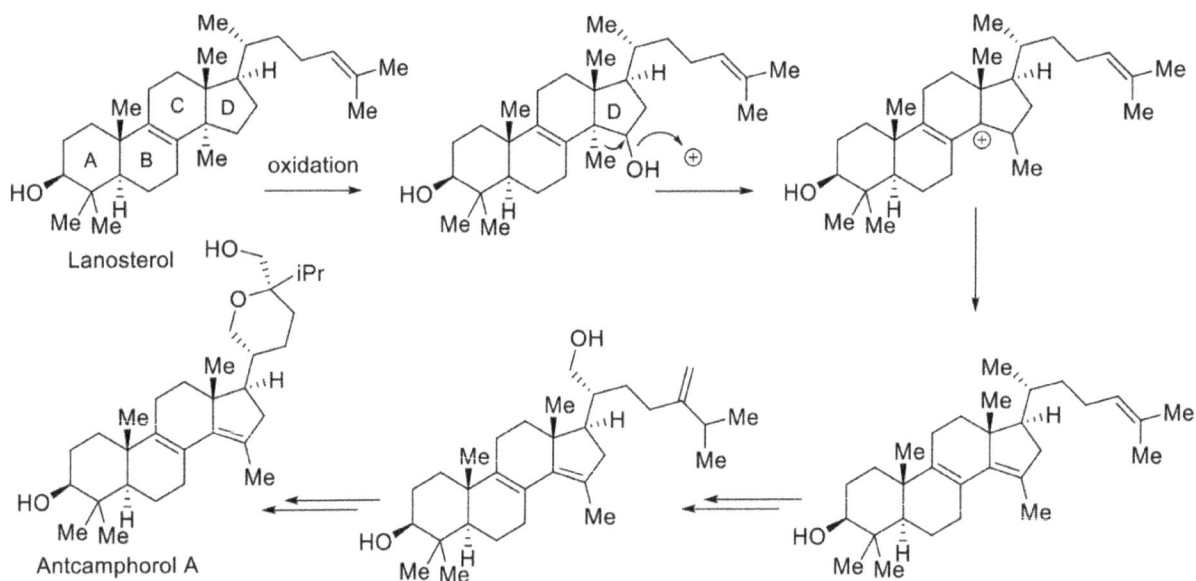

FIGURE 9.9 Biosynthesis of antcamphorol A.

9.2.2 Biosynthesis of Pentacyclic Triterpenoid

The major pentacyclic triterpenoid scaffolds (hopane, oleanane, ursane and lupine) are derived from β-amyrin, α-amyrin and lupeol. Squalene is the precursor for producing all types of pentacyclic triterpenoids. First of all, squalene is oxidized into 2,3-oxidosqualene in presence of squalene epoxidase. The conversion of 2,3-oxidosqualene into pentacyclic triterpenes involves most complex enzymatic reactions (Figure 9.13) [25].

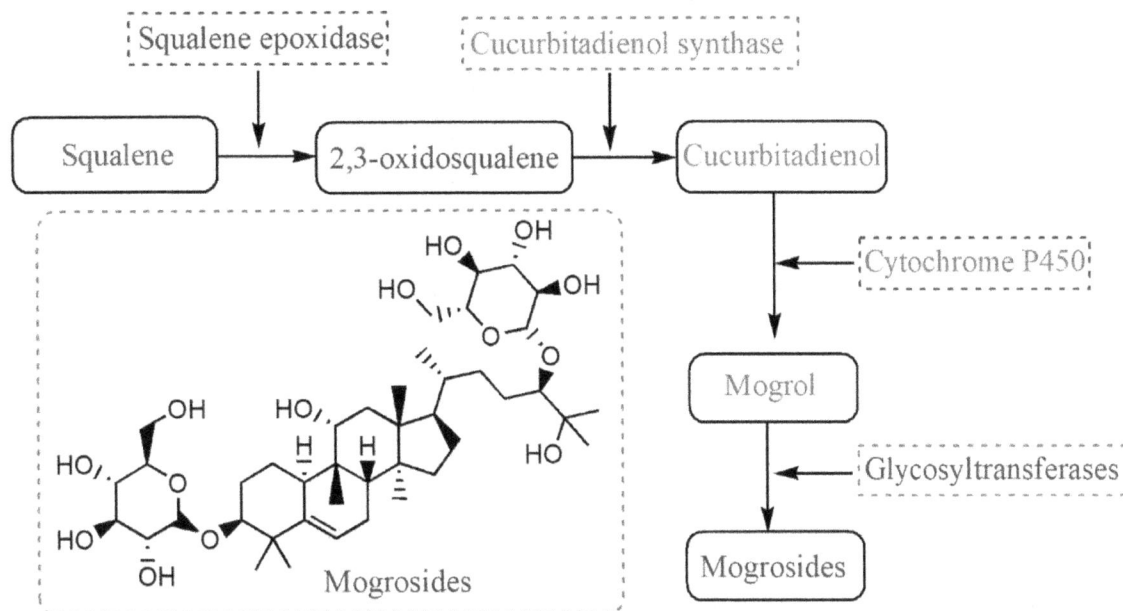

FIGURE 9.10 Synthesis of mogrosides.

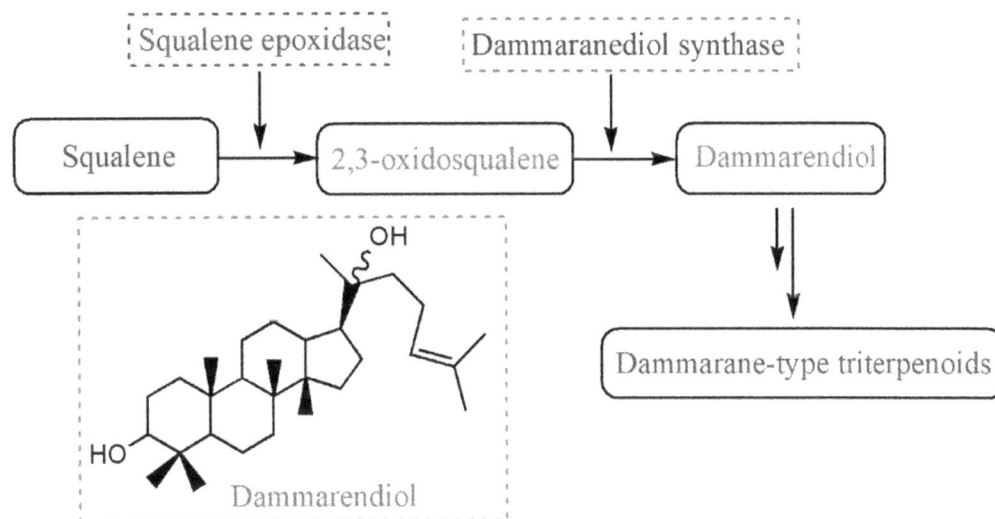

FIGURE 9.11 Synthesis of dammarane-type triterpenoids.

9.2.2.1 Biosynthesis of Hopanoids

Hopanoids are the pentacyclic triterpenoid lipids that occur in bacteria (*Bradyhrizobium*). These are synthesized from isopentenyl units that are produced in a biosynthetic route leading to furnish isopentenyl diphosphate. Six C_5 units are joined to produce squalene that acts as the immediate precursor for the synthesis of hopanoid [26]. The hopane skeleton is generated from squalene in presence of squalenehopene cyclase (Figure 9.14). Hopanoids consist of six isoprene units (C_5H_8) and their biosynthetic pathway can be divided into the following stages [27].

- Synthesis of the C5-precursor isopentenyl diphosphate (IPP) and its isomerization to dimethylallyl diphosphate (DMPP).

- Elongation of DMPP by head-to-tail condensation of two IPP to farnesyl diphosphate (C15).
- Reductive condensation of two farnesyl diphosphates, head-to-head arrangement to squalene (C30).
- Cyclization of squalene to hopene.
- Elongation reactions of hopene to hopanoids.

9.2.2.2 Biosynthesis of Oleanane and Ursane Triterpenoids

Oleanane and ursane pentacyclic triterpenoids are secondary metabolites of plants. Oleanolic acid is the oleanane-type triterpenoids with molecular formula $C_{30}H_{48}O_3$. Chemically it is 3β-hydroxy-olean-12-en-28-oic acid. Similarly glycyrrhetinic acid is an oleanane-type triterpenoid with the molecular

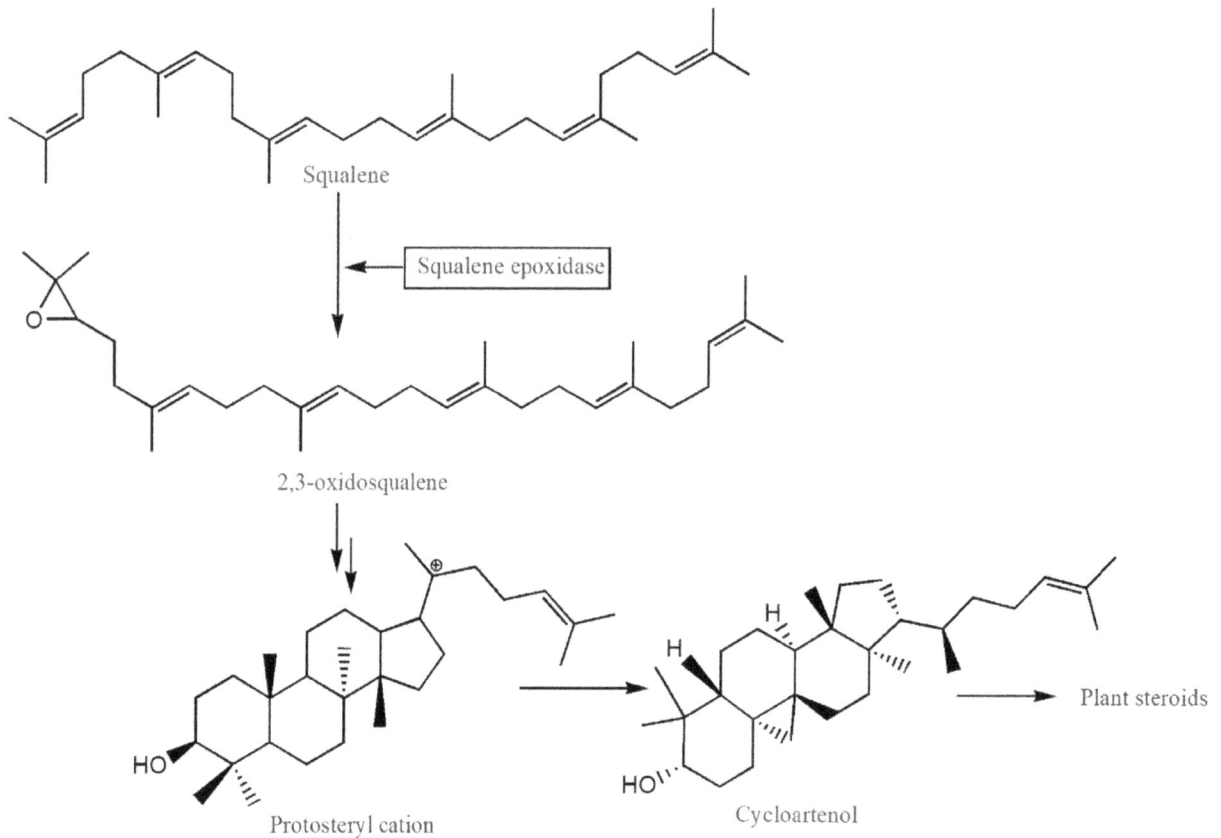

FIGURE 9.12 Biosynthesis of cycloartane.

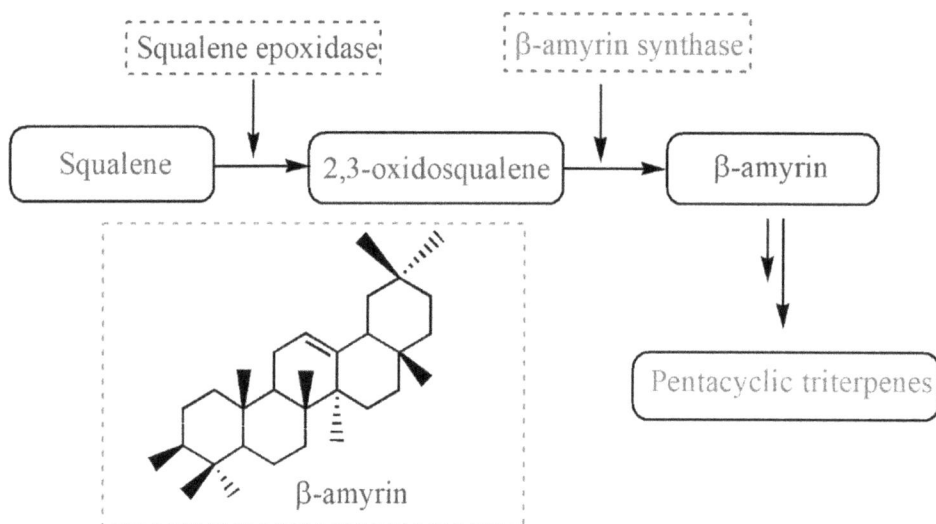

FIGURE 9.13 Biosynthesis of pentacyclic triterpenoid.

formula $C_{30}H_{46}O_4$. Chemically it is 3β-hydroxy-11-oxo-18β-olean-12-en-30-oic acid and ursolic acid is the most available ursane triterpenoid with molecular formula $C_{30}H_{48}O_3$. Chemically, it is 3β-hydroxy-urs-12-en-28-oic acid (Figure 9.15) [28].

Oleanane and ursane pentacyclic triterpenoids are obtained via the MVA pathway. It involves the generation of mevalonate from HMGCoA in presence of 3-hydroxy-3-methylglutaryl-CoA reductases (HMG1). Further mevalonate is transformed

into isopentenyl diphosphate and dimethylallyl diphosphate which are converted to farnesyl diphosphate by a farnesyl phosphate synthase (ERG20). Two molecules of farnesyl diphosphate are then converted to squalene by the action of enzyme squalene synthase (ERG9). Squalene is further transformed into 2,3-oxidosqualene in presence of squalene epoxidase (ERG1). Finally, the subsequent synthesis of pentacyclic triterpenoids involves plant genes encoding amyrin synthase, CYP450 and CYP450 reductase. Genes encoding β-amyrin

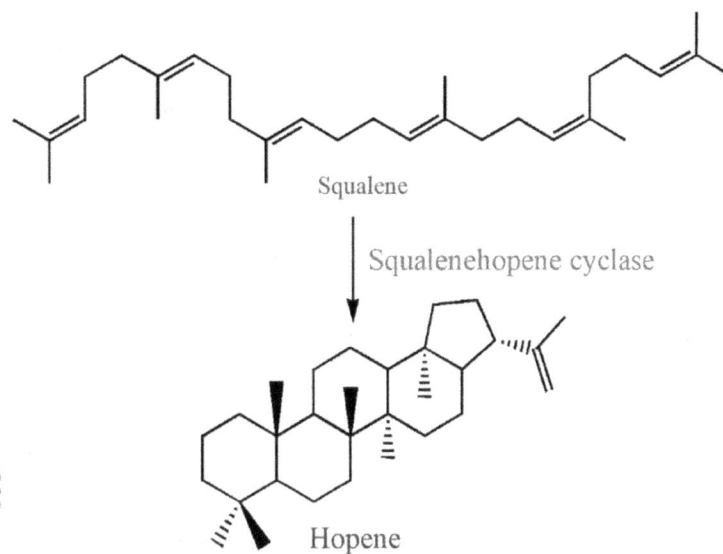

FIGURE 9.14 Biosynthesis of hopanoids.

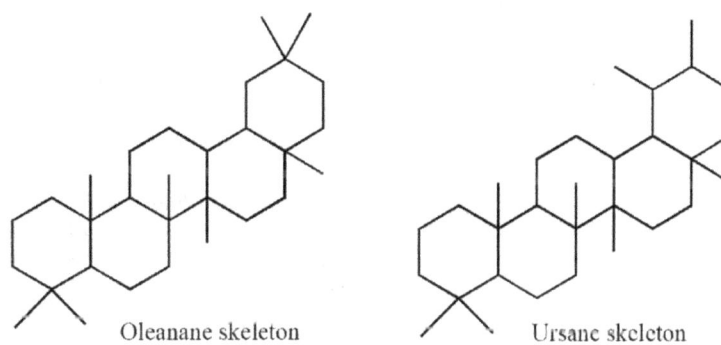

FIGURE 9.15 Structures of oleanane and ursane pentacyclic triterpenoids skeleton.

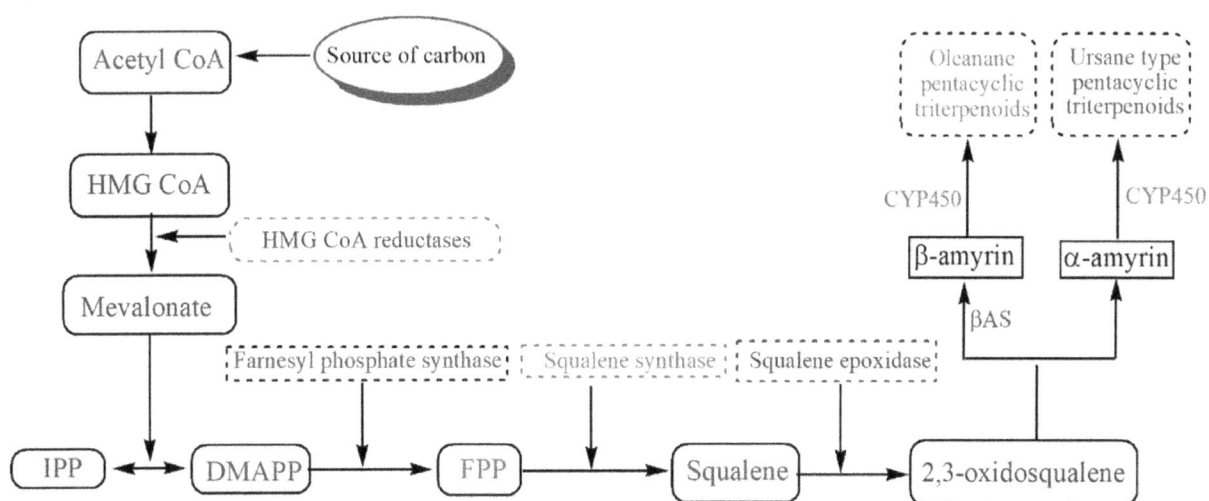

FIGURE 9.16 Biosynthesis of oleanane and ursane triterpenoids via MVA pathway.

synthase (*β*AS) that catalyzes the formation of *β*-amyrin from 2,3-oxidosqualene. *β*-Amyrin acts as a precursor of oleanane pentacyclic triterpenoids. Similarly, genes encoding CYP450 enzymes catalyze the subsequent conversion of *α*-amyrin to ursane-type pentacyclic triterpenoids (Figure 9.16) [29]. It was reported that these triterpenoids exhibit several biological activities such as antiviral, anti-inflammatory, antimicrobial and antitumor [30].

9.2.2.3 Biosynthesis of Taraxerol

Taraxerol is an oleanane-type pentacyclic triterpenoid with the molecular formula $C_{30}H_{50}O$. It is widely present in various higher plants such as *Taraxacum officinale, Alnus glutinosa* and *Litsea dealbata*. Chemically, it is (3β)-D-friedoolean-14-en-3-ol. Biosynthesis of taraxerol occurs through mevalonate (MVA) pathway in the cytosol of plants in which dimethylallyl diphosphate (DMAPP) and isopentyl pyrophosphate (IPP) are produced first followed by the synthesis of squalene. Squalene is the major precursor for the synthesis of triterpenoids (taraxerol) catalyzed by taraxerol synthase (Figure 9.17) [31].

9.2.2.4 Biosynthesis of Lupeol

Lupeol is a lupane-type pentacyclic triterpene with molecular formula $C_{30}H_{50}O$. It is found in Ficus septica, Diospyros morrisiana. The biosynthesis of lupeol starts with the condensation of two molecules of FPP to produce squalene in the presence of squalene synthase. Squalene is then undergoes oxidation to produce 2,3-oxidosqualene by the action of membrane-bound enzymes like monooxygenase and squalene epoxidase. Finally, oxidosqualene undergoes cyclization to produce lupeol in presence of lupeol synthase (Figure 9.18) [32].

9.2.2.5 Biosynthesis of Betulinic Acid

Betulinic acid is a lupane-type pentacyclic triterpene with the molecular formula $C_{30}H_{48}O_3$. Chemically, it is 3β, hydroxy-lup-20(29)-en-28-oic acid. In yeast, the biosynthetic precursors of betulinic acid include IPP and dimethylallyl pyrophosphate (DMAPP), which are derived from the acetyl-CoA via mevalonic acid (MVA) pathway. Two molecules of FPPs undergo condensation to produce squalene catalyzed by squalene synthase (ERG9). Further, squalene is converted to 2,3-oxidosqualene catalyzed by squalene

epoxidase (ERG1). 2,3-oxidosqualene is catalyzed by lupeol synthase (LUS) to produce lupeol followed by three sequential oxidations catalyzed by lupeol C-28 oxidases (cytochrome P450 monooxygenases) that leads to the generation of betulin, betulinic aldehyde and betulinic acid (Figure 9.19) [33].

9.2.2.6 Biosynthesis of Friedelin

Friedelin and its derivatives are commonly called friedelane triterpenoids. Friedelin is a pentacyclic triterpenoid with the chemical formula $C_{30}H_{50}O$. It is the most abundant triterpene found in Azima tetracantha, Orostachys japonica and Quercus stenophylla. It is also found in the roots of the Cannabis plant. Friedelin is a triterpenoid ketone produced biosynthetically after several rearrangements of the dammarenyl cation (Figure 9.20). It has been found that Friedelin exhibits anti-inflammatory, anti-diabetic and gastroprotective activities [34,35].

9.2.2.7 Biosynthesis of Celastrol

Celastrol is a pentacyclic triterpenoid isolated from *Tripterygium wilfordii* of family Celastraceae. The biosynthesis of celastrol starts with cycloisomerization of 2,3-oxidosqualene in presence of oxidosqualene cyclase and continues with oxidative decoration of the 30-carbon triterpenoid core structure by cytochrome P450 enzymes. In the case of biosynthesis of celastrol, the core triterpenoid scaffold is supposed to be friedelin that is synthesized in presence of friedelin synthases (FRSs). The first step of the pathway involves cyclization of the common triterpenoid precursor 2,3-oxidosqualene in presence of friedelin synthases to furnish friedelin. Further, several enzyme systems (CYP712K1, CYP712K2 or CYP712K3) catalyze the successive oxygenations of friedelin to form polpunonic acid via 29-hydroxy-friedelin and 29-oxo-friedelin. Dashed arrows represent multiple enzymatic steps involved in the biosynthesis of celastrol (Figure 9.21) [36].

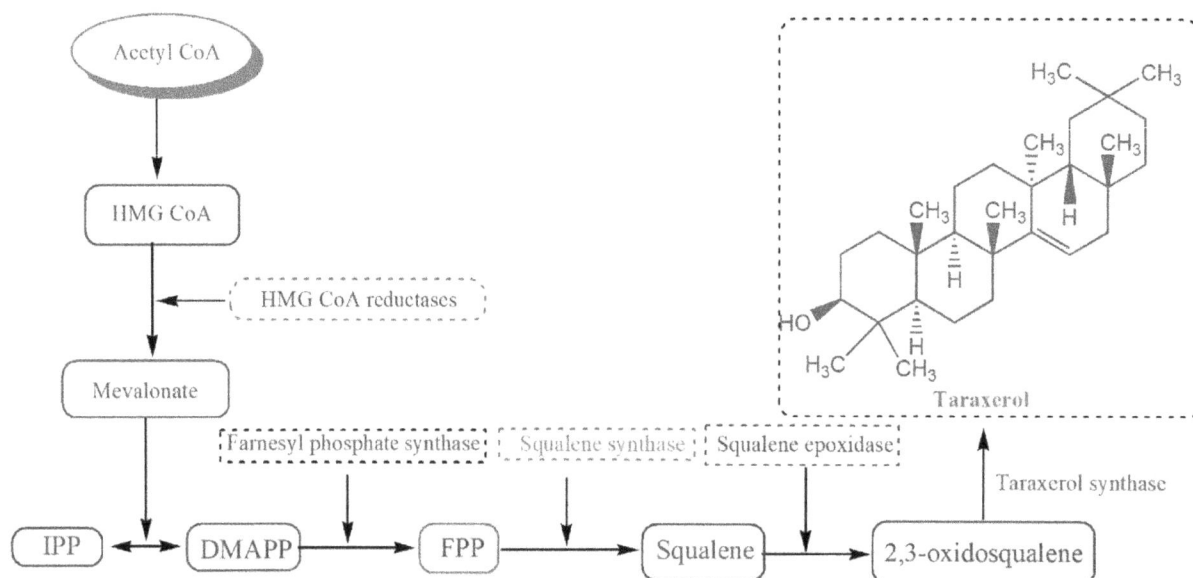

FIGURE 9.17 Biosynthetic pathway of taraxerol.

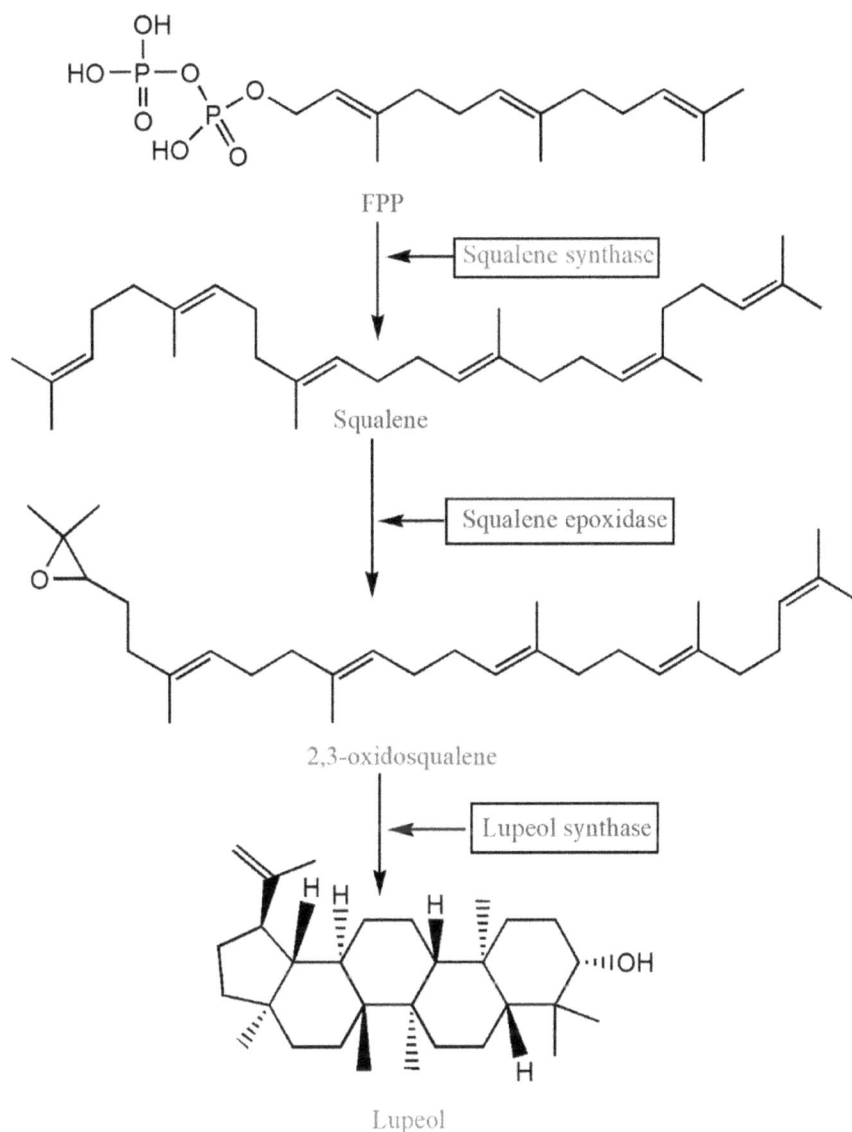

FIGURE 9.18 Biosynthetic route of lupeol.

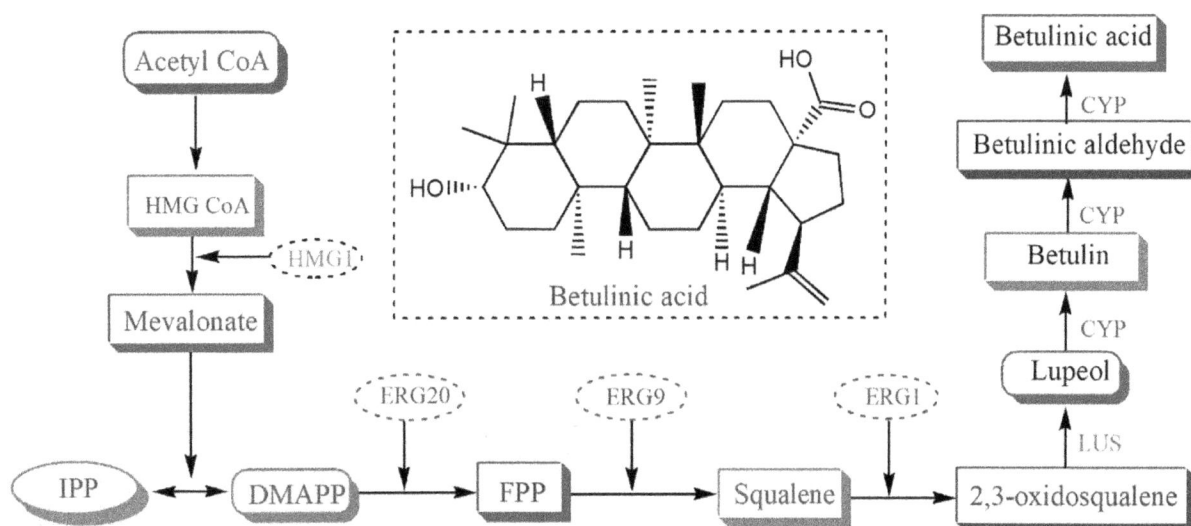

FIGURE 9.19 Biosynthesis of betulinic acid.

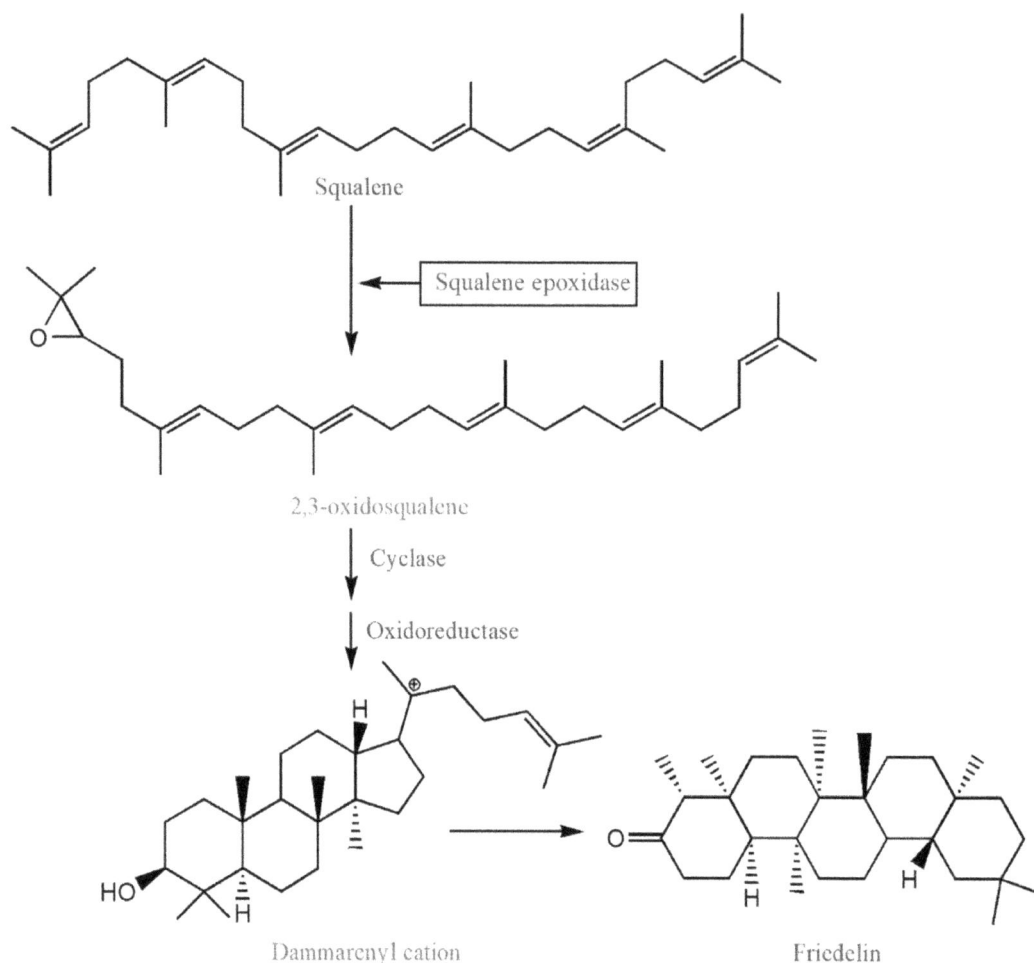

FIGURE 9.20 Biosynthesis of friedelin via dammarenyl cation.

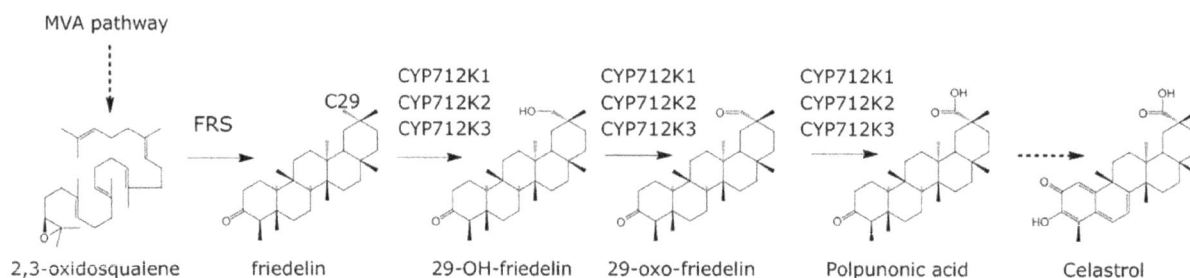

FIGURE 9.21 Biosynthetic pathway of celastrol.

9.2.2.8 Biosynthesis of Oleanolic Acid

Oleanolic acid is a pentacyclic triterpenoid with the molecular formula $C_{30}H_{48}O_3$. Chemically it is 3β-hydroxyolean-12-en-28-oic acid. The precursor molecule 2,3-oxidosqualene is synthesized in the cytosol of plant cell from IPP derived from the mevalonate pathway. The subsequent cyclization of 2,3-oxidosqualene forms the branch-point between the primary sterol and the secondary triterpenoid metabolism that leads to the generation of oleanolic acid (Figure 9.22). Oleanolic acid and its derivatives exhibit various pharmacological activities such as antioxidant, hepatoprotective effects, anti-inflammatory and anticancer activities [37].

9.2.2.9 Biosynthesis of Glycyrrhizin

Glycyrrhizin is a triterpenoid saponin with the molecular formula $C_{42}H_{62}O_{16}$. It is extracted from the root of *Glycyrrhiza uralensis*, *Glycyrrhiza glabra* and *Glycyrrhiza inflate*. The biosynthesis of glycyrrhizin involves the initial cyclization of 2,3-oxidosqualene to produce triterpene skeleton (β-amyrin) in presence of β-amyrin synthase (β-AS) followed by a series of oxidative reactions at positions C_{11}, C_{30} and glycosyl transfers to the C_3 hydroxyl group. Glycosylation is catalyzed by UDP glycosyltransferases (UGTs) (Figure 9.23) [38].

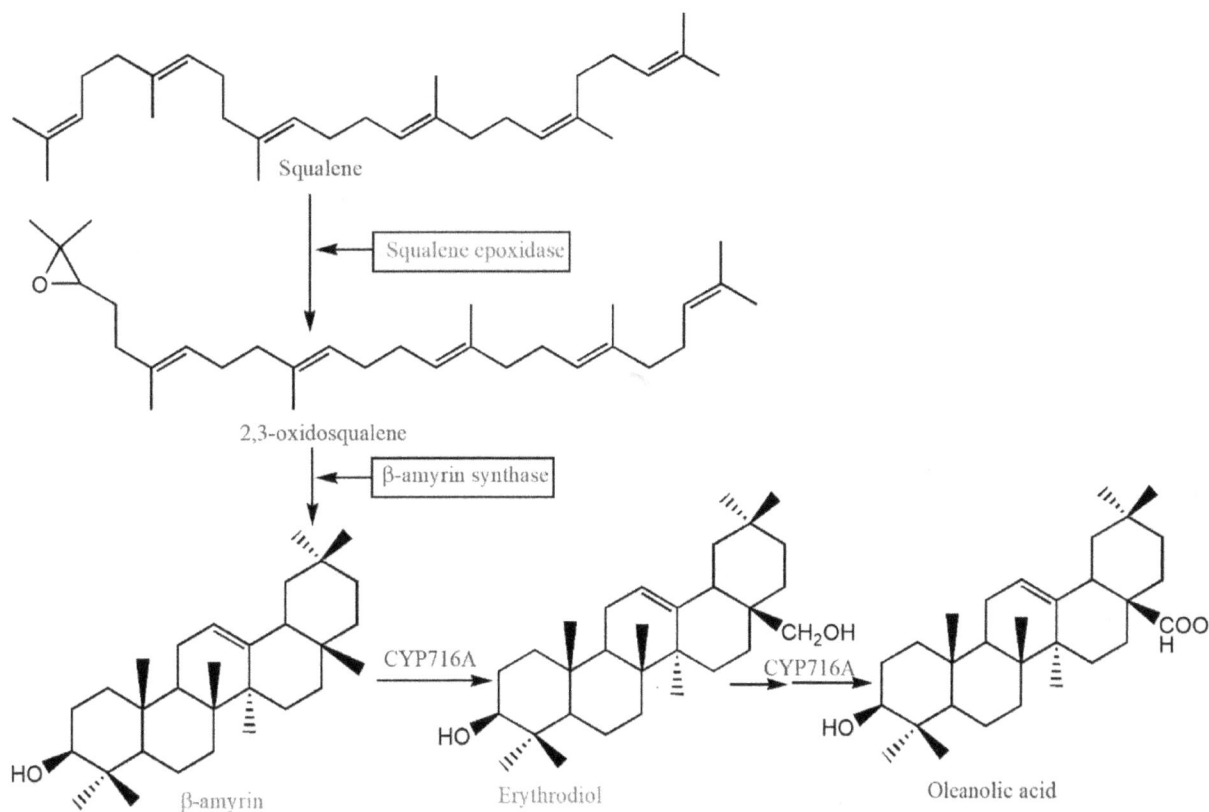

FIGURE 9.22 Oleanolic acid biosynthetic pathway.

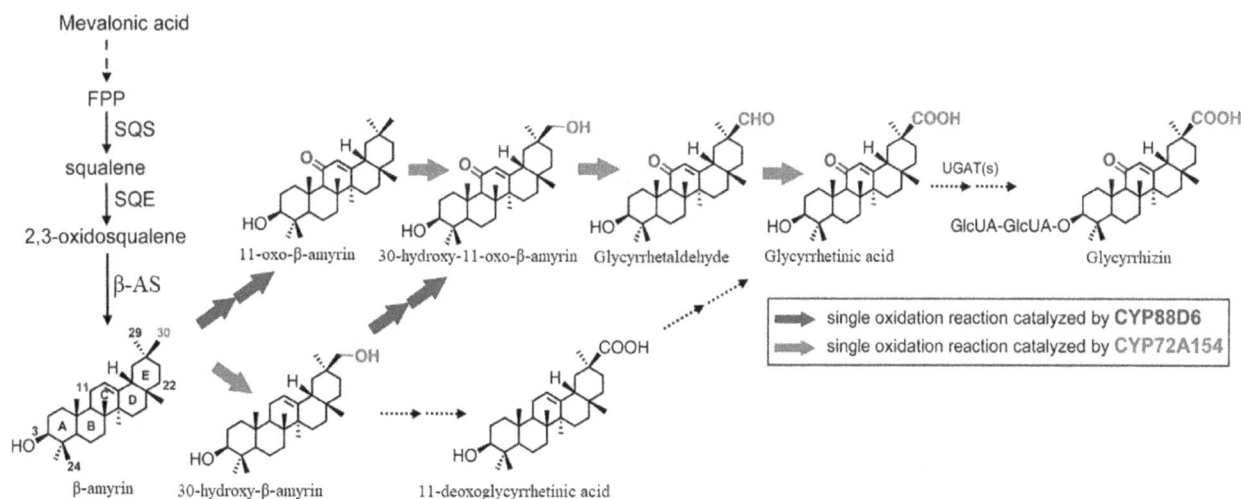

FIGURE 9.23 Biosynthesis of glycyrrhizin from 2,3-oxidosqualene.

9.3 Medicinal Uses of Triterpenoids

Triterpenoids posses several medicinal properties such as anticancer, antioxidant, antipyretic, hepatoprotective, cardiotonic, sedative, analgesic, anti-inflammatory, etc [39]. Triterpenoids with anti-diabetic activity include Astragaloside, Cucurbitacin, Diosgenin, Ginsenoside, Amyrin, Asiatic acid, Avicin, Betulinic acid, Escin, Glycyrrhizin, Oleanolic acid, etc. Similarly, triterpenoids used for the treatment of cardiovascular disorders

include Astragaloside, Cucurbitacin, Diosgenin, Ginsenoside, Gypenoside, Oleandrin and Betulinic acid (Table 9.2, Figure 9.24) [40–42].

9.3.1 Triterpenoids as Anticancer Agents

Triterpenoids exhibit antitumor activity due to their ability to block nuclear factor-κB (NF-κB) activation, inhibit signal transducer, induce apoptosis and activate transcription and angiogenesis [43]. Several triterpenoids that display antitumor activities

TABLE 9.2

List of Triterpenoids with Medicinal Properties

Triterpenoids	Medicinal Properties
Astragaloside, cucurbitacin, diosgenin, ginsenoside, amyrin, asiatic acid, avicin, betulinic acid, escin, glycyrrhizin, oleanolic acid, platycodon D, ursolic acid, withanolide	Anti-diabetic
Astragaloside, cucurbitacin, diosgenin, ginsenoside, gypenoside, oleandrin, betulinic acid, escin, glycyrrhizin, lupeol, oleanolic acid, platycodon D, saikosaponins, ursolic acid and withanolide	Cardiovascular agent
Cucurbitacin, diosgenin, ginsenoside, amyrin, boswellic acid, celastrol, glycyrrhizin, lupeol, oleanolic acid, CDDO-ME, ursolic acid and withanolide	Treatment of arthritis
Diosgenin, gypenoside, betulinic acid, glycyrrhizin, oleanolic acid and ursolic acid	Treatment of atherosclerosis
CDDO-MA	Anti-Parkinson's disease
CDDO-MA and alpha-onocerin	Anti-Alzheimer
Ginsenoside, betulinic acid, escin, glycyrrhizin, platycodon D, momordin, oleanolic acid and ursolic acid	Treatment of obesity
Oleanolic acid	Treatment of multiple sclerosis
Asiatic acid	Anti-depressant
Ursolic acid	Treatment of osteoporosis
Escin and asiatic acid	Treatment of cerebral ischemia
CDDO-MA	Treatment of memory loss

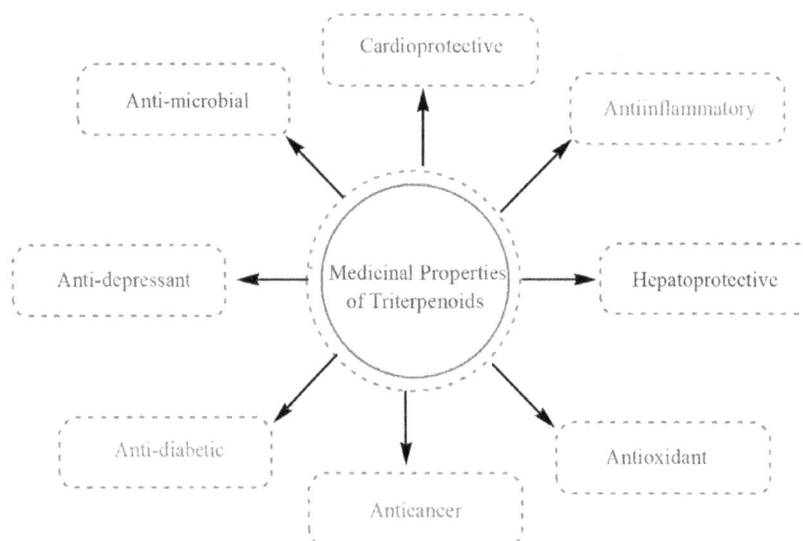

FIGURE 9.24 Medicinal uses of triterpenoids.

include celastrol, pristimerin, avicins, lupeol, betulinic acid, ursolic, oleanolic acid, etc. To enhance the efficacy of existing antitumor activity, some of the synthetic triterpenoid derivatives are also synthesized that include cyano-3,12-dioxooleana-1,9 (11)-dien-28-oic (CDDO), its methyl ester (CDDO-Me) and imidazolide (CDDO-Im) derivatives. Among these compounds, CDDO, CDDO-Me and betulinic acid exhibit promising antitumor activities and these derivatives are currently under phase I clinical studies (Table 9.3) [44].

9.3.1.1 Lupeol as Anticancer Agent

Lupeol is reported to exhibit anticancer activity against different types of cancers such as human prostate, breast, skin, liver and blood cancers. It is reported that lupeol displays anticancer activity in human cervical carcinoma (HeLa) cells. The inhibitory activity of lupeol is mediated through induction of S-phase cell cycle arrest and apoptosis. The effects of lupeol on human

non-small cell lung cancer (NSCLC) are due to suppression of the formation of NSCLC cells and also because of inhibiting the phosphorylation of epithelial growth factor receptor (EGFR).

TABLE 9.3

List of Triterpenoids in Clinical Trials

Triterpenoids	Disease State	Phase	Status
Betulinic acid	Dysplastic nervous syndrome	I/II	Ongoing
CDDO	Solid tumors	I	Completed
CDDO-ME	Liver disease	I/II	Terminated
CDDO-ME	Solid tumors	I	Terminated
Escin	Arm lymphedema	II	Completed
Ginsenoside	Breast cancer	II	Ongoing
Ginsenoside	Hypertension	II	Completed
Ginsenoside	Ischemic stroke	II/III	Completed
Glycyrrhizin	Hepatitis C	III	Ongoing
Glycyrrhetinic acid	Renal disease	II	Ongoing

Treatment of Lupeol on colorectal cancer cell lines (HCT116 and SW620) suppresses the migration of colorectal cancer cells by cytoskeleton RhoA-ROCK1 pathway inhibition that provides an anti-metastatic activity for CRC patients [45].

9.3.1.2 Limonoids as Anticancer Agent

Limonoids occur naturally only in plant species of the Rutaceae and Meliaceae plant families (Manners, 2007). Limonoids are highly oxygenated modified triterpenes that are biosynthesized from the acetate–mevalonate pathway in citrus fruits (Roy and Saraf, 2006). They occur in significant amounts as aglycones and glycosides in seeds and fruit tissues. Citrus fruits contain approximately 36 limonoid aglycones and 17 limonoid glucosides (Hasegawa and Miyake, 1996). Major citrus species accumulate limonin, nomilin, obacunone and deacetylnomilin. Limonoid aglycones that cause bitterness in numerous citrus fruits are converted into tasteless limonoid glucosides during fruit maturation. Limonoids have been shown to inhibit the growth of estrogen receptor-positive and -negative human breast cancer cells in culture. Nomilin has also been confirmed to significantly inhibit the growth of MCF-7 human breast cancer cells. Nomilin has also been shown to inhibit HIV-1 replication by inhibiting p-24 antigen activity and protease activity in cell systems. Maleastrones A and B are two limonoids isolated from *Malleastum* sp. collected from the Madagascar rain forest. They displayed significant cytotoxic activities against a panel of cancer cell lines including A2780, MDA-MB-435, HT-29, H522-T1 and U937 with IC_{50} values varying between 0.19 and 0.63 µM for the first compound and 0.20 and 0.49 µM for the second compound (Figure 9.25) [46].

9.3.1.3 Tirucallane as Anticancer Agent

Tirucallane is a tetracyclic triterpene. It is the 20S-stereoisomer of euphane and its derivatives, such as tirucalladienol. Jing et al. demonstrated the cytotoxic activities of new tirucallane-type triterpenoids (ficutirucins A-I). These compounds are isolated from the fruit of *Ficus carica*. The structures of these triterpenoids are established based on chemical and spectroscopic methods (IR, NMR, HR-ESIMS). All the isolated triterpenoids are evaluated for their cytotoxic activities against three human cancer cell lines such as MCF-7, HepG-2 and

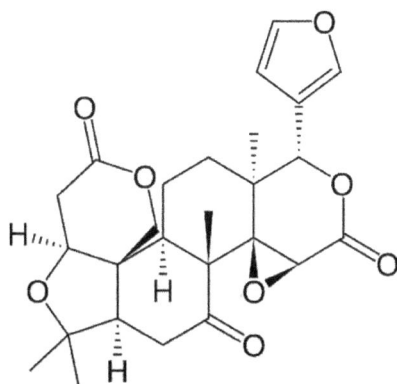

FIGURE 9.25 Molecular structure of limonoid.

U2OS. Evaluation study revealed that the tested compounds exhibit moderate cytotoxic activities with IC_{50} values of 11.67–45.61 µM against cancer cell lines (Figure 9.26, Table 9.4) [47].

9.3.1.4 Cycloartane Triterpene as Anticancer Agent

Cycloartane is a triterpene and chemically it is called 4,4,14-trimethyl-9,19-cyclo-5α, 9β-cholestane. Hassan et al. reported new cycloartane triterpene and other phytoconstituents from the aerial parts of *Euphorbia dendroides*. The newly isolated cycloartane-type triterpenoids are evaluated for *in vitro* cytotoxicity against different cell lines. It is observed that the tested compounds exhibit cytotoxic activities against different cancer cell lines such as human hepatoma cells (HepG2), human hepatocyte-derived carcinoma cells (Huh-7), human pancreatic cancer cells (KLM-1), human brain carcinoma cells (1321N1) and human cervical cancer cells (HeLa) with IC_{50} values of 20.67 ± 0.72, 16.24 ± 0.53, 22.59 ± 0.94, 25.99 ± 0.31 and 40.50 ± 3.14 µM, respectively as compared to 5-fluorouracil as positive control (Figure 9.27) [48].

9.3.1.5 Protostane Triterpene as Anticancer Agent

Protostane triterpene is a stereoisomer of the tetracyclic triterpene dammarane. Protostane triterpenes are isolated from the plants of genus *Alisma* and family Alismataceae. Both *in vitro* and *in vivo* studies were evaluated to identify their biological potentials such as antiviral properties against hepatitis B virus and HIV-I, antitumor activity, hepatoprotective activity and reversal of multi-drug resistance in cancer cells (Figure 9.28) [49].

Several protostane triterpenes isolated from Alismatis Rhizoma include alisol A, alisol B, alisol A 24acetate and alisol B 23acetate, alisol C 23-acetate, Alisol F and Alisol G. Evaluation study reports suggest that Alisol A induce cell apoptosis, arrest G1 phase of cell cycle, autophagy and generate intracellular reactive oxygen species (ROS) in MDAMB231 cells. Similarly, alisol A 24-acetate exhibit promising anti-cholesterolemic effects in an *in vivo* assay in which blood cholesterol levels are reduced by 61% in hypercholesterolmic rats (Figure 9.29) [50].

Protostane is a tetracyclic triterpene with the molecular formula of $C_{30}H_{50}O_5$. It occurs only in plants of genus Alisma. Li et al. reported a new protostane-type triterpenoid bearing an oxetane ring in the side chain, named called alisol W (Figure 9.30). It is isolated from the dried rhizome of *Alisma plantagoaquatica* subsp. *orientale*. This compound exhibits vasorelaxant activity and inhibition on 11 β-hydroxysteroid dehydrogenase type 1 (*11β*-HSD1). The inhibition of 11β-HSD1 indicates the potential therapeutic benefit to persons with metabolic syndrome and obesity-related disorders [51].

9.3.1.6 Ursolic Acid as Anticancer Agent

Ursolic acid (UA) inhibits the growth of breast cancer (BC) cells by modulating different signaling pathways such as PI3K/Akt/mTOR, ERK and EGFR. Shanmugam et al. demonstrated the antitumor potential of UA that arrests BC cells by inhibiting the growth of tumor, metastasis and angiogenesis (Figure 9.31) [52].

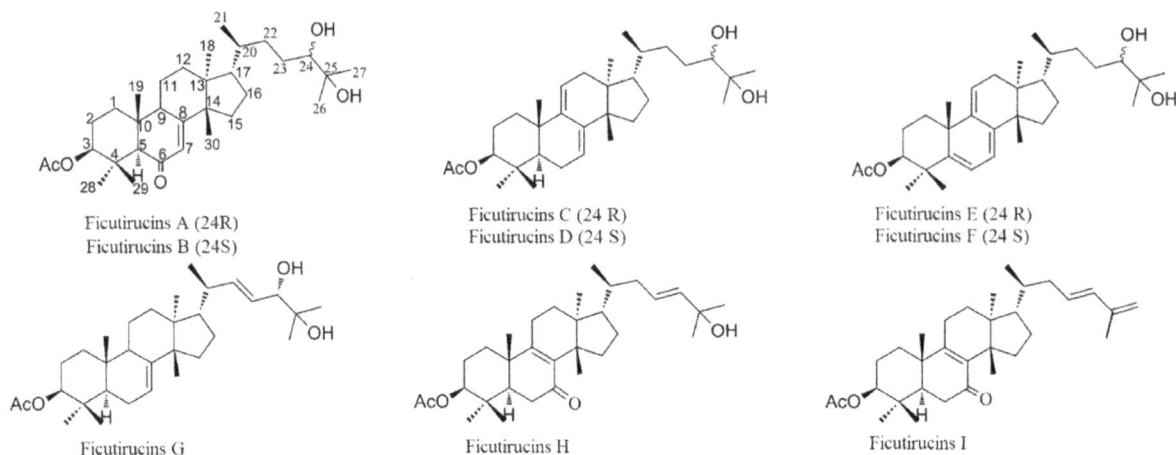

FIGURE 9.26 Structures of tirucallane-type triterpenoids (ficutirucins A-I).

TABLE 9.4

Tirucallane-Type Triterpenoids with IC_{50} Values for Anticancer Activity

Compounds	IC_{50} Values (µM)		
	MCF-7	HepG-2	U20S
Ficutirucins A	27.78 ± 2.21	**11.67 ± 0.09**	34.93 ± 3.28
Ficutirucins B	17.94 ± 0.33	17.76 ± 0.06	27.66 ± 0.17
Ficutirucins C	>50	42.02 ± 1.73	>50
Ficutirucins D	>50	>50	>50
Ficutirucins E	>50	>50	>50
Ficutirucins F	45.61 ± 1.34	24.68 ± 5.11	20.70 ± 1.99
Ficutirucins G	27.48 ± 0.57	>50	28.08 ± 0.26
Ficutirucins H	>50	>50	>50
Ficutirucins I	31.28 ± 1.14	23.19 ± 1.13	32.76 ± 1.00
cis-Platinum (positive controls)	9.69 ± 0.26	8.30 ± 0.58	6.19 ± 0.13

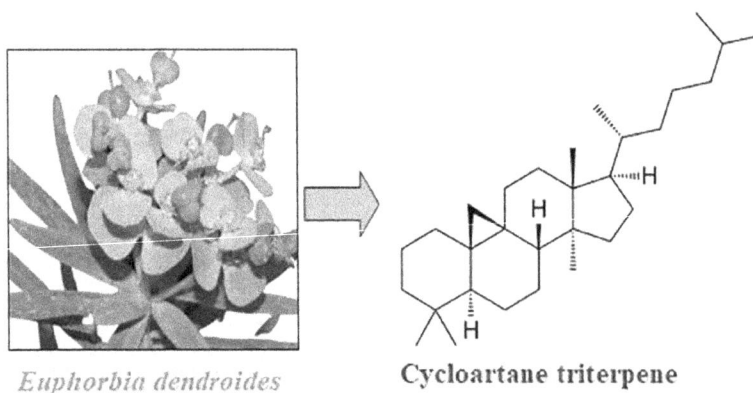

Euphorbia dendroides   **Cycloartane triterpene**

FIGURE 9.27 Structure of cycloartane triterpene.

UA induces apoptosis and arrest cell cycle, thereby reducing the tumor size. Another study report suggested that UA can inhibit the expression of cyclin D1 and CDK4 by regulating FoxM1 signaling pathway which is considered as a positive signal in the detection of cancer cells. UA has also down-regulated the expression of CDK4, FoxM1 and cyclin D1 thereby inducing apoptosis in MCF-7 BC cell lines. Yeh et al. reported that UA inhibits cancer cell invasion, migration and proliferation and cell colony formation in MDA-MB-231 [53]. In addition to this, UA significantly down-regulates the expression of u-PA and MMP-2 and up-regulates the PAI-1 and TIMP-2. Shishodia et al. demonstrated that UA can regulate the cytokines signaling cascade by inhibiting the TNF-induced stimulation of NF-kB (Figure 9.32) [54].

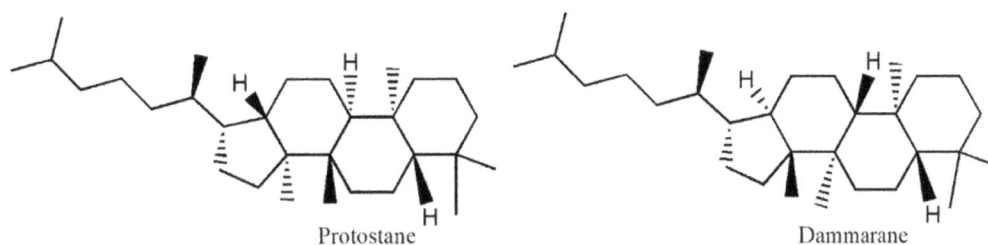

FIGURE 9.28 Structure of protostane triterpene.

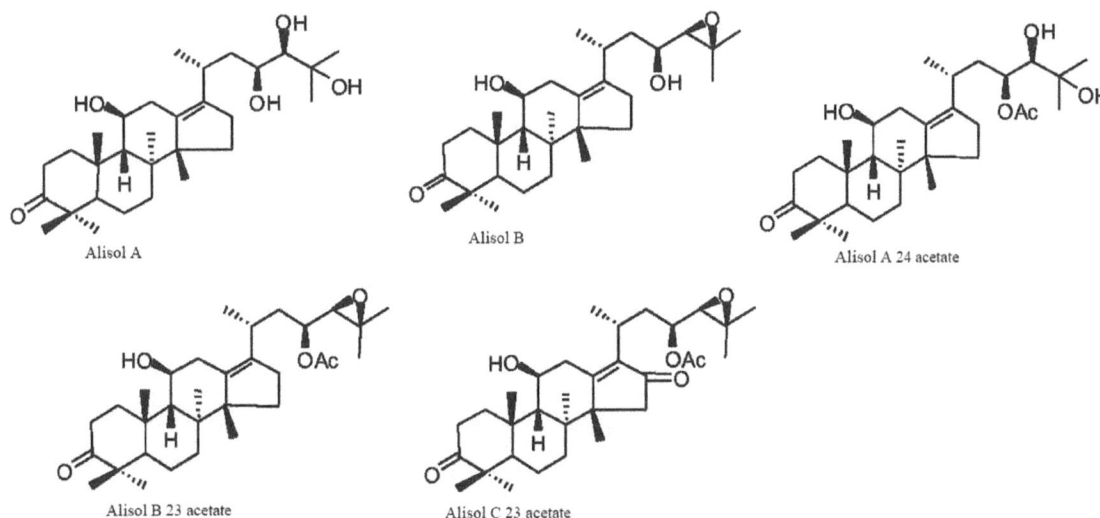

FIGURE 9.29 Structures of protostane triterpenes.

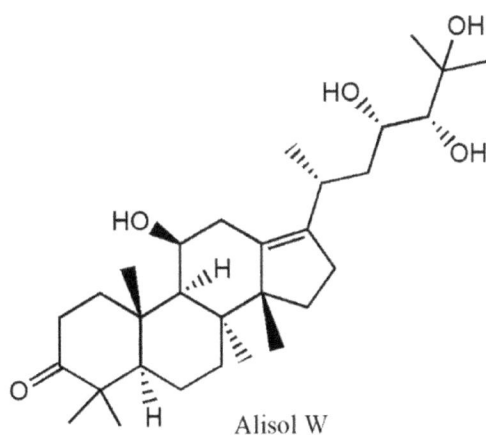

FIGURE 9.30 Structure of alisol W.

9.3.1.7 Betulinic Acid as Anticancer Agent

The anticancer activities of betulinic acid (BA) and betulin (BT) are due to its ability to induce apoptotic cell death in cancer cells by activating the mitochondrial pathway of apoptosis [55]. Both BA and BT are able to alter the cell cycle in human melanoma cell line, lung cell line and breast cancer cell lines [56]. BA induces cell cycle arrest in G_0/G_1 phase in human cervical cancer (Hela) cells and breast cancer. It may arrest MCF-7 cells and human cervical cancer cells in G_0/G_1 phase by increasing the expression of p53 and p21 or altering

the protein expression of related signaling pathways (PI3K/Akt, NF-κB signaling pathway). Several studies reported that BA and BT induce tumor cell apoptosis by inhibition of DNA topoisomerase (Figure 9.33) [57].

9.3.1.8 Ganoderic Acid D as Anticancer Agent

Ganoderic acid D is a lanostane-type tetracyclic triterpene with molecular formula $C_{30}H_{42}O_7$ (Figure 9.34). It is extracted from *Ganoderma lucidum*. Evaluation study reports suggest that it can suppress the proliferation of HeLa cells with IC_{50} of 17.3 μmol/L by inducing cell apoptosis and cell cycle arrest in G2/M phase [58]. Similarly, Ganoderiol F is another lanostane-type pentacyclic triterpene compound with molecular formula $C_{30}H_{46}O_3$ and is also extracted from *Ganoderma lucidum*. It is reported that Ganoderiol F prevents the proliferation of HepG2, Huh 7 and K562 cell lines with IC_{50} of 17, 8.5 and 4 μmol/L respectively [59].

9.3.1.9 Dehydrotrametenolic Acid as Anticancer Agent

Dehydrotrametenolic acid (DTA) is one of the lanostane-type tetracyclic triterpenes with molecular formula $C_{30}H_{46}O_3$ (Figure 9.35). It is extracted from *Poria cocos* of family Polyporaceae [59]. DTA can induce apoptosis of J82 cell through activation of the caspase-3 pathway and also arrest the cell cycle in the G2/M phase. It can also regulate the expression

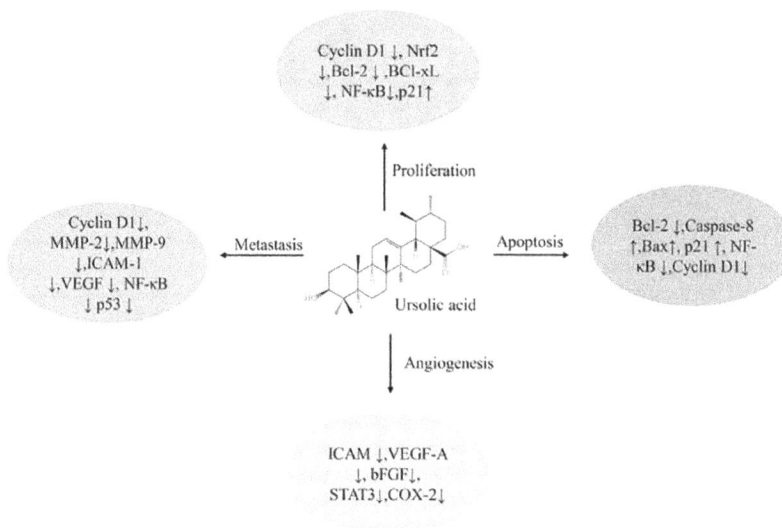

FIGURE 9.31 Molecular targets of ursolic acid.

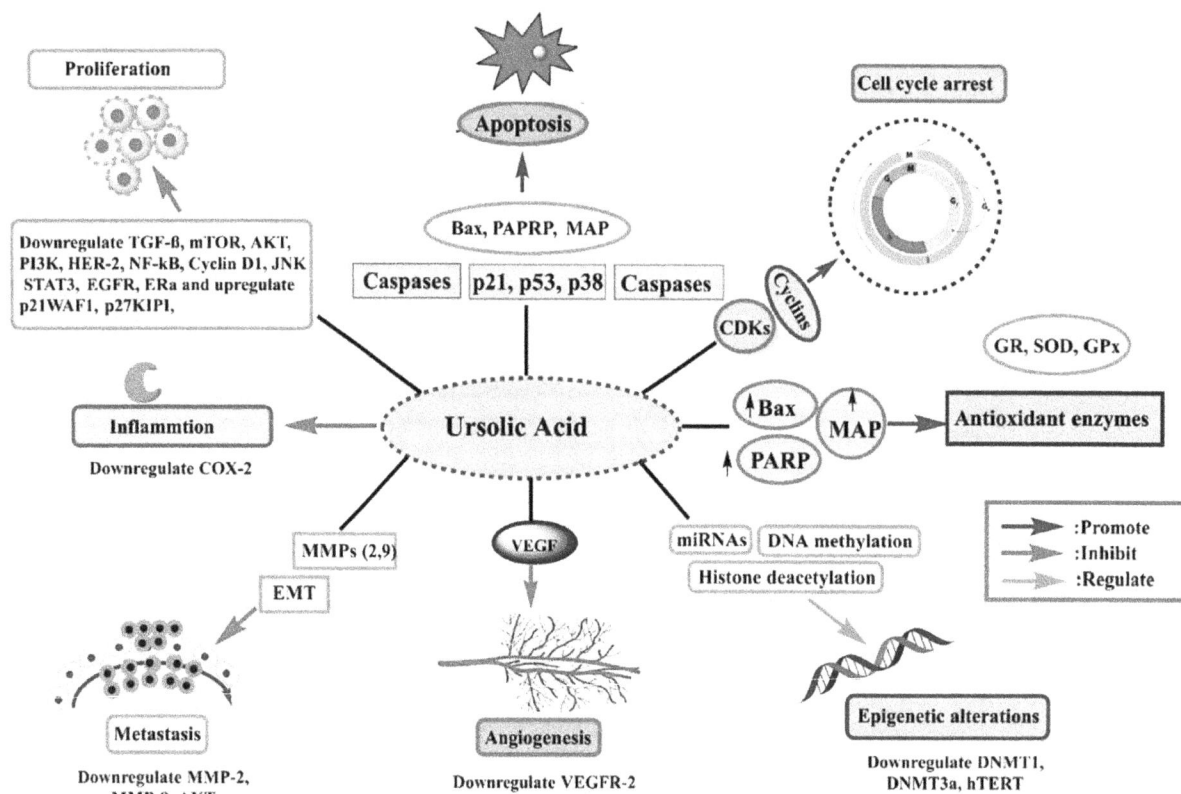

FIGURE 9.32 Molecular mechanism of action of ursolic acid for treatment of breast cancer.

of H-Ras, Akt and ERK. Other derivatives of dehydrotrametenolic acid isolated from *Poria cocos* include 25-hydroxyporicoic acid H, 16α, 25-dihydroxydehydroeburicoic acid, 5α, 8α-peroxydehydro-tumulosic acid and 15α-hydroxy dehydrotumulosic acid. These compounds are lanostane-type tetracyclic triterpenes with an inhibitory effect on Epstein-Barr Virus early antigen (EBV-EA) [60,61].

9.3.1.10 Impatienside A and Bivittoside D as Anticancer Agents

Impatienside A and bivittoside D are the lanostane-type tetracyclic triterpenoid glycosides that are isolated from sea cucumber *Holothuria impatien* (Figure 9.36). It is reported that impatienside A and bivittoside D exhibited significant

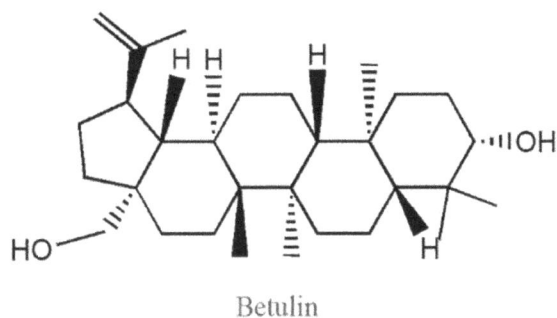

Betulin

FIGURE 9.33 Structure of betulin (BT).

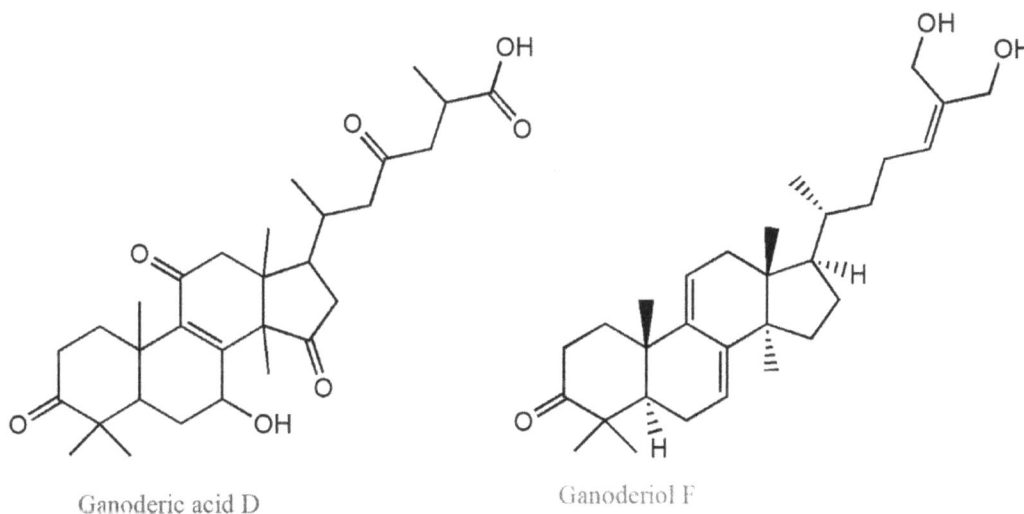

Ganoderic acid D

Ganoderiol F

FIGURE 9.34 Structure of ganoderic acid D.

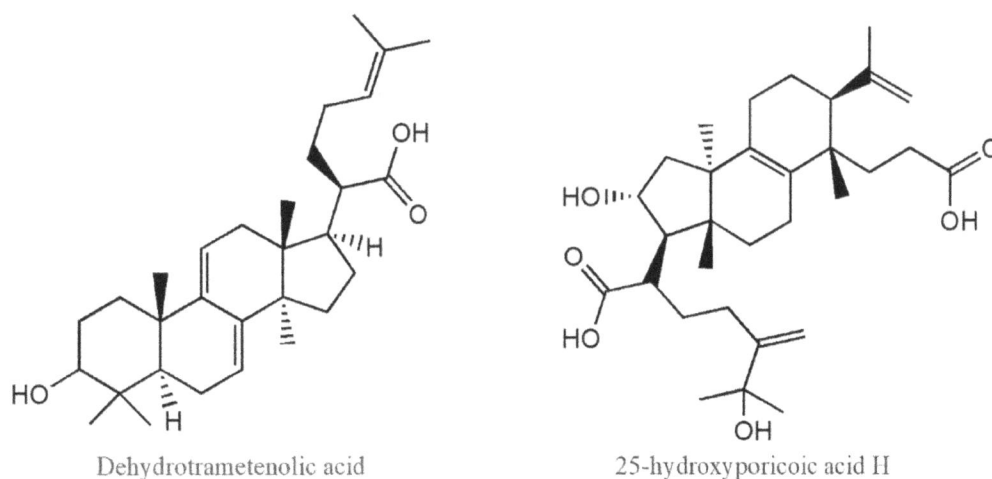

Dehydrotrametenolic acid

25-hydroxyporicoic acid H

FIGURE 9.35 Structure of dehydrotrametenolic acid (DTA).

cytotoxicity effects with an IC_{50} value of 0.25–1.9 µmol/L against tumor cell lines (HCT-116, A549, HepG2, DU145, MCF-7 and KB cells) as compared to the clinical applied anticancer drug etoposide [62].

9.3.1.11 Ananosic Acid B as Anticancer Agents

Ananosic acid B and Ananosic acid C are lanostane-type tetracyclic triterpenoids (Figure 9.37). These terpenoids are extracted from *Kadsura longepedunculata*. Evaluation study

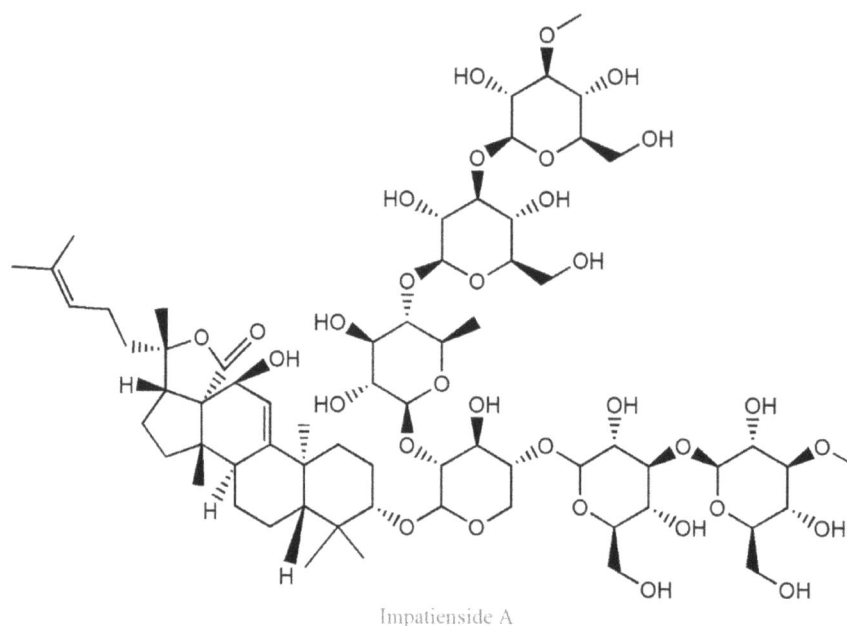

Impatienside A

FIGURE 9.36 Structure of impatienside A and bivittoside D.

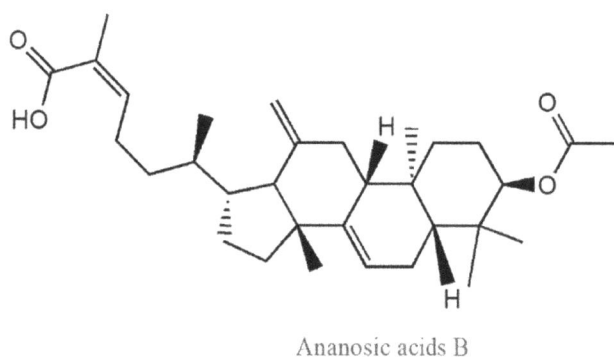

Ananosic acids B

FIGURE 9.37 Structure of ananosic acid B.

reported the antitumor proliferation activity of Ananosic acids B and Ananosic acids C on HeLa and HL-60 cells. Similarly, seco-coccinic acids A-E are the lanostane-type tetracyclic triterpenoids, extracted from the roots of *Kadsura coccinea* and also displayed anti-proliferative effects on HL-60 cell, all with IC_{50} values in the range of 6.8–42.1 µmol/L [63,64].

9.3.1.12 Daedaleasides as Anticancer Agents

Daedaleasides B-E are lanostane-type tetracyclic triterpenoids (Figure 9.38). These are isolated from the fruit of *Daedalea dickinsii*. Evaluation study reports suggest that these terpenoids can induce apoptosis in the HL-60 cell by inducing the target cell internucleosomal DNA fragmentation [65,66].

9.3.1.13 Inonotsuoxides as Anticancer Agent

Inonotsuoxides (A, B) are lanostane-type tetracyclic triterpenoids (Figure 9.39). These are isolated from sclerotia of *Inonotus obliquus*. These triterpenoids have potential antitumor promoting effects on Raji cells in patients with lymphoma. From the

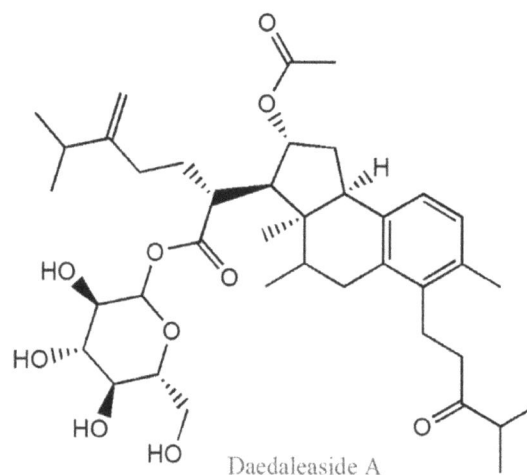

Daedaleaside A

FIGURE 9.38 Structure of daedaleaside A.

two-stage carcinogenesis test on mouse skin, it is revealed that this compounds exhibit significant antitumor activities [67].

Similarly, other lanostane-type tetracyclic triterpenoid isolated from *Inonotus obliquus* include lanosta-8,23E-diene-3b, 22R, 25-triollanosta-7,9,23E-triene-3b, 22R, 25-triol. The results of change in Epstein-Barr Virus (EBV) lymphoid stem cell viability by trypan blue staining reported that this triterpenoid exhibit antitumor activity. Inotodiol is another lanostane-type triterpenoid compound isolated from the sclerotia of *Inonotus obliquus*. The *in vivo* carcinogenesis test revealed that inotodiol exhibits the potent antitumor promoting activity. Further, inotodiol when treated on murine leukemia P388 cells, the proliferation of tumor was suppressed. The mechanism involved in this case is the activation of Caspase-3 and cell apoptosis (Figure 9.40) [68].

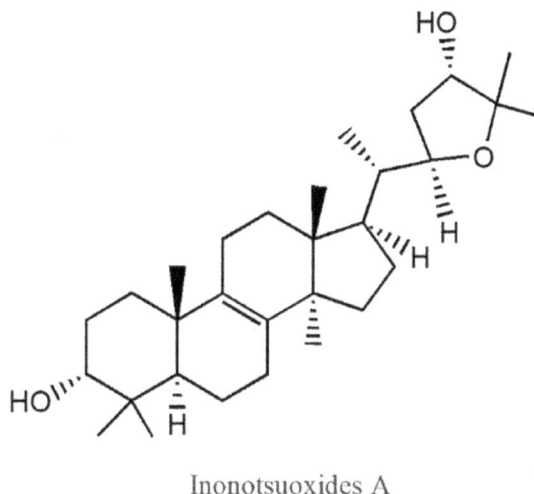

Inonotsuoxides A

FIGURE 9.39 Structure of inonotsuoxides A.

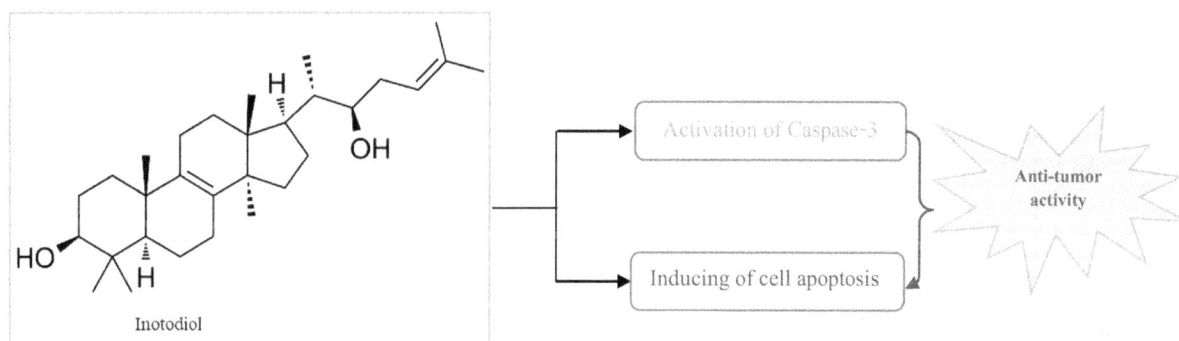

Inotodiol

FIGURE 9.40 Structure of inotodiol.

9.3.1.14 CDDO-Me as Anticancer Agent

CDDO-Me is an oleanane-type pentacyclic triterpenoids. Chemically, it is methyl-2-cyano-3,12-dioxooleana-1,9(11)-dien-28-oate (Figure 9.41). The treatment of CDDO-Me on pancreatic cancer MiaPaCa-2 and Pane-1 cells, prevent cell proliferation and induce cell apoptosis. The proliferation inhibition rate is found to be 7% and 16% when the drug concentration was 0.63 μmol/L and elevated to 84% and 80% with the concentration of 5 μmol/L [69].

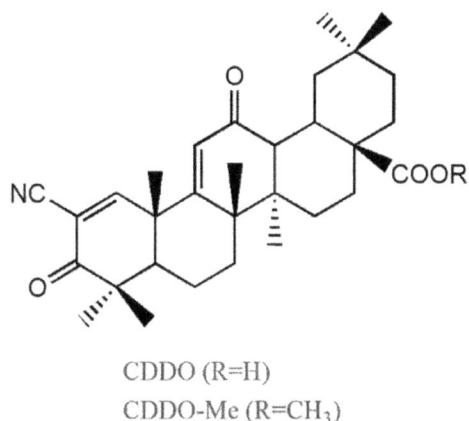

CDDO (R=H)
CDDO-Me (R=CH₃)

FIGURE 9.41 Structure of CDDO and CDDO-Me.

9.3.1.15 β-Escin as Anticancer Agent

β-Escin is a triterpenoids with anti-proliferative activity. It can inhibit the proliferation of HL-60 cells in drug concentration between 30 and 50 mg/mL. Apoptosis can be identified by AnnexinV-FITC analysis. Similarly, typical DNA apoptosis ladders can also be detected in DNA ladder electrophoresis which indicates β-escin can suppress the proliferation of HL-60 cells and induce apoptosis (Figure 9.42) [70].

9.3.1.16 Cimicifoetisides as Anticancer Agents

Cimicifoetisides A and B are triterpenes (Figure 9.43) isolated from the rhizome of *Cimicifuga foetida* (Ranunculaceae). These terpenes exhibit anti-proliferative activity and cytotoxic effects against tumor cells of rat and the human breast

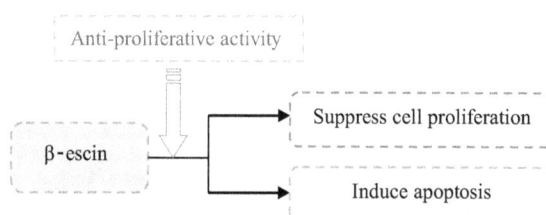

FIGURE 9.42 Mechanism of βescin as anticancer agent.

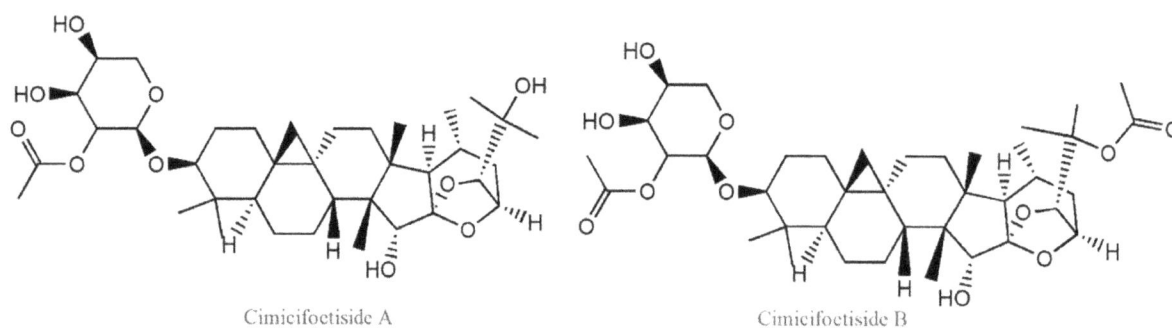

FIGURE 9.43   Structures of cimicifoetisides A and B.

cancer MDA-MB-A231 cell line, with IC_{50} values of 0.52 and 6.74 μmol/L by cimicifoetisides A and IC_{50} values of 0.19 and 10.21 μmol/L by cimicifoetisides B [71].

9.3.1.17 Nimbolide as Anticancer Agent

Nimbolide is a triterpenoid with the molecular formula $C_{27}H_{30}O_7$. It is isolated from the edible parts of the neem tree (*Azadirachta indica*). It can significantly inhibit the proliferation of HT-29 cells with a concentration of 2.5 mmol/L. Nimbolide also induces the cell cycle arrest of HT-29 cells (Figure 9.44) [72].

9.3.1.18 3-O-Acetyl-11-Keto-β-Boswellic Acid as Anticancer Agent

3-O-acetyl-11-keto-β-boswellic acid is the naturally occurring triterpenoid with the molecular formula $C_{32}H_{48}O_5$ (Figure 9.45). It is isolated from the gum resin exudate of the stem of plant *Boswellia serrate* of family frankincense. It can induce apoptosis

of tumor cells, when treated on PC-3 and LNCaP cell lines via activation of the death receptor DR-5 signaling pathway [73].

9.3.1.19 Asiatic Acid as Anticancer Agent

Asiatic acid (AA) is the ursane-type pentacyclic triterpenoid with molecular formula $C_{30}H_{48}O_5$ (Figure 9.46). It is isolated from the leaves of *Centella asiatica* [74]. It can induce apoptosis of human melanoma cells (SK-MEL-2) in which the mechanism of action may be due to an increase in the level of oxidative stress within the tumor cells and also activation of the intracellular mitochondrial apoptosis pathway [75]. In addition to this, AA can also induce the apoptosis of HepG2 cells in which the mechanism of action may be due to an increase in the intracellular Ca^{2+} levels and thereby activating the expression P53. Tran et al. reported the the synthesis of AA derivatives and evaluated their cytotoxicity against three cancer cell lines: KB (human carcinoma in the mouth), HepG2 (human hepatocellular carcinoma) and SK-LU-1 (human lung carcinoma). These compounds exhibited potential cytotoxic activity against three tested cell lines with IC_{50} values ranging from 0.67 to 37.39 μM [76].

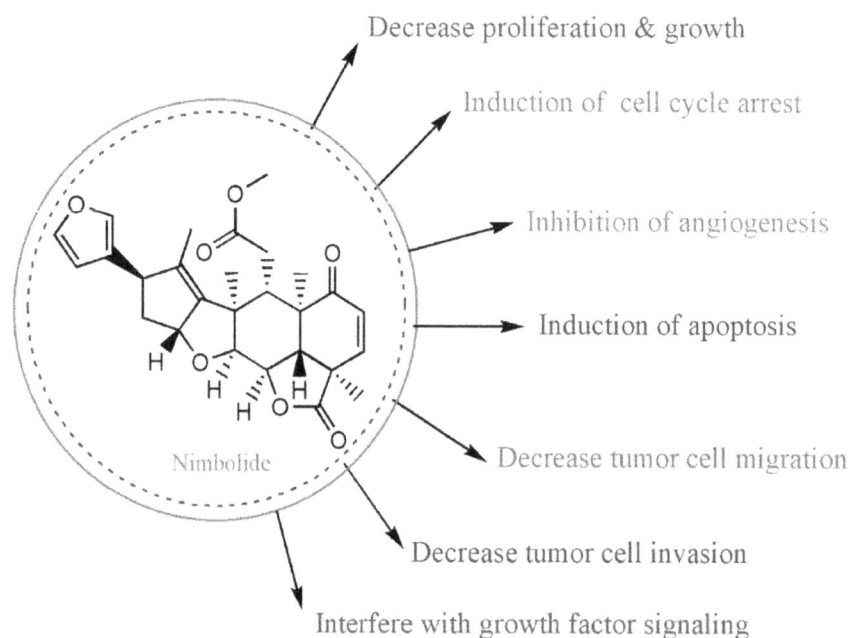

FIGURE 9.44   Molecular mechanism of nimbolide.

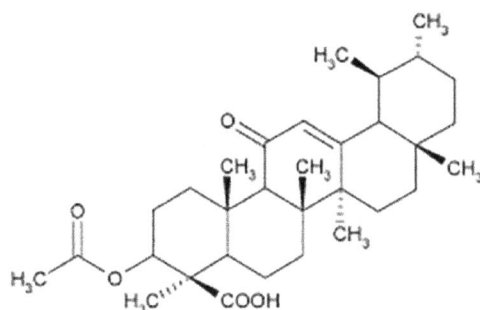

FIGURE 9.45 Structure of 3-O-acetyl-11-keto-β-boswellic acid.

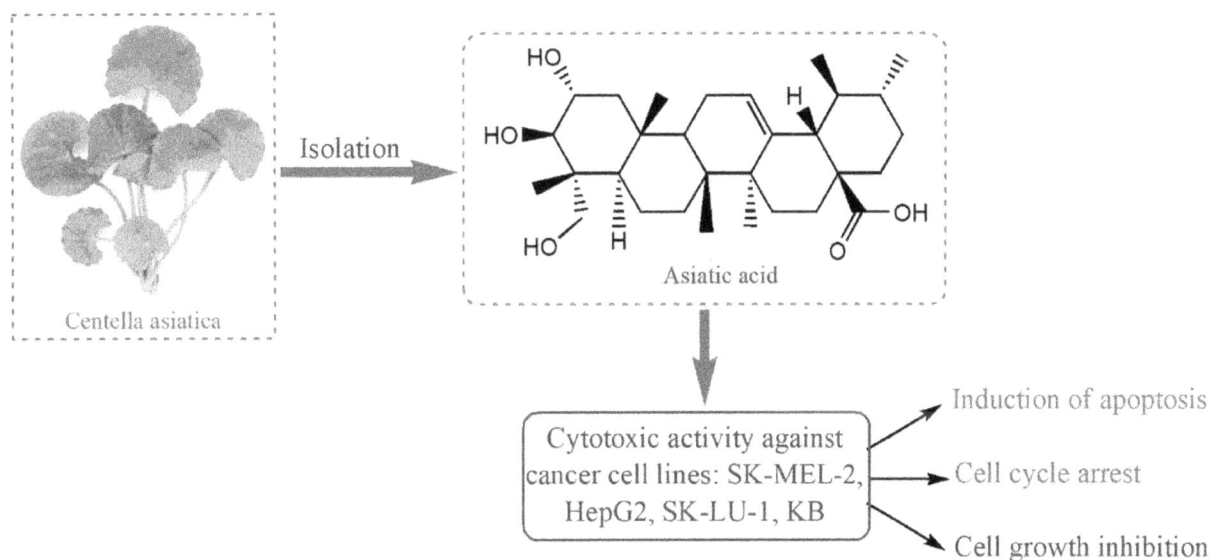

FIGURE 9.46 Molecular targets of asiatic acid as anticancer agent.

9.3.1.20 Celastrol as Anticancer Agent

Celastrol is the pentacyclic triterpenoid with the molecular formula $C_{29}H_{38}O_4$ (Figure 9.47). It exhibits potential cytotoxic activity on different human tumor cell lines, including A549, HCT-8, MCF-7, KB, with IC_{50} values of 0.21, 0.25, 0.23 and 0.20 ng/mL respectively. Zhu et al. reported the antitumor activity of celastrol by inhibition of proliferation, invasion and migration in cholangiocarcinoma via PTEN/PI3K/Akt pathway. Cholangiocarcinoma is the prime hepatobiliary malignancy that arise from the epithelial cells. Phosphatase and tensin homolog (PTEN) is a tumor suppressor that represents one of the most common targets for the genetic defect in human cancer [77].

9.3.1.21 Glycyrrhizin as Anticancer Agent

Glycyrrhizin is the chief constituent of *Glycyrrhiza glabra*. Glycyrrhizin is a glycosylated saponin that contains one molecule of glycyrretinic acid and two molecules of glucuronic acid. Ansari et al demonstrated the potential anti-proliferative and anticancer properties of Glycyrrhizin. It induces ROS dependent apoptosis and arrests cell cycle at G_0/G_1 phase in a dose-dependent manner in the case of HPV18+ human cervical cancer HeLa cell line in vitro. Glycyrrhizin also inhibits the growth of lung tumors in

PDX mice by downregulating the level of High Mobility Group Box 1 (HMGB1). HMGB1 is a nonhistone chromosomal protein that regulates the formation of nucleosome and gene transcription. HMGB1 promotes the invasion and metastasis of cancer cells. This mechanism involves the inhibition of the JAK/STAT signaling pathway by glycyrrhizin (Figure 9.48) [78,79].

9.3.2 Triterpenoids as Anti-Diabetic Agents

Diabetes mellitus is considered as a chronic metabolic disorder which is a serious health problem globally [80]. Several anti-diabetic drugs are available that focus on improving the sensitivity of insulin, increasing the production of insulin and decreasing the blood glucose level. Though a number of synthetic drugs are available, medicinal agents from natural sources play a major role in the treatment of diabetes due to the presence of suitable physicochemical properties. Several experiments are conducted to evaluate the anti-diabetic property of triterpenoids. Evaluation studies demonstrate that these compounds have several anti-diabetic mechanisms that include inhibition of enzymes involved in glucose metabolism, prevention of the development of insulin resistance and normalizing the plasma glucose and insulin levels (Figure 9.49). Triterpenoids are also found to have promising therapeutic

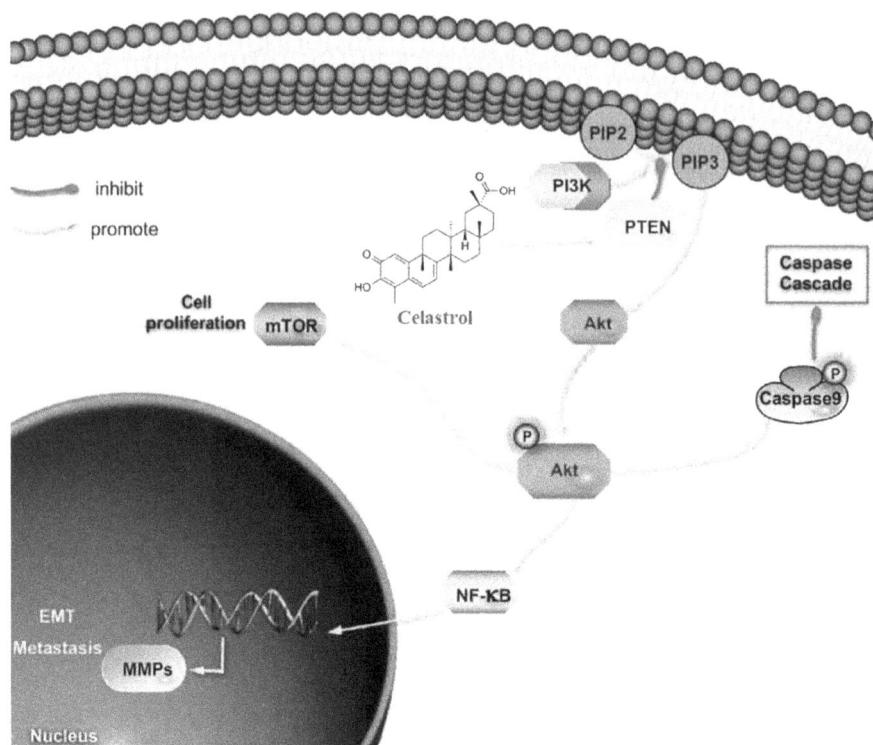

FIGURE 9.47 Cytotoxic effects of celastrol via PI3K/Akt/mTOR and NF-κB signaling pathways.

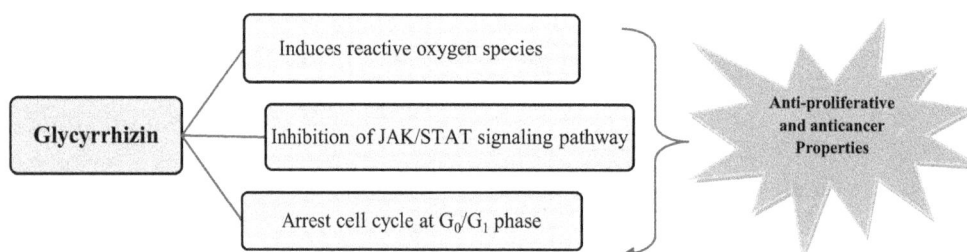

FIGURE 9.48 Molecular targets of glycyrrhizin as anticancer agent.

potentials to prevent diabetic complications. They exhibit strong antioxidant activity and also inhibit the formation of glycation end products that are associated with the pathogenesis of diabetic nephropathy, embryopathy and neuropathy or impaired wound healing [81].

9.3.2.1 Lupeol as Anti-Diabetic Agent

Lupeol exhibited its anti-diabetic activity by reducing hyperinsulinemia through the regulation of insulin receptor and GLUT 4 protein. Evaluation study reports the effects of Lupeol on enzymatic antioxidants superoxide dismutase (SOD), catalase (CAT) and non-enzymatic antioxidant (Vitamin C) in type-2 diabetic adult male rats and also decreases the levels of antioxidant enzymes (SOD, CAT and Vitamin C) in the liver of the type-2 diabetic rats. Lupeol is also used for the treatment of diabetes by inhibiting the α-glucosidase activity (Figure 9.50). It is reported to inhibit the activity of α-amylase enzyme responsible for the development of diabetes [82,83].

9.3.2.2 Oleanolic Acid as Anti-Diabetic Agent

Insulin resistance is considered as the hallmark of type 2 diabetes and also it is the major predictor of the onset of diabetes [84]. So, the treatment of OA (25 μmol/L) in the case of insulin-resistant HepG2 cells improves the insulin sensitivity by increasing the expression of insulin receptor substrate 1 (IRS-1) and glucose transporter 4 (GLUT-4) proteins via NF-κB. IRS1 is the key factor in the case of insulin signaling pathways whereas GLUT-4 is the primary glucose transporter in the case of skeletal muscle, adipose tissues and liver [85]. Hence, both IRS1 and GLUT-4 are called as therapeutic targets of Oleanolic acid in the management of diabetes. The administration of OA supplement (25 mg/kg/day) in adipose tissue of rats for 10 weeks also improves fructose-induced insulin resistance through the IRS-1/phosphatidylinositol 3-kinase/Akt pathway [86]. It is observed that ROS play a key role in diabetes mellitus and insulin resistance (Figure 9.51) [87,88].

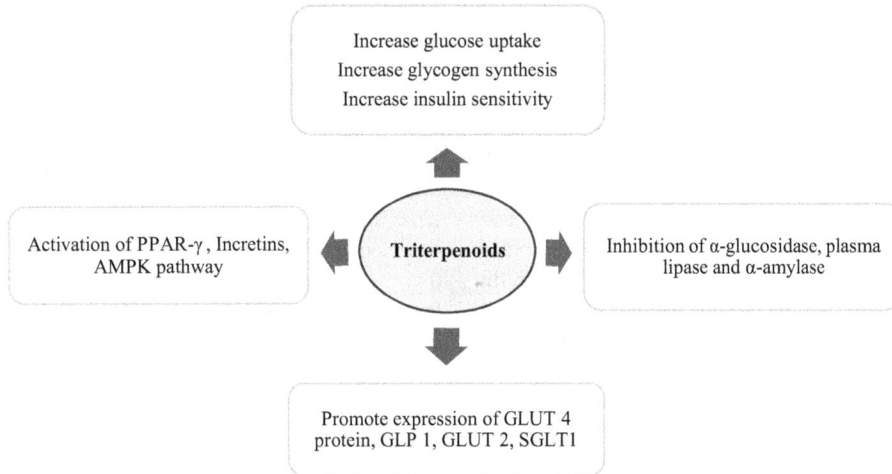

FIGURE 9.49 Mechanisms of actions of triterpenoids for anti-diabetic activity.

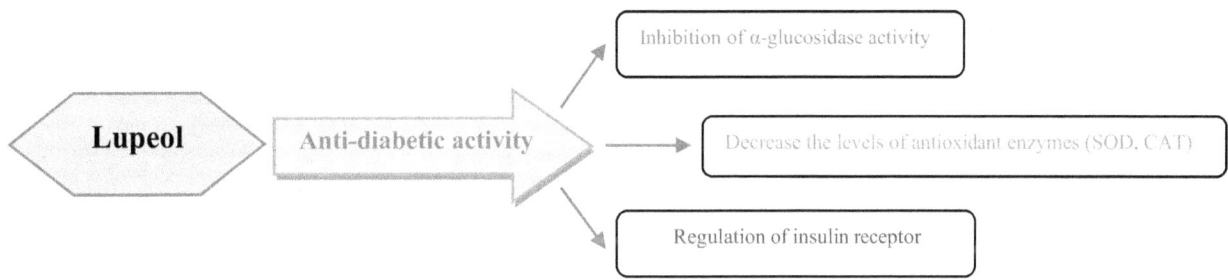

FIGURE 9.50 Molecular targets of lupeol as anti-diabetic agent.

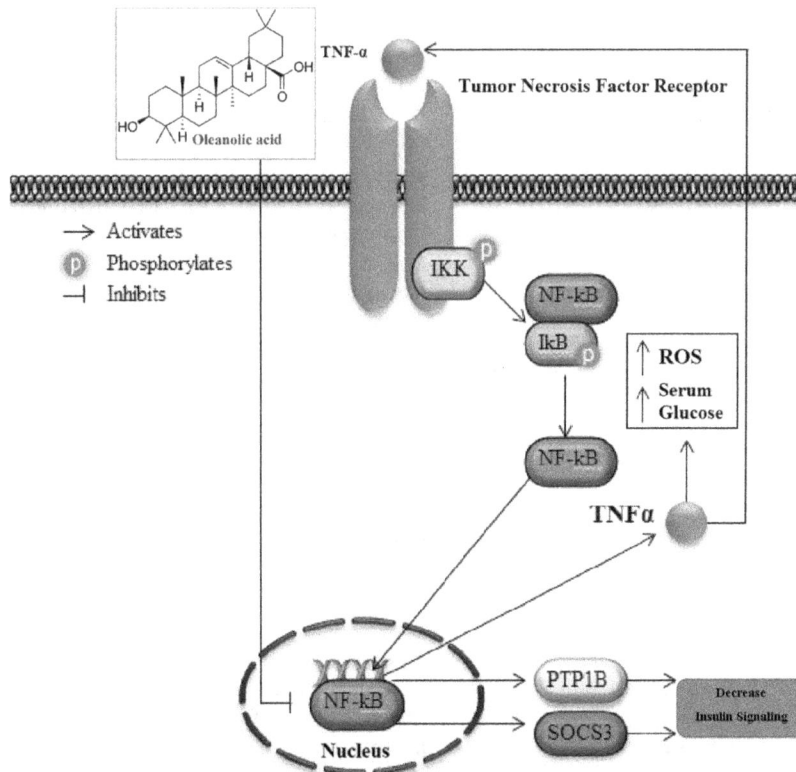

FIGURE 9.51 Mechanism of actions of oleanolic acid as antidiabetic agent.

9.3.2.3 Viburodorol A as Anti-Diabetic Agent

Viburodorol A is a triterpene present in *Viburnum odoratissimum* (Figure 9.52). Evaluation study report suggests that it exhibits antihyperglycemic activity due to stimulation of glucose absorption in insulin-resistant HepG2 cells [89]. Similarly, oleanolic acid and maslinic acid decrease the blood glucose level in experimental diabetes due to their inhibitory effects on the activity of α- amylase, α-glucosidase and sucrase, reduce plasma ghrelin concentrations, decreases in ghrelin, SGLT1, GLUT2, α-amylase and α-glucosidase expression in the gastrointestinal tract, reductions in MDA concentrations

and increases in SOD and GPx by comparison with the STZ diabetic control [90,91].

9.3.2.4 Ursolic Acid A as Anti-Diabetic Agent

The pentacyclic triterpenoid, UA exhibits potential anti-diabetic activity (Figure 9.53, Table 9.5). It is screened *in vitro* against α-glucosidase and the results showed that UA revealed significant inhibitory activity with the IC_{50} values of $0.71 \pm 0.27\,\mu M$. Further, the modification of structure at different positions of UA is an effective approach to generate a suitable ligand [92].

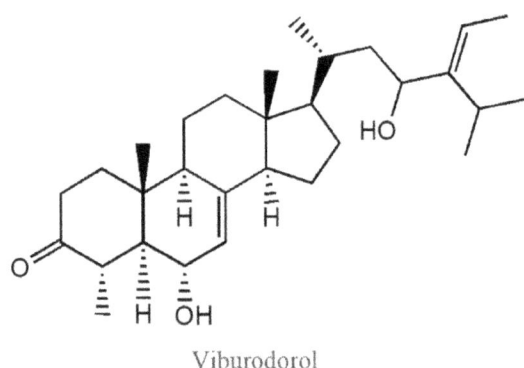

Viburodorol

FIGURE 9.52 Structure of viburodorol A.

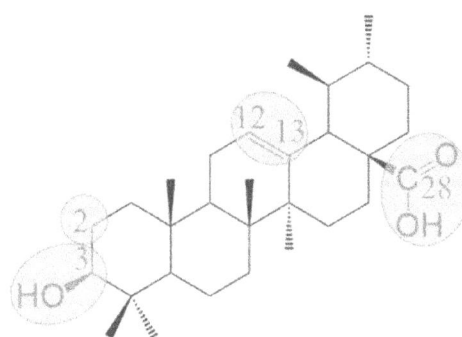

FIGURE 9.53 SAR study of ursolic acid.

TABLE 9.5

List of Triterpenoids with Anti-Diabetic Activity

Terpenoids	Biological Source	Mechanisms of Actions
Arjunolic acid	*Terminalia arjuna*	Reduces oxidative stress and decreases the level of serum pro-inflammatory TNF-α
Asiatic acid	*Centella asiatica*	Preserves pancreatic beta cell
Astragaloside	Astragaloside	Decreases in renal AGEs formation
Bacosine	*Bacopa monniera*	Decreases blood glucose levels by enhancing glucose utilization and protein glycation inhibition
Bartogenic acid	*Barringtonia racemosa*	Inhibits Pancreatic α-amylase
Corosolic acid	*Eriobotrya japonica*	Promotes 3H-glucose uptake, suppresses the differentiation and down-regulates the expression of PPAR-γ
Euscaphic acid	*Sanguisorba tenuifolia*	Regulates α-glucosidase inhibitory activity
Glycyrrhetinic acid	*Glycyrrhiza uralensis*	Improves glucose tolerance, Enhance insulin-stimulated glucose uptake through PPAR-γ activation and induced MRNA levels of insulin receptor
Gymnemic acid	*Gymnema sylvestre*	Reduces blood glucose, promotes insulin and inhibits GAPDH levels
Ginsenoside	*Panax ginseng*	Increases lucose uptake by activation of AMPK
Hodulcin	*Hovenia dulcis*	Inhibits sweetness
Ilexgenin A	*Ilex pubescens*	Suppresses of TXNIP
Lupenol	*Aegle marmelos*	Exerts Anti-diabetic action by β-cell protection
Maslinic acid	*Olea europaea*	Reduces blood glucose levels by inhibiting glycogen phosphorylase
Oleanolic acid	*Aralia elata*	Reduces blood glucose levels
Pistagremic acid	*Pistacia chinensis*	Activates α-glucosidase
Platycodin	*Platycodon grandiflorum*	Inhibits intracellular TG accumulation, down-regulation of PPAR-γ expression
Senegins	*Polygala senega*	Reduces blood glucose levels
3β-Taraxerol	*Mangifera indica*	Increases glycogen synthesis through the activation of PKB and activates glucose transport by inducing the activation of GLUT4.
Ursolic acids	*Phyllanthus amarus*	Inhibits pancreatic α-amylase
Ursane	*Rhododendron brachycarpum*	Helps to exert inhibitory activity against PTP 1B
Ziziphin	*Ziziphus jujuba*	Inhibits sweetness

9.3.3 Triterpenoids as Antimicrobial Agents

Kiplimo et al. demonstrated that triterpenoids (lupenyl acetate, oleanolic acid, α-amyrin, β-amyrin, β-amyrin acetate, friedelanone and friedelin acetate) isolated from *Vernonia auriculifera* exhibit antimicrobial activities. α- and β-Amyrin exhibit minimum inhibitory concentration (MIC) of 0.25 mg/mL against *Staphylococcus aureus, Bacillus subtilis, Enterococcus faecium* and *Staphylococcus saprophyticus* whereas lupenyl

acetate and oleanolic acid displayed MIC of 0.25 mg/mL against *Stenotrophomonas maltophilia* (Figure 9.54, Table 9.6) [93].

9.3.3.1 β-Amyrin as Antimicrobial Agents

Al-Homaidan et al. reported the antibacterial activity of β-amyrin isolated from *Laurencia microcladia*. The antibacterial activities of β-amyrin were performed using both

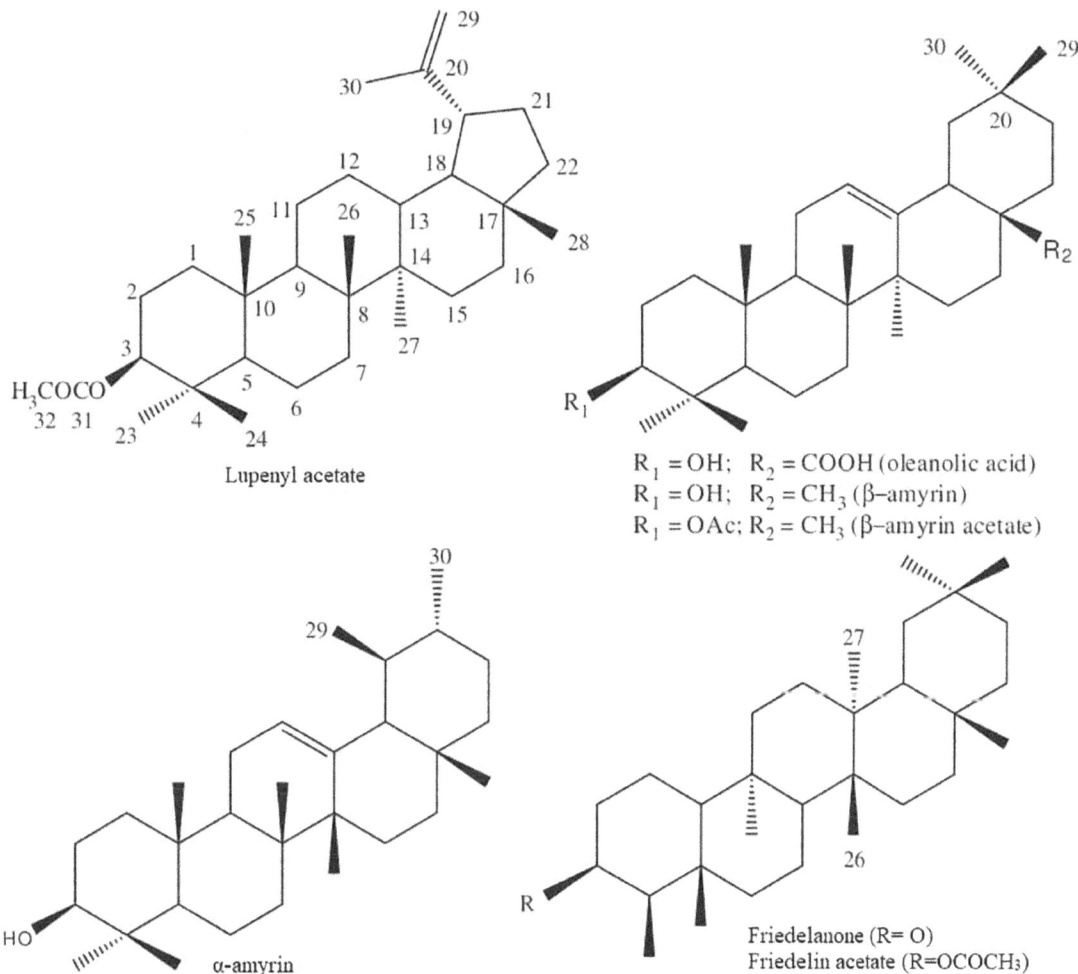

$R_1 = OH$; $R_2 = COOH$ (oleanolic acid)
$R_1 = OH$; $R_2 = CH_3$ (β-amyrin)
$R_1 = OAc$; $R_2 = CH_3$ (β-amyrin acetate)

Friedelanone (R= O)
Friedelin acetate (R=OCOCH₃)

FIGURE 9.54 Structures of triterpenoids isolated from *Vernonia auriculifera*.

TABLE 9.6

Antimicrobial Activities (MIC in mg/mL) of Compounds Isolated from *V. auriculifera*

Microorganisms	MIC (mg/mL)						
	Lupenyl Acetate	Oleanolic Acid	β-Amyrin	β-Amyrin Acetate	α-Amyrin	Friedelanone	Friedelin Acetate
S. aureus	1	0.5	0.25	1	0.25	1	1
B. subtilis	1	0.5	0.25	1	0.25	0.25	1
E. faecium	0.5	1	0.25	0.5	0.25	1	1
S. epidermidis	1	0.25	0.5	1	0.5	0.5	0.5
S. saprophyticus	0.25	1	0.25	1	0.25	1	1
E. coli	0.12	1	0.12	0.5	0.12	1	0.5
K. neumoniae	1	0.5	0.5	1	0.5	1	1
P. aeruginosa	1	1	1	1	1	1	1

gram-positive bacteria (*Bacillus subtilis* NCTC 1040 and *Staphylococcus aureus* NCTC 7447) and Gram-negative bacteria (*Salmonella typhi* ATCC 19430, *Escherichia coli*, NCTC 10416 and *Pseudomonas aeruginosa* ATCC 10145a). Evaluation results revealed that β-amyrin exhibits significant antibacterial activity [94]. The minimum inhibitory concentration (MIC) of β-amyrin is presented in Table 9.7.

9.3.3.2 Betulinic Acid as Antimicrobial Agents

Mutai et al. reported the antimicrobial activity of pentacyclic triterpenes isolated from *Acacia mellifera*. Three triterpenoids are isolated from this plant including (20*S*)-oxolupane-30-al-(20*R*)-oxolupane-30-al and BA. These compounds are evaluation for their antimicrobial activity against various pathogens of both bacterial and fungal strains such as *Staphylococcus aureus*, *Escherichia coli*, *Cryptococcus neoformans*, *Trichophyton mentagrophyte*, *Candida krusei*, *Microsporum gypseum* and *Sacharomyces cerevisiae*. From the evaluation results, it is reported that (20*S*)-oxolupane-30-al, (20*R*)-oxolupane-30-al and BA exerted antibacterial activity against *S. aureus* as presented in Table 9.8 [95].

9.3.4 Triterpenoids as Anti-Inflammatory Agents

Inflammation is a complex biological system of vascular tissues injured by harmful stimuli, including impaired cells, irritants, or disease-producing agents. Inflammation is categorized into two types such as acute and chronic type. Acute inflammation is initiated with minimal stimuli that result due to the elevation of migration of plasma and leukocytes from the blood into the injured cells. Chronic inflammation leads to the dynamic movement of some types of cells existing at the site of inflammation. NF-κB plays a vital role as transcription factor in mediating inflammation. NF-κB is activated by several factors such as stress, cytokines, ROS, ultraviolet (UV) rays and pathogens. Similarly,

Prostaglandins (PGs) are the inflammatory mediators that are produced from arachidonic acid by constitutive COX-1 and inducible COX-2 [96].

9.3.4.1 Avicins as Anti-Inflammatory Agents

Avicins are the triterpenoid saponins. Structurally, Avicins are of different types such as Avicins A, D, G (Figure 9.55) [97]. These are isolated from the seed pods of *Acacia victoriae* of family Leguminosae. The anti-inflammatory activity of avicins is mediated through AMPK activation, mTOR inhibition and autophagy induction. It is reported that Avicin D inhibits the NF-κB signaling and activate the NF-E2-related factor 2 (Nrf2) pathway. Similarly, Avicin G potentially blocks the motion of NF-κB-binding DNA [98].

9.3.4.2 Ginsenosides as Anti-Inflammatory Agents

Ginsenosides are triterpene saponins primarily extracted from *Panax ginseng* (Figure 9.56) [99]. It can be divided into two major groups such as 20(S)-protopanaxadiol (e.g. Rb_1, Rb_2, Rc and Rd) and 20(S)-protopanaxatriol (e.g. Re, Rg_1, Rg_2 and Rb_3). Ginsenosides suppress NF-κB signaling pathways [100]. It also decreases the production of interleukin-1 (IL-1) from LPS-treated RAW264.7 cells with sustaining cell viability. Similarly, Rb_1blocks IFN-γ induces JAK/STAT and ERK signaling pathways and downstreams transcription factors thereby expressing iNOS gene [101].

9.3.4.3 Escin as Anti-Inflammatory Agent

Escin is a pentacyclic triterpene with the molecular formula $C_{55}H_{86}O_{24}$ (Figure 9.57). It exists in two isomeric forms such as α and β [102]. It is the active constituent of *Aesculus hippocastanum*. It prevents the inflammatory edema *in vivo* by reproducing the initial exudative phase, including paw edema induced by irritative agents [103]. Escin also inhibit

TABLE 9.7

Minimum Inhibition Concentration of the Isolated β-Amyrin

Concentration (mg/mL)	Diameter of Inhibition Zone (mm)	
	Staphylococcus aureus (NCTC 7447)	*Salmonella typhi* (ATCC 19430)
10	26	20
5	20	16
2.5	14	11

TABLE 9.8

Antimicrobial Activities of Triterpenoids

Test Compounds	Diameter of Inhibition Zone (mm)	
	Staphylococcus aureus (Bacterial Strain)	*Microsporum gypseum* (Fungal Strain)
20*S*-Oxolupane-30-al	10	21
20*R*-Oxolupane-30-al	10	-
Betulinic acid	9	-
Chloramphenical	19	-
Fluconazole	-	0.25

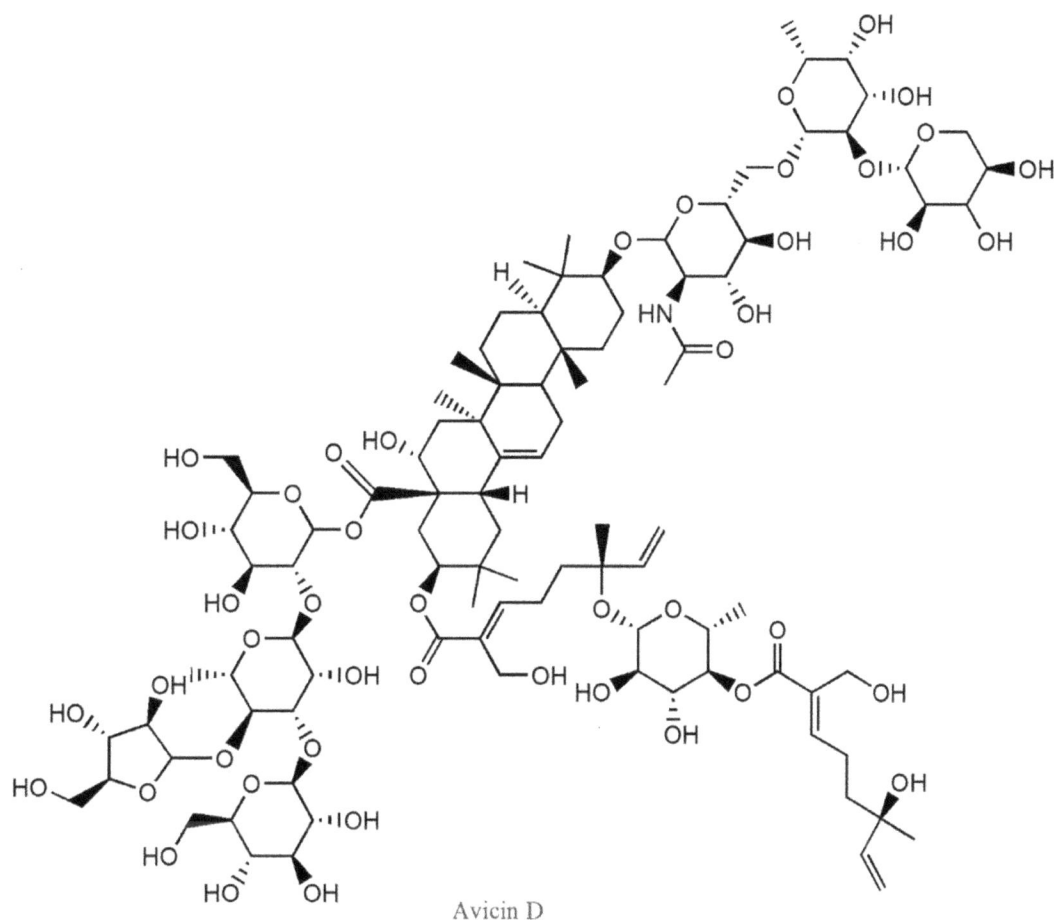

FIGURE 9.55 Structure of avicins D.

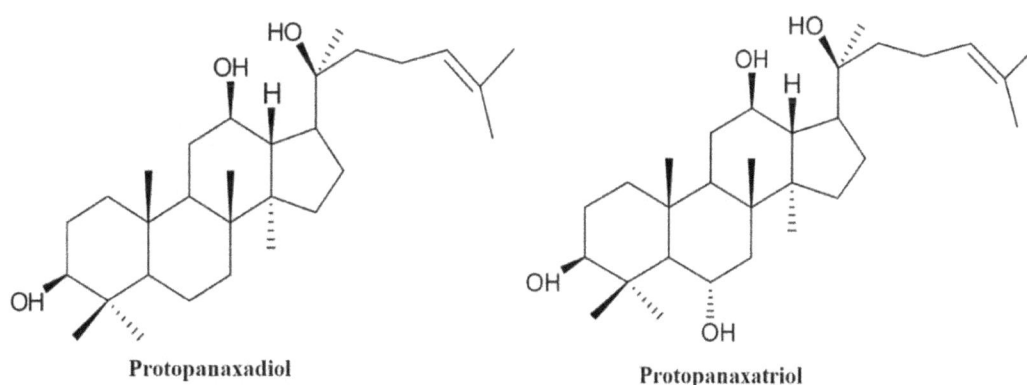

Protopanaxadiol

Protopanaxatriol

FIGURE 9.56 Chemical structure of ginsenosides.

the acetic acid-induced increase in capillary permeability and adhesion formation in an animal model. It was reported that escin down-regulates the expression of ICAM-1 and E-selectin protein and blocks the migration and adhesiveness of neutrophils [104].

9.3.4.4 Glycyrrhizin as Anti-Inflammatory Agent

Glycyrrhizin exhibits significant anti-inflammatory activities including inhibition of LPS/D-galactosamine-induced liver damage and prevention of free fatty acid-induced hepatic lipotoxicity [105]. Several evaluation studies have reported that glycyrrhizin displayed protective effects by inhibiting the production of NO, PGE2, TNF-α, IL-6, IL-1β and ROS and also reducing the expression of pro-inflammatory genes (iNOS and COX-2) and significantly block the activation of transcription factors (NF-κB, PI3K p110δ and p110γ) [106]. It is also reported that glycyrrhizin can inhibit NF-κB binding activity in CCl4 and ethanol-induced chronic inflammatory liver injury (Figure 9.58) [107,108].

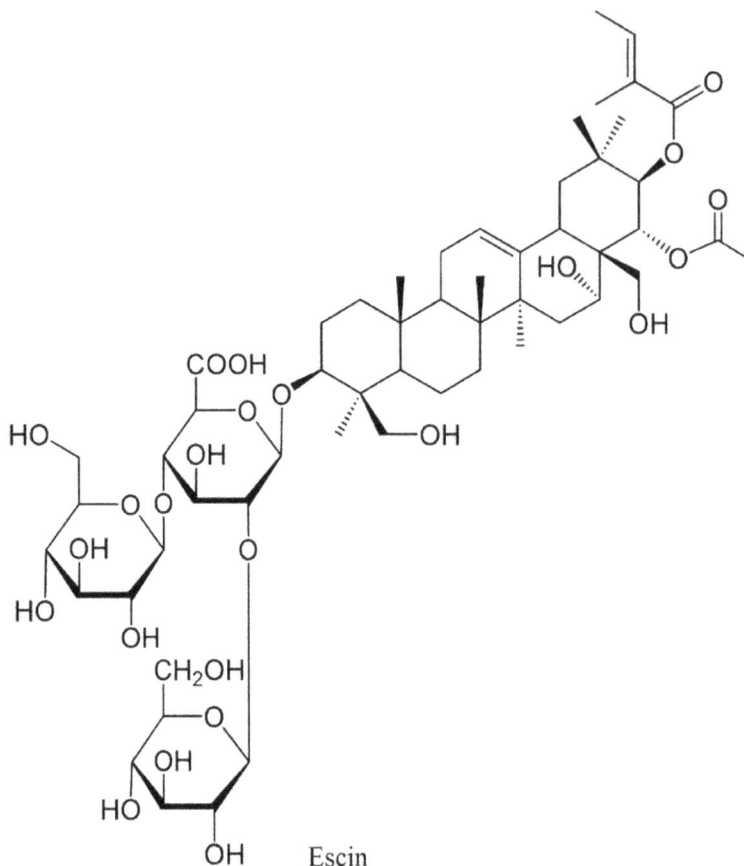

FIGURE 9.57　Chemical structure of escin.

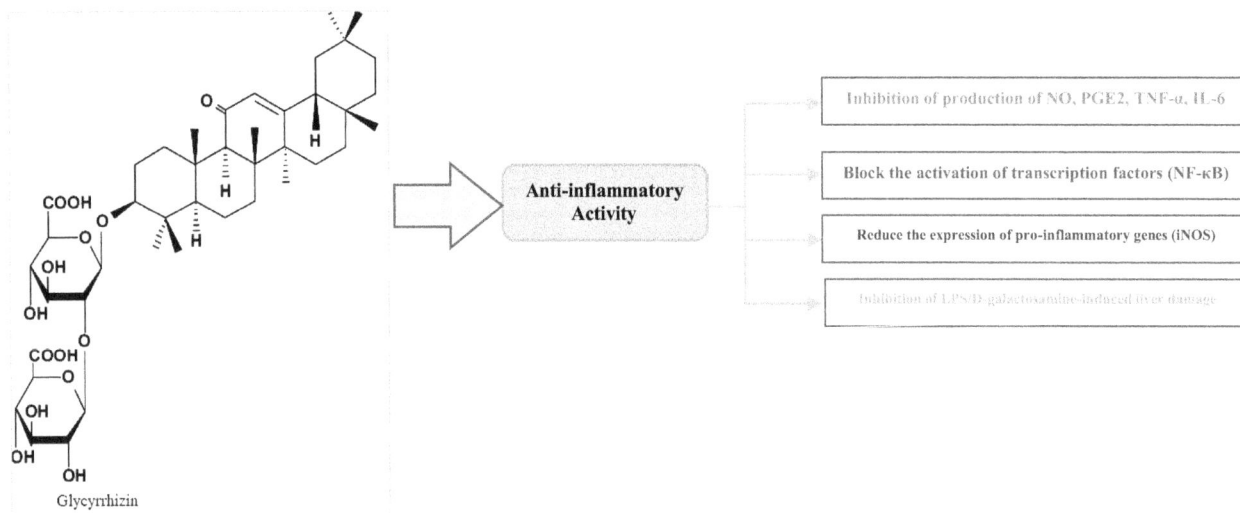

FIGURE 9.58　Molecular targets of glycyrrhizin as anti-inflammatory agent.

9.3.4.5 *Lupeol as Anti-Inflammatory Agent*

Several studies have revealed that lupeol abolishes the Akt-dependent pathways and also suppresses the NF-κB signaling, such as DNA binding of NF-κB complex, phosphorylation of IκBα protein and NF-κB-dependent reporter gene activity [109]. In addition to this, the anti-inflammatory activity of lupel can be evaluated from the observation that lupeol induces significant inhibition of PGE2 production in A23187-triggered macrophages (Figure 9.59) [110]. Geetha et al. demonstrated that Lupeol is used for the treatment of arthritis through the depletion of inflammation *in vivo*. This significant effect of lupeol for the treatment of arthritic inflammation in mice is mainly due to the modulation of the immune system and generation of inflammatory factors [111].

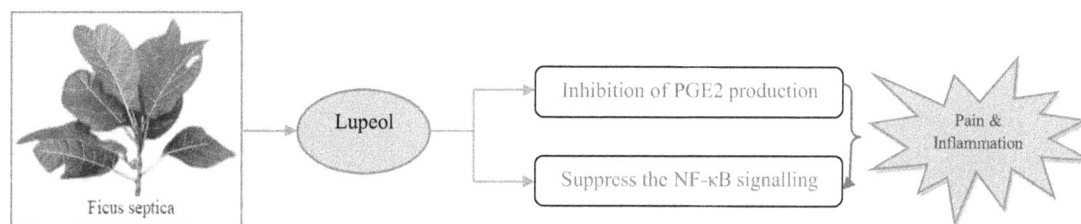

FIGURE 9.59 Molecular targets of lupeol as anti-inflammatory agent.

9.3.4.6 Oleanolic Acid as Anti-Inflammatory Agent

The anti-inflammatory properties of oleanolic acid (OA) are exerted by inhibiting the release of liposaccharide (LPS) mediated high mobility group box 1 (HMGB1) and cell adhesion molecules (CAMs) expression [112]. HMGBI is a protein which up-regulates the pro-inflammatory cytokines in inflammatory disorders. Lee et al. evaluated that OA diminishes the LPS-induced pro-inflammatory responses by the down-regulation of the expression of nuclear factor-κB (NF-κB) and tumor necrosis factor-α (TNF-α) by both in vivo and in vitro studies. Both NF-κB and TNF-α- are considered as biomarkers of inflammation [113]. Similarly, Gupta et al. demonstrated the inhibitory effect of OA on carrageenan-induced rat paw edema and formaldehyde-induced arthritis. OA and its derivatives are widely used as anti-inflammatory and immunosuppressive agents because it decreases the levels of pro-inflammatory cytokines, namely TNF-α, IL-1β and IL-6 [114].

9.3.4.7 Ursolic Acid as Anti-Inflammatory Agent

UA is natural pentacyclic triterpenoid with molecular formula $C_{30}H_{48}O_3$ and is structurally known as 3β-hydroxyurs-12-en-28-oic acid. Suh et al. demonstrated that UA decreases the production of iNOS and COX-2 via suppression of NF-κB in RAW264.7 mouse macrophages [115]. Similarly, Subbaramaiah et al suggested that the treatment with UA induces the suppression of TPA-mediated COX-2 protein and synthesis of PGE2 in human mammary epithelial cells. Several studies reported that UA abolishes the TPA-mediated activation of protein kinase C, extracellular signal regulated kinase 1/2 (ERK1/2), c-Jun NH2-terminal kinase 1/2

(JNK1/2) and p38 MAPK. From *in vitro* and *in vivo* studies, it is reported that UA has a variety of mechanism of action including reduction in expression of lipopolysaccharide (LPS)-triggered pro-inflammatory mediators in RAW264.7 cells and of 12-O-tetradecanoyl phorbol-13-acetate (TPA)-induced skin tumor promotion. In addition to this, UA abolishes the induction of COX-2, lipoxygenase, iNOS, MMP-9 and blocks the apoptotic gene that indicates UA potentiates its effects *via* down-regulation of NF-κB (Figure 9.60) [116].

9.3.4.8 Platycodin D as Anti-Inflammatory Agent

Platycodin D is an oleanane-type triterpenoid saponin with the molecular formula $C_{57}H_{92}O_{28}$ (Figure 9.61) It is isolated from the root of *Platycodon grandiflorum*. It exhibits anti-inflammatory activity *via* modulation of the NF- κB/IκB-α pathway [117]. Several reports suggest that platycodin D modulates NF-κB signaling pathways in the human keratinocyte cell line. In addition to this, platycodin D suppresses the expression of COX-2 by TPA which results in inhibition of PGE2 production in rat peritoneal macrophages [118].

9.3.4.9 Saikosaponins as Anti-Inflammatory Agents

Saikosaponins are the oleanane-type triterpenoid glycosides. It is isolated from *Bupleurum* species of family Umbelliferae (Figure 9.62). Saikosaponins represent a group of compounds including Saikosaponin A, Saikosaponin B, Saikosaponin B2, Saikosaponin E, Saikosaponin G, Saikosaponin I, Saikosaponin N, Saikosaponin U. Saikosaponin D suppresses the translocation of NF-κB and phosphorylation of IκBα which is related to the inhibition of T cell activation and apoptosis of cancer cells. *In vitro* studies suggest that the anti-inflammatory activity of

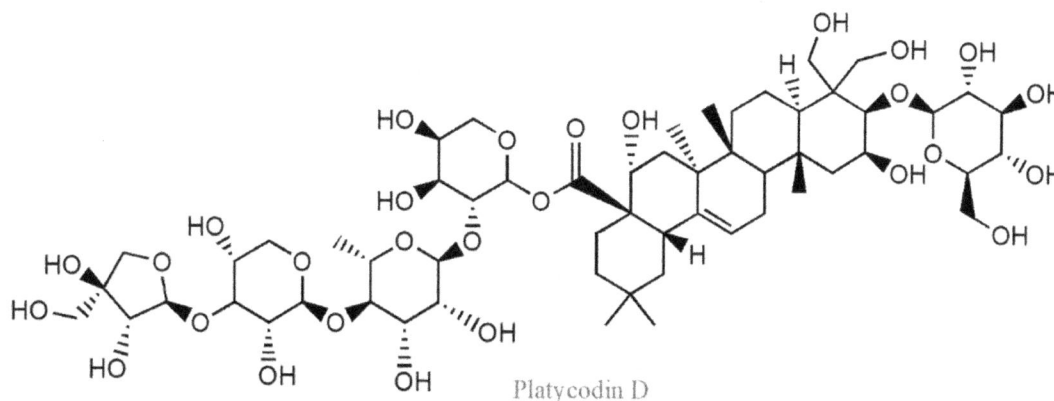

FIGURE 9.61 Structure of platycodin D.

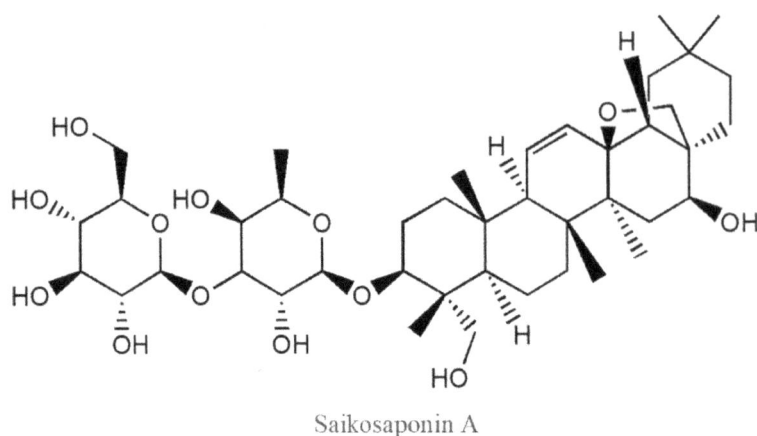

Saikosaponin A

FIGURE 9.62 Structure of saikosaponin A.

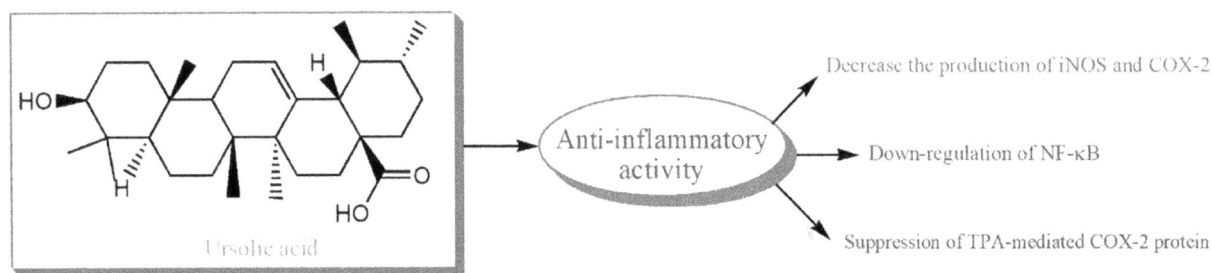

FIGURE 9.60 Molecular targets of ursolic acid for anti-inflammatory activity.

saikosaponins is exerted by suppression of ear and paw edema of mouse induced by phorbol 12-myristate 13-acetate (PMA), inhibition of cyclooxygenase (COX) and lipoxygenase (LOX) production as presented in Table 9.9 [119].

9.3.5 Triterpenoids as Antioxidant

Antioxidant activity is considered as one of the major protective effects produced by triterpenoids. It is involved in the prevention of carcinogenesis, aging, ischemia, neurodegenerative and cardiovascular disorders. ROS play a vital role in

TABLE 9.9

Molecular Targets of Triterpenoids for Anti-Inflammatory Activity [120–123]

Name of Triterpenoid	Type of Triterpenoid	Biological Sources	Molecular Targets
Avicin	Oleanane	*Acacia victoriae*	Inhibition of NF-κB DNA binding
Boswellic acid	Ursane	*Boswellia serrata* *Boswellia carteri*	Inhibition of NF-κB pathway
Celastrol	Oleanane	*Tripterygium wilfordii*	Inhibition of MAPK, JAK/STAT, Akt, COX-2, iNOS and NF-κB pathways
Escin	Oleanane	*Aesculus hippocastanum*	Inhibition of MAPK, TLR2, COX-2, ICAM-1, iNOS and NF-κB pathways
Ginsenosides	Dammarane	*Panax ginseng*	Inhibition of MMP-9 expression, MAPK, JAK/STAT, NF-κB pathways
Glycyrrhizin	Oleanane	*Glycyrrhiza glabra*	Inhibition of TNF-α, IL-6, IL-1β, COX-2, iNOS and NF-κB pathways
Lupeol	Lupane	*Mangifera indica*	Inhibition of Akt and NF-κB pathways
Oleanolic acid	Oleanane	*Arctostaphyllos uvaursi*	Inhibition of COX-2, iNOS and NF-κB pathways
Platycodon D	Oleanane	*Platycodon grandiflorum*	Inhibition of COX-2, iNOS and NF-κB pathways
Saikosaponins	Oleanane	*Bupleurum falcatum*	Inhibition of TNF-α, IL-6, IL-1β, COX-2, iNOS and NF-κB pathways
Ursolic acid	Ursane	*Calluna vulgaris*	Inhibition of MAPK, TNF-α, IL-6, IL-1β, TLR4, ICAM-1, COX-2, iNOS and NF-κB pathways

case of antioxidant activity. ROS is continuously produced by the body's normal use of oxygen during respiration and cell-mediated immune functions. ROS is also called as free radicals that include superoxide ($\cdot O_2^-$), hydrogen peroxide (H_2O_2), peroxynitrite ($\cdot ONOO^-$) and hydroxyl ($\cdot OH^-$). At physiological concentrations, ROS is necessary for normal cell function. If ROS is not scavenged effectively by cellular constituents, then the excess ROS can react with several biomolecules such as DNA, lipids and proteins that initiate the peroxidation of membrane lipids which leads to the accumulation of lipid peroxides and finally damage DNA and proteins [124]. It is observed that OA exhibits its antioxidant property by decreasing ROS level, inhibiting lipid peroxidation or stimulating the cellular antioxidant defenses [125].

9.3.5.1 *Fulgic Acids as Anti-Oxidants*

Fulgic acid A and Fulgic acid B are ursane type of triterpenoids. These are isolated from the roots of *Potentilla fulgens*.

Other triterpenoids isolated from the roots of *Potentilla fulgens* include UA, Euscaphic acid and Corosolic acid (Figure 9.63). All of these triterpenoids are evaluated for their antioxidant activity based on DPPH (2,2-diphenyl-1-picrylhydrazyl) and ABTS [2,2'-azinobis-(3-ethylbenzothiazoline-6-sulfonate)] assay methods. Tested compounds exhibit promising antioxidant activities in comparison with standard drug (Ascorbic acid) as shown in Table 9.10 [126].

9.3.5.2 *Multiflorane Triterpenoid as Anti-Oxidants*

Liu et al. reported a new multiflorane triterpenoid and two cucurbitane triterpenoids that are isolated from the stems of Momordica charantia (Figure 9.64). The structures of the newly isolated compounds are conformed by spectroscopic studies (IR, NMR, LC-MS). These compounds are evaluated for their antioxidant activity using the ABTS assay method. Evaluation study reports suggest that the tested triterpenoids exhibit free radical cation scavenging activity with IC_{50} values of 268.5 ± 7.9,

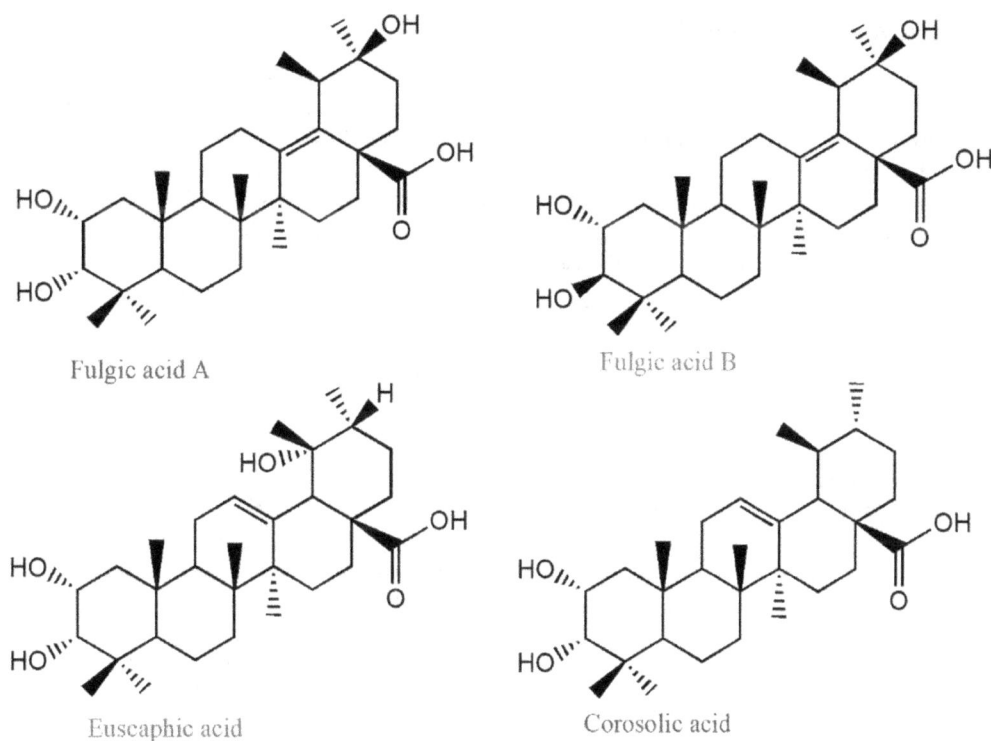

FIGURE 9.63 Structures of fulgic acid A, B, euscaphic acid and corosolic acid.

TABLE 9.10

Antioxidant Activity of *P. fulgens* Isolated Compounds

Test Compounds	IC_{50} Value ((μg/mL)	
	DPPH·± SD	ABTS+·± SD
Fulgic acid A	9.8 ± 0.5	19.9 ± 0.8
Fulgic acid B	11.6 ± 0.7	23.3 ± 0.3
Ursolic acid	5.3 ± 0.2	7.8 ± 0.4
Euscaphic acid	21.0 ± 0.3	30.2 ± 0.6
Corosolic acid	32.8 ± 0.5	38.7 ± 0.2
Ascorbic acid	3.6 ± 0.7	6.2 ± 0.2

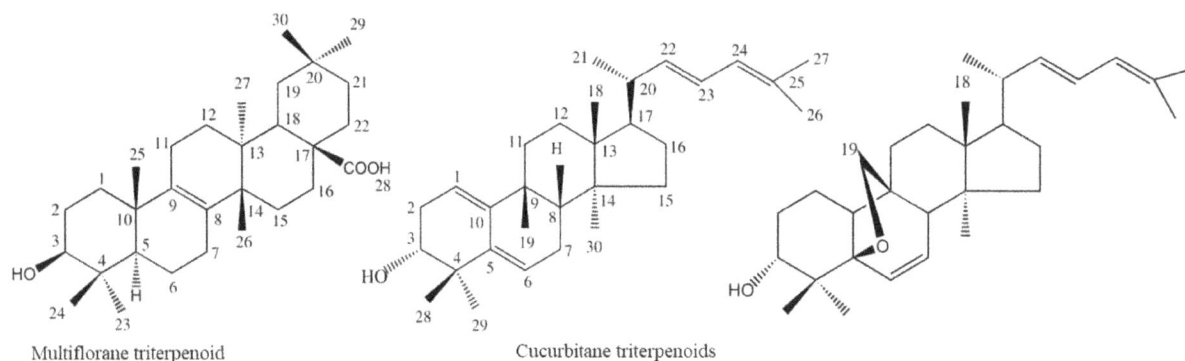

Multiflorane triterpenoid   Cucurbitane triterpenoids

FIGURE 9.64 Structures of multiflorane triterpenoids.

352.1 ± 11.5 and 458.9 ± 13.0 μM, respectively. Similarly, the inhibitory effect of these compounds was observed on xanthine oxidase activity with IC_{50} values of 142.3 ± 30.2, 36.8 ± 20.5 and 124.9 ± 8.3 μM respectively [127].

9.3.5.3 Madecassic Acid as Anti-Oxidants

Madecassic acid is a pentacyclic triterpenoid with the molecular formula $C_{30}H_{48}O_6$ (Figure 9.65). It is the active constituent present in *Centella asiatica*. *In vitro* experimental study indicates that Madecassic acid prevents H_2O_2-induced oxidative stress and apoptosis in human periodontal ligament fibroblast (hPDLF) cells. It represents a cellular model to identify the effect of Madecassic acid against inflammatory disease by reducing intracellular ROS production and maintaining mitochondrial membrane potential [128].

9.3.6 Triterpenoids as Hepatoprotective

Liver disorder is one of the leading causes of death worldwide. There is a remarkable advancement in the development of safe and effective treatment, but still potential hepatoprotective agents are urgent needed for patient care. So, natural products are a really good source of lead compounds. Hence, evaluation results report that there are large numbers of natural triterpenoids that possess promising hepatoprotective effects [129].

9.3.6.1 Triterpenoids Saponis as Hepatoprotective

Wang et al. reported seven new triterpenoid saponins (schekwangsiensides A-G) and a new triterpenoid (schekwangsienin).

These compounds are isolated from the aerial parts of *Schefflera kwangsiensis*. The structures of these compounds are interpreted based on their spectroscopic and chemical data [130]. The structure of schekwangsienside A is identified as 3β-O-(β-D-glucofuranosylurono-6,3-lactone)-12-en-28-oic acid. Similarly, the structure of schekwangsienside B is elucidated as 3β-O-hydroxyolean-12-en-28-oic acid 28-O-[6-O-((3R)-3-hydroxyl-3-methylglutaryl)-β-D-glucopyranosyl]-ester. The structure of schekwangsienside C is identified as 3β-O-hydroxyolean-12-en-28-oic acid 28-O-[β-D-xylopyranosyl-(1→2)-β-D-glucopyranosyl] ester. Chemically, schekwangsienside D is called as 3β-O-[α-L-rhamnopyranosyl-(1→2)-β-D-glucopyranosyl-(1→2)-β-D-glucuronopyranosyl] olean-12-en-28-oic acid 28-O-(β-D-glucopyranosyl)ester. Further, these compounds are evaluated for their hepatoprotective activities against D-galactosamine-induced HL-7702 cell damage. The hepatoprotective effects of the test compounds are determined by MTT colorimetric assay in HL-7702 cells [131].

9.3.6.2 Ceanothic Acid and Zizybrenalic Acid as Hepatoprotective

Ceanothic acid and zizybrenalic acid are the pentacyclic triterpenes isolated from *Ziziphus oxyphylla* Edgew (Figure 9.66). Hepatoprotective activity exhibited by zizybrenalic acid is more promising than ceanothic acid which is detected from the decrease in carbon tetrachloride (CCl_4)-induced elevation of serum biomarkers (ALP, ALT, TB and TP). ALT is present in the cytosol of hepatocytes and the damage to the liver is detected by elevated ALT level due to its leakage from hepatocytes into the blood. Similarly, the

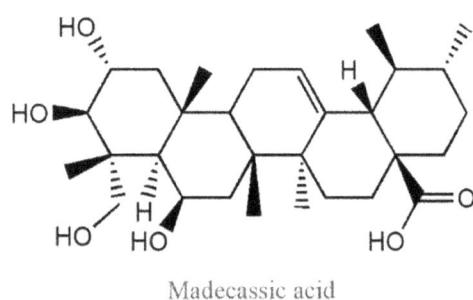

Madecassic acid

FIGURE 9.65 Structure of madecassic acid.

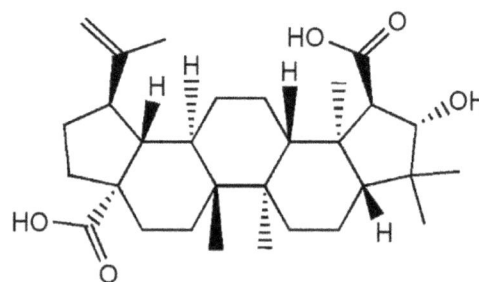

Ceanothic acid

FIGURE 9.66 Structure of ceanothic acid.

obstruction in the flow of bile and damage to hepatocytes is detected from an elevated level of ALP in serum. Ceanothic acid and zizybrenalic acid may inhibit the leakage of hepatocyte enzymes and exhibit hepatoprotection by membrane stabilizing activity. Oxidative stress is involved in hepatotoxicity development which occurs due to excessive formation of ROS in mitochondria subsequently damaging the macromolecules including dysfunction of proteins, DNA damage and lipid peroxidation [132].

9.3.7 Triterpenoids as Cardioprotective Agent

Cardiovascular complications are more common in the case of diabetic patients. Further, stress-induced hyperglycemia in the case of non-diabetic patients is associated with acute myocardial infarction that causes higher mortality rate in hospitalized patients. So, it has been investigated that the treatment of OA improves the heart function in streptozotocin-induced diabetic rats [133].

9.3.7.1 Lupeol as Cardioprotective Agent

Varalakshmi et al. reported the cardioprotective effect of pentacyclic triterpene (c) on cyclophosphamide-induced oxidative stress. Cyclophosphamide (CP) is an alkylating agent which is commonly used in cancer chemotherapy that causes cardiotoxicity to normal cells. Lupeol is the pentacyclic triterpene, isolated from the stem bark of Crataeva nurvala. It is investigated for its cardioprotective effect against CP-induced toxicity. The cardioprotective effect of lupeol (50 mg/kg body weight for 10 days orally) is observed from the significant reversal of the above alterations induced by CP. These observations signify the antioxidant property of triterpenes and their cytoprotective action against CP-induced cardiotoxicity [134].

9.3.7.2 Betulinic Acid as Cardioprotective Agent

Myocardial ischemia and reperfusion injury lead to cell death. Xia et al. demonstrated the cardioprotective effect of betulinic acid (BA) on myocardial ischemia reperfusion injury in rats. It is reported that BA ameliorates myocardial ischemia/reperfusion injury in rats by inhibiting the release of LDH and CK, suppressing myocyte apoptosis. LDH and CK are called markers of cellular necrosis [135].

9.3.8 Triterpenoids as Antiviral Agents

Viral infections are serious human diseases with a high rate of mortality. Antiviral drugs are the class of medication used for the treatment of viral infections. Most of the antivirals are target-specific but broad-spectrum antivirals are effective against wide range of viruses such as human immunodeficiency virus (HIV) infections, herpes simplex virus (HSV), varicella-zoster virus (VZV), cytomegalovirus (CMV), hepatitis B virus (HBV), hepatitis C virus (HCV), respiratory syncytial virus (RSV), human papillomavirus (HPV) and influenza virus [136]. Antiviral agents generally target different stages of the viral life cycle. The target stages involved in the viral life cycle include viral attachment to host cell, uncoating, synthesis of viral mRNA, translation of mRNA, maturation of new viral proteins, replication of viral DNA and RNA etc [137]. Herbal drugs mainly triterpenoids are a rich source of antiviral agents [138].

Wei et al. synthesized dammarane-type triterpene derivatives and evaluated their antiviral activity by inhibiting both HIV-1 and HCV proteases. Evaluation results report that these triterpenoids are potential inhibitors of HIV-1 protease with $IC_{50} < 10\,\mu M$. These compounds inhibit HCV protease with $IC_{50} < 10\,\mu M$ (Table 9.11) [139].

The pentacyclic triterpenes such as OA, UA, BA and glycyrrhetic acid inhibit the HIV infection. Ginseng roots mainly

TABLE 9.11

List of Antiviral Triterpenes [141–143]

Triterpenes	Biological Sources	Class	Virus	Cell Lines	IC_{50} or EC_{50}
Celastrol	*Tripterygium regelii* (Celastraceae)	OT	CoV	-	$10.3 \pm 0.2\,\mu M$
3β-Friedelanol	*Euphorbia neriifolia* (Euphorbiaceae)	OT	CoV	MRC-5	-
Lyonifoloside A	*Lyonia ovalifolia* (Ericaceae)	CAT	CVB3	Vero	$11.1 \pm 1.98\,\mu M/L$
Beesioside I	*Souliea vaginata* (Ranunculaceae)	CAT	HIV	MT-4	$2.32 \pm 0.46\,\mu M$
Lyonifoloside A	*Lyonia ovalifolia* (Ericaceae)	CAT	IV	MDCK	$11.1\,\mu M/L$
Camellenodiol 3-O-β-D-glucuronopyranoside	*Camellia japonica* (Theaceae)	-	PEDV	Vero	-
Oleanolic acid	*Fadogia tetraquetra*	OT	SFV	HK21	-
Ursolic acid	*Calluna vulgaris*	OT	CMV	GPEL	$6.8\,\mu g/mL$
Paracaseolin A	*Sonneratia paracaseolaris* (Sonneratiaceae)	-	-	MDCK	$28.4\,\mu g/mL$
Ganoderic acid T-Q	*Ganoderma lingzhi*	-	-	-	$1.2 \pm 1.0\,\mu M$

contain ginsenosides that consist of dammarane-type triterpenes in which protopanaxadiol and protopanaxatriol are present as aglycones. These aglycones are converted to panaxadiol and panaxatriol under strong acid conditions. These compounds inhibit HIV-1 and HCV PRs as well as the human aspartyl and serine proteases renin and trypsin. Panaxadiol and panaxatriol have the same moiety in ring A but panaxatriol possesses one more hydroxyl group in ring B. So, panaxatriol is called as 6α-hydroxy-PD. Panaxadiol possess two hydroxyl (–OH) groups at positions C_3 and C_{12} in the rings A and C respectively. Panaxatriol contains three hydroxyl (–OH) groups at positions C_3, C_6, C_{12} in the rings A, B and C respectively. The structure-activity relationships (SAR) study helps to produce suitable antiviral drugs (Figure 9.67) [140].

9.3.9 Miscellaneous

a. Glycyrrhizin

Glycyrrhizin is a triterpenoid glycoside that contains one glycyrrhetinic acid as aglycone part and two molecules of d-glucuronic acid. Glycyrrhizin undergoes hydrolysis in vivo to produce glycyrrhetinic acid that is responsible for its pharmacological properties such as anti-ulcer, anti-inflammatory, anti-diabetic, immunomodulator, antitussive, antioxidant, antitumor, hepatoprotective and antiviral (Figure 9.68) [144].

b. Boswellic acids

Boswellic acids (BA) are the series of pentacyclic terpenoids that are produced by gum resin of plants *Boswellia papyrifera*, *Boswellia serrata* and *Boswellia carteri* of Burseraceae family. There are different types of BAs identified including α and β-boswellic acids, acetylated α and β-boswellic acids, 11-keto-β-boswellic acid and 3-O-acetyl-11-keto-β-boswellic acid (Figure 9.69) [145].

Boswellic acids exhibit diverse pharmacological activities including immunomodulatory, anti-inflammatory, expectorant, anti-asthmatic, antiseptic, anxiolytic, anti-neurotic, analgesic, tranquilizing and antibacterial effects [145]. Boswellic acids can modulate various molecular targets such as enzymes, growth factors, kinases, transcription factors and receptors that allow to stimulate apoptosis and cell cycle arrest. It can also inhibit several signaling pathways related to cell survival, proliferation and metastasis [146]. Among these molecular targets, NF-κB and Akt play a major role in progression of cancer that regulates the cancer cell proliferation, survival, invasion, metastasis and high mortality of patients. The in vivo study is performed to determine the effect of BA by injecting a sensitization liquid (0.15 mL aluminum hydroxide gel at 88.67 mg/mL and 0.05 mg ovalbumin) intraperitoneally in an asthma model. The evaluation results revealed that BA minimizes the symptoms by abrogating p-STAT6 followed by a reduction in GATA3 expression. Similarly, a study on the effect of BAs in bovine serum albumin (BSA)-induced arthritis reported that the oral administration, BAs (25, 50 and 100 mg/kg/day) remarkably decreases the leucocyte population and inhibits its infiltration into the knee joint as well as the pleural cavity in a BSA-injected knee (Figure 9.70, Table 9.12) [147].

c. Celastrol

The pentacyclic triterpenoid, Celastrol is found to exhibit anti-inflammatory activity. It generally inhibits the IKK-NF-κB-signaling pathway and activates the heat shock response *via* heat shock factor-1 (HSF-1). Recent studies have reported that celastrol suppresses inflammatory cytokines, such as NF-κB, IL-1β and TNF-α. Lee et al. reported that celastrol

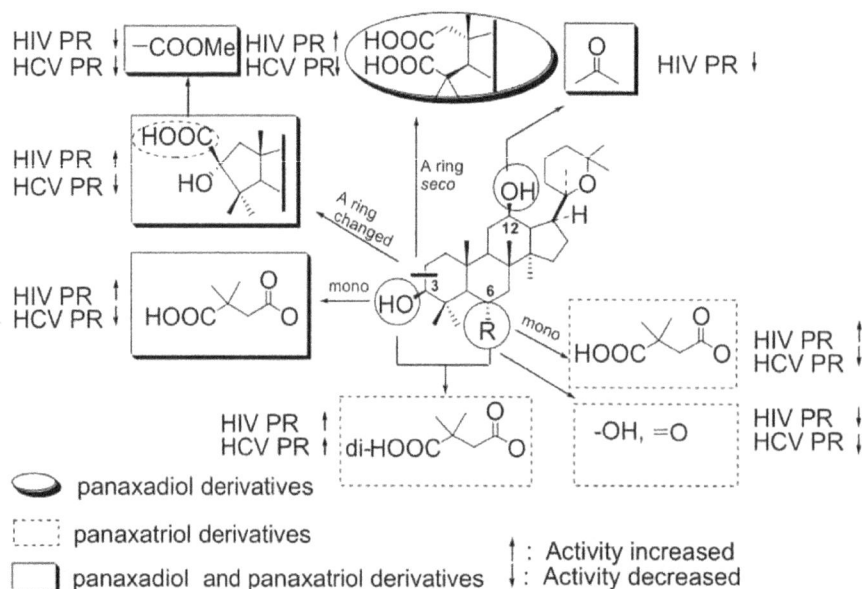

FIGURE 9.67 SAR study of dammarane triterpene derivatives against HIV and HCV PRs.

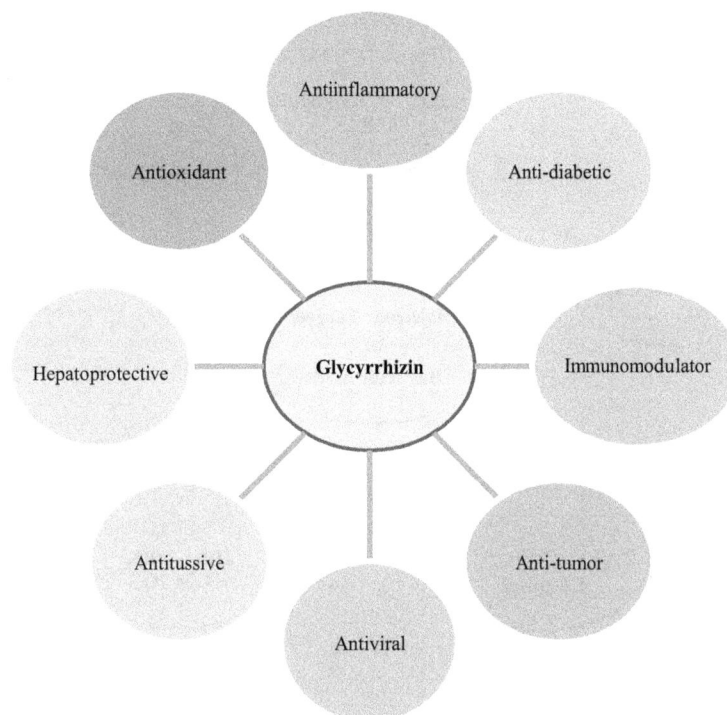

FIGURE 9.68 Pharmacological properties of glycyrrhizin.

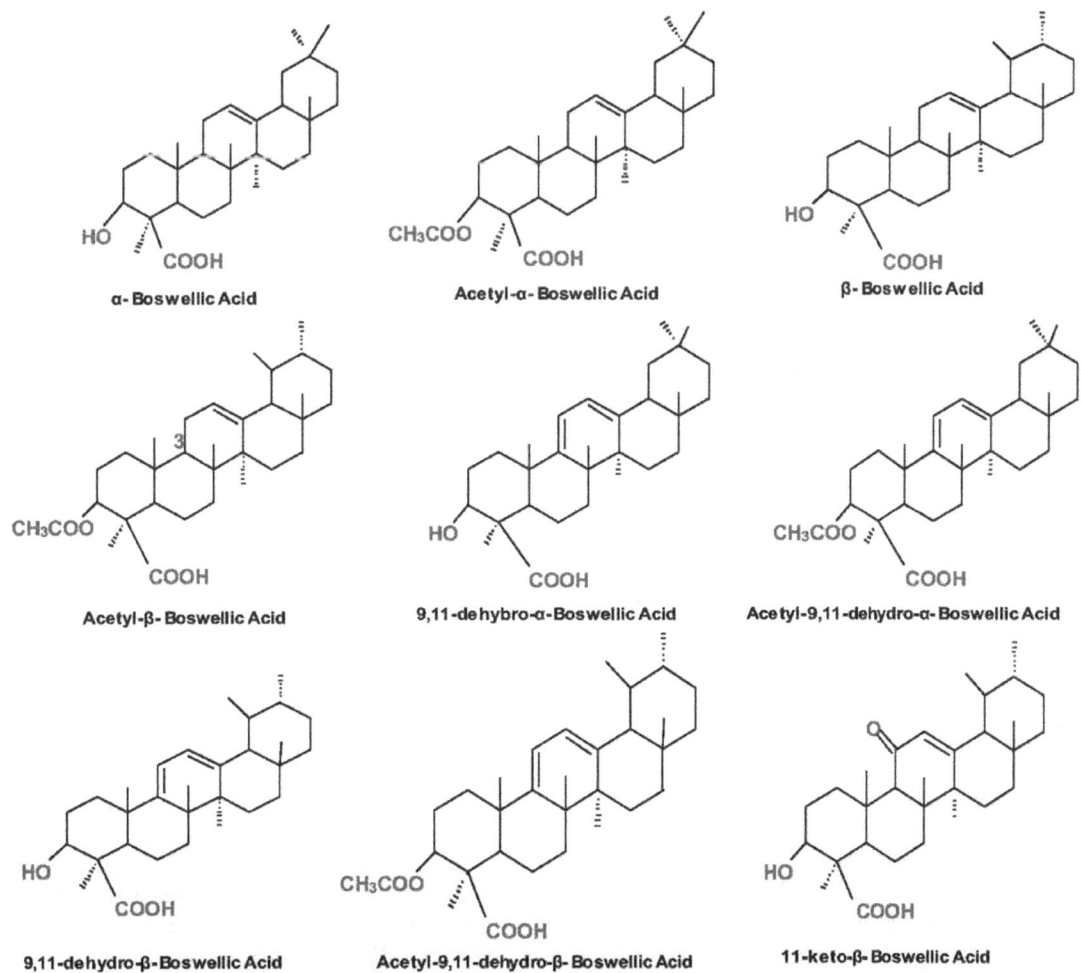

FIGURE 9.69 Structures of different types of boswellic acids.

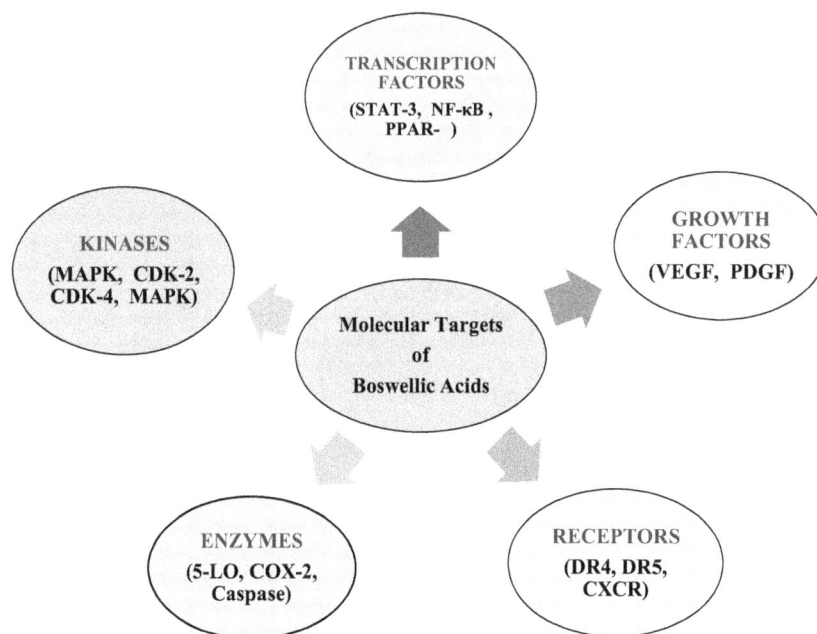

FIGURE 9.70   Molecular targets of boswellic acids.

TABLE 9.12

Biological Activity of Boswellic Acid

Diseases	Mechanism of Action
Arthritis	↓ Infiltration of leucocytes
Asthma	↓ Expression of pSTAT6 and GATA3
Atherosclerosis	↓ NF-κB activity
Brain cancer	↓ Apoptosis
Colon cancer	↓ Cyclin D1 and E, CDK 2 and 4
Lung cancer	↓ Apoptosis
Melanoma	↓ Topoisomerase II and MMPs
Neuroblastoma	↓ PARP cleavage, ↑ Apoptosis
Pancreatic cancer	↓ p-mTOR
Parkinson's disease	↓ Inflammatory markers
Diabetes	↓ Cytokine burst and blood glucose
Gastric ulcer	↓ Biosynthesis of leukotrienes

mainly decrease the activity of IKKα and IKKβ in a dose-dependent manner, both *in vitro* and *in vivo*. Celastrol is found to have different target molecules for several disorders such as inflammation, diabetes, obesity, cancer and neuronal disorders [148].

Pinna et al. demonstrated that celastrol effectively decreases the levels of pro-inflammatory cytokines, namely TNF-α, IL-1β, IL-6 and IL-8 in biopsies of the inflamed intestinal mucosa of patients with Crohn's disease [149].

Wu et al. reported that celastrol inhibits the migration and invasion of human chondrosarcoma cells with a maximum inhibition at 5 mg/L of celastrol. This is due to suppression of the cell proliferation regulating inhibitor of protein phosphatase 2 A/c-MYC (CIP2 A/c-MYC) signaling pathway, down-regulation

of the expression of CIP2 A protein, which results not only in decreased tumor invasion, but also decrease the cell proliferation and induce apoptosis [150].

Kim et al. demonstrated the anti-diabetic activity of celastrol by treating insulin-resistant diabetic mice with celastrol for a duration of 2 months and reported that this compound significantly decreases the fasting plasma glucose and hemoglobin A1c levels (HbA1c). It was observed by a lower mean fasting plasma glucose in celastrol-treated diabetic mice in comparison with non-treated diabetic mice, with values of 16.2 and 32.1 mmol respectively after 8 weeks of treatment [151].

Celastrol exhibits its anti-obesity activity through (1) inhibition of the generation of ROS (2) increase in the activity of antioxidant enzymes and (3) inhibiting the activity of the enzyme NADPH oxidase. The combination of these mechanisms of actions is associated with a decrease in the level of low-density lipoprotein (LDL) and triglyceride in obese mice. Leptin is a hormone produced by adipose tissue. It is considered as the hallmark of obesity as it is responsible for suppressing food intake and regulating the expenditure of energy. Liu et al. exhibited that celastrol reduces the body weight of leptin-resistant mice up to 45% through decrease in the consumption of food [152].

Celastrol is identified as a potential agent to slow down the neurodegenerative disorder such as Alzheimers disease and Parkinson's disease in animal models. It is reported that Celastrol nduces and modulates the activity of heat shock proteins (HSPs) in the cerebral cortex thereby exerting a neuroprotective effect. Chow et al. have demonstrated that celastrol induces the expression of a variety of HSP classes such as HSP27 and HSP32 in glial cells and HSP70 in neurons in rat models (Figure 9.71) [153,154].

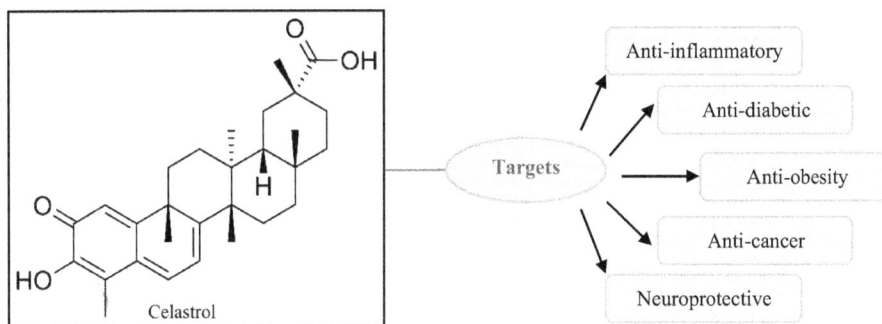

FIGURE 9.71 Biological properties of celastrol.

d. Quassinoids

Quassinoids are highly oxygenated and degraded triterpenes lactones [155]. These compounds are isolated as bitter principles from the plant *Quassia amara* of family Simaroubaceae. Quassinoids can be categorized into different groups based on the presence of basic skeletons C_{18}, C_{19}, C_{20}, C_{22} and C_{25} (Figure 9.72). Several quassinoids exhibit diverse biological properties such as anti-inflammatory, inseticidal, herbicidal, antiparasitic, antimalarial and anticancer [156].

e. Lanosterol

Lanosterol is used for the treatment of variety of diseases such as hepatitis, hypertension, neurasthenia, chronic bronchitis, leukopenia and cancer. Lanosterol synthase (LSS) is the rate-limiting enzyme for the biosynthesis of cholesterol. The inhibition of cholesterol biosynthesis from lanosterol results in the inhibition of cell cycle progression that leads to the induction of cell differentiation. Further, lanosterol is a potential herbal drug for the treatment of cataracts that can reverse the aggregation of intracrystalline proteins. The exposure of UV-B radiation induces oxidative injury that results in crystalline denaturation and apoptosis in lens epithelial cells. So, LSS might play a protective role during the early stages of this process [157,158].

9.4 Conclusion

Triterpenes are a diverse group of natural products. These are secondary metabolites of the plants. Squalene and 2,3-oxidosqualene are considered as the key precursors for the biosynthesis of triterpenes with diverse chemical structures.

Oxidosqualene cyclase-mediated cyclization of 2,3-oxidosqualene is the principal step for the biosynthesis of triterpenes in plants. Biosynthesis of terpenes is mediated through MVA or MEP pathway. Triterpenoids are categorized into different types such as acyclic, monocyclic, bicyclic, tricyclic, tetracyclic and pentacyclic triterpenes based on their structural features. The structures of these triterpenoids are elucidated on the basis of chemical and spectroscopic methods (FT-IR,^{1}H-NMR,^{13}C-NMR, HR-ESIMS). Triterpenoids have been used widely due to their several pharmacological activities anticancer, antidiabetic, antimicrobial, antiviral analgesic and anti-inflammatory effects. Extensive studies are conducted on terpenoids to determine their molecular targets and mechanism of actions.

Abbreviations

Ads	Alzheimer's disease
AMPK	5′ AMP-activated protein kinase
CAT	Cycloartane triterpene
CoV	Coronavirus
COX-2	Cyclooxygenase-2
CrtE	Geranylgeranyl pyrophosphate synthase
CVB3	Coxsackie B3
DMAPP	Dimethylallyl diphosphate
DXP	Deoxyxylulose 5-phosphate
FPP	Farnesyl diphosphate
EBV	Epstein-Barr virus
G3P	Glyceraldehyde 3-phosphate
GPP	Geranyl diphosphate
GPEL	Guinea pig embryo lung fibroblasts
HIV	Human immunodeficiency viruses
IKK	Inhibitor of nuclear factor kappa-B kinase subunit
IL-6	Interleukin-6
IPP	Isopentenyl diphosphate
Ipi	Isopentenyl diphosphate delta isomerase
IV	Influenza virus
MAPK	Mitogen-activated protein kinase
MMP-9	Matrix metalloproteinase-9
mTOR	Mammalian target of rapamycin
MEP	Methylerythritol 4-phosphate
Nrf2	NF-E2-related factor 2
NF-κB	Nuclear factor-kappa B
OT	Oleanane triterpene

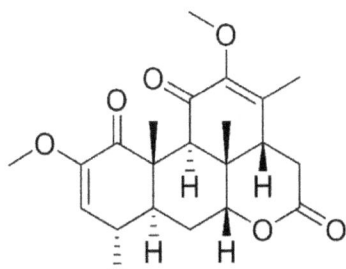

FIGURE 9.72 Molecular structure of quassinoids.

PEDV	Porcine epidemic diarrhea virus
PGs	Prostaglandins
PSPP	Presqualene diphosphate
ROS	Reactive oxygen species
SFV	Semliki forest virus
Sqs	Squalene synthase
STAT	Signal transducer and activator of transcription
STAT1	Signal transducers and activators of transcription 1
STAT3	Signal transducer and activator of transcription 3

REFERENCES

[1] Mahato, S. B.; Nandy, A. K.; Roy, G. Triterpenoids. *Phytochemistry.* **1992**, 31(7), 2199–2249. doi: 10.1016/0031-9422(92)83257-y.

[2] Muffler, K.; Leipold, D.; Scheller, M. C.; Haas, C.; Steingroewer, J.; Bley, T.; Ulber, R. Biotransformation of triterpenes. *Process Biochemistry*, **2011**, 46(1), 1–15.

[3] Sandeep; Ghosh, S. Triterpenoids: Structural diversity, biosynthetic pathway, and bioactivity. *Studies in Natural Products Chemistry*, **2020**, 411–461. doi: 10.1016/b978-0-12-819483-6.00012-6.

[4] Jäger, S.; Trojan, H.; Kopp, T.; Laszczyk, M. N.; Scheffler, A. Pentacyclic triterpene distribution in various plants–rich sources for a new group of multi-potent plant extracts. *Molecules*, **2009**, 14, 2016–2031.

[5] Kvasnica, M.; Urban, M.; Dickinson, N. J.; Sarek, J. Pentacyclic triterpenoids with nitrogen-and sulfur-containing heterocycles: Synthesis and medicinal significance. *Natural Product Reports*, 2015. doi: 10.1039/C5NP00015G.

[6] Pyka, A. Detection progress of selected drugs in TLC. *BioMed Research International*, **2014**, 732078. doi: 10.1155/2014/732078.

[7] Waksmundzka-Hajnos, M.; Sherma, J.; Kowalska, T. *Thin Layer Chromatography in Phytochemistry || TLC of Triterpenes (Including Saponins).* **2008**. doi: 10.1201/9781420046786.ch20.

[8] Castola, V.; Bighelli, A.; Casanova, J. Direct qualitative and quantitative analysis of triterpenes using ^{13}C-NMR spectroscopy exemplified by dichloromethanic extracts of cork. *Applied Spectroscopy*, **1999**, 53, 344–350.

[9] Xu, C.; Wang, B.; Pu, Y.; Tao, J.; Zhang, T. Techniques for the analysis of pentacyclic triterpenoids in medicinal plants. *Journal of Separation Science*, **2017**, 1–57. doi: 10.1002/jssc.201700201.

[10] Gill, B. S.; Kumar, S.; Navgeet. Triterpenes in cancer: Significance and their influence. *Molecular Biology Reports*, **2016**, 43, 881–896. doi: 10.1007/s11033-016-4032-9.

[11] Yadav, V. R.; Prasad, S.; Sung, B.; Kannappan, R.; Aggarwal, B. B. Targeting inflammatory pathways by triterpenoids for prevention and treatment of cancer. *Toxins*, **2010**, 2, 2428–2466.

[12] Safayhi, H.; Sailer, E. Anti-inflammatory actions of pentacyclic triterpenes. *Planta Medica*, **1997**, 63, 487–493.

[13] Zhang, J.; Yamada, S.; Ogihara, E.; Kurita, M.; Banno, N.; Qu, W.; Feng, F.; Akihisa, T. Biological activities of triterpenoids and phenolic compounds from *Myrica cerifera* Bark. *Chemistry & Biodiversity*, **2016**, 13(11), 1601–1609. doi: 10.1002/cbdv.201600247.

[14] Okada, K. The biosynthesis of isoprenoids and the mechanisms regulating it in plants. *Bioscience, Biotechnology, and Biochemistry*, **2011**, 75, 1219–1225.

[15] Phillips, D. R.; Rasbery, J. M.; Bartel, B.; Matsuda, S. P. Biosynthetic diversity in plant triterpene cyclization. *Current Opinion in Plant Biology*, **2006**, 9, 305–314.

[16] Thimmappa, R.; Geisler, K.; Louveau, T.; O'Maille, P.; Osbourn, A. Triterpene biosynthesis in plants. *Annual Review of Plant Biology*, **2014**, 65(1), 225–257. doi: 10.1146/annurev-arplant-050312-120229.

[17] Noushahi, H. A.; Khan, A. H.; Noushahi, U. F., et al. Biosynthetic pathways of triterpenoids and strategies to improve their biosynthetic efficiency. *Plant Growth Regulation*, **2022**. doi: 10.1007/s10725-022-00818-9.

[18] Sawai, S.; Uchiyama, H.; Mizuno, S., et al. Molecular characterization of an oxidosqualene cyclase that yields shionone, a unique tetracyclic triterpene ketone of *Aster tataricus*. *FEBS Letters*, **2011**. doi: 10.1016/j.febslet.2011.02.037.

[19] Rohmer, M. The discovery of a mevalonate-independent pathway for isoprenoid biosynthesis in bacteria, algae and higher plants. *Natural Product Reports*, **1999**, 16(5), 565–574. doi: 10.1039/a709175c.

[20] Pan, J.-J.; Bugni, T. S.; Poulter, C. D. Recombinant squalene synthase. Synthesis of cyclopentyl non-head-to-tail triterpenes. *The Journal of Organic Chemistry*, **2009**, 74(19), 7562–7565. doi: 10.1021/jo9014547.

[21] Ling, T.; Boyd, L.; Rivas, F. Triterpenoids as reactive oxygen species modulators of cell fate. *Chemical Research in Toxicology*, **2022**, 35(4), 569–584. doi: 10.1021/acs.chemrestox.1c00428.

[22] Dai, L.; Liu, C.; Zhu, Y., et al. Functional characterization of cucurbitadienol synthase and triterpene glycosyltransferase involved in biosynthesis of mogrosides from *Siraitia grosvenorii*. *Plant Cell Physiology*, **2015**, 56, 1172–1182. doi: 10.1093/pcp/pcv043.

[23] Li, D.; Zhang, Q.; Zhou, Z.; Zhao, F.; Lu, W. Heterologous biosynthesis of triterpenoid dammarenediol-II in engineered *Escherichia coli*. *Biotechnology Letters*, **2016**, 38(4), 603–609. doi: 10.1007/s10529-015-2032-9.

[24] Ohyama, K.; Suzuki, M.; Kikuchi, J., et al. Dual biosynthetic pathways to phytosterol via cycloartenol and lanosterol in *Arabidopsis*. *Proceedings of the National Academy of Sciences of the United States of America*, **2009**, 106, 725–730. doi: 10.1073/pnas.0807675106.

[25] Niemann, G. J. Biosynthesis of pentacyclic triterpenoids in leaves of *Ilex aquifolium* L. *Planta*, **1985**, 166(1), 51–56.

[26] Kannenberg, E. L.; Perzl, M.; Härtner, T. The occurrence of hopanoid lipids in *Bradyrhizobium* bacteria. *FEMS Microbiology Letters*, **1995**, 127, 255–262.

[27] Kannenberg, E.; Poralla, K. Hopanoid biosynthesis and function in bacteria. *Naturwissenschaften*, **1999**, 86, 168–176. doi: 10.1007/s001140050592.

[28] Thimmappa, R.; Geisler, K.; Louveau, T.; O'Maille, P.; Osbourn, A. Triterpene Biosynthesis in plants. *Annual Review of Plant Biology*, **2014**, 65, 225–257.

[29] Luchnikova, N. A.; Grishko, V. V.; Ivshina, I. B. Biotransformation of oleanane and ursane triterpenic acids. *Molecules*, **2020**, 25(23), 5526. doi: 10.3390/molecules25235526.

[30] Jesus, J. A.; Lago, J. H.; Laurenti, M. D.; Yamamoto, E. S.; Passero, L. F. Antimicrobial activity of oleanolic and ursolic acids: An update. *Evidence-Based Complementary and Alternative Medicine*, **2015**, 620472. doi: 10.1155/2015/620472.

[31] Mus, A. A.; Goh, L. P. W.; Marbawi, H.; Gansau, J. A. The biosynthesis and medicinal properties of taraxerol. *Biomedicines*, **2022**, 10, 807. doi: 10.3390/biomedicines10040807.

[32] Qiao, W.; Zhou, Z.; Liang, Q., et al. Improving lupeol production in yeast by recruiting pathway genes from different organisms. *Scientific Reports*, **2019**, 9, 2992. doi: 10.1038/s41598-019-39497-4.

[33] Jin, C. C.; Zhang, J. L.; Song, H., et al. Boosting the biosynthesis of betulinic acid and related triterpenoids in *Yarrowia lipolytica* via multimodular metabolic engineering. *Microbial Cell Factories*, **2019**, 18, 77. doi: 10.1186/s12934-019-1127-8.

[34] Betu Kasangana, P.; Stevanovic, T. Chapter 1: Studies of pentacyclic triterpenoids structures and antidiabetic properties of *Myrianthus* genus. *Studies in Natural Products Chemistry*, **2021**, 68, 1–27.

[35] Alves, T. B.; Souza-Moreira, T. M.; Valentini, S. R.; Zanelli, C. F.; Furlan, M. Friedelin in *Maytenus ilicifolia* is produced by friedelin synthase isoforms. *Molecules*, **2018**, 23, 700. doi: 10.3390/molecules23030700.

[36] Hansen, N. L.; Miettinen, K.; Zhao, Y., et al. Integrating pathway elucidation with yeast engineering to produce polpunonic acid the precursor of the anti-obesity agent celastrol. *Microbial Cell Factories*, **2020**, 19, 15. doi: 10.1186/s12934-020-1284-9.

[37] Lu, C.-Z.; Zhang, C.; Zhao, F.; Li, D.; Lu, W. Biosynthesis of ursolic acid and oleanolic acid in *Saccharomyces cerevisiae*. *AIChE Journal*, **2018**. doi: 10.1002/aic.16370.

[38] Ohyama, S. H.; Sawai, K. S.; Mizutani, M.; Ohnishi, T.; Sudo, H.; Akashi, T.; Aoki, T.; Saito, K.; Muranaka, T. Licorice β-amyrin 11-oxidase, a cytochrome P450 with a key role in the biosynthesis of the triterpene sweetener glycyrrhizin. *Proceedings of the National Academy of Sciences of the United States of America*, **2008**, 105, 14204–14209.

[39] Bishayee, A.; Ahmed, S.; Brankov, N.; Perloff, M. Triterpenoids as potential agents for the chemoprevention and therapy of breast cancer. *Frontiers in Bioscience*, **2011**, 16(3), 980–996. doi: 10.2741/3730.

[40] Setzer, W. N.; Setzer, M. C. Plant-derived triterpenoids as potential antineoplastic agents. *Mini-Reviews in Medicinal Chemistry*, **2003**, 3, 540–556.

[41] Smith-Kielland, I.; Domish, J. M.; Malterud, K. E.; Hvistendahl, G.; Rømming, C.; Bøckman, O. C.; Kolsaker, P.; Stenstrøm, Y; Nordal, A. Cytotoxic triterpenoids from the leaves of *Euphorbia pulcherrima*. *Planta Medica*, **1996**, 62, 322–325.

[42] Dzubak, P.; Hajduch, M.; Vydra, D.; Hustova, A.; Kvasnica, M.; Biedermann, D.; Sarek, J. Pharmacological activities of natural triterpenoids and their therapeutic implications. *Natural Product Reports*, **2006**, 23(3), 394. doi: 10.1039/b515312n.

[43] Petronelli, A.; Pannitteri, G.; Testa, U. Triterpenoids as new promising anticancer drugs. *Anti-Cancer Drugs*, **2009**, 20(10), 880–892. doi: 10.1097/cad.0b013e328330fd90.

[44] Zhang, W.; Men, X.; Lei, P. Review on anti-tumor effect of triterpene acid compounds. *The Journal of Cancer Research and Therapeutics*, **2014**, 10(Suppl S1), 14–19.

[45] Saleem, M. Lupeol, a novel anti-inflammatory and anti-cancer dietary triterpene. *Cancer Letters*, **2009**, 285(2), 109–115. doi: 10.1016/j.canlet.2009.04.033.

[46] Nagini, S. Neem limonoids as anticancer agents: Modulation of cancer hallmarks and oncogenic signaling. *Enzymes*, **2014**, 36, 131–147. doi: 10.1016/B978-0-12-802215-3.00007-0.

[47] Jing, L.; Zhang, Y. M.; Luo, J. G.; Kong, L. Y. Tirucallane-type triterpenoids from the fruit of *Ficus carica* and their cytotoxic activity. *Chemical and Pharmaceutical Bulletin*, **2015**, 63(3), 237–243. doi: 10.1248/cpb.c14-00779.

[48] Hassan, A. R.; Ashour, A.; Amen, Y.; Nagata, M.; El-Toumy, S. A.; Shimizu, K. A new cycloartane triterpene and other phytoconstituents from the aerial parts of *Euphorbia dendroides*. *Natural Product Research*, **2020**, 1–9. doi: 10.1080/14786419.2020.1800693.

[49] Zhao, M.; Gödecke, T.; Gunn, J.; Duan, J.-A.; Che, C.-T. Protostane and fusidane triterpenes: A mini-review. *Molecules*, **2013**, 18, 4054–4080. doi: 10.3390/molecules18044054.

[50] Wang, C.; Feng, L.; Ma, L.; Chen, H.; Tan, X.; Hou, X.; Song, J.; Cui, L.; Liu, D.; Chen, J.; et al. Alisol A 24-acetate and alisol B 23-acetate induced autophagy mediates apoptosis and nephrotoxicity in human renal proximal tubular cells. *Frontiers in Pharmacology*, **2017**, 8, 172.

[51] Li, H.-M.; Liu, D.; Dai, W.-F.; Chen, X.-Q.; Li, R.-T. A new protostane-type triterpenoid from *Alisma plantago-aquatica* subsp. orientale (Sam.) Sam. *Natural Product Research*, **2018**, 1–6. doi: 10.1080/14786419.2018.1519710.

[52] Shanmugam, M. K.; Dai, X.; Kumar, A. P.; Tan, B. K.; Sethi, G.; Bishayee, A. Ursolic acid in cancer prevention and treatment: molecular targets, pharmacokinetics and clinical studies. *Biochemical Pharmacology*, **2013**, 85(11), 1579–1587.

[53] Yeh, C. T.; Wu, C. H.; Yen, G. C. Ursolic acid, a naturally occurring triterpenoid, suppresses migration and invasion of human breast cancer cells by modulating c-Jun N-terminal kinase, Akt and mammalian target of rapamycin signalling. *Molecular Nutrition & Food Research*, **2010**, 54(9), 1285–1295.

[54] Shishodia, S.; Majumdar, S.; Banerjee, S.; Aggarwal, B. B. Ursolic acid inhibits nuclear factor-κB activation induced by carcinogenic agents through suppression of IκBα kinase and p65 phosphorylation: correlation with down-regulation of cyclooxygenase 2, matrix metalloproteinase 9, and cyclin D1. *Cancer Research*, **2003**, 63(15), 4375–4383.

[55] Pisha, E.; Chai, H.; Lee, I. S.; Chagwedera, T. E.; Farnsworth, N. R.; Cordell, G. A.; Beecher, C. W. W.; Fong, H. H. S.; Kinghorn, A. D.; Brown, D. M.; Wani, M. C.; Wall, M. E.; Hieken, T. J.; Das Gupta, T. K.; Pezzuto, J. M. Discovery of betulinic acid as a selective inhibitor of human melanoma that functions by induction of apoptosis. *Nature Medicine*, **1995**, 1, 1046–1051.

[56] Zhang, D. M.; Xu, H. G.; Wang, L.; Li, Y. J.; Sun, P. H.; Wu, X. M.; Wang, G. J.; Chen, W. M.; Ye, W. C. Betulinic acid and its derivatives as potential antitumor agents. *Medicinal Research Reviews*, **2015**, 35, 1127–1155.

[57] Kim, D. S. H. L; Pezzuto, J. M.; Pisha, E. Synthesis of betulinic acid derivatives with activity against human melanoma. *Bioorganic & Medicinal Chemistry Letters*, **1998**, 8, 1707–**1712**.

[58] Yue, Q. X.; Cao, Z. W.; Guan, S. H.; Liu, X. H.; Tao, L.; Wu, W. Y., et al. Proteomics characterization of the cytotoxicity mechanism of ganoderic acid D and computer-automated estimation of the possible drug target network. *Molecular & Cellular Proteomics*, **2008**, 7, 949–961.

[59] Chang, U. M.; Li, C. H.; Lin, L. I.; Huang, C. P.; Kan, L. S.; Lin, S. B. Ganoderiol F, a ganoderma triterpene, induces senescence in hepatoma HepG2 cells. *Life Sciences*, **2006**, 79, 1129–1139.

[60] Rios, J. L. Chemical constituents and pharmacological properties of *Poria cocos*. *Planta Medica*, **2011**, 77, 681–691.

[61] Kang, H. M.; Lee, S. K.; Shin, D. S.; Lee, M. Y.; Han, D. C.; Baek, N. I., et al. Dehydrotrametenolic acid selectively inhibits the growth of H-ras transformed rat2 cells and induces apoptosis through caspase-3 pathway. *Life Sciences*, **2006**, 78, 607–613.

[62] Sun, P.; Liu, B. S.; Yi, Y. H.; Li, L.; Gui, M.; Tang, H. F., et al. A new cytotoxic lanostane-type triterpene glycoside from the sea cucumber *Holothuria impatiens*. *Chemistry & Biodiversity*, **2007**, 4, 450–457.

[63] Chen, Y. G.; Hai, L. N.; Liao, X. R.; Qin, G. W.; Xie, Y. Y.; Halaweish, F. Ananosic acids B and C, two new 18 (13->12)-abeo-lanostane triterpenoids from *Kadsura ananosma*. *Journal of Natural Products*, **2004**, 67, 875–877.

[64] Wang, N.; Li, Z.; Song, D.; Li, W.; Fu, H.; Koike, K., et al. Lanostane-type triterpenoids from the roots of *Kadsura coccinea*. *Journal of Natural Products*, **2008**, 71, 990–994.

[65] Yoshikawa, K.; Kouso, K.; Takahashi, J.; Matsuda, A.; Okazoe, M.; Umeyama, A., et al. Cytotoxic constituents of the fruit body of *Daedalea dickisii*. *Journal of Natural Products*, **2005**, 68, 911–914.

[66] Lei, P.; Zhang, W.; Men, X. Review on anti-tumor effect of triterpene acid compounds. *Journal of Cancer Research and Therapeutics*, **2014**, 10(5), 14. doi: 10.4103/0973-1482.139746.

[67] Nakata, T.; Yamada, T.; Taji, S.; Ohishi, H.; Wada, S.; Tokuda, H., et al. Structure determination of inonotsuoxides A and B and in vivo anti-tumor promoting activity of inotodiol from the sclerotia of Inonotus obliquus. *Bioorganic & Medicinal Chemistry*, **2007**, 15, 257–264.

[68] Nomura, M.; Takahashi, T.; Uesugi, A.; Tanaka, R.; Kobayashi, S. Inotodiol, a lanostane triterpenoid, from Inonotus obliquus inhibits cell proliferation through caspase-3-dependent apoptosis. *Anticancer Research*, **2008**, 28, 2691–2696.

[69] Deeb, D.; Gao, X.; Liu, Y.; Kim, S. H.; Pindolia, K. R.; Arbab, A. S., et al. Inhibition of cell proliferation and induction of apoptosis by oleanane triterpenoid (CDDO-Me) in pancreatic cancer cells is associated with the suppression of hTERT gene expression and its telomerase activity. *Biochemical and Biophysical Research Communications*, **2012**, 422, 561–567.

[70] Niu, Y. P.; Wu, L. M.; Jiang, Y. L.; Wang, W. X.; Li, L. D. Beta-escin, a natural triterpenoid saponin from Chinese horse chestnut seeds, depresses HL-60 human leukaemia cell proliferation and induces apoptosis. *Journal of Pharmacy and Pharmacology*, **2008**, 60, 1213–1220.

[71] Sun, L. R.; Qing, C.; Zhang, Y. L.; Jia, S. Y.; Li, Z. R.; Pei, S. J., et al. Cimicifoetisides A and B, two cytotoxic cycloartane triterpenoid glycosides from the rhizomes of *Cimicifuga foetida*, inhibit proliferation of cancer cells. *Beilstein Journal of Organic Chemistry*, **2007**, 3, 3.

[72] Roy, M. K.; Kobori, M.; Takenaka, M.; Nakahara, K; Shinmoto, H.; Tsushida, T. Inhibition of colon cancer (HT-29) cell proliferation by a triterpenoid isolated from *Azadirachta indica* is accompanied by cell cycle arrest and up-regulation of p21. *Planta Medica*, **2006**, 72, 917–923.

[73] Li, W.; Liu, J.; Fu, W., et al. 3-O-acetyl-11-keto-β-boswellic acid exerts anti-tumor effects in glioblastoma by arresting cell cycle at G2/M phase. *Journal of Experimental & Clinical Cancer Research*, **2018**, 37(1), 132. doi: 10.1186/s13046-018-0805-4.

[74] Hsu, Y. L.; Kuo, P. L.; Lin, L. T.; Lin, C. C. Asiatic acid, a triterpene, induces apoptosis and cell cycle arrest through activation of extracellular signal-regulated kinase and p38 mitogen-activated protein kinase pathways in human breast cancer cells. *Journal of Pharmacology and Experimental Therapeutics*, **2005**, 313, 333–344.

[75] Lee, Y. S.; Jin, D. Q.; Kwon, E. J.; Park, S. H.; Lee, E. S.; Jeong, T. C., et al. Asiatic acid, a triterpene, induces apoptosis through intracellular Ca^{2+} release and enhanced expression of p53 in HepG2 human hepatoma cells. *Cancer Letters*, **2002**, 186, 83–91.

[76] Tran Van, L.; Vo Thi, Q. N.; Tran Van, C., et al. Synthesis of asiatic acid derivatives and their cytotoxic activity. *Medicinal Chemistry Research*, **2018**, 27, 1609–1623. doi: 10.1007/s00044-018-2176-y.

[77] Zhu, B.; Wei, Y. Antitumor activity of celastrol by inhibition of proliferation, invasion, and migration in cholangiocarcinoma via PTEN/phosphatidylinositol 3-kinase (PI3K)/Akt pathway. *Cancer Medicine*, **2019**, 9(2), 783–796. doi: 10.1002/cam4.2719.

[78] Farooqui, A.; Khan, F.; Khan, I.; Ansari, I. A. Glycyrrhizin induces reactive oxygen species-dependent apoptosis and cell cycle arrest at G0/G1 in HPV18+ human cervical cancer HeLa cell line. *Biomedicine & Pharmacotherapy*, **2018**, 97, 752–764.

[79] Jube, S.; Rivera, Z. S.; Bianchi, M. E., et al., Cancer cell secretion of the DAMP protein HMGB1 supports progression in malignant mesothelioma. *Cancer Research*, **2012**, 72(13), 3290–3301.

[80] Pozzilli, P.; Pieralice, S. Latent autoimmune diabetes in adults: Current status and new horizons. *Endocrinology and Metabolism*, **2018**, 33(2): 147–159.

[81] Nazaruk, J.; Borzym-Kluczyk, M. The role of triterpenes in the management of diabetes mellitus and its complications. *Phytochemistry Reviews: Proceedings of the Phytochemical Society of Europe*, **2015**, 14(4), 675–690. doi: 10.1007/s11101-014-9369-x.

[82] Pushpanjali, G., et al. Effect of lupeol on enzymatic and non-enzymatic antioxidants in type-2 diabetic adult male Wistar rats. *Drug Invention Today*, **2019**, 12, 5.

[83] Gupta, R., et al. Evaluation of antidiabetic and antioxidant potential of lupeol in experimental hyperglycaemia. *Natural Product Research*, **2012**, 26(12), 1125–1129.

[84] Taylor, R. Insulin resistance and type 2 diabetes. *Diabetes*, **2012**, 61, 778–779.

[85] Hanley, A. J.; Williams, K.; Gonzalez, C.; D'Agostino, R. B.; Wagenknecht, L. E.; Stern, M. P.; Haffner, S. M. Prediction of type 2 diabetes using simple measures of insulin resistance. *Diabetes*, **2003**, 52, 463–469.

[86] Li, M.; Han, Z.; Bei, W.; Rong, X.; Guo, J.; Hu, X. Oleanolic acid attenuates insulin resistance via NF-κB to regulate the IRS1-GLUT4 pathway in HepG2 cells. *Evidence-Based Complementary and Alternative Medicine*, **2015**, 643102.

[87] Li, Y.; Wang, J.; Gu, T.; Yamahara, J.; Li, Y. Oleanolic acid supplement attenuates liquid fructose-induced adipose tissue insulin resistance through the insulin receptor substrate-1/phosphatidylinositol 3-kinase/Akt signaling pathway in rats. *Toxicology and Applied Pharmacology*, **2014**, 277, 155–163.

[88] Wang, X.; Li, Y. L.; Wu, H.; Liu, J. Z.; Hu, J. X.; Liao, N.; Peng, J.; Cao, P. P.; Liang, X.; Hai, C. X. Antidiabetic effect of oleanolic acid: A promising use of a traditional pharmacological agent. *Phytotherapy Research*, **2011**, 25(7), 1031–1040. doi: 10.1002/ptr.3385.

[89] Jun, Z. M.; Xing, W. Y.; Jing, J. Z.; Xia, L.; Li, L. D.; Xiao, L. S.; Gang, X. Sterols and terpenoids from *Viburnum odoratissimum*. *Natural Products and Bioprospecting*, **2014**, 4, 175–180.

[90] Zeng, X. Y.; Wang, Y. P.; Cantley, J.; Iseli, T. J.; Molero, J. C.; Hegarty, B. D.; Kraegen, E. W.; Ye, Y.; Ye, M. Oleanolic acid reduces hyperglycemia beyond treatment period with Akt/FoxO1-induced suppression of hepatic gluconeogenesis in type-2 diabetic mice. *PLoS One*, **2012**, 7, 1–12.

[91] Jose, M. C.; Angeles, G.; Teresa, D.; Mirela, R.; Jose, A. C. Biochemical basis of the antidiabetic activity of oleanolic acid and related pentacyclic triterpenes. *Diabetes*, **2013**, 62, 1791–1799.

[92] Wu, P. P.; Zhang, B. J.; Cui, X. P. Synthesis and biological evaluation of novel ursolic acid analogues as potential α-glucosidase inhibitors. *Scientific Reports*, **2017**, 7, 45578. doi: 10.1038/srep45578.

[93] Kiplimo, J. J.; Koorbanally, N. A.; Chenia, H. Triterpenoids from *Vernonia auriculifera* Hiern exhibit antimicrobial activity. *African Journal of Pharmacy and Pharmacology*, **2011**, 5(8), 1150–1156.

[94] Abdel-Raouf, N.; Al-Enazi, N. M.; Al-Homaidan, A. A.; Ibraheem, I. B. M.; Al-Othman, M. R.; Hatamleh, A. A. Antibacterial β-amyrin isolated from *Laurencia microcladia*. *Arabian Journal of Chemistry*, **2015**, 8(1), 32–37. doi: 10.1016/j.arabjc.2013.09.033.

[95] Mutai, C.; Bii, C.; Rukunga, G.; Ondicho, J.; Mwitari, P.; Abatis, D.; Vagias, C.; Roussis, V.; Kirui, J. Antimicrobial activity of pentacyclic triterpenes isolated from *Acacia mellifera*. *African Journal of Traditional, Complementary, and Alternative Medicines*, **2009**, 6(1), 42–48.

[96] Ríos, J. L.; Recio, M. C.; Maáñez, S.; Giner, R. M. Natural triterpenoids as anti-inflammatory agents. *Studies in Natural Products Chemistry*, **2000**, 93–143. doi: 10.1016/s1572-5995(00)80024-1.

[97] Jayatilake, G. S.; Freeberg, D. R.; Liu, Z.; Richheimer, S. L.; Blake Nieto, M. E. Isolation and structures of avicins D and G: In vitro tumor-inhibitory saponins derived from *Acacia victoriae*. *Journal of Natural Products*, **2003**, 66, 779–783.

[98] Haridas, V.; Arntzen, C. J.; Gutterman, J. U. Avicins, a family of triterpenoid saponins from *Acacia victoriae* (Bentham), inhibit activation of nuclear factor kappa B by inhibiting both its nuclear localization and ability to bind DNA. *Proceedings of the National Academy of Sciences of the United States of America*, **2001**, 98, 11557–11562.

[99] Peng, D.; Wang, H.; Qu, C., et al. Ginsenoside Re: Its chemistry, metabolism and pharmacokinetics. *Chinese Medicine*, **2012**, 7, 2. doi: 10.1186/1749-8546-7-2.

[100] Lu, J. M.; Yao, Q.; Chen, C. Ginseng compounds: An update on their molecular mechanisms and medical applications. *Current Vascular Pharmacology*, **2009**, 7(3), 293–302.

[101] Hofseth, L. J.; Wargovich, M. J. Inflammation, cancer, and targets of ginseng. *Journal of Nutrition*, **2007**, 137(1 Suppl), 183S–185S.

[102] Patlolla, J. M. R.; Rao, C. V. Anti-inflammatory and Anticancer Properties of β-Escin, a *Triterpene* saponin. *Current Pharmacology Reports*, **2015**, 1, 170–178. doi: 10.1007/s40495-015-0019-9.

[103] Arnould, T.; Janssens, D.; Michiels, C.; Remacle, J. Effect of aescine on hypoxia induced activation of human endothelial cells. *European Journal of Pharmacology*, **1996**, 315, 227–233.

[104] Wang, T.; Fu, F.; Zhang, L.; Han, B.; Zhu, M.; Zhang, X. Effects of escin on acute inflammation and the immune system in mice. *Pharmacological Reports*, **2009**, 61, 487–494.

[105] Asl, M. N.; Hosseinzadeh, H. Review of pharmacological effects of *Glycyrrhiza* sp. and its bioactive compounds. *Phytotherapy Research*, **2008**, 22, 709–724.

[106] Yoshida, T.; Abe, K.; Ikeda, T.; Matsushita, T.; Wake, K.; Sato, T.; Sato, T.; Inoue, H. Inhibitory effect of glycyrrhizin on lipopolysaccharide and D-galactosamine-induced mouse liver injury. *European Journal of Pharmacology*, **2007**, 576, 136–142.

[107] Wang, C. Y.; Kao, T. C.; Lo, W. H.; Yen, G. C. Glycyrrhizic acid and 18β-glycyrrhetinic acid modulate lipopolysaccharide-induced inflammatoryresponse by suppression of NF-κB through PI3K p110δ and p110γ inhibitions. *Journal of Agricultural and Food Chemistry*, **2011**, 59, 7726–7733.

[108] Wu, X.; Zhang, L.; Gurley, E.; Studer, E.; Shang, J.; Wang, T.; Wang, C.; Yan, M.; Jiang, Z.; Hylemon, P. B.; Sanyal, A. J.; Pandak, W. M., Jr.; Zhou, H. Prevention of free fatty acid-induced hepatic lipotoxicity by 18β-glycyrrhetinic acid through lysosomal and mitochondrial pathways. *Hepatology*, **2008**, 47, 1905–1915.

[109] Saleem, M.; Afaq, F.; Adhami, V. M.; Mukhtar, H. Lupeol modulates NF-κB and PI3K/Akt pathways and inhibits skin cancer in CD-1 mice. *Oncogene*, **2004**, 23, 5203–5214.

[110] Lee, T. K.; Poon, R. T. P.; Wo, J. Y.; Ma, S.; Guan, X. Y.; Myers, J. N.; Altevogt, P.; Yuen, A. P. W. Lupeol suppresses cisplatin induced nuclear factor-κB activation in head and neck squamous cell carcinoma and inhibits local invasion and nodal metastasis in an orthotopic nude mouse model. *Cancer Research*, **2007**, 67, 8800–8809.

[111] Fernandez, M. A.; de las Heras, B.; Garcia, M. D.; Saenz, M. T.; Villar, A. New insights into the mechanism of action of the anti-inflammatory triterpene lupeol. *Journal of Pharmacy and Pharmacology*, **2001**, 53, 1533–1539.

[112] Ayeleso, T. B.; Matumba, M. G.; Mukwevho, E. Oleanolic acid and its derivatives: Biological activities and therapeutic potential in chronic diseases. *Molecules*, **2017**, 22(11), 1915. doi: 10.3390/molecules22111915.

[113] Lee, W.; Yang, E. J.; Ku, S. K.; Song, K. S.; Bae, J. S. Anti-inflammatory effects of oleanolic acid on LPS-induced inflammation in vitro and in vivo. *Inflammation*, **2013**, 36, 94–102. doi: 10.1007/s10753-012-9523-9.

[114] Gupta, M. B.; Bhalla, T. N.; Gupta, G. P.; Mitra, C. R.; Bhargava, K. P. Antiinflammatory activity of natural products (I) triterpenoids. *European Journal of Pharmacology*, **1969**, 6, 67–70.

[115] Suh, N.; Honda, T.; Finlay, H. J.; Barchowsky, A. Novel triterpenoids suppress inducible nitric oxide synthase (iNOS) and inducible cyclooxygenase (COX-2) in mouse macrophages. *Cancer Research*, **1998**, 58, 717–723.

[116] Subbaramaiah, K.; Michaluart, P.; Sporn, M. B.; Dannenberg, A. J. Ursolic acid inhibits cyclooxygenase-2 transcription in human mammary epithelial cells. *Cancer Research*, **2000**, 60, 2399–2404.

[117] Kim, Y. P.; Lee, E. B.; Kim, S. Y.; Li, D.; Ban, H. S.; Lim, S. S.; Shin, K. H.; Ohuchi, K. Inhibition of prostaglandin E2 production by platycodin D isolated from the root of *Platycodon grandiflorum*. *Planta Medica*, **2001**, 67, 362–364.

[118] Ahn, K. S.; Noh, E. J.; Zhao, H. L.; Jung, S. H.; Kang, S. S.; Kim, Y. S. Inhibition of inducible nitric oxide synthase and cyclooxygenase II by *Platycodon grandiflorum* saponins *via* suppression of nuclear factor-κB activation in RAW 264.7 cells. *Life Sciences*, **2005**, 76, 2315–2328.

[119] Bermejo-Benito, P.; Abad-Martínez, M. J.; Silván-Sen, A. M. In vivo and in vitro anti-inflammatory activity of saikosaponins. *Life Sciences*, **1998**, 63, 1147–1156.

[120] Connolly, J. D.; Hill, R. A. Triterpenoids. *Natural Product Reports*, **2010**, 27, 79–132.

[121] Karin, M.; Cao, Y.; Greten, F. R.; Li, Z. W. NF-kappa B in cancer: from innocent by stander to major culprit. *Nature Reviews Cancer*, **2002**, 2, 301–310.

[122] Hayden, M. S.; Ghosh, S. Signaling to NF-kappaB. *Genes Development*, **2004**, 18, 2195–2224.

[123] Sharma, M. L.; Bani, S.; Singh, G. B. Anti-arthritic activity of boswellic acids in bovine serum albumin (BSA)-induced arthritis. *International Journal of Immunopharmacology*, **1989**, 11, 647–652.

[124] Shahidi, F. Antioxidants. In: *Handbook of Antioxidants for Food Preservation*, **2015**, 1–14. doi: 10.1016/b978-1-78242-089-7.00001-4.

[125] Krishnaiah, D.; Sarbatly, R.; Bono, A. Phytochemical antioxidants for health and medicine a move towards nature. *Biotechnology and Molecular Biology Reviews*, **2007**, 2, 97–104.

[126] Choudhary, A.; Mittal, A. K.; Radhika, M.; Tripathy, D.; Chatterjee, A.; Banerjee, U. C.; Singh, I. P. Two new stereoisomeric antioxidant triterpenes from *Potentilla fulgens*. *Fitoterapia*, **2013**, 91, 290–297. doi: 10.1016/j.fitote.2013.09.008.

[127] Liu, C.-H.; Yen, M.-H.; Tsang, S.-F.; Gan, K.-H.; Hsu, H.-Y.; Lin, C.-N. Antioxidant triterpenoids from the stems of *Momordica charantia*. *Food Chemistry*, **2010**, 118(3), 751–756.

[128] Valdeira, A. S. C.; Ritt, D. A.; Morrison, D. K.; McMahon, J. B.; Gustafson, K. R.; Salvador, J. A. R. Synthesis and biological evaluation of new madecassic acid derivatives targeting ERK cascade signaling. *Frontiers in Chemistry*, **2018**, 6, 1–20. doi: 10.3389/fchem.2018.00434.

[129] Xu, G. B.; Xiao, Y. H.; Zhang, Q. Y.; Zhou, M.; Liao, S. G. Hepatoprotective natural triterpenoids. *European Journal of Medicinal Chemistry*, **2018**, 145, 691–716. doi: 10.1016/j.ejmech.2018.01.011.

[130] Wang, Y.; Zhang, C. L.; Liu, Y. F.; Liang, D.; Luo, H.; Hao, Z. Y.; Chen, R. Y.; Yu, D. Q. Hepatoprotective triterpenoids and saponins of *Schefflera kwangsiensis*. *Planta Medica*, **2014**, 80(2–3), 215–222. doi: 10.1055/s-0033-1360179.

[131] Abdullah Khan, M. A.; Ahmad, W.; Ahmad, M.; Adhikari, A.; Ibrar, M.; Asif, M. Antioxidant, antinociceptive, anti-inflammatory, and hepatoprotective activities of pentacyclic triterpenes isolated from *Ziziphus oxyphylla* Edgew. *Drug and Chemical Toxicology*, **2021**, 1–12.

[132] Mapanga, R. F.; Rajamani, U.; Dlamini, N.; Zungu-Edmondson, M.; Kelly-Laubscher, R.; Shafiullah, M.; Wahab, A.; Hasan, M. Y.; Fahim, M. A.; Rondeau, P.; Bourdon, E.; Essop, M. F. Oleanolic acid: A novel cardioprotective agent that blunts hyperglycemia-induced contractile dysfunction. *PLoS One*, **2012**, 7(10), e47322. doi: 10.1371/journal.pone.0047322.

[133] Sudharsan, P. T.; Mythili, Y.; Selvakumar, E.; Varalakshmi, P. Cardioprotective effect of pentacyclic triterpene, lupeol and its ester on cyclophosphamide-induced oxidative stress. *Human & Experimental Toxicology*, **2005**, 24(6), 313–318. doi: 10.1191/0960327105ht530oa.

[134] Xia, A.; Xue, Z.; Li, Y., et al. Cardioprotective effect of betulinic acid on myocardial ischemia reperfusion injury in rats. *Evidence-Based Complementary and Alternative Medicine*, **2014**, 2014, 573745. doi: 10.1155/2014/573745.

[135] Paintsil, E.; Cheng, Y. C. Antiviral agents. *Encyclopedia of Microbiology*. **2009**, 223–257. doi: 10.1016/B978-012373944-5.00178-4.

[136] Wu, J. J.; Pang, K. R.; Huang, D. B. Advances in antiviral therapy. *Dermatologic Clinics*. **2005**, 23, 313–322.

[137] Xiao, S.; Tian, Z.; Wang, Y.; Si, L.; Zhang, L; Zhou, D. Recent progress in the antiviral activity and mechanism study of pentacyclic triterpenoids and their derivatives. *Medical Research Review*, **2018**, 38(3), 951–976. doi: 10.1002/med.21484.

[138] Wei, Y.; Ma, C. M.; Hattori, M. Synthesis of dammarane-type triterpene derivatives and their ability to inhibit HIV and HCV proteases. *Bioorganic & Medicinal Chemistry*, **2009**, 17(8), 3003–3010. doi: 10.1016/j.bmc.2009.03.019.

[139] Darshani, P.; Sen Sarma, S.; Srivastava, A. K., et al. Antiviral triterpenes: A review. *Phytochemistry Reviews*, **2022**. doi: 10.1007/s11101-022-09808-1.

[140] Yang, J. L.; Ha, T. K. Q.; Dhodary, B., et al. Oleanane triterpenes from the flowers of *Camellia japonica* inhibit porcine epidemic diarrhea virus (PEDV) replication. *Journal of Medicinal Chemistry*, **2015**, 58, 1268–1280.

[141] Zhu, Q.; Bang, T. H.; Ohnuki, K., et al. Inhibition of neuraminidase by *Ganoderma triterpenoids* and implications for neuraminidase inhibitor design. *Scientific Reports*, **2015**, 5, 1–9.

[142] Wu, H.; Ma, G.; Yang, Q., et al. Discovery and synthesis of novel beesioside I derivatives with potent anti-HIV activity. *European Journal of Medicinal Chemistry*, **2019**, 166, 159–166.

[143] Graebin, C. S. The pharmacological activities of glycyrrhizinic acid ("glycyrrhizin") and glycyrrhetinic acid. *Sweeteners*, **2017**, 245–261. doi: 10.1007/978-3-319-27027-2_15.

[144] Roy, N. K.; Parama, D.; Banik, K.; Bordoloi, D.; Devi, A. K.; Thakur, K. K.; Padmavathi, G.; Shakibaei, M.; Fan, L.; Sethi, G.; Kunnumakkara, A. B. An update on pharmacological potential of boswellic acids against chronic diseases. *International Journal of Molecular Sciences*, **2019**, 20(17), 4101. doi: 10.3390/ijms20174101.

[145] Gupta, I.; Gupta, V.; Parihar, A.; Gupta, S.; Ludtke, R.; Safayhi, H.; Ammon, H. P. Effects of *Boswellia serrata* gum resin in patients with bronchial asthma: Results of a double blind, placebo-controlled, 6-week clinical study. *European Journal of Medical Research*, **1998**, 3, 511–514.

[146] Syrovets, T.; Buchele, B.; Krauss, C.; Laumonnier, Y.; Simmet, T. Acetyl-boswellic acids inhibit lipopolysaccharide-mediated TNF-α induction in monocytes by direct interaction with IkB kinases. *Journal of Immunology*, **2005**, 174, 498–506.

[147] Lee, J. H.; Koo, T. H.; Yoon, H. Inhibition of NF-κB activation through targeting I κB kinase by celastrol, a quinine methide triterpenoid. *Biochemical Pharmacology*, **2006**, 72, 1311–1321.

[148] Pinna, G. F.; Fiorucci, M.; Reimund, J. M.; Taquet, N.; Arondel, Y.; Muller, C. D. Celastrol inhibits pro-inflammatory cytokine secretion in Crohn's disease biopsies. *Biochemical and Biophysical Research Communications*, **2004**, 322(3), 778–786.

[149] Wu, J.; Ding, M.; Mao, N.; Wu, Y.; Wang, C.; Yuan, J.; Miao, X.; Li, J.; Shi, Z. Celastrol inhibits chondrosarcoma proliferation, migration and invasion through suppression CIP2A/c-MYC signaling pathway. *Journal of Pharmacological Sciences*, **2017**, 134(1), 22–28.

[150] Kim, J. E.; Lee, M. H.; Nam, D. H.; Song, H. K.; Kang, Y. S.; Lee, J. E.; Kim, H. W.; Cha, J. J.; Hyun, Y. Y.; Han, S. Y.; Han, K. H.; Han, J. Y.; Cha, D. R. Celastrol, an NF-kappaB inhibitor, improves insulin resistance and attenuates renal injury in db/db mice. *PLoS One*, **2013**, 8(4), e62068.

[151] Sethi, G.; Ahn, K. S.; Pandey, M. K.; Aggarwal, B. B. Celastrol, a novel triterpene, potentiates TNF-induced apoptosis and suppresses invasion of tumor cells by inhibiting NF-κB-regulated gene products and TAK1-mediated NF-κB activation. *Blood*, **2007**, 109, 2727–2735.

[152] Liu, J.; Lee, J.; Salazar Hernandez, M. A.; Mazitschek, R.; Ozcan, U. Treatment of obesity with celastrol. *Cell*, **2015**, 161(5), 999–1011.

[153] Chow, A. M.; Tang, D. W.; Hanif, A.; Brown, I. R. Induction of heat shock proteins in cerebral cortical cultures by celastrol. *Cell Stress and Chaperones*, **2013**, 18(2), 155–160.

[154] Guo, Z.; Vangapandu, S.; Sindelar, R. W.; Walker, L. A.; Sindelar, R. D. Biologically active quassinoids and their chemistry: Potential leads for drug design. *Current Medicinal Chemistry*, **2005**, 12(2), 173–190. doi: 10.2174/0929867053363351.

[155] Kawada, K.; Kim, M.; Watt, D. S. Synthesis of quassinoids. A review. *Organic Preparations and Procedures International*, **1989**, 21(5), 521–618. doi: 10.1080/00304948909356425.

[156] Zhao, L.; Chen, X. J.; Zhu, J.; Xi, Y. B.; Yang, X.; Hu, L. D., et al. Lanosterol reverses protein aggregation in cataracts. *Nature*, **2015**, 523, 607–611.

[157] Shanmugam, P. M.; Barigali, A.; Kadaskar, J., et al. Effect of lanosterol on human cataract nucleus. *Indian Journal of Ophthalmology*, **2015**, 63(12), 888–890. doi: 10.4103/0301-4738.176040.

10

Synthesis and Medicinal Uses of Sesquiterpenoids and Sesterpenoids

Biswa Mohan Sahoo*, **Bimal Krishna Banik†**, **and Abhishek Tiwari‡**

CONTENTS

10.1 Introduction 331
    10.1.1 Classification of Sesquiterpenoids 332
    10.1.2 Biosynthesis of Sesquiterpene 332
    10.1.3 Chemistry and Medicinal Uses of Sesquiterpenoids 332
        10.1.3.1 Acyclic Sesquiterpenoid 332
        10.1.3.2 Monocyclic Sesquiterpenoid 334
        10.1.3.3 Bicyclic Sesquiterpenoid 336
        10.1.3.4 Tricyclic Sesquiterpenoid 339
        10.1.3.5 Miscellaneous 339
    10.1.4 Chemistry and Medicinal Uses of Sesquiterpenoid Lactones 339
        10.1.4.1 Artemisinin 340
        10.1.4.2 Eudesmanolide 340
        10.1.4.3 Germacranolides 340
        10.1.4.4 Guaianolides 342
        10.1.4.5 Tenulin 343
    10.1.5 Chemistry and Medicinal Uses of Sesquiterpenols 343
    10.1.6 Chemistry and Medicinal Uses of Sesterpenes 345
10.2 Conclusion 355
Abbreviation 355
References 356

10.1 Introduction

Sesquiterpenes (C_{15} terpenoids) are the class of terpenes which consist of three isoprene units. Plant families, known to be principal producers of sesquiterpene volatiles, include Lamiaceae, Geraniaceae, Rutaceae, Myrtaceae, Asteraceae, Cannabaceae and Gingeraceae. A number of highly functionalized, nonvolatile sesquiterpenes of plant origin have demonstrated highly specific biological activity and have therefore been investigated for their clinical potential. Examples of sesquiterpene include α-Zingiberene (*Zingiber officinalis*), α-Farnesene (*Eucalyptus globules*) [1–3].

Another class of compounds contains various characteristic features such as α-methylene γ-lactone system, α, β-unsaturated carbonyls and epoxides. These are chemically distinct from the sesquiterpenoids and are collectively known as sesquiterpenoid lactones (Figure 10.1). The specific and vital nucleophiles present in the enzyme system such as thiol and amino moieties help to augment faster and reactive approach toward receptor sites by these sesquiterpenoid lactones. Thus, these compounds exhibit marked and pronounced biological activities such as anti-inflammatory, antimicrobial, antioxidant, antitumor and antimalarial. Examples of sesquiterpene lactone include Tenulin (*Helenium amarum*), Thapsigargin (*Thapsia garganica*), Artemisinin (*Artemisia annua*), Germacranolide (*Geranium macrorrhizum*), Eudesmanolide (*Magnolia obovata*), Matricarin (*Achillea vermacularis*) and Guaianolide (*Guaiacum officinale*) [4].

Sesquiterpenols possess fifteen carbon atoms (C_{15}) in their structure and exhibit diverse therapeutic effects such as antifungal, antiviral, antinociceptive, insecticidal, analgesic and anti-inflammatory. Sesquiterpenols are derived from Sesquiterpenes by addition of hydroxyl (-OH) group. Examples:

* Roland Institute of Pharmaceutical Sciences, Berhampur affiliated to Biju Patnaik University of Technology (BPUT), Rourkela, Odisha, India. Corresponding author e-mail address: drbiswamohansahoo@gmail.com.

† Department of Mathematics and Natural Sciences, College of Sciences and Human Studies, Prince Mohammad Bin Fahd University, Al Khobar, Kingdom of Saudi Arabia. Corresponding author e-mail address: bimalbanik10@gmail.com.

‡ Faculty of Pharmacy, Pharmacy Academy, IFTM University, Lodhipur Rajput, Moradabad-244102, Uttar Pradesh, India.

DOI: 10.1201/9781003008682-10

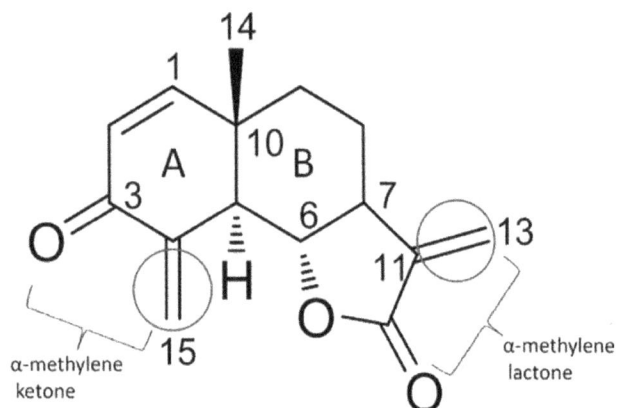

FIGURE 10.1 Structure of sesquiterpenoid lactones.

TABLE 10.1

Classification of Sesquiterpenoids

Types of Sesquiterpenes	Name of Sesquiterpenes
Acyclic	α-Farnesene
	β-Farnesene
Monocyclic	Abscisic acid
	α-Bisabolene
	α-Curcumene
	Humulene
	α-Zingiberene
Bicyclic	α-Bergamotene
	β-caryophyllene
	Cadinene
	Germacrene-A
	Guaiazulene
Tricyclic	Santalene

Farnesol (*Santalum spicatum*), α-Bisabolol (*Matricaria recutita*), Patchoulol (*patchouli cablin*) and Nerolidol (*Melaleuca quinquenervia*) [5].

Sesterpenes (C_{25} terpenoids) are the class of terpenes which consist of five isoprene units with molecular formula $C_{25}H_{40}$. Sesterpenes and their derivatives are collectively known as sesterterpenoids. These are ubiquitous secondary metabolites in fungi, marine organisms and plants. Their structural diversity includes carbotricyclic ophiobolanes, polycyclic anthracenones, polycyclic furan-2-ones and polycyclic hydroquinones. Examples of sesterterpenoids include Merochlorin-A (*Streptomyces sp.*), Ophiobolin-A (*Drechslera gigantean*), Secoemestrin-D (*Emericella sp.*), Manoalide (*Luffariella variabilis*), Luffariellolide (*Luffariella sp.*), Hippolide-A (*Hippospongia lachne*), Palauolol (*Thorectandra sp.*), Kohamaic acid-A (*Ircinia spp.*) and Petrosaspongiolides (*Petrosaspongia nigra*). Furthermore, most of them possess promising biological activities such as antibacterial, antifungal, anti-inflammatory, antioxidant and cytotoxicity [6].

10.1.1 Classification of Sesquiterpenoids

Sesquiterpenoids are categorized into four classes such as acyclic, monocyclic, bicyclic and tricyclic, as summarized in Table 10.1. Examples include α-Farnesene (*Eucalyptus globules*), α-bisabolene in black pepper (*Piper nigrum*) and β-caryophyllene in ylang ylang (*Cananga odorata*) and Santalene (*Santalum album*) (Table 10.2) [7,8].

10.1.2 Biosynthesis of Sesquiterpene

Sesquiterpenes are derived from farnesyl diphosphate (FDP) and can be cyclized to produce diverse chemical structures. The reaction of geranyl pyrophosphate with isopentenyl pyrophosphate results in the 15-carbon farnesyl pyrophosphate (FPP) (Figure 10.2), which is an intermediate in the biosynthesis of sesquiterpenes such as farnesene [9].

Cyclic sesquiterpenes are more common than cyclic monoterpenes because of the increased chain length and additional double bond in the sesquiterpene precursors. In addition to

common six-membered ring systems such as found in zingiberene (a constituent of the oil from ginger), cyclization of one end of the chain to the other end can lead to macrocyclic rings such as humulene. Cadinenes contain two fused six-membered rings, whereas caryophyllene contains nine-membered ring fused to a cyclobutane ring. Vetivazulene and guaiazulene are aromatic bicyclic sesquiterpenoids. With the addition of a third ring, the possible structures become increasingly varied. Examples include longifolene, copaene and the alcohol patchoulol. Sesquiterpene biosynthesis normally begins in the cytosol via the C_{15} prenyl diphosphate FDP, which itself is derived from mevalonic acid and the MVA pathway. Sesquiterpene synthases act on FDP to yield different types of sesquiterpene hydrocarbons [10].

10.1.3 Chemistry and Medicinal Uses of Sesquiterpenoids

10.1.3.1 Acyclic Sesquiterpenoid

10.1.3.1.1 Farnesene

Farnesene is an acyclic sesquiterpene olefin, which occurs in a wide range of both plant (*Eucalyptus globulus*) and animal. There are two isomers of Farnesene such as α-Farnesene and β-farnesene, which differ by the location of one double bond. Chemically, α-Farnesene is 3,7,11-trimethyl-1,3,6,10-dodecatetraene, whereas β-farnesene is 7,11-dimethyl-3-methylene-1,6,10-dodecatriene. The alpha form can exist as four stereoisomers that differ about the geometry of two of its three internal double bonds (the stereoisomers of the third internal double bond are identical). The beta isomer exists as two stereoisomers about the geometry of its central double bond. (E)-β-farnesene is the major component of pollen odor in *Lupinus* and stimulates pollination behavior in bumblebees. It also mimics the action of juvenile hormone III in some insects. The substrate for (E)-β-farnesene synthase is farnesyl diphosphate (FDP), which is a ubiquitous isoprenoid intermediate involved in cytoplasmic phytosterol biosynthesis (Figure 10.3). The ionization of the enzyme-bound nerolidyl diphosphate intermediate and proton elimination can also produce (E)-β-farnesene [11].

TABLE 10.2

List of Sesquiterpenoids and Their Biological Properties

Sesquiterpenes	Biological Source	Structure	Biological Properties
α-Zingiberene	*Zingiber officinalis*		Antioxidant, increases glutathione production and reduces inflammation
α-Farnesene	*Eucalyptus globulus*		Effective against cariogenic bacteria that cause tooth decay
α-Bisabolene	*Piper nigrum*		Acts as pheromones in different insects
β-Caryophyllene	*Cananga odorata*		Anti-toothache
Humulene	*Humulus lupulus*		Anti-inflammatory
α-Bergamotene	*Citrus bergamia*		Anti-inflammatory
α-Curcumene	*Curcuma longa*		Anti-inflammatory
δ-Cadinene	*Juniperuscommunis*		Antioxidant
Germacrene-A	*Lamium purpureum*		Antimicrobial and insecticidal
Guaiazulene	*Oil of guaiac and chamomile oil*		Component of skin care products
Abscisic acid	*Lemon, Avocado*		Plant growth hormone
Santalene	*Sandalwood oil*		Flavor and fragrance agent

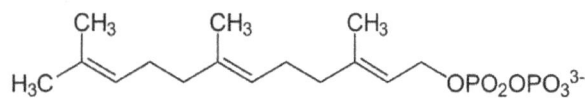

FIGURE 10.2 Farnesyl pyrophosphate.

10.1.3.2 Monocyclic Sesquiterpenoid

10.1.3.2.1 Abscisic Acid

Abscisic acid (ABA) is an isoprenoid plant growth hormone and structurally related to sesquiterpenes. There are four forms of ABA such as (+) and (−) or (S) and (R) (Figure 10.4). Among these, S-(+) enantimorph occurs naturally in plants. But the racemic mixture of ABA is available commercially. The mirror image pairs in the mixture form a more stable crystal structure with a higher melting point of 190°C as compared to natural ABA with melting point of 161°C.

Germination of seeds and growth of excised embryos are inhibited by ABA. In nongrowing organs, the function of ABA has been reported to inhibit synthesis of protein and nucleic acids, so as to induce abscission and senescence. ABA enhances the tolerance level of plants to various kinds of stresses caused by environmental or biotic factors. It also plays a major role in development and maturation of seeds.

ABA is synthesized in the plastidal 2-C-methyl-D-erythritol-4-phosphate (MEP) pathway. It is formed from the mevalonic acid-derived precursor farnesyl diphosphate (FDP), the C_{15} backbone of ABA [12]. FDP is the first sesquiterpenoid (C_{15}) compound, and it was reported that it produces directly ABA after oxidation and cyclization (Figure 10.5) [12].

10.1.3.2.2 α-Bisabolene

Bisabolenes are group of closely related natural chemical compounds, which are classified as sesquiterpenes. There are three isomers of bisabolene such as α-, β- and γ-bisabolene, which differ by the positions of the double bonds (Figure 10.6). Bisabolenes are produced from farnesyl pyrophosphate (FPP) and are present in the essential oils of bisabol. These compounds act as pheromones in different insects such as stink bugs [13].

10.1.3.2.3 α-Curcumene

α-Curcumene is a sesquiterpene. Chemically, it is 2-methyl-6-p-tolyl-2-heptene in which one of the hydrogens at position

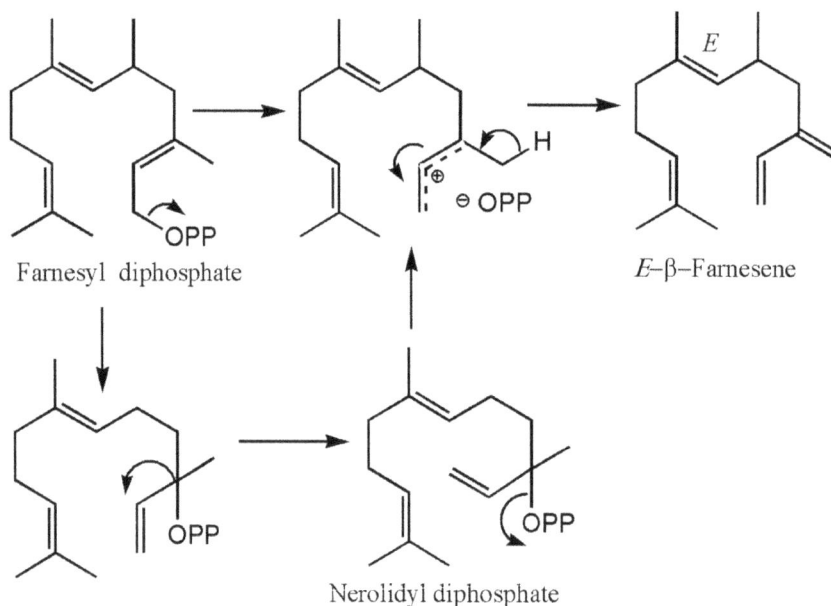

FIGURE 10.3 Proposed mechanism for the formation of (E)-β-farnesene from FDP.

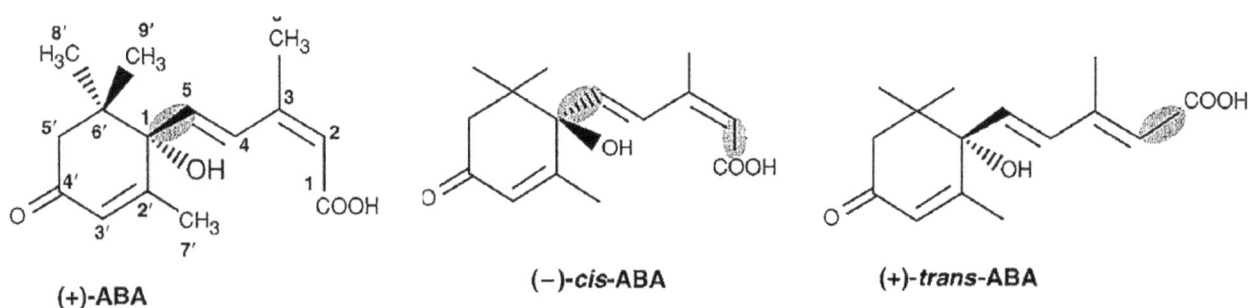

FIGURE 10.4 Structure of abscisic acid and its isomers.

Method-I

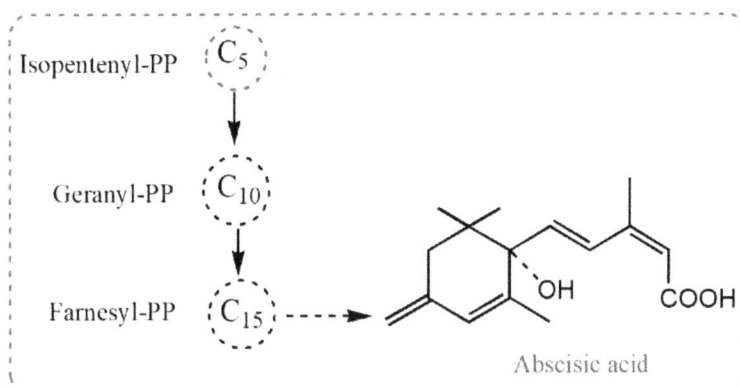

Method-II

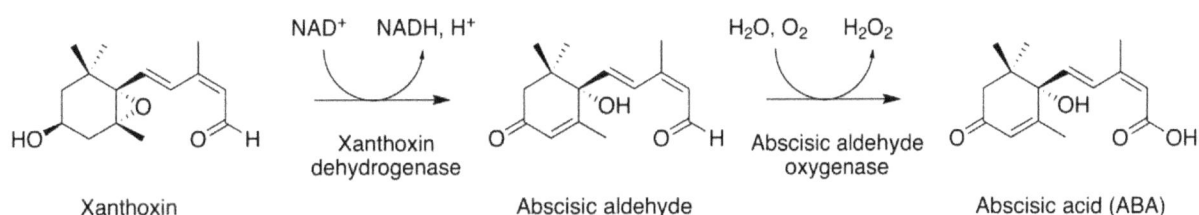

FIGURE 10.5 Synthesis of abscisic acid via Method-I & II.

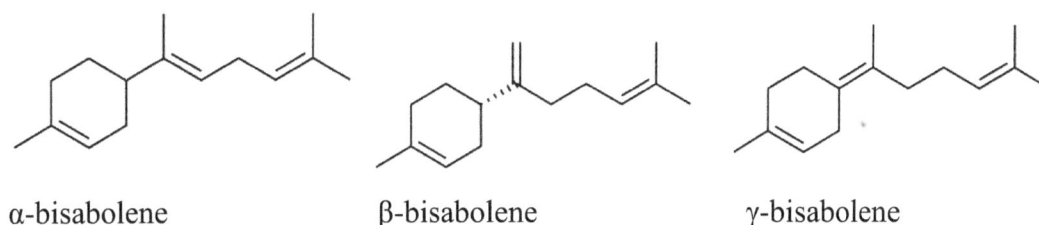

α-bisabolene β-bisabolene γ-bisabolene

FIGURE 10.6 Chemical structure of bisabolene isomers.

6 is substituted by p-tolyl group. It exhibits anti-inflammatory activity. The mode of action may be due to inhibition of the nitric oxide synthase derived NO and cyclooxygenase-2 derived prostaglandin-E_2 (PGE_2) production along with reduction of the LPS-induced nuclear translocations and trans activities of NFκB.

Silva et al. reported antimicrobial evaluation of sesquiterpene α-curcumene and its synergism with Imipenem. α-Curcumene was isolated from the fresh aerial parts of *Senecio selloi* Spreng. DC. Its biological activity was investigated against bacteria, yeasts and an alga by using microdilution method. The strongest effect was manifested against *Saccharomyces cerevisiae* with estimated values of MIC and MFC 0.8 mg/mL. The α-curcumene synergism in the concentrations of 1mM and 5mM, respectively, with selected antibiotics such as ciprofloxacin, imipenem, ceftazidim and a combination of amoxicillin and clavulanic acid was evaluated against *Staphylococcus aureus, Escherichia coli, Enterobacter*

cloacae, Staphylococcus haemolyticus and *Klebsiella pneumoniae* by the disk diffusion assay (Tables 10.3 and 10.4). The results represented that the occurrence of synergism of α-curcumene with imipenem against the clinical isolate *E. cloacae* with a significance level of $p > 0.05$ (Table 10.5) [14].

10.1.3.2.4 Humulene

Humulene is also known as α-humulene or α-caryophyllene. It is a naturally occurring monocyclic sesquiterpene with

TABLE 10.3

MIC Values of the α-Curcumene against Bacteria

Microorganisms		MIC (mg/mL)	MBC (mg/mL)
Gram-positive bacteria	*S. aureus*	13	13
	B. subtilis	13	13
Gram-negative bacteria	*E. coli*	13	13
	P. aeruginosa	13	13

molecular formula $C_{15}H_{24}$. It possesses 11-membered ring and consists of three isoprene units containing three nonconjugated C=C double bonds, two of them being triply substituted and one being doubly substituted. Chemically, it is 2,6,6,9-tetramethyl-1,4–8-cycloundecatriene. It was first found in the essential oils of *Humulus lupulus* (hops). Humulene is the ring-opened isomer of β-caryophyllene. Humulene was studied for potential anti-inflammatory effects. It can produce ROS that leads to an antineoplastic effect by inducing apoptosis. Interestingly, Humulene was found to increase secretion of IL-8, a chemokine with various functions, including promoting angiogenesis that leads to wound healing [15].

Humulene is derived from farnesyl diphosphate (FPP). The formation of humulene from FPP is catalyzed by sesquiterpene synthesis enzymes. Humulene can also be synthesized

TABLE 10.4

MIC Values of the α-Curcumene against Yeasts and an Algae

Microorganisms		MIC (mg/mL)	MFC (mg/mL)
Yeast	*C. albicans*	6.4	6.4
	S. cerevisiae	0.8	0.8
	C. glabrata	>6.4	ND
Algae	*P. zopfii*	>6.4	ND

ND, not determined.

TABLE 10.5

Evaluation of the Synergism of α-Curcumene with Antibiotics by the Disk Diffusion Against *E. cloacae* (A Clinical Isolate)

Antibiotics	Diameter of Zone of Inhibition (mm)		
	Control	1 mM	5 mM
Imipenem	24[a]	24[a]	27[b]
Ciprofloxacin	R	R	R
Ceftazidime	R	R	R
Combination of amoxicillin and clavulanic acid	R	R	R

Different letters superscripts are significantly different ($p < 0.05$).
R, resistant.

by using a combination of four-component assembly and palladium-mediated cyclization as summarized in Figure 10.7) [16,17].

10.1.3.2.5 Zingiberene

Essential oil of ginger (*Zingiber officinale*) contains active constituents such as terpenes including zingiberene, zingiberol and zingiberone. Zingiberene is a monocyclic sesquiterpene, the predominant constituent of the oil of ginger *Zingiber officinale*. Chemically, Zingiberene is 2-Methyl-5-(6-methylhept-5-en-2-yl) cyclohexa-1,3-diene. It is formed in the isoprenoid pathway from farnesyl pyrophosphate (FPP). FPP undergoes a rearrangement to give nerolidyl diphosphate. After the removal of pyrophosphate, the ring closes leaving a carbocation on the tertiary carbon attached to the ring. Then, 1,3-hydride shift takes place to produce more stable allylic carbocation. The final step in the formation of zingiberene is the removal of the cyclic allylic proton and consequent formation of a double bond. Zingiberene synthase is the enzyme responsible for catalyzing the reaction forming zingiberene (Figure 10.8).

It inhibits the generation of β-amyloid proteins and their aggregates that lead to decrease in the formation of the plaques in Alzheimer's disease. It also acts as carminative to reduce pain and discomfort caused by stomach gas. Similarly, Gingerol down-regulates the anti-apoptotic protein Bcl-2 and enhances the pro-apoptotic protein Bax, while gingerdione is an effective antitumor agent in human leukemia cells by inducing G1 arrest via down-regulation of cyclin D2, cyclin E and cdc25A and the up-regulation of CDK1 and p15. Gingerdione also decreases Bcl-2 accumulation and activates caspase-3 cleavage [18].

10.1.3.3 Bicyclic Sesquiterpenoid

10.1.3.3.1 α-Bergamotene

Bergamotene is a sesquiterpene. It is a constituent of oils of carrot (*Daucus carota*), bergamot (*Citrus bergamia*), lime (*Citrus aurantifolia*), citron (*Citrus medica*) and cottonseed oil (*Gossypium hirsutum*). It consists of bicyclo[3.1.1]hept-2-ene skeleton substituted at positions 2 and 6 by methyl groups

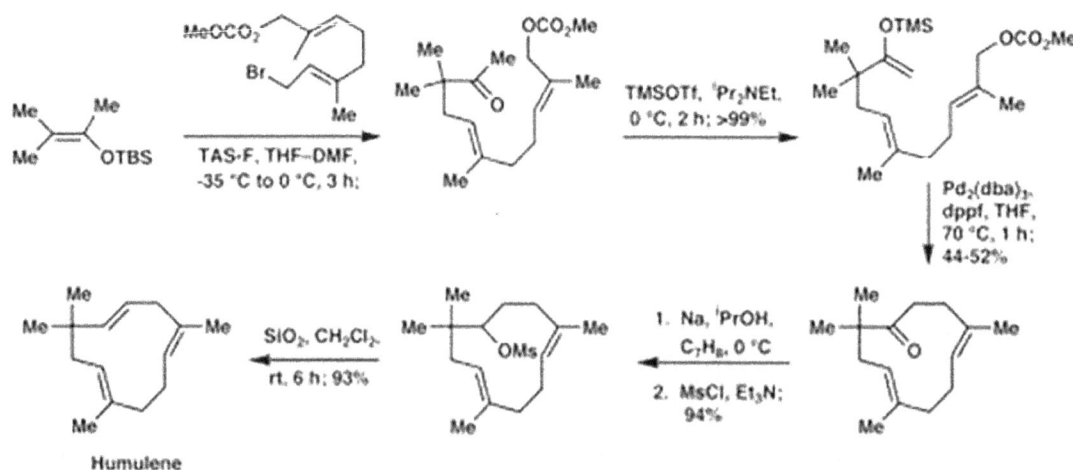

FIGURE 10.7 Synthetic route of humulene.

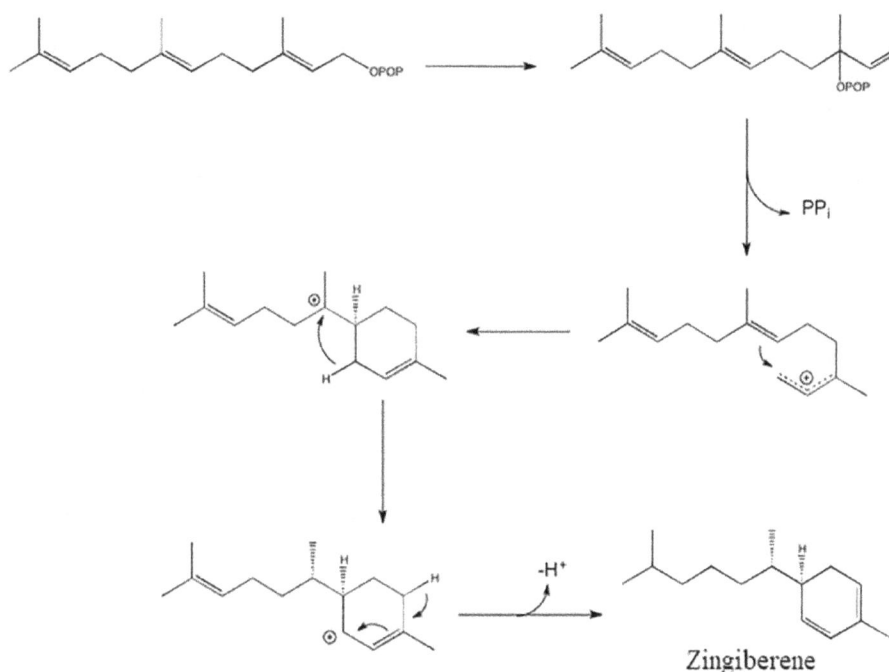

FIGURE 10.8 Biosynthesis of zingiberene.

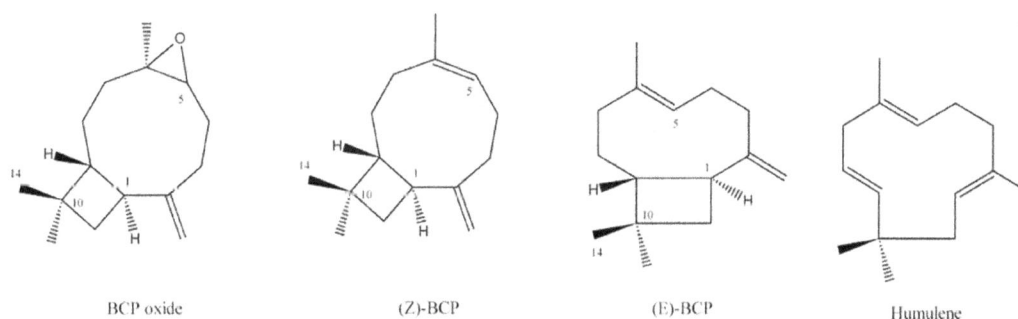

FIGURE 10.9 Structures of β-caryophyllene and its oxide.

and at position 6 by a 4-methylpent-3-en-1-yl group. There are two structural isomers, α-bergamotene and β-bergamotene, which differ only by the position of double bond. Both of these isomers have stereoisomers, the most common of which are known as the cis and trans-isomers (or endo- and exo-isomers). α-Bergamotene is present in the essential oils of bergamot, lime and cottonseed. Bergamotenes are biosynthesized from farnesyl pyrophosphate via a variety of enzymes such as exo-alpha-bergamotene synthase, (+)-endo-beta-bergamotene synthase and (–)-endo-alpha-bergamotene synthase. *β*-trans-Bergamotene acts as a precursor for biosynthesis of fumagillin, ovalicin and related antibiotics. It possesses strong anti-inflammatory property [19,20].

10.1.3.3.2 β-caryophyllene

β-Caryophyllene (BCP) and β-caryophyllene oxide (BCPO) are the natural bicyclic sesquiterpenes (Figure 10.9), present in essential oil of medicinal plants such as guava (*Psidium guajava*) and oregano (*Origanum vulgare* L.). Both BCP and BCPO possess significant anticancer activities which affect

the growth and proliferation of several cancer cells. BCP is a phytocannabinoid with strong affinity toward cannabinoid receptor type 2 (CB$_2$), but not cannabinoid receptor type 1 (CB$_1$). So, the selective activation of CB$_2$ may be considered as novel strategy for treatment of pain. Similarly, BCPO alters several key pathways of cancer development, such as mitogen-activated protein kinase (MAPK), PI3K/AKT/mTOR/S6K1 and STAT3 pathways. On the other hand, BCP isomer, α-humulene, exhibited significant antiproliferative activities against cancer cells (Table 10.6) [21].

10.1.3.3.3 δ-Cadinene

δ-Cadinene is a bicyclic sesquiterpene. Chemically, it is 4-isopropyl-1,6-dimethyl-decahydronaphthalene (Figure 10.10). Due to the presence of double bond and stereo centers, cadinene is categorized into four classes such as α, β, γ and δ-cadinene [22].

The formation of δ-cadinene involves a sequence of reaction in which a preliminary isomerization step (to nerolidyl diphosphate) is required to permit the ionization-dependent cyclization to the macrocycle, followed by 1,3-hydride shift, closure of the second ring and deprotonation to the bicyclic product. A small amount of δ-cadinene is produced by the recombinant synthase from FDP (Figure 10.11). It is interesting in light of the abundance of this bicyclic olefin in the sesquiterpene fraction of peppermint oil and the efficient production of this olefin in oil gland extracts. These observations suggest that an additional and distinct δ-cadinene synthase must operate in peppermint [23].

Kundu et al. repotted the antioxidant potential of essential oil and cadinene sesquiterpenes of *Eupatorium adenophorum.*

TABLE 10.6

Anticancer Activities (IC$_{50}$) of BCPO and BCP

Sesquiterpenes	Concentration (µg/mL)	Cell Lines
BCPO	0.87	HepG2
(*Psidium cattleianum*)	2.98	HeLa
	2.77	AGS
	3.69	SNU-1
	6.03	SNU-16
BCP	3.9	HCT 116
(*Aquilaria crassna*)	5.5	PANC-1
	12.9	HT-29
	19.4	ME-180
	21.3	PC3
	21.5	K562

The antioxidant activities of the essential oil and cadinenes were evaluated by using 2,2-diphenyl-1-picrylhydrazyl (DPPH) and the ferric reducing ability assay (FRAP), together with three antioxidant standards, i.e. ascorbic acid, tert-butyl-4-hydroxy toluene (BHT) and gallic acid. The antioxidant activity of essential oil was compared with the standards. The essential oil and the cadinene sesquiterpene-rich extract of *E. adenophorum* have exhibited potential antioxidant activities [24].

10.1.3.3.4 Germacrene-A

Germacrenes are a class of volatile organic hydrocarbons, specifically, sesquiterpenes It is formed from farnesyl diphosphate (FDP) by germacrene-A synthase (GAS). Chemically, it is (1E, 5E, 8S)-1,5-dimethyl-8-(prop-1-en-2-yl)cyclodeca-1,5-diene. Germacrene has five isomers such as Germacrene-A, B, C, D and E (Figure 10.12). These compounds possess antimicrobial and insecticidal properties [25].

10.1.3.3.5 Guaiazulene

Guaiazulene is a bicyclic sesquiterpene. It is the constituent of essential oils of guaiac and chamomile oil. It is the derivative of azulene and is chemically known as 1,4-dimethyl-7-isopropylazulene. It shows antioxidant/pro-oxidant activity by inhibiting lipid peroxidation and effectively scavenges hydroxyl radicals and DPPH [26].

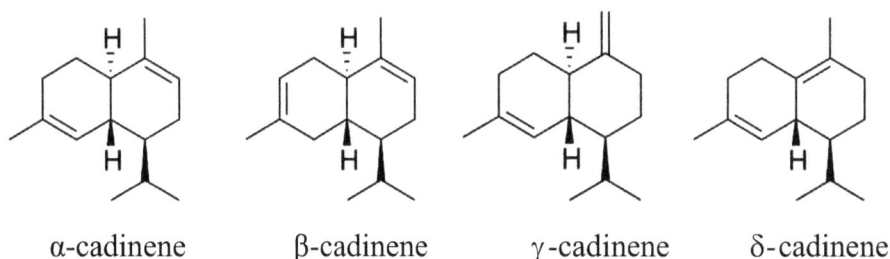

α-cadinene   β-cadinene   γ-cadinene   δ-cadinene

FIGURE 10.10   Structures of δ-cadinene.

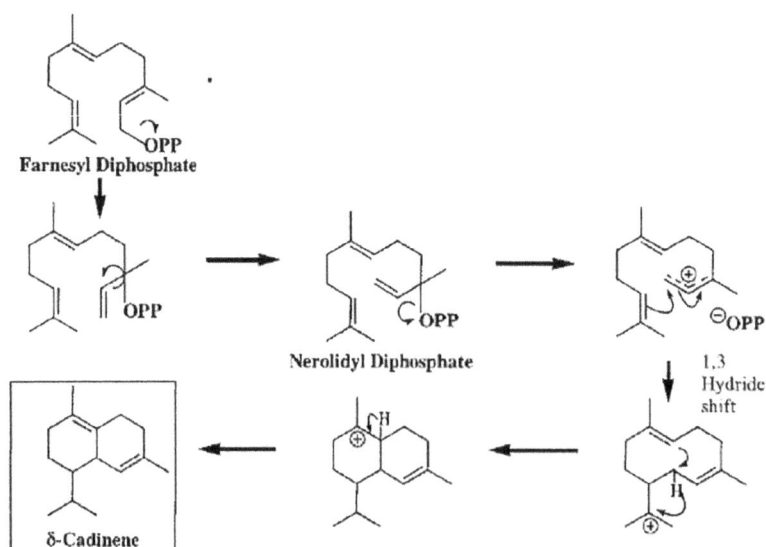

FIGURE 10.11   Proposed mechanism for the formation of δ-cadinene from FDP.

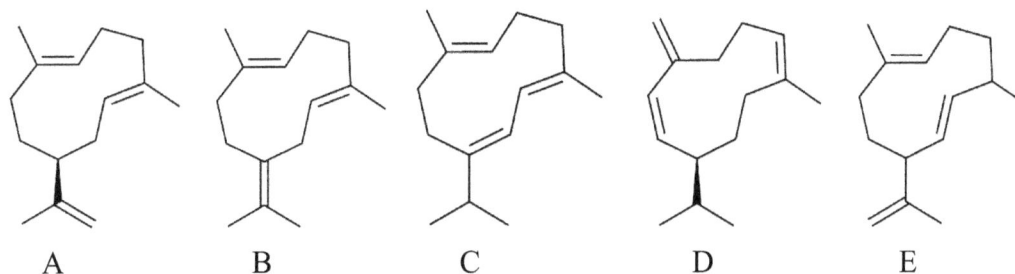

FIGURE 10.12 Structures of isomers of germacrene.

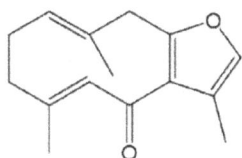

FIGURE 10.13 Structure of furanodiene.

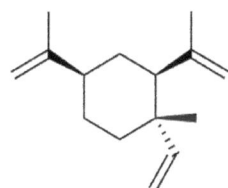

FIGURE 10.14 Structure of β-elemene.

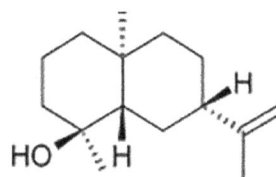

Isointermedeol

FIGURE 10.15 Structure of isointermedeol.

10.1.3.4 Tricyclic Sesquiterpenoid

10.1.3.4.1 Santalene

Santalene is the constituent of sandalwood oil (*Santalum album*). Chemically, it is 1,7-dimethyl-7-(4-methyl-3-pentenyl)-tricyclo[2.2.1.0(2,6)]heptanes. Santalene is present in two iso-form such as α- and β-santalene. Santalene sesquiterpene is the precursor of santalol which is an important perfumery ingredient with sweet-woody and balsamic odor. Alpha-santalene synthase is an enzyme with systematic name (2E, 6E)-farnesyl diphosphate lyase (cyclizing, (+)-α -santalene-forming). This enzyme catalyzes the following chemical reaction [27,28].

(2E, 6E)-farnesyl diphosphate ↔ (+)-α-santalene + diphosphate

10.1.3.5 Miscellaneous

Furanodiene, a sesquiterpene extracted from the essential oil of the rhizome of *Curcuma wenyujin* which inhibits the growth of uterine cervical (U14) tumors in mice (Figure 10.13). The rhizome and the aerial part of *Aristolochia mollissima* exhibit significant cytotoxic activity in human cervix carcinoma cell line HeLa [29].

Edris reported that *β*-elemene (Figure 10.14), a sesquiterpene from *Nigella sativa*, could inhibit the growth of laryngeal cancer cells by activating caspase-3 cleavage and decreasing the accumulation of eukaryotic initiation factors eIF-4E and 4G, basic fibroblast growth factor (bFGF) and vascular epithelial growth factor (VEGF) [30].

Singh et al. investigated the essential oil from a lemon grass variety of *Cymbopogon flexuosus* (CFO) and its major chemical constituent sesquiterpene isointermedeol (Figure 10.15) for their ability to induce apoptosis in human leukemia HL-60 cells. CFO and ISO inhibited the cell proliferation with IC_{50} of ~30 and 20 g/mL, respectively in 48 h. Both CFO and ISO also activated the apical death receptors TNFR1, DR4 and caspase-8 activity. Simultaneously, both increased the expression of mitochondrial cytochrome-c protein with its concomitant release to cytosol leading to caspase-9 activation. It was suggested that both CFO and ISO are involved in intrinsic and extrinsic pathways of apoptosis [31].

10.1.4 Chemistry and Medicinal Uses of Sesquiterpenoid Lactones

Sesquiterpene lactones (SLs) are a class of sesquiterpenoids which contain a lactone ring. These are colorless, lipophilic compounds. Sesquiterpene lactones can be divided into several classes such as germacranolides, heliangolides, guaianolides, pseudoguaianolides, hypocretenolides and eudesmanolides. SLs found in plants of the family Asteraceae, Umbelliferae and Magnoliaciae. Sesquiterpene lactones isolated from plants of Asteraceae family play a major role in human health [32]. These are used as pharmaceutical agents due to their potential for the treatment of cardiovascular disorders and cancer. Various mechanisms of actions are proposed for the reduction of inflammation and tumorigenesis at therapeutic levels in humans [33,34]. Sesquiterpene lactones have been found to sensitize tumor cells in comparison to conventional drug treatments. Some of the sesquiterpenoid lactones are antimicrobial which disrupts the cell wall of fungi and invasive bacteria (Table 10.7).

TABLE 10.7

List of Sesquiterpenoid Lactones and Their Properties

Sesquiterpenoid Lactones	Biological Source	Biological Properties
Artemisinin	*Artemisia annua*	Antimalarial
Eudesmanolide	*Magnolia obovata*	Anti-inflammatory
Germacranolides	*Geranium macrorrhizum*	Leishmanicidal activity
Guaianolides	Guaiacum officinale	Treatment of rheumatism and cancer
Parthenolide	*Tanacetum parthenium*	Antipyretic agent
Tenulin	*Helenium amarum*	Anticancer
Matricarin	*Achillea vermacularis*	Anti-inflammatory

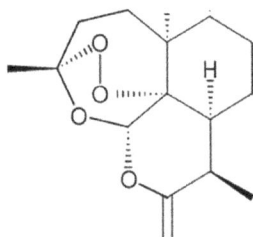

FIGURE 10.16   Structure of artemisinin.

10.1.4.1 Artemisinin

Artemisinin is a sesquiterpene lactone with an endoperoxide ring (Figure 10.16) which imparts its antimalarial properties [35]. Its mechanism of action is based on the degradation of the endoperoxide bridge in a heme-dependent process to form carbon-centered radicals which then alkylate multiple targets including heme and proteins. The involvement of heme plays a key role to improve the specificity of artemisinin toward *Plasmodium* species. The cause of toxicity to the parasite is not clear but it may cause interference with the conversion of heme to hemozoin [36–39].

The early stages of artemisinin biosynthesis in glandular trichomes of *Artemisia annua* of family Asteraceae were observed at the biochemical and molecular levels. The active principle artemisinin was isolated and identified in 1972. Currently, efforts are made to focus on the regulation of this pathway to improve the production of artemisinin. The pathway begins with the conversion of Farnesyl diphosphate (FDP) to amorpha-4,11-diene by amorpha-4,11-diene synthase (ADS). Biochemical characterization and metabolite profiling of plant tissue indicated the presence of this olefin as well as its oxygenated derivatives artemisinic alcohol and aldehyde, dihydroartemisinic alcohol and aldehyde and dihydroartemisinic acid. Following the cloning and expression of Cytochrome P450 71AV1 (CYP71AV1), which converts amorpha-4,11-diene to artemisinic alcohol and aldehydes [11–13]. Zhang et al. characterized the artemisinic aldehyde Δ11(13) reductase (Double bond reductase 2, DBR2) which mainly reduces artemisinic aldehyde to dihydroartemisinic aldehyde. This substrate is then converted to dihydroartemisinic acid by aldehyde dehydrogenase-1 (Aldh-1). From dihydroartemisinic acid, the remaining steps involve the photooxidative formation of the endoperoxide ring (Figure 10.17) [40–45].

10.1.4.2 Eudesmanolide

Eudesmanolide is isolated from *Magnolia obovata*. It is chemically known as 3-oxo-5αH, 8βH-eudesma-1,4(15), 7(11)-trien-8,12-olide. It was observed to reduce inflammation (Figure 10.18).

10.1.4.2.1 Santamarine

Santamarine (STM) is a sesquiterpene lactone (Figure 10.19) and component of *Magnolia grandiflora* and *Ambrosia confertiflora*. It was found to possess antimicrobial, antifungal, antibacterial, anti-inflammatory and anticancer activities. STM inhibits growth and induces apoptosis in A549 lung adenocarcinoma cells through the induction of oxidative stress. It also induces oxidative stress by promoting reactive oxygen species (ROS) generation, depleting intracellular glutathione (GSH) and inhibiting thioredoxin reductase (TrxR) activity in a dose-dependent manner. Further, the mechanistic study demonstrated that STM induces apoptosis by modulation of Bax/Bcl-2 expressions, disruption of mitochondrial membrane potential, activation of caspase-3 and cleavage of PARP in a dose-dependent manner. In addition to this activity, STM inhibited the constitutive and inducible translocation of NF-κBp65 into the nucleus. IKK-16 (I-κB kinase inhibitor) enhanced the STM-induced apoptosis, indicating that STM induces apoptosis in A549 cells at least in part through NF-κB inhibition [46].

10.1.4.2.2 Paludolactone

Paludolactone is a type of eudesmanolide lactone (Figure 10.20), obtained from *Wedelia Paludosa* Dc. (*Acmela Brasiliensis*). The analgesic effect of paludolactone was observed by using writhing test. It was administered through the intraperitoneal route. It was found more potent than some well-known analgesic and anti-inflammatory drugs [47].

10.1.4.3 Germacranolides

Germacranolides are a group of natural chemical compounds classified as sesquiterpene lactones (Figure 10.21). They are found in a variety of plant sources and isolated from *Geranium macrorrhizum*. Muschietti et al. reported Germacranolide-type sesquiterpene lactones from *Smallanthus sonchifolius* with promising activity against *Leishmania mexicana* and *Trypanosoma cruzi* [48].

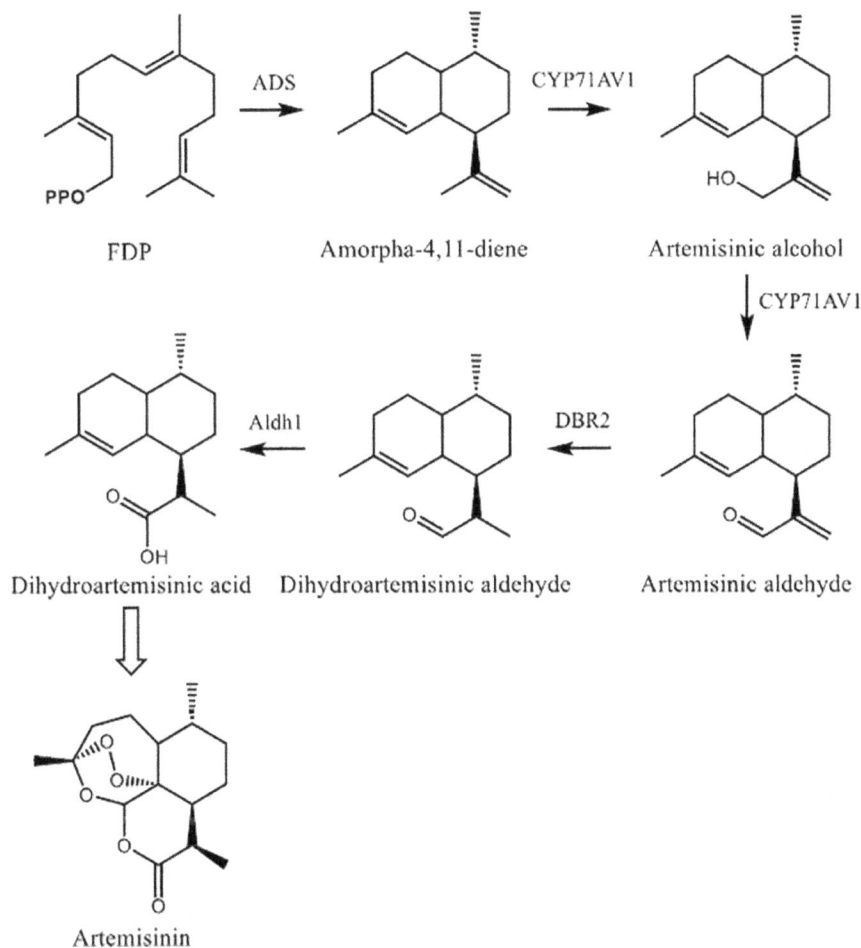

FIGURE 10.17 Steps involved in biosynthesis of the sesquiterpene lactone artemisinin in *A. annua*.

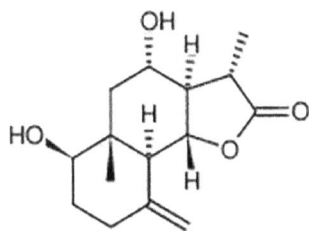

FIGURE 10.18 Structure of eudesmanolide.

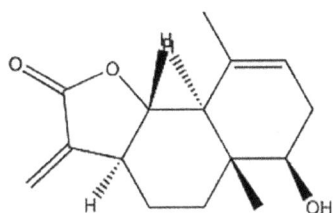

FIGURE 10.19 Structure of santamarine.

10.1.4.3.1 Lactuside

Lactuside is isolated from the roots of *Lactuca laciniata* (Figure 10.22). It is present in two forms such as Lactuside-A and Lactuside-B. Ping et al. reported that Lactuside-B

decreases the aquaporin-4 (AQP4) and caspase-3 mRNA expression in the hippocampus and striatum following cerebral ischemia-reperfusion injury in rats. The mechanism by which Lactuside-B recoverd ischemic brain injury may be associated with changes in AQP4 and caspase-3 mRNA expression in the hippocampus and the striatum [49].

10.1.4.3.2 Parthenolide

Parthenolide is a sesquiterpenoid lactone with the chemical name 4,5α-epoxy-6-β-hydroxy-germacra-1(10), 11(13)-dien-12-oic acid γ-lactone. It has an additional epoxide bridge between 4-and 5α-positions (Figure 10.23).

Parthenolide is obtained from the leaves of *Tanacetum parthenium* (L.) of family Asteraceae. It is also obtained from *Chrysanthemum parthenium* (L.) and *Magnolia grandiflora* (L.) of family Compositae and Magnoliaceae respectively. It is commonly known as feverfew and has employed as an effective febrifuge or antipyretic agent. It was found to act as a serotonin antagonist thereby causing the inhibition of the release of serotonin from blood platelets [50].

10.1.4.3.3 Elephantopin

Elephantopin is a natural chemical compound extracted from the plant *Elephantopus elatus* of family Compositae. It is a sesquiterpene lactone with a germacranolide skeleton

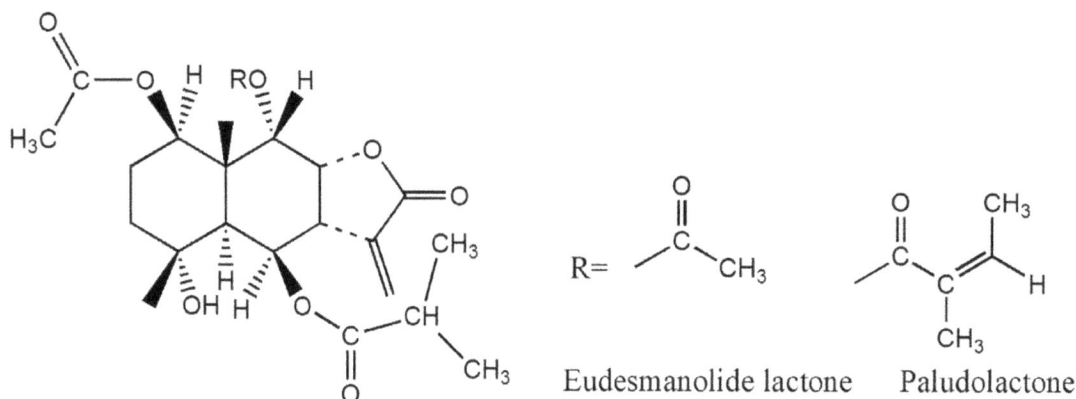

FIGURE 10.20   Structure of and eudesmanolide lactone paludolactone.

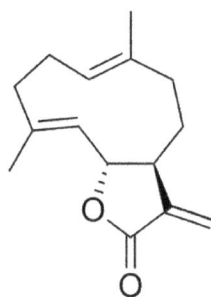

FIGURE 10.21   Structure of germacranolide.

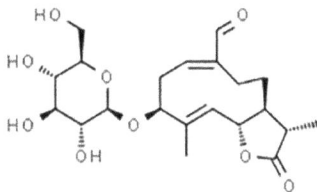

FIGURE 10.22   Structure of lactuside.

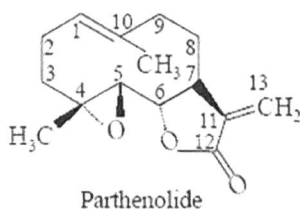

Parthenolide

FIGURE 10.23   Structure of parthenolide.

(Figure 10.24), which contains two lactone rings and an epoxide functional group [51].

10.1.4.4 Guaianolides

Guaianolides are a large group of sesquiterpene lactones of chemotaxonomic as well as medicinal importance. The ability of the thapsigargins isolated from the genus *Thapsia* (Apiaceae) to inhibit an intracellular calcium pump and thereby inducing apoptosis has encouraged intensive research in order to develop a new drug [52].

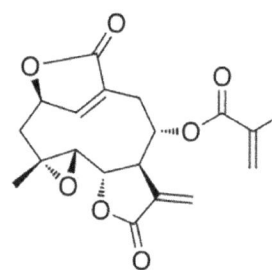

FIGURE 10.24   Structure of elephantopin.

10.1.4.4.1 Thapsigargin

Thapsigargin is a highly functionalized 6,12-guianolide sesquiterpene lactone which is produced by *Thapsia garganica* of family Apiaceae. Medicinally, it is used for the treatment of rheumatism, sterility and cold. Patkar et al. reported its histamine-releasing properties in the late1970s and its protein target in mammalian cells was later identified as the sarcoplasmic reticulum Ca^{2+}/ATPase pump (SERCA) [52,53]. Along with its properties as a SERCA inhibitor, it has potential to induce apoptosis. SERCA is responsible for pumping cytosolic Ca^{2+} into the sarcoplasmic reticulum in myocytes during muscle contraction or relaxation cycles [54]. Therefore, Thapsigargin-mediated inhibition of SERCA blocks muscle relaxation by preventing the removal of Ca^{2+} ions from the cytosol [55]. Thapsigargin induces endoplasmic reticulum stress generally, and one consequence of this is the inhibition of autophagosome fusion with the lysosome during autophagy which results in apoptosis. Thus, its ability to disrupt Ca^{2+} homeostasis in mammalian cells suggested a potential role in cancer therapy. Though it is highly cytotoxic but lacks specificity for tumor cells. So, the experiment was carried out for further development of anticancer drug [56].

Denmeade et al. successfully coupled thapsigargin to a prostate specific antigen (PSA) effectively to generate a prostate cancer-targeted prodrug [57,58]. Mipsigargin is a thapsigargin-based variant of this PSA-coupling approach which targets solid tumors. It is now in phase-II clinical trials for the treatment of hepatocellular carcinoma [59].

The sesquiterpene synthase TgTPS2 converts FDP into the sesquiterpene alcohol epi-kunzeaol (Figure 10.25). The screening of *T. garganica* cytochrome P450s by transient expression

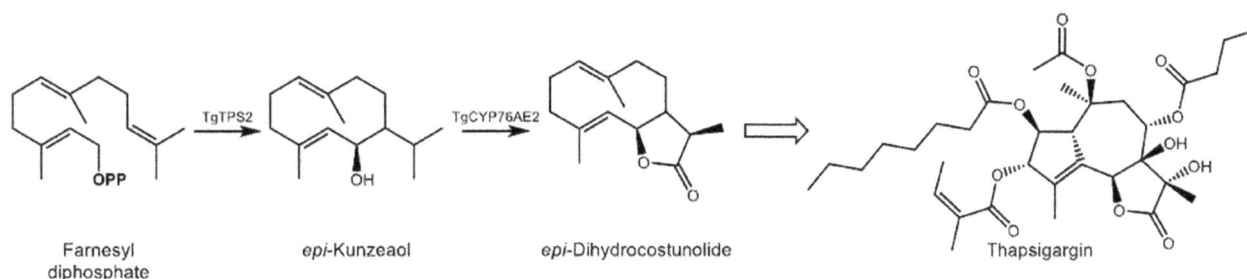

FIGURE 10.25 Biosynthetic steps for synthesis of thapsigargin in *T. garganica.*

in *Nicotiana benthamiana* followed by mass spectrometry analysis demonstrated that only one enzyme, TgCYP76AE2, was capable of transforming epi-kunzeaol into a product, identified as the germacrenolide sesquiterpenes lactone epi-dihydrocostunolide [60]. Andersen et al. also reported detection of transcripts for TgTPS2 and TgCYP76AE2 in the epithelial cells lining secretory ducts of T. garganica root sections [61].

10.1.4.4.2 Thapsigargicin

Thapsigargicin-induced histamine release from mesentery, lung and heart mast cells of the rat at concentrations from 0.1 μM but provoked only release from the corresponding guinea-pig cells in the concentration-range 0.16–1.6 μM. Thapsigargin increased the cytoplasmic free calcium level in intact human blood platelets at concentrations from 3.0 nM [62].

10.1.4.4.3 Matricarin

Matricarin is a sesquiterpene lactone with the molecular formula $C_{17}H_{20}O_2$ (Figure 10.26). It was isolated from Achillea vermacularis (Compositae). It is composed of a seven-membered ring which adopts a chair conformation, fused to two five-membered rings, one of which is essentially planar and the other exhibits an envelope conformation. It exhibits anti-inflammatory and antipasmodic therapeutic properties.

10.1.4.5 Tenulin

Tenulin is a chief constituent of sesquiterpene lactone (Figure 10.27), isolated from *Helenium amarum*. It was first identified by Clark in 1939. It has been reported to exert cytotoxic activity through inhibition of DNA synthesis and cellular enzymatic activity. Tenulin and isotenulin inhibit P-glycoprotein function and overcome multidrug resistance in cancer cells. It also disrupts the growth and development of insect larvae. The possible binding site of tenulin and isotenulin on P-gp was investigated through kinetic mechanism analyses with different P-gp substrates. P-gp has two substrate-binding sites, the H-site and the R-site, identified as preferentially binding Hoechst-33342 and rhodamine-123, respectively [63].

The molecular docking results showed that tenulin and isotenulin bind similarly on the NBD (ATPase binding site), indicating that the inhibition of drug efflux probably is related to ATP binding to P-gp. Consequently, when tenulin or isotenulin (0.1–1 μM) was combined with 200 μM verapamil in an ATPase assay, the consumption of ATP was decreased as

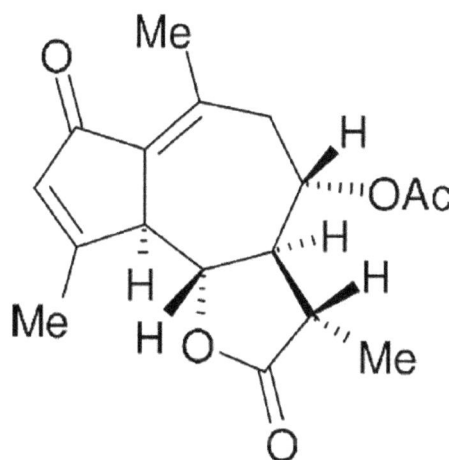

FIGURE 10.26 Structure o matricarin.

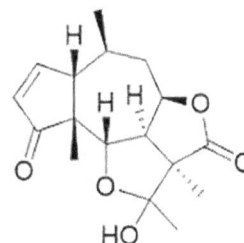

FIGURE 10.27 Structure of sesquiterpene lactone (Tenulin).

compared to verapamil treatment alone. This result demonstrated that tenulin and isotenulin likely bind to the same site as verapamil on P-gp ATPase [64].

10.1.5 Chemistry and Medicinal Uses of Sesquiterpenols

Sesquiterpenol is a sesquiterpene with an attached alcohol group. These are found in appreciable amounts in oils of ginger, carrot, valerian, patchouli and vetiver. Sesquiterpene alcohols are found in various essential oils such as Bisabolol, Farnesol, Nerolidol, Patchoulol etc [65]. The various pharmacological actions include anti-inflammatory, and stimulant to the liver and glands (Table 10.8).

Bisabolol is also known as levomenol. It is natural monocyclic sesquiterpene alcohol (Figure 10.28). It is colorless

TABLE 10.8

List of Sesquiterpenols and Their Properties

Sesquiterpenols	Biological Source	Biological Properties
α-Bisabolol	*Matricaria recutita*	Analgesic and Anti-inflammatory
Farnesol	*Santalum spicatum*	Bacteriostatic
Nerolidol	*Melaleuca quinquenervia*	Insecticidal and ovicidal effect
Patchoulol	*Patchouli cablin*	Antiviral and Anti-inflammatory,

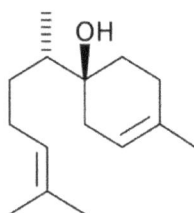

FIGURE 10.28 Structure of bisabolol.

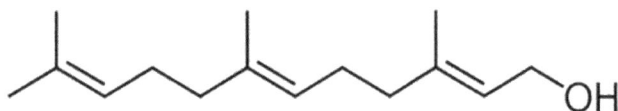

FIGURE 10.29 Structure of farnesol.

viscous oil and is the primary constituent of the essential oil present in *Matricaria recutita* and *Myoporum crassifolium*. Its enantiomer, α-(+)-bisabolol, is present naturally but synthetic bisabolol is a racemic mixture i.e. α-(±)-bisabolol. It is having anti-irritant, anti-inflammatory and antimicrobial properties [66].

Farnesol is natural acyclic sesquiterpene alcohol with the molecular formula $C_{15}H_{26}O$. Chemically, it is (2E, 6E)-3,7,11-trimethyldodeca-2,6,10-trien-1-ol (Figure 10.29). It is a colorless liquid, hydrophobic and thus insoluble in water.

It is isolated from *Santalum spicatum*. Farnesol is produced from 5-carbon isoprene compounds in both plants and animals. Geranyl pyrophosphate reacts with isopentenyl pyrophosphate and produces farnesyl pyrophosphate (C_{15}), which is an intermediate in the biosynthesis of sesquiterpenes such as farnesene (Figure 10.30). The oxidation of farnesene provides sesquiterpenoids alcohol (farnesol).

It is bacteriostatic and non-irritant agent. It is used in the formulation of deodorants in cosmetic products because of its

antibacterial activity. It also enhances the amphotericin-B and caspofungin activity against Candida. It acts as a chemopreventative and antitumor agent [67].

Nerolidol is a naturally occurring sesquiterpene alcohol found in the essential oils of several types of plants and flowers. Chemically, it is 3,7,11-trimethyl-1,6,10-dodecatrien-3-ol. It is also known as peruviol and penetrol. Nerolidol possesses four different isomeric forms which consist of two enantiomers and two geometric isomers (Figure 10.31). The existence of these isomeric forms is due to the presence of a double bond at the C-6 position and the asymmetric center at the C-3 position. The two geometric isomers of nerolidol include cis and trans-nerolidol, which differ in the geometry about the central double bond [68].

It exhibits various pharmacological activities including antioxidant, antifungal, anticancer and antimicrobial activity etc. Due to its hydrophobic nature, nerolidol is easily permeable across the plasma membrane and can interact with intracellular proteins. So, it has high cytotoxic potential and can disrupt the membrane. U.S. Food and Drug Administration (FDA) has also permitted the use of nerolidol as a food flavoring agent [68].

Nerolidol is synthesized as an intermediate in the production of (3E)-4,8-dimethy-1,3,7-nonatriene (DMNT), a herbivore-induced volatile which protects plants from herbivore damage. The overall biosynthesis mechanism of nerolidol is illustrated in Figure 10.32 as described by Bouwmeester et al [69].

Nigmatov et al. described the overall chemical synthesis of Nerolidol as illustrated in Figure 10.33. At first, nerolidol was synthesized as an intermediate in the chemical synthesis of geranyl esters from linalool. The process starts with the treatment of linalool with diketene or ethyl acctoacetate by the Carroll reaction to yield a mixture of (E) and (Z)-geranylacetone. The addition of acetylene to both (E) and (Z)-geranylacetone led to the production of (E) and (Z)-dehydronerolidol, respectively.

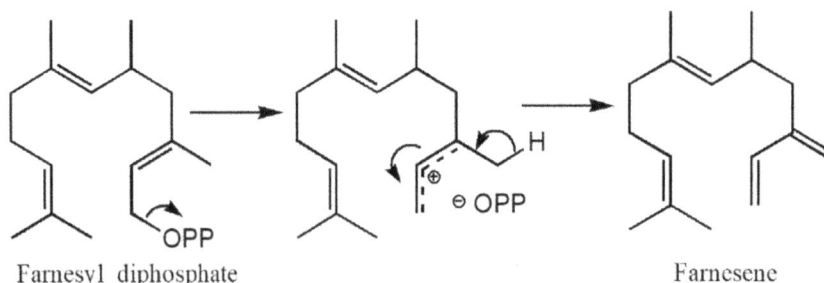

Farnesyl diphosphate   Farnesene

FIGURE 10.30 Biosynthesis of sesquiterpenes (farnesene).

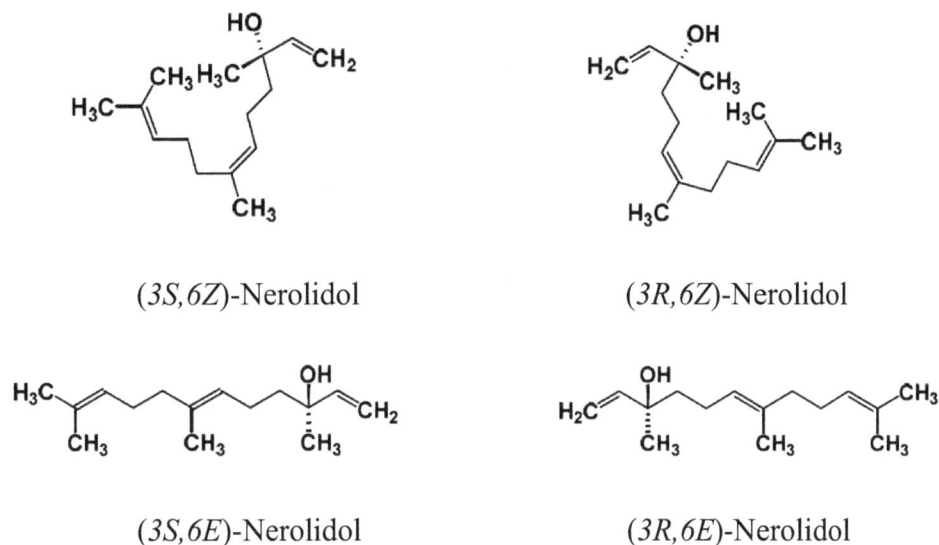

(*3S,6Z*)-Nerolidol

(*3R,6Z*)-Nerolidol

(*3S,6E*)-Nerolidol

(*3R,6E*)-Nerolidol

FIGURE 10.31 Chemical structures of isomers of nerolidol.

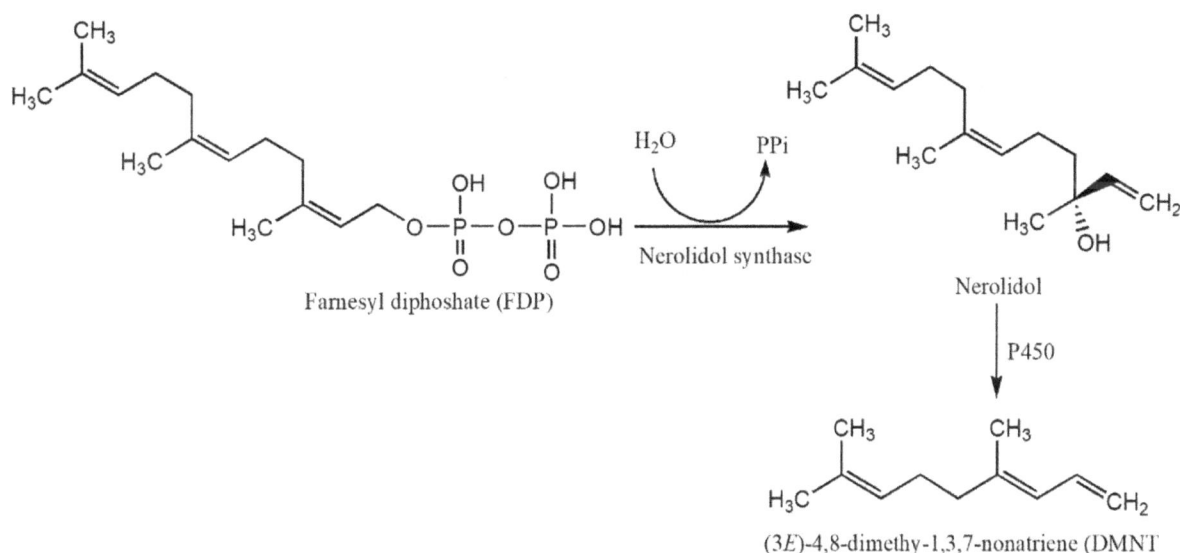

Farnesyl diphoshate (FDP)

Nerolidol synthase

Nerolidol

P450

(*3E*)-4,8-dimethy-1,3,7-nonatriene (DMNT

FIGURE 10.32 Synthetic route of nerolidol.

Finally, these compounds were selectively hydrogenated to trans- and cis-nerolidol, respectively by using a Lindlar catalyst [70].

Patchoulol or patchouli alcohol is a sesquiterpene alcohol (Figure 10.34). Patchouli alcohol was first isolated in 1869 by Gal and its chemical composition was later correctly formulated as $C_{15}H_{26}O$ by Montgolfier [71]. It is found in *Patchouli cablin* (patchouli) and is effective against influenza. It has also anti-inflammatory properties. Patchoulol is used in the synthesis of the chemotherapeutic drug (Taxol) [72].

10.1.6 Chemistry and Medicinal Uses of Sesterpenes

Sesterterpenes are a group of pentaprenyl terpenoids. These compounds possess C_{25} carbon frameworks derived from five isoprene units. These compounds exhibit various biological activities such as antibacterial, antifungal, cytotoxic, antimalarial and anti-inflammatory activities.

Sesterterpenoids originate from geranyl-farnesyl-pyrophosphate (GFPP) and are present mainly in fungi and marine organisms (Figure 10.35). Some specific metabolites are produced through various diverse cyclizations of the linear C_{25} precursor including cytotoxic agents like Ophiobolin-A and Bilosespene-A. Bilosespens A and B are the two novel sesterpenes, isolated from the Red Sea sponge *Dysidea cinerea*. These terpenes are cytotoxic to some of the human cancer cells [73,74].

Recently, investigations are carried out to determine the structural, chemical and biological properties of various sesterterpenoids. It was found that these compounds exhibit a broad spectrum of biological activities against bacteria, fungi and nematodes (Table 10.9).

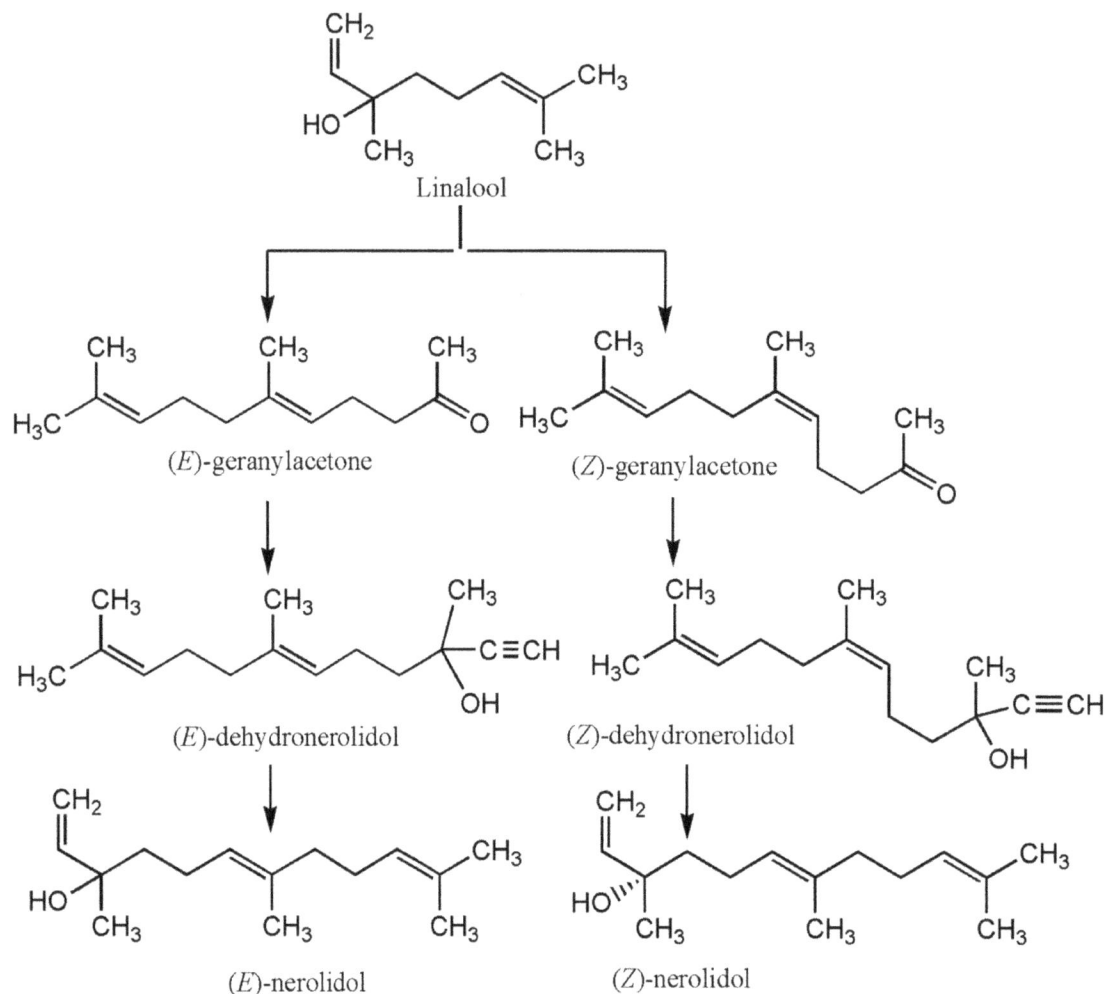

FIGURE 10.33 Synthetic pathway of nerolidol in industry.

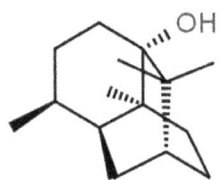

FIGURE 10.34 Chemical structure of patchoulol.

Ophiobolins (Ophs) are a group of tricarbocyclic sesterterpenoids and their structures contain tricyclic 5-8-5 carbotricyclic skeleton. The molecular weight of these molecules ranges from 338 to 432 and are produced by fungi mainly belong to the genus *Bipolaris* and *Aspergillus*. The first member of this family named as ophiobolin-A (Figure 10.36), was isolated from the pathogenic plant fungus *Ophiobolus miyabeanus* by Nakamura and Ishibashi in 1958 [75].

Ophiobolin is categorized into Ophiobolin-A (*Drechslera gigantean*), Ophiobolin-B (*Bipolaris oryzae*), Ophiobolin-C

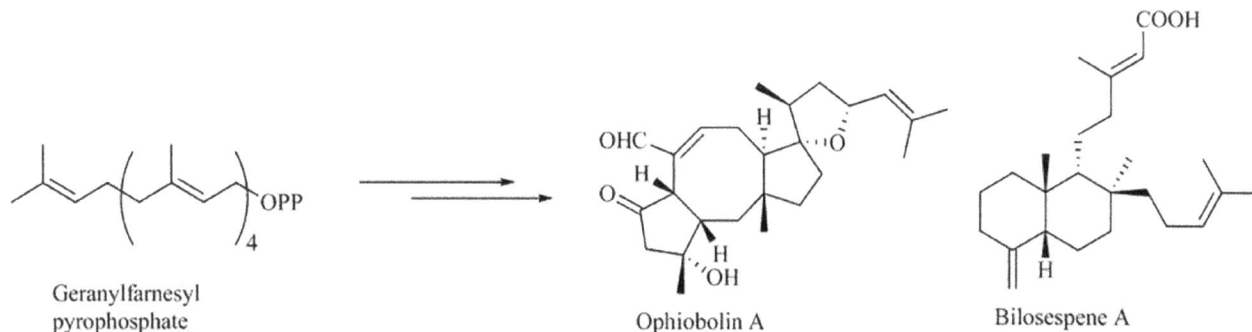

FIGURE 10.35 Synthesis of sesterpenes from GFPP (Ophiobolin-A and Bilosespene-A).

TABLE 10.9

List of Sesterpenes and Their Biological Source with Structure

Sesterpenes	Biological Source
Merochlorin-A	*Streptomyces sp.*
Ophiobolin-A	*Drechslera gigantea*
Secoemestrin-D	*Emericella sp.*
Manoalide	*Luffariella variabilis*
Luffariellolide	*Luffariella sp.*
Hippolide-A	*Hippospongia lachne*
Palauolol	*Thorectandra sp.*
Kohamaic acid-A	*Ircinia spp.*
Petrosaspongiolides	*Petrosaspongia nigra*

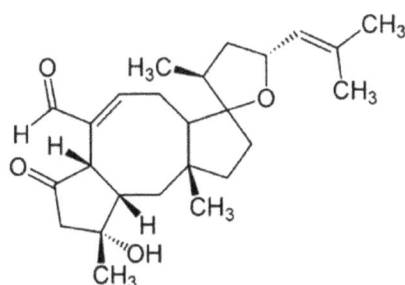

FIGURE 10.36 Structure of tricarbocyclic sesterterpenoids (Ophiobolin-A).

(*Aspergillus section Usti*), Ophiobolin K (*Aspergillus section Usti*), Ophiobolin T (*Ulocladium sp.*), Ophiobolin O (*Aspergillus ustus*). The majority of Ophs were discovered from the fungi genera of *Bipolaris* and *Aspergillus*. For example, genus *Bipolaris* produces ophiobolins A-B, I-J and L-M, while genus *Aspergillus* produces ophiobolins C, F-H, K, N and U-W (Table 10.10).

These bioactive molecules exhibited antibacterial activity against *S. aureus*, *M. intracellulare*, *B. subtilis*, MRSA, BCG and *M. smegmatis* (Table 10.11). The antibacterial activities are assessed on concentration/inhibitory diameter (μM/mm). Compound **5** displayed significant antibacterial activity against *B. subtilis*, MRSA and BCG, when compared with compound **3** and **4**, which demonstrated that the presence of hydroxyl group at C-6 position improve the antibacterial activity. Regarding antifungal activity, compounds **1**, **7**, **27** and **36** suppressed the growth of T. mentagrophytes at concentration of 60 μM, but there was no significant inhibition on *C. neoformans* and *C. albicans* [76].

Ophs also presented nematocidal and trypanocidal activities (Table 10.12). Compound 10 displayed the best nematocidal activities against the *Caenorhabditis elegans* at 5μM (LD$_{50}$ value) than other tested Ophs (11, 12, 33, 34, 37, 38). Similarly, compounds **37** and **38** also exhibited trypanocidal activities against *Trypanosoma cruzi* [77].

Ophs act as promising drug candidates due to the developments of their antiproliferative activities against a vast number of cancer cell lines, multidrug resistance (MDR) cells and

TABLE 10.10

Sources of Ophiobolins (Ophs)

Number	Sesterterpenoids	Biological Source
1	Ophiobolin-A	*Bipolaris oryzae*
2	6-epi-ophiobolin-A	*Penicillium patulum*
3	3-anhydroophiobolin-A	*Cochliobolus heterostrophus*
4	3-anhydro-6-epiophiobolin-A	*Drechslera gigantea*
5	3-anhydro-6-hydroxy-Ophiobolin-A	*Bipolaris oryzae*
6	Ophiobolin A lactone	*Polyangium cellulosum*
7	Ophiobolin B	*Bipolaris oryzae*
8	3-anhydro-6-epiophiobolin-B	*Cochliobolus heterostrophus*
9	Ophiobolin-B lactone	*Pseudomonas aeruginosa*
10	Ophiobolin-C	*Mollisia sp.*
11	6-epi-ophiobolin C	*Cochliobolus heterostrophus*
12	18-dihydroophiobolin C	*Cochliobolus heterostrophus*
13	Ophiobolin-D	*Cephalosporium caerulens*
14	Ophiobolin-E	*Drechslera gigantea*
15	Ophiobolin-F	*Cochliobolus heterostrophus*
16	Ophiobolin-G	*Aspergillus ustus*
23	Ophiobolin H	Aspergillus ustus
27	Ophiobolin I	Polyangium cellulosum
30	Ophiobolin J	Drechslera oryzae
33	Ophiobolin K	Emericella variecolor
35	Ophiobolin L	Cochliobolus heterostrophus
37	Ophiobolin M	Cochliobolus heterostrophus
39	6-epi-ophiobolin N	Emericella variecolor
40	Ophiobolin O	Aspergillus sp.

TABLE 10.11

Antibacterial Activities of Ophs

Number	Antibacterial Activity						
	E. coli	*S. aureus*	MRSA	*M. intracellulare*	*B. subtilis*	BCG	*M. smegmatis*
1	>62	62/7–12	NA	62/3–6	NA	NA	NA
3,4	NA	>100	>100	NA	>100	>100	NA
5	NA	31	31	NA	31	31	NA
7	>62	62/3–6	NA	>62	NA	NA	NA
16	>2.7×10³	NA	NA	NA	136/10	NA	NA
17	NA	NA	>250	NA	>250	125/NR	68/NR
18	NA	NA	>250	NA	>250	125/NR	NA
23	>2.6×10³	>1.6×10³	NA	NA	682/16	NA	NA
24	1.6×10³/10	>1.6×10³	NA	NA	NA	NA	NA

TABLE 10.12

Nematocidal Activity of Ophs

Number	Species	Concentration (LD$_{50}$, µM)
10	*C. elegans*	5
11	*C. elegans*	130
12	*C. elegans*	25
33	*C. elegans*	26
34	*T. cruzi*	9.6
37	*C. elegans*	13
38	*C. elegans*	130

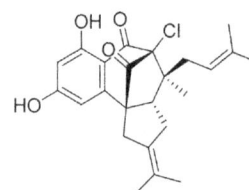

FIGURE 10.37 Structure of merochlorin-A.

TABLE 10.13

Cytotoxic Activities of Ophs

Number	Cell Line	IC$_{50}$ (µM)
1	U373-MG	0.87±0.01
2	HeLa	14
3	K562	40±5.5
4	A549	30±1
5	K562	4.1±0.50
7	CLL	2×10⁻³
10	CLL	8×10⁻³
16	P388	25
17	HepG2	0.37±0.03
18	HepG2	2.2±0.18

cancer stem cells (CSCs) (Table 10.13). The cytotoxic activities of Ophs against various cell lines are presented as IC$_{50}$ values by MTT assay [78].

Merochlorin-A is the first complex chlorinated resorcinol-containing metabolite (Figure 10.37) isolated from a marine-derived actinomycete strain CNH189 through a screening effort against clinically resistant bacteria. It was identified as an agent to act against a number of Gram-positive strains such as methicillin-resistant Staphylococcus aureus (MRSA, bacterium). Cytotoxicity study was performed in tissue culture assays at 24 and 72 hours against human HeLa cervical and L929 mouse sarcoma cell lines. A compound is considered as cytotoxic if there is 50% reduction in absorbance at 490 nm due to the reduced MTS ((3-(4,5-dimethylthiazol-2-yl)-5-(3-carboxymethoxyphenyl)-2-(4-sulfophenyl)-2H-tetrazolium) reagent in live cells. It was observed at 150 and 4 µM in HeLa cells at 24 and 72 hours, respectively, and at 80 and 9 µM in L929 cells at 24 and 72 hours, respectively [79].

TABLE 10.14

Cytotoxicity Data for Secoemestrin D

Compound	Cell Lines						
	NCI-H460	SF-268	MCF-7	PC-3M	MDA-MB-231	CHP-100	WI-38
Secoemestrin-D	0.15±0.01	0.06±0.01	0.14±0.01	0.17±0.01	0.06±0.01	0.10±0.01	0.24±0.01
Doxorubicin	0.05±0.01	0.31±0.04	0.35±0.08	0.27±0.03	0.68±0.0.05	0.79±0.10	NT

CHP-100, human neuroblastoma; MCF-7, human breast cancer; MDA-MB-231, human breast adenocarcinoma; NCI-H460, human non-small cell lung cancer; NT, Not tested; PC-3M, metastatic human prostate adenocarcinoma; SF-268, human CNS cancer (glioma); WI-38, human lung fibroblast. Results are expressed as IC$_{50}$ value±standard deviation in µM. Doxorubicin and DMSO were used as positive and negative controls.

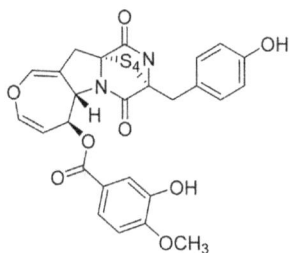

FIGURE 10.38 Structure of secoemestrin-D.

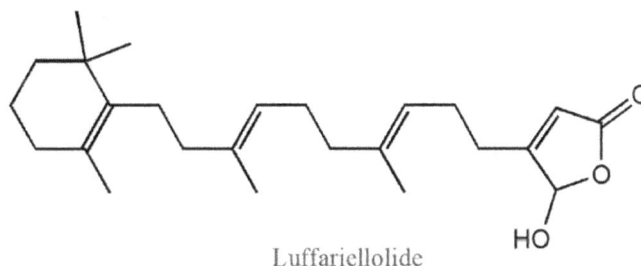

Manoalide

FIGURE 10.39 Structure of sesterterpenoid (Manoalide).

Secoemestrin-D exhibit significant cytotoxicity with IC_{50} ranging from 0.06 to 0.24 μM (Table 10.14) and moderate selectivity toward human glioma (SF-268) and metastatic breast adenocarcinoma (MDA-MB-231) cell lines. Chemically, it is epitetrathio-dioxo piperizine analog (Figure 10.38) [80].

Manoalide, a sesterterpenoid (Figure 10.39) was initially isolated as an antibacterial metabolite from *Luffariella variabilis* sponge in 1980. Later, it was found to be a potent inhibitor of phospholipase-A_2 (PLA_2) with an IC_{50} value of 1.7 μmol/L. So, it possesses anti-inflammatory property [81].

Luffariellolide is a sesterterpene (Figure 10.40) and isolated from the Palauan sponge *Luffariella sp*. It is having useful anti-inflammatory properties. It is a partially reversible inhibitor of phospholipase-A_2 (PLA_2) [82,83].

Aplysinoplides A-C (Figure 10.41), were isolated from sponge, *Aplysinopsis digitata*. Aplysinoplides-A and B are closely related to luffariellolide due to the aliphatic bridge joining the 5-hydroxyfuran-2-one residue with the trimethylcyclohexene moiety. Aplysinoplide-C is similar to luffariolide with the different functionalization of trimethylcyclohexene residue. Aplysinoplides A-C exhibited cytotoxic activity against P388 mouse leukemia cells with the IC_{50} values of ~1, ~1 and ~26 μM, respectively. Aplysinoplides-A and B did not inhibit

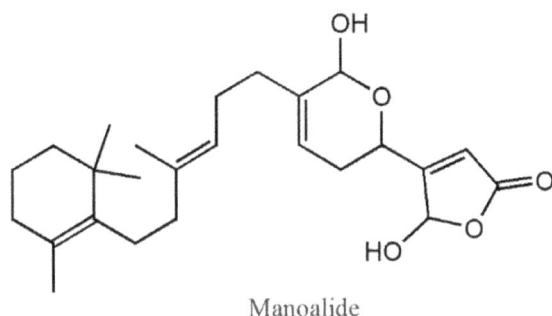

Luffariellolide

FIGURE 10.40 Structure of luffariellolide.

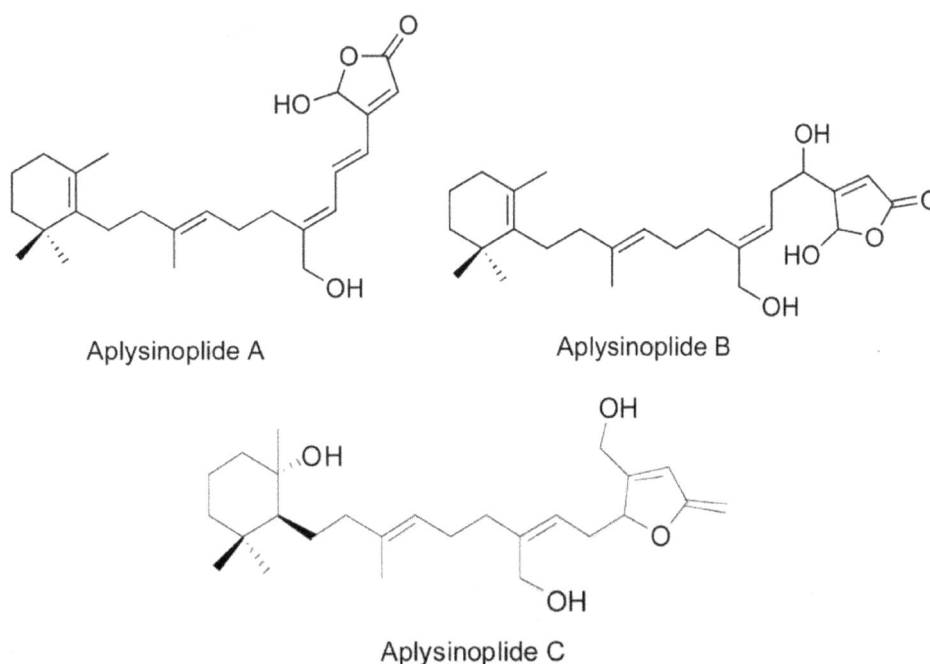

Aplysinoplide A

Aplysinoplide B

Aplysinoplide C

FIGURE 10.41 Structures of aplysinoplides (A, B, C).

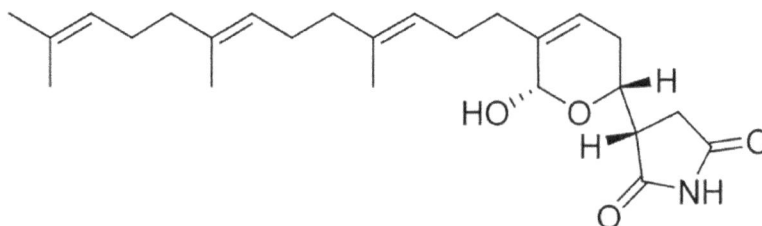

FIGURE 10.42 Structure of hippolide-A.

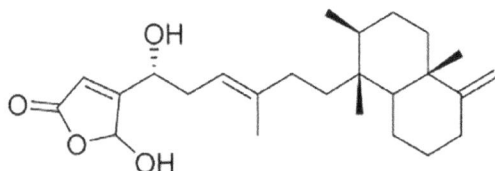

FIGURE 10.43 Structure of palauolol.

TABLE 10.15

IC_{50} Values of KA-A on the Activities of DNA Polymerases

DNA Polymerases	IC_{50} Value (μM)
Sea urchin DNA polymerase-α	6.1±0.8
Sea urchin DNA polymerase-β	30.3±2.1
Human DNA polymerase-ε	5.8±0.4
Human DNA polymerase-γ	14.9±2.2

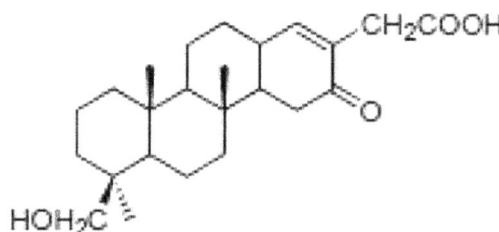

FIGURE 10.44 Structure of kohamaic acid-A.

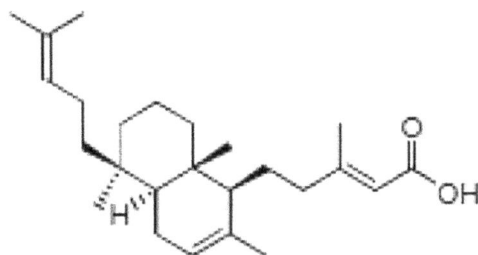

FIGURE 10.45 Structure of petrosaspongiolide.

TABLE 10.16

Effect of Petrosaspongiolides M-R on Secretory PLA_2

Petrosaspongiolides	Molecular Formula	$IC_{50}(\mu$M) Bee Venom	Human Synovial
M	$C_{25}H_{36}O_4$	0.6	1.6
N	$C_{29}H_{42}O_8$	N.D.	N.D.
P	$C_{25}H_{38}O_5$	N.D.	3.8
Q	$C_{27}H_{40}O_7$	N.D.	N.D.
R	$C_{25}H_{36}O_5$	N.D.	N.D.

ND, Not determined

bovine pancreatic PLA_2 at concentrations up to of 100 μM, in comparison with manoalide which exhibited potent activity in a parallel experiment [84].

Hippolide-A is isolated from the Sea sponge *Hipposspongia lachne* (Figure 10.42). It exhibited cytotoxic activity against A549, HeLa and HCT-116 cell lines with IC_{50} values of 5.22×10^{-2}, 4.80×10^{-2} and 9.78 μM respectively. It also exhibited moderate PTP_1B inhibitory activity with an IC_{50} value of 23.81 μM. In addition, Hippolide-A demonstrated weak anti-inflammatory activity, with IC_{50} values of 61.97 and 40.35 μM for PKCγ and PKCα respectively [85].

Palauolol was obtained as clear oil from *Fascaplysinopsis sp.* with molecular formula $C_{25}H_{38}O_4$ (Figure 10.43). It is a secondary alcohol which on dehydration produces palauolide. So, Palauolol is a biosynthetic precursor of palauolide. Palauolol is having antimicrobial activity against *S. aureus* and *B. subtilis*. Both palauolide (85% inhibition @ 0.8 μg/mL) and palauolol (82% inhibition @ 0.8 μg/mL) inactivate bee venom PLA_2 [86].

Kohamaic acid-A (KA-A) is isolated as a novel sesterterpenic acid from marine sponge *Ircinia sp* (Figure 10.44). It was screened as an inhibitor of cleavage of fertilized sea urchin eggs. It was also found that it inhibited DNA polymerases. DNA polymerase is associated with genomic DNA replication and its repair in eukaryotic cells. The IC_{50} values of KA-A were found to be 5.8–14.9 μM for mammalian DNA polymerases (Table 10.15). Therefore, the LD_{50} value of KA-A was slightly higher than the IC_{50} value *in-vitro* for DNA polymerases [87].

Petrosaspongiolides are isolated from marine sponge *Petrosaspongia nigra* and named as petrosaspongiolides M-R (Figure 10.45). Petrosaspongiolide-M and N are white amorphous solid with molecular formula $C_{25}H_{36}O_4$ and $C_{29}H_{42}O_8$ respectively [88]. All of these compounds inhibited phospholipase-A_2 (PLA_2) by blocking these enzymes irreversibly present human synovial and bee venom with IC_{50} values in the micromolar range (Table 10.16).

Miguel et al. reported the effects of Petrosaspongiolide-M as a novel Phospholipase-A_2 (PLA_2) inhibitor *in-vitro* and *in-vivo*, on acute and chronic inflammation. Petrosaspongiolide M (Figure 10.46) was isolated from the Caledonian marine sponge *Petrosaspongia nigra*. It showed selectivity toward secretory PLA_2 versus cytosolic PLA_2 with potency on the human synovial enzyme. Petrosaspongiolide-M decreased carrageenan-induced paw edema in mice after the oral administration of 5, 10, or 20 mg/kg. The inflammatory response of arthritis was reduced by petrosaspongiolide-M, which also inhibited leukotriene-B_4 levels in serum and PGE_2 levels in paw homogenates. Petrosaspongiolide-M also reduced the production of eicosanoids derived from the COX and 5-LO pathways in an acute inflammatory response (Table 10.17) [89].

Results are expressed as mean $\pm$ S.E.M. ($n = 6$–12). * $P < 0.05$, ** $P < 0.01$ compared with control group. Petrosaspongiolide M, manoalide and reference inhibitors were tested at 10 mM. N.D. Not determined. Drugs were incubated with a cytosolic fraction of U937 cells (cPLA2), high-speed supernatants from human neutrophil homogenates (5-LO), microsomal fraction from human monocytes treated with LPS (COX-2), microsomal fraction from human platelets (COX-1)

Scalaranes are sesterterpenoids, characterized by the typical tetracyclic carbon skeleton (Figure 10.47). These are mainly isolated from sponges. The first scalarane was isolated from a *Mediterranean* sponge in 1980 [90].

Sesterstatins-4 and 5 (Figure 10.48), are new pentacyclic furanosesterterpenes belonging to the scalarane subgroup. These are isolated from the marine black sponge *Hyrtios erecta* incorporating a benzoperhydro-phenanthrene ring system joined to a furan ring. These compounds inhibited the growth of various cancer cell lines *in-vitro* which include murine P338 leukemia and human BXPC-3 pancreas, RPMI-7951 melanoma, U-251 central nervous system (CNS), KAT-4 and SW-1736 thyroid, NCI-H460 NSCLC, FADU pharynx and DU-145 prostate cancer models with the IC_{50} growth inhibitory concentrations ranging between 4–13 and 5–26 μM, respectively [91].

Phyllophenones-D and E are considered as scalarane sesterterpenoids (Figure 10.49). These compounds were isolated from the sponge *Phyllospongia foliascens*. Phyllophenones-D possesses a cyclopent-2-enone ring joined to the D ring of the scalarane ring system, hydroxyl group at C_{12} and a tri-substituted double bond between C_{16} and C_{17} while Phyllophenones-E incorporates the same functionalities in the scalarane skeleton as well as two hydroxyl and the ethanoyl groups at C_{12}, C_{18} and C_{17} respectively. Phyllofenone-D has a rare α, β-unsaturated ketone ring, whereas phyllofenone-E possesses also a relatively rare 25-norscalarane framework. Phyllofenone-D exhibited cytotoxic activity against P388 leukemia cell line with the IC_{50} growth inhibitory concentration of 17 μM. In contrast, it was found that phyllofenone-E was inactive [92].

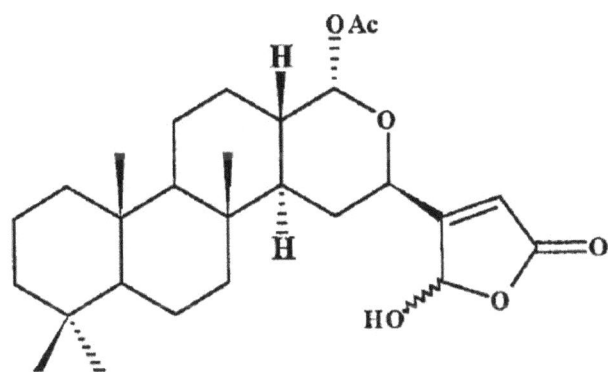

FIGURE 10.46 Chemical structure of petrosaspongiolide-M.

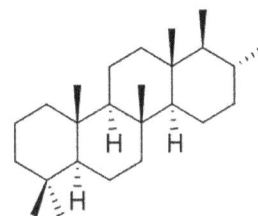

FIGURE 10.47 Chemical structure of scalarane.

TABLE 10.17

Effect of Petrosaspongiolide-M on cPLA2, 5-LO, COX-2, COX-1 Activities *in-vitro*

Compounds	cPLA$_2$ (pmol AA/mg min)	5-LO (ng LTB4/mL)	COX-2 (ng PGE2/mL)	COX-1 (ng PGE2/mL)
Petrosaspongiolide-M	5.7±0.7	9.8±1.3[b]	2.1±0.4	6.5±0.7
Indomethacin	N.D.	N.D.	1.1±0.1[a]	0.9±0.2[b]
Control	5.6±0.6	18.4±0.8	1.7±0.1	6.7±0.8
Manoalide	3.1±0.7[b]	12.0±1.0[b]	1.3±0.3	6.6±0.4

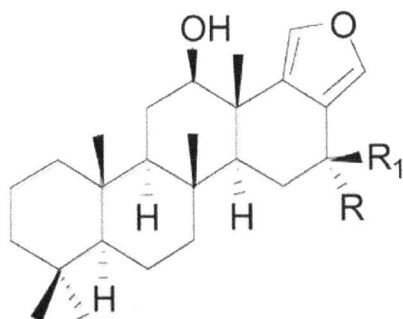

R=OH, R$_1$=H, Sesterstatin 4
R=H, R$_1$=OH, Sesterstatin 5

FIGURE 10.48 Chemical structure of sesterstatins-4 and 5.

Phyllophenones-D   Phyllophenones-E

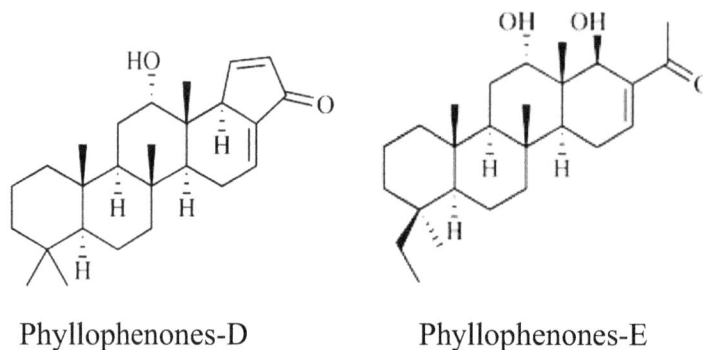

FIGURE 10.49   Chemical structure of scalarane sesterterpenoids.

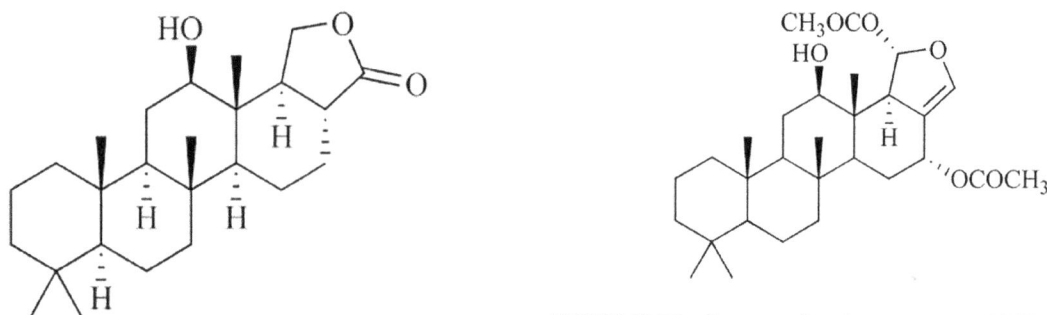

FIGURE 10.50   Structure of hyppospongide-B.

FIGURE 10.52   Structure of marine sesterterpenoid (Heteronemin).

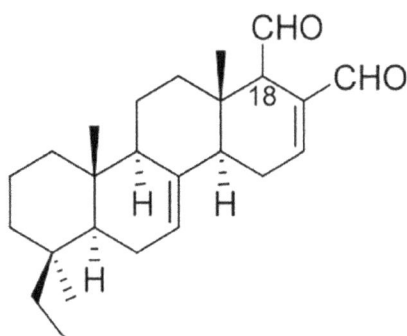

Cacospongionolide F

FIGURE 10.53   Structure of cacospongionolide-F.

FIGURE 10.51   Structure of mooloolabenes A (18R) and B (18S).

Hyppospongide-B is a scalarane sesterterpenoid which was isolated from the sponge *Hipponspongia sp*. It contains a 4,5-dihydrofuran ring joined to the D-ring of the scalarane ring system (Figure 10.50). It exhibited cytotoxic activity against DLD-1 and HCT-116 human colon, T-47D human breast and K-562 human leukemia cancer cell lines with the IC$_{50}$ growth inhibitory concentrations of >10 μM [93].

Mooloolabenes (A-E) are isolated from *Hyatella intestinalis* (Figure 10.51). These are scalarane norsesterpenoids which exhibited cytotoxic activity against P388 mouse leukemia cell line. Mooloolabenes A and B were found most potent with the IC$_{50}$ growth inhibitory concentrations of ~2 and ~3 μM respectively [94].

Heteronemin is a marine sesterterpenoid-type natural product (Figure 10.52). It was isolated from the sponge *Hyrtius sp*.

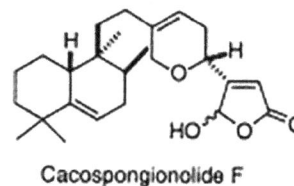

It possesses 4,5-dihydrofuran moiety joined to the D ring of the scalarane ring system and two methoxycarbonyl groups at C$_{16}$ and C$_5$. It exhibited a potent cytotoxic effect against LNcap and PC3 prostate cancer cells with IC$_{50}$ 1.4 and 2.7 μM after 24 h respectively. Heteronemin also induced apoptotic cell death and inhibited TNF-α-induced NF-kappa-B activation through proteasome inhibition in chronic myelogenous leukemia cells [95].

Cacospongionolide-F is a bioactive cacospongionolide-related sesterterpene with the molecular formula C$_{25}$H$_{36}$O$_4$ (Figure 10.53). It was isolated from the Northern Adriatic sponge *Fasciospongia cavernosa* [96]. Cacospongionolide-F is highly active (LC$_{50}$=0.17 ppm) in the *Artemia salina* bioassay, moderately toxic (LC$_{50}$=0.7 ppm) in a fish lethality assay and highly active against the gram-positive bacteria, *Bacillus subtilis* (MIC=0.78 μg/mL) and *Micrococcus luteus* (MIC=0.78 μg/mL) [97,98].

Snapper et al. reported the total synthesis of the anti-inflammatory marine sponge metabolites (+)-cacospongionolide-B and E. Cacospongionolide-E differs from cacospongionolide-B only in the position of its olefin in the decalin ring system.

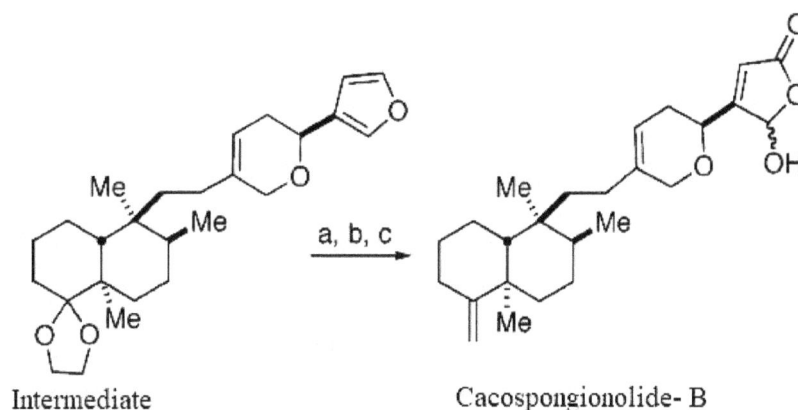

Intermediate   Cacospongionolide- B

Reaction conditions: (a) 1N HCl/THF (1:2), rt (90%). (b) $Ph_3P=CH_2$, DMSO, 75°C (84%). (c) O_2, Rose bengal, *i*-Pr_2NEt, 150 W. tungsten lamp, CH_2Cl_2, -78 °C (69%).

FIGURE 10.54 Scheme for synthesis of cacospongionolide-B.

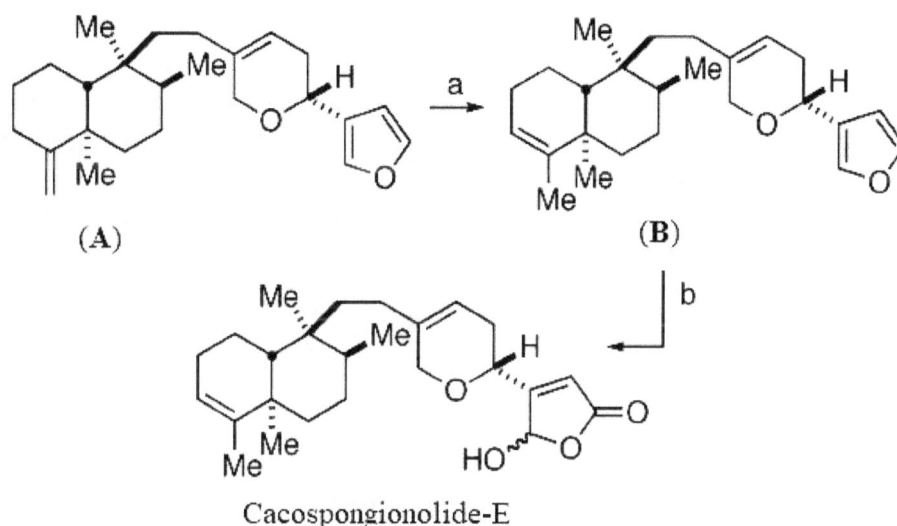

(A)   (B)

Cacospongionolide-E

Reaction conditions: (a) $RhCl_3\cdot H_2O$ in a mixture of EtOH/CHCl$_3$, 70 °C (70%). (b) O_2, Rose Bengal, *i*-Pr_2NEt, CH_2Cl_2, -78 °C (54%).

FIGURE 10.55 Scheme for synthesis of cacospongionolide-E.

On the basis of biological studies, this region may be responsible for inhibiting PLA$_2$ activity. Cacospongionolide-B is reported to suppress the expression of inflammatory enzymes by inhibiting the nuclear factor-κB (NF) activation [99]. The synthesis of cacospongionolide-B involves the reaction of the following intermediate in the prescribed reaction conditions as depicted in Figure 10.54.

Reaction conditions: (a) 1N HCl/THF (1:2), rt (90%). (b) Ph$_3$P=CH$_2$, DMSO, 75°C (84%). (c) O$_2$, Rose bengal, *i*-Pr$_2$NEt, 150 W. tungsten lamp, CH$_2$Cl$_2$, −78°C (69%).

The synthesis of cacospongionolide-E (**2**) was accomplished in two steps from the furanyl intermediate (**A**) (Figure 10.55). The exo-cyclic olefin of (**A**) was isomerized to the more stable internal position utilizing RhCl$_3$H$_2$O in a mixture of EtOH/

CHCl$_3$ (1:1). Heating the mixture to 70°C for 6 days provided the desired olefin isomerization product (**B**) in 70% yield with no detectable pyran olefin isomerization by^{1}H-NMR. Adduct (**B**) was converted to cacospongionolide-E with 54% yield by applying photooxidation procedure.

Reaction conditions: (a) RhCl$_3$·H$_2$O in a mixture of EtOH/ CHCl$_3$, 70°C (70%). (b) O$_2$, Rose Bengal, *i*-Pr$_2$NEt, CH$_2$Cl$_2$, −78°C (54%).

Arai et al. reported selective growth inhibition of cancer cells by Furospinosulin-1 (Figure 10.56) under hypoxia condition. Chemically, it is a Furanosesterterpene and isolated from an Indonesian marine sponge (*Dactylospongia elegans*). It exhibited selective antiproliferative activity against DU145 human prostate cancer cells under hypoxic conditions in

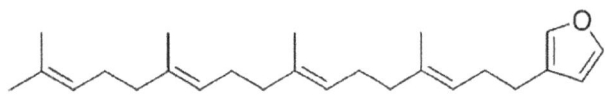

FIGURE 10.56 Chemical structure of furospinosulin-1.

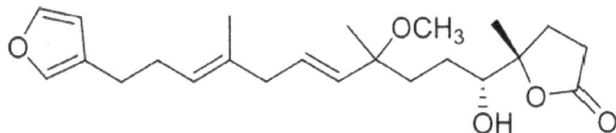

FIGURE 10.57 Structure of nor-sesterterpenoid (Irciformonin-I).

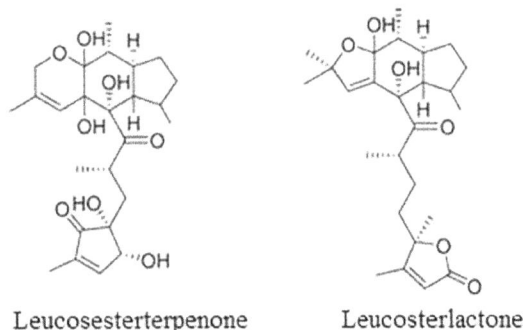

Leucosesterterpenone   Leucosterlactone

FIGURE 10.58 Structure of tetracyclic sesterterpenoids.

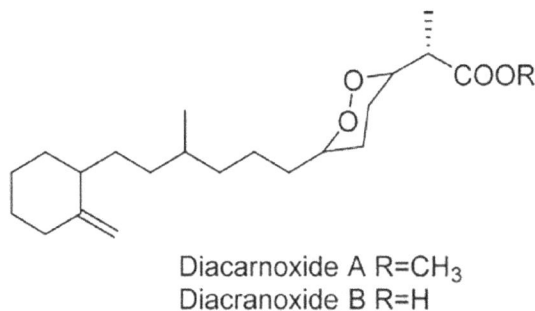

Diacarnoxide A R=CH₃
Diacranoxide B R=H

FIGURE 10.59 Structure of diacarnoxides-A and B.

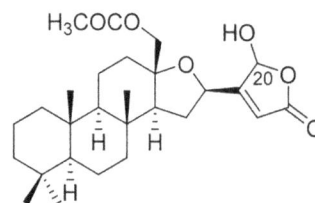

FIGURE 10.60 Structure of pentacyclic sesterterpene (Lintenolide).

concentrations ranging from 1 to 100 µm. Furospinosulin-1 also demonstrated antitumor activity at 10–50 mg/kg oral administration in a mouse model inoculated with sarcoma S180 cells. Mechanistic analysis revealed that furospinosulin-1 suppresses the transcription of the insulin-like growth factor-2 gene (IGF-2), which is selectively induced under hypoxic conditions through the prevention of the binding of nuclear proteins to the Sp1 consensus sequence in the IGF-2 promoter region [100].

Irciformonin-I is a nor-sesterterpenoid, containing a furanone and furan rings joined by means of an unsaturated and methoxylated bridge (Figure 10.57). It was isolated from *Ircinia formosana* (sponge). This compound was found to be toxic as it demonstrated the inhibition of peripheral blood mononuclear cell proliferation. 15-O-Acetylirciformonin-B and 10-O-acetylirciformonin-B are C_{22} furanosesterterpenoids, which differ from irciformonin-I by the functionalization of the bridge linking the furanone and furan rings. Both compounds were isolated from *Ircinia spp.* and exhibited cytotoxic activity against K562 human leukemia, DLD-1 human colon carcinoma, HepG2 human hepatocarcinoma and Hep3B human hepatocarcinoma cancer cell lines. 15-acetylirciformonin-B was found to be the most potent compound against these four cancer cell lines with the IC_{50} growth inhibitory concentrations ranging between 0.03 and ~5 µM [101,102].

Leucosesterterpenone and leucosterlactone (Figure 10.58) are considered as tetracyclic sesterterpenoids. These are isolated from Himalayan plant *Leusceptrum canum*. Leucosesterterpenone inhibited fibroblast growth factor-2-dependent proliferation, migration, chemotaxis and tube

formation with small and large vessel endothelial cells, whereas leucosterlactone was completely inactive. These two compounds formed a complex with the fibroblast growth factor-2 receptor-1 with the IC_{50} inhibitory concentrations of ~1 and 132 µM, respectively with simultaneous down-regulation of ERK1/2 phosphorylation. Leucosesterterpenone is less hydrophobic than leucosterlactone, which may contribute to its increased activity and thus potential application as an anti-angiogenic agent [103].

Diacarnoxides-A and B (Figure 10.59) are sesterterpenoid derivatives of propanoic acid and its methyl ester. Diacarnoxide-A is the methyl ester of diacarnoxides-B. These compounds were isolated from *Diacarnus levii* (sponge). They exhibited cytotoxic properties under hypoxic conditions against five human cancer cell lines including two prostate (DU145 and PC-3) and three breast (MCF-7, MDA-MB-231 and T47D) cancer models. Diacarnoxide-B inhibited cancer cell proliferation under normoxic versus hypoxic conditions with IC_{50} ratios of ~7/5 (DU145), ~17/13 (PC-3), ~14/7 (MCF-7), ~14/6 (MDA-MB-231) and ~6/6 (T47D) µM [104].

Lintenolides is a pentacyclic sesterterpenes (Figure 10.60). These were isolated from the Caribbean sponge *Cacospongia cf. linteiformis*. Lintenolides-A and G possess a perhydrophenanthrene ring system fused with a furanyl moiety and differ by the methoxycarbonyl versus hydroxyl group at C_{13}. But in the case of Lintenolide-E, the perhydrophenanthrene ring system is fused with a pyranyl moiety. These compounds exhibited in vitro cytotoxicity against WEHI-164 murine fibrosarcoma, J774-murine monocyte/macrophage, P388-murine leukemia and GM7373-bovine endothelial cell lines with the IC_{50} values in the range of 0.2–4 µM [105].

Bolivianine is a novel sesterpene with an unprecedented skeleton (Figures 10.61 and 10.62). It is isolated from the trunk bark of *Hedyosmum angustifolium* (Chloranthaceae). Whereas, isobolivianine is an isomer of Bolivianine produced

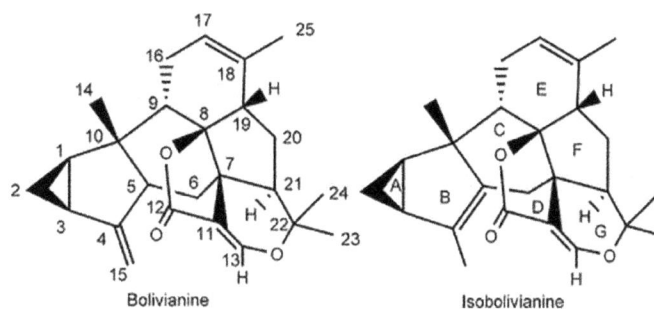

FIGURE 10.61 Structure of bolivianine and its isomer.

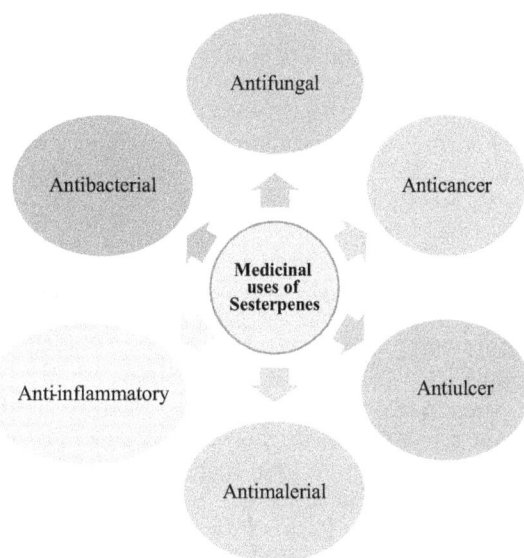

FIGURE 10.62 Medicinal uses of sesterpenes.

under acidic conditions. The structure and stereochemistry of these compounds are determined on the basis of chemical and spectroscopic data [106].

10.2 Conclusion

The development of highly functionalized natural terpenoids was punctuated by the rapid production of terpene synthase and cytochrome P450 families. It results in chemical diversification of the natural products through the introduction of different types of functional groups into the olefinic backbone of terpene hydrocarbons. Both the cytosolic MVA pathway and the plastid localized MEP pathway contribute to generate terpene skeletons that lead to functionalized and biologically active secondary metabolites. The emergence of generating highly potent and medicinally useful plant terpenes may exploit the natural similarities between human and insect physiology as well as a higher compatibility with proteins compared to synthetic compounds. Efforts made to devlop new approaches to elucidate the complex biochemical pathways and also to optimize the screening strategies for sesquiterpenoids. This will undoubtedly improve the discovery rate and availability of sesquiterpenoids for clinical use. Similarly, sesterterpenoids

represent attractive molecular architectures to facilitate the binding of drug molecules specifically to receptors or proteins involved in tumorigenesis. The recent advance in the synthesis of sesterterpenoids creates a new opportunity in their utilization for the development of novel therapeutic regimes.

Abbreviation

ABA	Abscisic acid
ADS	Amorpha-4,11-diene synthase
Aldh1	Aldehyde dehydrogenase-1
AQP4	Aquaporin-4
ATP	Adenosine triphospate
BCP	β-caryophyllene
BCPO	β-caryophyllene oxide
bFGF	basic fibroblast growth factor
BHT	Butyl-4-hydroxy toluene
CYP71AV1	Cytochrome P450 71AV1
°C	Degree Celsius
CB1	Cannabinoid receptor type-1
CFO	*Cymbopogon flexuosus*
CH$_2$Cl$_2$	Dichloromethane
CHCl$_3$	Trichloro methane
CNS	Central nervous system
COX	Cyclooxygenase
CSCs	Cancer stem cells
DBR2	Double bond reductase-2
DMNT	(3E)-4,8-dimethy-1,3,7-nonatriene
DMSO	Dimethyl sulfoxide
DNA	Deoxyribonucleic acid
DPPH	α, α-diphenyl-β-picrylhydrazyl
EtOH	Ethanol
FDA	Food and Drug Administration
FDP	Farnesyl diphosphate
FPP	Farnesyl pyrophosphate
FRAP	Ferric reducing ability assay
GAS	Germacrene-A synthase
GFPP	Geranyl-farnesyl-pyrophosphate
GSH	Glutathione
H$_2$O	Water
HCl	Hydrochloric acid
IC50	Half maximal inhibitory concentration
IGF-2	Insulin-like growth factor-2 gene
ISO	Isointermedeol
kg	Kilogram

LC50	Lethal concentration required to kill 50% of the population.
LD50	Lethal Dose (50%)
MAPK	Mitogen-activated protein kinase
MDR	Multidrug resistance
MFC	Minimum fungicidal concentration
mRNA	Messenger RNA
MTT	(3-(4,5-dimethylthiazol-2-yl)-2,5-diphenyltetrazolium bromide
mg	Milligram
MIC	Minimum (or minimal) inhibitory concentration
Ml	Milliliter
MEP	2-C-methyl-D-erythritol-4-phosphate
MRSA	Methicillin-resistant *Staphylococcus aureus*
MVA	Mevalonic acid
NF	Nuclear factor
NMR	Nuclear magnetic resonance
NO	Nitric oxide
O_2	Oxygen
Ophs	Ophiobolins
PGE_2	Prostaglandin-E_2
PLA_2	Phospholipase A_2
PSA	Prostate specific antigen
ROS	Reactive oxygen species
SERCA	Sarcoplasmic reticulum Ca^{2+}/ATPase pump
STM	Santamarine
TNF	Tumor necrosis factor
TrxR	Thioredoxin reductase
µg	Microgram
µM	Micromolar
rt	Room temperature
THF	Tetrahydrofuran
VEGF	Vascular epithelial growth factor

REFERENCES

1. Sharifi-Rad, J.; Sureda, A.; Tenore, G.; Daglia, M.; Sharifi-Rad, M.; Valussi, M.; Tundis, R.; Sharifi-Rad, M.; Loizzo, M.; Ademiluyi, A. Biological activities of essential oils: From plant chemoecology to traditional healing systems. *Molecules*, **2017**, 22, 70.

2. Lombard, J.; Moreira, D. Origins and early evolution of the mevalonate pathway of isoprenoid biosynthesis in the three domains of life. *Molecular Biology and Evolution*, **2010**, 28, 87–99.

3. Bouvier, F.; Rahier, A.; Camara, B. Biogenesis, molecular regulation and function of plant isoprenoids. *Progress in Lipid Research*, **2005**, 44, 357–429.

4. Gershenzon, J.; Dudareva, N. The function of terpene natural products in the natural world. *Nature Chemical Biology*, **2007**, 3, 408.

5. Bohlmann, J.; Meyer-Gauen, G.; Croteau, R. Plant terpenoid synthases: Molecular biology and phylogenetic analysis. *Proceedings of the National Academy of Sciences of the United States of America* **1998**, 95, 4126–4133.

6. Buckle, J. Chapter 3: Basic plant taxonomy, basic essential oil chemistry, extraction, biosynthesis, and analysis. In: *Clinical Aromatherapy*, 3rd Ed., United Kingdom: Churchill Living Stone, **2015**, 37–72.

7. Chizzola R. *Regular Monoterpenes and Sesquiterpenes (Essential Oils), Natural Products.* Berlin Heidelberg: Springer, 2013, 2973–3008.

8. Osbourn, A. E.; Lanzotti, V. *Plant-Derived Natural Products-Synthesis, Function and Application.* New York: Springer; 2009. 3–50.

9. Cordell G. A. Biosynthesis of sesquiterpenes. *Chemical Reviews* **1976**, 76, 425–460.

10. Davis, E. M.; Croteau, R. Cyclization enzymes in the biosynthesis of monoterpenes, sesquiterpenes, and diterpenes. *Topics in Current Chemistry.* **2000**, 209, 53–95.

11. Isolation and bacterial expression of a sesquiterpene synthase cDNA clone from peppermint (*Mentha piperita*, L.) that produces the aphid alarm pheromone (E)-β-farnesene. *Proceedings of the National Academy of Sciences of the United States of America*, **1997**, 94, 12833–12838.

12. Srivastava, L. M. Abscisic Acid, One of the Key Determinants of in Vitro Shoot Differentiation from Cotyledons of Vigna radiata. In: Srivastava, L. M. (ed.). *Plant Growth and Development Hormones and Environment*, Elsevier Science, Atlanta, GA, 2002, 217–231.

13. Spakowicz, D. J.; Strobel, S. A. Biosynthesis of hydrocarbons and volatile organic compounds by fungi: bioengineering potential. *Applied Microbiology and Biotechnology*, **2015**, 99(12), 4943–4951.

14. Silva, G. N. S. D.; Pozzatti, P.; Rigatti, F.; Hörner, R.; Alves, S. H.; Mallmann, C. A.; Heinzmann, B. M, Antimicrobial evaluation of sesquiterpene α-curcumene and its synergism with Imipenem. *The Journal of Microbiology, Biotechnology and Food Sciences*, **2015**, 4(5) 434–436.

15. Hartsel, J. A.; Eades, J.; Hickory, B.; Makriyannis, A. Cannabis sativa and Hemp. In: Gupta, R. C. (ed.). *Nutraceuticals: Efficacy, Safety and Toxicity*, Academic Press, Elsevier, Cambridge, 2016, 735–754.

16. Passos, G. F., Fernandes, E. S., da Cunha, F. M., Ferreira, J., Pianowski, L. F., Campos, M. M. and Calixto, J. B. Anti-inflammatory and anti-allergic properties of the essential oil and active compounds from Cordia verbenacea. *Journal of Ethnopharmacology*, **2007**, 110(2), 323–333.

17. Fernandes E.S.; Passos G.F.; Medeiros R.; da Cunha F.M.; Ferreira J.; Campos M.M.; Pianowski L.F.; Calixto J.B. Anti-inflammatory effects of compounds alpha-humulene and (−)-trans-caryophyllene isolated from the essential oil of Cordia verbenacea. *European Journal of Pharmacology*, **2007**, 569(3), 228–236.

18. Rani K. Cyclisation of farnesyl pyrophosphate into sesquiterpenoids in ginger rhizomes (*Zingiber officinale*). *Fitoterapia.* **1999**, 70(6), 568–574.

19. Cane, D. E.; McIlwaine, D. B.; Harrison, P. H. M. Bergamotene biosynthesis and the enzymic cyclization of farnesyl pyrophosphate. *Journal of the American Chemical Society*, **1989**, 111(3), 1152–1153.

20. Cane, D. E.; McIlwaine, D. B. The biosynthesis of ovalicin from β-trans-bergamotene. *Tetrahedron Letters*, **1987**, 28(52), 6545–6548.

21. Fidyt, K.; Fiedorowicz, A.; Strządała, L.; Szumny, A. β-caryophyllene and β-caryophyllene oxide-natural compounds of anticancer and analgesic properties. *Cancer Medicine* **2016**, 5(10), 3007–3017.

22. Borg Karlson, A. K.; Norin, T.; and Talvitie, A. Configurations and conformations of torreyol (δ-cadinol), α-cadinol, T-muurolol and T-cadinol. *Tetrahedron*, **1981**, 37(2), 425–430.

23. Lee, J. H.; Yang, H. Y.; Lee, H. S.; Hong, S. K. Chemical composition and antimicrobial activity of essential oil from cones of Pinus koraiensis. *The Journal of Microbiology and Biotechnology*, **2008**, 18, 497–502.

24. Kundu, A.; Saha, S.; Walia, S.; Ahluwalia, V.; Kaur, C. Antioxidant potential of essential oil and cadinene sesquiterpenes of *Eupatorium adenophorum, Toxicological & Environmental Chemistry*, **2013**, 95(1), 127–137.

25. Bohlmann, F.; Gupta, R. Six cadinene derivatives from *Ageratina adenophora. Phytochemistry*, **1981**, 20, 1432–1433.

26. Fraga, B. M. Natural sesquiterpenoids. *Natural Product Reports*, **1988**, 5(5), 497–521.

27. Wang, Y.-C.; Wen, M.-L.; Li, M.-G.; Zhao, J.-Y.; Han, X.-L. Progress in biosynthesis of santalene and santalol. *Chinese Journal of Biotechnology*, **2018**, 34, 862–875.

28. Jones, C. G.; Moniodis, J.; Zulak, K.G.; Scaffidi, A.; Plummer, J.A.; Ghisalberti, E.L.; Barbour, E.L.; Bohlmann, J. Sandalwood fragrance biosynthesis involves sesquiterpene synthases of both the terpene synthase (TPS)-a and TPS-b subfamilies, including santalene synthases. *The Journal of Biological Chemistry*, **2011**, 286(20), 17445–17454.

29. Sun, X. Y.; Zheng, Y. P.; Lin, D. H.; Zhang, H.; Zhao, F.; Yuan, C. S. Potential anti-cancer activities of Furanodiene, a Sesquiterpene from *Curcuma wenyujin. The American Journal of Chinese Medicine*, **2009**, 37, 589–596.

30. Edris, A. E. Anti-cancer properties of Nigella spp. essential oils and their major constituents, thymoquinone and beta-elemene. *Current Clinical Pharmacology*, **2009**, 4, 43–46.

31. Kumar, A.; Malik, F.; Bhushan, S.; Sethi, V. K.; Shahi, A. K.; Kaur, J.; Taneja, S. C.; Qazi, G. N.; Singh, J. An essential oil and its major constituent isointermedeol induce apoptosis by increased expression of mitochondrial cytochrome-c and apical death receptors in human leukaemia HL-60 cells. *Chemico-Biological Interactions*, **2008**, 171(3), 332–347.

32. Nakagawa, M.; Ohno, T.; Maruyama, R.; Okubo, M.; Nagatsu, A.; Inoue, M.; Tanabe, H.; Takemura, G.; Minatoguchi, S., Fujiwara, H. Sesquiterpene lactone suppresses vascular smooth muscle cell proliferation and migration via inhibition of cell cycle progression. *Biological and Pharmaceutical Bulletin*, **2007**, 30(9), 1754–1757.

33. Nishimura, K.; Miyase, T.; Ueno, A.; Noro, T.; Kuroyanagi, M.; Fukushima, S. Sesquiterpene lactones from Lactuca laciniata. *Phytochemistry*, **1986**, 25(10), 2375–2379.

34. Fischer, N. H.; Olivier, E. J.; Fischer H.D. The biogenesis and chemistry of sesquiterpene lactones. *Fortschritte der Chemie organischer Naturstoffe*, 1979, 38, 47–390.

35. Meshnick, S. R.; Taylor, T. E.; Kamchonwongpaisan, S. Artemisinin and the antimalarial endoperoxides: From herbal remedy to targeted chemotherapy. *Microbiology Review*, **1996**, 60, 301–315.

36. Nosten, F.; White, N. J. Artemisinin-based combination treatment of *Falciparum* Malaria. *American Journal of Tropical Medicine and Hygiene*, **2007**, 77, 181–192.

37. Meshnick, S. R. Artemisinin: Mechanisms of action, resistance and toxicity. *International Journal for Parasitology*, **2002**, 32, 1655–1660.

38. O'neill, P. M.; Barton, V. E.; Ward, S. A. The molecular mechanism of action of artemisinin-The debate continues. *Molecules*, **2010**, 15, 1705–1721.

39. Golenser, J.; Waknine, J. H.; Krugliak, M.; Hunt, N. H.; Grau, G. E. Current perspectives on the mechanism of action of artemisinins. *International Journal for Parasitology*, **2006**, 36, 1427–1441.

40. Bouwmeester, H. J.; Wallaart, T. E.; Janssen, M. H. A.; van Loo, B.; Jansen, B. J. M.; Posthumus, M. A.; Schmidt, C. O.; De Kraker, J.-W.; König, W. A.; Franssen, M. C. R. Amorpha-4,11-diene synthase catalyses the first probable step in artemisinin biosynthesis. *Phytochemistry*, **1999**, 52, 843–854.

41. Mercke, P.; Bengtsson, M.; Bouwmeester, H. J.; Posthumus, M. A.; Brodelius, P. E. Molecularcloning, expression, and characterization of amorpha-4,11-diene synthase, a key enzyme of artemisinin biosynthesis in *Artemisia annua* L. *Archives of Biochemistry and Biophysics*, **2000**, 381, 173–180.

42. Bertea, C.; Freije, J.; Van derWoude, H.; Verstappen, F.; Perk, L.; Marquez, V.; De Kraker, J.-W.; Posthumus, M.; Jansen, B.; De Groot, A. Identification of intermediates and enzymes involved in the early steps of artemisinin biosynthesis in *Artemisia annua. Planta Medica*, **2005**, 71, 40–47.

43. Teoh, K. H.; Polichuk, D. R.; Reed, D. W.; Nowak, G.; Covello, P. S. *Artemisia annua* L.(Asteraceae) trichome-specific cDNAs reveal CYP71AV1, a cytochrome P450 with a key role in the biosynthesis of the antimalarial sesquiterpene lactone artemisinin. *FEBS Letters*, **2006**, 580, 1411–1416.

44. Zhang, Y.; Teoh, K. H.; Reed, D. W.; Maes, L.; Goossens, A.; Olson, D. J.; Ross, A. R.; Covello, P. S. The molecular cloning of artemisinic aldehyde D11 (13) reductase and its role in glandular trichome-dependent biosynthesis of artemisinin in *Artemisia annua. Journal of Biological Chemistry*, **2008**, 283, 21501–21508.

45. Teoh, K. H.; Polichuk, D. R.; Reed, D. W.; Covello, P. S. Molecular cloning of an aldehydes dehydrogenase implicated in artemisinin biosynthesis in *Artemisia annua. Botany* **2009**, 87, 635–642.

46. El-Feraly, F. S.; Chan, Y. Isolation and characterization of the sesquiterpene lactones costunolide, parthenolide, costunolide diepoxide, santamarine, and reynosin from *Magnolia grandiflora* L. *Journal of Pharmaceutical Sciences*, **1978**, 67(3), 347–350.

47. Cechinel Filho, V.; Block, L. C.; Yunes, R. A.; Delle Monache F. Paludolactone: A new eudesmanolide lactone from *Wedelia paludosa* DC. (*Acmela brasiliensis*). *Natural Product Research*, **2004**, 18(5), 447–451.

48. Ulloa, J. L.; Spina, R.; Casasco, A.; Petray, P. B.; Martino, V.; Sosa, M.A.; Frank, F.M.; Muschietti, L. V. Germacranolide-type sesquiterpene lactones from *Smallanthus sonchifolius* with promising activity against *Leishmania mexicana* and *Trypanosoma cruzi. Parasites & Vectors*, **2017**, 10, 2–10.

49. Li, P. F., Zhan, H. Q., Li, S. Y., Liu, R. L., Yan, F. L., Cui, T. Z., Yang, Y. P., Li, P.; Wang, X. Y. Lactuside B decreases aquaporin-4 and caspase-3 mRNA expression in the hippocampus and striatum following cerebral ischaemia-reperfusion injury in rats. *Experimental and Therapeutic Medicine*, **2014**, 7, 675–680.

50. Tiuman T. S., Ueda-Nakamura T., Garcia Cortez D. A., Dias Filho B. P., Morgado-Díaz J. A., de Souza W., Nakamura C. V. Antileishmanial activity of parthenolide, a sesquiterpene lactone isolated from *Tanacetum parthenium*. *Antimicrobial Agents and Chemotherapy*, **2005**, 49(1), 176–182.

51. Kupchan, S. M.; Aynehchi, Y.; Cassady, J. M.; Mc Phail, A. T.; Sim, G. A.; Shnoes, H. K.; Burlingame, A. L. The isolation and structural elucidation of 2 novel sesquiterpenoid tumor inhibitors from *Elephantopus elatus*. *Journal of the American Chemical Society*, **1966**, 88(15), 3674–3676.

52. Simonsen, H. T.; Weitzel, C.; Christensen, S. B. Guaianolide sesquiterpenoids: Pharmacology and biosynthesis. In: Ramawat, K., Mérillon, JM. (eds) *Natural Products*, Springer, Berlin, Heidelberg, 2013, 3069–3098.

53. Ali, H., Christensen, S. B., Foreman, J. C., Pearce, F. L., Piotrowski, W., Thastrup, O. The ability of thapsigargin and thapsigargicin to activate cells involved in the inflammatory response. *British Journal of Pharmacology*, **1985**, 85(3), 705–712.

54. Patkar, S. A.; Rasmussen, U.; Diamant, B. On the mechanism of histamine release induced by thapsigargin from Thapsia garganica L. *Agents Actions*, **1979**, 9, 53–57.

55. Thastrup, O.; Cullen, P.J.; Drobak, B.K.; Hanley, M.R.; Dawson, A.P. Thapsigargin, a tumor promoter, discharges intracellular Ca^{2+} stores by specific inhibition of the endoplasmic reticulum Ca^{2+}-ATPase. *Proceedings of the National Academy of Sciences of the United States of America*, **1990**, 87, 2466–2470.

56. Sagara, Y.; Fernandez-Belda, F.; De Meis, L.; Inesi, G. Characterization of the inhibition of intracellular Ca^{2+} transport ATPases by thapsigargin. *Journal of Biological Chemistry*, **1992**, 267, 12606–12613.

57. Denmeade, S. R.; Jakobsen, C. M.; Janssen, S.; Khan, S. R.; Garrett, E. S.; Lilja, H.; Christensen, S. B.; Isaacs, J. T. Prostate-specific antigen-activated thapsigargin prodrug as targeted therapy for prostate cancer. *Journal of the National Cancer Institute*, **2003**, 95, 990–1000.

58. Jiang, S.; Chow, S.; Nicotera, P.; Orrenius, S. Intracellular Ca^{2+} signals activate apoptosis in thymocytes: Studies using the Ca^{2+}-ATPase inhibitor thapsigargin. *Experimental Cell Research*, **1994**, 212, 84–92.

59. Ganley, I. G.; Wong, P.-M.; Gammoh, N.; Jiang, X. Distinct autophagosomal-lysosomal fusion mechanism revealed by thapsigargin-induced autophagy arrest. *Molecular Cell*, **2011**, 42, 731–743.

60. Mahalingam, D.; Peguero, J.; Cen, P.; Arora, S. P.; Sarantopoulos, J.; Rowe, J.; Allgood, V.; Tubb, B.; Campos, L. A Phase II, Multicenter, single-arm study of mipsagargin (G-202) as a second-line therapy following sorafenib for adult patients with progressive advanced hepatocellular carcinoma. *Cancers*, **2019**, 11, 833.

61. Pickel, B.; Drew, D. P.; Manczak, T.; Weitzel, C.; Simonsen, H. T.; Ro, D. K. Identification and characterization of a kunzeaol synthase from Thapsia garganica: Implications for the biosynthesis of the pharmaceutical thapsigargin. *Biochemical Journal*, **2012**, 448, 261.

62. Andersen, T. B.; Martinez-Swatson, K. A.; Rasmussen, S. A.; Boughton, B. A.; Jorgensen, K.; Andersen-Ranberg, J.; Nyberg, N.; Christensen, S. B.; Simonsen, H. T.

63. Chang, Y. T.; Wang, C. C. N.; Wang, J. Y.; Lee, T. E.; Cheng Y. Y.; Morris-Natschke S. L.; Lee, K. H.; Hung C. C. Tenulin and isotenulin inhibit P-glycoprotein function and overcome multidrug resistance in cancer cells. *Phytomedicine*, **2019**, 53, 252–262.

64. Clark, E.P. The constituents of certain species of Helenium. II. Tenulin. *Journal of the American Chemical Society*, **1939**, 61, 1836–1840.

65. Kamatou, G. P.; Viljoen, A. M. A review of the application and pharmacological properties of α-bisabolol and α-bisabolol-rich oils. *Journal of the American Oil Chemists Society*, **2010**, 87(1), 1–7.

66. Alves Ade, M.; Goncalves, J.; Cruz J.; Araujo D. Evaluation of the sesquiterpene (–)-alpha-bisabolol as a novel peripheral nervous blocker. *Neuroscience Letters*, **2010**, 472(1), 11–15.

67. Joo, J. H; Jetten A. M. Molecular mechanisms involved in farnesol-induced apoptosis. *Cancer Letters*, **2009**, 287(2), 123–135.

68. Chan, W. K.; Tan, L. T. H.; Chan, K. G.; Lee, L. H.; Goh, B. H. Nerolidol: A sesquiterpene alcohol with multi-faceted pharmacological and biological activities. *Molecules*, **2016**, 21(5), 529.

69. Bouwmeester, H. J.; Verstappen, F. W.; Posthumus, M. A.; Dicke, M. Spider mite-induced (3S)-(E)-nerolidol synthase activity in cucumber and lima bean. The first dedicated step in acyclic c11-homoterpene biosynthesis. *Plant Physiology*. **1999**, 121, 173–180.

70. Nigmatov, A. G.; Serebryakov, é. P.; Yanovskaya, L. A. Improved method for the isolation of geranyl esters of (4E/Z, 8E)- and (4E/Z, 8Z)-farnesylacetic acid. *Pharmaceutical Chemistry Journal*. **1987**, 21, 529–533.

71. Kiyohara, H.; Ichino, C.; Kawamura, Y.; Nagai, T.; Sato, N.; Yamada, H. Patchouli alcohol: *in-vitro* direct anti-influenza virus sesquiterpene in *Pogostemon cablin* Benth. *The Journal of Natural Medicines-Tokyo*, **2012**, 66, 55–61.

72. Li, Y.; Xian, Y.; Ip, S.; Su, Z.; Su, J.; He, J. Antiinflammatory activity of patchouli alcohol isolated from pogostemonis herba in animal models. *Fitoterapia*, **2011**, 82(8): 1295–1301.

73. Rudi, A.; Yosief, T.; Schleyer, M.; Kashman, Y. Bilosespens A and B: Two novel cytotoxic sesterpenes from the marine sponge *Dysidea cinerea*. *Organic Letters*, **1999**, 1, 471–472.

74. Au, T. K.; Chick P. C.; Leung P. C. The biology of ophiobolins. *Life Science*, **2000**, 67, 733–742.

75. Pak, C. L.; William, A. T.; Wang, J. H.; Carl, L. T. Role of calmodulin inhibition in the mode of action of ophiobolin A. *Plant Physiology*, **1985**, 77, 303–308

76. Tian, W.; Deng, Z.; Hong, K. The biological activities of sesterterpenoid-type ophiobolins. *Marine Drugs*, **2017**, 15, 229.

77. Leung, P. C.; Taylor, W. A.; Wang, J. H.; Tipton, C. L. Ophiobolin A, A natural product inhibitor of calmodulin. *Journal of Biological Chemistry*, **1984**, 259, 2742–2747.

78. Zhu, W. M.; Lu, Z. Y.; Hong, K.; Miao, C. *The Preparation and Application of Sesterterpenoids Compounds Ophiobolins*. CN Patent 101591314, 2 December 2009.

Localization and in-vivo characterization of *Thapsia garganica*: CYP76AE2 indicates a role in thapsigargin biosynthesis. *Plant Physiology*, **2017**, 174, 56.

79. Sakoulas, G.; Nam, S.-J.; Loesgen, S.; Fenuical, W.; Gensen, P. R.; Nizet, V.; Hensler, M. Novel bacterial metabolite mer-hochlorin: A demonstrates *in-vitro* activities against multi-drug resistant methicillin-resistant Staphylococcus aureus. *PLoS One*, **2012**, 7, e29439.

80. Xu, Y.-m.; Espinosa-Artiles, P.; Liu, M. X.; Arnold, A. E.; Gunatilaka, A. A. L. Secoemestrin D: A cytotoxic epitetrathiodioxopiperizine and emericellenes A–E, five sesterterpenoids from *Emericella* sp. AST0036, a fungal endophyte of *Astragalus lentiginosus*. *Journal of Natural Products*, **2013**, 76(12), 2330–2336.

81. Soriente, A.; De Rosa, M. M.; Scettri, A.; Sodano, G.; Terencio, M. C.; Payá, M.; Alcaraz, M. J. Manoalide. *Current Medicinal Chemistry*, **1999**, 6(5), 415–431.

82. Albizati, K. F.; Holman, T.; Faulkner, D. J.; Glaser, K. B.; Jacobs, R. S. Luffariellolide, an anti-inflammatory sester-terpene from the marine spongeLuffariella sp., *Experientia*, **1987**, 43, 949–950.

83. Tsuda, M., Shigemori, H., Ishibashi, M., Sasaki, T. Luffariolides A-E, new cytotoxic sesterpenes from the okinawan marine sponge *Luffariella* sp. *The Journal of Organic Chemistry*, **1992**, 57, 3503–3507.

84. Ueoka, R.; Nakao, Y.; Fujii, S.; van Soest, W.M.; Matsunaga, S. Aplysinoplides A-C, cytotoxic sesterterpenes from the marine sponge *Aplysinopsis digitata*. *Journal of Natural Products*, **2008**, 71, 1089–1091.

85. Piao, S. J.; Zhang, H. J.; Lu, H. Y.; Yang, F.; Jiao, W. H.; Yi, Y. H.; Chen, W. S.; Lin, H. W. Hippolides A-H, acyclic mano-alide derivatives from the marine sponge *Hippospongia lachne*, *Journal of Natural Products*, **2011**, 74(5), 1248–1254.

86. Schmidt, E. W.; Faulkner, D. J. A new anti-inflammatory sesterterpene from the sponge *Fascaplysinopsis* sp. from Palau. *Tetrahedron Letters*, **1996**, 37(23), 3951–3954.

87. Mizushina, Y.; Murakami, C.; Yogi, K.; Ueda, K.; Ishidoh, T.; Takemura, M.; Perpelescu, M.; Suzuki, M.; Oshige, M.; Yamaguchi, T.; Saneyoshi, M. Kohamaic acid-A: A novel sesterterpenic acid, inhibits activities of DNA polymerases from deuterostomes. *Biochimica et Biophysica Acta (BBA)-Proteins and Proteomics*, **2003**, 1648(1–2), 55–61.

88. Randazzo, A.; Debitus, C.; Minale, L.; García Pastor, P.; Alcaraz, M. J.; Payá, M.; Gomez-Paloma, L. Petrosaspongiolides M-R: New potent and selective phos-pholipase-A_2 inhibitors from the new caledonian marine sponge *Petrosaspongia nigra*. *Journal of Natural Products*, **1998**, 61(5), 571–575.

89. Garcia-Pastor, P.; Randazzo, A.; Gomez-Paloma, L.; Alcaraz, M. J.; Paya, M. Effects of petrosaspongiolide M: A novel phospholipase-A_2 inhibitor, on acute and chronic inflammation. *The Journal of Pharmacology and Experimental Therapeutics*, **1999**, 289(1), 166–172.

90. González, M. A. Scalarane sesterterpenoids. *Current Bioactive Compounds*, **2010**, 6, 178–210.

91. Pettit, G. R.; Tan, R.; Melody, N.; Cichacz, Z. A.; Herald, D. L.; Hoard, M. S.; Pettit, R. K.; Chapuis, J. C. Antineoplastic agents 397: isolation and structure of sesterstatins 4 and 5 from *Hyrtios erecta*. *Bioorganic & Medicinal Chemistry Letters*, **1998**, 8, 2093–2098.

92. Zhang, H. J.; Fang, H. F.; Yi, Y. H.; Lin H. W. Scalarane sesterpenes from the Chinese spongia *Phyllospongia folia-scens*. *Helvetica Chimica Acta*, **2009**, 92, 762–767.

93. Chang, Y. C.; Tseng, S. W.; Liu, L. L.; Chou, Y.; Ho, Y.S.; Lu, M. C.; Su J. H. Cytotoxic sesterterpenoids from a sponge *Hyppospongia* sp. *Marine Drugs*, **2012**, 10, 987–997.

94. Somerville, M.J.; Hooper, J. N. A.; Garson, M. J. Mooloolabenes A-E, norsesterterpenes from the Australian sponge *Hyattella intestinalis*. *Journal of Natural Products*, **2006**, 69, 1587–1590.

95. Schumacher, M.; Cerella, C.; Eifes, S.; Chateuvieux, S.; Morceau, F.; Jaspars, M.; Dicato M.; Diederich, M. Heteronemin, a spongean sesterterpene, inhibits TNF alpha-induced NF-kB activation through proteasome inhibition and induces apoptotic cell death. *Biochemical Pharmacology*, **2010**, 79, 610–622.

96. Zhang, C.; Liu, Y. Targeting cancer with sesterterpenoids: The new potential antitumor drugs. *Journal of Natural Medicines*, **2015**, 69, 255–266.

97. De Rosa, S.; Crispino, A.; De Giulio, A.; Iodice, C.; Amodeo, P.; Tancredi, T. A New Cacospongionolide deriv-ative from the Sponge *Fasciospongia cavernosa*. *Journal of Natural Products*, **1999**, 62(9), 1316–1318.

98. De Rosa, S.; De Stefano, S.; Zavodnik, N. Cacospongionolide: A new antitumoral sesterpene, from the marine sponge *Cacospongia mollior*. *The Journal of Organic Chemistry*, **1988**, 53, 5020–5023.

99. Cheung, A. K., Murelli, R., & Snapper, M. L. Total syntheses of (+)- and (–)-cacospongionolide-B, cacospongionolide-E, and related analogues. Preliminary study of structural fea-tures required for phospholipase A_2 Inhibition. *The Journal of Organic Chemistry*, **2004**, 69(17), 5712–5719.

100. Arai, M.; Kawachi, T.; Seitawan, A.; Kobayashi, M. Hypoxia-selective growth inhibition of cancer cells by furospino-sulin-1, a furanosesterpene isolated from an Indonesian marine sponge. *ChemMedChem*, **2010**, 5, 1919–1926.

101. Shen, Y.-C.; Lo, K.-L.; Lin, Y.-C.; Khalil, A. T.; Kuo Y.-K.; Shih P.-S. Novel linear C_{22}-sesterterpenoids from sponge *Ircinia formosana*. *Tetrahedron Letters*, **2006**, 47, 4007–4010.

102. Su, J.-H.; Tseng, S.-W.; Lu, M.-C.; Liu, L.-L.; Chou, Y.; Sung, P.-J. Cytotoxic C21 and C22 terpenoid-derived metabolites from the sponge Ircinia sp. *Journal of Natural Products*, **2011**, 74, 2005–2009.

103. Hussain, S.; Slevin, M.; Matou, S.; Ahmed, N.; Iqbal Choudhary, M.; Ranjit, R.; West, D.; Gaffney, J. Anti-angiogenic activity of sesterterpenes; natural product inhib-itors of FGF-2-induced angiogenesis. *Angiogenesis*, **2008**, 11, 245–256.

104. Dai, J.; Liu, Y.; Zhou Y.-D.; Nagle, D. G. Hypoxia-selective antitumor agents: norsesterterpene peroxides from the marine sponge *Diacarnus levii* preferentially suppress the growth of tumor cells under hypoxia conditions. *Journal of Natural Products*, **2007**, 70, 648–651.

105. Carotenuto, A.; Fattorusso, E.; Lanzotti, V.; Magno, S.; Carnuccio, R.; Iuvone, T. Antiproliferative sesterterpenes from the caribbean sponge *Cacospongia cf. linteiformis*. *Comparative Biochemistry & Physiology*, **1998**, 119C, 119–123.

106. Acebey, L.; Sauvain, M.; Beck, S.; Moulis, C.; Gimenez, A.; Jullian, V. Bolivianine, a new sesterpene with an unusual skeleton from *Hedyosmum angustifolium*, and its isomer, isobolivianine. *Organic Letters*, **2007**, 9(23), 4693–4696. Doi: 10.1021/ol7015725.

11

Terpenes as Useful Drugs

Biswa Mohan Sahoo*, Bimal Krishna Banik†, and Abhishek Tiwari‡

CONTENTS

11.1 Introduction 361
    11.1.1 Terpenes as Antimalarial Drugs 363
    11.1.2 Terpenes as Anthelmintic Drugs 365
    11.1.3 Terpenes as Anti-inflammatory Drugs 367
    11.1.4 Terpenes as Anti-HIV Drugs 369
    11.1.5 Terpenes as Anti-Alzheimer Drugs 376
    11.1.6 Terpene as Antimicrobial Drugs 378
    11.1.7 Terpenes as Anticancer Drugs 379
    11.1.8 Terpenes as Antioxidants 382
    11.1.9 Immunomodulatory Activity 382
    11.1.10 Terpenes as Hepatoprotective 382
    11.1.11 Terpenes as Anti-Diabetic Drugs 383
11.2 Miscellaneous 386
11.3 Conclusion 392
Abbreviation 392
Acknowledgment 392
References 392

11.1 Introduction

Terpene is a hydrocarbon isolated from turpentine, volatile oil obtained from pine trees. Terpenoids are modified terpenes with changed positions of methyl groups or added oxygen atoms. Terpenoids are volatile substances and are mainly responsible for the smell of plants and flowers. These are common constituents of different plant groups namely citrus, conifers and eucalyptus and are usually present in their leaves and fruits. The basis of classification of naturally occurring terpene is the number of carbon atoms present and most of them possess the general formula of $(C_5H_8)n$, where 'n' differ among the various groups. So, the classification of terpenes can be made rationally based on the presence of isoprene units in their structures [1].

Some of the common properties of terpenoids include most of these compounds are colorless and fragrant liquids at room temperature and lighter than water. Some terpenoids are solid like camphor. Terpenoids are generally insoluble in water but can be dissolved in organic solvents. Structurally, terpenoids are open-chained or cyclic unsaturated compounds with one or more than one double bond. Chemically, terpenoids are liable for oxidation by oxidizing agents and when exposed thermally, can decompose to isoprene. Terpenoids also undergo polymerization and dehydrogenation to produce long-chain products. Based on the cycles present in the terpenoids, these are further classified into acyclic, monocyclic, bicyclic, tricyclic and tetracyclic terpenoids [2].

Natural monoterpenes such as limonene, carvone, carveol, pyrethrin etc have been reported to exert their antimicrobial or anticancer effect through various mechanisms such as apoptosis up-regulation, influencing the posttranslational modification or prenylation of some physiologically important protein molecules and blocking the repolarization of sodium channel of neurons of target tissue (Table 11.1) [3].

Most of the terpenes are found to possess various biological activities such as anti-diabetics, antimicrobial, digestive stimulant, anticancer, anti-Alzheimer, anthelmintic, antimalarial, antioxidant, immunomodulators, hepatoprotective, anti-HIV, analgesic and anti-inflammatory (Figure 11.1) [4].

---
* Roland Institute of Pharmaceutical Sciences, Berhampur affiliated to Biju Patnaik University of Technology (BPUT), Rourkela, Odisha, India. Corresponding author e-mail address: drbiswamohansahoo@gmail.com.
† Department of Mathematics and Natural Sciences, College of Sciences and Human Studies, Prince Mohammad Bin Fahd University, Al Khobar, Kingdom of Saudi Arabia. Corresponding author e-mail address: bimalbanik10@gmail.com.
‡ Faculty of Pharmacy, Pharmacy Academy, IFTM University, Lodhipur Rajput, Moradabad-244102, Uttar Pradesh, India.

TABLE 11.1

List of Terpenes with Diverse Therapeutic Effects

Terpenes	Type	Source	Therapeutic Effects
α-Pinene	Monoterpene	Rosemary	Anticancer Anti-inflammatory
β-Myrcene	Monoterpene	Cannabis	Sedative Analgesic Anti-inflammatory
Carvone	Monoterpene	Caraway	Anticancer Anti-inflammatory Anti-diabetic Antimicrobial
Limonene	Monoterpene	Citrus fruit	Anticancer Anti-anxiety Antioxidant
Linalool	Monoterpene alcohol	Lavender	Antimicrobial Antifungal
β-Caryophyllene	Sesquiterpene	Clove	Anti-inflammatory Antimicrobial Antioxidant
Abietane	Diterpene	Plectranthus	Anti-inflammatory Antioxidant Antitumor

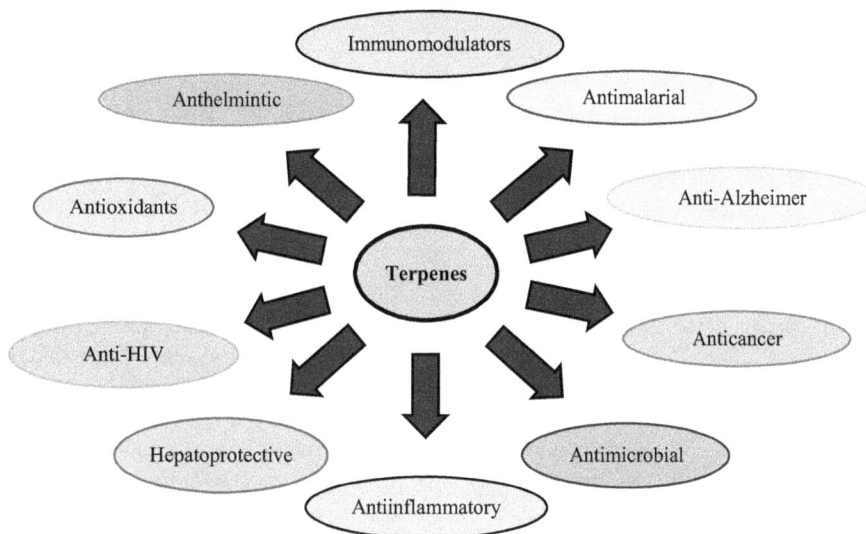

FIGURE 11.1 Terpenes as useful drugs.

11.1.1 Terpenes as Antimalarial Drugs

Malaria is a mosquito-borne infectious disease and major public health problem worldwide. Due to the emergence of malaria parasite strains resistant to conventional medicinal agents, it was necessary to find out new effective drug molecules [5]. So, herbal drugs like terpenes are evaluated for their antimalarial activity *in vitro* and *in vivo*. Malaria is caused by single-celled microorganisms of the *Plasmodium* group (*P. falciparum, P. vivax, P. ovale, P. malariae* and *Plasmodium knowlesi*). This disease is mainly spread by an infected female Anopheles mosquito. The cycle begins with the injection of infective sporozoites of salivary glands of Anopheles mosquito into the human host bloodstream [6].

Artemisinin is a sesquiterpene lactone found in *Artemisia annua* (Asteraceae) (Figure 11.2). It is used as an antimalarial drug with specific activity against the malaria parasite, *Plasmodium falciparum*. Artemisinin and its analogs can kill all stages of the life cycle of malarial parasites and

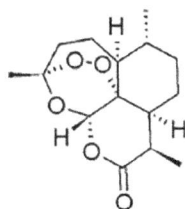

FIGURE 11.2 Structure of sesquiterpene lactone (Artemisinin).

are currently the main treatment against malaria worldwide. The antiparasitic activity is due to the presence of endoperoxide bridge which is eventually cleaved by the Fe^{2+} ion and forms radical ion intermediates which bind to the proteins like sarcoplasmic reticulum $Ca^{2+}ATPase$ and thereby killing the organism [7].

Nerolidol is a sesquiterpene alcohol present as an essential oil in several plants (ginger, jasmine, lavender, *Cannabis sativa* and lemon grass). There are two isomers of nerolidol such as cis and trans which differ in the geometry about the central double bond (Figure 11.3). Chemically, it is 3,7,11-trimethyl-1,6,10-dodecatrien-3-ol. The antimalarial activity of nerolidol was investigated in a mouse model of malaria. Mice were infected with Plasmodium berghei ANKA and were treated with 1,000 mg/kg/dose nerolidol in two doses administered by both oral and inhalation routes. Treatment with nerolidol by both routes inhibited parasitaemia by >99% and >80% for oral and inhalation routes respectively relative to untreated controls until 14 days after infection. Parasitaemia in mice treated by oral and inhalation routes showed similar trends on days 4, 7, 11, 14, 18 and 21 after infection ($p > 0.05$) and significant differences relative to untreated mice throughout the experiment ($p < 0.0001$) (Table 11.2). The toxicity of nerolidol administered by either route was not significant while genotoxicity was observed only at the highest dose tested. These results indicate that the combined use of nerolidol and other drugs targeting different points of the same isoprenoid pathway may be an effective treatment for malaria [8].

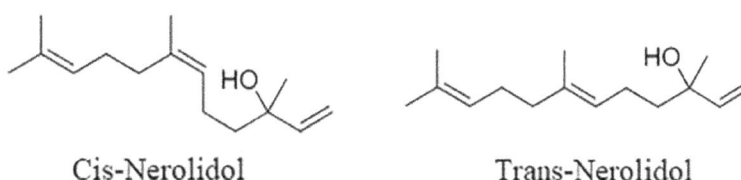

Cis-Nerolidol   Trans-Nerolidol

FIGURE 11.3 Structure of Nerolidol isomers.

TABLE 11.2

Antimalarial Activity of Nerolidol against *P. berghei* ANKA

Nerolidol (1,000 mg/kg)	Inhibition of Parasite Growth (%)					
	Day 4	**Day 5**	**Day 11**	**Day 14**	**Day 18**	**Day 21**
Inhalation route	57	93	60	80	78	44
Oral route	68	89	87	99	98.7	71.5

Bakuchiol is a meroterpene in the class of terpenophenol. In 1966, Mehta et al. first isolated Bakuchiol from the seeds of *Psoralea corylifolia*. The first complete synthesis of Bakuchiol has been described in 1973. Chemically, Bakuchiol is 4-[(1*E*, 3*S*)-3-ethenyl-3,7-dimethylocta-1,6-dienyl]phenol (Figure 11.4). It is made up of styryl moiety in conjunction with a monoterpene [9]. *In-vitro* anti-plasmodial activity was evaluated against chloroquine-sensitive *P. falciparum* NF-54 strain. It was demonstrated that the hexane extract of *P corylifolia* seeds showed good anti-plasmodial activity with IC_{50} of 18.3 ± 0.38 µg/mL. But the purified product from the hexane extract, bakuchiol exhibited improved anti-plasmodial activity with IC_{50} of 3.1 ± 0.35 µg/mL or 12.1 ± 0.92 µM in dose-dependent manner. Bakuchiol did not show hemolytic activity at higher concentration (3,900 µM). Bakuchiol showed synergistic interaction with the two most clinically used antimalarial drugs (chloroquine and quinine). The results of this study indicated that bakuchiol inhibits *P. falciparum* growth, inducing oxidative stress leads to mitochondrial membrane depolarization, DNA fragmentation and also acts synergistically. Hence, a continue efforts are needed to develop new antimalarial drugs for combating resistance to malaria [10,11]. The results of *in-vitro* drug interaction between bakuchiol and chloroquine or quinine are summarized in Tables 11.3 and 11.4 respectively.

Kayembe et al. reported the extraction of eleven terpenes from *Cassia alata* and *Ocimum gratissimum* leaves and screened for antimalarial activity against *Plasmodium falciparum in-vitro*. The results of the antimalarial activities of isolated terpenes showed that the four terpenes from Cassia alata (TCA1-TCQ4) and five terpenes from Ocimum gratissimum (TOG1, TOG2, TOG5, TOG6 and TOG7) are the very active against Plasmodium with all their IC_{50} below 1 µg/mL whereas TOG3 and TOG4 are active with IC_{50} values < 4 µg/mL (Table 11.5) [12].

Mayer et al. reported the antimalarial activity of a cis-terpenone. The oxidized derivative of hydroxy-cis terpenone (OHCT) is a synthetic molecule which is not toxic to cultured human liver cells at concentrations as high as 60 µM and inhibits the activity of cytochrome P450s that metabolize many drugs. *In-vitro* assay was carried out to determine the activity of OHCT against chloroquine-sensitive and -resistant strains of *Plasmodium falciparum* and *P. falciparum* clone that is partially resistant to artemisinin. OHCT was effective against all intraerythrocytic stages of *P. falciparum* and exhibited activity *in-vitro* against both chloroquine-sensitive and -resistant strains of *P. falciparum* as well as a *P. falciparum* clone that is partially resistant to artemisinin at nanomolar concentrations. However, OHCT exhibited potent activity against gametocytes, the form which is transmitted by mosquitoes and essential for the spread of malaria. OHCT displayed a significant inhibitory activity with an IC_{50} in the range of 3.9–93 nM depending on the parasite clone tested. The time course of action of OHCT (1 µM) was investigated with the *P. falciparum* clone Dd2 [13]. The parasitaemia level decreased by 50% following treatment with OHCT for 8 h, suggesting OHCT has a fast mechanism of action as shown in Tables 11.6–11.8.

Vanderwal et al. reported the antimalarial properties of Kalihinol analogs. Kalihinol is a member of isocyanoterpene family (Figure 11.5). The isocyanoterpenes (ICTs) are a group of sponge-derived polycyclic terpenoids bearing isonitrile functional group. The antimalarial activity was carried out by using both drug-sensitive and resistant *P. falciparum* strains [14].

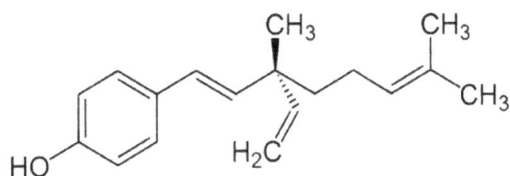

FIGURE 11.4 Structure of Bakuchiol.

TABLE 11.3

Interaction between Chloroquine and Bakuchiol

Combination	Ratio of Drugs		Chloroquine Mean FIC_{50}	Bakuchiol Mean FIC_{50}	Σ FICs	Interaction
	Chloroquine	**Bakuchiol**				
1	5	0	1.0	0.0	1.0	Additive
2	4	1	0.6	0.9	1.5	Additive
3	3	2	0.3	0.6	0.9	Synergistic
4	1	1	0.6	0.8	1.4	Additive
5	2	3	0.5	0.7	1.2	Additive
6	1	4	0.6	0.7	1.3	Additive
7	0	5	0.0	1.0	1.0	Additive

TABLE 11.4

Interaction between Quinine and Bakuchiol

Combination	Ratio of Drugs		Quinine Mean FIC_{50}	Bakuchiol mean FIC_{50}	Σ FICs	Interaction
	Quinine	**Bakuchiol**				
1	5	0	1.0	0.0	1.0	Additive
2	4	1	0.5	0.9	1.4	Additive
3	3	2	0.4	0.8	1.2	Synergistic
4	1	1	0.6	0.9	1.5	Additive
5	2	3	0.3	0.7	1.0	Additive
6	1	4	0.2	0.7	0.9	Additive
7	0	5	0.0	1.0	1.0	Additive

TABLE 11.5

Antimalarial Activity of Isolated Terpenes

Plant	Terpene	IC_{50} (μg/mL)
Cassia alata	TCA1	0.94
	TCA2	0.23
	TCA3	0.44
	TCA4	0.52
Ocimum gratissimum	TOG1	0.32
	TOG2	0.27
	TOG3	1.41
	TOG4	3.96
	TOG5	0.44
	TOG6	0.65
	TOG7	0.52
Quinine	-	0.1

TABLE 11.6

Inhibition of Plasmodium Survival by OHCT (nM)

	Plasmodium falciparum Clones			
	Chloroquine Resistant Clones		**Chloroquine-Sensitive Clone**	**Artemisinin Resistant Clone**
Test Compounds	**Dd2Nm**	**FCR3**	**HB3**	**78 GR**
OHCT-IC_{50}	85 ± 36.1	93 ± 48.7	38 ± 9.7	3.9 ± 0.5
OHCT-IC_{90}	422 ± 27.4	407 ± 150	523 ± 310	17.4 ± 0.9
Chlor-IC_{50}	85 ± 22.5	181 ± 92.8	-	-
Chlor-IC_{90}	546 ± 69.1	524 ± 150.3	< 20	-
Artem-IC_{50}	-	-	-	4.2 ± 0.4
Artem-IC_{90}	-	-	-	18.1 ± 1.0

* *P. falciparum* clones, two chloroquine resistant clones, Dd2Nm and FCR3, one chloroquine-sensitive clone, HB3 and the artemisinin resistant clone 78 GR were incubated with five concentrations of OHCT for 48 h. Percent parasitaemia in 9,500 infected human erythrocytes was estimated by microscopic observation of Giemsa-stained blood smears and by measurement of p(LDH). IC_{50} and IC_{90} values (nM) represent the mean $\pm$ SE of three experiments (vehicle and five duplicate concentrations per experiment).

11.1.2 Terpenes as Anthelmintic Drugs

Anthelmintics or antihelminthics are group of antiparasitic drugs which expel parasitic worms and other internal parasites from the body by either stunning or killing them without causing any damage to the host. There is an urgent need for developing new therapeutics for parasitic helminthic diseases affecting 1.5–2 billion people worldwide due to the threat of resistance development to existing treatments and also due to incomplete efficacies. The terpene thymol was successfully used to cure hookworm infections in the 1900s [15,16].

Horton et al. reported the anthemintic activity of dysinin-type sesquiterpenes (isolated from the marine sponge Dysidu herbarea Keller (Dysideidae)). Among which, (−) ennantiomers were active against the helminthes namely *Nippostrongylus brasiliensis, Nipostrongylus dubias* and *Hypselodorzs nana* whereas (+) ennantiomers were inactive. So, it was observed that the structural pattern and optical activity of terpenoids also influence their anthelmintic activity [17].

Ronaldo et al. reported the anthelmintic activity of the natural compound (+)-Limonene epoxide against *Schistosoma mansoni*.

TABLE 11.7

Time Course of OHCT Action

Time (min)	Parasitaemia (%)
30	8.6
60	8.5
120	8.6
240	8.1
360	6
480	4.7
Vehicle (0.1% DMSO)	8.9
Media	9.1

[a] The *P. falciparum* clone Dd2 (CQR) was treated with 0.5 μM OHCT for the times shown. Parasite viability was measured by the lactate dehydrogenase assay.

TABLE 11.8

OHCT (nM) Inhibits Growth of *P. falciparum* Gametocytes[a]

Test Compounds	Stage I and II		Stage III and IV	
	IC$_{50}$	IC$_{90}$	IC$_{50}$	IC$_{90}$
OHCT	22 ± 3.4	174 ± 5.4	36 ± 9.0	944 ± 92.0
Chloroquine	394 ± 127.0	1520 ± 27.5	226 ± 49.3	2109 ± 85.2

[a] IC$_{50}$ and IC$_{90}$ values (nM) represent the mean ± SE of duplicate experiments, each performed in duplicate (vehicle and 8 concentrations per experiment.

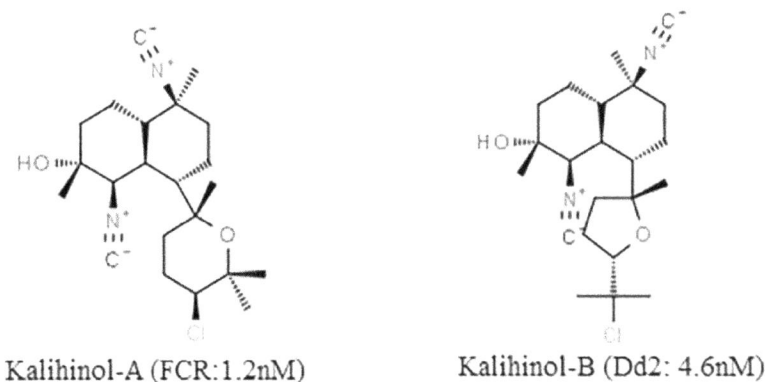

Kalihinol-A (FCR:1.2nM)   Kalihinol-B (Dd2: 4.6nM)

FIGURE 11.5 IC$_{50}$ values of Kalihinol against drug resistant strains of *P. falciparum*.

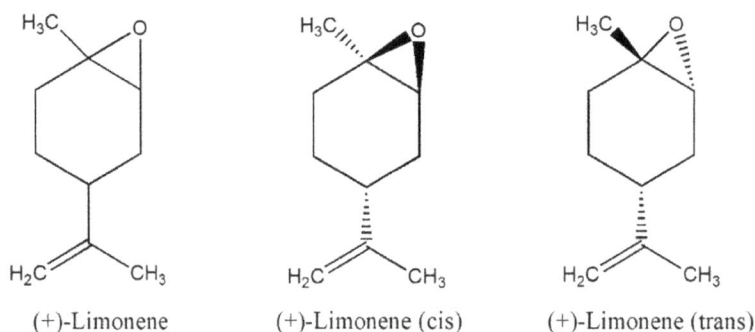

(+)-Limonene   (+)-Limonene (cis)   (+)-Limonene (trans)

FIGURE 11.6 Chemical structure of (+)-limonene epoxide.

(+)-Limonene epoxide is a mixture of cis and trans isomers present in many plants (Figure 11.6). The effects of (+)-limonene epoxide at 25 μg/mL against *S. mansoni* adult worms were similar to those of the positive control (praziquantel), with the reduction in motility and death of all worms after 120 h (Table 11.9) [18,19].

TABLE 11.9

In-vitro Effects of (+)-Limonene Epoxide against 49-Day-Old Adult *Schistosoma mansoni*

| Group | Period of Incubation (h) | Dead Worms (%)[a] | Motor Activity Reduction (%)[a] | | Worms with Tegumental Alterations (%)[a] | |
			Slight	Significant	Partial	Extensive
Control[b]	24	0	0	0	0	0
	48	0	0	0	0	0
	72	0	0	0	0	0
	96	0	0	0	0	0
	120	0	0	0	0	0
Praziquantel (10 µg/mL)	24	100	0	100	20	80
	48	100	0	100	20	80
	72	100	0	100	20	80
	96	100	0	100	20	80
	120	100	0	100	20	80
(+)-Limonene epoxide (12.5 µg/mL)	24	0	0	0	0	0
	48	0	0	0	0	0
	72	0	0	0	0	0
	96	0	20	0	0	0
	120	0	60	0	0	0
(+)-Limonene epoxide (25 µg/mL)	24	0	0	0	0	0
	48	0	0	0	0	0
	72	0	60	60	80	0
	96	0	40	100	40	60
	120	100	0	0	0	100
(+)-Limonene epoxide (50 µg/mL)	24	0	40	0	20	0
	48	0	40	60	40	0
	72	80	0	100	0	100
	96	100	0	100	0	100
	120	100	0	100	0	100
(+)-Limonene epoxide (75 µg/mL)	24	100	0	100	0	100
	48	100	0	100	0	100
	72	100	0	100	0	100
	96	100	0	100	0	100
	120	100	0	100	0	100

[a] Percentages relative to the 20 worms investigated.
[b] RPMI 1640.

11.1.3 Terpenes as Anti-inflammatory Drugs

Inflammation is a protective mechanism initiated by the organism to fight against the injurious stimulus and promote the healing process. Inflammation can be classified into acute or chronic type. Acute inflammation is the initial response which is characterized by the increased movement of plasma and innate immune system cells such as neutrophils and macrophages from the blood into the injured tissues. Chronic inflammation is a progressive change in the type of cells present at the site of the inflammatory reaction and is characterized by simultaneous destruction and healing of the injured tissue. So, anti-inflammatory drugs are used effectively to inhibit the progress of inflammation without interfering with normal homeostasis. Nonsteroidal anti-inflammatory drugs (NSAIDs) and glucocorticoids are used clinically as anti-inflammatory drugs but all suffer from adverse toxicity, gastrointestinal tract irritation and liver dysfunction [20,21]. Hence, Natural products always remain an area of interest as these compounds are toxic as well as cost-effective with less immune response.

A number of herbal drugs containing terpenes have been identified to target inflammatory mediators such as prostaglandins (PGs) and thromboxane (TX). Terpenoids (monoterpenoids, diterpenoids, sesquiterpenes, triterpenes and tetraterpenes) are isolated from plant origin which significantly exhibits anti-inflammatory activities in different animal models (Table 11.10) [22].

The labdane type of diterpenoid was isolated in ethanolic extract of *Ocimum labiatum* and anti-inflammatory effect was determined by using the cytometric bead array (CBA) technique. The noncytotoxic concentration (25 µg/mL) of *O. labiatum* extract significantly ($p < 0.05$) inhibited the production of pro-inflammatory cytokines such as IL-2, IL-4, IL-6 and IL-17A. Geraniol (GOH) is a monoterpenoid and the main ingredient of coriander oil. It has been shown to possess anti-inflammatory activity. Linalool is a naturally occurring terpenoid and was isolated from *Lindera umbellata*. It suppress the LPS-induced pro-inflammatory cytokine production such as that of NO, IL-6 and TNF-α in a dose-dependent manner (Figure 11.7) [23,24].

Terpene alkaloid from *Fagraea racemosa* inhibited the production of prostaglandin E2 in 3T3 murine fibroblasts (IC$_{50}$~5.1 µM). Shizukaols B and D are the lindenane type sesquiterpenoid dimer. These compounds were isolated from the whole plant of *Chloranthus serratus*. The compounds showed

TABLE 11.10

Anti-inflammatory Activity of Terpenes

Terpenes	Structure	Biological Source	Mechanism of Action
Thymol		Oregano	Diminishing the release of inflammatory mediators such as prostanoids, interleukins and leukotrienes.
Carnosic acid		Rosemary	Inhibits the formation of pro-inflammatory leukotrienes and 5-LOX
Carnosol		Rosemary	Inhibits the formation of pro-inflammatory leukotrienes and 5-LOX
Rosmanol		Rosemary	Inhibits the activation of NF-kB and STAT3.
Lupeol		Rhizophora mucronata	Inhibiting the prostaglandins (PG) synthesis
Andrographolide		Andrographis paniculata	Inhibits the expression of TNF-α, IL-6 and IL-1β
Filifolinone		Heliotropium filifolium	Increases the expression level of pro-inflammatory and anti-inflammatory cytokines

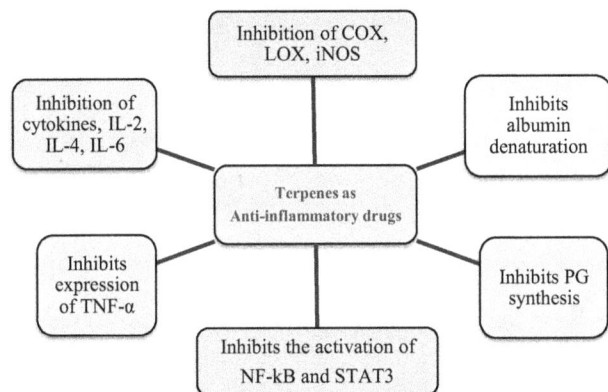

FIGURE 11.7 Molecular targets of terpenes as anti-inflammatory drugs.

significant anti-inflammatory activities with IC_{50} values of 0.22, 0.15 and 7.22 µM respectively. Thymol is chemically known as 2-isopropyl-5-methylphenol. It exhibited potent anti-inflammatory activity by diminishing the release of inflammatory mediators viz. prostanoids, interleukins and leukotrienes [25–27].

11.1.4 Terpenes as Anti-HIV Drugs

HIV represents the human immunodeficiency virus that damages the body's immune system. If HIV infection is not treated properly then it leads to acquired immunodeficiency syndrome (AIDS). There are two types of HIV such as HIV-1 and HIV-2. HIV-1 is transmitted more easily as compared to HIV-2. There are different classes of anti-HIV drugs based on their mechanism of action (MOA) and resistance profiles. These drugs include nucleoside-analog reverse transcriptase inhibitors (NNRTIs), non–nucleoside reverse transcriptase inhibitors (NNRTIs), integrase inhibitors, protease inhibitors (PIs) and fusion inhibitors etc. NRTIs are the first group of drugs approved by the FDA. NRTIs are administered as prodrugs that require host cell entry and phosphorylation by cellular kinases before exerting the antiviral effect. The resistance to NRTIs is mediated by several mechanisms such as ATP-dependent pyrophosphorolysis, reversal of chain termination and increase in the discrimination between the native

deoxyribonucleotide substrate and the inhibitor. So, the plant-derived terpenoids and their analogs are used for the treatment of HIV infections. Several terpenoids (Betulinic acid, Dryopteric acid A, Panaxodiol, Panaxotriol, Glycyrrhizic acid, Maslinic acid, Moronic acid, Noroleananes, Oleanolic acid, Papyriogenin A and Rosamultin) exhibit anti-HIV activity via inhibition of HIV-1 reverse transcriptase and HIV-1-protease (Table 11.11).

Chang et al. found that dehydroandrographolide succinic acid monoester (DASM), prepared from andrographolide, can inhibit HIV-1 (IIIB isolate) in H9 cells (Figure 11.8). The average minimal inhibitory concentration of DASM was found to be 4.6 µM and the mean nontoxic concentration was 250 µM. It also inhibited three HIV-1 strains (IIIB, LAV and UCD123) and one HIV-2 strain (U937) in activated human blood mononuclear cells (PBMC) [28].

Paris et al. discovered a diterpene from *Rosmarinus afficinalis* of family Labiatae as potent anti-HIV protease agent. Carnosic acid produced 90% inhibition of recombinant HIV-1 protease at 0.24 µM. It also exhibited potent anti-HIV-1 (IIIB and RF strains) activity in MT-2 cells with $EC_{90} = 0.96$ µM and cytotoxic to MT-2 cells with $IC_{90} = 1.08$ µM. It was found that the benzylic CH_2 and free carboxylic groups in carnosic acid played vital roles in potentiating the anti-protease activity. Subsequently, three carnosic acid derivatives were synthesized such as rosmanol, 7-o-methylrosmanol and 7-o-ethylrosmanol (Figure 11.9). It was found that these compounds were less potent in the protease assay with EC_{90} values of 1.73, 4.16 and 4.54 µM, respectively [29].

Gustafson et al. reported the isolation of antisane type diterpene [ent-3S-Hydroxy-atis-16(17)-en-1,14-dione] from *Homalanthus nutans* of family Euphorbiaceae and a kaurane diterpenoid [ent-15-Oxo-kaur-16-en-19-oic acid] from *Chrysobalanus icaco* of family Chrysobalanaceae as anti-HIV drugs (Figure 11.10). These compounds exhibited EC_{50} values of 19 and 1.58 µM respectively. However, both compounds were cytotoxic to H9 cells at concentrations 2 and 4 fold of each EC_{50} respectively [30].

Two new kaurane-type terpenoids such as neotripterifordin and tripterifordin were isolated as anti-HIV principles from the roots of *Tripterygium wilfordii* of family Celastraceae

TABLE 11.11

List of Terpenoids with Anti-HIV Activity

Terpenoids	Biological Sources	Mechanism of Action
Acutissimatriterpenes D and E	*Phylanthus acutissimius*	Inhibition of HIV-1 reverse transcriptase
Alnustic acid methyl ester	*Alnus firma*	Inhibition of HIV-1-protease
Betulinic acid	*Cratoxylum arborescens*	Inhibition of HIV-1 reverse transcriptase
Dryopteric acid A and B	*Dryopteris crassirhizoma*	Inhibition of HIV-1-protease
Dammaranes (Panaxodiol, Panaxotriol)	*Panax ginseng*	Inhibition of HIV-1-protease
Glycyrrhizic acid	*Glycyrrhiza glabra*	Inhibition of HIV-1-protease
Maslinic acid	*Myrceugenia euosma*	Inhibition of HIV-1-protease
Moronic acid	*Stauntonia obovatifoliola*	Inhibition of HIV-1-protease
Noroleananes, oleananes		Inhibition of HIV-1-protease
Oleanolic acid	*Syzygium samarangense*	Inhibition of HIV-1-protease
Papyriogenin A	*Tetrapanax papyriferus*	Inhibition of HIV-1-protease
Rosamultin	*Rosa rugose*	Inhibition of HIV-1-protease

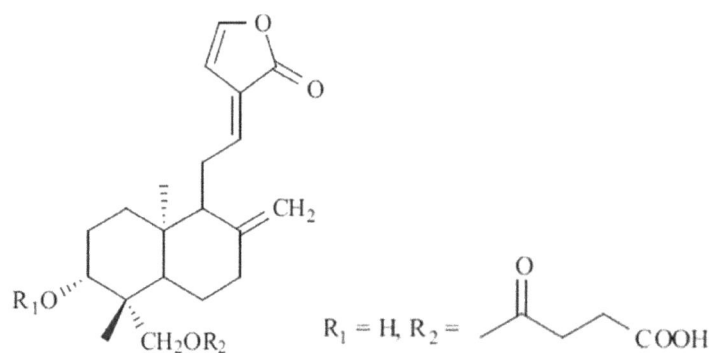

FIGURE 11.8 Structure of dehydroandrographolide succinic acid monoester (DASM).

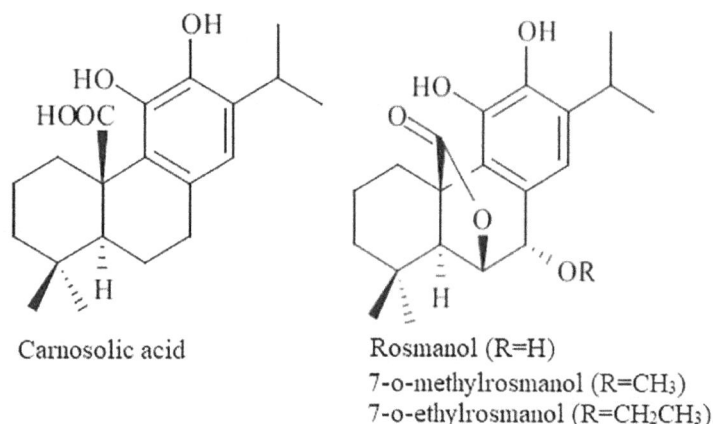

Carnosolic acid

Rosmanol (R=H)
7-o-methylrosmanol (R=CH₃)
7-o-ethylrosmanol (R=CH₂CH₃)

FIGURE 11.9 Structure of three carnosic acid derivatives.

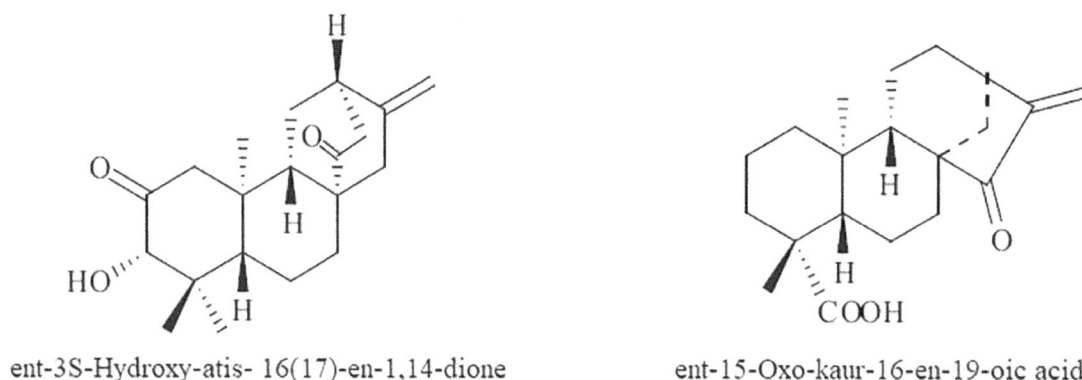

ent-3S-Hydroxy-atis- 16(17)-en-1,14-dione

ent-15-Oxo-kaur-16-en-19-oic acid

FIGURE 11.10 Structure of antisane type and kaurane diterpenoid.

(Figure 11.11). The former compound exhibited 50% inhibition of HIV-1 (IIIB isolate) replication at 0.025 μM in H9 lymphocytes whereas the latter showed the same inhibition at 3.1 μM. In addition, the more potent neotripterifordin also had a higher therapeutic index (TI: IC₅₀/EC₅₀ = 125) as compared to that of tripterifordin (TI = 5). Accordingly, the location of the lactone ring and the stereochemistry of the OH group affected both activity and toxicity [31,32].

Wu and Chang et al. isolated 27 kauranes from the fruits of *Annona squamosa* and *A. glabra* of Annonaceae and found anti-HIV activity with two compounds. 16β, 17-Dihydroxy-*ent*-kauran-19-oic acid (**a**) and methyl-16α-hydro-19-al-*ent*-kauran-17-oate (**b**) (Figure 11.12) produced 50% inhibition of

HIV-1 (IIIB isolate) replication at 2.38 and 15 μM respectively in H9 lymphocytes [33,34].

Gustafson and Gulakowski et al. discussed the anti-HIV activity of prostratin, a non-tumor promoting phorbol ester from *Homalanthus nutans* of family Euphorbiaceae in numerous virus strains and cell lines (Figure 11.13). Prostratin inhibited virus (HIV-1 RF strain) replication at concentrations ranging from 0.1 to 25 μM in both chronically infected CEM-SS and C-8166 cells. In two acutely infected cell systems such as monocytic cell line U937 and freshly isolated monocyte or macrophage cell culture, prostratin also produced cytoprotection at concentrations between 1 and 25 μM [35].

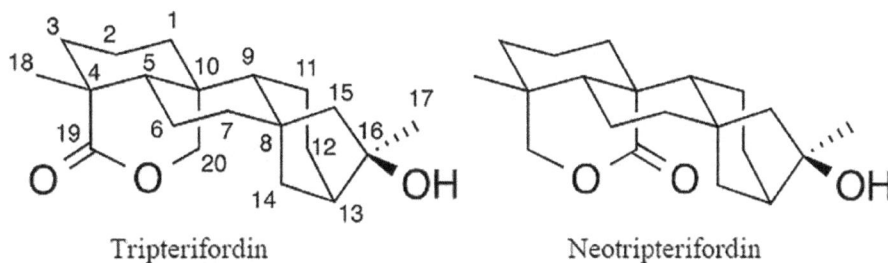

FIGURE 11.11 Structure of kaurane-type terpenoids.

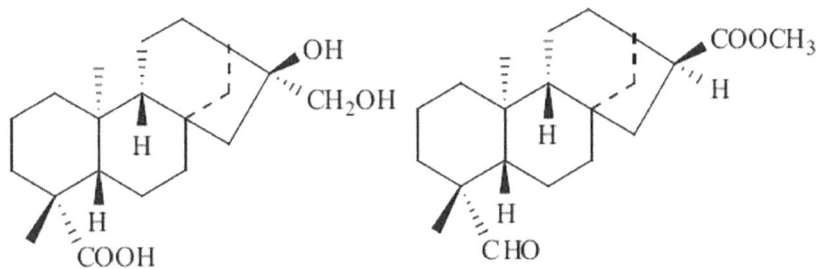

FIGURE 11.12 Structure of kaurane-type compounds (**a,b**).

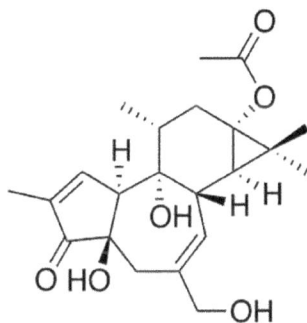

FIGURE 11.13 Structure of prostratin.

Erickson et al. isolated another novel phorbol ester from *Excoecaria agallocha of* family Euphorbiaceae. 12-deoxy-phorbol-13-(3E, 5E-decadienoate) (Figure 11.14) displayed potent HIV inhibition in H9 cells with an EC$_{50}$ of 6 nM and TI greater than 333 and also it is cytostatic at higher concentrations in H9 uninfected cells similarly to prostratin [36].

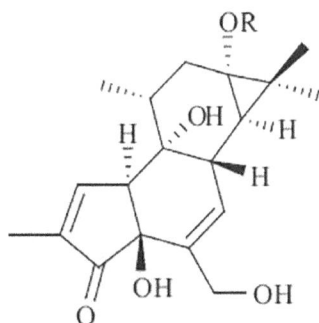

$$R = COCH_2 - (CH = CH)_2 - CH_2CH_2CH_2CH_3$$

Glycyrrhizin (GL) is a triterpene glycosides, isolated from *Glycyrrhiza radix* (Leguminosae) and was first cited as an anti-HIV agent in 1987 (Figure 11.15). Ito et al. reported that GL completely reduce HIV-1 (IIIB strain) plaque formation at a concentration of 0.6 mM in MT-4 cells in a dose-dependent manner. It showed 50% inhibition at 0.15 mM and a TI of 17.3. Similarly, GL exhibited an anti-cytopathic effect and antiviral antigen expression in MT-4 cells at 0.3 and 0.6–1.2 mM respectively. It was observed that the combination of GL with AZT produces additive cytoprotection against HIV-1 (IIIB strain) in MT-4 cells [37].

Nakashima et al. found that glycyrrhizin sulfate (GLS) was slightly more potent than GL against HIV-1 (IIIB strain) activity in MT-4 cells. GLS inhibited the viral cytopathic effect and

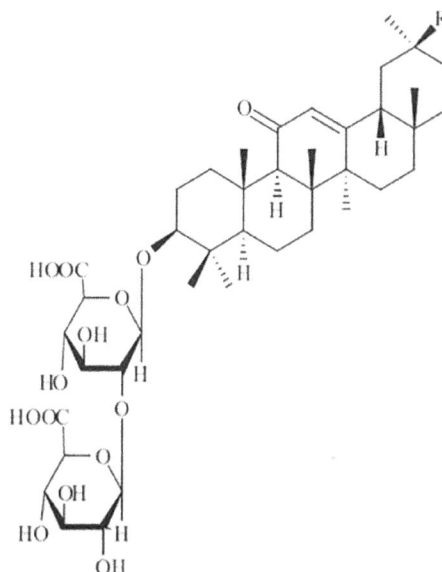

FIGURE 11.14 Structure of 12-deoxyphorbol-13-(3E,5E-decadienoate)

FIGURE 11.15 Structure of glycyrrhizin (R= COOH).

antigen expression at 0.184 and 0.736 mM respectively. It also prevented 90% of giant cell formation and suppressed 50% of HIV-induced plaque formation at 0.04 mM [38]. Tochikura et al. also reported that GLS produced these effects in three other HIV-1 isolates (LAV, ARV and YU-6) and one HIV-2 isolate (GH-1) at 0.736 mM [39].

Hirabayashi et al. reported three GL analogs (**a–c**) as anti-HIV agents in 1991 (Figure 11.16). The most potent compound (**b**) inhibited more than 90% of HIV-1 induced cytopathogenicity at 0.16 mM, while the compound (**a**) and (**c**) produced only 50% suppression at the same concentration. In addition to this, compound 16 was effective in decreasing antigen expression whereas compound (**a**) and (**c**) were not active at the same concentration. It was proposed that the presence of carboxylic acid at position 20 in the case of (**b**) and GL is essential for anti-HIV-1 potency. In contrast, the α, β-unsaturated ketone in the C-ring of GL and the hetero-annular diene of the C/D ring in (**b**) were interchangeable [40].

Abdallah et al. reported that anti-HIV activity of Astragalus species. Astragaloside II is the major component of *A. spino-susn* of family Leguminosae (Figure 11.17). It inhibited HIV replication in H9 lymphocytes (EC$_{50}$ = 25 μM) but it was not active *in vivo* [41].

Konoshima et al. examined the fruit extracts of *Gleditsia japonica* (Leguminosae) and *Gymnocladus chinesis* (Leguminosae) and isolated two known anti-HIV triterpenoid saponins such as gleditsia saponin-C and gymnocladus saponin-G (Figures 11.18 and 11.19) [42].

Gleditsia saponin-C prevented 50% of HIV-1 (IIIB isolate) replication in acutely infected H9 lymphocytes at 1.1 μM, while the Gymnocladus saponin-G showed the same effect at 2.7 μM. In addition to this, the structure of Gleditsia saponin-C had a slightly better TI: 8.9 than Gymnocladus saponin-G of TI: 5.2. Alkaline hydrolysis of Gleditsia saponin-C and Gymnocladus

saponin-G yielded less active or inactive prosapogenins. This finding led to the assumption that monoterpenyl groups and sugar moieties in Gleditsia saponin-C and Gymnocladus saponin-G were crucial for the anti-HIV potency. Accordingly, acetylation or methylation of the inactive aglycone echinocystic acid gave active products **a–c** (Figure 11.20), which showed comparable results with EC$_{50}$ of 3.1, 5.1, 2.3 μM, respectively.

Chen et al. first reported the friedelane terpenoid called salaspermic acid (Figure 11.21) as an anti-HIV-1 drug. It was isolated from the roots of *Tripterygium wilfordii* (Celastraceae).

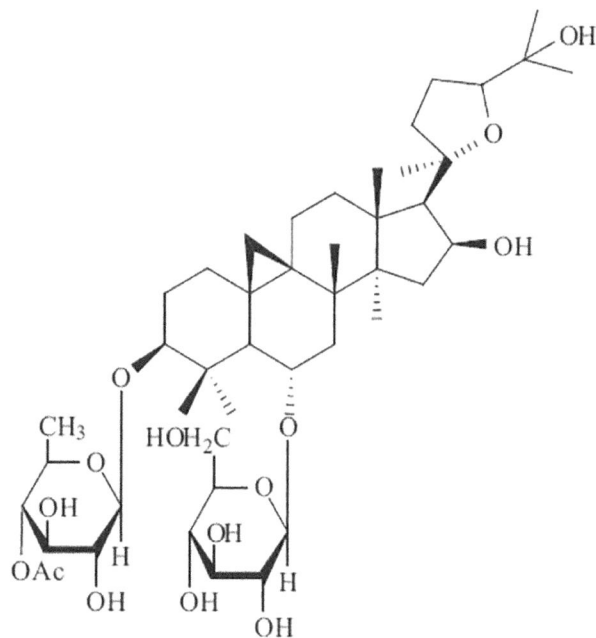

FIGURE 11.17   Structure of astragaloside II.

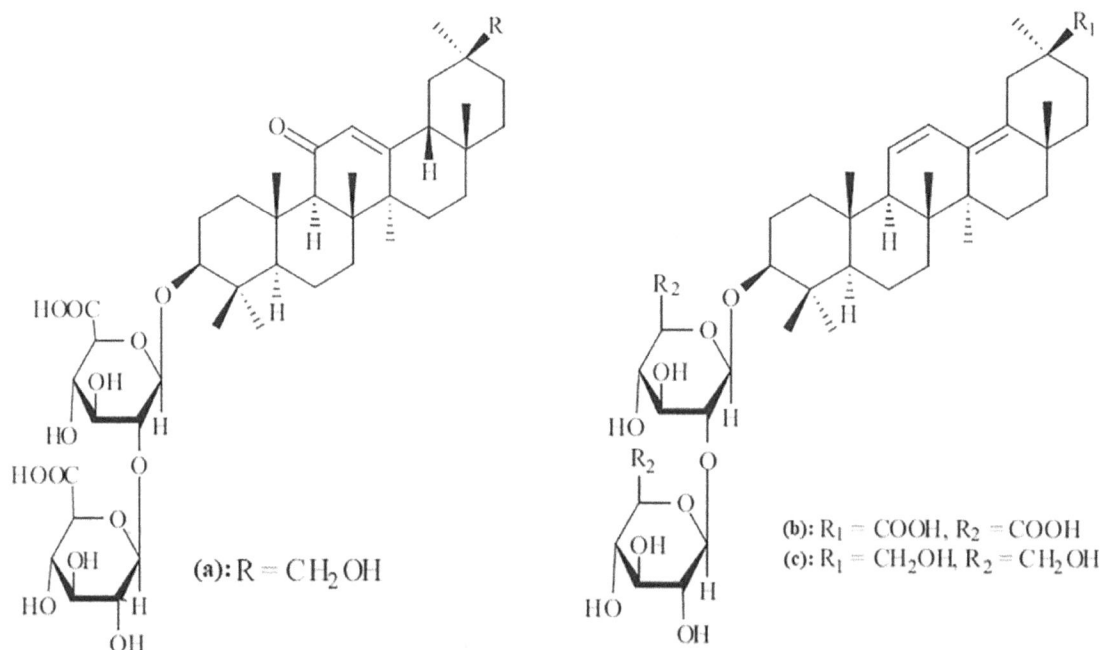

FIGURE 11.16   Structure of glycyrrhizin analogs (**a–c**).

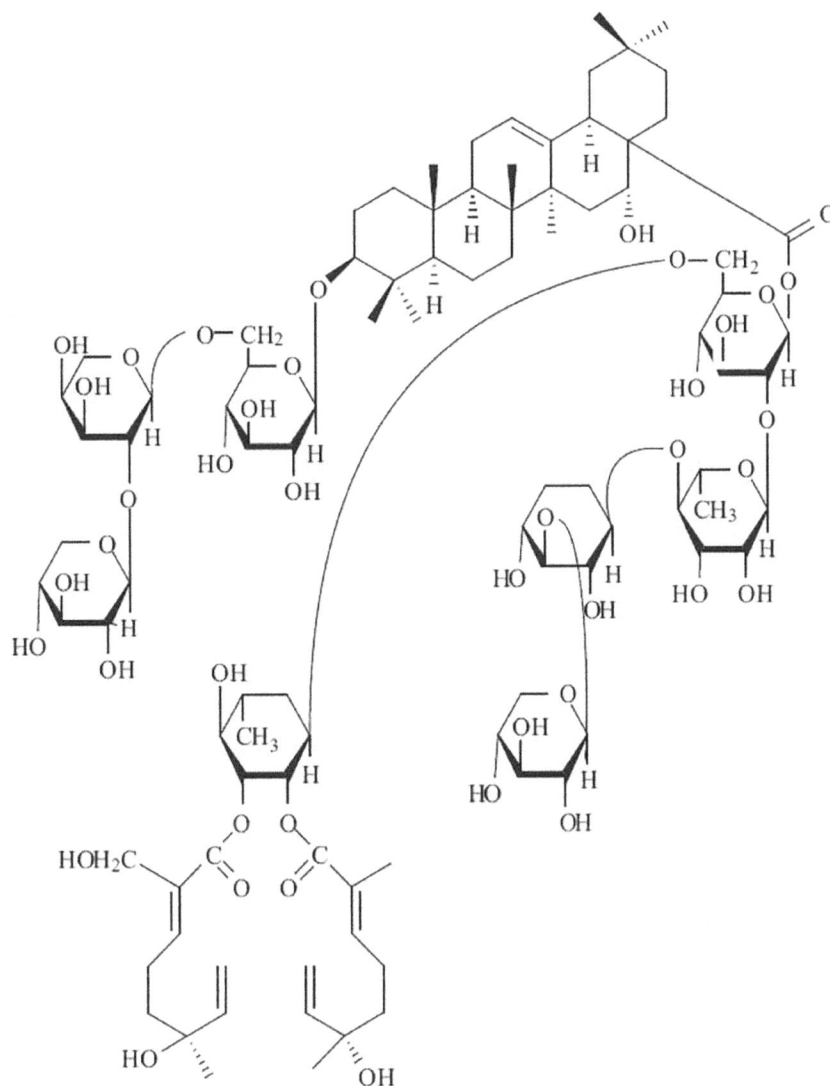

FIGURE 11.18 Structure of gleditsia saponin-C.

This compound exhibited EC_{50} and TI values of 10 µM and 5.3 respectively against the HIV-1 IIIB strain in H9 lymphocytes and showed moderate inhibitory activity against the recombinant reverse transcriptase [43].

Recently, Kuo et al. reported two new friedelin-type triterpenes isolated from stems of Celastrus hindsii (Celastraceae), which are structurally related to **a–e** (Figure 11.22). Celasdin-B (**f**) and maytenfolone-A (Figure 11.23) showed significant anti-HIV (IIIB strain) activity with EC_{50} values in H9 cells of 1.75 and 3.8 µM, respectively [44].

Four pentacyclic triterpenes such as maprounic acid (a), maprounic acid acetate (b), 1β-hydroxymaprounic 3-p-hydroxybenzoate (c) and 2α-hydroxymaprounic acid 2,3-bis-*p*-hydroxybenzoate (**d**) were isolated from *Maprounea africana* of family Euphorbiaceae (Figure 11.24). Compounds **a–d** together with hydrolyzed products e and f did not inhibit HIV-2 RT *in-vitro*. In contrast, compounds **c–f** showed potent inhibition against HIV-1 RT using poly(rA)-oligo(dT) as a template-primer, with EC_{50} values of 3–5 µM [45].

Lanostane (C_{31}) type triterpene called suberosol was isolated from the ethanolic extract of the stems and leaves of *Polyalthia*

suberosa of family Annonaceae (Figure 11.25). It displayed HIV-1 (IIIB strain) inhibition in H9 lymphocytes with an EC_{50} of 6.6 µM and a TI of 6.7 [46].

Xu et al. reported the isolation of five triterpenes from *Geum japonicum* (Rosaceae). Five compounds such as 2α, 19α-dihydroxy-3-oxo-12-ursen-28-oic acid, maslinic acid, epipomolic acid, tormentic acid (Figures 11.26 and 11.27) and ursolic acid produced 72%, 100%, 42%, 49% and 85% inhibition respectively against HIV-1 protease (PR) at a concentration of 17.9 µg/mL [47].

Ma et al. also reported the isolation of ursolic acid (Figure 11.28) and its malonate from the stems of *Cynomorium songaricum* (Cynomoriaceae). These compounds inhibited HIV-PR with IC_{50} values of 8 and 6 µM respectively [48].

Quere et al. hypothesized that rigid pentacyclic compounds could be promising HIV-1 PR inhibitors. They carried out a database search of non-peptide PIs based on a pharmacophore scaffold and analyzed the ideal compounds by computerized docking along with an established PR inhibition assay. Ursolic acid, oleanolic acid and betulinic acid (Figure 11.29) matched the hydrophobic pocket of a relaxed PR monomer and

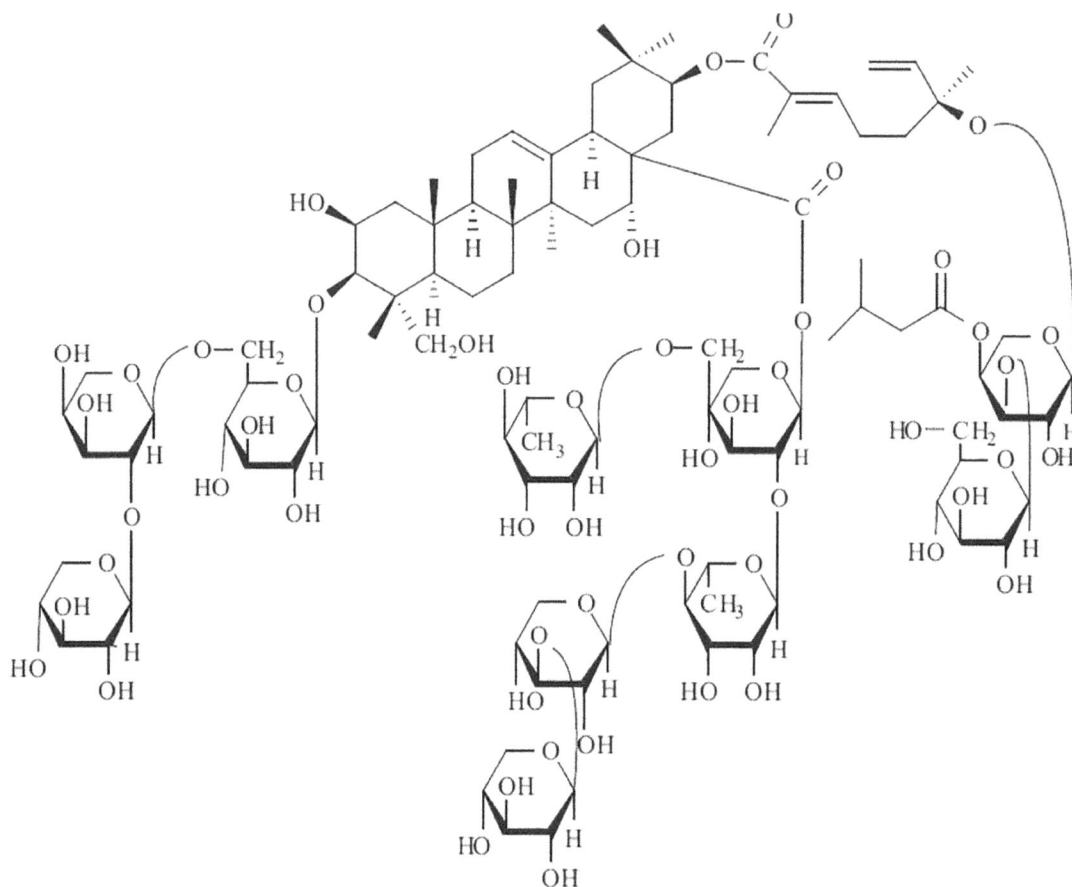

FIGURE 11.19 Structure of gymnocladus saponin-G.

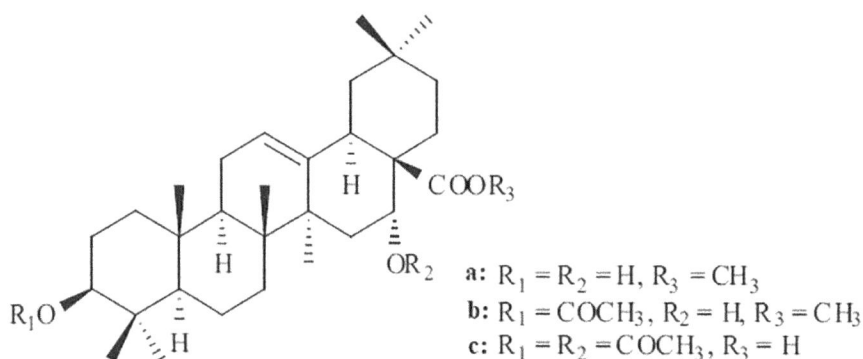

a: $R_1 = R_2 = H, R_3 = CH_3$
b: $R_1 = COCH_3, R_2 = H, R_3 = CH_3$
c: $R_1 = R_2 = COCH_3, R_3 = H$

FIGURE 11.20 Structure of active products **a–c**.

prevented dimerization in the PR inhibition assay with Ki values of 3.4, 1 and 2.5 μM, respectively. The rigid structure of these triterpenes coupled with the inhibitor protease complex stability encouraged further chemical modification in order to reach nanomolar inhibitory concentrations [49,50].

Recently, Kashiwada et al. extracted four known triterpene acids (asiantic acid, arjunolic acid, ursolic acid and oleanolic acid) from *Syzygium claviflorum* of family Myrtaceae. Among these triterpenes, ursolic acid, oleanolic acid and pomolic acid (Figure 11.30) demonstrated strong anti-HIV IIIB activity with EC$_{50}$ values of 4.4, 3.7 and 3.0 μM, respectively, and TI values of 3.3, 12.8 and 16.6, respectively [51].

Two lupanes such as Betulinic acid and platanic acid were isolated from *Syzygium claviflorum* (Myrtaceae). These compounds were first reported to possess antiviral activity in 1994. Betulinic acid and platanic acid also reduced HIV IIIB reproduction by 50% in H9 lymphocytes at concentrations of 1.4 and 6.5 μM, respectively. Alphitolic acid (Figure 11.31) was isolated from *Rosa woodsii* (Rosaceae), which has an additional hydroxy group at the C-2 position of betulinic acid. It displayed lower anti-HIV activity with EC$_{50}$ of 42.3 μM [52,53].

Soler et al. discovered the potent betulinic acid derivative RPR103611 based on rationale design (Figure 11.32). After adding an aminoalkanoic acid at the C-28 position of betulinic

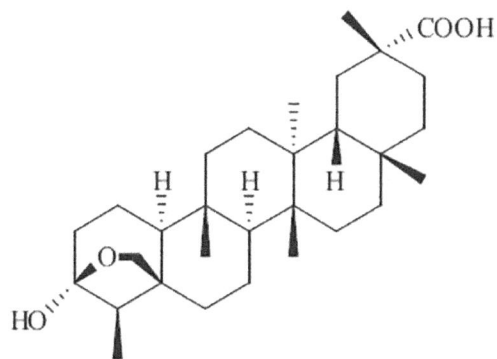

FIGURE 11.21 Structure of salaspermic acid.

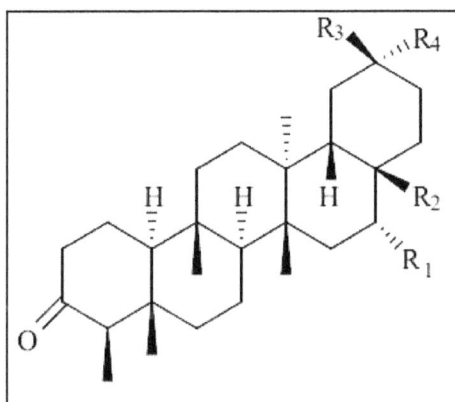

a: R_1 = H, R_2 = R_4 = CH_2OH, R_3 = CH_3

b: R_1 = H, R_2 = R_3 = CH_2OH, R_4 = CH_3

c: R_1 = H, R_2 = CH_2OH, R_3 = R_4 = CH_3

d: R_1 = H, R_2 = COOH, R_3 = R_4 = CH_3

e: R1 = OH, R_2 = R_3 = R_4 = CH_3

f: R_1 = OH, R_2 = CH_2OH, R_3 = R_4 = CH_3

FIGURE 11.22 Structure of compounds (**a–e**) and Celasdin-B (**f**).

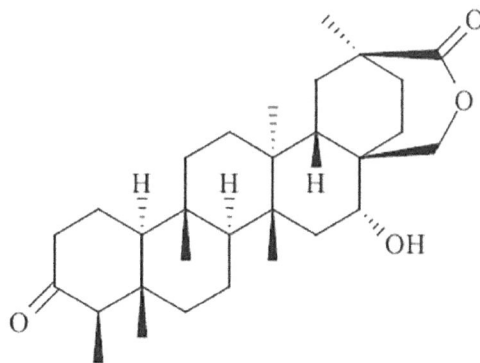

FIGURE 11.23 Structure of maytenfolone-A.

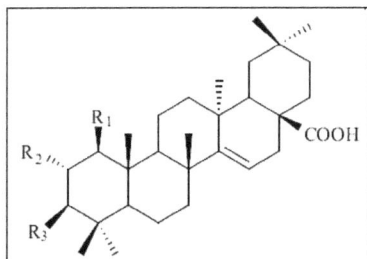

Maprounic acid (**a**): R1=H, R_2= H, R_3=OH
Maprounic acid acetate (**b**): R_1=H, R_2=H, R_3= $OCOCH_3$,
1β-Hydroxymaprounic 3-*p*-hydroxybenzoate
(**c**): R_1=OH, R_2=H, R_3=OCO(*p*-OH)C_6H_4, 2α-hydroxy-
maprounic acid-2,3-bis-*p*-hydroxybenzoate
(**d**): R_1=H, R_2=OCO(*p*-OH)C_6H_4, R_3= OCO(*p*-OH)C_6H_4,
1β-Hydroxymaprounic acid
(**e**): R_1=OH, R_2=H, R_3=OH, 2α-Hydroxy-maprounic acid
(**f**): R_1= H, R_2=OH, R_3=OH

FIGURE 11.24 Structure of pentacyclic triterpenes.

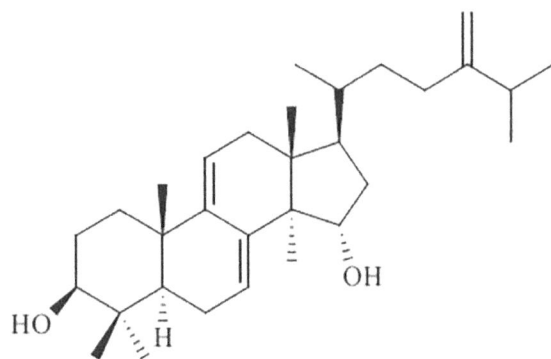

FIGURE 11.25   Structure of Suberosol.

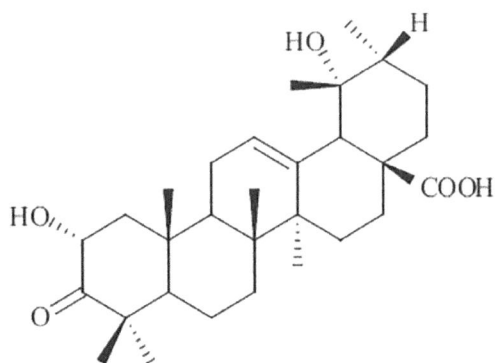

FIGURE 11.26   Structure of 2α,19α-dihydroxy-3-oxo-12-ursen-28-oic acid.

acid, they performed a systematic study to determine the optimal chain length. It was observed that this compound exhibited significant anti-HIV-1 IIIB/LAI action in CEM and MT-4 cells at a methylene length between 7 and 11. RPR103611 also produced significant cytoprotection against HIV-1 LAI/IIIB in certain cell lines (CEM-4, MT-4, H9 and U-937) with EC_{50} values ranging from 0.03 to 0.08 μM and TI values in excess of 90 [54–56].

Nigranoic acid is a triterpenoid isolated from *Schisandra sphaerandra* of family Schizandraceae (Figure 11.33). Chemically, it is 3,4-secocycloarta-4(28), 24-(Z)-diene substituted by carboxy groups at positions 3 and 26. It was found to possess HIV-1 reverse transcriptase activity [57].

Chen et al. also isolated four triterpene lactones (lancilactones A, B and C, and kadsulactone A) from an ethanol extract of *Kadsura lancilimba* (Schizandraceae). Lancilactone-C (Figure 11.34) showed potent anti-HIV (IIIB isolate) activity with $EC_{50}=3.01$ μM and TI>71.4. It was proposed that the opened ring-A in lancilactone C and the resulting free carboxylic acid might play role for producing this activity [58].

11.1.5 Terpenes as Anti-Alzheimer Drugs

Alzheimer's disease is a progressive disorder. It is the most common cause of dementia which creates continuous decline in memory, thinking and behavior. It also refers to chronic neurodegenerative disease which usually starts slowly and gradually worsens over time. Medications are available which can reduce and control some behavioral symptoms Acetyl cholinesterase inhibitors (AChEIs), were first approved by the

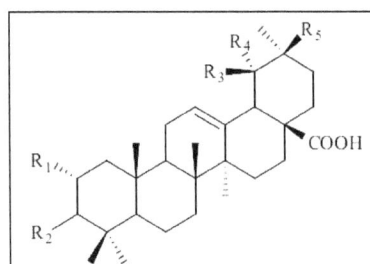

Maslinic Acid: R₁=α-OH, R₂=β-OH, R₃=H, R₄=H, R₅=CH₃
Maslinic Acid: R_1=α-OH, R_2=β-OH, R_3=H, R_4=H, R_5=CH₃
Epipomolic Acid: R_1=H, R_2=α-OH, R_3=CH₃, R_4=OH, R_5=H
Tormentic Acid: R_1=α-OH, R_2=β-OH, R_3=CH₃, R_4=OH, R_5=H

FIGURE 11.27   Structure of maslinic acid, epipomolic acid and tormentic acid.

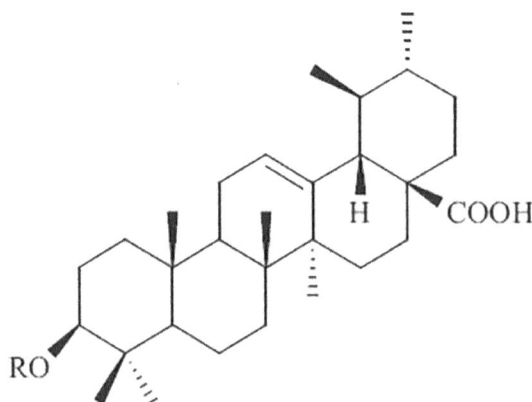

FIGURE 11.28   Structure of ursolic acid (R = H).

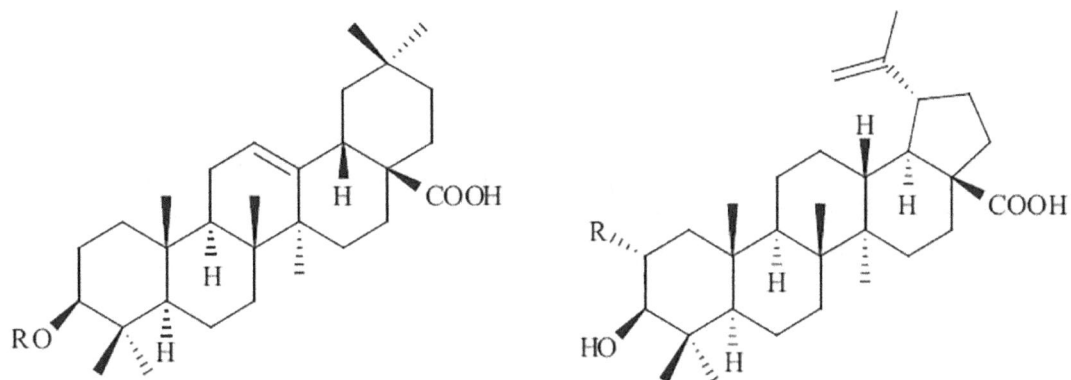

FIGURE 11.29 Structure of oleanolic acid (R = H) and betulinic acid (R = H).

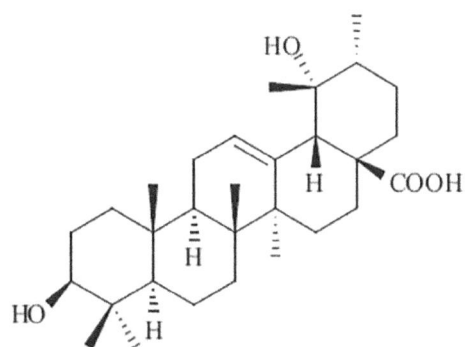

FIGURE 11.30 Structure of pomolic acid (R = H).

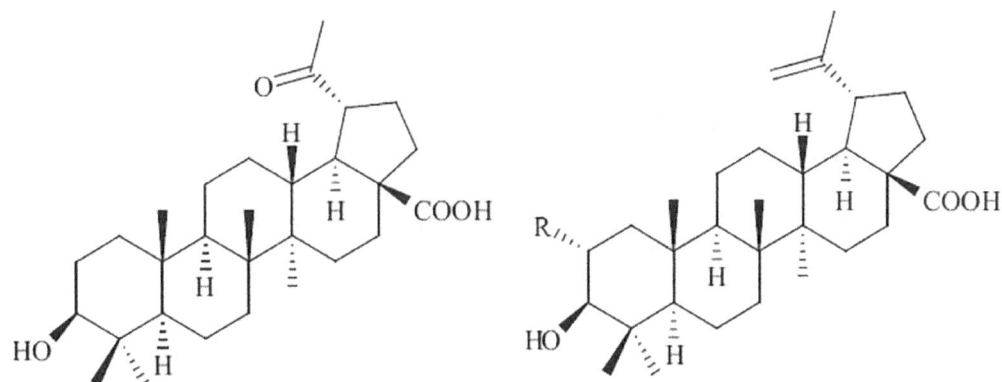

FIGURE 11.31 Structure of platanic acid and alphitolic acid (R = OH).

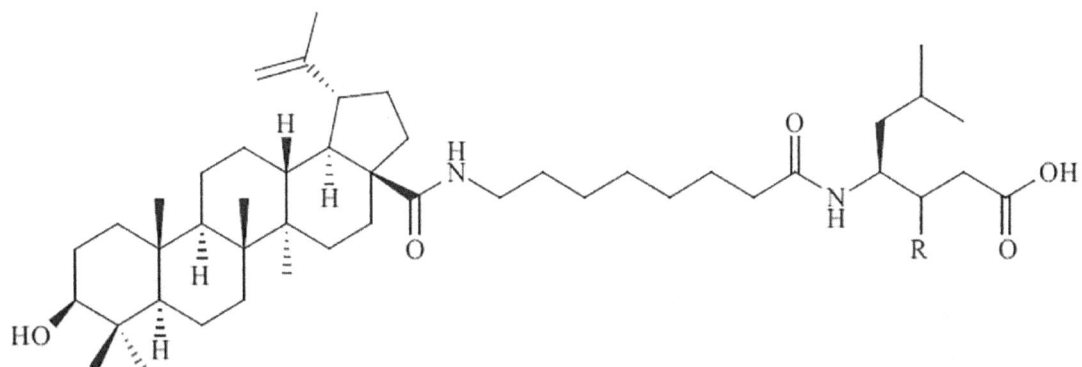

FIGURE 11.32 Structure of betulinic acid derivative RPR103611 [R=S (OH)].

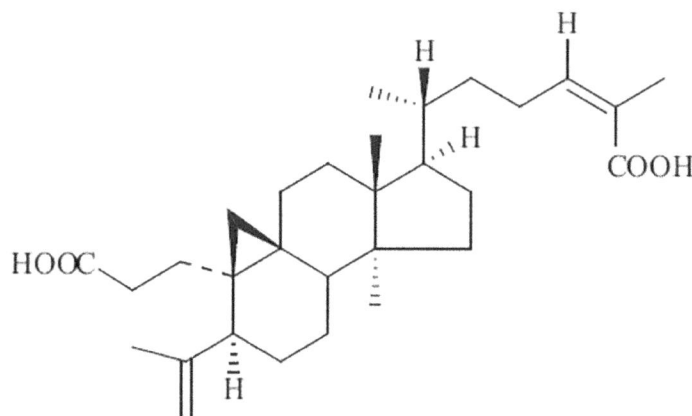

FIGURE 11.33 Structure of nigranoic acid.

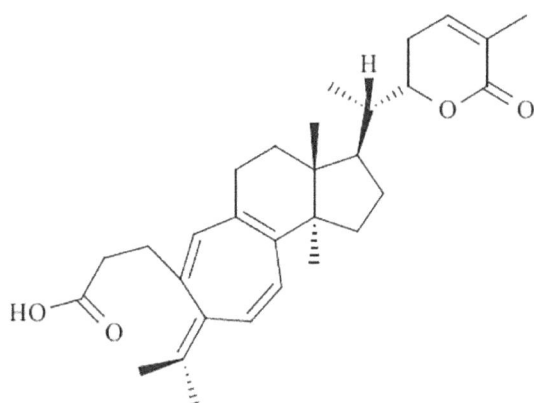

FIGURE 11.34 Structure of lancilactone-C.

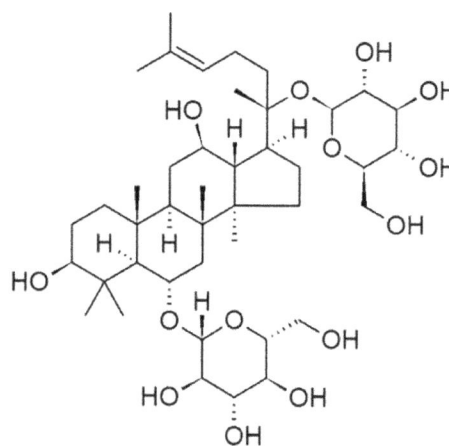

FIGURE 11.35 Structure of ginsenoside.

U.S. Food and Drug Administration (FDA) in 1995 based on clinical trials showing modest symptomatic benefit on cognitive, behavioral and global measures. In 2004, FDA approved memantine, an NMDA antagonist, for the treatment of dementia symptoms in moderate to severe AD. Memantine may be co-administered with an AChEI for clinical practice.

New therapeutic agents are discovered which can block the disease-inducing mechanisms. So, various efforts have been made to discover anti-AD agents from natural sources. Terpenoids such as ginsenosides, gingkolides and canabinoids are isolated and identified as potential anti-AD agents (Figure 11.35). Some other terpenoids including iridoid glycoside, oleanolic acid, tenuifolin, cryptotanshinone and ursolic acid are under investigation for their *in-vitro* and *in-vivo* animal studies [59–61].

11.1.6 Terpene as Antimicrobial Drugs

Antimicrobial drugs are chemical entities of natural or synthetic origin that inhibit the growth or kill the microorganisms including bacteria, fungi, protozoa and viruses. But the antimicrobial resistance (AMR) is one of the major problems to human health worldwide. So, the terpenoids (α-Pinene, Limonene, Myrcene, Geraniol, Linalool, Nerol and Terpineol) constitute several phyto-constituents with promising antimicrobial activities.

Sclareolide is a sesquiterpene lactone which is derived from various plant sources such as *Salvia sclarea* and *Salvia yosgadensis* (Figure 11.36). Sclareolide demonstrated good antibacterial activity against common human pathogens, including *Staphylococcus aureus*, *Pseudomonas aeruginosa*, *Escherichia coli* and *Enterococcus faecalis* [62].

Wang et al. reported antibacterial activity of seven terpenoids against one Gram-positive bacterium (*Staphylococcus aureus*) and two gram-negative bacteria (*Escherichia coli* and *Salmonella enterica*). MIC_{50} values ranging from 0.420 to 0.747 mg/mL were obtained for the seven terpenoids against *Staphylococcus aureus*, 0.421–0.747 mg/mL for *Escherichia coli* and 0.420–1.598 mg/mL for *Salmonella enterica* as presented in Table 11.12. The minimum inhibitory concentration (MIC_{50}) was determined using the agar dilution method.

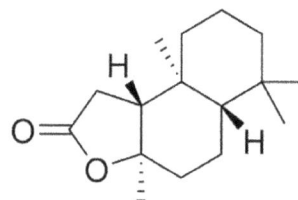

FIGURE 11.36 Structure of sclareolide.

At a concentration of 0.421 mg/mL, limonene showed the strongest antibacterial activity among the three terpenoids without a hydroxyl group–α-pinene, limonene and myrcene. For all seven terpenoids, the concentration of 3.432 mg/mL was determined to be the MBC as shown in Table 11.13.

MBC was defined as the lowest concentration of the samples wherein no visible turbidity or colonies were observed. For the results of MIC_{50} and MBC values, Gram-positive bacteria were observed to be slightly more sensitive than gram-negative bacteria. This is due to the hydrophilic cell wall structure of gram-negative bacteria. Gram-negative bacterial cell wall is mainly composed of lipo-polysaccharide constituent, which blocks the penetration of hydrophobic components of terpenoids and that's why Gram-positive bacteria are found to be more sensitive to the seven terpenoids [63].

Nakano et al. first isolated Copalic acid from *Hymenea courbaril* resin samples. It has been demonstrated that Copalic acid possesses significant antimicrobial activity against *B. subtilis*, *S. aureus* and *S. epidermidis*. Copalic acid was the most active diterpene, displaying a very promising MIC value of 3.1 g/mL against the key pathogen (*Porphyromonas gingivalis*) involved in infectious disease. Souza et al. reported that copalic acid was active against the microorganisms responsible for dental caries such as *Streptococcus salivarius, S. sobrinus, S. mutans, S. mitis, S. sanguinis* and *Lactobacillus casei* [64].

Paiva et al. isolated the kaurenoic acid from oleoresin of *C. langsdorffii a*nd reported its anti-inflammatory activity. It was observed that kaurenoic acid prevented tissue damage in the rat model of acetic acid colitis. It has been reported also that kaurenoic acid, isolated from *C. paupera* oleoresin, displayd antibacterial activity against *B. subtilis, S. aureus* and *S. epidermidis* [65].

Souza et al. reported that kaurenoic and polyalthic acids were capable of inhibiting the rhodamine 6G efflux in *Saccharomyces cerevisiae* with Pdr5p enzyme (protein that causes multiple drug resistance) [66].

Hardwickiic acid was isolated from the stem bark of *Irvingia gabonensis*. It inhibited the growth of several bacteria and fungus species using dilution methods. It exhibited significant antibacterial activity against *B. subtilis, S. aureus* and *Mycobacterium* [67].

11.1.7 Terpenes as Anticancer Drugs

Cancer is a group of diseases which involve abnormal cell growth with the potential to invade or spread to other tissue and organs of the body. It is the uncontrolled growth of abnormal cells anywhere in a body. The abnormal cells are called as cancer cells, malignant cells or tumor cells. Various anticancer drugs are available for the treatment of cancer such as alkylating agents (cisplatin, chlorambucil, procarbazine and carmustine), antimetabolites (methotrexate, cytarabine and gemcitabine), anti-microtubule agents (vinblastine and paclitaxel), topoisomerase inhibitors (etoposide and doxorubicin), cytotoxic agents (bleomycin and mitomycin). But these drugs exhibit severe adverse effects like hair loss, nausea, vomiting, anemia, etc. To reduce the side effects associated with above drugs, naturally occurring phytochemicals such as terpenes have shown enormous potential for the prevention and treatment of several cancers [68].

TABLE 11.12

Minimum Inhibitory Concentrations (MIC) of Terpenoids

| Terpenoids | MIC(mg/mL) | | |
	Escherichia coli	*Salmonella enterica*	*Staphylococcus aureus*
α-Pinene	0.686	0.686	0.420
Limonene	0.421	0.421	0.421
Myrcene	0.635	1.598	0.635
Geraniol	0.445	0.445	0.445
Linalool	0.686	0.420	0.420
Nerol	0.441	0.441	0.441
Terpineol	0.747	0.467	0.747

TABLE 11.13

Minimum Inhibitory Concentrations (MIC) of Terpenoids

| Terpenoids | MBC(mg/mL) | | |
	Escherichia coli	*Salmonella enterica*	*Staphylococcus aureus*
α-Pinene	3.432	1.716	1.716
Limonene	1.682	0.673	1.682
Myrcene	3.176	3.176	3.176
Geraniol	1.778	0.711	1.778
Linalool	1.716	0.686	1.716
Nerol	1.762	0.705	1.762
Terpineol	1.868	0.747	1.868

The spectrum of activity of terpenoids as anticancer agents is doubtlessly the most explored area of research in the last few decades. Monocyclic monoterpenes like D-limonene and perillyl alcohol are shown to inhibit the development of liver, skin, mammary, colon, lung, prostate and pancreatic carcinomas in a dose-dependent manner. However, the metabolites of D-limonene such as perillic acid, dihydroperillic acid, limonene-1,2-diol and the oxygenated derivative of D-limonene like carvone have also shown to have potent antitumor activities (Figure 11.37) [69].

Monoterpenes such as carveol, uroterpenol and sobrerol have shown activity against human mammary carcinomas. Carvone has also been shown to reduce pulmonary adenoma and forestomach tumor formation. In an *in-vitro* antitumor study, betulinic acid has been shown to induce apoptosis of several human tumor cells, whereas ursolic acid and oleanolic acid reduced leukemia cell growth and inhibited the proliferation of several transplantable tumors in animals. Recent study reported a new class of monoterpenes designated as miliusanes which was isolated from *Miliusa sinensis*. Differenet typs of miliusanes such a smiliusol, miliusate and miliusane-1 demonstrated potent cytotoxic activity against a panel of cancer cell lines. It was considered that the effects of carotenoids on colon and prostate cancers are mainly attributed to lutein, β-carotene, zeaxanthin and lycopene etc [70].

Carotenoids are the tetraterpenoids that impart characteristic color to the fruits and vegetables like pumpkins, carrots, corn, watermelon, tomatoes etc. These are yellow, orange and red-colored organic pigments. Lycopene is one of the naturally occurring carotenoids. The possible mechanisms of lycopene include cell cycle arrest, modulation of immune function and induction of apoptotic cell death. Lycopene also inhibits the production of reactive oxygen species (ROS) and decreases the phosphorylation of extracellular signal-regulated kinase (ERK) that results in inhibition of cancer cell growth (Figure 11.38) [71].

Halomin is a halogenated acyclic monoterpene derivative isolated from the red alga *Portieria hornemnnii*. It is effective against renal, brain, colon and lung cancer cell lines etc. Inhibition of cell proliferation and induction of apoptosis has been proposed as the underlying mechanism of the antitumor effects of β-elemene. In addition, several studies have indicated that β-elemene enhances the cytotoxic effect of radiation *in vitro* and *in vivo*. Li et al. suggested that β-elemene can enhance lung (A549) cell radiosensitivity through the enhancement of DNA damage and the suppression of DNA repair. The sesquiterpenes such as β-caryophyllene and caryophyllene

oxide were isolated from the oil of *Calocedrus formosana* leaves. These compounds presented antitermitic activity and antifungal activity against *L. sulphureus* [72].

Goren et al. reported that β-caryophyllene exhibited antimicrobial activity against *E. coli, S. aureus, K. pneumonia, P. aeruginosa and C. albicans* and caryophyllene oxide showed activity only for *C. albicans*. Zheng et al. also reported the isolation of compounds β-caryophyllene, β-caryophyllene oxide and α-humulene from *copaiba oleoresins* and presented the significant activity as inducers of the detoxifying enzyme glutathione S-transferase in mouse liver and small intestine [73].

δ-Cadinene inhibited the growth of *Streptococcus mutans* (cariogenic bacteria) and Propionibacterium acnes (bacteria responsible for acne). Pérez-Lopez et al. performed a bioassay of the essential oil obtained from the fruit of *Schinus molle* against *S. pneumonia* resistant to conventional antibiotics that lead to the identification of δ-cadinene as the principal active constituent with MIC of 31.25 g/mL [74].

Asiatic acid is a pentacyclic triterpene isolated from the Brahmi plant, *Centella asiatica*. It was successfully tested for its apoptotic effects in HepG2 cells. This triterpene decreased the viability of HepG2 cells mediated by an increase in intracellular calcium levels, which led to the increase in expression of the tumor suppressor gene p53 [75].

Overproduction of prostaglandins is considered ad mediator of inflammation and carcinogenic process. So, the inhibitors of prostaglandin biosynthetic enzyme cyclooxygenase (COX) have played a key role of anti-inflammatory and cancer chemopreventive agents. Curdione is one of the major sesquiterpene compounds (Figure 11.39) isolated from rhizome of *Curcuma zedoaria* with the inhibitory effect on the production of prostaglandin E2 in lipopolysaccharide (LPS)-stimulated mouse macrophage RAW 264.7 cells in a concentration-dependent manner with IC_{50} of 1.1 pM [76].

Taxol is also known as paclitaxel, a diterpenoid first isolated from the bark of *Taxus brevifolia* (Figure 11.40). In 1982, it was approved by the FDA as a medicine for the treatment of various types of cancer such as carcinoma (ovary, breast, lung, head, neck, bladder and cervix) and melanoma. The anticancer activity of taxol is based on the inhibition of mitosis where it targets tubulin causing difficulty with the spindle assembly, cell division and also chromosome segregation. Recently, taxol has been reported to be useful in treating neurodegenerative diseases such as Alzheimer's disease [77,78].

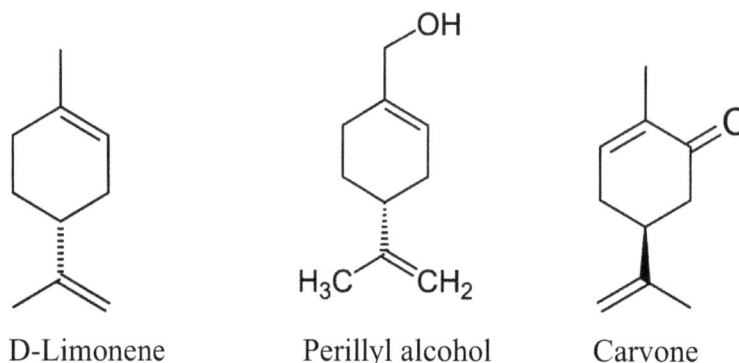

FIGURE 11.37 Structure of monocyclic monoterpenes.

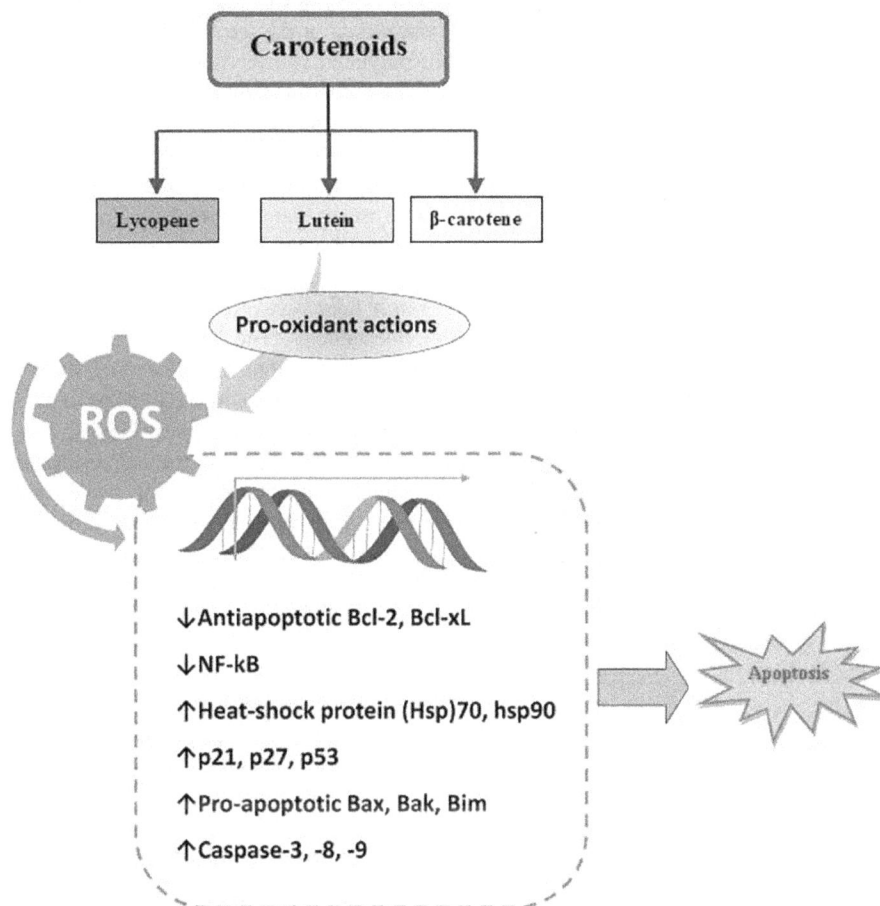

FIGURE 11.38 Molecular mechanism of carotenoids.

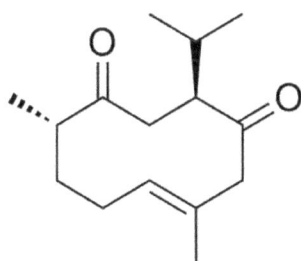

FIGURE 11.39 Chemical structure of curdione.

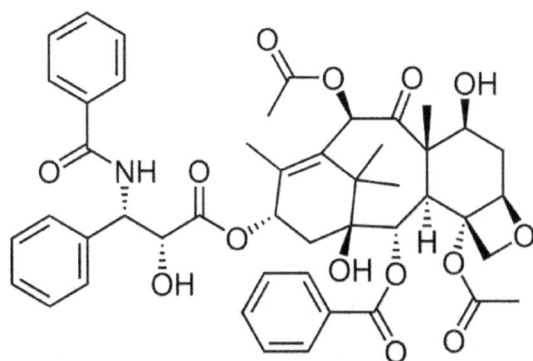

FIGURE 11.40 Chemical structure of Taxol.

Furanodiene is a sesquiterpene that exhibits hepatoprotective, anti-inflammatory, antioxidant and anticancer activities (Figure 11.41). The anti-inflammatory activity of furanodiene is comparable with that of commonly used anti-inflammatory agent (Indomethacin). Furanodiene inhibits the growth of HeLa, Hep-2, HL-60, PC-3, SGC-7901, MCF-7, MDA-MB-231 and HT-1080 cancer cell lines. Furanodiene inhibits the growth of sarcoma 180 tumors in mice *in-vivo*. The survival of mice was prolonged after intraperitoneal treatment with 10 and 30 mg/kg furanodiene once a day for 7 days. Furanodiene retards proliferation of HepG2 cancer cell through G2/M cell cycle and induces apoptosis. Furanodiene is also significantly enhances the sub-G1 peak in human leukemia HL-60 cells, which is accompanied by the activation of the caspase-3, caspase-8 and caspase-9 cascade [79–81].

Curcumol is one of the major components of *Rhizoma Curcumae*. It is a guaiane-type sesquiterpenoid hemiketal

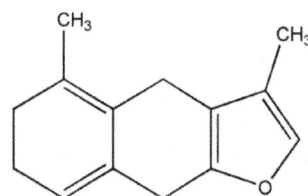

FIGURE 11.41 Structure of furanodiene.

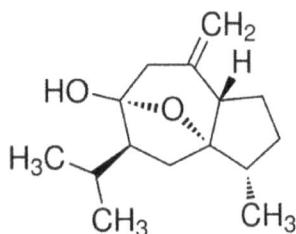

FIGURE 11.42 Structure of curcumol.

(Figure 11.42). It was first isolated in the year 1965 and its absolute stereostructure was determined in 1984. It exhibits anti-hepatic fibrosis activity via down-regulation of the transforming growth factor-β1 and cytochrome P450 in HSC-T6 cells. Curcumol also inhibits the proliferation of MCF-7, MM231, HeLa and OV-UL-2 cancer cells. It was reported that a dose of 50 mg/mL of curcumol significantly inhibits the total RNA synthesis in MCF-7, MM-231 and HeLa cells. Curcumol inhibits the proliferation and induces the apoptosis of nasopharyngeal carcinoma CNE-2 cells. For this, the mechanism involved is related to the down-regulation of NF-kB [82–86].

11.1.8 Terpenes as Antioxidants

Antioxidants are drug substances which protect the cells from being oxidized by free radicals. Reactive oxygen species (ROS) are highly reactive toxic molecules and induce oxidative diseases such as aging, arteriosclerosis, cancer, Alzheimer's disease and Parkinson's disease etc. DPPH free radical and reducing power assays are used to evaluate the antioxidant activities. Both are spectrophotometric methods for determining antioxidant activity. It is based on electron transfer reactions which visually rely on the reduction of colored oxidants. The DPPH free radical is characterized as a stable free radical, which is widely accepted as a method for estimating the free radical scavenging activity of antioxidant. The reducing power of antioxidants is a key indicator for potential antioxidant activity. The antioxidant effect exponentially increases as a function of the reducing power which indicates that the antioxidant properties are related with the development of reducing power. In the reducing power assay, the presence of reluctant or antioxidant causes the reduction of the Fe^{3+} or ferricyanide complex to its ferrous form [87–89].

Various terpenoids such as neral, citral, citronellal, menthone and isomenthone showed strong antioxidant activity. The essential oils with good radical scavenging activity arranged in the order clove > cinnamon > nutmeg > basil > oregano > thyme. The diterpenoid, 7-epi-candicandiol, inhibited the lipid peroxidation at IC_{50} of 43.1 mg/mL [90].

The reducing power activity of the seven terpenoids was determined by using L-ascorbic acid as a standard. Among seven terpenoids, α-pinene showed the highest reducing power at 213.7 ± 5.27 μg/mL, while myrcene showed the lowest reducing power at 21.59 ± 1.14 μg/mL. The reducing power of remaining terpenoid was in the order of limonene (133.48 ± 6.22 μg/mL) > nerol (79.15 ± 3.75 μg/mL) > terpinol (73.92 ± 3.34 μg/mL) > geraniol (73.78 ± 3.94 μg/mL) > lin-

alool (43.97 ± 1.09 μg/mL). However, the DPPH free radical scavenging of the terpenoids was found to be lower than that of butylated hydroxytoluene (BHT) which is known to be a strong reducing agent [91].

Wei and Shibamoto proposed that terpenes and terpenoids that contribute to the antioxidant activity of essential oils include α-terpinene, β-terpinene and β-terpinolene, 1,8-cineole, menthone, isomenthone, thymol and eugenol. Eugenol is a monoterpenoid and is present as major constituent of clove essential oil (*Syzygium aromaticum*). It was found to have inhibitory activity against lipid peroxidation by interfering with chain reactions of free radicals. The inhibitory activity of eugenol was about fivefold higher than that observed for α-tocopherol. Thymol and carvacrol the main constituents of thyme oil are observed to produce strong antioxidant activity. The antioxidant activity of caraway oil may be due to the presence of active constituents like linalool, carvacrol, anethole and estragole [92].

Wang et al. reported antioxidant activities of seven terpenoids such as α-pinene, limonene, myrcene, geraniol, linalool, nerol and terpineol. Among all seven terpenoids, α-pinene exhibited the strongest DPPH free radical scavenging effect at 12.57 ± 0.18 mg/mL. This was followed by limonene at 13.35 ± 0.26 mg/mL and nerol at 26.08 ± 2.28 mg/mL. Myrcene (40.83 ± 1.68 mg/mL) showed the lowest scavenging activity. The determined IC_{50} values were in the order BHT < α-pinene < limonene < nerol < terpinol < geraniol < linalool < myrcene [93].

11.1.9 Immunomodulatory Activity

Immunity system refers to the defense system of the body to protect from invading agents. This system is able to generate varieties of cells and molecules capable of recognizing and eliminating various types of foreign and undesirable agents. Modulation of the immune system denotes to any change in the immune response which can involve induction, expression, amplification or inhibition of any part or phase of the immune response. Thus, immunomodulatory is a drug substance used for its effect on the immune system. There are generally of two types immunomodulators such as immunosuppressants and immunostimulators based on their effects [94].

Carvone (100 μmole/kg), limonene (100 μmole/kg) and perillic acid (50 μmole/kg) were found to increase the total white blood cells (WBC) count in mice. Perillic acid is produced by metabolism of R-(+)-limonene in presence of the yeast *Yarrowia lipolytica*. The immunomodulatory effect of carotenoids extracted from carrot was also assessed in rats by the analysis of immune parameters in the blood. There are significant increases in monocyte neutrophils, percentage lymphocytes and platelet count in the group which received carotenoid complex in comparison with the control group. These results suggest that carotenoids have immunomodulatory effects (Table 11.14). *β*-Carotene is able to enhance cell-mediated immunity in case of the elderly patients [95].

11.1.10 Terpenes as Hepatoprotective

Alqasoumi et al. reported the isolation of six diterpenoids such as 4-epi-abietol, ferruginol, hinokiol, sugiol, Z-communic acid

TABLE 11.14

Terpenoids as Immunomodulators

Terpenoids	Phytochemical/ Biological Source	Structure	Mechanism of Action
Andrographolide (Labdane diterpenoid)	*Andrographis paniculata*		Enhances the expression of IL-2 Inhibition of NO in endotoxin stimulated macrophages
Boswellic acid (Pentacyclic triterpenoids)	*Bowella serrata*		Inhibition of mast cell degranulation
Ursolic acid (Pentacyclic triterpenoids)	*Mirabilis jalapa*		Activates intracellular killing effect of macrophages during Mycobacterium tuberculosis infection

and hinokiol-1-one-3β, 12-dihydroxyabieta-8,11,13-triene-1-one from petroleum ether fraction of *Juniperus procera* (Figure 11.43 and Table 11.15). The hepatoprotective activity was evaluated through the quantification of biochemical parameters and confirmed using histopathology analysis. These compounds showed significant activity as hepatoprotective when investigated against carbon tetrachloride-induced liver injury. Among these compounds, 4-epi-abietol and sugiol were most effective in reducing the elevated liver enzymes as indication for liver protection [96,97].

11.1.11 Terpenes as Anti-Diabetic Drugs

Diabetes is considered as chronic metabolic disorder worldwide and around 285 million people are affected by this disease. The number of diabetic patients is expected to double in the next two decades. So, anti-diabetic are mainly focused is to augment the production of insulin, sensitivity of insulin receptors and reduce the blood glucose level. But the utility of these drug therapies has been obstructed due to their side effects and poor efficacy and drug resistance. Hence natural products like terpenoids have been screened for their anti-diabetic property so as to minimize the side effects and improve efficacy [98].

There are several approaches for the treatment of diabetes such as (1) Insulin or insulin-mimetic, (2) Enhancement of insulin resistance, (3) Inhibitors of hepatic glucose production, (4) Agents which inhibit glucose uptake, (5) Insulin sensitizers.

α-Glucosidase and α-amylase are the enzymes that hydrolyze oligosaccharides (sucrose) into absorbable monosaccharides (glucose) thereby increasing the blood glucose level [99]. The inhibition of these enzymes prevents the quick increase of postprandial hyperglycemia which plays a vital role in the treatment of diabetes mellitus (DM). Hyptadienic acid is a triterpenoid with the molecular formula $C_{30}H_{46}O_4$ (Figure 11.44). It is isolated from the root of *Potentilla fulgens*. It is found to have inhibitory activity against α-glucosidase thereby it is proved to be a potential candidate for the management of diabetes. Similarly, pentacyclic triterpene acetates isolated from the stem bark of *Fagara tessmannii* exhibit significant inhibition against α-glucosidase [100]. New triterpenoids isolated from the roots of *Salacia hainanensis* have displayed inhibitory action against α-glucosidase as compared to the control (Acarbose). Eugenol is an alcoholic terpene that inhibits pancreatic α-amylase, lipase as well as ACE activity in in-vitro [101]. Similarly, sesquiterpene lactones isolated from the *Smallanthus macroscyphus* reduce the blood glucose level by inhibiting intestinal enzymes.

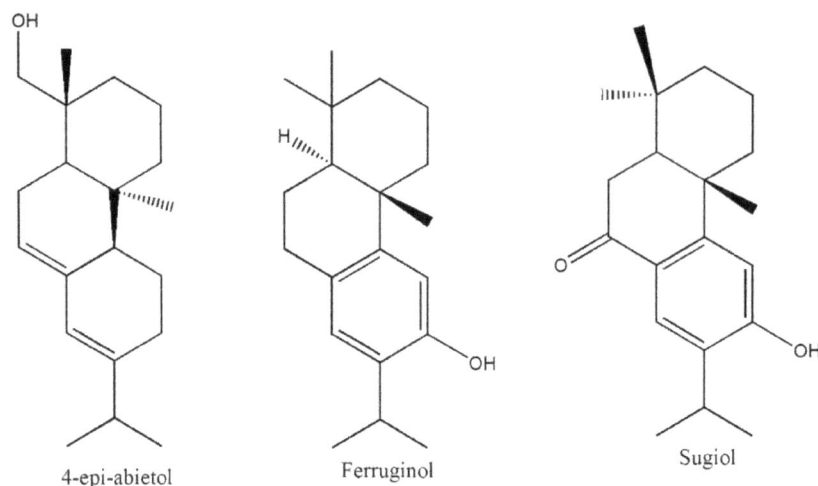

FIGURE 11.43　Structure of diterpenoids.

TABLE 11.15

List of Terpenes as Hepatoprotective Drugs

Chemical Class	Name of Compound	Source	Family
Diterpene	Acanthoic acid	Acanthopanax koreanum	Araliaceae
Diterpene	Neoandrographolide	Andrographis paniculata	Acanthaceae
Triterpene	α-Amyrin and β-Amyrin	Protium heptaphyllum	Burseraceae
Triterpene	γ-Amyrone	Sedum sarmentosum	Crassulaceae
Sesquiterpene	Torilin and Torilolone	Cnidium monnieri	Apiaceae

Corosolic acid is the pentacyclic triterpenes with the molecular formula $C_{30}H_{48}O_4$ (Figure 11.45). It is isolated from the leaves of *Lagerstroemia speciosa*. It exerts potential inhibition against α-glucosidase with an IC_{50} value of 3.53 μg/mL [102].

Further, the progress of diabetic complications is due to oxidative stress produced by ROS. So, terpenes with antioxidant properties can be applied to reduce oxidative stress which is beneficial for the treatment of diabetes. Safranal is a monoterpene with the molecular formula $C_{10}H_{14}O$ (Figure 11.46). Chemically, it is 2,6,6-trimethylcyclohexa-1,3-diene-1-carbaldehyde. It is present in the essential oil of saffron. It is reported that safranal is a degradation product of the carotenoid zeaxanthin via the

intermediate picrocrocin. It exhibits a protective effect against oxidative damage in diabetic rats. The administration of safranal to diabetic rats through intraperitoneal route decreases blood glucose by improving the cellular antioxidant defence mechanism in a dose-dependent manner [103].

Lupeol is a pentacyclic triterpenoid with the molecular formula $C_{30}H_{50}O$ (Figure 11.47). It is found in strawberry, olive,

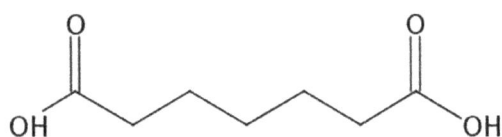

FIGURE 11.44　Structure of hyptadienic acid.

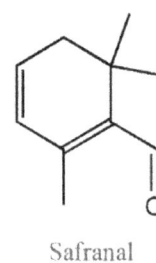

Safranal

FIGURE 11.46　Structure of safranal.

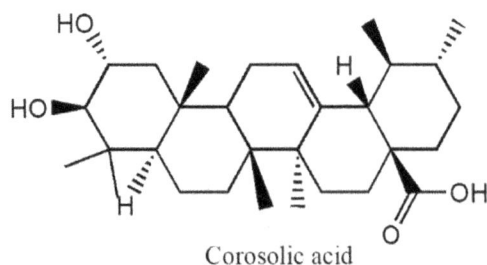

Corosolic acid

FIGURE 11.45　Structure of corosolic acid.

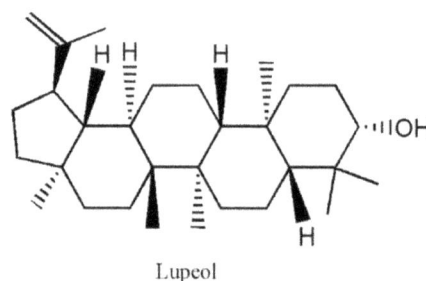

Lupeol

FIGURE 11.47　Structure of lupeol.

mangoes and grapes. Evaluation study revealed that Lupeol decreases the oxidative stress and also restores the antioxidant defense system in the liver of Swiss albino mice. As it decreases the level of ROS level, it plays a major role in the treatment of diabetes [104].

Deutschlander et al. reported the anti-diabetic property of Betulin. It is a pentacyclic triterpenoid with the molecular formula $C_{30}H_{50}O_2$ (Figure 11.48). It is isolated from the root bark of *Euclea undulate*. It is found to exhibit a significant decrease in blood glucose level as compared to standard drug (glibenclamide). The MOA of Betulin mainly involves the inhibition of transcription factor associated with the activation of genes involved in the biosynthesis of cholesterol, fatty acids and triglycerides. The down-regulation of target genes results in a decrease in cellular lipid levels that leads to improved insulin resistance [105].

Momordicoside is a cucurbitane-type triterpenoid glycosides with molecular formula $C_{36}H_{58}O_8$ (Figure 11.49). It is isolated from bittermelon (*Momordica charantia*). It stimulates GLUT4 translocation with increased activity of AMP-activated protein [106]. It also enhances the oxidation of fatty acid and glucose disposal in both insulin-sensitive and insulin-resistant mice during the Oral glucose tolerance test (OGTT). These results reveal that Momordicoside can be used as a potential therapeutic for the treatment of diabetes and obesity [107].

Stevioside is a diterpene glycoside with the molecular formula $C_{38}H_{60}O_{18}$ (Figure 11.50). It is a natural sweetener extracted from the leaves of *Stevia rebaudiana* (Bertoni) [108]. The antihyperglycemic activity of Stevioside may be due to an increase in glycolysis and suppression of gluconeogenesis by inducing genes involved in glycolysis or inhibiting the action of ATP phosphorylation and NADH oxidase activity in liver mitochondria [109]. It also exhibits antihyperglycemic, insulinotropic, blood pressure-lowering and glucagonostatic effect in type II diabetic rats [110].

Costunolide is a naturally occurring sesquiterpene lactone with the molecular formula $C_{15}H_{20}O_2$ (Figure 11.51). Chemically, it is (3aS, 6E, 10E, 11aR)-6,10-dimethyl-3-methylidene-3a, 4,5,8,9,11a-hexahydrocyclodeca[b]furan-2-one. It is isolated from the rhizomes of *Costus speciosus*. It enhances the glucose absorption capacity in the muscle and adipose tissue which produce an insulin-like effect on the peripheral tissues. It also stimulates the regeneration of remaining β-Cells [111].

Bassic acid is a naturally occurring triterpenoid with the molecular formula $C_{30}H_{46}O_5$ (Figure 11.52). It is isolated from the root bark of *Bumelia sartorum*. It stimulates the hypoglycemic activity mediated through enhanced secretion of insulin from the pancreatic β-cells [112].

Palbinone is a triterpene with the molecular formula $C_{22}H_{30}O_4$ (Figure 11.53). It is isolated from Moutan Cortex (Paeonia suffruticosa, Paeoniaceae). It stimulates glucose uptake and glycogen synthesis via activation of AMPK in insulin-resistant human HepG2 Cells [113]. Palbinone effectively reduces the retinal inflammation and oxidative stress resulting from high glucose in STZ-induced diabetic rats. It also activates the antioxidative Nrf2 pathway and inhibits NLRP3 inflammasome formation during diabetic retinopathy (Figure 11.54 and Table 11.16) [114].

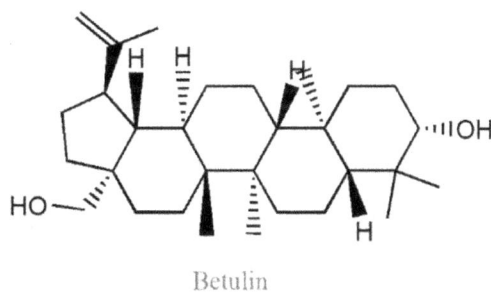

Betulin

FIGURE 11.48 Structure of betulin.

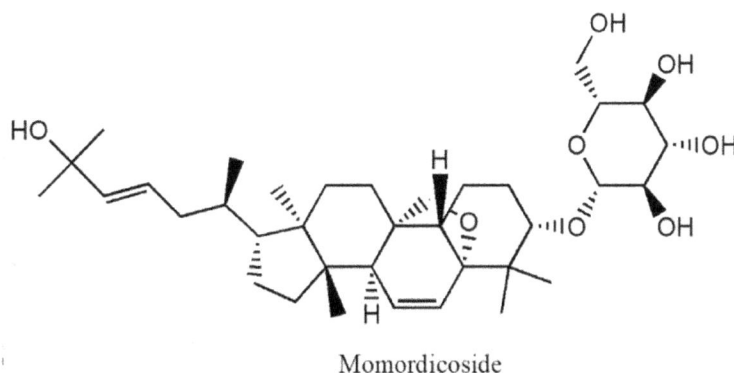

Momordicoside

FIGURE 11.49 Structure of momordicoside.

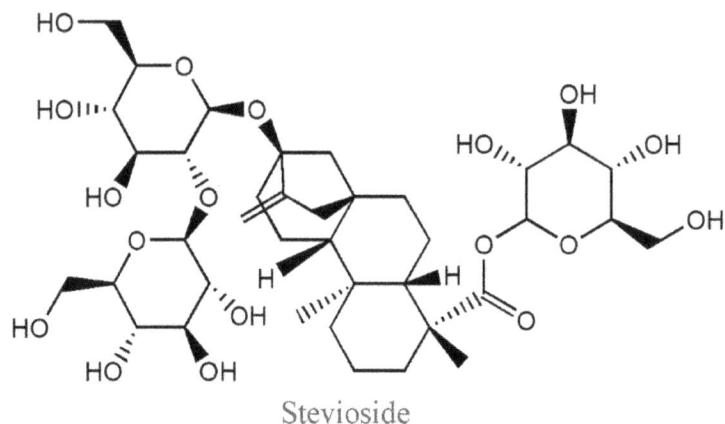

Stevioside

FIGURE 11.50   Structure of stevioside.

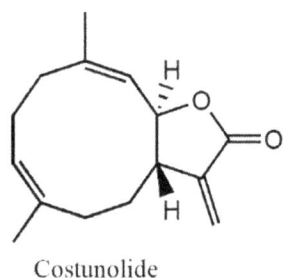

Costunolide

FIGURE 11.51   Structure of costunolide.

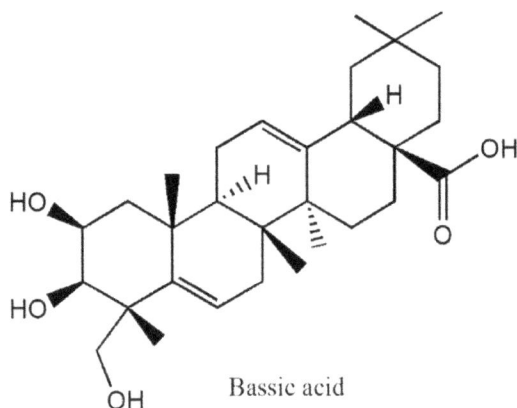

Bassic acid

FIGURE 11.52   Structure of bassic acid.

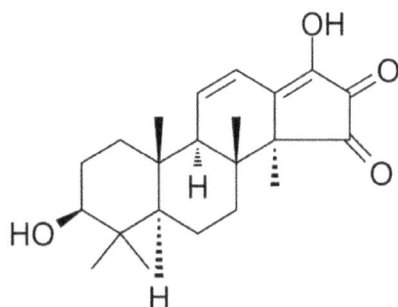

FIGURE 11.53   Structure of palbinone.

11.2 Miscellaneous

Furanodienone is one of the main bioactive constituents of *Rhizoma Curcumae* that exhibits anti-inflammatory activity (Figure 11.55). It was observed that Furanodienone inhibits COX-2 more as compared to COX-1. Recent studies show that furanodienone may be a potential lead drug compound for breast cancer therapy, especially against estrogen receptor-α (ER-α)-positive breast cancer. It also causes the arrest of G1 cell cycle in BT474 cells and induces apoptosis in SKBR3 cells. Furanodienone interferes with epidermal growth factor receptor (EGFR) and human epidermal growth factor receptor-2 (HER-2) signaling in treated cells [115].

α- and β-amyrin, a triterpene mixture isolated from the trunk of medicinal plant, *Protium heptaphyllum* (Figure 11.56). It provides hepatoprotection against acetaminophen-induced hepatotoxicity. It results in diminution of oxidative stress and inhibition of cytochrome-P450 activity [116].

Lupeol is a pentacyclic triterpenoid in which the hydrogen at 3β-position is substituted by a hydroxy group (Figure 11.57). It occurs in the skin of lupin seeds as well as in the latex of fig trees. Lupeol reduced the cell proliferation of both keratinocytes and fibroblasts. It was investigated to determine the effects of lupeol (0.1, 1, 10 and 20 μg/mL) on *in-vitro* wound healing assays and signaling mechanisms in human neonatal for skin keratinocytes and fibroblasts. It increased *in-vitro* wound healing in keratinocytes and promoted the contraction of dermal fibroblasts in the collagen gel matrix. The underlying mechanism for the positive effect of lupeol on wound healing may involve the activation of PI3k/Akt and p38 MAPK, suppression of NF-κB signaling and Keratin 16, as well as the cytoprotective effects of MMP-2 and Tie-2 [117].

Gulacti et al. reported some novel terpenoids with potential anti-Alzheimer activity from *Nepeta obtusicrena*. Dichloromethane extract of *Nepeta obtusicrena* afforded a diterpenoid and a triterpenoid in addition to two known triterpenoids such as oleanolic acid and ursolic acid. Purification of the diterpenoid was carried out by HPLC and its structure was elucidated as 14α-acetoxy-6-oxo-abieta-7-ene and structure of the triterpenoid was elucidated as 2α, 3β, 19α, 24-tetrahydroxy-11-oxo-olean-12-ene. Both of the novel terpenes were obtained from nature for the first time and named as obtusicrenone and nemrutolone respectively (Figure 11.58). Anticholinesterase (anti-Alzheimer) and

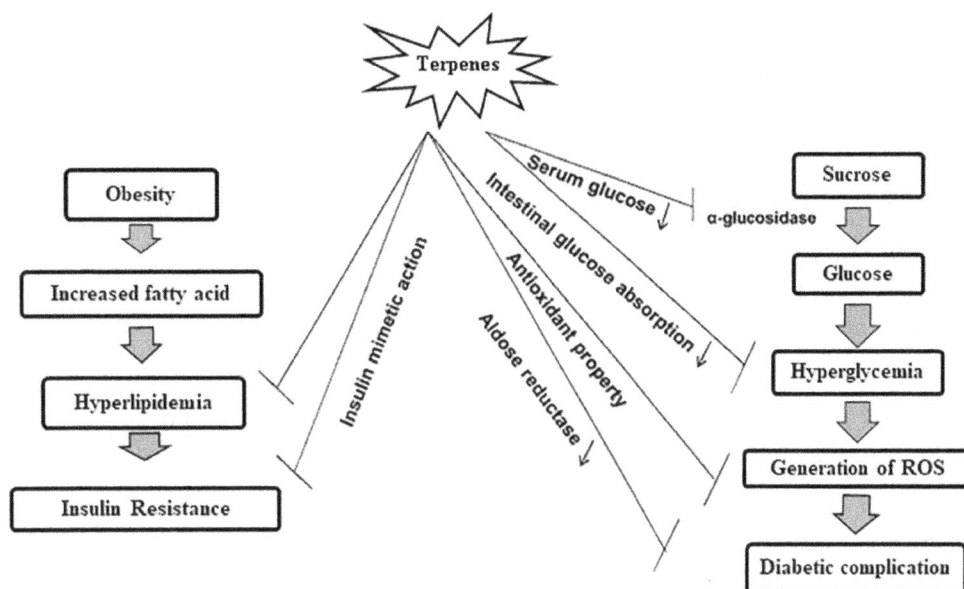

FIGURE 11.54 Targets of terpenes for anti-diabetic activity.

TABLE 11.16

List of Terpenes as Anti-diabetic Agents

Terpenes	Source	Effect
Bassic acid	Root bark of *Bumelia sartorum*	Enhance secretion of insulin
D-limonene	Citrus fruits	Decrease plasma glucose, increase activity of glycolytic enzymes
Stevioside	Leaves of *Stevia rebaudiana*	Insulinotropic, Glucagonostatic
Rebaudioside	Leaves of *Stevia rebaudiana*	Insulinotropic
Lupeol	Mango	Decrease oxidative stress
Palbinone	Moutan cortex	Induce glucose uptake via AMPK pathway
Betulin	Root bark of *Euclea undulata*	Inhibition of α-glucosidase
Lactucain A	*Lactuca indica*	Lowering of plasma glucose
Tinosporaside	*Tinospora cordifolia*	Anti-hyperglycemic
Urosolic acid	Leaves of *Rosmarinus officinalis*	Anti-hyperglycemic

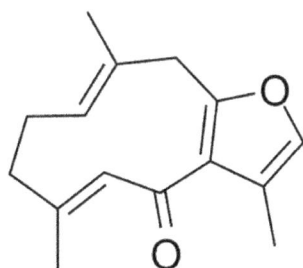

FIGURE 11.55 Structure of furanodienone.

antioxidant activities [DPPH free radical scavenging activity, ABTS cation radical scavenging activity, lipid peroxidation inhibitory activity, CUPRAC (Cupric Reducing Antioxidant Capacity)] of the dichloromethane and methanol extracts and the isolated above four terpenoids were investigated (Tables 11.17–11.19). Both of the extracts and the isolated four terpenoids exhibited high anticholinesterase activity mainly against acetylcholinesterase (AChE) enzyme. None of the samples tested demonstrated high antioxidant activity. BHT (Butylated hydroxytoluene), BHA (Butylated hydroxyanisole) and α-tocopherol

were used as standards for DPPH free radical scavenging activity assay and ABTS cation radical scavenging activity assay [118].

Valdir et al. reported the chemistry and biological activities of terpenoids from Copaiba (*Copaifera spp.*) Oleoresins. Copaiba oleoresins are exuded from the trunks of trees of the *Copaifera species* of family Leguminosae). This oleoresin is a solution of diterpenoids mainly mono- and di-acids, solubilized by sesquiterpene hydrocarbons. The sesquiterpenes and diterpenes (labdane, clerodane and kaurane skeletons) are different for each Copaifera species and have been linked to several reported biological activities such as antitumor, antimicrobial, anti-inflammatory and antioxidant activities as presented in Table 11.20 [119].

The diterpenes most commonly isolated from copaiba oleoresins are copalic, polyalthic, hardwickiic, kaurenoic and ent-kaurenoic acids, together with their derivatives 3-hydroxy-copalic, 3-acetoxy-copalic and ent-agathic (Figure 11.59).

The main sesquiterpenes present in copaiba oleoresins are β-caryophyllene, caryophyllene oxide, α-humulene, δ-cadinene, α-cadinol, α-cubebene, α- and β-selinene, β-elemene, α-copaene, trans-α-bergamotene and β-bisabolene (Figure 11.60).

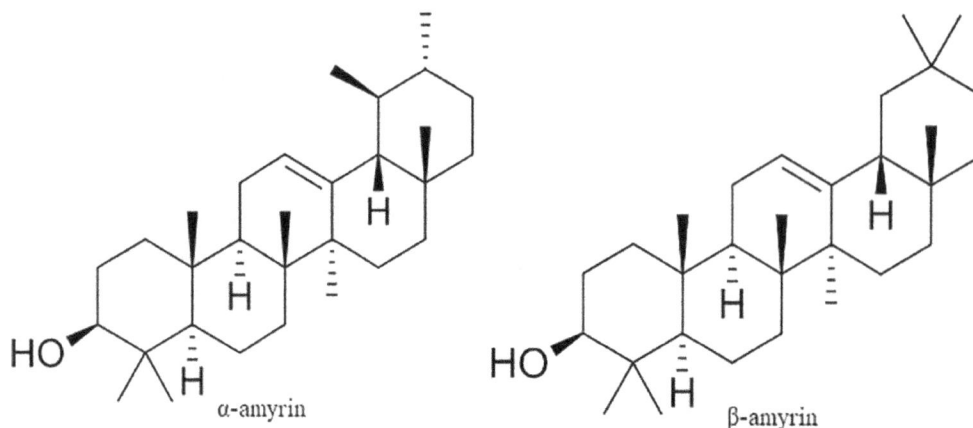

FIGURE 11.56 Structure of α- and β-amyrin.

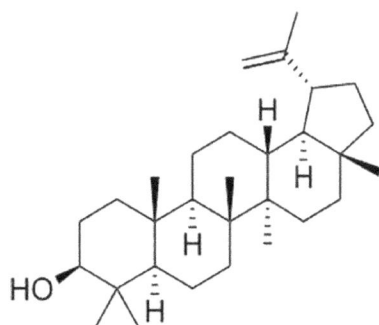

FIGURE 11.57 Structure of lupeol.

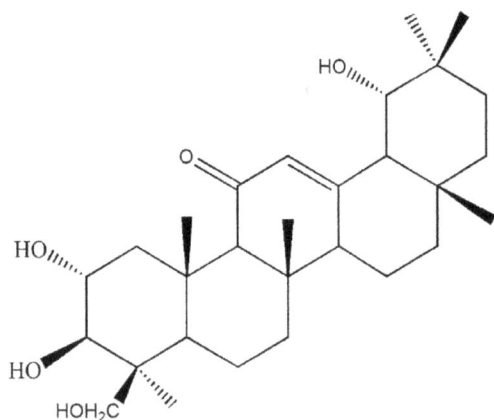

FIGURE 11.58 Structure of nemrutolone.

TABLE 11.17

Anticholinesterase Activity of the Extracts at 200 μg/mL

Samples	Inhibition % against AChE
NOD[a]	71.03
NOM[b]	73.59
Galantamine	74.12

[a] NOD, dichloromethane extract of *Nepeta obtusicrena*.

[b] NOM, methanol extract of *Nepeta obtusicrena*.

TABLE 11.18

IC_{50} Values of the Samples (μg/mL)

Samples	DPPH	ABTS	Anti-AChE
NOD[a]	NA[c]	NA	-
NOM[b]	162.21	69.16	-
Obtusicrenone	NA	173.15	13.78
Nemrutolone	NA	116.67	54.36
Oleanolic acid	NA	NA	27.88
Ursolic acid	NA	NA	36.09
BHA	16.19	12.66	-
BHT	74.59	18.35	-
α-Tocopherol	33.09	19.35	-
Galantamine	-	-	5.01

[a] NOD, dichloromethane extract of *Nepeta* obtusicrena.

[b] NOM, methanol extract of *Nepeta* obtusicrena; [c]NA: Not active ($IC_{50} \geq 200$).

Braga et al. detected Kaurenoic acid in copaiba oleoresins in 1998 and isolated it from *C. cearensis* using ion-exchange chromatography. Various pharmacological activities were conducted by using kauran acid, to determine its uterine muscle relaxant, anti-inflammatory, bactericidal and cytotoxicity effects. Cunha et al. isolated the diterpene, kaurenoic acid from *C. langsdorffii* oleoresin and reported the uterine relaxant effects. Kaurenoic acid exerts the relaxant effect mainly through blocking of calcium channel and opening of ATP-sensitive potassium channels [120].

Excisanin-A is a diterpenoid, extracted from *Isodon Macrocalyxin* (Figure 11.61). It was found to be an effective and potent tumor growth inhibitor in BALB/c nude mice implanted with Hep3B cells. A daily intraperitoneal (ip) injection of 10–20 mg/kg of excisanin-A in this xenografted mouse model and the subsequent analysis of the tumor samples after sacrifice established the inhibition of the AKT signaling pathway in tumor cells [121].

Paeoniflorin is a pinane monoterpene glycoside with the chemical formula of $C_{23}H_{28}O_{11}$ and molecular weight of 480.50 (Figure 11.62). Paeoniflorin is isolated from the plants of *Paeoniaceae* family (*Paeonia lactiflora*). It exerts several biological activities including anticancer, antioxidative stress, antiplatelet aggregation, vasodilator and anti-inflammatory

TABLE 11.19

CUPRAC Assay Results of Extracts Terpenoids

Samples	10 μg/mL	25 μg/mL	50 μg/mL	100 μg/mL
NOD[a]	0.13	0.17	0.22	0.30
NOM[b]	0.13	0.21	0.32	0.49
Obtusicrenone	0.11	0.19	0.26	0.41
Nemrutolone	0.14	0.20	0.28	0.36
Oleanolic acid	0.90	0.18	0.24	0.31
Ursolic acid	0.12	0.16	0.23	0.30
BHA	1.32	1.74	2.04	2.57
BHT	1.51	1.84	2.24	2.69
α-Tocopherol	0.32	0.49	0.75	1.47

[a] NOD, Dichloromethane extract of *Nepeta* obtusicrena.
[b] NOM, Methanol extract of *Nepeta obtusicrena*.

TABLE 11.20

Biological Activities Tested in Different Species of *Copaifera oleoresins*

Species	Biological Activity
C. reticulata Ducke	Anti-inflammatory, antimicrobial, antinociceptive, anxiolytic, antileishmanial, teratogenicity and embriotoxicity
C. paupera (Herzog) Dwyer	Antimicrobial, antileishmanial
C. multijuga Hayne	Anti-inflammatory, antitumor, anti-nociceptive, antimicrobial and anti-leishmanial
C. langsdorffii Desf.	Antileishmanial, antioxidant, insecticide, Antimutagenic, anti-inflammatory and wound Healing
C. lucens Dwyer	Antimicrobial, antileishmanial

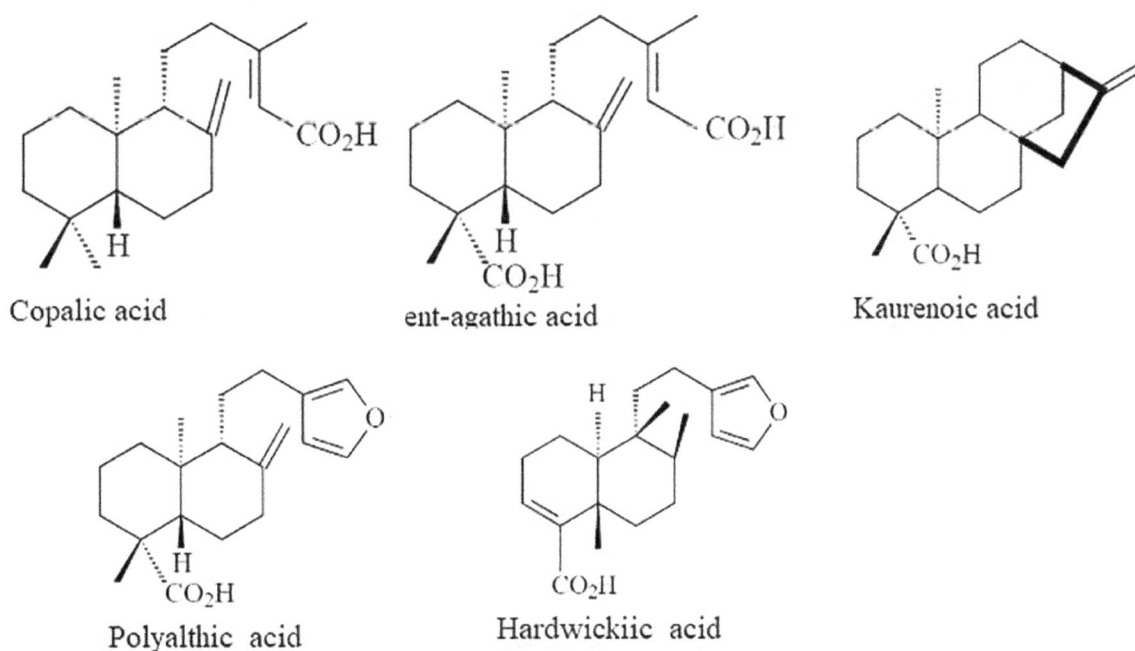

FIGURE 11.59 Structure of diterpenes detected in copaiba oleoresin.

etc. Evaluation study suggested that Paeoniflorin exhibits significant anticancer activities on various cancers such as liver cancer, gastric cancer, breast cancer, lung cancer, pancreatic cancer, colorectal cancer, bladder cancer and leukemia. The *in-vitro* study revealed that Paeoniflorin with concentration of 0.5–2.0 mg/mL activates the activity of Caspase-3 and inhibits the phosphorylation of p65 and IκB, thereby inducing apoptosis of cancer cell HepG2 [122].

Yang et al. demonstrated that Paeoniflorin exhibits potential anticancer effect through inhibiting the pathways MMP-9 and ERK while increasing apoptosis of pancreatic cancer cell panc-1. Similarly, Tsuboi et al. observed that Paeoniflorin induces the expressions of mitogen-activated protein kinase (MAPK) family protein ERK, C-Jun N-terminal kinase (JNK), p38 and then induces apoptosis of leukemia Jurkal cell, via decreasing MMP, activating Caspase. Qi et al.

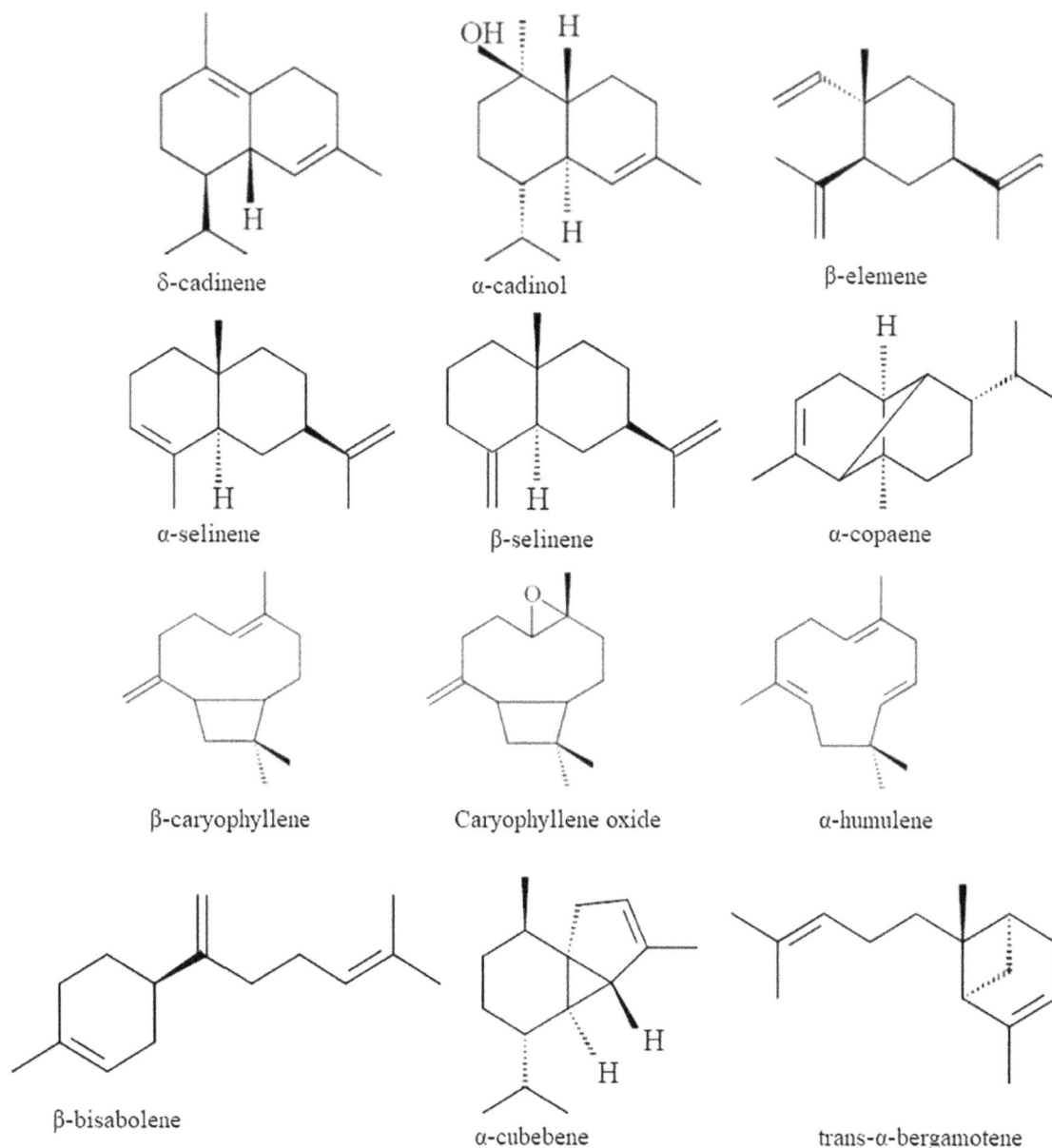

FIGURE 11.60 Structure of sesquiterpenes detected in copaiba oleoresin.

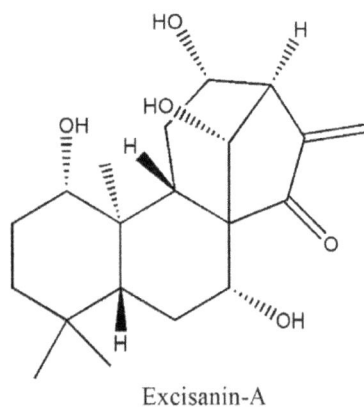

FIGURE 11.61 Structure of excisanin-A.

applied Paeoniflorin (5–40 μmol/L) to activate NF-κB pathway and found that Paeoniflorin induces apoptosis of cell A549 (Figure 11.63) [123].

Ganoderic acids (A, E, H, K, M, X, Y, Z) are the class of triterpenoids with molecular formula $C_{30}H_{44}O_7$ (Figure 11.64). It is isolated from Ganoderma mushrooms. Chemically, it is (2R, 6R)-6-[(5R, 7S, 10S, 13R, 14R, 15S, 17R)-7,15-dihydroxy-4,4,10,13,14-pentamethyl-3,11-dioxo-2,5,6,7,12,15,16,17-octahydro-1H-cyclopenta[a]phenanthren-17-yl]-2-methyl-4-oxoheptanoic acid. It has been reported to exhibit antinociceptive, antioxidative, cytotoxic, hepatoprotective and anticancer activities. Ganoderic acids can inhibit inflammation via activation of farnesoid-X-receptor (FXR). FXR plays a vital role as neuroprotective in multiple sclerosis [124].

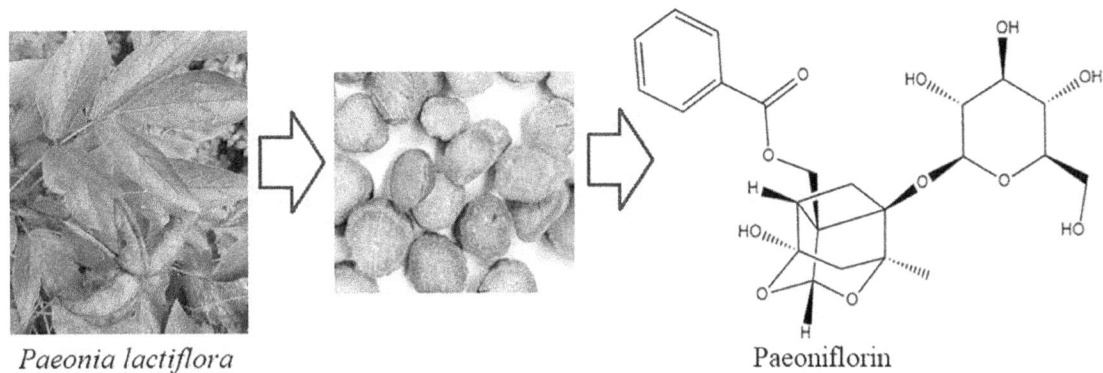

FIGURE 11.62 Structure of paeoniflorin.

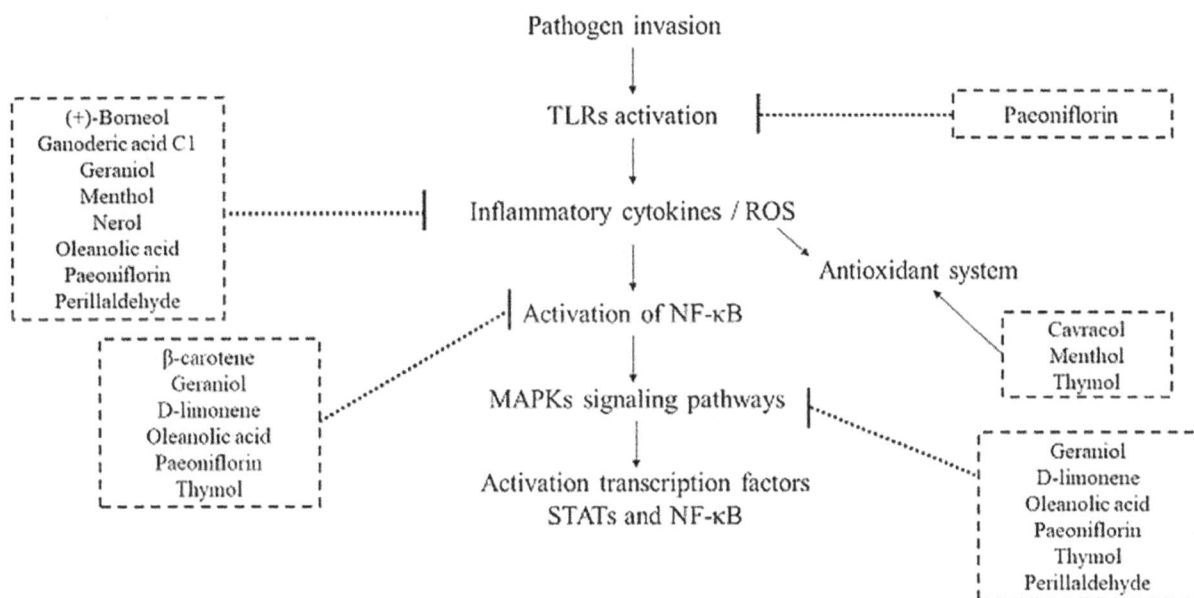

FIGURE 11.63 Targets of terpenes for anticancer and antioxidant activity.

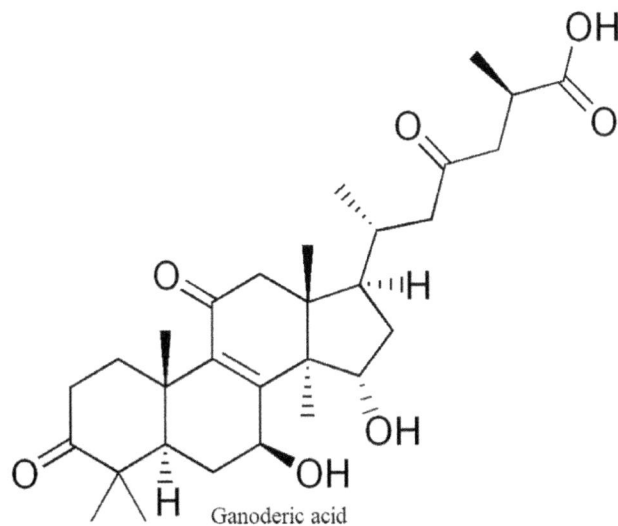

FIGURE 11.64 Structure of ganoderic acid.

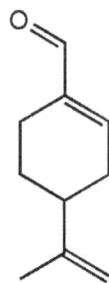

Perillaldehyde

FIGURE 11.65   Structure of perillaldehyde.

Perillaldehyde (PAE) is a naturally occurring monocyclic terpenoid with the chemical formula $C_{10}H_{14}O$ (Figure 11.65). It is abundantly present in the herb *perilla frutescence* and *Zingiber mioga*. Chemically, it is 4-prop-1-en-2-ylcyclohex-ene-1-carbaldehyde. It has been reported to exert different pharmacological activities such as anti-inflammatory, antioxidative and anti-fungal [125].

11.3 Conclusion

Natural product-based therapeutics play a major role for the discovery of terpenoids and their derivatives which have huge variety of pharmacological actions and therapeutic potentials. The wide array of structures and functionalities that have evolved in natural terpenoids provide an excellent pool for generating new drug molecules for the treatment of various disorders in human beings. Terpenoids are successfully used clinically as anti-diabetic, antimicrobial, digestive stimulant, anticancer, anti-Alzheimer, anthelmintic, antimalarial, antioxidant, immunomodulator, hepatoprotective, anti-HIV and anti-inflammatory agents. The use of bioinformatics, molecular databases and analytical experimental tools have helped greatly in analyzing the chemical, molecular mechanisms and functional studies about different types of terpenes. Hence, most of the terpenoids are in the discovery pipeline to determine their various MOA for other biological activities.

Abbreviation

ABTS	2,2'-azino-bis(3-ethylbenzothiazoline-6-sulfonic acid)
AChE	Acetylcholinesterase
AChEI	Acetylcholinesterase inhibitor
AD	Alzheimer Disease
ATP	Adenosine triphosphate
BHA	Butylated hydroxyanisole
BHT	Butylated hydroxytoluene
CBA	Cytometric bead array
COX	Cyclooxygenase
CUPRAC	Cupric Reducing Antioxidant Capacity
DASM	Dehydroandrographolide succinic acid monoester
DMSO	Dimethyl sulfoxide
DNA	Deoxyribonucleic acid

DPPH	2,2-Diphenyl-1-picrylhydrazyl
EC50	Half maximal effective concentration
EGFR	Epidermal growth factor receptor
FDA	Food and Drug Administration
FIC50	Fractional inhibitory concentrations (FICs) at the IC_{50} level
GL	Glycyrrhizin
GOH	Geraniol
HER-2	Growth factor receptor-2
HIV	Human immunodeficiency viruses
HPLC	High Performance Liquid Chromatography
IC_{50}	Half maximal inhibitory concentration
ICTs	Isocyanoterpenes
IL	Interleukin
ip	Intraperitoneal
LOX	Lipoxygenase
LPS	Lipopolysaccharide stimulated
MAPK	Mitogen-activated protein kinase
MBC	Minimum bactericidal concentration
mg	Milligram
MIC50	Minimum Inhibitory Concentration required to inhibit the growth of 50% of organisms.
MIC90	Minimum Inhibitory Concentration required to inhibit the growth of 90% of organisms
mL	Milliliter
μM	Micrometer
MMP-2	Matrix metalloproteinase-2
NF-κB	Nuclear factor kappa-light-chain-enhancer of activated B cells
NMDA	N-methyl-D-aspartate
NO	Nitric oxide
NSAIDs	Nonsteroidal anti-inflammatory drugs
OHCT	Oxidized derivative of hydroxy-cis terpenone
PGs	Prostaglandins
PR	Protease
RNA	Ribonucleic acid
ROS	Reactive oxygen species
RT	Reverse Transcriptase
TI	Therapeutic Index
TNF-α	Tumor necrosis factor alpha
TX	Thromboxane
WBC	White blood cells

Acknowledgment

The authors are grateful to Roland Institute of Pharmaceutical Sciences and Prince Mohammad Bin Fahd University.

REFERENCES

1. Gershenzon, J.; Dudareva, N. The function of terpene natural products in the natural world. *Nature Chemical Biology*, **2007**, 3(7), 408–414.
2. Wang, G.; Tang, W.; Bidigare, R.R. Zhang, L.; Demain, A.L. Natural products drug discovery and therapeutic medicine. Eds, *Terpenoids as Therapeutic Drugs and Pharmaceutical Agent*. Totowa, NJ: Humana Press, 2005, 197–227.

3. Zwenger, S.; Basu, C. Plant terpenoids: Applications and future potentials. *Biotechnology and Molecular Biology Reviews*, **2008**, 3(1), 1–7.

4. Canter, P. H.; Thomas, H.; Ernst, E. Bringing medicinal plants into cultivation: Opportunities and challenges for biotechnology. *Nature Chemical Biology*, **2005**, 23, 180–185.

5. Miller, L. H., Ackerman, H. C.; Su, X.; Wellems, T. E. Malaria biology and disease pathogenesis: Insights for new treatments. *Nature Medicine*, **2013**, 19, 156–167.

6. Aguiar, A. C., Rocha, E. M., Souza, N. B., Franca, T. C., Krettli, A. U. New approaches in antimalarial drug discovery and development: A review. *Memórias do Instituto Oswaldo Cruz*, **2012**, 107, 831–845.

7. O'Neill, P.M.; Barton, V.E.; Ward, S.A. The molecular mechanism of action of artemisinin-the debate continues. *Molecules*, **2010**, 15, 1705–1721.

8. Saito, A. Y.; Marin Rodriguez, A. A.; Menchaca Vega, D. S.; Rodrigo A. C.; Sussmann, E. A.; Kimura, A. M. Katzin, Antimalarial activity of the terpene nerolidol. *International Journal of Antimicrobial Agents*, **2016**, 48(6), 641–646.

9. Mehta, G. N. U. R., Nayak, U. R. and Dev, S. Bakuchiol, a novel monoterpenoid. *Tetrahedron Letters*, **1966**, 7(38), 4561–4567.

10. Banerji, A; Chintalwar, G. Biosynthesis of bakuchiol, A meroterpene from *Psoralea corylifolia*. *Phytochemistry*, **1983**, 22(9), 1945–1947.

11. Cheema, H. S.; Singh, A.; Pal, A.; Bhakuni, R. S.; Darokar, M. P. Antimalarial activity, mechanism of action and drug interaction study of active constituent isolated from the dried seeds of *Psoralea corylifolia*. *International Journal of Pharmacognosy and Phytochemical Research*, **2017**, 9(3), 364–375.

12. Kayembe, J. S.; Taba, K. M.; Ntumba, K.; Kazadi, T. K. *In vitro* Antimalarial Activity of 11 Terpenes Isolated from *Ocimum gratissimum* and *Cassia alata* Leaves. Screening of Their binding affinity with Haemin. *Journal of Plant Studies*, **2012**, 1(2), 168–172.

13. Ghislaine Mayer, D. C. Bruce, M., Kochurova, O., Stewart, J. K. and Zhou, Q. Antimalarial activity of a *cis*-terpenone. *Malaria Journal*, **2009**, 8, 139.

14. Daub, M. E.; Prudhomme, J.; Ben Mamoun, C.; Le Roch, K. G.; Vanderwal, C. D. Antimalarial properties of simplified kalihinol analogues. *ACS Medicinal Chemistry Letters*, **2017**, 8, 355–360.

15. Utzinger, J.; Keiser, J. Schistosomiasis and soil-transmitted helminthiasis: common drugs for treatment and control. *Expert Opinion on Pharmacotherapy*, **2004**, 5, 263–285.

16. Caffrey, C. R. Chemotherapy of schistosomiasis: Present and future. *Current Opinion in Chemical Biology*, **2007**, 11, 433–439.

17. Horton, P.; Inman, W. D.; Crew, P. Enantiomeric relationships and anthelmintic activity of dysinin derivatives from Dyszdea marine sponges. *Journal of Natural Products*, **1990**, 53,143–151.

18. Doenho, M. J.; Cioli, D.; Utzinger, J. Praziquantel: Mechanisms of action, resistance and new derivatives for schistosomiasis. *Current Opinion in Infectious Diseases*, **2008**, 21, 659–667.

19. de Moraes, J.; Almeida, A. A.; Brito, M. R.; Marques, T. H.; Lima, T. C.; de Sousa, D. P.; Nakano, E.; Mendonça, R.Z.; Freitas, R.M. Anthelmintic activity of the natural compound (+)-limonene epoxide against *Schistosoma mansoni*. *Planta Medica*, **2013**, 79, 253–258.

20. Charles, A. D. Anti inflammatory agents: present and future. *Cell*, **2010**, 140, 935–950.

21. Rainsford, K. D. Profile and mechanisms of gastrointestinal and other side effects of nonsteroidal anti-inflammatory drugs (NSAIDs). *The American Journal of Medicine*, **1999**, 107, 27S–35S.

22. De las Heras, B., Hortelano S., Molecular basis of the anti-inflammatory effects of terpenoids. *Inflammation & Allergy - Drug Targets*, **2009**, 8(1), 28–39.

23. Kapewangolo, P.; Omolo, J. J.; Bruwer, R.; Fonteh, P.; Meyer, D. Antioxidant and anti-inflammatory activity of *Ocimum labiatum* extract and isolated labdane diterpenoid. *Journal of Inflammation Research*, **2015**, 12, 4.

24. Jayachandran, M.; Chandrasekaran, B.; Namasivayam, N. Geraniol attenuates fibrosis and exerts anti-inflammatory effects on diet induced atherogenesis by NF-?B signaling pathway. *European Journal of Pharmacology*, **2015**, 762, 102–111.

25. Maeda, H.; Yamazaki, M.; Katagata, Y. Kuromoji (*Lindera umbellata*) essential oil inhibits LPS-induced inflammation in RAW 264.7 cells. *Bioscience, Biotechnology, and Biochemistry*, **2013**, 77(3), 482–486.

26. Prakash, V. Terpenoids as source of anti-inflammatory compounds. *Asian Journal of Pharmaceutical and Clinical Research*, **2017**, 10(3), 68–76.

27. Zhang, M.; Iinuma, M.; Wang, J. S.; Oyama, M.; Ito, T.; Kong, L. Y. Terpenoids from *Chloranthus serratus* and their anti-inflammatory activities. *Journal of Natural Products*, **2012**, 75, 694–698.

28. Chang, R. S.; Ding, L.; Chen, G. Q.; Pan, Q. C.; Zhao, Z. L.; Smith, K. M. Dehydroandrographolide succinic acid monoester as an inhibitor against the human immunodeficiency virus. *Proceedings of the Society for Experimental Biology and Medicine*, **1991**, 197, 59–66.

29. Paris, A.; Strukelj, B.; Renko, M.; Turk, V.; Pukl, M.; Umek, A.; Korant, B. D. Inhibitory effect of carnosic acid on HIV-1 protease in cell-free assays. *Journal of Natural Products*, **1993**, 56, 1426–1430.

30. Gustafson, K. R.; Munro, M. H. G.; Blunt, J. W.; Cardellina, J. H. II; McMahon, J. B.; Gulakowski, R. J.; Cragg, G. M.; Cox, P. A.; Brinen, L. S.; Clardy, J.; Boyd, M. R. HIV Inhibitory natural products. 3. Diterpenes from *Homalanthus acuminatus* and *Chrysobalanus icaco*. *Tetrahedron*, **1991**, 47, 4547–4554.

31. Chen, K.; Shi, Q.; Fujioka, T.; Zhang, D. C.; Hu, C. Q.; Jin, J. Q.; Kilkuskie, R. E.; Lee, K. H. Anti-AIDS Agents, 4. Tripterifordin, a novel anti-HIV principle from *Tripterygium wilfordii*: Isolation and structural elucidation. *Journal of Natural Products*, **1992**, 55, 88–92.

32. Chen, K.; Shi, Q.; Fujioka, T.; Nakano, T.; Hu, C. Q.; Jin, J. Q.; Kilkuskie, R. E.; Lee, K. H. Anti-AIDS agents, XIX. Neotripteriofordin, a novel anti-HIV principle from *Tripterygium wilfordii*: Isolation and structural elucidation. *Bioorganic & Medicinal Chemistry*, **1995**, 3, 1345–1348.

33. Wu, Y. C.; Hung, Y. C.; Chang, F. R.; Cosentino, L. M.; Wang, H. K.; Lee, K. H. Identification of Ent-16β, 17-dihydroxykauran-19-oic acid as an Anti-HIV principle and isolation of the new diterpenoids annosquamosins A and B from *Annona squamosa*. *Journal of Natural Products*, **1996**, 59, 635–637.

34. Chang, F. R.; Yang, P. Y.; Lin, J. Y.; Lee, K. H.; Wu, Y.C. Bioactive kaurane diterpenoids from annona glabra. *Journal of Natural Products*, **1998**, 61, 437–439.

35. Gustafson, K. R.; Cardellina, J. H. II; McMahon, J. B.; Gulakowski, R. J.; Ishitoya, J.; Szallasi, Z.; Lewin, N. E.; Blumberg, P. M.; Weislow, O. S.; Beutler, J. A.; Buckheit Jr, R. W.; Cragg, G. M.; Cox, P. A.; Bader, J. P.; Boyd, M. R. A Nonpromoting phorbol from the samoan medicinal plant *Homalanthus nutans* inhibits cell killing by HIV-1. *Journal of Medicinal Chemistry*, **1992**, 35, 1978–1986.

36. Erickson, K. L.; Beutler, J. A.; Cardellina, J. H. II; McMahon, J. B.; Newman, J. D.; Boyd, M. R. A Novel phorbol ester from *Excoecaria agallocha*. *Journal of Natural Products*, **1995**, 58, 769–772.

37. Ito, M.; Sato, A.; Hirabayashi, K.; Tanabe, F.; Shigeta, S.;Baba, M.; De Clercq, E.; Nakashima, H.; Yamamoto, N. Mechanism of inhibitory effect of glycyrrhizin on replication of human immunodeficiency virus (HIV). *Antiviral Research*, **1988**, 10, 289–298.

38. Nakashima, H.; Matsui, T.; Yoshida, O.; Isowa, Y.; Kido, Y.; Motoki, Y.; Ito, M.; Shigeta, S.; Mori, T.; Yamamoto, N. A New anti-human immunodeficiency virus substance, glycyrrhizin sulfate; endowment of glycyrrhizin with reverse transcriptase-inhibitory activity by chemical modification. *Japanese Journal of Cancer Research*, **1987**, 78, 767–771.

39. Tochikura, T. S.; Nakashima, H.; Yamamoto, N. Antiviral agents with activity against human retroviruses. *JAIDS Journal of Acquired Immune Deficiency Syndromes*, **1989**, 2, 441–447.

40. Hirabayashi, K.; Iwata, S.; Matsumoto, H.; Mori, T.; Shibata, S.; Baba, M.; Ito, M.; Shigeta, S.; Nakashima, H.; Yamamoto, N. Antiviral activities of glycyrrhizin and its modified compounds against human immunodeficiency virus type 1 (HIV-1) and herpes simplex virus type 1 (HSV-1) *in Vitro. Chemical and Pharmaceutical Bulletin,* **1991**, 39, 112–115.

41. Abdallah, R. M.; Ghazy, N. M.; El-Sebakhy, N. A.; Pirillo, A.; Verotta, L. Astragalosides from egyptian *Astragalus spinosus* Vahl. *Pharmazie* **1993**, 48, 452–454.

42. Konoshima, T.; Yasuda, I.; Kashiwada, Y.; Cosentino, L. M.; Lee, K. H. Anti-AIDS Agents, 21. Triterpenoid Saponins as Anti-HIV Principles from Fruits of *Gleditsia japonica* and *Gymnocladus chinensis*, and a Structure activity Correlation. *Journal of Natural Products*, **1995**, 58, 1372–1377.

43. Chen, K.; Shi, Q.; Kashiwada, Y.; Zhang, D. C.; Hu, C. Q.; Jin, J. Q.; Nozaki, H.; Kilkuskie, R. E.; Tramontano, E.; Cheng, Y. C.; McPhail, D. R.; McPhail, A. T.; Lee, K. H. Anti-AIDS Agents, 6. Salaspermic acid, an Anti-HIV principle from *Tripterygium wilfordii*, and the structure activity correlation with its related compounds. *Journal of Natural Products*, **1992**, 55, 340–346.

44. Kuo, Y. H.; Kuo, L. M. Anti-tumor and Anti-AIDS triterpenes from *Celastrus hindsii*. *Phytochemistry* **1997**, 44, 1275–1281.

45. Jiang, Y.; Ng, T. B.; Wang, C. R.; Zhang, D.; Cheng, Z. H.; Liu, Z. K.; Qiao, W. T.; Geng, Y. Q.; Li, N.; Liu, F. Inhibitors from natural products to HIV-1 reverse transcriptase, protease and integrase. *Mini-Reviews in Medicinal Chemistry*, **2010**, 10(14), 1331–1344.

46. Li H.-Y., Sun N.-J., Kashiwada Y., Sun L., Snider J. V., Cosentino L. M. Anti-AIDS agents, 9. suberosol, a new C31 lanostane-type triterpene and anti-HIV principle from *Polyalthia suberosa. Journal of Natural Products*, **1993**, 56 1130–1133.

47. Xu, H. X. Zeng, F. Q.; Wan, M.; Sim, K. Y. Anti-HIV triterpene acids from *Geum japonicum. Journal of Natural Products*, **1996**, 59, 643–645.

48. Ma, C.; Nakamura, N.; Miyasho, H.; Hattori, M.; Shimotohno, K. Inhibitory effects of constituents from *Cynomorium songaricum* and related triterpene derivatives on HIV-1 protease. *Chemical and Pharmaceutical Bulletin*, **1999**, 47, 141–145.

49. Quere, L.; Wenger, T.; Schramm, H. J. Triterpenes as potential dimerization inhibitors of HIV-1 protease. *Biochemical and Biophysical Research Communications*, **1996**, 227, 484–488.

50. Sun, I. C.; Wang, H. K.; Kashiwada, Y.; Shen, J. K.; Cosentino, L. M.; Chen, C. H.; Yang, L. M.; Lee, K. H. Anti-AIDS agents. 34. Synthesis and structure-activity relationships of betulin derivatives. *Journal of Medicinal Chemistry*, **1998**, 41, 4648–4657.

51. Kashiwada, Y.; Wang, H. K.; Nagao, T.; Kitanaka, S.; Yasuda, I.; Fujioka, T.; Yamagishi, T.; Cosentino, L. M.; Kozuka, M.; Okabe, H.; Ikeshiro, Y.; Hu, C. Q.; Yeh, E.; Lee, K. H. Anti AIDS agents, 30. Anti-HIV activity of oleanolic acid, pomolic acid, and structurally related triterpenoids *Journal of Natural Products*, **1998**, 61, 1090–1095.

52. Fujioka, T.; Kashiwada, Y.; Kilkuskie, R. E.; Cosentino, L. M.; Ballas, M. L.; Jiang, J. B.; Janzen, W. P.; Chen, I. S.; Lee, K. H. Anti-AIDS agents, 11. Betulinic acid and platanic acid as anti-HIV principles from *Syzygium claviflorum*, and the Anti-HIV activity of structurally related triterpenoids. *Journal of Natural Products*, **1994**, 57, 243–247.

53. Kashiwada, Y.; Hashimoto, F.; Cosentino, L. M.; Chen, C. H.; Garrett, P. E.; Lee, K. H. Betulinic acid and dihydrobetulinic acid derivatives as potent anti-HIV agents. *Journal of Medicinal Chemistry*, **1996**, 39, 1016–1017.

54. Soler, F.; Poujade, C.; Evers, M.; Carry, J. C.; Henin, Y.; Bousseau, A.; Huet, T.; Pauwels, R.; De Clercq, E.; Mayaux, J. F.; Le Pecq, J. B.; Dereu, N. Betulinic acid derivatives: A new class of specific inhibitors of human immunodeficiency virus type-1 entry. *Journal of Medicinal Chemistry*, **1996**, 39, 1069–1083.

55. Evers, M.; Poujade, C.; Soler, F.; Ribeill, Y.; James, C.; Lelievre, Y.; Gueguen, J. C.; Reisdorf, D.; Morize, I.; Pauwels, R.; De Clercq, E.; Henin, Y.; Bousseau, A.; Mayaux, J. F.; Le Pecq, J. B.; Dereu, N. Betulinic acid derivatives: A new class of human immunodeficiency virus type-1 specific inhibitors with a new mode of action. *Journal of Medicinal Chemistry*, **1996**, 39, 1056–1068.

56. Mayaux, J. F.; Bousseau, A.; Pauwels, R.; Huet, T.; Henin, Y.; Dereu, N.; Evers, M.; Soler, F.; Poujade, C.; De Clercq, E.; Le Pecq, J. B. Triterpene derivatives that block entry of human immunodeficiency virus type-1 into cells. *Proceedings of the National Academy of Sciences of the United States of America*, **1994**, 91, 3564–3568.

57. Sun, H. D.; Qiu, S. X.; Lin, L. Z.; Wang, Z. Y.; Lin, Z. W.; Pengsuparp, T.; Pezzuto, J. M.; Fong, H. H. S.; Cordell, G. A.; Farnsworth, N. R. Nigranoic acid, a triterpenoid

from *Schisandra sphaerandra* that inhibits HIV-1 reverse transcriptase. *Journal of Medicinal Chemistry*, **1996**, 59, 525–527.

58. Chen, D. F.; Zhang, S. X.; Wang, H. K.; Zhang, S. Y.; Sun, Q. Z.; Cosentino, L. M.; Lee, K. H. Novel anti-HIV lancilactone C and related triterpenes from *Kadsura lancilimba*. *Journal of Natural Products*, **1999**, 62, 94–97.

59. Lleó, A.; Greenberg, S. M.; Growdon, J. H. Current pharmacotherapy for Alzheimer's disease. *Annual Review of Medicine*, **2006**, 57, 513–33.

60. Akhondzadeh, S.; Abbasi, S. H. Herbal medicine in the treatment of Alzheimer's disease. *American Journal of Alzheimer's Disease & Other Dementias*, **2006**, 21(2), 113–118.

61. Yoo, K. Y.; Park, S.Y. Terpenoids as potential anti-alzheimer's disease therapeutics. *Molecules*, **2012**, 17, 3524–3538.

62. Hayet, E.; Fatma, B.; Souhir, I.; Waheb, F. A.; Abderaouf, K.; Mahjoub, A.; Maha, M. Antibacterial and cytotoxic activity of the acetone extract of the flowers of *Salvia sclarea* and some natural products. *Pakistan Journal of Pharmaceutical Sciences*, **2007**, 20(2):146–148.

63. Wang, C. Y.; Chen, Y. W.; Hou, C. Y. Antioxidant and antibacterial activity of seven predominant terpenoids. *International Journal of Food Properties*, **2019**, 22(1), 230–238.

64. Nakano, T.; Djerassi, C. Terpenoids. XLVI. Copalic acid. *The Journal of Organic Chemistry*, **1961**, 26, 167–173. Souza, A. B.; Souza, M. G. M.; Moreira, M. A.; Moreira, M. R.; Furtado, N. A. J. C.; Martins, C. H. G.; Bastos, J. K.; Santos, R. A.; Heleno, V. C. G.; Ambrosio, S. R. Antimicrobial evaluation of diterpenes from *Copaifera langsdorffii* oleoresin against periodontal anaerobic bacteria. *Molecules*, **2011**, 16, 9611–9619.

65. Paiva, L. A. F.; Gurgel, L. A.; Silva, R. M.; Tomé, A. R.; Gramosa, N. V.; Silveira, E. R.; Santos, F. A.; Rao, V. S. N. Anti-inflammatory effect of kaurenoic acid, a diterpene from *Copaifera langsdorffii* on acetic acid-induced colitis in rats. *Vascular Pharmacology*, **2003**, 39, 303–307.

66. Souza, P. A.; Rangel, L. P.; Oigman, S. S.; Elias, M. M.; Ferreira-Pereira, A.; Lucasa, N. C.; Leitão, G. G. Isolation of two bioactive diterpenic acids from *Copaifera glycycarpa* oleoresin by high-speed counter-current chromatography. *Phytochemical Analysis*, **2010**, 21, 539–543.

67. Mcchesneya, J. D.; Clark, A. M.; Silveira, E. R. Antimicrobial diterpenes of *Croton sonderzanus*, 1-Hardwickic and 3,4-secotrachylobanoic acids. *Journal of Natural Products*, **1991**, 54, 1625–1633.

68. Shi, W.; Gould, M. N. Induction of cytostasis in mammary carcinoma cells treated with the anticancer agent perillyl alcohol. *Carcinogenesis*, **2002**, 23, 131–142.

69. Gould, M. N. Cancer chemoprevention and therapy by monoterpenes. *Environmental Health Perspectives*, **1997**, 105, 977–979.

70. Zou, L.; Liu, W.; Yu, L. β-Elemene induces apoptosis of K562 leukemia cells. *Chinese Journal of Oncology*, **2001**, 23, 196–198.

71. Li, L. J.; Zhong, L. F.; Jiang, L. P.; Geng, C. Y.; Zou, L. J. β-elemene radiosensitizes lung cancer A549 cells by enhancing DNA damage and inhibiting DNA repair. *Phytotherapy Research*, **2011**, 25, 1095–1097.

72. Sun, Y.; Liu, G.; Zhang, Y.; Zhu, H.; Ren, Y.; Shen, Y. M. Synthesis and *in vitro* anti-proliferative activity of β-elemene monosubstituted derivatives in HeLa cells mediated through arrest of cell cycle at the G1 phase. *Bioorganic & Medicinal Chemistry*, **2009**, 17, 1118–1124.

73. Rogerio, A. P.; Andrade, E. L.; Leite, D. F. P.; Figueiredo, C. P.; Calixto, J. B. Preventive and therapeutic anti-inflammatory properties of the sesquiterpene α-humulene in experimental airways allergic inflammation. *British Journal of Pharmacology*, **2009**, 158, 1074–1087.

74. Pérez-López, A.; Cirio, A. T.; Rivas-Galindo, V. M.; Aranda, R. S.; Torres, N. W. Activity against Streptococcus pneumoniae of the essential oil and δ-cadinene isolated from Schinus molle fruit. *Journal of Essential Oil Research*, **2011**, 23, 25–28.

75. Lee, Y. S.; Jin, D. Q.; Kwon, E. J.; Park, S. H.; Lee, E. S.; Jeong, T. C.; Nam, D. H.; Huh, K.; Kim, J. A. Asiatic acid, a triterpene, induces apoptosis through intracellular Ca^{2+} release and enhanced expression of p53 in HepG2 human hepatoma cells. *Cancer Letters*, **2002**, 186, 83–91.

76. Oh, O. J.; Min, H. Y.; Lee, S. K. Inhibition of inducible prostaglandin E2 production and cyclooxy-genase-2 expression by curdione from *Curcuma zedoaria*. *Archives of Pharmacal Research*, **2007**, 30(10), 1226–1239.

77. Hezari, M.; Croteau, R. Taxol biosynthesis: An update. *Planta Medica* **1997**, 63, 291–295.

78. Priyadarshini, K.; Keerthi Aparajitha, U. Paclitaxel against cancer: A short review. *Medical Chemistry* **2012**, 2, 139–141.

79. Wang, E.; Li, X.; Sun, Y.; Tai, X.; Li, W.; Guo, T.; Induction of apoptosis by furanodiene in HL60 leukemia cells through activation of TNFR1 signaling pathway. *Cancer Letters*, **2008**, 271, 158–166.

80. Sun, X. Y.; Zheng, Y.P.; Lin, D.H.; Zhang, H.; Zhao, F.; Yuan, C. S. Potential anti-cancer activities of Furanodiene, a Sesquiterpene from *Curcuma wenyujin*. *The American Journal of Chinese Medicine*, **2009**, 37, 589–596.

81. Zhong, Z. F.; Li, Y. B.; Wang, S. P.; Tan, W.; Chen, X. P., Chen, M. W., Wang, Y. T. Furanodiene enhances tamoxifen-induced growth inhibitory activity of ERa- positive breast cancer cells in a PPARg independent manner. *Journal of Cellular Biochemistry* **2012**, 113, 2643–2651.

82. Makabe, H.; Maru, N.; Kuwabara, A.; Kamo, T.; Hirota, M. Anti-inflammatory sesquiterpenes from *Curcuma zedoaria*. *Natural Product Research* **2006**, 20, 680–685.

83. Tanaka, K.; Kuba, Y.; Ina, A.; Watanabe, H.; Komatsu, K., Prediction of cyclooxygenase inhibitory activity of curcuma rhizome from chromatograms by multivariate analysis. *Chemical and Pharmaceutical Bulletin* **2008**, 56, 936–940.

84. Lou, Y.; Zhang, H.; He, H.; Peng, K.; Kang, N.; Wei, X.; Li, X.; Chen, L.; Yao, X.; Qiu, F. Isolation and identification of phase-1 metabolites of curcumol in rats. *Drug Metabolism and Disposition* **2010**, 38, 2014–2022.

85. Xu, L.; Bian, K.; Liu, Z.; Zhou, J.; Wang, G. The inhibitory effect of the curcumol on women cancer cells and synthesis of RNA. *Tumor*, **2005**, 25, 570–572.

86. Zhang, W.; Wang, Z.; Chen, T. Curcumol induces apoptosis via caspases-independent mitochondrial pathway in human lung adeno carcinoma ASTC-a-1 cells. *Medical Oncology*, **2011**, 28, 307–314.

87. Halliwell, B. Antioxidants and human disease: A general introduction. *Nutrition Reviews*, **1997**, 55, 49–52.

88. Pham-Huy, L. A.; He, H.; Pham-Huy, C.; Radicals, F. Antioxidants in disease and health. *International Journal of Biomedical Science*, **2008**, 4(2), 89–96.

89. Kedare, S. B.; Singh, R. P. Genesis and development of DPPH method of antioxidant assay. *Journal of Food Science and Technology*, **2011**, 48(4), 412–422.

90. Edris, A. E. Pharmaceutical and therapeutic potentials of essential oils and their individual volatile constituents: A review. *Phytotherapy Research*, **2007**, 21, 308–323.

91. Miguel, M. G. Antioxidant and anti-inflammatory activities of essential oils: A short review. *Molecules.* **2010**, 15, 9252–9287.

92. Wei, A.; Shibamoto, T. Antioxidant/lipoxygenase inhibitory activities and chemical compositions of selected essential oils. *Journal of Agricultural and Food Chemistry*, **2010**, 58, 7218–7225.

93. Wang, C. Y., Chen, Y. W.; Hou, C. Y. Antioxidant and antibacterial activity of seven predominant terpenoids. *International Journal of Food Properties*, **2019**, 22(1), 229–237.

94. Shantilal, S.; Vaghela, J. S.; Sisodia, S. S. Review on immunomodulation and immunomodulatory activity of some medicinal plant. *European Journal of Biomedical and Pharmaceutical sciences*, **2018**, 5(8), 163–174.

95. Raphael, T.J.; Kuttan, G. Immunomodulatory activity of naturally occurring monoterpenes carvone, limonene, and perillic acid. *Immunopharmacol Immunotoxicol*, **2003**, 25(2), 285–94.

96. Alqasoumi, S. I., Abdel-Kader, M. S. Terpenoids from Juniperus procera with hepatoprotective activity. *Pakistan Journal of Pharmaceutical Sciences*, **2012**, 25(2), 315–22.

97. Dey, P., Saha, M. R., Sen, A. Hepatotoxicity and the present herbal hepatoprotective scenario. *International Journal of Green Pharmacy*, **2013**, 7, 265–73.

98. Putta, S.; Yarla, N. S.; Kilari, E. K.; Surekha, C.; Aliev, G.; Divakara, M. B.; Santosh, M. S.; Ramu, R.; Zameer, F.; Mn, N. P.; Chintala, R.; Rao, P. V.; Shiralgi, Y.; Dhananjaya, B. L. Therapeutic potentials of triterpenes in diabetes and its associated complications. *Current Topics in Medicinal Chemistry*, 2016, 16(23), 2532–2542.

99. Kumar, D.; Ghosh, R.; Pal, B. C. α-Glucosidase inhibitory terpenoids from *Potentilla fulgens* and their quantitative estimation by validated HPLC method. *Journal of Functional Foods*, 2013, 5, 1135–1141. Doi: 10.1016/j.jff.2013.03.010.

100. Mbaze, L.M.; Poumale, H. M.; Wansi, J. D.; Lado, J. A.; Khan, S. N.; Iqbal, M. C.; Ngadjui, B. T.; Laatsch, H. Alpha-glucosidase inhibitory pentacyclic triterpenes from the stem bark of *Fagara tessmannii* (Rutaceae). *Phytochemistry*, **2007**, 68, 591–595. Doi: 10.1016/j.phytochem.2006.12.015.

101. Mnafgui, K.; Kaanich, F.; Derbali, A.; Hamden, K.; Derbali, F.; Slama, S.; Allouche, N.; Elfeki, A. Inhibition of key enzymes related to diabetes and hypertension by Eugenol *in vitro* and in alloxan-induced diabetic rats. *Archives of Physiology and Biochemistry*, **2013**, 1–9. Doi: 10.3109/13813455.2013.822521.

102. Hou, W.; Li, Y.; Zhang, Q.; Wei, X.; Peng, A.; Chen, L.; Wei, Y. Triterpene acids isolated from Lagerstroemia speciosa leaves as α-glucosidase inhibitors. *Phytotherapy Research*, **2009**, 23, 614–618. Doi: 10.1002/ptr.2661.

103. Samarghandian, S.; Borji, A.; Delkhosh, M. B., Samini, F. Safranal treatment improves hyperglycemia, hyperlipidemia and oxidative stress in streptozotocin induced diabetic rats. *Journal of Pharmaceutical Sciences*, **2013**, 16, 352–362. Doi: 10.18433/j3zs3q.

104. Prasad, S.; Kalra, N.; Shukla, Y. Hepatoprotective effects of lupeol and mango pulp extract of carcinogen induced alteration in Swiss albino mice. *Molecular Nutrition & Food Research*, **2007**, 51, 352–359. Doi: 10.1002/mnfr.200600113.

105. Deutschländer, M. S.; Lall, N.; Van de Venter, M.; Hussein, A.A. Hypoglycaemic evaluation of a new triterpene and other compounds isolated from Eucleaundulata Thunb. var. myrtina (Ebenaceae) root bark. *Journal of Ethnopharmacology*, **2011**, 133, 1091–1095. Doi: 10.1016/j.jep.2010.11.038.

106. Güdr, A. Influence of total anthocyanins from bitter melon (momordica charantia linn.) as antidiabetic and radical scavenging agents. *Iranian Journal of Pharmaceutical Research*, **2016**, 15:301–309.

107. Tan, M. J.; Ye, J. M.; Turner, N.; Hohnen-Behrens, C.; Ke, C. Q.; Tang, C. P.; Chen, T.; Weiss, H. C.; Gesing, E. R.; Rowland, A.; James, D. E. Antidiabetic activities of triterpenoids isolated from bitter melon associated with activation of the AMPK pathway. *Chemical Biology*, **2008**, 15, 263–273. Doi: 10.1016/j.chembiol.2008.01.013.

108. Geuns, J. M. Stevioside. *Phytochemistry*, **2003**, 64(5):913–921. Doi: 10.1016/s0031-9422(03)00426-6.

109. Jeppesen, P. B., Gregersen, S., Poulsen, C. R., Hermansen, K. Stevioside acts directly on pancreatic β cells to secrete insulin: actions independent of cyclic adenosine monophosphate and adenosine triphosphate sensitive K+ channel activity. *Metabolism*, **2000**, 49, 208–214. Doi: 10.1016/s0026-0495(00)91325-8.

110. Jeppesen, P. B.; Gregersen, S.; Alstrup, K. K.; Hermansen, K. Stevioside induces antihyperglycemic, insulinotropic and glucagonostatic effects *in vivo*: Studies in the diabetic GotoKakizaki (GK) rats. *Phytomedicine*, 2002, 9, 9–14. Doi: 10.1078/0944-7113-00081.

111. Eliza, J.; Daisya, P.; Ignacimuthu, S.; Duraipandiyan, V. Normoglycemic and hypolipidemic effect of costunolide isolated from *Costus speciosus* (Koen ex. Retz.)Sm. in streptozotocin-induced diabetic rats. *Chemico-Biological Interactions*, **2009**, 179, 329–334. Doi: 10.1016/j.cbi.2008.10.017.

112. Naik, S. R.; Barbosa Filho, J. M.; Dhuley, J. N.; Deshmukh, V. Probable mechanism of hypoglycemic activity of bassic acid, a natural product isolated from *Bumelia sartorum*. *Journal of Ethnopharmacology*, **1991**, 33, 3744. Doi: 10.1016/0378-8741(91)90158-a.

113. Ha do, T.; Tuan, D. T.; Thu, N. B.; Nhiem, N. X.; Ngoc, T. M.; Yim, N.; Bae K. Palbinone and triterpenes from Moutan Cortex (*Paeonia suffruticosa*, Paeoniaceae) stimulate glucose uptake and glycogen synthesis via activation of AMPK in insulin-resistant human HepG2 cells. *Bioorganic & Medicinal Chemistry Letters*, **2009**, 19(19), 5556–5559. Doi: 10.1016/j.bmcl.2009.08.048.

114. Shi, Q.; Wang, J.; Cheng, Y.; Dong, X.; Zhang, M.; Pei, C Palbinone alleviates diabetic retinopathy in STZ-induced rats by inhibiting NLRP3 inflammatory activity. *Biochemical and Molecular Toxicology*, **2019**, e22489. Doi: 10.1002/jbt.22489.

115. Li, Y. W.; Zhu, G. Y.; Shen, X. L.; Chu, J. H.; Yu, Z. L.; Fong, W. F. Furanodienone induces cell cycle arrest and apoptosis by suppressing EGFR/HER2 signaling in HER2-overexpressing human breast cancer cells. *Cancer Chemotherapy and Pharmacology*, **2011**, 68(5), 1315–1323.

116. Oliveira, F. A.; Chaves, M. H.; Almeida, F. R.; Lima, R. C. Jr; Silva, R. M.; Maia, J. L.; Brito, G. A.; Santos, F. A.; Rao, V. S. Protective effect of alpha- and beta-amyrin, a triterpene mixture from Protium heptaphyllum (Aubl.) March. trunk wood resin, against acetaminophen-induced liver injury in mice. *Journal of Ethnopharmacology*, **2005**, 98(1–2), 103–108.

117. Pereira Beserra, F.; Xue, M.; Maia, G. L. D. A.; Leite Rozza, A.; Helena Pellizzon, C.; Jackson, C. J. Lupeol, a pentacyclic triterpene, promotes migration, wound closure, and contractile effect *in vitro*: Possible involvement of PI3K/Akt and p38/ERK/MAPK pathways. *Molecules*, **2018**, 23, 1–17.

118. Yilmaz, A.; Boga, M.; Topcu, G. Novel terpenoids with potential anti-Alzheimer activity from *Nepeta obtusicrena*. *Records of Natural Products*, **2016**, 10(5), 530–541.

119. Leandro, L. M., de Sousa Vargas, F., Barbosa, P. C. S., Neves, J. K. O., Da Silva, J. A.; da Veiga-Junior, V. F. Chemistry and biological activities of terpenoids from copaiba (*Copaifera spp.*) oleoresins. *Molecules*, **2012**, 17, 3866–3889.

120. Cunha, K. M. A.; Paiva L. A.; Santos, F. A.; Gramosa, N. V.; Silveira, E. R.; Rao, V. S. Smooth muscle relaxant effect of kaurenoic acid, a diterpene from *Copaifera langsdorffii* on rat uterus *in vitro*. *Phytotherapy Research*, **2003**, 17, 320–324.

121. Deng, R.; Tang, J.; Xia, L. P.; Li, D. D.; Zhou, W. J.; Wang, L. L.; Feng, G. K.; Zeng, Y. X.; Gao, Y. H.; Zhu, X. F. ExcisaninA, a diterpenoid compound purified from Isodon Macrocalyxin D, induces tumor cells apoptosis and suppresses tumor growth through inhibition of PKB/AKT kinase activity and blockade of its signal pathway. *Molecular Cancer Therapeutics*, **2009**, 8, 873–882.

122. Niu, K.; Liu, Y.; Zhou, Z.; Wu, X.; Wang, H.; Yan J. Antitumor effects of paeoniflorin on hippo signaling pathway in gastric cancer cells. *Journal of Clinical Oncology*, **2021**, 2021, 4724938. Doi: 10.1155/2021/4724938.

123. Yang, J.; Ren, Y.; Lou, Z.-G.; Wan, X.; Weng, G.-B.; Cen, D. Paeoniflorin inhibits the growth of bladder carcinoma via deactivation of STAT3. *Acta Pharmaceutica*, **2018**, 68(2), 211–222. doi: 10.2478/acph-2018-0013.

124. Cao, F.-R.; Feng, L.; Ye, L.-H.; Wang, L.-S.; Xiao, B.-X.; Tao, X.; Chang, Q. Ganoderic acid a metabolites and their metabolic kinetics. *Frontiers in Pharmacology*, 2017, 8. Doi: 10.3389/fphar.2017.00101.

125. Chen, L.; Qu, S.; Yang, K.; Liu, M.; Li, Y. X.; Keller, N. P.; Zeng, X.; Tian, J. Perillaldehyde: A promising antifungal agent to treat oropharyngeal candidiasis. *Biochemical Pharmacology*, **2020**, 180, 114201. Doi: 10.1016/j. bcp.2020.114201.

12

Terpenoids: A Brief Survey of Naturally Occurring Terpenes Molecules

Bimal Krishna Banik*, Abhishek Tiwari†, and Biswa Mohan Sahoo‡

CONTENTS

12.1 Introduction 399
12.2 Sources 400
12.3 Classification 400
12.4 Terpenoids as Chemotherapeutics 441
12.5 Classification Terpenoids 442
12.6 Mechanism of Action of Selected Terpenoids against Cancer Progression 442
12.7 Monoterpenoids 442
12.8 Menthol 447
12.9 Auraptene 448
12.10 D-Limonene 448
12.11 Perillic Acid 448
12.12 Ascaridole 448
12.13 Carvacrol 449
12.14 Sesquiterpenoids and Sesquiterpene Lactones 449
12.15 Artesunate and Artemisinin 449
12.16 β-Elemene 449
12.17 Diterpenoids 449
12.18 Crocetin 450
12.19 Triterpenoids 450
12.20 Betulinic Acid 450
12.21 Lupeol 450
12.22 Conclusion 450
References 451

12.1 Introduction

The study of natural products derived from plants, animals, or microbes is what sets this field apart. The products could be medically beneficial or harmful. Natural product chemistry investigates the natural source, the mechanisms by which the source biosynthetically creates the product, the processes by which the product can be extracted from the source, and the methods by which the product can be identified. This research lay the framework for pharmacological testing of a potentially helpful natural substance or biochemical analysis of a natural toxin. Courses in natural products and medicinal chemistry, chemistry, botany, and microbiology, as well as pharmacology and pharmaceutics, would be the focus of a natural products chemistry curriculum.

Terpenoids are a class of chemicals found primarily in plants, though a few terpenoids have been isolated from other sources. Essential oils, which are volatile oils extracted from the sap and tissues of some plants and trees, are mostly composed of mono- and sesqui-terpenoids. Since the beginning of time, essential oils have been utilised in perfumery. The non-steam volatile di- and tri-terpenoids are extracted from plant and tree gums and resins. Carotenoids are a category of chemicals that includes tetraterpenoids. The most important polyterpenoid is rubber.

* Department of Mathematics and Natural Sciences, College of Sciences and Human Studies, Prince Mohammad Bin Fahd University, Al Khobar, Kingdom of Saudi Arabia. bimalbanik10@gmail.com; bbanik@pmu.edu.sa.

† Faculty of Pharmacy, Pharmacy Academy, IFTM University, Lodhipur Rajput, Moradabad-244102, Uttar Pradesh, India. abhishekt1983@gmail.com.

‡ Roland Institute of Pharmaceutical Sciences, Berhampur affiliated to Biju Patnaik University of Technology (BPUT), Rourkela, Odisha, India.

CH₃ ... (chemical structure diagram) **Monoterpenoid**

(chemical structure diagram) **Sesquiterpenoids**

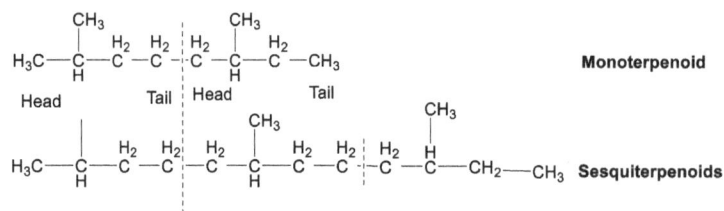

FIGURE 12.1 Structure of Monoterpennoids and Sesquiterpenoids

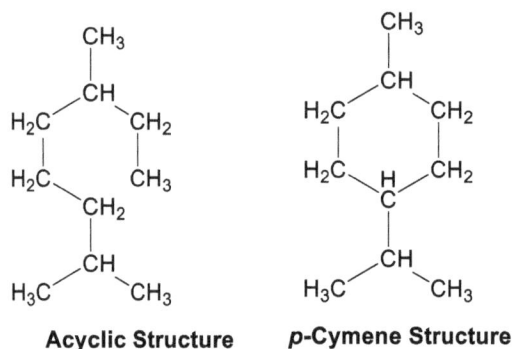

Acyclic Structure   ***p*-Cymene Structure**

FIGURE 12.2 Acyclic and cyclic terpenes.

The term "terpene" was first used to designate a combination of isomeric hydrocarbons with the Molecular Formula $C_{10}H_{16}$ found in turpentine and numerous essential oils extracted from plant and tree sap and tissues. At the time, oxygenated derivatives such as alcohols, aldehydes, ketones, and so on were referred to as "camphors." The phrases "terpenes" and "camphors" were combined to form the term "terpenoids." The following is the contemporary definition: "It encompasses plant-derived hydrocarbons with the general formula $(C_5H_8)_n$, as well as their oxygenated, hydrogenated, and dehydrogenated derivatives." The molecular formula of most natural terpenoid hydrocarbons is $(C_5H_8)_n$, and the value of n is used to classify them. There are also oxygenated derivatives of each class that occur naturally, which are mostly alcohols, aldehydes, or ketones, in addition to terpenoid hydrocarbons.

Isoprene is one of the results of the thermal disintegration of practically all terpenoids, leading to the hypothesis that the skeletal structures of all naturally existing terpenoids can be built up of isoprene units; this is known as the isoprene rule, and it was initially proposed by Wallach [1].

In addition, Ingold (1925) noted that in natural terpenoids, the isoprene units were connected "head to tail" (the head being the branched end of isoprene) [2].

The special isoprene rule can be used to describe the divisibility into isoprene units and their head-to-tail union. Open chain monoterpenoids and sesquiterpenoids have the following carbon skeletons (Figure 12.1):

Monocyclic terpenoids have a six-membered ring, and the monoterpenoid open chain can only give rise to one monocyclic monoterpenoid structure, namely the *p*-Cymene structure. The following structures demonstrate this, with the acyclic structure written in the traditional "ring form." *p*-Cymene is the source of the majority of monocyclic monoterpenoids found in nature (Figure 12.2) [3].

12.2 Sources

Terpenes are a class of naturally occurring chemicals, most of which are found in plants, although a few have been derived from other sources (Table 12.1).

12.3 Classification

On the basis of isoprene units, the terpenoids are classified and enlisted in Table 12.1 and Figure 12.3.

Terpenoids are classified in the following categories:

A. **Hemi and Monoterpenes**
   a. **Hemiterpenes:** Terpenes containing only one isoprene (2-methyl-1,3-butadiene) moiety are called as hemiterpenes i.e. Prenol, 3-methyl-2-buten-1-ol occurs in ylang-ylang oil and oil of hops from *Humulus lupulus* etc.
   b. **Monoterpenes:** Terpenes possessing at least two isoprene ($C_{10}H_{16}$) units are called Monoterpenes. A few examples are enlisted in Figure 12.4. Monoterpenes can also be categorised into **three** types:
      i. Acyclic monoterpenes
      ii. Monocyclic monoterpenes
      iii. Bicyclic monoterpenes
   i. **Acyclic monoterpenes:** These terpenes do not possess any cyclic structure i.e citronellal, petitgrain, E-myrcene, Perillene, (*E, Z*)-isomers geranial and neral etc as shown (Figure 12.3)

TABLE 12.1

List of Terpenoids with Their Plant Sources, Synnonyms and Uses [7–26]

S. No	Compound	Source	Synonyms of Plant	Uses
1.	Thymol	**Plant Name:** *Trachyspermum copticum* **Family:** Apiaceae (Umbelliferae)	Bishop's weed, Carum seed or Carum ajowan	Used in cholera, Powerful antiseptic and germicide, carminative.
2.	Eugenol	**Plant Name:** Pimenta dioic **Family:** Myrtaceae	Allspice tree, Pimento tree	Flatulent indigestion and externally for neuralgic or rheumatic pain. Also possess anaesthetic, analgesic, antioxidant, antiseptic, carminative, muscle relaxant, rubefacient, and stimulant.
3.	Benzaldehyde (95%), Prussic acid (3%)	**Plant Name:** Prunus dulcis var. amara **Family:** Rosaceae	Amygdalus communis, Amygdalus dulcis, Prunus amygdalus, Prunus communis	A 'fixed' oil commonly known as 'sweet almond oil' is made by pressing the kernels from both the sweet and bitter almond trees. Unlike the essential oil, this fixed oil does not contain any benzaldehyde or prussic acid and has many medical and cosmetic uses. It is used as a laxative, for bronchitis, coughs, heartburn and for disorders of the kidneys, bladder, and biliary ducts. It helps relieve muscular aches and pains, softens the skin, and promotes a clear complexion. Also used as Anaesthetic, antispasmodic, narcotic, vermifuge. Bitter almond oil is no longer used for internal medication.
4.	Ambrettolide and ambrettolic acid	**Plant Name:** *Abelmoschus moschatus* **Family:** Malvaceae	Annual hibiscus, Galu Gasturi, muskdana, Musk mallow, Musk okra, Ornamental okra, Rose mallow,	It is used in Aromatherapy improving circulation, muscles and joints: Cramp, fatigue, muscular aches and pains, poor circulation. Also used in anxiety, depression, nervous tension and stress-related conditions. Used for flavouring alcoholic and soft drinks as well as some foodstuffs, especially confectionery.
5.	Caryophyllene, cadinene and cadinol	**Plant Name:** *Amyris balsamifea* **Family:** Rutaceae	Rhus arborescens, Toxicodendron arborescens, Amyris funckiana, Elemifera balsamifera, Kuntze. Schimmelia oleifera	Aromatherapy/home use Perfume. As a cheap substitute for East Indian sandalwood in perfumes and cosmetics, although it does not have the same rich tenacity; chiefly employed as a fixative in soaps. Limited application in flavouring work, especially liqueurs.
6.	Phellandrene, Pinene, Limonene, Linalol and Borneol	**Plant Name:** *Angelica archangelica* **Family:** Apiaceae (Umbelliferae)	Celery, Norwegian angelica	Dull and congested skin, irritated conditions, psoriasis, Accumulation of toxins, arthritis, gout, rheumatism, water retention, Bronchitis, coughs, Anaemia, anorexia, flatulence, indigestion, Fatigue, migraine, nervous tension and stress-related disorders, Colds.
7.	Trans-anethole (80%–90%).	**Plant Name:** *Illicium verum* **Family:** Illiciaceae	Star anise	Antiseptic, carminative, expectorant, insect repellent, stimulant
8.	Trans-anethole (75%–90%).	**Plant Name:** *Pimpinella anisum* **Family:** Apiaceae (Umbelliferae	Anise	Antiseptic, antispasmodic, carminative, diuretic, expectorant, galactagogue, stimulant, stomachic.

(Continued)

TABLE 12.1 (*Continued*)

List of Terpenoids with Their Plant Sources, Synnonyms and Uses [7–26]

S. No	Compound	Source	Synonyms of Plant	Uses
9.	Thymohydroquinone dimethyl ether (80% approx.), Isobutyric ester of phlorol (20% approx.)	**Plant Name:** *Arnica montana* **Family:** Asteraceae (Compositae)	Mountain tobacco, Mountain arnica	Stimulant, anti-inflammatory, vulnerary.
10.	Ferulic acid, Valeric, Vanillin	**Plant Name:** *Ferula asa-foetida* **Family:** Apiaceae (Umbelliferae)	Asafetida, *F. narthex*	Antispasmodic, carminative, expectorant, hypo-tensive, stimulant. Animals are repelled by its odour.
11.	Citral, Citronellol, Eugenol, Geraniol, Linalyl acetate	**Plant Name:** *Melissa officinalis* **Family:** Lamiaceae (Labiatae)	Lemon balm	Antidepressant, antihistaminic, antispasmodic, bactericidal, carminative, cordial, diaphoretic, emmenagogue, febrifuge, hypertensive, insect-repellent, nervine, sedative, stomachic, sudorific, tonic, uterine, vermifuge.
12.	Pinene and phellandrene	**Plant Name:** *Abies balsamea* **Family:** Pinaceae	Canada balsam	Antiseptic (genito-urinary, pulmonary), antitussive, astringent, cicatrisant, diuretic, expectorant, purgative, regulatory, sedative (nerve), tonic, vulnery
13.	Caryophyllene	**Plant Name:** *Copaifera officinalis* **Family:** Fabaceae (Leguminosae)	Balsam Copaiba, Copahu Balsam,	The oleoresin is used in pharmaceutical products especially cough medicines and diuretics. The oil and crude balsam are extensively used as a fixative and fragrance component in all types of perfumes, soaps, cosmetics and detergents. The crude is also used in porcelain painting.
14.	Benzyl cinnamate and Cinnamyl cinnamate and cinnamein	**Plant Name:** *Myroxylon balsamum var. pereirae* **Family:** Fabaceae (Leguminosae)	Tolu tree, tolu balsam tree, *Myroxylon toluiferum*	Anti-inflammatory, antiseptic, balsamic, expectorant, parasiticide, stimulant; promotes the growth of epithelial cells
15.	Eugenol and vanillin	**Plant Name:** *Myroxylon balsamum var. balsamum* **Family:** Fabaceae (Leguminosae)	Myroxylon pereirae, Peruvian balsam	Antitussive, antiseptic, balsamic, expectorant, stimulant
16.	Linalol, Cineol, Camphor, Eugenol, limonene and citronellol	**Plant Name:** *Ocimum basilicum* **Family:** Lamiaceae (Labiatae)	Basil, Sweet basil	The oil is employed in high class fragrances, soaps and dental products; used extensively in major food categories especially meat products and savories.
17.	Cineol (30%–50%), pinene, linalol, terpineol acetate, methyl eugenol.	**Plant Name:** *Laurus nobilis* **Family:** Lauraceae	Bay laurel, bay tree	Antirheumatic, antiseptic, bactericidal, diaphoretic, digestive, diuretic, emmenagogue, fungicidal, hypotensive, sedative, stomachic
18.	Eugenol (up to 56%), myrcene, chavicol and, in lesser amounts	**Plant Name:** *Pimenta racemosa* **Family:** Myrtaceae	Bay rum tree, West Indian bay tree, bay-berry	Analgesic, anticonvulsant, antineuralgic, antirheumatic, antiseptic, astringent, expectorant, stimulant, tonic (for hair).
19.	Coniferyl cinnamate, sumaresinolic acid, traces of styrene, vanillin and benzaldehyde.	**Plant Name:** *Styrax benzoin* **Family:** Styracaceae	Benzoin tree, Styrax benzoin	Anti-inflammatory, antioxidant, antiseptic, astringent, carminative, cordial, deodorant, diuretic, expectorant, sedative, styptic, vulnerary
20.	Linalyl acetate (30%–60%), linalol (11%–22%) sesquiterpenes, terpenes, alkanes and furocoumarins (including bergapten, 0.30%–0.39%).	**Plant Name:** *Citrus bergamia* **Family:** Rutaceae	Citrus aurantium subsp. bergamia.	Analgesic, anthelmintic, antidepressant, antiseptic (pulmonary, genito-urinary), antispasmodic, antitoxic, carminative, digestive, diuretic, deodorant, febrifuge, laxative, parasiticide, rubefacient, stimulant, stomachic, tonic, vermifuge, vulnery

(Continued)

TABLE 12.1 (*Continued*)
List of Terpenoids with Their Plant Sources, Synnonyms and Uses [7–26]

S. No	Compound	Source	Synonyms of Plant	Uses
21.	Methyl salicylate (98%) identical to wintergreen oil.	**Plant Name:** *Betula lenta* **Family:** Betulaceae	*B. capinefolia*, cherry birch, southern birch, mahogany birch, mountain mahogany.	Analgesic, anti-inflammatory, antipyretic, antirheumatic, antiseptic, astringent, depurative, diuretic, rubefacient, tonic. Limited use as a counterirritant in anti-arthritic and antineuralgic ointments and analgesic balms. Limited use as a fragrance component in cosmetics and perfumes; extensively used as a flavouring agent, especially 'root beer', chewing gum, toothpaste, etc. (usually very low-level use)
22.	Betulenol, sesquiterpenes (phenol, cresol, xylenol, guaiacol, creosol, pyrocatechol, pyrobetulin)	**Plant Name:** *Betula alba* **Family:** Betulaceae	B. alba var. pubescens, B. odorata, B. pendula,	Anti-inflammatory, antiseptic, cholagogue, diaphoretic, diuretic, febrifuge, tonic.
23.	Cymene, ascaridole, cineol, linalol	**Plant Name:** *Peumus boldus,* **Family:** Monimiaceae	Boldu boldus, Boldoa fragrans, boldus, boldu.	Antiseptic, cholagogue, diaphoretic, diuretic, hepatic, sedative, tonic, urinary demulcent
24.	d-Borneol, 35% terpenes: pinene, camphene, dipentene, d-borneol, terpineol	**Plant Name:** *Dryobalanops aromatic* **Family:** Dipterocarpaceae	*D. camphora*, Borneo camphor, East Indian camphor, Baros camphor, Sumatra camphor, Malayan camphor	Mildly analgesic, antidepressant, antiseptic, antispasmodic, antiviral, carminative, rubefacient, stimulant of the adrenal cortex, tonic (cardiac and general).
25.	Eugenol, triacontane, phenols, ethyl alcohol and ethyl formate	**Plant Name:** *Boronia megastigma* **Family:** Rutaceae	Brown boronia.	Aromatic. The absolute is used in high-class perfumery work, especially florals. Used in specialized flavour work, especially rich fruit products
26.	Capryllic acid, phenols, aliphatics, terpenes, esters, scoparin and sparteine, as well as wax, etc.	**Plant Name:** *Spartium junceum* **Family:** Fabaceae (Leguminosae).	*Genista juncea*, genista, weavers broom, broom (absolute), genet (absolute).	Antihaemorrhagic, cardioactive, diuretic, cathartic, emmenagogue, narcotic, vasoconstrictor.
27.	Diosphenol (25%–40%), limonene, menthone	**Plant Name:** *Agothosma betulina* **Family:** Rutaceae	*Barosma betulina*, short buchu, mountain buchu, bookoo, buku, bucco	Antiseptic (especially urinary), diuretic, insecticide
28.	Nerolidol (80%), farnesol, bisabolol	**Plant Name:** *Myrocarpus fastigiatus* **Family:** Fabaceae (Leguminosae)	Cabureicica, 'Baume de Perou brun'	Antiseptic, balsamic, cicatrisant
29.	Cadinene, cadinol, p-cresol, guaiacol	**Plant Name:** *Juniperus oxycedrus* **Family:** Cupressaceae	Juniper tar, prickly cedar, medlar tree, prickly juniper	Analgesic, antimicrobial, antipruritic, antiseptic, disinfectant, parasiticide, vermifuge
30.	Cineol (14%–65%), terpineol, terpinyl acetate, pinene, nerolidol	**Plant Name:** *Melaleuca cajeputi* **Family:** Myrtaceae	Cajuput, white tea tree, white wood, swamp tea tree, punk tree	Mildly analgesic, antimicrobial, antineuralgic, antispasmodic, antiseptic (pulmonary, urinary, intestinal), anthelmintic, diaphoretic, carminative, expectorant, febrifuge, insecticide, sudorific, tonic.
31.	Citral, nerol, citronellol, limonene and geraniol. metatabilacetone (3%–5%)	**Plant Name:** *Calamintha officinalis* **Family:** Lamiaceae (Labiatae)	C. *clinopodium, Melissa calaminta*, calamint, common calamint, mill mountain,	Anaesthetic (local), antirheumatic, antispasmodic, astringent, carminative, diaphoretic, emmenagogue, febrifuge, nervine, sedative, tonic
32.	Beta-asarone 80%, calamene, calamol, calamenene, eugenol and shyobunones.	**Plant Name:** Acorus calamus var. angustus **Family:** Araceae	*Calamus aromaticus*, sweet flag, sweet sedge, sweet root	Anticonvulsant, antiseptic, bactericidal, carminative, diaphoretic, expectorant, hypotensive, insecticide, spasmolytic, stimulant, stomachic, tonic, vermifuge. The oil of calamus is reported to have carcinogenic properties.

(Continued)

TABLE 12.1 (*Continued*)

List of Terpenoids with Their Plant Sources, Synnonyms and Uses [7–26]

S. No	Compound	Source	Synonyms of Plant	Uses
33.	Cineol, pinene, terpineol, menthol, thymol. Brown camphor contains safrol (80%).	**Plant Name:** *Cinnamomum camphora* **Family:** Lauraceae	*Laurus camphora*, true camphor, hon-sho, laurel camphor, gum camphor, Japanese camphor, Formosa camphor.	Anti-inflammatory, antiseptic, antiviral, bactericidal, counter-irritant, diuretic, expectorant, stimulant, rubefacient, vermifuge. Acne, inflammation, oily conditions, spots; also for insect prevention (flies, moths, etc).
34.	Caryophyllene, benzyl acetate, benzyl alcohol, farnesol, terpineol, borneol, geranyl acetate, safrol, linalol, limonrne, methyl salicylate	**Plant Name:** *Cananga odorata* **Family:** Annonaceae	C. odoratum var. macrophylla	Antiseptic, antidepressant, aphrodisiac, hypotensive, nervine, sedative, tonic.
35.	Mainly carvone (50%–60%), limonene (40%), carveol, dihydrocarveol, dihydrocarvone, pinene, phellandrene.	**Plant Name:** *Carum carvi* **Family:** Apiaceae (Umbelliferae)	*Apium carvi*, carum, caraway fruits.	Antihistaminic, antimicrobial, antiseptic, aperitif, astringent, carminative, diuretic, emmenagogue, expectorant, galactagogue, larvicidal, stimulant, spasmolytic, stomachic, tonic, vermifuge.
36.	Terpinyl acetate, cineol limonene, sabinene, linalol, linalyl acetate, pinene, zingiberene	**Plant Name:** *Elettaria cardamomum* **Family:** Zingiberaceae	*Elettaria cardomomum var. cardomomum*, cardomom, cardamomi, cardomum, mysore cardomom	Antiseptic, antispasmodic, aphrodisiac, carminative, cephalic, digestive, diuretic, sialogogue, stimulant, stomachic, tonic (nerve).
37.	Pinene, carotol, daucol, limonene, bisabolene, elemene, geraniol, geranyl acetate, caryophyllene	**Plant Name:** *Daucus carota* **Family:** Apiaceae (Umbelliferae)	Wild carrot, Queen Anne's lace, bird's nest.	Anthelmintic, antiseptic, carminative, depurative, diuretic, emmenagogue, hepatic, stimulant, tonic, vasodilatory and smooth muscle relaxant.
38.	Cymene, diterpene, limonene, caryophyllene, terpineol and eugenol	**Plant Name:** *Croton eluteria* **Family:** Euphorbiaceae	Cascarilla, sweetwood bark, sweet bark, Bahama cascarilla, aromatic quinquina, false quinquina	Astringent, antimicrobial, antiseptic, carminative, digestive, expectorant, stomachic, tonic.
39.	Cinnamic aldehyde (75%–90%) methyl eugenol, salicylaldehyde and methyl-salicylaldehyde	**Plant Name:** *Cinnamomum cassia* **Family:** Lauraceae	*C. aromaticum, Laurus cassia*, Chinese cinnamon, false cinnamon, cassia cinnamon, cassia lignea.	Antidiarrhoeal, anti-emetic, antimicrobial, astringent, carminative, spasmolytic
40.	Benzyl alcohol, methyl salicylate, farnesol, geraniol and linalol	**Plant Name:** *Acacia farnesiana* **Family:** Mimosaceae	*Cassia ancienne*, sweet acacia, huisache, popinac, opopanax.	Antirheumatic, antiseptic, antispasmodic, aphrodisiac, balsamic, insecticide
41.	Atlantone, caryophyllene, cedrol, cadinene	**Plant Name:** *Cedrus atlantica* **Family:** Pinaceae	Atlantic cedar, Atlas cedar, African cedar, Moroccan cedarwood (oil), libanol (oil).	Antiseptic, antiputrescent, antiseborrheic, aphrodisiac, astringent, diuretic, expectorant, fungicidal, mucolytic, sedative (nervous), stimulant (circulatory), tonic.
42.	Cedrene, cedrol, thujopsene, sabinene	**Plant Name:** *Juniperus ashei* **Family:** Cupressaceae	*J. mexicana*, mountain cedar, Mexican cedar, rock cedar, Mexican juniper.	Antiseptic, antispasmodic, astringent, diuretic, expectorant, sedative (nervous), stimulant (circulatory).
43.	Cedrene (up to 80%), cedrol (3%–14%) cedrenol	**Plant Name:** *Juniperus virginiana*	Cupressaceae red cedar, eastern red cedar, southern red cedar, Bedford cedarwood (oil).	Abortifacient, antiseborrhoeic, antiseptic (pulmonary, genito-urinary), antispasmodic, astringent, balsamic, diuretic, emmenagogue, expectorant, insecticide, sedative (nervous), stimulant (circulatory).
44.	Limonene (60%), apiol, selinene, santalol, sedanolide and sedanolic acid anhydride	**Plant Name:** *Apium graveolens* **Family:** Apiaceae (Umbelliferae)	Celery fruit	Anti-oxidative, antirheumatic, antiseptic (urinary), antispasmodic, aperitif, depurative, digestive, diuretic, carminative, cholagogue, emmenagogue, galactagogue, hepatic, nervine, sedative (nervous), stimulant (uterine), stomachic, tonic (digestive).

(Continued)

TABLE 12.1 (Continued)

List of Terpenoids with Their Plant Sources, Synnonyms and Uses [7–26]

S. No	Compound	Source	Synonyms of Plant	Uses
45.	Chamazulene, farnesene, bisabolol oxide, en-yndicycloether	**Plant Name:** *Matricaria recutica* **Family:** Asteraceae (Compositae)	*M. chamomilla*, camomile, blue chamomile, matricaria, Hungarian chamomile, sweet false chamomile, single chamomile, chamomile blue (oil).	Analgesic, anti-allergenic, anti-inflammatory, antiphlogistic, antispasmodic, bactericidal, carlminative, cicatrisant, cholagogue, digestive, emmenagogue, febrifuge, fungicidal, hepatic, nerve sedative, stimulant of leucocyte production, stomachic, sudorific, vermifuge, vulnerary
46.	Lavender, lavandin, vetiver, cedarwood, oakmoss, labdanum, olibanum and artemisia oils.	**Plant Name:** *Ormenis multicaulis* **Family:** Asteraceae (Compositae)	*O. mixta*, *Anthemis mixta*, Moroccan chamomile	Antispasmodic, cholagogue, emmenagogue, hepatic, sedative.
47.	Angelic and tiglic acids (approx. 85%), pinene, farnesol, nerolidol, chamazulene, pinocarvone, cineol,	**Plant Name:** *Chamaemelum nobile* **Family:** Asteraceae (Compositae)	*Anthemis nobilis*, camomile, English chamomile, garden chamomile, sweet chamomile, true chamomile	Analgesic, anti-anaemic, antineuralgic, antiphlogistic, antiseptic, antispasmodic, bactericidal, carminative, cholagogue, cicatrisant, digestive, emmenagogue, febrifuge, hepatic, hypnotic, nerve sedative, stomachic, sudorific, tonic, vermifuge, vulnerary
48.	Chavicol, also 1-allyl-2, 4-dimethoxybenzene and anethole.	**Plant Name:** *Anthriscus cerefolium* **Family:** Apiaceae (Umbelliferae)	*A. longirostris*, garden chervil, salad chervil.	Aperitif, antiseptic, carminative, cicatrisant, depurative, diaphoretic, digestive, diuretic, nervine, restorative, stimulant (metabolism), stomachic, tonic. EXTRACTION Essential oil by steam distillation from seeds or fruit
49.	Eugenol (80%–96%), eugenol acetate, cinnamaldehyde (3%), benzyl benzoate, linalol, safrol, cinnamaldehyde (40%–50%), eugenol (4%–10%), benzaldehyde, cuminaldehyde, pinene, cineol, phellandrene, furfurol, cymene, linalol,	**Plant Name:** *Cinnamomum zeylanicum* **Family:** Lauraceae	*C. verum*, *Laurus cinnamomum*, Ceylon cinnamon, Seychelles cinnamon, Madagascar cinnamon, true cinnamon, cinnamon leaf (oil), cinnamon bark (oil).	Anthelmintic, antidiarrhoeal, antidote (to poison), antimicrobial, antiseptic, antispasmodic, antiputrescent, aphrodisiac, astringent, carminative, digestive, emmenagogue, haemostatic, orexigenic, parasiticide, refrigerant, spasmolytic, stimulant (circulatory, cardiac, respiratory), stomachic, vermifuge
50.	Geraniol (up to 45%) citronellal (up to 50%), geranyl acetate, limonene, camphene	**Plant Name:** *Cymbopogon nardus* **Family:** Poaceae (Gramineae)	*Andropogon nardus*, Sri Lanka citronella, Lenabatu citronella.	Antiseptic, antispasmodic, bactericidal, deodorant, diaphoretic, diuretic, emmenagogue, febrifuge, fungicidal, insecticide, stomachic, tonic, vermifuge.
51.	Eugenol, eugenyl acetate, caryophyllene	**Plant Name:** *Syzygium aromaticum* **Family:** Myrtaceae	Eugenia aromatica, E. caryophyllata, E. caryophyllus.	Anthelmintic, antibiotic, anti-emetic, antihistaminic, antirheumatic, antineuralgic, antioxidant, antiseptic, antiviral, aphrodisiac, carminative, counter-irritant, expectorant, larvicidal, spasmolytic, stimulant, stomachic, vermifuge.
52.	linalol (55%–75%), decyl aldehyde, borneol, geraniol, carvone, anethole	**Plant Name:** *Coriandrum sativum* **Family:** Apiaceae (Umbelliferae)	Coriander seed, Chinese parsley	Analgesic, aperitif, aphrodisiac, anti-oxidant, anti-rheumatic, antispasmodic, bactericidal, depurative, digestive, carminative, cytotoxic, fungicidal, larvicidal, lipolytic, revitalizing, stimulant (cardiac, circulatory, nervous system), stomachic
53.	Di-hydrocostus lactone (50%), Costols, caryophyllene, selinene, oleic acids	**Plant Name:** *Saussurea costus* **Family:** Asteraceae (Compositae) 5.	Lappa, Aucklandia costus, Aplotaxis lappa, A. auriculata.	Antiseptic, antispasmodic, antiviral, bactericidal, carminative, digestive, expectorant, febrifuge, hypotensive, stimulant, stomachic, tonic

(Continued)

TABLE 12.1 (*Continued*)

List of Terpenoids with Their Plant Sources, Synnonyms and Uses [7–26]

S. No	Compound	Source	Synonyms of Plant	Uses
54.	Caryophyllene, cadinene, cubebene, sabinene	**Plant Name:** *Piper cubeba* **Family:** Piperaceae	*Cubeba officinalis*, cubeba, tailed pepper, cubeb berry, false pepper.	Antiseptic (pulmonary, genito-urinary), antispasmodic, antiviral, bactericidal, carminative, diuretic, expectorant, stimulant.
55.	Cuminaldehyde, pinenes, terpinenes, cymene, phellandrene, myrcene and limonene, farnesene, caryophyllene	**Plant Name:** *Cuminum cyminum* **Family:** Apiaceae (Umbelliferae)	*C. odorum*, cummin, roman caraway.	Anti-oxidant, antiseptic, antispasmodic, antitoxic, aphrodisiac, bactericidal, carminative, depurative, digestive, diuretic, emmenagogue, larvicidal, nervine, stimulant, tonic
56.	Pinene, camphene, sylvestrene, cymene, sabinol,	**Plant Name:** *Cupressus sempervirens* **Family:** Cupressaceae	Italian cypress, Mediterranean cypress.	Antirheumatic, antiseptic, antispasmodic, astringent, deodorant, diuretic, hepatic, styptic, sudorific, tonic, vasoconstrictive
57.	Coumarin (1.6%)	**Plant Name:** *Carphephorus odoratissimus* **Family:** Asteraceae (Compositae)	Trilisa odoratissima, Liatris odoratissima, Frasera speciosa, hound's tongue, deer's tongue	Antiseptic, demulcent, diaphoretic, diuretic, febrifuge, stimulant, tonic
58.	Carvone (30%–60%), limonene, phellandrene, eugenol, pinene limonene, pinene	**Plant Name:** *Anethum graveolens* **Family:** Apiaceae (Umbelliferae)	Peucedanum graveolens, Fructus anethi, European dill, American dill.	Antispasmodic, bactericidal, carminative, digestive, emmenagogue, galactagogue, hypotensive, stimulant, stomachic.
59.	Alantolactone (helenin), isolactone, dihydroisalantolactone, dihydralantolactone, alantic acid, azulene	**Plant Name:** *Inula helenium* **Family:** Asteraceae (Compositae)	*Helenium grandiflorum, Aster officinalis, A. helenium*, inula, scabwort, alant, horseheal, yellow starwort, elf dock, wild sunflower, velvet dock, 'essence d'aunee'	Anthelmintic, anti-inflammatory, antiseptic, antispasmodic, antitussive, astringent, bactericidal, diaphoretic, diuretic, expectorant, fungicidal, hyperglycaemic, hypotensive, stomachic, tonic Elecampane root is the richest source of inulin.
60.	Phellandrene, Elemol, elemicin, terpineol, carvone and terpinolene	**Plant Name:** *Canarium luzonicum* **Family:** Burseraceae	*C. commune*, Manila elemi, elemi gum, elemi resin, elemi (oleoresin).	Antiseptic, balsamic, cicatrisant, expectorant, fortifying, regulatory, stimulant, stomachic, tonic.
61.	Cineol (70%–85%), pinene, limonene, cymene, phellandrene, terpinene, aromadendrene,	**Plant Name:** *Eucalyptus globulus var. globulus* **Family:** Myrtaceae tree, stringy bark.	Gum tree, southern blue gum, Tasmanian blue gum, fever	Analgesic, antineuralgic, antirheumatic, antiseptic, antispasmodic, antiviral, balsamic, cicatrisant, decongestant, deodorant, depurative, diuretic, expectorant, febrifuge, hypoglycaemic, parasiticide, prophylactic, rubefacient, stimulant, vermifuge, vulnerary.
62.	Cineol (70%–85%), pinene, limonene, cymene, phellandrene, terpinene, aromadendrene,	**Plant Name:** *Eucalyptus citriodora* **Family:** Myrtaceae	Lemon-scented gum, citron-scented gum, scented gum tree, spotted gum, 'boabo'.	Antiseptic, antiviral, bactericidal, deodorant, expectorant, fungicidal, insecticide
63.	Piperitone (40%–50%), phellandrene (20%–30%), camphene, cymene, terpinene and thujene.	**Plant Name:** *Eucalyptus dives var. Type* **Family:** Myrtaceae	Broad-leaf peppermint, blue peppermint, menthol-scented gum.	Cuts, sores, ulcers etc Arthritis, muscular aches and pains, rheumatism, sports injuries Asthma, bronchitis, catarrh, coughs, throat and mouth infections Colds, fevers, 'flu, infectious illness, e.g. measles Headaches, nervous exhaustion, neuralgia, sciatica. **Other Uses:** Little used medicinally these days except in deodorants, disinfectants, mouthwashes, gargles and in veterinary practice. 'Piperitone' rich oils are used in solvents. Employed for the manufacture of thymol and menthol (from piperitone).

(Continued)

TABLE 12.1 (*Continued*)

List of Terpenoids with Their Plant Sources, Synnonyms and Uses [7–26]

S. No	Compound	Source	Synonyms of Plant	Uses
64.	Anethole (50%–60%), limonene, phellandrene, pinene, anisic acid, anisic aldehyde, camphene, limonene	**Plant Name:** *Foeniculum vulgare* **Family:** Apiaceae (Umbelliferae)	F. officinale, F. capillaceum, Anethum foeniculum,	Aperitif, anti-inflammatory, antimicrobial, antiseptic, antispasmodic, carminative, depurative, diuretic, emmenagogue, expectorant, galactagogue, laxative, orexigenic, stimulant (circulatory), splenic, stomachic, tonic, vermifuge
65.	Santene, pinene, limonene, bornyl acetate, lauraldehyde	**Plant Name:** *Abies alba* **Family:** Pinaceae (oil), fir needle (oil).	A. pectinata, whitespruce, European silver fir, edeltanne, weisstanne, templin (cone oil), Strassburg or Vosges turpentine	Analgesic, antiseptic (pulmonary), antitussive, deodorant, expectorant, rubefacient, stimulant, tonic.
66.	Pinene, dipentene, limonene, thujene, phellandrene, cymene, myrcene, terpinene	**Plant Name:** *Boswellia carteri* **Family:** Burseraceae	Olibanum, gum thus.	Anti-inflammatory, antiseptic, astringent, carminative, cicatrisant, cytophylactic, digestive, diuretic, emmenagogue, expectorant, sedative, tonic, uterine, vulnerary
67.	Pinene, cineol, eugenol and sesquiterpenes.	**Plant Name:** *Alpinia officinarum* **Family:** Zingiberaceae	*Radix galanga minoris, Languas officinarum,* galanga, small galangal, Chinese ginger, ginger root, colic root, East Indian root	Antiseptic, bactericidal, carminative, diaphoretic, stimulant, stomachic.
68.	Pinene, cadinol, cadinene and myrcene	**Plant Name:** *Ferula galbaniflua* **Family:** Apiaceae (Umbelliferae)	F. gummosa, galbanum gum, galbanum resin, 'bubonion'	Analgesic, anti-inflammatory, antimicrobial, antiseptic, antispasmodic, aphrodisiac, balsamic, carminative, cicatrisant, digestive, diuretic, emmenagogue, expectorant, hypotensive, restorative, tonic.
69.	Benzyl acetate, phenyl acetate, linalol, linalyl acetate, terpineol and methyl anthranilate	**Plant Name:** *Gardenia jasminoides* **Family:** Rubiaceae	G. grandiflora, G. radicans, florida, gardinia, Cape jasmine, common gardenia.	**Use:** Perfume. **Other Uses:** Employed in high-class perfumery, especially oriental fragrances
70.	Allicin, allylpropyl disulphide, diallyl disulphide, diallyl trisulphide, citral, geraniol, linalol, phellandrene	**Plant Name:** *Allium sativum* **Family:** Amaryllidaceae or Liliaceae	Common garlic, allium, poor man's treacle!	Amoebicidal, anthelmintic, antibiotic, antimicrobial, antiseptic, antitoxic, antitumour, antiviral, bactericidal, carminative, cholagogue, hypocholesterolemic, depurative, diaphoretic, diuretic, expectorant, febrifuge, fungicidal, hypoglycaemic, hypotensive, insecticidal, larvicidal, promotes leucocytosis, stomachic, tonic.
71.	Citronellol, geraniol, linalol, isomenthone, menthone, phellandrene, sabinene, limonene	**Plant Name:** *Pelargonium graveolens* **Family:** Geraniaceae	Rose geranium, pelargonium	Antidepressant, antihaemorrhagic, anti-inflammatory, antiseptic, astringent, cicatrisant, deodorant, diuretic, fungicidal, haemostatic, stimulant (adrenal cortex), styptic, tonic, vermifuge, vulnerary.
72.	Gingerin, gingenol, gingerone, zingiberine, linalol, camphene, phellandrene, citral, cineol, borneol,	**Plant Name:** *Zingiber officinale* **Family:** Zingiberaceae.	Common ginger, Jamaica ginger	Analgesic, anti-oxidant, antiseptic, antispasmodic, antitussive, aperitif, aphrodisiac, bactericidal, carminative, cephalic, diaphoretic, expectorant, febrifuge, laxative, rubefacient, stimulant, stomachic, tonic.
73.	Limonene (90%), cadinene, paradisiol, neral, geraniol, citronellal, sinensal	**Plant Name:** *Citrus x paradisi* **Family:** Rutaceae	C. racemosa, C. maxima var. racemosa, shaddock (oil).	Antiseptic, antitoxic, astringent, bactericidal, diuretic, depurative, stimulant (lymphatic, digestive), tonic.
74.	Guaiol (42%–72%), bulnesol, bulnesene, guaiene, patchoulene, guaioxide.	**Plant Name:** *Bulnesia sarmienti* **Family:** Zygophyllaceae	Champaca wood (oil), 'palo santo'	Anti-inflammatory, anti-oxidant, antirheumatic, antiseptic, diaphoretic, diuretic, laxative

(Continued)

TABLE 12.1 (*Continued*)

List of Terpenoids with Their Plant Sources, Synnonyms and Uses [7–26]

S. No	Compound	Source	Synonyms of Plant	Uses
75.	Nerol, neryl acetate (30%–50%), geraniol, pinene, linalol, isovaleric aldehyde, sesquiterpenes, furfurol and eugenol	**Plant Name:** *Helichrysum angustifolium* **Family:** Asteraceae (Compositae) SYNOYNMS	Immortelle, everlasting, St John's herb	Anti-allergenic, anti-inflammatory, antimicrobial, antitussive, antiseptic, astringent, cholagogue, cicatrisant, diuretic, expectorant, fungicidal, hepatic, nervine.
76.	Humulene, myrcene, caryophyllene and farnesene	**Plant Name:** *Humulus lupulus* **Family:** Moraceae.	Common hop, European hop, lupulus	Anodyne, an aphrodisiac, antimicrobial, antiseptic, antispasmodic, astringent, bactericidal, carminative, diuretic, emollient, oestrogenic properties, hypnotic, nervine, sedative, soporific. **Extraction:** Essential oil by steam distillation from the dried cones or catkins, known as 'strobiles'. (An absolute is also produced by solvent extraction for perfumery use.)
77.	Allyl isothiocyanate (75%), with phenylethyl isothiocyanate	**Plant Name:** *Armoracia rusticana* **Family:** Brassicaceae (Cruciferae)	Cochlearia armoracia, A. lapathifolia, red cole, raifort	Antibiotic, antiseptic, diuretic, carminative, expectorant, laxative (mild), rubefacient, stimulant.
78.	Phenylethyl alcohol, benzaldehyde, cinnamaldehyde, benzyl alcohol, benzoic acid, benzyl acetate, benzyl benzoate, eugenol, methyl eugenol and hydroquinone	**Plant Name:** *Hyacinthus orientalis* **Family:** Liliaceae *Scilla nutans*, bluebell		Antiseptic, balsamic, hypnotic, sedative, styptic
79.	Pinocamphone, isopinocamphone, estragole, borneol, geraniol, limonene, thujone, myrcene, caryophyllene	**Plant Name:** *Hyssopus officinalis* **Family:** Lamiaceae (Labiatae) SYNONYM	Azob.	Astringent, antiseptic, antispasmodic, antiviral, bactericidal, carminative, cephalic, cicatrisant, digestive, diuretic, emmenagogue, expectorant, febrifuge, hypertensive, nervine, sedative, sudorific, tonic (heart and circulation), vermifuge, vulnerary
80.	Pilocarpine, isopilocarpine, pilocarpidine, methyl nonyl ketone, dipentene	**Plant Name:** *Pilocarpus jaborandi* **Family:** Rutaceae	*Pernambuco jaborandi, P. pennatifolius,* iaborandi, jamborandi, arrudo do mato, arruda brava, jamguaraddi, juarandi	Antiseptic, diaphoretic, emmenagogue, galactagogue, stimulant (nerve)
81.	Benzyl acetate, linalol, phenylacetic acid, benzyl alcohol, farnesol, methyl anthranilate, cis-jasmone, methyl jasmonate, among others.	**Plant Name:** *Jasminum officinale* **Family:** Oleaceae	Jasmin, jessamine, common jasmine, poet's jessamine	Analgesic (mild), antidepressant, anti-inflammatory, antiseptic, antispasmodic, aphrodisiac, carminative, cicatrisant, expectorant, galactagogue, parturient, sedative, tonic (uterine).
82.	Pinene, myrcene, sabinene, limonene, cymene, terpinene, thujene, camphene,	**Plant Name:** *Juniperus communis* **Family:** Cupressaceae SYNONYM	Common juniper	Antirheumatic, antiseptic, antispasmodic, antitoxic, aphrodisiac, astringent, carminative, cicatrisant, depurative, diuretic, emmenagogue, nervine, parasiticide, rubefacient, sedative, stomachic, sudorific, tonic, vulnerary.
83.	Camphene, sabinene, myrcene, phellandrene, l imonene, cymene, cineol, borneol, nerol, geraniol, fenchone	**Plant Name:** *Cistus ladaniferus* **Family:** Cistaceae	Cistus (oil), gum cistus, ciste, cyste (absolute), labdanum gum, ambreine, European rock rose.	Antimicrobial, antiseptic, antitussive, astringent, balsamic, emmenagogue, expectorant, tonic.

(Continued)

TABLE 12.1 (*Continued*)

List of Terpenoids with Their Plant Sources, Synnonyms and Uses [7–26]

S. No	Compound	Source	Synonyms of Plant	Uses
84.	Linalyl acetate (30%–32%), linalol, cineol, camphene, pinene	**Plant Name:** *Lavandula x intermedia* **Family:** Lamiaceae (Labiatae).	Lavandula hybrida, L. hortensis, bastard lavender	Analgesic, anticonvulsive, antidepressant, antimicrobial, antirheumatic, antiseptic, antispasmodic, antitoxic, carminative, cholagogue, choleretic, cicatrisant, cordial, cytophylactic, deodorant, diuretic, emmenagogue, hypotensive, insecticide, nervine, parasiticide, rubefacient, sedative, stimulant, sudorific, tonic, vermifuge, vulnerary.
85.	Cineol, camphor (40%–60%), linalool, linalyl acetate.	**Plant Name:** *Lavandula latifolia* **Family:** Lamiaceae (Labiatae)	*L. spica*, aspic, broad-leaved lavender, lesser lavender, spike.	Analgesic, anticonvulsive, antidepressant, antimicrobial, antirheumatic, antiseptic, antispasmodic, antitoxic, carminative, cholagogue, choleretic, cicatrisant, cordial, cytophylactic, deodorant, diuretic, emmenagogue, hypotensive, insecticide, nervine, parasiticide, rubefacient, sedative, stimulant, sudorific, tonic, vermifuge, vulnerary.
86.	Linalol, lavandulol, lavandulyl acetate, terpineol, cineol, limonene, ocimene, caryophyllene.	**Plant Name:** *Lavandula angustifolia* **Family:** Lamiaceae (Labiatae)	*L. vera, L. officinalis*, garden lavender, common lavender	Analgesic, anticonvulsive, antidepressant, antimicrobial, antirheumatic, antiseptic, antispasmodic, antitoxic, carminative, cholagogue, choleretic, cicatrisant, cordial, cytophylactic, deodorant, diuretic, emmenagogue, hypotensive, insecticide, nervine, parasiticide, rubefacient, sedative, stimulant, sudorific, tonic, vermifuge, vulnerary.
87.	Limonene, terpinene, pinenes, sabinene, myrcene, citral, linalol, geraniol, octanol, nonanol, citronellal, bergamotene,	**Plant Name:** *Citrus limon* **Family:** Rutaceae	*C. limonum*, cedro oil.	Anti-anaemic, antimicrobial, antirheumatic, antisclerotic, antiscorbutic, antiseptic, antispasmodic, antitoxic, astringent, bactericidal, carminative, cicatrisant, depurative, diaphoretic, diuretic, febrifuge, haemostatic, hypotensive, insecticidal, rubefacient, stimulates white corpuscles, tonic, vermifuge.
88.	Citral (65%–85%), myrcene (12%–25%), dipentene, methylheptenone, linalol, geraniol, nerol, citronellol, farnesol	**Plant Name:** *Cymbopogon citratus* **Family:** Poaceae (Gramineae)	*Andropogon citratus, A. schoenathus*, West Indian lemongrass, Madagascar lemongrass, Guatemala lemongrass. *Andropogon flexuosus, Cymbopogon flexuosus*, East Indian lemongrass, Cochin lemongrass, native lemongrass, British India lemongrass, 'vervaine Indienne' or France Indian verbena.	Analgesic, antidepressant, antimicrobial, anti-oxidant, antipyretic, antiseptic, astringent, bactericidal, carminative, deodorant, febrifuge, fungicidal, galactagogue, insecticidal, nervine, sedative (nervous), tonic
89.	Limonene, pinenes, camphene, sabinene, citral, cymene, cineols and linalol	**Plant Name:** *Citrus aurantifolia* **Family:** Rutaceae	*C.medica var. acida, C. latifolia*, Mexican lime, West Indian lime, sour lime.	Antirheumatic, antiscorbutic, antiseptic, antiviral, aperitif, bactericidal, febrifuge, restorative, tonic
90.	Mainly linalol, linalyl acetate.	**Plant Name:** *Bursera glabrifolia* **Family:** Burseraceae	*B. delpechiana*, Mexican linaloe, 'copal limon'	Anticonvulsant, anti-inflammatory, antiseptic, bactericidal, deodorant, gentle tonic
91.	Farnesol	**Plant Name:** *Tilia vulgaris* **Family:** Tiliaceae	*T. europaea*, lime tree, common lime, lyne, tillet, tilea	Astringent (mild), antispasmodic, bechic, carminative, cephalic, diaphoretic, diuretic, emollient, nervine, sedative, tonic

(Continued)

TABLE 12.1 (*Continued*)

List of Terpenoids with Their Plant Sources, Synnonyms and Uses [7–26]

S. No	Compound	Source	Synonyms of Plant	Uses
92.	Citral (up to 85%).	**Plant Name:** *Litsea cubeba* **Family:** Lauraceae	*L. citrata*, 'may chang', exotic verbena, tropical verbena.	Antiseptic, deodorant, digestive, disinfectant, insecticidal, stimulant, stomachic
93.	Mainly phthalides (70%) butylidene, dihydrobutylidene, butylphthalides and ligostilides.	**Plant Name:** *Levisticum officinale* **Family:** Apiaceae (Umbelliferae)	Angelica levisticum, Ligusticum	Antimicrobial, antiseptic, antispasmodic, diaphoretic, digestive, diuretic, carminative, depurative, emmenagogue, expectorant, febrifuge, stimulant (digestive), stomachic
94.	Limonene, methyl methylanthranilate, geraniol, citral, citronellal,	**Plant Name:** *Citrus reticulata* **Family:** Rutaceae	C. *nobilis*, C. *madurensis*, C. *unshiu*, C. *deliciosa*, European mandarin, true mandarin, tangerine, satsuma	Antiseptic, antispasmodic, carminative, digestive, diuretic (mild), laxative (mild), sedative, stimulant (digestive and lymphatic), tonic
95.	Calendulin	**Plant Name:** *Calendula officinalis* **Family:** Asteraceae (Compositae)	Calendula, marygold, marybud, gold-bloom, pot marigold, hollygold, common marigold, poet's marigold.	Antihaemorrhagic, anti-inflammatory, antiseptic, antispasmodic, astringent, diaphoretic, cholagogue, cicatrisant, emmenagogue, febrifuge, fungicidal, styptic, tonic, vulnerary.
96.	Terpinenes, terpineol, sabinenes, linalol, carvacrol, linalyl acetate, ocimene, cadinene, geranyl acetate, citral, eugenol,	**Plant Name:** *Origanum majorana* **Family:** Lamiaceae (Labiatae)	*Marjorana hortensis*, knotted marjoram.	Chilblains, bruises, ticks, Arthritis, lumbago, muscular aches and stiffness, rheumatism, sprains, strains, Asthma, bronchitis, coughs, Colic, constipation, dyspepsia, flatulence, Amenorrhoea, dysmenorrhoea, leucorrhoea, Colds, Headache, hypertension, insomnia, migraine, nervous tension and stress-related conditions. The oil and oleoresin are used as fragrance components in soaps, detergents, cosmetics and perfumes. Employed in most major food categories, especially meats, seasonings and sauces, as well as soft drinks and alcoholic beverages such as vermouths and bitters.
97.	Pinenes	**Plant Name:** *Pistacia lentiscus* **Family:** Anacardiaceae	Mastick tree, mastick, mastix, mastich, lentisk	Antimicrobial, antiseptic, antispasmodic, astringent, diuretic, expectorant, stimulant
98.	Coumarins – melilotic acid, orthocoumaric acid	**Plant Name:** *Melilotus officinalis* **Family:** Fabaceae (Leguminosae)	Common melilot, yellow melilot, white melilot, corn melilot, melilot trefoil, sweet clover, plaster clover, sweet lucerne, wild laburnum, king's clover, melilotin (oleoresin).	Anti-inflammatory, antirheumatic, antispasmodic, astringent, emollient, expectorant, digestive, insecticidal (against moth), sedative
99.	Palmic aldehyde, enanthic acid, anisic acid, acetic acid and phenols.	**Plant Name:** *Acacia dealbata* **Family:** Mimosaceae	A. decurrens var. dealbata, Sydney black wattle.	Antiseptic, astringent
100.	Menthol (70%–95%), menthone (10%–20%), pinene, menthyl acetate, isomenthone, thujone, phellandrene, piperitone and menthofuran.	**Plant Name:** *Mentha arvensis* **Family:** Lamiaceae (Labiatae)	Field mint, Japanese mint	Anaesthetic, antimicrobial, antiseptic, antispasmodic, carminative, cytotoxic, digestive, expectorant, stimulant, stomachic
101.	Menthol (29%–48%), menthone (20%–31%), menthyl acetate, menthofuran, limonene, pulegone, cineol, among	**Plant Name:** *Mentha piperita* **Family:** Lamiaceae (Labiatae)	Brandy mint, balm mint.	Analgesic, anti-inflammatory, antimicrobial, antiphlogistic, antipruritic, antiseptic, antispasmodic, antiviral, astringent, carminative, cephalic, cholagogue, cordial, emmenagogue, expectorant, febrifuge, hepatic, nervine, stomachic, sudorific, vasoconstrictor, vermifuge.

(Continued)

TABLE 12.1 (*Continued*)

List of Terpenoids with Their Plant Sources, Synonyms and Uses [7–26]

S. No	Compound	Source	Synonyms of Plant	Uses
102.	L-carvone (50%–70%), dihydrocarvone, phellandrene, limonene, menthone, menthol, pulegone, cineol, linalol, pinenes	**Plant Name:** *Mentha spicata* **Family:** Lamiaceae (Labiatae)	M. *viridis*, common spearmint, garden spearmint, spire mint, green mint, lamb mint, pea mint, fish mint.	Anaesthetic (local), antiseptic, antispasmodic, astringent, carminative, cephalic, cholagogue, decongestant, digestive, diuretic, expectorant, febrifuge, hepatic, nervine, stimulant, stomachic, tonic.
103.	Thujone, cineol, pinenes and dihydromatricaria ester,	*Artemisia vulgaris* **Family:** Asteraceae (Compositae)	Armoise, wild wormwood, felon herb, St John's plant.	Anthelmintic, antispasmodic, carminative, choleretic, diaphoretic, diuretic, emmenagogue, nervine, orexigenic, stimulant, stomachic, tonic (uterine, womb), vermifuge.
104.	Allyl isothiocyanate (99%).	**Plant Name:** *Brassica nigra* **Family:** Brassicaceae (Cruciferae) *Sinapsis nigra, B. sinapioides*, black mustard.		Aperitif, antimicrobial, antiseptic, diuretic, emetic, febrifuge, rubefacient (produces blistering of the skin), stimulant
105.	Heerabolene, limonene, dipentene, pinene, eugenol, cinnamaldehyde, cuminaldehyde, cadinene,	**Plant Name:** *Commiphora myrrha* **Family:** Burseraceae	*Balsamodendrom myrrha*, gum myrrh, common myrrh, hirabol myrrh, myrrha.	Anticatarrhal, anti-inflammatory, antimicrobial, antiphlogistic, antiseptic, astringent, balsamic, carminative, cicatrisant, emmenagogue, expectorant, fungicidal, revitalizing, sedative, stimulant (digestive, pulmonary), stomachic, tonic, uterine, vulnerary.
106.	Cineol, myrtenol, pinene, geraniol, linalol, camphene, among others	**Plant Name:** *Myrtus communis* **Family:** Myrtaceae SYNONYM	Corsican pepper	Anticatarrhal, antiseptic (urinary, pulmonary), astringent, balsamic, bactericidal, expectorant, regulator, slightly sedative.
107.	Quercetin.	**Plant Name:** *Narcissus poeticus* **Family:** Amaryllidaceae.	Pinkster lily, pheasant's eye, poet's narcissus	Antispasmodic, aphrodisiac, emetic, narcotic, sedative.
108.	Cineol (50%–65%), terpineol, pinene, limonene, citrene, terebenthene, valeric ester, acetic ester, butyric ester	**Plant Name:** *Melaleuca viridiflora* **Family:** Myrtaceae	M. quinquenervia, 'gomenol'	Analgesic, anthelmintic, anticatarrhal, antirheumatic, antiseptic, antispasmodic, bactericidal, balsamic, cicatrisant, diaphoretic, expectorant, regulator, stimulant, vermifuge
109.	Camphene, pinene, dipentene, sabinene, cymene, geraniol, borneol, linalol, terpineol, myristicin (4%–8%), safrol, elemincin	**Plant Name:** *Myristica fragrans* **Family:** Myristicaceae	M. *officinalis, M. aromata, Nux moschata*, myristica (oil), mace (husk), macis (oil).	Analgesic, anti-emetic, anti-oxidant, antirheumatic, antiseptic, antispasmodic, aphrodisiac, carminative, digestive, emmenagogue, gastric secretory stimulant, larvicidal, orexigenic, prostaglandin inhibitor, stimulant, tonic
110.	Evernic acid, dusnic acid, some atranorine, chloratronorine.	**Plant Name:** *Evernia prunastri* **Family:** Usneaceae	Mousse de chêne, treemoss.	Antiseptic, demulcent, expectorant, fixative
111.	Dipropyl disulphide, methylpropyl disulphide, dipropyl trisulphide, methylpropyl trisulphide and allylpropyl disulphide.	**Plant Name:** *Allium cepa* **Family:** Liliaceae	Common onion, Strasburg onion.	Anthelmintic, antimicrobial, antirheumatic, antiseptic, antisclerotic, antispasmodic, antiviral, bactericidal, carminative, depurative, digestive, diuretic, expectorant, fungicidal, hypocholesterolaemic, hypoglycaemic, hypotensive, stomachic, tonic, vermifuge.
112.	Bisabolene sesquiterpene alcohols.	**Plant Name:** *Commiphora erythraea* **Family:** Burseraceae	*C. erythraea var. glabrascens*, bisabol myrrh, sweet myrrh.	Antiseptic, antispasmodic, balsamic, expectorant

(*Continued*)

TABLE 12.1 (*Continued*)

List of Terpenoids with Their Plant Sources, Synnonyms and Uses [7–26]

S. No	Compound	Source	Synonyms of Plant	Uses
113.	Limonene, myrcene, camphene, pinene, ocimene, cymene	**Plant Name:** *Citrus aurantium var. amara* **Family:** Rutaceae	*C. vulgaris, C. bigaradia,* Seville orange, sour orange bigarade (oil).	Anti-inflammatory, antiseptic, astringent, bactericidal, carminative, choleretic, fungicidal, sedative (mild), stomachic, tonic
114.	Linalol, linalyl acetate, limonene, pinene, nerolidol, geraniol, nerol, methyl anthranilate, indole, citral, jasmone	**Plant Name:** *Citrus aurantium var. amara* **Family:** Rutaceae	*C. vulgaris, C. bigaradia,* orange flower, neroli, neroli bigarade	Antidepressant, antiseptic, antispasmodic, aphrodisiac, bactericidal, carminative, cicatrisant, cordial, deodorant, digestive, fungicidal, hypnotic (mild), stimulant (nervous), tonic (cardiac, circulatory).
115.	Bergapten, auraptenol	**Plant Name:** *Citrus sinensis* **Family:** Rutaceae	C. aurantium var. dulcis, C. aurantium var. sinensis, China orange, Portugal orange	Antidepressant, anti-inflammatory, antiseptic, bactericidal, carminative, choleretic, digestive, fungicidal, hypotensive, sedative (nervous), stimulant (digestive and lymphatic), stomachic, tonic.
116.	Carvacrol, thymol, cymene, caryophyllene, pinene, bisabolene, linalol, borneol, geranyl acetate, linalyl acetate, terpinene.	**Plant Name:** *Origanum vulgare* **Family:** Lamiaceae (Labiatae)	European oregano, wild marjoram, common marjoram, grove marjoram, joy of the mountain, origanum (oil).	Analgesic, anthelmintic, antirheumatic, antiseptic, antispasmodic, antitoxic, antiviral, bactericidal, carminative, choleretic, cytophylactic, diaphoretic, diuretic, emmenagogue, expectorant, febrifuge, fungicidal, parasiticide, rubefacient, stimulant, tonic
117.	Carvacrol, thymol, cymene, caryophyllene, pinene, limonene, linalol, borneol, myrcene, thujone, terpinene	**Plant Name:** *Thymus capitatus* **Family:** Lamiaceae (Labiatae)	T. *capitans, Coridothymus capitatus, Satureja capitata, Thymbra capitata,* oreganum (oil), Israeli oreganum (oil), Cretan thyme, corido thyme, conehead thyme, headed savory, thyme of the ancients.	
118.	Myristic acid, alpha-irone and oleic acid.	**Plant Name:** *Iris pallida* **Family:** Iridaceae	Orris root, iris, flag iris, pale iris, orris butter (oil).	Dried Root – antidiarrhoeal, demulcent, expectorant. Fresh Root – diuretic, cathartic, emetic.
119.	Geraniol, farnesol, geranyl acetate, methyl heptenone, citronellol, citral, dipentene and limonene.	**Plant Name:** Cymbopogon martinii var. martinii **Family:** Graminaceae	*Andropogon martinii, A. martinii var. motia,* East Indian geranium, Turkish geranium, Indian rosha, motia.	Antiseptic, bactericidal, cicatrisant, digestive, febrifuge, hydrating, stimulant (digestive, circulatory), tonic
120.	Apiol, Myristicin, tetramethoxyally-benzene, pinene, menthatriene, pinene and carotel	**Plant Name:** *Petroselinum sativum* **Family:** Apiaceae (Umbelliferae)	P. hortense, Apium petroselinum, Carum petroselinum, common parsley, garden parsley	Antimicrobial, antirheumatic, antiseptic, astringent, carminative, diuretic, depurative, emmenagogue, febrifuge, hypotensive, laxative, stimulant (mild), stomachic, tonic (uterine).
121.	Patchouli alcohol (40%), pogostol, bulnesol, nor patchoulenol, bulnese, patchoulene	**Plant Name:** *Pogostemon cablin* **Family:** Lamiaceae (Labiatae)	P. *patchouly,* patchouly, puchaput.	Antidepressant, anti-inflammatory, anti-emetic, antimicrobial, antiphlogistic, antiseptic, antitoxic, antiviral, aphrodisiac, astringent, bactericidal, carminative, cicatrisant, deodorant, digestive, diuretic, febrifuge, fungicidal, nervine, prophylactic, stimulant (nervous), stomachic, tonic.
122.	Pulegone, menthone, iso-menthone, octanol, piperitenone trans-iso-pulegone.	**Plant Name:** *Mentha pulegium* **Family:** Lamiaceae (Labiatae).	Pulegium, European pennyroyal, pudding grass	Antiseptic, antispasmodic, diaphoretic, carminative, digestive, emmenagogue, insect repellent, refrigerant, stimulant.
123.	Thujene, pinene, camphene, sabinene, carene, myrcene, limonene, phellandrene.	**Plant Name:** *Piper nigrum* **Family:** Piperaceae	Piper, pepper	Analgesic, antimicrobial, antiseptic, antispasmodic, antitoxic, aperitif, aphrodisiac, bactericidal, carminative, diaphoretic, digestive, diuretic, febrifuge, laxative, rubefacient, stimulant (nervous, circulatory, digestive), stomachic, tonic

(*Continued*)

TABLE 12.1 (*Continued*)

List of Terpenoids with Their Plant Sources, Synnonyms and Uses [7–26]

S. No	Compound	Source	Synonyms of Plant	Uses
124.	Linalyl acetate, geranyl acetate, linalol, nerol, terpineol, geraniol, nerolidol, farnesol, limonene	**Plant Name:** *Citrus aurantium var. amara* **Family:** Rutaceae.	*C.bigaradia*, petitgrain bigarade (oil), petitgrain Paraguay (oil). See also *bitter orange*	Antiseptic, antispasmodic, deodorant, digestive, nervine, stimulant (digestive, nervous), stomachic, tonic.
125.	Limonene, pinenes, phellandrene, dipentene, camphene, myrcene and bornyl acetate	**Plant Name:** *Pinus mugo var. pumilio* **Family:** Pinaceae	*P. mugo, P. montana, P. pumilio*, mountain pine, Swiss mountain pine, pine needle (oil).	Analgesic, antimicrobial, antiseptic, antitussive, antiviral, balsamic, diuretic, expectorant, rubefacient.
126.	Terpineol, estragole, fenchone, fenchyl alcohol and borneol	**Plant Name:** *Pinus palustris* **Family:** Pinaceae	Longleaf yellow pine, southern yellow pine, pitch pine, pine (oil).	Analgesic (mild), antirheumatic, antiseptic, bactericidal, expectorant, insecticidal, stimulant.
127.	Pinenes, carene, dipentene, limonene, terpinenes, myrcene, ocimene, camphene, sabinene;	**Plant Name:** *Pinus sylvestris* **Family:** Pinaceae	Forest pine, Scots pine, Norway pine, pine needle (oil).	Antimicrobial, antineuralgic, antirheumatic, antiscorbutic, antiseptic (pulmonary, urinary, hepatic), antiviral, bactericidal, balsamic, cholagogue, choleretic, deodorant, diuretic, expectorant, hypertensive, insecticidal, restorative, rubefacient, stimulant (adrenal cortex, circulatory, nervous), vermifuge.
128.	Citronellol (18%–22%), phenyl ethanol (63%), geraniol and nerol (10%–15%), stearopten (8%), farnesol (0.2%–2%)	**Plant Name:** *Rosa centifolia* **Family:** Rosaceae	Rose maroc, French rose, Provence rose, hundred-leaved rose, Moroccan otto of rose (oil), French otto of rose (oil), rose de mai (absolute or concrete).	Antidepressant, antiphlogistic, antiseptic, antispasmodic, anti-tubercular agent, antiviral, aphrodisiac, astringent, bactericidal, choleretic, cicitrisant, depurative, emmenagogue, haemostatic, hepatic, laxative, regulator of appetite, sedative (nervous), stomachic, tonic (heart, liver, stomach, uterus).
129.	Mainly citronellol (34%–55%), geraniol and nerol (30%–40%), stearopten (16%–22%), phenyl ethanol (1.5%–3%), farnesol (0.2%–2%)	**Plant Name:** *Rosa damascena* **Family:** Rosaceae	Summer damask rose, Bulgarian rose, Turkish rose (Anatolian rose oil), otto of rose (oil), attar of rose (oil).	Abdominal and chest pains, strengthening the heart, menstrual bleeding, digestive problems and constipation.
130.	Pinenes, camphene, limonene, cineol, borneol with camphor, linalol, terpineol, octanone, bornyl acetate	**Plant Name:** *Rosmarinus officinalis* **Family:** Lamiaceae (Labiatae)	*R. coronarium*, compass plant, incensier	Analgesic, antimicrobial, anti-oxidant, antirheumatic, antiseptic, antispasmodic, aphrodisiac, astringent, carminative, cephalic, cholagogue, choleretic, cicatrisant, cordial, cytophylactic, diaphoretic, digestive, diuretic, emmenagogue, fungicidal, hepatic, hypertensive, nervine, parasiticide, restorative, rubefacient, stimulant (circulatory, adrenal cortex, hepatobiliary), stomachic, sudorific, tonic (nervous, general), vulnerary.
131.	Linalol (90%–97%), Cineol, terpineol, geraniol, citronellal, limonene, pinene	**Plant Name:** *Aniba rosaeodora* **Family:** Lauraceae	*A. rosaeodora var. amazonica*, bois de rose, Brazilian rosewood.	Mildly analgesic, anticonvulsant, antidepressant, anti-microbial, antiseptic, aphrodisiac, bactericidal, cellular stimulant, cephalic, deodorant, stimulant (immune system), tissue regenerator, tonic.
132.	Methyl nonyl ketone (90%).	**Plant Name:** *Ruta graveolens* **Family:** Rutaceae	Garden rue, herb-of-grace, herbygrass.	Antitoxic, antitussive, antiseptic, antispasmodic, diuretic, emmenagogue, insecticidal, nervine, rubefacient, stimulant, tonic, vermifuge.

(*Continued*)

TABLE 12.1 (*Continued*)

List of Terpenoids with Their Plant Sources, Synnonyms and Uses [7–26]

S. No	Compound	Source	Synonyms of Plant	Uses
133.	Linalyl acetate (75%), linalol, pinene, myrcene and phellandrene.	**Plant Name:** *Salvia sclarea* **Family:** Lamiaceae (Labiatae)	Clary, clary wort, muscatel sage, clear eye, see bright, common clary, clarry, eye bright.	Anticonvulsive, antidepressant, antiphlogistic, antiseptic, antispasmodic, aphrodisiac, astringent, bactericidal, carminative, cicatrisant, deodorant, digestive, emmenagogue, hypotensive, nervine, regulator (of seborrhoea), sedative, stomachic, tonic, uterine.
134.	Thujone (42%), cineol, borneol, caryophyllene	**Plant Name:** *Salvia officinalis* **Family:** Lamiaceae (Labiatae)	Garden sage, true sage, Dalmatian sage	Anti-inflammatory, antimicrobial, anti-oxidant, antiseptic, antispasmodic, astringent, digestive, diuretic, emmenagogue, febrifuge, hypertensive, insecticidal, laxative, stomachic, tonic
135.	Camphor (34%), cineol (up to 35%), limonene (up to 41%), camphene (up to 20%), pinene (up to 20%)	**Plant Name:** *Salvia lavendulaefolia* **Family:** Lamiaceae (Labiatae)	Lavender-leaved sage	Antidepressant, anti-inflammatory, antimicrobial, antiseptic, antispasmodic, astringent, carminative, deodorant, depurative, digestive, emmenagogue, expectorant, febrifuge, hypotensive, nervine, regulator (of seborrhoea), stimulant (hepatobiliary, adrenocortical glands, circulation), stomachic, tonic (nerve and general).
136.	Santalols, Santene, teresantol, borneol, santalone, tri-cyclo-ekasantalal,	**Plant Name:** *Santalum album* **Family:** Santalaceae	White sandalwood, yellow sandalwood, East Indian sandalwood, sandalwood Mysore, sanders-wood, santal (oil), white saunders (oil), yellow saunders (oil).	Antidepressant, antiphlogistic, antiseptic (urinary and pulmonary), antispasmodic, aphrodisiac, astringent, bactericidal, carminative, cicatrisant, diuretic, expectorant, fungicidal, insecticidal, sedative, tonic.
137.	Santolinenone	**Plant Name:** *Santolina chamaecyparissus* **Family:** Asteraceae (Compositae)	*Lavandula taemina*, cotton lavender	Antispasmodic, antitoxic, anthelmintic, insecticidal, stimulant, vermifuge
138.	Safrole (80%–90%), pinenes, phellandrenes, asarone, camphor, thujone, myristicin and menthone,	**Plant Name:** *Sassafras albidum* **Family:** Lauraceae	*S. officinale, Laurus sassafras, S. variifolium*, common sassafras, North American sassafras, sassafrax.	Antiviral, diaphoretic, diuretic, carminative, pediculicide (destroys lice), stimulant
139.	Sabinol, sabinyl acetate, terpinene, pinene, sabinene, decyl aldehyde, citronellol, geraniol, cadinene and dihydrocuminyl alcohol.	**Plant Name:** *Juniperus sabina* **Family:** Cupressaceae	Sabina cacumina, savin (oil).	Powerful emmenagogue, rubefacient, stimulant
140.	Carvacrol, pinene, cymene, camphene, limonene, phellandrene and borneol	**Plant Name:** *Satureja hortensis* **Family:** Lamiaceae (Labiatae)	Satureia hortensis, Calamintha hortensis, garden savory	Anticatarrhal, antiputrescent, antispasmodic, aphrodisiac, astringent, bactericidal, carminative, cicatrisant, emmenagogue, expectorant, fungicidal, stimulant, vermifuge
141.	Carvacrol, cymene and thymol, with lesser amounts of pinenes, limonene, cineol, borneol and terpineol	**Plant Name:** *Satureja montana* **Family:** Lamiaceae (Labiatae) 5.	Obovata, Calamintha montana	See summer savory
142.	Phellandrene, also caryophyllene, pinene and carvacrol	**Plant Name:** *Schinus molle* **Family:** Anacardiaceae	Peruvian pepper, Peruvian mastic, Californian pepper tree.	Antiseptic, antiviral, bactericidal, carminative, stimulant, stomachic.
143.	Pinene, linalol, borneol, terpineol, geraniol, eugenol and methyl eugenol, among others.	**Plant Name:** *Asarum canadense* **Family:** Aristolochiaceae	Wild ginger, Indian ginger	Anti-inflammatory, antispasmodic, carminative, diuretic, diaphoretic, emmenagogue, expectorant, febrifuge, stimulant, stomachic

(Continued)

TABLE 12.1 (*Continued*)

List of Terpenoids with Their Plant Sources, Synnonyms and Uses [7–26]

S. No	Compound	Source	Synonyms of Plant	Uses
144.	Bornyl acetate, isobornyl valerianate, borneol, patchouli alcohol, terpinyl valerianate, terpineol, eugenol and pinenes,	**Plant Name:** *Nardostachys jatamansi* **Family:** Valerianaceae	Nard, 'false' Indian valerian root (oil).	Anti-inflammatory, antipyretic, bactericidal, deodorant, fungicidal, laxative, sedative, tonic.
145.	Pinenes, limonene, bornyl acetate, tricyclene, phellandrene, myrcene, thujone, dipentene and cadinene	**Plant Name:** *Tsuga canadensis* **Family:** Pinaceae	*Pinus canadensis, Abies canadensis*, spruce, eastern hemlock, common hemlock, hemlock (oil), spruce (oil), fir needle (oil).	Antimicrobial, antiseptic, antitussive, astringent, diaphoretic, diuretic, expectorant, nervine, rubefacient, tonic.
146.	Styrene with vanillin, phenylpropyl alcohol, cinnamic alcohol, benzyl alcohol, ethyl alcohol,	**Plant Name:** *Liquidambar orientalis* **Family:** Hamamelidaceae	*Balsam styracis*, oriental sweetgum, Turkish sweetgum, asiatic styrax, styrax, storax, liquid storax.	Anti-inflammatory, antimicrobial, antiseptic, antitussive, bactericidal, balsamic, expectorant, nervine, stimulant
147.	Mainly tagetones, Ocimene, myrcene, linalol, limonene, pinenes, carvone, citral, camphene, valeric acid and salicylaldehyde	**Plant Name:** *Tagetes minuta* **Family:** Asteraceae (Compositae)	*T. glandulifera*, tagette, taget, marigold, Mexican marigold, wrongly called 'calendula' (oil).	Anthelmintic, antispasmodic, bactericidal, carminative, diaphoretic, emmenagogue, fungicidal, stomachic
108.	Thujone (66%–81%), camphor, borneol,	**Plant Name:** *Tanacetum vulgare* **Family:** Asteraceae (Compositae)	*Chrysanthemum vulgare, C. tanacetum*, buttons, bitter buttons, bachelor's buttons, scented fern, cheese.	Anthelmintic, anti-inflammatory, antispasmodic, carminative, diaphoretic, digestive, emmenagogue, febrifuge, nervine, stimulant, tonic, vermifuge
108.	Estragole (70%), capillene, ocimene, nerol, phellandrene, thujone and cineol,	**Plant Name:** *Artemisia dracunculus* **Family:** Asteraceae (Compositae)	Estragon (oil), little dragon, Russian tarragon.	Anthelmintic, antiseptic, antispasmodic, aperitif, carminative, digestive, diuretic, emmenagogue, hypnotic, stimulant, stomachic, vermifuge.
108.	Terpinene-4-ol (up to 30%), cineol, pinene, terpinenes, cymene, sesquiterpenes, sesquiterpene alcohols	**Plant Name:** *Melaleuca alternifolia* **Family:** Myrtaceae	Narrow-leaved paperbark tea tree, ti-tree, ti-trol, melasol.	Anti-infectious, anti-inflammatory, antiseptic, antiviral, bactericidal, balsamic, cicatrisant, diaphoretic, expectorant, fungicidal, immuno-stimulant, parasiticide, vulnerary
108.	Thujone (60%), fenchone, camphor, sabinene, pinene	**Plant Name:** *Thuja occidentalis* **Family:** Cupressaceae	Swamp cedar, white cedar, northern white cedar, eastern white cedar, tree of life, American arborvitae, cedarleaf (oil).	Antirheumatic, astringent, diuretic, emmenagogue, expectorant, insect repellent, rubefacient, stimulant (nerve, uterus and heart muscles), tonic, vermifuge
108.	Thymol and carvacrol (60%), cymene, terpinene, camphene, borneol, linalool, can also contain geraniol, citral, thuyanol	**Plant Name:** *Thymus vulgaris* **Family:** Lamiaceae (Labiatae)	*T. aestivus, T. ilerdensis, T. webbianus, T. valentianus*, French thyme, garden thyme, red thyme (oil), white thyme (oil).	Anthelmintic, antimicrobial, anti-oxidant, antiputrescent, antirheumatic, antiseptic (intestinal, pulmonary, genito-urinary), antispasmodic, antitussive, antitoxic, aperitif, astringent, aphrodisiac, bactericidal, balsamic, carminative, cicatrisant, diuretic, emmenagogue, expectorant, fungicidal, hypertensive, nervine, revulsive, rubefacient, parasiticide, stimulant (immune system, circulation), sudorific, tonic, vermifuge.
108.	Coumarin (20%–40%)	**Plant Name:** *Dipteryx odorata* **Family:** Leguminosae	*Coumarouna odorata*, tonquin bean, Dutch tonka bean.	Insecticidal, narcotic, tonic (cardiac).
108.	Methyl benzoate, methyl anthranilate, benzyl alcohol, butyric acid, eugenol, nerol, farnesol, geraniol	**Plant Name:** *Polianthes tuberosa* **Family:** Agavaceae	Tuberosa, tubereuse	Narcotic. Used in high class perfumes, especially of an oriental, floral or fantasy type. Occasionally used for flavouring confectionery and some beverages

(Continued)

TABLE 12.1 (*Continued*)

List of Terpenoids with Their Plant Sources, Synnonyms and Uses [7–26]

S. No	Compound	Source	Synonyms of Plant	Uses
108.	Mainly tumerone (60%), ar-tumerone, atlantones, zingiberene, cineol, borneol, sabinene and phellandrene,	**Plant Name:** *Curcuma longa* **Family:** Zingiberaceae	C. *domestica, Amomoum curcuma,* curcuma, Indian saffron, Indian yellow root, curmuma (oil).	Analgesic, anti-arthritic, anti-inflammatory, antioxidant, bactericidal, cholagogue, digestive, diuretic, hypotensive, insecticidal, laxative, rubefacient, stimulant
108.	Mainly alphapinene (50%), betapinene (25%–35%), carene (20%–60%)	**Plant Name:** *Pinus palustris* and other *Pinus* species **Family:** Pinaceae	Terebinth, therebentine, gum thus, gum turpentine, turpentine balsam, spirit of turpentine (oil).	Analgesic, antimicrobial, antirheumatic, antiseptic, antispasmodic, balsamic, diuretic, cicatrisant, counter-irritant, expectorant, haemostatic, parasiticide, rubefacient, stimulant, tonic, vermifuge.
108.	Bornyl acetate, isovalerate, caryophyllene, pinenes, valeranone, ionone, eugenyl isovalerate, borneol, patchouli alcohol, valerianol	**Plant Name:** *Valeriana fauriei* **Family:** Valerianaceae	V. *officinalis, V. officinalis var. angustifolium, V. officinalis var. latifolia,* European valerian, common valerian, Belgian valerian, fragrant valerian, garden valerian.	Anodyne (mild), antidandruff, diuretic, antispasmodic, bactericidal, carminative, depressant of the central nervous system, hypnotic, hypotensive, regulator, sedative, stomachic
108.	Vanillin (1.3%–2.9%) hydroxybenzaldehyde, acetic acid, isobutyric acid, caproic acid, eugenol, furfural	**Plant Name:** *Vanilla planifolia* **Family:** Orchidaceae	V. *fragrans,* common vanilla, Mexican vanilla, Bourbon vanilla, Reunion vanilla.	Balsamic, Used in pharmaceutical products as a flavouring agent. Used as a fragrance ingredient in perfumes, especially oriental types. Widely used to flavour tobacco and as a food flavouring, mainly in ice cream, yoghurt and chocolate
108.	Citral (30%–35%), nerol, geraniol,	**Plant Name:** *Aloysia triphylla* **Family:** Verbenaceae	A. citriodora, Verbena triphylla, Lippia citriodora, L. triphylla, verbena, herb Louisa.	Antiseptic, antispasmodic, carminative, detoxifying, digestive, febrifuge, hepatobiliary stimulant, sedative (nervous), stomachic
108.	Vetiverol, vitivone, vetivenes	**Plant Name:** *Vetiveria zizanoides* **Family:** Poaceae (Gramineae)	*Andropogon muricatus,* vetivert, khus khus	Antiseptic, antispasmodic, depurative, rubefacient, sedative (nervous system), stimulant (circulatory, production of red corpuscles), tonic, vermifuge
108.	Nonadienal, parmone, hexyl alcohol, quercitin	**Plant Name:** *Viola odorata* **Family:** Violaceae	English violet, garden violet, blue violet, sweet-scented violet.	Analgesic (mild), anti-inflammatory, antirheumatic, antiseptic, decongestant (liver), diuretic, expectorant, laxative, soporific, stimulant (circulation).
108.	Methyl salicylate, formaldehyde, gaultheriline.	**Plant Name:** *Gaultheria procumbens* **Family:** Ericaceae	Aromatic wintergreen, checkerberry, teaberry, gaultheria (oil).	Analgesic (mild), anti-inflammatory, antirheumatic, antitussive, astringent, carminative, diuretic, emmenagogue, galactagogue, stimulant
108.	Ascaridole (60%–80%), cymene, limonene, terpinene, myrcene	**Plant Name:** Chenopodium ambrosioides var. anthelminticum **Family:** Chenopodiaceae	*C. anthelminticum,* American wormseed, chenopodium, Californian spearmint, Jesuit's tea, Mexican tea, herb sancti mariae, Baltimore (oil).	Anthelmintic, antirheumatic, antispasmodic, expectorant, hypotensive
108.	Thujone (71%), azulenes, terpenes	**Plant Name:** *Artemisia absinthium* **Family:** Asteraceae (Compositae)	Common wormwood, green ginger, armoise, absinthium (oil).	Anthelmintic, choleretic, deodorant, emmenagogue, febrifuge, insect repellent, narcotic, stimulant (digestive), tonic, vermifuge.
108.	Azulene (51%), pinenes, caryophyllene, borneol, terpineol, cineol, bornyl acetate, camphor, sabinene, thujone	**Plant Name:** *Achillea millefolium* **Family:** Asteraceae (Compositae)	Milfoil, common yarrow, nosebleed, thousand leaf – and many other country names.	Anti-inflammatory, antipyretic, antirheumatic, antiseptic, antispasmodic, astringent, carminative, cicatrisant, diaphoretic, digestive, expectorant, haemostatic, hypotensive, stomachic, tonic.
108.	Methyl benzoate, methyl salicylate, methyl para-cretol, benzyl acetate, eugenol, geraniol, linalool, pinene, cadinene	**Plant Name:** *Cananga odorata var. genuina* **Family:** Annonaceae	*Unona odorantissimum,* flower of flowers.	Aphrodisiac, antidepressant, anti-infectious, antiseborrhoeic, antiseptic, euphoric, hypotensive, nervine, regulator, sedative (nervous), stimulant (circulatory), tonic

S. No.	No. of Isoprene Units	Number of Carbon Atoms	Class	Attachments
(1)	1	5	Hemiterpenoids (C_5H_8)	Head to tail
(2)	2	10	Monoterpenoids ($C_{10}H_{16}$)	Head to tail
(3)	3	15	Sesquiterpenoids ($C_{15}H_{24}$)	Head to tail
(4)	4	20	Diterpenoids ($C_{20}H_{32}$)	Head to tail
(5)	5	25	Sesterterpenoids ($C_{25}H_{40}$)	Head to tail
(6)	6	30	Triterpenoids ($C_{30}H_{48}$)	Tail to tail
(7)	8	40	Tetraterpenoids (carotenoids) ($C_{40}H_{64}$)	Tail to tail
(8)	> 8	> 40	Polyterpenoids (C_5H_8	Head to tail

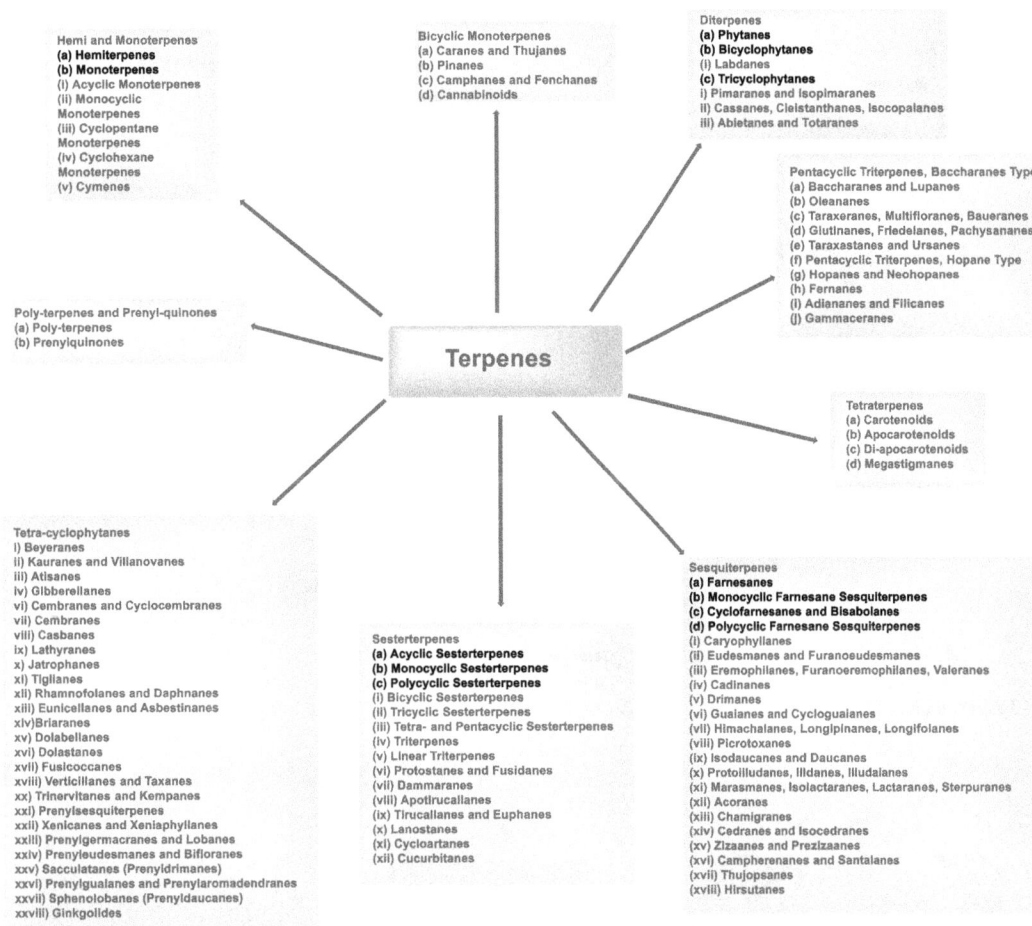

FIGURE 12.3 Classification of terpenes.

ii. **Monocyclic monoterpenes:** Terpenes that possess one cyclic structure are called as monocyclic monoterpenes i.e. Cyclopropane and Cyclobutane, Cinerins, jasmolins, pyrethrins, Chrysanthemol, cinerolone as shown in Figure 12.5.

a. **Cyclopentane Monoterpenes**

Monoterpenes that possess at least one cyclopentane ring is called as Cyclopentane Monoterpenes. These also possess iridoides (cyclopentanopyran) nucleus. Examples are 1-acetyl-4-isopropenylcyclopentene, approx. 200 cyclopentane monoterpenes, (+)-Iridomyrmecin etc as shown in Figure 12.6. (-)-Loganin and (-)-secologanin comes in seco-iridoids glucoside categories. Yellowish (+)-jasmolactone A and (+)-Oleuropein is the first seco-iridoid to be isolated.

b. **Cyclohexane Monoterpenes**

Monoterpenes possess at least one cyclo-hexane ring are called as Cyclohexane monoterpenes. Examples include *cis-trans*-isomers of *p*-menthane, *Trans-p*-menthane, *o*- and *m*-isomers are rarely occurring rearrangement products of *p*-menthane, Limonene, D-phellandrene, *p*-Menthan-3-ol, (*R*)- (+)-E-Phellandrene, etc., as shown in Figure 12.7.

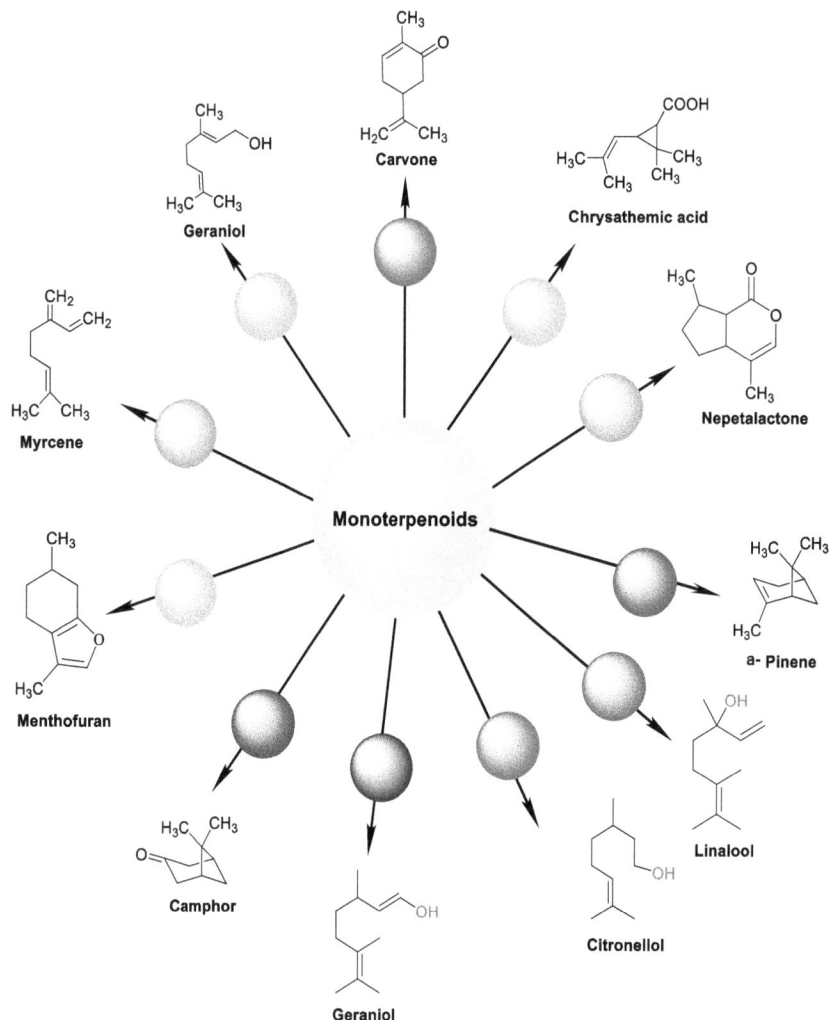

FIGURE 12.4 Few examples of monoterpenes.

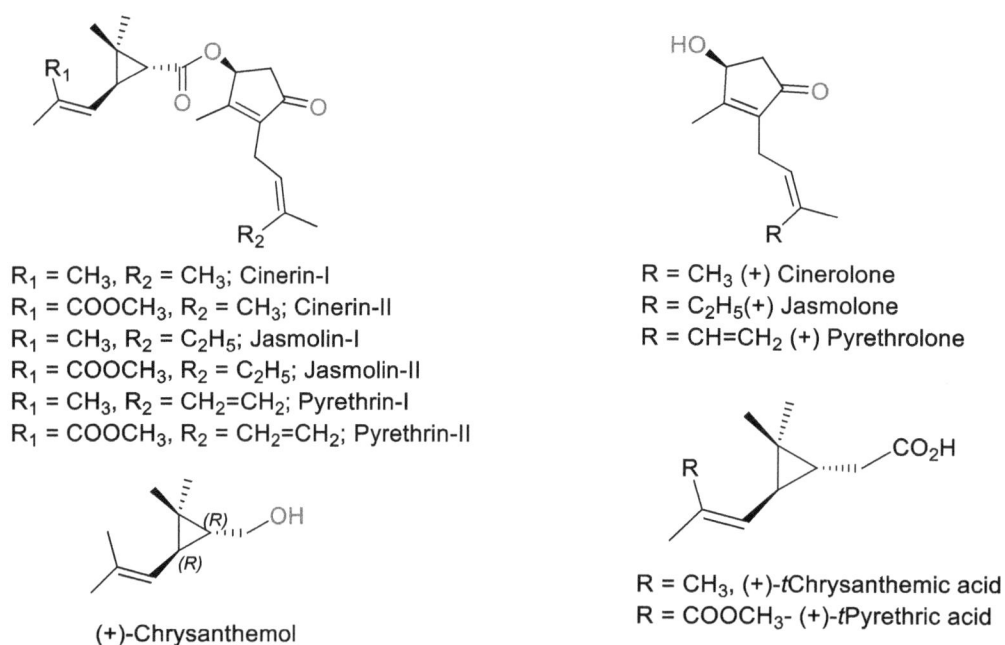

R₁ = CH₃, R₂ = CH₃; Cinerin-I
R₁ = COOCH₃, R₂ = CH₃; Cinerin-II
R₁ = CH₃, R₂ = C₂H₅; Jasmolin-I
R₁ = COOCH₃, R₂ = C₂H₅; Jasmolin-II
R₁ = CH₃, R₂ = CH₂=CH₂; Pyrethrin-I
R₁ = COOCH₃, R₂ = CH₂=CH₂; Pyrethrin-II

R = CH₃ (+) Cinerolone
R = C₂H₅(+) Jasmolone
R = CH=CH₂ (+) Pyrethrolone

(+)-Chrysanthemol

R = CH₃, (+)-*t*Chrysanthemic acid
R = COOCH₃- (+)-*t*Pyrethric acid

FIGURE 12.5 Monocyclic monoterpenes.

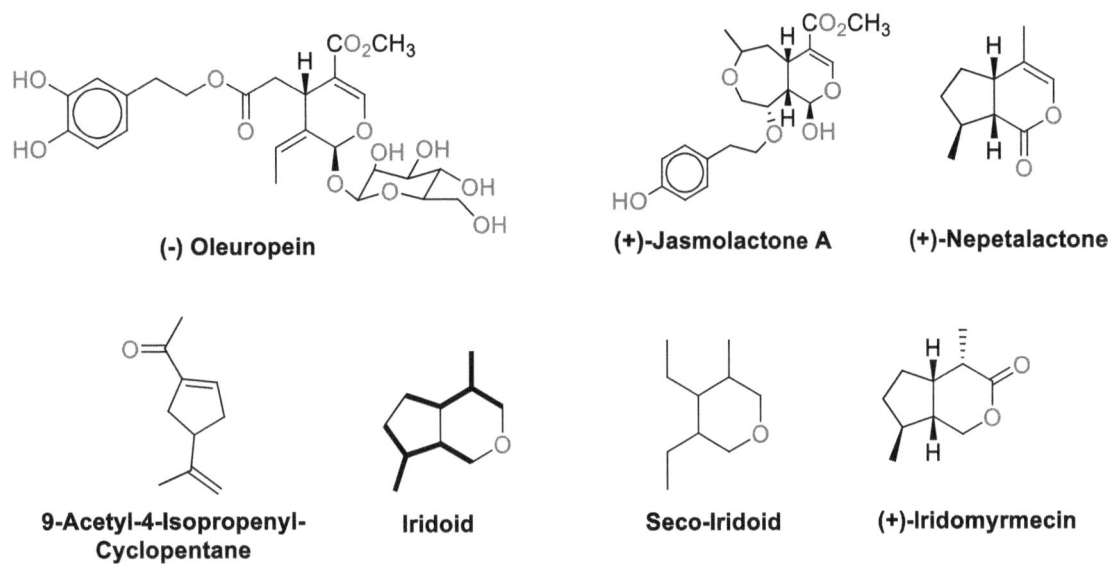

(-) Oleuropein

(+)-Jasmolactone A

(+)-Nepetalactone

9-Acetyl-4-Isopropenyl-Cyclopentane

Iridoid

Seco-Iridoid

(+)-Iridomyrmecin

FIGURE 12.6 Cyclopentane monoterpenes.

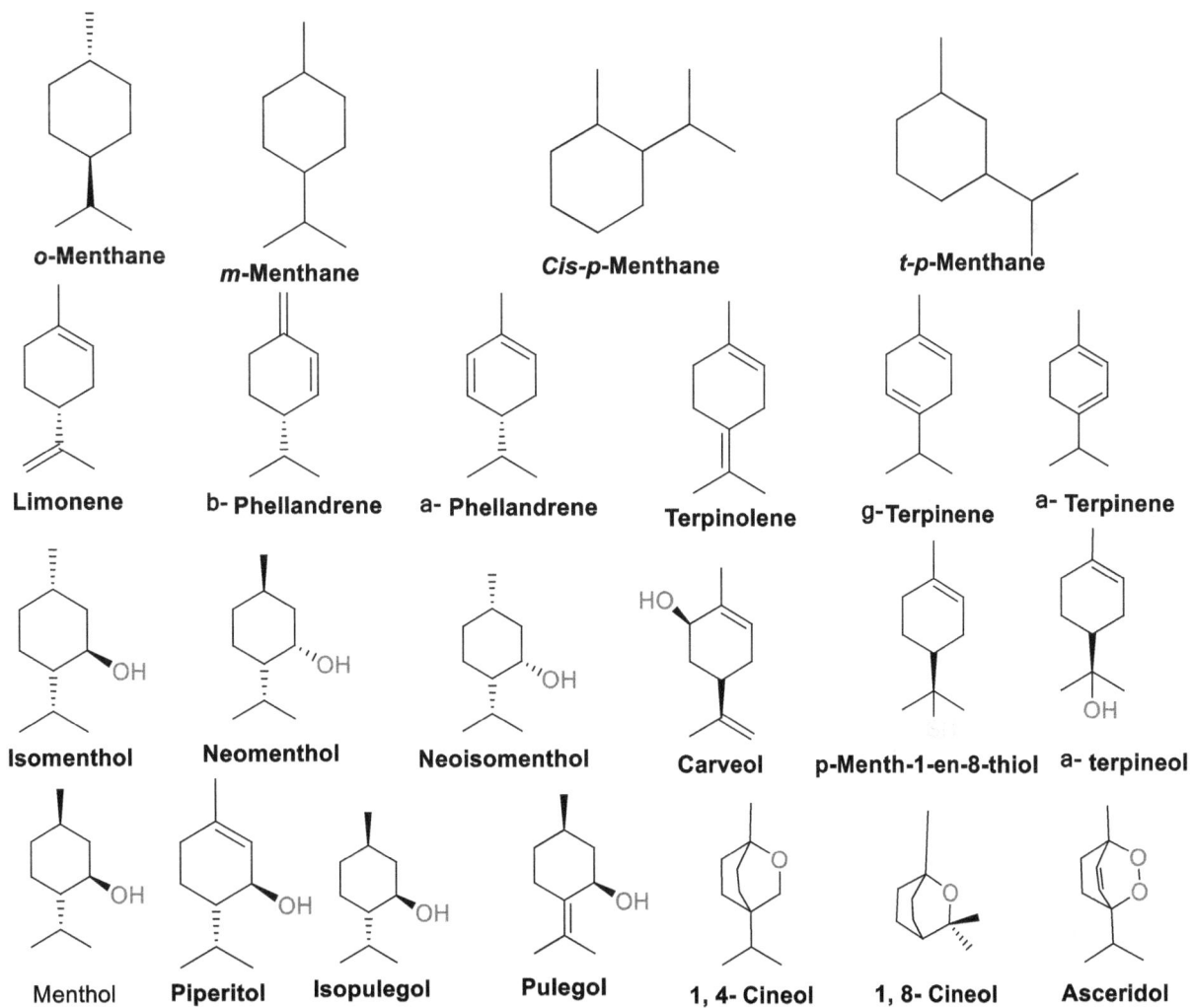

o-Menthane

m-Menthane

Cis-p-Menthane

t-p-Menthane

Limonene

b- Phellandrene

a- Phellandrene

Terpinolene

g-Terpinene

a- Terpinene

Isomenthol

Neomenthol

Neoisomenthol

Carveol

p-Menth-1-en-8-thiol

a- terpineol

Menthol

Piperitol

Isopulegol

Pulegol

1, 4- Cineol

1, 8- Cineol

Asceridol

FIGURE 12.7 Cyclohexane monoterpenes.

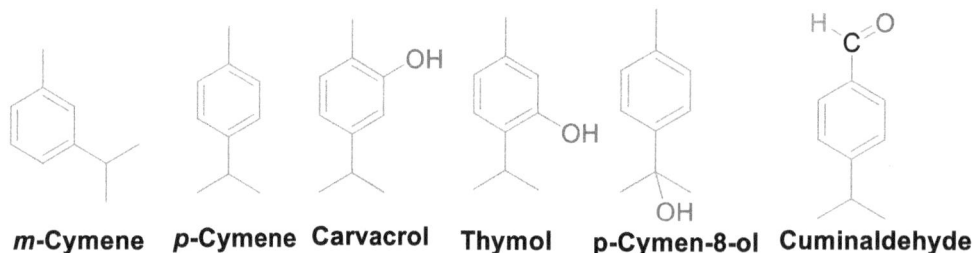

FIGURE 12.8 Cymenes.

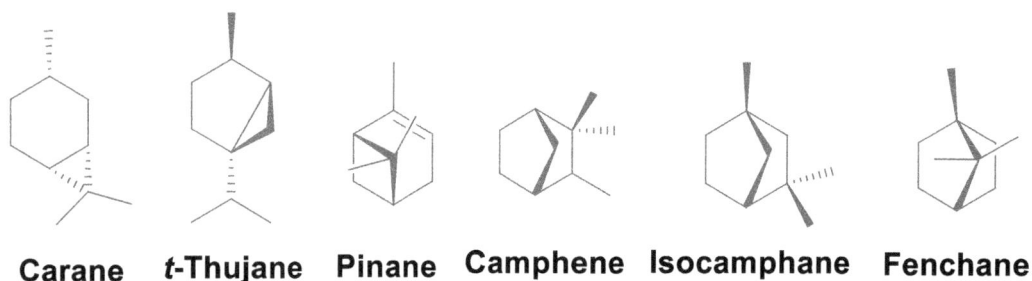

FIGURE 12.9 Bicyclic monoterpenes.

Other examples includes (-)-pulegol, (+)-piperitol, carvol, p-methy-1-en-8-thiol, 1,4-cineol, ascaridole etc.

c. Cymenes

Cymenes are benzenoid menthanes, *o*-isomer has not yet been found in nature. Examples include are *m*-Cymene (*Ribes nigrum*, Saxifragaceae) whereas *p*-cymene in (cinnamon, cypress, eucalyptus, thyme, turpentine and others). Carvacrol (marjoram, origanum, summer savoy and thyme). Thymol (*Thymus vulgaris*, labiatae). *p*-Cymen-8-ol (*Hylotrupes bajulus* (Cerambycodae). Cuminaldehyde (eucalyptus and myrrh) etc as shown in Figure 12.8.

iii. Bicyclic Monoterpenes

As the name denotes, these possess two rings condensed together and are called as Bicyclic cyclopropanes. Examples are carane and thujane, bicyclic cyclobutane pinane, bicyclic heptanes such as camphane, isocamphane and fenchane are the most important skeletons of naturally occurring bicyclic monoterpenes (Figure 12.9).

a. Caranes and Thujanes

(+)-3-Carene is a constituent of turpentine oil also present in *Pinus longifolia*, juniper (*Juniperus*) and *Citrus*. Carboxylic acids derived from carane and carene such as (+)-chaminic acid are found in *Chamaecyparis nootkatensis* (Cupressaceae).

Thujane derivatives such as (+)-3-Thujanones (thujone and its 4-epimer isothujone), Thujol [(+)-thujan-3D-ol] its 4-epimer (+)-isothujol

are found in *Artemisia*, *Juniperus* and *Thuja* species are most abundant in plant family. (+)-4(10)-Thujene, better known as (+)- sabinene, occurs in the oil of savin obtained from fresh tops of *Juniperus sabina* (Cupressaceae).

b. Pinanes

Turpentine oil obtained on large-scale possess more than 70% pinene. Enantiomers of both regio-isomers are found in many other conifers (Pinaceae). (+)-*trans*-Verbenol (oil of turpentine) and (+)- verbenone, a pinenone (bark beetles), regioisomeric (-)-pinocarvone, pinocarveol, (+)-myrtenal (Figure 12.10).

c. Camphanes and Fenchanes

Naturally occurring camphanes like borneols, isoborneols possess (*exo* OH and 2-camphanone (2-bonanone) referred as Camphor. It also includes (+)-Borneol and Iso-borneol (Figure 12.11).

d. Cannabinoids

The Indian *Cannabis sativa* var. *indica* (Moraceae, Cannabaceae) belongs to the cannabinoids. These are benzopyrans derived biogenetically from the monoterpene p-menth-1-ene and a phenol [3,4].

B. Sesquiterpenes

Terpenes consist of three isoprene units ($C_{15}H_{24}$) may be linear, cyclic, bicyclic, or tricyclic forms. Lactone rings are another kind of sesquiterpenes. Examples are artemisinin, Farnesanes, Sinensal, (*S*)-2,3-Dihydrofarnesol, Nerolidol etc.

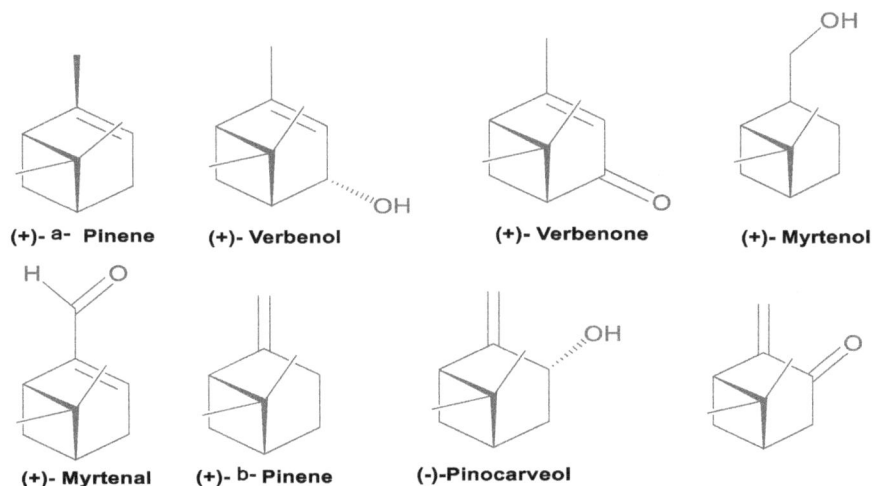

FIGURE 12.10 Pinanes.

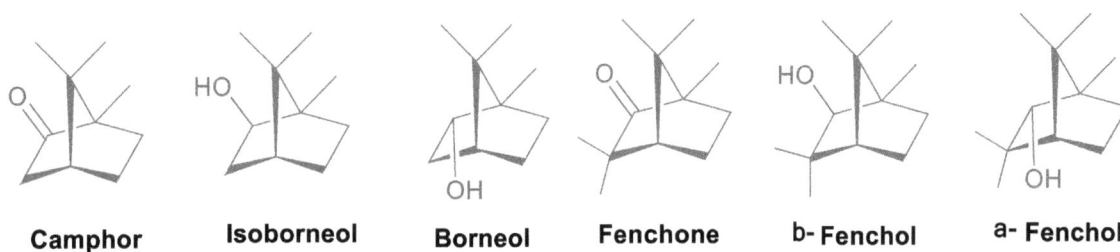

FIGURE 12.11 Camphanes and Fenchanes.

a. **Farnesanes**

The *(E, E)*-isomer of D-farnesene is a component of the flavors and natural coatings of apples, pears and other fruits. Associated with *(E)*-farnesene, it also occurs in several ethereal oils, for example, those of camomile, citrus, and hops.

b. **Monocyclic Farnesane Sesquiterpenes**

Farnesane containing one cyclic ring is known as monocyclic farnesane sesquiterpenes i.e humulane, caryophyllane etc.

c. **Cyclofarnesanes and Bisabolanes**

Abscisic acid (cabbage-leaves, potatoes, roses and cotton-young fruits) is secreted through fungus *Cercospora cruenta*. Bisabolanes represents a more eminent class of monocyclic sesquiterpenes (Figure 12.12).

d. **Polycyclic Farnesane Sesquiterpenes**

i. **Caryophyllanes**

Examples: Humulane, Caryophyllene. E-Caryophyllene is a mixture of its *cis* isomer iso-caryophyllene present in clove oil (up to 10%), *Szygium aromaticum* (Figure 12.12).

ii. **Eudesmanes and Furanoeudesmanes**

Examples: E-eudesmol, *epi*-Jeudesmol, (+)-E-Costus acid and (+)-E-costol (Figure 12.13).

Several structural lactones derivatives derived from eudesmane occur in *Artemisia* species (Asteraceae). These include various 3-oxo-12,6-eudesmanolides such as (+)-santonane from the flowers of *A. pauciflora*, taurin from *A. taurica* and anti-helmintic but toxic santonines.

iii. **Eremophilanes, Furanoeremophilanes, Valeranes**

A methyl shift from C-10 to C-5 in eudesmane leads to the basic skeleton of more than 150 eremophilane and furanoeremophilane derivatives isolated so far from higher plants. Whereas, valeranes rising from shifting of the CH_3 group C-15 in eudesmane from C-4 to C-5, in contrast, very rarely occur. Examples valerenones (*Valeriana officinalis*) and (*Nardostachys jatamansi* (Valerianaceae), Eremophilanes (*Eremophila mitchelli*), eremofortins and other derivatives of eremophilanes (Figure 12.13).

iv. **Cadinanes**

(-)-4,9-Cadinadiene (D-cadinene) from the oil of hops (*Humulus lupulus*, Cannabaceae) as well as (-)-3,9-cadinadiene, known as E-cadinene, widely spread in plants, spicy smelling and isolated from the oil of cade obtained by distillation of the wood of

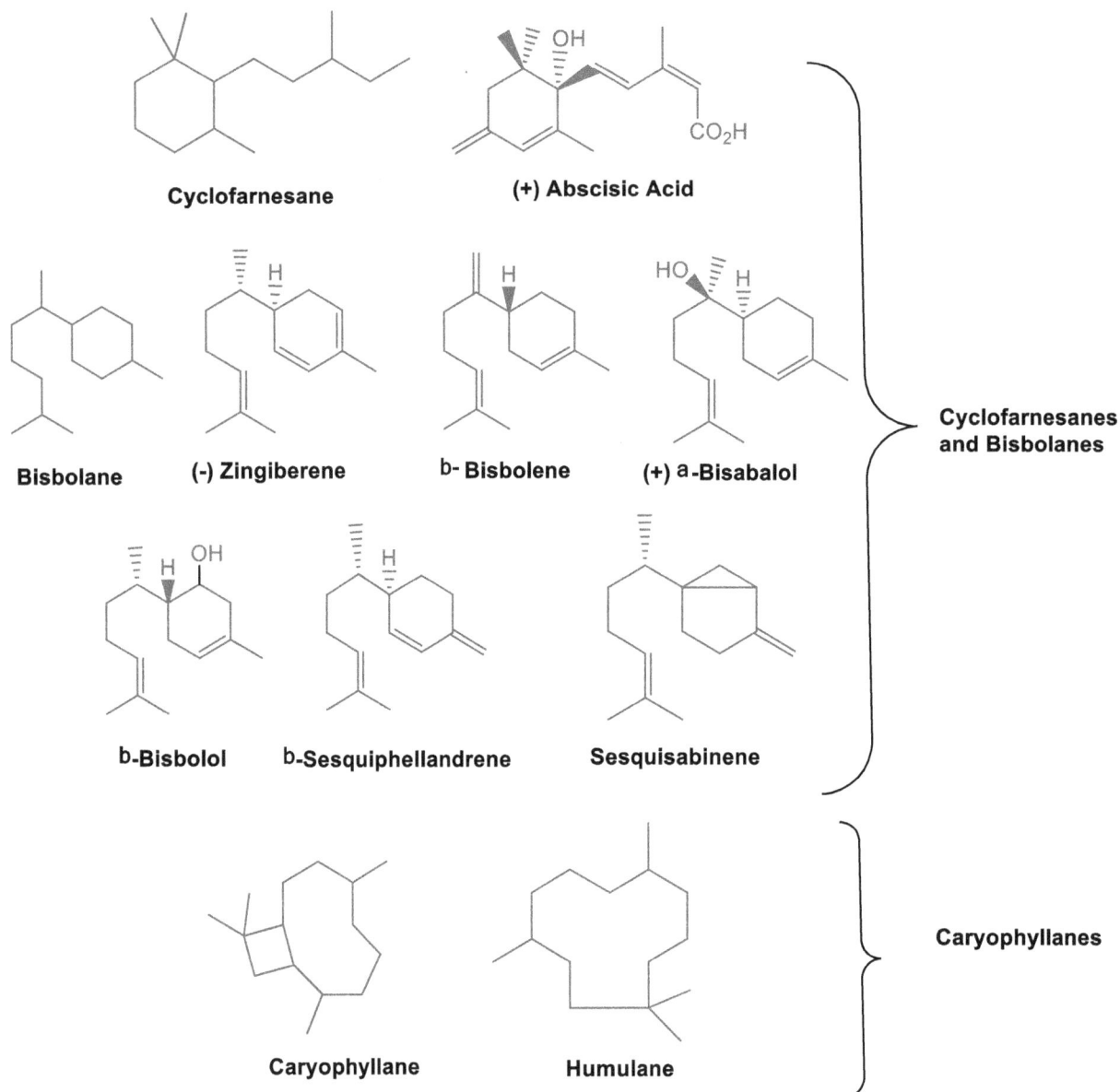

FIGURE 12.12 Cyclofarnesanes and Bisabolanes and Caryophyllanes.

Mediterranean juniper *Juniperus oxycedru* (Cupressaceae), exemplify the cadinanes. Berries of juniper species *Juniperus communis* and *J. oxycedrus* contain bulgaranes such as (0)-4,9-bulgaradiene (E1-bulgarene) and (-)-4(15), 10(14)-bulgaradiene (Figure 12.13).

v. **Drimanes**

The parent hydrocarbon 5D, 8D, 9E, 10E-drimane with *trans*-decalin as core structure occurs in paraffin oil. Examples are drimenol, drimenine, (+)-8-drimen-7-one, and 6,14,15-trihydroxy-8-drimen-12,11-olide, 11,15-Nordrimanes (Figure 12.14).

vi. **Guaianes and Cycloguaianes**

(+)-1(5), 6-Guaiadiene and its 4E-stereoisomer occur in *Balsamum tolutanum*

(*Myroxylon balsamum*), ()-1(5), 11-, ()-1(10), 11-and ()-1(10), 7(11)-guaiadiene (*Guajacum officinale),* patchouli oil (*Pogostemon patchouli)*are examples (Figure 12.14).

vii. **Himachalanes, Longipinanes, Longifolanes**

Examples are D-himachalene and himachalol are constituents of cedar wood oil (*Cedrus deodara*) (Figure 12.14).

viii. **Picrotoxanes**

Sesquiterpene alkaloids with picrotoxane skeleton such as (-)-dendrobin obtained from *Dendrobium nobile*. (-)-Picrotoxinin is obtained from *Anamirta cocculus* (syn. *Menispermum cocculus*, Menispermaceae); picrotoxin possesses CNS and respiratory stimulant and antidote to barbiturates.

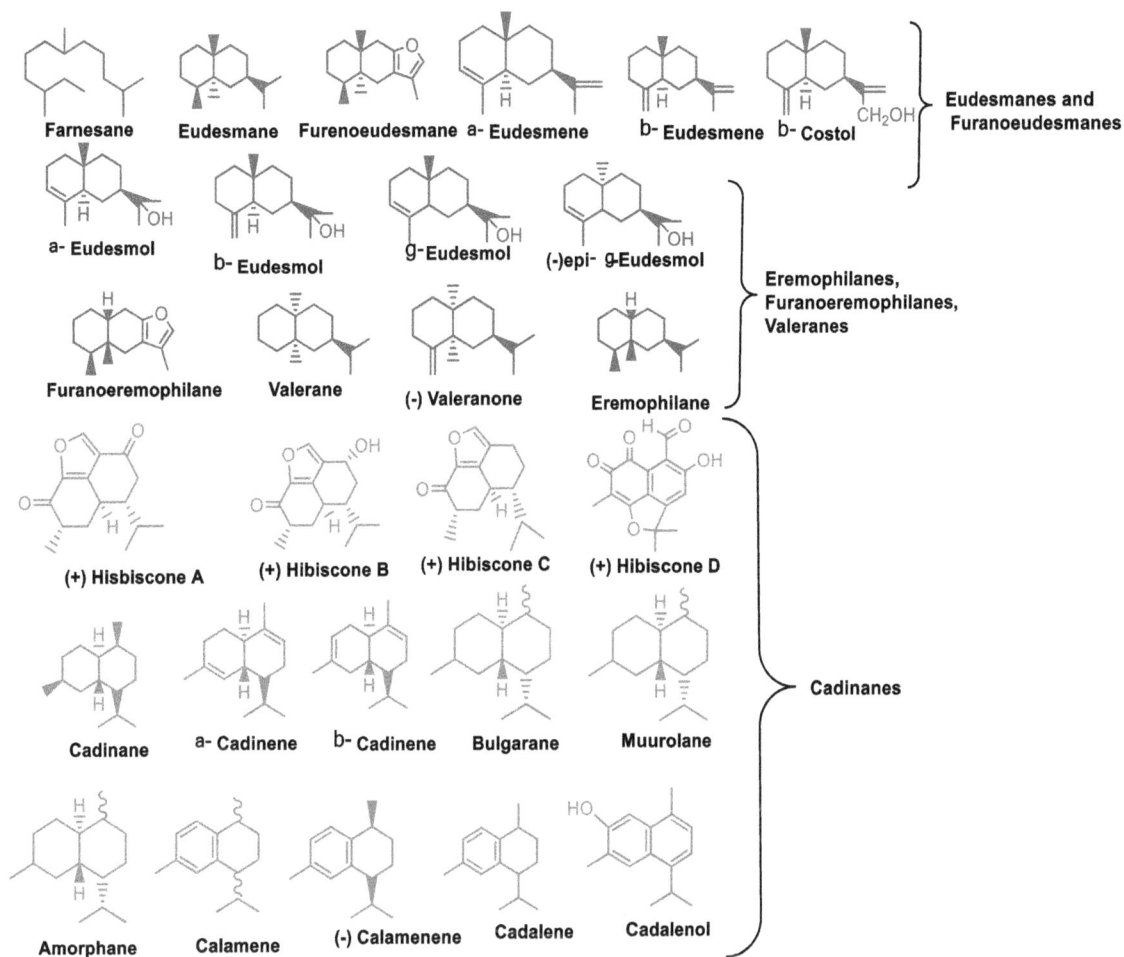

FIGURE 12.13 Eudesmanes, Furanoeudesmanes, Eremophilanes, Furanoeremophilanes, Valeranes and Cadinanes.

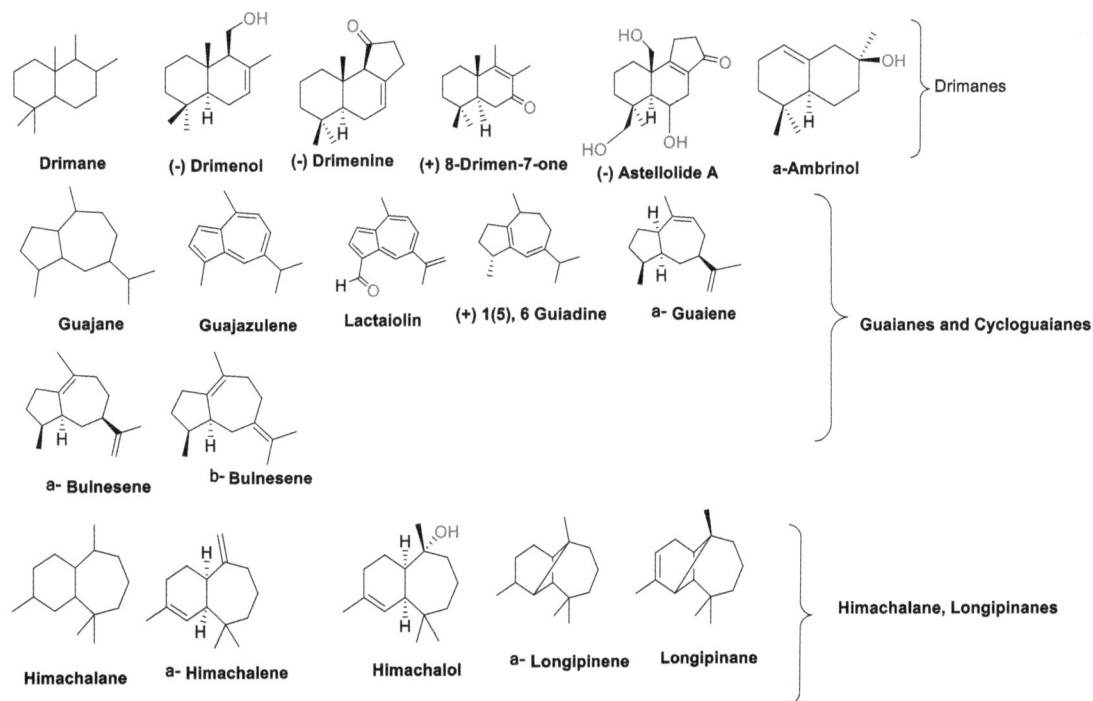

FIGURE 12.14 Drimanes, Guaianes, Cycloguaianes and Himachalanes, Longipinanes, Longifolanes.

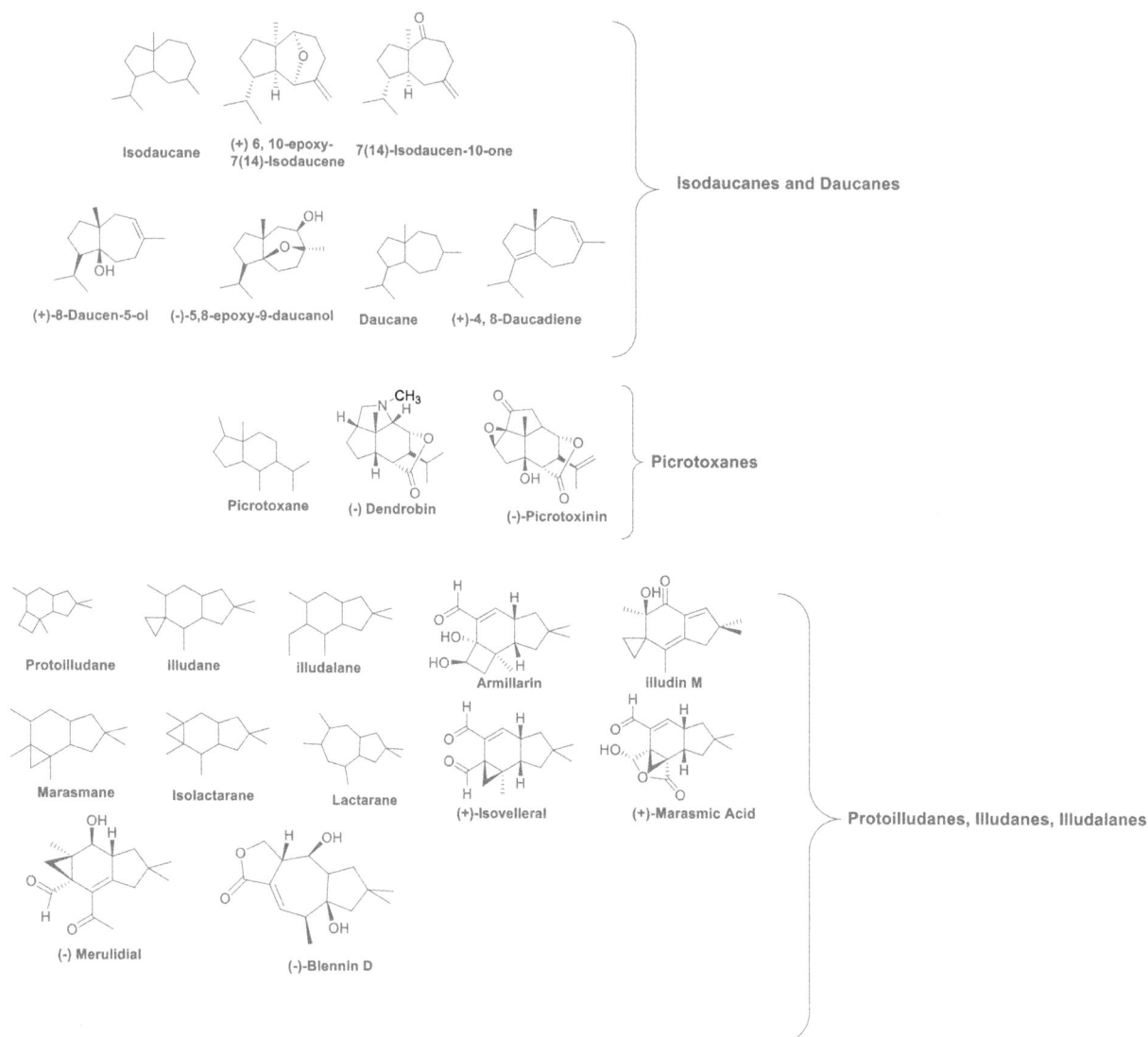

FIGURE 12.15   Isodaucanes, Daucanes, Picrotoxanes, Protoilludanes, Illudanes and Illudalanes.

ix. **Isodaucanes and Daucanes**

Isodaucanes such as (+)-6,10-epoxy-7(14)-isodaucene and 7(14)-isodaucen-10-one are constituents of the oil of sage from *Salvia sclarea* (Labiatae). Examples are (+)-4,8- daucadiene, (+)-8-daucen-5-ol and ()-5,8-epoxy-9-daucanol (Figure 12.15).

x. **Protoilludanes, Illudanes, Illudalanes**

seco-Illudane, also referred to as illudalane; protoilludane formally converts into spirocyclic illudane (Figure 12.15).

xi. **Marasmanes, Isolactaranes, Lactaranes, Sterpuranes**

Bond formation from C_1 to C_{11}, C_2 to C_9, C_3 to C_6 and disconnection of the C_4-C_5 bond in farnesane formally leads to marasmane. Isolactarane arises from the latter by cleavage of the C_3–C_4 and connection of the C_5–C_7 bond which, on its part, formally expands to

lactarane by migration of C3 from C6 to C-4 involving disconnection of the C5–C7 bond [3,5].

xii. **Acoranes**

The name of this class of sesquiterpenes stems from the *Acorus* species. (-)-4-Acoren-3-one, for example, has been isolated from *Acorus calamus* (Calamus, Araceae) and from the carrot *Daucus carota* (Umbelliferae). (+)-3,7(11)-Acoradiene is a constituent of juniper *Juniperus rigida*; its enantiomer occurs in *Chamaecyparis nootkatensis* (Cupressaceae) (Figure 12.16).

xiii. **Chamigranes**

More than 50 naturally occurring representatives with halogens as substituents are predominantly isolated from algae 2. Examples include (-)-10-bromo-1,3-, 7(14)-chamigratrien-9-ol (obtusadiene) and

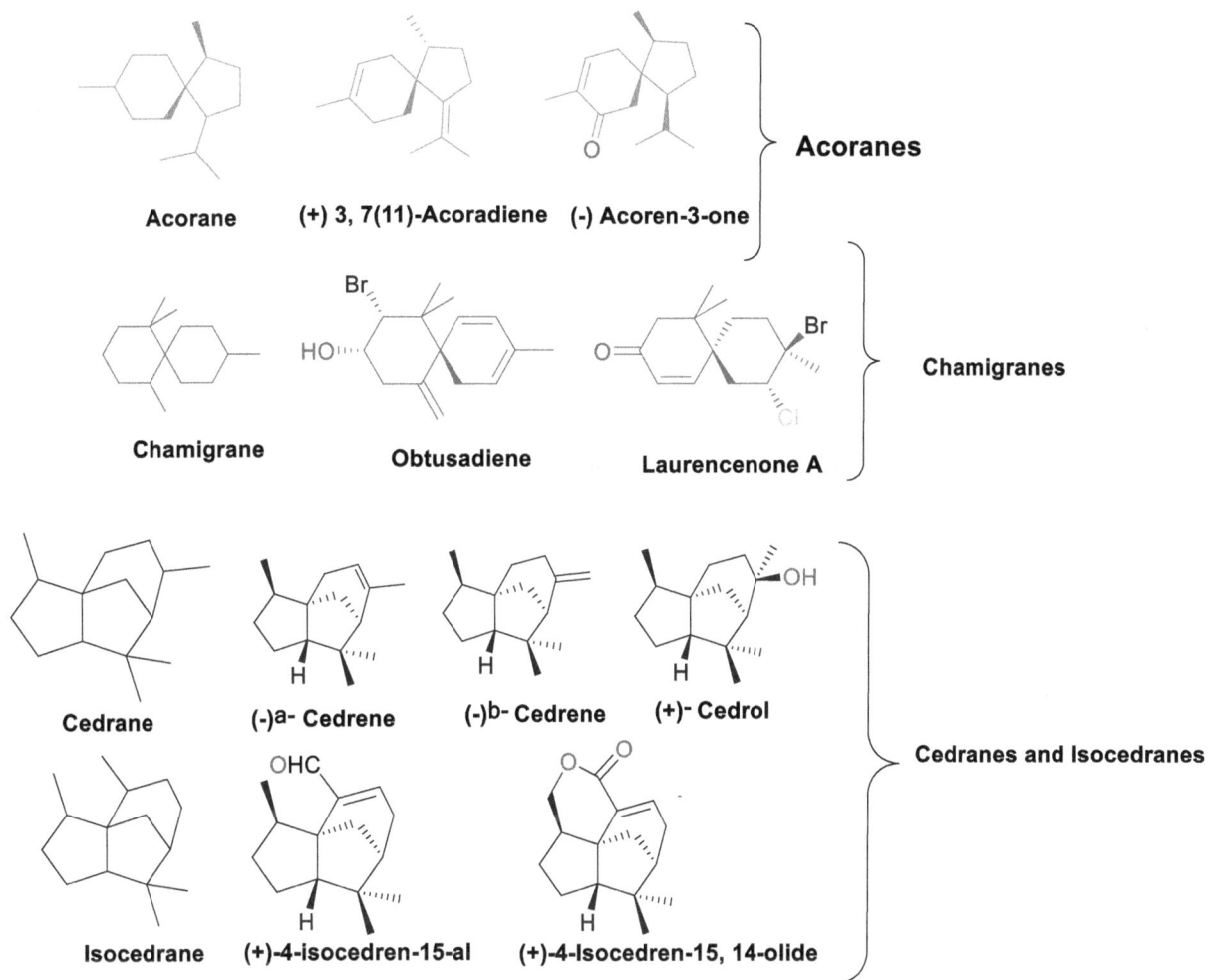

FIGURE 12.16 Acroanes, Chamigranes, Cedranes and Isocedranes.

3-bromo-2D-chloro-7-chamigren-9-one (laurencenone A) from the red alga *Laurencia obtusa*. (-)-Chamigra-3,7(14)-diene found in *Chamaecyparis taiwanensis* (Cupressaceae) is one of the rare chamigrenes of plant origin (Figure 12.16).

xiv. **Cedranes and Isocedranes**

Cedranes are formally derived from farnesane by bond formation between C-1 and C-6, C-2 and C-11 as well as C-6 and C-10. Cedrane is formally converted into isocedrane by migration of the C-15 methyl group from C-3 to C-5 (Figure 12.16).

xv. **Zizaanes and Prezizaanes**

The oil of vetiver from *Vetiveria zizanoides* (Poaceae), prezizaanes and zizaanes (+)-prezizaene, (-)-7-prezizaanol, (+)-6(13)-zizaene, (+)-6(13)-zizaen-12-ol, also known as khusimol (Figure 12.17).

xvi. **Campherenanes and Santalanes**

(-)-Campherenol and (-)-campherenone (*Cinnamomum camphora*). An additional bond from C-2 to C-4 of campherenane

formally leads to tricyclic Dsantalane, the basic skeleton of some constituents found in sandalwood oil 18 (oil of santal). Examples include (+)-*(Z)*-D-santalol with a slight woody odor similar to cedar wood oil and (+)-*(E)*-D-santalal, with a strong woody odor (Figure 12.17).

xvii. **Thujopsanes**

(-)-3-Thujopsene and (+)-15-nor-4-thujopsen-3-one (mayurone) from hiba oil. ()-3-Thujopsene does not shape the odor, but is, in addition to ()-D-cedrene and (+)-cedrol, one of the chief constituents (up to 25%) of the Texan oil of cedar wood from *Juniperus virginiana* (Cupressaceae) (Figure 12.17) [3,6].

xviii. **Hirsutanes**

Hirsutene is a hydrocarbon arising biogenetically from cyclization of humulene and isolated from cultures of the fungus *Coriolus consors* (Basidomycetae). Epoxidized derivatives such as hirsutic acid produced by the fungus *Stereum hirsutum* (Basidomycetae)

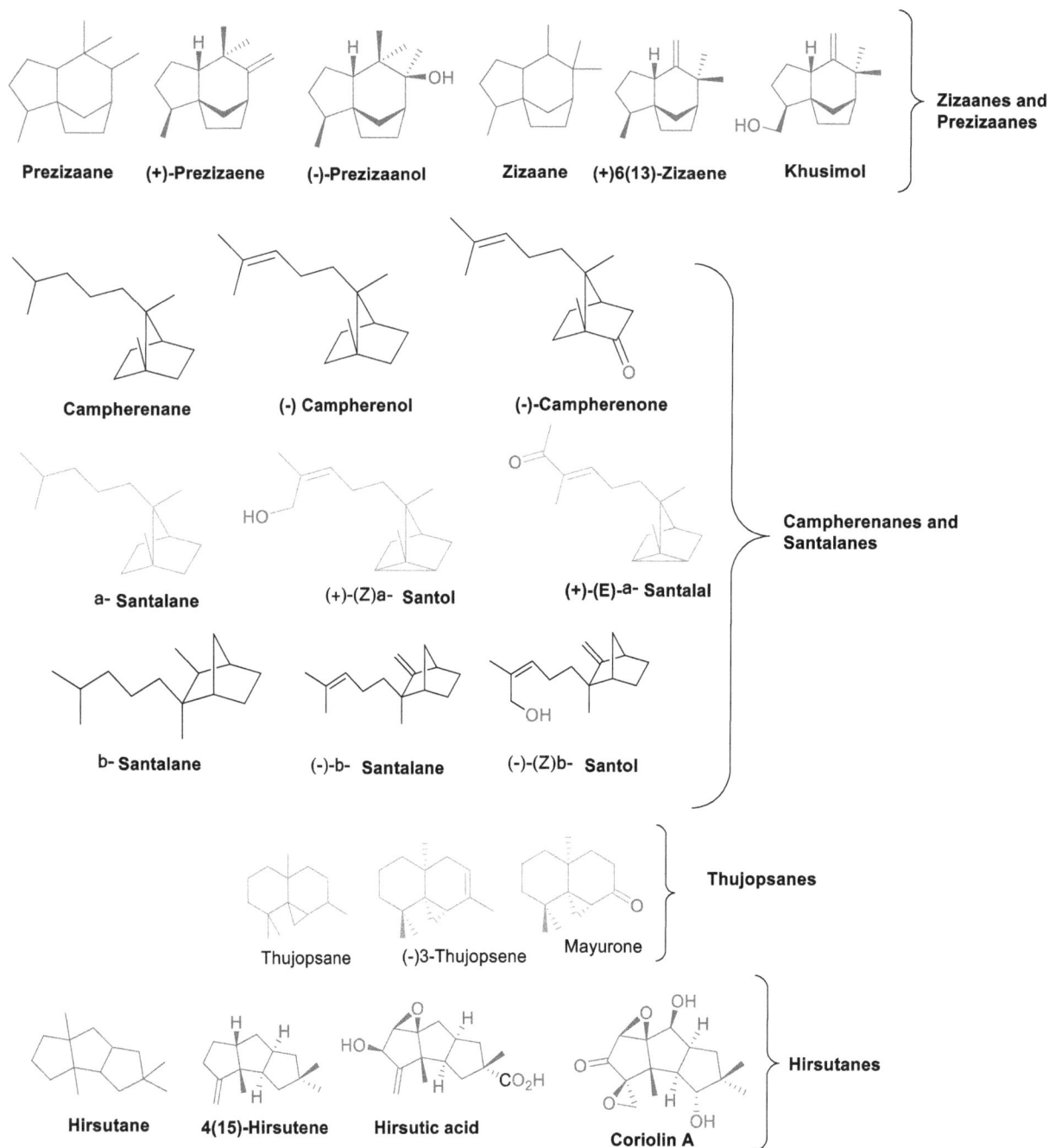

FIGURE 12.17   Zizaanes, Prezizaanes, Campherenanes, Santalanes, Thujopsanes and Hirsutanes.

growing on the dead wood of broad-leaved trees, and coriolin A including two acylated derivatives (B, C) isolated from *Coriolus consors*, display antibiotic and antineoplastic activity (Figure 12.17).

C. **Diterpenes**

Diterpenoids are a diverse class of chemical components with a molecular formula of $C_{20}H_{32}$ and four isoprene units that can be found in a variety of natural sources. The biological activity of this class of chemicals included anti-inflammatory, antibacterial,

anticancer, and antifungal properties. Grayanotoxin, forskolin, eleganolone, marrubenol, and 14-deoxyandrographolide are some of the diterpenes that have cardiovascular action. Diterpenes of the kaurane and pimarane types are biologically active metabolites isolated from the roots and leaves of various plants. These are categorised as:

a. **Phytanes**

Approx. 5,000 naturally occurring acyclic and cyclic diterpenes derived from phytane. The (3R, 7R, 11R)-enantiomer of phytane has been found in meteorites, oil slate, other sediments and, last but

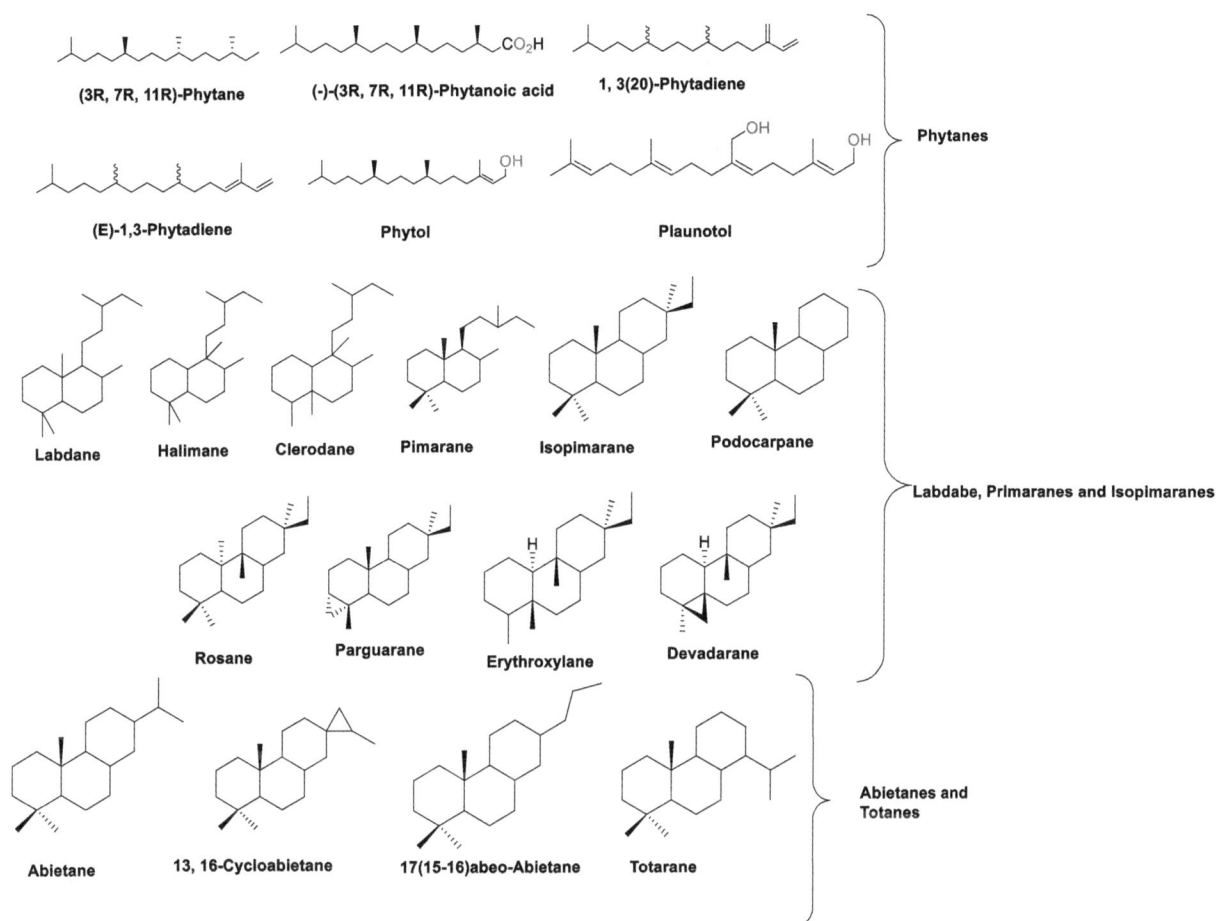

FIGURE 12.18   Phytanes, Labdanes, Pimaranes, Isopimaranes, Abietanes and Totaranes.

not least, in the human liver. 1,3(20)-Phytadiene is obtained from *Nicotiana tabacum* (Solanaceae); (E)-1,3-phytadiene and its (Z)-isomer are found in zooplankton (Figure 12.18).

b. **Bicyclophytanes**

 i. **Labdanes**

  More than 500 labdanes predominantly isolated from higher plants are known. They represent 8,11-10,15-cyclophytanes that contain the decalin bicycle as a core structure, which also defines the usually accepted ring numbering. The name labdane stems from *Cistus labdaniferus*: (Cistaceae) has a pleasant smell like ambergris and contains not only D-pinene but also labdan-8D, 15-diol and 8Ehydroxylabdan-15-oic acid (Figure 12.18) [3,7].

c. **Tricyclophytanes**

 i. **Pimaranes and Isopimaranes**

  Pimaranes and isopimaranes 2 are 13,17-cyclolabdanes with the perhydrophenanthrene basic skeleton. Rearranged and cyclized pimaranes include the rosanes (shift of the methyl group C-20 from C-10 to C-9), the parguaranes (3,18-cyclopimaranes), the erythroxylanes (shift of the methyl group C-19 of rosane from C-4 to C-5), and the devadaranes (4,19-cycloerythroxylanes). Podocarpanes are formally derived from pimaranes by omitting the carbon atoms 15–17 (15,16,17-trinorpimaranes)

 ii. **Abietanes and Totaranes**

  In plants, however, they emerge from the cyclization of geranylgeranyl diphosphate. Related parent diterpene hydrocarbons include 13,16- cycloabietanes, 17(15–16)-*abeo*-abietanes in which the methyl group C-17 has shifted from C-15 to C-16, and totaranes. The latter formally arise from abietane when the isopropyl group migrates from C-13 to C-14 (Figure 12.18).

 iii. **Cassanes, Cleistanthanes, Isocopalanes**

  Cassane is the basic skeleton of cassaic acid, the parent diterpenoid of alkaloids from the bark of *Erythrophleum guinese* and other *Erythrophleum* species (Fabaceae). *Erythrophleum* alkaloids such as cassaidine, cassain and cassamine are 2-(N, N-dimethylamino)ethyl esters of cassaic acid derivatives (Figure 12.19).

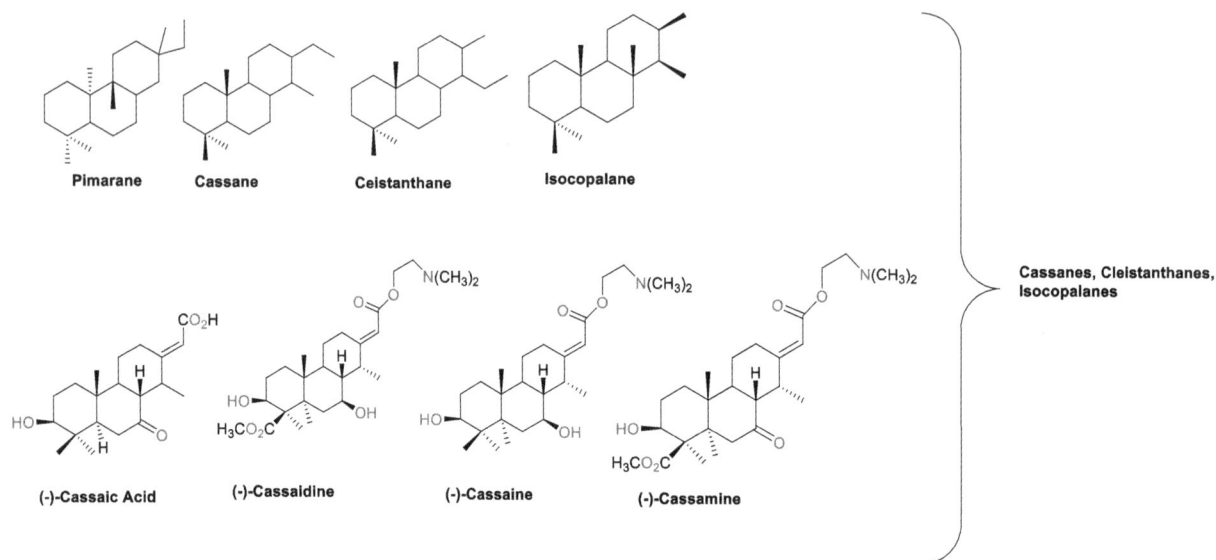

FIGURE 12.19 Cassanes, Cleistanthanes, Isocopalanes.

iv. **Tetra-cyclophytanes**

Tetracyclophytanes with individual ring numbering are, for the most part, derived from the skeleton of pimarane. Beyerane as an example is an 8,16-cyclopimarane, from which kaurane formally arises by migration of the methyl group C_{17} from position C_{13} to C_{16} (Figure 12.20).

i. **Beyeranes**

The name beyerane stems (*Beyeria leschenaultia*) possess (+)-17-*O*-cinnamoyl-15-beyerene-3,17,19-triol, also referred to as the cinnamic acid ester of beyerol. (+)-15-Beyerene as the parent hydrocarbon is one of the constituents of *Erythroxylon monogynum* (Araliaceae); its () enantiomer is found in the ethereal oils of some conifers such as *Thujopsis dolabrata* and *Cupressus macrocarpa* (Cupressaceae). (+)-15-Beyeren-3-one occurs in the Tambooti wood from *Spirostachys africana*, and (+)-7-hydroxy-15-beyeren-19-oic acid was isolated from *Stevia aristata* (Asteraceae) (Figure 12.20).

ii. **Kauranes and Villanovanes**

More than 200 kauranes are reported to exist in plants. The parent hydrocarbon ()-kaurane occurs in *Aristolochia triangularis*, and its 16D-stereoisomer in various sediments. (+)-16,17 Dihydroxy-9(11)-kauren-19-oic acid is one of the constituents of roasted coffee. ()-1,7,14-Trihydroxy-16-kauren-15-one, which has antibacterial and antineoplastic activities *in vitro*, was isolated with other kaurane derivatives from

various *Rabdosia* species. The fungus *Gibberella fujikuroi* produces not only the gibberellines but also (-)-16-kaurene, ()-17,18-dihydroxy-16-kauren-19,6Eolide and other kauranolides. Some rearranged furanokauranes such as ()-cafestol and ()-kahweol are the antiinflammatory constituents of coffee also present in the green coffee oil obtained from *Coffea arabica* (Rubiaceae) (Figure 12.20).

iii. **Atisanes**

Atisane is the basic skeleton of various diterpene alkaloids (aconitum-alkaloids) found in the plant families of Ranunculaceae and Garryaceae. (-)-Atisine, as a typical representative, was isolated from the Atis plant *Aconitum heterophyllum* (Ranunculaceae), which also contains ()-15,20-dihydroxy-16-atisen-19-oic acid as the lactone (19,20-olide). *Erythroxylon monogynum* and related species are reported to contain ()-16-atisene. Euphorbiaceae such as *Euphorbia acaulis* and *E. fidjiana* produce ()-16D, 17-dihydroxyatisan-3-one. (+)-13-Atisene-16E, 17-diol is known as serradiol due to its natural occurrence in *Sideritis serrata* (Labiatae) (Figure 12.21) [3,8].

iv. **Gibberellanes**

More than 60 gibberellanes isolated from higher plants and fungi play a vital role as plant growth hormones, and also regulate the degradation of chlorophyll as well as the formation

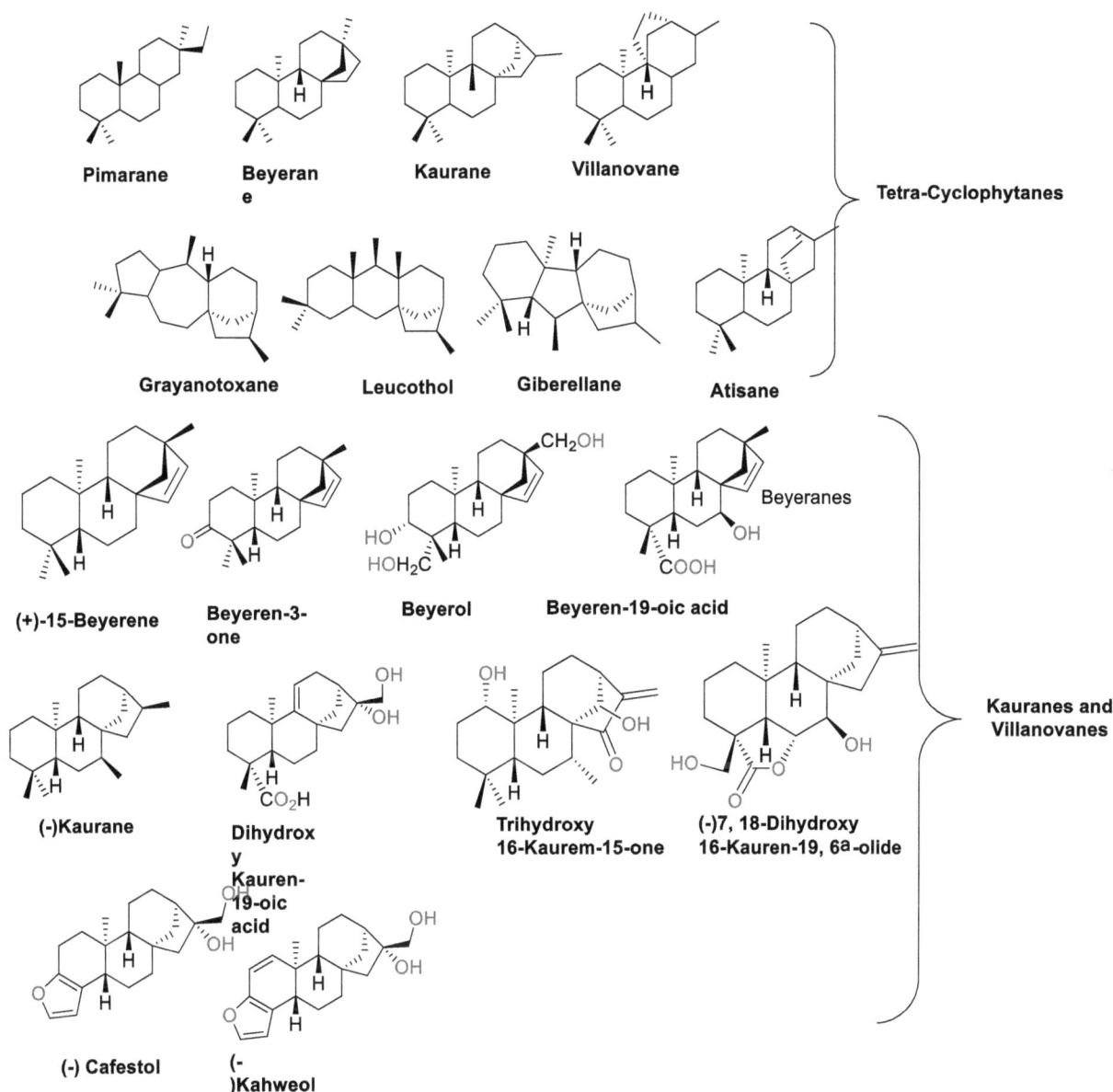

Pimarane **Beyerane** **Kaurane** **Villanovane** } **Tetra-Cyclophytanes**

Grayanotoxane **Leucothol** **Giberellane** **Atisane**

(+)-15-Beyerene **Beyeren-3-one** **Beyerol** **Beyeren-19-oic acid** Beyeranes

(-)Kaurane **Dihydroxy Kauren-19-oic acid** **Trihydroxy 16-Kaurem-15-one** **(-)7, 18-Dihydroxy 16-Kauren-19, 6ª-olide** } **Kauranes and Villanovanes**

(-) Cafestol **(-)Kahweol**

FIGURE 12.20 Tetra-cyclophytanes, Beyeranes, Kauranes and Villanovanes.

of fruits, and thus are used in agriculture. Large amounts of (+)-gibberellin A3, known as gibberellic acid, are isolated from the culture filtrate of the Japanese fungus *Gibberella fujikuroii* (Figure 12.21).

v. **Grayanotoxanes**

Leucothoe grayana (Ericaceae) defined the name of two other classes of tetracyclic diterpenes, the rare leucothols and the more abundant grayanatoxins. Examples include ()-leucothol C, 10(20), 15-grayanatoxadiene-3E, 5E, 6E-triol and 10(20)-grayanotoxene-2E, 5E, 6E, 14E, 16D-pentol as neurotoxic constituents of the leaves of *Leucothoe*

grayana (Ericaceae), 2,3-epoxygrayanotoxane-5E, 6E, 10D, 14E, 16E-pentol from *Rhododendron japonicum* (Ericaceae) (Figure 12.22).

vi. **Cembranes and Cyclocembranes**

Various bi- and tricyclic diterpenes are derived from the monocyclic membrane. Casbane, for example, is simply 2,15-cyclocembrane; another bond between C-6 and C-10 leads to lathyrane, from which the jatrophanes arise by opening the C_1-C_2 bond (Figure 12.23).

vii. **Cembranes**

More than 100 cembrane derivatives have been reported to occur in plants. Among those, Cembrene A is widely

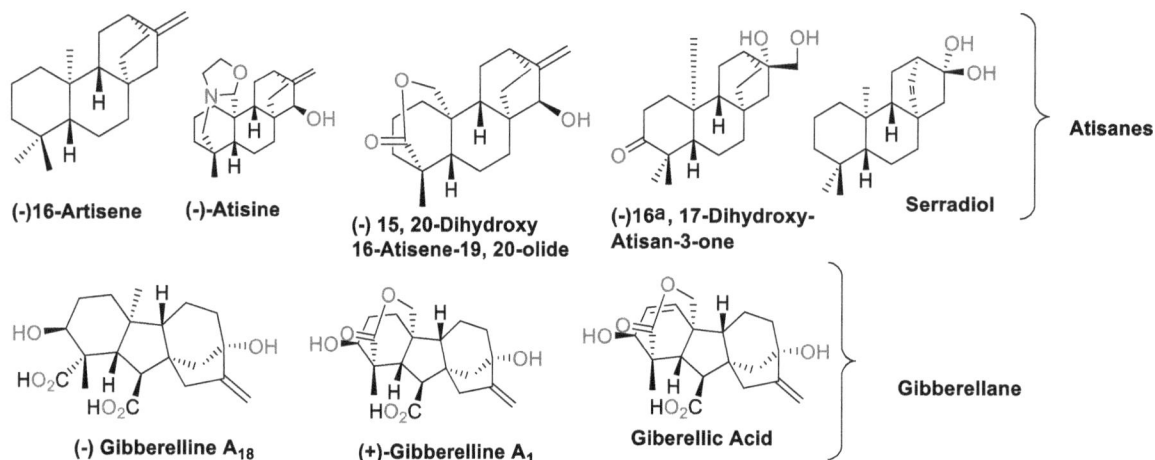

FIGURE 12.21   Atisanes and Gibberellanes.

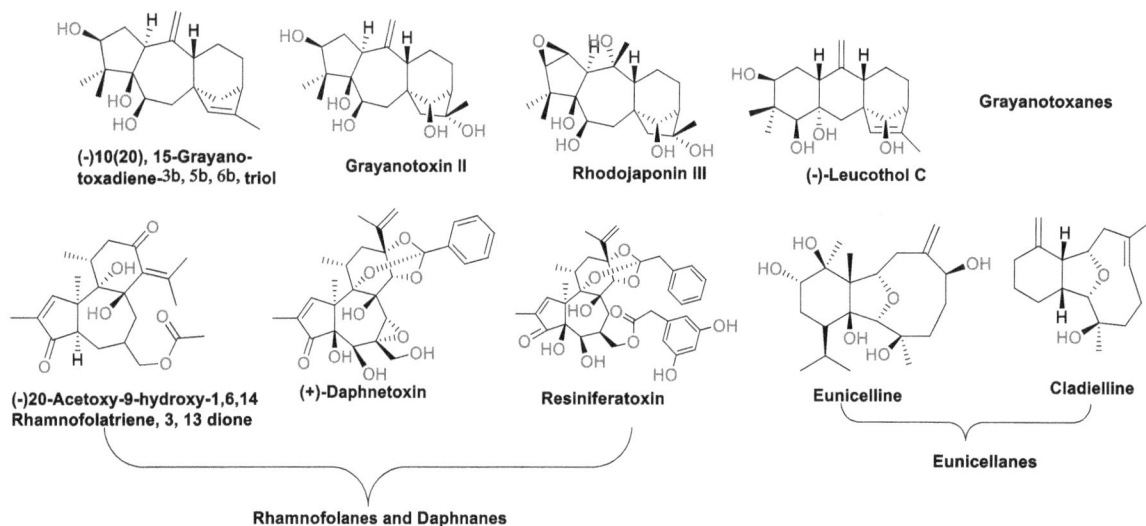

FIGURE 12.22   Grayanotoxanes.

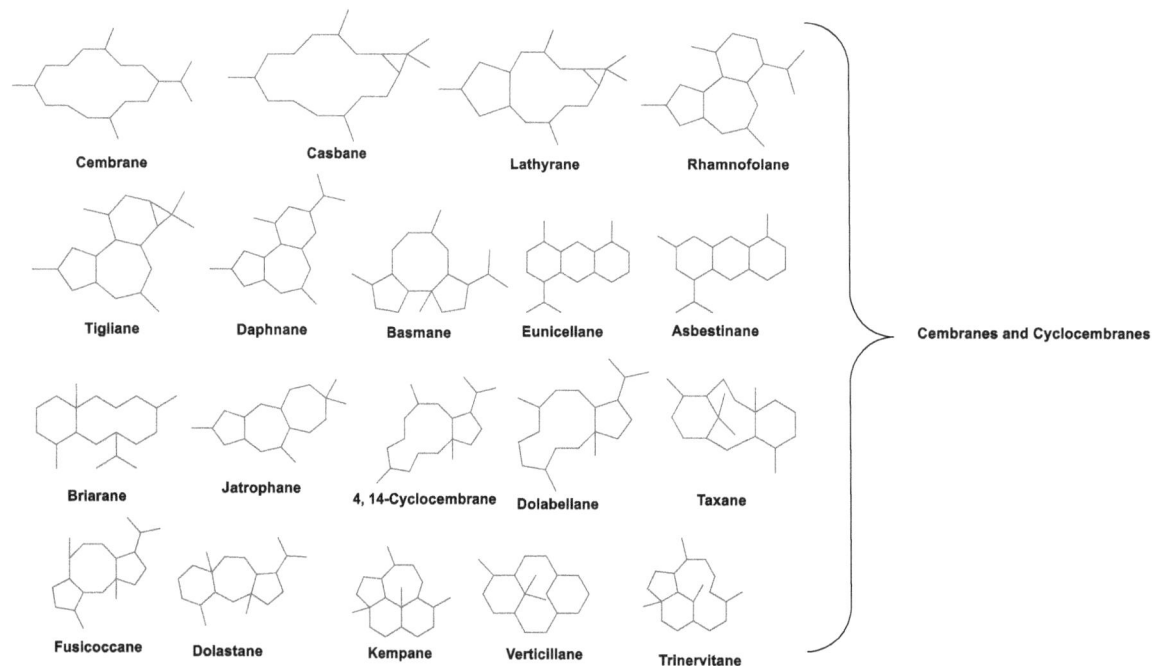

FIGURE 12.23   Cembranes and Cyclocembranes.

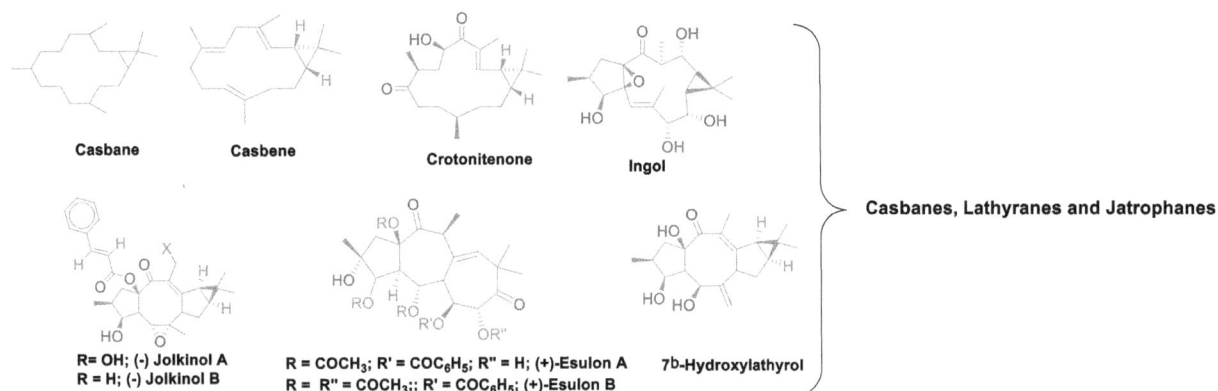

FIGURE 12.24 Cembranes.

spread among higher plants. Moreover, it serves as a pheromone of various termites and belongs, together with Serratol to the odorless constituents of incense (olibanum) from *Boswellia serrata* (Burseraceae). Some species of tobacco also contain cembranes; E-cembrenediol, more precisely (+)-2,7,11-cembratriene-4E, 6D-diol, from *Nicotiana tabacum* (Solanaceae) is an example.

viii. **Casbanes**

Casbanes are also rare in higher plants. Casbene, for example, occurs as an antifungal in the seed and sprout of *Ricinus communis* (Euphorbiaceae). (+)-Crotonitenone from *Croton nitens* (Euphorbiaceae) is another representative (Figure 12.24).

ix. **Lathyranes**

Lathyranes are, for the most part, isolated from Euphorbiaceae. Representatives include 7E-hydroxylathyrol from *Euphorbia lathyris*, the cinnamic acid esters denoted as jolkinols from *E. jolkini*, and ingol, a skin-irritating and antineoplastic hydrolyzate obtained from *E. ingens* and *E. kamerunica* (Figure 12.24).

x. **Jatrophanes**

The name jatrophane stems from *Jatropha gossypiifolia* (Euphorbiaceae) and possesses antineoplastic and anti-leukemic (+)-jatrophone. Various differently substituted jatrophanes are isolated from Euphorbiaceae, such as the esulones from *Euphorbia esula* and the euphornines from *E. helioscopia* and *E. maddeni* (Figure 12.24) [3,9].

xi. **Tiglianes**

Polyhydroxylated tiglianes esterified with linoleic and palmitic acid are

among the irritant and cocarcinogenic (tumor-promoting) constituents of various Euphorbiaceae. They occur in the purgative and counterirritant croton oil expressed from the seeds of *Croton tiglium* (Euphorbiaceae). Phorbol and isophorbol are obtained upon hydrolysis of these esters. Hydrolysis of prostratin, isolated from *Pimela prostrata*, yields 12-deoxyphorbol. Fatty acid esters of the latter occur in various Euphorbiaceae.

xii. **Rhamnofolanes and Daphnanes**

Rhamnofolanes such as (-)-20-acetoxy-9-hydroxy-1,6,14-rhamnofolatriene-3,13-dione from *Croton rhamnifolius* (Euphorbiaceae) and other constituents from various *Jatropha* species rarely occur in plants.

xiii. **Eunicellanes and Asbestinanes**

Eunicellanes and (infrequently) asbestinanes are diterpenes of marine origin, including eunicelline from the alga *Eunicella stricta* and cladielline found in some *Cladiella* species. Asbestinanes such as ()-asbestinine 2 are characteristic of the gorgonia *Briareum asbestinum* (Araceae).

xiv. **Briaranes**

Briaranes are found in marine organisms including corals, represented by the briantheines W and X from *Briareum polyanthes*. Brianthein X is reported to be an insecticide, while the closely related (-)-solenolide A isolated from *Solenopodium* species exhibits antiviral and anti-inflammatory activities. These briarane lactones are usually referred to as erythrolides (Figure 12.25) [3,10].

xv. **Dolabellanes**

Dolabellanes occur in various corals and algae. Cytotoxic and ichthyotoxic

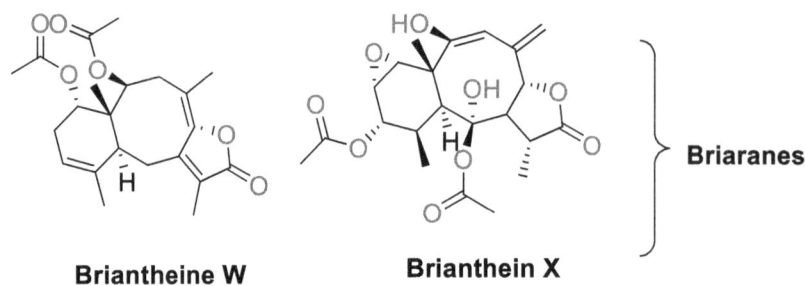

FIGURE 12.25 Briaranes.

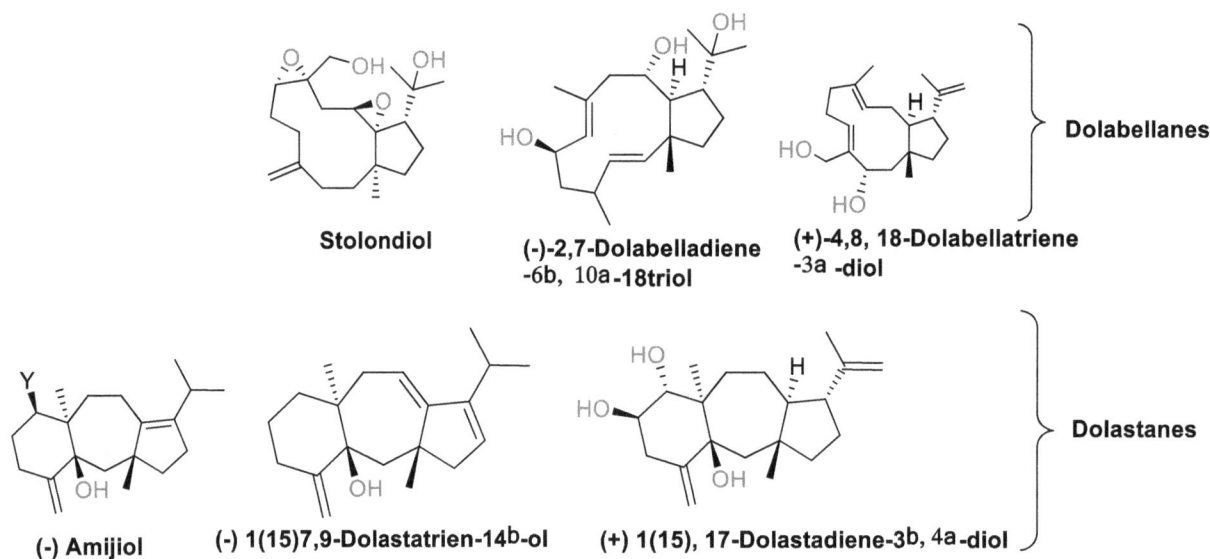

FIGURE 12.26 Dolabellanes and Dolastanes.

(poisonous for fish) due to (-)-stolondiol from *Clavularia* corals, (-)-2,6-dolabelladiene-6E, 10D, 18-triol from the herbivorous (plant-eating) sea rabbit *Dolabella california* and 4,8,18-dolabellatriene-3D, 16-diol and other dolabellanes isolated from various *Dictyota* brown algae are representative examples (Figure 12.26).

xvi. **Dolastanes**

Metabolites of various algae and corals also include dolastadienes and -trienes, exemplified by Amijiol, its isomer isoamijiol, and 1(15), 7,9-dolastatrien-14E-ol from brown algae *Dictyota linearis* and *D. cervicornis*, as well as 1(15), 17-dolastadiene-3D, 4E-diol from the soft coral *Clavularia inflata* (Figure 12.26).

xvii. **Fusicoccanes**

Fusicoccanes are metabolites of some fungi, liverworts and algae. Some of these and their glycosides act as growth regulators. (+)-Epoxydictymene from the brown alga *Dictyota dichotoma*, ()-fusicoplagin A from the liverwort *Plagiochila acanthophylla* and (+)-fusicoccin H from the fungus *Fusicoccum amygdali* are selected representatives (Figure 12.27).

xviii. **Verticillanes and Taxanes**

Verticillanes such as (+)-verticillol from the wood of *Sciadopitys verticillate* (Taxodiaceae) rarely occur as natural products. 8,19-Cycloverticillane, however, is the core skeleton of (+)-taxine A, the crystalline chief constituent of the poisonous alkaloid mixture taxine isolated from the needles but not the red berries of the European yew tree *Taxus baccata* (Taxaceae), causing gastrointestinal irritation as well as cardiac and respiratory failure. Various taxanes are also found in the alkaloid mixtures obtained from Taxaceae, exemplified by (-)-10-deacetylbaccatin from the needles and the bark of European *Taxus baccata* and ()-taxol 19 from the bark of Pacific *Taxus brevifolia* and *T. cuspidate* (Figure 12.27) [3,11].

xix. **Trinervitanes and Kempanes**

Trinervitanes including their *seco-*, methyl- and 11,15-cyclo-derivatives (also referred to as kempanes) belong to the defense pheromones 14–17 of various termite species, represented by 1(15), 8-trinervitadien-14E-ol and 1(15), 8-trinervitadiene-13D, 14E-diol from *Trinervitermes gratiosus* as well as 6,8-kempadiene-2D, 13D-diol (known as kempene 1) from *Nasutitermes kempae*.

xx. **Prenylsesquiterpenes**

About 200 diterpenes are reported in which an isoprenyl ("prenyl") residue extends one of the side chains of a sesquiterpene, consequently referred to as prenylsesquiterpenes (Figure 12.28).

xxi. **Xenicanes and Xeniaphyllanes**

Xeniaphyllanes (prenylcaryophyllane) incorporating the cylobutane ring of the parent caryophyllane and represented by isoxeniaphyllenol in some *Xenia* corals such as *X. macrospiculata*, are rarely found in other organisms. Xenicanes with opened cyclobutane ring more frequently occur

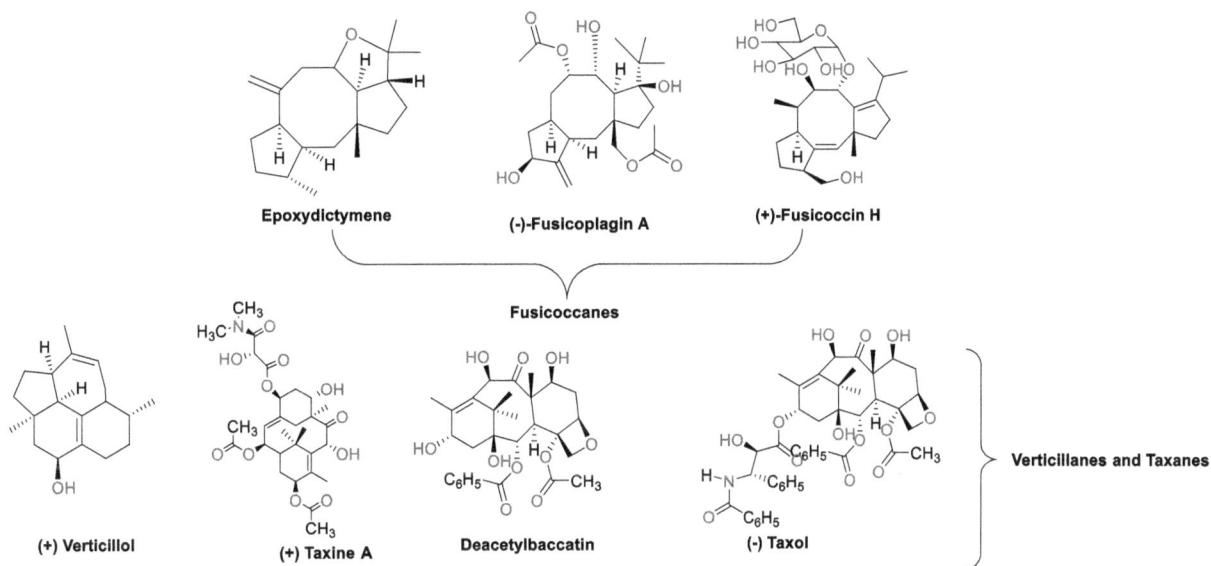

FIGURE 12.27 Fusicoccanes, Verticillanes and Taxanes.

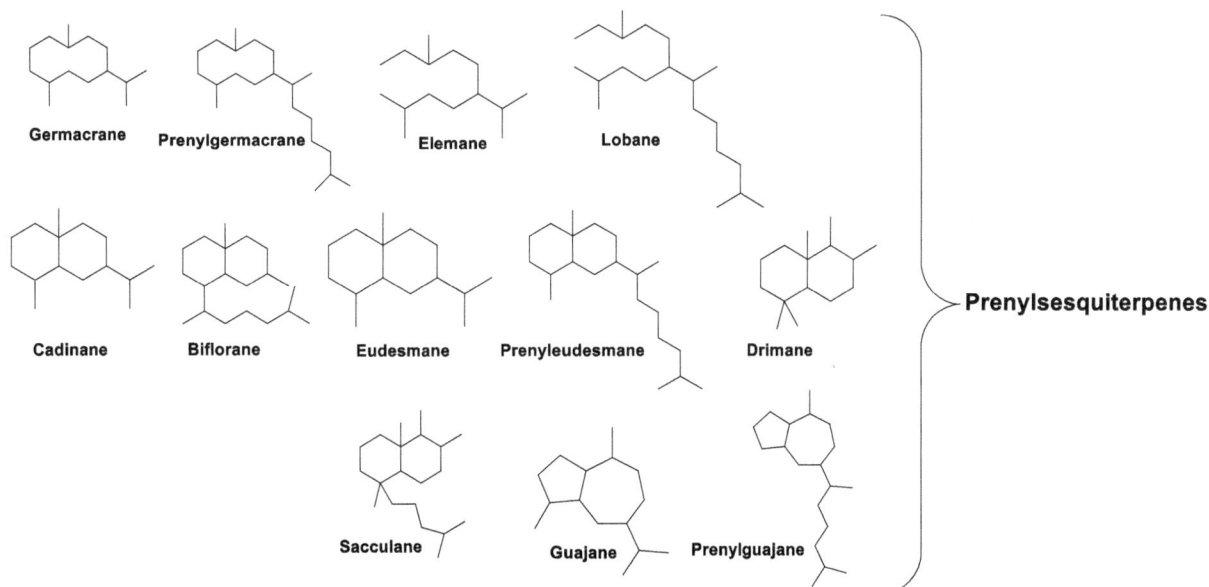

FIGURE 12.28 Prenylsesquiterpenes.

in algae and corals, for example, iso-dictyohemiacetal in the alga *Dictyota dichotoma*, xeniolit A and xeniaacetal in the corals *Xenia macrospiculata* and *Xenia crassa* (Figure 12.29).

Pachydictyon coriaceum. 17,18-Epoxy-8,10,13(15)-lobatriene and (+)-14,17-diacetoxy-8,10,13(15)-lobatrien-18-ol representing the prenylelemanes are isolated from several *Lobophytum* species (Figure 12.29).

xxii. **Prenylgermacranes and Lobanes**

Prenylgermacranes and prenyl-elemanes, usually known as lobanes and possibly arising from COPE rear-rangements of appropriate prenylger-macranes, are also metabolites of some marine organisms. Various derivatives of the prenylgermacrane dilophol, for example, occur in the algae *Dilophus ligulatus*, *Dictyota dichotoma* and

xxiii. **Prenyleudesmanes and Bifloranes**

The alga *Dictyota acutiloba* pro-duces dictyolene, one of the rarely abundant prenyleudesmanes. Prenyl-cadinanes are more frequently found not only in marine organisms but also in higher plants and insects. They are usually referred to as bifloranes and, when containing a benzenoid

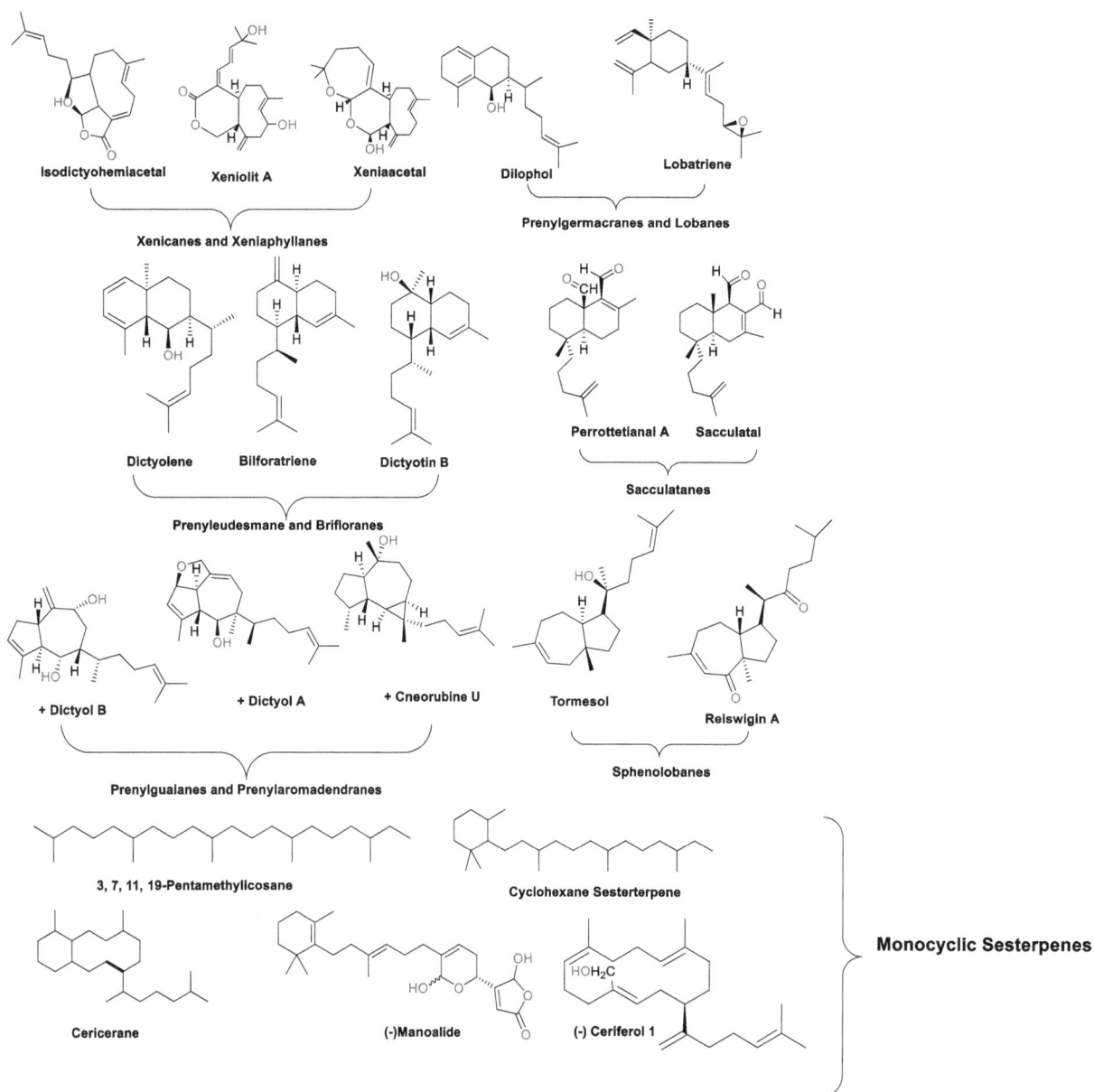

FIGURE 12.29 Xenicanes and Xeniaphyllanes, Prenylgermacranes and Lobanes, Prenyleudesmanes and Bifloranes, Sacculatanes, Prenylguaianes, Prenylaromadendranes, Sphenolobanes and Monocyclic Sesterterpenes.

ring, as serrulatanes. (-)-4,10(19), 15-Bifloratriene, for example, is secreted by the termite *Cubitermes umbratus* as a constituent of the defense pheromone cocktail (Figure 12.29) [3,12].

xxiv. **Sacculatanes (Prenyldrimanes)**

Sacculatanes (prenyldrimanes) occur in various liverwort species. *Porella perrottetiana*, for instance, protects itself against insects by producing the strongly bitter compounds 3E-hydroxy-7,14-sacculatadien-12,11-olide and perrottetianal A; the latter also inhibits the sprouting of rice; 18-hydroxy-7,16-sacculatadien-11,12-dial (sacculatal) and tumor-promoting 7,17 sacculatadiene-11,12-dial were found to be among the active substances (Figure 12.29).

xxv. **Prenylguaianes and Prenylaromadendranes**

Prenylguaianes occur predominantly in algae *Dictyota* and *Aplysia*, and less frequently in corals *Xenia*. Following their dominant origin, prenylguaianes are referred to as dictyols. (+)-Dictyol A and (+)-dictyol B from the brown alga *Dictyota dichotoma* are typical representatives. Prenylaromadendranes characterize the Cneoraceae and are consequently denoted as cneorubines (Figure 12.29).

xxvi. **Sphenolobanes (Prenyldaucanes)**

Prenyldaucanes predominantly found in liverworts and marine organisms are commonly referred to as sphenolobanes. (+)-3D, 4D-Epoxy-13(15), 16-sphenolobadiene-5D, 18-diol which inhibits sprouting and growth of rice, and its 5D-acetoxy derivative, both from the liverwort *Anastrophyllum minutum*, as well as (+)-3,17-sphenolobadien-13-ol (tormesol) from *Halimium viscosum*, are representatives.

xxvii. **Ginkgolides**

The leaves of the maiden's hair tree *Ginkgo biloba*, a survivor of the Ginkgoaceae genus wide-spread during the Mesozoikum era of the Earth's history and valued as a park tree, contain flavonoids, the tetracyclic sesquiterpene-lactone bilobalide A and various hexacyclic high-melting and bitter-tasting diterpene lactones known as the ginkgolides A-M, which are resistant towards acids 2. Bilobalide and ginkgolides carry a *t*-butyl group which rarely occurs in natural compounds.

D. **Sesterterpenes**

a. **Acyclic Sesterterpenes**

The acyclic representatives are derived from 3,7,11,15,19-pentamethylicosane C25H52. Their biogenesis primarily yields geranylfarnesyl-diphosphate, involving the known pathway. Sesterterpenes rarely occur in higher plants. 3,7,11,15,19-Pentamethylicosa-2,6-dien-1-ol as an example is found in the leaves of potatoes *Solanum tuberosum* (Solanaceae).

b. **Monocyclic Sesterterpenes**

The lactone (-)-manoalide isolated from the sponge *Luffariella variabilis* living in the Manoa valley of Oahu Island (Hawai) as one of the few examples is reported to have analgesic, anti-inflammatory and immunosuppressive properties (Figure 12.29).

c. **Polycyclic Sesterterpenes**

i. **Bicyclic Sesterterpenes**

The parent hydrocarbon skeletons of bicyclic sesterterpenes are derived predominantly from some sesquiterpenes. Diprenyldrimanes, for example, (+)- dysideapalaunic acid and related lactones, are partly reported to be natural inhibitors of protein phosphatase. Prenyldrimanes in higher plants are represented by ()- salvisyriacolide from *Salvia syriaca* and as (+)-salvileucolide methylester from *Salvia hypoleuca* (Labiatae) (Figure 12.30) [3,13].

ii. **Tricyclic Sesterterpenes**

Tricyclic sesterterpenes known to date incorporate tricyclic diterpenes as core structures which are prenylated at the side chains such as cheilanthanes and ophiobolanes, representing prenylisocopalanes and prenylfusicoccanes, respectively. (+)-Cheilanthatriol isolated from the fern *Cheilanthes farinosa* and the ophiobolins found in the phytopathogenic fungi *Ophiobolus miyabeanus*, *O. heterostrophus* and *Aspergillus ustus* are examples. Some of the ophiobolines exhibit antibacterial and phytotoxic activities (Figure 12.30).

iii. **Tetra- and Pentacyclic Sesterterpenes**

Various scalaranes, exemplified by (+)-scalarin in *Cacospongia scalaris*, (+)-desoxyscalarin in *Spongia officinalis*, and hyrtial in *Hyrtios erecta* with antiinflammatory activity (Figure 12.30).

iv. **Triterpenes**

Linear Triterpenes

Most of these are derived from squalane and squalene, with two farnesane units linked in a tail-to-tail manner (Figure 12.30).

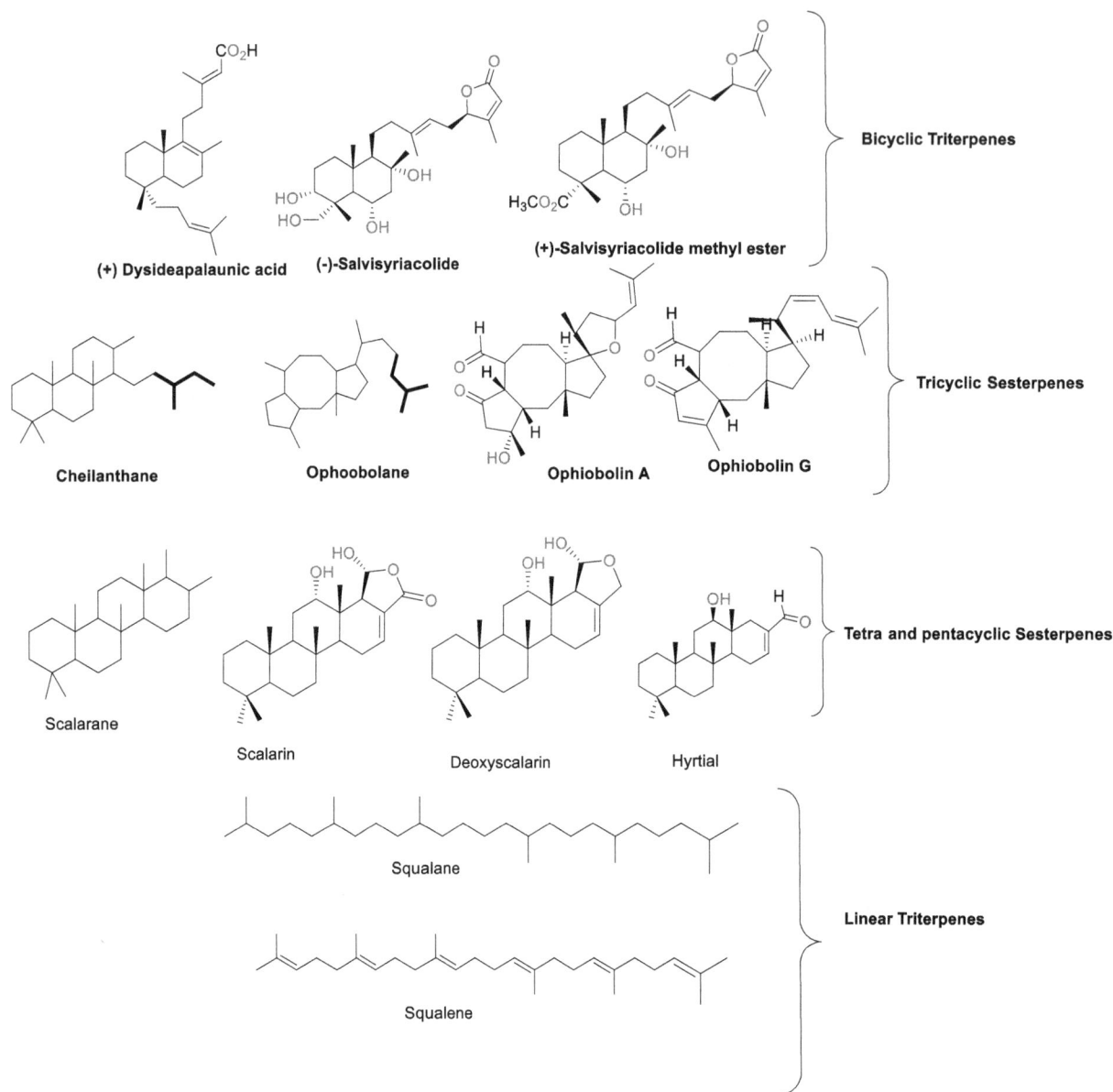

FIGURE 12.30 Bicyclic, Tricyclic Tetra and Pentacyclic Sesterterpenes & Linear Triterpenes.

v. **Tetracyclic Triterpenes, Gonane Type**

The cyclization of 2,3-epoxysqualene leads to the stereoisomers protostane and dammarane with the gonane tetracycle known from steroids. 29-Norprotostanes are referred to as fusidanes. Dammarane converts into apotirucallane involving migration of the methyl group C_{18} from C_{14} to C_{13} (Figure 12.31).

vi. **Protostanes and Fusidanes**

Protostane and their 29-nor-derivatives denoted as fusidanes are fungal metabolites. *Cephalosporium caerulens*, for instance, produces (+)-protosta-17(20)-*(Z)*-24- diene-3E-ol. Fusidanes such as helvolic acid from the *Helvola* mutant of the mold *Aspergillus fumigatus* and related structures are widely

used as antibacterials. Fusidic acid isolated from the fermentation broth of *Fusidium coccineum* and related tribes is an example (Figure 12.32) [3,14].

vii. **Dammaranes**

Dammaranes exemplified by (+)-dammara-20,24-diene-3E-20R-diol   and   its (+)-20S-diastereomer belong to the constituents of the yellowish-white gum damar, the resinous exudate of the southeast Asian damar tree *Shorea wiesneri* (Dipterocarpaceae) used in plasters, varnishes, and lacquers. Examples include (-)-dammar-24-ene-3E, 12E, 20S-triol and dammar-24-ene-3E, 6D, 12E, 20R-tetrol, better known as protopanaxatriol.

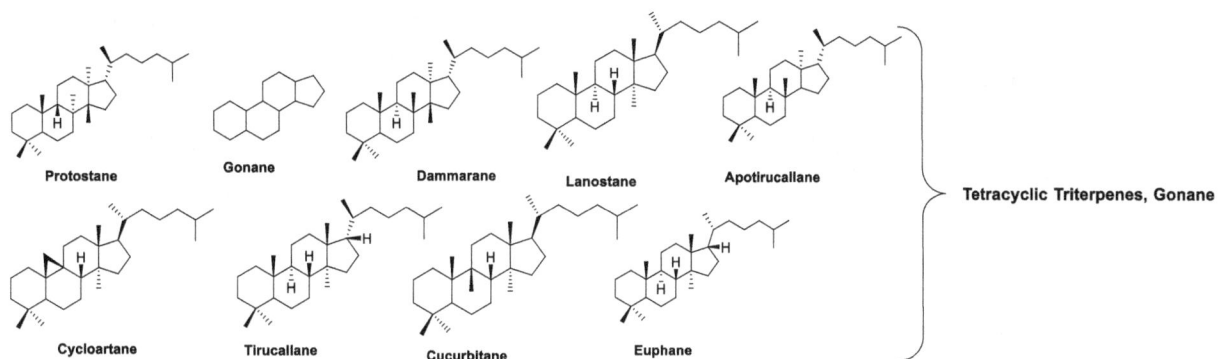

FIGURE 12.31 Tetracyclic Triterpenes.

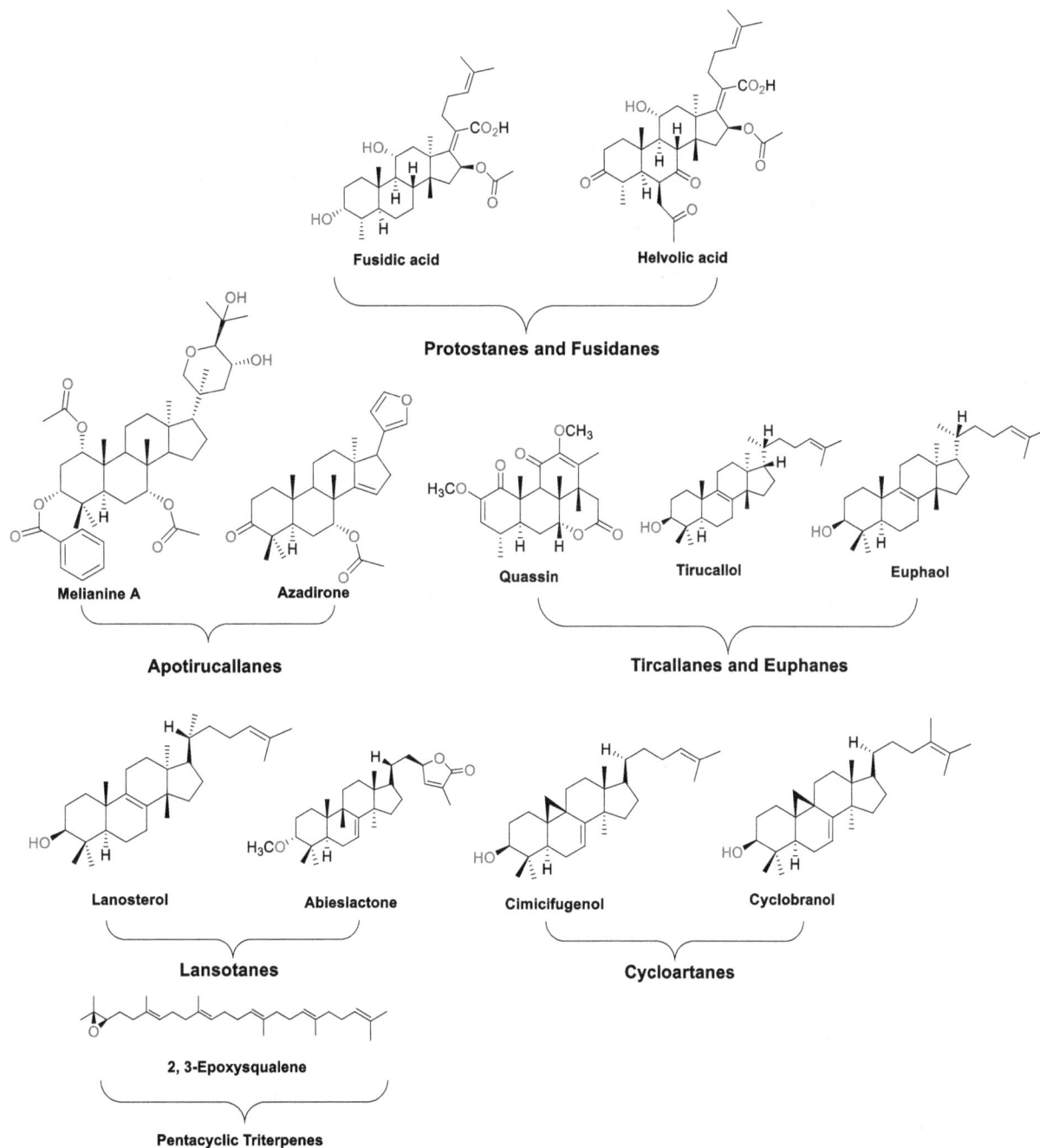

FIGURE 12.32 Protostanes, Fusidanes, Dammaranes, Apotirucallanes, Tirucallanes, Euphanes Lanostanes and Pentacyclic Triterpenes.

viii. **Apotirucallanes**

Apotirucallanes such as melianine A occur in Indian lilac *Azadirachta indica* (Meliaceae); infusions are reported to have insecticidal and slightly anaesthetic actions. Some tetranortriterpenes (C26) and other degradation products of triterpenes referred to as quassinoids (C20) originate from the apotirucallane tetracycle (Figure 12.32).

ix. **Tirucallanes and Euphanes**

Tirucallanes and the 20-epimeric euphanes are frequently found in Euphorbiaceae, and are exemplied by tirucallol and its (20*R*)-epimer euphol from *Euphorbia tirucalli* and related species (Figure 12.32).

x. **Lanostanes**

More than 200 naturally abundant lanostanes are reported 2. They are found as constituents of some higher plants and as fungal metabolites. (+)-Lanosta-8,24-dien-3E-ol, commonly known as lanosterol, (-)-Abieslactone from the bark of fir *Abies mariesii* (Pinaceae) is another natural lanostane derivative (Figure 12.32).

xi. **Cycloartanes**

(+)-Cycloartenol, also known as cyclobranol, from the fruits of *Strychnos nux vomica* (Loganiaceae), *Solanum tuberosum* (Solanaceae) and the seed of rice *Oryza sativa* (Poaceae), is one typical representative of more than 120 naturally abundant cycloartanes (Figure 12.32).

xii. **Cucurbitanes**

The name of about 50 naturally abundant cucurbitanes stems from Cucurbitaceae, the Latin term for cucurbitaceous plants such as cucumbers and pumpkins, known since antiquity for their beneficial and toxic properties. One of the most frequently isolated representatives is the bitter substance (+)-cucurbitacin B from *Phormium tenax* and *Ecballium elaterium* (Cucurbitaceae), also found in *Iberis* species (Cruciferae) [3,15].

a. **Pentacyclic Triterpenes, Baccharane Type**

Cyclization of 2,3-epoxysqualene into six-membered ring *D*, followed by *Wagner-Meerwein* rearrangements, builds up the intermediate cation of the tetracyclic triterpene 3E-hydroxybacchar-21-ene, which finally closes the fifth ring to the pentacyclic 3E-hydroxylupanium ion. Examples include 2,3-epoxysqualene, 3β-hydroxylupane cation, 3β-hydroxybacchar-21-ene etc.

1. **Baccharanes and Lupanes**

The first lupane triterpene was isolated from the skin of lupin seeds *Lupinus luteus* (Leguminosae) and is therefore referred

to as lupeol. Lupanes are also detected in cocoons of the silk worm *Bombyx mori*. Examples of other lupane derivatives are (+)-1,11-dihydroxy-20(29)-lupen-3-one from *Salvia deserta*, (+)-20(29)-lupene-3E, 11D-diol (+)-12,20(29)-lupadiene-3E, 27,28-triol. Although, Baccharane triterpenes (+)-12,21-baccharadiene (*Lemmaphyllum microphyllum* var. *obovatum)* are rarely found.

2. **Oleananes**

Approx. 300 oleananes are reported in plant kingdom. The basic moiety (+)-oleanane is isolated from petroleum, and also (+)-3E, 11D, 13E-triol derivatives from *Pistazia vera* (Anacardiaceae). Examples includes ()-priverogenin B (cowslip *Primula veris* (Primulaceae), (+)-soyasapogenol (*Glycine* species) (Leguminosae), *Quillaja* saponin (*Quillaja saponaria* (Rosaceae)).

3. **Taraxeranes, Multifloranes, Baueranes**

(+)-14-Taraxeren-3E-ol, one of 30 naturally abundant taraxane derivatives known as taraxerol obtained from lion's tooth *Taraxacum officinale* (Asteraceae). Examples include (-)-7-multifloren-3E-ol (multiflorenol), (-)-7-baueren-3E-ol as synonym Ilexol.

4. **Glutinanes, Friedelanes, Pachysananes**

(+)-5-Glutinen-3E-ol, also named (+)-alnusenol obtained from *Alnus glutinosa* (Betulaceae). Approx. five naturally abundant pachysananes are reported prominent one is (+)-16,21-pachysanadiene-3E, 28-diol obtained from *Pachysandra terminalis* (Buxaceae) [3,16].

5. **Taraxastanes and Ursanes**

Approx. 25 taraxastanes have been isolated, predominantly from Asteraceae. Examples include (+)-20-Taraxasten-3E-ol (taraxasterol), (+)-20(30)-taraxasten-3E, 16E-diol (arnidenediol) from lion's tooth *Taraxacum officinale* and *Arnica montana,* 20-taraxasten-3E, 16E-diol (faradiol) from *Arnica montana, Tussilago farfara, Senecio alpinus* and gold-bloom *Calendula officinalis* (all Asteraceae) etc.

Approx.150 ursanes from plant kingdom are reported namely (+)-3E-Hydroxyursan-28-oic acid, obtained from *Nerium oleander* (Apocynaceae), (+)-3E-Hydroxy-12-ursen-28-oic acid isolated from berries of bearberry *Arctostaphylos uva-ursi* also known as ursolic acid. It is also found in *Rhododendron* species, in cranberries *Vaccinum macrocarpon* (Ericaceae), (+)-3E, 19D-dihydroxy-12-ursen-28-oic acid (known as pomolic acid) extracted from apple wax coat.

6. **Pentacyclic Triterpenes, Hopane Type**

The basic hopane skeleton is formed through 2,7-, 6,11-, 10,15-, 14,19-, and 18,22-cyclization of the carbenium ion in fivefold chair conformation (as drawn) arising from regioselective protonation of the 2,3-double bond of squalene (but not of the 2,3-epoxide). Therefore, hopanes are not usually hydroxylated at 3-position. Methyl shifts at hopane nucleus leads to neohopane (C-28 methyl from C-18 to C-17), fernane (C-27-methyl from C-14 to C-13; C-26 methyl from C-8 to C-14), adianane (C-25 methyl from C-10 to C-9) and finally, filicane (C-24 methyl from C-4 to C-5) (Figure 12.33).

7. **Hopanes and Neohopanes**

Stereoisomers of hopane as the parent nucleus of approx. A total of hundred naturally occurring hopane triterpenes can be isolated from oil slate as well as petroleum. Hydroxylated hopanes such as (+)-6D, 22-hopanediol (some lichens) linked with (+)-22(29)-hopen-6D, 21E-diol occurs in roots of *Iris missouriensis* (Liliaceae). Neohopanes exemplified by (+)-12-neohopen-3E-ol from *Rhododendron linearifolium* (Ericaceae) rarely occur as

natural products. Other examples include Bacteriohopane-32,33,34,35-tetrol moderately substituting cholesterol from culture possess an unbranched polyhydroxylated C5-C6 alkyl chain attached to C-30 of the hopane skeleton (Figure 12.33) [3,17].

8 Fernanes

Various fernanes are reported in some ferns, e.g. (+)-7- and (-)-8- fernene as well as (-)-7,9(11)-fernadiene from *Adiantum monochlamys* and *A. pedatum* (Pteridaceae). Hydroxylated derivatives retigeric acid A in *Lobaria retigera* (Sticataceae) and in higher plants, e.g. (-)-7- fernen-3E-ol in *Rhododendron linearifolium* (Ericaceae) and (-)-3E, 11E-dihydroxy-8-fernen-7-one in *Euphorbia supina* (Euphorbiaceae) (Figure 12.33).

9 Adiananes and Filicanes

The Adiananes term is derived from genus *Adiantum* (Pteridaceae) from which (-)-5-adianene has been isolated. Examples includes (+)-5-adianen-3E-ol (Simiarenol) obtained from *Rhododendron simiarum* (Ericaceae) leaves, Filicanes are also isolated from ferns and 3-filicen-23-al (filicenal) from *Adiantum pedatum* (Figure 12.33).

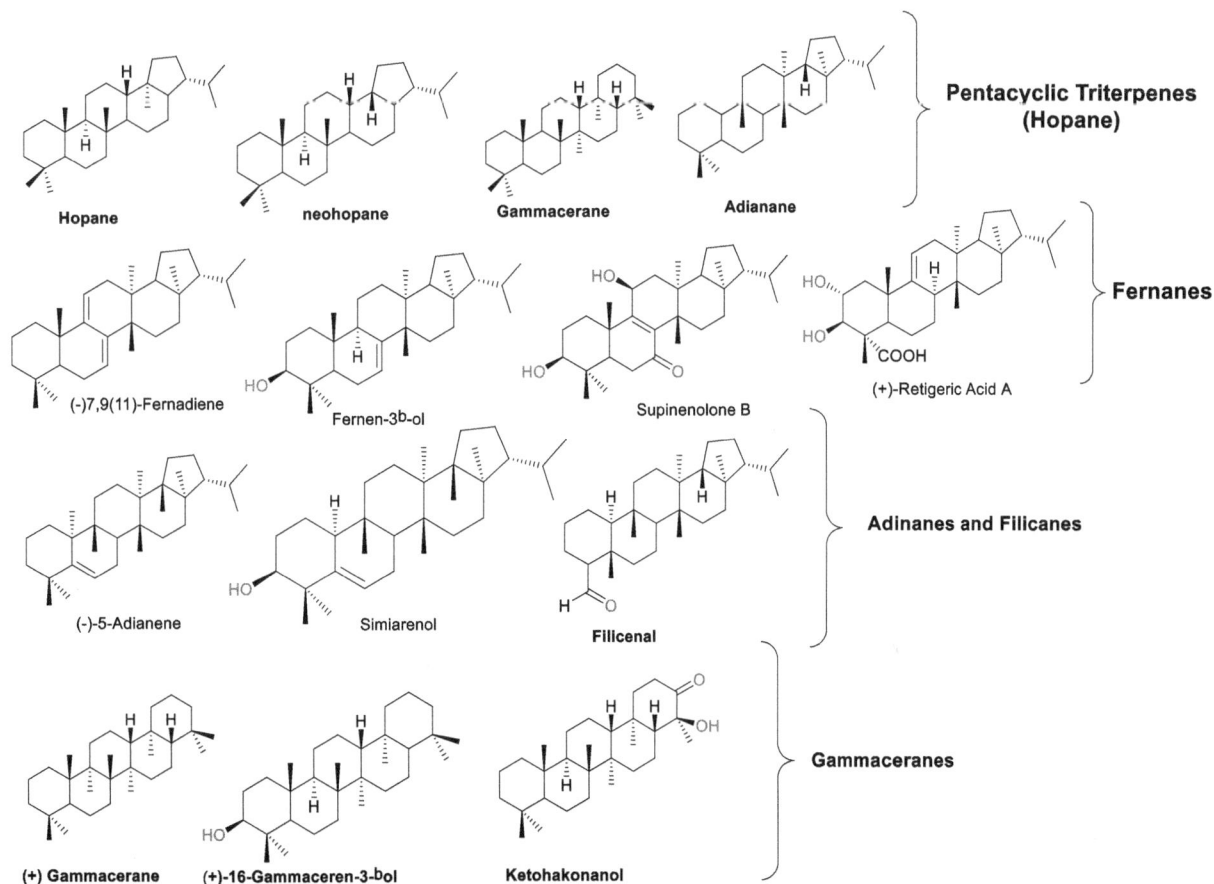

FIGURE 12.33 Pentacyclic Triterpenes, Hopanes, Neohopanes, Fernanes, Adiananes and Filicanes and Gammaceranes.

10 Gammaceranes

Minute amounts of dextrorotatory gammaceranes have been extracted from oil slate. 3-Hydroxy-derivatives i.e (+)-16-gammacerene-3E-ol are obtained from roots of *Picris hieracioides* (Asteraceae). 22E-Hydroxy-30-nor-gammaceran-21-one is found in the Japanese fern *Adiantum monochlamys* (Figure 12.33).

E. Tetraterpenes

These are naturally occurring terpenes that possess 8 isoprene moieties with structural formula $C_{40}H_{64}$. These can be categorized into following categories:

a. Carotenoids

This is the prominent group comprising approx. 200 natural tetraterpenes referred as carotenoids, all carotenoids represent structural alternates of E-carotene (*Daucus carota* (Umbelliferae)) and possess 11–12 conjugated CC double bonds. The most accepted name is Carotene. Examples include lycopene (acyclic red triterpene) from *Lycopersicon esculantum*, Solanaceae, Jcarotene red and light-sensitive obtained from (*Penicillium sclerotiorum*) also referred as carotene, the orange-yellow 7′, 8′-dihydro-derivative of corn (maize, *Zea mays*, Poaceae) (Figure 12.34).

b. Apocarotenoids

Terpenoids arising from carotenoids due to the separation of terminal substituents are called as apocarotenoids. Examples includes Ecarotenal (Orange peel), 8′-apo-E-caroten-8′-al from egg yolk (Figure 12.34) [3,18].

c. Di-apocarotenoids

8,8′-diapocarotenoids are orange-yellow crocetin from *Gardenia* species (Rubiaceae) and *Mimosa pudica* (Mimosaceae), the dimethyl-lester J-crocetin and the di-gentiobiose ester (+)-D-crocin in various *Crocus* species (Iridaceae). (+)-D-Crocin (yellow-red constituents of saffron) Rosafluin, systematically named 10,10′-diapocaroten-10,10′-diol, occurs in yellow rose flowers (Figure 12.34).

d. Megastigmanes

These are formed by condensation of 2-butyl-1,1,3-trimethylcyclohexane with E-carotene basic skeleton referred as Megastigmanes. Examples includes 2,6,6-trimethyl-2-cyclohexenone, 2,4,4-trimethyl-cyclohexene-3-carbaldehyde and 5,5,9-trimethyl-1-oxabicyclo[4.3.0]-3-nonen-2-one etc may also contribute to the fragrances of flowers. Typical megastigmanes include (R)-D- and E-ionone from *Viola odorata* (Violaceae), *Freesia refracta* and *Boronia megastigma* (Rutaceae), and (S)-Jionone from *Tamarindus indica* (Leguminosae). (-)-5,6- Epoxy-7-megastigmen-9-one is extracted

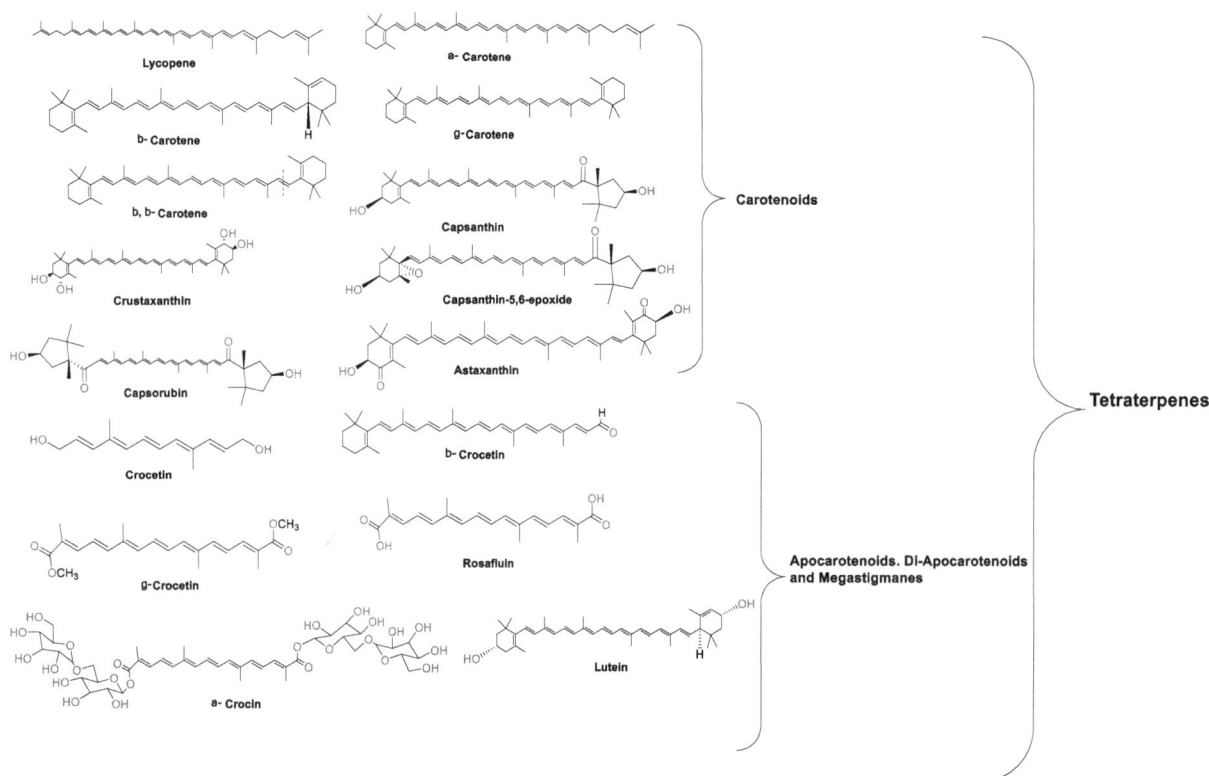

FIGURE 12.34 Carotenoids, Apocarotenoids, Di-apocarotenoids and Megastigmanes.

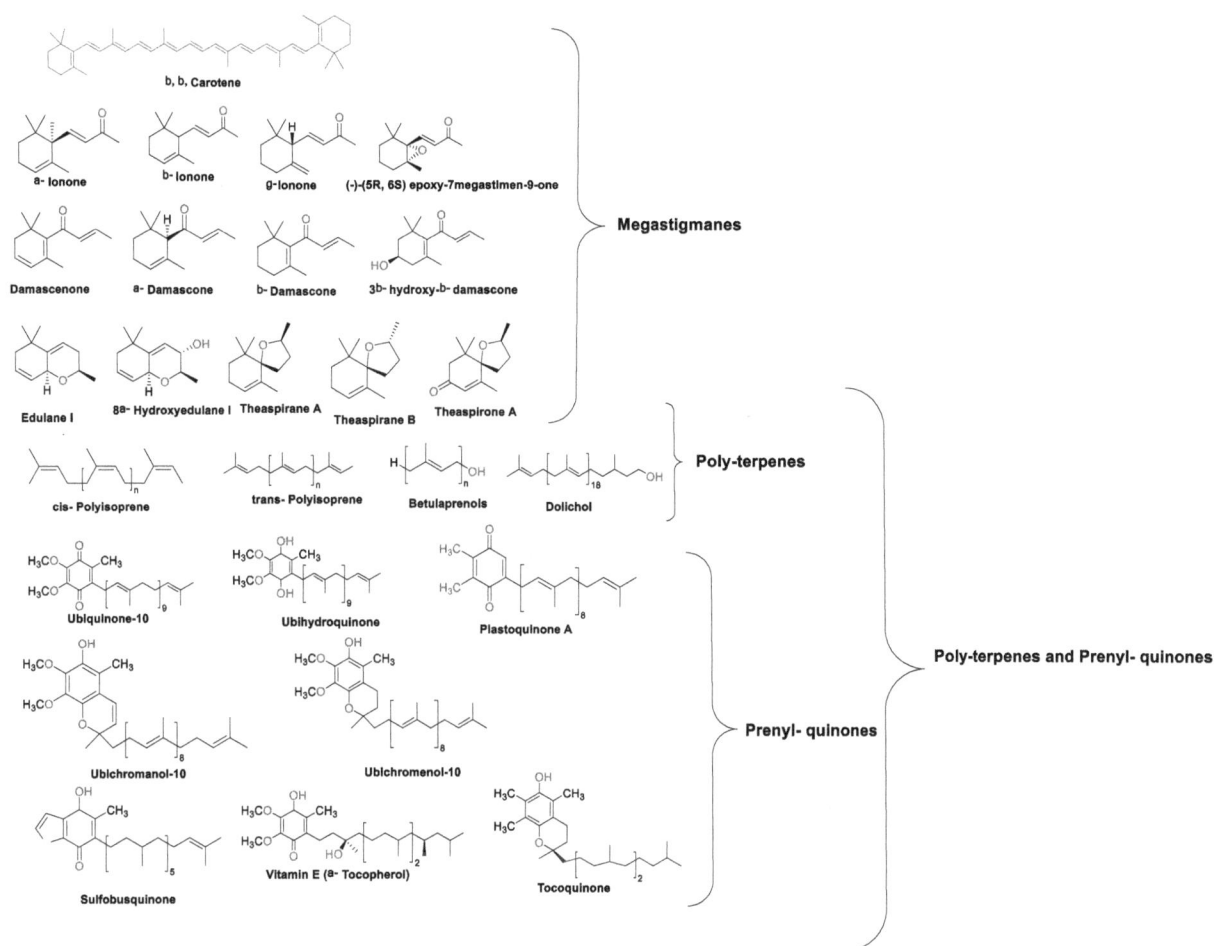

FIGURE 12.35 Megastigmanes, Poly-terpenes and Prenylquinones.

from the ethereal oils of carrots (Umbelliferae), tomatoes (Solanaceae), and tobacco (Solanaceae). Damascenone and damascones are fragrance-forming compounds derived from Bulgarian rose oil (*Rosa damascena*) (Rosaceae). Tobacco *Nicotiana tabacum* contains 3E-Hydroxydamascone (Solanaceae). The aroma of the passion flower, *Passiflora edulis*, is shaped by edulanes (Passifloraceae) (Figure 12.34).

F. Poly-Terpenes and Prenyl-Quinones

a. Poly-terpenes

Poly-terpene contains more than eight isoprene units i.e. natural rubber (caoutchouc) mainly obtained *Hevea brasiliensis* (Euphorbiaceae) latex from Brazil and south-eastern asia. *Trans*-polyisoprene is the main constituent of gutta-percha (*Palaquium gutta* and *P. oblongifolia*) (Sapotaceae). Various *trans*-oligoterpenols isolated from *Betula verrucosa* (Betulaceae) are known as betulaprenols, Betulaprenol-9 also found in tobacco (*Nicotiana tabacum* Solanaceae). Betulaprenol-11,12 are reported *Morus nigra* (Moraceae) leaves and silk-worms (*Bombyx mori*) faeces (Figure 12.35) [3,19].

b. Prenylquinones

It contains terpenyl groups containing up to ten isoprene units, capable of undergoing reductions to the corresponding hydroquinones and cyclizations to chromenols and chromanols. Various lipid-soluble prenyl-benzoquinones derived biogenetically from the amino acids phenylalanine and tyrosine occur in the cells of almost all aerobic organisms (bacteria, plants, animals), and are therefore referred to as ubiquinones. Vitamin E also known as (+)-D-tocopherol or 2-prenyl-3,4-dihydro-2*H*-1-benzopyran-6-ol, represents a prenyl-chromanol. (Figure 12.36). Vitamin-E also has anti-inflammatory and anti-rheumatic properties. It influences fertility (fertility vitamin); for example, vitamin E deficient food causes sterility in rats and prevents honey-bees from metamorphosing to the queen [19].

12.4 Terpenoids as Chemotherapeutics

Cancer is the biggest cause of death worldwide. The cancer burden is measured by the incidence and mortality rates. The International Agency for Research on Cancer (IARC) has

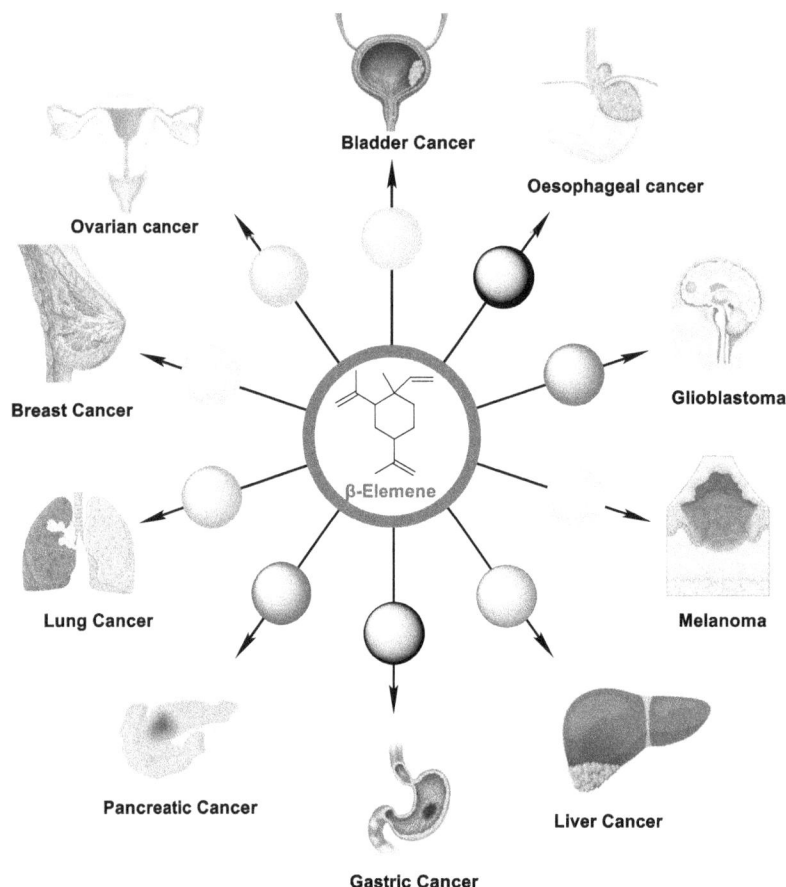

FIGURE 12.36   Activity of β-Elemene against different types of cancer.

issued an assessment on the global cancer burden based on ***Globocan*** 2020, a projection of cancer incidence and mortality (IARC). In 2020, there will be 19.3 million new cancer diagnoses globally, with around 10 million cancer-related deaths. The most common cancer in women was anticipated to be female breast cancer, with 2.3 million new diagnoses, second to lung, colorectal, prostate and stomach cancer (7.3%). Lung cancer was the biggest killer in terms of deaths, compensating 1.8 million casualties (18%), accompanied by colorectal (9.4%), liver (8.3%), stomach (7.7%), and female breast cancer (7.7%). Global cancer incidence is expected to hit 28.4 million in 2040, rising 47% from 2020 [3,20].

In 2018, worldwide cancer statistics anticipated 18.1 million additional cases and 9.6 million mortalities taking lung cancer as the leading cause of mortality in males, accompanied by stomach and liver cancer, and breast cancer seems to be the most prevalent ailment and the major cause of morbidity in females, pursued by lung and colorectal cancer. Esophageal, pancreatic, and prostate cancers had lower death rates than other different cancers [20].

Cancer is a disease characterised by genetic abnormalities that result in abnormal growth of cells locally or metastatically and this mutation may be induced by external or internal DNA damage. Damaged cells, under ordinary situations, would trigger a repair mechanism that included cell cycle block to avoid additional damage. P53 is a tumour suppressor gene that controls cell cycle and triggers mortality when necessary.

This method prevents mutations from being propagated along to succeeding cell generations. When this mechanism is disrupted, cancer cells rapidly mutate, leading to the creation of tumours. Several terpenoids have been shown to have apoptotic potential against cancer cells in these circumstances in many studies [3,21].

12.5 Classification Terpenoids

The classification of terpenoids is shown in Table 12.2.

12.6 Mechanism of Action of Selected Terpenoids against Cancer Progression

Table 12.3 shows the summery of various mechanisms of action of terpenoids against different types of Cancer.

12.7 Monoterpenoids

Thymol

Thymol suppresses tumor progression in acute promyelocytic leukaemia (HL-60) cells by depolarizing mitochondria and halting the cell cycle during the G0/G1 phase. Thymol anticancer potential is caused by the formation of ROS and the

TABLE 12.2

Classification of Terpenoids [3–24]

S. No.	Terpenoids	Examples
1.	Monoterpenoids	Thymol, Menthol, Auraptene, D-limonene, Perillic acid, Ascaridole, Carvacrol, Thymoquinone
2.	Sesquiterpenoids and Sesquiterpene lactones	Ambrosin, coronopilin, and dindol-01, Parthenolide, Costunolide, Dehydrocostuslactone, Helenalin, EM23, Artesunate & Artemisinin, β-Elemene
3.	Diterpenoids	Triptolide, Crocetin, Phytol
4.	Triterpenoids	Ursolic acid, Betulinic acid, Lupeol

TABLE 12.3

Summery of Various Mechanism of Action of Terpenoids against Different Types of Cancer [8–26]

Terpenoids	Cell cycle	Authophagy	Differentiation	Anti-angiogenesis	Anti-metastasis	Anti-MDR	Chemo prevention	MOA
D-limonene		√	√			√		Inhibition of HMG Co-A reductase and CoA synthesis
Cantharidine		√	√	√			√	Inhibition of serine
Artemisnin and derivatives		√					√	Cleavage of Iron or heme-mediated peroxide
Triptolide			√				√	Inhibition of XPB ATPase and transcription factor
Pseudolaric acid			√		√		√	Blockage of microtubules
Cucurbitacin							√	Interfere with F-actin and inhibition of STAT 3
Allisol			√	√	√		√	Inhibition of sarcoplasmic/ endoplasmic reticulum
Pachymic acid	√	√	√	√			√	Inhibition of DNS topoisomerases
Lycopene		√						Scavengers of ROS, Inhibition of MMP2
Thymol	√		√	√			√	caspase activation and apoptosis by boosting Bax protein expression while lowering Bcl-2 protein
Menthol	√	√				√	√	Reduction in topoisomerase-I and II, followed by DNA damage and cell death.
Auraptene	√	√		√		√	√	Inhibited the mTOR signalling pathway and its downstream actions, slowing cell proliferation
Perillic Acid	√	√	√	'	√		√	Promoted upregulation of p21 and downregulation of CDK1 and CDK4 in HCT116 colon cancer cells
Ascaridole	√			√	√		√	Exhibits cytotoxicity toward NER-deficient cells at much lower doses than repair-proficient cells

(Continued)

TABLE 12.3 (*Continued*)

Summery of Various Mechanism of Action of Terpenoids against Different Types of Cancer [8–26]

Terpenoids	Cell cycle	Authophagy	Differentiation	Anti-angiogenesis	Anti-metastasis	Anti-MDR	Chemo prevention	MOA
Carvacrol	√	√	√	√			√	boosted the formation of ROS and interacted with intracellular targets in the HL-60 human leukaemia cell line, HCT-8 human colon cancer cell line, SF-295 brain tumour cell line, and MDA-MB-435 human breast cancer cells
Artesunate and Artemisinin		√	√	√		√	√	down-regulation of proliferating cell nuclear antigen (PCNA) and Bcl-2 gene expression and up-regulation of Bax gene level
β-Elemene	√	√	√	√		√	√	Inhibited CDK1 by upregulating p27, reducing CDK1 phosphorylation on threonine, and decreasing cyclin B1 expression, leading in cyclin B1-CDK1 multiplex under expression and cell cycle arrest in the G2/M phase
Triptolide		√	√	√	√		√	Inhibits SENP1, c-Jun, and androgen receptors via targeting XPB and Rpb1 (triptolide targets), however further study is required.
Crocetin	√	√	√		√		√	Triggered mitochondrial death by altering mitochondrial membrane potential, resulting in Bax protein overexpression, caspase 3 activation, and Bcl-2 expression downregulation overexpression, caspase 3 activation, and Bcl-2 expression downregulation
Ursolic acid		√	√	√	√		v	Triggered cell death in MDA-MB-231 cells, mostly via the extrinsic apoptotic pathway, which included activation of the Fas receptor, caspase 8 and polymerase (PARP) breakage, up-regulation of Bax, down-regulation of Bcl-2, and cytochrome C release.
Betulinic acid	√		√	√	√	√	√	Inducing MMP to release cytochrome C from mitochondria into cytosol, leading to caspase cleavage, initiation of mitochondrial apoptosis pathway and nuclear degradation

(*Continued*)

TABLE 12.3 (*Continued*)

Summery of Various Mechanism of Action of Terpenoids against Different Types of Cancer [8–26]

Terpenoids	Cell cycle	Authophagy	Differentiation	Anti-angiogenesis	Anti-metastasis	Anti-MDR	Chemo prevention	MOA
Lupeol	√	√	√	√			√	Inhibits the growth of prostate cancer (PC3) and gallbladder cancer (GBC-SD) cells by arresting the cell cycle and targeting the PI3K and EGFR/MMP-9 signalling pathways.
D-Limonene		√	√			√		Inhibition of HMG Co-A reductase and CoA synthesis
Cantharidine		√	√	√			√	Inhibition of serine
Artemisnin and derivatives		√					√	Cleavage of Iron or heme-mediated peroxide
Triptolide			√				√	Inhibition of XPB ATPase and transcription factor
Pseudolaric acid			√		√		√	Blockage of microtubules
Cucurbitacin							√	Interfere with F-actin and inhibition of STAT 3
Allisol			√	√	√		√	Inhibition of sarcoplasmic/ endoplasmic reticulum
Pachymic Acid	√	√	√	√			√	Inhibition of DNS topoisomerases
Lycopene		√						Scavengers of ROS, Inhibition of MMP2
Thymol	√		√	√			√	caspase activation and apoptosis by boosting Bax protein expression while lowering Bcl-2 protein
Menthol	√	√				√	√	Reduction in topoisomerase-I and II, followed by DNA damage and cell death.
Auraptene	√	√		√		√	√	Inhibited the mTOR signalling pathway and its downstream actions, slowing cell proliferation
Perillic acid	√	√	√		√		√	Promoted upregulation of p21 and downregulation of CDK1 and CDK4 in HCT116 colon cancer cells
Ascaridole	√			√	√		√	Exhibits cytotoxicity toward NER-deficient cells at much lower doses than repair-proficient cells
Carvacrol	√	√	√	√			√	Boosted the formation of ROS and interacted with intracellular targets in the HL-60 human leukaemia cell line, HCT-8 human colon cancer cell line, SF-295 brain tumour cell line, and MDA-MB-435 human breast cancer cells

(Continued)

TABLE 12.3 (*Continued*)

Summery of Various Mechanism of Action of Terpenoids against Different Types of Cancer [8–26]

Terpenoids	Cell cycle	Authophagy	Differentiation	Anti-angiogenesis	Anti-metastasis	Anti-MDR	Chemo prevention	MOA
Artesunate and Artemisinin		√	√	√		√	√	Down-regulation of proliferating cell nuclear antigen (PCNA) and Bcl-2 gene expression and up-regulation of Bax gene level
β-Elemene	√	√	√	√		√	√	Inhibited CDK1 by upregulating p27, reducing CDK1 phosphorylation on threonine, and decreasing cyclin B1 expression, leading in cyclin B1-CDK1 multiplex under expression and cell cycle arrest in the G2/M phase
Triptolide		√	√	√	√		√	Inhibits SENP1, c-Jun, and androgen receptors via targeting XPB and Rpb1 (triptolide targets), however further study is required.
Crocetin	√	√	√		√		√	Triggered mitochondrial death by altering mitochondrial membrane potential, resulting in Bax protein overexpression, caspase 3 activation, and Bcl-2 expression downregulation overexpression, caspase 3 activation, and Bcl-2 expression downregulation
Ursolic acid		√	√	√	√		v	Triggered cell death in MDA-MB-231 cells, mostly via the extrinsic apoptotic pathway, which included activation of the Fas receptor, caspase 8 and polymerase (PARP) breakage, up-regulation of Bax, down-regulation of Bcl-2, and cytochrome C release.
Betulinic acid	√		√	√	√	√	√	Inducing MMP to release cytochrome C from mitochondria into cytosol, leading to caspase cleavage, initiation of mitochondrial apoptosis pathway and nuclear degradation
Lupeol	√	√	√	√			√	Inhibits the growth of prostate cancer (PC3) and gallbladder cancer (GBC-SD) cells by arresting the cell cycle and targeting the PI3K and EGFR/MMP-9 signalling pathways.

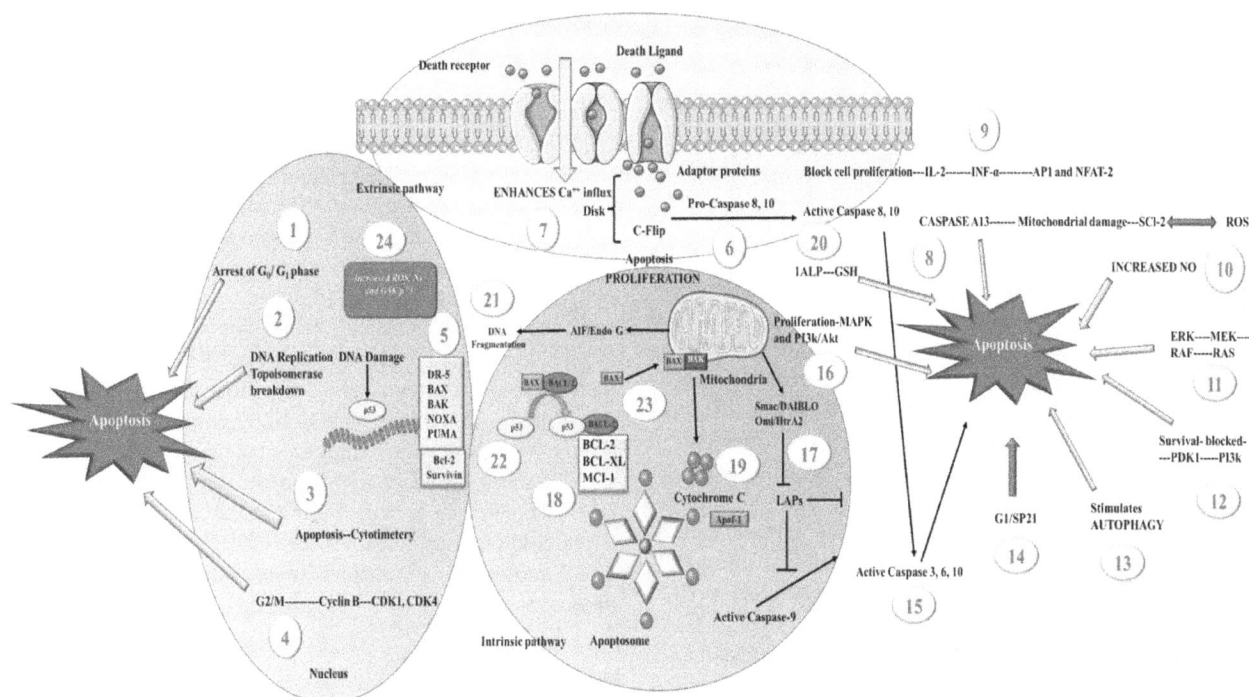

FIGURE 12.37 Various mechanisms involved in the chemotherapeutics of terpenoids like Thymquinone acts via pathways 15-24, Carbacrol acts via pathways 2, 3, 14, 16 and 19, D-Limonine acts via pathways via 1, 2, 3, 4, 8 and 14, Menthol acts via 1-4; Artesenate and Artemisnin acts via 1,2,18, 5, 22; β-elemene acts through 1,2, 5, 22, 16; Triptolide acts through 9,22, 17, 11,14; Crocetin 15,6,20, 18, 22; Ursolic acid acts through 18, 22, 19, 1,2; Betulinic acid acts through 23,24,19, 5′ Lupeol acts through 1,2, 15,14, 8.

breakdown of mitochondrial inner membrane. Thymol may possibly cause caspase activation and apoptosis by boosting Bax protein expression while lowering Bcl-2 protein. Thymol also produced lipid breakdown, mitochondrial depolarization, nucleolar segregation and apoptosis in Caco-2 colorectal cancer cells. Despite the fact that an electron microscopic investigation revealed that thymol is cytotoxic to cancer cells, further research is required to establish the mechanisms behind this cytotoxicity. Thymol exhibited a similar effect on AGC human gastric cancer cells, promoting cell death through the intrinsic apoptosis pathway and the production of reactive oxygen species (ROS). In bladder cancer cells (T24, SW780, J82), thymol triggered caspases and produced mortality. In a 7-day therapy study, thymol, generates an anti-apoptotic response through the Bcl-2/Bax pathway. Thymol-induced oxidative destruction was also identified in oral squamous carcinoma cells (OSCC) and cervical cancer cells in a xenograft nude mice model. Thymol inhibited the synthesis of interleukin-2 (IL-2) and interferon-gamma (IFN-Y) and acted as an immunomodulator on T-cell in human T lymphocyte Jurkat cells (JM).

The modes of action of thymol are assumed to be the down-regulation of activator protein 1 (AP-1) and nuclear factor of activated T-cell (NFAT-2) as transcription factors. Thymol increases phospholipase C-dependent Ca^{+2} release from ER and $Ca+2$ entry via $Ca+2$ channels, promotes apoptosis in MG63 human osteosarcoma cells, DBTRG-05MG human glioblastoma cells, and PC3 human prostate cancer cells. Additionally, the specific mechanism of cell death is unknown since thymol-induced Ca^{+2} secretion results in the generation

of reactive oxygen species (ROS). Figure 12.37 depicts the probable anticancer pathways of thymol [3,21].

12.8 Menthol

The inhibitory chemical mechanism of menthol in SNU-5 human gastric cancer cells prompted the decrease of topoisomerase I and II, followed by DNA damage and cell death, according to Lin et al. Targeting topoisomerase inhibition as a therapeutic technique might eventually be employed in conjunction with other clinical treatments.

Another study found that by targeting the reduction of Bcl-2 protein expression and suppression of p21, $\alpha_1$alpha, 25-dihydroxyvitamin D_3 $(1,25(OH)_2D_3)$ in combination with menthol may enhance the anticancer effect against prostate cancer cells. Furthermore, the 24-hydroxylase enzyme, which is a 1α, $25(OH)_2D_3$ catabolic enzyme, was shown to be up-regulated in various forms of cancer, resulting in the inactivation of 1α, $25(OH)_2D_3$ and suggesting a tumour escape mechanism. Furthermore, the action of menthol on metabolic enzymes might be a difficult strategy to take in medication development. Cancer has been shown to cause the outflow of calcium from bones into circulation, culminating in the activation of a calcium-permeable channel ($TRPM_8$) and cell death. Menthol caused mitochondrial membrane depolarization in bladder cancer cells (T_{24}) by raising Ca^{+2} levels through overexpression of $TRPM_8$, resulting in cancer cell death. The total anticancer molecular mechanism of menthol is shown in Figure 12.37.

12.9 Auraptene

Auraptene inhibits the cell cycle and induces apoptosis in SNU-1 human gastric cancer cells by disrupting mitochondrial membranes. Auraptene has been shown to boost p53 tumour suppressor protein levels while lowering cyclin D1 levels, thereby stopping cell development during the G1 phase and eventually leading to cell death. Auraptenβe also inhibited the mTOR signalling pathway and its downstream actions, slowing cell proliferation. Although auraptene's anticancer method promoted mTOR inhibition, indirect mTOR-rictor activation is still a mystery that may be investigated further. The synergistic activity of auraptene in conjunction with anticancer drugs (cisplatin, paclitaxel, 5-fluorouracil) was examined in KYSE30 esophageal carcinoma cells. Auraptene's association with the P-gp efflux pump boosted anticancer drug efflux transporters, increasing the medication's detrimental effects on cancer cells. Auraptene also reduces cancer cell malignancy by upregulating p53 and p21 overexpression and downregulating CSC markers including CD44 and BMI-1.

Auraptene also decreased cell proliferation, migration, and metastasis in human cervical cancer cells (HeLa cells) and ovarian cancer cells (A2780 cells). In these cell lines, auraptene was shown to have a potential anticancer impact via lowering the expression levels of matrix metalloproteinases (MMP-2 and MMP-9). Auraptene also inhibits the activity of the β-catenin T-cell factor (TCF) in colorectal cancer cells and reduces the overexpression of the c-Myc proto-oncogene. Auraptene treatment decreased cell growth and produced G2/M phase arrest in Caco-2 human colorectal carcinoma and DLD-1 colorectal adenocarcinoma cells. This monoterpenoid has been shown to decrease β-catenin/TCF activity in Caco-2 cells while increasing it in DLD-1 cells. This discrepancy in auraptene action shows that auraptene's cell growth reduction of colorectal cancer cells is not reliant on -catenin-TCF signalling targeting [3,22].

12.10 D-Limonene

D-immunomodulatory limonene's activity on macrophages was suggested to reduce cell proliferation by increasing nitric oxide (NO) generation in BALB/c mice with lymphoma cells, showing that D-limonene anticancer potency is connected to macrophage activation. Because D-limonene has been linked to immune reaction control through macrophages, future research should focus on the mechanisms behind macrophage activation and innate and specific immunity. D-limonene suppressed the proliferation of HL-60 leukaemia cells by downregulating Bcl-2 and changing the p53 gene while raising the expression of the pro-apoptotic Bax gene [3,23]. This drug stimulated autophagy in lung cancer cells by cleaving the pro-apoptotic agent Atg5 and interacting with the Bcl-XL transmembrane protein in mitochondria (A549 and H1299). As a consequence, apoptosis and cell death-related caspases were activated. Because it has been suggested that Atg5 may play a role in D-limonene-induced apoptosis, more research into the mechanism behind autophagy's function in

D-limonene-induced apoptosis is required. Furthermore, the synergistic impact of D-limonene and berberine was investigated in MGC803 human gastric cancer cells, where it increased intracellular ROS generation, facilitated mitochondrial damage, increased caspase 3 activity, decreased Bcl-2 protein expression, and accelerated cell death. Cell cycle suppression in the G1 and G2/M phases was generated by the interaction of D-synergistic limonene with berberine, as well as apoptosis induction through the intrinsic mitochondrial route. More research is needed, however, to completely understand the molecular interactions of combination medications. D-limonene inhibited the expression of Ras, Raf, MEK, and ERK1/2 in skin tumours. This anticancer mechanism was discovered in combination with Bcl-2 downregulation and Bax protein overexpression. Inhibition of the PI3K/AKT pathway has been found as a D-limonene anticancer strategy in LS174T colon cancer cells. Future research might look at different colon cancer cell lines for further information. Figure 12.37 summarises the molecular mechanism of D-anticancer limonene.

12.11 Perillic Acid

Perillic acid generated a dose-dependent cytotoxic response, cell cycle arrest, and apoptosis in non-small cell lung cancer (A549) cells, as well as a substantial increase in Bax, p21, and caspase 3 expression. This monoterpenoid, on the other hand, is considered to target a variety of signalling locations, including DNA repair inhibition, in addition to the apoptotic signalling pathway. As a consequence, further investigation into other possible mechanistic pathways is required. The anticancer effect of perillic acid promoted upregulation of p21 and downregulation of CDK1 and CDK4 in HCT116 colon cancer cells. Although this effect was expected in colon cancer cells, we only looked at the wild-type (p53) cell line in our research. Null p53 colon cancer cells might be employed in future studies [3,23].

12.12 Ascaridole

Oxidative DNA damage and G2/M phase cell cycle arrest in NER (nucleotide excision repair, the main technique to remove DNA damage)-deficient cell lines, such as XP3BE (XPC-deficient) and GM10902, were proposed in the investigation of ascaridole's anticancer mechanism (ERCC6 deficient). When comparing the cytotoxicity of ascaridole to that of repair-proficient cells, it was shown that ascaridole exhibited preferential cytotoxicity toward NER-deficient cells at much lower doses than repair-proficient cells. As a consequence, other mechanisms, such as synthetic lethal interactions, might be implicated. Because this effect was only seen at fatal dosages, the role of ascaridole-induced oxidative stress in cellular death warrants additional investigation. In the HL-60 human leukaemia cell line, HCT-8 human colon cancer, SF-295 brain tumour cell line, and MDA-MB-435 human breast cancer cells, ascoridole, a natural endoperoxide, increased the generation of ROS and interacted with intracellular targets.

12.13 Carvacrol

In the study of ascaridole's anticancer mechanism, oxidative DNA damage and G2/M phase cell cycle arrest in NER (nucleotide excision repair, the basic approach for removing DNA damage)-deficient cell lines, such as XP3BE (XPC-deficient) and GM10902, were postulated (ERCC6 deficient). When comparing the cytotoxicity of ascaridole to that of repair-proficient cells, it was shown that ascaridole had a preference for NER-deficient cells and did so at far lower levels than repair-proficient cells. Other processes, such as synthetic lethal interactions, may be implicated as a result. Because this impact was only seen at lethal doses, more research into the function of ascaridole-induced oxidative stress in cellular death is needed. Ascoridole, a natural endoperoxide, boosted the formation of ROS and interacted with intracellular targets in the HL-60 human leukaemia cell line, HCT-8 human colon cancer cell line, SF-295 brain tumour cell line, and MDA-MB-435 human breast cancer cells. Figure 12.37 depicts the anticancer molecular mechanism of carvacrol against several cancer cells [3,24].

12.14 Sesquiterpenoids and Sesquiterpene Lactones

Sesquiterpene Lactones

Sesquiterpene lactones are large secondary plant metabolites. Ambrosin, coronopilin and dindol-01, for example, have been found to produce DNA damage in MCF-10A, MCF-7, JIMT-1, and IICC1937 breast cancer cells by triggering cell cycle arrest in the S and G2 phases, which boosted p53 protein expression. A possible anticancer mechanism is the regulation of the NF-KB signalling pathway by sesquiterpene lactone in the extract of many Mexican Indian medicinal herbs. Sesquiterpene lactone extracted from Artemisia macrocephala has anticancer effects against 3T3, HeLa, and MCF-7 cells by generating ROS and targeting the sarcoplasmic reticulum calcium ATPases pump, NF-KB, p53 signalling pathway, angiogenesis, and metastasis. Sesquiterpene lactones have been found to have a significant anticancer activity in oxidative stress-mediated apoptosis via the control of NF-KB, STAT3, and the involvement of the p53 protein. The mechanism of action of some sesquiterpene lactones against cancer is discussed below, with a description of the mechanism of action displayed in Figure 12.37 [3,24].

12.15 Artesunate and Artemisinin

Artesunate, a sesquiterpene lactone, is an effective antimalaria drug. In H22 solid hepatic carcinoma cells, artesunate has been demonstrated to exhibit anticancer capabilities, with the mechanism of action being down-regulation of proliferating cell nuclear antigen (PCNA) and Bcl-2 gene expression and up-regulation of Bax gene level. The synergistic anticancer effects of artesunate in combination with allicin showed that osteosarcoma (MG-63 and U20S) cell growth was suppressed

through a mitochondrial apoptosis-dependent pathway, as shown by caspase 3/9 expression. Artesunate inhibited ferroptosis resistance and induced iron-dependent ROS-mediated ferroptosis in head and neck cancer cells (HNC) via the Nrf2 antioxidant response element (ARE) pathway, which inhibited ferroptosis resistance and induced iron-dependent ROS-mediated ferroptosis, according to another study. By reducing GSH and raising lipid ROS levels, artesunate triggered ferroptosis. In uveal melanoma cells, artesunate inhibits tumour growth by suppressing β-catenin accumulation and activating downstream target genes like c-Myc and CDK1. In an in vivo study, inhibition of β-catenin accumulation was also seen in colorectal cancer as a result of artesunate treatment, resulting in apoptotic activation. Artesunate's anticancer properties, as well as its safety, need further investigation using appropriate animal models to corroborate the findings and better understand the mechanism of action.

12.16 β-Elemene

Curcuma wenyujin, a Chinese herbal medicine, has another sesquiterpenoid, β-elemene, as one of its active components. This chemical has been investigated in human NSCCL cell lines (H460 and A549). The findings revealed that at the G2/M and S phases, β-elemene arrests the cell cycle. Indeed, β-elemene inhibited CDK1 by upregulating p27, reducing CDK1 phosphorylation on threonine, and decreasing cyclin B1 expression, leading to cyclin B1-CDK1 multiplex underexpression and cell cycle arrest in the G2/M phase. Furthermore, β-elemene has been demonstrated to up-regulate the expression of a checkpoint kinase (Chk2), resulting in CDC25C phosphorylation and CDK1 down-regulation, halting cells in the G2/M phase through a Chk2-dependent mechanism. β-Elemene also suppressed the production of cyclin A and the formation of the cyclin A-CDK2 complex in human lung cancer cells. When cells were exposed to β-elemene, however, CDK2 activation required phosphorylation on threonine 160, which was inhibited by up-regulated p27, resulting in S phase arrest and death. More study is required to determine how β-elemene affects the expression levels of cyclin-A, cyclin B1, CDC25C, and Bcl-2 genes, as well as whether β-elemene-induced cell cycle arrest is linked to apoptosis.

12.17 Diterpenoids

Triptolide

Triptolide induced senescence-like phenotypes in tumor-derived human prostate epithelial cells at modest dosages (1 ng/mL). Triptolide, on the other hand, induced apoptosis at higher dosages (50–100 ng/mL), with the p53 protein level rising in response. The molecular signalling pathway underlying the two types of cell death produced by low and high dosages of triptolide needs further research. According to another study, triptolide inhibited prostate cancer development in a xenograft model by lowering SUMO-specific protease 1 activity (SENP1), as well as androgen receptor expression and c-Jun transcription activity. Triptolide inhibits SENP1, c-Jun, and androgen receptors via targeting XPB and Rpb1 (triptolide targets), however further study is required. Triptolide activates caspase 3 and activates the apoptosis pathway in MDA-MB-231 triple-negative breast cancer cells, as well as regulating the autophagy signalling pathway by changing the amounts of LC3B-II autophagy protein and p62 receptor. More in vitro and animal-based research is required to understand and validate the influence of triptolide on numerous pathways in triple-negative breast cancer cells. Triptolide decreased SKOV/DDP cell metastasis by lowering adhesion-related proteins like integrin beta 1 (ITGB1) and inhibiting apoptotic agents including MMP-2 and MMP-9. According to in-vivo studies, triptolide enhanced the expression of IL-2 and TNF, which boosted the expression of NK cell-related proteins like CD16 and CD56 while decreasing the expression of proteins that promote blood vessel development like CD31 and CD105. This is reported that triptolide combined with cisplatin has a synergistic anti-tumour impact against epithelial ovarian cancer cells (EOC).

12.18 Crocetin

The anticancer activity of Crocetin on human esophageal squamous carcinoma (KYSE-150) cells was validated by targeting the PI_3K/AKT and $ERK_{1/2}$ signalling pathways, as well as p38 down-regulation and p53 and p21 protein overexpression.

These changes triggered mitochondrial death by altering mitochondrial membrane potential, resulting in Bax protein overexpression, caspase 3 activation, and Bcl-2 expression downregulation. However, it's important to investigate crocetin's anticancer properties in different cancer cell lines and verify the targets with inhibitors to validate the crocetin-pathway relationship. Crocetin in combination with cisplatin had a synergistic impact on KYSE-150 cells, resulting in increased cisplatin apoptotic activity. Crocetin also arrests KYSE-150 cells in the S phase, changes cell shape, and increases the expression of Bax and apoptotic genes, according to another research [3,25].

12.19 Triterpenoids

Ursolic Acid

Several investigations found that ursolic acid has a modulatory influence on the expression and function of mitochondrial enzymes, growth inhibition, and apoptosis in distinct in vitro cell lines (HaCaT, M4Beu, HuH7) and in vivo experimental cancer models. In a mouse model, the hepatoprotective effect of ursolic acid was studied. In mice, ursolic acid therapy protected liver mitochondria against damage caused by Ca^{2+} release from the mitochondrial matrix, cytochrome C, and apoptosis induction. This points to ursolic acid having a direct inhibitory impact on mitochondrial permeability transition, which slows cell proliferation through caspase-dependent or independent pathways. A mitochondrial permeability transition pore has been identified as a possible future target in the hunt for the precise mechanism of ursolic acid's hepatoprotective action. Furthermore, ursolic acid increased oxidative stress

and triggered autophagy in osteosarcoma cells (U-2OS and MG-63) to cause apoptosis. Ursolic acid also triggered cell death in MDA-MB-231 cells, mostly via the extrinsic apoptotic pathway, which included activation of the Fas receptor, caspase 8 and polymerase (PARP) breakage, up-regulation of Bax, down-regulation of Bcl-2, and cytochrome C release. Ursolic acid also caused caspase 9 cleavage, indicating that it promotes both intrinsic and extrinsic apoptotic pathways. Only one breast cancer cell line was investigated in this research, which is a flaw. Another research found that ursolic acid inhibited colon adenocarcinoma (SW480) cell growth by up-regulating the proteins p53, Bax, and p21, which resulted in caspase 3-dependent death [26].

12.20 Betulinic Acid

Several studies demonstrated that betulinic acid induces apoptosis in cancer cells via a direct effect on mitochondria. Betulinic acid exerts anticancer activity by inducing MMP to release cytochrome C from mitochondria into cytosol, leading to caspase cleavage, initiation of mitochondrial apoptosis pathway and nuclear degradation. Other reported mechanisms of action of betulinic acid include ROS generation, p38 and MAPKs activation, which cause apoptosis in UISO-Mel-1 human melanoma cells. The study suggested conducting clinical studies to explore the clinical outcomes of betulinic acid [27].

12.21 Lupeol

Grapes, cabbages, and olives contain lupeol, a pentacyclic triterpenoid found in culinary vegetables, fruits, and medicinal plants. Pitchai et al. discovered mitochondrial-mediated apoptosis in MCF-7 breast cancer cells treated with lupeol, as shown by anti-apoptotic proteins being down-regulated. Lupeol also inhibited the development of lung cancer (A549) cells by lowering anti-apoptotic protein mRNA expression. Several studies have demonstrated that lupeol inhibits the growth of prostate cancer (PC3) and gallbladder cancer (GBC-SD) cells by arresting the cell cycle and targeting the PI3K and EGFR/MMP-9 signalling pathways. Lupeol also suppressed the proliferation of melanoma cells (Mel-928, Mel-1241, Mel-1011) by interfering with the Wnt/-catenin pathway and disrupting the Wnt signalling pathway. Lupeol also suppressed the development of colorectal cancer cells (SW480 and HCT-116) by causing cell cycle arrest in the S phase as a consequence of inhibiting the agene and protein levels, according to the findings of that research. Although CCNA2 is an oncogenic gene that regulates colorectal cancer growth, cell migration, and cell cycle, lowering its expression does not completely explain cell invasion and migration. As a result, further research is needed to properly comprehend the underlying mechanism of action of lupeol [3,27].

12.22 Conclusion

This chapter provides comprehensive information regarding the natural occurring terpene molecules, sources, elaborated classification of mono, di and tri, tetra, ploy and sesterterpenes. This chapter also discuss the mechanism of action of some Terpenoids such as Thymol, Menthol, Auraptene, D-Limonene, Perillic Acid, Ascaridole, Carvacrol, Sesquiterpene Lactones, Artesunate and Artemisinin, β-Elemene, Triptolide, Crocetin, Ursolic acid, Betulinic acid and Lupeol against Cancer Progression.

REFERENCES

1. Wallach, O. Zur Kenntniss der Terpene und ätherischen Oele. *European Journal of Organic Chemistry*, **1887**, 238, 78–89.
2. Ingold, E. H. The Nature of the Alternating Effect in Carbon Chains. Part II. The Directing Influence of the a-Methoxyvinyl Group in Aromatic Substitution. *Journal of the Chemical Society*, **1925**, 127, 870.
3. Breitmaier, E. *Terpenes*. WILEY-VCH Verlag GmbH & Co. KGaA, Weinheim, 2006, pp:1–197. ISBN: 3-527-31786-4.
4. Aceret, T.L.; Sammarco, P.W.; Coll, J.C. Effects of Diterpenes Derived from the Soft Coral Sinularia flexibilis on the Eggs, Sperm and Embryos of the Scleractinian Corals Montipora digitata and Acropora tenuis. *Marine Biology*, **1995**, 122, 317–323.
5. Adolph, S.; Jung, V.; Rattke, J.; Pohnert, G. Wound Closure in the Invasive Green Alga Caulerpa taxifolia by Enzymatic Activation of a Protein Cross-Linker. *Angewandte Chemie International Edition*, **2005**, 44, 2806–2808.
6. Aimi, N.; Odaka, H.; Sakai, S.I.; Fujiki, H.; Suganuma, M.; Moore, R.E.; Patterson, G.M.L. Lyngbyatoxins B and C, Two New Irritants from Lyngbya majuscula. *Journal of Natural Products*, **1990**, 53, 1593–1596.
7. Akhter, S.; Nath, S.K.; Tse, C.M.; Williams, J.; Zasloff, M.; Donowitz, M. Squalamine, a Novel Cationic Steroid, Specifically Inhibits the Brush-Border Na+/H+ Exchanger Isoform NHE3. *American Journal of Physiology-Cell Physiology*, **1999**, 276, 136–144.
8. Amico, V.; Oriente, G.; Piattelli, M.; Tringali, C.; Fattorusso, E.; Magno, S.; Mayol, L. Caulerpenyne, an Unusual Sesquiterpenoid from the Green Alga Caulerpa prolifera. *Tetrahedron Letters*, 1978, 19, 3593–3596.
9. Avila, C. Natural Products of Opistobranch Molluscs: A Biological Review. *Oceanography and Marine Biology: An Annual Review*, **1995**, 33, 487–559.
10. Bakus, G.J.; Targett, N.M.; Schulte, B. Chemical Ecology of Marine Organisms: An Overview. *Journal of Chemical Ecology*, **1986**, 12, 951–987.
11. Barbier, P.; Guise, S.; Huitorel, P.; Amade, P.; Pesando, D.; Briand, C.; Peyrot, V. Caulerpenyne from Caulerpa taxifolia Has an Antiproliferative Activity on Tumor Cell Line SK-NSH and Modifies the Microtubule Network. *Life Sciences*, **2001**, 70, 415–429.
12. Bellan-Santini, D.; Arnaud, P.M.; Bellan, G.; Verlaque, M. The Influence of the Introduced Tropical Alga Caulerpa taxifolia, on the Biodiversity of the Mediterranean Marine Biota. *Journal of the Marine Biological Association of the United Kingdom*, **1996**, 76, 235–237.
13. Bergmann, W.; Johnson, T.B. The Chemistry of Marine Animals. I. The Sponge Microciona pralifera. *Zeitschrift für Physiologische Chemie*, **1933**, 222, 220–226.
14. Brust, A.; Garson, M.J. Advanced Precursors in Marine Biosynthetic Study. The Biosynthesis of Dichloroimines in the Tropical Marine Sponge Stylotella aurantium. *Tetrahedron Letters*, **2003**, 44, 327–330.
15. Cinel, B.; Roberge, M.; Behrisch, H.; Van Ofwegen, L.; Castro, C.B.; Andersen, R.J. Antimitotic Diterpenes from Erythropodium caribaeorum Test Pharmacophore Models for Microtubule Stabilization. *Organic Letters*, **2000**, 2, 257–260.
16. Ciomei, M.; Albanese, C.; Pastori, W.; Grandi, M.; Pietra, F.; D'Ambrosio, M.; Guerriero, A.; Battistini, C. Sarcodictyins: A New Class of Marine Derivatives with Mode of Action Similar to Taxol. *Proceedings of the American Association for Cancer Research*, **1997**, 38, 5.
17. Ciulla, T.A.; Criswell, M.H.; Danis, R.P.; Williams, J.I.; McLane, M.P.; Holroyd, K.J. Squalamine Lactate Reduces Choroidal Neovascularization in a Laser-Injury Model in the Rat. *Retina*, **2003**, 23, 808–814.
18. Coleman, A.C.; Kerr, R.G. Radioactivity-Guided Isolation and Characterization of the Bicyclic Pseudopterosin Diterpene Cyclase Product from Pseudopterogorgia elisabethae. *Tetrahedron*, **2000**, 56, 9569–9574.
19. Cvejic, J.H.; Rohmer, M. CO_2 as Main Carbon Source for Isoprenoid Biosynthesis via the Mevalonate-Independent Methylerythritol 4-Phosphate Route in the Marine Diatoms Phaeodactylum tricornutum and Nitzschia ovalis. *Phytochemistry*, **2000**, 53, 21–28.
20. D'Ambrosio, M.; Guerriero, A.; Pietra, F. Sarcodictyin A and Sarcodictyin B, Novel Diterpenoidic Alcohols Esterified by (E)-N(1)-Methylurocanic Acid. Isolation from the Mediterranean stolonifer Sarcodictyon roseum. *Helvetica Chimica Acta*, **1987**, 70, 2019–2027.
21. Dauben, W.G.; Thiessen, W.E.; Resnick, P.R. Cembrene, a Fourteen-Membered Ring Diterpene Hydrocarbon. *Journal of Organic Chemistry*, **1965**, 30, 1693–1698.
22. De Freitas, J.C.; Blankemeier, L.A.; Jacobs, R.S. In Vitro Inactivation of the Neurotoxic Action of b-Bungarotoxin by the Natural Product Manoalide. *Experientia*, **1984**, 40, 864–865.
23. Edwards, D.J.; Gerwick, W.H. Lyngbyatoxin Biosynthesis: Sequence of Biosynthetic Gene Cluster and Identification of a Novel Aromatic Prenyltransferase. *Journal of the American Chemical Society*, **2004**, 126, 11432–11433.
24. Faulkner, D.J. Marine Pharmacology. *Antonie van Leeuwenhoek*, **2000**, 77, 135–145.
25. Feller, M.; Rudi, A.; Berer, N.; Goldberg, I.; Stein, Z.; Benayahu, Y.; Schleyer, M.; Kashman, Y. Isoprenoids of the Soft Coral Sarcophyton glaucum: Nyaloide, a New Biscembranoid, and Other Terpenoids. *Journal of Natural Products*, **2004**, 67, 1303–1308.
26. König, G.M.; Wright, A.D.; Angerhofer, C.K. Novel Potent Antimalarial Diterpene Isocyanates, Isothiocyanates and Isonitriles from the Tropical Marine Sponge Cymbastela hooperi. *Journal of Organic Chemistry*, **1996**, 61, 3259–3267.
27. Hifzur Rahman Siddique, Mohammad Saleem. Beneficial health effects of lupeol triterpene: A review of preclinical studies. *Life Sciences*, **2011**, 88, (7–8): 285–293.

13

A Brief Account on Plant and Marine-Derived Terpenoids

Bimal Krishna Banik*, Abhishek Tiwari†, and Biswa Mohan Sahoo‡

CONTENTS

13.1 Microbial Synthesis of Terpenoids 455
13.2 Epoxidation 455
13.3 Baeyer–Villiger Oxidation 458
13.4 Kinetic Resolution 460
Conclusion 462
References 462

The isolation of natural products from marine sources accelerated rapidly beginning in the early 1970 and is only now hinting that a plateau may have been reached, as measured by the number of yearly publications. Chemical synthesis of marine natural products has climbed steadily since the late 1970s and has reached a plateau of around 60–70 per year for the past 5 years. The annual number of publications dealing with all aspects of marine natural products synthesis is considerably higher. The rapid growth in the breadth and depth of this field in a comparatively short period of time mirrors the growth and interests of the synthesis community at large. Synthesis chemists are stimulated primarily by compounds which possess potential biomedical importance and/or provocative structures, of which there is no shortage of either in the collective metabolites from marine sources. Continued growth in this area is inevitable and is a sign of the vigor and vitality of marine natural products chemistry today.

Terpenes comprise primary and secondary metabolites, all derived from the five-carbon isoprene entity. Combination and modifications of these isoprene units lead to a multitude of diverse structures with different chemical and biological properties. Terpenoids from higher plants are well studied and ethnopharmacologically applied for centuries. However, it was not until the early to mid-20th century that their marine counterparts were explored. Steroidal terpenoids were the first marine isoprenes to be discovered. Bergmann studied during the 1930s and the following decade sterols from various marine microorganisms. Later on, one of his students, Leon Ciereszko, was attracted by the odours of gorgonians and catalysed with his findings research in the field of marine terpenoid chemistry. To date a large number of marine terpenoid structures are known Terpenes from taxa occurring predominantly or exclusively in the oceans, such as certain algae and invertebrates are, due to their uniqueness, the most interesting

structures. Since the 1970s terpenoids of marine origin have been described in several general reviews devoted to marine natural products. Marine terpenoids are also listed in the Dictionary of Terpenoids and can be found in annual reviews on the isolation of new C_{10}, C_{15}, C_{20} and C_{30} isoprenoids and sterols. Additionally, there exists a wealth of literature dealing predominantly with marine isoprenoids covering aspects of marine monoterpenoids marine diterpenoids algal sesquiterpenoids marine sesquiterpenes marine triterpenoid oligoglycosides and marine sterols. Other authors highlighted the presence and significance of terpenoids occurring in marine invertebrates. A number of reviews dealing with the synthesis of marine terpenoids have been reported as well as theoretical considerations concerning the biosynthesis of marine isoprenoids. However, despite the many structures known and their ecological and pharmacological importance, only a few experimental biosynthetic studies on marine terpenoid compounds were performed. Examples are flexibilide (1), dihydroflexibilide (2), chloromertensene (3), sphonodictidine (4), 7-deacetoxy-olepupuane (5), litophynol (6), caulepenyne (7), oxytoxin (8), eleutherobin (9), and sarcodicity A-E (10–14) [1–3] (Figure 13.1).

Secondary metabolites produced by marine organisms in order to ensure their survival may be envisaged as 'optimized' by evolutionary mechanisms, e.g. mutation and natural selection, to specifically influence certain biological target structures, i.e. DNA, enzymes, receptors, and membranes. Since some molecular targets are highly conserved defensive toxins may also bind to therapeutically relevant target sites. In this way, the biological activity of marine terpenes, as illustrated by their ecological role, may also be exploited in terms of pharmacology [4,5].

Several biologically active terpenoids turned out to possess biomedical potential and are thus already in preclinical

---

* Department of Mathematics and Natural Sciences, College of Sciences and Human Studies, Prince Mohammad Bin Fahd University, Al Khobar, Kingdom of Saudi Arabia. bimalbanik10@gmail.com; bbanik@pmu.edu.sa.
† Faculty of Pharmacy, Pharmacy Academy, IFTM University, Lodhipur Rajput, Moradabad-244102, Uttar Pradesh, India. abhishekt1983@gmail.com.
‡ Roland Institute of Pharmaceutical Sciences, Berhampur affiliated to Biju Patnaik University of Technology (BPUT), Rourkela, Odisha, India.

DOI: 10.1201/9781003008682-13

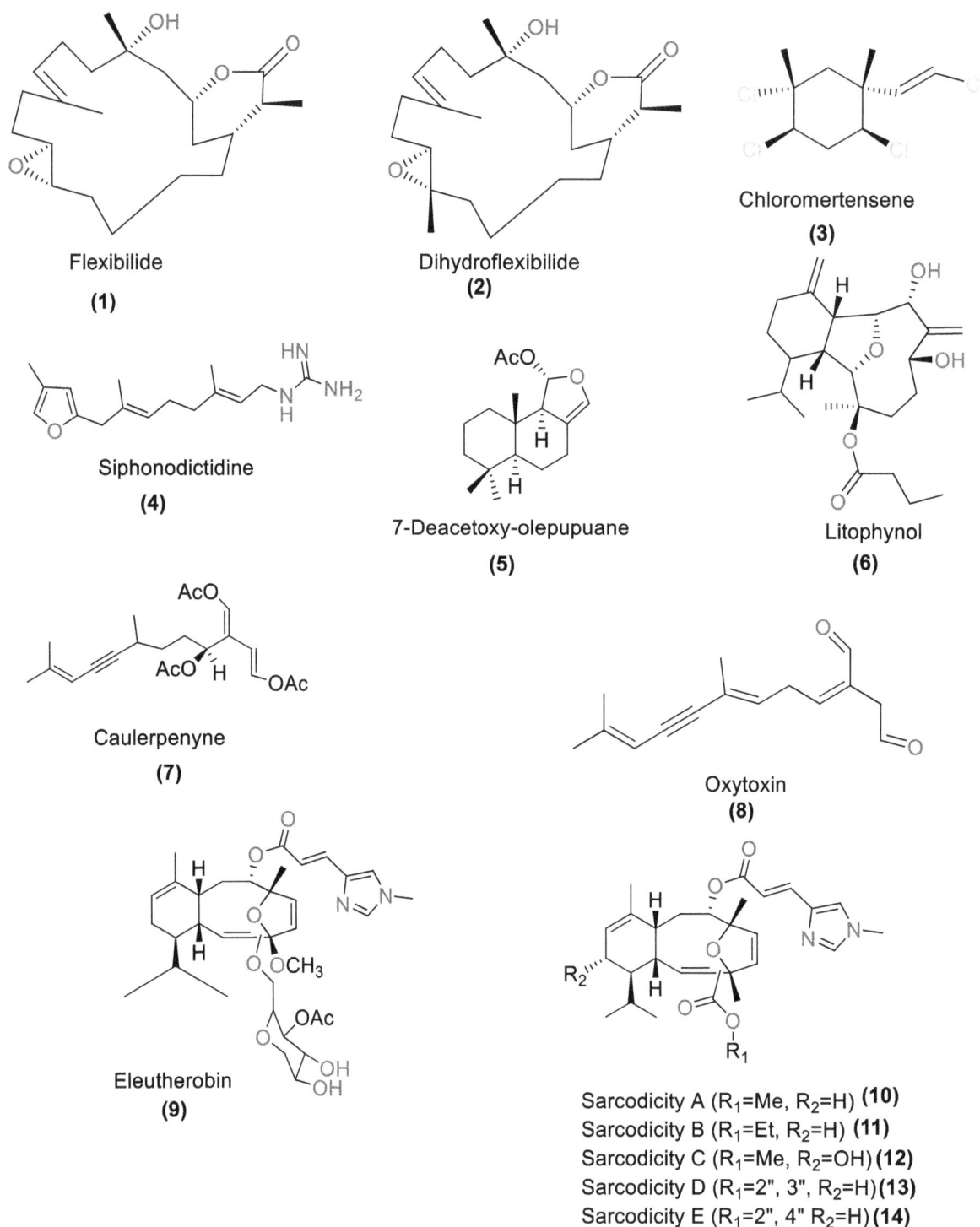

Flexibilide

(1)

Dihydroflexibilide

(2)

Chloromertensene

(3)

Siphonodictidine

(4)

7-Deacetoxy-olepupuane

(5)

Litophynol

(6)

Caulerpenyne

(7)

Oxytoxin

(8)

Eleutherobin

(9)

Sarcodicity A (R$_1$=Me, R$_2$=H) **(10)**
Sarcodicity B (R$_1$=Et, R$_2$=H) **(11)**
Sarcodicity C (R$_1$=Me, R$_2$=OH) **(12)**
Sarcodicity D (R$_1$=2", 3", R$_2$=H) **(13)**
Sarcodicity E (R$_1$=2", 4" R$_2$=H) **(14)**

FIGURE 13.1 Examples of Marine terpenes.

or clinical development. Eleutherobin is a microtubuline stabilizing compound in preclinical trials, which competes for the paclitaxel binding site on the microtubule polymer. Aside from paclitaxel, its mode of action is shared with the natural products epothilone A and B and discodermolide. Ojima et al. (1999) proved that these four natural products share a similar 3D structure, and suggested a common pharmacophore, necessary for binding to the paclitaxel site. Eleutherobin originates

from a soft coral of the genus Eleutherobia collected from Australian waters and has been re-isolated along with congeners from another octocoral from the Caribbean which can be kept by aquaculture [6–9].

The terpene nucleus of eleutherobin is bound to a sugar moiety and linked via an ester bond to a partial structure most probably derived from histidine. Total synthesis was achieved but a paucity of material currently hinders the development of the compound. The sarcodictyins possess a very similar structure and were already described in the 1980s by Pietra et al. without recognising the activity of these compounds. They were derived from the Mediterranean stolonifera sarcodictyin series. In 1997 the tubulin stabilizing activity of sarcodictyin was discovered. Consequently, this compound is now in preclinical development (Figures 13.2 and 13.3) [10–14].

Table 13.1 shows the occurrence and medicinal application of terpenoid natural compounds from seawater algae, sponges, and fungi.

Table 13.2 shows the different marine terpenoids with the method of modification to increase the yield.

13.1 Microbial Synthesis of Terpenoids

Stereoselective hydroxylation of terpenoids is biotransformation of (−)-menthol by *Rhizoctonia solani*. (−)-Menthol is a constituent of mint oil; however, its production from natural sources is too low to cover all the needs. It is also possible to synthesize it from (+)-citronellal, thymol or myrcene, which are quite easily available substrates. (−)-Menthol is commonly used as a food, cosmetic, and pharmaceutical additive due to its pleasant flavor and fragrance. Bio-transformations of (−)-menthol by *Aspergillus* species, *Penicillium* species and *Rhizoctania solani*. Scientists have shown that 3 out of 12 species of *R. solani* produced: (−)-6-hydroxymenthol, (+)-6,8-dihydroxymenthol, and (−)-1-hydroxymenthol are intermediates for syntheses of

compounds which can be applied as agents for control of snow blight disease (Scheme 13.1).

The second example of microbial hydroxylation is biotransformation of (−)-α-pinene by fungi. Microbial transformation of (-)-α-Pinene by *Botrytis cinerea* resulted in obtaining (1R, 3S, 5R)- trans-β-pinen-3-ol and (1S, 2R, 5S)-trans-7-oxoverbenol (Scheme 13.2) [15–18].

13.2 Epoxidation

There are many literature examples describing the preparation of terpenoid epoxides via bio-transformations. Over 60 fungi species were tested for their capability of transforming (R)-(+)- and (S)-(−)-limonene, which yielded trans and cis epoxides using solid-phase microextraction as the monitoring technique. Products of biotransformation of (R)-limonene by *Penicillium digitatum* were mainly (R)-(+)-α-terpineol and γ-terpinene. *cis*-Limonene oxide and trans-limonene oxide were the only side products in this reaction. These products have high repellent activity. Limonene oxide is a major constituent of essential oil of the repellent plant *Lippia javanica*. Another example of microbial epoxidation is biotransformation of unsaturated bicyclic γ-lactone. *Absidia cylindrospora* efficiently transforms 4,4,6-trimethyl-9-oxabicyclo [4.3.0] non-2-en-8-one to corresponding trans-epoxylactones (R_1, R_2=CH$_3$) (Scheme 13.3) [19,20].

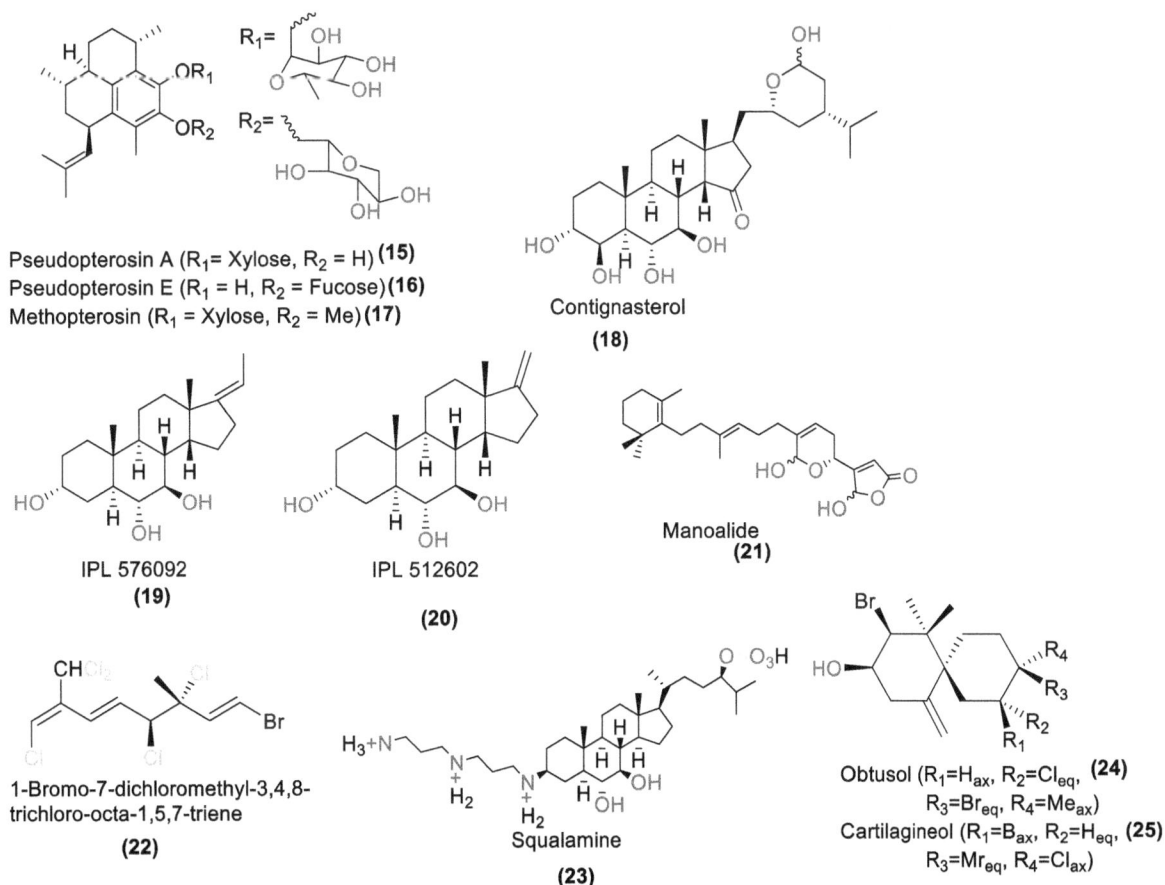

Pseudopterosin A (R_1= Xylose, R_2 = H) **(15)**
Pseudopterosin E (R_1 = H, R_2 = Fucose) **(16)**
Methopterosin (R_1 = Xylose, R_2 = Me) **(17)**

Contignasterol
(18)

IPL 576092
(19)

IPL 512602
(20)

Manoalide
(21)

1-Bromo-7-dichloromethyl-3,4,8-trichloro-octa-1,5,7-triene
(22)

Squalamine
(23)

Obtusol (R_1=H$_{ax}$, R_2=Cl$_{eq}$, **(24)**
R_3=Br$_{eq}$, R_4=Me$_{ax}$)
Cartilagineol (R_1=B$_{ax}$, R_2=H$_{eq}$, **(25)**
R_3=Mr$_{eq}$, R_4=Cl$_{ax}$)

FIGURE 13.2 Few more examples of Marine terpenes.

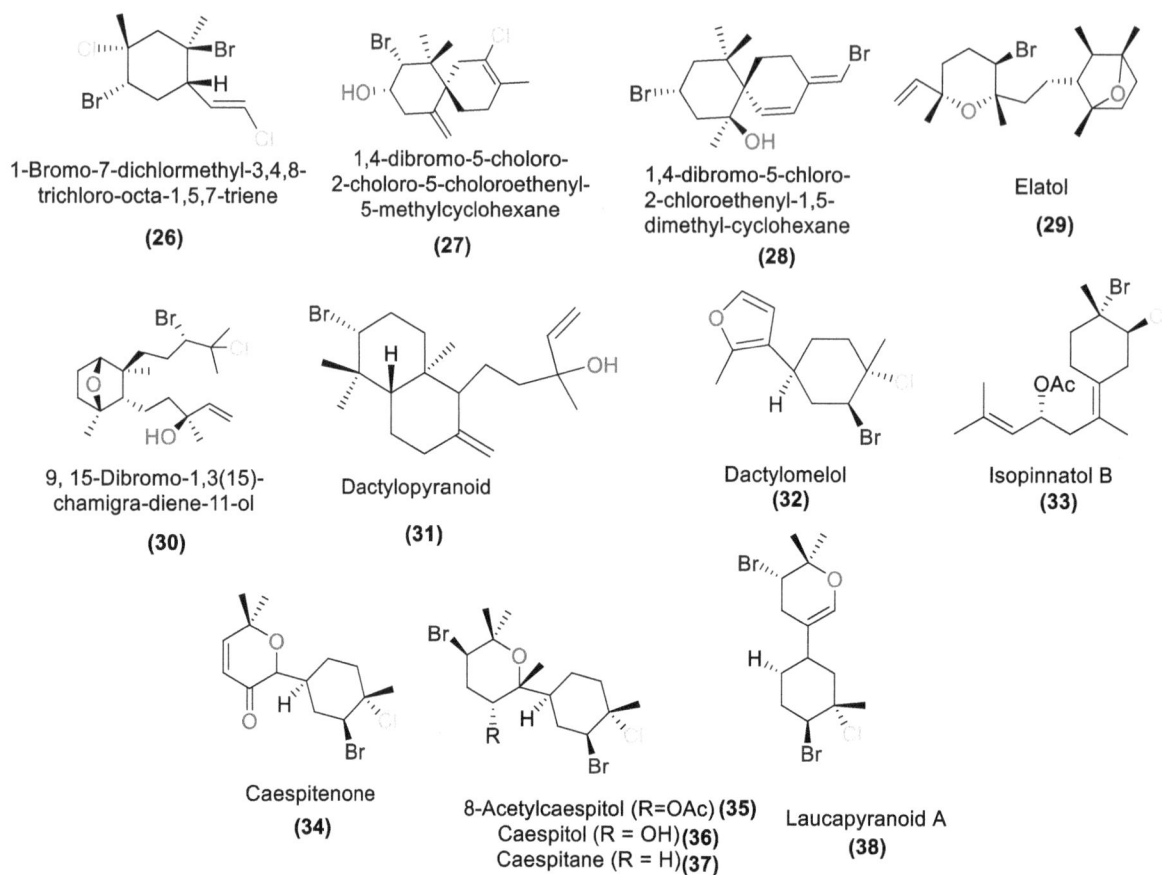

FIGURE 13.3   Some examples of Marine terpenes.

TABLE 13.1

Occurrence and Medicinal Application of Terpenoid Natural Compounds from Seawater Algae, Sponges, and Fungi

S. No.	Compounds	Source	Type of Terpenoids	Activity	Ref.
1.	(−) Dilopholide	Dilophus ligulatus Brown alga	Monocyclic monoterpene C20	Cytotoxic activity	[21]
2.	(−) Reiswigin	Epipolasis reiswigi sponge	Bicyclic diterpene C20	Antiviral activity, herpes simplex	[22]
3.	(+) Dictyol	*Dictyota dichotoma,Halimeda stuposa –Dictyota sp.* Brown alga	Bicyclic diterpene C20	Antimicrobial andcytotoxic activity	[23]
4.	(+)-Isoagatholactone	*Spongia officinalis* sponge	Tricyclic diterpene C20	Antimicrobial andcytostatic activity	[24]
5.	(−) Aplroseol-1	*Aplysilla rosea, Igernella notabilis, Dendrilla rosea* sponges	Tricyclic diterpene C20	Cytostatic activity	[24,25]
6.	(−) Manoalide	*Luffariella variabilis*sponge	Monocyclic sesterterpene C25	Analgesic, antiinflammatory and antibiotic activity, hepatitis C inhibitor	[26]
7.	(+) Scalarin	*Cacospongia (Scalarisponga) scalaris,Spongia virgultosa*and other *Ircinia*-genera sponges	Tetracyclic sesterterpene C25	Cytotoxic activity	[27]
8.	(−) Fusidic Acid	*Fusidium coccineum* and other *Fusidium sp.* fungi *Corchorusaestuans L.* herb in India	Tetracyclic triterpene acid C30	Bacteriostatic antibiotic	[27]

TABLE 13.2

Marine Terpenoids along with Method of Modification to Increase the Yield [15,29–43]

SN	Isoprenoids	Methods	Reference
1.	Zeaxanthin, β-carotene and lycopene	Translation of heterologous carotenoid genes on plasmids	[28]
2.	Zeaxanthin	Translation of heterologous carotenoid genes on different plasmids	[29]
3.	1hydroxyneurosporene, demethylspheroidene, 1'-HO-γ-carotene and 7,8-dihydrozeaxanthin	Translation of heterologous carotenoid genes	[30]
4.	Lycopene, β-carotene and phytoene	Translation of IPP isomerases from *P. rhodozyma, H. pluvialis* and S. cerevisiae	[31]
5.	Astaxanthin	Translation of native IPP isomerase and GGPP synthase from *A. fulgidus*	[32]
6.	Lycopene and ubiquinone (UQ-8)	Translation of heterologous genes from *B. subtilis* and *Synechocystis* encoding 1-deoxy-D-xylose-5-phosphate synthase	[33]
7.	Lycopene and zeaxanthin	Translation of native gene encoding D-1-deoxyxylulose 5-phosphate synthase	[34]
8.	Lycopene	Translation of the *idi* gene and the *dxs* gene or in combination	
9.	Taxadiene	Translation of DXP synthase	[35]
10.	Lycopene	Translation of *dxs* and *dxr* gene with three different promoters	[36]
11.	Limonene, carveol and carvone	Translation of heterologous genes and feeding of precursor	[37]
12.	Amorpha-4,11-diene	Translation of the genes from *S. cerevisiae* encoding the mevalonate pathway and amorpha-4,11-synthase	[38]
13.	Lycopene	Overexpression of genome-wide stoichiometric flux balance analysis	[39]
14.	Mevalonate pathway optimization for amorpha-4,11-diene	Balancing enzymatic activity of the heterologous expressed mevalonate pathway	[40]
15.	Lycopene	Overexpression targets identified by screening of genomic libraries	[41]
16.	8-Hydroxycadinene and artemisinic acid	Over-expression of plant cytochrome P450s followed by optimization	[42]

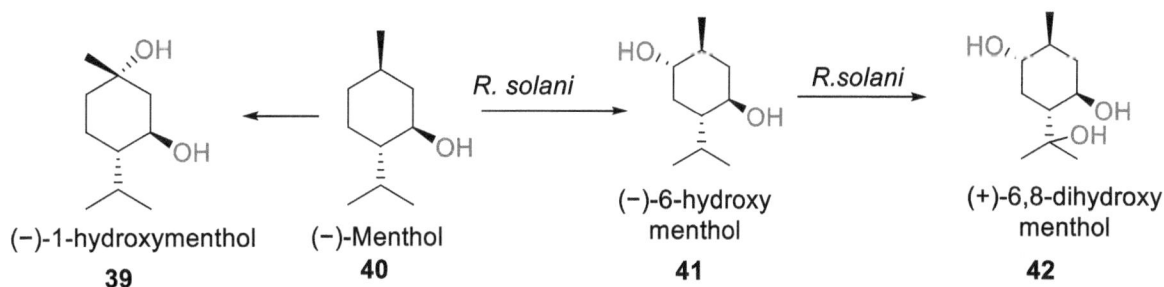

SCHEME 13.1 Microbial transformation of Menthol to hydroxy menthol [44].

(−)-1-hydroxymenthol **39**   (−)-Menthol **40**   (−)-6-hydroxy menthol **41**   (+)-6,8-dihydroxy menthol **42**

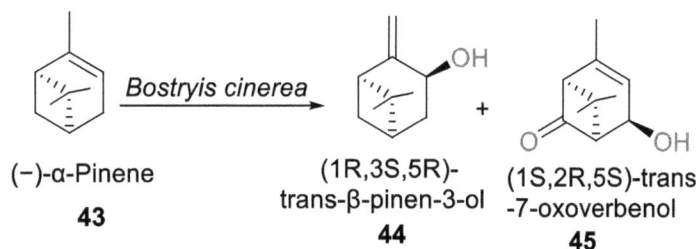

(−)-α-Pinene **43**   (1R,3S,5R)-trans-β-pinen-3-ol **44**   (1S,2R,5S)-trans-7-oxoverbenol **45**

SCHEME 13.2 Microbial transformation of (−)-α-pinene [44].

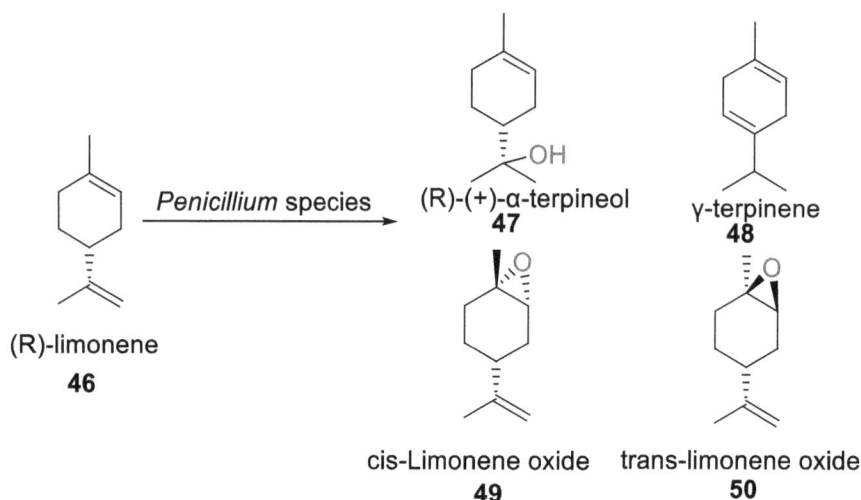

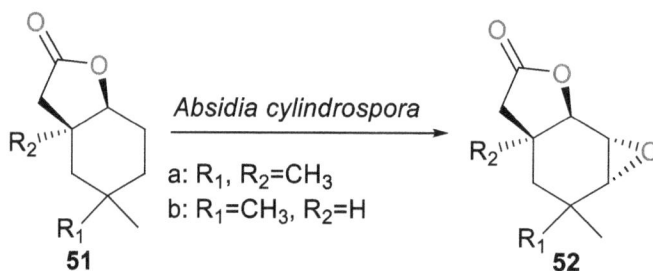

SCHEME 13.3 Microbial transformation of (R)-Limonene to terpinol, terpinene and its epoxides [44].

13.3 Baeyer–Villiger Oxidation

Baeyer–Villiger oxidation of linear and cyclic ketones to corresponding esters and lactones is an important reaction in the synthesis of biologically relevant compounds. Microorganisms that produce Baeyer–Villiger monooxygenases (BVMO) are able to oxidize ketones with high regio- and stereospecificity. An interesting example of microbial Baeyer–Villiger oxidation is biotransformation of (±)-dihydrocarvone by *Acinetobacter calcoaceticus*, resulting in two lactones. Monooxygenase from this species exhibits regio- and enantiospecifity (Scheme 13.4) [44–54].

Biotransformation of this terpene by *A. calcoaceticus* leads to the migration of secondary carbon atom during oxidation resulting in lactone. This compound is an intermediate in the synthesis of (3S, 6R)-3-methyl-6-(1-methylethyl)-9-decen-1-yl acetate which is an attractant for male *Aonidiella aurantii*—citrus fruits pest and can be applied with combination with other pest control agents. The next example of whole cells Baeyer–Villiger oxidation is biotransformation of (4R)-(–)-carvone by yeast *Trichosporum cutaneum* CCT 1903. The main product is hydrogenated lactone and side products–ketoacid, dihydrocarveol, and epoxydihydrocarveol. Biocatalytic Baeyer–Villiger oxidation can be performed also

by isolated enzymes. The most common sources of isolated BVMO are species of Acinetobacter, Nocardia (Norris) and, and Pseudomonas (Scheme 13.5) [51–53,55,56].

An example of selective bio-reduction of terpenes are bio-transformations of (–)-menthone 21 and (+)-pulegone to (1R, 3S, 4S)-(+)-neomenthol. Biocatalyst used for this transformation was *Hormonema* species. (+)-Neomenthol possesses antifungal activity against *Fusarium verticillioides* which causes bakanae disease in rice seedlings and Colletotrichum gloeosporioides–plants pathogenic fungus infesting apple and mango. Yeast *Hormonema species* was also applied for selective reduction of (4S)-isopiperitenone (Scheme 13.6).

Isopiperitenol is an intermediate in the preparation of (–)-menthol which production from natural sources. Another example of selective bio-reduction is biotransformation of (4R)-carvone. In case of biotransformation of (4R)-carvone the first step was hydrogenation of (-) Menthone to (1R, 4R)-dihydrocarvone and subsequent stereoselective reduction to (1R, 2S, 4R)- dihydrocarveol. In addition, when G. butleri had been used as biocatalysts, a small amount of side products were present. Dihydrocarveol is a flavoring food additive so environmentally friendly method of its production would be desirable (Scheme 13.7) [20].

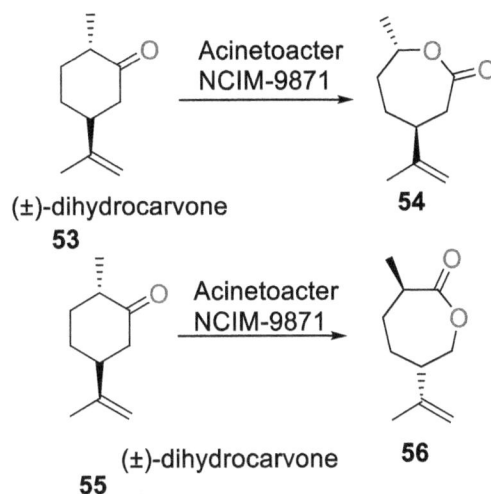

SCHEME 13.4 Microbial transformation of (±) Dihydrocarvone to seven-membered α- and β-lactone [44].

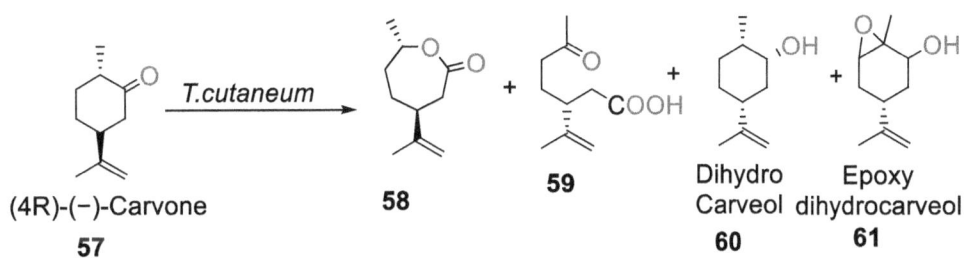

SCHEME 13.5 Microbial transformation of (4R)-(-)-Carvone [44].

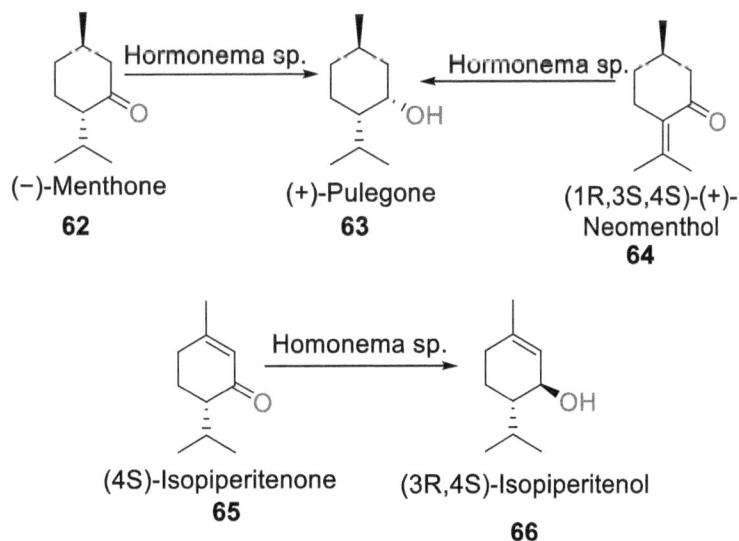

SCHEME 13.6 Microbial transformation (-)-Menthone and (4S)-Isopiperitenone [44].

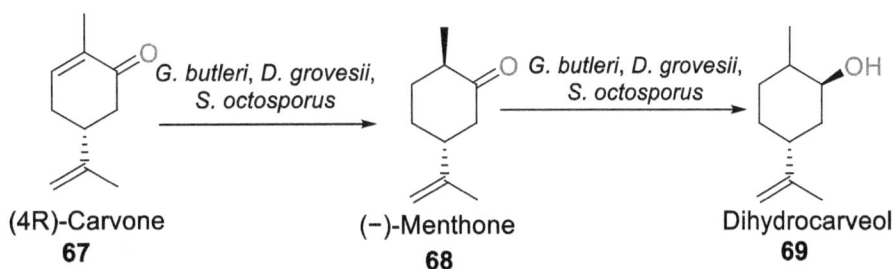

SCHEME 13.7 Microbial transformation of (4R)-Carvone [44].

The same biocatalysts were used for bioreduction of second isomer of carvone–(4S)-carvone. Biotransformation by *D. grovesii* resulted in the formation of allyl alcohol which possesses fragrance resembling that of mint and caraway. It is used in the cosmetic industry and as flavor additive in the food industry [57,58]. The same product was obtained in the case of using *G. butleri* and *S. octosporus* [7–13].

13.4 Kinetic Resolution

The same microorganism was used to resolve the diastereomeric mixture of carveol by selective oxidation. Biotransformation by *R. erythropolis* DCL14 revealed high stereoselectivity of dehydrogenase. As a result, two optically active products were obtained: (–)- (R)-carvone and (–)-cis-carveol (Schemes 13.8 and 13.9) [54,55,59].

Despite good results, commenced studies over largescale microbial transformations are focused generally on bio-transformations of small, nonpolar molecules. The challenge is to develop a system suitable for carrying out biotransformation of larger, more polar, and multifunctional compounds (Scheme 13.10) [60–62].

A variety of cyclic halogenated and oxygenated monoterpenoids from the red marine algae, the *Ochtodes crockeri* have been isolated. Many of these exhibit antifeedant, sedative and antibacterial properties. Bromination of 89, available from dimedone, followed by the addition of vinyl magnesium bromide affords bis-allylic alcohol 90 as a 1:1 mixture of diastereomers. Hydrolysis in aqueous acetone provides metabolites 91 and 91 in three steps. Chlorination of 97 by reaction with thionyl chloride gives the rearranged allylic chloride 95. Treatment with one equivalent of silver acetate and separation of the isomers affords acetate 96 which is converted to

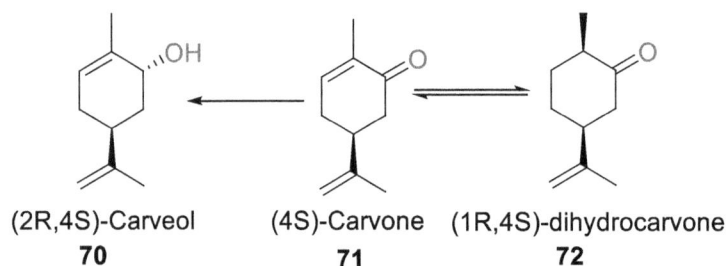

SCHEME 13.8   Microbial transformation (4S)-Carvone [44].

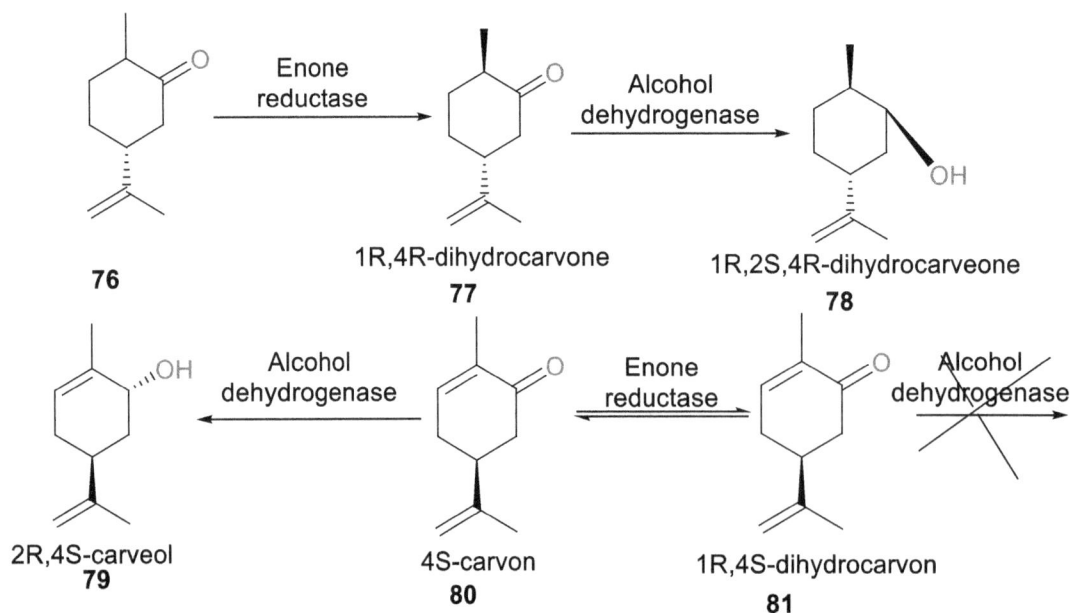

SCHEME 13.9   Microbial transformation (4R)-Carvone and 2R, 4S-Carveol [44].

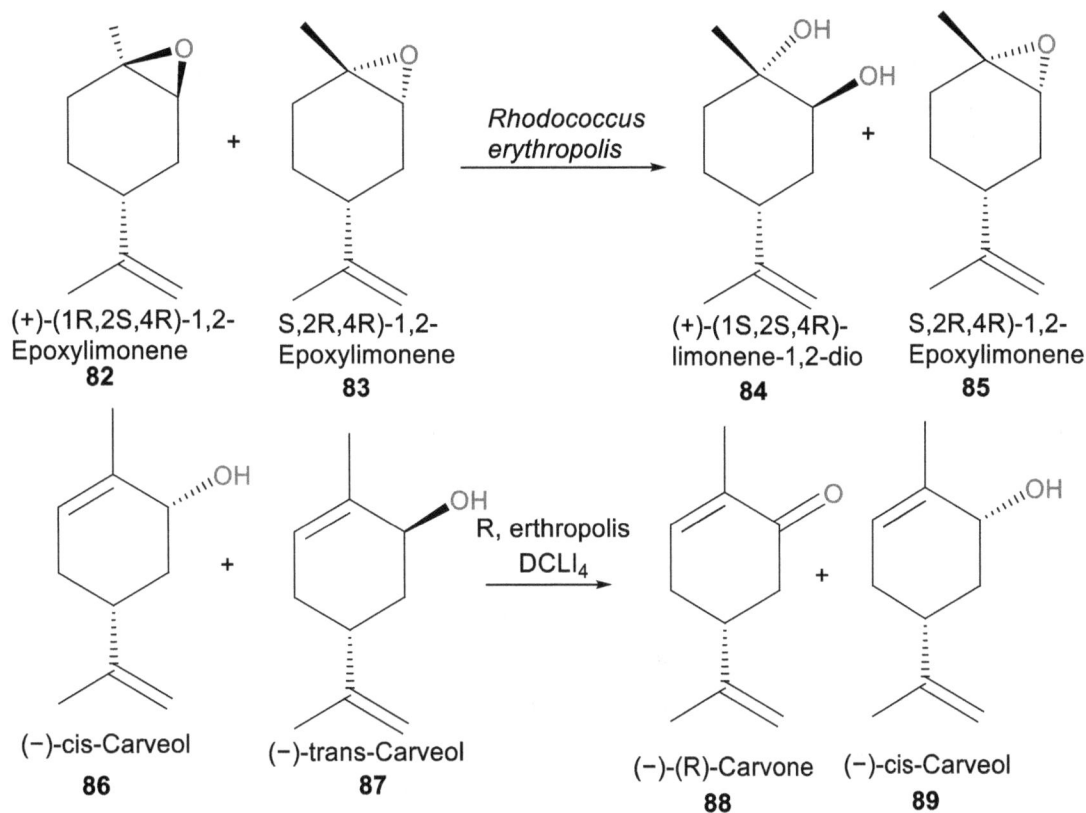

SCHEME 13.10 Microbial transformation of (+)-(1R, 2S, 4R)-1,2- Epoxylimonene and (-) cis- synthesis of Terpenoids from *Chondrococcus (Desmia)* and *Ochtodes* spp.

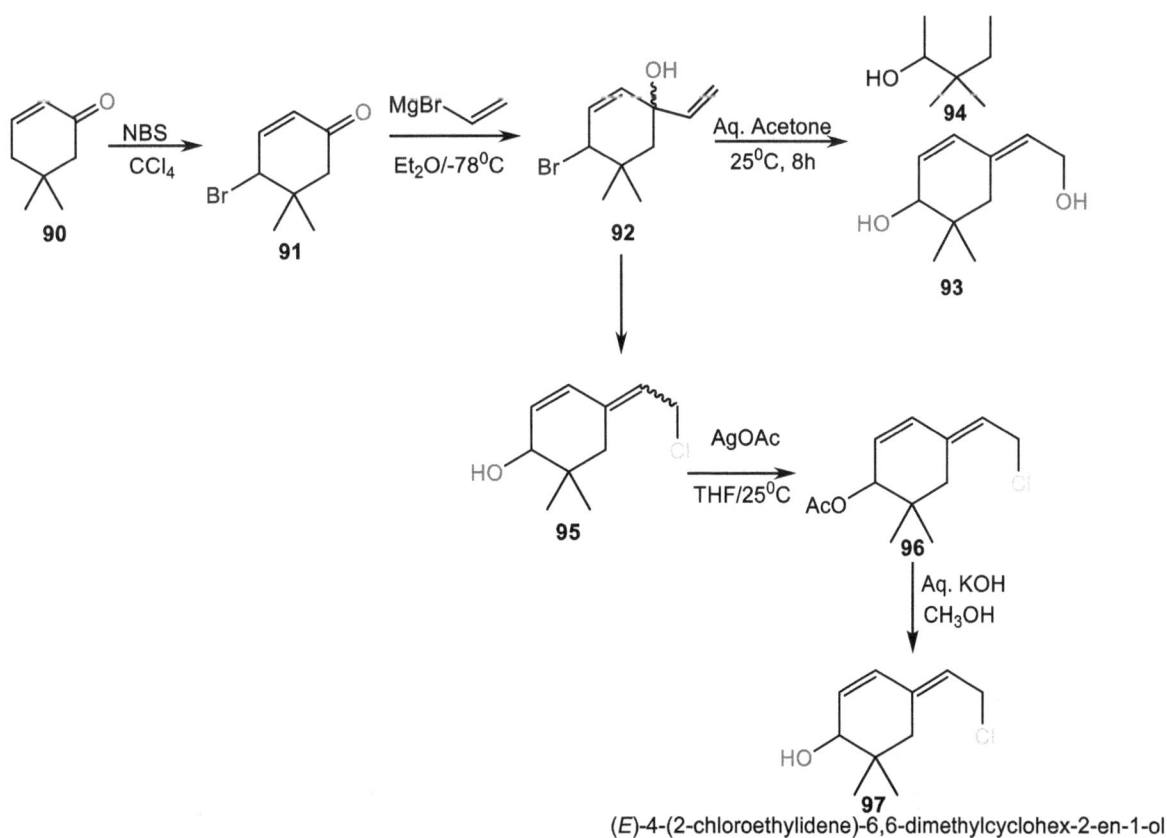

(+)-(1R,2S,4R)-1,2-
Epoxylimonene
82

S,2R,4R)-1,2-
Epoxylimonene
83

Rhodococcus erythropolis

(+)-(1S,2S,4R)-
limonene-1,2-dio
84

S,2R,4R)-1,2-
Epoxylimonene
85

(−)-cis-Carveol
86

(−)-trans-Carveol
87

R, erthropolis
DCLI$_4$

(−)-(R)-Carvone
88

(−)-cis-Carveol
89

90

NBS
CCl$_4$

91

MgBr
Et$_2$O/-78^{0}C

92

Aq. Acetone
25^{0}C, 8h

94

93

95

AgOAc
THF/25^{0}C

96

Aq. KOH
CH$_3$OH

97
(E)-4-(2-chloroethylidene)-6,6-dimethylcyclohex-2-en-1-ol

SCHEME 13.11 Synthesis of *Ochtodes crockery* metabolites.

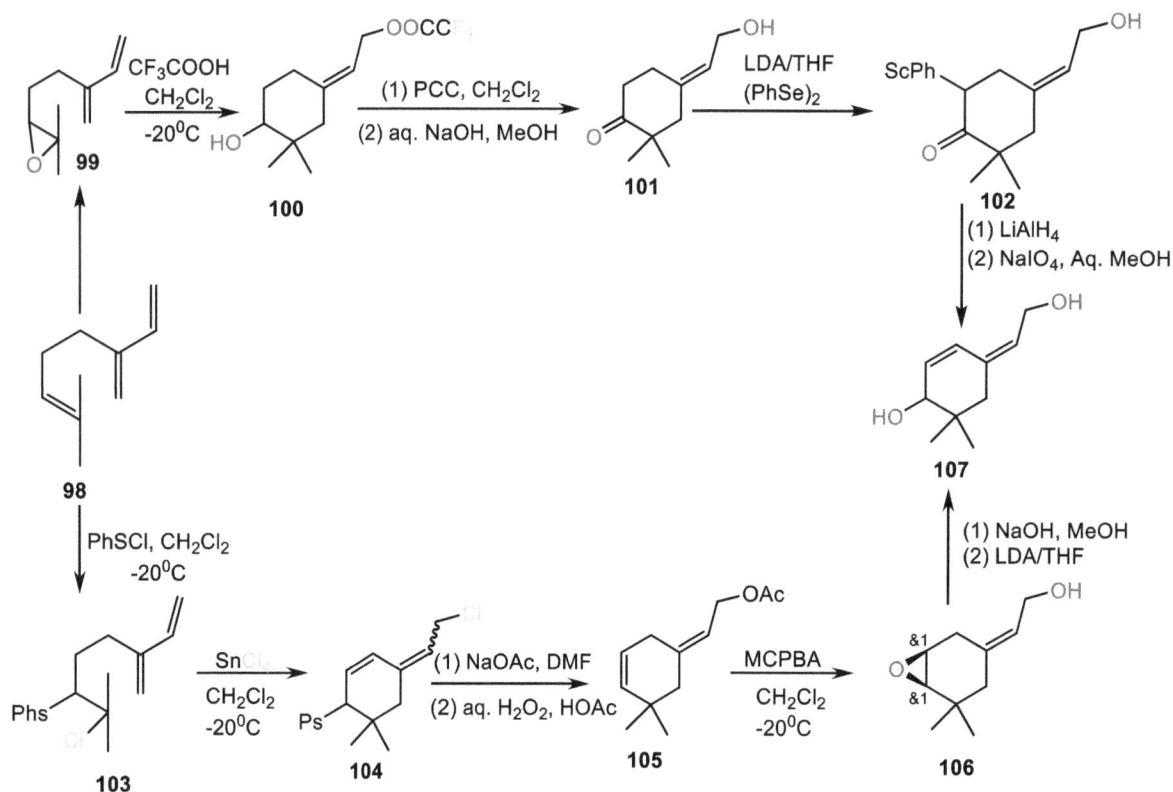

SCHEME 13.12 Synthesis of *Ochtodes crockeri* metabolite.

metabolite 97 by saponification, thus giving 97 in five steps (Scheme 13.11) [60–62].

Another *Ochtodes crockeri* metabolite 107 has been synthesized by two different routes by Masaki (Scheme 13.12). Cation-olefin cyclization of epoxide 99 gives trifluoroacetate 100. PCC oxidation gives 101 and saponification followed by phenylselenylation of the ketone yields 102. LAH reduction of the ketone 107. A second route to 107 involves the addition of phenylsulfenyl chloride to myrcene 98 to obtain 103. Treatment of 103 with tin (IV) chloride initiates cation olefin cyclization via episulfonium ion formation to produce allylic chloride 104 as a mixture of isomers. Displacement of the chloride with acetate and sulfoxide elimination provides 105. Oxidation of 105 to the epoxide 106 occurs upon treatment with MCPBA. Saponification of the acetate and eliminative opening of the epoxide with LDA proceeds with kinetic selectivity and provides the natural product 107 in seven steps [63].

Conclusion

This chapter summarizes the comprehensive information about terpenes from marine sources, e.g. seawater algae, sponges, and fungi their occurrence and medicinal applications, along with method of modification in order to enhance the yield. This chapter also comprises the microbial transformation of terpenoids i.e menthol to hydroxy menthol, microbial transformation of (R)-Limonene to terpinol, terpinene and its epoxides, Baeyer–Villiger oxidation, Microbial transformation of (±) Dihydrocarvone to seven-membered

α- and β-lactone, Menthone and (4S)-Isopiperitenone etc. It also includes Kinetic resolution, Microbial transformation (4R)-Carvoneand2R, 4S-Carveol etc.

REFERENCES

1. Coll, J.C.; Bowden, B.F.; Tapiolas, D.M.; Willis, R.H.; Djura, P.; Streamer, M.; Trott, L. The Terpenoid Chemistry of Soft Corals and Its Implications. *Tetrahedron*, **1985**, 41, 1085–1092.

2. Aceret, T.L.; Sammarco, P.W.; Coll, J.C. Effects of Diterpenes Derived from the Soft Coral Sinularia flexibilis on the Eggs, Sperm and Embryos of the Scleractinian Corals Montipora digitata and Acropora tenuis. *Marine Biology*, **1995**, 122, 317–323.

3. De Nys, R.; Coll, J.C.; Price, I.R. Chemically Mediated Interactions between the Red Alga Plocamium hamatum (Rhodophyta) and the Octocoral Sinularia cruciata (Alcyonacea). *Marine Biology*, **1991**, 108, 315–320

4. Leone, P.A.; Bowden, B.F.; Carroll, A.R.; Coll, J.C. Chemical Consequences of the Relocation of the Soft Coral Lobophytum compactum and Its Placement in Contact with the Red Alga Plocamium hamatum. *Marine Biology*, **1995**, 122,: 675–679

5. Sullivan, B.; Faulkner, J.D.; Webb, L. Siphonodictidine, a Metabolite of the Burrowing Sponge Siphonodictyon sp. That Inhibits Coral Growth. *Science*, **1983**, 221, 1175–1176.

6. Ojima, I.; Chakravarty, S.; Inoue, T.; Lin, S.; He, L.; Horwitz, S.B.; Kuduk, S.D.; Danishefsky, S.J. A Common Pharmacophore for Cytotoxic Natural Products That Stabilize Microtubules. *Proceedings of the National Academy of Sciences of the United States of America*, **1999**, 96, 4256–4261.

7. Lindel, T.; Jensen, P.R.; Fenical, W.; Long, B.H.; Casazza, A.M.; Carboni, J.; Fairchild, C.R. Eleutherobin, a New Cytotoxin That Mimics Paclitaxel (Taxol) by Stabilizing Microtubules. *Journal of the American Chemical Society*, **1997**, 119, 8744–8745.

8. Cinel, B.; Roberge, M.; Behrisch, H.; Van Ofwegen, L.; Castro, C.B.; Andersen, R.J. Antimitotic Diterpenes from Erythropodium caribaeorum Test Pharmacophore Models for Microtubule Stabilization. *Organic Letters*, **2000**, 2, 257–260.

9. Taglialatela-Scafati, O.; Deo-Jangra, U.; Campbell, M.; Roberge, M.; Andersen, R.J. Diterpenoids from cultured Erythropodium caribaeorum. *Organic Letters*, **2002**, 4, 4085–4088.

10. Pietras, R.J.; Weinberg, O.K. Antiangiogenic Steroids in Human Cancer Therapy. *Evidence-Based Complementary and Alternative Medicine*, **2005**, 2, 49–57.

11. Wessels, M.; Konig, G.M.; Wright, A.D. New Natural Product Isolation and Comparison of the Secondary Metabolite Content of Three Distinct Samples of the Sea Hare Aplysia dactylomela from Tenerife. *Journal of Natural Products*, **2000**, 63, 920–928.

12. Garson, M.J.; Simpson, J.S. Marine Isocyanides and Related Natural Products – Structure, Biosynthesis and Ecology. *Natural Product Reports*, **2004**, 21, 164–179.

13. Angerhofer, C.K.; Pezzuto, J.M.; König, G.M.; Wright, A.D.; Sticher, O. Antimalarial Activity of Sesquiterpenes from the Marine Sponge Acanthella klethra. *Journal of Natural Products*, **1992**, 55, 1787–1789.

14. Wright, A.D.; König, G.M.; Angerhofer, C.K.; Greenidge, P.; Linden, A.; Desqueyroux-Faundez, R. Antimalarial Activity: The Search for Marine-Derived Natural Products with Selective Antimalarial Activity. *Journal of Natural Products*, **1996**, 59, 710–716.

15. Adam, W.; Mitchell, C.M.; Saha-Moller, C.R. Steric and Electronic Effects in the Diastereoselective Catalytic Epoxidation of Cyclic Allylic Alcohols with Methyltrioxorhenium (MTO). *European Journal of Organic Chemistry*, **1999**, 4, 785–790.

16. Itoh, T.; Jitsukawa, K.; Kaneda, K.; Teranishi, S. Vanadium Catalyzed Epoxidation of Cyclic Allylic Alcohols. Stereoselectivity and Stereocontrol Mechanism. *Journal of the American Chemical Society*, **1979**, 101, 159–169.

17. de Faveri, G.; Ilyashenko, G.; Watkinson, M. Recent Advances in Catalytic Asymmetric Epoxidation Using the Environmentally Benign Oxidant Hydrogen Peroxide and Its Derivatives. *Chemical Society Reviews*, **2011**, 40, 1722–1760.

18. Farooq A, Tahara S, Choudhary I, Rahman A-U, Ahmed Z, Baser KH, Demirci F Biotransformation of (–)-α-Pinene by Botrytis cinerea. *Naturforsch*, **2002**, 57c, 303–306.

19. Lochyński, S.; Frąckowiak, B.; Librowski, T.; Czarnecki, R.; Grochowski, J.; Serda, P.; Pasenkiewicz-Gierula, M. Stereochemistry of Terpene Derivatives. Part 3: Hydrolytic Kinetic Resolution as a Convenient Approach to Chiral Aminohydroxyiminocaranes with Local Anaesthetic Activity. *Tetrahedron: Asymmetry*, **2002**, 13, 873–878.

20. Moniczewski, A.; Librowski, T.; Lochyński, S.; Strub, D. Evaluation of the Irritating Influence of Carane Derivatives and Their Antioxidant Properties in a Deoxyribose Degradation test. *Pharmaceutical Representative*, **2011**, 63, 120–129.

21. Bouaicha, N.; Pesando, D.; Puel, D.; Tringali, C. Cytotoxic Diterpenoids from the Brown Alga Dilophus ligulatus. *Journal of Natural Products*, **1993**, 56, 1747–1752.

22. (a) Kashman Y, Hirsch S, Koehn F, Cross S. Reiswigins A and B, Novel Antiviral Diterpenes from a Deepwater Sponge. *Tetrahedron Letters*, **1987**, 28, 5461–5464; (b) Snider BB, Yang K. Synthesis of (–)-Reiswigin A. Assignment of Absolute and Relative Configuration. *Tetrahedron Letters*, **1989**, 30, 2465–2468; (c) Snider BB, Yang K. Stero- and Enantiospecific Synthesis of (–)-reiswigin A and B. Assignment of Absolute and Relative Configuration. *The Journal of Organic Chemistry*, **1990**, 55, 4392–4399; (d) Kim D, Shin KJ, Kim IY, Park SW. A Total Synthesis of (–)-Reiswigin A via Sequential Claisen Rearrangement-Intramolecular Ester Enolate Alkylation. *Tetrahedron Letters*, **1984**, 35, 7957–7960.

23. a Gedara SR, Abdel-Halim OB, El-Sharkawy SH, Salama OM, Shier TW, Halim AF. Cytotoxic Hydrazulene Diterpenes from the Brown Alga Dictyota dichotoma. *Zeitschrift für Naturforschung*, **2003**, 58c, 17–22; (b) Siamopoulou P, Bimplakis A, Iliopoulou D, Vagias C, Cos P, Berghe DV, Roussis V. Diterpenes from Brown Algae Dictyota dichotoma and Dictyota linearis. *Phytochemistry*, **2004**, 65, 2025–2030; (c) Ovenden SPB, Nielson JL, Liptrot CH, Willis RH, Tapiolas DM, Wright AD, Motti CA. Update of Spectroscopic Data for 4-Hydroxydictyolactone and Dictyol E Isolated from Halimeda stuposa – Dictyota sp. Assemblage. *Molecules*, **2012**, 17, 2929–2938; (d) Abou-El-Wafa GSE, Shaaban M, Shaaban KA, El-Nagger MEE, Maier A, Fiebig HH, Laatsch H. Pachydictyols B and C: New Diterpenes from Dictyota dichotoma Hudson. *Marine Drugs*, **2013**, 11, 3109–3123.

24 González MA. Spongiane Diterpenoids. *Current Bioactive Compounds*, **2007**, 3, 1–36; (f) González MA. Total Syntheses and Synthetic Studies of Spongiane Diterpenes. *Tetrahedron*, **2007**, 64, 445–467.

25. (a) Schmitz FJ, Chang JS, Hossain MB, van der Helm D. Marine Natural Products: Spongiane Derivatives from the Sponge Igernella notabilis. *The Journal of OrganicChemistry*, **1985**, 50, 2862–2865;(b) Karuso P, Taylor WC. The Constituents of Marine Sponges II. The Isolation from Aplysilla rosea (dendroceratida) of (5R*, 7S*, 8R*, 9S*, 10R*, 13S*, 14S*, 15S*)-15,17-Epoxy-17-Hydroxy-16-Oxospongian-7-yl Butyrate (aply-reosol-1) and related Diterpenes (Aplyreosol-2 to-6). *Australian Journal of Chemistry*, **1986**, 39, 1629–1641; (c) Hambley TW, Taylor WC, Toth S. The Constituents of Marine Sponges VII: The absolute Stereochemistry of Aplyroseol-1. *Australian Journal of Chemistry*, **1997**, 50, 391–394.(d) Abad A, Arnó M, Marin ML, Zaragozá RJ. Spongian Pentacyclic Diterpenes. Stereoselectice Synthesis of Aplyroseol-1, Aplyroseol-2, and Deacetylaplyroseol-2. *Journal of the Royal Society of Chemistry: Perkin Transactions*, **1993**, 1, 1861–1867; (e) Arnó M, Betancur-Galvis L, González MA, Sierra J, Zaragozá RJ. Synthesis and Cytotoxic Activity of Novel C7-Functionalized Spongiane Diterpenes. *Bioorganic & Medicinal Chemistry*, **2003**, 11, 3171–3177.

26. (a) De Silva ED, Scheuer PJ. Manoalide, an Antibiotic Sesterterpenoid from the Marine Sponge Lufariella variabilis (Polajeff). *Tetrahedron Letters*, **1980**, 21,

1611–1614;(b) Soriente A, De Rosa M, Scettri A, Sodano G, Terencio MC, Paya M, Alcaraz MJ. Manoalide. *Current Medicinal Chemistry*, **1999**, 6, 415–431;(c) Folmer F, Jaspars M, Schumacher M, Dicato M, Diederich M. Marine Natural Products Targeting PhospholipaseA2. *Biochemical Pharmacology*, **2010**, 80, 1793–1800;(d) Hog DT, Webster R, Trauner D. Synthetic Approaches toward Seesterterpenoids. *Natural Product Research*, **2012**, 29, 752–779;(e) Salam KA, Furuta A, Noda N, Tsuneda S, Sekiguchi Y, Yamashita A, Moriishi K, Nakakoshi M, Tsubuki M, Tani H, Tanaka J, Akimitsu N. Inhibition of Hepatitis C Virus NS3 Helicase by Manoalide. *Journal of Natural Products*, **2012**, 75, 650–654.

27. (a) Fattorusso E, Magno S, Santacroce C, Sica D. Scalarin, a New Pentacyclic C-25 Terpenoid from the Sponge Cacospongia scalaris. *Tetrahedron*, **1972**, 28, 5993–5997;(b) Cimino G, De Stefano S, Minale L, Trivellone E. 12-Epi-Scalarin and 12-Epi-Deoxoscalarin, Sesterterpenes from the Sponge Spongia nitens. *Journal of the Royal Society of Chemistry: Perkin Transactions*, **1977**, 1, 1587–1593;(c) Cambie RC, Rickard CEF, Rutledge PS, Yang XS. Scalarolide and Scalarin, Sesterterpenes from Cacospongia and Ircinia Sponges. *Acta Crystallographica Section C*, **1999**, C55, 112–114;(d) Yang X, Shao Z, Zhang X. Sesterterpenes from the Sponge Dysidea sp. *Zeitschrift für Naturforschung*, **2010**, 65b, 625–627;(e) González MA. Scalarene Sesterterpenoids. *Current Bioactive Compounds*, **2010**, 6, 1–29;(f) Lee YJ, Lee JW, Lee DG, Lee HS, Kang JS, Yun J. Cytotoxic Sesterterpenoids Isolated from the Marine Sponge Scalrispongia sp. *International Journal of Molecular Sciences*, **2014**, 15, 20045–20053.

28. Misawa N, Nakagawa M, Kobayashi K, Yamano S, Izawa Y, Nakamura K, Harashima K. Elucidation of the Erwinia uredovora Carotenoid Biosynthetic Pathway by Functional Analysis of Gene Products Expressed in Escherichia coli. *Journal of Bacteriology*, **1990**, 172(12), 6704–6712. doi:10.1128/jb.172.12.6704-6712.1990.

29. Ruther, A., Misawa, N., Böger, P. et al. Production of Zeaxanthin in Escherichia coli Transformed with Different Carotenogenic Plasmids. *Applied Microbiology and Biotechnology*, **1997**, 48, 162–167. doi:10.1007/s002530051032.

30. Albrecht, U.; Sun, Z.S.; Eichele, G.; Lee, C.C. A Differential Response of Two Putative Mammalian Circadian Regulators, mper1and mper2, to Light. *Cell*, **1997**, 91(7), 1055–1064.

31. Kajiwara S, Fraser PD, Kondo K, Misawa N. Expression of an Exogenous Isopentenyl Diphosphate Isomerase Gene Enhances Isoprenoid Biosynthesis in Escherichia coli. *Biochemical Journal*, **1997**, 324(Pt 2), 421–426.

32. Wang, Q., Hasan, G., Pikielny, C.W. Preferential Expression of Biotransformation Enzymes in the Olfactory Organs of Drosophila melanogaster, the Antennae. *Journal of Biological Chemistry*, **1999**, 274(15), 10309–10315.

33. Harker M., Bramley P.M. Expression of Prokaryotic 1-Deoxy-D-Xylulose-5-Phosphatases in Escherichia coli Increases Carotenoid and Ubiquinone Biosynthesis. *FEBS Letters*, **1999**, 448(1), 115–119.

34. Matthews PD, Wurtzel ET. Metabolic Engineering of Carotenoid Accumulation in Escherichia coli by Modulation of the Isoprenoid Precursor Pool with Expression of Deoxyxylulose Phosphate Synthase. *Applied Microbiology and Biotechnology*, **2000**, 53(4), 396–400. doi:10.1007/s002530051632.

35. Wang Z, Haydon PG, Yeung ES. Direct Observation of Calcium-Independent Intercellular ATP Signaling in Astrocytes. *Analytical Chemistry*, **2000**, 72, 2001–2007.

36. Huang Q., Roessner C. A., Croteau R., Scott A. I. Engineering Escherichia coli for the Synthesis of Taxadiene, a Key Intermediate in the Biosynthesis of Taxol. *Bioorganic & Medicinal Chemistry*, **2001**, 9, 2237–2242. doi:10.1016/S0968-0896(01)00072-4.

37. Kim S. W., Keasling J. D. Metabolic Engineering of the Nonmevalonate Isopentenyl Diphosphate Synthesis Pathway in Escherichia coli Enhances Lycopene Production. *Biotechnology and Bioengineering*, **2001**, 72, 408–415. doi:10.1002/1097-0290(20000220)72:4<408::AID-BIT1003>3.0.CO;2-H.

38. Carter OA, Peters RJ, Croteau R. Monoterpene Biosynthesis Pathway Construction in Escherichia coli. *Phytochemistry*, **2003**, 64, 425–433. doi:10.1016/S0031-9422(03)00204-8.

39. Martin, V.J.; Pitera, D.J.; Withers, S.T.; Newman, J.D.; Keasling, J.D. Engineering a Mevalonate Pathway in Escherichia coli for Production of Terpenoids. Nature Biotechnology, 2003, 21, 796–802.

40. Alper H, Jin Y-S, Moxley JF, et al. Identifying Gene Targets for the Metabolic Engineering of Lycopene Biosynthesis in Escherichia coli, *Metabolic Engineering*, **2005**, 7, 155–164.

41. Pitera DJ, Paddon CJ, Newman JD, Keasling JD. Balancing a Heterologous Mevalonate Pathway for Improved Isoprenoid Production in Escherichia coli. *Metabolic Engineering*, **2007**, 9(2), 193–207. doi:10.1016/j.ymben.2006.11.002.

42. Jin, Y. S.; Stephanopoulos, G. Multi-Dimensional Gene Target Search for Improving Lycopene Biosynthesis in Escherichia coli. *Metabolic Engineering*, **2007**, 9, 337–347.

43. Chang, M.C.; Eachus, R.A.; Trieu, W.; Ro, D.K.; Keasling, J.D. Engineering Escherichia coli for Production of Functionalized Terpenoids Using Plant P450s. *Nature Chemical Biology*, **2007**, 3, 274–277.

44. Kuriata, R.A.; Strub, D.; Lochyński, S. Application of Microorganisms towards Synthesis of Chiral Terpenoid Derivatives. *Applied Microbiology and Biotechnology*, **2012**, 95, 1427–1436. doi:10.1007/s00253-012-4304-9.

45. Mihovilovic MD, Mueller B, Stanetty P Monooxygenasemediated Baeyer–Villiger Oxidations. *European Journal of Organic Chemistry*, **2002**, 22, 3711–3730.

46. Kamerbeek NM, Janssen DB, van Berkel WJH, Fraaije MW. Baeyer–Villiger Monooxygenases, an Emerging Family of Flavin Dependent Biocatalysts. *Advanced Synthesis & Catalysis*, **2003**, 345, 667–678.

47. Renz, M.; Meunier, B. 100 years of Baeyer–Villiger Oxidations. *European Journal of Organic Chemistry*, **1999**, 4, 737–750.

48. Ottolina, G.; Carrea, G.; Collonna, S.; Ruckemann, A. A Predictive Active Site Model of Cyclohexanone Monooxygenase Catalyzed Baeyer–Villiger Oxidation. *Tetrahedron: Asymmetry*, **1996**, 7, 1123–1136.

49. Anderson, R.J.; Adams, K.G.; Chinn, H.R.; Henrick, C.A. Synthesis of the Optical Isomers Of 3-Methyl-6-Isopropenyl-9-Decen-1-yl Acetate, a Component of the California Red Scale Pheromone. *Journal of Organic Chemistry*, **1980**, 45, 2229–2236.

50. Pinheiro, L.; Marsaioli, A.J. Microbial Monooxygenases Applied to Fragrance Compounds. *Journal of Molecular Catalysis B: Enzymatic*, **2007**, 44, 78–86.

51. Secundo, F.; Carrea, G.; Riva, S.; Battistel, E.; Bianchi, D. Cyclohexanone Monooxygenase Catalyzed Oxidation of Methyl Phenyl Sulfide and Cyclohexanones with Macromolecular NADP in a Membrane Reactor. *Biotechnology Letters*, **1993**, 15, 865–870.

52. Norris, D.B.; Trudgill, P.W. The Metabolism of Cyclohexanol by Nocardia globerula CL 1. *Biochemical Journal*, **1971**, 121, 363–370.

53. Alphand, V.; Furstoss, R.; Pedragosa-Moreau, S.; Roberts, S.M.; Willetts, A.J. Comparison of Microbiologically and Enzymatically Mediated Baeyer–Villiger Oxidation: Synthesis of Optically Active Caprolactones. *Journal of the Chemical Society, Perkin Transactions 1*, **1996**, 1, 1867–1872.

54. Faber, K. *Biotransformation in Organic Chemistry*. Heidelberg: Springer, 2000, 177–217.

55. Van Dyk, M.S.; van Rensburg, E.; Rensburg, I.P.B.; Moleleki, N. Biotransformation of Monoterpenoid Ketones by Yeasts and Yeast-Like Fungi. *Journal of Molecular Catalysis B: Enzymatic*, **1998**, 5, 149–154.

56 Roberts, S.M.; Wan, P.W. Enzyme-Catalyzed Baeyer–Villiger Oxidation. *Journal of Molecular Catalysis B: Enzymatic*, **1998**, 4, 111–136.

57. Serra, S.; Brenna, E.; Fuganti, C.; Maggioni, F. Lipase-Catalyzed Resolution of p-Menthan-3-ols Monoterpenes: Preparation of the Enantiomer-Enriched Forms of Menthol, Isopulegol, Trans- and Cispiperitol, and cis Isopiperitenol. *Tetrahedron: Asymmetry*, **2003**, 14, 3313–3319.

58. Carballeira, J.D.; Valmaseda, M.; Alvarez, E.; Sinisterra Gago, J.V. Gongronella butleri, Schizosaccharomyces octosporus and Diplogelasinospora grovesii: Novel Microorganisms Useful for the Stereoselective Reduction of Ketones. *Enzyme and Microbial Technology*, **2004**, 34, 611–623.

59. Robinson, D.E.J.E.; Bull, S.D. Kinetic Resolution Strategies Using Non-Enzymatic Catalysts. *Tetrahedron: Asymmetry*, **2003**, 14, 1407–1446.

60. van der Werf, M.J.; Orru, R.V.A.; Overkamp, K.M.; Swarts, H.J.; Osprian, I.; Steinreiber, A.; de Bont, J.A.M.; Faber, K. Substrate Specificity and Stereospecificity of Limonene-1,2-Epoxide Hydrolase from Rhodococcus erythropolis DCL14; an Enzyme Showing Sequential and Enantioconvergent Substrate Conversion. *Applied Microbiology and Biotechnology*, **1999a**, 52, 380–385.

61. Tecelao, C.S.R.; van Keulen, F.; da Fonseca, M.M.R. Development of a Reaction System for the Selective Conversion of (–)-Trans-Carveol to (–)-Carvone with Whole Cells of *Rhodococcus erythropolis* DCL14. *Journal of Molecular Catalysis B: Enzymatic*, 11, 719–724.

62. van der Werf, M.J.; van der Ven, C.; Barbirato, F.; Eppink, M.H.M.; de Bont, J.A.M.; van Berkel, J.H. Stereoselective Carveol Dehydrogenase from Rhodococcus erythropolis DCL14. *Journal of Biological Chemistry*, **1999b**, 274, 26296–26304.

63. Cikoš, A.M.; Jurin, M.; Rakovac, R.C.; Joki, S.; Jerkovi, I. Update on Monoterpenes from Red Macroalgae: Isolation, Analysis, and Bioactivity. *Marine Drugs*, **2019**, 17, 537; doi:10.3390/md17090537.

14

A Synthetic Overview of Terpenes and Nor Diterpenes

Bimal Krishna Banik[*]**, Abhishek Tiwari**[†]**, and Biswa Mohan Sahoo**[‡]

CONTENTS

14.1   Introduction 468
14.2   Vitamin A (Retinol Acetate) 468
     14.2.1   Retrosynthetic Route of Dehydrolinalool...468
     14.2.2   First Step: Synthesis of C-15 Wittig Salt from Dehydrolinalool..............................468
     14.2.3   Second Step: Synthesis of C5 Acetate..468
     14.2.4   Third Step: Synthesis of Vitamin A acetate...468
     14.2.5   Fourth Step: Synthesis of β-carotene 468
14.3   Synthetic Route of Cafestol 468
14.4   Retrosynthesis of (-)-Cafestol 468
     14.4.1   Step 1: Synthesis of diazoketoester 470
     14.4.2   Step 2: Synthesis of (-) Cafestol...472
14.5   Baccatin III as a Taxol Precursor 472
14.6   Baccatin Synthesis..472
     14.6.1   Step 1: Synthesis of sulfonylhydrazone..474
     14.6.2   Step 2: Synthesis of cyclohexenaldehyde 474
     14.6.3   Step 3: Synthesis of diastereomers 475
     14.6.4   Step 4: Synthesis of 7-O-triethylsilylbaccatin-III 475
14.7   Triterpenes..476
     14.7.1   Lupeol...476
14.8   Synthesis of (+) Lupeol 476
     14.8.1   Retro synthesis of 6-methoxy- a-tetralone from 476
           14.8.1.1   Step 1: Synthesis of substituted benzoate 476
           14.8.1.2   Step 2: Synthesis of allylketobenzoate 479
           14.8.1.3   Step 3: Synthesis of ketal ester 479
14.9   Diterpenoids..481
14.10 Classification...481
14.11 Gibberellanes..481
14.12 Alkaloids of C20-Diterpenoid...481
14.13 C20-Atisines Class 482
     14.13.1 C20-Denudatine Class...482
     14.13.2 C20-Hetidine Class 482
     14.13.3 C20-Vakognavine Class 483
     14.13.4 C20-Napelline Class..483
     14.13.5 C20-Anopterine Class 484
     14.13.6 C20-Rearranged Classes 484
     14.13.7 Bis-Diterpenoid Alkaloids..485
14.14 Conclusion 486
References..486

[*] Department of Mathematics and Natural Sciences, College of Sciences and Human Studies, Prince Mohammad Bin Fahd University, Al Khobar, Kingdom of Saudi Arabia. bimalbanik10@gmail.com; bbanik@pmu.edu.sa.

[†] Faculty of Pharmacy, Pharmacy Academy, IFTM University, Lodhipur Rajput, Moradabad-244102, Uttar Pradesh, India. abhishekt1983@gmail.com.

[‡] Roland Institute of Pharmaceutical Sciences, Berhampur affiliated to Biju Patnaik University of Technology (BPUT), Rourkela, Odisha, India.

DOI: 10.1201/9781003008682-14

14.1 Introduction

Terpenes are a type of natural product that includes chemicals having the formula (C5H8)n. These unsaturated hydrocarbons, which are made up of over 30,000 chemicals, are mostly generated by plants, notably conifers. Monoterpenes (C10), sesquiterpenes (C15), diterpenes (C20), and so on are the different types of terpenes. Alpha-pinene, a primary component of turpentine, is a well-known monoterpene. In this section we have discussed the synthesis of Vitamin A, Cafestol, Baccatin III and Lupeol.

14.2 Vitamin A (Retinol Acetate)

14.2.1 Retrosynthetic Route of Dehydrolinalool

In the course of a Wittig reaction between C15 Wittig salt and C5 acetate, the C-11, C-12 double bond of the diterpene retinol acetate is clearly detached. Retrosynthesis of the C15 salt yields dehydrolinalool as the starting reagent through Eionylidenethanol, vinylionol, ethynylionol, E-ionone and pseudoionone, with processes similar to those used in monoterpene commercial syntheses (Scheme 14.1). The C5 acetate is formed when 4,4-dialkoxy-3-methyl-1-buten-3-ol undergoes an allyl rearrangement. The latter is, of course, the result of hydrogenation of 4,4-dialkoxy-3-methyl-1-butin-3-ol, which is made possible by ethynylation of dialkoxyacetone, which is generated by oxidation of acetoneketal. Pommer developed the notion of a convergent commercial synthesis of retinol acetate (Scheme 14.1) [1].

Vitamin A synthesis involves three steps. The first step is the synthesis of C-15 Wittig Salt from Dehydrolinalool, then synthesis of C5 Acetate and the last synthesis of Vitamin A acetate. In last we have also discussed the Synthesis of β- carotene [1].

14.2.2 First Step: Synthesis of C-15 Wittig Salt from Dehydrolinalool

In order to accomplish the synthesis, methylacetoacetate is transesterified with racemic dehydrolinalool and the resulting dehydrolinaloylacetoacetate is subjected to *Carroll* decarboxylation to pseudoionone. This undergoes acid-catalyzed cyclization to E-ionone which is ethynylated to ethynylionone by ethyne in the presence of sodium hydroxide as the base. Ethynylionone undergoes partial catalytic hydrogenation to vinylionol using a deactivated catalyst. In the presence of hydrobromicacid, allyl rearrangement of vinylionol directly yields E-ionylidenebromoethane which reacts with triphenylphosphane to the crystallizing C15-Wittig salt (Scheme 14.2).

14.2.3 Second Step: Synthesis of C5 Acetate

To complete the synthesis, methylacetoacetate is transesterified with racemic dehydrolinalool and the resultant dehydrolinaloylacetoacetate is decarboxylated to pseudoionone

through Carroll decarboxylation. This is cyclized acidically to E-ionone, which is then ethynylated to ethynylionone by ethyne in the presence of sodium hydroxide as the base. Using a deactivated catalyst, ethynylionone is partially catalytically hydrogenated to vinylionol. The allyl rearrangement of vinylionol in the presence of hydrobromicacid produces E-ionylidenebromoethane, which combines with triphenylphosphane to form the crystallising C15-Wittig salt (Scheme 14.3).

14.2.4 Third Step: Synthesis of Vitamin A acetate

The C15 salt undergoes Wittig alkenylation with the C5 acetate in a nearly quantitative manner. Retinol acetate, a more stable form of vitamin A (retinol), is separated from n-hexane by recrystallization (Scheme 14.4) [2].

14.2.5 Fourth Step: Synthesis of β-carotene

E-carotene is produced via an intermolecular Mcmurry deoxygenative coupling of retinal (vitamin A aldehyde) (Scheme 14.5).

14.3 Synthetic Route of Cafestol

(-)-Cafestol, an anti-inflammatory diterpenoid found in coffee belongs to the class of rearranged kauranes.

14.4 Retrosynthesis of (-)-Cafestol

The 1,2-diol substructure of cafestol **32** is introduced via dihydroxylation of the exocyclic C-16-C-17 double bond in the precursor **33**, according to an evident design of its synthesis. The idea of Corey and coworkers **37** to use the ring strain of the cyclopropylmethylium ion 36 as a vehicle to build up the kauranepentacycle: thus, 36 is expected to generate the precursor 34 via carbenium ion 35. Additional double bonds in positions 5,6 and 11,12 of the precursor 34, the purposes of which are less obvious at first glance, arise from the idea of Corey and coworkers 37 to use the ring strain of the The intermediate carbenium ion 36 should be formed by protonation of the main alcohol 37 produced by reducing the carboxylic acid ester function in **38**. On the one hand, the extra keto functionin 38 should allow for the intramolecular [2+1]-cycloaddition synthesis of the carbene intermediate **39** necessary for cyclopropanation, as well as the Claisen creation of the Eketoester **40**. On the other hand, condensation with ester **41**. The ester **42**, on the other hand, might be obtained via alkylation of 1,3-cyclohexadien-5-carboxylate 43 with the electrophile halide 44. The tertiary alcohol **45** is a distinct precursor of the dihydrobenzofuran **44**, which may be obtained by alkylation of the tetrahydrobenzofuranone **46**, first in D-position to the carbonyl carbon, yielding **45** and then precisely there, yielding **44**. Finally, Afeist-Benary furan synthesis produces tetrahydrobenzofuranone **14** by mixing 1,3-cyclohexanedione 48 with chloroacetaldehyde **47** (Scheme 14.6).

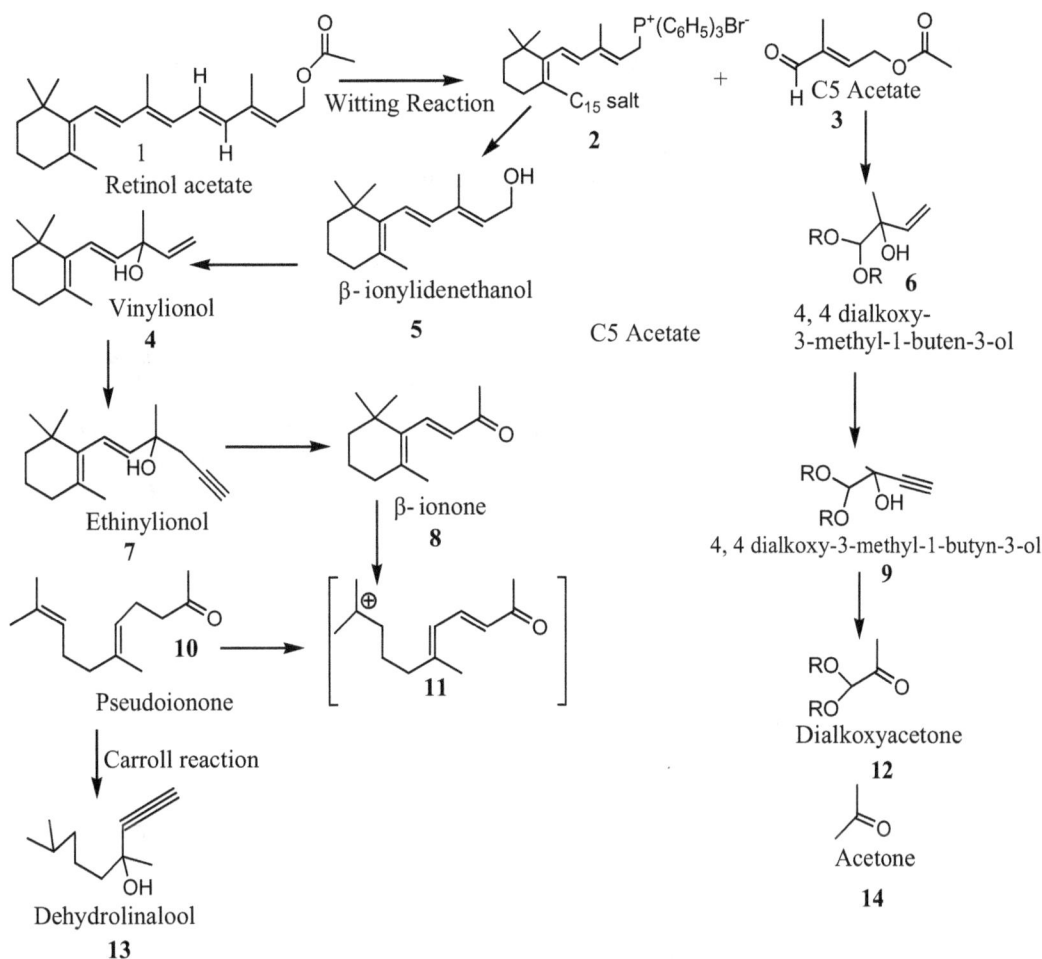

SCHEME 14.1 Retro Synthesis of Dehydrolinalool.

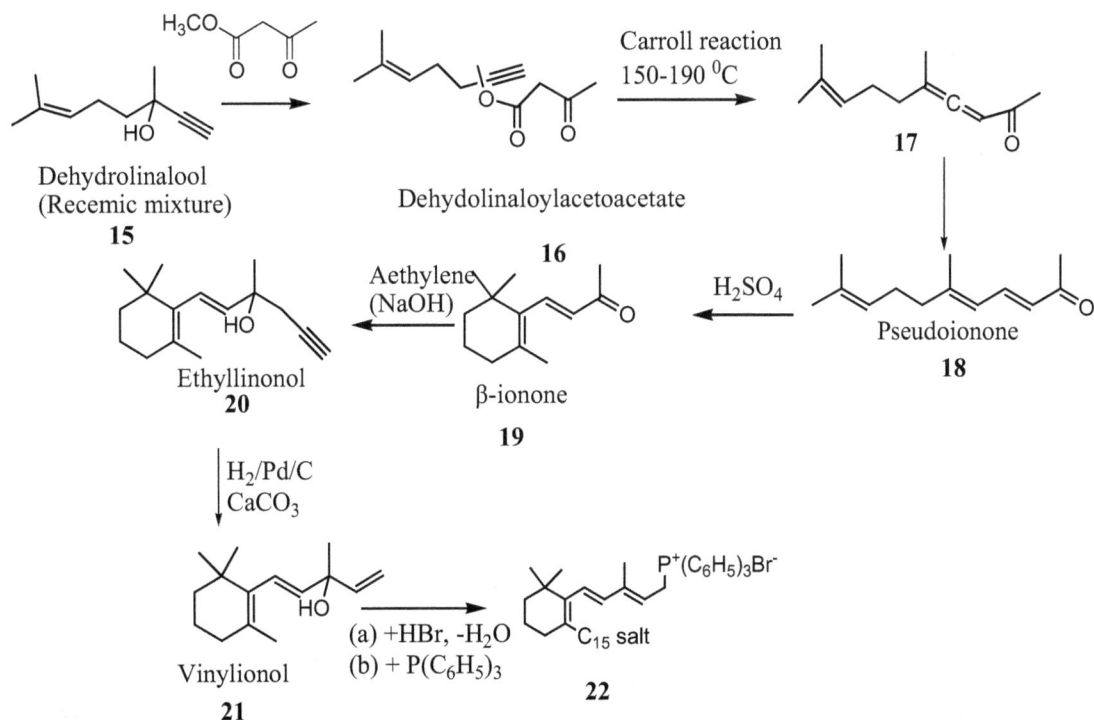

SCHEME 14.2 Synthesis of C-15 Wittig Salt.

Scheme 14.3 text/labels:

Acetylene (NaOH)

H₃COC / COCH₃ / O — **23** Dimethoxyacetone

4,4-Dimethoxy-3methyl-1-butyn-3-ol **24**

H₂/Pd/C / CaCO₃

C5 alcohol **25**

+ (CH₃CO)₂O / −CH₃CO₂H

C5 ester **26**

120 °C / Cu⁺⁺ / −2CH₃OH

C5 acetate **27**

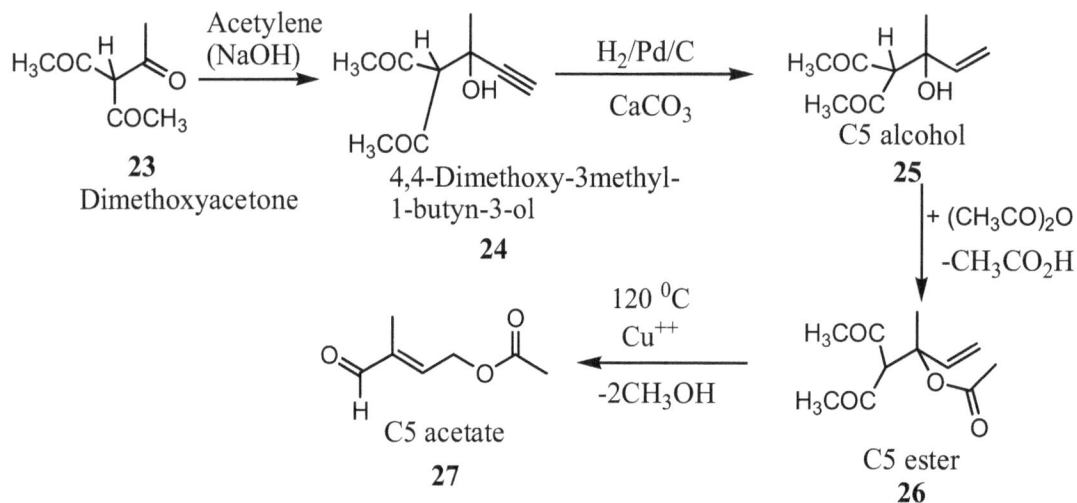

SCHEME 14.3 Synthesis of C5 Acetate.

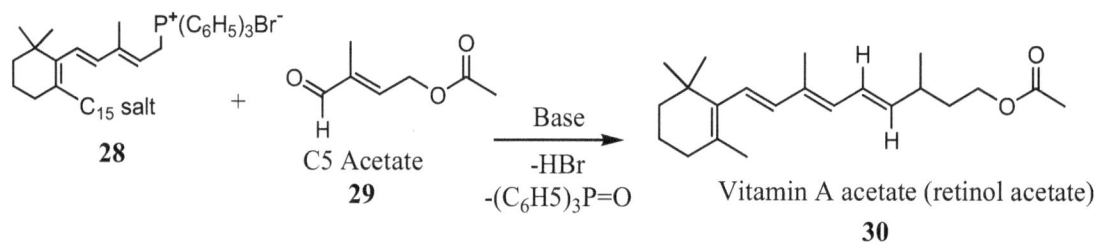

Scheme 14.4 text/labels:

P⁺(C₆H₅)₃Br⁻

C₁₅ salt **28**

+ C5 Acetate **29**

Base / −HBr / −(C₆H5)₃P=O

Vitamin A acetate (retinol acetate) **30**

SCHEME 14.4 Synthesis of Vitamin A acetate.

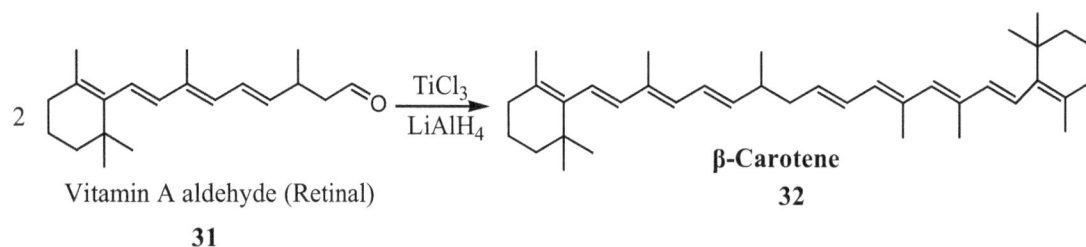

Scheme 14.5 text/labels:

2 · Vitamin A aldehyde (Retinal) **31**

TiCl₃ / LiAlH₄

β-Carotene **32**

SCHEME 14.5 Synthesis of β- carotene.

14.4.1 Step 1: Synthesis of diazoketoester

The synthesis of racemic cafestol described by Corey and associates starts with the major enol tautomer **48** of 1,3-cyclohexanedione, which cyclizes with chloroacetaldehyde **47** in an ethanol solution of sodium hydroxide to give tetra-hydrobenzofuranone **46**. This is deprotonated with lithium-diisopropylamide(LDA), then methylated in D-position to the carbonyl group with iodomethane, producing **45**. The carbonyl function is alkynylated bylithiatedtrimethylsilylethyne, which is immediately dehydrated to trimethylsilylenyne **49** in the presence of pyridinium p-toluenesulfonate (PPTS) and magnesium sulphate in benzene in a variant of the synthetic idea. Desilylation of potassium fluoride in dimethylsulfoxide generates ethyne **50**, which is transformed to aldehyde **51** after hydroboration with diisoamylborane(iA2BH) and oxidation with hydrogen peroxide. In the presence of triphenylphosphane and imidazole, sodium borohydride in methanol lowers **51** to the primary alcohol, which is then reacted with iodine to form the iodethyl molecule **43**. After deprotonation with LDA, the latter alkylates meth-ylcyclohexa-1,3-diene-5-carboxylic acid methyl ester **42** to the tricyclic ester **41**. The ester **41** is hydrolyzed with sodium hydro-xide to produce the E-ketoester **40**; 1,1′-carbonyldiimidazole transforms the carboxylate to the imidazolide, which is then reacted with D-lithiatedt-butylacetate. The diazoke-toester **52** is formed by reacting the Eketoester **40** with p-toluenesulfonylazide in a solution of diazabicyclo[5.4.0] undec-7-ene (DBU) in dichloromethane (Scheme 14.7) [2,3].

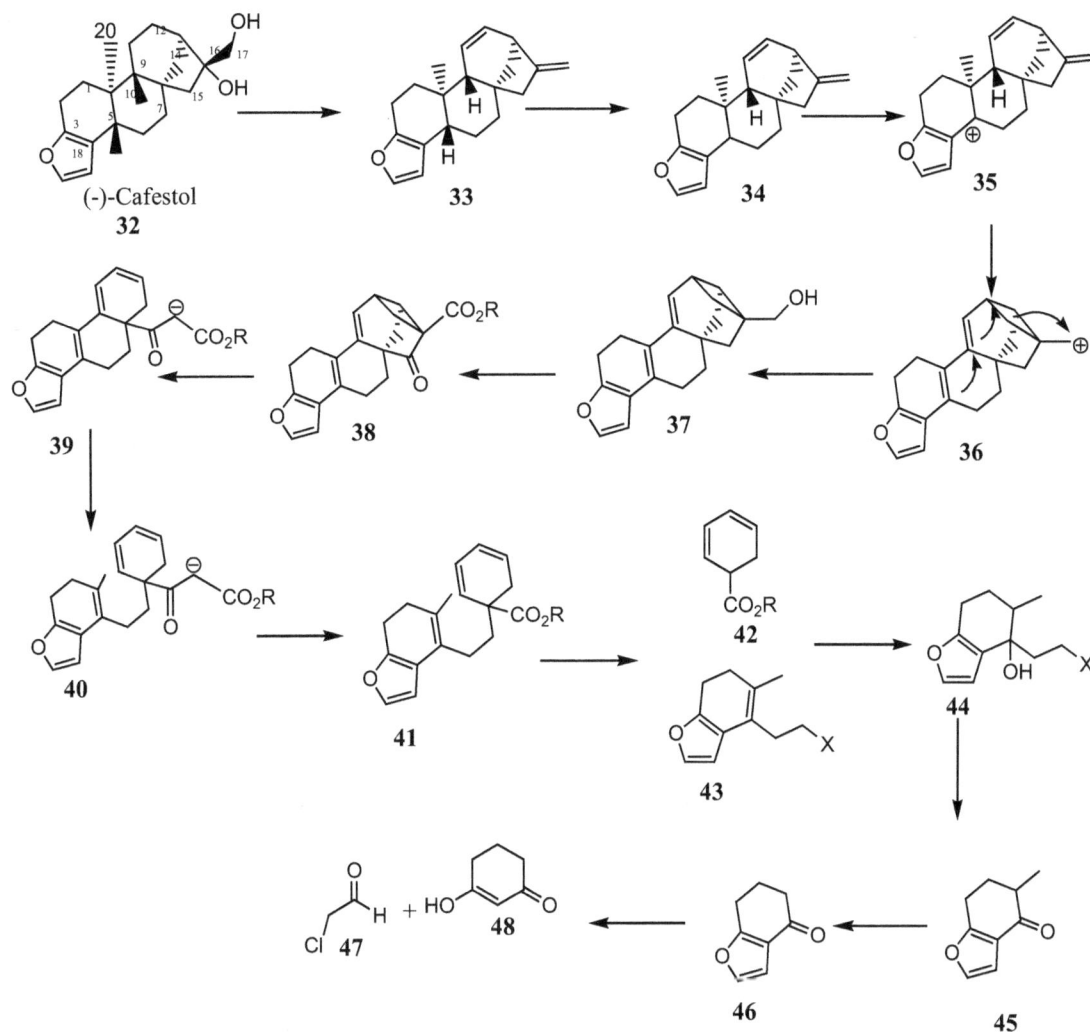

SCHEME 14.6 Retrosynthesis of (-)-Cafestol from cyclohexanedione and chloroacetaldehyde.

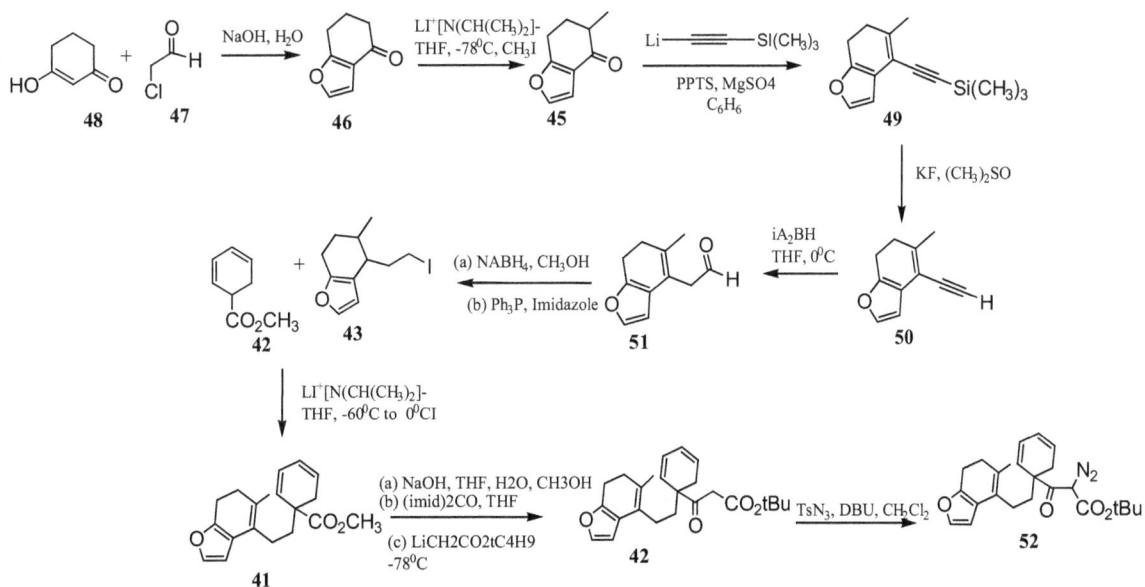

SCHEME 14.7 Synthesis of diazoketoester from cyclohexanedione and chloroacetaldehyde.

14.4.2 Step 2: Synthesis of (-) Cafestol

The diazoketoester **52** undergoes cyclopropanation to the unstable pentacycle **38** (major product) when added dropwise to a solution of copper(II)bis-salicylaldehyde-*t*-butylimine in toluene, involving the intermediate carbene **39**. Sodium borohydride reduces **38** in methanol solution to the secondary alcohol which is, for protection, converted to the benzylether **53** by *Williamson synthesis* involving deprotonation to the alcoholate by sodium hydride and O-alkylation by benzylbromide so that diisobutylaluminumhydride (DIBAH) reduces the *t*-butylester in **53** to the primary alcohol **54** in dichloromethane (Scheme 14.8).

Refluxing **54** with trifluoroacetic anhydride in 2,6-lutidine is the crucial reaction to the pentacycle **55**, which is unstable towards acids. The CC double bond in conjugation with the furan ring is reduced by sodium in liquid ammonia after lithium in ethanol deprotects **55** to the secondary alcohol **56**. The required trans-isomer **57** is mostly produced by working up in tetrahydrofuran (THF) and water. To eliminate the alcohol function that is no longer necessary, **57** is prepared for anallyl rearrangement by reacting it with methanesulfonyl chloride in the presence of triethylamine, resulting in the formation of the allyl iodide **58**, which may be regenerated in two steps: The cafestol precursor **33** is obtained by hydrazinolysis of **58** in a combination of dimethoxyethane and t-butylalcohol, which yields the allylhydrazine **59** after oxidation to the diimine involving a 1,5-sigmatropic hydrogen shift followed by nitrogen removal. The furan ring must be protected prior to catalytic hydrogenation by D-lithiation and subsequent reaction with triisopropylsilyltriflate (TIPSOTf) to form D-triisopropylsilylfuran **60**. Dihydroxylation of the exocyclic CC double bond with osmiumtetroxide in THF, catalytic hydrogenation of the C-11-C-12 double bond in the presence of rhodium and aluminium oxide and finally desilylation of the furan ring with hydrogen fluoride in a mixture of THF and acetonitrile are the final steps to racemic cafestol **32** [4].

14.5 Baccatin III as a Taxol Precursor

Baccatin III is found in Taxus baccata, a European yew tree. It is a desirable precursor in partial syntheses of (-)-taxol, which is extracted from the Pacific cousin Taxus brevifolius and utilized in the treatment of leukaemia and other cancers.

14.6 Baccatin Synthesis

They assumed that protection of the 1,2-diol by cyclocarbonate would be valid following deoxygenation of baccatin III in the allyl position (C-13) to the precursor **61** while constructing the synthesis. The alkene **63** is formed via retrosynthetic disconnection of the oxetane ring in cyclocarbonate **62**, with the vicinal alcohol functions concealed as acetone ketal; the oxetane ring is then predicted to close by adding the primary alcohol group to the C-5-C-6 double bond in **63**. A double bond in

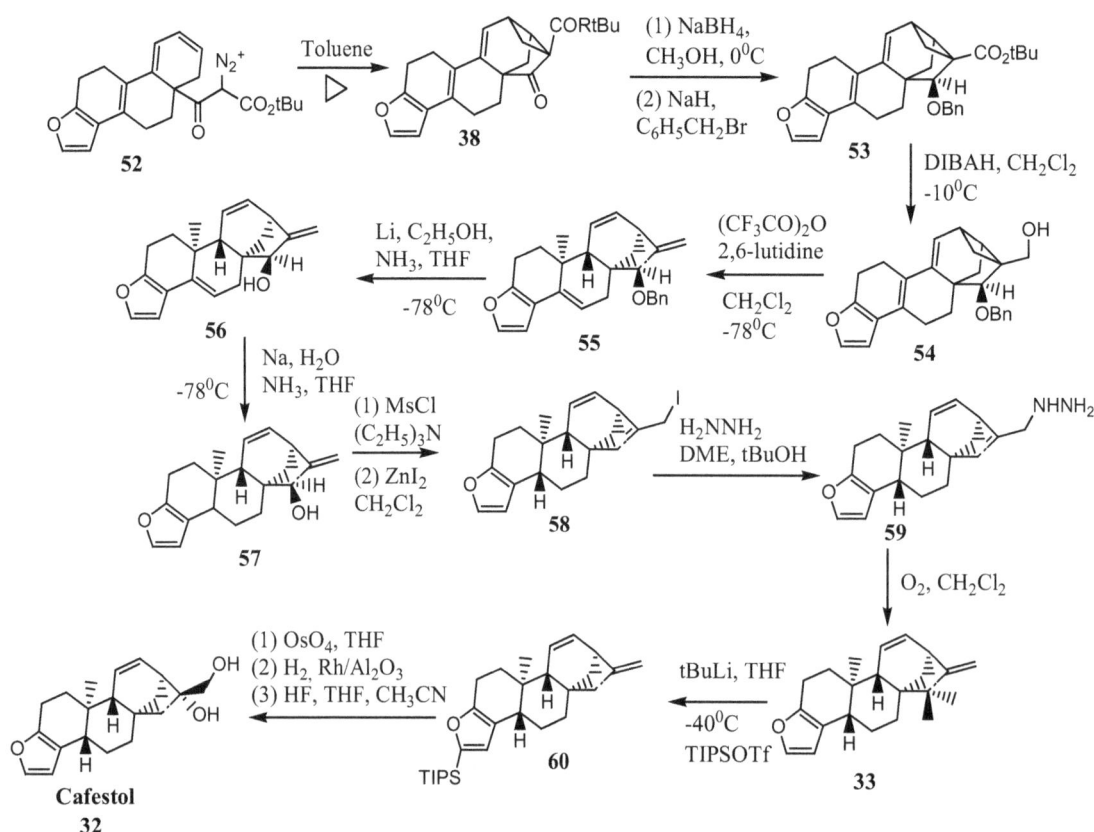

SCHEME 14.8 Synthesis of (-) Cafestol from diazoketoester.

64, which is shown to come from an intramolecular Mcmurry deoxygenative coupling of the dialdehyde **65**, which, in turn, is the product of oxidation of the (protected) primary alcohol functionalities in **66**, may subsequently be used to introduce functional groups in the 9,10-position.

The allylalkohol **66** appears to be the obvious precursor of the protected diol **65** and **66** is derived from a Shapiro coupling of cyclohexadienyllithium **67** with thecyclohexene-4-aldehyde **68**, which is the product of oxidation and subsequent diol protection of the bicyclic hydroxylactone **69**; the latter is derived from rearrangement of the Diels-A The sulfonylhydrazone of the ketone **70**, which is a Diels-Alder cycloadduct of the protected 3-hydroxymethyl-2,4-dimethyl-1,3-pentadiene **72** and ketene **71** as the dienophile, gives rise to the cyclohexadienyllithium **67** (Scheme 14.9) [4,5].

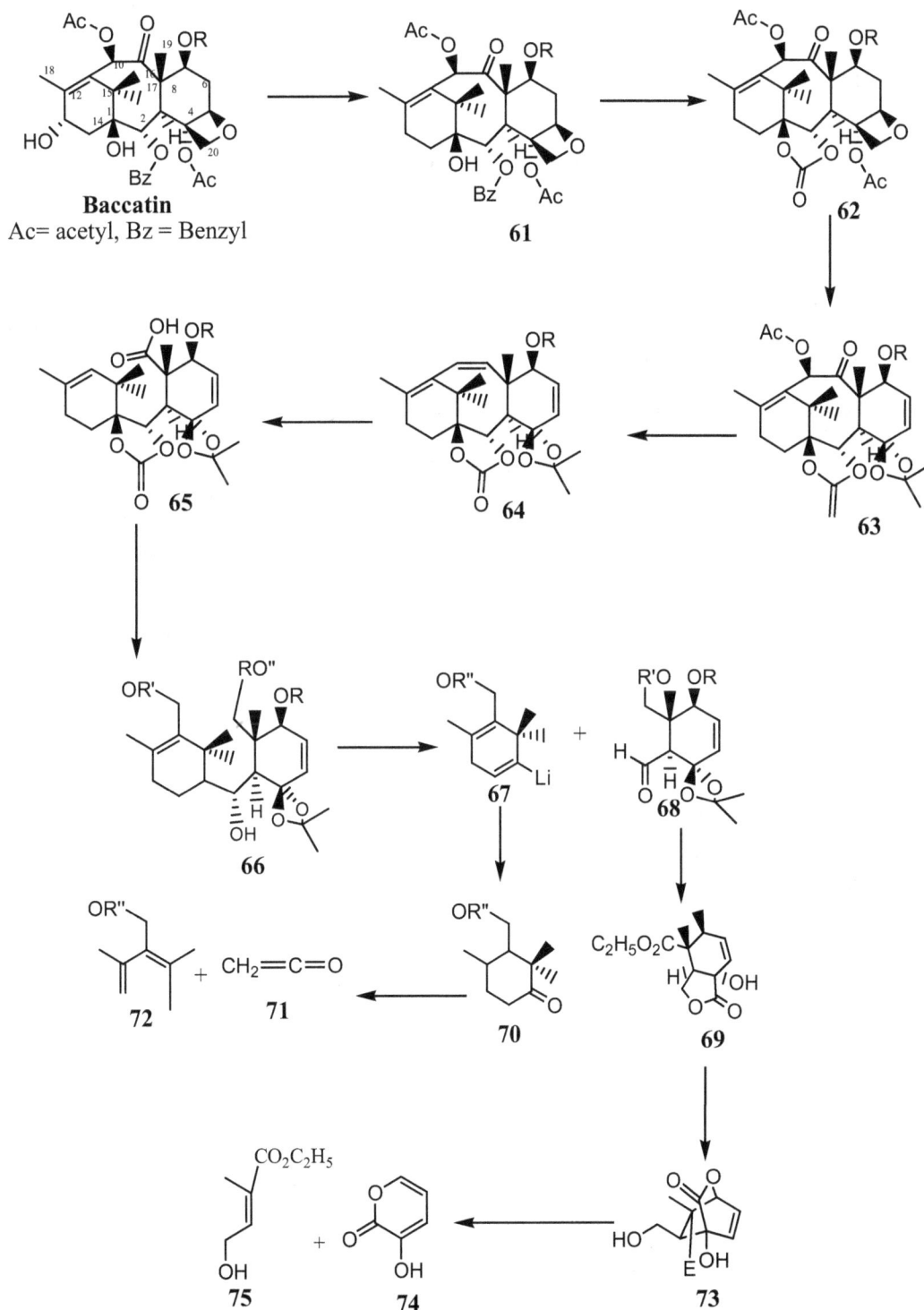

SCHEME 14.9 Retro Synthesis Baccatin to 3-hydroxy-2-pyrone and 4-hydroxy-2-methyl-2-butenoate.

14.6.1 Step 1: Synthesis of sulfonylhydrazone

A stable precursor of thecyclohexadienyllithium 67 is the aryl-sulfonylhydrazone 78. To make this starting reagent, 3-acetoxy-methyl-2,4-dimethyl-1,3-pentadiene 72 is treated with the ketene equivalent chloroacrylnitrile 71a in a Diels-Alder reaction. In t-butyl alcohol, cycloadduct 76 is predominantly hydrolyzed to the hydroxy ketone 77. The primary alcohol function of the inter-mediate 70 is protected by a subsequent reaction with t-butyldi-methylsilylchloride (TBSCl) and imidazole in dichloromethane, in which the keto carbonyl group is derivatized to the required arylsulfonylhydrazone 78 with 2,4,6-triisopropylphenylsulfonyl-hydrazide in tetrahydrofuran.

The allylalkohol 66 appears to be the obvious precursor of the protected diol 65, and 66 is derived from a Shapiro cou-pling of cyclohexadienyllithium 67 with the cyclohexene-4-aldehyde 68, which is the product of oxidation and subsequent diol protection of the bicyclic hydroxylactone 69; the latter is derived from rearrangement of the Diels-A The cyclohexa-dienyllithium 67 is derived from the ketone sulfonylhydra-zone 70, which is a Diels-Alder cycloadduct of the protected 3-hydroxymethyl-2,4-dimethyl-1,3-pentadiene 72 with ketene 71 as the dienophile. A stable precursor of the cyclohexadi-enyllithium 67 is the arylsulfonylhydrazone 78. To make this starting reagent, 3-acetoxymethyl-2,4-dimethyl-1,3-pentadi-ene 72 is treated with the ketene equivalent chloroacrylnitrile 71a in a Diels-Alder reaction. In t-butyl alcohol, the cycload-duct 76 is predominantly hydrolyzed to hydroxyketone 77. The primary alcohol function of the intermediate 70 is protected by a subsequent reaction with t-butyldimethylsilylchloride (TBSCl) and imidazole in dichloromethane, in which the keto carbonyl group is derivatized to the required arylsulfonylhy-drazone 78 with 2,4,6-triisopropylphenylsulfonylhydrazide in tetrahydrofuran (Scheme 14.10).

14.6.2 Step 2: Synthesis of cyclohexenaldehyde

A stable precursor of the cyclohexadienyllithium 67 is the aryl-sulfonylhydrazone 78. To make this starting reagent, 3-ace-toxymethyl-2,4-dimethyl-1,3-pentadiene 72 is treated with the ketene equivalent chloroacrylnitrile 71a in a Diels–Alder reaction. In t-butyl alcohol, cycloadduct 76 is predominantly hydrolyzed to the hydroxyketone 77. The primaryalcohol function of the intermediate 70 is protected by a subsequent reaction with t-butyldimethylsilylchloride (TBSCl) and imid-azole in dichloromethane, in which the keto carbonyl group is derivatized to the required arylsulfonylhydrazone 78 with 2,4,6-triisopropylphenylsulfonylhydrazide in tetrahydrofuran

The allylalkohol 66 appears to be the obvious precursor of the protected diol 65, and 66 is derived from a Shapiro cou-pling of cyclohexadienyllithium 67 with thecyclohexene-4-aldehyde 68, which is the product of oxidation and subsequent diol protection of the bicyclic hydroxylactone 69; the latter is derived from rearrangement of the Diels-A The cyclohexa-dienyllithium 67 is derived from the ketone sulfonylhydra-zone 70, which is a Diels-Alder cycloadduct of the protected 3-hydroxymethyl-2,4-dimethyl-1,3-pentadiene 12 with ketene 71 as the dienophile. A stable precursor of thecyclohexadienyl-lithium 67 is the arylsulfonylhydrazone 78. To make this start-ing reagent, 3-acetoxymethyl-2,4-dimethyl-1,3-pentadiene 72 is treated with the ketene equivalent chloroacrylnitrile 71a in a Diels-Alder reaction. In t-butyl alcohol, the cycloadduct 76 is predominantly hydrolyzed to the hydroxyketone 77. The pri-maryalcohol function of the intermediate 70 is protected by a subsequent reaction with t-butyldimethylsilylchloride (TBSCl) and imidazole in dichloromethane, in which the keto carbonyl group is derivatized to the required arylsulfonylhydrazone 78 with 2,4,6-triisopropylphenylsulfonylhydrazide in tetrahydro-furan (Scheme 14.11) [4,5].

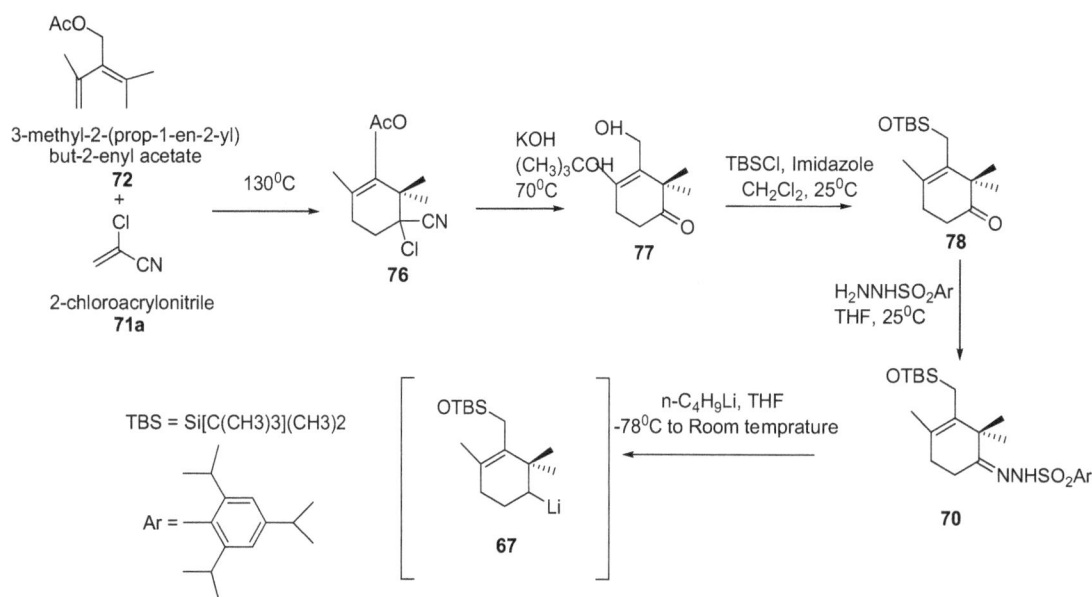

SCHEME 14.10   Synthesis of sulfonylhydrazone of the ketone.

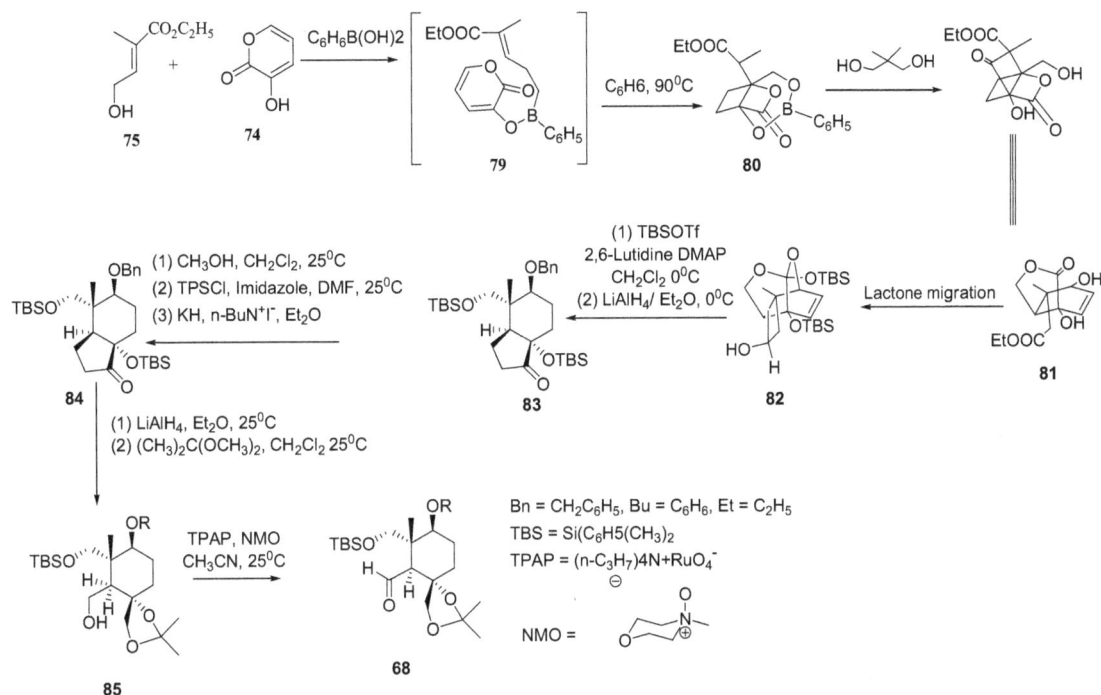

SCHEME 14.11 Synthesis of cyclohexenaldehyde.

14.6.3 Step 3: Synthesis of diastereomers

Following the sharpless epoxidation, the coupling product **66** has an allylic alcohol substructure that is epoxidized by t-butylhydroperoxide in benzene catalyzed by tiny quantities of vanadyl(IV)acetylacetonate. Lithiumaluminumhydride in ether reductively opens the oxirane ring, yielding the trans-diol **87**, which is then deprotonated to the trans-diolate by potassium hydride inhexamethylphosphoric acid triamide (HMPA) and reacts with phosgene to form the cyclocarbonate **88** for protection at room temperature. At room temperature, tetra-n-butylammoniumfluoride (TBAF) in THF cleaves both silyl protective groups, releasing the primary alcohol groups for oxidation to the dialdehyde **65** needed for intramolecular MCMURRY coupling by tetrapropylammoniumperruthenate (TPAP) and N-methylmorpholine-N-oxide (NMO) in dichloromethane. The *Mcmurry* reaction parameters are changed to produce the 9,10-diol **89**, which is more closely linked to the baccatin structure than the alkene. Excess titanium-(III)chloride in dimethoxyethane (DME) is used as a reducing agent, along with an excessive zinc-copper pair. This step produces the racemic cis-diol 89, which is reacted with (1S)-(-)-camphanic chloride in dichloromethane and triethylamine as the base to produce diastereomeric camphanates **90** in reasonable yields. Chromatography is used to separate diastereomers, and X-ray crystallography is used to identify them. The subsequent stages of the synthesis are carried out with the required dextrorotatory diol 89 produced by hydrolysis of its camphanate (Scheme 14.12) [5].

14.6.4 Step 4: Synthesis of 7-O-triethylsilylbaccatin-III

The alcohol functions of the *cis*-diol **89** clearly exhibit different reactivities: camphanic chloride selectively acylates the 9-OH group to **90**, while acetic anhydride indichloromethane acylates the allyl alcohol function 10-OH at room temperature in the presence of DMAP as acylation catalyst. There is no straightforward explanation of this advantageous selectivity of acylation, which enables oxidation of the free 9-OH group with TPAP and NMO in dichloromethane to the D-acetoxyketone **63**. In contrast, the selectivity of the subsequent hydroboration (BH3/THF) and oxidation (H₂O₂, NaHCO₃) at the sterically less hindered C-5-C-6 double bond in favor of the desired alcohol **91** is reasonable. The undesirable regioisomeric C-6-alcohol is a minor product. From the hydroxy functions deprotected by hydrolysis of the ketal with hydrogen chloride in methanol and water at room temperature, thesterically less-hindered primary alcohol group is acetylated selectively. After removal of the benzyl protective group by hydrogenation in ethyl acetate, 7-OH is protected once again by triethylchlorosilane, while 5-OH is activated with methane sulfonyl chloride. This enables C-20-OH, deprotected by potassium carbonate inmethanol and water, to substitute intramolecularly the C-5-mesylate in butanone containing tetra-*n*-butylammonium acetate, thus closing the oxetane ring in **94**. Acetylation of the tertiary 4-OH catalyzed by DMAP is achieved in spite of sterichindrance, and phenyllithium in THF opens the cyclocarbonate ring at −78°C to the benzoate **95**. In order to introduce the allyl alcohol function C-13-OH,

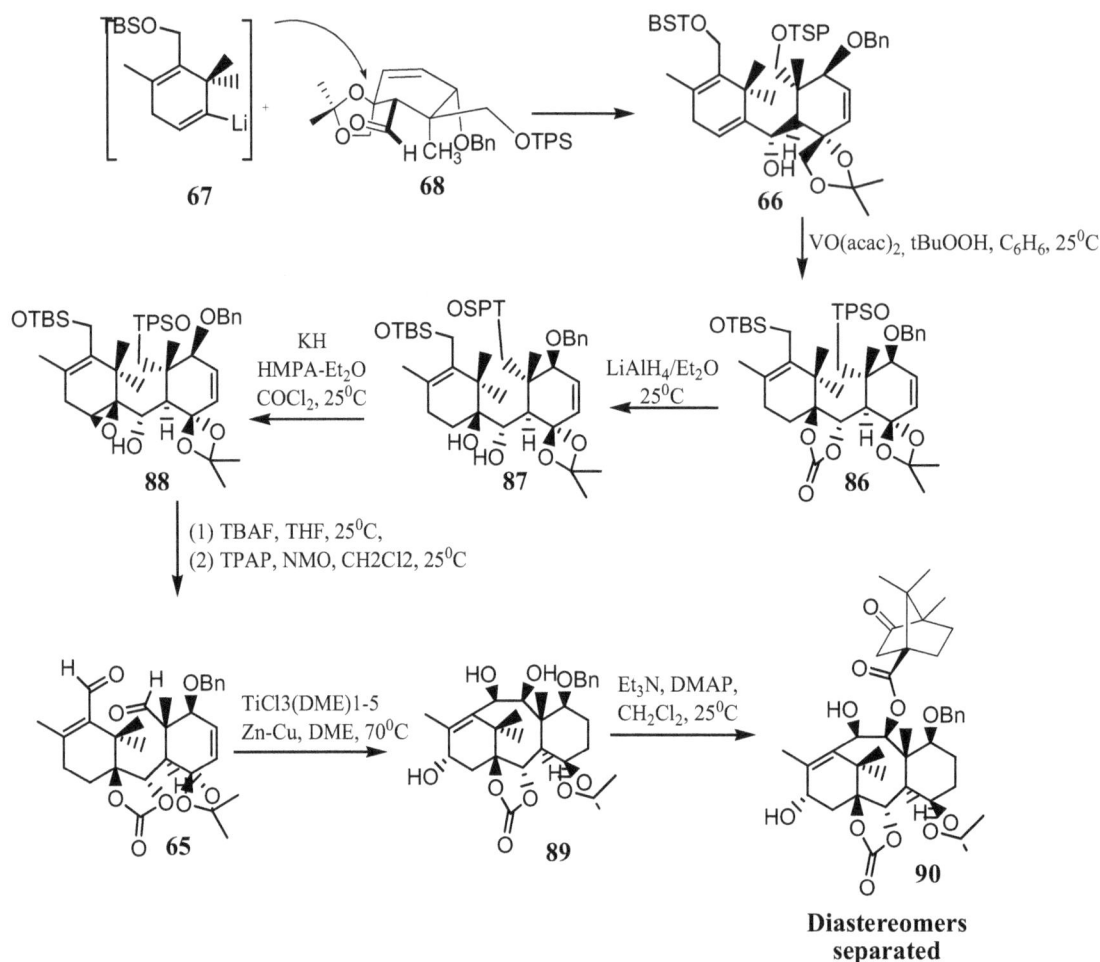

SCHEME 14.12 Synthesis of diastereomers.

pyridiniumchlorochromate (PCC) oxidizes in the allyl position in benzene solution, yielding the enone precursor. Sodium borohydride in methanol solution finally reduces the enone to the target 7-*O*-triethylsilylbaccatin III (Scheme 14.13) [4,5].

14.7 Triterpenes

14.7.1 Lupeol

Total triterpene syntheses with well-documented total syntheses are rare in the literature. G. Stork and associates [4,5] described the synthesis of racemic lupeol as an example. (+)-Lupeol is a pentacyclic triterpene that is found in abundance in plants.

14.8 Synthesis of (+) Lupeol

Knowing that genuine (+)- lupeol may be degraded to the pentacyclic ketal ester **96** was helpful in designing the synthesis. In refluxing 1,4-dioxane, excess methyllithium reacts with 1 to form the ethyleneketal of lupan-20-ol-3-one. The isopropenyl group is regenerated after dehydration with phosphorylchloride

in pyridine. Sodium borohydride in methanol converts the ketone to genuine (+)-lupeol after hydrolysis of the ketal. To summarise, the ketalester **96** will be an appealing target for a complete synthesis as a precursor of lupeol.

The cyclopentane ring E is predicted to close via an intramolecular nucleophilic substitution of OR- (tosylate) by the carbanion D to the methoxycarbonyl function in **97**, resulting in retrosynthetic disconnection of the ketal ester **96**. The aldehyde carboxylic acid is generated by ozonolysis of the enol **98** as a tautomer of the ketone **99**, and the main alcohol as a precursor of the methylether turns out to be the result of esterification and reduction of the aldehyde carboxylic acid (Scheme 14.14) [6,7].

14.8.1 Retro synthesis of 6-methoxy-a-tetralone from

14.8.1.1 Step 1: Synthesis of substituted benzoate

The oxidation of the secondary alcohol 100, which is a ketal derivative of the hydroxy-D, D-dimethylketone **101**, obviously results in the synthesis of substituted benzoate Ketone **99**. The latter is the result of methylation of the mono-D-methylketone **102**, which is achieved by hydrogenation of the CC

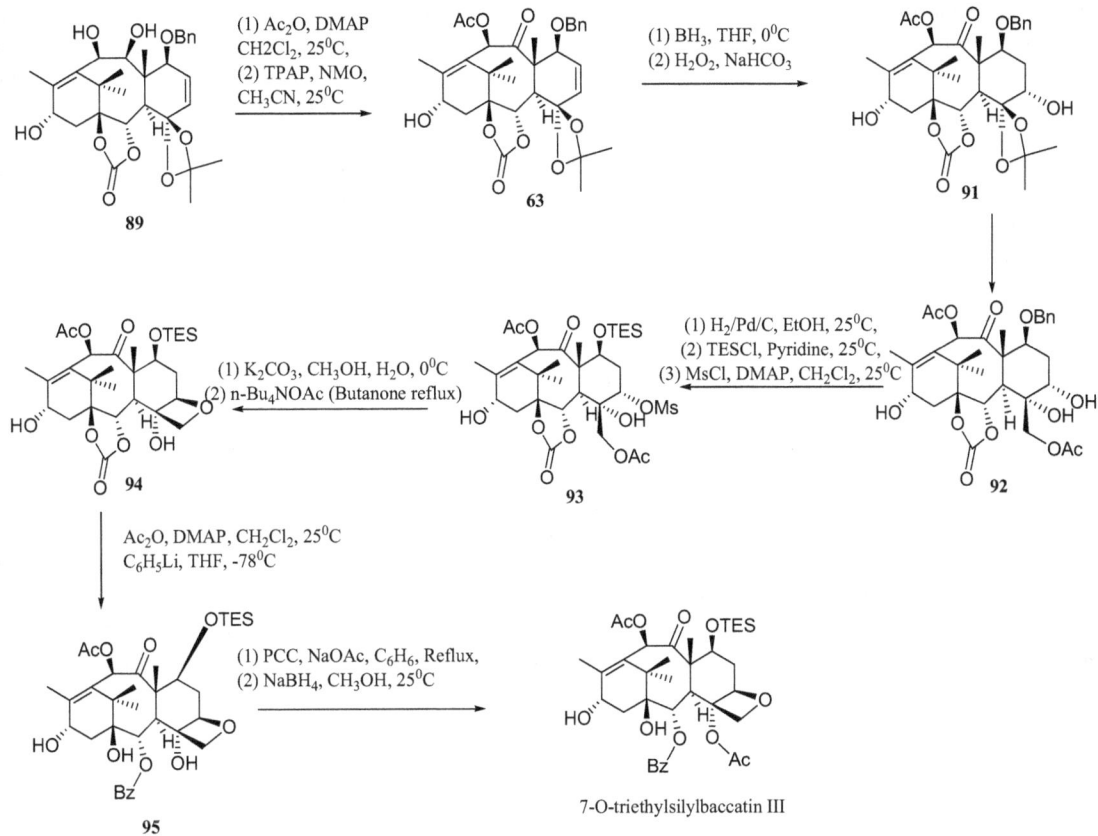

SCHEME 14.13 Synthesis of 7-O-triethylsilylbaccatin-III.

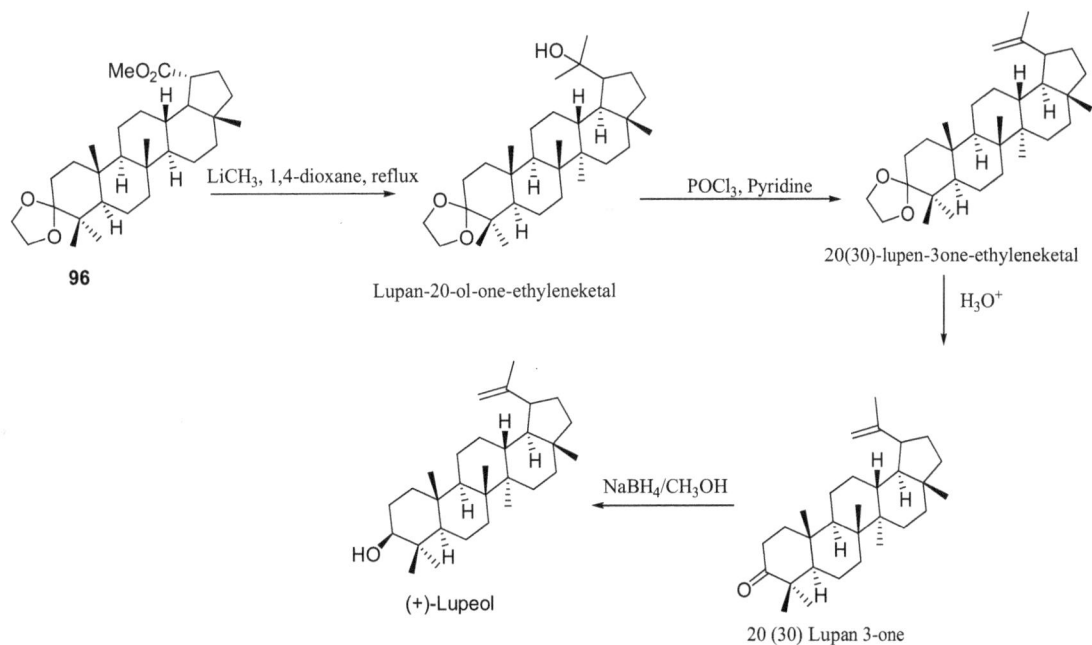

SCHEME 14.14 Synthesis of (+) Lupeol from ketal ester.

double bond in the enone **103**, which, in turn, is the result of intramolecular ***Knoevenagel*** alkenylation of the diketone **104**. The nucleophilic addition of ethylmagnesium halide to the electrophilic lactone carbonyl carbon is predicted to produce 104 from the G-keto acid **106** through the equivalent enol lactone **105**. The synthesis of different steroids 40 was effectively accomplished by adding alkyl magnesium halide to G-keto acids and the parent enol lactones, followed by intramolecular Knoevenagel alkenylation. The G-keto acid 106 is thought to be produced via allylation of the enone108 and hydroboration and oxidation of the terminal CC double bond in **107** [6]. Enone **108** is formed by adding ethyl magnesium halide to the enol lactone **110** of the G-keto acid **111**, resulting in intramolecular ***Knoevenagel*** alkenylation of the D, G-diketone **109**. The allyl compound **112** is generated before this keto acid by nucleophilic opening of the cyclopropane ring with hydride and methylation of the carbonyl in the D position in the cyclopropyl ketone **113**. The mesylate **114** of the primary alcohol **115**, generated by reduction of the aldehyde **116**, will be used to perform cyclopropanation, provided that the keto function is safeguarded as a ketal before reduction. By reducing the E-keto nitrile **117**, which is produced by nucleophilic addition of hydrogen cyanide to the electrophilic double bond of

the enone 118, the aldehyde is introduced through the imine. The latter is the result of a thermal oxy-Cope rearrangement of the intermediate tricyclic allylenolethercarbenium ion **119**, which is produced from enone **120**. This is the result of the arylmethylether **121** and its precursor 122 being reduced. 1,9,10,10a-Tetrahydro-7-methoxy-3(2H)-phenanthrone **122** can be readily obtained by anellation of 2-methyl-6-methoxy-D-tetralone **123**, which can be synthesized from commercial 6-methoxy-D-tetralone using a variety of methods, including D-methylation of the ketone using the enamine or the enolether; alternatively, the cyclization of 6-methoxy-E-methyl-Dtetralone 123 with 4-N, N-dimethylamino-2-butanone methiodide in the presence of potassium t-butanolate to 1,9,10,10a-tetrahydro-7-methoxy-3-methoxy-3-methoxy-3-methoxy-3-methoxy-3-methoxy-3-methoxy-3-methoxy-3-meth (2H) -phenanthrone **123** The octahydrophenanthrol **121** is obtained by reducing the enone CC double bond with sodium borohydride and then hydrogenating it with palladium and strontium carbonate as a little deactivated catalyst. Lithium in liquid ammonia is used to partially reduce the benzenoid ring to the enone. To preserve the hydroxy group throughout the ensuing synthesis steps, the enone is derivatized to benzoate **120** (Scheme 14.15) [8,9].

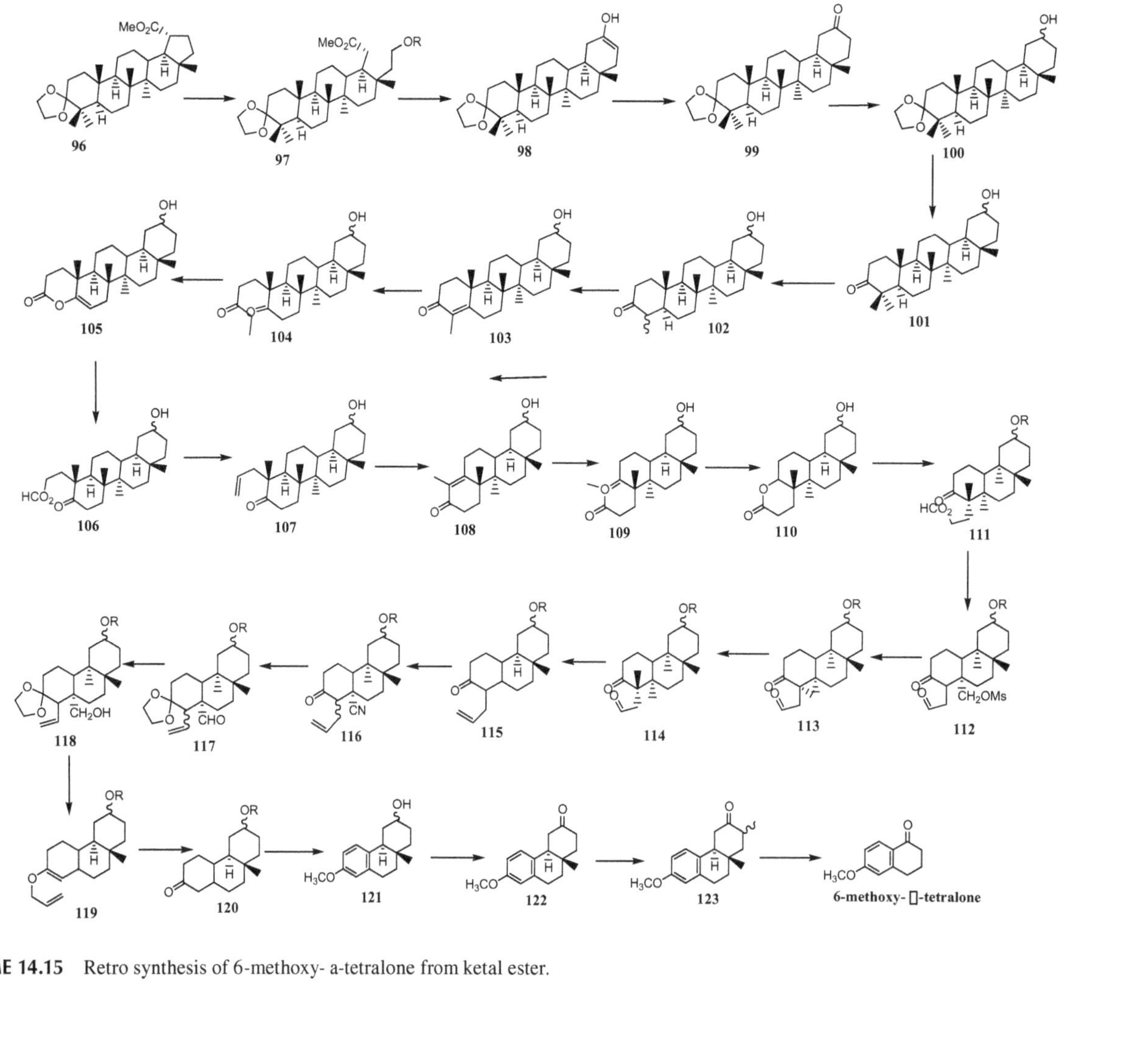

SCHEME 14.15 Retro synthesis of 6-methoxy- a-tetralone from ketal ester.

A combination of triallyl orthoformate and allyl alcohol in tetrahydrofuran (THF) at room temperature, catalyzed by p-toluenesulfonic acid, converts the benzoate **120** to the allylenolether intermediate; oxa-COPE rearrangement to the Dallylenone **118** proceeds by refluxing in pyridine. The cyanoketone **117** is formed by the nucleophilic addition of diethylaluminum cyanide to the enone in a combination of benzene and toluene, which is then reduced by lithiumaluminumhydride to the aldehyde **116** involving the intermediate imine after the keto function is protected as a 1,3-dioxolane. Sodium borohydride is employed to convert the aldehyde **116** to the main alcohol **115** if rebenzoylation is necessary due to benzoate hydrolysis during work-up (Scheme 14.16) [10,11].

14.8.1.2 Step 2: Synthesis of allylketobenzoate

When diluted hydrochloric acid is added to a THF solution, the mesylate **114** produced by mesylation of **115** with methanesulfonyl chloride in pyridine undergoes simultaneous deprotection and cyclopropanation. The cyclopropane ring opens to the methyl molecule with lithium in liquid ammonia after rebenzoylation with benzoyl chloride in pyridine, and the resultant intermediate undergoes direct reductive methylation by iodomethane in hexamethylphosphoric acid triamide D to the carbonyl function in **112**. The propionic acid **111** and the corresponding enol lactone **110** crystallise in the reaction mixture after hydroboration and oxidation of the terminal vinyl group. When ethylmagnesium bromide is mixed with diethylether and THF, the D-G-diketone **109** is formed, which quickly cyclizes to the enone108 during work-up in aqueous sodium hydroxide and methanol solution. The allylketobenzoate **107** is obtained by trapping the enolate of **108** with triallyl

orthoformate in the presence of ptoluenesulfonic acid at room temperature, followed by oxy-***Cope Rearrangement*** of the resultant allylenolether in refluxing pyridine and benzoyl chloride (Scheme 14.17).

14.8.1.3 Step 3: Synthesis of ketal ester

After rebenzoylation, a process of hydroboration, ***Jones oxidation***, and 1,2-addition of ethyl magnesium bromide yields the benzoate enone **103**, with the G-keto acid **106**, its enol lactone **105**, and the D-G-diketone **104** as intermediates. The geminal methyl group is introduced by reductive methylation of the enolateanion/carbanion of the resultant cycloalkanone **102** with iodomethane after reducing the CC double bond of the enone with lithium in liquid ammonia. The intended D, D-dimethylketone **101** must be isolated from various chemical byproducts using chromatography. Prior to oxidation of the secondary alcohol function of ring E to the ketone **99**, the keto function is preserved as a 1,3-dioxolane **100** in order to constrict the cyclohexane ring E in the pentacyclic ketoalcohol 101. Its enolate is trapped as enol acetate **98** with acetic anhydride after being produced with sodium and hexamethyldisilazane (HMDS) in THF. The tosylate ester **97** is obtained by ozonolysis of the enol acetate **98** in dichloromethane, reduction with sodium borohydride in sodium hydroxide at 0°C, smooth acidification with diluted aqueous acetic acid, esterification of the resultant carboxylic acid with diazomethane and tosylation. After adding HMDS and sodium to target **96** as the intended precursor of racemic lupeol, the latter cyclizes in a modified Dieckmann reaction when heated in benzene solution (Scheme 14.18) [10,11].

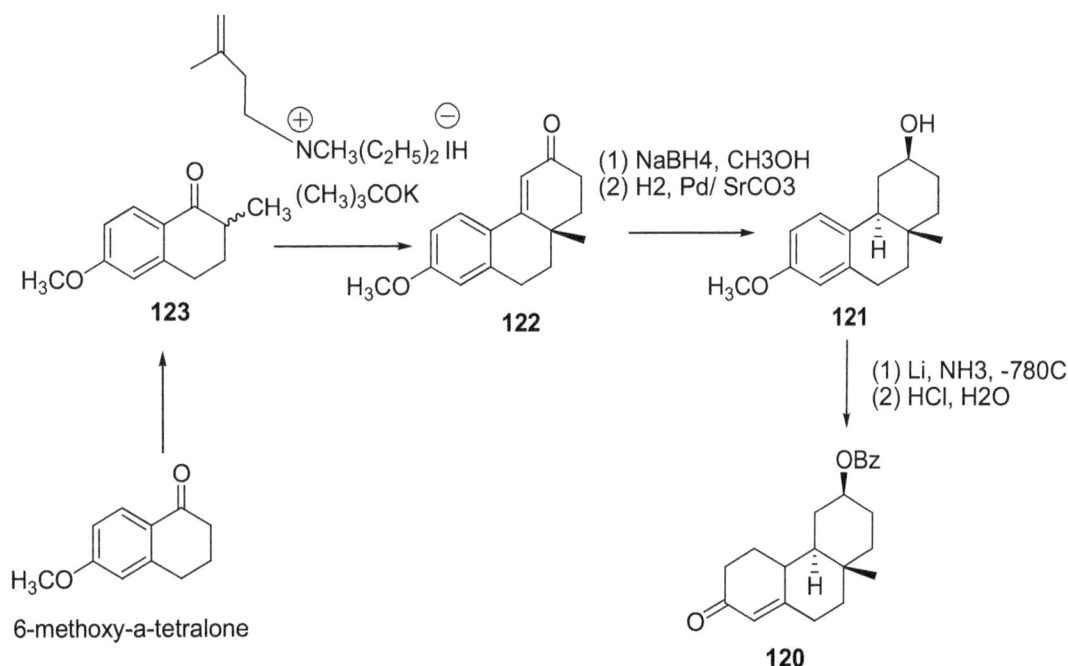

SCHEME 14.16 Synthesis of benzoate from 6-methoxy-αtetralone.

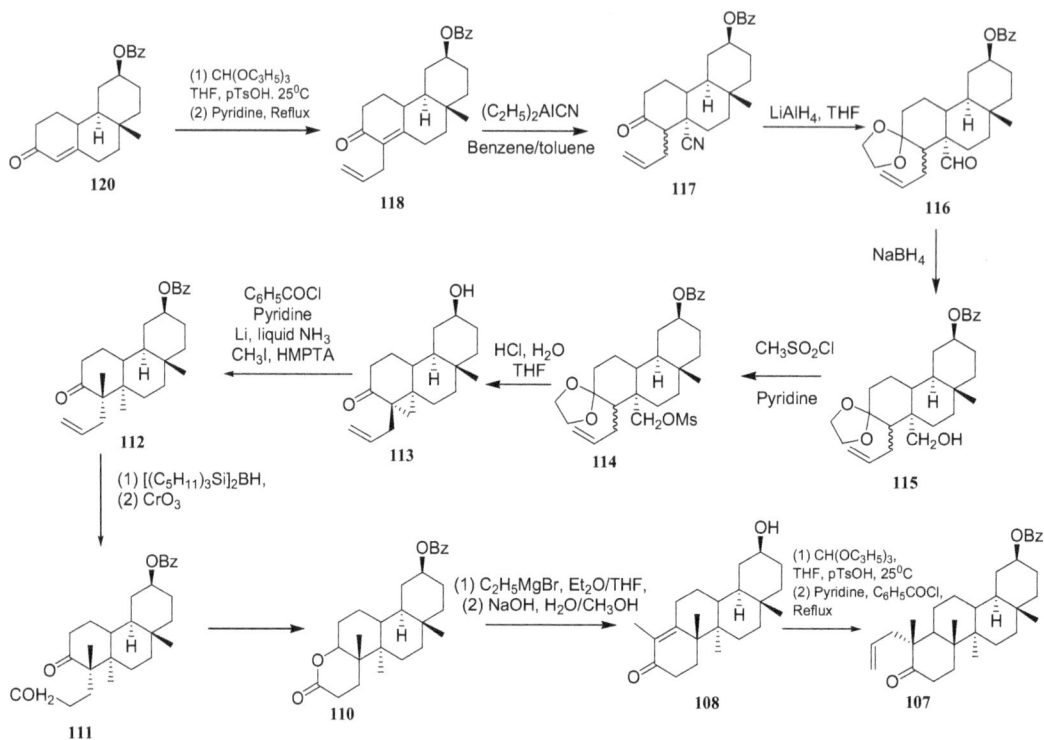

SCHEME 14.17 Synthesis of allylketobenzoate.

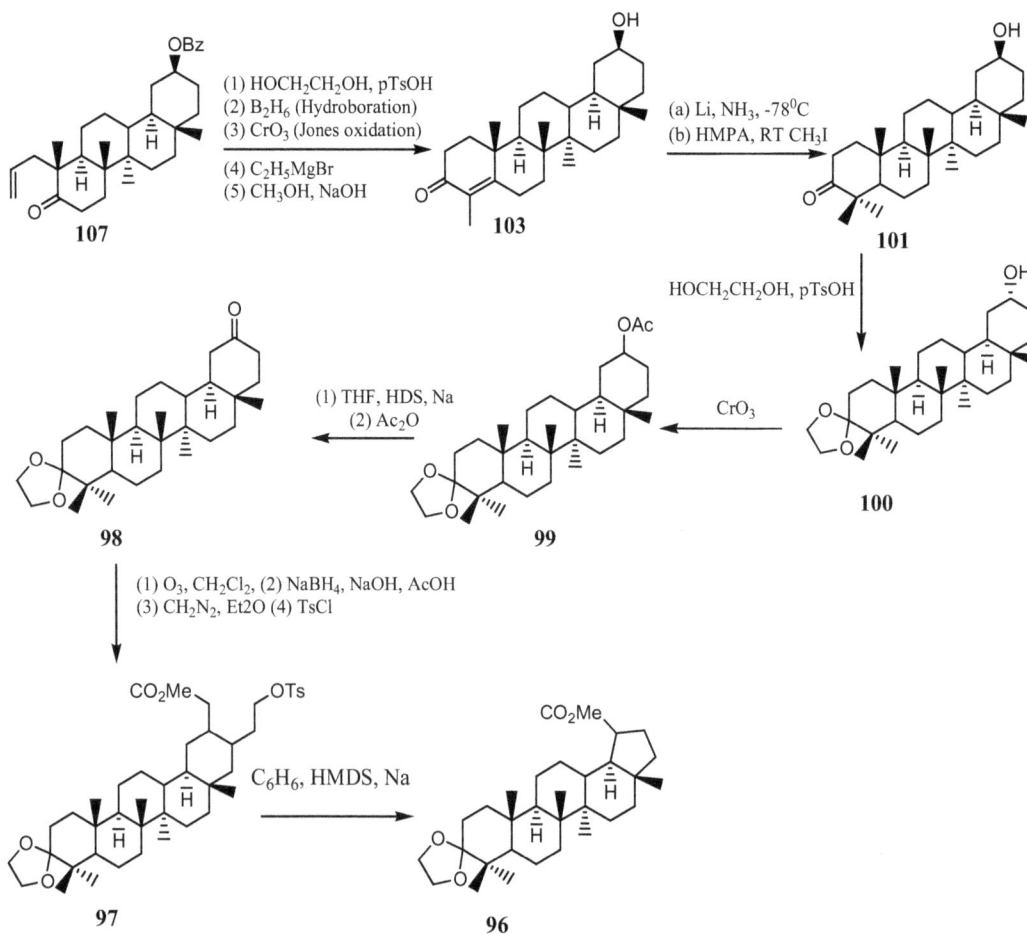

SCHEME 14.18 Synthesis of ketal ester.

14.9 Diterpenoids

Diterpenoids are compounds generated by a variety of natural plants that have considerable thematic challenges, bioactivity and shady toxicity. More than 1,500 Diterpenoids have been identified and characterized so far. Diterpenoids have a variety of pharmacological effects, including neurotropic, antibacterial, antitumor, hypotensive, analgesic, anti-inflammatory, muscle relaxant, antiarrhythmic and local anaesthetic [2]. Diterpenoids isolated from several Ranunculaceae plants, particularly the genera Delphinium, Aconitum and Consolida, are typically distinguishable and recognized as cytotoxic against cancer. Many scientific articles and reviews on phytochemistry, chemical reactions, compositional and botanical studies, and the biological activities of 'Diterpenoids' have been published, classifying these substances as structures containing either 18 or 19 or 20 cycled carbon atoms (C18, C19, C20), i.e., according to the number of contiguous carbon atoms that make up their central arrangement.

14.10 Classification

Diterpenoid alkaloids are heterocyclic systems containing -aminoethanol, methylamine, or ethylamine nitrogen atoms derived from the amination of tetra- or pentacyclic diterpenoids, and are classified into C18-, C19- and C20-diterpenoid alkaloids, respectively, based on their carbon skeleton configuration.

14.11 Gibberellanes

Too far, more than 60 gibberellanes have been identified from higher plants and fungi 2, the majority of which are C-20-norditerpenes. They are utilized in agriculture because they play an important function as plant growth hormones, as well as regulating the breakdown of chlorophyll and the development of fruits. The culture filtrate of the Japanese fungus *Gibberellafujikuroii* yields large quantities of (+)-gibberellin A3, also known as gibberellic acid (Figure 14.1 and structures **124–125**) [2].

Due to numerous gibberellins, it creates from geranylgeranyl-diphosphate through (-)-16(17)-kaurene, this fungus accelerates the development of rice seedlings excessively (Bakanae disease). Dihydrogibberellic acid, commonly known as (+)-gibberelline A1, is found in a variety of higher plants. The rabbit clover Lupinus luteus (Leguminosae) has an unripe seed that contains (-)-gibberelline A18 with a methyl group (C-20) [2].

14.12 Alkaloids of C20-Diterpenoid

C20 diterpenoids are more complicated than C18 and C19 diterpenoids. They are 20-carbon skeleton tetracyclic diterpenes with a Trans ring connecting C19 and C20. Atisine, denudatine, hetisine, hetidine, anopterine, napelline and vakognavine are the most common Diterpenoids -C20 identified from Delphinium (Figure 14.2 and structures **127–133**).

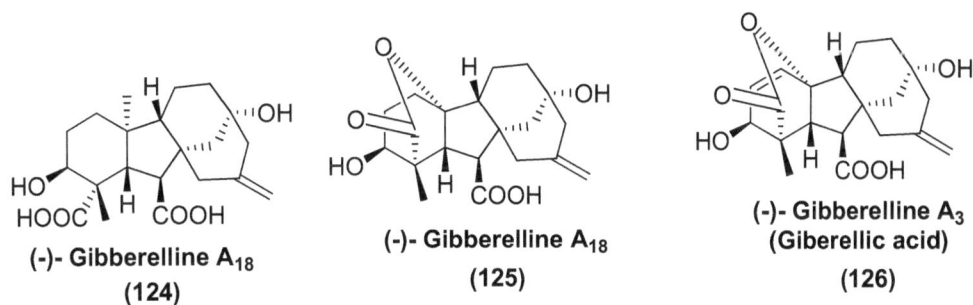

(-)- Gibberelline A$_{18}$
(124)

(-)- Gibberelline A$_{18}$
(125)

(-)- Gibberelline A$_3$
(Giberellic acid)
(126)

FIGURE 14.1 Structures of Gibberellanes.

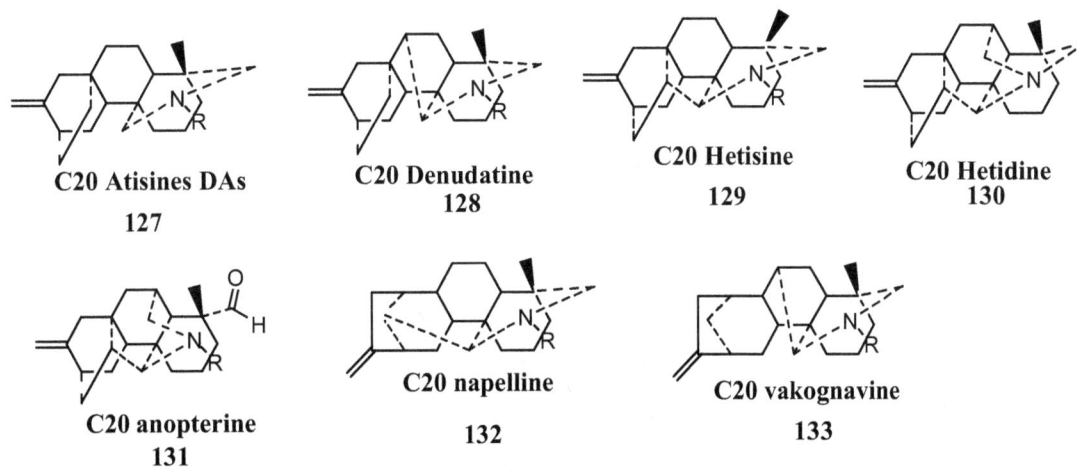

C20 Atisines DAs
127

C20 Denudatine
128

C20 Hetisine
129

C20 Hetidine
130

C20 anopterine
131

C20 napelline
132

C20 vakognavine
133

FIGURE 14.2 Structures of C20-Diterpenoid Alkaloids.

14.13 C20-Atisines Class

C20-Atisines Diterpenoids of the class Atisines have been isolated from several species of the genera Aconitum, Delphinium and Spiraea. Spirae japonica var. acuminata produces spirimines A and B (**134** and **135** in Figure 14.3). The methoxy group on C19 is present in DA-135, but the quaternary ammonium hydroxyethyl group is present in leucostomines A and B (**136** and **137** in Figure 14.3). The oxazolidine ring is revealed in compounds **138–140**, while the trimethyl-oxocyclohexyloxy group is revealed in compound **139**. Between C7 and C20, diterpenoids **141–143** have an O–C–N unit, while structure **143** has a carbonyl group at C15 [12].

14.13.1 C20-Denudatine Class

Except for Diterpenoids **144–145** derived from the complete herb *of Delphinium anthriscifolium* var. *Savatieri* [2,10,12],

the majority of denudatine Diterpenoids (compounds **144–149** in Figure 14.4) come from Aconitum spp.

The most significant C20-Diterpenoids members are hetisines (structures **150–154** in Figure 14.5). Structures **150–154** demonstrate that they are mostly isolated from Aconitum spp. and Delphinium spp. and include hydroxyl or methoxide groups in C6 and C3. On C6, diterpenoids **150–153** have an OH group, but compound **154** has a methoxide group. The majority of C20-Diterpenoids have an a-oriented OH group at C3, whereas compound **153** has a -oriented OH group. Compounds **154** isolated from the lateral roots of *A. Carmichaelii* had propionyloxy in C13 and 2-methyl butyryloxy moieties in C2, as well as a quaternary N base.

14.13.2 C20-Hetidine Class

Three compounds (**155–157** in Figure 14.6) make up the smallest category in the hetidine categorization. In all hetidine-Diterpenoids, the presence of the N=CH group, an

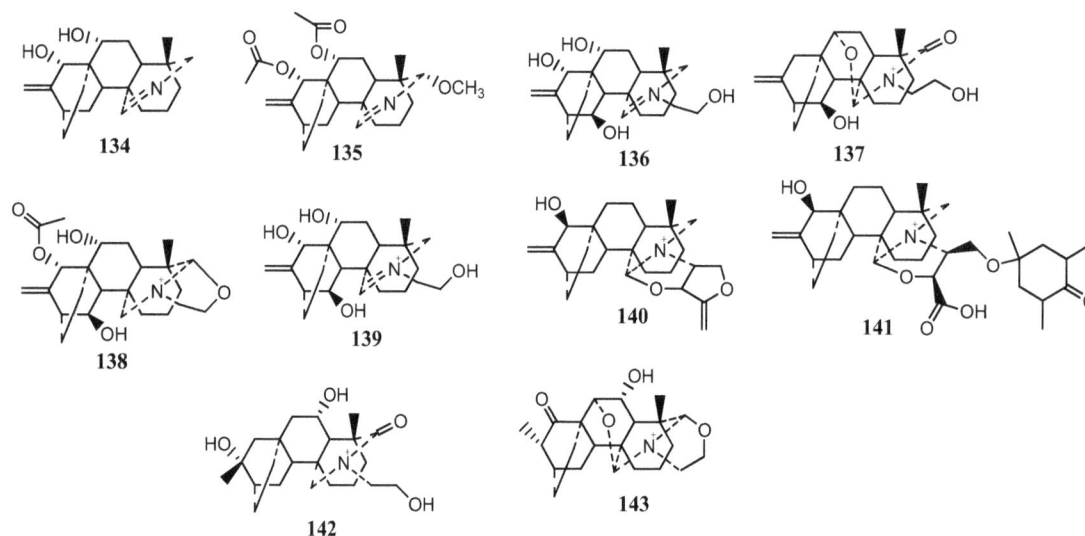

FIGURE 14.3 Structures of C20-Atrisines Diterpenoids.

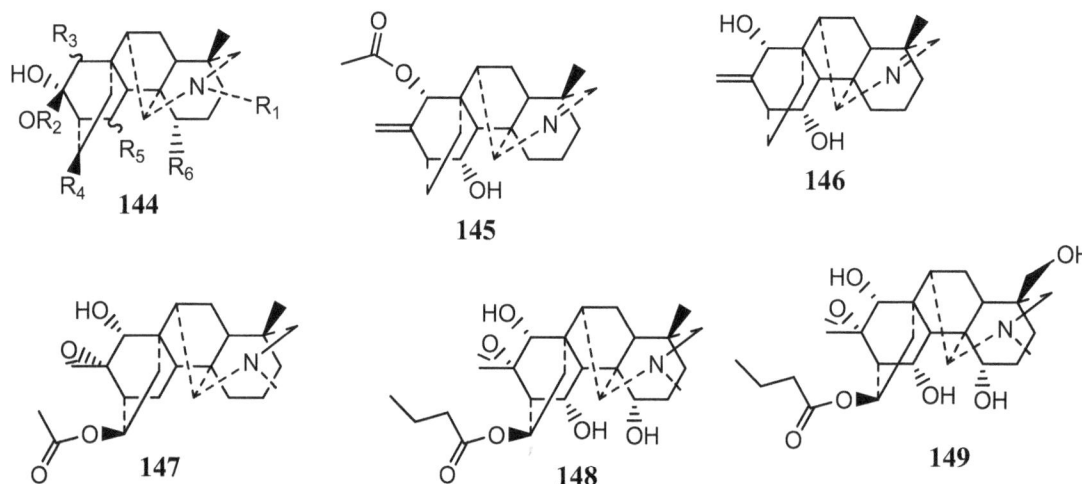

FIGURE 14.4 Structures of C20-Denudatine Class.

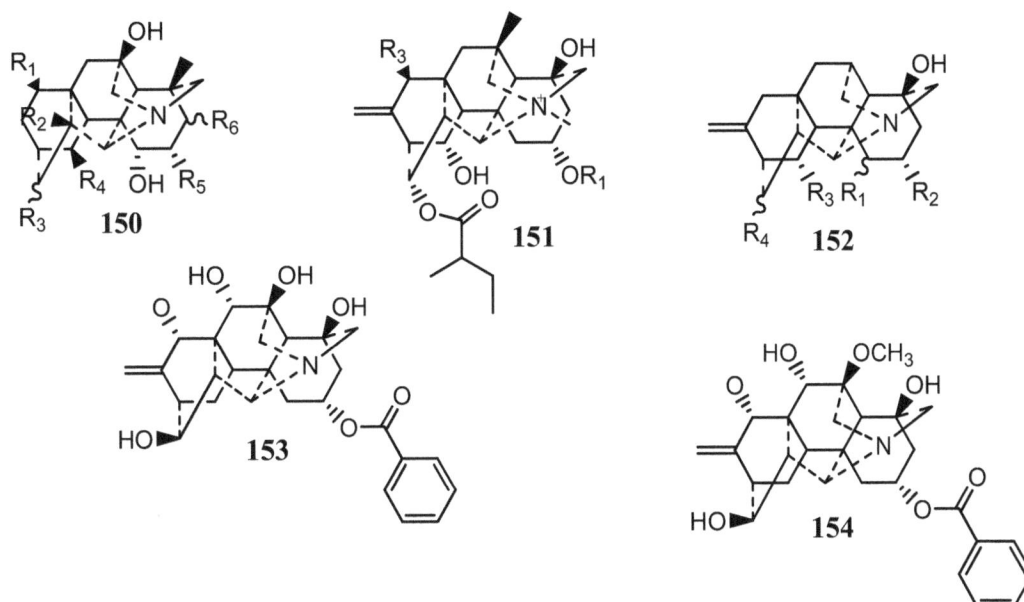

FIGURE 14.5 Structures of C20 Hetisine diterpenoids.

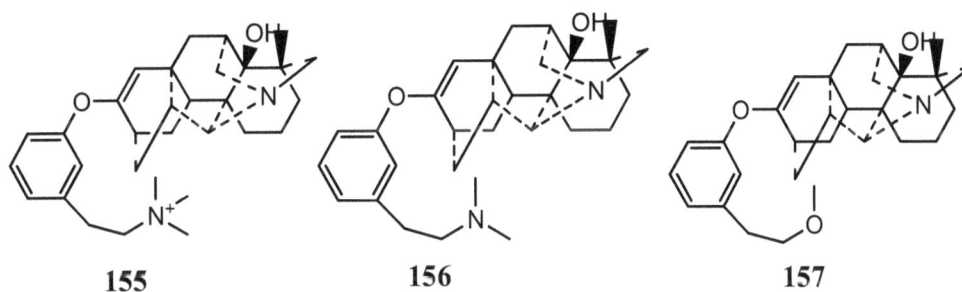

FIGURE 14.6 Structures of C20-Hetidine Class.

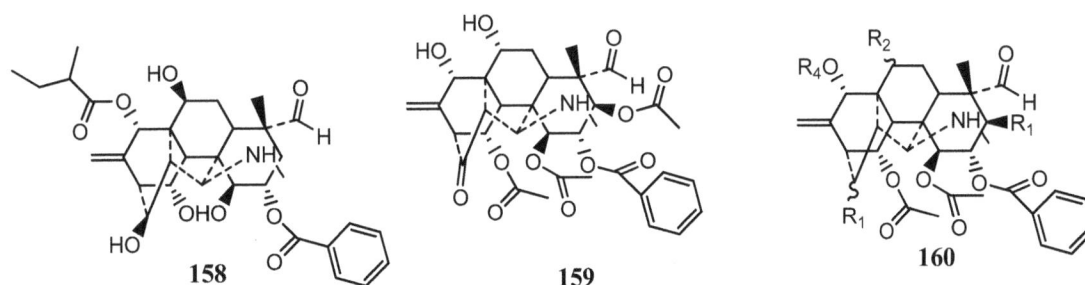

FIGURE 14.7 Structures of C20-Vakognavine diterpenoids.

endocyclic double bond, and a hydroxyl at C5 distinguishes it. In C17, Rotundifosine F (structure **155** in Figure 14.6) has a cardicine chloride, but DA **156** has a hordenine group in the same location, and DA **157** has a (2-methoxyethyl)-benzene ethanol moiety.

14.13.3 C20-Vakognavine Class

Aconitum and Delphinium produce the majority of vakognavine Diterpenoids (**158–159** in Figure 14.7). They have an uncommon double bond between C16 and C17, an aldehyde group in C19, and a distinctive N–Me group structurally [12].

14.13.4 C20-Napelline Class

This class contains a small number of alkaloids (**161–166** in Figure 14.8). DA 148 has an endocyclic double bond and an N=C19; DA 162 is a lactam fragment [13,14]. The class is completed by aconicarmichinium A trifluoroacetate, aconicarmichinium B trifluoroacetate and aconicarmichinium C chloride (**164–166**), all of which are derived from the alcohol iminium salts of *A. Carmichaelii.*

14.13.5 C20-Anopterine Class

This group of diterpenoids comes from the Anopterus/ Anopterusmacleayanus species (**167–169** in Figure 14.9). There are two hydroxyl groups, a N–Me, and an endocyclic double bond in all anopterine Diterpenoids. Only the substituent in C11 differs; compounds **167** and **168** have a hydroxymethyl butenoate group, whereas DA-169 has an O-benzoyl group.

14.13.6 C20-Rearranged Classes

They are novel C20-Diterpenoids with altered carbon skeletons (**170–174** in Figure 14.10). Isodonrubescens produces Kaurine A and B (**170, 171** in Figure 14.10). Instead of a 19,20, these two compounds have a 7,20-aza-ent-kaurane structure. A lactone between C11 and C16 is also found in the DA-170.

D. grandiflorum is the source of compound **172**. The connection between the N atom and C17 was open compared to the hetisine class skeleton due to the formation of a five-member ring with C4, C5, C6, C18 and the N atom. The roots of Delphinium trichophorum are used to make DA **174**. Its skeleton has a C-ring that is organized in a pentacyclic configuration rather than a hexacyclic one, as in the hetisane class. Oxygenated groups are found in almost all C20-diterpenoid alkaloids. C20-Diterpenoids, on the other hand, differ from C19-diterpenoid alkaloids in the following ways [11–15]:

i. As C19-Diterpenoids, most of them lack a methoxy group;
ii. Some alkaloids have an acetoxy or benzyloxy ester group, or both, but no other ester groups;
iii. Most C20-Diterpenoids have exocyclic methylene, and many of them have a secondary hydroxyl function in the allylic position;
iv. Few atisine and hetidine-

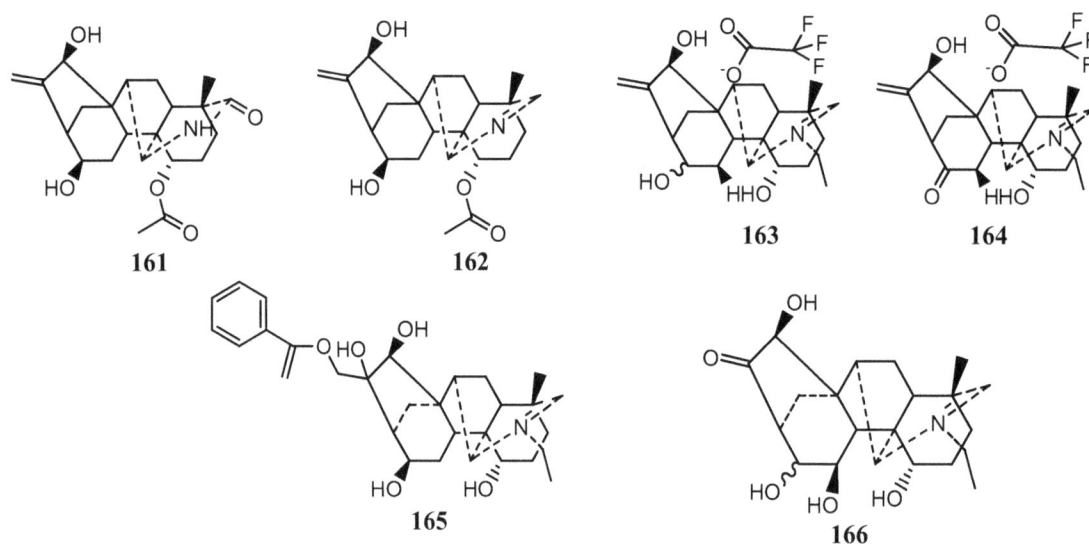

FIGURE 14.8 Structures of C20-Napelline Class.

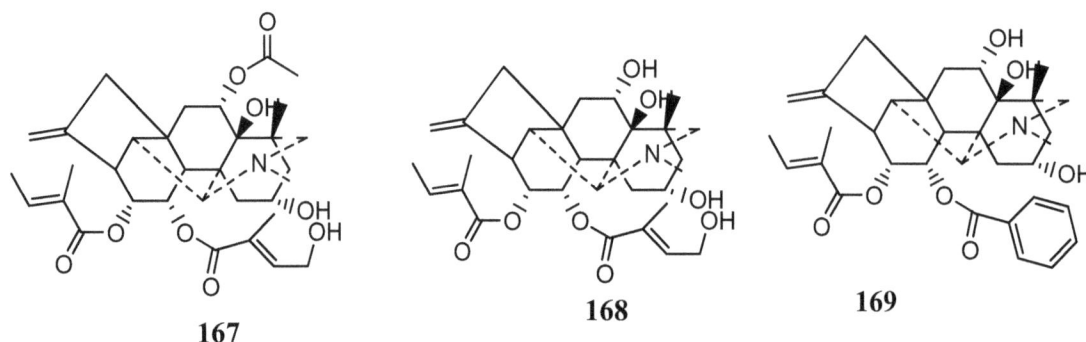

FIGURE 14.9 Structures of C20-Anopterine Class.

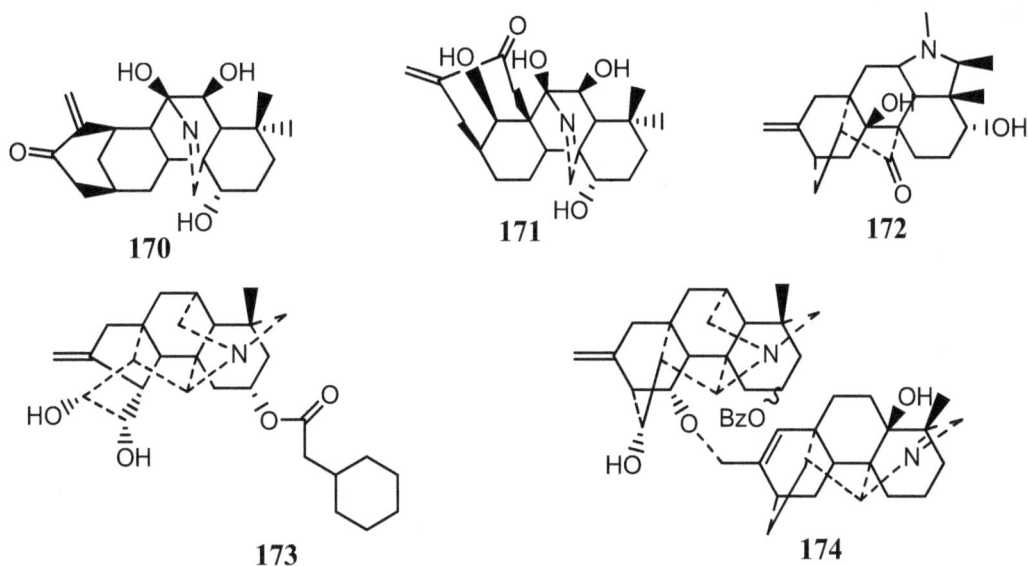

FIGURE 14.10 Structures of C20-Rearranged Classes.

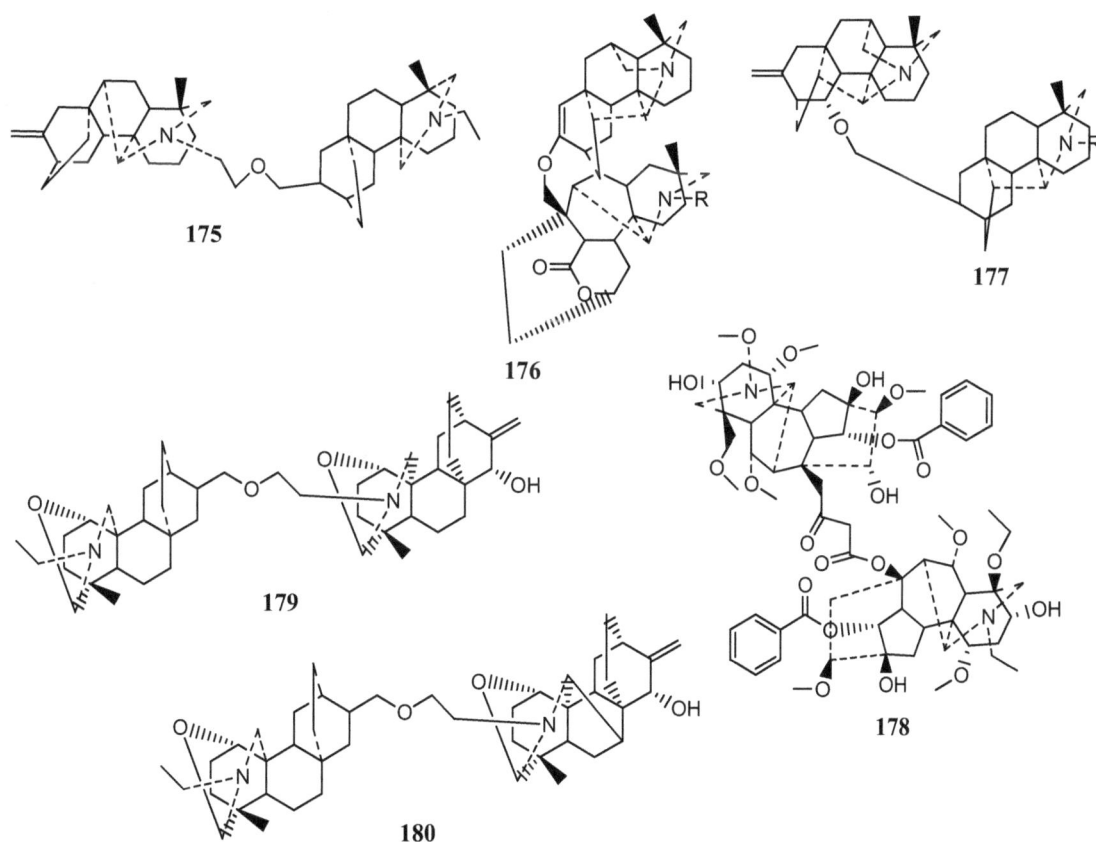

FIGURE 14.11 Structures of Bis-Diterpenoid Alkaloids.

14.13.7 Bis-Diterpenoid Alkaloids

Bis-Diterpenoids (**175–180** in Figure 14.11) are divided into three structural classes: hetidine–hetisine (**176** in Figure 14.11), heteratisine–hetidine and heteratisine–hetidine (**177** in Figure 14.11). The atisine–denudatine complex is made up of an atisine-type and a denudatine-type C20-DA with an O-ether connection between them. Hetidine–hetisine is a combination made up of a hetidine-type and a hetisine-type C20-DA with an oxygen atom connecting the two. Heteratisine–hetidine is a connection between a lactone-type C19-DA and a hetidine-type C20-DA [16–20]. Hetidine-type alkaloids have N, O-mixed acetal/ketal units.

14.14 Conclusion

This chapter provides comprehensive information regarding synthetic overview of terpenes and Nor-diterpenes. It includes Retrosynthetic route of Dehydro-linalool, Cafestol, Synthesis of Baccatin, Lupeol, retro synthesis of 6-methoxy-a-tetralone, substituted benzoate etc. It also includes diterpenes, Gibberellanes, C20 Diterpenoid, Atisines, Denudatine, Hetidine, Vakognavine and Bis-Diterpenoid alkaloids.

REFERENCES

[1]. Pommer, H. Viatmin A. *Angewandte Chemie*, **1960**, 72, 811, 911.

[2]. McMurry. Coupling β-Carotene from Vitamin-A-Aldehyde 36 D. Lenoir, Synthesis **1989**, 883.

[3]. Corey, E. J.; Wess, G.; Xiang, Y. B.; Singh, A. K. Cafestol (racemic). *Journal of the American Chemical Society*, **1987**, 109, 4717.

[4]. Nicolaou, K. C.; Guy, R. K. Die Eroberung von Taxol. *Angewandte Chemie*, **1995**, 107(34), 2247–2079.

[5]. Stork, G.; Uyeo, S.; Wakamatsu, T.; Grieco, P.; Labowitz, J. Triterpenes, Lupeol and Precursor. *Journal of the American Chemical Society*, **1971**, 93, 4945.

[6]. Stork, G.; Loewenthal, H. J. E.; Mukharji, P. C. A Total Synthesis of 1-Oxygenated Steroids. *Journal of the American Chemical Society*, 78, **1956**, 501.

[7]. Hesse, B.; Meier, H.; Zeeh, B. *Spectroscopic Methods in Organic Chemistry*, 7th Edition. Georg Thieme, Stuttgart, **2005**.

[8]. Joulain, D.; König, W. A. *The Atlas of Spectral Data of Sesquiterpene Hydrocarbons*. E. B. Verlag, Hamburg, **1999** (collection of spectroscopic data).

[9]. Friebolin, H. *Basic One- and Two-Dimensional NMR Spectroscopy*, 4th Edition. Wiley-VCH, Weinheim, **2004**.

[10]. Günther, H. *NMR-Spektroskopie*, 3rd Edition, Georg Thieme, Stuttgart, **1992**.

[11]. Braun, S.; Kalinowski, H.-O.; Berger, S. *200 and More Basic NMR Experiments, A Practical Course*, 3rd Edition, Wiley-VCH, Weinheim, **2004**.

[12]. Csupor, D.; Chang, F. R.; El-Shazly, M.; D'Auria, J. C. *Molecules*, **2021**, 26(13), 4103. doi: 10.3390/molecules26134103.

[13]. Sun, L.-M.; Huang, H.-L.; Li, W.-H.; Nan, Z.-D.; Zhao, G.-X.; Yuan, C.-S. ChemInform Abstract: Alkaloids from Aconitum barbatum var. puberulum. *Helvetica Chimica Acta*, **2009**, 40, 1126–1133.

[14]. Yu, H.-J.; Liang, T.-T. A New Alkaloid from the Roots of Aconitum carmichaeli Debx. *Journal of the Chinese Chemical Society*, **2012**, 59, 693–695.

[15]. Meng, X. H.; Jiang, Z. B.; Zhu, C. G.; Guo, Q. L.; Xu, C. B.; Shi, J. G. Napelline-type C20-diterpenoid alkaloid iminiums from an aqueous extract of "fu zi": Solvent-/base-/acid-dependent transformation and equilibration between alcohol iminium and aza acetal forms. *Chinese Chemical Letters*, **2016**, 27, 993–1003.

[16]. Ren, M.-Y.; Yu, Q.-T.; Shi, C.-Y.; Luo, J.-B. Anticancer Activities of C18-, C19-, C20-, and Bis-Diterpenoid Alkaloids Derived from Genus Aconitum. *Molecules*, **2017**, 22, 267.

[17]. Lin, L.; Chen, D.-L.; Liu, X.-Y.; Chen, Q.-H.; Wang, F.-P.; Yang, C.-Y. Bis-Diterpenoid Alkaloids from Aconitum tanguticum Var. trichocarpum. *Natural Product Communications*, **2009**, 4, 897–901.

[18]. Zhang, Z. T.; Chen, D. L.; Chen, Q. H.; Wang, F. P. Bis-Diterpenoid Alkaloids from Aconitum tanguticum var. trichocarpum. *Helvetica Chimica Acta*, **2013**, 96, 710–718.

[19]. Guo, Q.; Xia, H.; Wu, Y.; Shao, S.; Xu, C.; Zhang, T.; Shi, J. Structure, Property, Biogenesis, and Activity of Diterpenoid Alkaloids Containing a Sulfonic Acid Group from Aconitum carmichaelii. *Acta Pharmaceutica Sinica B*, **2020**, 10, 1954–1965.

[20]. Duan, X. Y.; Zhao, D. K.; Shen, Y. Two New bis-C20-Diterpenoid Alkaloids with Anti-Inflammation Activity from Aconitum bulleyanum. *Journal of Asian Natural Products Research*, **2019**, 21, 323–330.

15

Sesquiterpenes: A Chemical Synthesis and Biological Activity

Bimal Krishna Banik*, Abhishek Tiwari†, and Biswa Mohan Sahoo‡

CONTENTS

15.1 Introduction 488
15.2 Acyclic Sesquiterpenoids 488
    15.2.1 Farnesanes 488
15.3 Monocyclic Farnesane Sesquiterpenes 489
    15.3.1 Cyclofarnesanes and Bisabolanes 489
15.4 Monocyclic Farnesane Sesquiterpenes 489
    15.4.1 Cyclofarnesanes and Bisabolanes 489
    15.4.2 Germacranes and Elemanes 490
    15.4.3 Humulanes 491
15.5 Polycyclic Farnesane Sesquiterpenes 491
    15.5.1 Caryophyllanes 491
    15.5.2 Eudesmanes and Furanoeudesmanes 492
    15.5.3 Eremophilanes, Furanoeremophilanes, Valeranes 493
    15.5.4 Cadinanes 495
    15.5.5 Drimanes 497
    15.5.6 Guaianes and Cycloguaianes 497
15.6 Himachalanes, Longipinanes, Longifolanes 499
    15.6.1 Picrotoxanes 500
    15.6.2 Isodaucanes and Daucanes 501
    15.6.3 Protoilludanes, Illudanes, Illudalanes 501
    15.6.4 Marasmanes, Isolactaranes, Lactaranes, Sterpuranes 501
    15.6.5 Acoranes 502
    15.6.6 Chamigranes 503
    15.6.7 Cedranes and Isocedranes 503
    15.6.8 Zizaanes and Prezizaanes 504
    15.6.9 Campherenanes and Santalanes 504
    15.6.10 Thujopsanes 505
    15.6.11 Hirsutanes 505
15.7 Other Polycyclic Sesquiterpenes 506
    15.7.1 Pinguisanes 506
    15.7.2 Presilphiperfolianes, Silphiperfolianes, Silphinanes, Isocomanes 506
15.8 Important Synthesis of Sesquiterpenes 507
    15.8.1 Farnesene 507
    15.8.2 Farnesol 507
    15.8.3 Nerolidol 509
    15.8.4 E-Selinene 509
    15.8.5 Isocomene 510
    15.8.6 Cedrene 510
    15.8.7 Periplanone B 511
    15.8.8 Laurene 514
15.9 Conclusion 514
References 515

* Department of Mathematics and Natural Sciences, College of Sciences and Human Studies, Prince Mohammad Bin Fahd University, Al Khobar, Kingdom of Saudi Arabia. bimalbanik10@gmail.com; bbanik@pmu.edu.sa.

† Faculty of Pharmacy, Pharmacy Academy, IFTM University, Lodhipur Rajput, Moradabad-244102, Uttar Pradesh, India. abhishekt1983@gmail.com.

‡ Roland Institute of Pharmaceutical Sciences, Berhampur affiliated to Biju Patnaik University of Technology (BPUT), Rourkela, Odisha, India.

DOI: 10.1201/9781003008682-15

15.1 Introduction

The sesquiterpenoids, in general, form the higher boiling fraction of the essential oils; this provides their chief source. Sesquiterpenoid structure is built up of three isoprene units; this has been shown to be the case for the majority of the known sesquiterpenoids, but there are some exceptions.

The sesquiterpenoids are classified into four groups according to the number of rings present in the structure. If we use the isoprene rule, then when three isoprene units are linked (head to tail) to form an acyclic sesquiterpenoid hydrocarbon as shown in Figure 15.1, the latter will contain four double bonds. Each isoprene unit contains two double bonds, but one

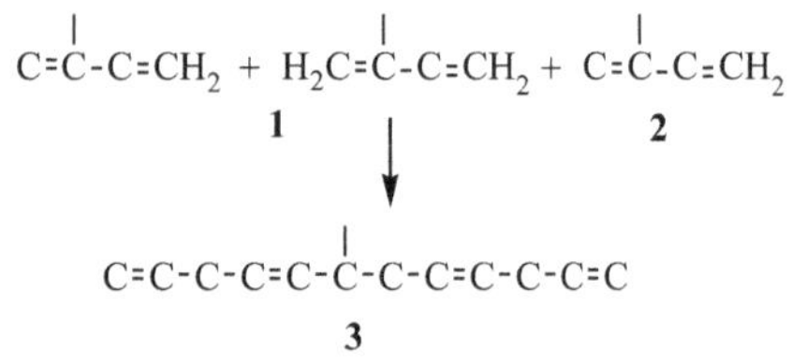

FIGURE 15.1 Acyclic sesquiterpenoid hydrocarbon (linked (head to tail) of three isoprene units).

TABLE 15.1

Different Classes of Sesquiterpenoids

Class of Sesquiterpenoids	Number of Double Bonds
Acyclic	4
Monocyclic	3
Bicyclic	2
Tricyclic	1

disappears for each pair that is connected, different classes of sesquiterpenoids are shown in Table 15.1 [1–3].

15.2 Acyclic Sesquiterpenoids

15.2.1 Farnesanes

2,6,10-Trimethyldodecane or farnesane, the parent compound of about 10,000 sesquiterpenes known to date [4] is found in the oil slate. The *(E, E)*-isomer of D-farnesene (4) is a component of the flavors and natural coatings of apples, pears and other fruits. Associated with *E*-farnesene (5), it also occurs in several ethereal oils, for example, those of camomile, citrus and hops. Aldehydes such as α- and β-sinensal (7, 8) derived from α- and β-farnesene contribute to the flavor of the oil of orange expressed from the fresh peel of ripe fruits of *Citrus sinensis* (Rutaceae); mandarin peel oil from *Citrus reticulata* and *C. aurantium* (Rutaceae) contains 0.2% of α-sinensal (7) with the smell of oranges. (*S*)-(+)-Nerolidol (9) in the oil of neroli obtained from orange flowers and found in many other flowers is used in perfumery, similar to farnesol (10) from *Acacia farnensiana* (Mimosaceae) and the oils of bergamot, hibiscus, jasmine and rose and pleasantly smelling blossoms such as lily of the valley. (*S*)-2,3-Dihydrofarnesol, known as terrestrol, is the marking pheromone of the male bumble bee *Bombus terrestris* (Figure 15.2).

The chains of some furanoid-farnesane derivatives are terminated by furan rings. Dendrolasin from sweet potatoes, also isolated from some marine snails, for example, is an alarm and defense pheromone of the ant *Dendrolasius fulginosus* [Dandrolasin (11)]. Sesquiro-sefuran (12) and longifolin (13) occur in the leaves of *Actinodaphne longifolia* (Figure 15.3) [5].

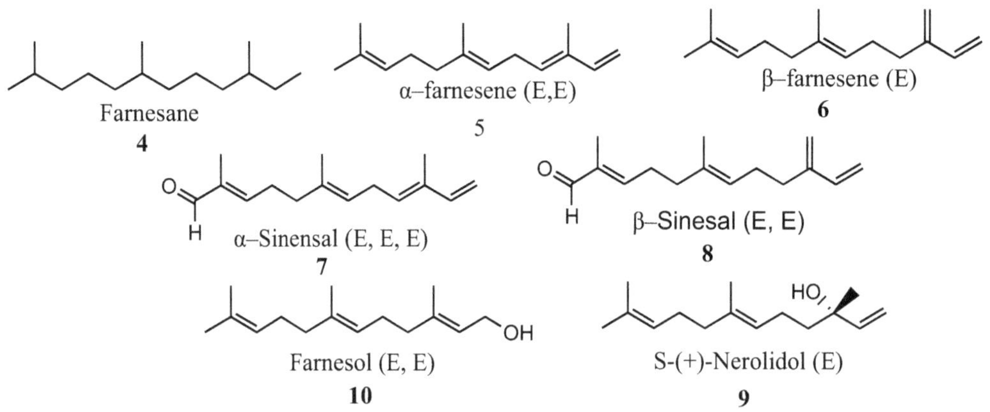

FIGURE 15.2 Farnesanes and its derivatives.

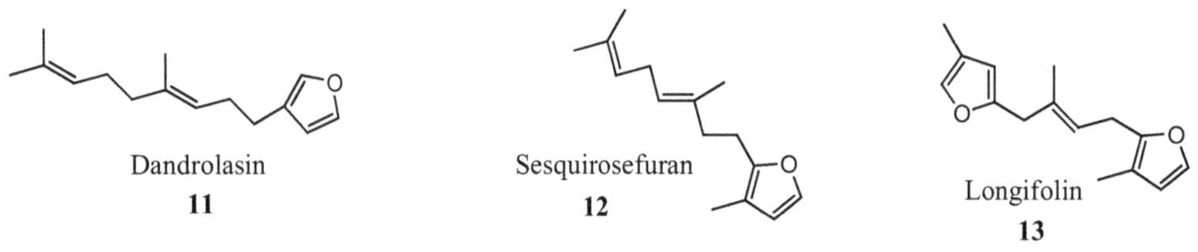

FIGURE 15.3 Farnesanes containing furan ring.

15.3 Monocyclic Farnesane Sesquiterpenes

15.3.1 Cyclofarnesanes and Bisabolanes

Cyclofarnesanes formally arise when carbon atoms C-6 and C-7 of farnesane close a ring. Abscisic acid, occurring in the leaves of cabbage, potatoes, roses and young fruits of cotton, and (+)-dihydroxy-J-ionylidene acetic acid produced by the fungus *Cercospora cruenta* which is anti-bacterially active, are examples. Abscisic acid acts as an antagonist of plant growth hormones and controls flowering, falling of fruits and shedding of leaves.

15.4 Monocyclic Farnesane Sesquiterpenes

15.4.1 Cyclofarnesanes and Bisabolanes

Cyclofarnesanes (**14**) formally arise when carbon atoms C-6 and C-7 of farnesane close a ring. Abscisic acid (**15**), occurring in the leaves of cabbage, potatoes, roses and young fruits of cotton, and (+)-dihydroxy-γ-ionylidene (**16**) acetic acid produced by the fungus *Cercospora cruenta* which is

anti-bacterially active, are examples. Abscisic acid acts as an antagonist of plant growth hormones and controls flowering, falling of fruits and shedding of leaves (Figure 15.4).

Formally, C-1 and C-6 of farnesane close a cyclohexane ring in the bisabolanes (17), which represent a more prominent class of monocyclic sesquiterpenes. Additional cyclizations increase the diversity (Figure 15.5).

Oil of ginger obtained from the rhizome of *Zingiber officinalis* (Zingiberaceae) consists predominantly of (-)-zingiberene (18) (20%–40%), E-sesquiphellandrene (22) and (+)-E-bisabolene (19). The latter also occurs in *Chamaecyparis nootkatensis* (Cupressaceae) and in the Sibirian pine tree *Pinus sibirica* (Pinaceae). (-)-D- (20) and (+)-E-bisabolol (21) are fragrant sesquiterpenes found in the essential oils of various plants; they also contribute to the odors of camomile and of bergamot oil from unripe fruits of *Citrus aurantium* var. *bergamia* (Rutaceae) growing in southern Italy. (-) E-Sesquiphellandrene (22) from *Curcuma longa* Sesquisabinene (23) from pepper *Piper nigrum* (Piperaceae), sesquithujene (24) from ginger *Zingiber officinalis* (Zingiberaceae) and sesquicarene (25) from *Schisandra chinensis* represent bicyclic bisabolanes [2,5].

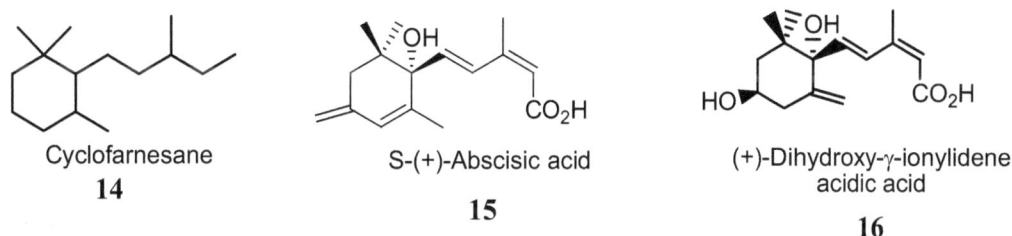

Cyclofarnesane
14

S-(+)-Abscisic acid
15

(+)-Dihydroxy-γ-ionylidene
acidic acid
16

FIGURE 15.4 Examples of Cyclofarnesanes.

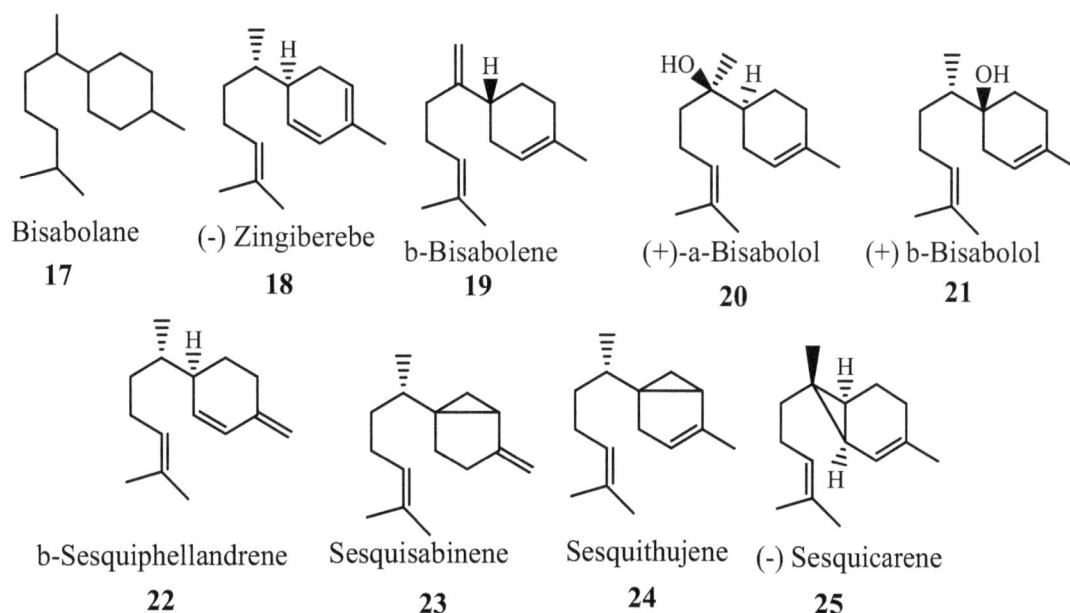

Bisabolane
17

(-) Zingiberebe
18

b-Bisabolene
19

(+)-a-Bisabolol
20

(+) b-Bisabolol
21

b-Sesquiphellandrene
22

Sesquisabinene
23

Sesquithujene
24

(-) Sesquicarene
25

FIGURE 15.5 Examples of Bisabolanes.

15.4.2 Germacranes and Elemanes

Germacranes (27, 28) formally result from ring closure of C-1 and C-10 of farnesane (26). 1(10), 4-Germacradienes such as 1(10), 4-germacradien-6-ol (30) present as a glycoside in *Pittosporum tobira* may undergo COPE rearrangements to elemadienes, exemplified by shyobunol (31) from the oils of galbanum and kalmus, so that some isolated elemane (29) derivatives are supposed to be artifacts arising from germacranes (Figure 15.6).

More than 300 naturally occurring germacranes are reported. Among these are beta-Elemenone (33) from *Rhododendron calostrotum*, germacrene B (38) [1(10)-*E*, 4-*E*, 7(11)-germacratriene] from the peel of *Citrus junos*, germacrene D (39) from bergamot oil (*Citrus bergamia*, Rutaceae) and germacrone (32) [1(10)-*E*, 4-*E*, 7(11)-germacratrien-8-one], derived from germacrene B (38), a pleasantly flowery to herby-smelling component isolated from the essential oil of myrrh (*Commiphora*

abyssinica, family-Burseraceae) as well as from ethereal oils of *Geranium macrorhyzum* (Geraniaceae) and *Rhododendron adamsii* (Ericaceae). Periplanones A-D (24, 25, 26, 37) act, in contrast to long-range pheromones, as close proximity sex excitants. They are found in the alimentary tract and excreta of the female American cockroach *Periplaneta americana* and cause the males to run and to perform the courtship display. Periplanones A and B, occurring in a ratio of 10:1, are 100 times more active than C and D (Figure 15.7) [6,7].

About 50 elemanes known to date comprise E-elemenone (33) from the oil of myrrh, representing the **COPE rearrangement** product of germacrone (32), (-)-bicycloelemene (40) from peppermint oils of various provenance (e.g. *Mentha piperita* or *Mentha arvensis*) and E-elemol (41) which is not only a minor component of Javanese oil of citronella but is also found in the elemi oil with an odor like pepper and lemon, expressed from the Manila elemi resin of the tree *Canarum luzonicum* (Burseraceae) (Figure 15.8) [6,7].

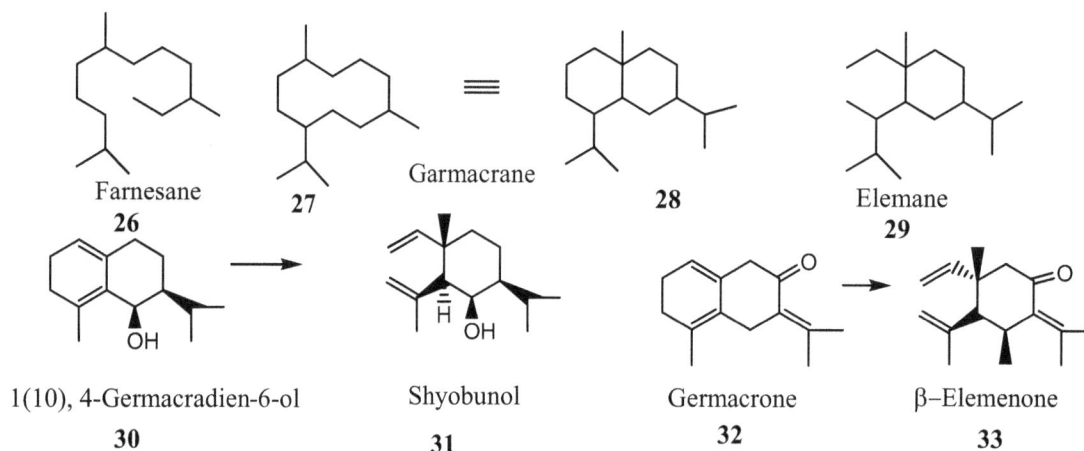

FIGURE 15.6 Examples of Germacranes.

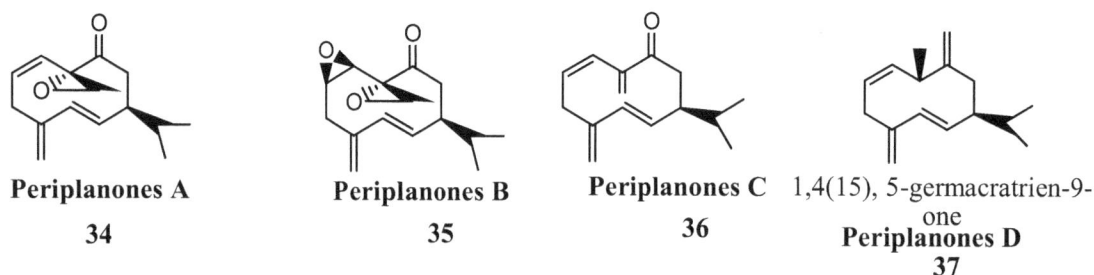

FIGURE 15.7 Examples of Periplanones.

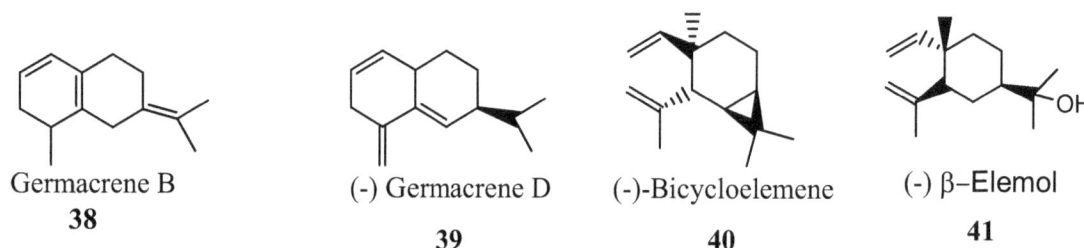

FIGURE 15.8 Examples of germacrene and Elemenes.

15.4.3 Humulanes

Ring closure of C-1 and C-11 of farnesane (42), not only formally but also in biogenesis *via* farnesyldiphosphate, produces the sesquiterpene skeleton of more than 30 naturally occurring humulanes (43). Regioisomeric D- (44) and E-humulene (45) occur in the leaves of *Lindera strychnifolia* (Lauraceae) (Figure 15.9).

Humulenes are, associated with epoxy humuladienes (46, 47) derived from D-humulene as well as (-)-humulol (48) and (+)-humuladienone (49), prominent constituents of the essential oils of hops (*Humulus lupulus*, Cannabaceae), cloves (*Caryophylliflos*, Caryophyllaceae) and ginger (*Zingiber zerumbeticum*, Zingiberaceae) (Figure 15.10) [2].

15.5 Polycyclic Farnesane Sesquiterpenes

15.5.1 Caryophyllanes

Approximately 30 naturally abundant caryophyllanes2 are derived from humulanes (50) in which C-2 and C-10 close a cyclobutane ring (Figure 15.11).

(-)-E-Caryophyllene (51, 52, 53) occurs as a mixture with its *cis* isomer iso-caryophyllene in the clove oil (up to 10%) from dried flower buds of cloves (*Caryophylliflos*, Caryophyllaceae), in the oil obtained from stems and flowers of *Szygium aromaticum* (Myrtaceae), as well as in the oils of cinnamon, citrus, eucalyptus, sage and thyme. Clove oil, with its pleasantly sweet, spicy and fruity odor, is used not only in perfumery

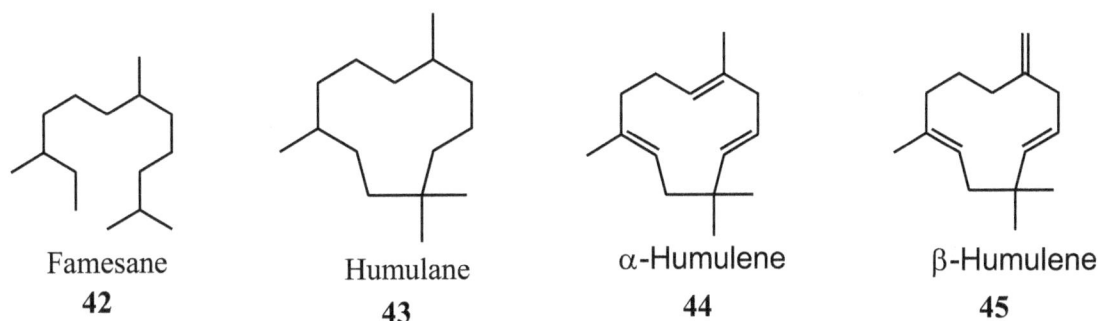

FIGURE 15.9 Examples of Humulanes.

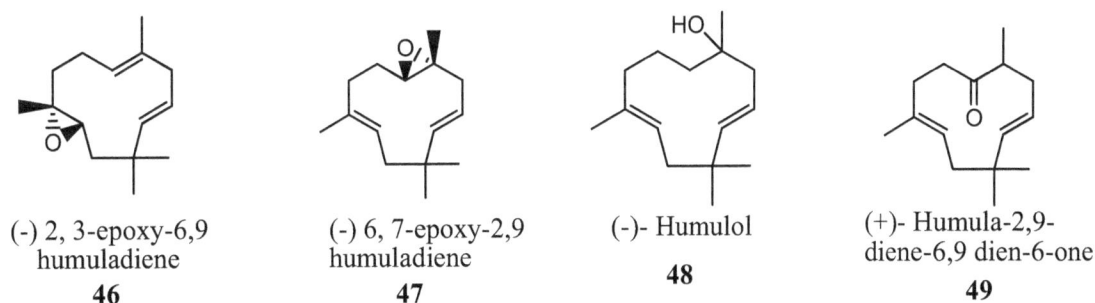

FIGURE 15.10 Few examples of Humulanes.

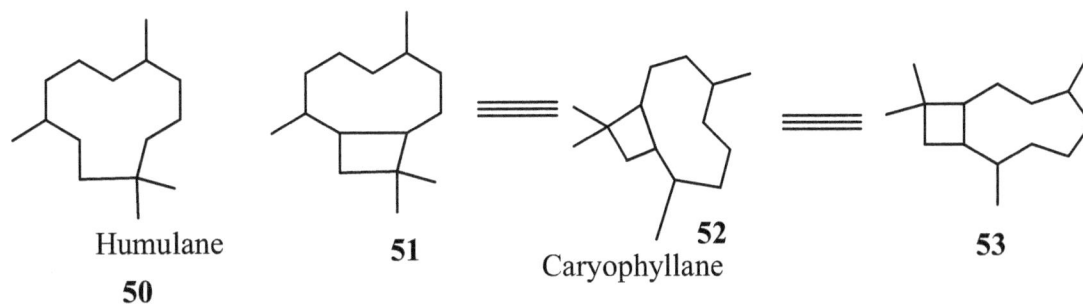

FIGURE 15.11 Examples of Caryophyllanes.

and for flavoring chewing gums, but also as a dental analgesic, carminative and counterirritant. Other representatives include (-)-6,7-epoxy-3(15)-caryophyllene from the leaves, flowers and stems of cloves, (+)-6-caryophyllen-15-al from the oil of sage (*Salvia sclarea*, Labiatae) and (-)-3(15), 7-caryophylladien-6-ol from Indian hemp *Cannabis sativa* var. *indica* (Cannabaceae) (Figure 15.12) [2,5].

15.5.2 Eudesmanes and Furanoeudesmanes

Carbon atoms C-1 and C-10 in addition to C-2 and C-7 of farnesane (58) link up to close the eudesmane (59, 60) bicyclic skeleton of sesquiterpenes with *trans*-decalin as a core structure with the corresponding numbering of the ring positions (Figure 15.13).

Well-known eudesmane (61) derivatives in flavors and fragrances include D- (62) and Eselinene (63) from the oils of *Cannabis sativa* var. *indica* (Moraceae), celery (*Apium graeveolens*, Umbelliferae) and hops (*Humulus lupulus*, Moraceae), (+)-D- (65) and (+)- E-eudesmol (65) from some oils of eucalyptus (*Eucalyptus macarthuri*), (-)-*epi*-Jeudesmol with its woody odor from the north African oil of geranium

(*Pelargonium odoratissimum* and allied species), and the almost odorless diastereomeric (+)-J-eudesmol from various ethereal oils. (+)-E-Costus acid and (+)-E-costol belong to the constituents of the essential oil obtained from the roots of *Saussurealappa*(Asteraceae) which is used to treat stomach ailments in Chinese and Japanese popular medicine (Figure 15.14) [2,5,8].

Several structural variants of lactones derived from eudesmane occur in *Artemisia*species (Asteraceae). These include various 3-oxo-12,6-eudesmanolides such as (+)-santonane (66) from the flowers of *A. pauciflora*, (-)-taurin (67) from *A. Taurica*, the antihelmintic but toxic santonines (68, 69) isolated from the dried unexpanded flowerheads of *A. Maritime* (contents up to 1.5% of D-santonin) and allied species (Figure 15.15).

Furanoeudesmanes such as (-)-furanoeudesma-1,3-diene, furanoeudesma-1,4-dien-6-one and furanoeudesma-1,4(15)-diene, known as (-)-lindestrene, belong to the sweetish balsamic-smelling constituents of the yellowish red gum-resin myrrh, used as a carminative and astringent and obtained from *Commiphora* species (e.g., *Commiphora abyssinica*, *C. molmol*, Burseraceae). Tubipofuran (70), a diastereomer

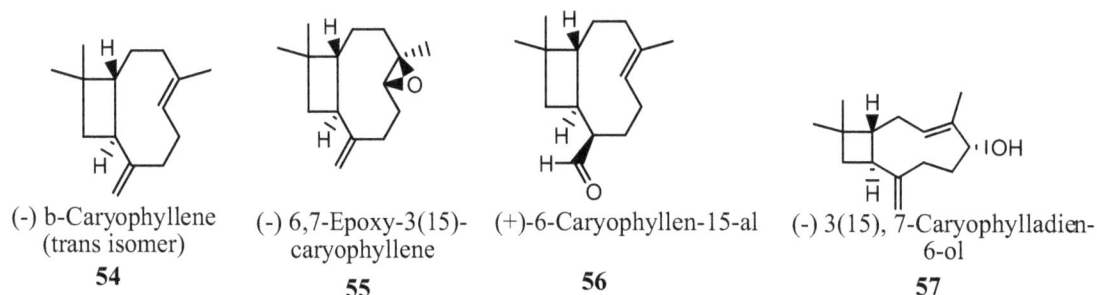

(-) b-Caryophyllene (trans isomer) **54**

(-) 6,7-Epoxy-3(15)-caryophyllene **55**

(+)-6-Caryophyllen-15-al **56**

(-) 3(15), 7-Caryophylladien-6-ol **57**

FIGURE 15.12 Few more examples of Caryophyllanes.

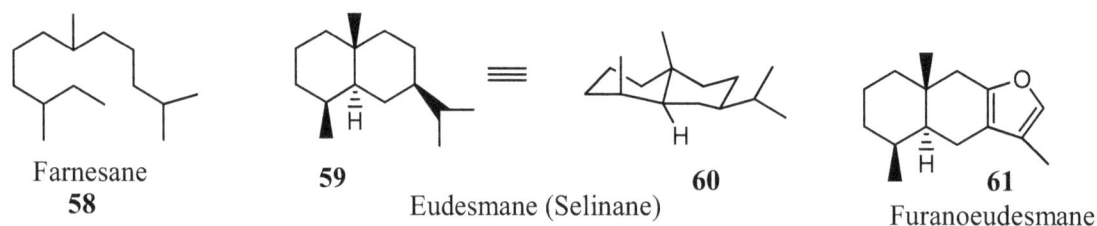

Farnesane **58**

59

Eudesmane (Selinane) **60**

Furanoeudesmane **61**

FIGURE 15.13 Examples of Eudesmanes.

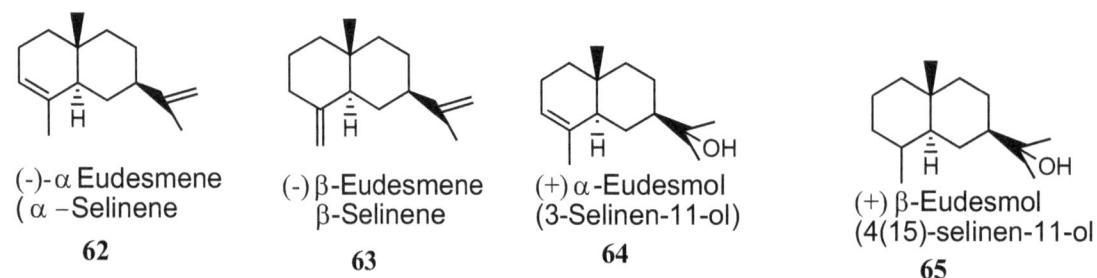

(-)-α Eudesmene (α –Selinene **62**

(-) β-Eudesmene β-Selinene **63**

(+) α-Eudesmol (3-Selinen-11-ol) **64**

(+) β-Eudesmol (4(15)-selinen-11-ol **65**

FIGURE 15.14 Examples of Furanoeudesmane.

of (-)-furanoeudesma-1,3-diene (71) isolated from *Tubipora musica* exhibits cyto- and ichthyotoxic activity (killing cells and fish, respectively), furanoeudesma-1,4 (15)-diene (72) from myrrh, furanoeudesma-1,4-diene-6-one (73) from *Commiphora myrrha* (Figure 15.16) [7,8].

15.5.3 Eremophilanes, Furanoeremophilanes, Valeranes

A methyl shift from C-10 to C-5 in eudesmane (74) leads to the basic skeleton of more than 150 eremophilane (75, 76) and furanoeremophilane (77) derivatives isolated so far from higher plants (Figure 15.17).

In contrast, valeranes (78, 79) arise from migration of the methyl group C-15 in eudesmane from C 4 to C-5, in contrast,

very rarely occur. Examples include the valerenones (80) from the roots of valerian *Valeriana officinalis* and from *Nardostachys jatamansi* (Valerianaceae).

The Australian tree *Eremophila mitchelli* gave its name to the eremophilanes with both methyl groups in E-positions of the decalin bicycle; the wood of *Eremophila mitchelli* contains various eremophiladienones (81, 82). Non-toxic metabolites of the fungus *Penicillium roqueforti* growing in some kinds of cheese and referred to as eremofortins (83, 84) are more prominent representatives of the eremophilanes (Figure 15.18) [2,9].

Eremophilanes with both methyl groups in D-positions of the decalin core structure are referred to as valencanes, exemplified by (-)-nootkatene (85) from the nootka cypress *Chamaecyparis nootkatensis* (Cupressaceae) and

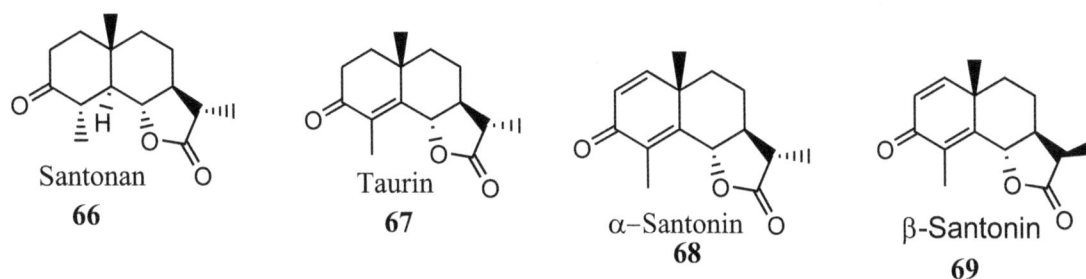

FIGURE 15.15 Few more examples of Santolins.

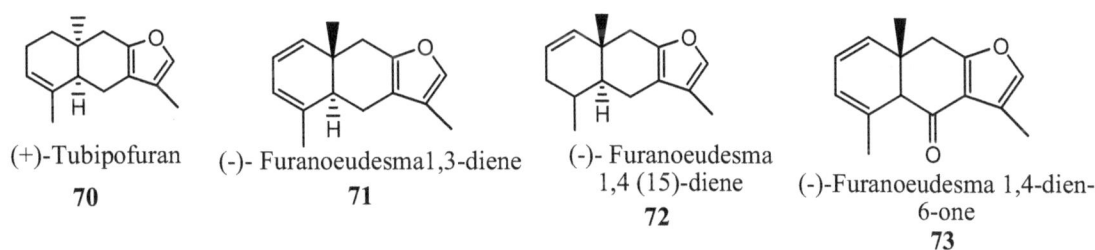

FIGURE 15.16 Few more examples of Furanoeudesmanes.

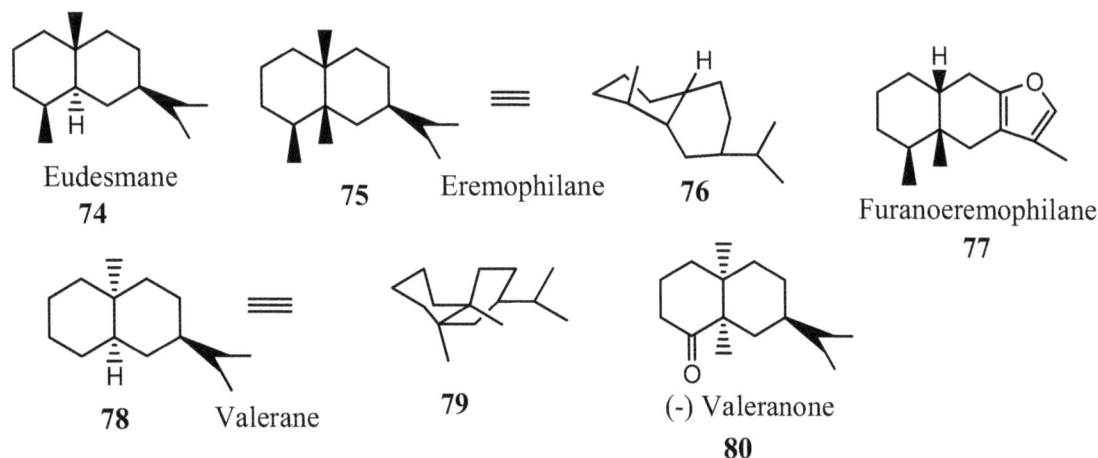

FIGURE 15.17 Examples of Eremophilanes.

(+)-11-eremophilen-2,9-dione (88) from the oil of grapefruit. Additional examples include (+)-valerianol [1(10)-eremophilen-11-ol] (86) from valerian (*Valeriana officinalis*) as well as nootkatone [1(10), 11-eremophiladien-2-one] from the oil of grapefruit (*Citrus paradisii*) and the nootka cypress which is added as a flavor to drinks. The regioisomeric (+)- isonootkatene [D-vetivone, 1(10), 7(11)-eremophiladien-2-one] (87) is a main constituent of the oil of vetiver distilled from the roots of tropical vetiver grass *Vetiveria zizanoides* (Poaceae) used in soap formulations and perfumery (Figure 15.19) [2,5,9,10].

More than 100 furanoeremophilanes are described as constituents in higher plants, chiefly in Senecio species (Asteraceae, formerly Compositae) (Figure 15.20).

Furanoeremophilan-9-one, a constituent of golden rag-wort (squaw weed, life root) from the dried plant of *Senecio aureus*, (+)-furanoeremophilen-3-one from *Senecio nemorensis*, (-)-1,10-epoxyfuranoeremophilane from *Senecio glastifolius*, and (-)-1,10 epoxy furanoeremophilan-6,9-dione from

Senecio smithii represent typical examples. Nardosinanes (94) such as (-)-kanshone A (95), including some structural variants and nardosinone (96), a 1,2-dioxolane (cyclic peroxide) isolated from *Nardostachys chinensis* (Valerianaceae) formally emerge from eremophilane by migration of the isopropylgroup from C-7 to C-6. Aristolanes (97) are 6,11-cycloeremophilanes. Some representatives are found in Aristolochiaceae, for example 9-aristolen-8-one (99) and 1(10)-aristolen-12-al (100) in *Aristolochia debilis*. 1(10)-Aristolene (98) is a constituent of the Gurjun balm flowing from the caves cut into the giant Dipterocarpaceae growing in Bengal upon setting a fire to these trees (Figure 15.21).

Ishwaranes (102) represent 7–11/10–12-bicycloeremophilanes (101). These rare tetracyclic sesquiterpenes are found in Aristolochiaceae, for example ishwarane (102) itself and ishwaranol (103) in the roots of *Aristolochia indica* and 3-ishwaranone (104) from *Aristolochia debilis* (Figure 15.22) [2,8].

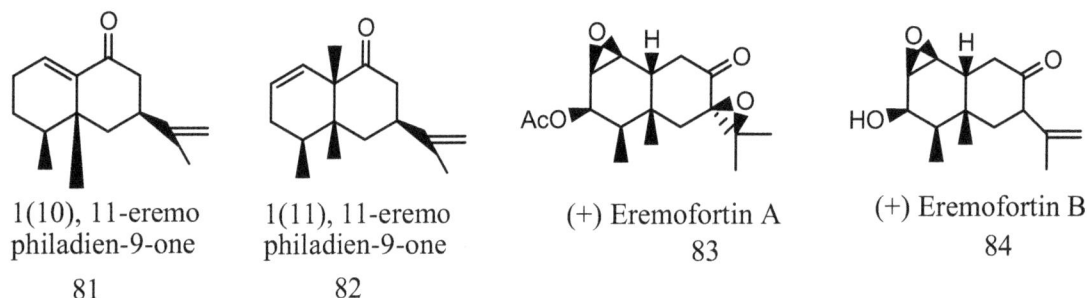

FIGURE 15.18 Examples of Furanoeremophilanes.

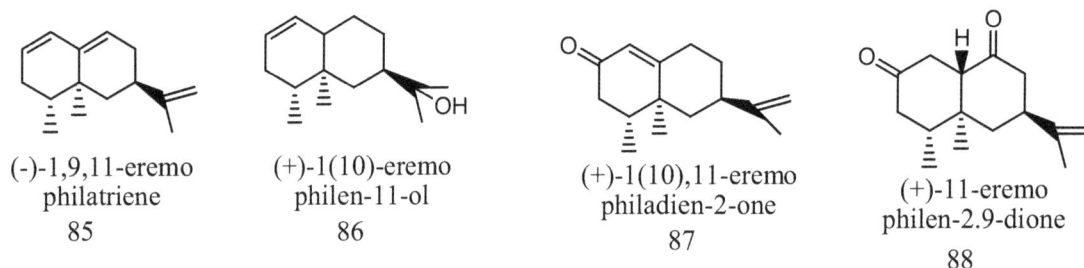

FIGURE 15.19 Examples of Valeranes.

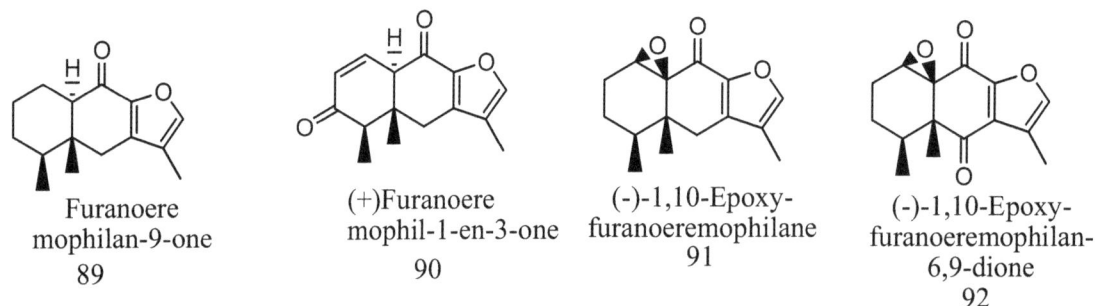

FIGURE 15.20 Few more examples of furanoeremophilanes.

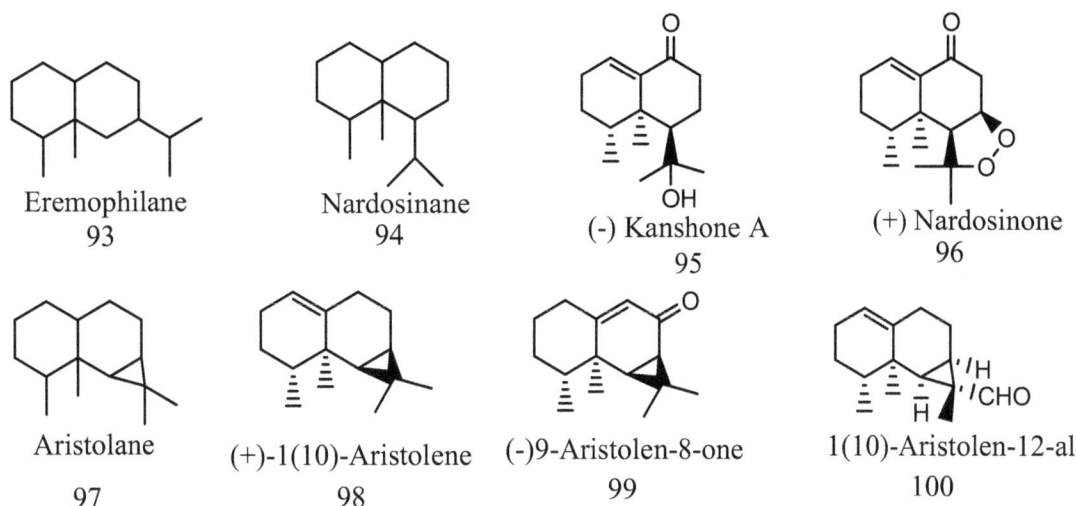

FIGURE 15.21 Examples of Eremophilanes and Aristolanes.

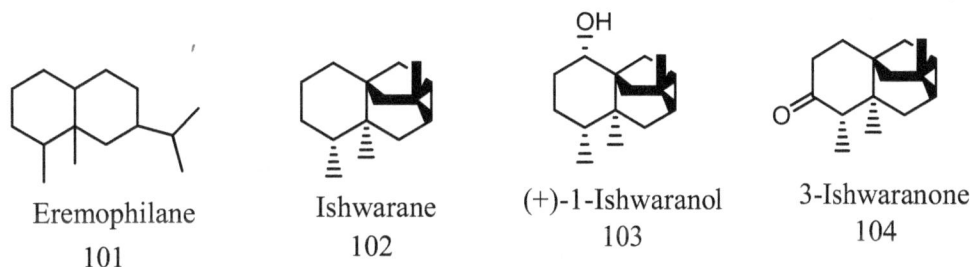

FIGURE 15.22 Examples of Eremophilanes and Ishwarane.

15.5.4 Cadinanes

More than 200 naturally abundant cadinanes (110) formally arise from ring closure of C-1 and C-6 as well as C-5 and C-10 of farnesane (109). The generally accepted numbering system, however, is not derived from farnesane, but from germacrane. Depending on the relative configuration at C-1, C-6 and C-7, the *trans* decalines cadinane and bulgarane (113) are distinguished from muurolane (116) and amorphane (119), each with the *iso*-propyl group in E- or D- position at C-7. Calamenenes (122) contain one benzenoid ring; cadalene (125) incorporates the naphthalene bicycle. (-)-4,9-Cadinadiene (D-cadinene, 111) from the oil of hops (*Humulus lupulus*, Cannabaceae) as well as (-)-3,9-cadinadiene (112), known as E-cadinene, widely spread in plants, spicy smelling and isolated from the oil of cade obtained by distillation of the wood of Mediterranean juniper *Juniperus oxycedru* (Cupressaceae), exemplify the cadinanes. Berries of juniper species *Juniperus communis* and *J. Oxycedrus* contain bulgaranes such as (-)-4,9-bulgaradiene (114) (El-bulgarene) and (-)-4(15), 10(14)-bulgaradiene (115). Muurolanes (116) include (+)-4(15), 10(14)-muuroladiene (117) (H-muurolene) from Swedish turpentine and ylang-ylang oil obtained by steam distillation of freshly picked flowers of the cananga tree *Cananga odorata* (Annonaceae) growing in Madagascar and the Philippine islands. They are pleasantly smelling and used in delicate perfumes. (-)-4,10(14)-Muuroladiene (118) (J-muurolene) also occurs in

the expectorant oil of pine needles (*Pinus silvestris,* Pinaceae). Amorphanes (119) are represented by (-)-4,11-amorphadiene (120) from *Viguiera oblongifolia*; its 12-carboxylic acid, also referred to as artemisic or qinghao acid (121), is isolated from *Artemisia annua* (Asteraceae) and exhibits antibacterial activity. Benzenoid (-)-(7S, 10S)-calamenene (122) is isolated from *Ulmus thomasii* (Ulmaceae) and (+)-3,8-calamenenediol (124) from *Heterotheca subaxillaris*. The naphthalene sesquiterpene cadalene (125) occurs in conifers, for example in the resin of fir *Abies sibirica* (Pinaceae). The wood of several trees contains 3-cadalenol (126). Cadalen-2,3-quinone (127) also known as mansonone C, is a constituent of *Mansonia altissima* and *Ulmus lactinata*. Hibiscones and various reddish-brown hibiscoquinones from *Hibiscus elatus* (Malvaceae) represent furanoid derivatives of cadinene (Figure 15.23) [2,8,9].

Antimalarials derived from 4,5-*seco*-cadinane (128) are found as constituents of the traditional Chinese medicinal herb *Artemisia annua* (Asteraceae), well-known as qinghao. Artemisinine, also referred to as qinghaosu (129), is a 3,6-per-oxide of the acylal formed by 4,5-*seco*-cadinane-5-aldehyde-12-oic acid. Dihydroqinghaosu (130) and the 11(13)-dehydro derivative artemisitene (131) are the active substances which, nowadays, are applied as semisynthetic esters and ethers (e.g. artemether, 130) to cure malaria. These peroxides probably eliminate singlet oxygen, which damages the membrane of the pathogens and disturbs their nucleic acid metabolism (Figure 15.24) [2,8,9].

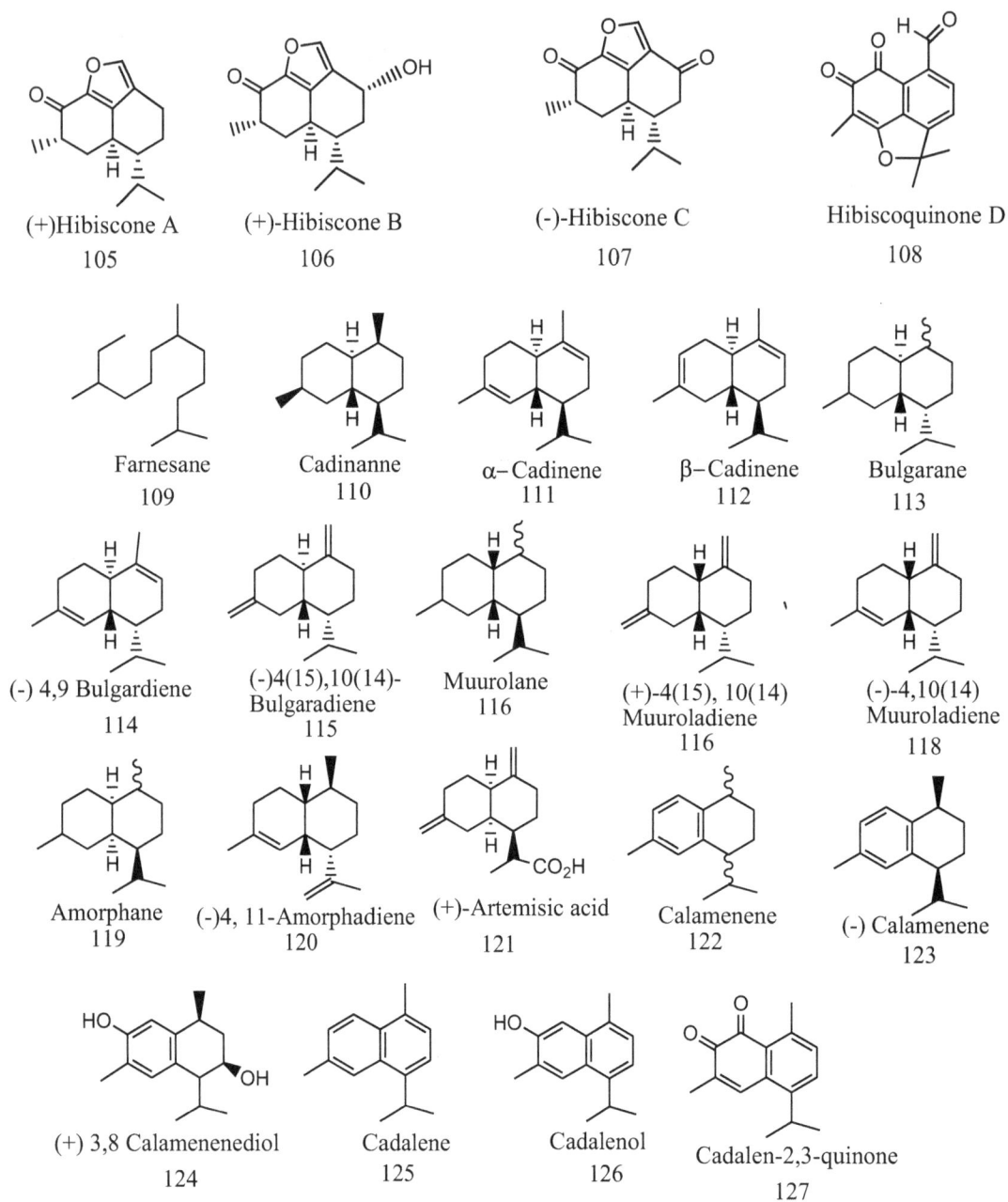

(+)Hibiscone A
105

(+)-Hibiscone B
106

(-)-Hibiscone C
107

Hibiscoquinone D
108

Farnesane
109

Cadinanne
110

α–Cadinene
111

β–Cadinene
112

Bulgarane
113

(-) 4,9 Bulgardiene
114

(-)4(15),10(14)-Bulgaradiene
115

Muurolane
116

(+)-4(15), 10(14)
Muuroladiene
116

(-)-4,10(14)
Muuroladiene
118

Amorphane
119

(-)4, 11-Amorphadiene
120

(+)-Artemisic acid
121

Calamenene
122

(-) Calamenene
123

(+) 3,8 Calamenenediol
124

Cadalene
125

Cadalenol
126

Cadalen-2,3-quinone
127

FIGURE 15.23 Examples of Cadinanes.

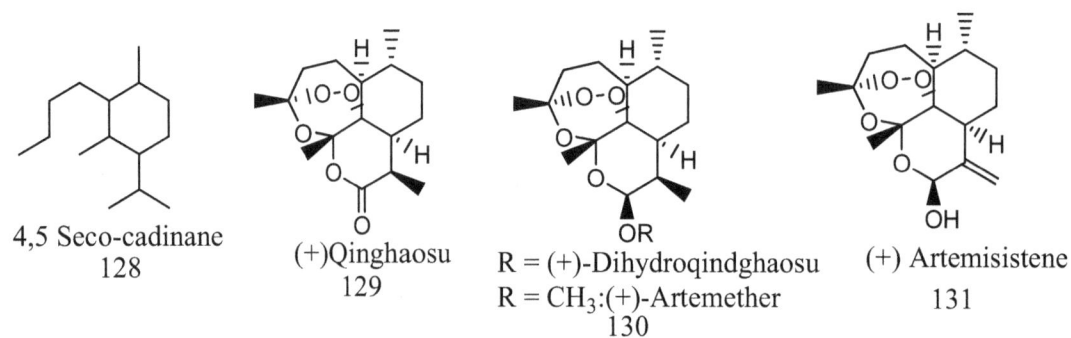

4,5 Seco-cadinane
128

(+)Qinghaosu
129

R = (+)-Dihydroqindghaosu
R = CH$_3$:(+)-Artemether
130

(+) Artemisistene
131

FIGURE 15.24 Examples of Cadinanes and Artemether.

15.5.5 Drimanes

Bond formation between C-2 and C-7 as well as C-6 and C-11 of farnesane (132) formally leads to the drimane (133) basic skeleton of sesquiterpenes. The accepted numbering system is derived from decalin and not from farnesane. The parent hydrocarbon 5D, 8D, 9E, 10E-drimane (134, 135) with *trans*-decalin as core structure occurs in paraffin oil (Figure 15.25) [2,11].

15.5.6 Guaianes and Cycloguaianes

Bond formation from C-1 to C-10 and C-2 to C-6 of farnesane (136) formally produces the bicyclic skeleton of more than 500 guaianes (137) isolated so far from higher plants with the numbering system adopted from that of decalin. Guaianes are also referred to as proazulenes because their naturally occurring derivatives frequently undergo dehydration to terpenoid azulenes (guaia-1,3,5,7,9-pentaenes) upon heating or steam distillation. Deep blue-violet oily guaiazulene (138) (guaia-1,3,5,7,9-pentaene) was obtained as an artifact upon work-up of the oils of camomile and guaiac wood from *Guajacum* species (Zygophyllaceae) is a well-known example. The milky

juice of the delicious fungus *Lactarius deliciosus* turns from orange to greenish upon damaging the fungal body when the genuine yellow 15-stearoyloxyguaia-1,3,5,7,9,11-hexaene is decomposed enzymatically to violet lactaroviolin (139) (guaia-1,3,5,7,9,11-hexaen-4-aldehyde) (Figure 15.26) [2,11].

(+)-1(5), 6-Guaiadiene (140) and its 4E-stereoisomer occur in *balsamum tolutanum*, a balm obtained from the tree *Myroxylon balsamum* (Leguminosae) which grows in the northern areas of South America α-guaiene (141), α- bulnesene (142) and β- bulnesene (143) are found among the constituents of guaiac wood oil from the tree *Guajacum officinale* (Zygophyllaceae) native in central America, and of pleasantly smelling patchouli oil obtained by steam distillation of fermented leaves of the patchouli shrub *Pogostemon patchouli* (Labiatae) cultivated for perfumery in tropical countries (Figure 15.27) [2,11].

The air-dried milky exudation of the roots of *Ferula galbaniflua* (Umbelliferae), collected in Iran and known as galbanum resin or gum galbanum, as well as guaiac wood oil, contain the tertiary alcohols guaiol (144), buinesol (145) and (+)-9-guaien-11-ol (146) with a spicy odor of leaves and wood. (-)-Kessoglycol (147), a guaiane tricycle, is found in the rhizome of valerian *Valeriana officinalis* (Valerianaceae) (Figure 15.28).

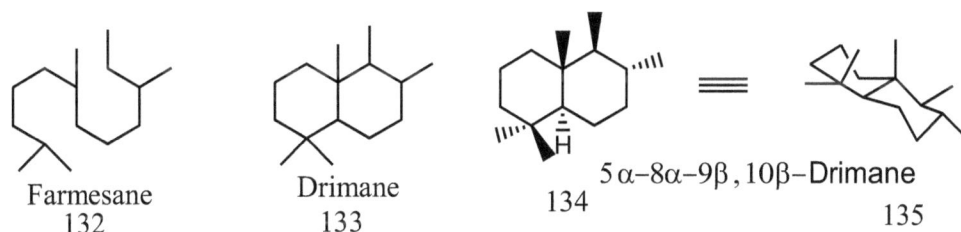

Farmesane 132	Drimane 133	134 5α–8α–9β, 10β–Drimane	135

FIGURE 15.25 Examples of Drimanes.

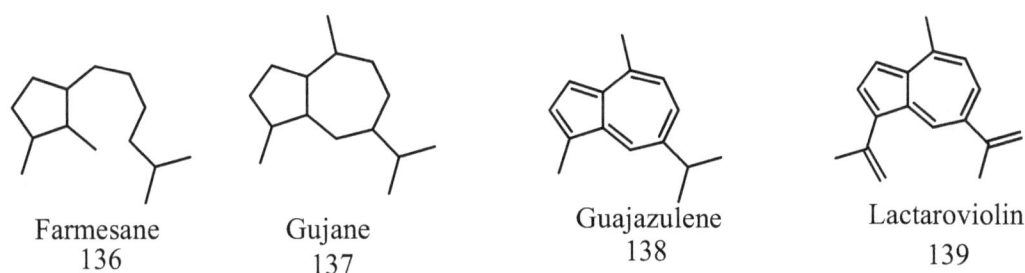

Farmesane 136	Gujane 137	Guajazulene 138	Lactaroviolin 139

FIGURE 15.26 Examples of Guaianes and Cycloguaianes.

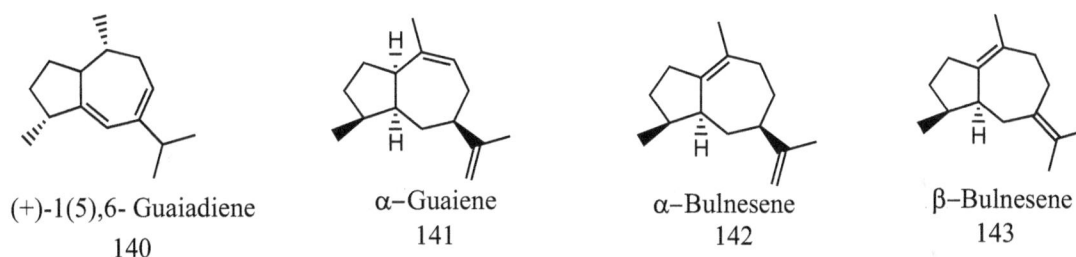

(+)-1(5),6- Guaiadiene 140	α–Guaiene 141	α–Bulnesene 142	β–Bulnesene 143

FIGURE 15.27 Few more examples of Guaianes and Cycloguaianes.

12,6-Guaianolides as a class of tricyclic sesquiterpene lactones 2,21 are found in a large variety of the plant family Asteraceae (formerly denoted as Compositae). Well-known examples include (-)-artabsin containing a cyclopentadiene partial structure and its ***DIELS-ALDER*** dimer absinthin as the chief bitter principle (bitterness threshold 1: 70,000) of wormwood (absinthium) *Artemisia absinthum*. This is used predominantly in perfumery but, because ingestion of the extracts may cause stupor, convulsions and even death, much less often it is applied to flavor alcoholic beverages (e.g., vermouth). (+)-Arglabin (148) from *Artemisia glabella*, tanaparthin-α-peroxide (149) from the daisy flower *Tanacetum parthenium*, achillicin (150) from yarrow (*Achillea millefolium*), matricin (151) and its 8-*O*-acetyl derivative from camomile *Matricaria chamomilla* (Asteraceae) are additional representatives (Figure 15.29) [2].

Pseudoguaianes (152) formally arise from guaianes by methyl shift from C-4 to C-5, represented by ambrosic acid (153) and the antineoplastic pseudoguaianolide ambrosin (154), both isolated from the herb *Ambrosia maritima* (Asteraceae) and other *Ambrosia* species. Helenalin (155) from the flowers of *Helenium autumnale* and *Arnica Montana* (Asteraceae) acts as an abortive, anti-inflammatory, antineoplastic, antirheumatic and antipyretic; external contact may cause skin irritations and sneezing, however, while ingestion may initiate vomiting, diarrhoea, vertigo and heart pounding up to circulatory collapse and death (Figure 15.30) [9,11].

6,11-Cycloguaianes are referred to as **aromadendranes** (156). Various aromadendrenes (157) and aromadendradienes (158) belong to the constituents of *balsamum tolutanum* from *Myroxylon balsamum* (Leguminosae). The oil of sage from *Salvia sclarea* (Labiatae), cultivated for perfumery in the

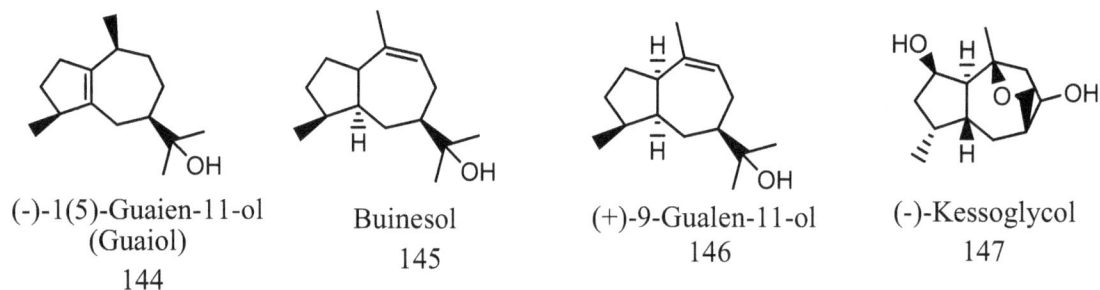

(-)-1(5)-Guaien-11-ol
(Guaiol)
144

Buinesol
145

(+)-9-Gualen-11-ol
146

(-)-Kessoglycol
147

FIGURE 15.28 Examples of Guaianes and its derivatives.

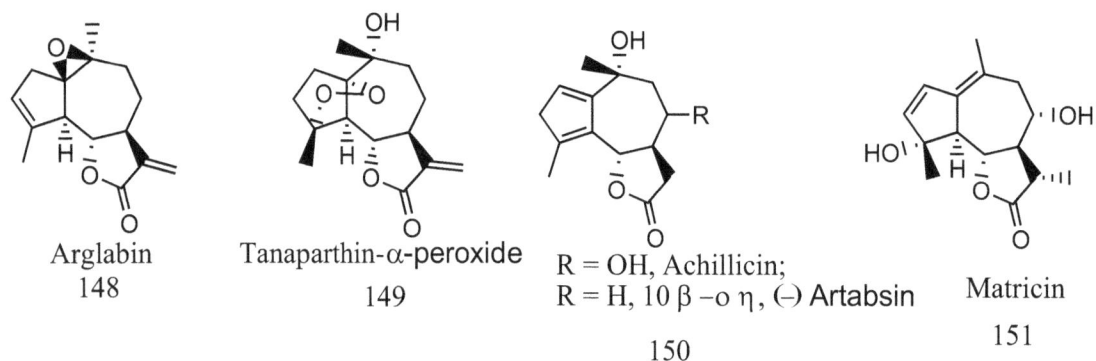

Arglabin
148

Tanaparthin-α-peroxide
149

R = OH, Achillicin;
R = H, 10 β –o η, (–) Artabsin
150

Matricin
151

FIGURE 15.29 Examples of Arglabin, matricin, absinthin and tanaparthin-α-peroxide.

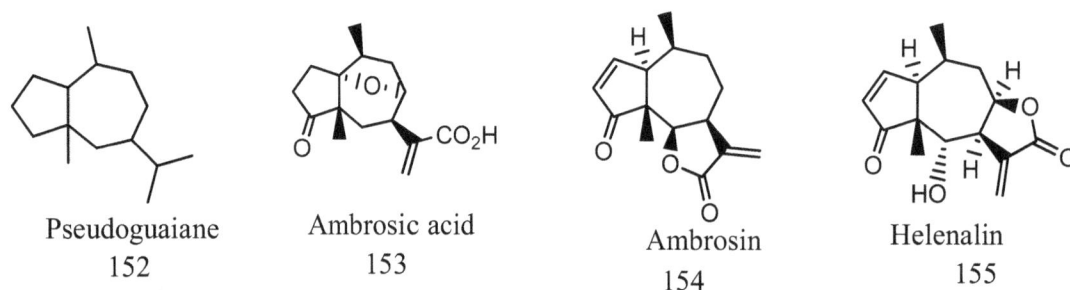

Pseudoguaiane
152

Ambrosic acid
153

Ambrosin
154

Helenalin
155

FIGURE 15.30 Examples of Pseudoguaianes.

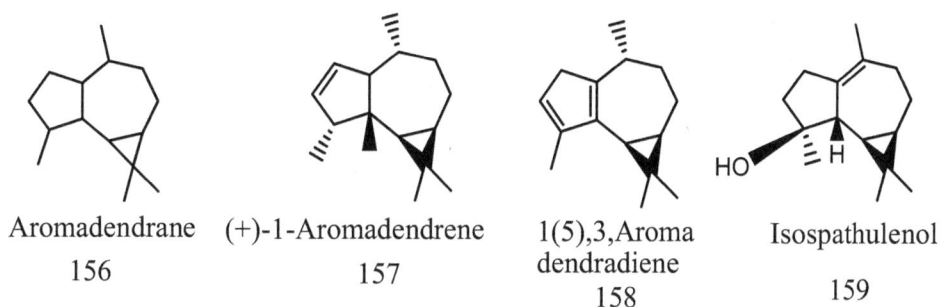

FIGURE 15.31 Examples of Aromadendranes.

Aromadendrane 156   (+)-1-Aromadendrene 157   1(5),3,Aroma dendradiene 158   Isospathulenol 159

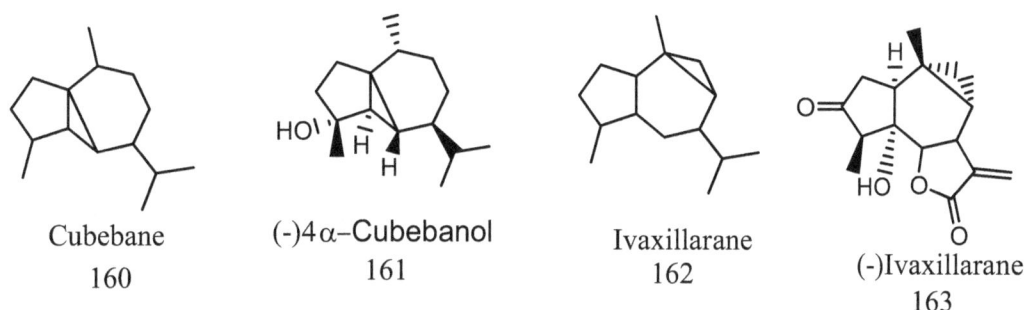

FIGURE 15.32 Examples of Cubebanes.

Cubebane 160   (-)4 α–Cubebanol 161   Ivaxillarane 162   (-)Ivaxillarane 163

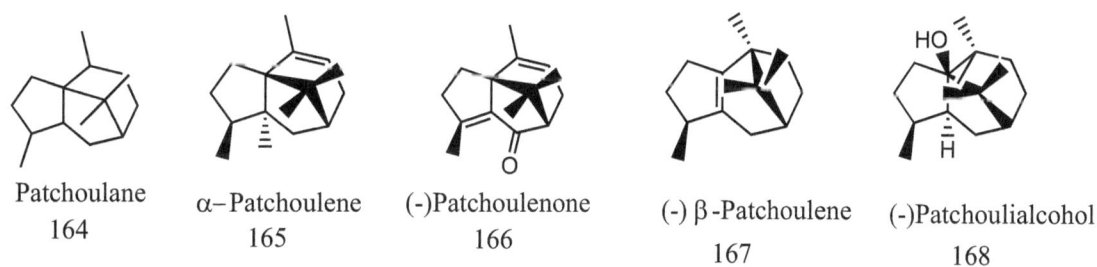

FIGURE 15.33 Examples of Patchoulanes.

Patchoulane 164   α–Patchoulene 165   (-)Patchoulenone 166   (-) β-Patchoulene 167   (-)Patchoulialcohol 168

south of Europe, contains (+)-1(10)-aromadendren-7-ol, also known as isospathulenol (159) (Figure 15.31) [2,5].

Cubebanes (160) and **ivaxillaranes** (162) represent 1,6- and 8,10-cycloguaianes, respectively. The fruits of Java pepper *Piper cubeba* (Piperaceae) contain (-)-4-cubebanol (161). The name ivaxillarane stems from *Iva axillaris* with (-)-ivaxillarin (163) as a constituent (Figure 15.32).

Patchoulanes (164) represent 1,11-cycloguaianes and rearranged derivatives of those, occurring in the the oils of cypress, guaiac wood and patchouli. Examples include D-patchoulene (165) and (-) patchoulenone (166) from the oil of cypress (*Cupressus sempervirens*, Pinaceae) as well as the rearranged derivatives E-patchoulene and (-)- patchoulialcohol (168) (patchoulol) from patchouli oil (Figure 15.33) [5].

Valerenanes is the common name of 8(7-6)-abeoguaianes, in which the seven-membered ring of guaiane (169) has contracted to cyclohexane, involving migration of C-8 from C-7

to C-6. Few valerenanes (170) are known to date, including valerenol (171), valerenal (172) and (-)-valerenoic acid (173), all of which belong to the constituents of the rhizome and roots of valerian *Valeriana officinalis* (Valerianaceae) (Figure 15.34) [10].

15.6 Himachalanes, Longipinanes, Longifolanes

Bonds between C-1 and C-6 as well as C-1 and C-11 formally convert farnesane (174) into the bicyclic skeleton of himachalane (175) with the numbering system adopted from that of farnesane. Several himachalanes such as D-himachalene (176) and himachalol (177) are constituents of the oil of cedar wood from *Cedrus deodara* (Pinaceae) (Figure 15.35).

2,7-Cyclohimachalanes are known as longipinanes (178). They occur in various oils of pine wood and some Asteraceae, exemplified by 3-longipinene from *Pinus* species (Pinaceae) and 3-longipinen-5-one from *Chrysanthemum vulgare* (Asteraceae). Longipinanes (179) are differentiated from longifolanes (182, 183) which formally and biogenetically also emerge from farnesane by cleaving the C-3-C-4 bond and closing the bonds C-1-C-6, C-2-C-4, C-3-C-7 and C-1-C-11 to the tricycle. Examples are the isomers longicyclene (184) and longifolene (185), widely spread in ethereal oils, the latter present to an extent of up to 20% in Indian turpentine oil which is produced commercially from the Himalayan pine *Pinus longifolia* (Pinaceae) for the synthesis of a widely used chiral hydroboration agent (Figure 15.36) [10].

15.6.1 Picrotoxanes

Farnesane (186) is formally converted into picrotoxane (187) by making the bonds C-3-C-7 and C-2-C-10 (numbering system of farnesane) also involving methyl migration from C-6 to C-13 (numbering system of picrotoxane). About 15 toxic sesquiterpene alkaloids with picrotoxane skeleton such as (-)-dendrobin (188) are among the constituents of the orchid *Dendrobium nobile* (Orchidaceae), the stems of which are used as an antipyretic and tonic in China and Japan. (-)-Picrotoxinin (189) is one of the bitter and ichthyotoxic (fish-killing) constituents of picrotoxin produced from the fruits and seed *Anamirta cocculus* (syn. *Menispermum cocculus*, Menispermaceae); picrotoxin is used as a CNS and respiratory stimulant as well as an antidote to barbiturates (Figure 15.37).

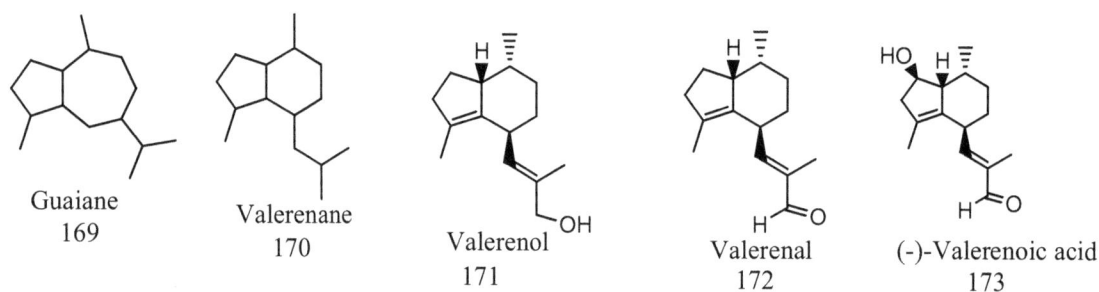

Guaiane
169

Valerenane
170

Valerenol
171

Valerenal
172

(-)-Valerenoic acid
173

FIGURE 15.34   Examples of Valerenanes.

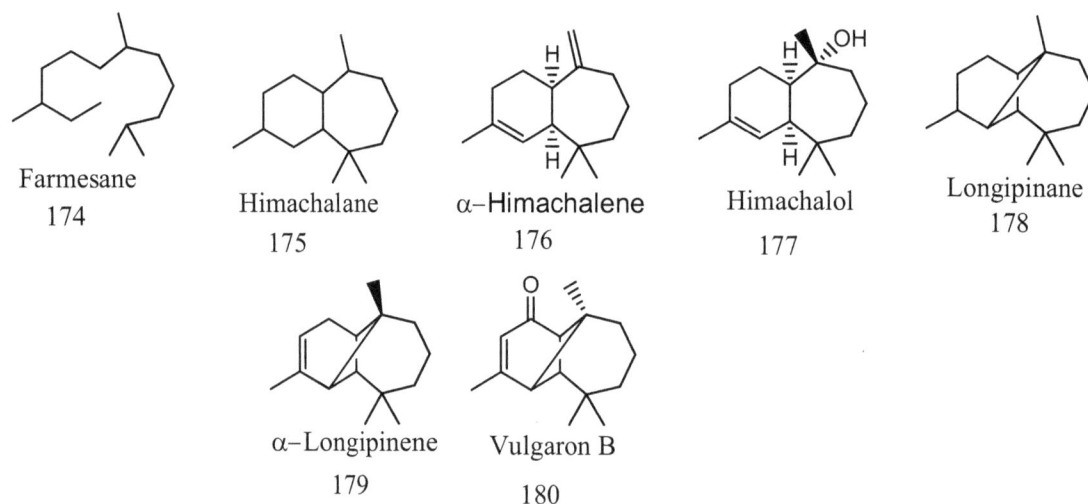

Farnesane
174

Himachalane
175

α–Himachalene
176

Himachalol
177

Longipinane
178

α–Longipinene
179

Vulgaron B
180

FIGURE 15.35   Examples of Himachalanes, Longipinanes, Longifolanes.

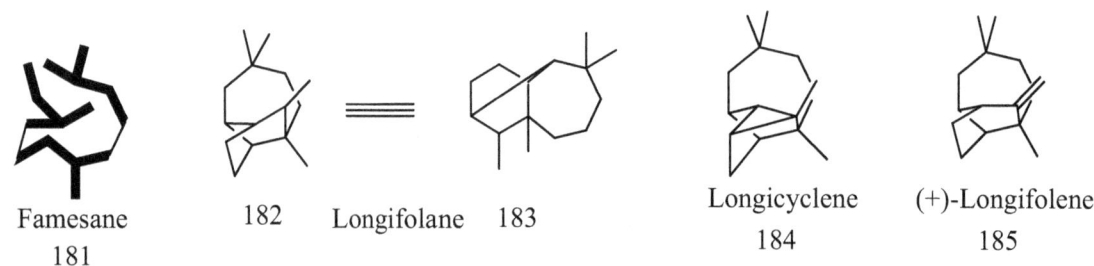

Famesane
181

182   Longifolane   183

Longicyclene
184

(+)-Longifolene
185

FIGURE 15.36   Few more examples of Cyclohimachalanes.

15.6.2 Isodaucanes and Daucanes

Bonds from C-1 to C-7 and from C-1 to C-10 in farnesane (190) formally build up the sesquiterpene skeleton of isodaucanes which are converted to daucanes by migration of one methyl group (C-14) from C-7 to C-8. Isodaucanes (191) such as (+)-6,10- isodaucen-10-one (192) and 7(14)-isodaucen-10-one (193) are constituents of the oil of sage from *Salvia sclarea* (Labiatae). The name daucane (194) stems from the carrot *Daucus carota* (Umbelliferae), from which (+)-4,8- daucadiene (195), (+)-8-daucen-5-ol (196) and (-)-5,8-epoxy-9-daucanol (197) have been isolated (Figure 15.38).

15.6.3 Protoilludanes, Illudanes, Illudalanes

The tricyclic skeleton of protoilludane arises formally from farnesane (198) by making bonds from C-1 to C-11, C-2 to C-9 and C-3 to C-6. Cleavage of the C-3C-4 bond results in the formation of *seco*-illudane, also referred to as illudalane; protoilludane (199) formally converts into spirocyclic illudane (200) by migration of C-4 from C-3 to C-6 (Figure 15.39).

Phytopathogenic fungi *Armillariellamellea* (Basidomycetae), for example, produce the antifungal (+)-armillarin (202) with protoilludan skeleton as an ester of *o*-orsellinic acid (2,4-dihydroxy-6-methylbenzoic acid). Two anti-tumor antibiotics (-)-illudin M (203) and S isolated from the poisonous and luminous fungus *Clitocybeilludens* (Basidomycetae), represent the illudanes. The six-membered ring of natural illudalanes is benzenoid for the most part; variously substituted derivatives occur in the fern *Pteridium aquilinum* (Polypodiaceae). Onitin (204) from *Onychiumauratum* and *O. Siliculosum* acts as a mild muscle relaxant.

15.6.4 Marasmanes, Isolactaranes, Lactaranes, Sterpuranes

Bond formation from C-1 to C-11, C-2 to C-9, C-3 to C-6 and disconnection of the C-4, C-5 bond in farnesane (205) formally leads to marasmane (206). Isolactarane (207) arises from the latter by cleavage of the C-3, C-4 and connection of the C-5, C-7 bond which, on its part, formally expands to

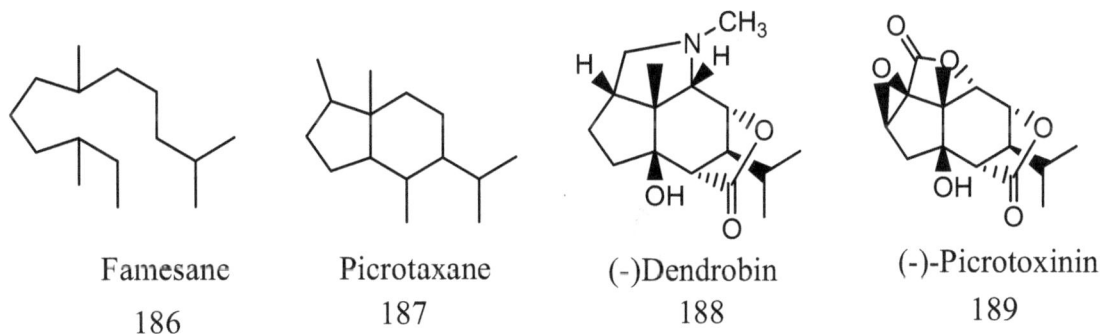

Famesane 186 Picrotaxane 187 (-)Dendrobin 188 (-)-Picrotoxinin 189

FIGURE 15.37 Examples of Picrotoxanes.

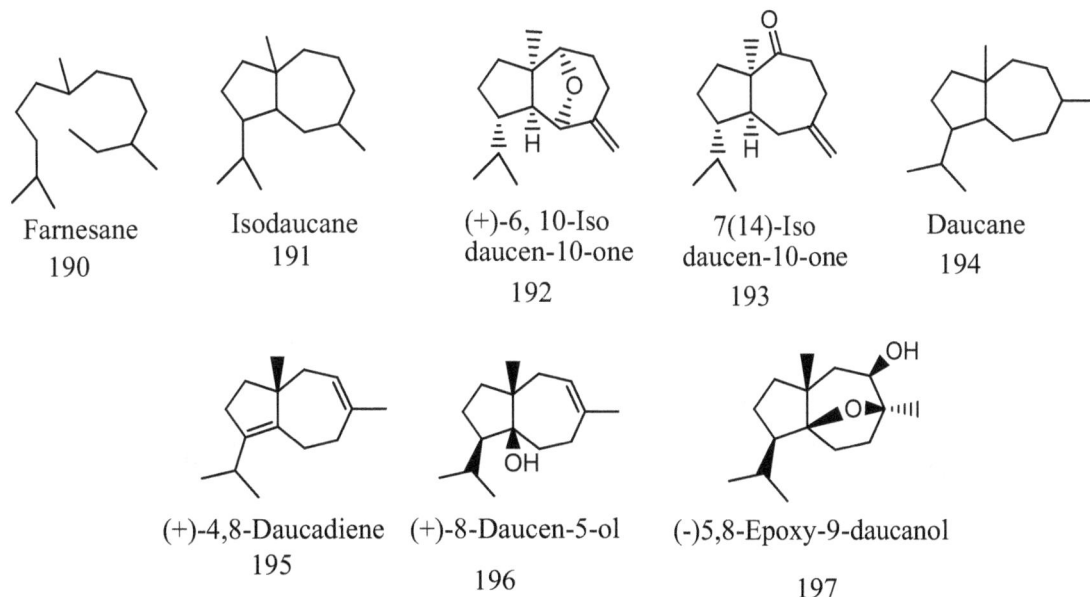

Farnesane 190 Isodaucane 191 (+)-6, 10-Iso daucen-10-one 192 7(14)-Iso daucen-10-one 193 Daucane 194

(+)-4,8-Daucadiene 195 (+)-8-Daucen-5-ol 196 (-)5,8-Epoxy-9-daucanol 197

FIGURE 15.38 Examples of Isodaucanes and Daucanes.

lactarane (208) by migration of C-3 from C-6 to C-4 involving disconnection of the C-5, C-7 bond. The names are derived from those of the fungal genera *Marasmius* and *Lactarius* (Figure 15.40).

(+)-Isovelleral (209) is a strong antibiotic with a marasmane skeleton isolated from the fungus *Lactarius vellereus* (Basidomycetae) and closely related species; due to the sharp taste, the fungus uses isovelleral (209) as an antifeedant against animals. Marasmic acid (210), an antibacterial and mutagenic constituent from *Marasmius conigenus* and other Basidomycetae, represents an acylal of a dialdehyde acid. Merulidial (211), a metabolite of the fungus *Meruliustremellosus* (Basidomycetae) with the isolactarane skeleton, acts as an antibacterial and antimycotic. Various lactarane derivatives referred to as blennins, such as the lactone (+)-blennin D (212), have been isolated from *Lactarius blennius*.

Culture of the phytopathogenic fungus *Stereum purpureum* (Basidomycetae), a parasite of some trees, produces unsaturated and hydroxylated sesquiterpenes with the tricyclic skeleton of sterpurane, formally arising from isolactarane by opening the C-5, C-6- and closing the C-4C-5 bond, or directly from farnesane by linking the bonds C-1C-11, C-2C-9 and C-4, C-7 (Figure 15.41).

15.6.5 Acoranes

Connecting the bonds C-1, C-6 and C-6, C-10 in farnesane (216) formally produces the spiro decane basic skeleton of acorane (217). The name of this class of sesquiterpenes stems from the *Acorus* species. (-)-4-Acoren-3-one (219), for example, has been isolated from *Acorus calamus* (Calamus, Araceae) and from the carrot *Daucus carota* (Umbelliferae). The oil of calamus (oil of sweet flag) from the rhizome of *Acorus calamus*

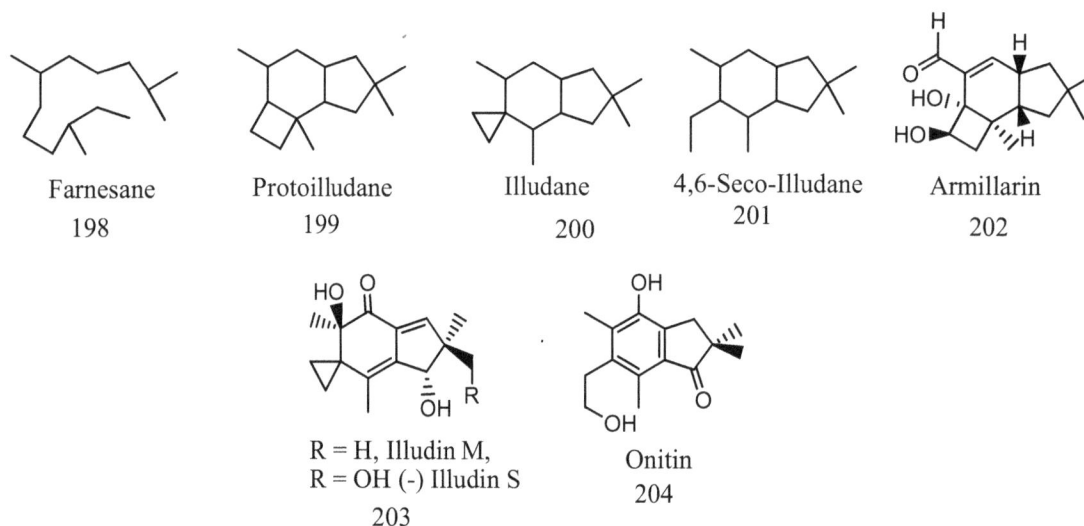

Farnesane
198

Protoilludane
199

Illudane
200

4,6-Seco-Illudane
201

Armillarin
202

R = H, Illudin M,
R = OH (-) Illudin S
203

Onitin
204

FIGURE 15.39 Examples of Protoilludanes, Illudanes, Illudalanes.

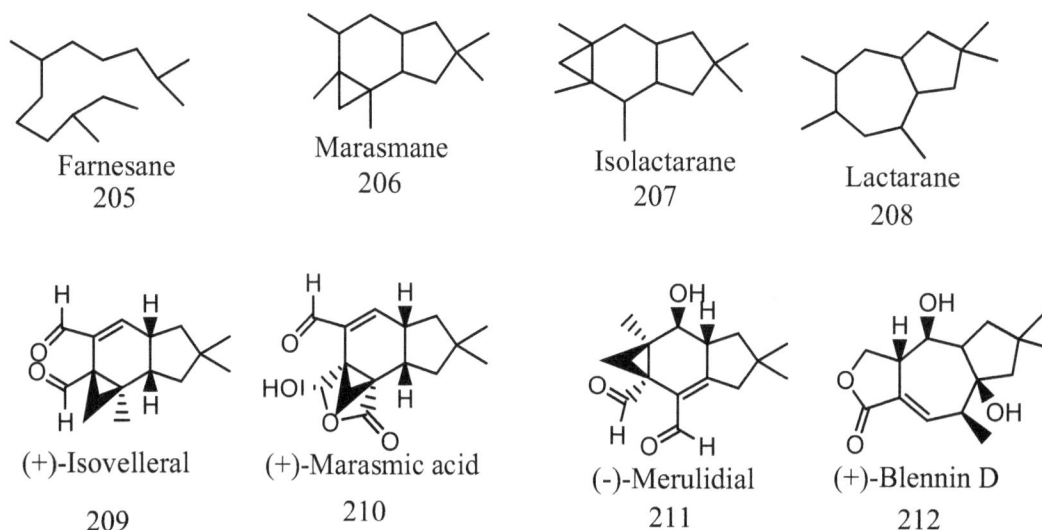

Farnesane
205

Marasmane
206

Isolactarane
207

Lactarane
208

(+)-Isovelleral
209

(+)-Marasmic acid
210

(-)-Merulidial
211

(+)-Blennin D
212

FIGURE 15.40 Examples of Marasmanes, Isolactaranes, Lactaranes.

with its warm and spicy odor and pleasant bitter taste is predominantly used in perfumery and as a minor (possibly carcinogenic) ingredient of vermouth, some flavored wines and liqueurs. (+)-3,7(11) Acoradiene (218) is a constituent of juniper *Juniperus rigida*; its enantiomer occurs in *Chamaecyparis nootkatensis* (Cupressaceae) (Figure 15.42) [2,5,10].

15.6.6 Chamigranes

A spiro[5,5]undecane skeleton characterizes the chamigranes (221) which formally originate from linking the bonds C-1C-6 and C-6C-11 of farnesane (220). More than 50 naturally occurring representatives with halogens as substituents are predominantly isolated from algae. Examples include (-)-10-bromo-1,3-, 7(14)-chamigratrien-9-ol (obtusadiene, 222) and 3-bromo-2D-chloro-7-chamigren-9-one (laurencenone A, 223) from the red alga *Laurencia obtusa*. (-)-Chamigra-3,7(14)-diene found in *Chamaecyparis taiwanensis* (Cupressaceae) is one of the rare chamigrenes of plant origin (Figure 15.43) [2].

15.6.7 Cedranes and Isocedranes

Cedranes are formally derived from farnesane (224) by the connection of bonds between C-1 and C-6, C-2 and C-11 as well as C-6 and C-10. Cedrane (225) is formally converted into isocedrane (229) by migration of the C-15 methyl group from C-3 to C-5 (Figure 15.44) [10].

Cedrane derivatives such as (-)-3-cedrene (D-cedrene, 226) and (+)-3(15)-cedrene (E-cedrene, 227) are widespread among *Juniperus* species (Cupressaceae). (-)-D-Cedrene (content up to 25%) and (+)-cedrol, 228 (content 20–40%) are the chief constituents of the oil of cedar wood used in perfumery and as an insect repellent, obtained from *Juniperus virginiana* growing in the south-east of USA. (+)-Cedrol shapes the pleasant woody balsamic odor of this volatile oil which intensifies upon acetylation of this sesquiterpenol. Isocedrane derivatives occur predominantly in *Jungia*species, exemplified by (+)-4-isocedren-15-al (230) in *Jungiamalvae folia* and (+)-4-isocedren-15,14-olide (231) and other isocedrenes in *Jungiastuebelii*.

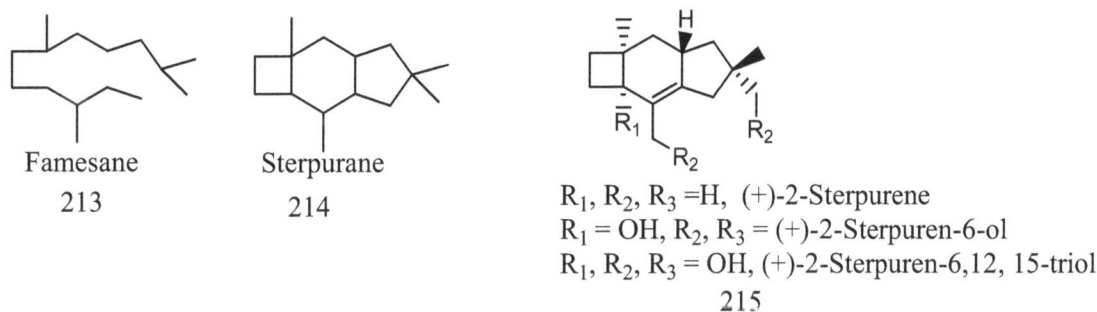

Famesane
213

Sterpurane
214

R$_1$, R$_2$, R$_3$ =H, (+)-2-Sterpurene
R$_1$ = OH, R$_2$, R$_3$ = (+)-2-Sterpuren-6-ol
R$_1$, R$_2$, R$_3$ = OH, (+)-2-Sterpuren-6,12, 15-triol
215

FIGURE 15.41 Examples of Famesane and Sterpuranes.

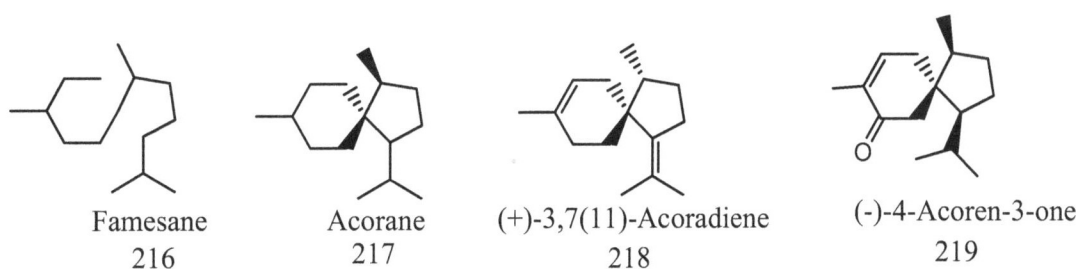

Famesane
216

Acorane
217

(+)-3,7(11)-Acoradiene
218

(-)-4-Acoren-3-one
219

FIGURE 15.42 Examples of Acoranes.

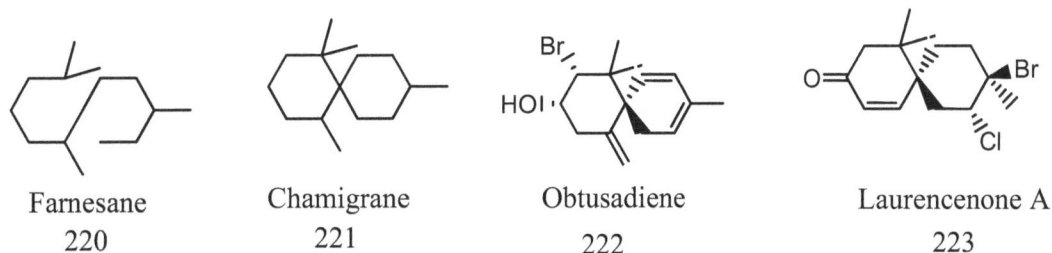

Farnesane
220

Chamigrane
221

Obtusadiene
222

Laurencenone A
223

FIGURE 15.43 Examples of Chamigranes.

15.6.8 Zizaanes and Prezizaanes

Bonds between C-2 and C-11, C-6 and C-10 as well as C-6 and C-15 of farnesane (232) formally link up the tricyclic sesquiterpene skeleton prezizaane, which converts into zizaane (236) by *methyl shift*. The oil of vetiver with its aromatic to harsh, woody odor, steam-distilled for perfumery from roots of vetiver grass *Vetiveria zizanoides* (Poaceae) which is grown chiefly in Haiti, India and Java, contains some prezizaanes (233) and zizaanes such as (+)-prezizaene (234), (-)-7-prezizaanol (235), (+)-6(13)-zizaene (237) and (+)-6(13)-zizaen-12-ol (238), also known as khusimol (Figure 15.45).

15.6.9 Campherenanes and Santalanes

Bonds from C-1 to C-6 and C-3 to C-7 of farnesane (239) formally build up the bicycle [2.2.1] heptane core structure of campherenane (240). (-)-Campherenol (241) and (-)-campherenone (242) are constituents found in the camphor tree *Cinnamomum camphora* (Lauraceae) and in the ethereal oil of freshly pressed lemon juice (Figure 15.46) [5,10].

An additional bond from C-2 to C-4 of campherenane formally leads to tricyclic Dsantalane, the basic skeleton of some constituents found in sandalwood oil 243 (oil of santal) with a woody, sweet, balsamic odor, used in perfumery and as a urinary anti-infective 244, obtained from dried heartwood of the tree *Santalum album* (Santalaceae) grown in east India. Examples include (+)-*(Z)*-D-santalol 245 with a slight woody odor similar to cedar wood oil and (+)-*(E)*-D-santalal 246, with a strong woody odor (Figure 15.47).

Cleavage of the C-3C-4 bond of farnesane 247 and connection of new bonds from C-2 to C-4, C-3 to C-7 and C-1 to C-4 lead to bicyclic E-santalane 248, which is another basic

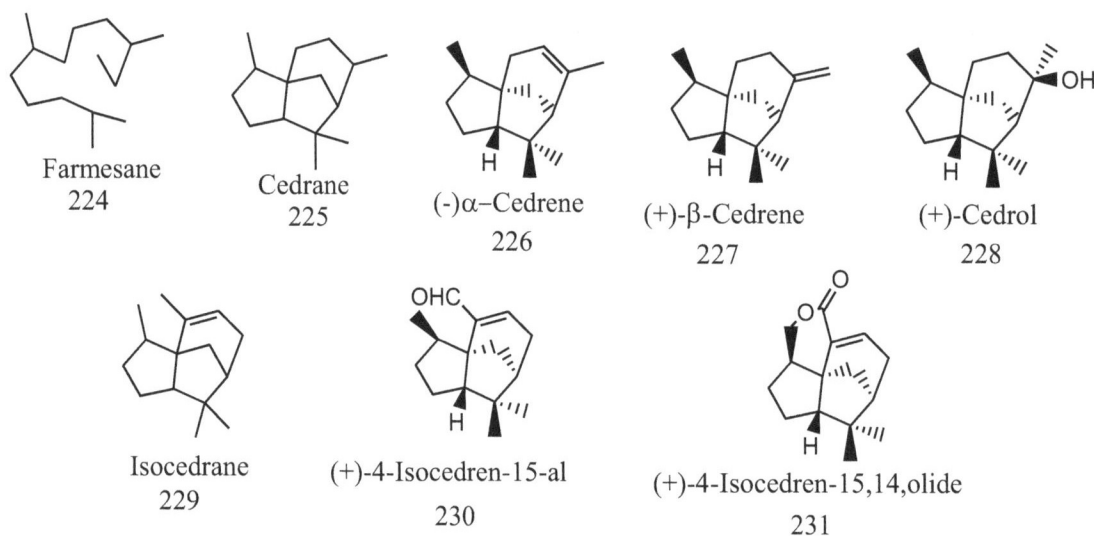

Farmesane
224

Cedrane
225

(-)α–Cedrene
226

(+)-β-Cedrene
227

(+)-Cedrol
228

Isocedrane
229

(+)-4-Isocedren-15-al
230

(+)-4-Isocedren-15,14,olide
231

FIGURE 15.44 Examples of Cedranes and Isocedranes.

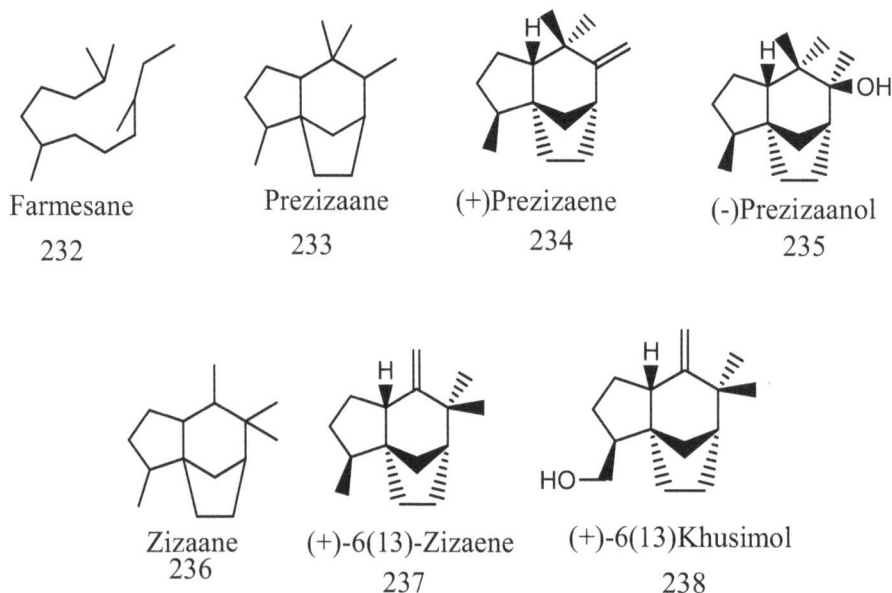

Farmesane
232

Prezizaane
233

(+)Prezizaene
234

(-)Prezizaanol
235

Zizaane
236

(+)-6(13)-Zizaene
237

(+)-6(13)Khusimol
238

FIGURE 15.45 Examples of Zizaanes and Prezizaanes.

skeleton of substances found in sandalwood. Examples are (-)-E-santalene 249 and (-) -(Z)-E-santalol 250 with a pleasant woody, slightly urinary odor used in perfumes and detergents (Figure 15.48) [10].

15.6.10 Thujopsanes

Farnesane 251 formally changes to thujopsane when the C-5C-6 bond cleaves and new bonds C-1C-6, C-2C-6, C-5C-7 and C-6C-11 connect. (-)-3-Thujopsene 252 and (+)-15-nor-4-thujopsen-3-one (Mayurone 254) from hiba oil obtained from the hiba live tree *Thujopsis dolabrata* (Cupressaceae) grown in Japan represent this small group of sesquiterpenes. (-)-3 Thujopsene 253 does not shape the odor, but is, in addition to (-)-D-cedrene and (+)-cedrol, one of the chief constituents

(up to 25%) of the Texan oil of cedar wood from *Juniperus virginiana* (Cupressaceae) (Figure 15.49) [2,5,10].

15.6.11 Hirsutanes

Connecting the bonds C-3C-7, C-2C-9 and C-1C-11 in farnesane 255 and subsequent shifts of the methyl groups C-14 and C-15 formally lead to the triquinane skeleton of hirsutanes (Hirsutane 257, Hirsutene 258, Hirsutic acid 259) occurring predominantly as fungal metabolites. 4(15)- Hirsutene 257 is a hydrocarbon arising biogenetically from cyclization of humulene and isolated from cultures of the fungus *Coriolusconsors* (Basidomycetae). Epoxidized derivatives such as hirsutic acid produced by the fungus *Stereumhirsutum* (Basidomycetae) growing on the dead wood of broad-leaved trees and coriolin A 260 including two

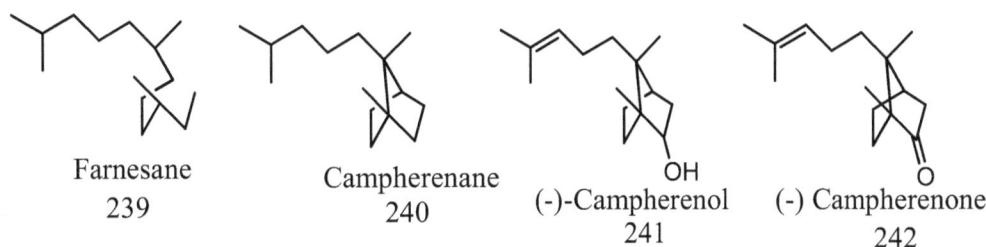

FIGURE 15.46 Examples of Campherenanes and Santalanes.

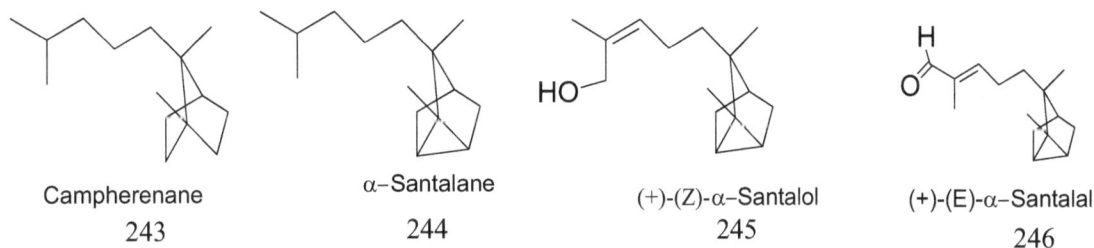

FIGURE 15.47 Few more Examples of Campherenanes and Santalanes.

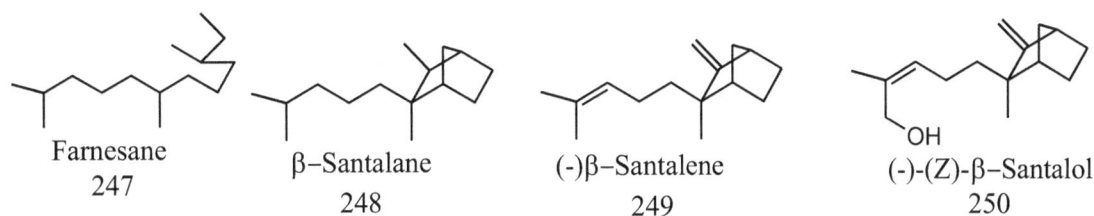

FIGURE 15.48 Examples of Farnesane, santalane and santalol.

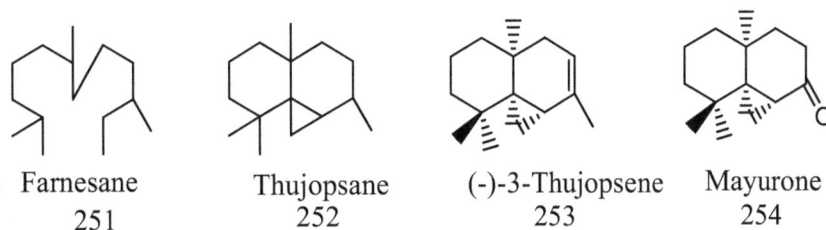

FIGURE 15.49 Examples of Thujopsanes.

acylated derivatives (B, C) isolated from *Coriolusconsors*, display antibiotic and antineoplastic activity (Figure 15.50) [2].

15.7 Other Polycyclic Sesquiterpenes

15.7.1 Pinguisanes

Unlike the sesquiterpenes presented so far, not all bonds between the isoprene units (boldface in the formula) within the pinguisane skeleton are attached in the usual head-to-tail manner. As a result, their structure (constitution) cannot be formally derived from cyclizations of farnesane 261. Moreover, one methyl group of the isopropyl residue has migrated from the side chain to the six-membered ring (Figure 15.51).

Pinguisanes 262 occur in various species of liverwort, exemplified by (-)-D-pinguisene 263 in *Porella platyphylla*. The name of this small group of sesquiterpenes stems from *Aneurapinguis*containing (+)-pinguisone 264; its bitter taste protects as an antifeedant against insects.

15.7.2 Presilphiperfolianes, Silphiperfolianes, Silphinanes, Isocomanes

Presilphiperfolianes represent an additional group of sesquiterpenes with structures that do not follow the isoprene rule and consequently are not deducible from farnesane 265 by formal cyclizations. Silphiperfoliane 266, as another structural variation, derives from presiphiperfoliane by migration of C-11 from C-7 to C-8; shifting C-9 from C-1 to C-8 results in the formation of silphinane 267. Another methyl shift from C-6 to C-7 leads from silphinane to the isocomane 268 skeleton with individual ring numbering. (-)-8E-Presilphiperfolanol from *Eriophyllumstaechadifolium* and *Fluorensiaheterolepsi*, (-)-5-silphiperfolene 273 as well as (-)-1-silphinene from the compass plant *Silphium perfoliatum* (Asteraceae) growing in the north west of the USA and defining the name, (+)-3-silphinenone from *Dugaldiahoopesii*, and (+)-arnicenone 274 from *Arnica parryi* (Asteraceae), exemplify this small group of tricyclic sesquiterpenes with the attractive triquinane skeleton.

Another triquinane hydrocarbon of natural origin referred to as (-)-isocomene merits special attention; it is a constituent of *Isocomawrigthii* (Asteraceae) growing in the south east of the USA, better known there as rayless golden rod and is highly toxic for cattle and sheep. It also occurs with the synonym berkheyaradulene in south African Asteraceae, such as in the roots of *Berkheya* and parts of *Senecio isatideus*. (-)-Modhephene 275 is an additional hydrocarbon isolated from these plants. Its exceptional [3.3.3] propellane parent skeleton arises formally from silphinane by migration of C-3 from C-4 to C-7 (Figure 15.52) [12,13].

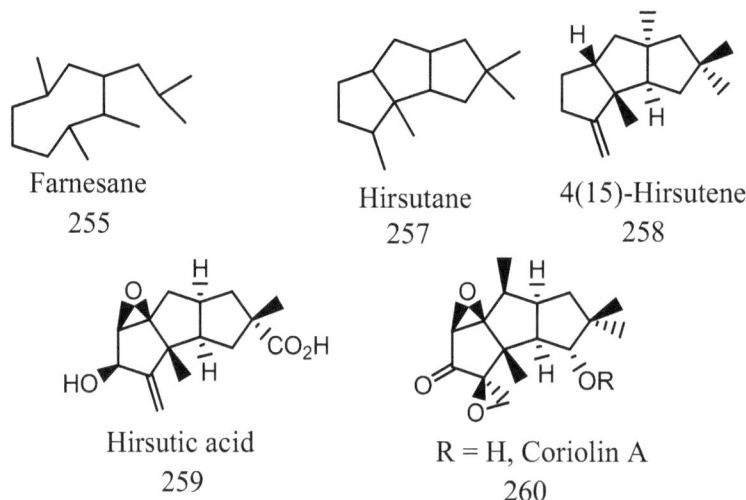

Farnesane
255

Hirsutane
257

4(15)-Hirsutene
258

Hirsutic acid
259

R = H, Coriolin A
260

FIGURE 15.50 Examples of Hirsutanes.

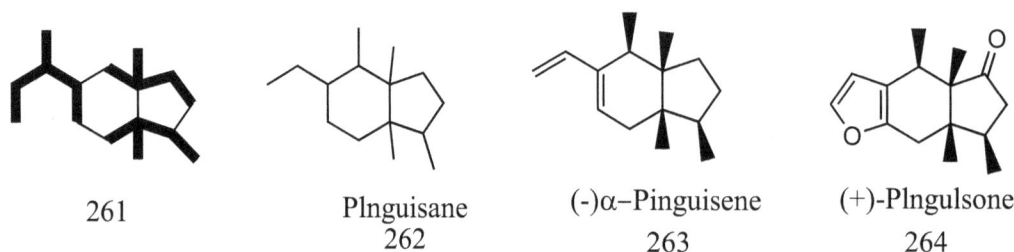

261

Plnguisane
262

(-)α–Pinguisene
263

(+)-Plngulsone
264

FIGURE 15.51 Examples of Pinguisanes.

15.8 Important Synthesis of Sesquiterpenes

15.8.1 Farnesene

Farnesene, $C_{15}H_{24}$, b.p. 128–130°C/12 mm.

This is obtained by the dehydration of farnesol with potassium hydrogen sulphate. This compound is the α-isomer 276, and it has now been shown that the β-isomer 277 occurs naturally (in oil of hops). β-Farnesene is also obtained by the dehydration of nerolidol (Figure 15.53) [12,13].

15.8.2 Farnesol

Farnesol 277, $C_{15}H_{26}O$, b.p. 120°C/0.3 mm.

This occurs in the oil of ambrette seeds and structure has also been elucidated. When oxidized with chromic acid,

farnesol (**278**) is converted into farnesal (**279**), $C_{15}H_{24}O$, a compound which behaves as an aldehyde. Thus farnesol is a primary alcohol. Conversion of farnesal into its oxime, followed by dehydration with acetic anhydride, produces cyanide (**280**) which, on hydrolysis with alkali, forms farnesenic acid (**281**), $C_{15}H_{24}O_2$, and a ketone, $C_{13}O_{22}O$ (282) (Scheme 15.1).

This ketone was then found to be dihydro-*pseudo*-ionone (geranylacetone, 283). In the formation of this ketone, two carbon atoms are removed from its precursor. This reaction is characteristic of α, β-unsaturated carbonyl compounds, and so it is inferred that the precursor, farnesenic acid (or its nitrile), is an α, β-unsaturated compound. Thus the foregoing facts may be formulated as follows, on the basis of the known structure of geranylacetone (Figure 15.54) [12,13].

A recent synthesis of farnesol 287 has been carried out by Corey et al. (1967) (Scheme 15.2) [14].

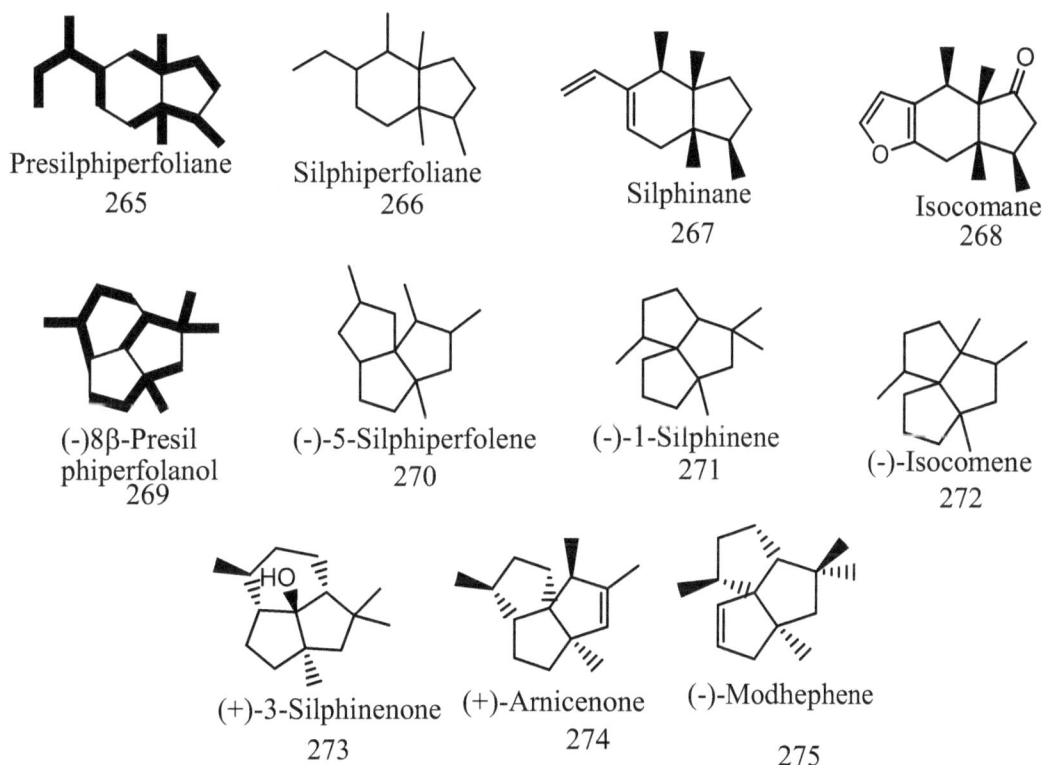

FIGURE 15.52 Examples of Presilphiperfolianes, Silphiperfolianes, Silphinanes, Isocomanes.

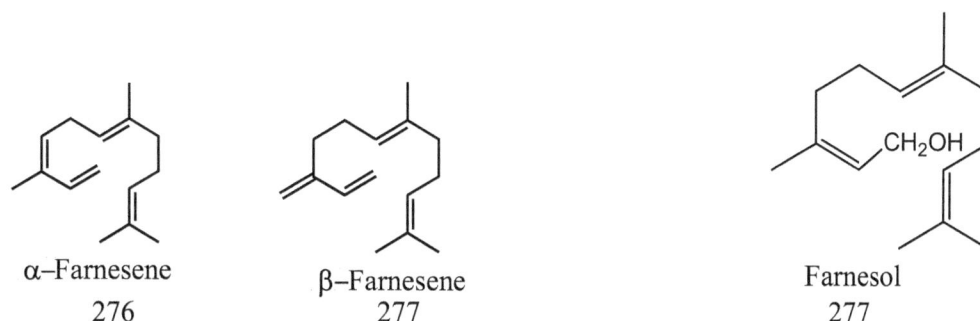

FIGURE 15.53 Isomers of Farnesene

FIGURE 15.54 Examples of Farnesol.

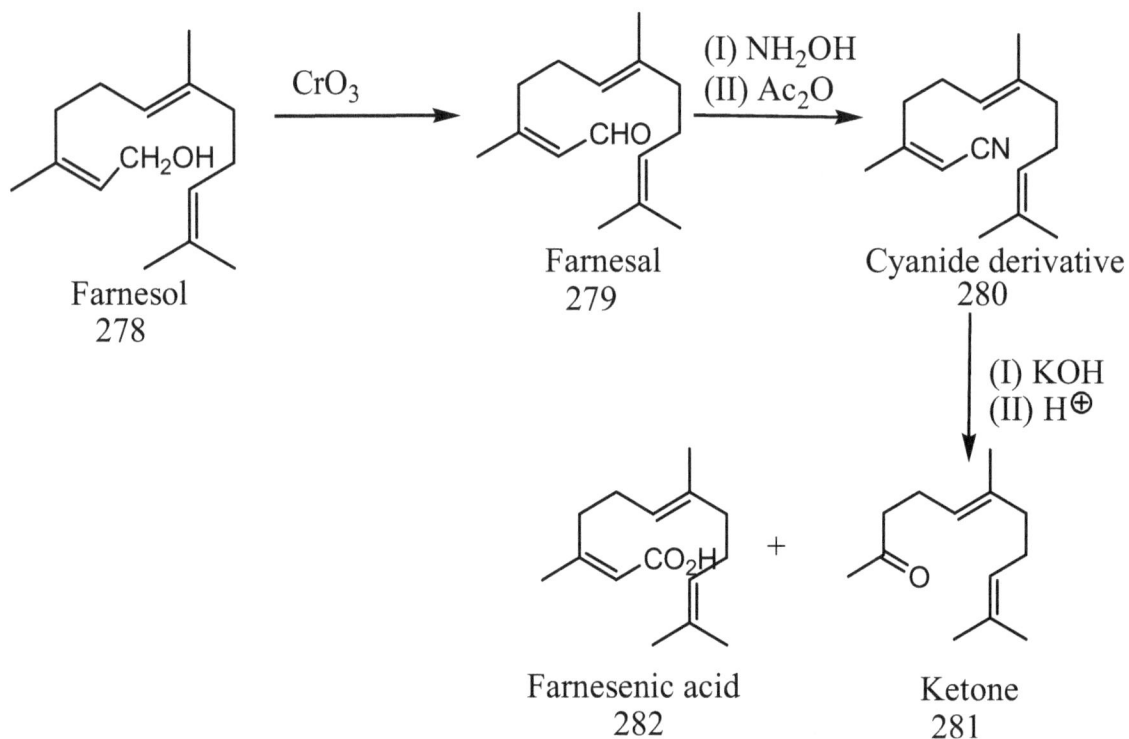

SCHEME 15.1 Synthesis of Farnesenic acid from Farnesol.

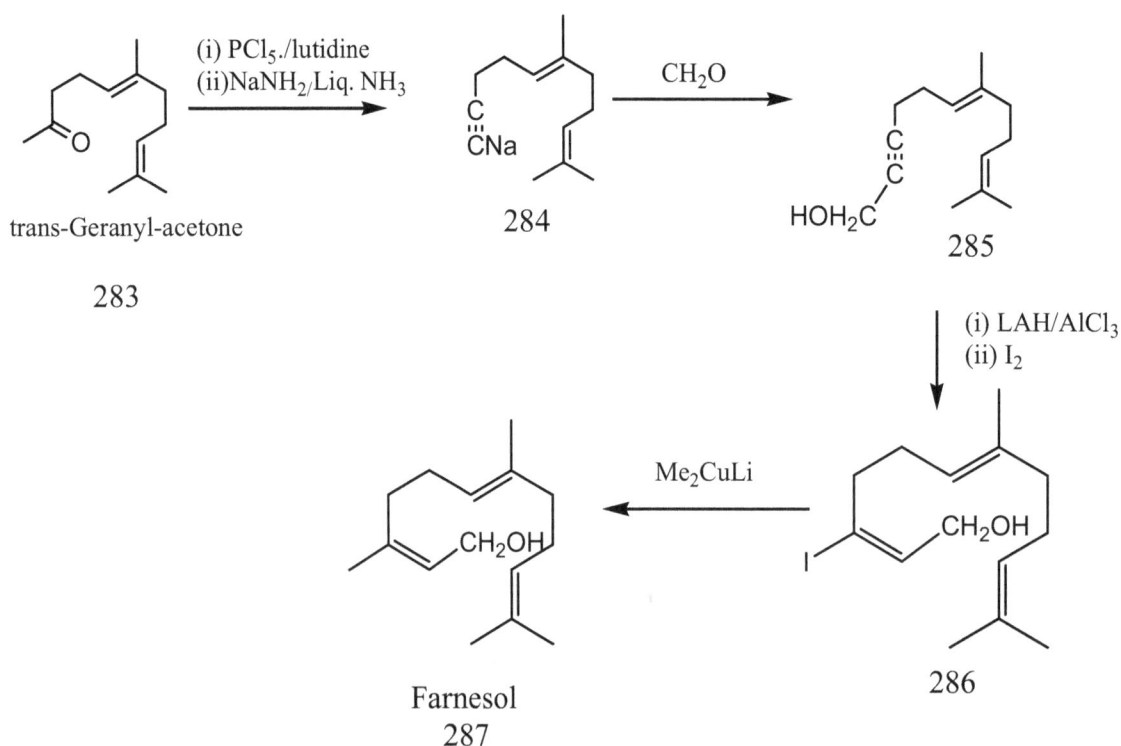

SCHEME 15.2 Synthesis of Farnesol from Geranyl acetone.

15.8.3 Nerolidol

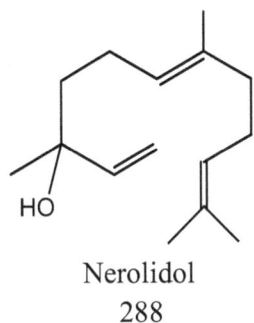

Nerolidol
288

Nerolidol 288, $C_{15}H_{26}O$, bp.125–127°C/4.5 mm. This occurs in the oil of neroli in the (+)- form. Nerolidol is isomeric with farnesol showed that the relationship between the two is the same as that between linalool and geraniol and confirmed the structure of nerolidol by synthesis (Scheme 15.3).

15.8.4 E-Selinene

The "right half" of the sesquiterpene (+)-E-selinene (as drawn below) includes (R)- (+)-limonene as a substructure. Retrosynthetic disconnection to (R)-(+)-limonene leads to the intermediate carbenium ions **1a** and **1b** via 15-nor-11-eudesmen-4-one (carbonyl alkenylation) and 15-nor-13-chloro-2-eudesmen-4-one (dehydrogenation, protective masking of the double bond in the side chain). These carbenium ions arise from (R)-9-chloro-p-menth-1-ene and the acylium ion **1c** (synthone) originating from 3-butenoic acid as reagent (synthetic equivalent). (R)-p-Menth-1-en-9-ol, on its part obtained by hydroboration and oxidation of (R)-(+)-limonene, turns out to be the precursor of the chloromenthene (Scheme 15.4).

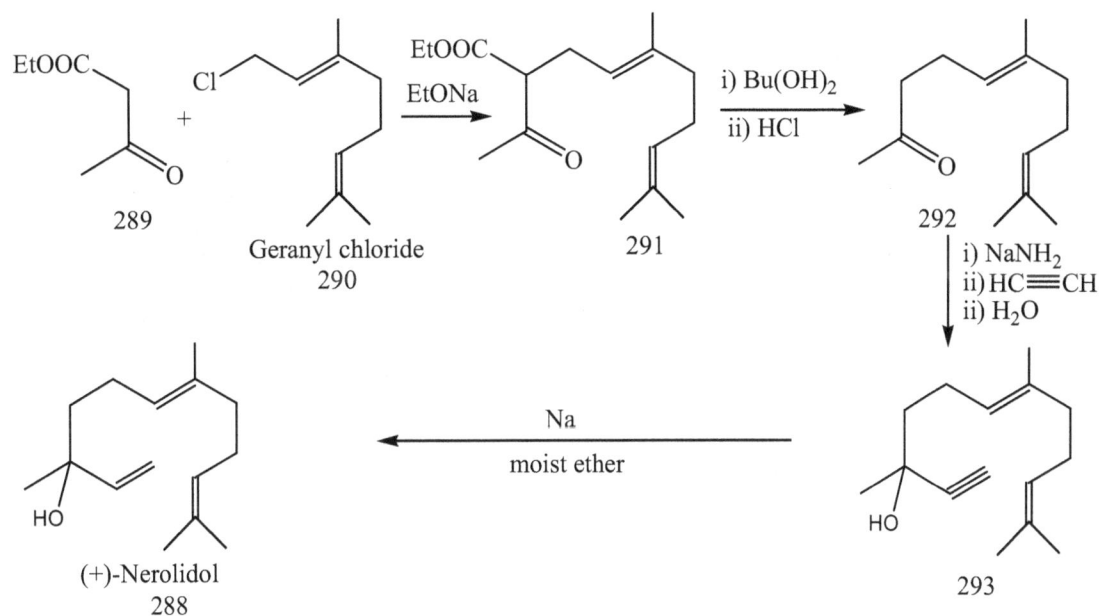

SCHEME 15.3 Synthesis of Nerolidol from Geranyl chloride.

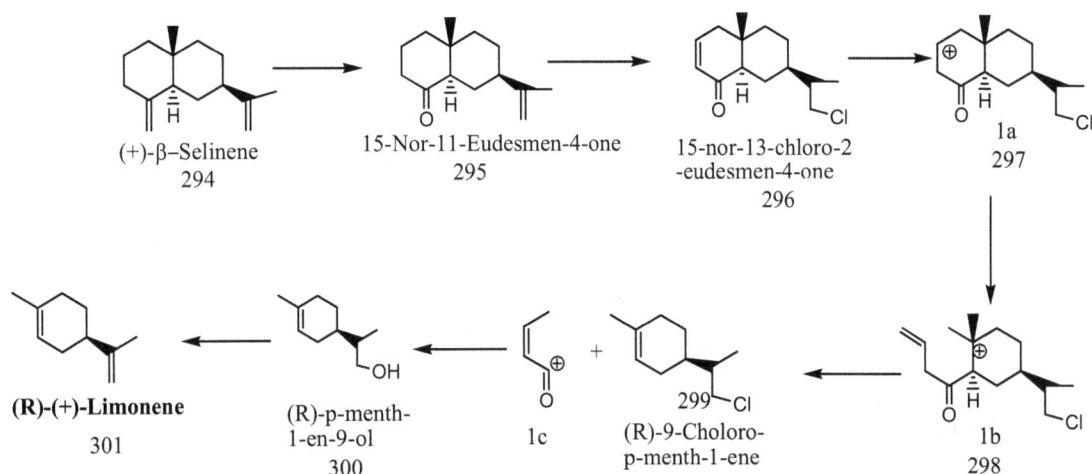

SCHEME 15.4 Synthesis of Limonene from Selinene.

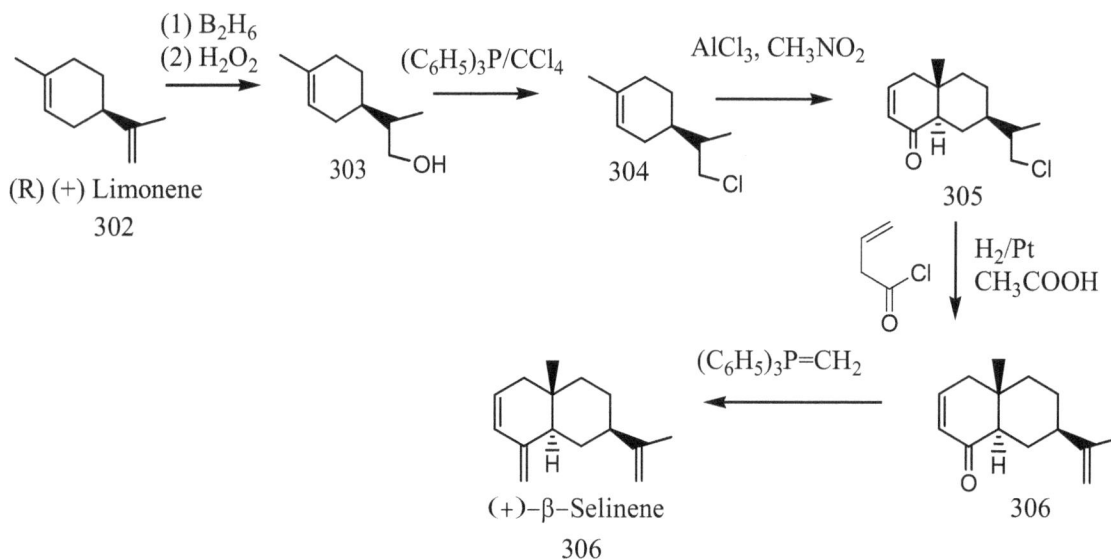

SCHEME 15.5 Synthesis of Selinene from Limonene.

Thus, the first step of a stereoselective synthesis of E-selinene reported by **MACKENZIE, ANGELO** and **WOLINSKI** involves hydroxylation of the C-8C-9- double bond of (R)-(+)-limonene by hydroboration with diborane and subsequent oxidation with hydrogen peroxide. The nucleophilic chlorination to (R)-9-chloro-pmenth-1-ene is achieved with tetrachloromethane and triphenylphosphane. 3-Butenoic acid chloride in the presence of aluminum chloride closes the ring to 15-nor-13-chloro-2-eudesmen-4-one. Catalytic hydrogenation and simultaneous dehydrochlorination in glacial acetic acid afford 15-nor-11-eudesmen-4-one as precursor, which is finally subjected to **WITTIG methylenation** to (+)-E-selinene (Scheme 15.5).

15.8.5 Isocomene

The retrosynthetic disconnection of isocomene leads primarily to the intermediate tertiary carbenium ion **308**, which may arise from the intermediate carbenium ion **309** by anionotropic 1,2-alkyl shift. The latter turns out to be the protonation product of the tricycle **310** containing an exocyclic CC-double bond which is generated by a **WITTIG-methylenation** of the tricyclic ketone **311**. The concept behind this is the formation of the cyclobutane ring in **311** by means of an intramolecular [2+2]-photocycloaddition of the 1,6-diene **312**. The enone substructure in **312** results from hydrolysis of the enolether and dehydration of the tertiary alcohol function in (6S)-1-alkoxy-2,4- dimethyl-3-(2-methyl-1-penten-5-yl) cyclohexene **313**. The tertiary alcohol **313** emerges from a nucleophilic alkylation of (6S)-3-alkoxy-2,6-dimethyl-2-cyclohexen-1-one **314** with metallated 5-halo-2-methyl-1-pentene obtained by **GRIGNARD reaction orlithiation**. The desired enantiomer **314** becomes feasible by methylation of the methylene group D to the carbonyl function of 3-alkoxy-2-methyl-2-cyclohexen-1-one **315**, followed by resolution of the racemic mixture (Scheme 15.6).

A synthesis reported by **PIRRUNG** 32 follows this concept. 3-Ethoxy-2-methyl-2- cyclohexen-1-one obtained from 1,3-cyclohexanedione in two steps involving C and O-alkylation is lithiated with lithiumdiisopropylamide (LDA) and methylated with iodomethane to racemic. The **GRIGNARD-reagent** prepared from 5-bromo-2-methyl-1-pentene alkylates to the unstable tertiary alcohol which is not isolated, reacting during workup of the reaction mixture with aqueous hydrochloric acid directly to the enone. This undergoes an intramolecular [2+2]-photocycloaddition to the tricyclic ketone upon irradiation with UV light (350nm) in *n*-hexane solution. Sterically forced by the methyl groups at the six-membered ring, the cyclobutane ring closes opposite to these methyl groups. A crystal structure confirms the relative configuration of the tricycle. Carbonyl methylenation with methylentriphenyl phosphorane, generated *in situ* from methyltriphenylphosphonium iodide in dimethylsulfoxide at 70°C, introduces the exocyclic double bond. The resulting alkene is protonated with *p*-toluenesulfonic acid in refluxing benzene solution to the intermediate carbenium ion. This undergoes a ring-expanding 1,2-alkyl shift to the carbenium ion, which deprotonates to racemic isocomene, as expected (Scheme 15.7).

15.8.6 Cedrene

Cedrene and cedrol are used as fragrances in perfumery and as insect repellants. Acid-catalyzed dehydration of cedrol, the product of a nucleophilic addition of methyllithium to the carbonyl function of 15-nor-3-cedrone, is a straightforward method to introduce the exocyclic 3(15)-double bond. An intramolecular opening of the cyclopropane ring in the tricyclic ketone **318** by the (nucleophilic) CC double bond permits the formation of the five-membered ring *B* in 15-nor-3-cedrone. [2+1]- Cycloaddition with diazomethane as carbene generator or the **SIMMONS-SMITH reaction** are suitable methods for cyclopropanation of the corresponding dienone precursor.

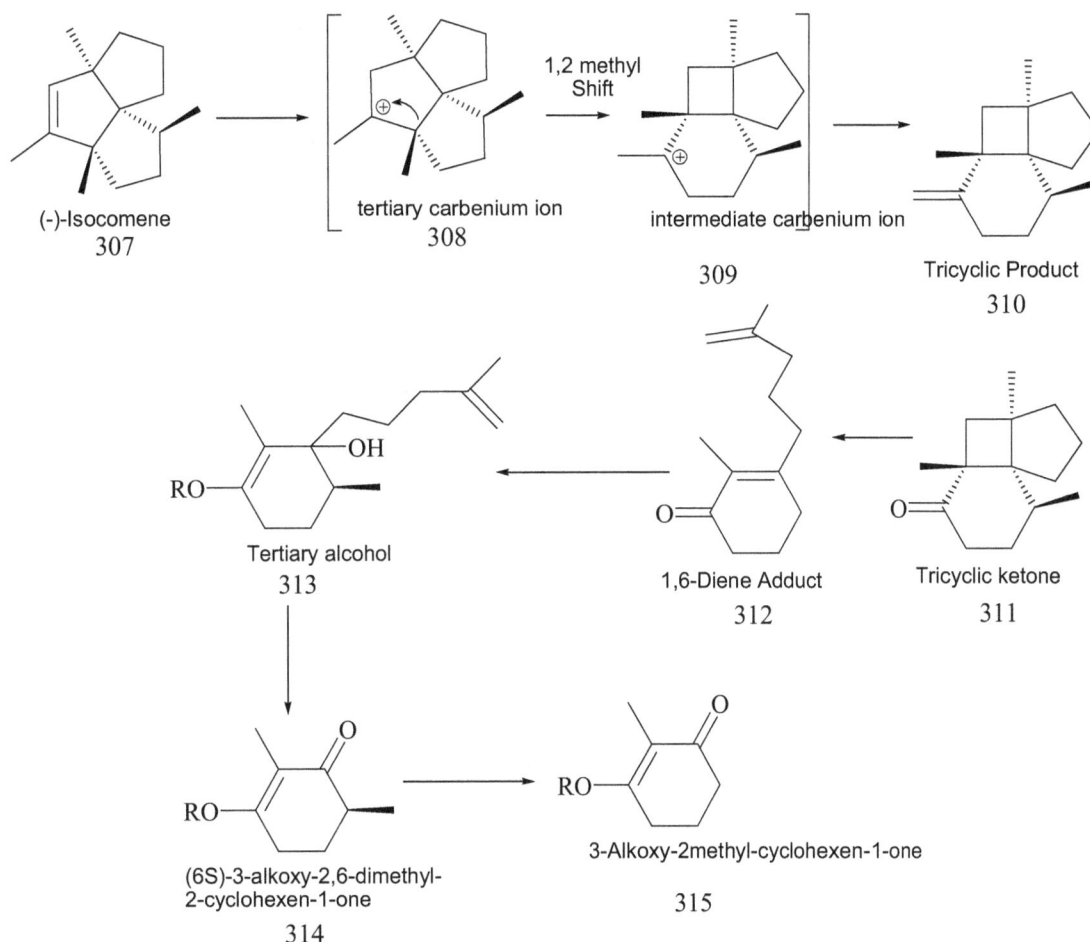

SCHEME 15.6 Synthesis of 3-alkoxy-2,6-dimethyl-2-cyclohexene-1-one.

In order to avoid the competing ring homologization of the cycloalkanone, the carbonyl function in **318** must be masked before cyclopropanation. The obvious way to do so is to use the secondary alcohol 2-cyclopentenol **3** as the precursor. Thus, the retrosynthesis leads from **318** *via* **319** to **320**. The latter turns out to be the reduction product of the cyclopentenone **321**, which is feasible by nucleophilic addition of lithiated 2-halo-6-methyl-5-heptene **322** with the enolether **323** of 1,3-cyclopentanedione (Scheme 15.8).

This is the concept of a synthesis of racemic cedrene reported by **E.J. COREY** and **R.D. BALANSON**. Lithiated 2-chloro-6-methyl-5-heptene **5** as carbon nucleophilealkylates 3-methoxy-2-cyclopentenone **323**; work-up of the reaction mixture in aqueous acidic solution converts the primarily formed alkylated 3-methoxy-2-cyclopentenol directly to the corresponding 2-cyclopentenone, which is reduced to the secondary allyl alcohol by diisobutylaluminumhydride. Regioselective cyclopropanation using the **SIMMONS-SMITH reaction** gives the diastereomeric bicyclo[3.1.0]hexanols **2**, which are oxidized to the ketones by chromium(VI)oxide in pyridine. The desired diastereomer rearranges to 15-nor-3-cedrone in the presence of the mixed anhydride from methanesulfonic and acetic acid as catalyst. Nucleophilic addition of methyllithium to the carbonyl function of affords cedrol, which dehydrates in the presence of formic acid to racemic cedrene (Scheme 15.9).

15.8.7 Periplanone B

Periplanone B is the most active sex pheromone found in the alimentary tract and excreta of the American cockroach *Periplaneta americana*. An elegant total synthesis of this germacrane sesquiterpene was achieved by *SCHREIBER* and *SANTINI*. Cyclodecatrienone 328 is an obvious precursor. One of the oxirane rings arises from epoxidation of the enone CC double bond, the other from [2+1]-cycloaddition of a carbene to the carbonyl bond of the enone. Oxidation of the methylene group introduces the additional carbonyl double bond. The CC double bond of the enone results from an elimination of HX in the D-X-substituted cyclodecadienone 329, which, on its part, is feasible by substitution of cyclodecadienone 329. An electrocyclic opening of the cyclobutene ring in 330 provides the 1,3-diene substructure in 329 (Scheme 15.10).

Enolization of the carbonyl function in 330 gives rise to a 1,5-dien-1-ol 331 which is nothing but the product of a *COPE-rearrangement* of the bicyclo[4.2.0]octanol 6332. The vinyl group in 333 is introduced by 1,2-addition of vinylmagnesiumhalide to thecarbonyl function of bicyclo[4.2.0]octanone333; the precursor 333 turns out to be the product of a [2+2]-photocycloaddition 334 of allene to the CC double bond of 4-*i*propyl-2-cyclohexenone 335.

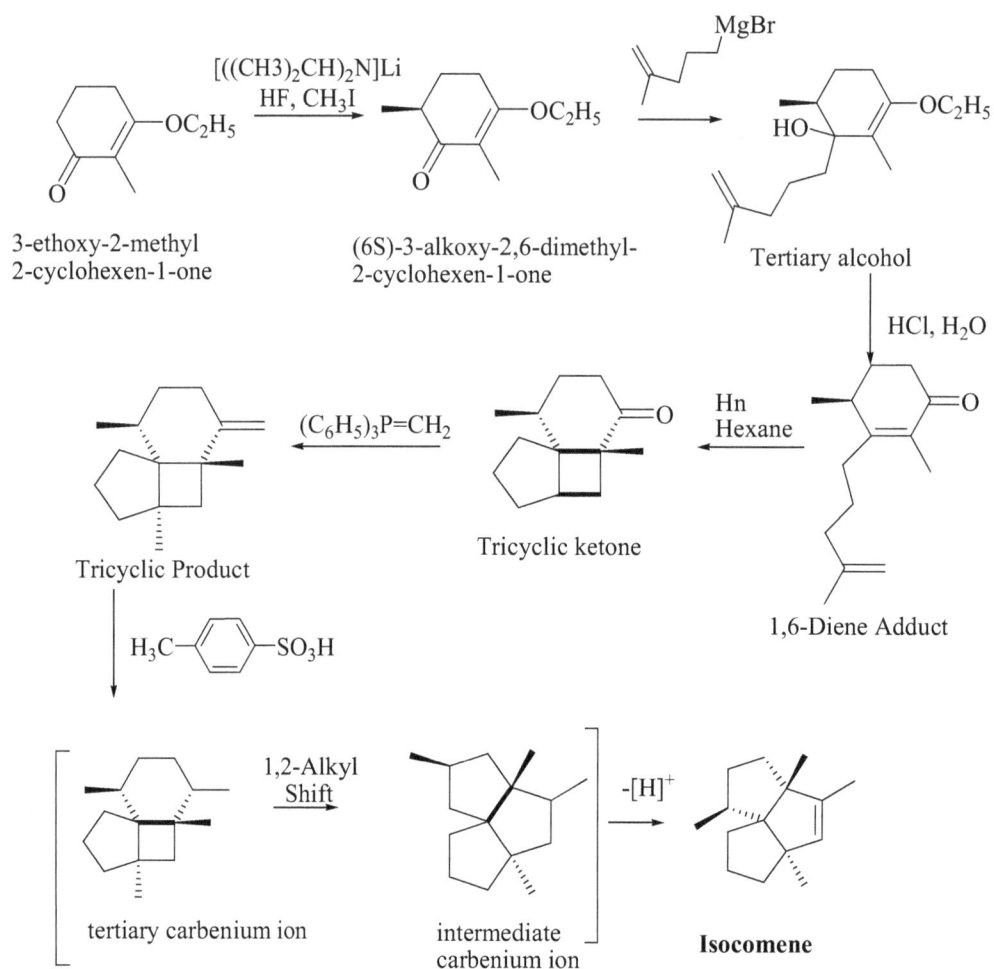

SCHEME 15.7 Synthesis of Isocomene from Tricyclic derivatives.

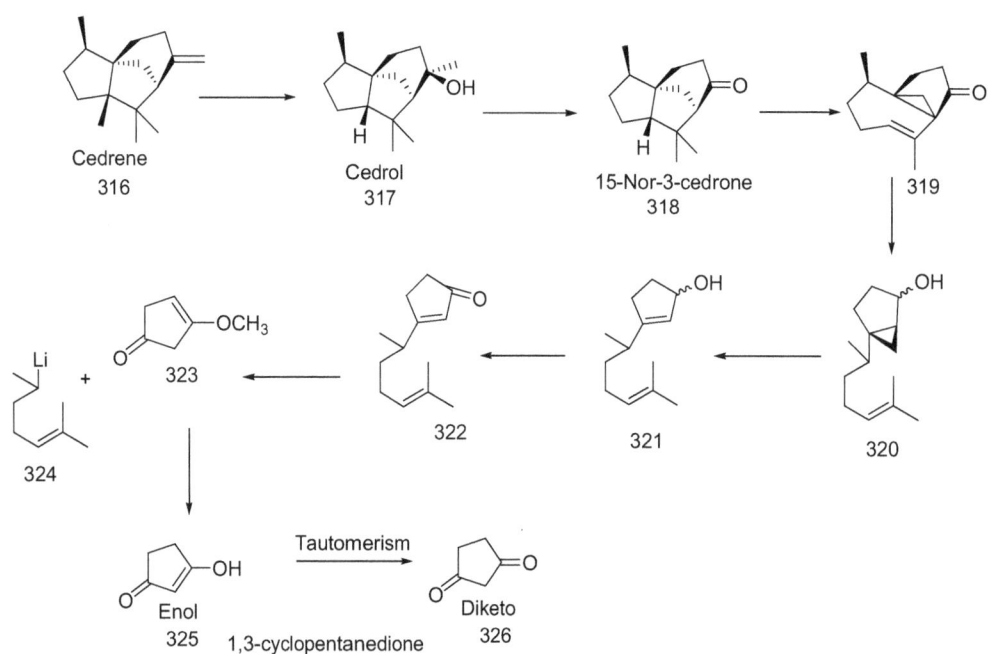

SCHEME 15.8 Synthesis of Diketo from Cedrene.

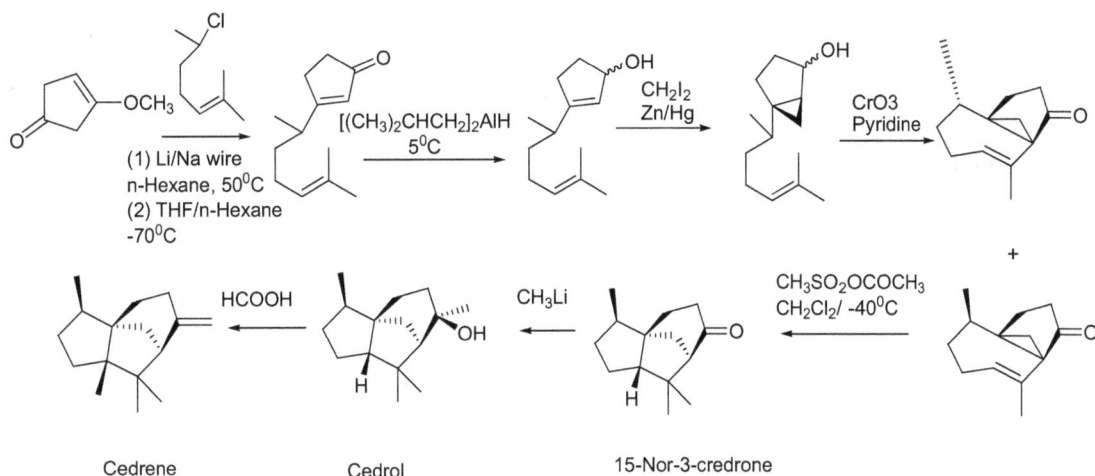

SCHEME 15.9 Synthesis of cedrene and derivatives.

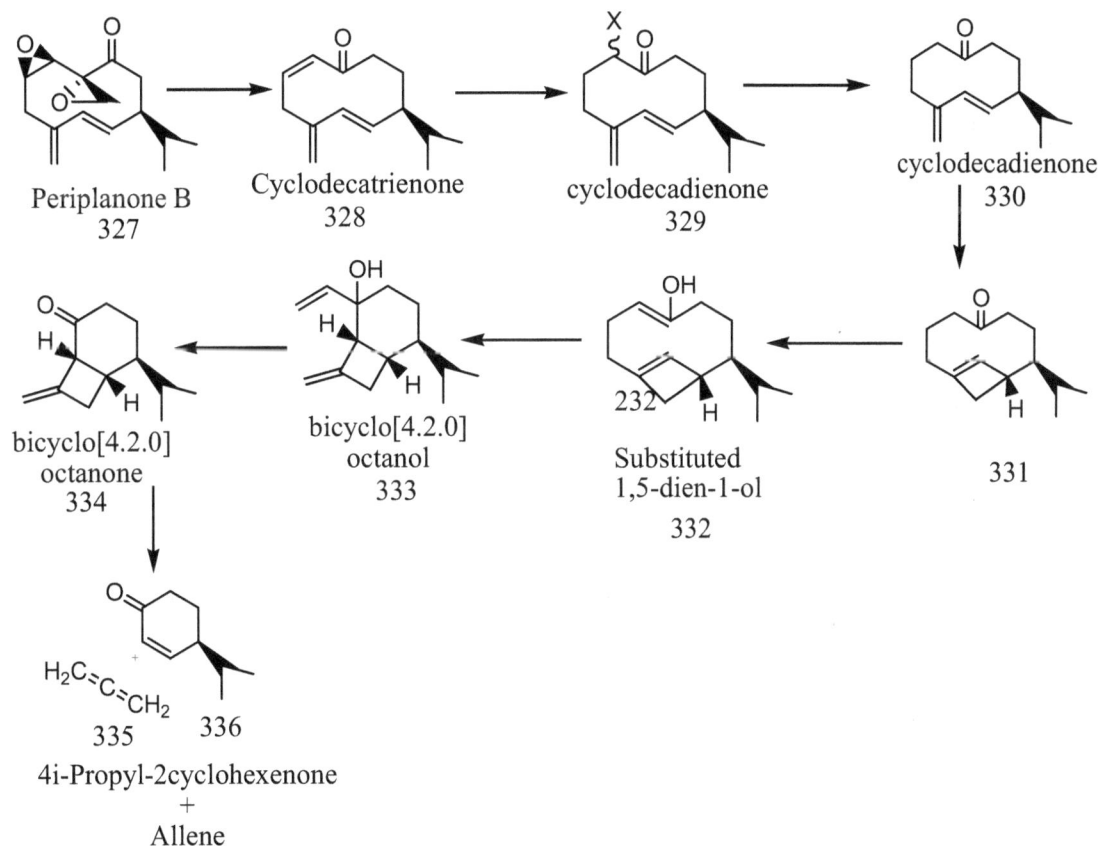

SCHEME 15.10 Synthesis of Propyl-2cyclohexanone from Periplanone.

In the first step of this synthesis 327, [2+2]-photocycloaddition of allene to racemic 4-*i*-propyl-2-cyclohexenone yields a mixture of the *anti*- and *syn*-cycloadducts 336 and 337, which do not have to be separated because the cyclodecadienone is built from both isomers in the course of subsequent procedures. The 1,2-addition of vinylmagnesiumbromide to the ketones 336 and 337 gives the diastereomeric alcohols 6a which

are deprotonated with potassium hydride and 18-crown-6 to the alcoholates 339, and these undergo oxa-COPE rearrangement to the bicyclic ketone with the expected 2:1 ratio of isomers. Thermally induced electrocyclic ring opening yields the desired *trans*-cyclohexadienone as a major product. Lithium-bis(trimethylsilyl)amide (LBTMSA) activates the enolate which is sulfenylated by diphenyldisulfide dioxide

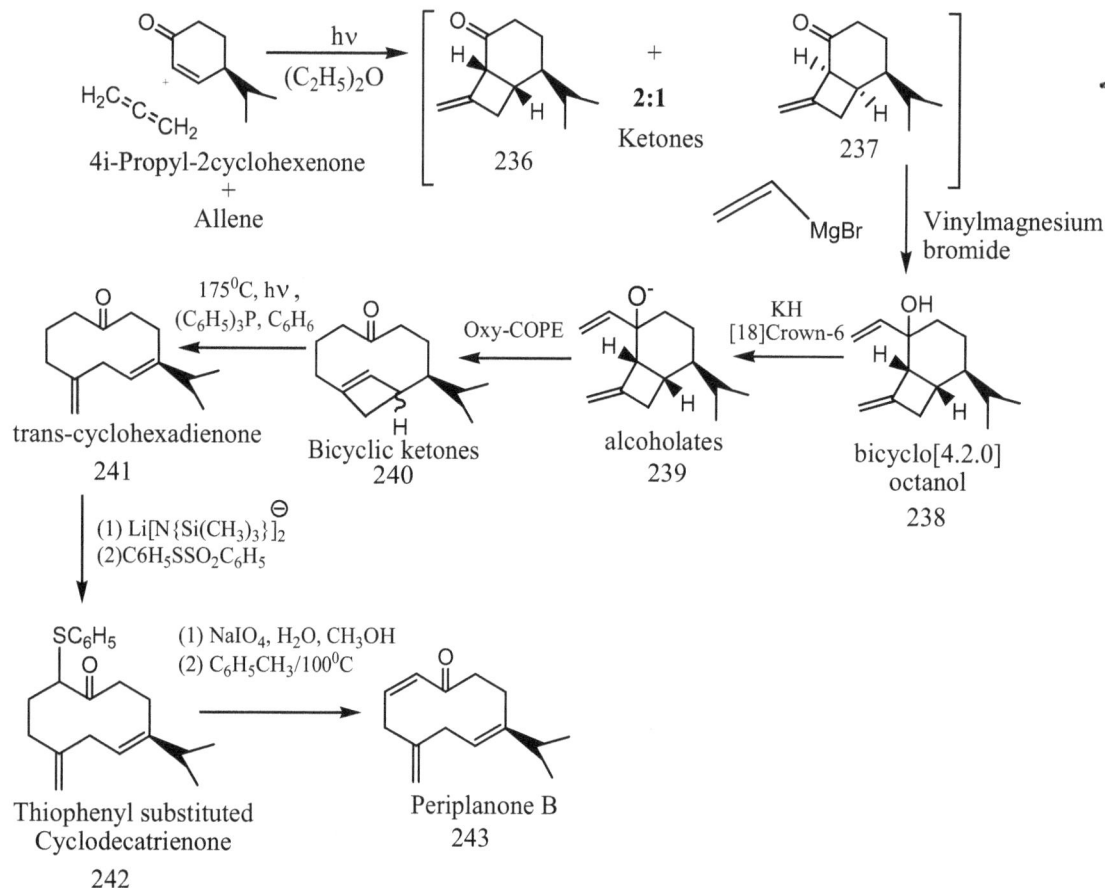

SCHEME 15.11   Synthesis of Periplanone B from Propy-2cyclohexenone.

(TROST´s reagent), predominantly to the desired regioisomer. Subsequent oxidation of the thioether with sodium periodate affords the intermediate phenylsulfone, which undergoes elimination to the cyclodecatrienone upon heating in toluene solution (Scheme 15.11).

Insertion of the oxirane ring in the 1,2-position is achieved by epoxidation of the electron-deficient enone CC double bond in with *t*-butylhydroperoxide. In order to introduce the carbonyl function at C-10, the 1,2-epoxycyclodecadienone is activated once again with LBTMSA to the enolate, which is converted by electrophilic addition of phenylselenylbromide to the phenylselenide as a masked ketone. Hydrogen peroxide oxidizes the selenide to the selenoxide which, upon acylation with acetic anhydride and sodium acetate in tetrahydrofuran, undergoes a selena-***PUMMERER rearrangement*** to the epoxycylodecadienedione involving the intermediates which are not isolated. A regioselective cycloaddition of dimethylsulfoniummethylide to the carbonyl double bond next to the first oxirane ring completes the last synthetic step to racemic periplanone B (Scheme 15.12) [11,12].

15.8.8 Laurene

The bicyclic sesquiterpene laurene (352) was first isolated from *Laurencia glandulijera* and has been found in several *Laurencia* species, including the marine red alga *Laurencia elata*. The major synthetic difficulty encountered in the preparation of laurene is the correct stereochemistry at the vicinal methyl groups. The key reaction in this strategy is a radical cyclization to form the cyclopentane ring with exocyclic methylene. Allylation of ester followed by LAH reduction, PCC oxidation and acetal formation provides the methyl substituted quaternary center in the form of acetal.

Hydroboration-oxidation with subsequent PCC oxidation converts the allyl moiety to an aldehyde, which is converted to an alkyne 48 via dibro-momethylene Wittig olefination and elimination. Hydrolysis of the acetal 48, the addition of methyl-magnesium iodide and conversion of the resulting alcohol to the xanthate provides the needed radical precursor. Treatment of49 with tri-n-butyltin hydride produces a 1: 1 mixture of laurene and epi-Iaurene in 12 steps. Slightly better control of diastereo selectivity in the synthesis of laurene (SO) is obtained by Taber (Scheme 15.13).

15.9 Conclusion

The present chapter provides comprehensive information regarding the sesquiterpenes, acyclic sesquiterpenes i.e Farnesanes, Monocyclic Farnesane Sesquiterpenes, Cyclofarnesanes and Bisabolanes, Germacranes and

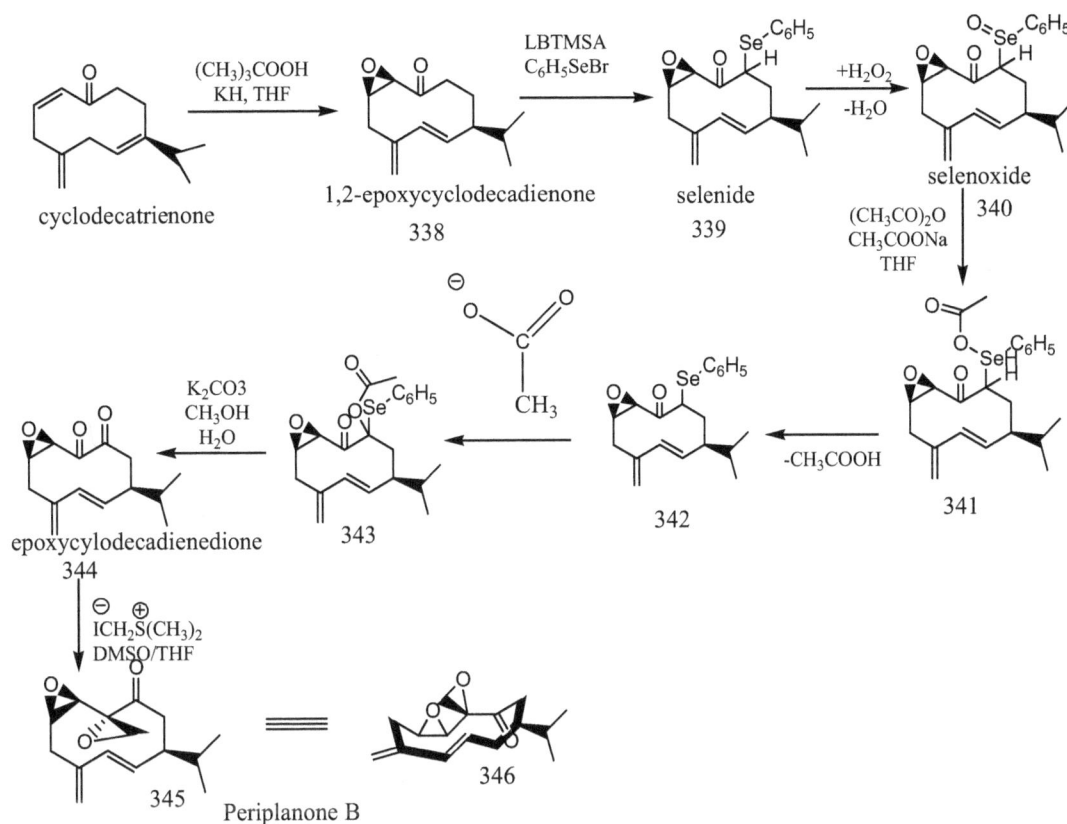

SCHEME 15.12 Synthesis of Periplanone B from Cyclodecatrienone.

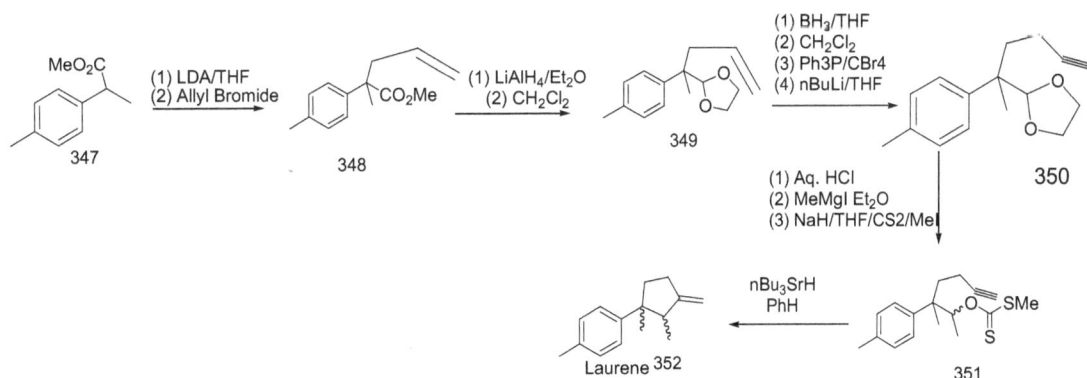

SCHEME 15.13 Synthesis of bicyclic sesquiterpene laurene.

Elemanes, Humulanes, Polycyclic Farnesane Sesquiterpenes i.e Caryophyllanes, Eudesmanes and Furanoeudesmanes, Eremophilanes, Furanoeremophilanes, Valeranes, Cadinanes, Drimanes, Guaianes and Cycloguaianes etc. It also explains about Marasmanes, Isolactaranes, Lactaranes, Sterpuranes, Acoranes, Chamigranes, Cedranes and Isocedranes, Zizaanes, Prezizaanes, Campherenanes and Santalanes, Hirsutanes and other polycyclic sesquiterpenes like Pinguisanes, Presilphiperfolianes, Silphiperfolianes, Silphinanes, Isocomanes etc. It also includes some important synthesis of Farnesene, Farnesol, Nerolidol, E-Selinene, Isocomene, Cedrene, Periplanone B, Laurene etc.

REFERENCES

1. Ruzicka, L. History of the Isoprene Rule. (Faraday Lecture). *Proceedings of the Chemical Society (London),* **1959**, 341.

2. Connolly, J. D.; Hill, R. A. *Dictionary of Terpenoids, Vol. 1: Mono- and Sesquiterpenoids, Vol. 2: Di- and higher Terpenoids, Vol. 3: Indexes.* Chapman & Hall, London, New York, Tokyo, Melbourne, Madras, 1991.

3. Newman, A. A. *Chemistry of Terpenes and Terpenoids.* Academic Press, London, New York, 1972.

4. Rücker, G. Sesquiterpene. *Angewandte Chemie,* **1973**, 85, 895; *Angewandte Chemie International Edition in English.* **1973**, 12, 793.

5. Ohloff, G. *Riechstoffe und Geruchsinn- Die molekulare Welt der Düfte.* Springer, Berlin, Heidelberg, New York, 1990; *Düfte – Signale der Gefühlswelt*, Verlag Helvetica Chimica Acta, Zürich, Wiley-VCH, Weinheim, 2004.

6. Jacobsen, M. *Insect Sex Pheromones.* Academic Press, London, New York, 1972.

7. Beroza, M. *Chemicals Controlling Insect Behaviour.* Academic Press, London, New York, 1970.

8. Sowndhararajan, K., Kim, S. Influence of Fragrances on Human Psychophysiological Activity: With Special Reference to Human Electroencephalographic Response. *Scientia Pharmaceutica*, **2016**, 84(4), 724–752.

9. Breitmaier, E. *Terpenes.* Wiley-VCH, Mörlenbach, 2007, 24–32,34.

10. *The Merck Index*, 13th Edition. Merck & Co., Inc., Whitehouse Station, NJ, 2001.

11. Kolodzej, H., Sesquiterpenlactone - Biologische Aktivitäten, Dtsch. *Apotheker Ztg*, **1993**, 133, 1795.

12. Singh, G. *Chemistry of Terpenoids and Carotenoids.* DPH, New Delhi, 2007, 82–100.

13. Stierle, D.B., Sims, J.J. Marine Natural Products XV: Polyhalogenated Cyclic Monoterpenes from the Red Alga Plocamium cartilagineum of Antarctica. *Tetrahedron*, 1979, 35,1261.

14. Corey, E. J.. General Methods for the Construction of Complex Molecules. *Pure and Applied Chemistry*, **1967**, 14(1), 9–38. https://doi.org/10.1351/pac196714010019

16

Sesquiterpenes Lactone: Biogenesis and Chemical Synthesis

Bimal Krishna Banik*, Abhishek Tiwari†, and Biswa Mohan Sahoo‡

CONTENTS

16.1   Introduction 518
16.2   Classification...520
16.3   Structure 520
16.4   Chemical Characteristics..521
16.5   Biosynthesis of Sesquiterpene Lactones 521
    16.5.1   Biosynthesis of the Isoprenoid Precursors in Plants 521
    16.5.2   Mevalonate (MVA) Pathway 521
    16.5.3   2-C-Methyl-D-erythritol 4-phosphate (MEP) Pathway 522
    16.5.4   Cross-Talk between MVA and MEP Pathways 522
16.6   Farnesyl Diphosphate Synthase..523
    16.6.1   Branch Point of Sesquiterpene Lactone Biosynthesis..523
    16.6.2   Sesquiterpene Lactone Pathway 523
    16.6.3   Hypothetical Pathway...524
16.7   Enzymes of Sesquiterpene Lactone Biosynthesis 525
    16.7.1   Biosynthesis of the Sesquiterpene Framework 525
    16.7.2   Biosynthesis of the Lactone Ring 525
    16.7.3   Function of the α-Methylene-γ-Lactone Group 526
    16.7.4   Sesquiterpene Lactone Synthesis 527
    16.7.5   Synthetic Strategy of Parthenolide...528
16.8   Parthenolide Semi-synthesis 530
    16.8.1   General Strategies for the Synthesis of Guaianolides...530
16.9   Synthesis of Guaianolides 531
    16.9.1   Semi-synthesis of Guaianolides 532
16.10 Costunolide and Its Derivatives..535
16.11 Parthenolide and Analogues 535
16.12 Artemisinin and Its Derivatives..535
16.13 Santonin and Its Analogues 535
16.14 Structure–Activity Relationship..537
16.15 Biological Activities 537
16.16 Role of Sesquiterpene Lactones 537
16.17 Pathways Affected by Sesquiterpene Lactones...537
16.18 IIR Signalling SLTS-Modified in Inflammation..540
16.19 Mechanism of Different STLs...540
    16.19.1 Santonin 540
    16.19.2 Artemisinin 540
    16.19.3 Parthenolide 541
    16.19.4 Costunolide..541

* Department of Mathematics and Natural Sciences, College of Sciences and Human Studies, Prince Mohammad Bin Fahd University, Al Khobar, Kingdom of Saudi Arabia. bimalbanik10@gmail.com; bbanik@pmu.edu.sa.

† Faculty of Pharmacy, Pharmacy Academy, IFTM University, Lodhipur Rajput, Moradabad-244102, Uttar Pradesh, India. abhishekt1983@gmail.com.

‡ Roland Institute of Pharmaceutical Sciences, Berhampur affiliated to Biju Patnaik University of Technology (BPUT), Rourkela, Odisha, India.

DOI: 10.1201/9781003008682-16

16.19.5 Dehydroleucodine...541
16.19.6 Helenalin 541
16.19.7 Thapsigargin..542
16.19.8 Arglabin...542
16.19.9 Cynaropicrin...543
Conclusion 543
References...543

16.1 Introduction

Since ancient times, herbal treatments, traditional folk medicine, and, in a wider sense, natural goods have been used to maintain human health. Natural origin chemicals are now being used as lead molecules in the development of medicines and functional foods. Derivatization of natural products by synthetic or organic alterations is still a realistic objective for the advancement of preventive and therapeutic medicine, according to reports. Natural-origin therapies with fewer side effects and high bioavailability have gotten a lot of interest as a consequence of promising findings. Plants with their varied secondary metabolites, which differ in chemical structure and physiological activity, present a potential collection of samples to study. Using natural compounds as templates for medicinal applications has been a hot issue in pharmacology during the past several decades. Natural products or their derivatives have been used to generate innovative and efficient treatment techniques in a variety of sectors, including cancer research, antimicrobial agent development and metabolic illness prevention.

Sesquiterpene lactones (STLs) are naturally occurring plant compounds that belong to the terpenoid family. STLs are present in a wide range of plant species, including edible and medicinal plants. A growing number of research are

reporting on the biological activities of STLs, as well as some important evaluations of their relevance, chemical variety and therapeutic potential.

STLs are a broad and diversified collection of physiologically active plant compounds found in a variety of plant families, including the Acanthaceae, Anacardiaceae, Apiaceae, Euphorbiaceae, Lauraceae, Magnoliaceae, Menispermaceae, Rutaceae, Winteraceae and Hepatideae [1], among others. However, the Compositae (Asteraceae) family has the most structures, with over 3,000 distinct forms described. Sesquiterpene lactones are a family of naturally occurring plant terpenoids that develop from the head-to-tail condensation of three isoprene units, followed by cyclization and oxidative transformation to generate a cis or trans-fused lactone. They are a major ingredient of essential oils. Pseudoguainolides (**1**), guaianalides (**2**), germanocranolides (**3**), eudesmanolides (**4**), heliangolides (**5**) and hyptocretenolides (**6**), for example, are categorized principally based on their carbocyclic skeletons (Figures 16.1 and 16.2). Costunolide (**7**), a germanacranoride linked to the ten-membered carbocyclic sesquiterpene germacrone, is the source of the suffix "*olide,*" which alludes to the lactone activity. SLs, on the other hand, have a variety of skeleton layouts. A single skeleton type of SL is produced by each plant species, which is concentrated predominantly in the leaves and flower crowns. Other examples are Tagitinin A (**8**), Tagitinin C (**9**), Cynaropicrin (**10**), Eupatoriopicrin (**11**), Parthenin (**12**),

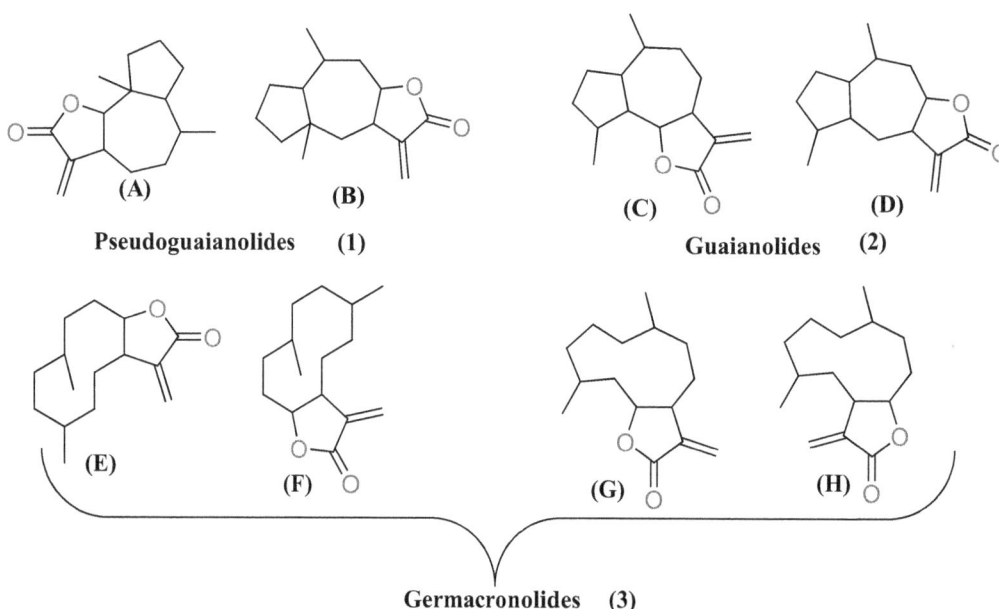

FIGURE 16.1 Basic skeletons of sesquiterpenes lactones.

Helenalin (13), Artemisnin (14), Vernodalin (15), Partholide (16), Dehydrocostus lactone (17), Vernolide (18), Hanphyllin (19), Alantolactone (20), 4,5 Epoxy, 4(5) α- dihydrosantonin (21), 4(5) α-Epoxy, 4,5 – dihydrosantonin (22), Neurolenin (23), Vernodal (24), 11, 13, dihydro vernodalin (25), Rupicolin (26), Ridentin (27), Artemisnin (14) [Dihydoartemisnin (28), Artemether (29), Artesunate (30), Sodium artesunate (31), Sodium Artelinate (32), Vernodalol (33), Neurolenin B (34), Ridentin (35) and Rupicolin A-8-acetate (36) [2].

The amount of SLs in a dry weight might range from 0.01% to 8%. Animals that have been inebriated by plants containing SLs have been known to die. They've been demonstrated to engage in a variety of biological functions.

A common characteristic of SLs is the presence of a γ-lactone ring (closed towards either C-6 or C-8) containing a α-methylene group in many situations. The introduction of hydroxyls or ester-ified hydroxyls, as well as an epoxide ring, are frequent altera-tions. Some SLs are glycosides, while others include halogen or sulphur atoms. The majority of SLs have shown cytotoxic

action (in vitro against KB and P388 leukaemia) as well as activ-ity against P388 leukaemia in vivo. Various cytotoxic SLs react with thiols, such as cysteine residues in the protein, through a fast *Michael* type of addition, according to structure-activity rela-tionship studies. Chemically, the, α, β-unsaturated carbonyl sys-tem found in SLs mediates these reactions. These findings back up the theory that SLs limit tumour development by selectively alkylating growth regulation biological macromolecules such important enzymes, which govern cell division and so hinder a range of cellular processes, leading to apoptosis [3].

Different numbers of alkylating structural components may explain differences in activity between individual SLs. Other parameters, such as lipophilicity, molecular shape and chemi-cal environment, as well as the target sulfhydryl, may impact sesquiterpene lactone activity. Figure 16.3 depicts a number of structurally different sesquiterpene lactones.

At first, sesquiterpene lactones were thought to be very cytotoxic. Chemical modifications have improved their bio-logical activities while lowering their cytotoxicity, therefore

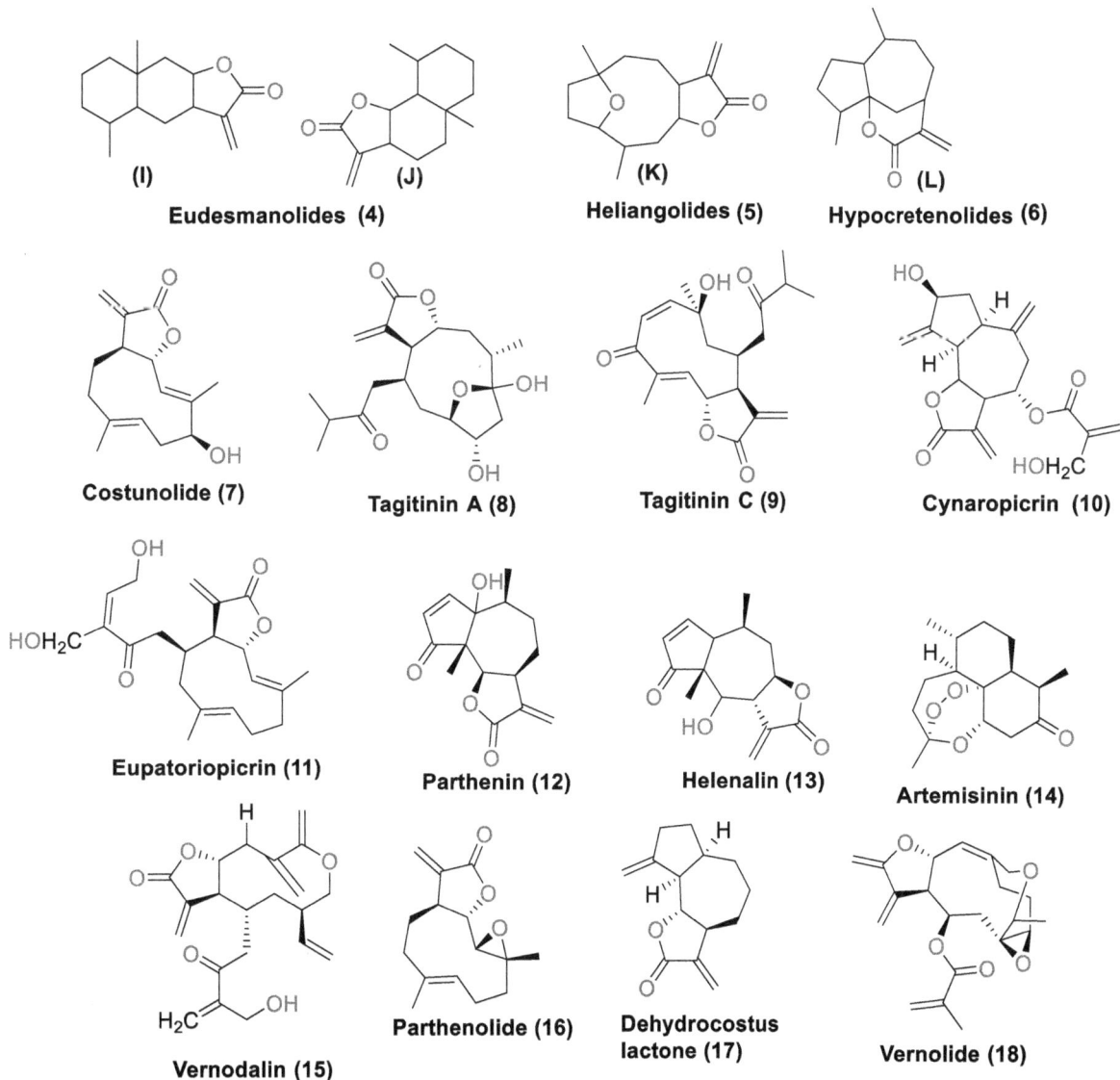

FIGURE 16.2 wide structures range of sesquiterpene lactones.

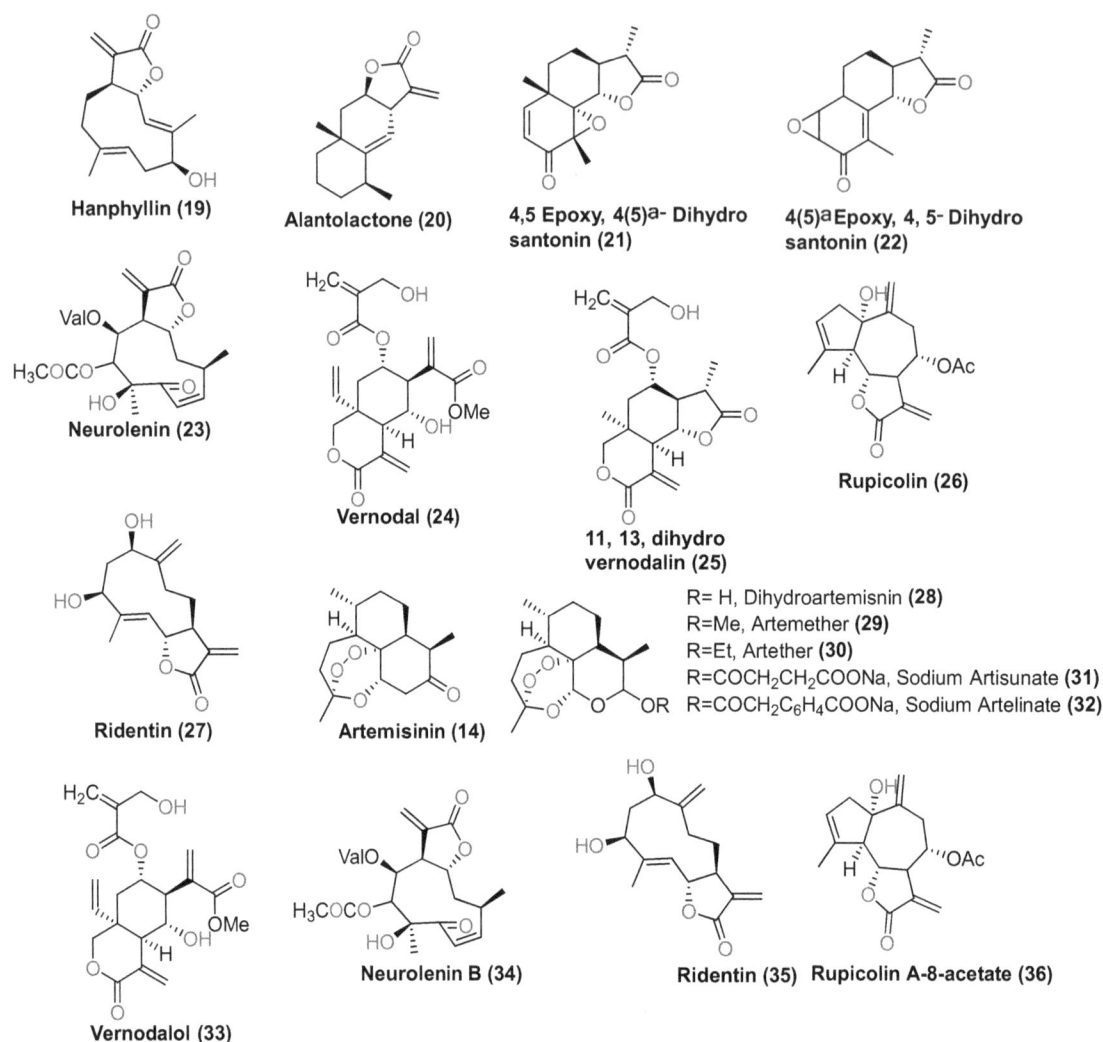

FIGURE 16.3 Diversity in structures of sesquiterpene lactones.

TABLE 16.1

Main Classes of STL

S. No.	Category	Size of Ring	Proto Type	Source
(1)	Guaianolides	5/7-fused bicyclic	Arglabin	*Artemisia myriantha*
(2)	Pseudoguaianolides	5/7-fused bicyclic	Helenalin	*Arnica montana*
(3)	Eudesmanolides	6/6-fused bicyclic	Alantolactone	*Inula helenium*
(4)	Carabranolide	6/3-tricyclic	Carabrol	*Carpesium faberi*
(5)	Xanthanolides	7-membered	Xanthiatin	Xanthium family
(6)	Germacranolides	10-membered	Parthenolide	*Tanacetum parthenium*

they've resurfaced as potential lead compounds. Artemisinin derivatives, artesunate and artemether are now in use, while clinical studies are underway for dimethylamino parthenolide, a parthenolide synthetic counterpart and mipsagargin, a prodrug derived from thapsigargin [4].

16.2 Classification

There are several species of STL and more than 8,000 chemicals have been discovered. The main classes of STL are shown in Table 16.1.

16.3 Structure

Sesquiterpenes are lipophilic molecules that are colorless. Plants synthesise three isoprene units through farnesyl pyrophosphate (FPP) in the endoplasmic reticulum (ER). Sesquiterpenes have a 15-carbon backbone and, although they vary in structure, the majority and most useful forms are cyclic, hence the emphasis of this study will be on these molecules. A large number of sesquiterpene synthases, combined with the fact that a single synthase may produce numerous products and that further modifications after sesquiterpene synthesis, such as oxidation and glycosylation, occur, resulting in a vast number of varied structures. Many similar synthases may produce the same products, in different ratios, which affect a plant's metabolite profile and can be used to classify closely related species or subspecies. The processes are tightly regulated in certain species where sesquiterpenes are generated as a stress response. For example, *Aquilaria sinensis* (Lour.) generates sesquiterpenes solely in reaction to herbivory and 26 unigenes coding for seven enzymes have been identified. The lactone ring of germacranolides type sesquiterpenes is formed by oxidation of the 3C side chain of germacranolides and guaianolides and eudesmanolides are produced from this. Sesquiterpene lactone biosynthesis is well studied, with thorough papers available, among other things. The most typical classes include pseudoguianolides (1), guaianolides (2), germacranolides (3) and eudesmanolides (4), with germacranolides being the most important in terms of their function in humans [5].

16.4 Chemical Characteristics

Sesquiterpene lactones have a fifteen-carbon (C15) backbone, which is mostly cyclic and may be modified to produce a variety of configurations. The existence of a -lactone ring closed toward either C-6 or C-8 is a distinguishing characteristic of STLs. In many circumstances, an exo-methylene group is linked to the carbonyl group in this -lactone. Because the lactone ring may be fused to the remaining skeleton in either a trans or cis configuration, the lactonization stereochemistry can be either or (trans- or cis-fused STLs). The most frequent arrangement is trans and the H-7 of STLs is usually oriented.

In certain STLs, the exocyclic methylene is reduced and the double bond is endocyclic as in artemisinin (14), matricin (37), achillin (38) and santonin (39) (Figure 16.4) [6].

Propionate, isobutyrate, methacrylate, isovalerate, epoxymethacrylate, 2-methylbutanoate, tiglate, angelate, senecioate, epoxyangelate, sarracinate, acetylsarracinate and other related residues are often found in STLs. There have only been a few glycosylated lactones or lactones with halogen or sulphur atoms in their structures identified. STLs also have a cyclopentenone moiety (dehydroleucodine (40), achillin (38)) and α, β unsaturated lactone ring (mikanolide (41), deoxyelephantopin (42) [7].

16.5 Biosynthesis of Sesquiterpene Lactones

16.5.1 Biosynthesis of the Isoprenoid Precursors in Plants

The mevalonate route (*cytosol*) and the 2-C-methyl-D-erythritol 4-phosphate pathway (*cytosol*) are two separate biosynthetic processes that coexist in plants and are responsible for the production of the universal terpenoid precursors Isopentenyl pyrophosphate (IPP) and dimethylallyl diphosphate (DMAPP) (*plastids*). On the other hand, the mevalonic acid (MVA) route, is found in yeast, whilst the Non-mevalonate (MEP) pathway is found in bacteria.

16.5.2 Mevalonate (MVA) Pathway

First, acetoacetyl CoA thiolase catalyzes the condensation of two acetyl-CoA molecules in the MVA (AACT). In the *second* enzymatic step, HMG-CoA synthase forms 3-hydroxy-3-methylglutaryl-CoA. (HMG-CoA). *Third*, HMG-CoA reductase (HMGR) uses two NADPH to convert HMG-CoA to mevalonate (MVA). HMGR is a membrane-bound ER protein that has been studied in numerous species and is assumed to be the rate-limiting step in the MVA pathway. Mevalonate

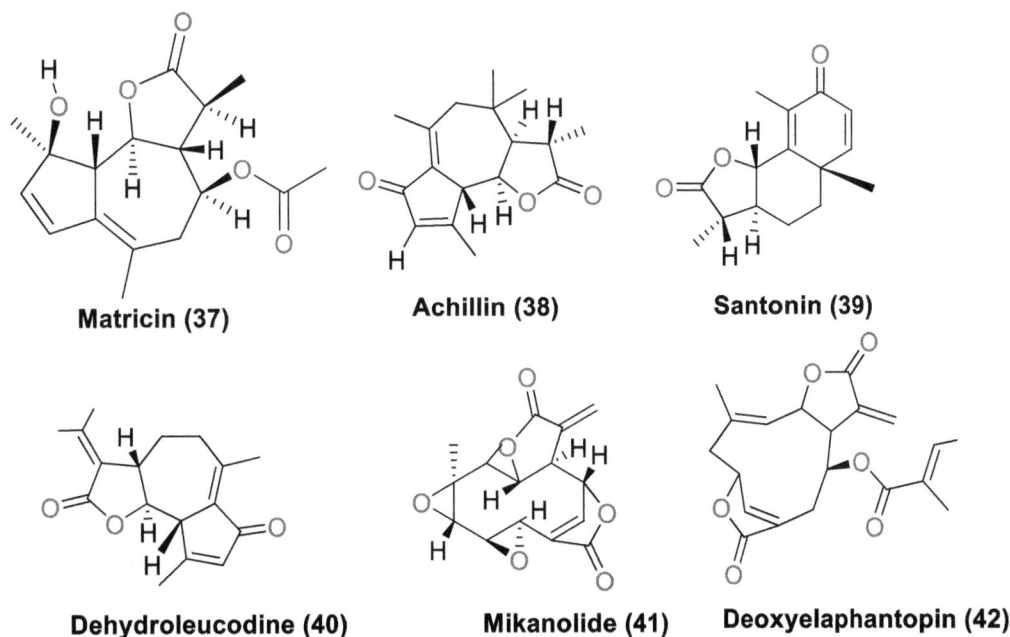

Matricin (37) **Achillin (38)** **Santonin (39)**

Dehydroleucodine (40) **Mikanolide (41)** **Deoxyelephantopin (42)**

FIGURE 16.4 Some examples of sesquiterpene lactones.

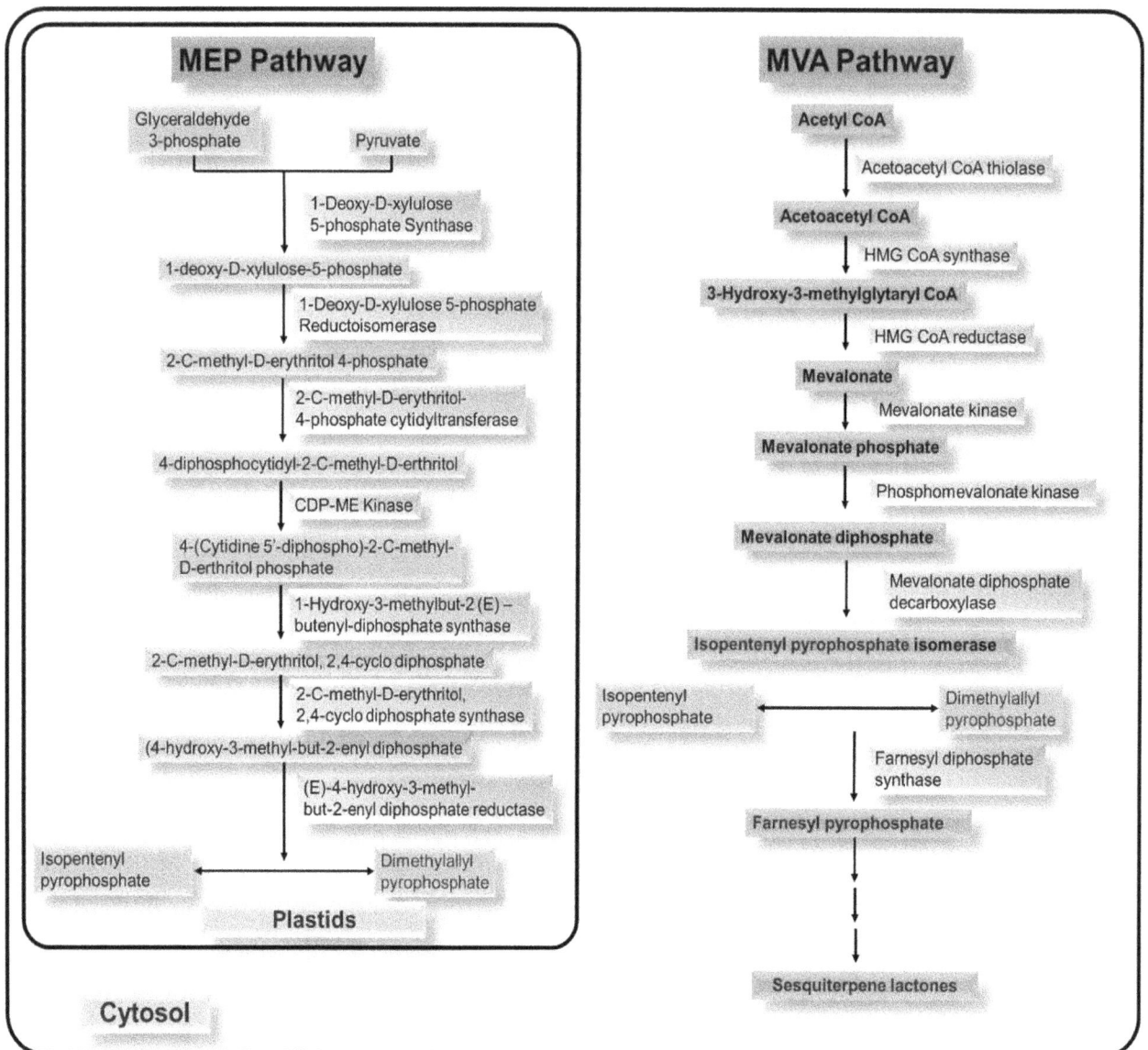

FIGURE 16.5 Biosynthetic pathway of sesquiterpene lactones.

is converted to IPP by mevalonate kinase, phosphomevalonate kinase and mevalonate diphosphate decarboxylase (MVD). IDI transforms IPP to DMAPP in cytosol, chloroplasts and mitochondria (Figure 16.5) [8].

16.5.3 2-C-Methyl-D-erythritol 4-phosphate (MEP) Pathway

There are seven enzymatic processes in the MEP pathway. The condensation of pyruvate and glyceraldehyde 3-phosphate (GA3P) to form 1-deoxy-D-xylulose 5-phosphate (DXP) is catalyzed by the deoxyxylulose phosphate synthase (DXS). The deoxy-D-xylulose 5-phosphate reductoisomerase (DXR) converts DXP to 2-C-methylD-erythritol 4-phosphate (MEP). In the MEP route, DXS and DXR are both considered rate-limiting enzymes. 2-C-methyl-D-erythritol 4-phosphate cytidylyltransferase (MCT), CDP-ME kinase (CMK), 2-C-methyl-D-erythritol 2,4-cyclodiphosphate synthase (MDS) and 1-hydroxy-2-methyl-2-(E)-butenyl-4-diphosphate synthase (HDS) convert MEP to (E (HDS). The (E)-4-hydroxy-3-methylbut-2-enyl diphosphate reductase (HDR) is responsible for the last step, which converts HMBPP into IPP and its isomer DMAPP [4,5,9].

16.5.4 Cross-Talk between MVA and MEP Pathways

The MVA route is thought to supply isoprenoid precursors for the production of sterols, sesquiterpenes and triterpenes, while the MEP pathway is thought to give precursors for the production of monoterpenes, diterpenes and carotenoids. In the synthesis of several terpenoids, despite the spatial isolation of both routes, an interchange of IPP and DMAPP between cytosol and plastids has been documented. Artemisinin integrated 14C-HMGCoA effectively; however, the competitive inhibitor mevinolin reduced up to 83% of the label incorporation into this secondary metabolite.

16.6 Farnesyl Diphosphate Synthase

16.6.1 Branch Point of Sesquiterpene Lactone Biosynthesis

Farnesyl phosphate (FPP) is the common precursor of all STLs, sterols, triterpenes and prenylated proteins and is made from IPP and DMAPP by farnesyl diphosphate synthase (FDS). FPP is a competitive route for isoprenoid precursors since it is a substrate of squalene synthase (SQS), which is the first step in the synthesis of sterols and brassinosteroids. FDS is mostly found in the cytosol and mitochondria of cells. FDS overexpression resulted in greater artemisinin content in A. annua than in non-transformed plants, demonstrating its significance as a rate-limiting step in this process.

16.6.2 Sesquiterpene Lactone Pathway

The cyclization of FPP, which is catalyzed by sesquiterpene synthases (STS), is the initial step in the production of STLs. STS are found mostly in the cytosol and are distinguished by their plasticity, demonstrating the ability to use diverse substrates. The enzyme Santalene synthase (SaSSy) creates a combination of santalenes (α, ß and epi-ß-santalene) and α-exo-bergamotene from sandalwood (Santalum album), which is utilized in the industry for its essential oil smell. At5g44630 and At 5g23960, two STPS-encoding genes in Arabidopsis thaliana, are thought to be responsible for at least 20 distinct STLs found in the

plant's floral volatiles. Germacrene A synthase (GAS), which converts FPP to germacrene A, is one of the most well-studied STS (GA). Many germacranolide-type STLs have GA as a precursor. A. annua, Barnadesia spinosa, sunflower and lettuce have been isolated by gene encoding GAS. The backbone skeleton for the production of costunolides and parthenolides, both of which have anticancer action and are of pharmaceutical relevance. GA is transformed to germacra-1(10), 4,11 (13)-trien-12-oic acid (GAA) by germacrene in a sequence of oxidation processes. A cytochrome P450-like enzyme called oxidase (GAO). A costunolide synthase then oxidises GAA to produce costunolide (7). A parthenolide synthase from feverfew (*T. parthenium*) was recently discovered; the enzyme catalyzes the epoxidation of costunolide molecules' C4–C5 double bonds to produce parthenolide (16), which possesses antimigraine and anticancer properties. As a result, parthenolide (16) production has become the second STL route to be entirely characterized. The antimalarial drug artemisinin, generated by *A. annua*, is the most significant and commercially valuable STL. The biosynthesis route for artemisinin was the first to be thoroughly characterized (Scheme 16.1). The enzyme amorpha-4,11-diene synthase (ADS) catalyzes the first step, which transforms FPP to amorpha-4,11-diene.

A cytochrome P450 monooxygenase (CYP71AV1) and a cytochrome P450 oxidoreductase (CPR) as the native redox partners hydroxylate amorpha-4,11-diene to yield artemisinic alcohol in the following reaction. An alcohol dehydrogenase (ADH1) converts the alcohol to artemisinic aldehyde. Zhang

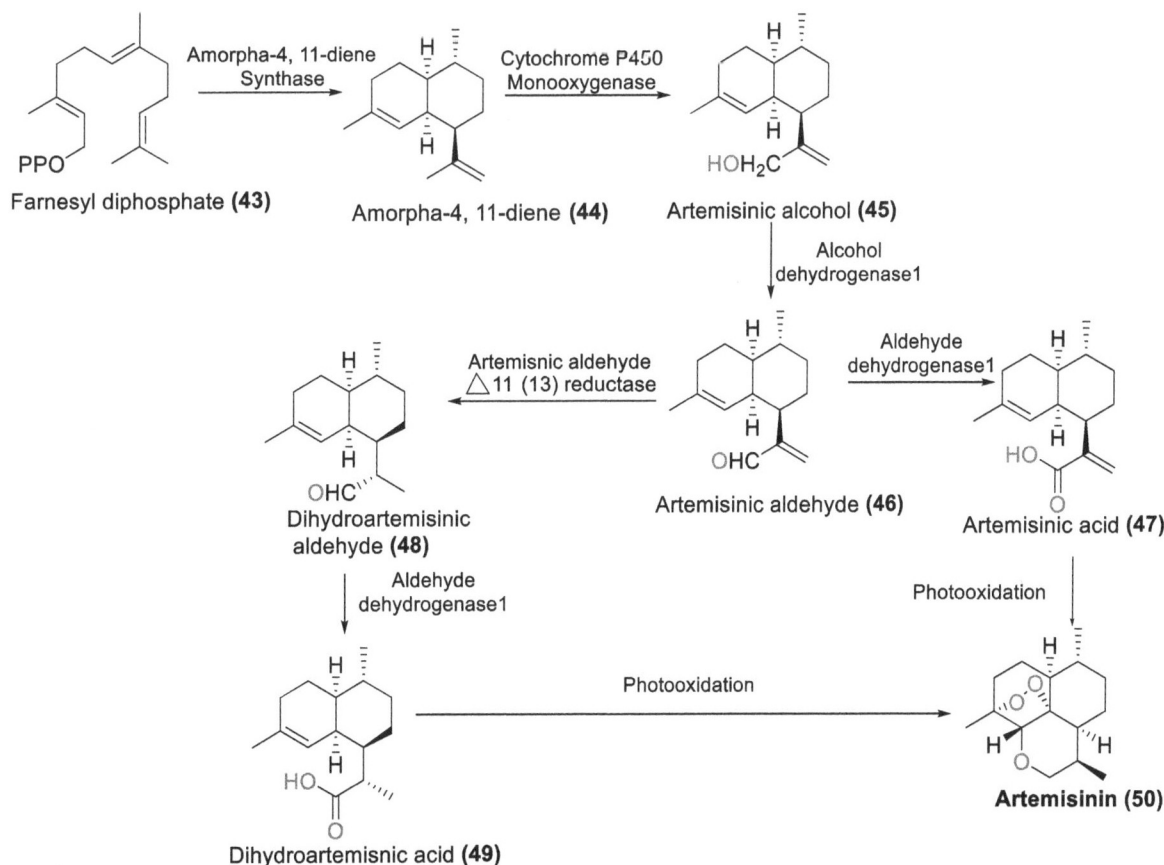

SCHEME 16.1 Biosynthetic pathway in *Artemisia annua*.

et al. discovered that the artemisinic aldehyde 11(13) reductase (DBR2) catalyzes the conversion of artemisinic aldehyde to dihydroartemisinic aldehyde, which is then transformed to dihydroartemisinic acid by an aldehyde dehydrogenase (ALDH1). Dihydroartemisinin acid is the first step in the production of artemisinin. Although this issue is still up for dispute, the reaction seems to be a non-enzymatic photooxidation mechanism. ALDH1 might possibly convert artemisinic aldehyde to artemisinic acid (AA) as an alternate route to artemisinin synthesis. (Scheme 16.1) [10].

16.6.3 Hypothetical Pathway

Despite the plethora of knowledge about Sesquiterpene lactones' structural features and biological activity, little is known about their production. The germacranolides are expected to be the source of the majority of sesquiterpene lactones and (+)-costunolide is thought to be the common intermediate of all germacranolide-derived lactones with a 6,7-lactone ring. Several publications have suggested a biosynthetic route for (+)-costunolide, as shown in Scheme 16.2 [10–14].

Farnesyl diphosphate (FPP), the common precursor of all sesquiterpenoids, is cyclized to the germacrene structure of germacrene A, or alternatively germacrene B, in this hypothesized route (Scheme 16.3). Germacrene A is hydroxylated at the isopropenyl side chain, then oxidized to produce germacrene carboxylic acid. Hydroxylation at the C6 position of germacrene carboxylic acid and subsequent lactonization would provide (+)-costunolide, while hydroxylation at the C8 position would yield (+)-inunolide, a putative intermediary of sesquiterpene lactones with a 7,8-lactone ring. The proposed (+)-costunolide Pathway is based mostly on the isolation of Cope rearrangement and cyclization products of the implicated germacrene intermediates from *Saussurea lappa*, a plant whose roots are exceptionally high in (+)-costunolide (7). However, there is no direct biochemical evidence (e.g., enzyme activity isolation) to support this proposed route [15].

It was discovered that eudesmanes are only created from germacrenes and germacrene 1,10-epoxides, whilst guaianes are only formed from 4,5-epoxides and 7-hydroxy-l(10), 6-cyclodienes after cyclization of a series of model compounds employing homogenized chicory roots as a catalyst (Scheme 16.3) [16,17].

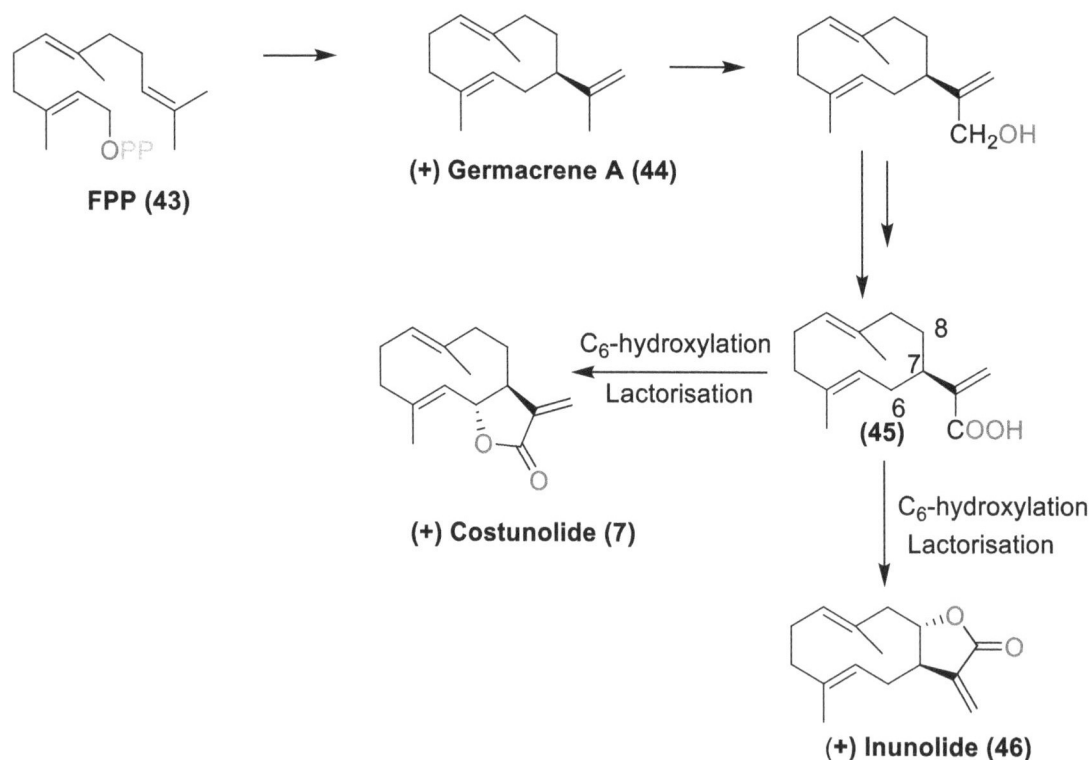

SCHEME 16.2   Proposed pathway for the biosynthesis of (+) Costunolide and (+) Inunolide.

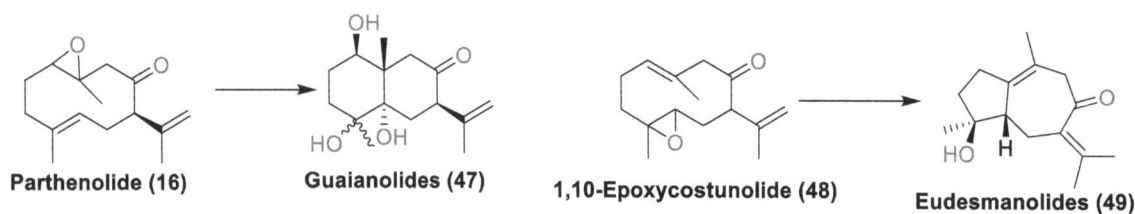

SCHEME 16.3   Cyclization of Germacrene from germacrene 1,10-epoxide and Eudesmanolides.

The role of epoxy-germacranolides information in the synthesis of guaiane and eudesmane lactones has been hypothesized many times in the literature, and it is based on biomimetic reactions: Guaianolides (**47**) would result from 4,5-epoxycostunolide (Parthenolide (**16**)), whereas eudesmanolides (**49**) would result from 1,10-epoxycostunolide (**48**) [13].

16.7 Enzymes of Sesquiterpene Lactone Biosynthesis

16.7.1 Biosynthesis of the Sesquiterpene Framework

Sesquiterpene synthases catalyze the manufacture of sesquiterpene olefins from farnesyl diphosphate (FPP, **43**). The same is presumably true for the biosynthesis of the sesquiterpene framework seen in sesquiterpene lactones, regardless of whether or not a germacradiene is involved. These enzymes are also known as sesquiterpene cyclases since most sesquiterpenes are cyclic and many have numerous ring configurations. Ionization of the diphosphate ester to Gene rate an allylic carbocation that may be attacked by the distal double bond initiates the enzymatic cyclization of FPP, producing the Macrocyclic trans-germacradienyl **50** or humulyl-cation **51** (Scheme 16.4). Before the reaction is halted by deprotonation of the carbocation or the capture of a nucleophile **52**, the carbocation may undergo further electrophilic cyclizations and rearrangements, such as hydride shifts and methyl migrations. Deprotonation of the trans-germacradienyl cation is all that is required to make germacrene A (**44**) or germacrene B (**53**) [18–21].

Ionization of the diphosphate ester of FPP (**43**), separation of charges **54** (Scheme 16.5), leading the formation of **55**, where ring expansion via **56** to **59**, which leads to the biosynthesis of daucanes and **60,** which leads to the biosynthesis of cadinanes, the driving force is stability, whereas the ring closure gives **58** leads the biosynthesis of biabolanes, santalanes and bergamotanes (Scheme 16.5)

16.7.2 Biosynthesis of the Lactone Ring

Biosynthesis of sesquiterpene lactones requires oxidising enzymes, although little is known about these enzymes. The conversion of 5-epi aristolochene (**61**) into capsidiol (**62**) in green pepper (*Capsicum annuum*) and into capsidiol (**62**) and debneyol (**63**) in tobacco (*Nicotiana tabacum*), and the conversion of vetispiradiene (**64**) into lubimin (6 (*Solanum tuberosum*). Far more is known about the oxidising enzymes of monoterpene biosynthesis, especially those involved in oxidising limonene to mint oil's monoterpene alcohols and ketones. Cytochrome P450 enzymes hydroxylate monoterpenes to alcohols, while NAD^+-dependent dehydrogenases oxidise to ketones/aldehydes. Further monoterpene skeleton changes include reduction, isomerization and conjugation (Scheme 16.6) [22].

The dehydrogenases of monoterpene biosynthesis are rather specific enzymes that oxidise an arrow range of related monoterpene alcohols.

In Geranylgeranyl diphosphate pathway, the oxidation of geranylgeranyl diphosphate (**66**) converted into abietadiene (**67**), which is oxidized to abietadienol (**68**) by cytochrome P450 and then into abietadienal (**69**) and finally to abietic acid (**70**) (Scheme 16.7) [23–25].

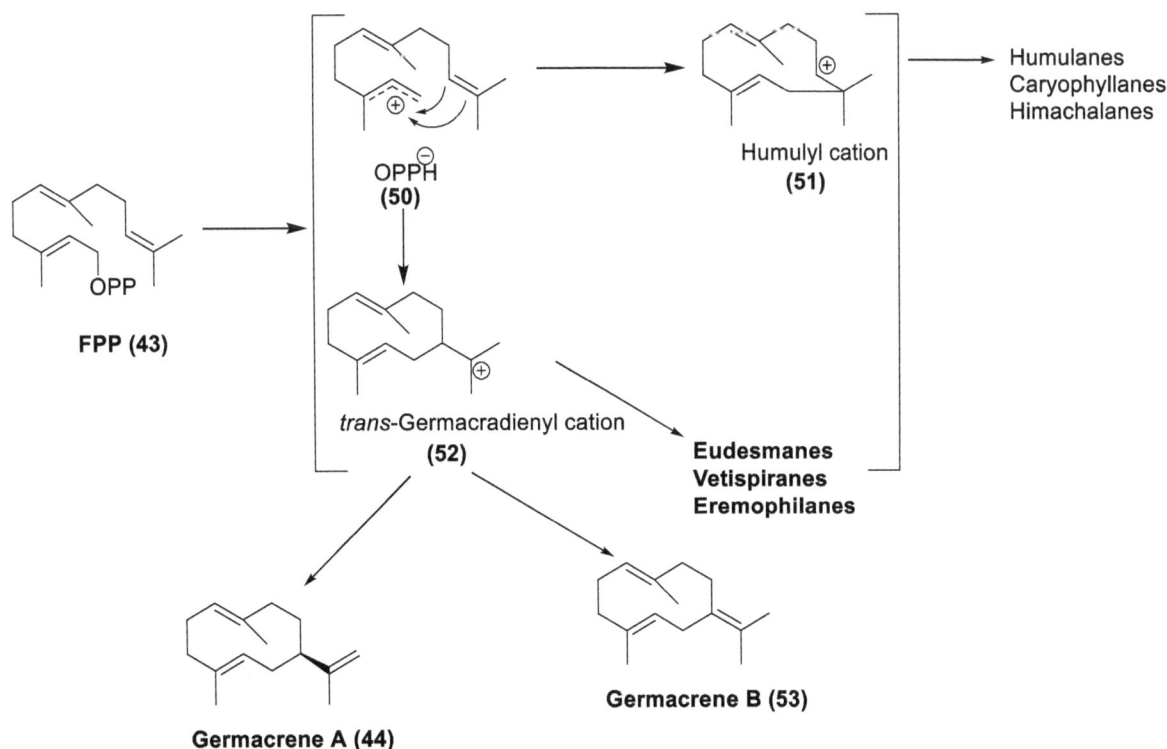

SCHEME 16.4 Sesquiterpene synthase cyclizes Farnesyl diphosphate through humulyl-o-trans-germacardienyl cation.

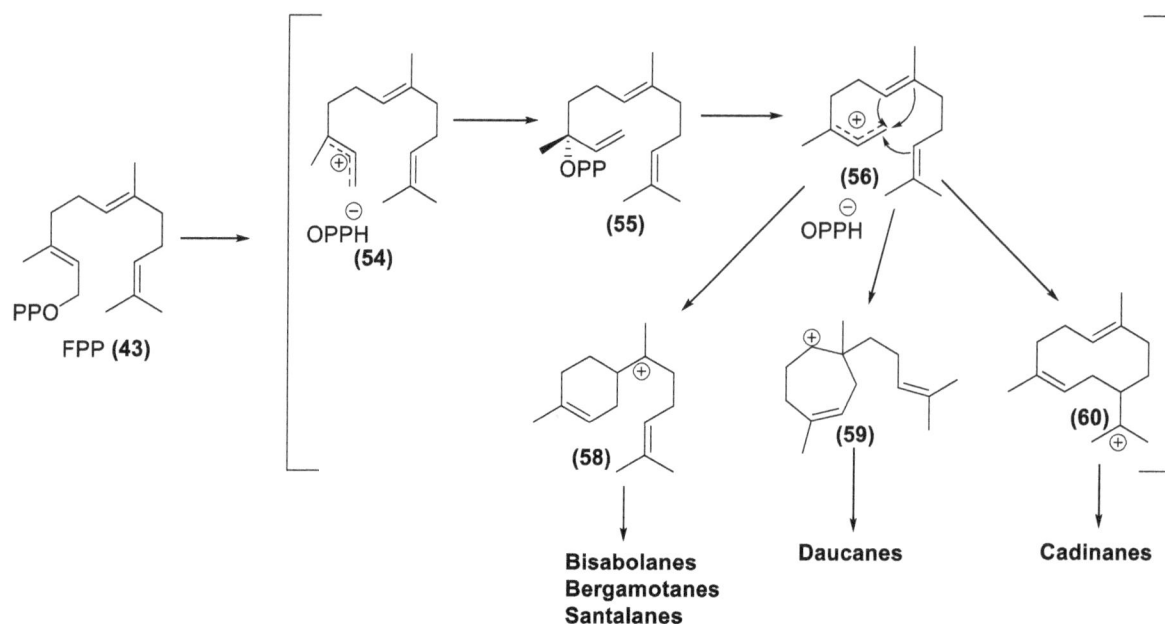

SCHEME 16.5 Cyclization of Farnesyl diphosphate by sesquiterpene synthase via neryl diphosphate.

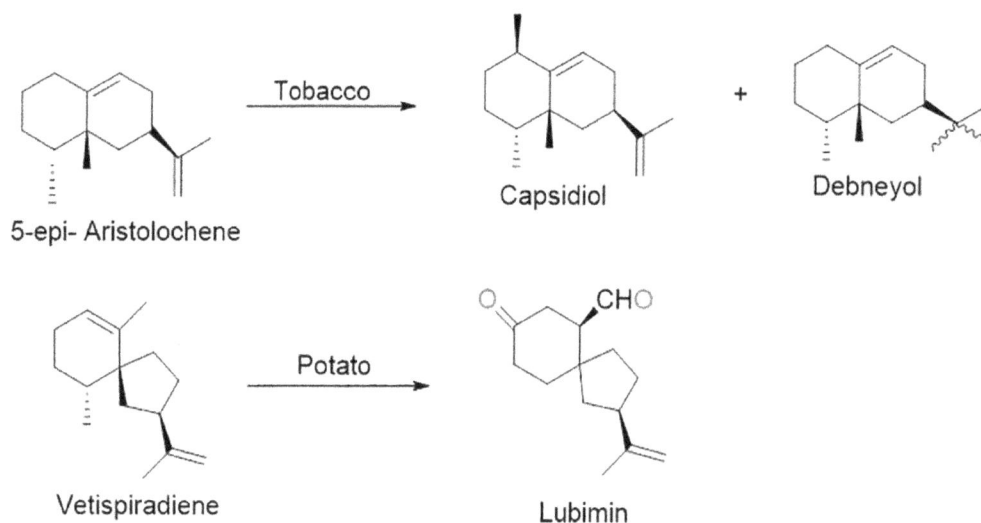

SCHEME 16.6 Oxidation of Sesquiterpenes in Solanaceae.

In conclusion, for the biosynthesis of the lactone ring, the literature suggests that the postulated hydroxylations of the germacrene intermediates at the C6-and C8-position are catalyzed by cytochrome P450 enzymes. Though, further oxidation of the hydroxyl group at C_{12} into a carboxylic acid might either completely be catalyzed by dehydrogenases or also involve cytochromeP450enzymes.

16.7.3 Function of the α-Methylene-γ-Lactone Group

The α-methylene-γ-lactone group (αMγL), an oxygen-containing ring structure with a carbonyl group, is the group mainly responsible for biological effects in humans such as tumour therapy and blood pressure reduction. This is due to

its alkylation power operating on transcription factors and enzymes in the human body, creating steric and chemical alterations and compromising the target's capacity to operate properly. Lactonization occurs when C12 is oxidized to generate the carbonyl function, followed by hydroxylation at C6 or C8, resulting in the development of the ring structure. This ring is responsible for, and is often thought to be crucial for, the cytotoxic effects of its contained compounds, with other functional groups in the structure just changing the ring's potency via steric and chemical impacts. The αMγL unit, which exerts its impact via alkylation of thiol groups present in proteins, is responsible for most of the functioning of sesquiterpene lactones [26–28].

This explains the predominance of contact dermatitis caused by Asteraceae members, as well as cell wall destruction when

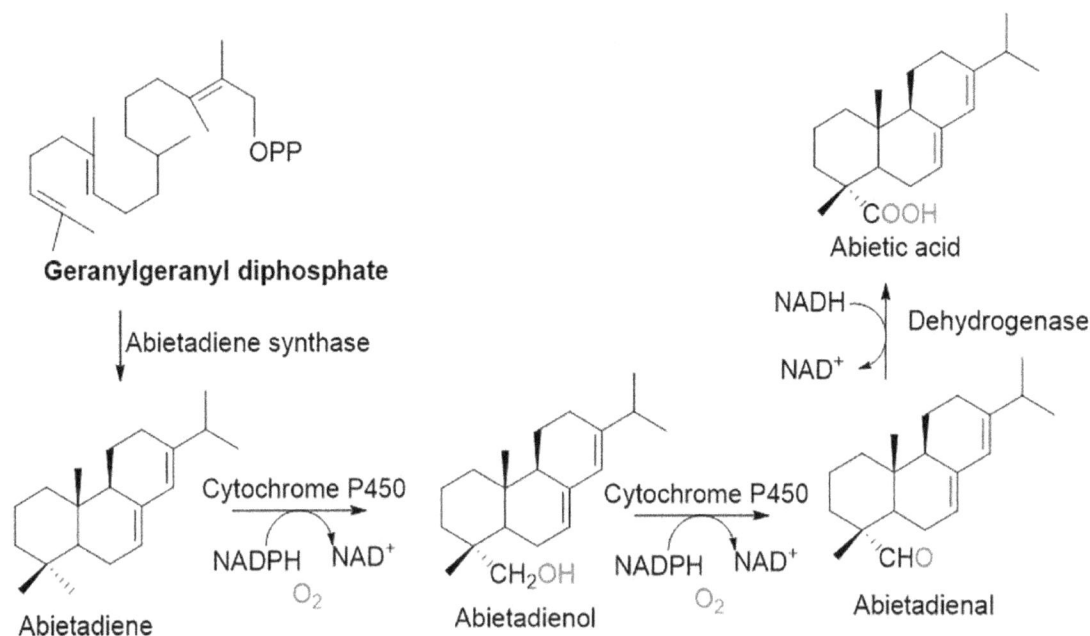

SCHEME 16.7 Pathways for the biosynthesis of abietic acid Geranylgeranyl diphosphate (GGPP).

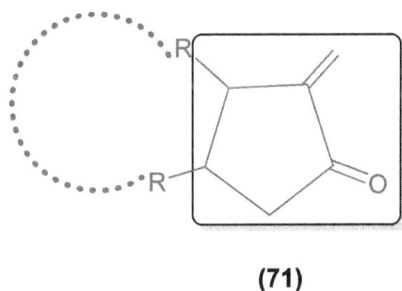

(71)

FIGURE 16.6 General structure of α-methylene-γ-lactone.

used as an antibacterial agent, which will be described more below. The αMγL group is also in charge of Michael reaction changes, which result in a variety of consequences such as gene expression control by activating and deactivating transcription factors, modifying expression with the effect of sensitising cancer cells and other activities. On the basis of the efficacy of sesquiterpene lactones containing, -unsaturated ketones instead of α-methyl-γ-lactones in relation to tumour cell death induction, it has been suggested that the functional unit is the unsaturated carbonyl O=C–C=CH₂ moiety present within the αMγL group, rather than the lactone itself, but more recent studies have moved away from this focus. The same research discovered that compounds having both an αMγL and an, -unsaturated ketone were the most cytotoxic. Other functional groups are thought to boost the activity of the αMγL group by chemical or steric processes. Because heliangolides have higher conformational flexibility than other structures, Macias et al. discovered that they are more effective at slowing plant growth than guaianolides. Large residues are known to inhibit activity, while smaller flexible germacranolides and compounds containing an OH or O-acyl group next to the αMγL are likely to be more efficient. The lactone's cis or trans

arrangement has not been shown to have a substantial influence on cytotoxicity [27–31]. The antimalarial effects of artemisinin are attributed to the peroxide bond, which is considered to trigger the production of ROS in the body, which may work in a variety of ways to kill or incapacitate the Plasmodium parasite that causes malaria (Figure 16.6).

16.7.4 Sesquiterpene Lactone Synthesis

In the synthesis of sesquiterpene lactones with diverse skeleton topologies, the generation of the α-methylene- γ-lactone moiety and the assembly of the core skeleton are two important processes. Various schemes will go over the various methods for creating the ten-membered germacrene carbocycle and the hydroazulene skeleton that is characteristic of the guaiane sesquiterpenoid.

Germacranolides are a subclass of germacrene sesquiterpene lactones with a ten-membered carbocyclic skeleton fused to a γ-lactone. Synthetic Strategies for the Germacranolide Skeleton in General Germacranolides are a subclass of germacrene sesquiterpene lactones with a ten-membered carbocyclic skeleton fused to a γ-lactone. Germacrene intermediates are utilized to create a wide range of cyclic terpenes, which have been proven to have a variety of biological roles. Because germacrene precursors are plentiful, this important intermediate can be converted into other sesquiterpene subclasses including eudemanes and guaianes. Despite decades of research into the whole synthesis of germacranolides, the synthesis of the ten-membered carbocyclic core has remained a challenge. The cyclodecadiene core is broken down when exposed to acidic, basic and high-temperature conditions, resulting in cyclized and rearranged fragmented products. Furthermore, at room temperature, germacranolides exist as conformer isomers, creating a challenge for purification and analysis [32,33]. There are four phases in the typical synthesis technique for germacrene lactones:

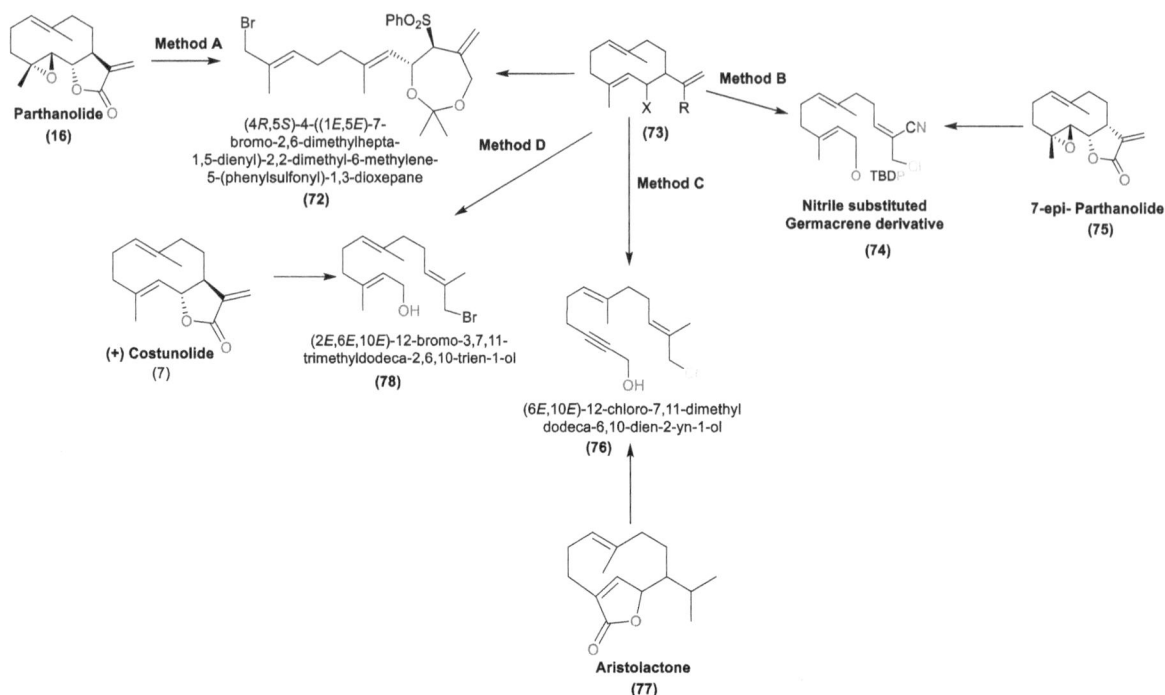

SCHEME 16.8   Synthetic strategy for germacranolides' ten-membered core skeleton.

a. Medium-sized ring synthesis through direct synthesis,

b. Regio- and stereoselective placement of the (E) and (Z) bonds on a ten-membered ring system,

c. The introduction of functional groups in a highly selective manner and

d. Synthesis of the -methylene-lactone group with efficiency.

The synthesis of germacranolides' ten-membered core skeleton has been described using a number of approaches, as shown in Scheme 16.8. **Method A** was used to make parthenolide (**16**), which starts with intramolecular alkylation of a sulfone derivative (**72**) and proceeds through the sesquiterpene lactone biosynthetic pathway. **Method B**, which begins with an intramolecular **Barbier-type** reaction involving compound (**73**) and finishes with the -methylene—lactone nitrile germacrene system **74**, was used to synthesise 7-epi-parthenolide (**75**). The germacrene skeleton was generated in the synthesis of aristolactone (**77**) using **Method C**, which involves the cyclization of the chloro alcohol precursor (**76**). Finally, in **Method D**, the bromo alcohol cyclization precursor was used to build the important ten-membered ring skeleton (**78**), which was used to make costunolide (**7**). In a different method, this crucial intermediate allows access to a diverse family of members, including selinanes, guaianes and elements [15].

16.7.5 Synthetic Strategy of Parthenolide

Parthenolide (**16**) is the active ingredient identified from fever few, a traditional herbal. It belongs to the sesquiterpene lactone class of natural compounds. In the presence of normal hematopoietic stem cells, parthenolide has been shown to

target leukaemia stem cells with remarkable selectivity. The biosynthetic route of sesquiterpene lactones was used to create parthenolide in an asymmetric total synthesis [34–39].

The starting components for this synthetic pathway are an aldehyde and a, a β, γ-unsaturated chiral sulfonylamide, from which the desired product is synthesized via a sequence of chemical transformations [40–42]. The ten-membered carbocyclic ring intermediate of parthenolide (**16**) was synthesized using approach A, as indicated in Scheme 16.9. Analyzing retrosynthetic process, compound **16** was predicted to be made from the 6,7-trans-germacrane ring system (**79**), which could be made from the sulfone (80) via an intermolecular -alkylation. The aldol reaction between aldehyde (**82**) and β, γ—unsaturated sulfonyl amide (**83**) yielded sulfone (**80**), which may be controlled by the aldol reaction [43,44].

Scheme 16.10 shows the procedures involved in obtaining parthenolide (**16**). The initial step was to make the unsaturated sulfonyl amide (**84**), which was done in two steps via a *Horner–Wadsworth–Emmons* reaction with a ketone and diethylphosphonoacetic acid. In the presence of titanium tetrachloride ($TiCl_4$) and di-isopropylethyl amine (i-Pr_2NEt) in dichloromethane (CH_2Cl_2), the aldol reaction between compound (**85**) and aldehyde (**86**) yielded compound (**87**) with the expected 6,7-stereochemistry as the primary result. The acetonide was obtained by selectively cleaving the tert-butyldimethylsylyl ether (TBS) protecting group of compound (**87**) with HCl in ethanol at 0°C followed by 2-methoxypropene treatment (**88**). A reduction reaction was then used to produce the thioether product (**89**), which was then treated with diphenyl disulfide/tri-n-butylphosphine. Compound (**89**) was oxidized in tert-butanol (t-BuOH) and pyridine using hydrogen peroxide/ammonium heptamolybdate ($H_2O_2/(NH_4)_6Mo_7O_{24}$) (**90**). Tetrabutylammonium Zuoride was used to remove the

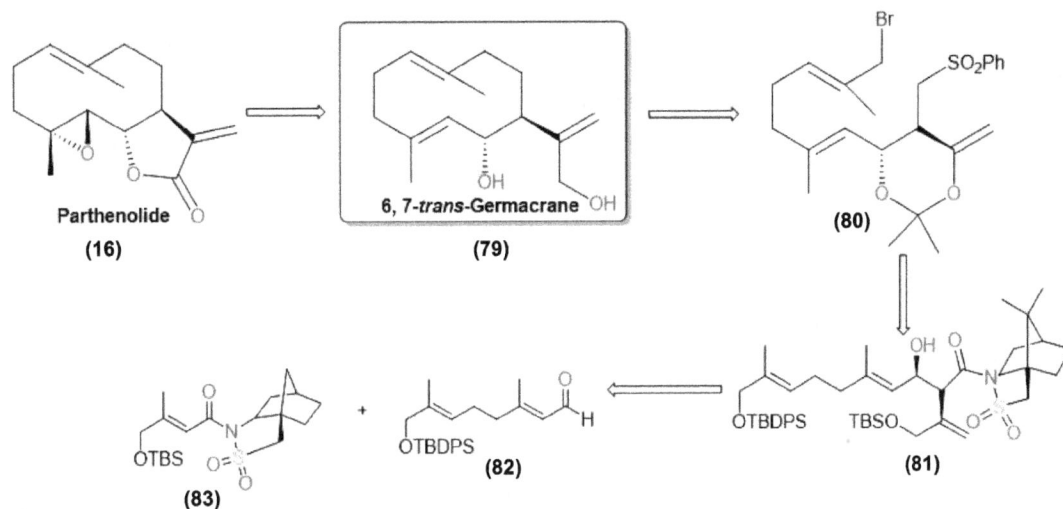

SCHEME 16.9 Retro-analysis to the synthesis of parthenolide

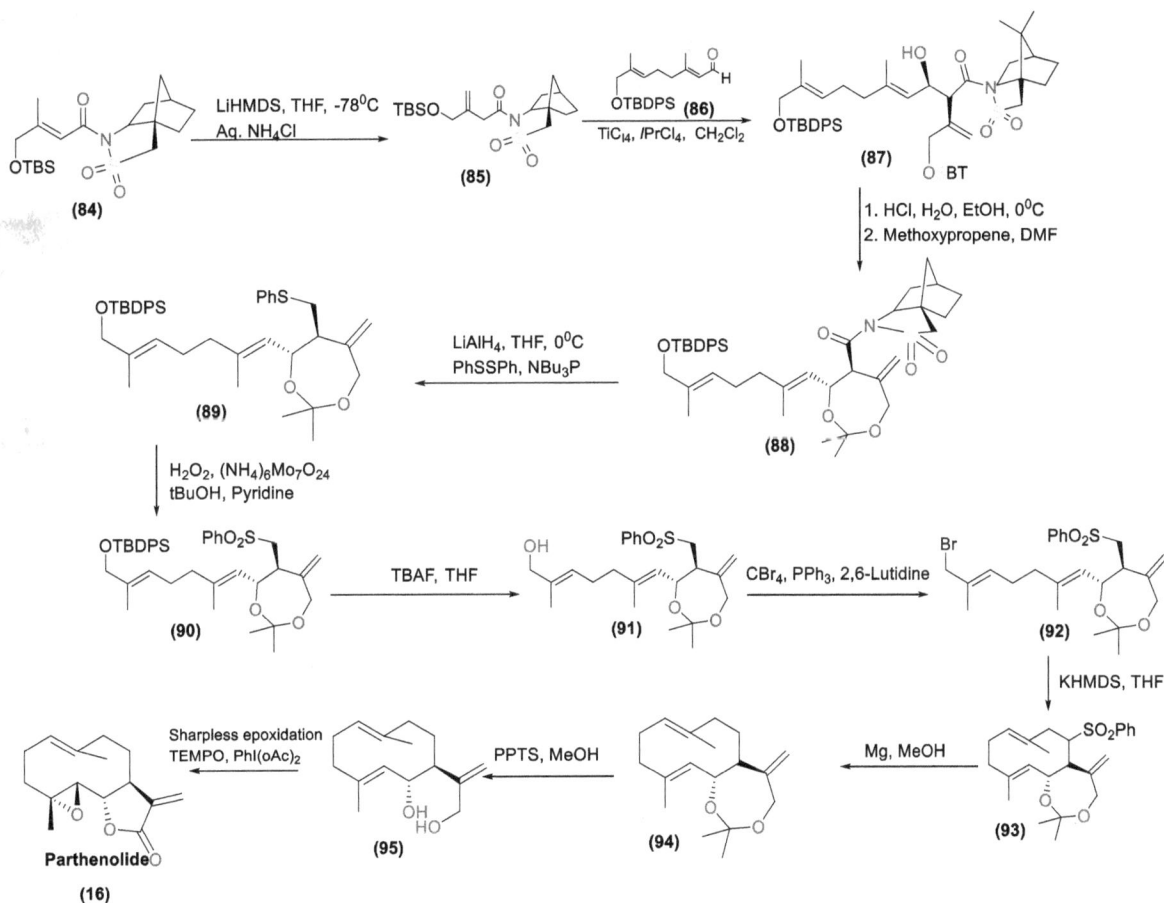

SCHEME 16.10 Synthesis of Parthenolide.

tert-butyldiphenylsilyl ether (TBDPS) group from chemical (**90**), resulting in alcohol (**91**).

At 0°C, tetrabromomethane (CBr$_4$), triphenylphosphine (PPh$_3$) and 2,6-lutidine were used as bases to convert alcohol (**91**) to its corresponding brominated molecule (**87**). The intended cyclized product (**93**) was obtained by treating

this chemical (**92**) with four equivalents of potassium bis-(trimethylsilyl) amide (KHMDS). After that, the sulfone moiety on the cyclized product (**93**) was eliminated by adding magnesium/methanol (Mg/MeOH), resulting in product (**94**). The use of pyridinium p-toluenesulfonate in methanol as a reagent to remove the acetonide group of (**94**) and form the required ten-membered carbocyclic germacrene ring intermediate has proven to be successful (**95**). Parthenolide (**16**) was obtained by Sharpless epoxidation of diol (**95**) followed by oxidation with 2,2,6,6-tetramethyl-1-piperidinyloxy (TEMPO) and (diacetoxyiodo)benzene (PhI(OAc)$_2$) [45,46].

16.8 Parthenolide Semi-synthesis

Large and complex compounds obtained from natural sources are frequently used as starting materials in the semi-synthesis strategy. When the precursor molecule comprises a structurally complex component that is either too expensive or too difficult to synthesise via complete synthesis, this approach comes in handy. The simplest technique for synthesising a complicated natural product is to start with molecules that already have the requisite germacranolide skeleton and then synthesise the target molecule through a sequence of chemical changes. Parthenolide (**16**) is synthesized using a protection-free technique using the natural product costunolide (**7**) as a starting material (Scheme 16.11). Because this germacranolide is easily extracted from the roots of *Saussurea lappa*, costunolide (**7**) has been discovered to be an excellent substrate for the synthesis of parthenolide. To obtain the crucial germacrane intermediate, costunolide (**7**) is treated with diisobutylaluminum hydride (DIBAL) in toluene at room temperature (**96**). The protected primary alcohol (**97**) was obtained in good yield

by treating (**96**) with tert-butyldimethylsilyl chloride (TBSCl). Compound (**98**) was obtained via selective epoxidation of the C4–C5 bond of compound (**97**) in CH$_2$Cl$_2$ at room temperature with titanium isopropoxide (Ti(Oi-Pr)$_4$), (-)-diisopropyl D-tartrate (D-(-)-DIPT) and tert-butyl hydroperoxide (TBHP). Parthenolide (**16**) was obtained by deprotecting compound (**99**) and then oxidising it with *TEMPO* and PhI(OAc)$_2$.

16.8.1 General Strategies for the Synthesis of Guaianolides

Guaianolides are a class of biologically active sesquiterpenes that includes a wide range of compounds. They're frequently employed to create new active compounds as scaffolds. The primary skeletal backbone of guaianolides is made up of a 5,7,5-ring system that may be found in the guaian-6,12-olide (**100**) and guaian-8,12-olide (**101**) structures. The skeleton corresponding to the seco-guaianolides (**102**), in which a C–C single bond is broken in one of the rings, is the least common (Figure 16.7).

Ring rearrangement or enlargement, as well as intramolecular and intermolecular cycloaddition, have all been developed as synthetic techniques for the synthesis of the hydroazulene skeleton. One of the most frequent synthetic procedures for generating the hydroazulene skeleton of guaianolides is to use rearrangement processes. The synthesis involves two important steps:

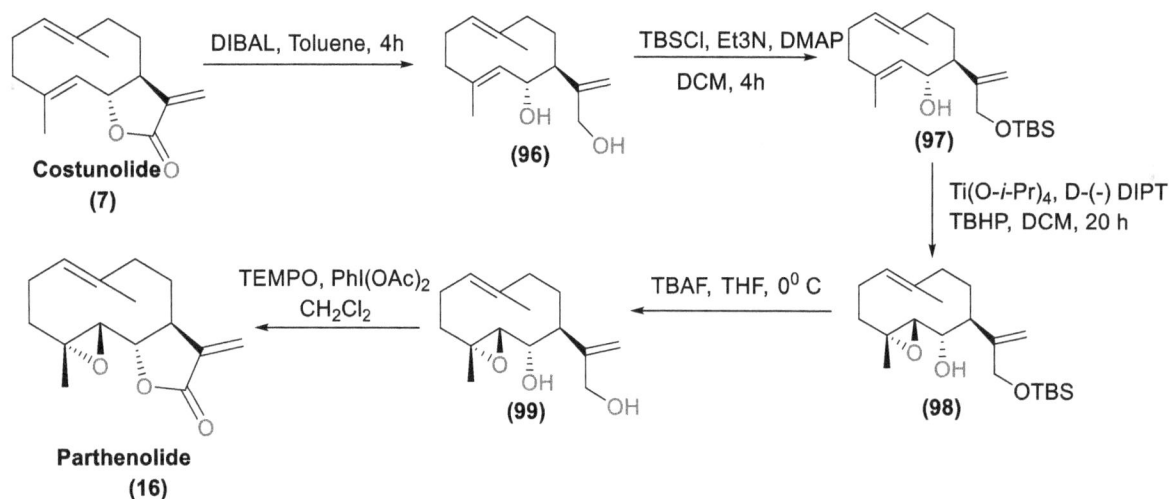

SCHEME 16.11   Semi-synthesis of Parthenolide.

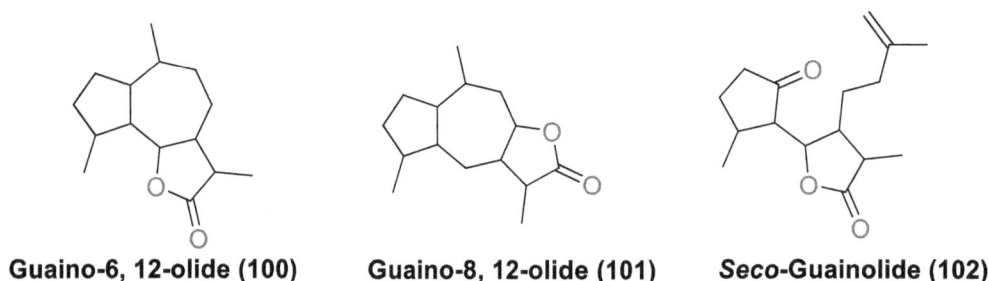

Guaino-6, 12-olide (100)   **Guaino-8, 12-olide (101)**   ***Seco*-Guainolide (102)**

FIGURE 16.7   Central skeletal of guaianolides.

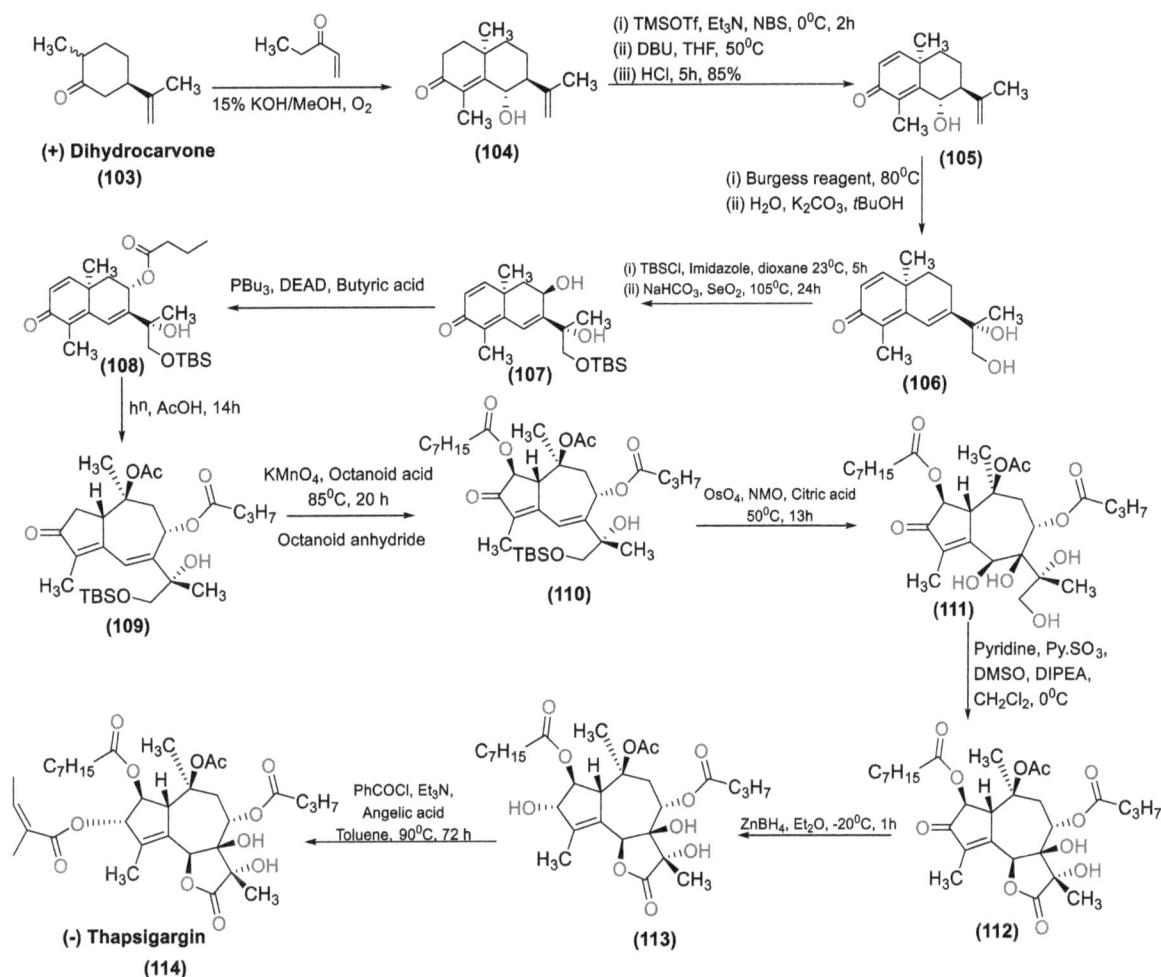

SCHEME 16.12 Synthesis of (-) Thapsigargin.

1. The use of the traditional photosantonin rearrangement.
2. Adding multiple oxygen atoms to the guaiano-lide skeleton to achieve scalable total synthesis of (-)-thapsigargin (**114**). (Scheme 16.12) [47].

Thapsigargin (**114**), a powerful inhibitor of the sarco-endoplasmic reticulum Ca^{+2} ATP-dependent pump protein, has shown to be a promising contender in a variety of medical fields.

The *Robinson* annulation and subsequent -hydroxylation between (+)-dihydrocarvone (**103**) and ethyl vinyl ketone yield decalin, which is used to generate the target molecule's skeleton carbons (**104**). The dienone (**105**) was obtained from a one-pot gramme scale bromination/ elimination sequence of (**104**). The chemo- and diastereoselective dihydroxylation of the terminal olefin with AD-mix followed by the treatment of compound (**105**) with the *Burgess* reagent yielded diol (**106**) in a good yield. The allylic alcohol (**107**) was diastereoselectively synthesized by first selectively protecting the main alcohol and then in situ allylic C–H oxidation with selenium dioxide (SeO$_2$). The *Mitsunobu inversion* with butyric acid was then used to install the butyrate with the required stereochemical configuration at C–8. The ring enlargement was achieved by using a Hg lamp to irradiate (**108**) in glacial acetic acid. The essential guaianolide skeleton intermediate (**109**) had good

stereoselectivity thanks to this gram-scale technique. The oxidation of (**109**) with potassium permanganate (KMnO$_4$) in the presence of octanoic acid and octanoic anhydride in toluene under reflux conditions resulted in the octanoylated enone (**110**). The tetra-ol was then produced using *Upjohn's* modified process, which uses citric acid at 50°C (**111**). To allow for the lactonization of (**111**), *Parikh–Doering* conditions were used (**112**). The reduction of (**113**) with zinc borohydride was followed by acylation with angelic anhydride and benzoyl chloride as the final step of their entire synthesis. In 11 stages, the ultimate product was (-) thapsigargin (**114**) [17–19].

16.9 Synthesis of Guaianolides

A novel method for synthesising highly oxygenated 6,12-guaianolide (**120**) derivatives is shown in Scheme 16.13. To obtain the allenyl ester, the monoprotected butynediol undergoes a *Johnson–Claisen* rearrangement (**115**). After that, the ester (**115**) is treated with methoxyethyl amine hydrochloride and *i*-PrMgCl to produce the *Weinreb* amide, which is subsequently converted to the alkynone (**116**) by treating it with ethyl magnesium bromide. The alkynoate is then formed by reducing the carbonyl group of alkylynone with

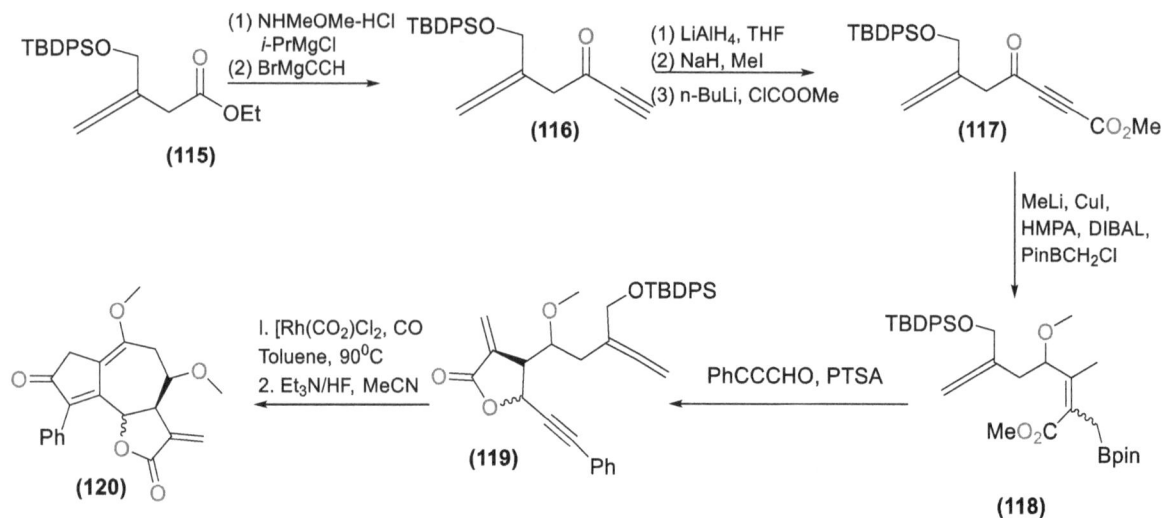

SCHEME 16.13 Synthesis of 6, 12 Guainolide derivatives.

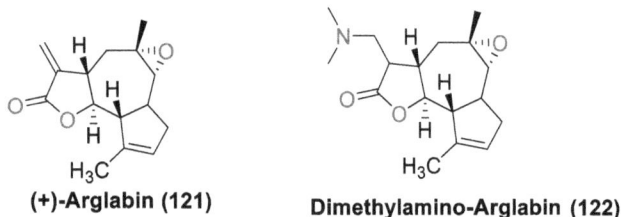

FIGURE 16.8 Few naturally occurring sesquiterpene lactones.

lithium aluminium hydride, forming the appropriate methyl ether, deprotonating the terminal alkyne with n-butyl lithium and finally adding chloromethyl ester (**117**). Alkyl boronates were produced by reacting (**117**) with diisopropylaluminum hydride (DIBAL), copper (I) iodide (CuI), methyl lithium and ClCH$_2$BP (**118**). After heating the alkyl boronate (**118**) mixture with 3-phenylpropiolaldehyde for an allylboration/lactonization process, a complex mixture was produced. After treatment with p-toluenesulfonic acid, the lactone (**119**) was produced (PTSA). The cyclocarbonylation product (**120**) was obtained after treatment with rhodium biscarbonyl chloride dimer and elimination of the tert-butyldiphenylsilyl ether (TBDPS) (Scheme 16.13) [49,50].

Arglabin (**121**) is another important member of one of the largest classes of naturally occurring sesquiterpene lactones (Figure 16.8). Artemisia glabella produces this natural product, which has been shown to be a strong farnesyl transferase inhibitor with promising anticancer activity and cytotoxicity against human tumour cell lines. Arglabin (**121**) has been effectively utilized for the treatment of colon, breast, ovarian and lung malignancies after being transformed into its dimethylamine hydrochloride adduct (**122**).

The first enantioselective synthesis of arglabin (**121**) utilising furan derivatives as starting materials is shown in Scheme 16.14. Starting with methyl-2-furoate, a two-step procedure utilising Cu1-catalyzed asymmetric cyclopropanation followed by ozonolysis yielded diastereo- and enantiomerically pure cyclopropanecarbaldehyde (**123**) in its diastereo- and enantiomerically

pure form. A selective methyl cuprate addition and Ni(II)-catalyzed cross-coupling with trimethylsilyl-methylenemagnesium chloride were used to make the chiral trans-substituted allylsilane (**124**) from furfuryl alcohol. When these two compounds were mixed, (**125**) was formed with excellent stereocontrol, with the carbonyl group of (**123**) being attacked by allylsilane (**124**) from the face opposite its methyl group, as per the *Felkin-Anh* paradigm. The addition of a base triggered saponification of the labile oxalic ester in (**125**) and subsequent lactonization provided the lactone-aldehyde (**126**). Compound (**126**) was subjected to a *Hosomi–Sakurai* allylation with 2-methylallylsilane followed by acylation to produce diene (**127**) of diastereomers. A ring-closing metathesis in the presence of a Grubbs second-generation catalyst produced the required guaianolide skeleton (**128**). PMB deprotection was used to make the homoallyl alcohol (**129**). Using catalytic quantities of vanadyl acetylacetonate (VO(acac)$_2$) and t-butyl hydroperoxide (TBHP) as the stoichiometric oxidants and the free hydroxyl group as a directing group, the desired -epoxide (**129**) was obtained. In the presence of pyridine, exposure of epoxide (**130**) to tri-Zuoromethanesulfonic anhydride (Tf$_2$O) produced alkene (**131**) as a single regioisomer. The deoxygenated product was obtained through acetate deprotection using the *Barton–McCombie* technique (**132**). The exo-methylene group, which is responsible for sesquiterpene lactones' biological activity, is introduced in the final stage of this synthesis. A dimethylamino arglabin derivative was obtained by alkylation of (**132**) with *Eschenmoser's* salt (**133**). The target compound (+)-arglabin (**121**) was obtained by quaternization with methyl iodide and removal of trimethylamine [51,52].

16.9.1 Semi-synthesis of Guaianolides

The abundant natural substance parthenolide (**16**) has recently been used to develop a biomimetic semi-synthesis of arglabin (**121**) (Scheme 16.15). Micheliolide (**134**) was obtained by treating parthenolide (**16**) with p-toluenesulfonic acid (p-TSA) in high quantities. The required -epoxide (**135**) was obtained as a single stereoisomer by epoxidating the double bond on the

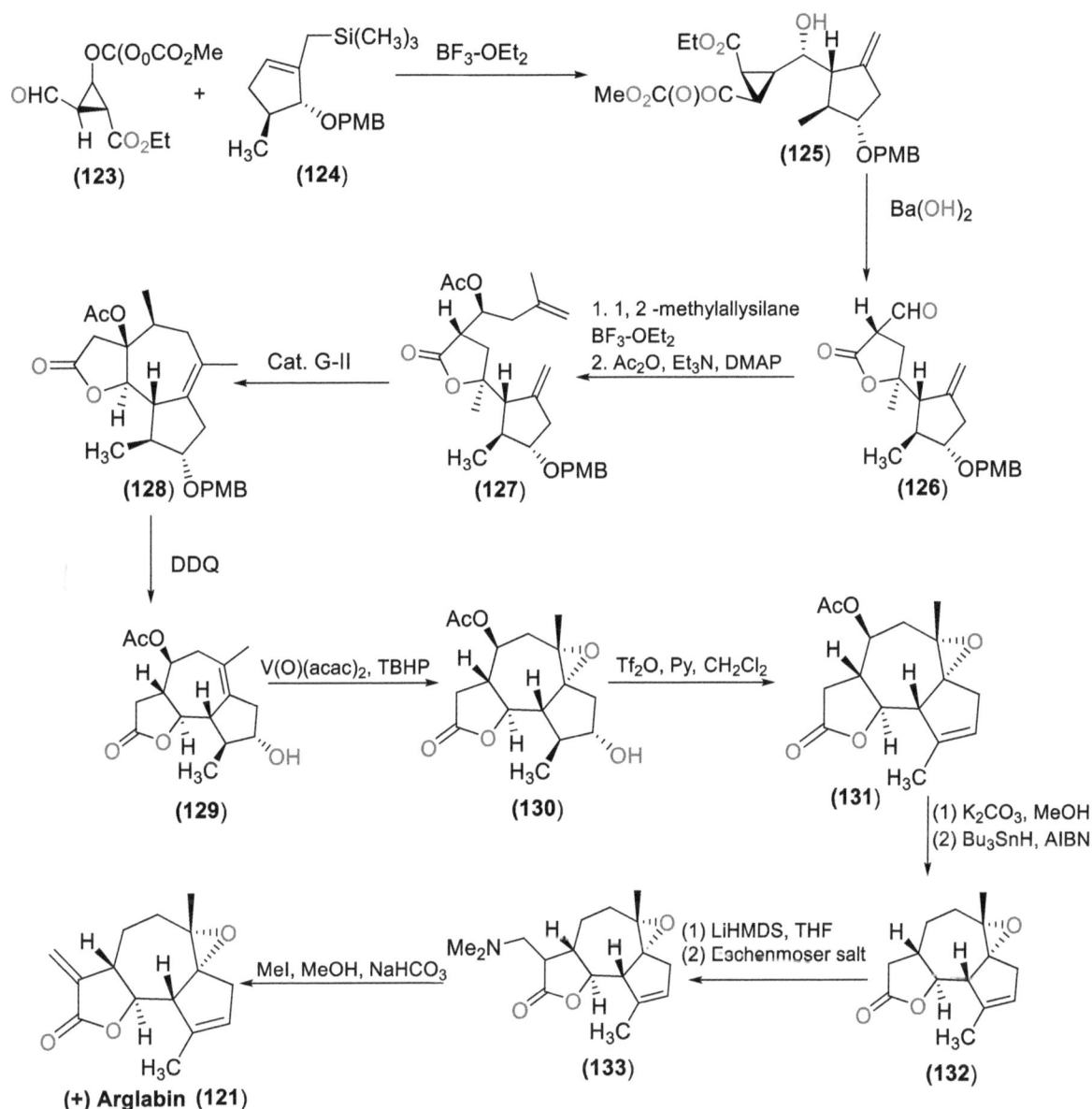

SCHEME 16.14 Total synthesis of arglabin.

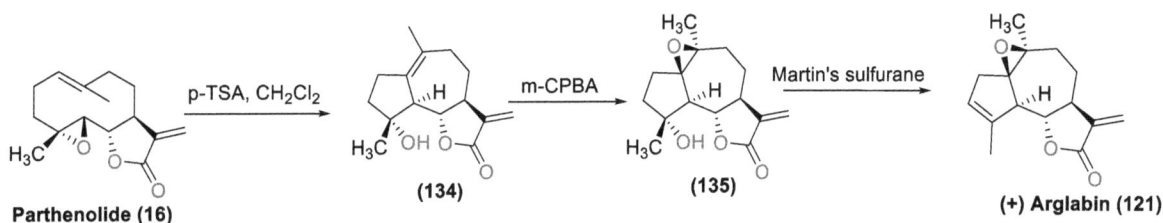

SCHEME 16.15 Synthesis of arglabin from parthenolide.

seven-member ring of chemical (**134**) with m-chloroperbenzoic acid (*m*-CPBA). Because epoxidation from the top face must overcome the steric effects of the upward methyl group, this reaction exhibits a strong stereoselectivity. In the epoxidation of homoallylic alcohols, the hydroxyl substituent can also act as a guiding group. The dehydration of (**135**) with *Martin's sulfurane* in CH$_2$Cl$_2$ yielded arglabin in the next step (**121**) [53–55].

Ludartin (**136**) was also used as the starting point for the semi-synthesis of arglabin (**121**) (Scheme 16.16). The stereoselective ring opening of the epoxide in ludartin (**136**) was achieved using BF$_3$•Et$_2$O in a 1:1 dioxane-water mixture, yielding compound (**137**) with good stereoselectivity. The epoxidation of its C(1)10 double bond with *m*-CPBA produced a variety of diastereomers, which were isolated using column

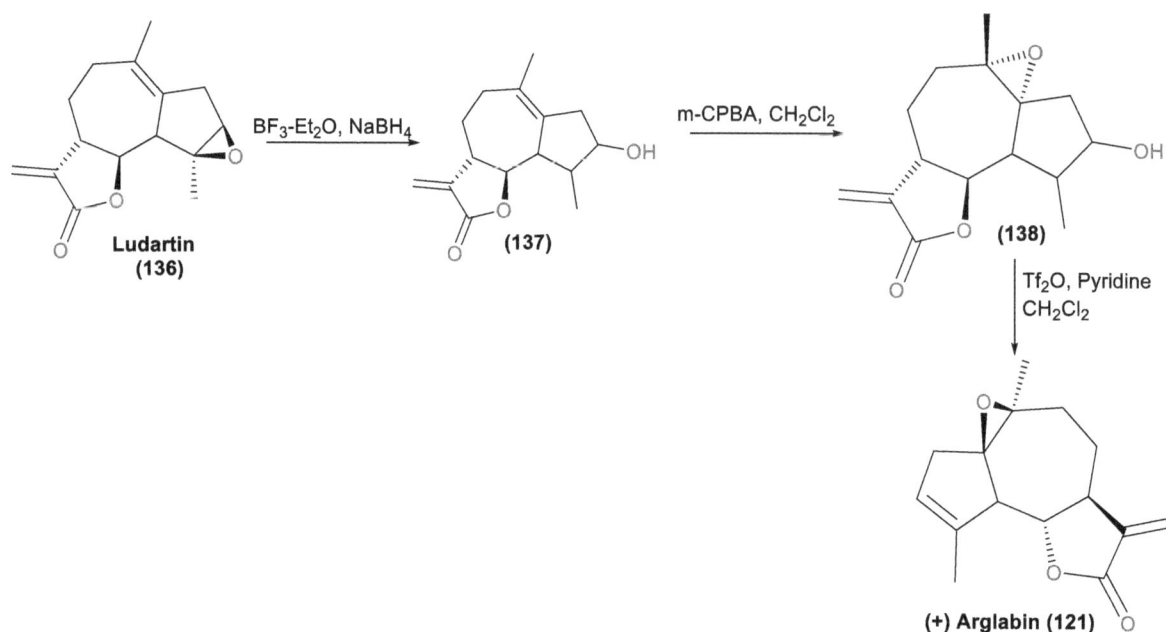

SCHEME 16.16 Synthesis of arglabin from ludartin.

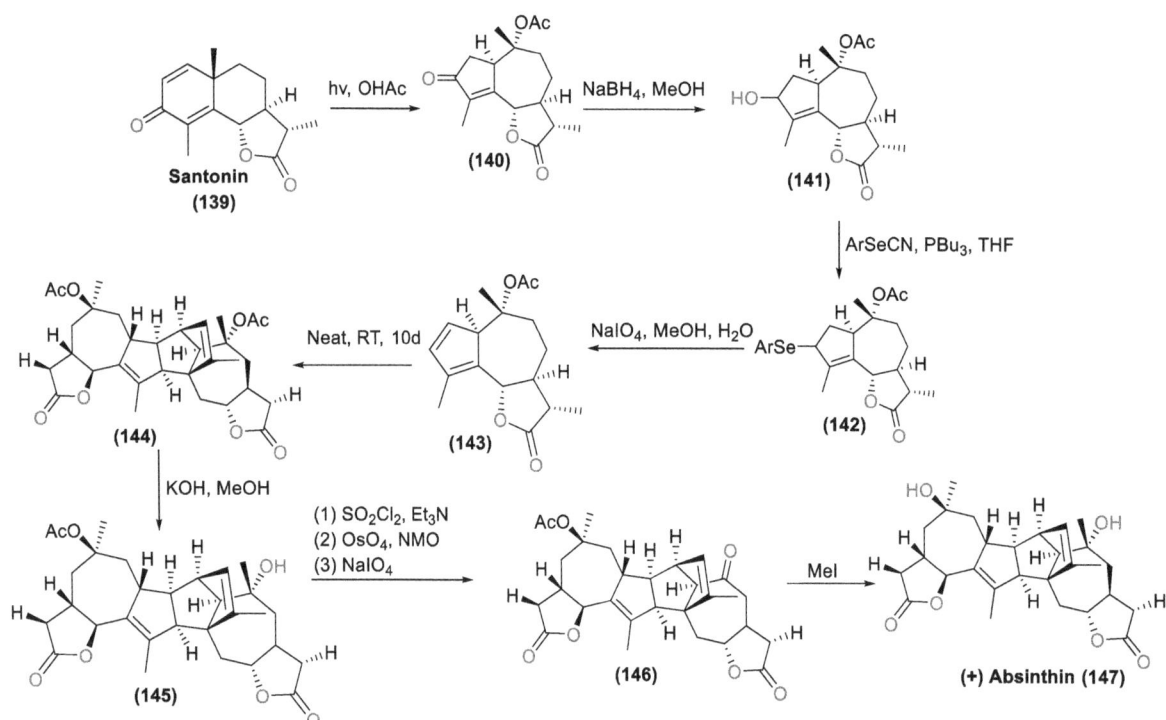

SCHEME 16.17 Synthesis of absinthin from santonin.

chromatography to produce compound (**138**). Finally, in argla-bin, dehydration with Tf$_2$O and pyridine yielded the required trisubstituted alkene product (**121**).

The synthesis of (+)-Absinthin (**147**) as a dimeric guaianolide from *Arthemisia absinthium* is shown in Scheme 16.17 [56]. The O-acetylisophotosantonic lactone was produced by photolysis with a Hg lamp using santonin (**139**) in acetic acid as a start-ing material (**140**). The diastereomeric alcohol combination was

obtained by reducing the enone carbonyl with NaBH$_4$ (**141**). The *Mitsunobu* aryl selenylation of (**141**) resulted in the synthesis of selenides (**142**), which were then treated with NaIO$_4$ to produce substituted cyclopentadiene (**143**). The *Diels-Alder* cycloaddi-tion of (**143**) yielded compound (**144**) with strong stereoselec-tivity as well as regioselectivity. Epi-absinthin was obtained by saponification of (**144**) with a methanolic potassium hydroxide solution followed by acidification with HCl (**145**). Diol (**145**)

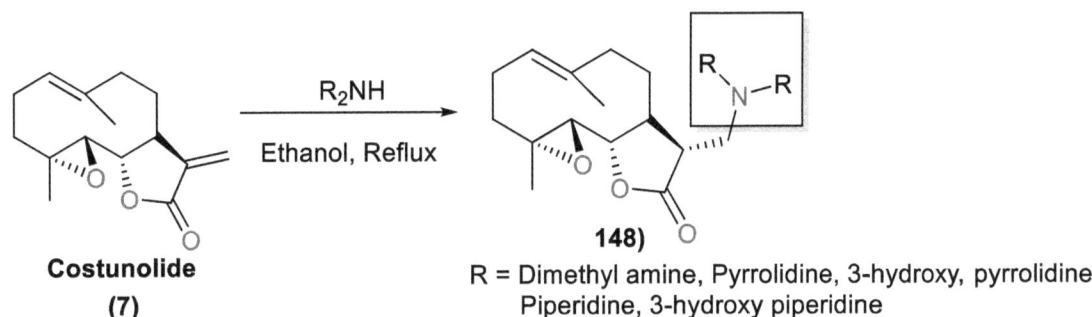

SCHEME 16.18 Synthesis of Amine derivative of costunolide.

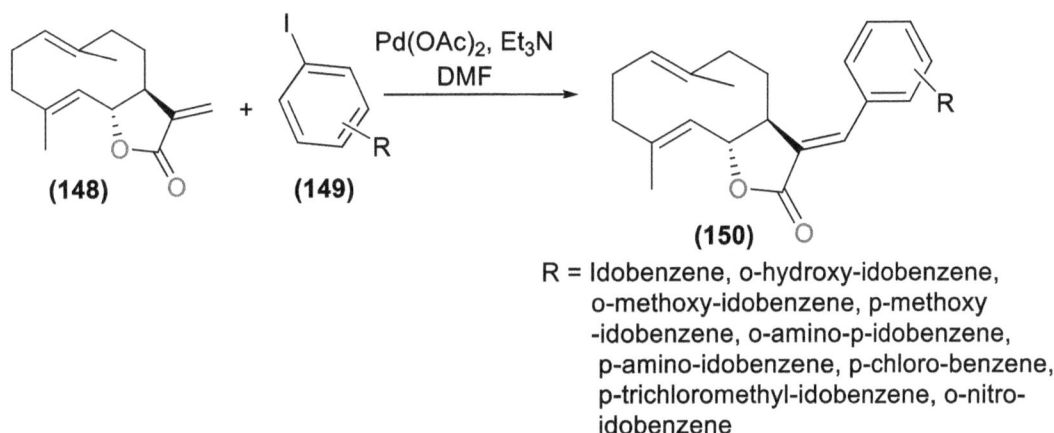

SCHEME 16.19 Synthesis of palladium-catalyzed derivative of Parthenolide.

was transformed to diketone (**146**) by oxidative degradation because the two alcohols had the incorrect configuration. The methyl groups were installed chemo- and stereo-selectively to finish the synthesis of (+)-Absinthin (**147**) [57–58].

16.10 Costunolide and Its Derivatives

Costunolide analogues (**148**, Scheme 16.18) have been synthesized after arylation of the α-methylene-γ-lactone under standard *Heck reaction* conditions as shown in Scheme 16.18 [59].

16.11 Parthenolide and Analogues

The palladium-catalyzed arylation reaction of parthenolide with aryl iodide (**149**) derivatives is shown in Scheme 16.19 [59–61].

16.12 Artemisinin and Its Derivatives

The sesquiterpene lactone Artemisinin **14** has peroxide isolated from *Artemisia annua*. The World Health Organization has decided to adopt the artemisinin class of compounds as the preferred basis for treating infections with *Plasmodium falciparum* strains, cerebral malaria and malaria in children due to a rise in resistance to most of the drugs currently used to treat malaria. Artemisinin amino derivatives (**151–152**) with various

lipophilic moieties and substituents have been produced in a novel class. An ethyl ether linker at the C-10 of artemisinin was used to incorporate aliphatic, alicyclic and aromatic amine groups into the synthesis (Figure 16.9). Artemisinin derivatives (**153–155**) have antiviral and anticancer activity [62–71].

16.13 Santonin and Its Analogues

Researchers have been focusing on *seco*-guaianolides in the last few years in order to develop new bioactive chemicals that might be used as therapeutic leads. The synthesis of *seco*-guaianolides is shown in Scheme 16.20. At low temperatures and in the presence of filter solutions, Ni(II) and Co(II), acetic acid (AcOH) and water, α-santonin (**139**) were converted into the guaianolide isophotosantonin (**156**). On molecule (**156**) the dehydration of the alcohol was catalyzed by acid, resulting in the diene (**157**). Seco-guaianolide (**158**) was produced by oxidising (**157**) with ozone and dimethyl sulphide.

Artemisia santonica has produced α-santonin (139), a sesquiterpene lactone with a eudesmane structure. α-santonin has been utilized as an anthelminthic in the past, and investigations have shown that it possesses important biological qualities including antipyretic, anti-inflammatory and fungicidal [72,73]. Furthermore, the sesquiterpene lactone α-santonin (139) has been changed to include the necessary α-methylene-γ-lactone, and these derivatives were shown to have a significant cytotoxic effect on cancer cells [74].

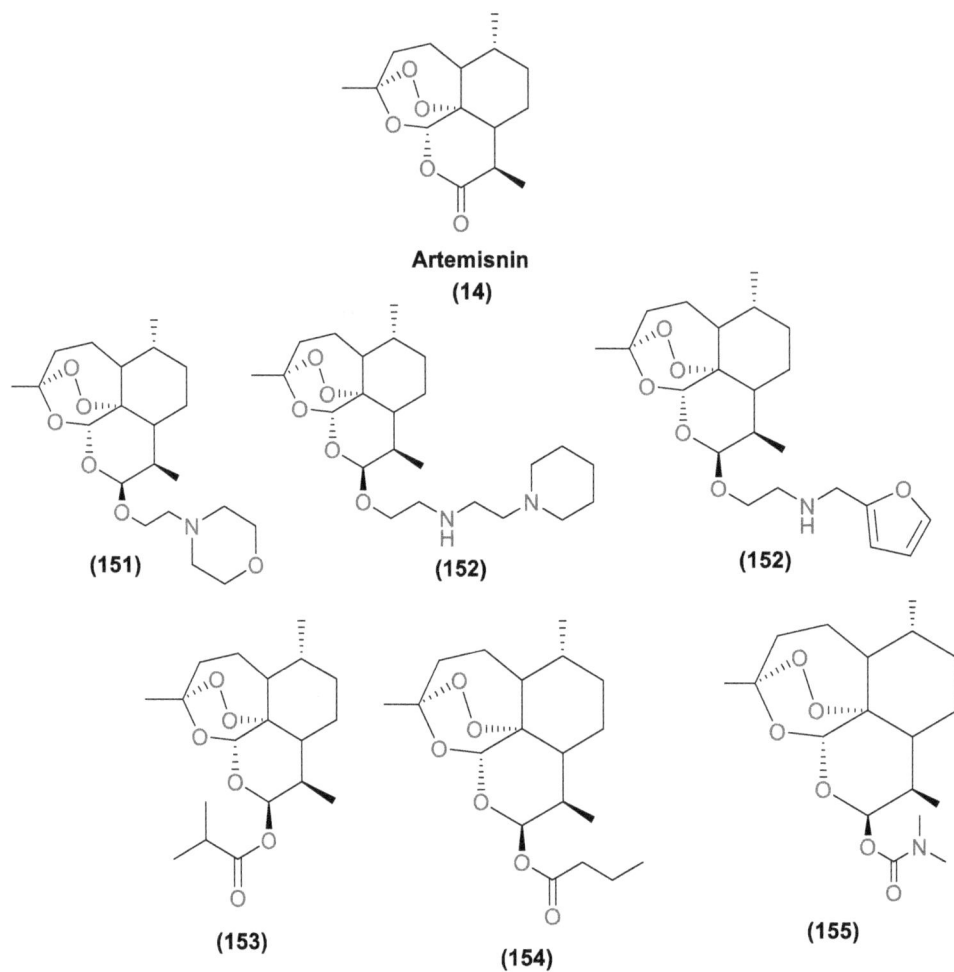

Artemisnin
(14)

(151)   **(152)**   **(152)**

(153)   **(154)**   **(155)**

FIGURE 16.9   Artemisinin and its derivatives.

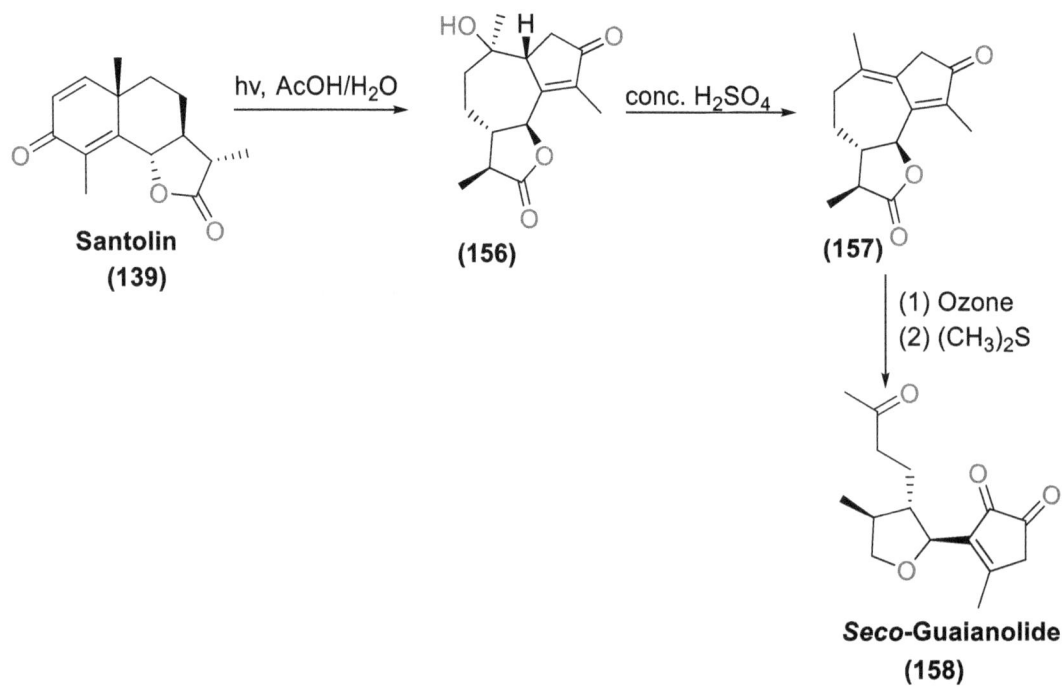

Santolin
(139)

hv, AcOH/H₂O

(156)

conc. H₂SO₄

(157)

(1) Ozone
(2) (CH₃)₂S

***Seco*-Guaianolide**
(158)

SCHEME 16.20   Synthesis of seco-Guainolide from santolin.

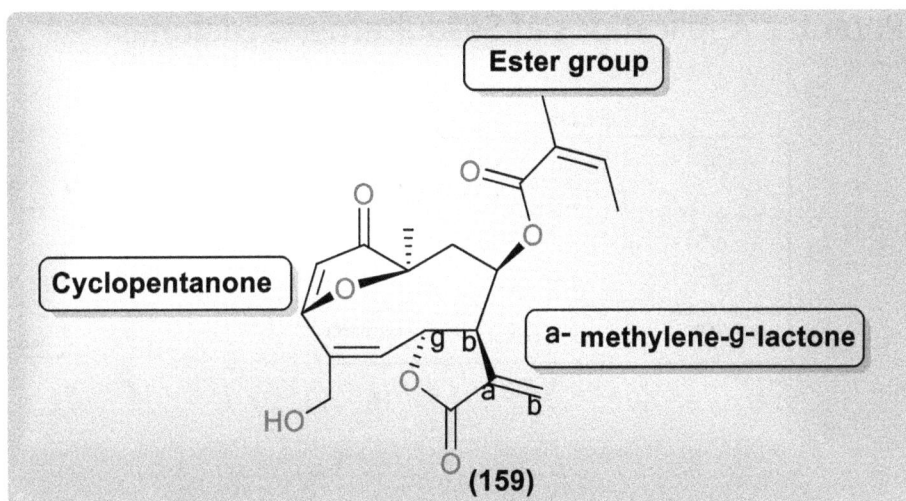

FIGURE 16.10 Structure-activity relationship of sesquiterpene lactones containing α,β-unsaturated cyclopentenone with an ester (angelate) group.

16.14 Structure–Activity Relationship

The anti-inflammatory, analgesic, cytotoxic, antitumorigenic and other therapeutic characteristics of α, β- or α, β, γ-unsaturated carbonyl moieties, such as those found in α-methylene-γ-lactone and, α, β-unsaturated cyclopentenone are dependent on them. STLs are architecturally diverse, and the bulk of them might be linked to a single synthase, resulting in a plethora of products with a wide range of effects. The cyclopentanone ring with oxygen bridge, connected to the ester moiety, and in the α-methylene-γ-lactone are the minimum structural requirements for sesquiterpene lactones, as illustrated in Figure 16.10 [75,76].

16.15 Biological Activities

Sesquiterpene lactones' bitter taste can also be caused by polar moieties and lipophilic centres. Bitterness is caused by C11-C13 exocyclic double bond. Sesquiterpene lactones inhibit enzymes with important thiol groups, including phosphofructokinase, glycogen synthase, DNA polymerase and thymidylate synthase. Sesquiterpene lactones hinder DNA synthesis by interfering with the DNA-template. Sesquiterpene lactones may be cytotoxic and anti-tumor because they impede DNA synthesis and thiol-containing enzymes.

Artemisinin is the sole western medicine sesquiterpene lactone. Parthenolide (**16**) is a migraine medication. Sesquiterpene lactones are commonly credited with traditional medicine's effectiveness. Figure 16.11 shows sesquiterpene lactone pharmacological applications [77–79].

16.16 Role of Sesquiterpene Lactones

The following drugs (Figure 16.12) are included in Figure 16.13 and their role in different signalling pathways is discussed [80,81].

16.17 Pathways Affected by Sesquiterpene Lactones

Figure 16.13 shows the essential components of 4 pathways that inhibit and stimulate STLs: JAK-STAT, PI3K-Akt, MAPK and NF-B. The different pathways are as follows:

a. MAPK Pathway

The mammalian mitogen-activated protein kinases (MAPK) family has 3 members: Signal-reduced kinase (ERK), p38 and c-Jun NH_2-terminal kinase (JNK). MAPK-kinase phosphorylates MAPK, activating it for cell proliferation, apoptosis, inflammation and differentiation. SLTS increased/decreased/unchanged/had no effect on phosphorylation of ERK, p38 and JNK of MAPK. Induction, inhibition and differentiation did not affect phosphorylation extent. SLTS also affects MAPK-dependent c-FOS expression. Few research reporting inhibitory effects of dehydro-costus lactone on differentiation say it does not impact ERK phosphorylation and other studies say SLTS does not influence JNK phosphorylation [82–84].

b. NFκB Pathway

Its 5 genes (NFB1, NFB2, RelA, c-Rel and RelB) encode 7 proteins (p50, p105, p52, p100, p65, c-Rel, RelB). Classical/alternative routes stimulate this pathway; activated NF-κB acts as a transcription factor in the nucleus. Differentiation, survival and inflammation may affect these pathways. SLTS alter the NFκB pathway, decreasing phosphorylation and inhibiting/stimulating differentiation. In all studies testing SLTS's impact, downstream NFATc1 expression was lowered. NF-κB may be altered by SLTS as a universal response, not specifically related to differentiation. Other pathways may be more effective at phosphorylating and activating these pathways.

Potential Applications of Sesquiterpene Lactones

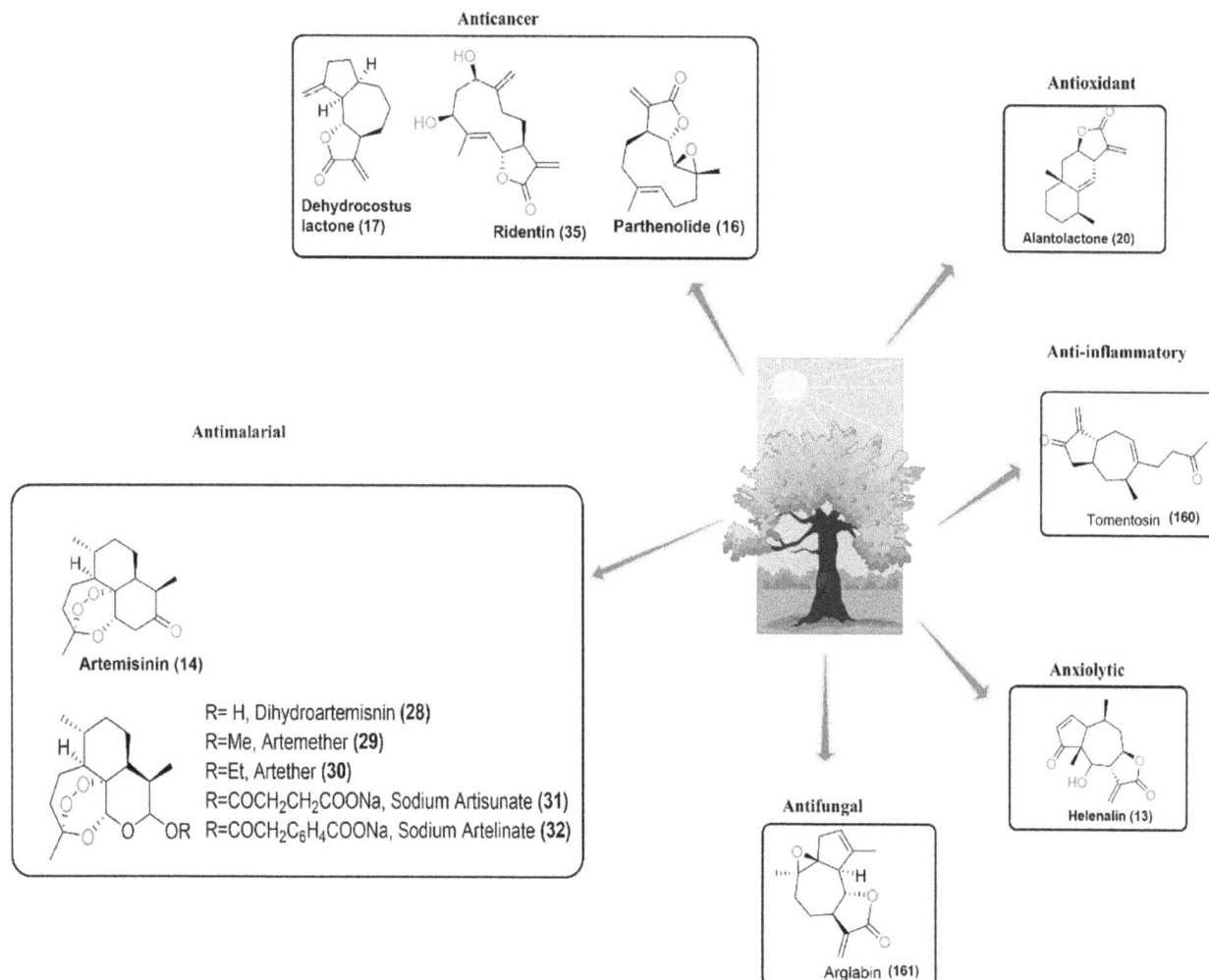

FIGURE 16.11 Potential applications of sesquiterpene lactones.

SLTS structural abnormalities, cell function and therapy requirements also limit discussion [85].

c. JAK-STAT Pathway

This family contains JAK1, JAK2, JAK3, TYK2 and 7 proteins, STAT1, STAT2, STAT3, STAT4, STAT5a, STAT5b and STAT6. This depends on cell types and signalling receptors. Ligand interaction with JAK-binding transmembrane receptors causes multimerization and transphosphorylation. Activated JAKs phosphorylate STATs, which dimerize and act as transcription factors in the nucleus. Activated STATs may impact cell-cycle progression, inhibition and lipid metabolism. SLTS affects this route, according to Figure 16.13 [86,87].

d. Phosphatydilinositol-3-kinase- Akt Pathway (PI3K-Akt)

Growth factors and regulators boost these. Triggering PI3K forms phosphatidylinositol-3,4,5-trisphosphate, which transfers and activates AKT at the plasma membrane. Activated PI3K/AKT pathway may affect cell cycle progression, apoptosis and differentiation. PI3K/AKT pathway can indirectly alter cellular differentiation by interacting with NF-κB (through IKK activation) and JAK/STAT pathways. The PI3K-AKT pathway responds to 3 SLTS (atractylenolide (I, II), 11,13-dihydro-[11R]-dehydroleucodine, iso-alantolactone, etc.) through AKT. Three inhibiting/stimulating lactones reduce AKT phosphorylation.

e. PPARγ and C/EBPα Pathway

SLTS lowered PPARγ and C/EBPα expression, but zalzuzanin had no effect. SLTS inhibits these pathways and C/EBPα. This route and C/EBPα are involved in adipogenesis and lipid metabolism, which sesquiterpene lactones block. Sesquiterpene lactones may impact PPARγ and C/EBPα in other cells, including cancer cells.

f. Miscellaneous Pathways

Atractrynolide induced differentiation by amplifying SHH and GLI expression, suggesting a role

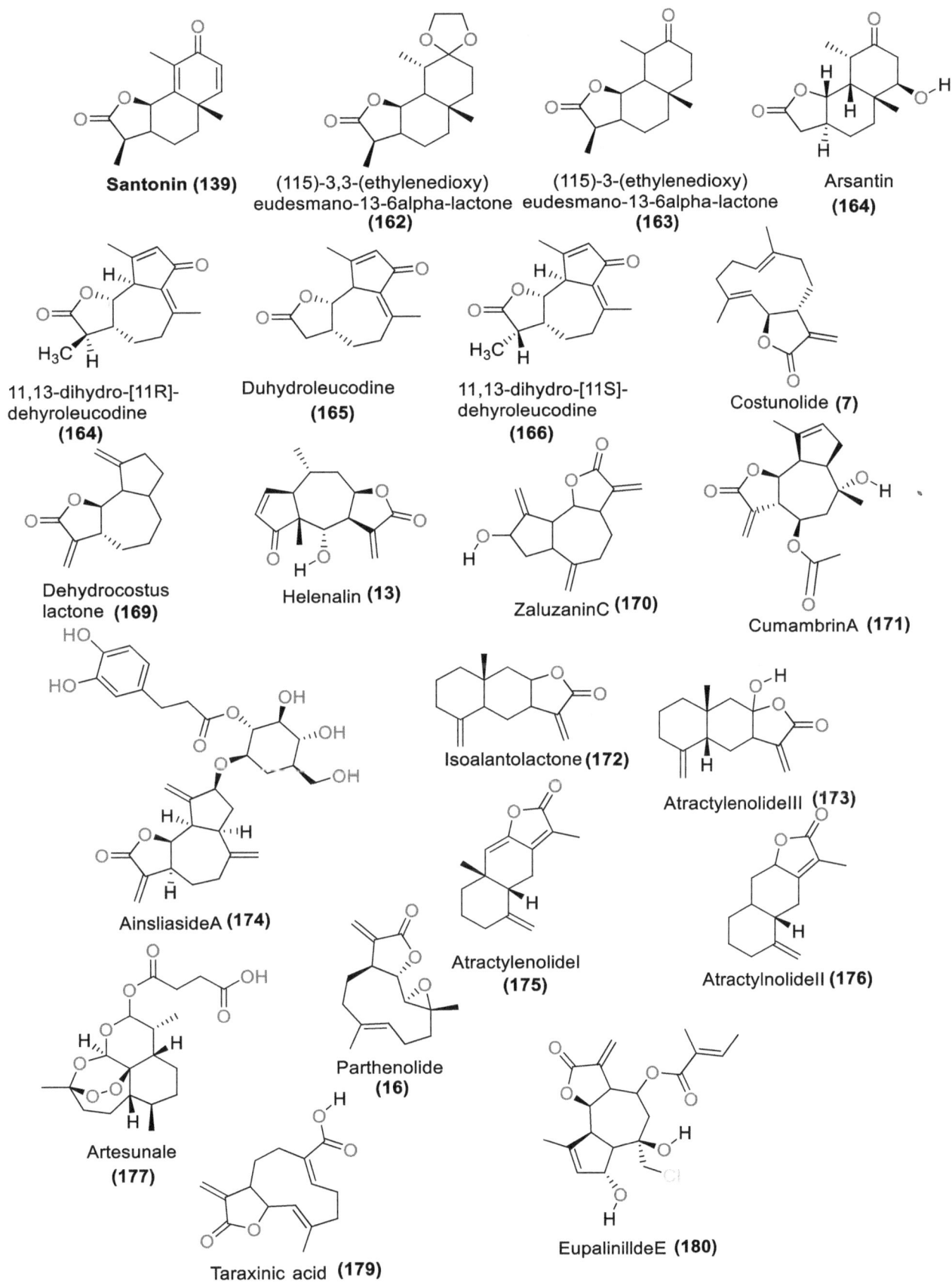

Santonin (139)

(115)-3,3-(ethylenedioxy)
eudesmano-13-6alpha-lactone
(162)

(115)-3-(ethylenedioxy)
eudesmano-13-6alpha-lactone
(163)

Arsantin
(164)

11,13-dihydro-[11R]-
dehyroleucodine
(164)

Duhydroleucodine
(165)

11,13-dihydro-[11S]-
dehyroleucodine
(166)

Costunolide **(7)**

Dehydrocostus
lactone **(169)**

Helenalin **(13)**

ZaluzaninC **(170)**

CumambrinA **(171)**

AinsliasideA **(174)**

Isoalantolactone **(172)**

AtractylenolideIII **(173)**

Atractylenolidel
(175)

AtractylnolideII **(176)**

Artesunale
(177)

Parthenolide
(16)

Taraxinic acid **(179)**

EupaliniIldeE **(180)**

FIGURE 16.12 Some more examples of sesquiterpene lactones.

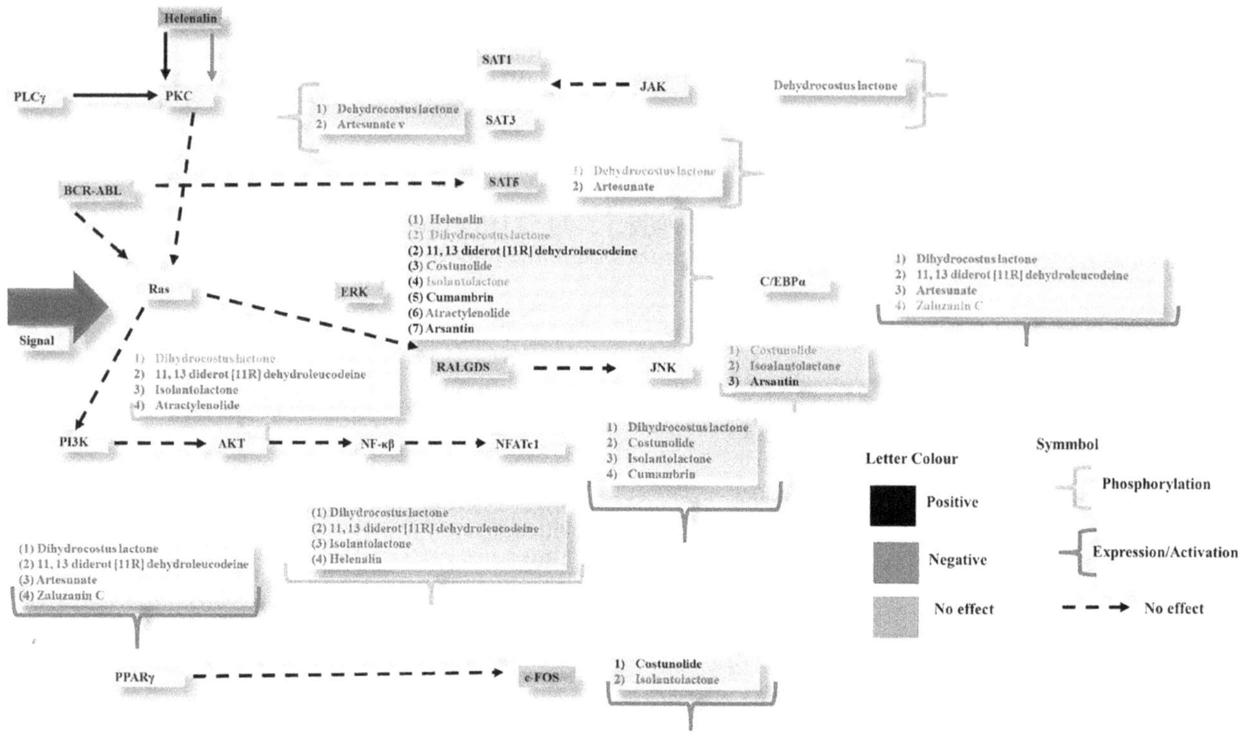

FIGURE 16.13 Pathways affected by sesquiterpene lactones.

for the Sonic-Hedgehog signalling pathway. SLTS decrease the expression of Fas involved in apoptosis, increase the expression of p27 and p21 and decrease the expression of c-Myc involved in the cell cycle.

16.18 IIR Signalling SLTS-Modified in Inflammation

It's a physiological response to damaging stimuli, infections, damaged cells/toxins, carried out by immune cells such monocytes/macrophages, lymphocytes, neutrophils and dendritic cells. Acute and chronic inflammatory reactions contribute to tissue homeostasis and resolve acute inflammation, respectively. NFκB, MAPK and JAK-STAT pathways are triggered in cells that come into touch with external substances, such as epidermal keratinocytes, lung and intestinal epithelial cells and blood leukocytes. Non-immune signal channels influence the inflammatory process in immune cells (like macrophages). Nuclear factor B (NF-B) signalling depends on transcription factors including p50 and RelA (p65), which govern gene expression that favors an inflammatory and immunological response. The typical NF-kB is triggered by external stimuli at the cell surface, resulting in twofold phosphorylation of IkB in the cytoplasm by IKK, leading to its destruction and the release of RelA and P50 (Figure 16.14) [87]. These two proteins reach the nucleus and influence immunological response, cell death, cell-cycle progression, inflammation and oncogenesis. NF-kB activation affects macrophage activity, T lymphocyte chemoattraction, neutrophil survival and dendritic cell maturation. Parthenolide is a potent anti-inflammatory medication

that can deactivate NF-kB at 5 M. The drug's high binding selectivity allows it to make a covalent bond with the NF-kB RelA subunit's cysteine, resulting in alkylation and inhibition of DNA binding (Figure 16.14).

Abbreviations: The inflammatory immune response=IIR; Tumor necrosis factor-alpha =TNF-α; Interleukin 1 beta =l1β; Inducible nitric oxide synthase=Inos

16.19 Mechanism of Different STLs

16.19.1 Santonin

One of the first STLs identified was santonin (**139**), a eudesmanolide found in Artemisia santonica (1830). This STL has been used as an ascaricide, to eliminate all types of worms and to prevent urine retention and enuresis due to atony or other causes. Because of its hazardous properties, it was no longer used in pharmaceuticals. It has been shown to have anti-inflammatory, antipyretic and analgesic properties. Numerous chemical changes to the santonin structure have been made in trying to improve its antiproliferative and cell differentiation activities in leukaemia cells, as well as its antimalarial action. Other guaianolides and eudesmanolides may be synthesized using this molecule as a starting point. It was one of the earliest STLs to have its structure revealed.

16.19.2 Artemisinin

Artemisinin (14) is a unique molecule because its molecular structure includes an endoperoxide ring. Artemisia annua was used to isolate this STL (Asteraceae). In Chinese traditional

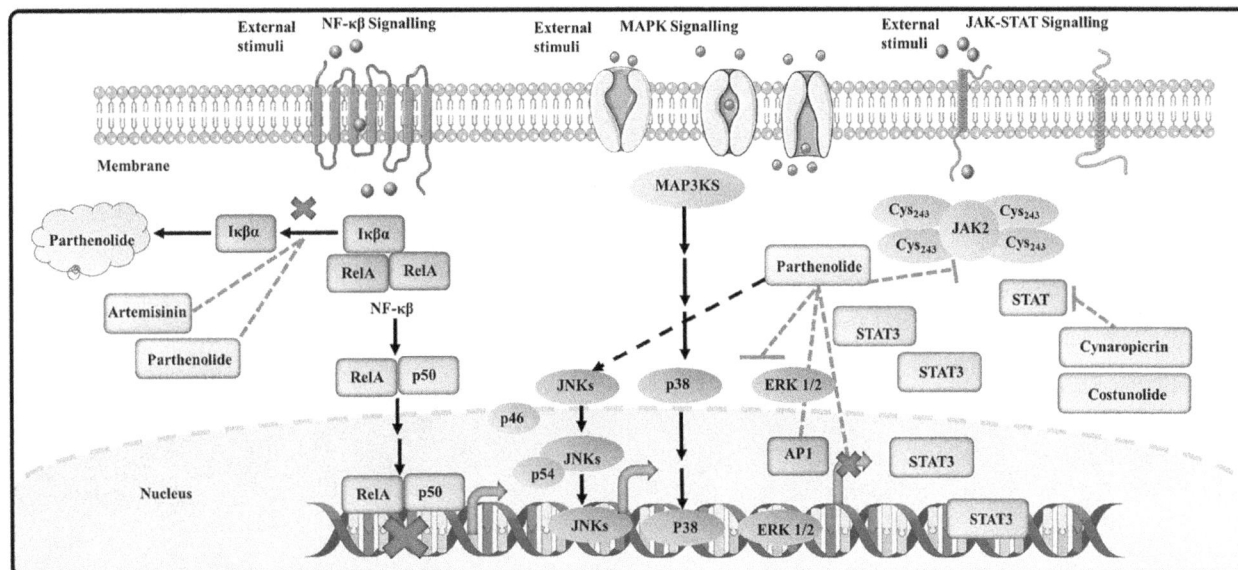

FIGURE 16.14 Signalling pathways involved in IIR modulated through sesquiterpene lactones.

medicine, the aerial portions of this plant have been employed as a febrifuge. Artemisinin and its derivatives are being utilized as antimalarials against Plasmodium falciparum, which is resistant to chloroquine. Leishmanicidal and anticancer actions have also been documented for this STL. In human hepatocellular carcinoma cells (SMMC-7721) and other cell lines, the latter trait is owing to its ability to suppress cell development and cause apoptosis. Artemisinin also has anti-istosomal properties. There has also been evidence of activity against *Helicobacter pylori*. In phase I and II studies, artemisinin derivatives are being tested for lupus nephritis, breast, colorectal and lung malignancies.

16.19.3 Parthenolide

Feverfew's active ingredient is parthenolide (**16**). (Tanacetum parthenium, Asteraceae). It's a traditional herbal remedy that's been used to cure migraines, fevers and arthritis for generations. This STL exhibits an antiproliferative effect against a variety of cancer cells, including melanoma, breast, colon and lung cancer, as well as leukaemia. The ability of a chemical to induce apoptosis in cancer cells offers a unique therapeutic option for the treatment of cancer and inflammation-related diseases. Parthenolide has been shown to have antiprotozoal (against Trypanosoma cruzi and Leishmania spp.) and anti-inflammatory properties.

16.19.4 Costunolide

The germacranolide-type STL costunolide (**6**) is found in the roots of Saussurea lappa, a traditional Chinese medicinal plant with anticancer and anti-inflammatory effects. Other plant species that contain this chemical include Magnolia sp., Laurus nobilis and Costus speciosus, among others. It has antidiabetic and antioxidant properties, as well as anti-inflammatory, anti-ulcerogenic, anticlastogenic and possible anticancer properties. Costunolide inhibits cell proliferation by causing apoptosis via a variety of methods, including the formation of reactive oxygen

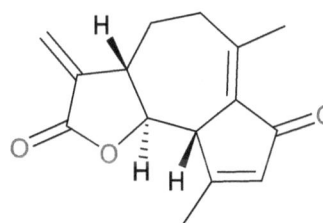

Dehydroleucodine (181)

FIGURE 16.15 Structure of dehydroleucodine.

species (ROS) and cell cycle arrest. Lung carcinoma, colon, bladder and platinum-resistant ovarian cancer; hepatoma and leukemic cells are all targets for this STL [86,87].

16.19.5 Dehydroleucodine

In an animal model, dehydroleucodine (**181,** Figure 16.15), an STL derived from Artemisia douglasiana, has cytotoxic effect against human leukaemia cells and inhibits the development of melanoma cells. It protects the stomach mucosa by reducing inflammation and gastrointestinal ethanol-induced damage, as shown in vivo studies. *T. cruzi* infective forms and *Leishmania mexicana* promastigotes are both inhibited by this chemical. This chemical has also been shown to have antimicrobial efficacy against *Pseudomonas aeruginosa* multi-resistant strains.

16.19.6 Helenalin

Helenalin (**13**) is a guaianolide STL that has been isolated from *Arnica montana* and other Asteraceae plants. Antimicrobial, cytotoxic, hepatoprotective, anti-inflammatory, antioxidant and antimicrobial activities have all been observed against Staphylococcus aureus. It possesses cardiotonic action and influences steroidogenesis in rat adrenocortical cells. There have also been reports of trypanocidal effects.

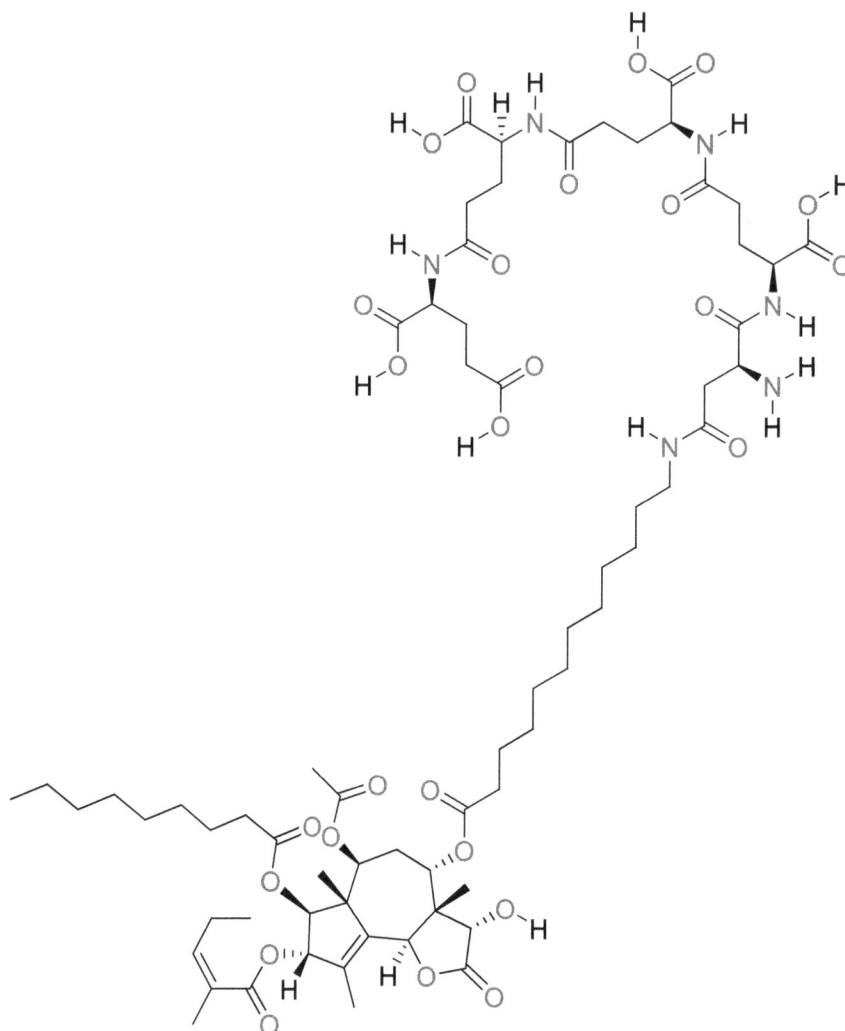

Mipsagargin (182)

FIGURE 16.16 Structure of mipasagargin.

16.19.7 Thapsigargin

The guaianolide STL Thapsigargin (**114**) was isolated from *Thapsia garganica* (Apiaceae). Hippocrates, Theophrastus, Dioscorides and Plinius all identified this Mediterranean medicinal plant as a skin irritant, effective for pulmonary ill-ness, catarrh and fever, as well as the treatment of rheumatic symptoms. Between 1980 and 1985, thapsigargin (**114**) was extracted from the fruits and roots in the hunt for the skin-irri-tant principle and its structure and absolute configuration were identified. This substance was shown to be a powerful hista-mine liberator as well as a cocarcinogen that promoted skin cancer in mice. Nonetheless, the discovery of thapsigargin's capacity to block the sarco-endoplasmic reticulum calcium ATPase (SERCA) pump piqued people's curiosity. Apoptosis is caused by the blockage of this pump, which results in a high concentration of calcium in the cytosol. Several analogues of thapsigargin have been discovered, as well as a prodrug called mipsagargin (**182**). In patients with solid tumours, Mipsagargin (**182,** Figure 16.16) has shown adequate tolerability and a posi-tive pharmacokinetic profile.

16.19.8 Arglabin

Arglabin (**121**) is a guaianolide-type STL identified for the first time from the plant Artemisia glabella, which grows in Kazakhstan. It may be found in the sections of the body that are above ground (leaves, bud flowers and stems). Later, it was discovered that arglabin (**121**) was contained in A. myriantha, a well-known plant used in Chinese traditional medicine. Arglabin has anticancer efficacy against a variety of tumour cell lines. Many derivatives have been discovered and those with bromine and chlorine atoms as well as an epoxy group on the C(3)=C(4) double bond seem to have higher anticancer activity. One of these compounds, dimethylamino arglabin has been used to treat lung, liver and ovarian malignancies and is now being studied in phase I and II clinical studies. In the Russian Federation, Kazakhstan, Uzbekistan, Tajikistan, the Kirghiz Republic and Georgia, this STL has been patented and registered as an anticancer medica-tion. Arglabin, unlike artemisinin, thapsigargin derivatives and parthenolide (16), functions as an anticancer agent via a distinct mechanism. It works by inhibiting the enzyme farnesyl trans-ferase, which has been linked to the development of malignant

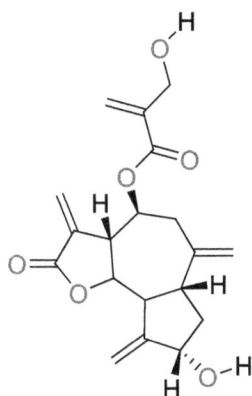

Cynaropicrin (183)

FIGURE 16.17 Structure of cynaropicrin.

tumours. Other biological actions of this chemical include an inhibitory impact on the influenza A virus, the ability to reestablish the generation of cytokines and other anti-inflammatory mediators in in vivo models of inflammation (carrageenan, histame and formalin models) and immunomodulatory activity. have shown that arglabin lowers inflammation in pancreatic -cells in vivo and in the INS-1 cell line in vitro, indicating that it might be a novel promising molecule for the treatment of inflammation and type 2 diabetes.

16.19.9 Cynaropicrin

The bitter principle of Cynara scolymus is **cynaropicrin (183**, Figure 16.17). It's a guaianolide-type STL that's been isolated from Saussurea lappa as well. Both Trypanosoma brucei and Trypanosoma cruzi are inhibited by this chemical. It contains antispasmodic and antiphotoaging properties, as well as cytotoxic action on leukemic cell lines [87].

Conclusion

This chapter provides comprehensive information about sesquiterpene lactone biogenesis and chemical synthesis. It also includes the classification, structure, chemical characteristics, Biosynthesis of Sesquiterpene Lactones, Sesquiterpene Lactone Pathway, Biosynthesis of the Sesquiterpene Framework, Biosynthesis of the lactone ring, Function of the α-Methylene-γ-Lactone Group, Synthetic strategy of Parthenolide, Sesquiterpene Lactone, General strategies for the synthesis of Guaianolides, semi-synthesis of Guaianolides, Artemisinin and Its Derivatives etc. It also discusses mechanism of action different sesquiterpene lactones.

REFERENCES

1. (a) Robles, M.; Aregullin, M.; West, J.; Rodriguez, E. Recent Studies on the Zoo Pharmacognosy, Pharmacology and Neurotoxicology of Sesquiterpene Lactones. *Planta Medica*, **1995**, 61, 199.

2. Modzelewska, A.; Sur, S.; Kumar, S. K.; Khan, S. R. Sesquiterpenes: Natural Products That Decrease Cancer Growth. *Current Mediclinical Chemistry - Anticancer Agents*, **2005**, 5, 477.

3. Chen, H. C.; Chou, C. K.; Lee, S. D.; Wang, J. C.; Yeh, S. F. Active Compounds from Saussurea lappa Clarks That Suppress Hepatitis B Virus Surface Antigen Gene Expression in Human Hepatoma Cells. *Antiviral Research*, **1995**, 27, 99.

4. Vranová, E.; Coman, D.; Gruissem, W. Structure and Dynamics of the Isoprenoid Pathway Network. *Molecular Plant*, **2012**, 5(2):318–333.

5. Vranová, E.; Coman, D.; Gruissem, W. Network Analysis of the MVA and MEP Pathways for Isoprenoid Synthesis. *Annual Review of Plant Biology*, **2013**, 64, 665–700.

6. Chaturvedi D. Sesquiterpene lactones: Structural diversity and their biological activities. In: *Opportunity, Challenge and Scope of Natural Products in Medicinal Chemistry*, Research Signpost, Trivandrum, India, 2011, 313–334.

7. Coricello, A.; El-Magboub, A., Luna, M.; Ferrario, A.; Haworth, I. S.; Gomer, C. J., Aiello, F.; Adams, J. D. Rational Drug Design and Synthesis of New α-Santonin Derivatives as Potential COX-2 Inhibitors. *Bioorganic & Medicinal Chemistry Letters*, **2018**, 28(6), 993–996. doi: 10.1016/j.bmcl.2018.02.036.

8. Gou, J.; Fuhua, H.; Huang, C.; Zhang, Y.; Chen, F. Discovery of a Non-Stereoselective Cytochrome P450 Catalyzing Either 8α- or 8β-Hydroxylation of Germacrene A Acid from the Chinese Medicinal Plant, Inula hupehensis. *The Plant Journal*, **2018**, 93, 92–106.

9. Tholl, D. Biosynthesis and Biological Functions of Terpenoids in Plants. *Advances in Biochemical Engineering/Biotechnology*, **2015**, 148, 63–106.

10. Teoh, K. H.; Polichuk, D. R.; Reed, D. W.; et al. *Artemisia annua* L. (Asteraceae) Trichome-Specific cDNAs Reveal CYP71AV1, a Cytochrome P450 with a Key Role in the Biosynthesis of the Antimalarial Sesquiterpene Lactone Artemisinin. *FEBS Letters*, **2006**, 580, 1411–1416.

11. Teoh, K. H.; Polichuk, D. R.; Reed, D. W.; et al. Molecular Cloning of an Aldehyde Dehydrogenase Implicated in Artemisinin Biosynthesis in *Artemisia annua*. *Botany*, **2009**, 87, 635–642.

12. Zhang, Y.; Teoh, K. H.; Reed, D. W.; et al. The Molecular Cloning of Artemisinic Aldehyde D-11(13) Reductase and Its Role in Glandular Trichome-Dependent Biosynthesis of Artemisinin in *Artemisia annua*. *Journal of Biological Chemistry*, **2008**, 283, 21501–21508.

13. Wen, W.; Yu, R. Artemisinin Biosynthesis and Its Regulatory Enzymes: Progress and Perspective. *Pharmacognosy Reviews*, **2011**, 5(10), 189–194.

14. Turconi, J.; Griolet, F.; Guevel, R.; et al. Semisynthetic Artemisinin, the Chemical Path to Industrial Production. *Organic Process Research & Development*, **2014**, 18, 417–422.

15. Liu, Q.; Majdi, M.; Cankar, K.; Goedbloed, M.; Charnikhova, T.; Verstappen, F. W.; de Vos R. C.; Beekwilder, J.; van der Krol, S.; Bouwmeester, H. J. Reconstitution of the Costunolide Biosynthetic Pathway in Yeast and Nicotiana benthamiana. *PLoS One*, **2011**, 6(8), e23255. doi: 10.1371/journal.pone.0023255.

16. Fischer, N. H.; Olivier, E. J.; Fischer, H. D. The biogenesis and chemistry of sesquiterpene lactones. In Herz W, Grisebach H, Kirby GW (eds.) *Progress in the Chemistry of Organic Natural Products*, Vol. 38. Springer Verlag, Wien—New York, 1979, 47– 320.

17. Buckingham, J. *Dictionary of Natural Products on CD-ROM, Version 9:2.* Chapman & Hall, London, 2001.

18. Piet, D. P.; Minnaard, A. J.; van der Heyden, K. A.; Franssen, M. C. R.; Wijnberg, J. B. P. A.; de Groot Ae. Biotransformation of (±)-4,8-dimethylcyclodeca-3(E), 7(E)-dien-ip-ol and (+)-Hedycaryol by Cichorium intybus. *Tetrahedron*, **1995**, 51, 243–254.

19. Cane, D. E. Enzymatic Formation of Sesquiterpenes. *Chemical Reviews*, **1990**, 90, 1089–1103.

20. McCaskill, D.; Croteau, R. Prospects for the bioengineering of isoprenoid biosynthesis. In Berger R (ed.) *Biotechnology of Aroma Compounds (Advances in Biochemical Engineering Biotechnology)*. Springer Verlag, Berlin, 1997, 107–147.

21. Croteau, R.; Kutchan, M.; Lewis, N. G. Natural products (secondary metabolites). In Buchanan B, Gruissem W, Jones R (eds.) *Biochemistry & Molecular Biology of Plants*. American Society of Plant Physiologists, Rockeville, 2000, 1250–1318.

22. McConkey, M. E.; Gershenzon, J.; Croteau, R. B. Developmental Regulation of Monoterpene Biosynthesis in the Glandular Trichomes of Peppermint. *Plant Physiology*, **2000**, 122, 215–223.

23. Dhavalikar, R. S.; Rangachari, P. N.; Bhattacharyya, P. K. Microbiological Transformations of Terpenes: Part DC—Pathways of Degradation of Limonene in a Soil Pseudomonad. *Indian Journal of Biochemistry*, **1966**, 3, 158–164.

24. Baker, F. C.; Mauchamp, B.; Tsai, L. W.; Schooley, D. W. Farnesol and Farnesal Dehydrogenases in Corpora Allata of the Tobacco Hornworm Moth, *Manduca sexta. Journal of Lipid Research*, **1983**, 24, 1586–1594.

25. Funk, C.; Croteau, R. Diterpenoid Resin acid Biosynthesis in Conifers: Characterisation of Two Cytochrome P450-Dependent Monooxygenases and an Aldehyde Dehydrogenase Involved in Abietic Acid Biosynthesis. *Archives of Biochemistry and Biophysics*, **1994**, 308, 258–266.

26. Yang, Y. I.; Kim, J. H.; Lee, K. T.; et al. Costunolide Induces Apoptosis in Platinum-Resistant Human Ovarian Cancer Cells by Generating Reactive Oxygen Species. *Gynecologic Oncology*, **2011**, 123(3), 588–596.

27. Long, J.; Zhang, S. F.; Wang, P. P.; et al. Total Syntheses of Parthenolide and Its Analogues with Macrocyclic Stereocontrol. *Journal of Medicinal Chemistry* **2014**, 57, 7098–7112.

28. Marshall, J. A.; Lebreton, J.; Dehoff, B. S. et al. Stereoselective Total Synthesis of Aristolactone and Epiaristolactone Via [2,3] Wittig Ring Contraction. *Journal of Organic Chemistry*, **1987**, 52, 3883–3889.

29. Shibuya, H, Ohashi K, Kawashima K et al. Synthesis of Costunolide, an Antitumor Germacranolide, from E, E-Farnesol by Use of a Low-Valent Chromium Reagent. *Chemical Letters*, **1986**, 1, 85–86.

30. Foo, K.; Usui, I.; Gootz, D. C.; et al. Scalable, Enantioselective Synthesis of Germacrenes and Related Sesquiterpenes Inspired by Terpene Cyclase Phase Logic. *Angewandte Chemie International Edition*, **2012**, 51, 11491–11495.

31. Chen, L. X.; Zhu, H. J.; Wang, R.; et al. Ent-Labdane Diterpenoid Lactone Stereoisomers from Andrographis paniculata. *Journal of Natural Products*, **2008**, 71, 852–855.

32. Mang, C.; Jakupovic, S.; Schunk, S.; et al. Natural Products in Combinatorial Chemistry: An Andrographolide-Based Library. *Journal of Combinatorial Chemistry*, **2006**, 8, 268–274.

33. Reynolds, A. J.; Scott, A. J.; Turner, C. I.; et al. The Intramolecular Carboxyarylation Approach to Podophyllotoxin. *Journal of the American Chemical Society*, **2003**, 125, 12108–12109.

34. Siedle, B.; Garcia-Pineres, A. J.; Murillo, R.; et al. Quantitative Structure – Activity Relationship of Sesquiterpene Lactones as Inhibitors of the Transcription Factor NF-kappa B. *Journal of Medicinal Chemistry*, **2004**, 47, 6042–6054.

35. Kummer, D. A., Brenneman, J. B., Martin, S. F.; et al. Application of a Domino Intramolecular Enyne Metathesis/ Cross Metathesis Reaction to the Total Synthesis of (+)-8-Epi-Xanthatin. *Organic Letters*, **2005**, 7, 4621–4623.

36. Merten, J.; Hennig, A.; Schwab, P.; et al. A Concise Sultone Route to Highly Oxygenated 1,10-Seco-Eudesmanolides – Enantioselective Total Synthesis of the Antileukemic Sesquiterpene Lactones (−)-Eriolanin and (−)-Eriolangin. *European Journal of Organic Chemistry*, **2006**, 5, 1144–1161.

37. Merten, J.; Frohlich, R.; Metz, P. et al. Enantioselective Total Synthesis of the Highly Oxygenated 1,10-Seco-Eudesmanolides Eriolanin and Eriolangin. *Angewandte Chemie International Edition*, **2004**, 43, 5991–5994.

38. Hirose, T.; Miyakoshi, N.; Mukai, C.; et al. Total Synthesis of (+)-Achalensolide Based on the Rh(I)-Catalyzed Allenic Pauson-Khand-Type Reaction. *Journal of Organic Chemistry*, **2008**, 73, 1061–1066.

39. Kalidindi, S.; Jeong, W. B.; Schall, A.; et al. Enantioselective Synthesis of Arglabin. *Angewandte Chemie International Edition*, **2007**, 46, 6361–6363.

40. Nakamura, T.; Tsuboi, K.; Oshida, M.; et al. Total Synthesis of (−)-Diversifolin. *Tetrahedron Letters*, **2009**, 50, 2835–2839.

41. Mihelcic, J.; Moeller, K. D.; Anodic Cyclization Reactions: The Total Synthesis of Alliacol A. *Journal of the American Chemical Society*, **2003**, 125, 36–37.

42. Justicia, J.; de Cienfuegos, L. A.; Estevez, R. E.; et al. Ti-Catalyzed Transannular Cyclization of Epoxygermacrolides. Synthesis of Antifungal (+)-Tuberiferine and (+)-Dehydrobrachylaenolide. *Tetrahedron*, **2008**, 64, 11938–11943.

43. Yang, Z. J.; Ge, W. Z.; Li, Q. Y.; et al. Syntheses and Biological Evaluation of Costunolide, Parthenolide, and Their Fluorinated Analogues. *Journal of Medicinal Chemistry*, **2015**, 58, 7007–7020.

44. Guzman, M. L.; Li, X. J.; Corbett, C.; et al. Mechanisms Controlling Selective Death of Leukemia Stem Cells in Response to Parthenolide. *Blood*, **2005**, 106, 141a–141a.

45. Long, J.; Ding, Y. H.; Wang, P. P.; et al. Protection-Group-Free Semisyntheses of Parthenolide and its Cyclopropyl Analogue. *Journal of Organic Chemistry*, **2013**, 78, 10512–10518.

46. Zhang, Q.; Cai, D. F.; Liu, J. H. Matrix Solid-Phase Dispersion Extraction Coupled with HPLC-Diode Array Detection Method for the Analysis of Sesquiterpene Lactones in Root of Saussurea lappa CB Clarke. *Journal of Chromatography B: Analytical Technologies in the Biomedical and Life Sciences*, **2011**, 879, 2809–2814.

47. Chu, H.; Smith, J. M.; Felding, J.; et al. Scalable Synthesis of (−)-Thapsigargin. *ACS Central Science*, **2017**, 3, 47–51.

48. Andrews, S. P.; Ball, M.; Wierschem, F.; et al. Total Synthesis of Five Thapsigargins: Guaianolide Natural Products Exhibiting Sub-Nanomolar SERCA Inhibition. *Chemistry*, **2007**, 13, 5688–5712.

49. Grillet, F.; Huang, C. F.; Brummond, K. M.; et al. An Allenic Pauson-Khand Approach to 6,12-Guaianolides. *Organic Letters*, **2011**, 13, 6304–6307.

50. Wen, B.; Hexum, J. K.; Widen, J. C.; et al A Redox Economical Synthesis of Bioactive 6,12-Guaianolides. *Organic Letters*, **2013**, 15, 2644–2647.

51. Ivanescu, B., Miron, A., Corciova, A. Sesquiterpene Lactones from Artemisia Genus: Biological Activities and Methods of Analysis. *Journal of Analytical Chemistry*, **2015**, 2015, 247685–247696.

52. Shaikenov, T. E.; Adekenov, S. M.; Williams, R. M.; et al. Arglabin-DMA, a Plant Derived Sesquiterpene, Inhibits Farnesyltransferase. *Oncology Reports*, **2001**, 8, 173–179.

53. Zhai, J. D.; Li, D. M.; Long. J.; et al. Biomimetic Semisynthesis of Arglabin from Parthenolide. *Journal of Organic Chemistry*, **2012**, 77, 7103–7107.

54. Roboz, G. J.; Guzman, M. Acute myeloid Leukemia Stem Cells: Seek and Destroy. *Expert Review of Hematology*, **2009**, 2, 663–672.

55. Peese, K. New Agents for the Treatment of Leukemia: Discovery of DMAPT (LC-1). *Drug Discovery Today*, **2010**, 15, 322–322.

56. Lone, S. H.; Bhat, K. A. Hemisynthesis of a Naturally Occurring Clinically Significant Antitumor Arglabin from Ludartin. *Tetrahedron Letters*, **2015**, 56, 1908–1910.

57. Zhang, W. H.; Luo, S. J.; Fang, F.; et al. Total Synthesis of Absinthin. *Journal of the American Chemical Society*, **2005**, 127, 18–19.

58. Sun, C. M.; Syu, W. J.; Don, M. J.; et al. Cytotoxic Sesquiterpene Lactones from the Root of Saussurea lappa. *Journal of Natural Products*, **2003**, 66, 1175–1180.

59. Han, C. H.; Barrios, F. J.; Riofsky, M. V.; et al. Semisynthetic Derivatives of Sesquiterpene Lactones by Palladium-Catalyzed Arylation of the Alpha-Methylene-Gamma-Lactone Substructure. *Journal of Organic Chemistry*, **2009**, 74, 7176–7179.

60. Kwok, B. H.; Koh, B. D.; Ndubiusi, M. I.; et al. The Sesquiterpene Lactone Parthenolide Binds and Inhibits IKK Beta. *Molecular Biology of the Cell*, **2001a**, 12, 271a.

61. Kwok, B. H. B.; Koh, B.; Ndubiusi, M. I.; et al. The Anti-Inflammatory Natural Product Parthenolide from the Medicinal Herb Feverfew Directly Binds to and Inhibits I kappa B kinase. *Chemical Biology*, **2001b**, 8, 759–766.

62. Yeung, S.; Pongtavornpinyo, W.; Hastings, I. M.; et al. Antimalarial Drug Resistance, Artemisinin-Based Combination Therapy, and the Contribution of Modeling to Elucidating Policy Choices. *American Journal of Tropical Medicine and Hygiene*, **2004**, 71, 179–186.

63. Ploypradith, P. Development of Artemisinin and Its Structurally Simplified Trioxane Derivatives as Antimalarial Drugs. *Acta Tropica*, **2004**, 89, 329–342.

64. Krishna, S.; Uhlemann, A. C.; Haynes, R. K.; et al. Artemisinins: Mechanisms of Action and Potential for Resistance. *Drug Resistance Updates*, **2004**, 7, 233–244.

65. Duthaler, U.; Huwyler, J.; Rinaldi, L.; et al. Evaluation of the Pharmacokinetic Profile of Artesunate, Artemether and their Metabolites in Sheep Naturally Infected with Fasciola hepatica. *Veterinary Parasitology*, **2012**, 186, 270–280.

66. Chadwick, J.; Jones, M.; Chadwick, A. E.; et al. Design, Synthesis and Antimalarial/Anticancer Evaluation of Spermidine Linked Artemisinin Conjugates Designed to Exploit Polyamine Transporters in Plasmodium falciparum and HL-60 Cancer Cell Lines. *Bioorganic & Medicinal Chemistry*, **2010**, 18, 2586–2597.

67. Calas, M.; Cordina, G.; Bompart, J.; et al. Antimalarial Activity of Molecules Interfering with Plasmodium falciparum Phospholipid Metabolism. Structure-Activity Relationship Analysis. *Journal of Medicinal Chemistry*, **1997**, 40, 3557–3566.

68. Calas, M.; Ancelin, M. L.; Cordina, G.; et al. Antimalarial Activity of Compounds Interfering with Plasmodium falciparum Phospholipid Metabolism: Comparison between Mono- and Bisquaternary Ammonium Salts. *Journal of Medicinal Chemistry*, **2000**, 43, 505–516.

69. Cloete, T. T.; Breytenbach, J. W.; de Koch, C.; et al. Synthesis, Antimalarial Activity and Cytotoxicity of 10-Aminoethylether Derivatives of Artemisinin. *Bioorganic & Medicinal Chemistry*, 20, **2012**, 4701–4709.

70. Efferth, T.; Kah, S.; Paulus, K.; et al. Phytochemistry and Pharmacogenomics of Natural Products Derived from Traditional Chinese Medicine and Chinese Materia Medica with Activity Against Tumor Cells. *Molecular Cancer Therapeutics*, **2008a**, 7, 152–161.

71. Blazquez, A. G.; Fernandez-Dolon, M.; Sanchez Vicente, L.; et al. Novel Artemisinin Derivatives with Potential Usefulness Against Liver/Colon Cancer and Viral Hepatitis. *Bioorganic & Medicinal Chemistry*, **2013**, 21, 4432–4441.

72. Singh, B.; Srivastava, J. S.; Khosa, R. L.; et al. Individual and Combined Effects of Berberine and Santonin on Spore Germination of Some Fungi. *Folia Microbiologica*, **2001**, 46, 137–142.

73. Arantes, F. F. P.; Barbosa, L. C. A.; Alvarenga, E. S.; et al. Synthesis and Cytotoxic Activity of Alpha-Santonin Derivatives. *European Journal of Medicinal Chemistry*, **2009**, 44, 3739–3745.

74. Wu, Z.; Wang, C.; Huang, M.; Tao, Z.; Yan, W.; Du, Y. Naturally Occurring Sesquiterpene Lactone-Santonin, Exerts Anticancer Effects in Multi-Drug Resistant Breast Cancer Cells by Inducing Mitochondrial Mediated Apoptosis, Caspase Activation, Cell Cycle Arrest, and by Targeting Ras/Raf/MEK/ERK Signalling Pathway. *Medical Science Monitor: International Medical Journal of Experimental and Clinical Research*, **2019**, 25, 3676–3682.

75. Ohhini, M.; Yoshimi, N.; Kawamori, T.; Ino, N.; Hirose, Y.; Tanaka, T.; Yamahara, J.; Miyata, H.; Mori, H. Inhibitory Effects of Dietary Protocatechuic Acid and Costunolide on 7, 12 Dimethylbenz[a]anthracene Induced Hamster Check Pouch Carcinogenesis. *Japanese Journal of Cancer Research*, **1997**, 88, 111.

76. Choi, J. H.; Ha, J.; Park, J. H.; Lee, J. Y.; Lee, Y. S.; Park, H. J.; Choi, J. W.; Masuda, Y.; Nakaya, K.; Lee, K. T. Costunolide Triggers Apoptosis in Human Leukemia U937 Cells by Depleting Intracellular Thiols. *Japanese Journal of Cancer Research*, **2002**, 93, 1327.

77. Lee, M. G.; Lee, K. T.; Chi, S. G., Park, J. H. Costunolide Induces Apoptosis by ROS-Mediated Mitochondrial Permeability Transition and Cytochrome C Release. *Biological and Pharmaceutical Bulletin*, **2001**, 24, 303.

78. Park, H. J.; Kwon, S. H.; Han, Y. N.; Choi, J. W.; Miyamoto, K.; Lee, S. H.; Lee, K. T. Apoptosis-Inducing Costunolide and a Novel Acyclic Monoterpene from the Stem Bark of Magnolia sieboldii. *Archives of Pharmacal Research*, **2001**, 24, 342.

79. Jeong, S. J.; Itokawa, T.; Shibuya, M.; Kuwano, M.; Ono, M.; Higuchi, R.; Miyamoto, T. Costunolide, a Sesquiterpene Lactone from Saussurea lappa, Inhibits the VEGFR KDR/Flk-1 Signaling Pathway. *Cancer Letters*, **2002**, 187, 129.

80. Bocca, C.; Gabriel, L.; Bozzo, F.; Miglietta, A. *Chemico-Biological Interactions*, **2004**, 147, 79.

81. Knoght, D. W. A Sesquiterpene Lactone, Costunolide, Interacts with Microtubule Protein and Inhibits the Growth of MCF-7 Cells. Feverfew: Chemistry and Biological Activity. *Natural Product Reports*, **1995**, 12, 271.

82. Garcia-Pineres, A. J.; Castro, V.; Mora, G.; Schmidt, T. J.; Strunck, E.; Pahl, H. L.; Merfort, I. Cysteine 38 in p65/NF-kappaB Plays a Crucial Role in DNA Binding Inhibition by Sesquiterpene Lactones. *Journal of Biological Chemistry*, **2001**, 276, 39713.

83. Subota, R.; Szwed, M.; Kasza, A.; Bugno, M.; Kordula, T. Parthenolide Inhibits Activation of Signal Transducers and Activators of Transcription (STATs) Induced by Cytokines of the IL-6 Family. *Biochemical and Biophysical Research Communications*, **2000**, 267, 329.

84. Zunino, S. J.; Ducore, J. M.; Storms, D. H. Parthenolide Induces Significant Apoptosis and Production of Reactive Oxygen Species in High-Risk Pre-B Leukemia Cells. *Cancer Letters*, **2007**, 254, 119.

85. Bedoya, L. M.; Abad, M. J.; Bermejo, P. The Role of Parthenolide in Intracellular Signalling Processes: Review of Current Knowledge. *Current Signal Transduction Therapy*, **2008**, 3, 82.

86. Sülsen, V. P.; Martino, V. S. *Sesquiterpene Lactone*. Springer, Argentiana, 2018, 332–340. doi: 10.1007/978-3-319-78274-4.

87. Hohmann, M. S.; Longhi-Balbinot, D. T.; Guazelli, C. F.; et al. Sesquiterpene lactones: structural diversity and perspectives as anti-inflammatory molecules. In: Atta-ur-Rahman (ed.) *Studies in Natural Products Chemistry*, Vol. 49. Elsevier, Amsterdam, 2016, 243–264.

17

An Overview of Lactarane: A New Class of Bio-Active Molecules

Bimal Krishna Banik[*]**, Abhishek Tiwari**[†]**, and Biswa Mohan Sahoo**[‡]

CONTENTS

17.1 Introduction..547
17.2 Biosynthesis of Lactarane and Marasmane Sesquiterpenes...547
17.3 Isolation of Lactarane and Marasmane Sesquiterpenes 548
17.4 Pharmacological Applications of Lactarane and Marasmane Sesquiterpenes...........................549
17.5 Synthetic Approaches toward Lactarane and Marasmane Sesquiterpenes 550
17.6 Ring Enlargements..550
17.7 Cyclizations..553
Conclusion 555
References..555

17.1 Introduction

Men have been fascinated by mushrooms and toadstools for a long time. They come in a wide range of colors and shapes. As far as fairy tales and hallucinations go, mushrooms haven't just sparked our imaginations. They're the spore-producing fruit bodies of fungi. For example, people know that some species have good taste and can be dangerous. The first has even led to farming and profiting from it. This is just one example of how mushrooms can be good for us. We've made a lot of medicines based on structures that were found in mushrooms.

Besides sugars, amino acids, fatty acids and nucleotides, which have a well-known and important role, living organisms are also known to make organic compounds that have a less important function, called secondary metabolites. The idea that secondary metabolites in mushrooms are to blame for the effects above led to a lot of research into their isolation, characterization and biological activity [1]. Compounds from mushrooms in the *Lactarius* genus have been found in a lot of different ways. Terpenes, which are also called terpenoids, isopentenoids, or isoprenoids, make up most of our body. It is made up of multiples of the C_5-unit isopentenyl pyrophosphate. (2) There are seven types of terpenes: Hemiterpenes (C_5): This repeating unit is also used to classify them. Monoterpenes

(C_{10}): This repeating unit is also used to classify them. Sesquiterpenoid C_{15}: This repeating unit is also used to classify them (C_{30}).

Most of the terpenes from Lactarius are sesquiterpenes, and the majority of them have either the **lactarane** skeleton (**1**) or the **marasmane** skeleton (**2**), but not both at the same time (Figure 17.1). A lot of sesquiterpenes found in nature come from mushrooms in the Lactarius and Russula groups.

17.2 Biosynthesis of Lactarane and Marasmane Sesquiterpenes

As a general rule, the way to make **lactarane** and **marasmane** sesquiterpenes is to make farnesyl pyrophosphate (**3**) to humulene (**5**) by connecting the two carbon C_2-C_3 that make up the molecule (Scheme 17.1). Further proton-induced ring closure of (**6**) yields the *protoilludane* carbocation, which via subsequent cyclobutane ring contraction rearranges to the cyclopropylcarbinyl carbocation (**7**). Finally, the rearrangement of cyclopropylcarbinyl carbocation (**7**) results in the seven-membered ring formation of the **lactarane** skeleton (**8**) [2–4].

Isolation of *two protoilludanes* (**6, 7**) from **Lactarius violascens** shows that this biosynthetic route can be used [5]. Another thing we know for sure is that in some mushrooms

[*] Department of Mathematics and Natural Sciences, College of Sciences and Human Studies, Prince Mohammad Bin Fahd University, Al Khobar, Kingdom of Saudi Arabia. bimalbanik10@gmail.com; bbanik@pmu.edu.sa.

[†] Faculty of Pharmacy, Pharmacy Academy, IFTM University, Lodhipur Rajput, Moradabad-244102, Uttar Pradesh, India. abhishekt1983@gmail.com.

[‡] Roland Institute of Pharmaceutical Sciences, Berhampur affiliated to Biju Patnaik University of Technology (BPUT), Rourkela, Odisha, India.

DOI: 10.1201/9781003008682-17

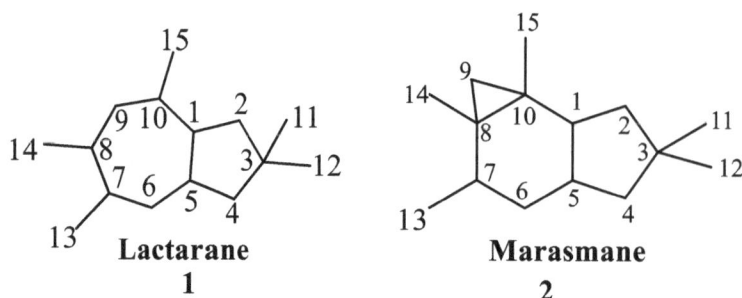

FIGURE 17.1   Carbon skeleton of lactarane and marasmane.

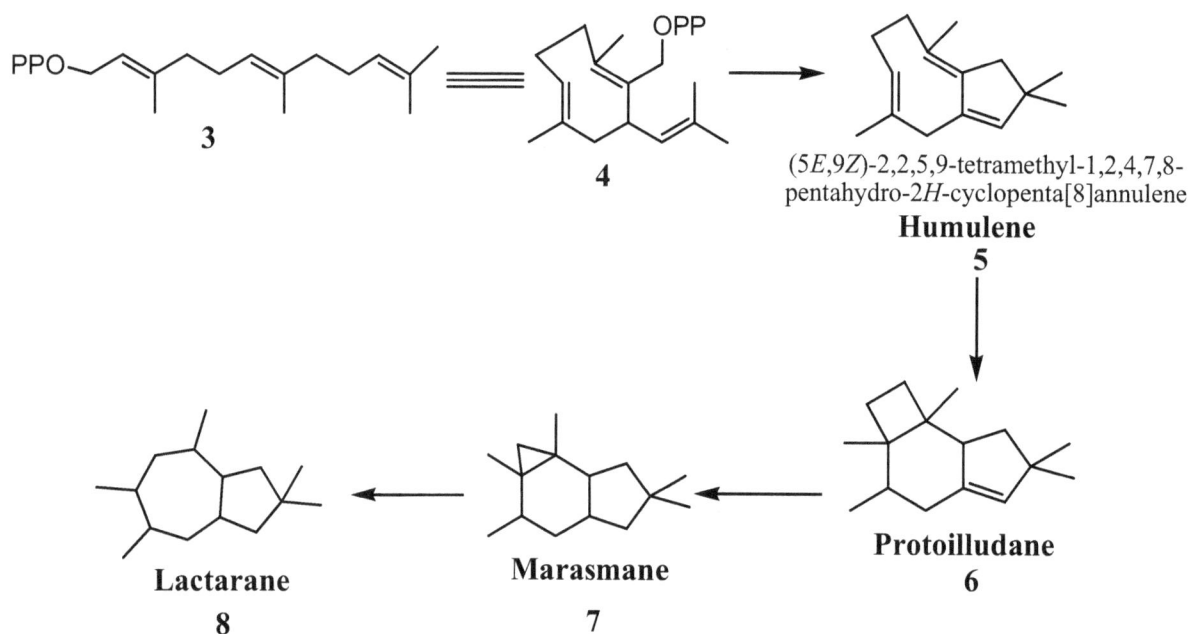

SCHEME 17.1   Synthesis of *Lactarane* sesquiterpenes from farnesyl pyrophosphate.

from the *Lactarius* genus, marasmanes are the precursors to lactaranes. *Velutinal* (**9**, R=H) is a type of sesquiterpene from the marasmane family that is made up of fatty acid derivatives. In young fruit bodies, this can make up to 6% of the dry weight of the fruit. Fruit bodies that have been cut or damaged release enzymes that make the velutinal esters in them turn into lactaranes, marasmanes and other things. Schemes 17.2 and 17.3 show how *Lactarius vellereus* might make stearoylvelutinal (**9**, where R is $CO(CH_2)_{16}CH_3$) [6,7]. First, the hydrolysis of the labile ester group in **9** (R=$CO(CH_2)_{16}CH_3$) may happen. This is because, in carefully taken extracts of *Lactarius vellereus*, "free" velutinal (**9**, R=H) can be found [6,7].

Enol (**10**) is made when an enzyme helps β-elimination of the epoxide in (**9**). Then, the marasmane isovelleral (**11**) is made (Scheme 17.2). A possible way to get lactaranes starts with an enzyme breaking open the epoxide ring. This results in the cyclopropylcarbinyl cation **13**, which then moves to the more energetically favorable carbocation **14** right away (Scheme 17.3). The C_9 to C_{10} hydride shift makes this cation more stable. At the same time, the allylic proton at C_{13}

is removed. Velleral (**16**), piperdial (**17**), or epi-piperdial lactaranes are the lactaranes that are formed (**18**).

17.3 Isolation of Lactarane and Marasmane Sesquiterpenes

Lactaranes are named after **Lactarius vellereus**, the plant from which the first lactarane was found. When one cut into the flesh of a mushroom from the genus *Lactarius*, one might get a milky substance. The marasmanes are named after *Marasmus conigenus*, which was the source of the first compound with this skeleton that could be found. In the past, people have talked about how to get lactaranes and marasmanes out of nature and figured out how they work [4,8]. Lactaranes and marasmanes, which come from *Lactarius*, are the subject of the most recent reviews. These mushrooms include *Russulas*, *Marasmius*, *Lentinellus* and other fungi that are very close to each other [8,9].

All of the lactaranes found in *Lactarius* have oxidized C_{13} and C_{14} positions, which are often connected to each other,

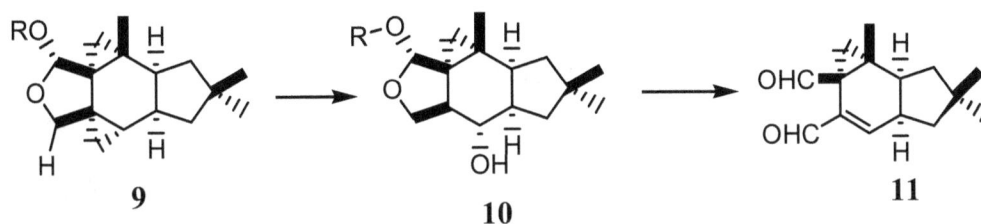

SCHEME 17.2 Synthesis of velutinal and its derivatives.

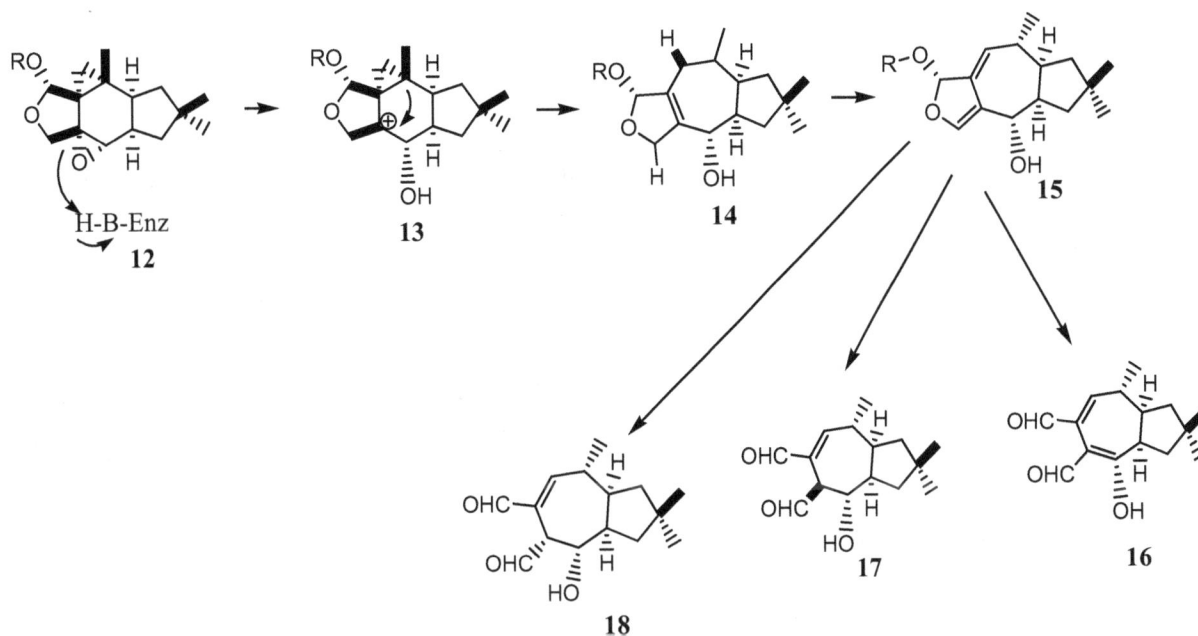

SCHEME 17.3 Formation of Velleral (**16**), piperdial (**17**), or epi-piperdial lactaranes (**18**).

forming a *furan* or *lactone* ring. Many lactaranes have unsaturated bonds (up to **three**) in the seven-membered ring that makes them up. Most of the time, there are *hydroxyl* groups in places such as C_6, C_7 and C_{10}. Sometimes, C_6 and C_{10} are linked together by an ether bridge. If you look at *furoscrobiculin D*, you'll see that the junction of the 5,7-member ring system is always *cis*. It is also found in the 5,6-membered ring system of naturally occurring *marasmanes*, where the cis ring junction can be found. This is true for both of the places where marasmane is oxidized, C_{13} and C_{14}. This makes them part of another ring system, like lactone or cyclic (hemi)acetal. Often, marasmanes have an unsaturated bond at the C_6 and C_7 positions. Sometimes, an epoxide or diol moiety is there, too, but it isn't always. The unfunctionalized C_{15} methyl group is found in almost every marasmane and lactarane that comes from *Lactarius*. Marasmanes from other genera usually have an oxidized C_{15} atom.

Some of the lactaranes and marasmanes that have been found are thought to be the result of the labile velutinal ester (**12**) being broken down or purified on a column [7,10]. Another way to make artifacts is to let them oxidize in the air [10]. However, there isn't a clear line between what is a natural product and what is an artifact.

17.4 Pharmacological Applications of Lactarane and Marasmane Sesquiterpenes

Lactarane and marasmane sesquiterpenes have been studied a lot in the past about how they affect the body [8]. Anti-feedant activity, or the ability to stop predators from eating without killing them, has been found in many of the compounds that have been isolated or made in the lab. Most of these antifeedant studies were done on pests that live in storage. The opossum, which is a natural fungivore, has also been shown to be a target. Most of the time, compounds with more hydroxyl groups are less active. Isovelleral (**11**) and velleral (**16**) are two of the strongest antifeedants on the market. As in these compounds and piperdial (**17**) and epi-piperdial (**18**), the unsubstituted dialdehyde moiety thought to be behind both of these things and the pungent taste of certain *Lactarius* species that aren't edible is thought to be the reason why. Also, the unfilled dialdehyde functionality found in other biologically active but biogenetically different terpenes is thought to be responsible for other activities, such as mutagenic and antibacterial and antifungal activities. It's possible that the velutinal esters (**12**) that make up most of the marasmanes and lactaranes are made

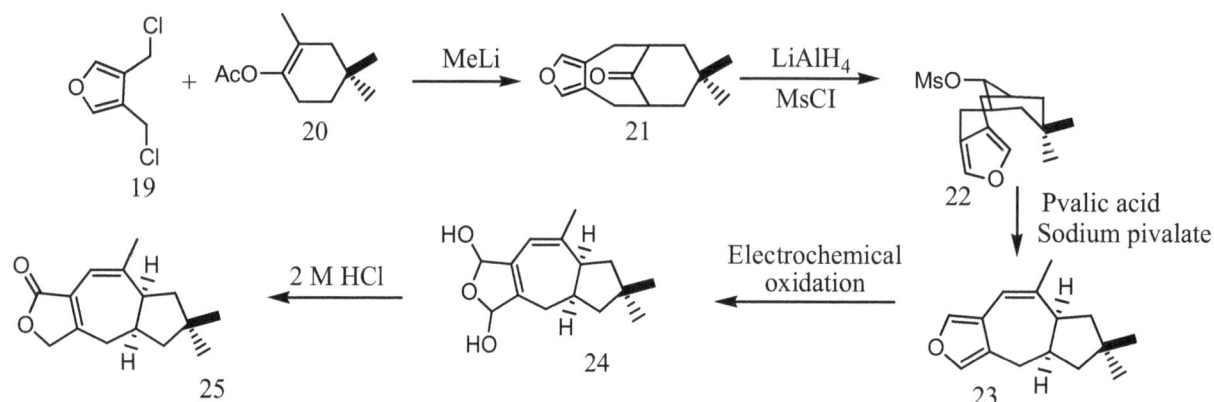

SCHEME 17.4   First total synthesis of Lactarane

when the mushroom's flesh is hurt. As soon as the mushroom is damaged, the velutinal esters that make up the mushroom's cell walls are exposed to the mushroom's cellular content and quickly enzymatically changed into the biologically active iso-velleral (8) or other unsaturated dialdehydes (e.g. **17, 18**). After a few minutes or hours, these compounds are changed by the mushroom into substances that are less toxic, preventing long-term contact with its own antifungal.

17.5 Synthetic Approaches toward Lactarane and Marasmane Sesquiterpenes

Scientists Daniewski and Vidari [8] reviewed various synthe-ses, and some syntheses of molecules strongly resembling the natural compounds are included.

For the synthesis of Lactarane sesquiterpenes, the usual method used is the ring expansion, which helped to make the 5,7-membered system. As part of the second group, cycloaddition and intermolecular cyclizations were used to do the main thing.

17.6 Ring Enlargements

As far back as 1975, **Froborget** and his coworkers wrote about the *first total synthesis* of a *lactarane* [11]. Even though the reported stereo-structure of the target molecule pyrovellerolac-tone (25) was wrong, the authors were able to get a sample of [12] that was exactly like the natural product. Later, the structure of **25** was changed again [13]. For the sake of clarity, the structures in Scheme 17.4 show the correct stereo- and regiochemistry. It starts with the cycloalkylation of **20** with 19 to make tricyclic **21**. One product was made when the carbonyl group in 21 was reduced. This product, mesylate **22**, was then used to make mesylate **23**. In the most important step of the synthesis, sodium pivalate was added to **22** to make **23** in a low yield. Electrochemical oxidation of **23** made dihydroxy-dihydrofuran **24** and some other material that had been oxidized too much. Hydrolysis of **24** made pyrovel-lerolactone (**25**) in a very small amount [13,14].

There were more efficient ways to get to 21 years later, but the same people also made velleral (**34**) and vellerolactone (**32**) in the same way [13]. [2+2] cycloaddition of enamine **28** with

a difunctionalized acetylene (**29**) was the key step in these syntheses. Then, the rings were heated to make them bigger (Scheme 17.5). **29** were used, depending on the substitution pat-tern of the target molecule that was being used. Deamination of compound **29** with diborane led to **30**, respectively, but the C_{10} allylic methyl group didn't turn into a different methyl group. The target molecules were made after more functional group changes.

In 1984, Christensen et al. wrote about how to make the basic core of the lactarane skeleton (**37**). (Scheme 17.6) [15]. Starting with the known diketone 35, dissolving metals and then mesylating them led to compound **36**, which when treated with TfOH changed into compound **37**. It took some research to figure out how to change the functional groups on **37**, but no natural product came out of this method.

In 1992, Thompson and Heathcock published the total syn-thesis of Hydrolactarorufin A (**39**) and lactarorufin A (**40**) [16]. Both compounds were obtained upon acid-catalyzed ring expansion of marasmane derivative **38** (Scheme 17.7).

Besides the categorization presented in this paragraph, a fur-ther subdivision can be made. This classification is based on the strategy used for the construction of the ring system before enlargement. Several, including the following syntheses, used a Diels-Alder approach to build up the initial ring system.

In 1994, Ogino et al. published their first synthesis of furo-scrobiculin B (**50**) [17], in which a furan ring transfer (FRT) reaction was used for the construction of a furan-annulated decalin system, which in turn was converted to a C_{15} nor-lactarane via a pinacol-type rearrangement (Scheme 17.8). Starting from dimedone (**41**), compound **42** was prepared in six steps. Basic treatment of **42** resulted in the generation of allene **43**, which underwent an intramolecular Diels-Alder reaction to intermediate **44**. Subsequent eliminative opening of the ether-bridge in **44** yielded the compounds **45** and **46** (ratio 1:3). It should be mentioned that no Diels-Alder reac-tion was observed when alkyne **42** was treated under neutral conditions, which indicated that the formation of allene **43** was essential for the FRT reaction. Compounds **45** and **46** were separately converted to 47, which via a two-step procedure gave **48** in low yield. Pinacol rearrangement of **48** gave **49** as a 5:2 mixture of diastereoisomers in high yield [18]. After sepa-ration, the major diastereoisomer was converted to furoscro-biculin B (**50**) and C_{10} epi-furoscrobiculin B (ratio 7:2).

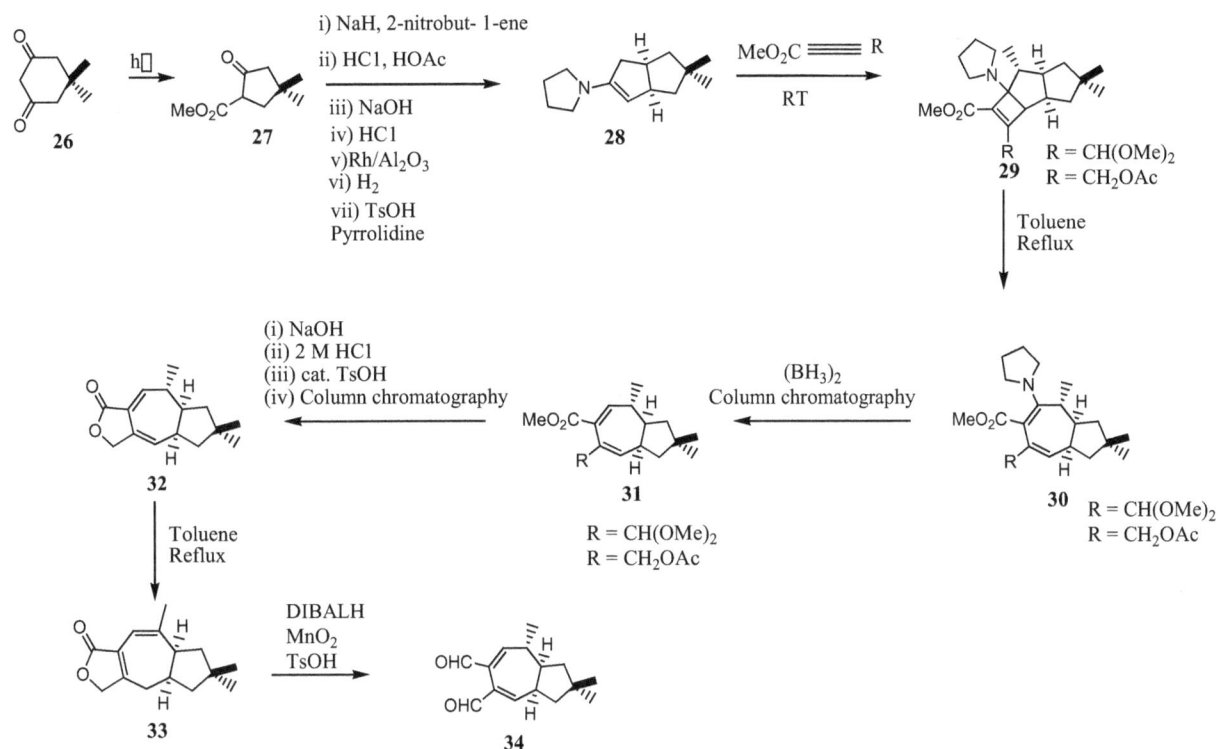

SCHEME 17.5 Synthesis of Velleral

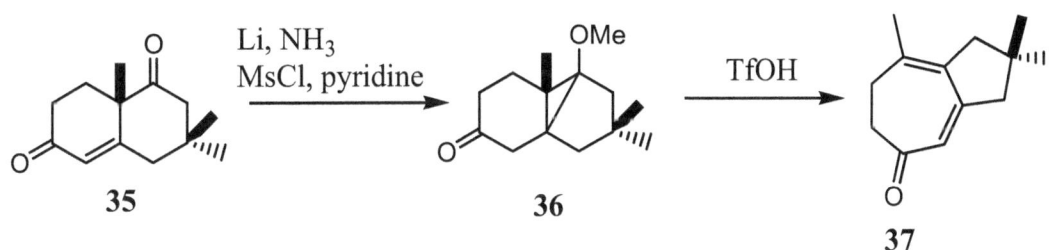

SCHEME 17.6 Construction of basic skeleton of lactarane.

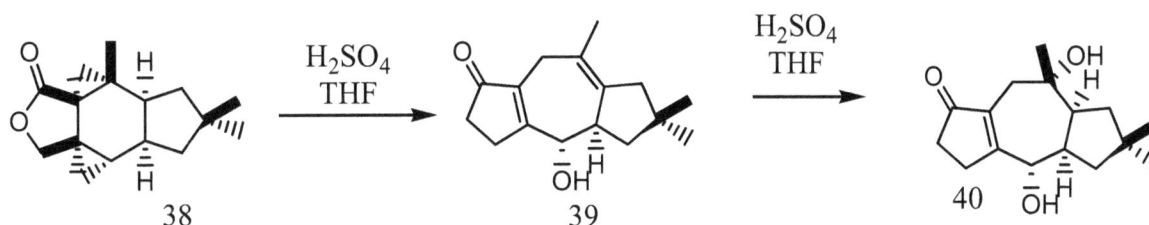

SCHEME 17.7 Hydrolactarorufin A.

In 1997, the same group made a slightly different version of furoscrobiculin B (**57**) [19]. In this method, the starting material, enone **51**, was changed in nine steps to **52**, which is the starting material for the FRT reaction (Scheme 17.9). **51** were formed when **52** was treated with tBuOK in tBuOH at 80°C. This was thought to be caused by the steric hindrance of the germinal methyl group in the intramolecular Diels-Alder reaction. Minor isomer **51** had to be deprotected first.

Then the major isomer **51** had to be deprotected, oxidized and selectively reduced with $Zn(BH_4)_2$ through a chelated intermediate of **51**. Pinacol rearrangement of **53** led to compound **54**, which after methylation led to the unnatural epifuroscrobiculin B as the main product after the methylation. To make furoscrobiculin B (**57**), **55** was first isomerized to enone **56**, then **57** was made in the same way as shown in Scheme 17.8 [17,18].

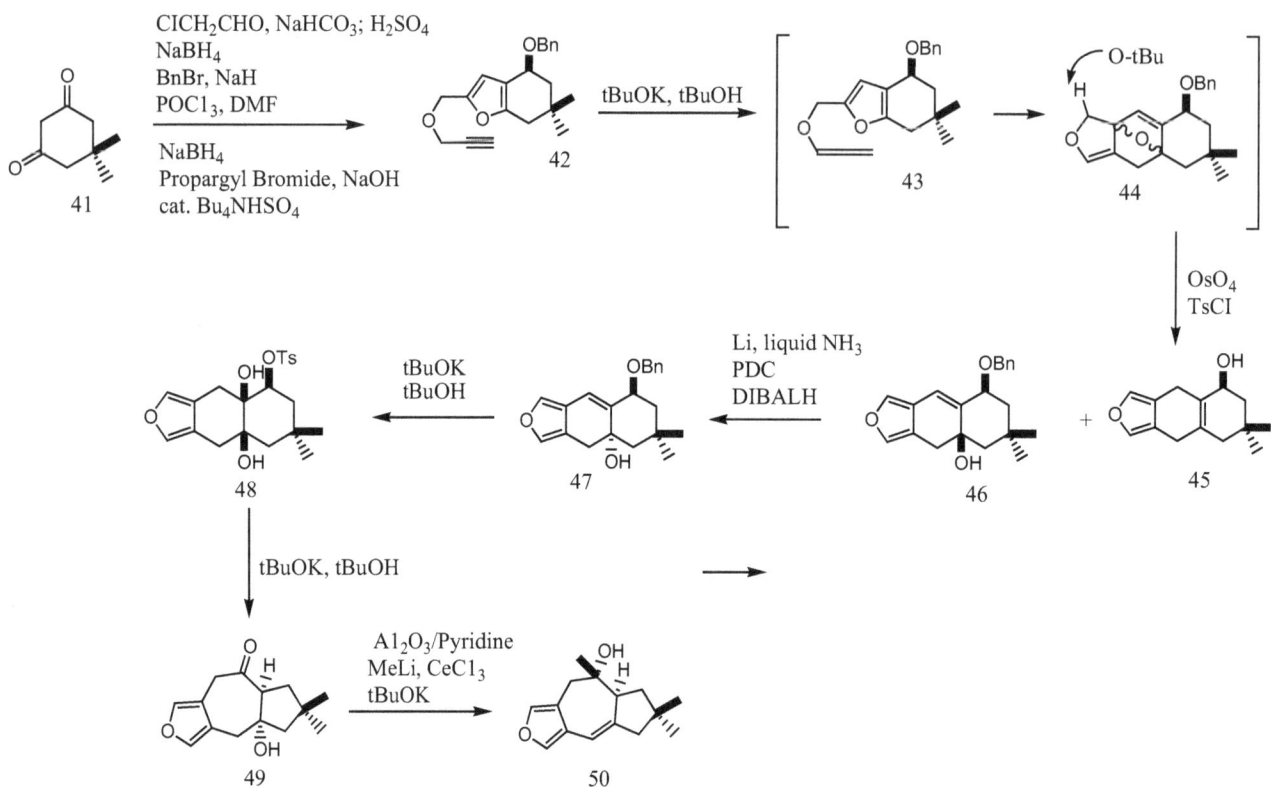

SCHEME 17.8 Furoscrobiculin B.

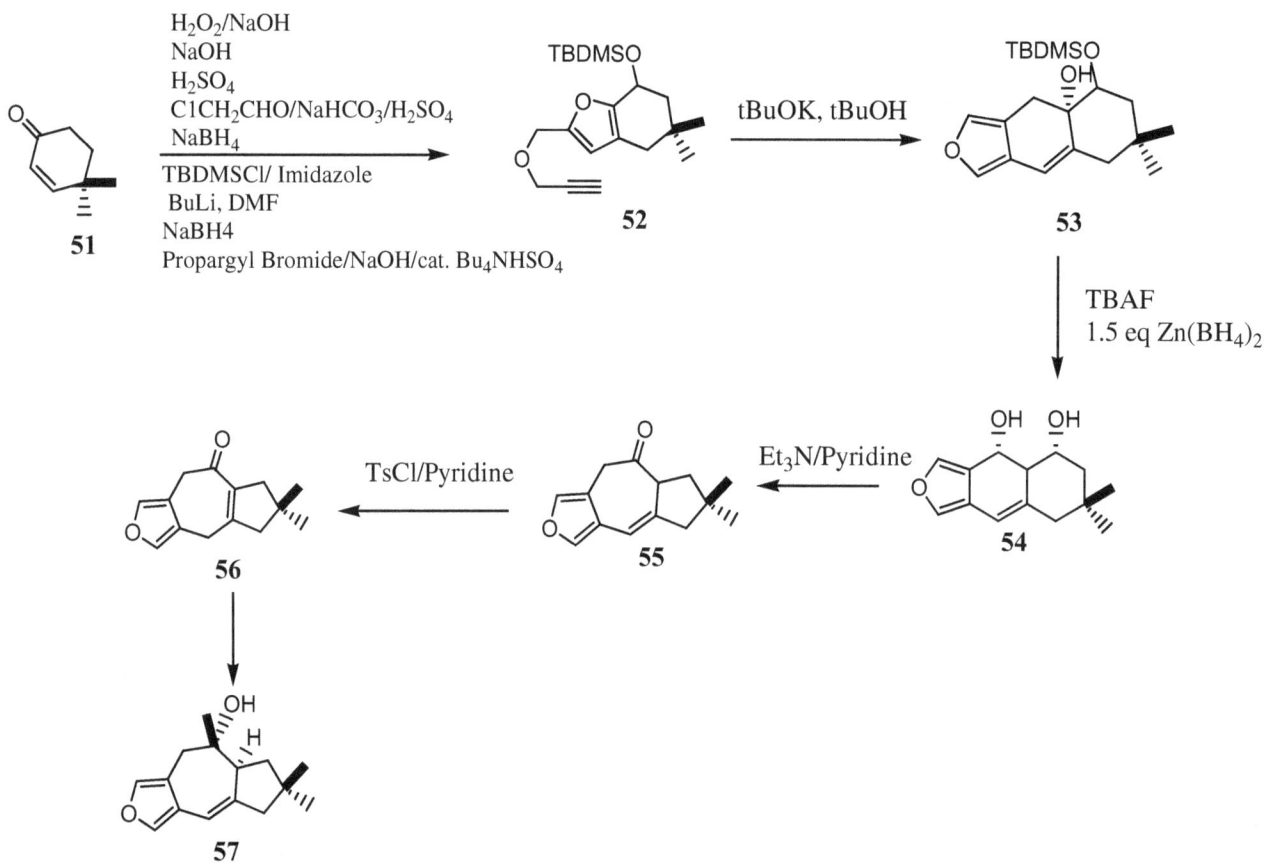

SCHEME 17.9 Furoscrobiculin B.

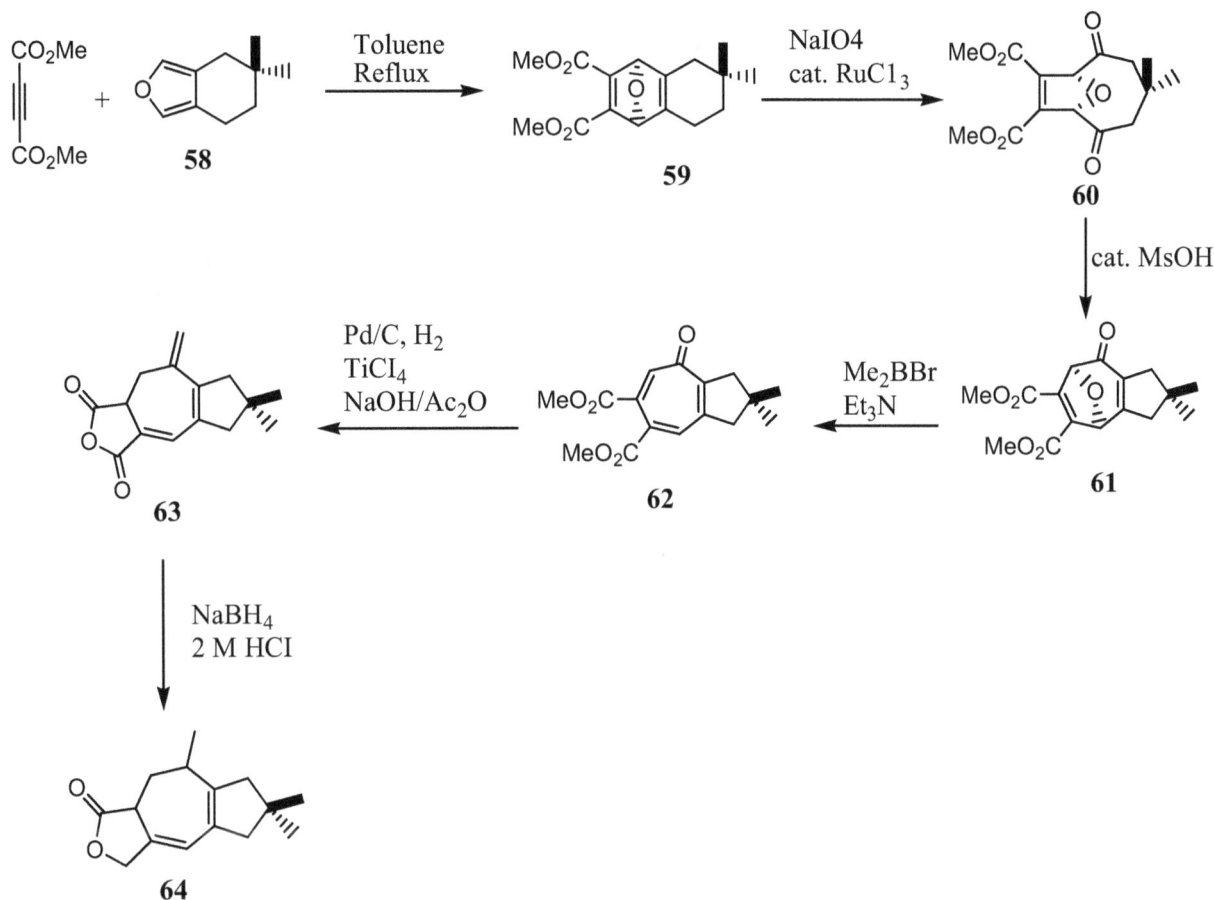

SCHEME 17.10 (10)-5(6)-bisanhydrolactarorufin A.

In 1997, Wockenful and other people wrote about how to make 1 (10)-5(6)-bisanhydro lactarorufin A [19,20] (**64**). For the lactarane skeleton, a Diels-Alder reaction of furan **58** with dimethyl acetylenedicarboxylate was used. Then the most electron-rich double bond in **59** was broken by oxidation (Scheme 17.10). When aldol condensation took place inside **60**, the geminal dimethyl group played an important role in the process. With $(CH_3)_2BBr$, the ether bridge in **61** was broken down. Then, the C_8-C_9 double bond in **61** was hydrogenated only in **62**. The methylation of the carbonyl group at C_{10} with 'theiNysted reagent was preferred to Witting olefination because it would not make the ester group at C_{15} turn into a different type of ester C_8. The diester moiety was turned into anhydride 63 under standard conditions. Then, $NaBH_4$ was used to reduce the anhydride **63** to 1(10)-5(6)-bisanhydrolactarorufin A (**64**). As the authors said, it was a little surprising that only the conjugated carbonyl group in **63** was reduced in the presence of an unconjugated carbonyl.

The prelactarane was made by Tochtermannet and others before this work [12,21]. Then, it didn't work out to make this compound into a **lactarane** further down the line.

17.7 Cyclizations

In 1989, Price and Schore came up with two slightly different ways to make furanether B (71) [22]. The first method

started with compound **65**, which was made from 2-methylfuran and tetrabromoacetone in a well-known way [23]. The ketone in 65 had to be reduced to avoid problems with identifying a carbonyl function that was added later in the process (Scheme 17.11) [22,24]. It was made from **66** (a 2:1 mixture of exo and endo) with the help of octa carbonyldicobalt, which did the Pauson-Khand reaction. Stereoselective: Only exo isomers were made in the reaction, but the regioselectivity (anti:syn = 1.5:1) was moderate, so only exo isomers were made. The four isomers in the reaction mixture were split into two pairs: the first one had the alcohols **67**, and the second one had the alcohols **68**. Both pairs were converted to the same intermediate **69** on their own. A mixture of different geometric isomers was made when **69** was formylated and then butanethiol was added. Treating **70** with trimethylsulfoniummethylsulfate made an epoxide, which changed when it was at room temperature. The rearranged product was then aromatized, and the result was furanether (**71**) [22–24].

The second method was given by Price and Schore, who came up with a more efficient way to get furanether B (**77**). The furan moiety was added earlier in the reaction sequence to stop isomers from being made at the same time (Scheme 17.11). Starting with compound **72**, the furan ring in compound **73** was added in the same way as in the first method. In this case, the Pauson-Khand cycloaddition only made **74** and **75**. This mixture was turned into **76**, and about 25% of

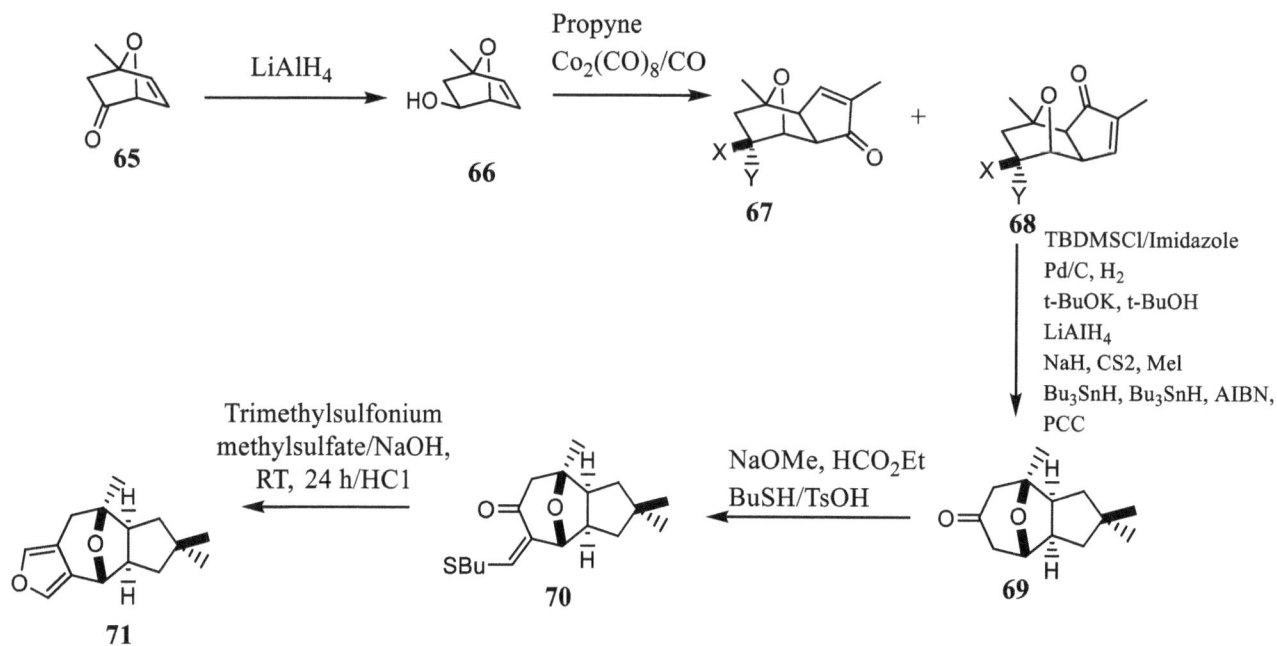

SCHEME 17.11   Synthesis of furanether B using trimethylsulfoniummethylsulfate

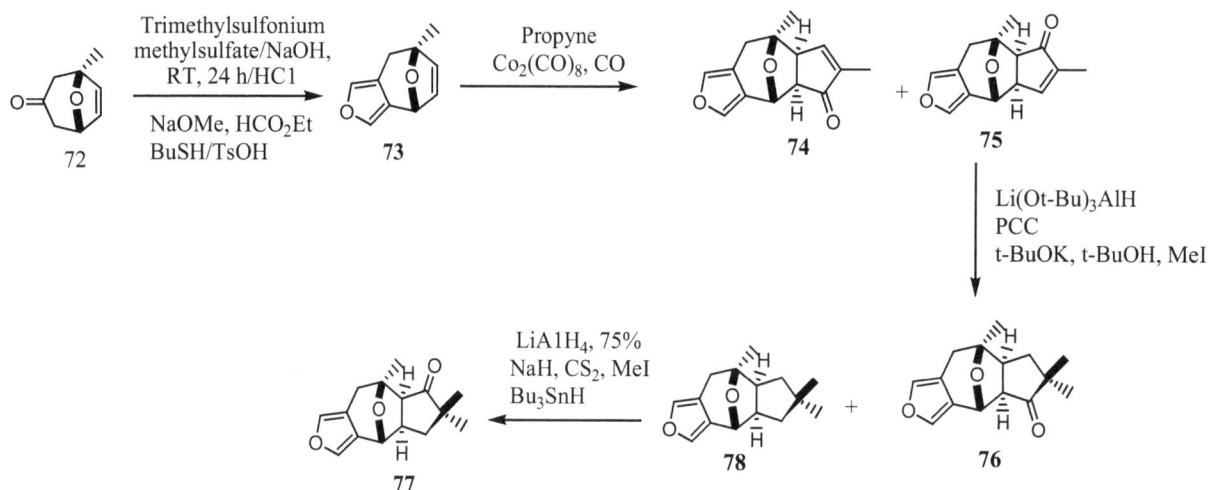

SCHEME 17.12   Synthesis of furanether B using LiAlH4

it was made with the ether bridge broken (Scheme 17.12). The mixture of **75** and **76** was treated with LiA1H4 and then deoxygenated, and the only thing that came out was furanether B (**77**).

In 1995, Molander et al. [24] wrote about another way to make furanether B (**86**). They used a [3+4] annulation reaction to make the core of the lactarane skeleton. They made the easily equilibrated compound **81** from **79** in the way shown in Scheme 17.13. [3+4] There was a good amount of

lactarane-like compound made when **81** was mixed with **82**. Decarboxylation proved that **83** had the same relative stereochemistry as **83**, which is what we already knew [22]. To get butanolide **85**, you had to change the acid in compound **83** into its triflate form and then reduce the ester group. Then, you used a palladium catalyst to make carbonylation. FuranetherB was made by reducing **85** with DIBALH (Di-isobutyl aluminum hydride) in CH_2Cl_2 to lactol and then dehydrating it (**86**) [25].

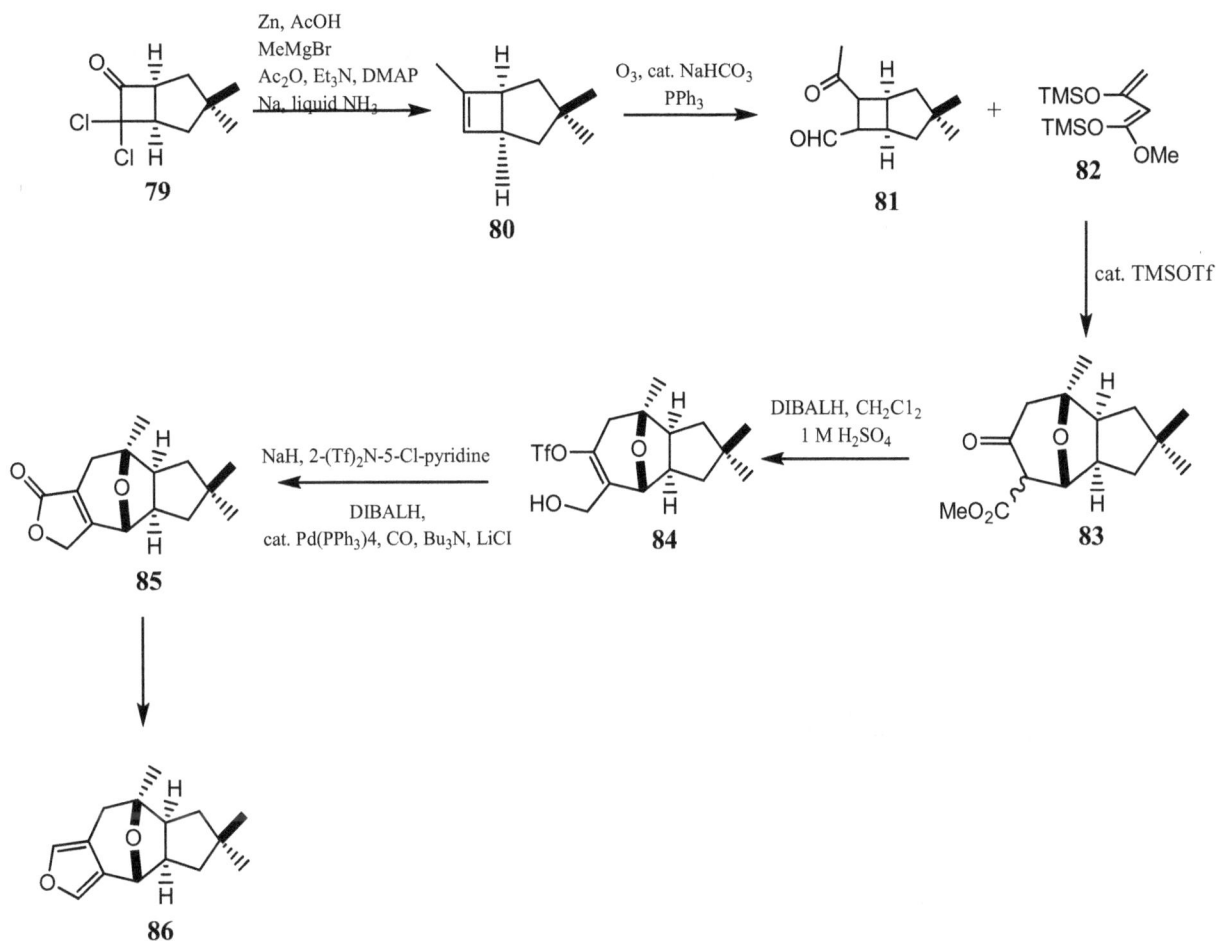

SCHEME 17.13 Synthesis of furanether B.

Conclusion

This chapter includes about biosynthesis and pharmacological applications of lactarane and marasmane sesquiterpenes. It includes the Isolation of Lactarane and Marasmane sesquiterpenes describes numerous processes like ring enlargements, Cyclizations and synthesis of Furanether B and Furoscrobiculin B etc. This chapter also includes the synthesis of Hydrolactarorufin A from marasmane derivative.

REFERENCES

1. Connolly, J. D.; Hill, R. A. *Dictionary of Terpenoids* (Vol. I). Chapman & Hall, London, **1991**.
2. Banthorpe, D. V. *Natural Products: Their Chemistry and Biological significance.* Longman Scientific & Technical: Harlow, **1994**.
3. Vidari, G.; Garlaschelli, L.; Rossi, A.; Vita Finzi, P. *Tetrahedron Letters*, **1998**, 39, 1957–1960.
4. Ayer, W. A.; Browne, L. M. *Tetrahedron*, **1981**, 37, 2199.
5. Sterner, T. Allergic Disease among Adolescents, IgE-sensitisation, Health-Related Quality of Life and physical activity. [Doctoral Thesis (compilation), Department of Clinical Sciences, Malmö]. Lund University, Faculty of Medicine. 2020.
6. Hanson, T. Examining the response of top marine predators to ecological change using stable isotope proxies. Ph. D. Thesis, Lund University, Faculty of Medicine. 2012.
7. Sterner, O.; Bergman, R.; Kihlberg, J.; Oluwadiya, J.; Wickberg, B.; Vidari, G.; De Bernardi, M.; De Marchi, F.; Fronza, G.; Vita Finzi, P. Basidiomycete sesquiterpenes: the Silica Gel Induced Degradation of Velutinal Derivatives. *Journal of Organic Chemistry*, **1985**, 50(7), 950–953.
8. (a) Vidari, G.; Vita Finzi, P. *Sesquiterpenes and Other Secondary Metabolites of Genus Lactarius (Basidiomycetes): Chemistry and Biological Activity* (Vol. 17). Elsevier, Amsterdam, **1995**. (b) Ayer, W. A.; Brandt, E. V.; Coetzee, J.; Daniewski, W. M.; Ferreira, D.; Malan, E.; Trifonov, L. S.; Vidari, G.. Progress in the Chemistry of Organic Natural Products. Springer-Verlag Wien 1999. https://doi.org/10.1007/978-3-7091-6366-5.
9. Garlaschelli, L.; Toma, L.; Vidari, G.; Colombo, D. Conformational studies and stereochemical assignments of the lactarane sesquiterpenes furoscrobiculin D and blennin D. *Tetrahedron*, **1994**, 50, 1211.
10. Sterner, O; Bergman, R; Kihlberg, J; Wickberg, B. The sesquiterpenes of Laetarius vellereus and Their Role in a Proposed Chemical Defense System. *Journal of Natural Products*, **1985**, 48, 279–288.

11. De Bernardi, M.; Garlaschelli, L.; Toma, L.; Vidari, G.; Vita-Finzi, P. The Chemical Basis of Hot-Tasting and Yellowing of the Mushrooms Lactarius chrysorrheus and L. scrobiculatus. *Tetrahedron*, **1993**, 49(7), 1489–1504.

12. Froborg, J.; Magnusson, G.; Thoren, S. Fungal Extractives. IX. Synthesis of the Velleral Skeleton and a Total Synthesis of Pyrovellerolactone. *Journal of Organic Chemistry*, **1975**, 40, 1595.

13. Magnusson, G.; Thoren, S. Fungal Extractives III. Two Sesquiterpenes Lactones from Lactrarius *Acta Chemica Scandinavica*, **1973**, 27, 1572.

14. Danielewski, M.; Matuszewska, A.; Szeląg, A.; Sozański, T. The Impact of Anthocyanins and Iridoids on Transcription Factors Crucial for Lipid and Cholesterol Homeostasis. *International Journal of Molecular Sciences*, 2021, 22, 6074.

15. Christensen, J. R.; Reusch, W. Preparation and Evaluation of a Perhydroazulene Intermediate for Lactarane Sesquiterpene Synthesis. *Canadian Journal of Chemistry*, **1984**, 62, 1954.

16. Thompson, S. K.; Heathcock, C. H. Total Synthesis of Some Marasmane and Lactarane Sesquiterpenes. *Journal of Organic Chemistry*, **1992**, 57, 5979–5989. DOI: 10.1021/jo00048a036.

17. Ogino, T.; Kurihara, C.; Baba, Y.; Kanematsu, K. Molecular Structures and Dimensions. *Journal of the Chemical Society, Chemical Communications*, **1994**, 1979.

18. Malagòn, O.; Porta, A.; Clericuzio, M.; Gilardoni, G.; Gozzini, D.; Vidari, G. Structures and Biological Significance of Lactarane Sesquiterpenes from the European mushroom Russula nobilis. *Phytochemistry*, **2014**, 107, 126–134. https://doi.org/10.1016/j.phytochem.2014.08.018

19. Evans, P.A.; Inglesby, P.A.; Kilbride, K. A Concise Total Synthesis of Pyrovellerolactone Using a Rhodium-Catalyzed [(3 + 2) + 2] Carbocyclization Reaction. *Organic Letter*, **2013**, 15(8), 1798–1801. doi: 10.1021/ol400165x.

20. Seki, M., Sakamoto, T., Suemune H., Kanematsu K. Total Synthesis of (±)-Furoscrobiculin B. *Journal of the Chemical Society, Perkin Transactions 1*, **1997**, 1707–1714.

21. Wilson, S. R.; Turner, R. B. Studies in Sesquiterpene Synthesis. The Marasmic Acid Skeleton. *Journal of Organic Chemistry*, **1973**, 38, 2870.

22. Tochtermann, W.; Bruhn, S.; Meints, M.; Wolff, C.; Peters, E.-M.; Peters, K.; Schnering von, H. G. Synthesis of 2(3)-8(9)-bis-anhydro-lactarorufin A *Tetrahedron*, **1995**, 51, 1623.

23. Segneanu, A.; Velciov, S. M. ; Olariu, S.; Florentina Cziple, F.; Damian, D.; Grozescu, I. Bioactive Molecules Profile from Natural Compounds. In T. Asao, & M. Asaduzzaman (Eds.), *Amino Acid - New Insights and Roles in Plant and Animal*. IntechOpen, London. 2017. https://doi.org/10.5772/intechopen.68643.

24. Molander, G. A.; Carey, J. S. Total Synthesis of Furanether B. An Application of a [3 + 4] Annulation Strategy. *Journal of Organic Chemistry*, **1995**, 60, 4845–4849.

25. Chen, I.T.; Baitinger, I.; Schreyer, L.; Trauner, D. Total Synthesis of Sandresolide B and Amphilectolide. *Organic Letters*, **2014**, 16(1):166–169. doi: 10.1021/ol403156r.

18

Terpenes: A Potential Chiral Pool for the Synthesis of Medicinally Privileged Bioactive Molecules

Bimal Krishna Banik*, **Abhishek Tiwari**†, and **Biswa Mohan Sahoo**‡

CONTENTS

18.1   Introduction 558
18.2   Diastereomers 558
18.3   Enantiomers (Chirality) 558
18.4   Chiral Terpenoids Synthesis 560
    18.4.1   Myrcene 560
    18.4.2   Citral 560
    18.4.3   Stereochemistry of Citral 562
18.5   Geraniol 562
    18.5.1   Linalool 563
    18.5.2   Lavandulol 563
    18.5.3   Citronellal 563
    18.5.4   Citronellol 563
    18.5.5   α-Terpineol 563
18.6   Synthesis of α-Terpineol Starts with p-Toluic Acid 564
    18.6.1   Carvone 564
    18.6.2   Carvone Oxidation and Reduction 564
    18.6.3   Limonene 566
    18.6.4   Menthol 566
    18.6.5   Menthone 567
18.7   Synthesis of Menthone and Menthol 567
    18.7.1   Pinene 568
    18.7.2   Total synthesis of α-Pinene 568
18.8   The Bornane (Camphene)-Norbornane (Isocamphane) Group 568
18.9   Camphor 568
18.10 Synthesis of Camphor (Haller, 1896) 570
18.11 Farnesene 570
18.12 Farnesol 570
18.13 Nerolidol 570
18.14 Metabolism Reactions of Chiral Terpenoids 572
18.15 Camphene 572
18.16 Camphor 572
18.17 Carvacrol 573
18.18 Carvone 573
18.19 1,4-Cineole 573
18.20 1,8-Cineole 574
18.21 Citral 575
18.22 Citronellal 575

* Department of Mathematics and Natural Sciences, College of Sciences and Human Studies, Prince Mohammad Bin Fahd University, Al Khobar, Kingdom of Saudi Arabia. bimalbanik10@gmail.com; bbanik@pmu.edu.sa.

† Faculty of Pharmacy, Pharmacy Academy, IFTM University, Lodhipur Rajput, Moradabad-244102, Uttar Pradesh, India. abhishekt1983@gmail.com.

‡ Roland Institute of Pharmaceutical Sciences, Berhampur affiliated to Biju Patnaik University of Technology (BPUT), Rourkela, Odisha, India.

DOI: 10.1201/9781003008682-18

18.23   Fenchone 575
18.24   Geraniol 575
18.25   Limonene 576
18.26   Linalool 576
18.27   Linalyl Acetate 576
18.28   Menthol 576
18.29   Myrcene 577
18.30   Pinene 577
18.31   Pulegone 578
18.32   α-Terpineol 578
18.33   α- and β-Thujone 579
18.34   Thymol 579
18.35   Caryophyllene 580
18.36   Farnesol 580
18.37   Longifolene 581
18.38   Patchouli Alcohol 581
18.39   Chiral Synthesis of Selected Terpenoids 582
         18.39.1   Synthesis of (+) Apiosporamide 582
18.40   Synthesis of (+)- Neosymbioimine 583
Conclusion 586
References 586

18.1 Introduction

Before contemporary chromatographic techniques, scientists often failed to separate isomer mixtures or mistook them for a single chemical. Isomers are molecules with the same molecular formula, such as a simple monoterpene with 10 carbons and numerous hydrogens and a hydrogen number that represents the amount of double bonds and cyclization (number of rings). $C_{10}H_{16}$ can represent hundreds of isomers of limonene. These "constitutional" isomers are not extremely similar; hence they are only recognized when contextually appropriate, i.e. when they are similar enough to be classed together, yet differ by atom-to-atom bonds. Alpha- and beta-pinene are constitutional isomers because they have the same atom-to-atom connection but distinct double bonds [1–5]. An example of a 'regioisomer' is 3-methoxy-4,5-hydroxybenzoic acid, contrasted to 4-methoxy-3,5-hydroxybenzoic acid. In both cases, the methoxy group is at position 3. These compounds are structural isomers because they are so similar. Stereoisomers are more closely related than constitutional isomers since they contain the same chemical formula and atom attachments, including double bond positions, but the atoms in bonds at a 'stereo-centre' point in different directions. Not all molecules have 'stereo-centres' or 'chiral centres,' but those that do are called stereoisomers. The hierarchy of isomers is shown in Figure 18.1 [6,7].

18.2 Diastereomers

Some plant species, including *Pelargonium crispum*, *Cymbopogon citratus* and Citrus x limon, contain a high concentration of diastereomers known collectively as citral, which contribute to their characteristic lemony aroma. In 1948, Ernest Guenther invented the term citral (sometimes spelled

lemonal) to describe the compound found in Citrus limon that is responsible for the fruit's signature lemon scent.

In the past, citral was thought to be a unique terpene aldehyde. However, recent research has disproved this theory after more than a decade. Historically, trans- and cis-citral were distinguished by the names "citral a" and "citral b," respectively; these terms have now been replaced by the more modern geranial (α-citral) and neral (β-citral), respectively. Specifically, these are stereoisomers known as cis and trans diastereomers. One class of diastereomers, induced by the geometry of the double bond, includes the cis and trans isomers. Citral a and citral b are two examples of double bond diastereomers, with the Z-(cis) and E-(trans) prefixes used to indicate the orientation of the priority groups surrounding the double bond. Some examples are Z, E-Nepetalactone (1), E, Z-Nepetalactone (2), γ-Cadinene (3), γ-Muurolene (4), (+)-Lpomeamarone (5) and (−) Ngaione (6) as shown in Figure 18.2 [8].

Priority groups with a Z- or cis designation are those that are geographically adjacent to one another, whereas priority groups with an E- or trans designation are those that are geographically distant from one another. Priority groups are defined by the Cahn, Ingold and Prelog priority rules, which rank elements in the order of their atomic numbers (O>N>C>H).

18.3 Enantiomers (Chirality)

To distinguish between the two possible spatial arrangements of atoms in a molecule, resulting in a mirror image form of the molecule, stereoisomers are defined in terms of 'chirality,' where a prefix like l- or d- stands for the Greek letters (−) or (+), respectively. The Greek root for the word chiral is kheir, which means "hand," hence enantiomers are analogized to

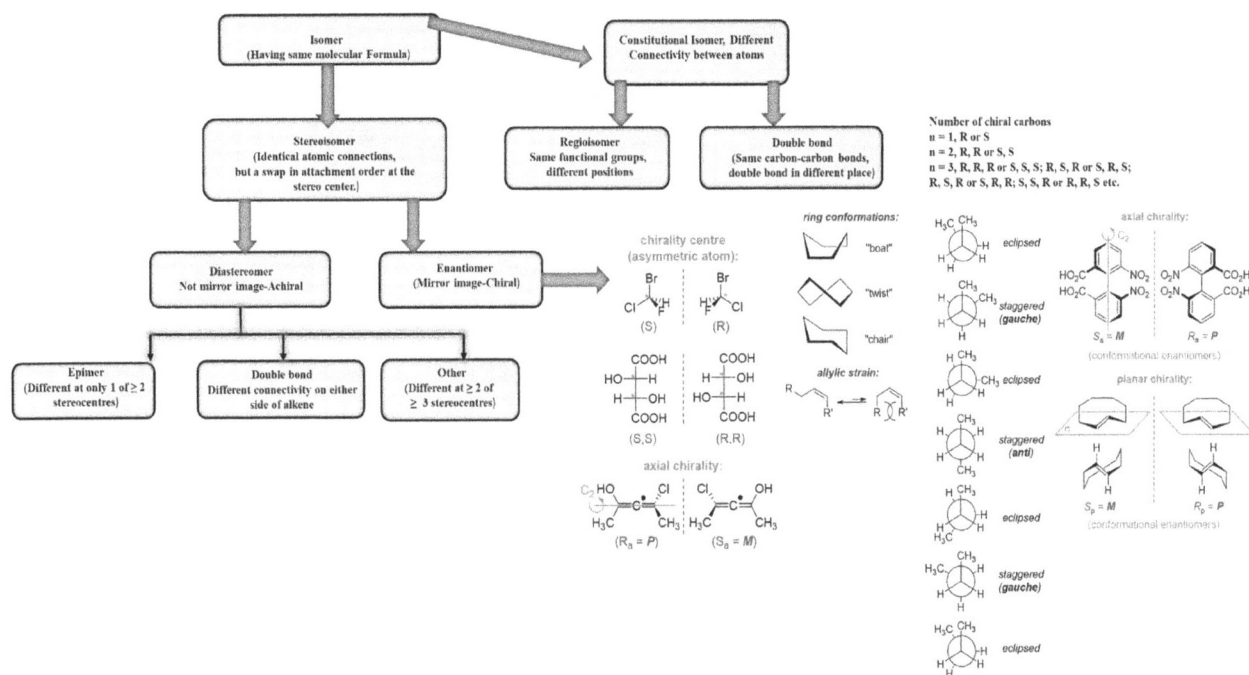

FIGURE 18.1 Isomer hierarchy in the nomenclature of organic chemistry.

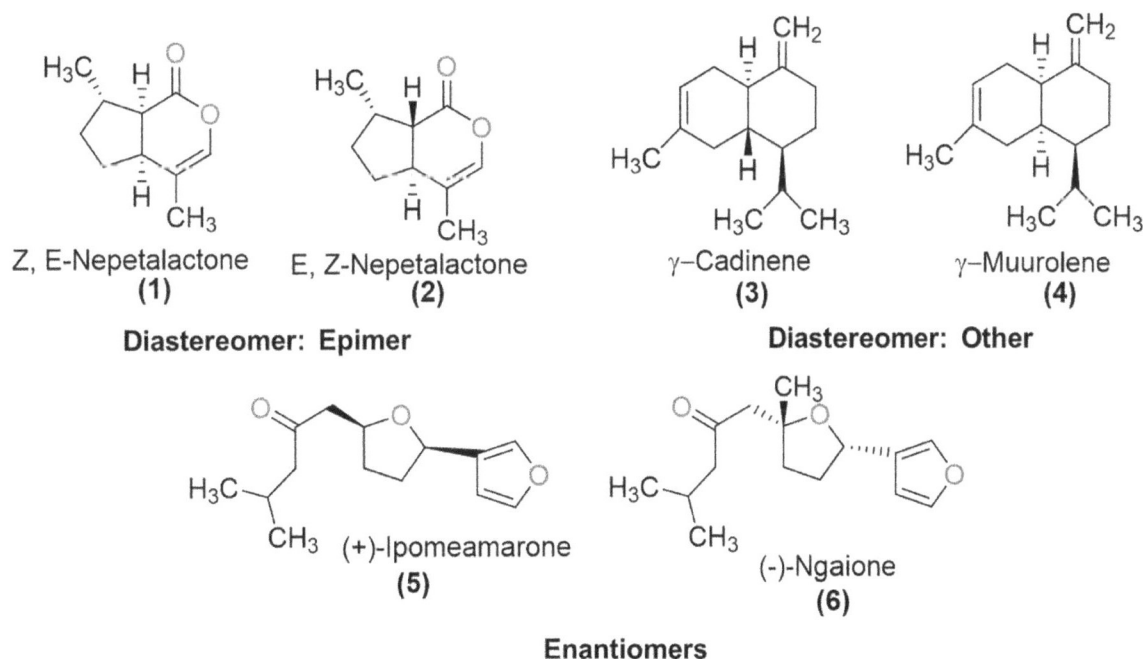

Z, E-Nepetalactone **(1)**   E, Z-Nepetalactone **(2)**

Diastereomer: Epimer

γ–Cadinene **(3)**   γ–Muurolene **(4)**

Diastereomer: Other

(+)-Ipomeamarone **(5)**

(-)-Ngaione **(6)**

Enantiomers

FIGURE 18.2 Examples of diastereomers and enantiomers.

non-superimposable left and right hands. The chiral carbon is a carbon atom whose substituents are bonded in a slightly different order than those of other carbon atoms. A chiral carbon is one with four distinct substituents. All chiral carbons in a molecule must be the mirror image of its enantiomer for the molecule to be an enantiomer; otherwise, the molecule will be a diastereomer due to the differing spatial configurations

at its n-1 chiral centres. The examples are (−) Citmellol (**7**), (−) Citronellal (**8**), Citronellene (**9**), (−) Carvone (**10**), (+) Carvone (**11**), (−) Dihydro-carvone (**12**), (+) Pulegone (**13**), (−)-Pulegone (**14**), (−) Perillaldehyde (**15**), α-Phellandrene (**16**), (−)-Menthone (**17**), Napatalactone (**18**) and Santonin (**19**) (Figure 18.3). Few chiral terpenoids, their sources and uses are shown in Table 18.1 [9,10].

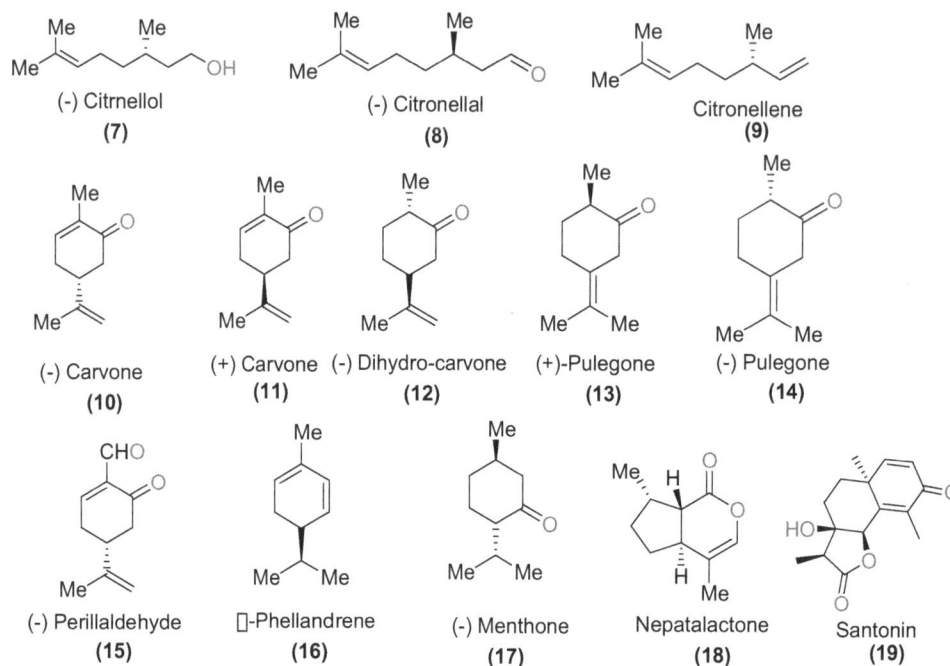

FIGURE 18.3 Example of enantiomers.

18.4 Chiral Terpenoids Synthesis

18.4.1 Myrcene

Verbena and bay oils contain this acyclic monoterpenoid hydrocarbon (i.e. terpene). Myrcene is an open-chain molecule with three double bonds that can be converted to a decane ($C_{10}H_{22}$) by catalytic hydrogenation (using platinum). To add to this, there are now two conjugated double bonds in myrcene because it has formed an adduct with maleic anhydride. The fact that optical exaltation is observed in myrcene lends credence to this conjugation. In addition to acetone, formaldehyde and $C_5H_6O_3$ (a ketodialdehyde), myrcene ozonolysis also yields succinic acid, carbon dioxide and chromic acid. Assigning structure 20(a) to myrcene provides an explanation for these findings. Terpenoid chemistry typically involves depicting the carbon skeleton as a ring 20(b) [11,12].

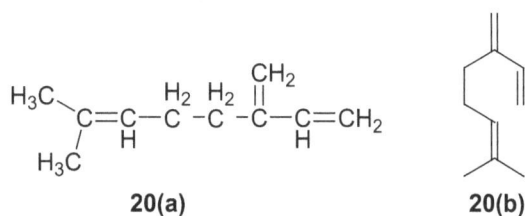

20(a)   **20(b)**

The systematic name of the compound is obtained by the use of the rule for acyclic polyenes. Therefore, 7-methyl-3-methyleneocta-1,6-diene is the correct chemical formula for myrcene. The ozonolysis and subsequent oxidation of the ketoaldehyde are depicted in the diagram below (Scheme 18.1) [12–14].

The formation of an alcohol upon hydration of myrcene (in the presence of suphuric acid) and the subsequent formation of citral upon oxidation provide more evidence for this proposed structure for myrcene. The structure of this compound

is known and its formation is in accord with the structure given to myrcene [15,16].

18.4.2 Citral

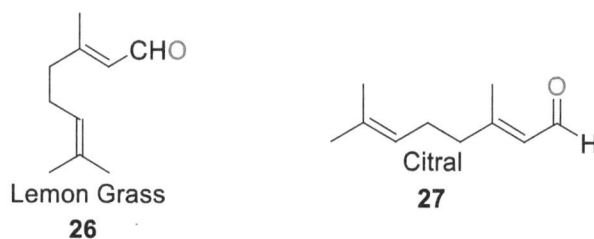

Given that the structures of the majority of the other acyclic monoterpenoids are derived from citral, this chemical deserves top billing within its genus ($C_{10}H_{16}O$). Lemon grass oil contains citral at concentrations between 60% and 80%. Lemon-scented citral is a liquid that goes by the name "citral." The presence of an oxo group in citral was previously established (it produces an oxime, for example). Citral is converted to p-cymene when heated with potassium hydrogen sulphate (28, 29). (Semmler, 1891). Using this reaction, Semmler was able to pinpoint where the citral methyl and isopropyl groups were located, confirming his earlier hypothesis that citral was acyclic (28) (two isoprene units joined head to tail).

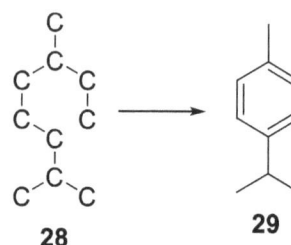

TABLE 18.1

Sources and Various Uses of Chiral Terpenoids [11–14]

S. No.		Natural Source	Category of Terpene	Application in Medicine
	Coenzyme Q,	vegetables specially legumes,	Acyclic polyisoprene	Heart failure and cancer
	1, 8 Cineol	*Eucalyptus globulus* and species	Monocyclic terpene C10	Expectorant, Bronchial asthma, anti-ulcer
	(–) Menthol	*Mentha piperita* and *arvensis*	Monocyclic terpene C10	Antibacterial, anti-ulcer, antispasmodic, antiseptic
	(–) Cis Carveol	*Mentha spicata*	Monocyclic terpene	Breast and prostate cancer
	(–) Zingiberene	*Zingiber officinalis*	Monocyclic sesquiterpene C15	Cosmetics industry, antiulcer, antiviral anticancer agent
	(–)-α-Bisabolol	*Matricaria chamomilla*	Monocyclic sesquiterpene C15	Cosmetic, Perfumery, antiulcer and anti-leishmaniasis
	(+) β-Eudesmol	*Atractylodes lancea, Pterocarpus santalinus Ginkgo biloba*	Bicyclic sesquiterpene C15	Anti-dementia, antiangiogenic which induces apoptosis
	(–)-α-Santonin	*Artemisia maritima* and species.	Bicyclic sesquiterpene C15	anthelmintic
	(–) Matricin	*Matricaria recutita, Chamomilla recutita*	bicyclic sesquiterpene C15	Anti-inflammatory, bowel disorders
	(+) Artemisinin	*Artemisia annua,*	Monocyclic sesquiterpene lacton C15	Antimalarial
	(–) Englerin A	*Phyllanthus engleri*	Bicyclic sesquiterpene C15	Renal cancer
	(–) Ginkgolide	*Ginkgo biloba* ginkgo tree in China	Same as above	Alzheimer's, Ischemic and haemorrhagic stroke, anti-inflammatory
	(–) Merrilactone	*Illicium merrillianum*	Same as above	Alzheimer's and Parkinson's diseases
	(–) Sclareol	*Salvia sclarea*	Bicyclic diterpene alcohol C20	Perfumery, Leukemia, breast and colon cancer,
	(–) Forskolin	*Plectranthus barbatus* (*Coleus forskohlii*) shrub in Brazil, Africa and China	Same as above	Heart failure, Psoriasis, autoimmune disorders
	(–) Acanthoic acid	*Acanthopanax koreanum*	Tricyclic diterpene C20	Anti-inflammatory agent, Promoter of apoptosis
	(–) 7-Oxo-13α-Hydroxyabiet-8(14)-en-18-onic acid	*Pinus sp*	Same as above	Cytotoxic activity
	(–)-Taxol	*Taxis brevifolia*	Same as above	Cytostatic agent
	(+) Resiniferatoxin	*Euphorbia resinifera*	Same as above	Anti-inflammatory, cancer
	(–)-Trigocherrin	*Trigonostemon cherrieri*	Same as above	Chikungunya-fever
	(+) Ingenol	Euphorbia peplus	Tetracyclic diterpeneC20	Actinic keratoses
	(+) Cucurbitacin B	*Cucurbitaceae sp.*	Same as above C30	Anticancer, hepatoprotective anti-inflammatory drug
	(–) Betulin	*Betula sp.*	Pentacyclic triterpene alcohol C30	Anticancer, anti-HIV, antibacterial and antiviral effect
	(–) Avicin D	Acacia victoriae	Saponin C30	Anticancer agents, initiators of cell apoptosis
	β-Carotene	*Solanum lycopersicum, Daucus carota,*	Bicyclic carotenoid C40	Age-related macular degeneration

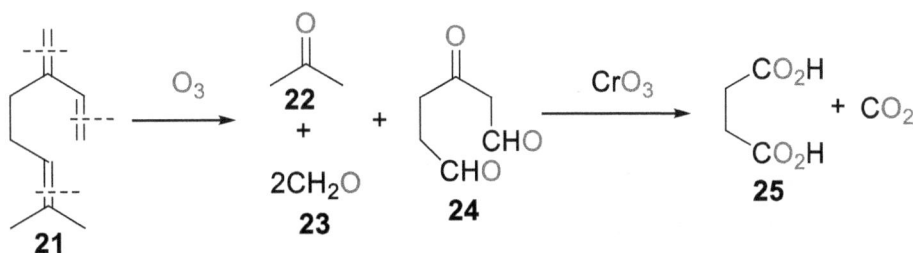

SCHEME 18.1 Ozonolysis and oxidation of myrcene.

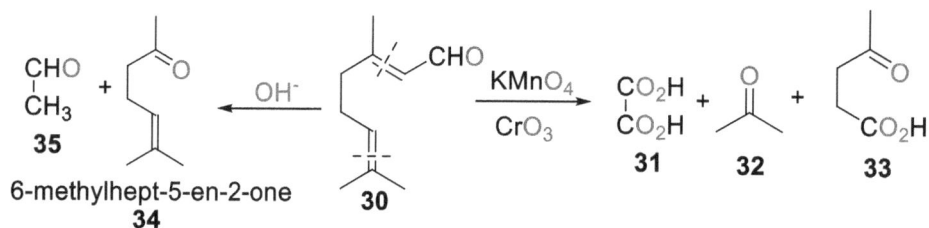

SCHEME 18.2 Oxidation of geranial.

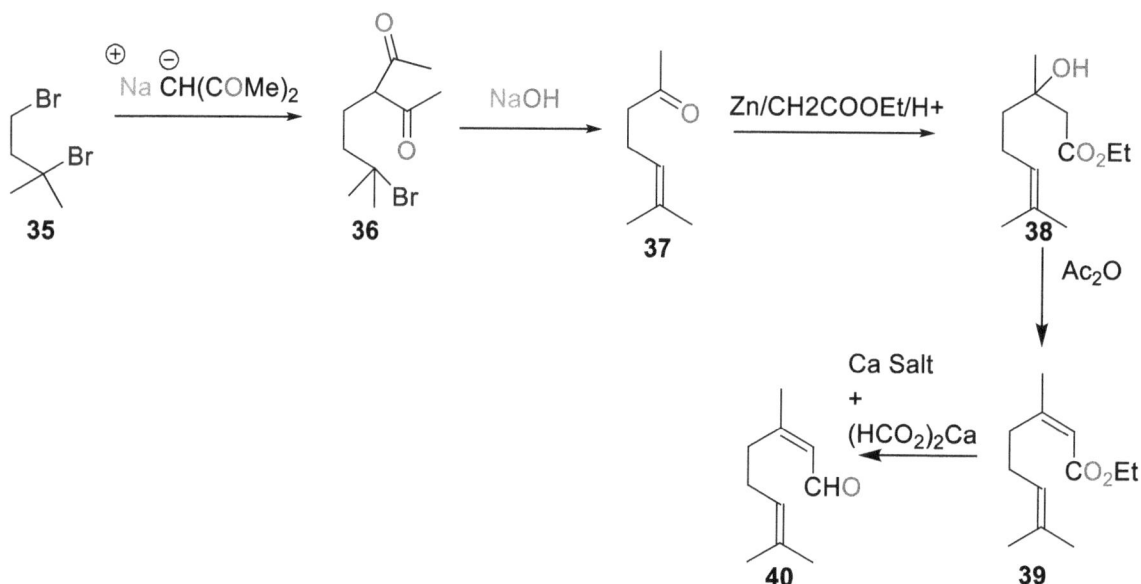

SCHEME 18.3 Synthesis of geranial.

Since no carbon is lost during oxidation to the acid, the oxo group in citral must be an aldehyde. Sodium amalgam can convert citral to the alcohol geraniol ($C_{10}H_{18}O$), while silver oxide oxidises it to geranic acid ($C_{10}H_{16}O_2$). When citral is exposed to alkaline permanganate and then chromic acid, three different acids are produced: acetone, oxalic and laevulic (Scheme 18.2) [12,17].

Verley's research provides additional evidence for the structure by showing that citral is converted to 6-methylhept-5-en-2-one (34) and acetaldehyde when treated with aqueous potassium carbonate and to oxalic acid (31), acetone (32) and 2-keto-pentanoic acid when treated with potassium permanganate (33). Assuming (30) suffers cleavage at the, -double bond, which is a common reactivity of, -unsaturated oxo compounds to alkaline reagents, provides a simple explanation for the generation of these products. Citral's structure was verified via the synthesis of methylheptenone (37), its subsequent conversion into geranic ester, and finally, its realization via the heating of a combination of the calcium salts of geranic and formic acids (Scheme 18.3) [12,18,19].

18.4.3 Stereochemistry of Citral

Citral can exist as two geometrical isomers, as evidenced by its formula. With regard to the methylene group in the main chain, the functional group (aldehyde) is either trans or cis. Citral-a (also

known as geranial) has a boiling point of 118°C–119°C/20mm. while citral-b (also known as neral) has a boiling point of 117°C–118°C/20mm. Both isomers are found in nature; for example, citral produces two semicarbazones. From analyzing the ring closures of the matching alcohols, the structures of these two forms have been determined (geraniol) [12,17,20].

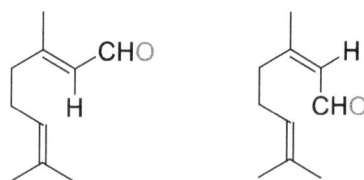

trans-form (or E-),   *cis*-form (or Z), Citral-b
Citral-a, Geraniol   Neral
42   **43**

18.5 Geraniol

Geraniol, is found in many essential oils, particularly rose oil. Hence, geraniol and nerol are geometrical isomers. Nerol occurs naturally in various essential oils, *e.g.*, oil of neroli, bergamot, etc.; its b.p. is 225°C–226°C. Geraniol has been assigned the *trans* configuration and nerol the *cis* on the fact

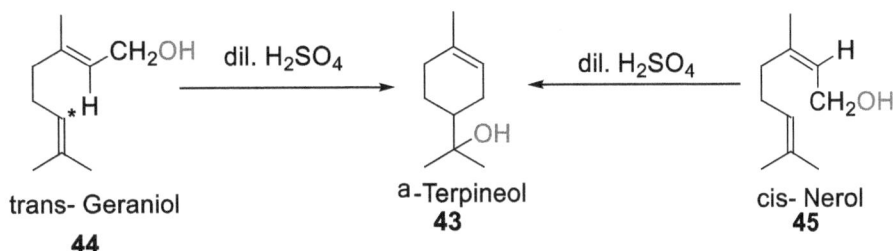

SCHEME 18.4 Reaction of from trans-geraniol.

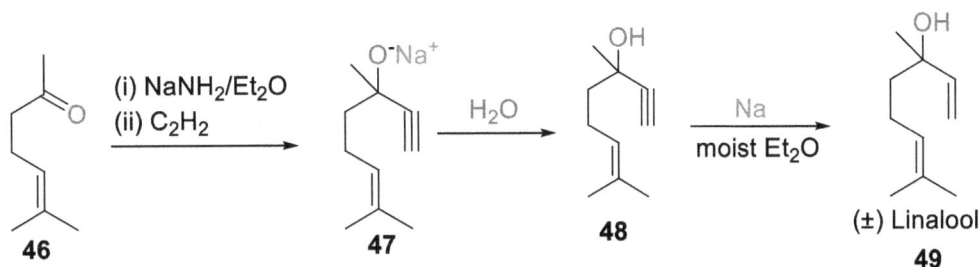

SCHEME 18.5 Synthesis of (±) linalool.

that cyclization to α-terpineol by means of dilute sulphuric acid takes place about nine times as fast with nerol as it does with geraniol; this faster rate with nerol is due to the proximity of the alcoholic groups to the carbon (*) which is involved in the ring formation (Scheme 18.4) [12,21,22].

18.5.1 Linalool

Linalool, an optically active molecule, can be found in two different forms: the (−)-form is found in rose oil, and the (+)-form is found in orange oil. Synthesis of linalool by treating the sodium derivative of methylheptenone with acetylene, followed by partial reduction of the triple bond, validated the structure of linalool (Scheme 18.5).

18.5.2 Lavandulol

Both free and ester forms of this compound can be found in French lavender oil. Because it deviates from the special isoprene rule in that its two isoprene units are not connected head to tail, this acyclic monoterpenoid is of particular interest.

18.5.3 Citronellal

The citronella oil contains an optically active component. Citronellol, $C10H_{20}O$, is the alcohol formed when citronellal is reduced with sodium amalgam, while citronellic acid, **(52)** $C_{10}H_{18}O_2$, is formed when citronellol is oxidized. When citronellal is reacted with chromic acid, 3-methyladipic acid and acetone are produced (Scheme 18.6).

18.5.4 Citronellol

There is a (−)-form of this in rose and geranium oils. Degradation by oxygen (to produce acetone and 3-methyladipic

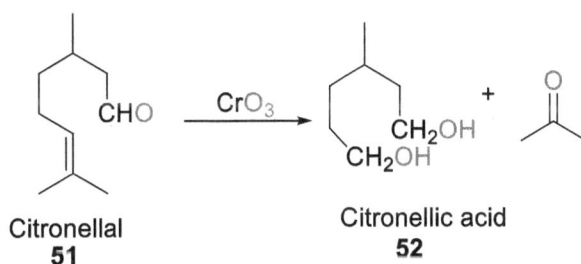

SCHEME 18.6 Synthesis of citronellic acid.

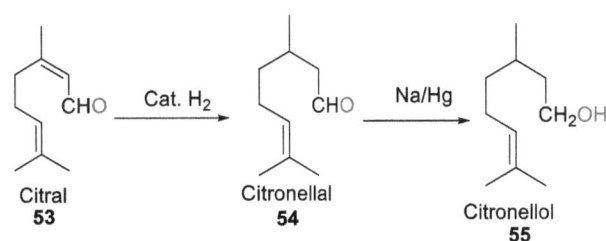

SCHEME 18.7 Synthesis of citronellol.

acid) and a series of other reactions elucidated its molecular make-up (Scheme 18.7).

18.5.5 α-Terpineol

The (+) and (−) forms of this optically active monoterpenoid occur in nature; the racemic modification has a melting point of 35°C and exists as a solid. $C_{10}H_{18}O$ is the chemical formula for α-terpineol, and its reactions demonstrate that the oxygen atom is bonded to a tertiary alcoholic group. When two bromine atoms are attached to α-terpineol, it becomes a compound with a single double bond. The chemical formula for the parent (saturated) hydrocarbon from which α-terpineol is derived is $C_{10}H_{20}$. This is consistent with C_nH_{2n}, the generic formula for the (monocyclic) cycloalkanes, and so confirms that α-terpineol is a monocyclic molecule.

When a-terpineol is cooked in sulfuric acid, a little amount of p-cymene is produced. This, along with the hypothesis that a-terpineol is monocyclic, suggests that the p-cymene skeleton must be present in a-terpineol. From this, we can deduce that -terpineol is likely p-menthane with one double bond and a tertiary alcoholic group. With the help of graded oxidation, Wallach was able to determine where exactly these functional groups were situated.

The (+), (−), and (±)-forms of this optically active monoterpenoid occur in nature; the racemic modification has a melting point of 35°C and exists as a solid. $C_{10}H_{18}O$ is the chemical formula for α-terpineol, and its reactions demonstrate that the oxygen atom is bonded to a tertiary alcoholic group. When two bromine atoms are attached to -terpineol, it becomes a compound with a single double bond. The chemical formula for the parent (saturated) hydrocarbon from which α-terpineol is derived is $C_{10}H_{20}$. This is consistent with C_nH_{2n}, the generic formula for the (monocyclic) cycloalkanes, and so confirms that α-terpineol is a monocyclic molecule.

When α-terpineol is cooked in sulfuric acid, a little amount of p-cymene is produced. This, along with the hypothesis that a-terpineol is monocyclic, suggests that the p-cymene skeleton must be present in a-terpineol. From this, we can deduce that -terpineol is likely p-menthane with one double bond and a tertiary alcoholic group. With the help of graded oxidation, Wallach (1893, 1895) was able to determine where exactly these functional groups were situated (Scheme 18.8) [12,23–26].

18.6 Synthesis of α-Terpineol Starts with *p*-Toluic Acid

Perkin, jr., Meldrum, and Fisher successfully synthesized α-terpineol from p-toluic acid (Scheme 18.9).

A much simpler synthesis of α-terpineol has been carried out by Alder and Vogt this makes use of the Diels-Alder reaction, using isoprene and methyl vinyl ketone as the starting materials (Scheme 18.10) [12,27].

Two other terpineols are also known: β- and γ-terpineol; the latter occurs naturally.

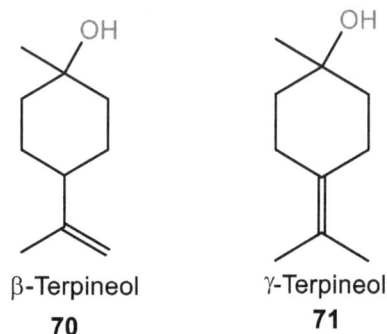

β-Terpineol **70** γ-Terpineol **71**

18.6.1 Carvone

This is found in both the optically active and the racemic modifications in a number of essential oils, including those of spearmint and caraway. Carvone's structure is predicated in large part on the fact that it can be synthesized from α-terpineol in the following manner (Scheme 18.11):

It follows Markownikoff's rule that adding nitrosyl chloride to α-terpineol (72) results in α-terpineol nitrosochloride (73) (the chlorine is the negative part of the addendum). The nitrosochloride in question undergoes a spontaneous rearrangement to an oxime molecule (74). By adding sodium ethoxide to (74), product (75), which, upon heating in dilute sulphuric acid, loses a water molecule while hydrolyzing the oxime to provide carvone (76). Therefore, carvone is p-menth-6,8-dien-2-one, as suggested by the reactions.

18.6.2 Carvone Oxidation and Reduction

The oxidation and reduction of carvone is shown in Scheme 18.12 [28,12].

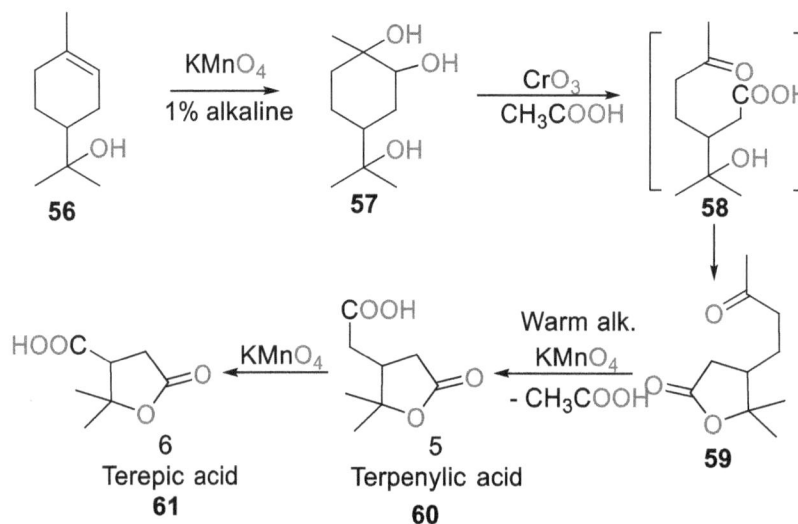

SCHEME 18.8 Synthesis of terepic acid.

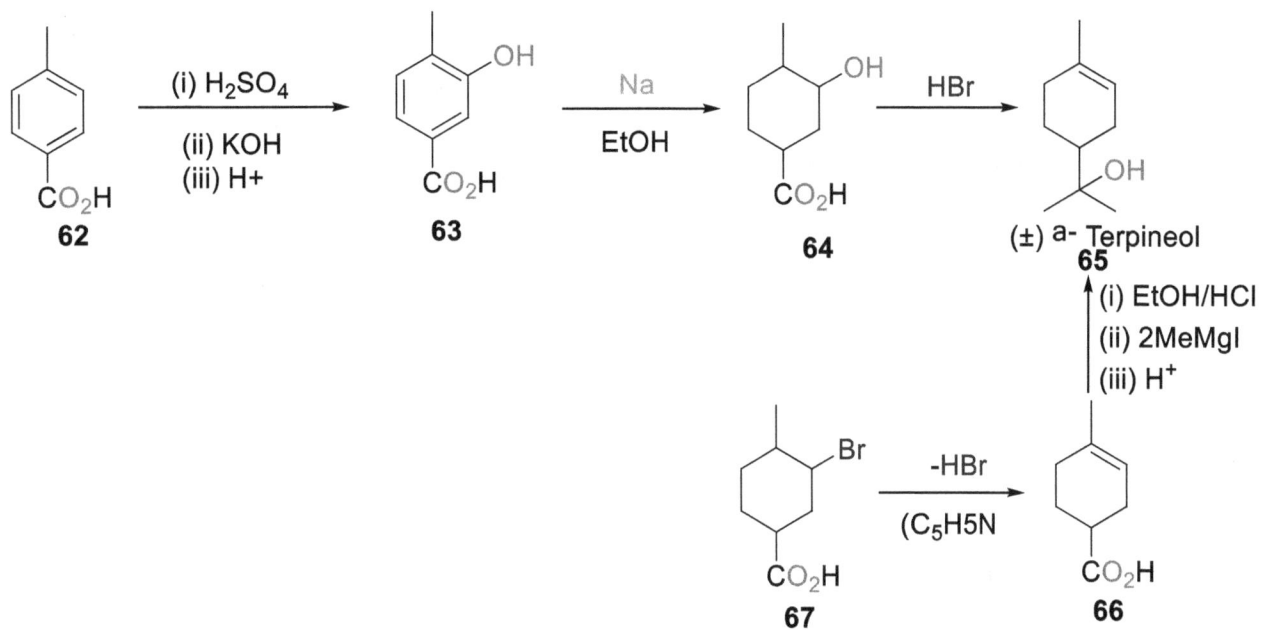

SCHEME 18.9 Synthesis of α-terpineol starts with *p*-toluic acid.

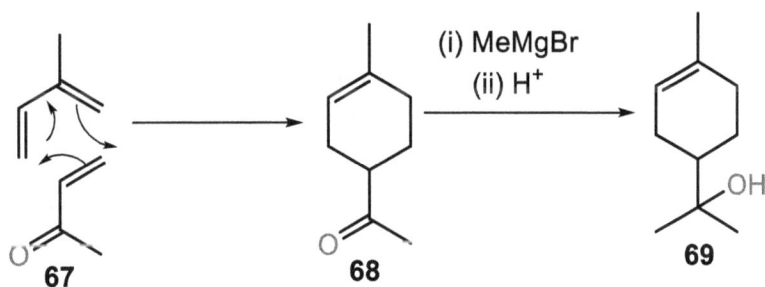

SCHEME 18.10 Diel's- Alder reaction of α-terpineol.

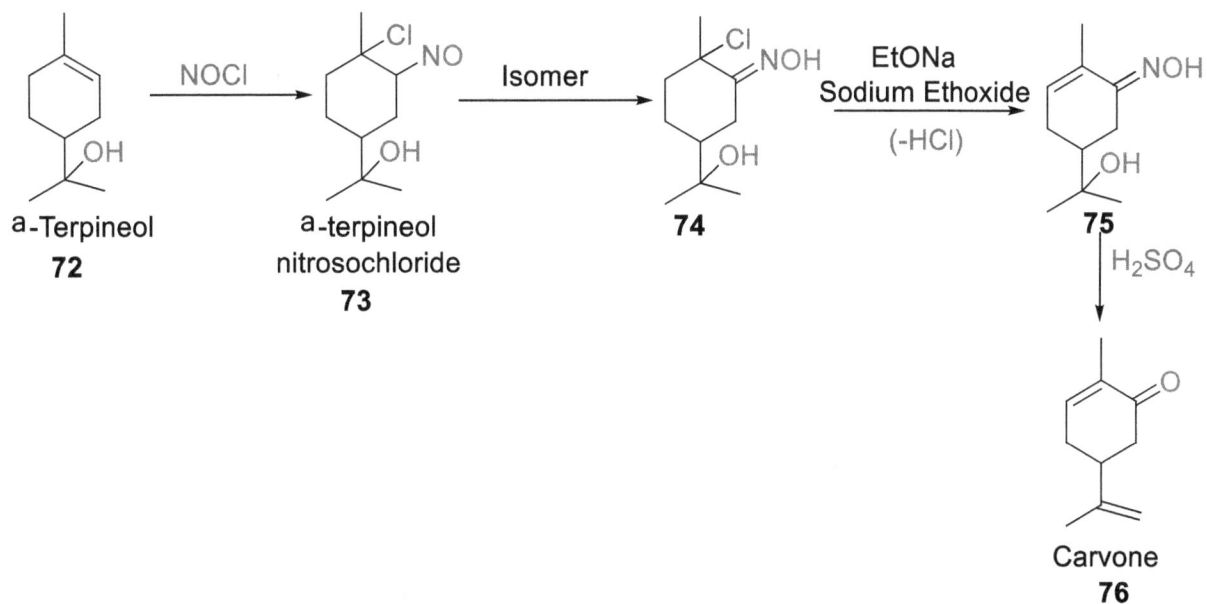

SCHEME 18.11 Synthesis of carvone from α-terpineol.

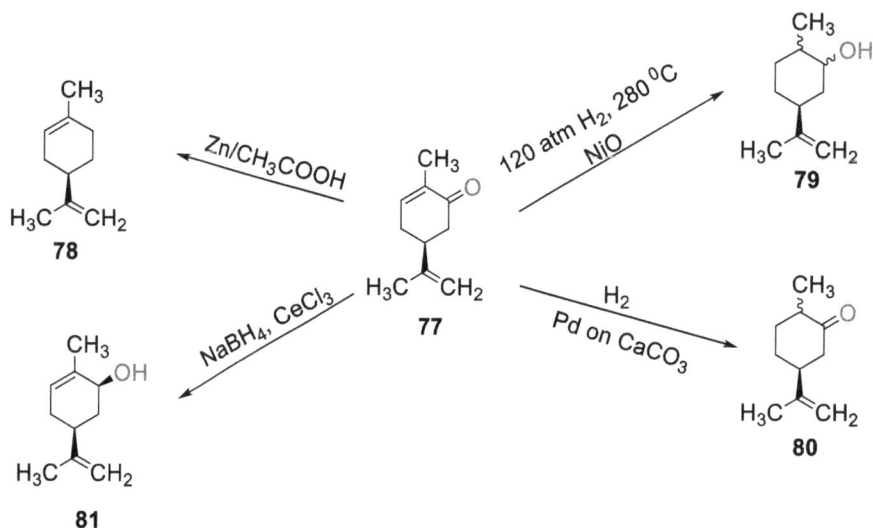

SCHEME 18.12 Oxidation and reduction of carvone.

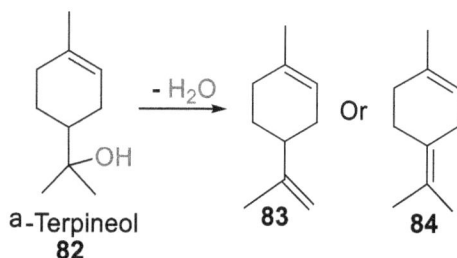

SCHEME 18.13 Dehydration of α-terpineol.

18.6.3 Limonene

This is optically active; the (+)-form occurs in lemon and orange oils, the (−)-form in peppermint oil, and the (−)-form in turpentine oil. At roughly 250°C, the optically active forms are racemized to create the racemic modification. The racemic modification is also known as dipentene; this term was given to the inactive form before its link to the active form (limonene) was established. Two double bonds are present in limonene because of the addition of four bromine atoms. (+)-Limonene may be made by dehydrating (+)-α-terpineol with potassium hydrogen sulphate, and limonene (or dipentene) may be turned into a-terpineol on shaking with dilute sulfuric acid (Scheme 18.13) [29].

The location of one double bond in limonene, as well as its carbon skeleton, are known. For position 8, the following chemical reactions provide supporting evidence:

Limonene $\xrightarrow{\text{NOCl}}$ Limonene nitrosyl chloride $\xrightarrow[\text{EtOH}]{\text{KOH}}$ Carvoxime

Since the structure of carvoxime is known, it therefore follows that (88) must have one double bond in position 8; thus, the above reactions may be written as (Scheme 18.14) [30]:

18.6.4 Menthol

Methol is a compound with optical activity; the (−)-form is the only one found in nature (peppermint oil, for example). Saturated chemical (−)-menthol has a melting point of 43°C;

its reactions, such as the formation of esters, demonstrate that its oxygen atom has an alcoholic functional character. The alcoholic group in menthol is secondary because it is oxidized to menthone, a ketone. Also, p-menthane is produced through hydrogen iodide reduction, suggesting that menthol likely contains this carbon skeleton. Finally, since menthol is produced upon reduction of (+)-pulegone and the structure of pulegone is known to be (89), it follows that menthol must be (90).

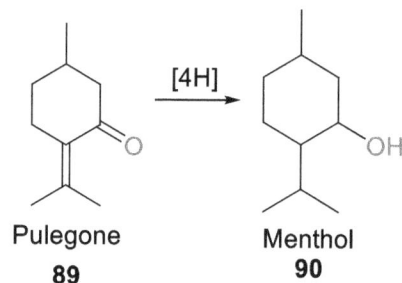

Examining the structure of menthol reveals three distinct chiral centres (91, 92, 93 and 94), allowing for the theoretical existence of eight optically active forms (four racemic modifications). The following are the known configurations of all eight enantiomers (the horizontal lines represent the plane of the cyclohexane ring).

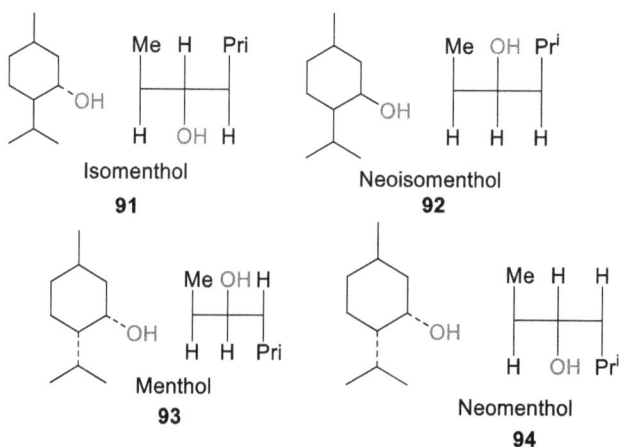

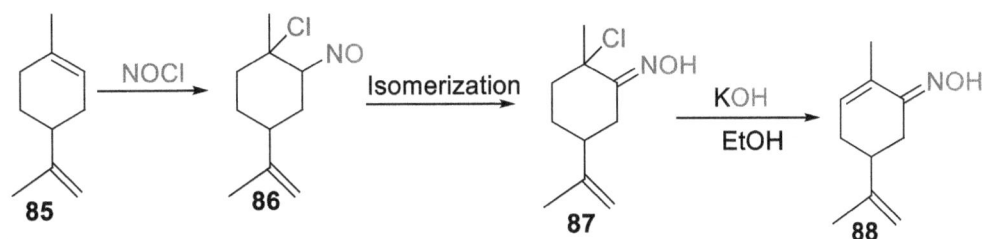

SCHEME 18.14 Synthesis of carvoxime.

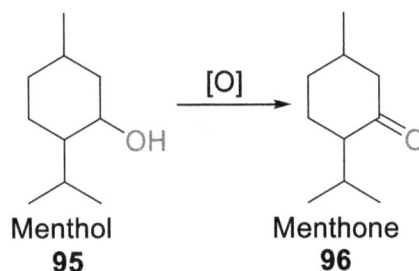

SCHEME 18.15 Oxidation of menthol to menthone.

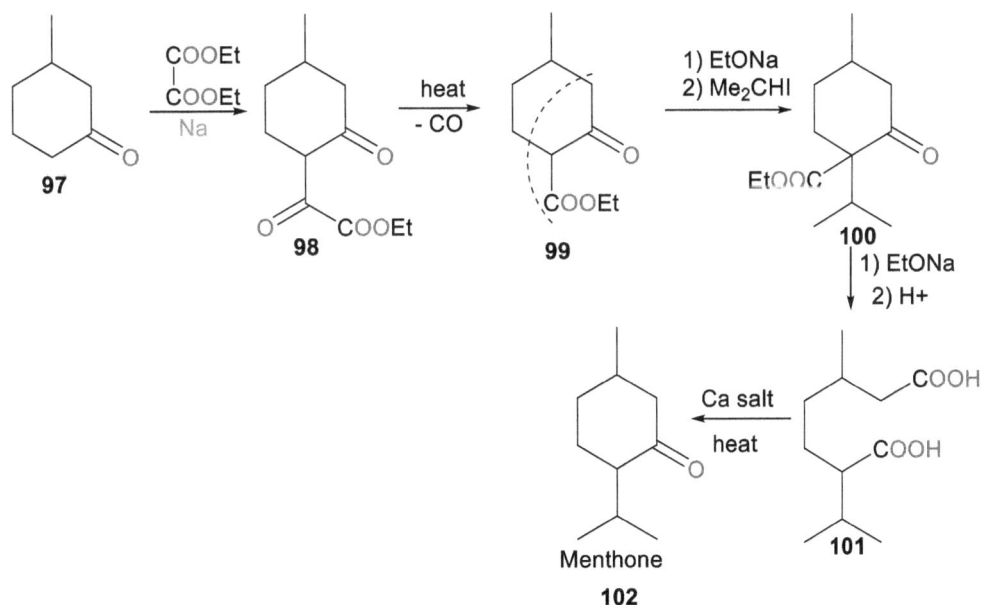

SCHEME 18.16 Synthesis of menthone.

18.6.5 Menthone

(−)-Menthone may be easily synthesized from (−)-menthol and chromic acid, and it is found naturally in peppermint oil (Scheme 18.15).

18.7 Synthesis of Menthone and Menthol

Menthone was first synthesized in 1907 by Kotz and Schwartz, who used the following method to distil the calcium salt

of 2-isopropyl-5-methylpimelic acid. The ethyl ester of 4-methylcyclohexan-2-one-1-carboxylic acid can be obtained through the condensation of 3-methylcyclohexanone and ethyl oxalate in the presence of salt, followed by heating the resulting keto ester under decreased pressure. This latter compound was obtained by treating it with EtONa and then adding isopropyl iodide; the resulting isopropyl cyclohexanone derivative was boiled with EtONa and then acidified to get 2-isopropyl-5-methylpimelic acid. Heating the equivalent dicarboxylic acid on the calcium salt produces menthone (Scheme 18.16).

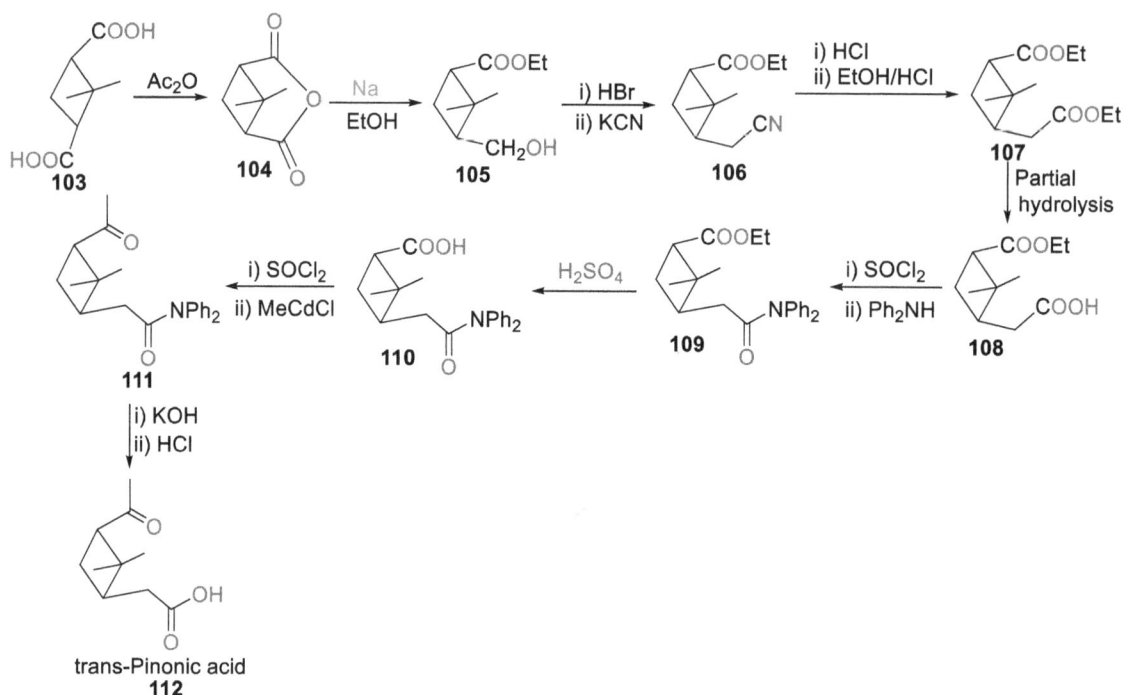

SCHEME 18.17 Synthesis of trans-pinonic acid.

The reduction of menthone using $NaBH_4$ in alcohol or pulegone in the presence of reduced catalyst gave the corresponding menthol [12,31,32].

18.7.1 Pinene

The precursor to all other members of this family, this chemical is synthesized from either alpha- or beta-pinene via catalytic hydrogenation using nickel or platinum. There are two pairs of enantiomers for each of pinene's two geometric isomers, cis and trans. α-Pinene, boiling point of 156°C. In the family of pinanes, this one stand head and shoulders above the rest. All turpentine oils contain both the (+)- and (–)-forms of this compound.

18.7.2 Total synthesis of α-Pinene

The α-pinene has been synthesized starting from trans-pinonic acid which obtained from 2,2-dimethyl-1,3-cyclobutandicarboxylic acid as described in the following Scheme 18.17.

As a result of all of these steps, a combination of alpha and delta-pinene is produced. Since this was discovered by the synthesis of nitrosyl chloride, we know that one of the pinenes is a, but we don't know which one. Pinenes react with diazoacetic ester to yield pyrazoline derivatives, which then breakdown into cyclopropane derivatives upon heating on their own or in the presence of cupper powder [12,33] (Scheme 18.18).

The oxidation of the molecules produced by treating α-pinene and δ-pinene with this method yielded 1-methylcyclopropane-1,2,3-tri-carboxylic acid and cyclopropane-1,2,3-tricarboxylic acid, respectively. Similar structures to those of α- and δ-pinenes can be seen in these products [12,34] (Scheme 18.19).

18.8 The Bornane (Camphene)-Norbornane (Isocamphane) Group

Bornane (camphene), $C_{10}H_{18}$. This is a synthetic compound and may be prepared from camphor, e.g.,

i. By first converting camphor to a variety of borneols, then to bornyl iodides, and lastly to bornane, the process begins (Scheme 18.20).

ii. Camphor may also be converted into bornane by means of Wolff-Kishner reduction. Bornane is solid, m.p. 156°C; it is optically inactive (Scheme 18.21) [12,35].

18.9 Camphor

The camphor tree, native to Formosa and Japan, is a natural example of this phenomenon. It has a melting point of 180°C and is optically active, with the (+)- and (–)-forms occurring in nature with racemic camphor being the most common synthetic form.

Bredt based his formula off of the aforementioned information as well as the following: (a) oxidising camphor with nitric acid yields camphoric acid, $C_{10}H_{16}O_4$; and (b) oxidising camphoric acid (or camphor) with nitric acid yields camphoronic acid, $C_9H_{14}O_6$. To demonstrate the connections among camphor, camphoric acid and camphoronic acid, Bredt proposed the following reactions based on the assumptions that (138) represented the structure of camphoric acid and (139) represented the structure of camphor (140) (Scheme 18.22) [12,36].

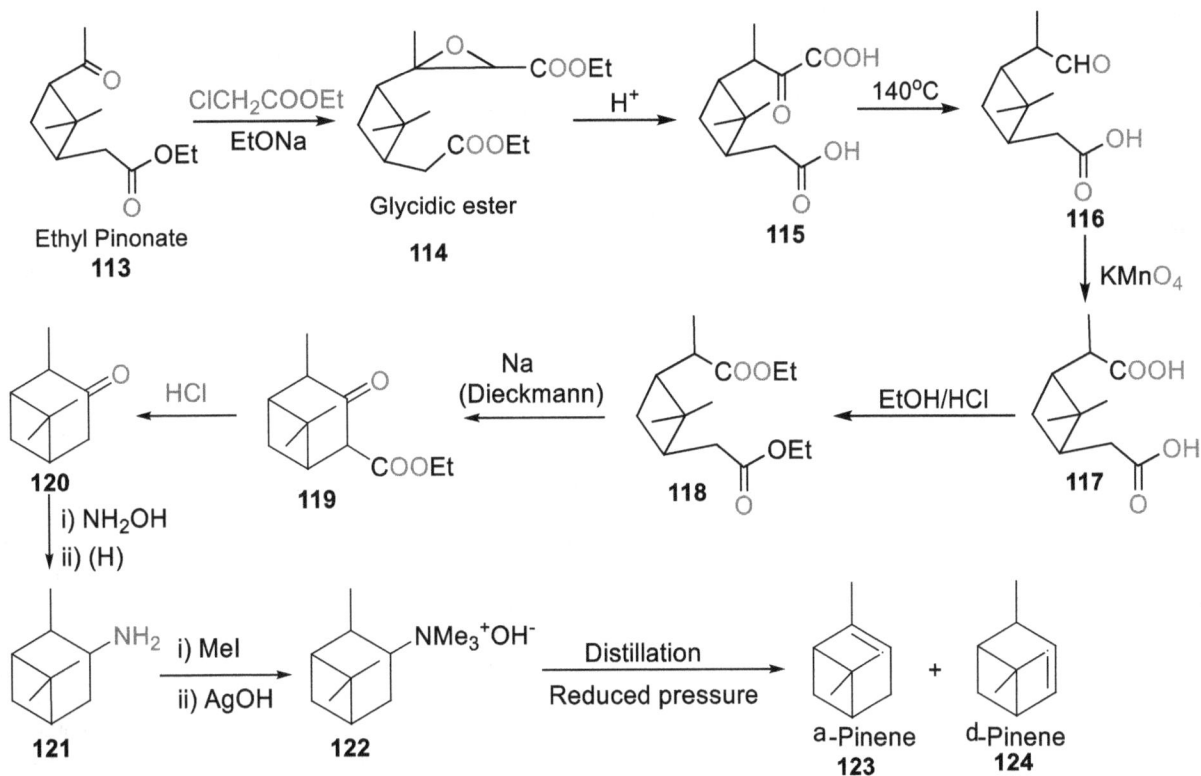

SCHEME 18.18 Synthesis of α-pinene and δ-pinene.

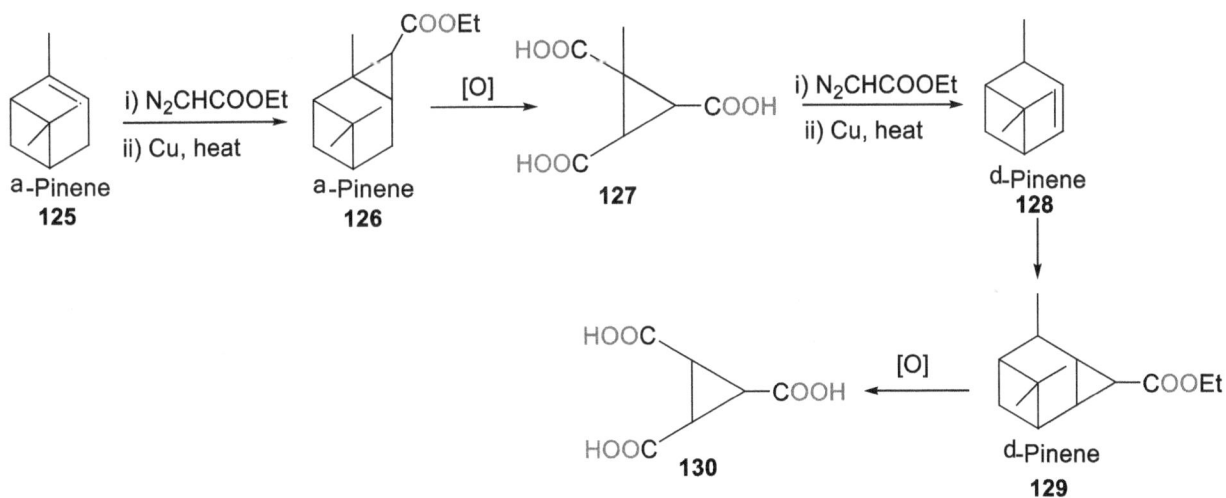

SCHEME 18.19 Oxidation of α-pinene.

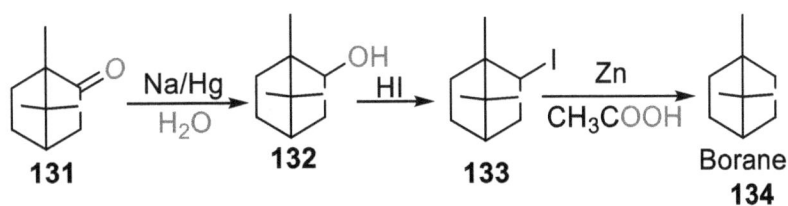

SCHEME 18.20 Reduction of camphor.

18.10 Synthesis of Camphor (Haller, 1896)

Haller started with camphoric acid created by the oxidation of camphor, but since the acid was made later by Komppa, we now have a whole synthesis of camphor (Scheme 18.23).

18.11 Farnesene

This is the product of farnesol dehydration with potassium hydrogen sulphate. This molecule has been identified as

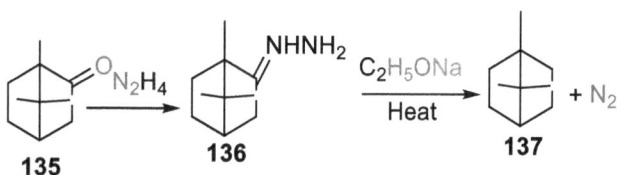

SCHEME 18.21 Wolff-Kishner reduction of camphor.

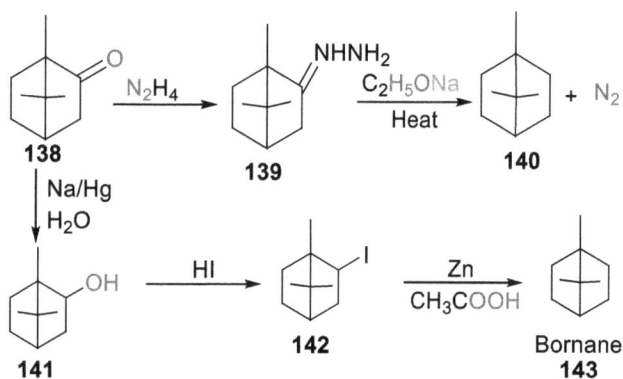

SCHEME 18.22 Synthesis of camphor.

the β-isomer, and it has been proved that the -isomer exists naturally (in oil of hops), with the structure indicated above having been ascribed. Dehydration of nerolidol also yields α-farnesene [12].

18.12 Farnesol

Farnesol, occurs in the oil of ambrette seeds. Kerschbaum provided the following explanation of its internal structure. An aldehyde, farnesal (150), is produced from farnesol (149) by oxidation with chromic acid. This means that farnesol is a main alcohol. Cyanide (151) is formed from farnesal oxime via dehydration with acetic anhydride; upon hydrolysis with alkali, farnesenic acid (152) is formed along with a ketone (153, 154) (Scheme 18.24) [12].

Dihydro-pseudo-ionone was ultimately identified as the ketone in question (geranylacetone). This ketone is formed by eliminating two carbons from its precursor. Since this reaction is typical of, α, β-unsaturated carbonyl compounds, we can deduce that farnesenic acid (or its nitrile) is a, α, β-unsaturated chemical as well. Based on what is already known about the chemical structure of geranylacetone, the foregoing can be expressed as follows. A recent synthesis of farnesol has been carried out by Corey et al. (Scheme 18.25) [12,37].

18.13 Nerolidol

The oil of neroli contains the (+)-form of nerolidol. Nerolidol and farnesol are isomers, and Ruzicka validated the structure of nerolidol through synthesis, demonstrating that their connection is analogous to that of linalool and geraniol (Scheme 18.26) [12,38].

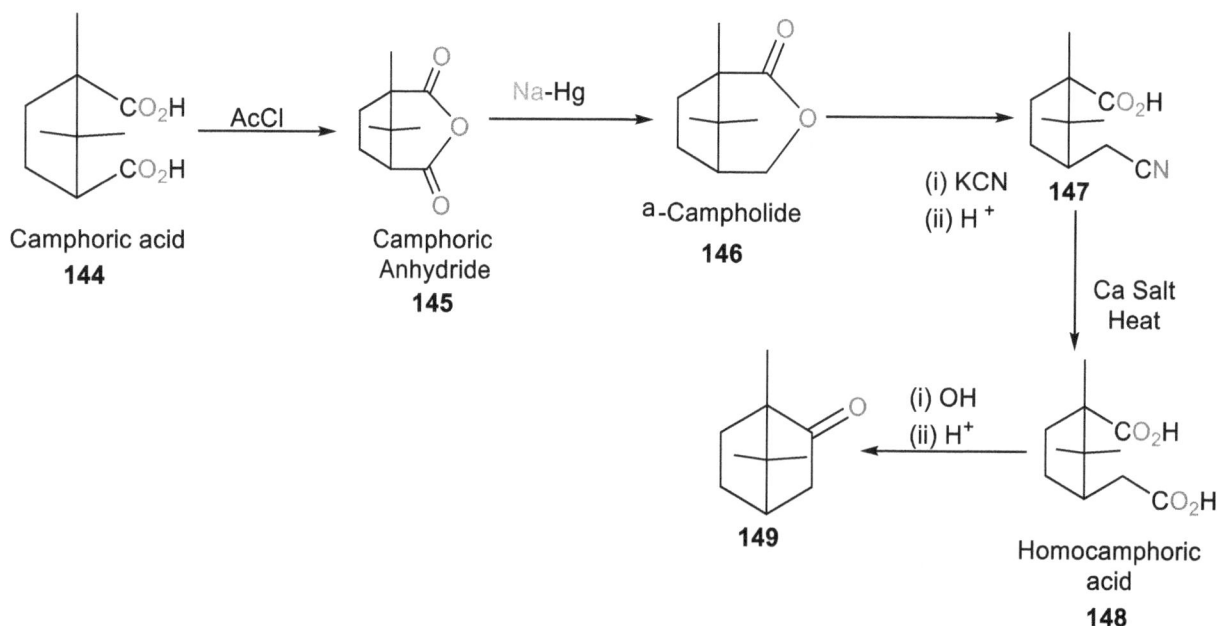

SCHEME 18.23 Synthesis of camphor.

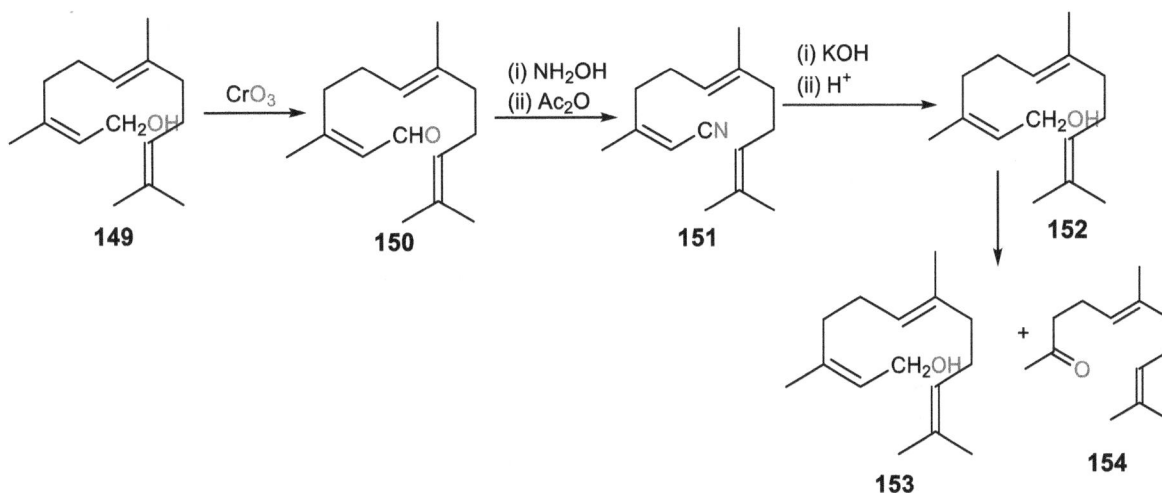

SCHEME 18.24 Metabolism of Farnesol.

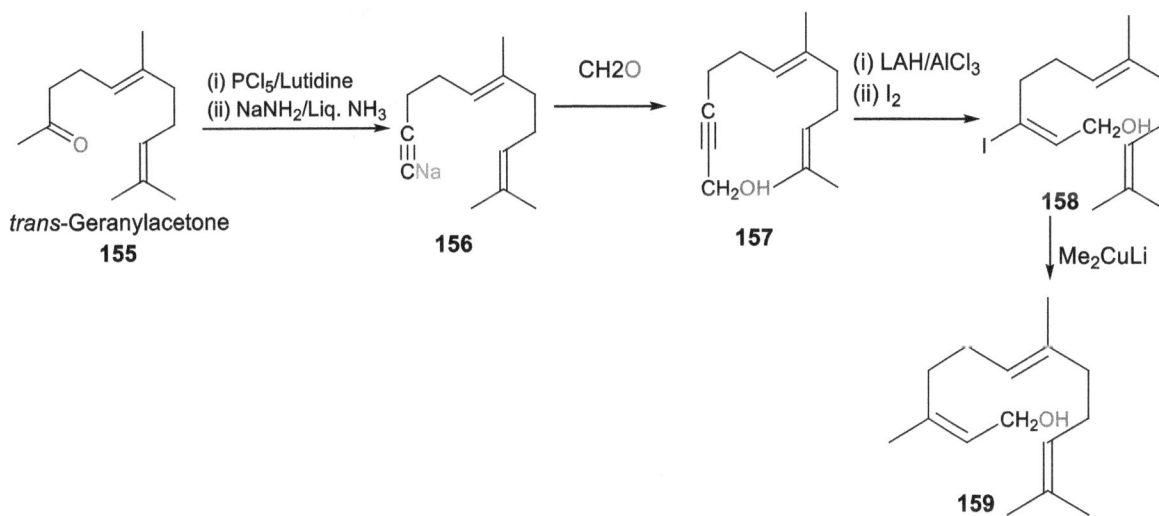

SCHEME 18.25 Metabolism of *trans*-geranylacetone.

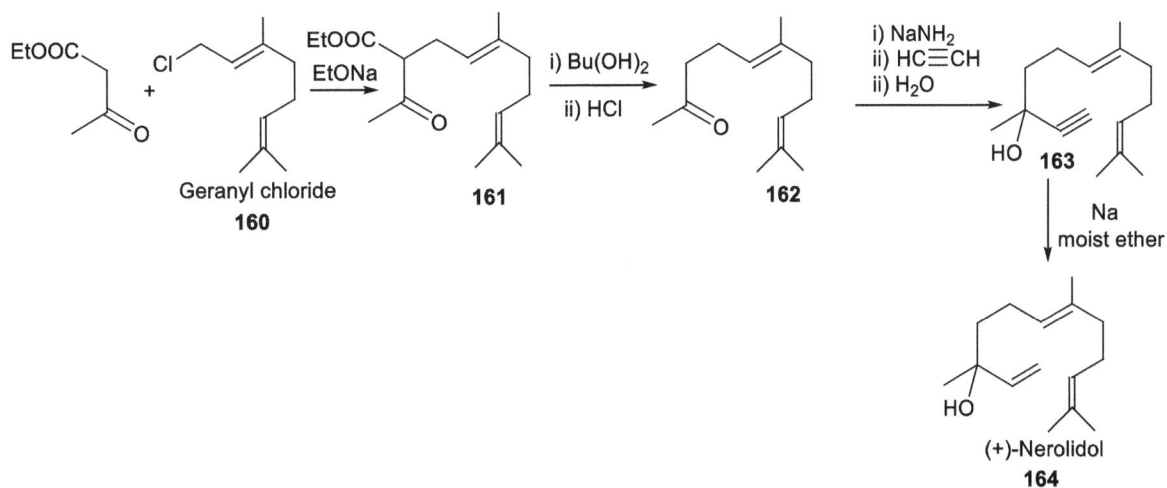

SCHEME 18.26 Metabolism of nerolidol.

18.14 Metabolism Reactions of Chiral Terpenoids

Terpenoids are the primary chemical components of essential oils extracted from plants. Food, fragrance, and the medicinal industries all make extensive use of them because of their appealing aromas and flavours. Terpenoids have a wide variety of medicinal uses in both modern and traditional medicine, including analeptic, antibacterial, antifungal, anticancer, and sedative effects. A lack of information about their biotransformation in humans belies their widespread use in industry. However, terpenoids' metabolism can result in the synthesis of novel biotransformation products with distinct structural features and, frequently, altered flavour and biological activity. All terpenoids are readily absorbed orally, absorbed via the skin, or inhaled, and they frequently reach detectable blood quantities in humans. However, several enzymes can quickly convert these chemicals into molecules that are more easily dissolved in water. Almost every organ or tissue in the body has some capacity for drug metabolism, but the liver is where this process occurs most efficiently. Phase I and Phase II reactions are the two main types of metabolic biotransformation. To a large extent, oxidation, reduction and hydrolysis mediated by cytochrome P450 (CYP) enzymes are of interest in phase I. When a product from Phase I undergoes Phase II, it is totally changed to have a high-water solubility. This is done by attaching already highly water-soluble endogenous entities such as sugars (glucuronic acids) or salts (sulphates) to the Phase I intermediate and generating a Phase II final product.

Many terpenoids can be eliminated from plants without going through both Phases I and II.

18.15 Camphene

Camphene occurs naturally in a wide variety of plants, although it is most abundant in the essential oil of *Thymus vulgaris's* leaves and flowers. Coughs and respiratory tract infections can be effectively treated with medicines containing camphene due to its expectorant, spasmolytic and antibacterial characteristics. Camphene is broken down into two diastereomeric glycols, as shown in Scheme 18.27 [12,39].

18.16 Camphor

The cinnamomum camphora tree grows in both Southeast Asia and North America, and its wood is used to make camphor, a bicyclic monoterpene. Additionally, it is a significant component of common sage's volatile oil (*Salvia officinalis*). Camphor is used commercially as a moth repellent and as a preservative in pharmaceuticals and cosmetics. It crystallises into white, fatty masses that have a strong camphoraceous odour. The primary hydroxylation products of d- and l-camphor were 5-endo- and 5-exo-hydroxycamphor, however camphor is extensively metabolized in dogs, rabbits and rats. Furthermore, a trace quantity of 3-endo-hydroxycamphor was detected (Scheme 18.28). Three- and five-bornane groups can be further reduced to 2,5-bornanedione. Reducing camphor to

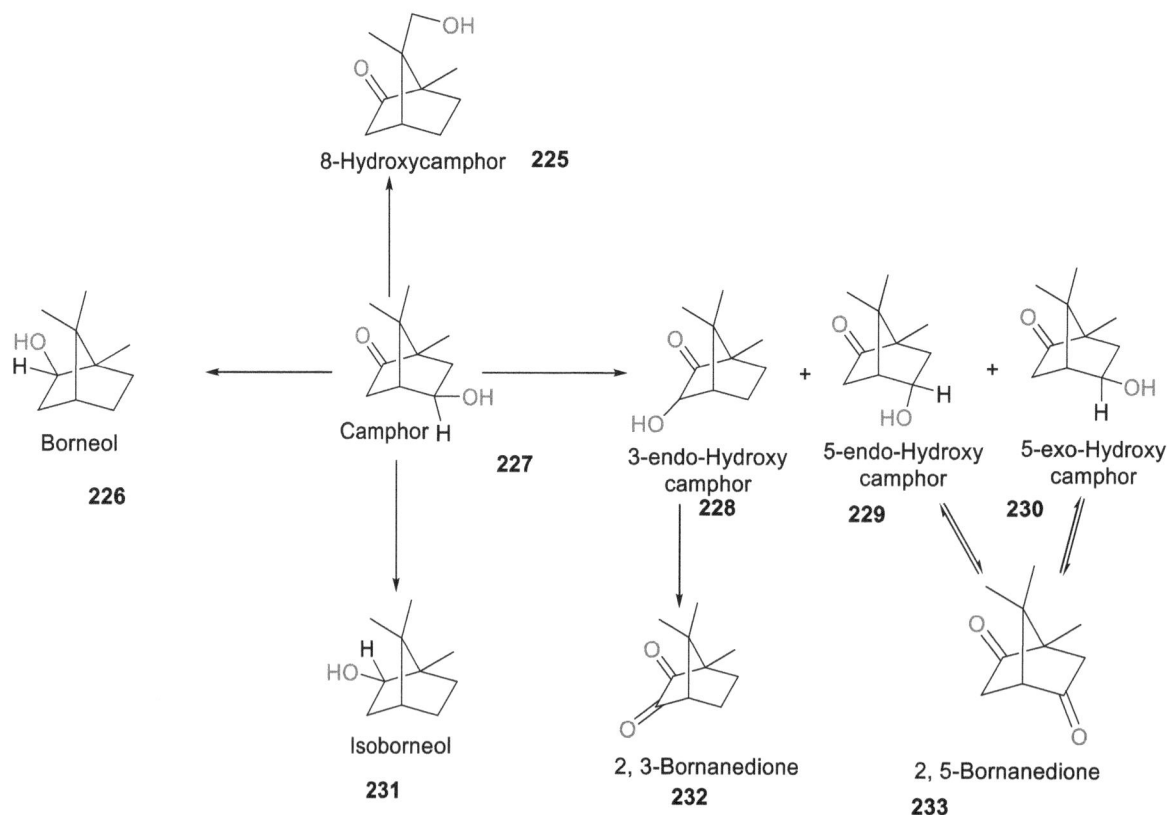

SCHEME 18.27 Metabolism of camphene.

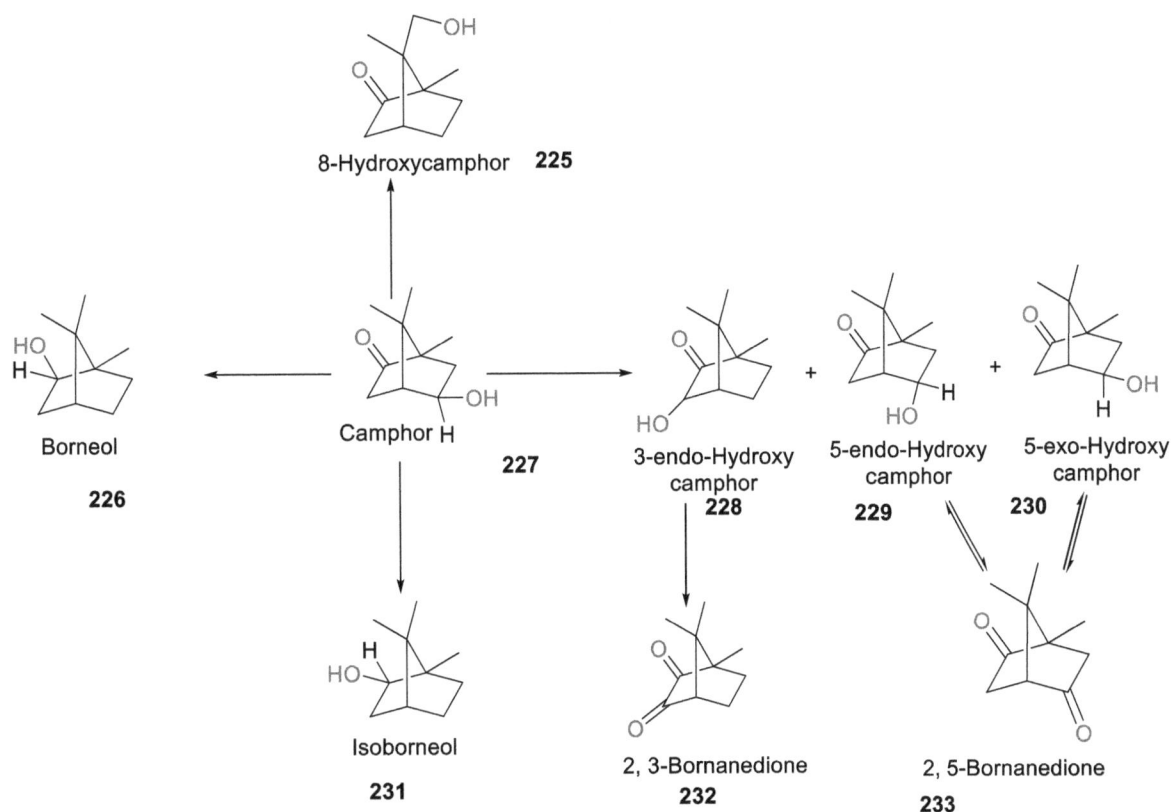

SCHEME 18.28 Metabolism of camphor

borneol and isoborneol are two of the many minor biotransformation stages. It is interesting to note that in a Phase II reaction, all hydroxylated camphor metabolites are further conjugated with glucuronic acid. Camphor, according to animal studies, should be heavily metabolized in humans as well. Indeed, hydroxylation to 5-exo-hydroxycamphor was also detected in a fairly recent work using human liver microsomes. Camphor's oral administration to humans also suggests the development of other metabolites, including borneol, isoborneol and other glucuronides, which are detected in the blood and urine of the test subjects [12,40].

18.17 Carvacrol

Carvacrol, a monoterpenic alcohol, is abundant in the essential oil of *Thymus vulgaris*, a plant native to the Mediterranean region and parts of southern Europe. It has been claimed that carvacrol is employed as a flavouring agent in a variety of meals and drinks (Scheme 18.29).

18.18 Carvone

There are two enantiomers of the monoterpene ketone carvone found in nature, designated (R)-(−)- and (S)-(+). Oil from spearmint leaves (*Mentha spicata* var. crispa) contains roughly 50% of (R)-(−)-carvone in addition to other terpenes, whereas essential oil from caraway seeds (Carum carvi) primarily

consists of (S)-(+)-carvone (Wichtel, 2002). Both enantiomers smell and taste different, and they each have their own applications in the culinary, fragrance and medicinal industries. Toothpaste, mouthwash and chewing gum often contain significant levels of (R)-(−)-carvone due to its minty flavour and taste. To improve flavour, (S)-(+)-carvone is commonly utilized in the food and fragrance industries due to its signature caraway aroma. (S)-(+)-carvone is employed as a stomachic and carminative in several drug preparations for its spasmolytic activity. As an added bonus, both (S)-(+)- and (R)-(−)-carvone are used in aromatherapy massage treatments for nerve tension and numerous skin problems. The (R)-(−)-carvone undergoes stereoselective metabolism to (4R, 6S)-(−)-carveol and (4R, 6S)-(−)-carveol glucuronide, whereas (S)-(+)-carvone undergoes nonselective metabolism. For both enantiomers, plasma metabolites were undetectable. By the addition to carveol, dihydrocarveol, carvonic acid, dihydrocarvonic acid and uroterpenolone were all detectable in the urine samples [12,41].

18.19 1,4-Cineole

Lime (*Citrus aurantiifolia*) is rich in 1,4-cineole, a monoterpene cyclic ether and *Eucalyptus polybractea*, a terpene, has been utilized for decades as a flavouring and aroma agent. Recent in vitro and in vivo animal research have shown considerable biotransformation of this monoterpene, strongly implying biotransformation in the human body as well, although

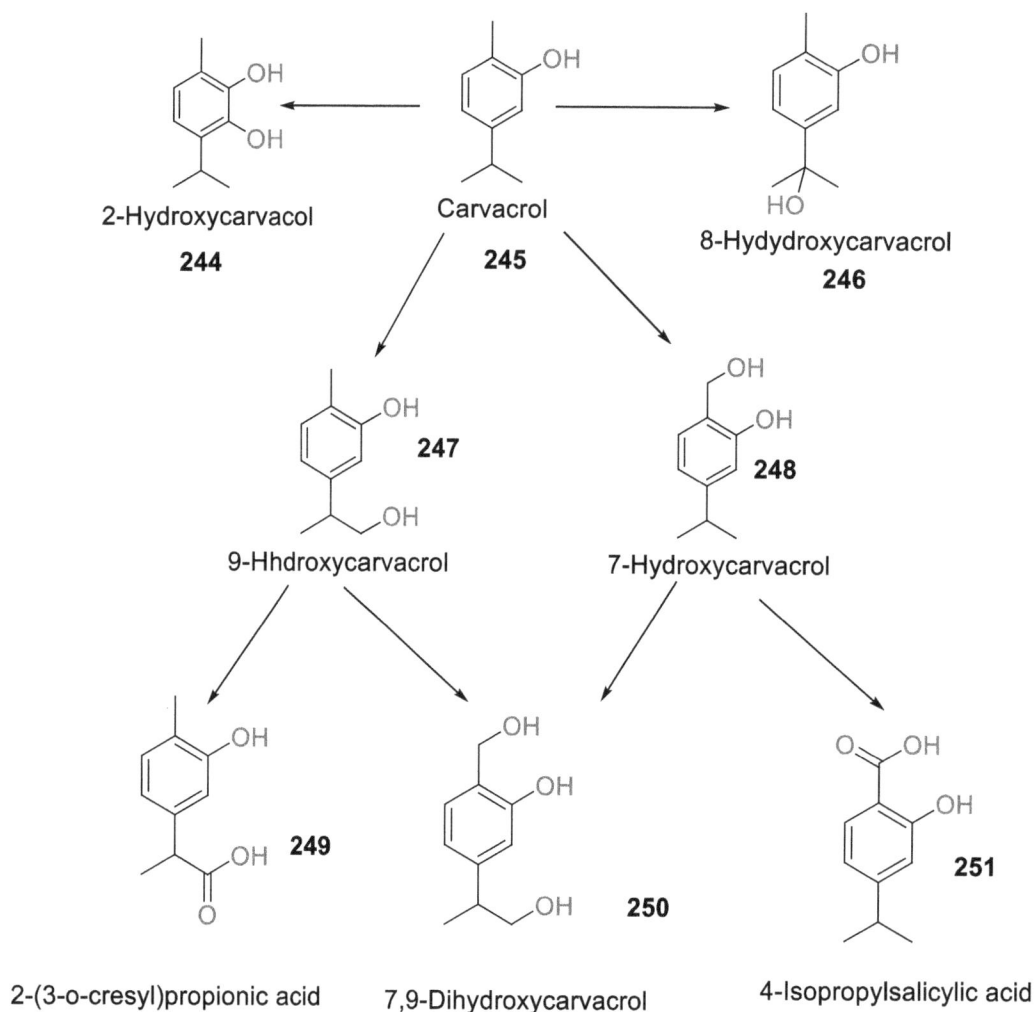

SCHEME 18.29   Metabolism of carvacrol.

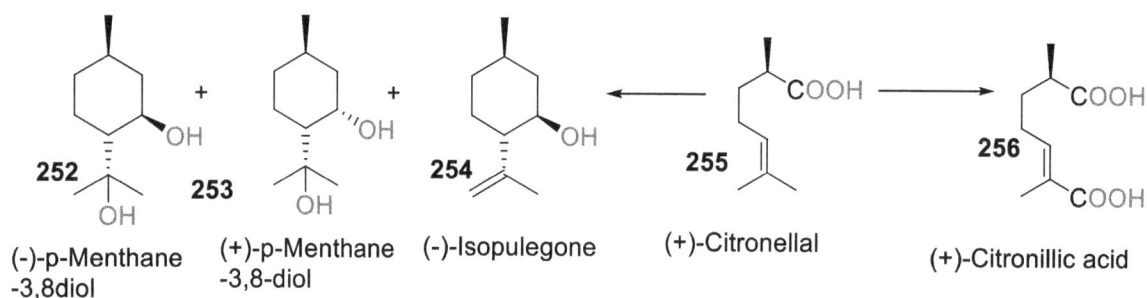

SCHEME 18.30   Metabolism of 1,4-cineole.

there are no in vivo data about the metabolism of 1,4- cineole in humans. Four neutral and one acidic metabolite, including 9-hydroxy-1,4-cineole, 3,8-dihydroxy-1,4-cineole, 8,9-dihydroxy-1,4-cineole, 1,4-cineole-8-en-9-ol and 1,4-cineole-9-carboxylic acid, were recovered from rabbit urine following oral administration. However, only 1,4-cineole 2 hydroxylation was detected using rat and human liver microsomes, suggesting species-related variations in 1,4-cineole metabolism (Scheme 18.30) [12,42].

18.20 1,8-Cineole

1,8-Cineole, commonly known as eucalyptol, is a monoterpene cyclic ether that is abundant in plants. It is present in particularly high amounts in the essential oil of *Eucalyptus polybractea*. It has several medicinal and cosmetic applications, including relief from coughs, sore muscles, neuropathy, rheumatism, asthma and urinary stones. 1,8-cineole is primarily

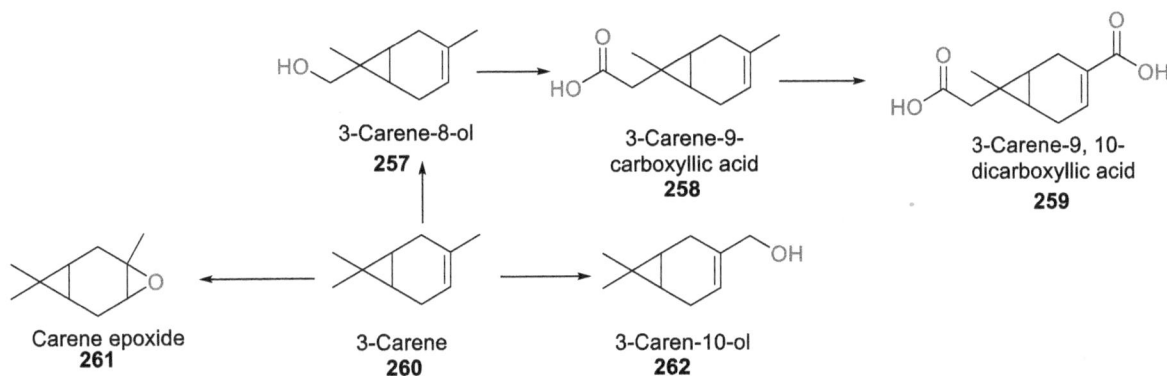

SCHEME 18.31 Metabolism of 1,8-cineole

transformed to 3-hydroxy-1,8-cineole by rat liver microsomes, then to 2- and finally to 9-hydroxycineole. Scheme 18.31 shows that only the 2-hydroxy- and 3-hydroxy products catalyzed by the isoenzyme CYP_3A_4 were detected in human liver microsome. Three healthy human volunteers were given a cold medication containing 1,8-cineole, and their urine was tested for the presence of both metabolites. Both metabolites can be measured in the urine to determine how much 1,8-cineole a person has consumed (Scheme 18.31) [12,42].

18.21 Citral

Isomers of geranial (E-3,7-dimethyl2,6-octadienal) and neral make up both natural and synthesized citral (Z-3,7-dimethyl-2,6-octadienal). The geranial isomer is typically the more abundant form in isomer blends. Citrus fruit essential oils (up to 5% in lemon oil) and other plants and herbs, including *Melissa officinalis*, lemongrass (70%–80%) and eucalyptus, contain it naturally. Since the early 1900s, citral has been widely employed in the food, cosmetic and detergent sectors due to its strong lemon fragrance and flavour. Citral is rapidly converted to various acids and a biliary glucuronide and expelled, as indicated by studies in rats; the majority of citral is eliminated by urine (48%–63%), followed by expired air (8%–17%) and faeces (7%–16%). The results of the isolation and identification of seven metabolites from urine: Acids 3-hydroxy-3,7-dimethyl-6-octenedioic and 3,8-dihydroxy-3,7-dimethyl-6-octenedioic E- and Z-3,7-dimethyl-2,6-octenedioic acid, 3,7-dimethyl-6-octenedioic acid and 3,9-dihydroxy-3,7-dimethyl-6-octenedioic acid are all derivatives of 3,7-dimethyl-2,6-octenedioic acid. No information about citral's metabolism in people is available at this time, either in vitro or in vivo. However, the aforementioned research in rats strongly suggests that citral undergoes substantial biotransformation [12,43].

18.22 Citronellal

Essential oil of Melissa officinalis contains the greatest levels of citronellal (a monocyclic monoterpene) (1%–20%). Although citronellal is rarely utilized in the food and flavour industries, it is an important component of many therapeutic products due

to its moderate sedative and stomachicum effects. It is widely utilized in patients, however there is a lack of information about its metabolism. Biotransformation of citronellal in rabbits was described in only one research. There are three neutral metabolites of (+)-citronellal that were isolated by Ishida et al. from rabbit urine: (−)-trans-menthane-3,8-diol, (+)-cis menthane-3,8-diol and (−)-isopulegol. As a result of regioselective oxidation of the aldehyde and dimethyl allyl groups, the acidic metabolite (6E)-3, 7-dimethyl-6-octene-1,8-dioic acid was produced.

18.23 Fenchone

The essential oil of *Foeniculum vulgare* has the highest concentration of the bicyclic monoterpene fancone. Fenchone, which has a camphoraceous aroma, is a flavouring and fragrance ingredient. The biotransformation of fenchone in human liver microsomes has recently been studied, with results showing the synthesis of 6-exohydroxyfenchone, 6-endo-hydroxyfenchone and 10-hydroxyfenchone. There is currently a lack of information regarding the metabolism of this chemical in human subjects. However, in vitro research employing human liver microsomes suggest that identical biotransformation products should also be identified in human blood or urine samples following dietary ingestion of food containing fenchone.

18.24 Geraniol

Concentrations of the monoterpene alcohol geraniol (12%–25%) are very high in the essential oil of the herb *Cymbopogon winteranus* Jowitt. Rose, palmarosa, citronella, geranium, lemon and many other essential oils contain trace amounts of this compound as well. In the perfume and flavouring industries, its rose-like aroma is highly valued. Direct oxidation of geraniol produces geranic acid and 3-hydroxycitronelic acid, while metabolic intermediates including 8-carboxygeraniol and 3,7-dimethyl-2,6-octenedioic acid (Hildebrandt's acid) are produced via the metabolism of geraniol. 8-hydroxygeraniol and 8-carboxygeraniol are both by products of selective oxidation of the C-8 of geraniol. 8-hydroxylation of geraniol is the initial step in the manufacture of indole alkaloids, which is why it also occurs in higher plants [12,44].

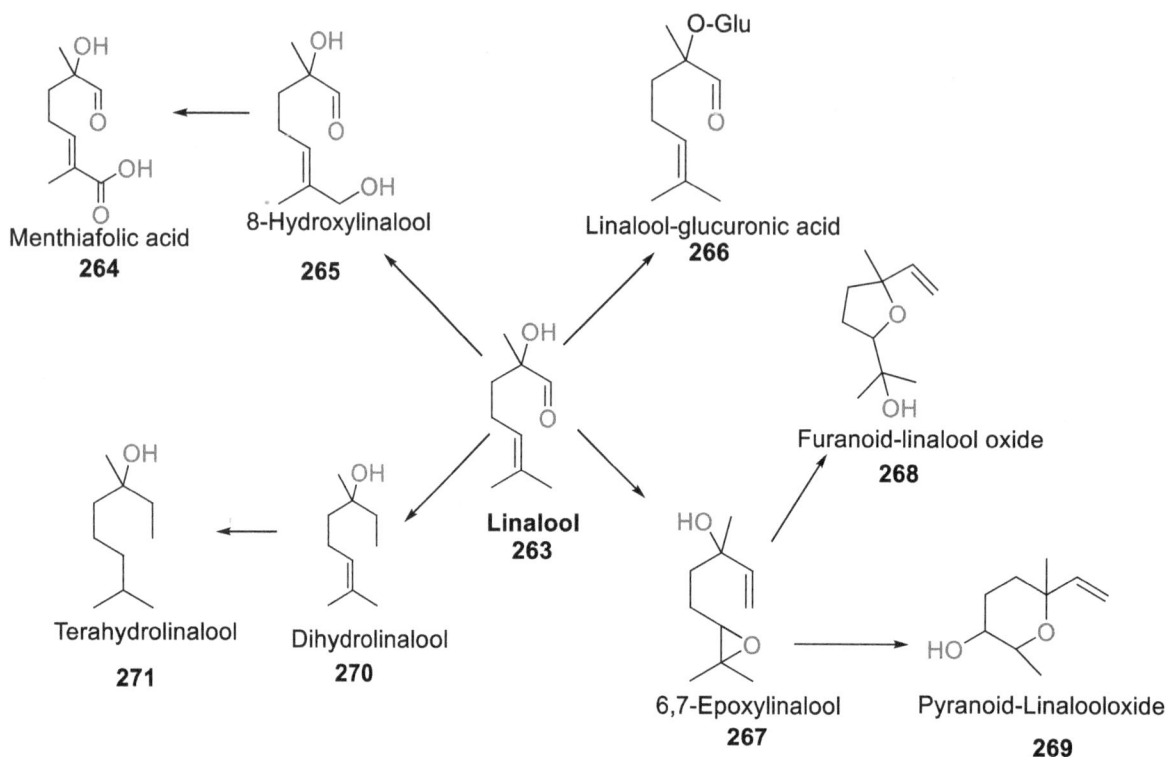

SCHEME 18.32 Metabolism of linalool.

18.25 Limonene

Orange peel (*Citrus aurantium* L. sp. aurantium) and other plants have been found to contain the monocyclic monoterpene (+)- and (−)-limonene enantiomers, which are widely employed as scents in consumer products and components of synthetic essential oils. The (+)-limonene isomer is found in higher concentrations in plants than either the racemic mixture or the (−)-limonene isomer. Previous research in rats and mice has established the chemo-preventive effects of (+) limonene. Since (+)-limonene is more useful in the food and fragrance industries, we will focus on its metabolism rather than that of (−)-limonene. Several laboratories have published detailed accounts of the biotransformation of (+)-limonene in vitro (using rat and human liver microsomes) and in vivo (using mice) (rat, mice, guinea, pigs, dogs, rabbits, human volunteers and patients). The main biotransformation products in humans include perillyl alcohol, perillic acid, p-mentha-1,8-dien-carboxylic acid (an isomer of perillic acid), cis-dihydroperillic acid, trans-dihydroperillic acid, limonene, 1,2-diol, limonene.

18.26 Linalool

Oils of Cajenne rosewood, Brazil rosewood, Mexican linaloe and coriander seed can be fractionally distilled and then rectified to produce linalool. Essential oil of *Ocimum basilicum* contains the greatest proportion of linalool (up to 75%). Pure linalool has a clean, pleasant, fruity fragrance with a hint of citrus that is often used in soap and detergent manufacturing.

Despite widespread use in the perfume business, we have no information on how linalool is metabolized by people. However, cytochrome P450 (CYP) isoenzymes in rats convert linalool to the more stable di- and tetrahydro-forms, as well as to 8-hydroxylinalool, which is then oxidized to 8-carboxylinalool (Scheme 18.32). Then, the glucuronide conjugates of CYP-derived metabolites are released [12,45].

18.27 Linalyl Acetate

Many different types of scent compounds have linalyl acetate as a component. Decorative cosmetics, fi ne perfumes, shampoos, toilet soaps and noncosmetic items like cleansers and detergents may all contain it as scents. Linalyl acetate is present in a variety of plants, however the essential oil of Citrus aurantium spp. aurantium has the highest concentration. Since linalyl acetate is an ester, it must be hydrolyzed in living organisms by carboxylesterases or esterases to yield linalool, which is subsequently metabolized to a wide variety of oxidative biotransformation products [46].

18.28 Menthol

Among the many active ingredients in mint oils, menthol is particularly prominent. Both peppermint oil (from *Mentha piperita*) and cornmint oil (from *Mentha arvensis*) are easily derived from the plant using steam distillation. It is commonly used to add flavour to meals and oral pharmaceutical preparations like cough drops and toothpaste because to its pleasant,

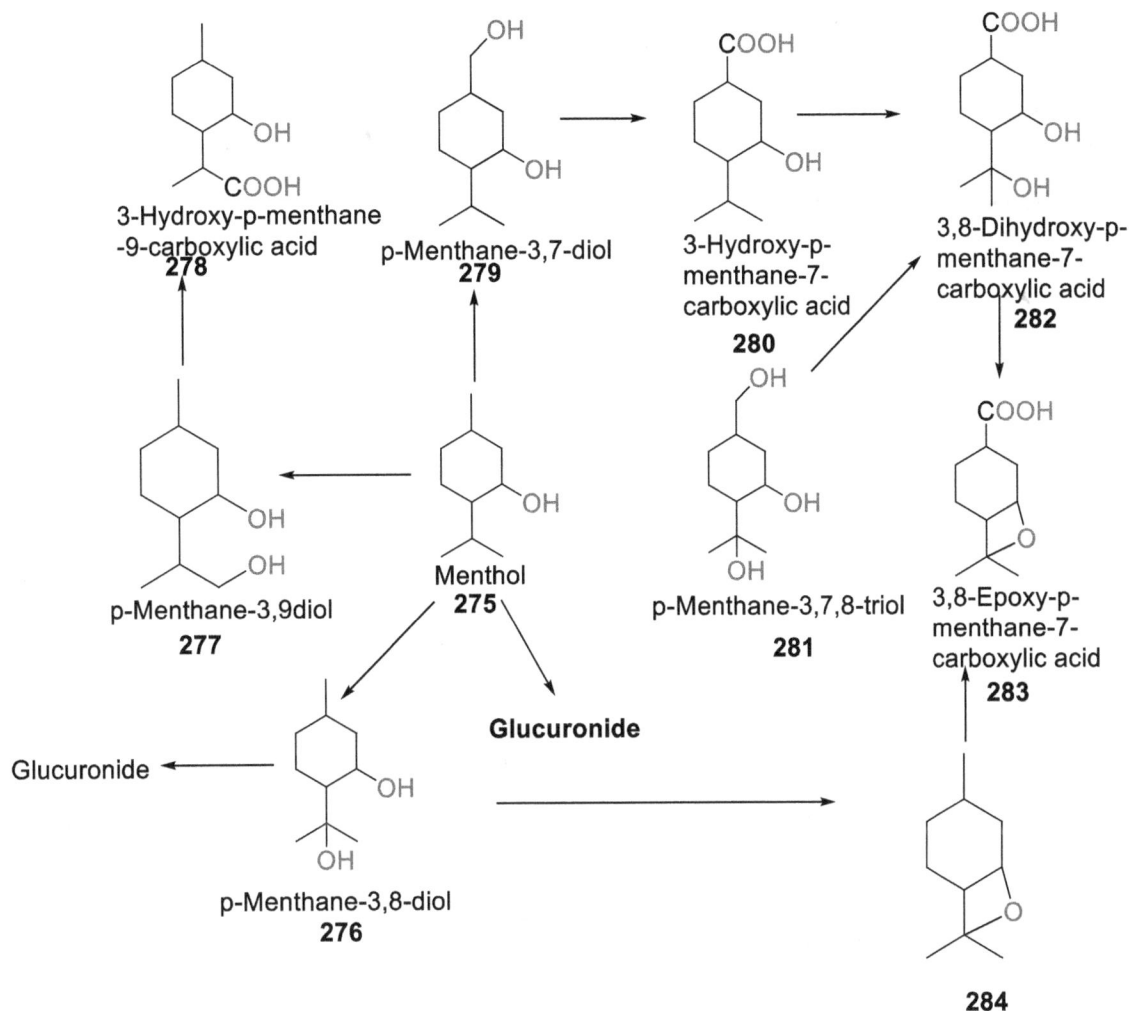

SCHEME 18.33 Metabolism of menthol.

traditional minty odour and taste. Volunteers' plasma and urine only contained menthol glucuronide, indicating that the compound was swiftly metabolized after oral treatment. Contrary to expectations, transdermal administration was required to detect unconjugated menthol. In rats, however, hydroxylation of the C-7 methyl group as well as the C-8 and C-9 isopropyl moiety forms a number of mono- and dihydroxymenthols and carboxylic acids, some of which are excreted in part as glucuronic acid conjugates. Derivatives of menthol that have been mono- or dihydroxylated are additional metabolites (Scheme 18.33). Again, menthol glucuronide was the primary metabolite in rats, just as it is in humans. Derivatives of Menthone are shown in Scheme 18.34 [12,47].

18.29 Myrcene

Humulus lupulus, essential oil, which is used to make beer and wine, primarily consists of myrcene. *Humulus lupulus*, is also commonly employed as a moderate sedative in various pharmaceutical preparations for the treatment of insomnia. Researchers have identified many metabolites in rabbit urine following oral administration, including two glycols that may

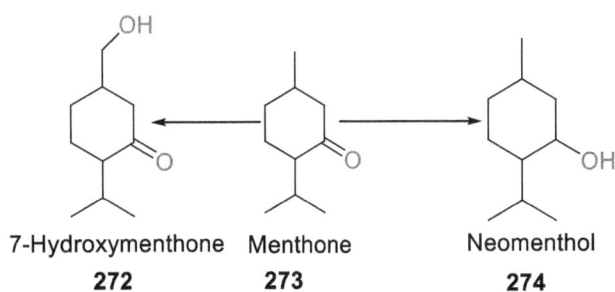

SCHEME 18.34 Metabolism of menthone.

have been produced as intermediates following the hydration of their respective epoxides. In the acidic environment of rabbit stomachs, myrcene was converted to limonene, which could be a step in the synthesis of uroterpenol.

18.30 Pinene

Both α- and β-pinene, with slightly different structural isotopes, occur naturally. Both isoforms have significant roles in pine resin, as their names imply. It's interesting to note that a-pinene is more prevalent in European pines whereas b-pinene is more abundant in North American pines. They are abundant in plant life and can be found in the resin of many different types of conifers. One of the highest concentrations of a- and b-pinene is in the essential oil of the fruit of juniper (*Juniperis communis*) with a total content of about 80% of these isoforms. Additionally, a-pinene can be found in the racemic mixture of eucalyptus oil and in the essential oil of rosemary (*Rosmarinus officinalis*). Both α- and β-pinene are used commercially to make gin and other spirits. Recently collected findings from the analysis of human urine following occupational exposure to sawing fumes further reveal that cis- and trans-verbenol are further hydroxylated to diols. The primary urinary metabolite of a-pinene in rabbits is trans-verbenol; the minor biotransformation products are myrtenol and myrtenic acid. The primary urinary metabolites of b-pinene, in rabbits, however, is cis-verbenol indicating stereoselective hydroxylation [47].

18.31 Pulegone

(R)-(+)- Essential oils from various types of mint contain pulegone, a monoterpene ketone. Pennyroyal is the common name for two different species of mint, *Hedeoma pulegoides* and *Mentha pulegium*. (S)-(−) Essential oils containing pulegone are quite rare. Pennyroyal oil has many uses, including as a flavouring in food and drink, a part of perfumes and as a flea repellent. The pennyroyal herb has a long history of medical use, including as an abortion and menstrual inducer. Penny royal oil may have caused toxicity to the central nervous system, stomach ulcers, liver and kidney failure, lung toxicity and death if it was taken in large enough quantities. Pennyroyal

oil purchased from health food stores was discovered to contain over 80% pulegone, a terpene that was found to be hepatotoxic and pneumotoxic in rats. Several research have been conducted to examine the metabolism of (R)-(+)-pulegone in response to its toxic effects being found in animal models and people. Pulegone undergoes selective 10-position oxidation, yielding 10-hydroxypulegone, at safe doses. It has also been found in very small amounts in urine samples, suggesting that it may be metabolized to menthone. No pulegol was detected in the urine samples, which suggests that pulegone may also undergo carbonyl group first reduction. Because of this, pulegol is efficiently converted to menthol or reorganized into 3-p-menthen-8-ol [47].

18.32 α-Terpineol

Cajuput oil, pine oil and petit grain oil were all discovered to contain α-terpineol, a monocyclic monoterpene tertiary alcohol that has been isolated from a variation of a-terpineol. However, tee tree essential oil contained the highest percentage of a-terpineol (up to 30%) (*Melaleuca alternifolia*). Due to its similarity to the scent of lilac, a-terpineol finds widespread application in the creation of fragrances, cosmetics, soaps and even disinfectants. When given to rats orally, α-terpineol is converted to p-menthane-1,2,8-triol via an epoxide intermediate. Two primary pathways for a-terpineol biotransformation in rats are allylic methyl oxidation and 1,2-double bound reduction. The alcohol p-ment-1-ene-7,8-diol could not be extracted from the urine samples, despite the fact that allylic oxidation of C-1 methyl appears to be the predominant pathway. This molecule is probably stored and can easily undergo additional oxidation to form oleuropeic acid (Scheme 18.35) [12,48].

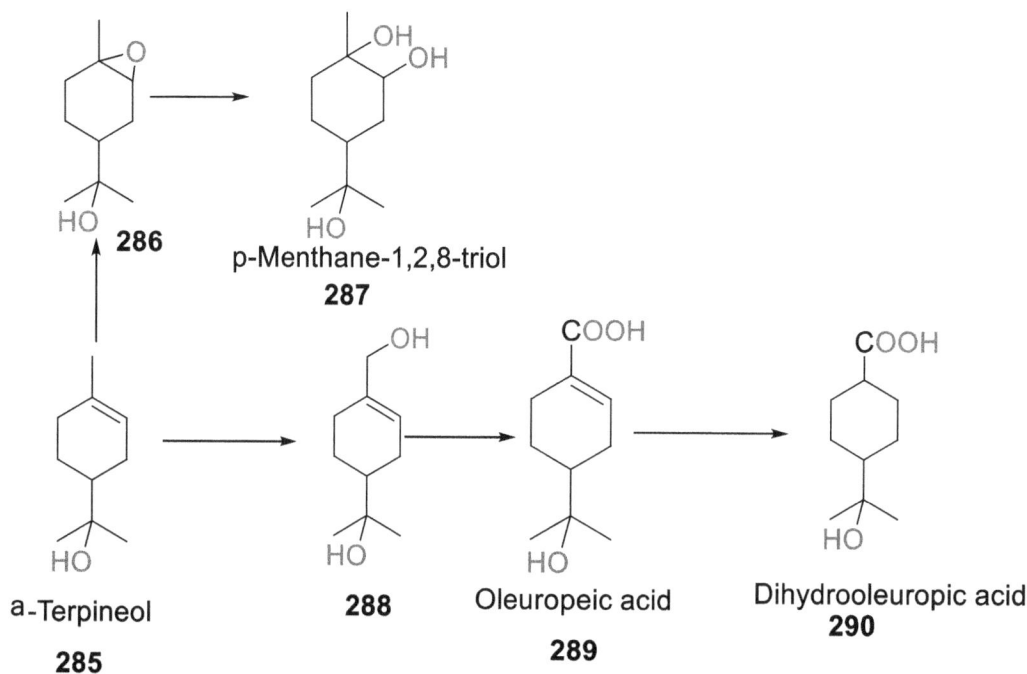

SCHEME 18.35 Metabolism of α-terpineol.

18.33 α- and β-Thujone

They are both bicyclic monoterpenes, but a- and b-thujone are distinguished from one another by the stereochemistry of their C-4 methyl group. The isomer ratio is plant-specific, with b-thujone being more abundant in wormwood oil and a-thujone being more abundant in cedar leaf oil. Herbal remedies, essential oils, meals, flavourings and beverages frequently contain these as well. α-Thujone is the main constituent in absinthe, a strong liquor that was widely used in Europe during the 19th century. Rat, mouse and human liver microsomes substantially metabolise α- and β-thujone to six hydroxythujones and three dehydrothujones and in vivo rat and mouse models were consistent with these in vitro observations (Scheme 18.36). Hydroxylation of methyl substituents at position 8 or 10 and subsequent dehydration may give rise to the dehydro derivatives. Numerous metabolites undergo further glucuronic acid conjugation in Phase II processes. Biotransformation following human consumption of α- and β-thujone should mirror that observed in vitro and in vivo [12,49].

18.34 Thymol

Natural sources of thymol are the bee balms (*Monarda fistulosa* and *Monarda didyma*). Larger levels of thymol exist in the essential oil of thyme (Thymus vulgaris) with concentrations up to 70%. Thymol is the major constituent in modern commercial mouthwash formulas as it displays antibacterial qualities. This could explain why thyme is used to treat sore throats and other similar infections in herbal therapy. It is notable that thymol is also utilized as flavor addition in a lot of meals and beverages. Despite substantial oxidation of the methyl and isopropyl groups and their subsequent excretion in large quantities unaltered or as their glucuronide and sulphate conjugates following oral administration to rats. This resulted in the production of derivatives of benzyl alcohol and 2-phenylpropanol

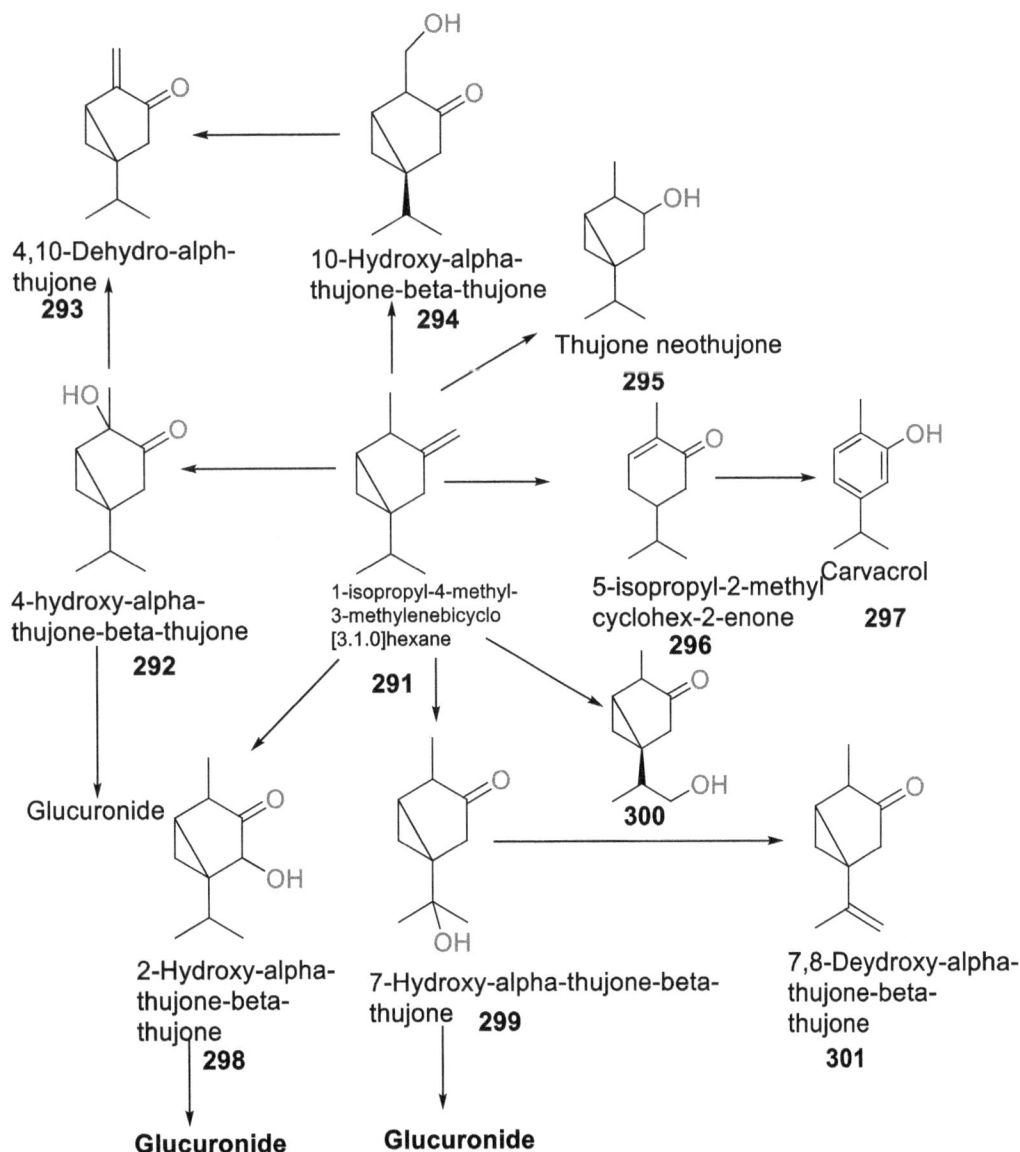

4,10-Dehydro-alph-
thujone
293

10-Hydroxy-alpha-
thujone-beta-thujone
294

Thujone neothujone
295

4-hydroxy-alpha-
thujone-beta-thujone
292

1-isopropyl-4-methyl-
3-methylenebicyclo
[3.1.0]hexane
291

5-isopropyl-2-methyl
cyclohex-2-enone
296

Carvacrol
297

Glucuronide

300

2-Hydroxy-alpha-
thujone-beta-
thujone
298

7-Hydroxy-alpha-thujone-beta-
thujone **299**

7,8-Deydroxy-alpha-
thujone-beta-
thujone
301

Glucuronide

Glucuronide

SCHEME 18.36 Synthesis of α- and β-thujone.

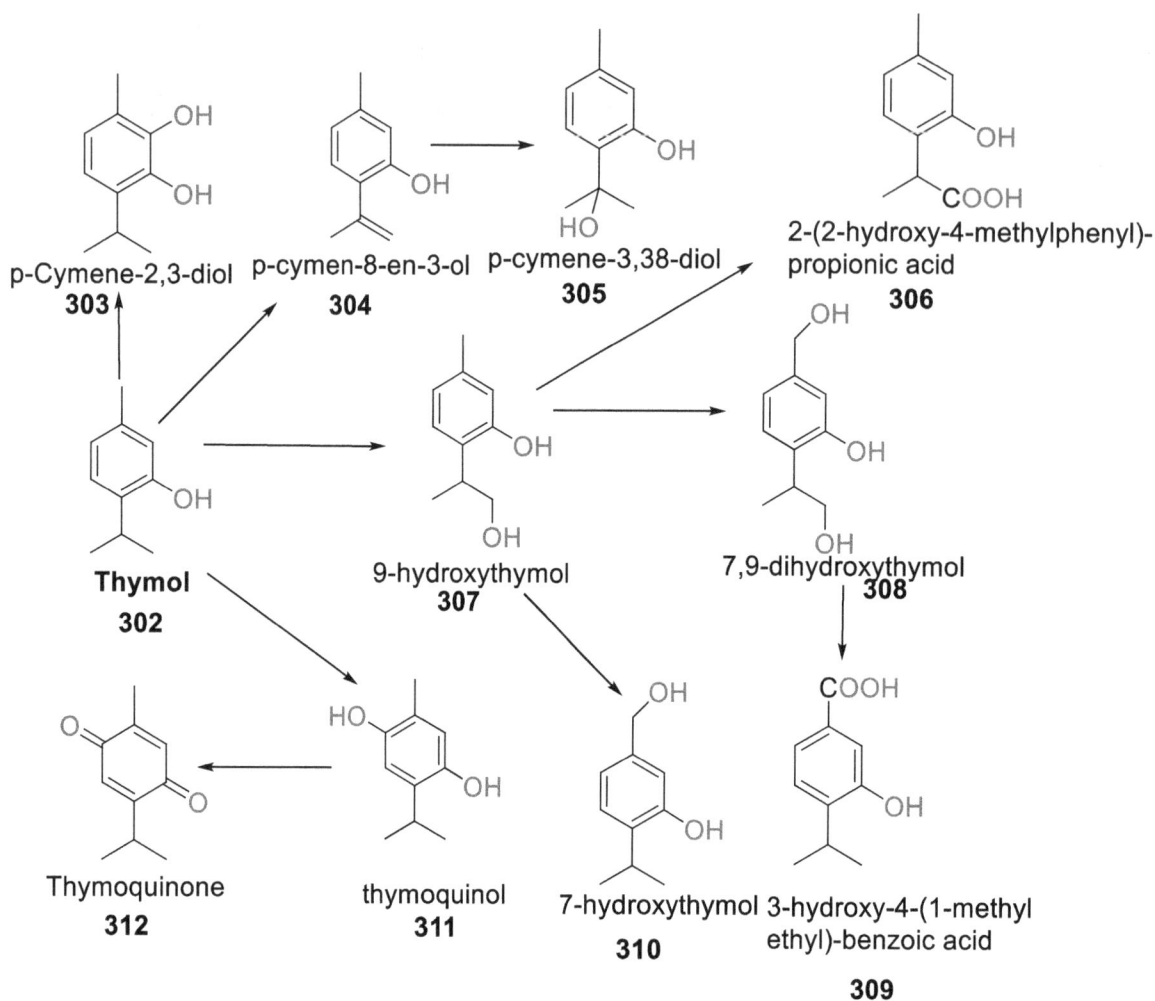

SCHEME 18.37   Metabolism of thymol.

and their corresponding carboxylic acids. Ring hydroxylation was merely a mild reaction (Scheme 18.37) [12,49].

18.35 Caryophyllene

The primary sesquiterpene found in hops is caryophyllene, which is added to soaps and scents for cosmetic purposes. Fresh hops stored in aerobic environments quickly oxidise. The primary by product of this process, (−)-caryophyllene-5,6-oxide, has a significant impact on beer quality. The mild sedative effects of hops are also ascribed to (−)-caryophyllene in herbal medicine. Additionally, it exhibits in vitro cytotoxicity against breast cancer cells. The primary metabolite [10S-(−)-14-hydroxycaryophyllene-5,6-oxide] and the minor biotransformation product cayrophyllene-5,6-oxide-2,12-diol were produced during the biotransformation of (−)-β-caryophyllene in rabbits. The diepoxide intermediate provides a simple explanation for how the minor metabolites are formed. Thus, it is hypothesized that the epoxide underwent regioselective hydroxylation. It is unknown at this time whether (−)-caryophyllene uses the same metabolic pathway in humans. However, after oral administration, significant

biotransformation in humans is strongly recommended based on in vivo findings from rabbits.

18.36 Farnesol

Hops contain the main sesquiterpene caryophyllene, which is added to soaps and fragrances for cosmetic purposes. In aerobic settings, fresh hops quickly oxidise. This process's main by product, (−)-caryophyllene-5,6-oxide, has a big impact on the taste of the beer. In herbal medicine, (−)-caryophyllene is also credited with the mild sedative properties of hops. Furthermore, it has in vitro cytotoxicity against breast cancer cells. The biotransformation of (−)-b-caryophyllene in rabbits resulted in the production of the main metabolite [10S-(−)-14-hydroxycaryophyllene-5,6-oxide] and the minor biotransformation product cayrophyllene-5,6-oxide-2,12-diol. The formation of the minor metabolites is simply explained by the diepoxide intermediate. As a result, it is assumed that regioselective hydroxylation occurred on the epoxide. Whether (−)-caryophyllene follows the same metabolic route in humans is currently unknown. However, based on in vivo results from rabbits, considerable

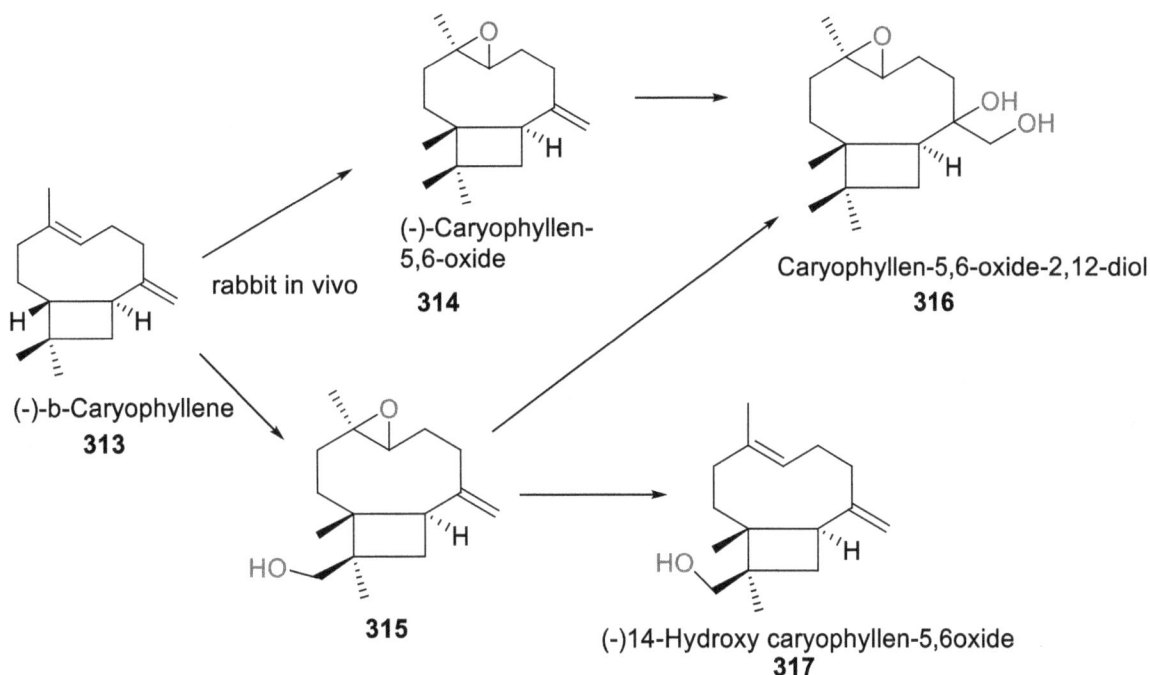

SCHEME 18.38 Metabolism of farnesol.

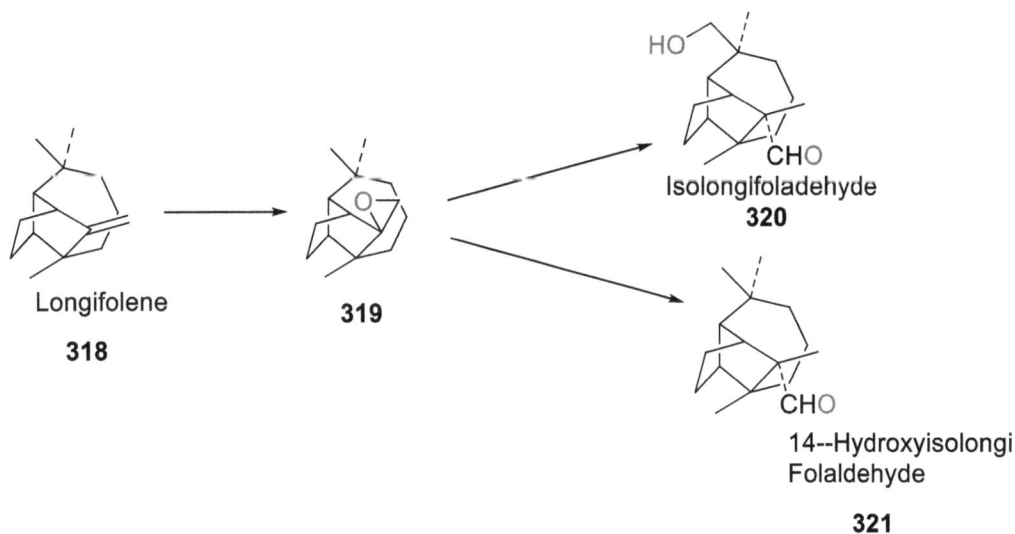

SCHEME 18.39 Metabolism of longifolene.

biotransformation in humans is strongly advised following oral treatment (Scheme 18.38) [12,50].

18.37 Longifolene

A naturally occurring oily liquid hydrocarbon with the common name "longifolene" is mostly present in some pine resins, particularly that of the Pinus longifolia tree, after which it is named. In addition to pines, longifolene is a key component of clove. Longifolene is employed in the food business because of its pleasant scent. Longifolene is metabolized by rabbits in the

following way: the exo-methylene group on the endo-face is attacked to create its epoxide, which is then isomerized to create a stable endo-aldehyde. Then this endo-aldehyde is quickly hydroxylated by cytochrome P450 (Scheme 18.39).

18.38 Patchouli Alcohol

In addition to pines, longifolene is a key component of clove. Longifolene is employed in the food business because of its pleasant scent. Longifolene is metabolized by rabbits in the following way: the exo-methylene group on the endo-face is

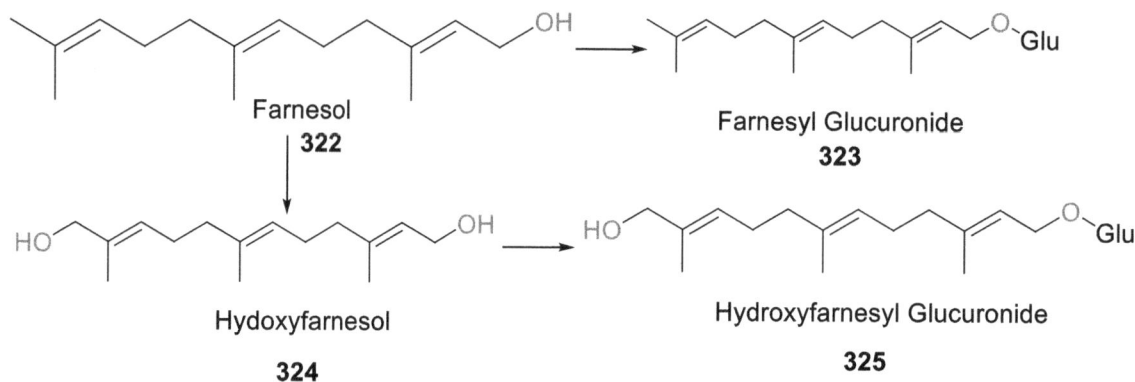

SCHEME 18.40 Metabolism of patchouli alcohol.

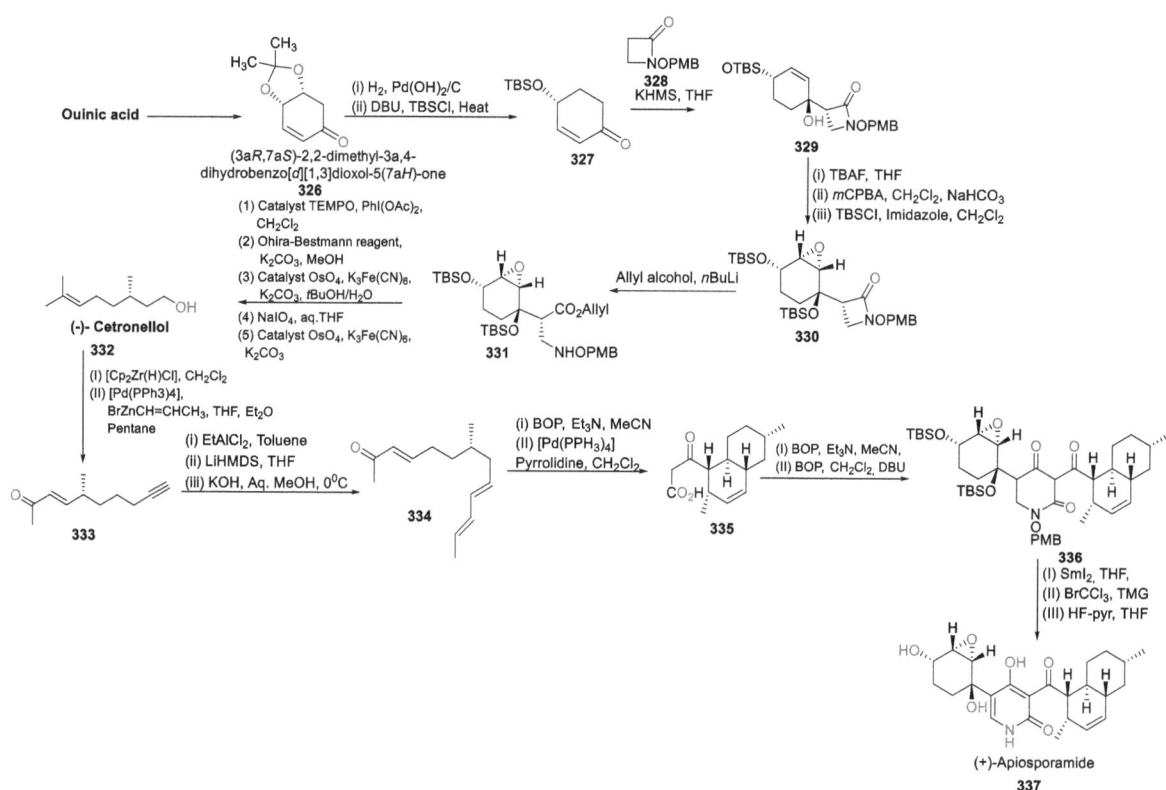

SCHEME 18.41 Synthesis of (+) apiosporamide from citronellol.

attacked to create its epoxide, which is then isomerized to create a stable endo-aldehyde. Then this endo-aldehyde is quickly hydroxylated by cytochrome P450 (Scheme 18.40).

18.39 Chiral Synthesis of Selected Terpenoids

18.39.1 Synthesis of (+) Apiosporamide

This is an example of the intramolecular Diels-Alder (IMDA) reaction and its transannular modification. So, in order to create the antibiotic alkaloid (+)-apiosporamide, two advanced intermediates from the chiral carbon pool are combined: amino acid 330, which comes from quinic acid and trans-decalin 334, which comes from citronellol, 332. In the first step, quinic acid was

used to make cyclohexenone 326, which was then deoxygenated to produce 1.2. Deprotonated β-lactam 328 was added, resulting in adduct 329, which was removed from the mixture and stereoselectively transformed into epoxide 330. Alyl alcohol was used to open the lactam in order to produce the necessary amino acid derivative 331. Citronellol, 332 was transformed into alkyne 333 in a parallel process, and this alkyne was then employed to produce the Diels-Alder precursor 334 through a Negishi-type chain elongation. High endo-selectivity IMDA was used to produce the trans-decalin ketone, which was then carboxylated to produce acid 335. The allyl-protecting group was taken out after coupling with 331 and carbonyl activation came next. Keto lactam 336 was produced via Dieckmann ring closure and then deprotected and aromatized to produce 337 (Scheme 18.41) [12,51].

18.40 Synthesis of (+)- Neosymbioimine

The marine alkaloid (S)-(−)-citronellol, 337 was employed in the complete synthesis of (+)-neosymbioimine. An endo-IMDA of triene 343, allowed access to the critical octalene structure 344. It was consequently transformed into aldehyde 339, which underwent a-hydroxylation and olefination to produce enoate 340. Triene 343 was produced by two more olefination stages, and it underwent a thermal IMDA reaction to produce decalin 344 with a 95% stereoselectivity. 345 was produced by stereoselective methylation and chain elongation to the nitrile. Regular functional group modification produced the compound ketal 346, which was then cyclized to produce imine 347. Following di-sulfonation and selective mono-hydrolysis of one sulfonate ester, demethylation of the aryl OMe ethers produced 348 (Scheme 18.42) (+) Citronellal, 349 has been utilized in IMDA reactions, producing the fungus metabolite hirsutellone B. (Scheme 18.42). Thus, (R)-citronellal, 349 was transformed into epoxy vinyl iodide 351, which was then combined with stannane 352 in a Stille-cross coupling to produce polyene 352. In the presence of Et$_2$AlCl, the stannane engages

in a Sakurai-type cyclization with the epoxide to produce the cyclohexane derivative 354, which is then subjected to a tandem IMDA reaction in which two extra rings are annulated. Thus, stereoselective synthesis of the tricyclic intermediate 355 was accomplished. The transformation of 356 into sulfone 359 was the next subgoal.

After the aryl methyl group was functionalized, the aldehyde 356 produced via Mukaiyama etherification and reduction of the ester was employed to extend the sidechain and produce the ketone 357. The introduction of a primary iodide and a thioaetate produced the intermediate 358, which under basic circumstances was cyclized to the thioether. Sulfone was produced by oxidation 359. The (Z)-cycloolefin was produced by the Ramberg-Backlund ring contraction and converted to the b-keto ester by carboxylation. Alcohol 360 was produced stereoselectively using Sharpless asymmetric dihydroxylation (AD) of the olefin and regioselective Barton-McCombie deoxygenation of the diol. After the alcohol was converted to a ketone, it was heated with NH$_3$ to produce 361 through an amidation-epimerization cyclization cascade of C-17 (Scheme 18.43) [12, 52].

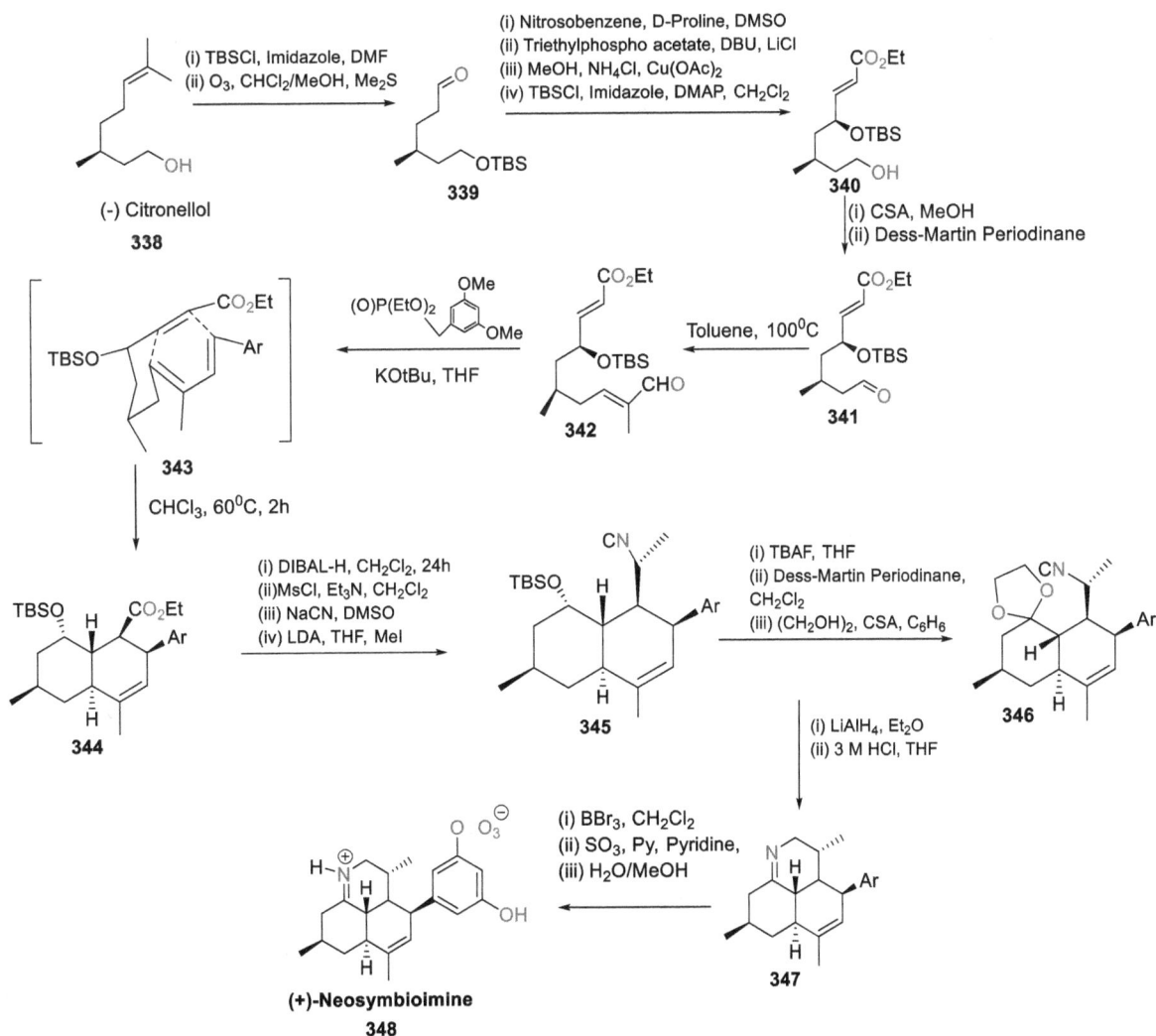

SCHEME 18.42 Synthesis of (+)- neosymbioimine from (+) citronellol.

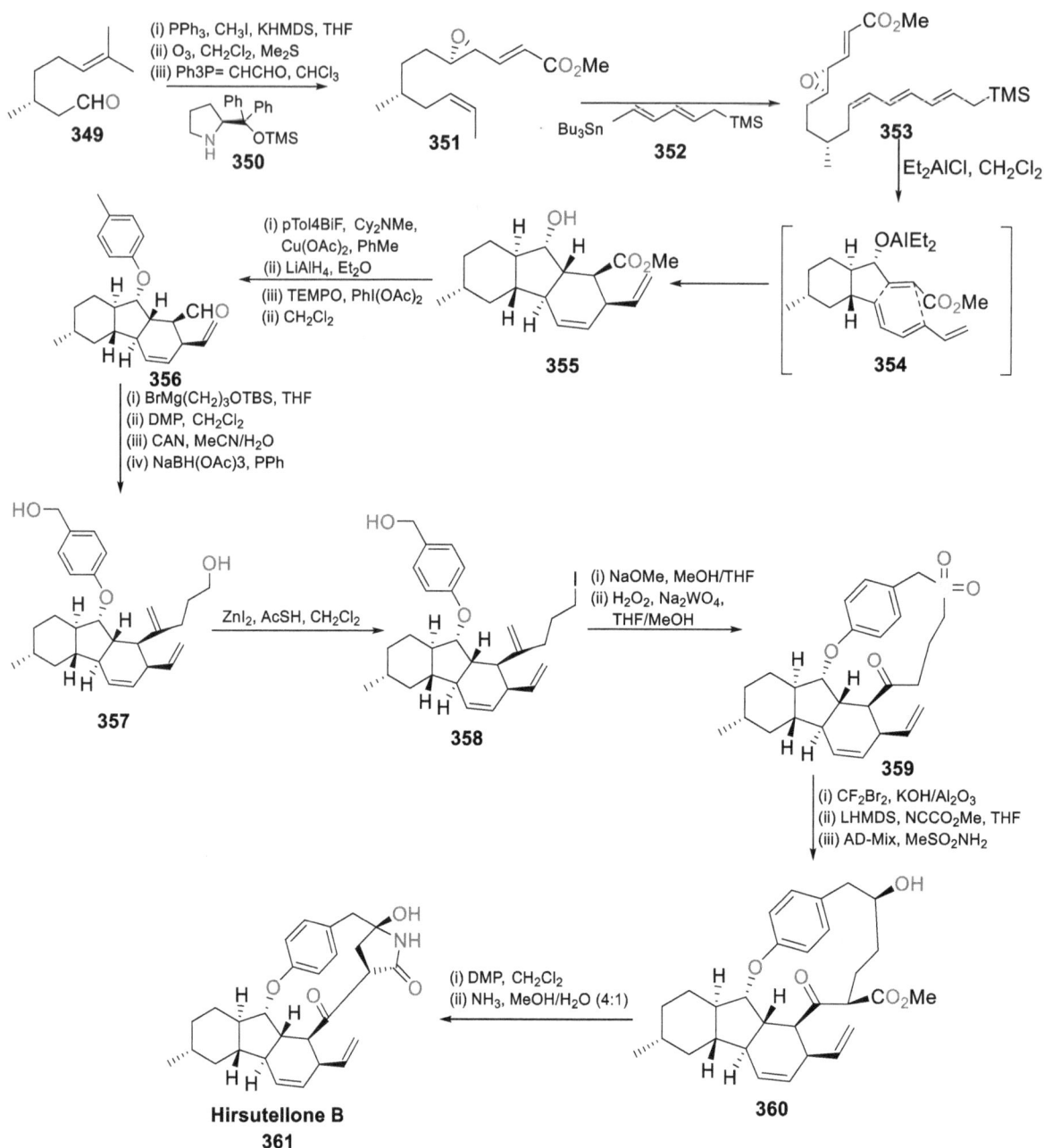

SCHEME 18.43   Synthesis of hirsutellone B from (+)-citronellal.

Diels-Alder addition of an electron-rich diene is a simple annulations technique. Thus, under Lewis's acid catalysis, (R)-carvone 362 was treated with diene 363 to produce the Diels-Alder adduct 364 that was stereo-selectively influenced by the large isopropenyl sidechain in the synthesis of the cell adhesion inhibitor (+)-Peribysin E, 371 (Scheme 18.43). A Saegusa oxidation produced the enone double bond. Protection of the more reactive ketone was necessary since the unprotected one underwent thioketal and Wittig reactions, which resulted in alcohol 365 upon reduction of the intermediate aldehyde. Further reduction to a methyl group was followed by the deprotection of the thioketal and oxidation of the isopropenyl appendage to produce methyl ketone (366). Baeyer-Villiger oxidation to the acetate Scheffer-Weitz (371) and α-iodination produced

vinyl iodide (367), which was used in a Suzuki coupling with vinyl boronic ester to produce diene (368) (Scheme 18.44) [12].

The marine diterpene (+)-chatancin, 380 is actually synthesized through this crucial transition (Scheme 18.44). Citronellol, 372 was first transformed into bromide 3.1, which was then combined with the anion of sulfone 374 to begin the synthesis. Desulfonation was followed by oxidation to produce aldehyde 375, which developed into an alkynoic ester and a keto sulfone at opposite ends. A powerful Michael acceptor was produced through Lindlar hydrogenation and oxidation, and it was employed in macrocyclization to create the TADA precursor 378. The Diels-Alder substrate 379 was produced by heating in the presence of water, and it spontaneously cycled into chatancin, 380 (Scheme 18.45) [12, 35, 52, 53].

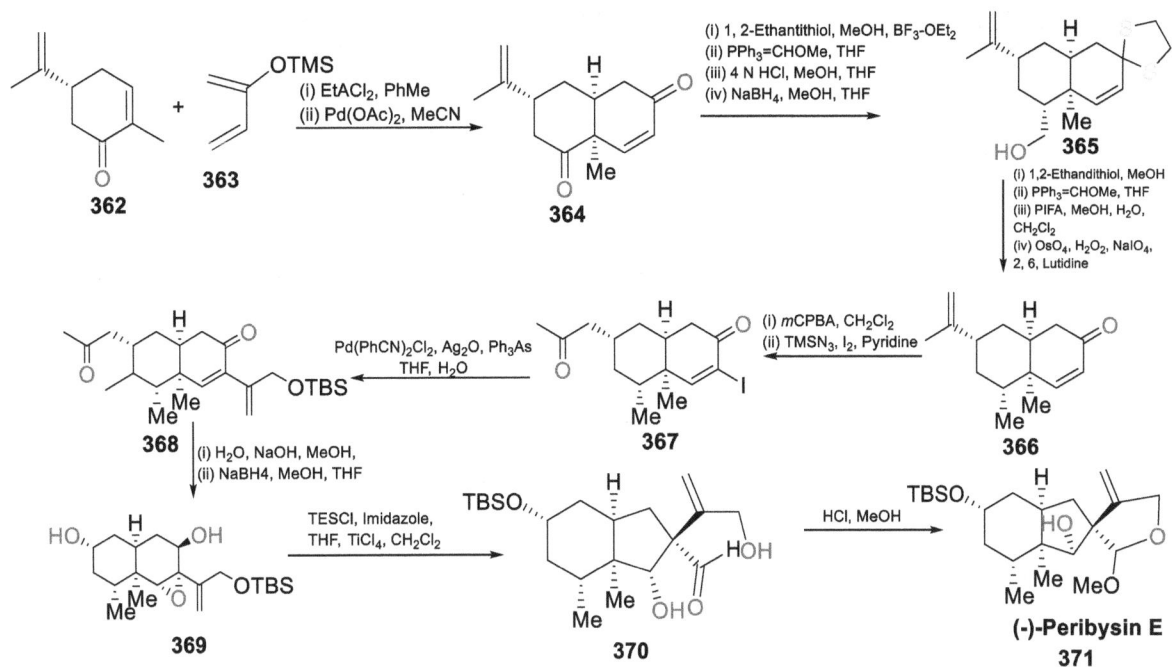

SCHEME 18.44 Synthesis of (–) peribysin E from (–) carvone.

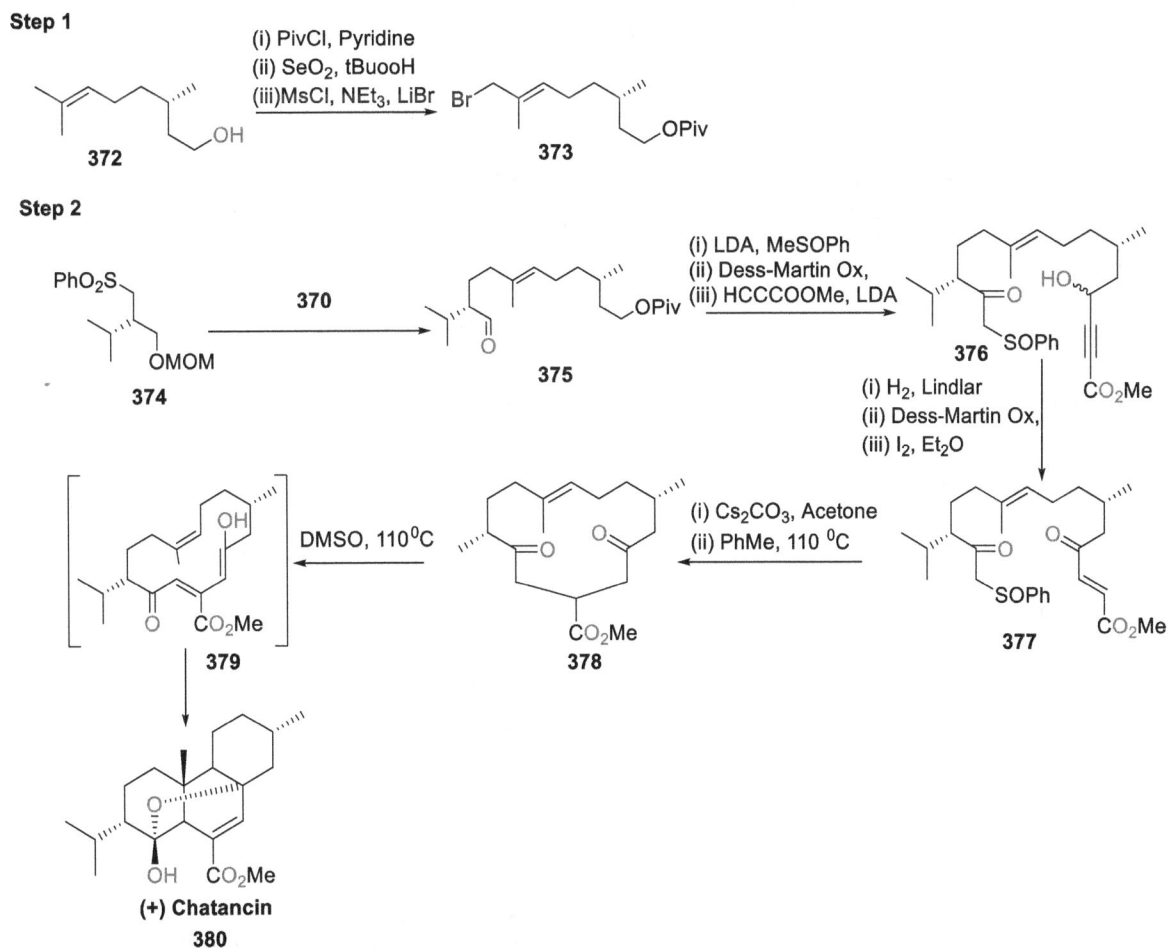

SCHEME 18.45 Synthesis of (+) chatancin from (+) citronellol.

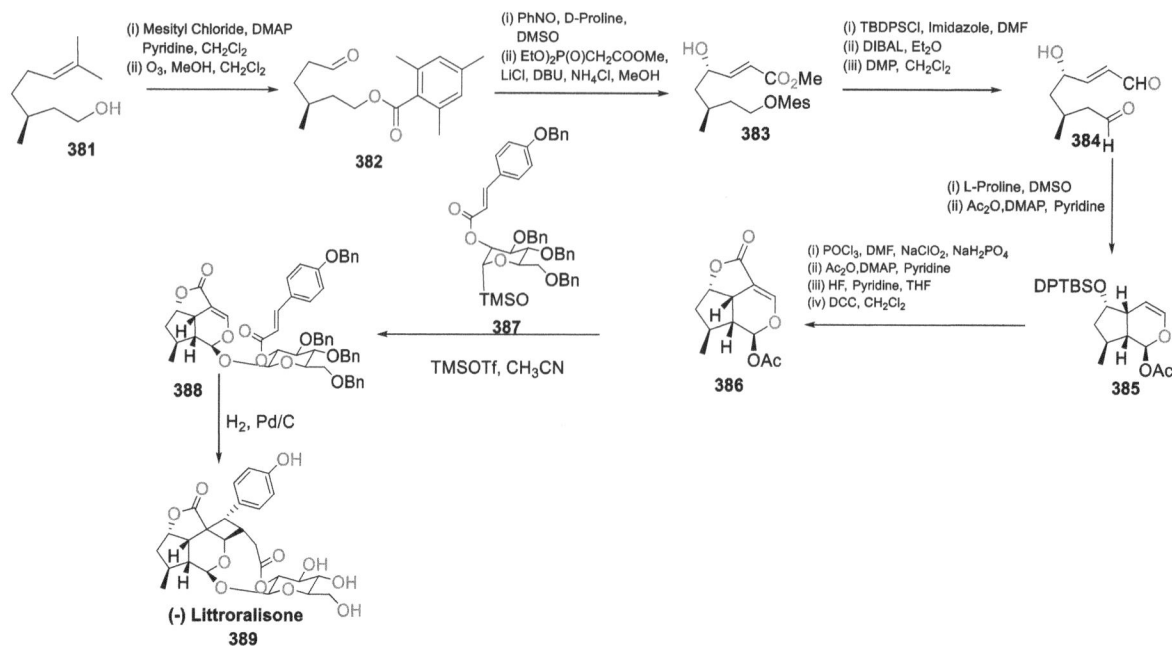

SCHEME 18.46 Synthesis of (–)-litteralisone from (–) citronellol.

The polycyclic neurotrophic iridoid (–)-litteralisone was made using an organocatalytic intramolecular Michael addition (Scheme 18.45). After Keck-Masamune olefination, citronellol, 381 was transformed into ester aldehyde 382, which underwent an organocatalytic a-hydroxylation to produce hydroxy enoate 383. 383 was converted into di-aldehyde 384, which served as the substrate for a Michael addition that was catalyzed by proline and produced glycoside 385 after acetylation. The creation of lactone 386 by Vilsmeyer formylation and Pinnick oxidation allowed for glycosidation with 1-O-TMS-b-D-glucose tetraacetate (387). After obtaining an amorphously pure diene 388, the cyclobutane ring was produced by photocycloaddition. The product of hydrogenolysis was (–)-litteralisone 389 (Scheme 18.46) [12,54].

Conclusion

This chapter describes about the terpenes, diastereomers, enantiomers, sources and various uses of Chiral terpenoids, Chiral terpenoids synthesis of Myrcene, Citral, Geraniol, Linalool, Lavandulol, Citronellal, Citronellol, α-Terpineol, Synthesis of α-terpineol starts with p-toluic acid, Carvone, Limonene, Menthol, Menthone, Synthesis of Menthone and Menthol, Total synthesis of α-pinene, bornane (camphene)-norbornane (isocamphane), Camphor, Synthesis of camphor, Farnesene, Farnesol, Nerolidol, etc. It also describes the metabolism reactions of Chiral terpenoids.

REFERENCES

[1] Breitmaier, E. *Terpenes: Flavors, Fragrances, Pharmaca, Pheromones*; Wiley-VCH, Weinheim, 2006.

[2] Rude, M. A.; Schirmer, A. New Microbial Fuels: A Biotech Perspective. *Current Opinion in Microbiology*, **2009**, 12, 274–281.

[3] Wilbon, P. A.; Chu, F.; Tang, C. Progress in Renewable Polymers from Natural Terpenes, Terpenoids, and Rosin. *Macromolecular Rapid Communications*, **2013**, 34, 8–37.

[4] Blaser, H.-U. The Chiral Pool as a Source of Enantioselective Catalysts and Auxiliaries. *Chemical Reviews*, **1992**, 92, 935–952.

[5] Ager, D., (ed.), *Handbook of Chiral Chemicals* (2nd edition). CRC Press, Taylor & Francis, Boca Raton, FL, 2005.

[6] Hanessian, S. *Total Synthesis of Natural Products: The Chiron Approach*. Pergamon Press, Oxford, 1983.

[7] Nakata, M. Monosaccharides as Chiral Pools for the Synthesis of Complex Natural Compounds. In Fraser-Reid, B. O.; Tatsuta, K., Thiem, J. (eds.), *Glycoscience* (2nd Edition). Springer-Verlag, Berlin, 2017.

[8] Sadgrove, N. J.; Padilla-González, G. F.; Phumthum, M. Fundamental Chemistry of Essential Oils and Volatile Organic Compounds, Methods of Analysis and Authentication. *Plants*, **2022**, 11, 789. https://doi.org/10.3390/plants11060789.

[9] Sadgrove, N. J. Comparing Essential Oils from Australia's 'Victorian Christmas Bush' (*Prostanthera lasianthos* Labill., Lamiaceae) to Closely Allied New Species: Phenotypic Plasticity and Taxonomic Variability. *Phytochemistry*, **2020**, 176, 112403.

[10] Brooks, W. H.; Guida, W. C.; Daniel, K. G. The Significance of Chirality in Drug Design and Development. *Current Topics in Medicinal Chemistry*, **2011**, 11, 760–770.

[11] Ochoa-Villarreal, M.; Howat, S.; Hong, S.; Jang, M. O.; Jin, Y. W.; Lee, E. K.; Loake, G. J. Plant Cell Culture Strategies for the Production of Natural Products. *BMB Reports*, **2016**, 49, 149–158.

[12] Sell, C. Chapter 5. Chemistry of essential oils. In Baser, K. H. C.; Buchbauer, G. (eds.). *Handbook of Essential Oils: Science, Technology, and Applications*. CRC Press, Taylor and Francis Group, London, 2010.

[13] Zhao, L.; Chang, W.-c.; Xiao, Y.; Liu, H.-w.; Liu, P. Methylerythritol Phosphate Pathway of Isoprenoid Biosynthesis. *Annual Review of Biochemistry*, **2013**, 82, 497–530.

[14] Santos-Sánchez, N. F.; Salas-Coronado, R.; Hernández-Carlos, B.; Villanueva-Cañongo, C. Shikimic acid pathway in biosynthesis of phenolic compounds. In Soto-Hernández, M.; García-Mateos, R.; Palma-Tenango, M. (eds.). *Plant Physiological Aspects of Phenolic Compounds* (Vol. 1). InTech Open, London, 2019.

[15] Pandey, A. V.; Tekwani, B. L.; Singh, R. L.; et al. Artemisinin, an Endoperoxide Antimalarial, Disrupts the Hemoglobin Catabolism and Heme Detoxifi Cation Systems in Malarial Parasites. *Journal of Biological Chemistry*, **1999**, 274(27), 19383–19388.

[16] Trombetta, D.; Francesco, C.; Sarpietro, M. G.; et al. Mechanisms of Antibacterial Action of Three Monoterpenes. *Antimicrobial Agents Chemotherapy*, **2005**, 49(6), 2474–2478.

[17] Berk, Z. Morphology and chemical composition. In: *Citrus Fruit Processing*. Academic Press, San Diego, CA, 2016, 9–54.

[18] Maarse, H. *Volatile Compounds in Foods and Beverages*. Marcel Dekker Inc., New York, 1991.

[19] Zeng, S.; Kapur, A.; Patankar, M. S.; Xiong, M. P. Formulation, Characterization, and Antitumor Properties of Trans- and Cis-Citral in the 4T1 Breast Cancer Xenograft Mouse Model. *Pharmaceutical Research*, **2015**, 32(8), 2548–2558.

[20] Anonymous. Ntp technical report on the toxicology and carcinogenesis studies of citral (microencapsulated) in F344/N rats and B6c3f1 mice, **2003**. https://ntp.niehs.nih.gov/ntp/htdocs/lt_rpts/tr505.pdf.

[21] Kozak, J. A.; Dake, G. R. *Angewandte Chemie International Edition*, 2008, 47, 4221–4223.

[22] Chadha, A.; Madyastha, M. K. Metabolism of Geraniol and Linalool in the Rat and Effects on Liver and Lung Microsomal Enzymes. *Xenobiotica*, **1984**, 14, 365–374.

[23] Asakawa, Y.; Ishida, T.; Toyota, M.; Takemoto, T. Terpenoid Biotransformation in Mammals IV. Biotransformation of (+)-Longifolene, (-)-Caryophyllene, (-)-Caryophyllene Oxide, (-)-Cyclocolorenone, (+)-Nootkatone, (-)-Elemol, (-)-Abietic and (+)-Dehydroabietic Acid in Rabbits. *Xenobiotica*, **1986**, 16, 753–767.

[24] Asakawa, Y.; Tair, Z.; Takemoto, T. X-Ray Crystal Structure Analysis of 14-Hydroxycaryophyllene Oxide, a New Metabolite of (-)-Caryophyllene, in Rabbits. *Journal of Pharmaceutical Sciences*, **1981**, 70, 710–711.

[25] Asakawa, Y.; Toyota, M.; Ishida, T. Biotransformation of 1,4-Cineole, a Monoterpene Ether. *Xenobiotica*, **1988**, 18, 1129–1134.

[26] Lawrence, W.T. Synthesis and Preparation of Terebic and Terpenylic acids. *Journal of the Chemical Society Transactions*, **1899**, 75, 527–533.

[27] Alder, K.; Vogt, W. Zur Kenntnis Der Diensynthese Mit Unsymmetrischen Addenden-Die Diensynthesen Des Piperylens, 1, 3-Dimethyl-Butadiens Und Des 1, 1, 3-Trimethyl-Butadiens Mit Acrylsaure Und Mit Acrolein. *Annalen Der Chemie-Justus Liebig*, **1949**, 564(2), 120–137.

[28] Simonsen, J. L. *The Terpenes* (Vol. 1, 2nd edition). Cambridge University Press, Cambridge, 1953, 394–408.

[29] E. Von Rudloff. Gas–Liquid Chromatography of Terpenes: Part II. The Dehydration Products of α-Terpineol. *Canadian Journal of Chemistry*, **1961**, 39(1), 1–12. https://doi.org/10.1139/v61-001.

[30] Chen, J.; Lu, M.; Jing, Y.; Dong, J. The Synthesis of L-Carvone and Limonene Derivatives with Increased Antiproliferative Effect and Activation of ERK Pathway in Prostate Cancer Cells. *Bioorganic & Medicinal Chemistry*, **2006**, 14(19), 6539–6547. doi: 10.1016/j.bmc.2006.06.013.

[31] Kotz, C. M.; Briggs, J. E.; Grace, M. K.; Levine, A. S.; Billington, C. J. Divergence of the Feeding and Thermogenic Pathways Influenced by NPY in the Hypothalamic PVN of the Rat. *The American Journal of Physiology-Regulatory, Integrative and Comparative Physiology*, **1998**, 275, R471–R477.

[32] Kotz, C. M.; Levine, A. S.; Billington, C. J. Effect of Naltrexone on Feeding, Neuropeptide Y and Uncoupling Protein Gene Expression During Lactation. *Neuroendocrinology*, **1997**, 65, 259–264.

[33] Lloyd, W. D.; Hedrick, G. W. Preparation and Beckmann Rearrangement of Pinonic Acid Oxime. *Industrial & Engineering Chemistry Product Research and Development*, **1963**, 2(2), 143–144.

[34] Alice Upshur, M.; Vega, M. M.; Gray Bé, A.; Chase, H. M.; Zhang, Y. Synthesis and Surface Spectroscopy of α-Pinene Isotopologues and Their Corresponding Secondary Organic Material. *Chemical Science*, **2019**, 10, 8390–8398.

[35] Ishida, T.; Toyota, M.; Asakawa, Y. Terpenoid Biotransformation in Mammals. V. Metabolism of (+)-Citro nellal, (+/-)-7-Hydroxycitronellal, Citral, (-)-Perillaldehyd, (-)-Myrtenal, Cuminaldehyde, Thujone, and (+/-)-Carvone in Rabbits. *Xenobiotica*, **1989**, 19, 843–855.

[36] Hong, Y. J.; Tantillo, D. J. Quantum Chemical Dissection of the Classic Terpinyl/Pinyl/Bornyl/Camphyl Cation Conundrum – The Role of Pyrophosphate in Manipulating Pathways to Monoterpenes. *Organic & Biomolecular Chemistry*, **2010**, 8, 4589–4600.

[37] Corey, E. J.; Ha, D.-C. Total Synthesis of Venustatriol. *Tetrahedron Letters*, **1988**, 29(26), 3171–3174.

[38] Chan, W. K.; Tan, L. T.; Chan, K. G.; Lee, L. H.; Goh, B. H. Nerolidol: A Sesquiterpene Alcohol with Multi-Faceted Pharmacological and Biological Activities. *Molecules*, **2016**, 21(5), 529. doi: 10.3390/molecules21050529.

[39] Vallianou, I.; Hadzopoulou-Cladaras, M. Camphene, a Plant Derived Monoterpene, Exerts Its Hypolipidemic Action by Affecting SREBP-1 and MTP Expression. *PLoS One*, **2016**, 11(1), e0147117. https://doi.org/10.1371/journal.pone.0147117.

[40] Willetts, A.; Masters, P.; Steadman, C. Regulation of Camphor Metabolism: Induction and Repression of Relevant Monooxygenases in *Pseudomonas putida* NCIMB 10007. *Microorganisms*, **2018**, 6(2), 41. doi: 10.3390/microorganisms6020041.

[41] Krause, S. T.; Crocoll, C.; Boachon, B.; Förster, C.; Leidecker, F. The Biosynthesis of Thymol, Carvacrol, and Thymohydroquinone in Lamiaceae Proceeds Via Cytochrome P450s and a Short-Chain Dehydrogenase. *PNAS*, **2021**, 118(52), e2110092118.

[42] Duisken, M.; Sandner, F.; Blömeke, B.; Hollender, J. Metabolism of 1,8-Cineole by Human Cytochrome P450 Enzymes: Identification of a New Hydroxylated Metabolite. *Biochimica et Biophysica Acta*, **2005**, 1722(3), 304–311. doi: 10.1016/j.bbagen.2004.12.019.

[43] Tak, J. H.; Isman, M. B. Metabolism of Citral, the Major Constituent of Lemongrass Oil, in the Cabbage Looper, Trichoplusia Ni, and Effects of Enzyme Inhibitors on Toxicity and Metabolism. *Pesticide Biochemistry and Physiology*, **2016**, 133, 20–25. doi: 10.1016/j.pestbp.2016.03.009.

[44] Chadha, A.; Madyastha, K. M. Metabolism of Geraniol and Linalool in the Rat and Effects on Liver and Lung Microsomal Enzymes. *Xenobiotica*, **1984**, 14(5), 365–374. doi: 10.3109/00498258409151425. PMID: 6475100.

[45] Milanos, S.; Elsharif, S. A.; Janzen, D.; Buettner, A.; Villmann, C. Metabolic Products of Linalool and Modulation of GABAA Receptors. *Frontiers in Chemistry*, **2017**, 5, 46. doi: 10.3389/fchem.2017.00046.

[46] Wichtel, M. *Teedrogen und Phytopharmaka*. Wissenschaftliche Verlagsgesellschaft, Stuttgart, 2002.

[47] Kjonaas, R.; Martinkus-Taylor, C.; Croteau, R. Metabolism of Monoterpenes: Conversion of l-Menthone to l-Menthol and d-Neomenthol by Stereospecific Dehydrogenases from Peppermint (*Mentha piperita*) Leaves. *Plant Physiology*, **1982**, 69(5), 1013–1017. doi: 10.1104/pp.69.5.1013.

[48] Madyastha, K. M.; Renganathan, V. Metabolism of Alpha-Terpineol by *Pseudomonas incognita*. *Canadian Journal of Microbiology*, **1984**, 30(12), 1429–1436. doi: 10.1139/m84-228.

[49] Sirisoma, N. S.; Höld, K. M.; Casida, J. E. α- and β-Thujones (Herbal Medicines and Food Additives): Synthesis and Analysis of Hydroxy and Dehydro Metabolites. *Journal of Agricultural and Food Chemistry*, **2001**, 49(4), 1915–1921.

[50] Schooley, D. A.; Goodman, W. G.; Cusson, M.; Gilbert, L. I. The Juvenile Hormones. *Reference Module in Life Sciences* **2019**, 55–103.

[51] Alfatafta, A. A.; Gloer, J. B.; Scott, J. A.; Malloch, D. Apiosporamide, a New Antifungal Agent from the Coprophilous Fungus *Apiospora montagnei*. *Journal of Natural Products*, **1994**, 57(12), 1696–702. doi: 10.1021/np50114a012.

[52] Varseev, G. N.; Maier, M. E. Enantioselective Total Synthesis of (+)-Neosymbioimine. *Organic Letters*, **2007**, 9(8), 1461–1464.

[53] Nicolaou, K. C.; Sarlah, D.; Wu, T. R.; Zhan, W. Total Synthesis of Hirsutellone B. *Angewandte Chemie International Edition*, **2009**, 48(37), 6870–6874. doi: 10.1002/anie.200903382.

[54] Zhao, Y. M.; Maimone, T. J. Short, Enantioselective Total Synthesis of Chatancin. *Angewandte Chemie International Edition*, **2015**, 54(4), 1223–1226. doi: 10.1002/anie.201410443.

Terpenes: A Unique Source of Oil and Fragrance in Industry

Bimal Krishna Banik*, Abhishek Tiwari†, and Biswa Mohan Sahoo‡

CONTENTS

19.1 Introduction 590
19.2 Classification of Volatile Oil 590
19.3 Volatile Oil Extraction 590
19.4 Sources of Natural Essential Oil 591
    19.4.1 Hydrodistillation 592
        19.4.1.1 Mechanism of Distillation 593
    19.4.2 Essential Oil Extraction by Expression 595
    19.4.3 Essential Oil Extraction with Cold Fat (Enfleurage) 595
    19.4.4 Enfleurage and Defleurage 596
    19.4.5 Hot Maceration Process 596
    19.4.6 Modern (Non-Traditional) Methods of Extraction of Essential Oils 596
        19.4.6.1 Headspace Trapping Techniques 596
        19.4.6.2 Static Headspace Technique 596
        19.4.6.3 Vacuum Headspace Technique 596
        19.4.6.4 Dynamic Headspace Technique 596
    19.4.7 Solid Phase Micro-Extraction (SPME) 596
    19.4.8 Supercritical Fluid Extraction (SFE) 597
    19.4.9 Simultaneous Distillation Extraction (SDE) 597
    19.4.10 Microwave Distillation 597
    19.4.11 Controlled Instantaneous Decomposition (CID) 598
    19.4.12 Thermo-Micro Distillation 598
    19.4.13 Molecular Spinning Band Distillation 598
19.5 Worldwide Current Trends in Essential Oils Market 598
19.6 Industrial Applications 598
    19.6.1 Food and Food Preservation 599
    19.6.2 Nano-Formulation Incorporating Essential Oil 600
    19.6.3 Role of Essential Oils as Natural Cosmetic Preservatives 600
    19.6.4 Antimicrobial Effects of Essential Oils 602
    19.6.5 Current Trends on the Application of EOs in the Food Industry 603
    19.6.6 Role of Essential Oils in Perfumes 604
19.7 Analysis of Essential Oils 608
19.8 Legislation of EO'S in Food 608
Conclusion 609
References 609

* Department of Mathematics and Natural Sciences, College of Sciences and Human Studies, Prince Mohammad Bin Fahd University, Al Khobar, Kingdom of Saudi Arabia. bimalbanik10@gmail.com; bbanik@pmu.edu.sa.

† Faculty of Pharmacy, Pharmacy Academy, IFTM University, Lodhipur Rajput, Moradabad-244102, Uttar Pradesh, India. abhishekt1983@gmail.com.

‡ Roland Institute of Pharmaceutical Sciences, Berhampur affiliated to Biju Patnaik University of Technology (BPUT), Rourkela, Odisha, India.

19.1 Introduction

Since the beginning of civilization, plants and plant-derived natural products have been employed for medicinal, religious and cosmetic purposes. As a means of discovering novel chemical entities as leads, one of the most rapidly developing disciplines of inquiry examines natural plant products. It has been established that bioactive phytochemical and/or bionutrients contained in medicinal plants have a crucial role in the prevention of chronic diseases such as cancer, diabetes and coronary heart diseases.

These chemicals are produced by plants as a kind of self-defense, but they also protect plants from disease and injury, as well as contribute to the plant's color, aroma and flavor. Essential oils (EOs), which may also be considered a sort of secondary metabolites, contribute to the therapeutic properties of aromatic plants. Due to their aroma, essential oils have traditionally been linked with food flavorings, cosmetics and perfumes. However, research indicates that volatile monoterpene components have substantial promise for treating and preventing human diseases. Figure 19.1 shows various sources of essential oils.

In recent years, plant EOs have gained growing prominence in phytomedicine and aromatherapy, and their widespread use has attracted the interest of basic researchers, particularly in regard to their antibacterial, antioxidant and anticancer capabilities. The bulk of EOs are complex natural combinations of terpene and phenylpropanoids (benzene derivatives) that are responsible for their biological effects. Terpenes and essential oils (EOs) appear to be the same thing at first glance: they both originate from plants, are aromatic and are often used for the same purpose. Due to their similarities, many individuals mistakenly assume that they are identical. Figure 19.2 depicts the various heterogenous groups present in Essential oils [1,2].

19.2 Classification of Volatile Oil

Table 19.1 shows how volatile oils are categorized according to the functional groups they contain.

19.3 Volatile Oil Extraction

Essential oils are utilized in several consumer products, including detergents, soaps, toiletries, cosmetics, medicines, fragrances, confectionary food items, soft drinks, distilled alcoholic beverages (hard drinks) and pesticides. The global production and use of essential oils and fragrances are expanding at a rapid rate. Essential to improving the overall yield and quality of essential oil is production technology. Traditional techniques for processing essential oils continue to be utilized in many regions of the world and are of significant importance. The most frequent and traditional techniques include water distillation, water

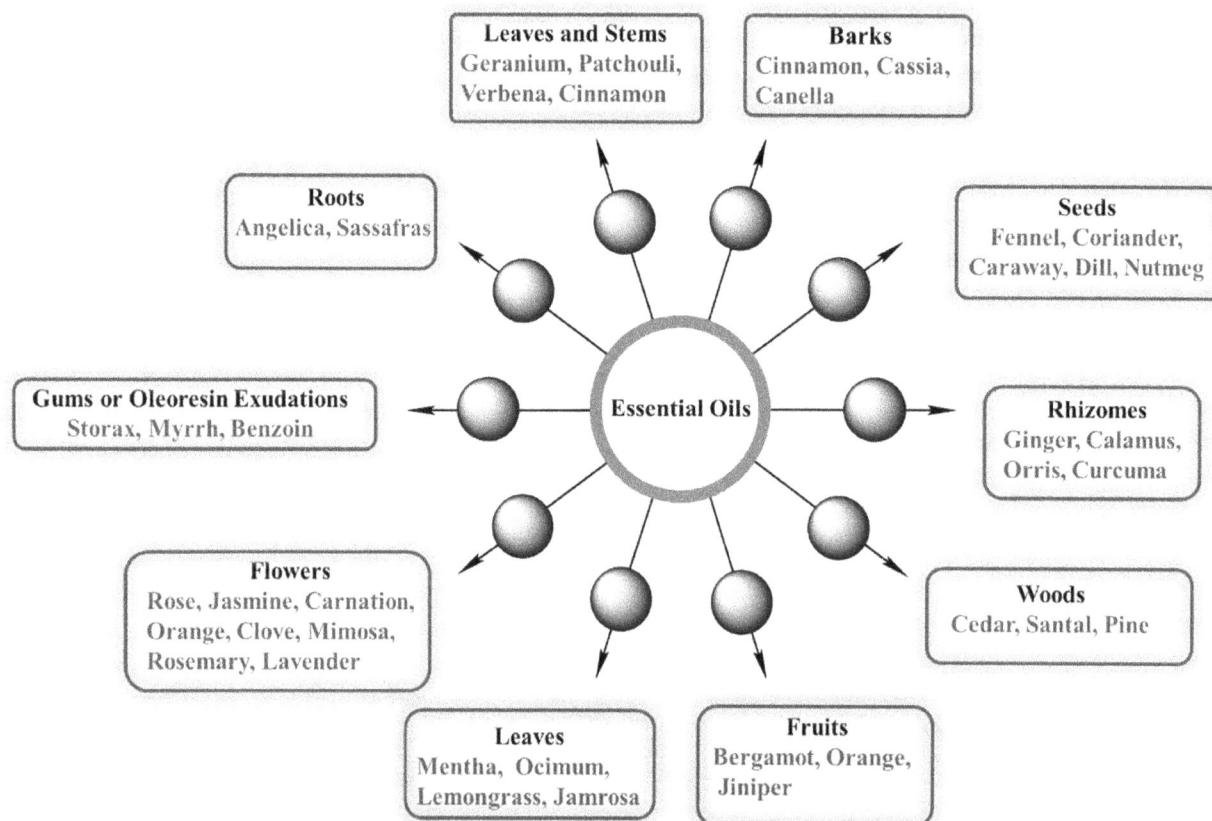

FIGURE 19.1 Sources of essential oils.

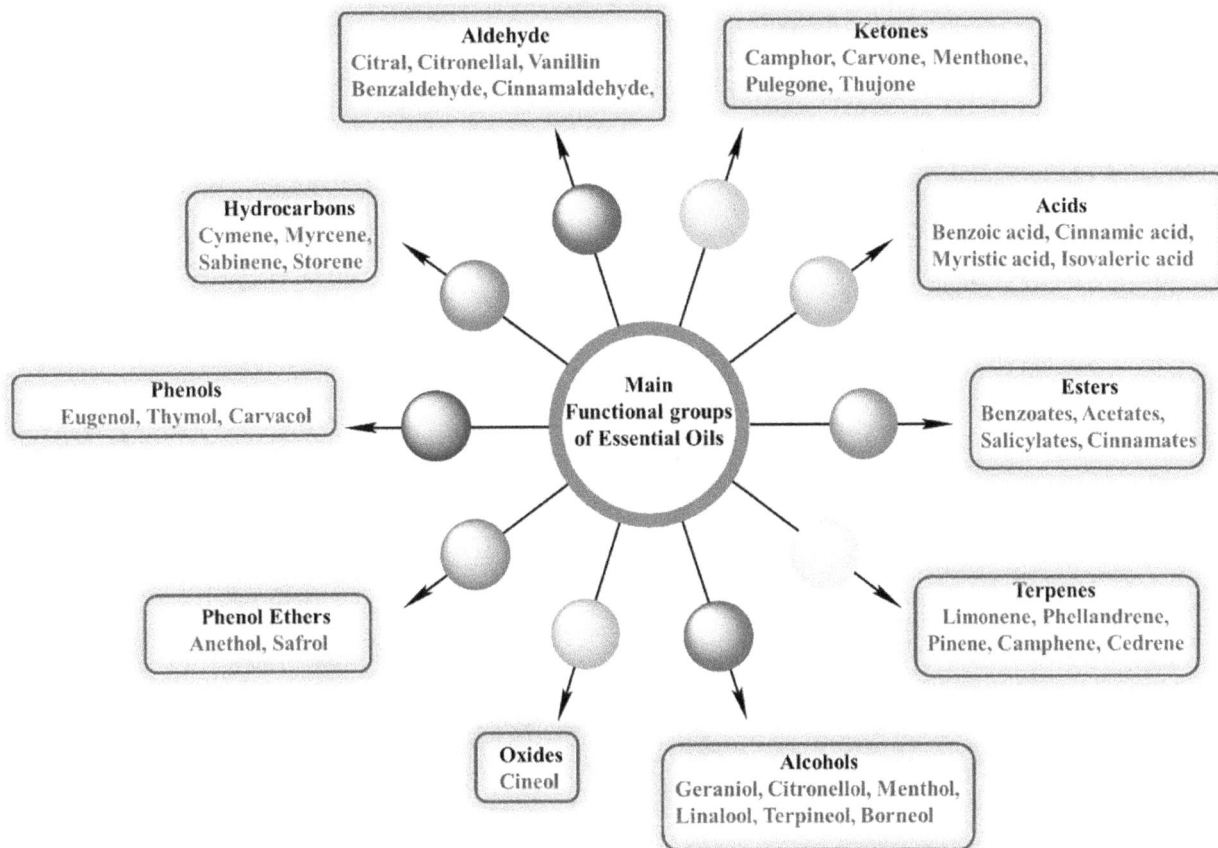

FIGURE 19.2 Heterogeneous chemical group present in essential oils.

TABLE 19.1

Different Constituents of Essential Oils

S. No.	Functional Groups	Examples
(1)	Hydrocarbons	Turpentine oil
(2)	Alcohols	Peppermint oil, Pudina, Sandalwood oil, etc.
(3)	Aldehydes	Cymbopogon sp., Lemongrass oil, Cinnamon Cassia and Saffron
(4)	Ketones	Camphor, Caraway and Dill, Jatamansi, Fennel, etc.
(5)	Phenols	Clove, Ajowan, Tulsi, etc.
(6)	Phenolic ethers	Nutmeg, Calamus, etc.
(7)	Oxides	Eucalyptus, Cardamom and Chenopodium oil
(8)	Esters	Valerian, Rosemary oil, Garlic, Gaultheria oil, etc.

and steam distillation, steam distillation, cohobation, maceration and enfleurage. When oil production from distillation is low, maceration is an adaptive alternative. Distillation techniques are suited for powdered almonds, rose petals and rose flowers, but solvent extraction is ideal for costly, fragile and thermally unstable substances such as jasmine, tuberose and hyacinth. The preferred method for extracting citronella oil from plant matter is water distillation (Figure 19.3) [1,2].

19.4 Sources of Natural Essential Oil

Essential oils are typically extracted from one or more plant parts, including flowers (e.g. rose, jasmine, carnation, clove, mimosa, rosemary and lavender), leaves (e.g. mint, Ocimum spp., lemongrass and jamrosa), leaves and stems (e.g. geranium, patchouli, petitgrain and verbena), bark (e.g. cinnamon, cassia and canella), wood (e.g. (e.g. balsam of Peru, Myroxylon balsamum, storax, myrrh and benzoin).

4.1 Methods of Producing Essential Oils.

The essential oils industry has devised terminology to differentiate three forms of hydrodistillation: water distillation, water and steam distillation and direct steam distillation. These words were invented by Von-Rechenberg and have since become standard in the essential oil market. All three techniques are subject to the same distillation of two-phase systems-related theoretical issues. The primary distinction lies in the material handling procedures.

Some volatile oils cannot be distilled without decomposition and are thus often extracted by expression or other mechanical processes (lemon oil and orange oil). In certain places, citrus oil is often extracted by rolling the fruit over a trough lined with sharp projections that are long enough to penetrate the epidermis and pierce the oil glands in the outer section of the peel (Ecuelle method). A pressing motion on the fruit extracts

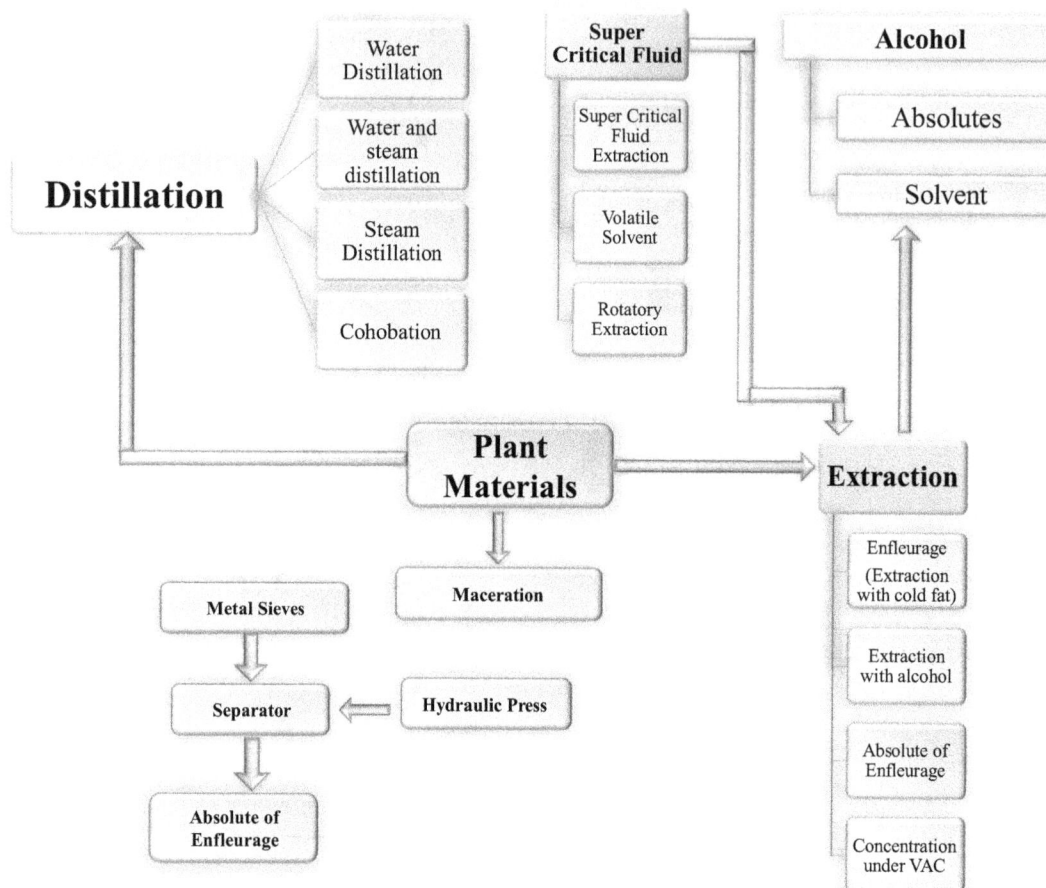

FIGURE 19.3 Extraction methods or essential oils.

the oil from the glands and a fine spray of water washes the oil from the mashed peel as the juice is collected through a tube that runs through the fruit's center.

Through centrifugation, the resultant oil-water emulsion is separated. In a variant of this method, the fruit's skin is removed prior to oil extraction. Frequently, the volatile oil concentration of fresh plant parts (flower petals) is so low that oil extraction using the aforementioned techniques is not commercially viable. On glass plates, an odorless, flavorless, fixed oil or fat is applied in a thin layer. The flower petals are laid on the fat for a few hours, and then the oil petals are continuously removed and replaced with a new layer of petals. After the fat has absorbed the maximum amount of scent, the oil may be extracted using alcohol. This technique, known as enfleurage, was once widely employed in the manufacture of fragrances and pomades.

In the current perfume business, the vast majority of essential oils are extracted utilizing volatile solvents such as petroleum ether and hexane. The primary benefit of extraction over distillation is the ability to keep a constant temperature (about 50°C) throughout the process. As a consequence, extracted oils have a more natural aroma than distilled oils, which may have experienced chemical changes due to the high temperature. This characteristic is very important to the perfume industry, although the conventional distillation method is less expensive than the extraction approach [3–5].

In destructive distillation, volatile oil is distilled in the absence of air. When Pinaceae or Cupressaceae wood or resin is burned without air, breakdown occurs and a number of volatile chemicals are pushed out. The remaining material is charcoal. The condensed volatile matter often separates into two layers: an aqueous layer comprising wood naptha (methyl alcohol) and pyroligneous acid (crude acetic) and a tarry liquid consisting of pine tar, juniper tar, or other tars depending on the type of wood utilized. This dry distillation is often carried out in retorts, and if the wood is chopped or coarsely powdered and the heat is delivered swiftly, the output is frequently around 10% of the wood weight employed.

19.4.1 Hydrodistillation

In order to separate essential oils using hydrodistillation, aromatic plant material is packed into a still with a suitable amount of water, which is then heated to a boil; alternatively, the plant charge is steamed with live steam. Under the action of hot water and steam, the essential oil is released from the plant tissue's oil glands. The water-oil vapor combination is condensed by indirect cooling with water. From the condenser, distillate runs into a separator, where oil and distillate water are separated mechanically (Figure 19.4) [3–5].

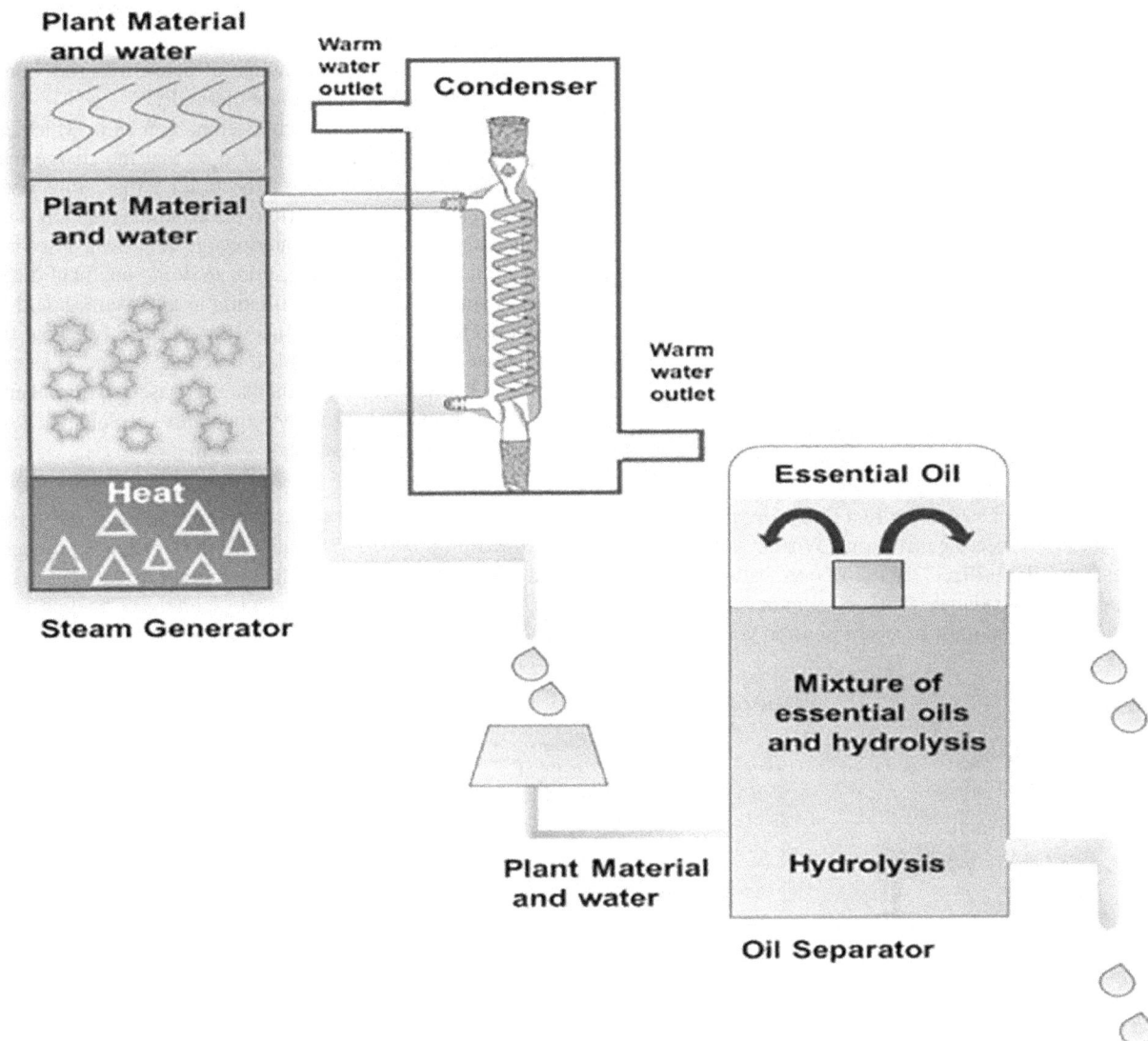

FIGURE 19.4 Hydro-distillation method.

19.4.1.1 Mechanism of Distillation

Hydrodistillation of plant material involves the following main physicochemical processes:

i. Hydrodiffusion
ii. Hydrolysis
iii. Decomposition by heat

19.4.1.1.1 Hydrodiffusion

Hydrodiffusion is the diffusion of essential oils and hot water through plant membranes. The steam does not truly permeate the dry cell membranes during steam distillation. Therefore, dry plant material can only be exhausted with dry steam after all volatile oil has been liberated from the oil-bearing cells by first comminuting the plant material thoroughly. However, when plant material is saturated with water, the exchange of vapors inside the tissue is dependent on the tissue's permeability when swelled. Plant cell membranes are nearly impervious to volatile oils. Therefore, in the real process, at the temperature of boiling water, a portion of the volatile oil dissolves in the water within the glands, and this oil-water solution permeates, via osmosis, the inflated membranes until it reaches the outside surface, where it is vaporized by passing steam.

Another facet of hydrodiffusion is that the rate of oil evaporation is not affected by the volatility of the oil's components, but rather by their water solubility. Therefore, the water-soluble, high-boiling components of oil in plant tissue distil before the water-insoluble, low-boiling components. Since hydrodiffusion rates are sluggish, it takes longer to distil uncomminuted material than comminuted material [3–5].

19.4.1.1.2 Hydrolysis

Hydrolysis is described in the current context as a chemical interaction between water and certain components of essential oils. In the presence of water, especially at high temperatures, esters tend to react with water to generate acids and alcohols. Esters are components of essential oils. In addition, as this is a time-dependent process, the rate of hydrolysis relies on how long oil and water remain in contact. This is one of water distillation's downsides.

19.4.1.1.3 Three Types of Hydrodistillation

Three are three types of hydrodistillation for isolating essential oils from plant materials:

1. Water distillation
2. Water and steam distillation
3. Direct steam distillation

19.4.1.1.3.1 Water Distillation Immersion in water followed by boiling through direct fire, steam jacket, closed steam jacket, closed steam coil, or open steam coil are included in these procedures. This technique is distinguished by the direct contact of boiling water and plant matter (Figure 19.5).

When the still is heated by direct fire, proper care must be taken to avoid overheating the charge. When a steam jacket or closed steam coil is utilized, the risk of overheating is reduced; with open steam coils, this risk is eliminated. With open steam, however, care must be taken to avoid condensation of water within the still. Consequently, the still should be well insulated. The plant material in the still must be stirred while the water boils; otherwise, clumps of dense material would sink to the bottom and deteriorate thermally.

Certain plant materials, such as cinnamon bark, which are rich in mucilage, must be powdered so that the charge may easily disperse in water; when the water temperature rises, the mucilage will be leached from the ground cinnamon. This significantly increases the water-charge mixture's viscosity, allowing it to char. Therefore, prior to doing any field distillation, a small-scale water distillation in glassware should be conducted to monitor any changes that occur throughout the distillation process. It is possible to calculate the production of oil from a known weight of plant material based on this laboratory experiment. The laboratory apparatus recommended for trial distillations is the Clevenger system [5].

19.4.1.1.3.2 Traditional Method of Producing Attar Using Hydro-distillation Floral attars are the distillates produced from the hydrodistillation of flowers (such as saffron, marigold, rose, jasmine and pandanus) in sandal wood oil or other base materials such as paraffin. Because the blooms must be

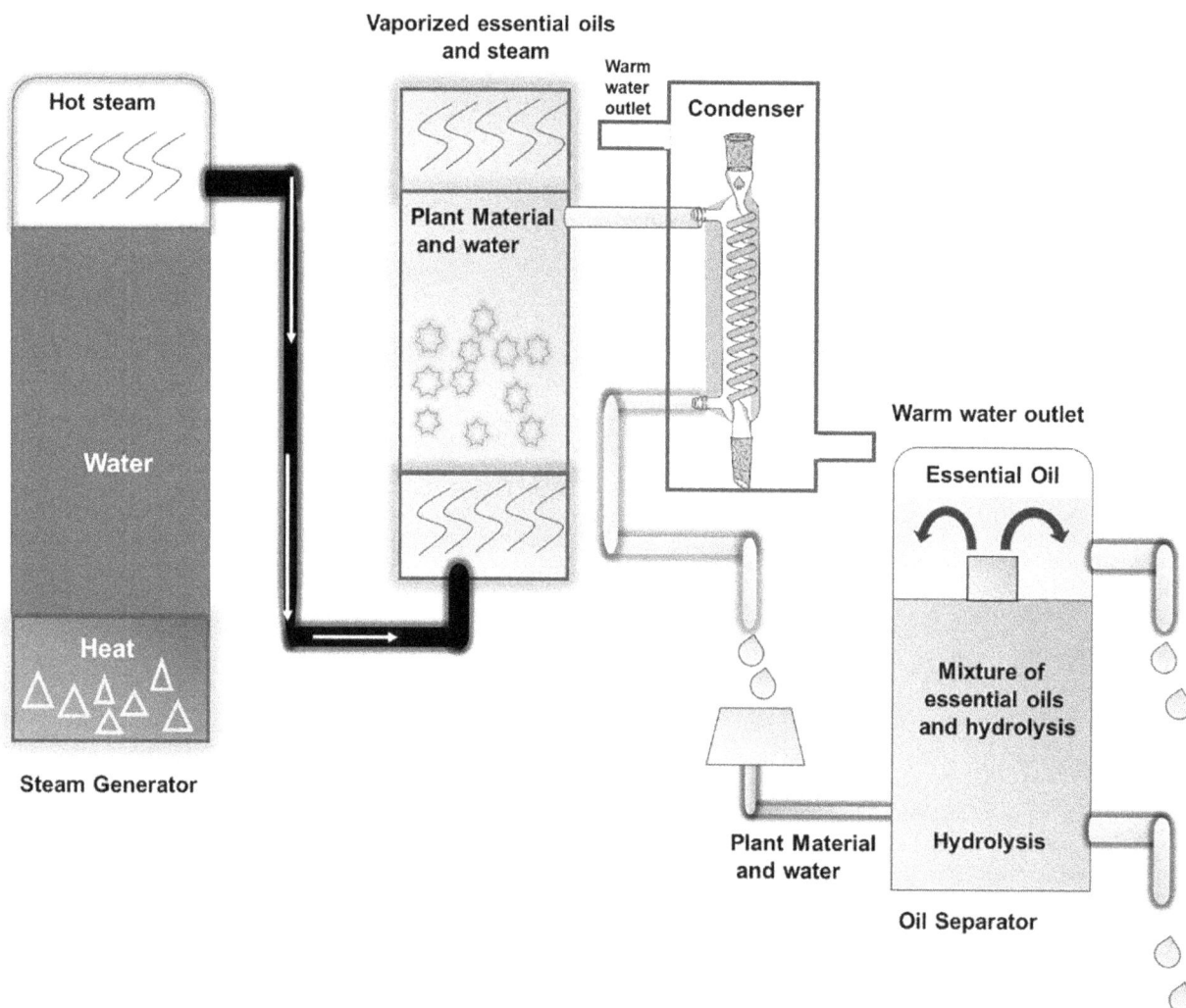

FIGURE 19.5 Water distillation assembly.

treated shortly after gathering, attar production takes place in isolated areas. The apparatus and equipment used to produce attar are lightweight, flexible, simple to repair and relatively efficient. Keeping these realities in mind, the traditional "deg and bhapka" method has been employed for generations with the following traditional implements.

- Deg (still)
- Bhapka (receiver)
- Chonga (bamboo condenser)
- Traditional bhatti (furnace)
- Gachchi (cooling water tank)
- Kuppi (leather bottle)

19.4.1.1.3.3 Water and Steam Distillation In water and steam distillation, the steam can be generated either in a satellite boiler or within the still, however, the plant material is kept separate. Similar to water distillation, water and steam distillation is a common practice in rural regions. Furthermore, it does not need much greater capital expenditures than water distillation. Also, the apparatus is similar to that used in water distillation, except the plant material is suspended on a perforated grid above the boiling water. In fact, it is usual for those who do water distillation to advance to water and steam distillation.

Cohobation

Cohobation is a process that can only be utilized during water or water and steam distillation. After the oil has been extracted from the distillate water, the water is returned to the still so that it can be reboiling. Its purpose is to limit the loss of oxygenated components, especially phenols, which dissolve to some degree in distillate water. For most oils, oil loss through solution in water is less than 0.2%, but for phenol-rich oils, the quantity of oil dissolved in the distillate water ranges from 0.2% to 0.7%. As this substance is continually re-vaporized, condensed and re-vaporized, any dissolved oxygenated components will encourage hydrolysis and deterioration of themselves or other oil components. Similarly, the likelihood of deterioration is increased if an oxygenated component is consistently exposed to a direct heat source or the side of a still that is substantially hotter than 100°C.

It is not advised until the oxygenated components of the distillate are exposed to temperatures no greater than 100° C. In steam and water distillation, plant material cannot come in direct contact with the fire source beneath the still. However, the walls of the still are efficient heat conductors, therefore still notes can also be generated from the thermal degradation processes of plant material touching the walls of the still [6].

19.4.1.1.3.4 Direct Steam Distillation It is the process of distilling plant matter using steam generated outside of the still in a satellite steam generator, sometimes known as a boiler. The plant material is supported on a perforated grid above the steam input, as in water and steam distillation. Controlling the quantity of steam produced by satellites is a significant advantage of this technology. As a result of the satellite boiler's steam generation, the plant material is heated to a maximum of

100°C and should thus not incur thermal deterioration. Steam distillation is the most commonly acknowledged method for the large-scale manufacturing of essential oils. It is an industry-wide standard in the flavor and fragrance supply industry.

19.4.2 Essential Oil Extraction by Expression

Expression, also known as cold pressing, is solely used to obtain citrus oils. The term expression refers to any physical action that breaks or crushes the essential oil glands in the peel to release the oil. An ancient technique, notably prevalent in Sicily (spugna method), included slicing citrus fruits in half and removing the pulp with a spoon-knife (known as a rastrello). The oil was extracted from the peel by pressing it against a hard item of baked clay (concolina) put beneath a big natural sponge, or by bending the peel into the sponge. The absorbed oil emulsion was extracted by pressing the sponge into the concolina or another container. According to reports, oil produced using this procedure includes more of the fruit's aroma than oil made using any other method.

Equaling, also known as the scodella method, employs a shallow bowl of copper (or occasionally brass) with a hollow center tube; the equaling instrument resembles a shallow funnel. The bowl is fitted with blunt-tipped brass prongs across which the entire citrus fruit is rolled by hand until all of the oil glands have ruptured. The oil and aqueous cell contents are allowed to drip down the hollow tube and into a container, where the oil is separated by decantation.

Citrus fruits are fed from a hopper into the machine's abrasive shell during the Pelatrice process. The fruits are rotated against an abrasive shell by a slow-moving Archimedian screw whose surface rasps the fruit surfaces, causing certain important oil cavities on the peel to break and release their oil-water emulsion. In Italy, most bergamot oil and some lemon oil are produced in this manner.

A copper chain is dragged by two horizontal ribbed rollers to form the sfumatrice apparatus. The peels are passed through these rollers, where they are squeezed and twisted in order to extract their oil [6,7].

19.4.3 Essential Oil Extraction with Cold Fat (Enfleurage)

Enfleurage is presently only practiced on a wide scale in the French province of Grasse, with the probable exception of a few rare cases in India, where the procedure has remained archaic. The success of enfleurage is largely determined by the quality of the fat base used. The corps must be prepared with the utmost care. It must be nearly odorless and of the appropriate consistency. If the corps is excessively rigid, the flowers will not have adequate contact with the fat, reducing its absorption capacity and leading to a below-average output of floral oil. Alternatively, if it is excessively soft, it will tend to swallow the flowers, causing the fatigued ones to adhere; when removed, the flowers will retain the clinging fat, leading to a significant loss of corpse. Therefore, the consistency of the corps must be such that it provides a semi-hard surface from which the spent blooms may be readily extracted. Enfleurage is performed in cool cellars, and each producer must prepare the corps according to the prevailing temperature in the cellars throughout the flower harvesting months.

19.4.4 Enfleurage and Defleurage

Each enfleurage facility is outfitted with tens of thousands of so-called chassis, which are used to transport the fat corps during the process. The chassis is a rectangular wooden frame. At the onset of the enfleurage process, the fat corps is applied with a spatula to both sides of a glass plate held by the frame. When stacked, the chassis form airtight chambers, with a layer of fat on both the top and bottom of each glass plate. Every morning throughout the harvest, newly gathered flowers are delivered, cleansed of impurities such as leaves and stems and then placed by hand on the fat layer on each glass dish. Never use dew- or rain-soaked blossoms, as any trace of moisture may cause the corpse to go rotten. The chassis is then stacked and stored for at least 24 h, depending on the type of flowers. These rest in direct touch with one fat layer (the bottom one), which functions as a direct solvent, while the second fat layer (beneath the glass plate of the chassis above) absorbs only the volatile aroma emitted by the flowers.

19.4.5 Hot Maceration Process

This technique shortens the traditional enfleurage process by immersing petals in molten fat heated to 45°C–60°C for 1–2 h, depending on the plant type. After each soaking, the fat from the petals is collected and purified. After 10–20 immersions, discarded flowers and water are cleansed of fat. The maceration absolute is produced by extracting and concentrating fat-containing oil under low pressure. It is typically used for exceptionally delicate flowers, such as lily-of-the-valley, whose physiological capabilities are lost shortly after harvesting.

19.4.6 Modern (Non-Traditional) Methods of Extraction of Essential Oils

Traditional techniques for extracting essential oils have been addressed, and these techniques are the most used on a commercial basis. With the advancement of technology, however, new techniques have been developed that may not be widely used in the commercial production of essential oils, but are valuable in certain circumstances, such as the production of expensive essential oils in their natural state without altering their thermosensitive components or the extraction of essential oils for microanalysis. These techniques consist of:

19.4.6.1 Headspace Trapping Techniques

Because they enable the direct measurement of the headspace gas above the heated samples, where the chemicals responsible for the scent recognized by the human nose are transferred, they are commonly utilized in wine aroma analyses. The static headspace approach is suited for the study of aromatic compounds with the highest concentration, whereas dynamic headspace (purge and trap) techniques are required for the analysis of trace levels. Typically, either cryo or sorbent traps

are employed to retain volatiles. Utilizing the same adsorbents as SPE, the retention of water and ethanol is decreased with sorbent traps, which are often preferred in practice. The trapped volatiles are retrieved by extraction with tiny amounts of solvent or by thermal desorption, which can be performed in the chromatographic injector, allowing for one-step analysis of the whole sample. However, the material can be fractionated and hence injected into many chromatographic cycles [6,7].

19.4.6.2 Static Headspace Technique

Extraction of volatiles from diverse food matrices using static headspace is a simple, convenient and often automated process. It is also utilized for the monitoring of volatile lipid oxidation products, among other uses. However, the greatest drawback of this technology is its limited sensitivity, particularly in comparison to solid-phase microextraction and dynamic headspace

19.4.6.3 Vacuum Headspace Technique

Using four distinct isolation procedures, the volatile components of yellow passion fruit were extracted. The most typical and representative extract was produced by vacuum headspace sampling followed by liquid-liquid extraction of the aqueous phase. This vacuum headspace concentration was pre-fractionated using silica gel medium-pressure adsorption chromatography.

19.4.6.4 Dynamic Headspace Technique

Three distinct Thymus species (*T. zygis* subsp. gracilis (Boiss.) Morales, *T. x citriodorus* (Pers.) Schreb., and *T. fontqueri* (Jalas) Molero e Vias) were hydrodistilled and supercritical fluid CO_2 extracted, respectively, to yield natural essential oils and extracts. In addition, the olfactory fingerprint of the plant materials was evaluated by the collection of volatiles using the dynamic headspace-thermal desorption approach. To achieve this objective, the odor space was described by computing odor values and employing a previously validated olfactory family model for the measurement and categorization of volatile odorants. By combining gas chromatography and mass spectrometry, the organoleptic characteristics and chemical makeup of the extracted oils were determined [8].

19.4.7 Solid Phase Micro-Extraction (SPME)

SPME is an efficient sample preparation approach for combining many activities, including sample collection, extraction, analyte enrichment and isolation from sample matrices; it has been used to extract analytes from gaseous, liquid and solid samples. SPME is a nonexhaustive approach based on the equilibrium partitioning of analytes between the sample matrix and the extraction phase. The SPME method includes two fundamental steps: (1) the partitioning of analytes between the extraction phase and the sample matrix and (2) the subsequent desorption of concentrated extracts into analytical equipment. Due to the fact that sampling, extraction, preconcentration and sample introduction into an analytical instrument may be conducted in a single step, SPME procedures yield quantitative or semiquantitative data [8].

19.4.8 Supercritical Fluid Extraction (SFE)

Extraction using supercritical fluids (SFE) has been utilized as an alternate technique to extract carotenoids from food samples. This is a sophisticated separation method based on the increased solvating power of gases above their critical point. CO_2 is the favored gas due to its lower critical temperature, nontoxic and nonflammable nature, inexpensive price and high purity (Figure 19.6). It may be utilized at temperatures between 40°C and 80°C and pressures between 35 and 70 MPa. SFE performs selective separation of carotenoids in a single step, hence eliminating processing at excessive temperatures. This makes it an ideal environment for extracting heat-sensitive carotenoids. In addition, the removal of organic solvents results in the absence of chemical residues in the extracted substances. SFE is compatible with supercritical fluid chromatography since both techniques may utilize the same mobile phase and apparatus, hence facilitating the evolution of extraction and separation techniques. This method has been successfully employed to isolate lycopene from other carotenoids in tomato fruits.

19.4.9 Simultaneous Distillation Extraction (SDE)

Widespread application of SDE to the measurement of volatile components in tea samples. Low water temperature in the circulating system resulted in improved recoveries and repeatability of volatile or semi-volatile and heat-stable components than SPME or HSME.

19.4.10 Microwave Distillation

In recent years, Microwave-Assisted Techniques (MATs) have been introduced as a new process design and operation for the extraction of essential oils, representing a viable alternative to conventional old-type methods of distillation used routinely for the isolation of essential oils from herbs, flowers and spices prior to gas chromatographic analysis. The innovation of the process is in the use of a microwave heating source to generate a combination of boiling solvent and raw plant material that has settled on top (or drenched inside). Several kinds of distillation procedures are examined in terms of energy efficiency, speed, product output, cleanliness and product quality. Results indicate the efficacy of MATs, which permit extraction of essential oils in shorter extraction periods (up to nine times quicker) utilizing "greener" techniques and produce a higher-quality essential oil with enhanced sensory and antioxidant capabilities.

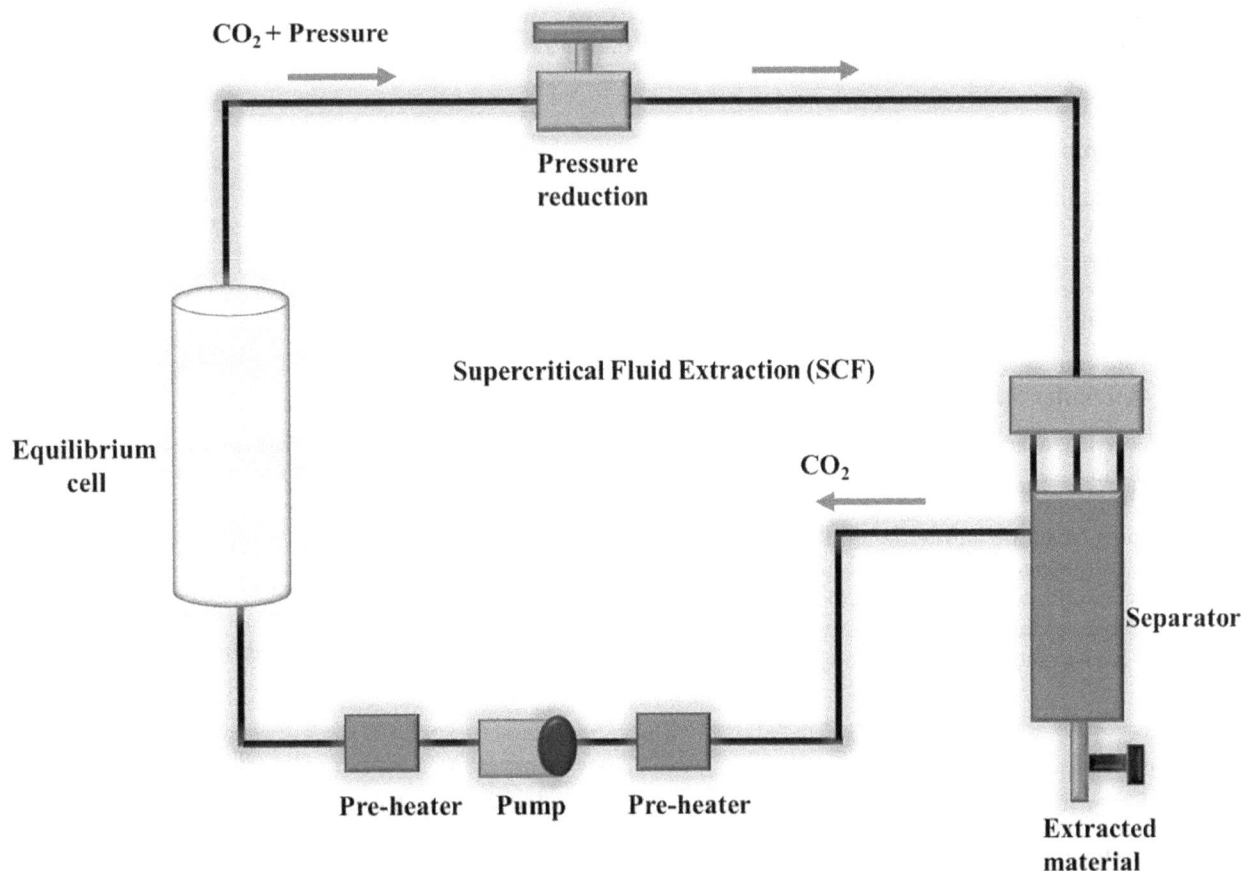

FIGURE 19.6 Solid phase micro-extraction technique.

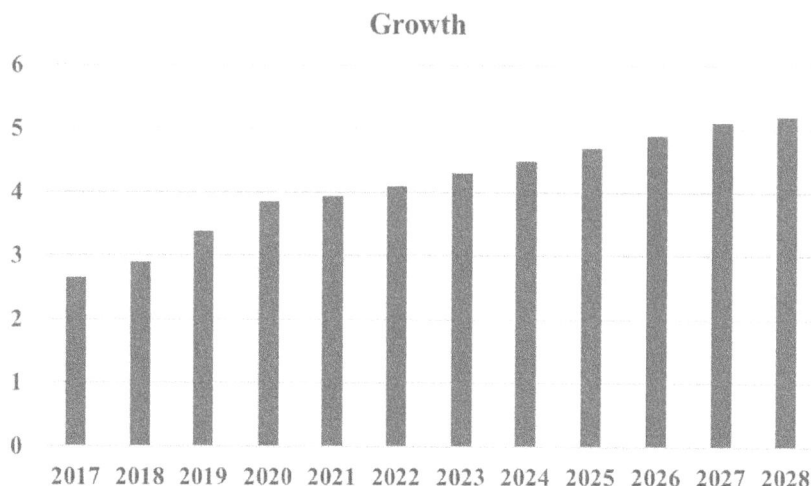

FIGURE 19.7 Expected increasing demand of essential oil in Europe from 2017 to 2028.

19.4.11 Controlled Instantaneous Decomposition (CID)

The procedure comprises exposing the partly humidified plant material to a steam pressure ranging from 0.5 to 3 bar for a brief period of time, followed by a quick decompression to a vacuum (about 15 mbar) for 200 ms each time. Auto-vaporization of plant material creates vapor with a mechanical strength that ruptures the oil cells. Due to pressure differences between the extractor and vacuum chamber, oil-rich vapor is promptly drawn into the vacuum chamber at each opening of the pressure valve, where it immediately condenses.

19.4.12 Thermo-Micro Distillation

It involves the utilization of Seal, Indicator silica, Glass tube (Pasteur pipette), Oven (220°C), Sample, Glass-wool and TLC Plate. The primary characteristic of this method is the employment of a TAS oven in which the temperature is kept low in order to extract thermolabile components.

19.4.13 Molecular Spinning Band Distillation

Using a revolving helical band, spinning band distillation generates a large number of theoretical plates. The bands can be constructed of either Teflon or metal. Below 225°C, Teflon spinning bands are utilized for distillation. For higher temperature distillations where Teflon would become brittle, metal bands are utilized. Spinning band distillation results in a highly effective separation in a short distillation column with low column hold up and low-pressure drop [6–8].

19.5 Worldwide Current Trends in Essential Oils Market

Figure 19.7 depicts the expected market of essential oil rising tremendously in Europe in different fields. The graph showed the expected growth of market in US $ during the period of 2017–2028.

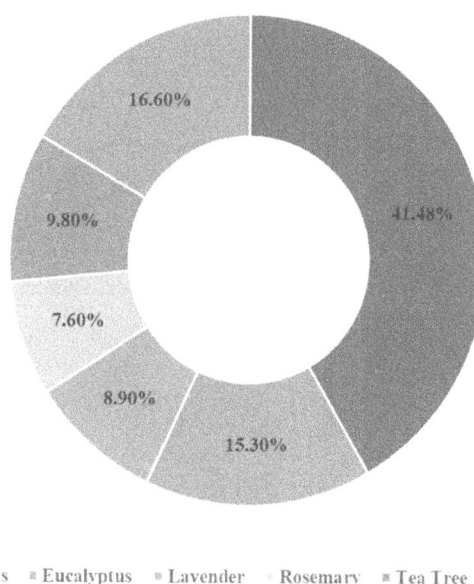

FIGURE 19.8 Worldwide Market share of different essential oils like Citrus, Eucalyptus, Lavender, Rosmary, tea tree and others.

Figure 19.8 depicts the share of different essential oil in the above market like major share has been taken by citru sand eucalyptus oil followed by lavender, rosemary, tea tree, etc.

19.6 Industrial Applications

It includes essential oils used as flavors and fragrances, in the pharmaceutical and medicinal industry, in alternative medicines, in the cosmetic industry, in the food industry, etc., are discussed below.

19.6.1 Food and Food Preservation

Increasing emphasis has been paid to the use of essential oils (EOs) in food systems as natural inhibitors or biopreservatives, mostly due to consumer concerns around artificial preservatives.

Due to the dangers associated with synthetic preservatives, essential oils are gaining popularity as natural additions for extending the shelf life of food goods. Synthetic additives help minimize food deterioration; however, the current generation is extremely health-conscious and favors natural goods over synthetic ones because of their possible toxicity and other problems [9]. Therefore, the extraction of essential oils from various plant organs and their use in meals is one of the most significant developing technologies. Essential oils are a rich source of a variety of bioactive chemicals with antioxidative and antibacterial capabilities, thus their application in food preservation is highly advantageous. Although essential oils have been demonstrated to be a viable alternative to chemical preservatives, they have unique constraints that must be addressed prior to their use in food systems. Low water solubility, high volatility and a strong odor are the primary characteristics that make food applications challenging.

Several scientific studies have demonstrated that EOs are efficient against a wide variety of food-contaminating germs.

In this context, Friedman et al. [10] evaluated the activity of 96 EOs and 23 EO compounds against Campylobacter jejuni, Escherichia coli O157:H7, Listeria monocytogenes and Salmonella enterica. Table 19.2 summarizes the action of plant EOs against common dietary bacteria.

The use of EOs to reduce food spoiling and mycotoxin generation caused by fungus has also been investigated. Several EOs and their constituents have been discovered to be effective against a variety of food-borne fungus responsible for the degradation of food items (Table 19.3).

Moreover, lipid oxidation is unquestionably a significant source of food degradation, which shortens shelf life and diminishes the quality of meals and food items. Included in this category are meat products that are susceptible to oxidation and have limited oxidation stability, resulting in the loss of their tissue, flavor, aroma and color. Similarly, edible oils are susceptible to the oxidation process and can create undesirable flavor and taste as well as develop hazardous chemicals in these items. In order to manage and prevent oxidation, antioxidants are essential for extending the shelf life of these items. In recent years, a number of studies have demonstrated the effectiveness of EOs as natural antioxidants, hence demonstrating their enormous potential in the food business.

TABLE 19.2

Essential Oils Active Against Pathogens Used in Food Preservation

S. No.	Essential Oil	Major Constituents	Effective against Fungus	Therapeutic Potential
(1)	*C. citratus*	Geranial (41.3%), Neral (33%), Myrvene (10.4%)	*A. ochraceus* *A. oryzae, A. fumigatus, A. parasiticus*	Restrict the growth
(2)	*O. majorana*	Terpene- 4-ol (28.98%), α-terpeniol (16.75%)	*A. flavus*	Antifungal and anti-afltoxigenic
(3)	*C. citratus*	Citrate (62.85%)	*P. expansum*	The highest inhibition was found to be at 750 µl for 21 days and inhibition increases as concentration increases.
(4)	*G. fragrantisiuma*	Methyl salicylate (98%)	*A. flavus*	Antifungal and Antiafltoxigenic, conservation of millets
(5)	*L. nobilis*	1,8 cineole (35.5%)	*Same as above*	Enhances shelf life by controlling food spoilage
(6)	*Cinnamom spp.*	Cinnamaldehyde (89.51%)	*A. niger, B. cinerea, p. digitatum*	Postharvest preservation of vegetable and fruits
(7)	*O. sanctum, O. basilicum, O. canum*	Eugenol (53.39%), Estragole (43.72%), Camphor (28.72%)	*B. spp, Penicillum spp, P. granati*	Showed synergistic potential in preservation of seeds storage.
(8)	*C. verum, C. citrates, O. vulgare*	Euginol (56.9%), Geranial (41.4%), Carvacrol (51.5%)	*Botrytis spp, Penicillum spp, P. granati*	Chitosan coated particle used in management of fungal infection.
(9)	*Z. officinalis*	α-zingiberene (23.85%), Geranial (14.16%)	*Zygosaccharo-myces*	Showed significant antifungal activity
(10)	*Illicium verum*	Anethole (89.12%)	*A. flavus*	Antifungal, anti-oxidant and Antiafltoxigenic,
(11)	*M. fragrans*	Elemicin (27.08%), Myristicine (21.29%), Thujanol (18.55%)	*A. flavus*	Protect rice against fungal infections and aflatoxins B1 secretion
(12)	*T. leptobotrys, satureioides, broussonnetii, and riatarum*	Carvacrol (76.94%), Borneol (27.71%), Camphor (46.17%),	*P. digitatum, P. italicum, Geotricum citri-aurantii*	Post harvest preservation of citrus against fungal pathogens
(13)	*C. longa*	Ar-tumerone (35.17%), Tumerone (11.93%)	*A. flavus*	Protection of maize
(14)	*Ocimum gratissimum*	Methyl cinnamate (48.29%), γ-terpinine (26.08%)	Same as above	Fungal, aflatoxin mediated biodegradation
(15)	*Callistemon lanceolatus*	1,8-cineole (56%)	Same as above	Antifungal

TABLE 19.3

Industrial Usage of Essential Oils, Major Food Product and Effectivity Against Bacteria

S. No.	Product	Ingredients	Bacteria	Major Active Ingredients	Effectivity
1.	Lighvan cheese	*M. spicata*	*Listeria monocytogenes* Lighvan	Carvone (78.76%)	Limonene (11.5%)
2.	*S. aureus*	Sardine fish	*C. aurantium*	Limonene (77.37%)	Inhibit the growth *S. aureus* in Sardine.
3.	*L. monocytogenes*	Minced beef	*Satureja Montana Pistacia lentiscus*	b-myrcene (15.18%) Carvacrol (29.19%)	Effective *against. monocytogenes*
4.	*L. monocytogenes S. typhimurium*	Chicken breast filet	*Zingiber officinale*	a-zingiberene (24.96%) b-sesquiphellandrene	Preserve the poultry meat
5.	*L. monocytogenes* Iranian white	Cheese	*M. pulegium*	Pulegone (36.68%), Piperitenone (16.88%)	Inhibit the growth of *L. monocytogenes*
6.	*B. subtilis*	Bread	*Coriandrum sativum*	b-linalool (66.07%)	Antibacterial activity against *B. subtilis*
7.	*L. monocytogenes, S. enteritidis, Y. enterocolitica, E. coli* and *Pseudomonas spp.*	Vegetables	*Thymus vulgaris Rosmarinus officinalis*	1,8-cineole (45.27%), Borneol (12.94%) p-cymene (39.18%), Thymol (25.05%)	Reduce the viable counts of bacteria in vegetables
8.	*E. coli S. choleraesuis L. monocytogenes S. aureus*	Leafy vegetables	*C. citratus*	Neral (50%) Geranial (35%)	More effective against tested bacteria
9.	*S. aureus, E. coli, Bacillus cereus* and *S. typhimurium*	Cream-filled cakes and pastries	*Same as above*	Nerial (39.0%), Geranial (33.3%)	Effective antimicrobial agent
10.	*Escherichia coli* O157:H7 *L. monocytogenes*	Iranian white cheese	*Bunium persicum*	Cuminaldehyde (11.4%)	Decreased bacterial growth and prolonged the shelf life during 45 days.

19.6.2 Nano-Formulation Incorporating Essential Oil

Nano-formulations available on the market are produced by industries using a variety of polymers, including chitoan, b-cyclodextrin, Hydroxypropyl, zein, Chitosan/k-carageenan, Polyvinylalcohol crosslinked glutaraldehyde, Sodium casein and maltodextrin, Soybean lecithin, Nanoprecipitation, etc., as showed in Table 19.4 [11–13].

19.6.3 Role of Essential Oils as Natural Cosmetic Preservatives

The safety of chemical preservatives is currently being questioned by a significant number of consumers. Traditional preservatives frequently irritate the skin and cause allergic responses. Antimicrobial essential oils (EOs) were proposed to replace synthetic preservatives in response to the increasing demand for natural and preservative-free cosmetics. The antibacterial activity of essential oil is contingent on the composition, concentration and interactions of its principal active components. At a minimal concentration, effective preservatives should exhibit a broad range of antibacterial action. Due to their synergistic effect, formulations including both types of preservatives—essential oil and a synthetic one—have been evaluated and offered as a compromise that permits a reduction in the concentration of both components. Although the vast majority of essential oils are considered harmless, a

few of them may pose a risk of contact allergy or phototoxicity. A well-balanced risk-benefit evaluation of essential oils is one of the greatest difficulties facing scientists and health policymakers.

The effectiveness of preservatives in cosmetic compositions is tested using a challenge test in accordance with the requirements of the European Pharmacopoeia. The challenge test is a standard procedure that involves artificial contamination of cosmetics with a predetermined number of viable bacteria and fungi (105–106 viable cells ml⁻¹ or g⁻¹ of product) and periodic removal of samples at fixed intervals for counting viable microorganisms present in the formulation during the test. The challenge test utilizes the bacterial strains *Staphylococcus aureus*, *Pseudomonas aeruginosa* and *Escherichia coli*, as well as the fungal species *Aspergillus niger* and *Candida albicans*.

Early research on the effectiveness of Thymus vulgaris essential oil in two cream formulations demonstrated a poor impact of preservation in the challenge test. In spite of the high content of thyme oil, the required conditions were not met for *Aspergillus niger* (3%). Oil samples exhibited a significant quantity of thymol (38.6%–43%), but a comparatively low concentration of carvacrol (9.8% and 2.2%). The authors hypothesized that the ideal ratio of both phenolic oil ingredients would enhance antifungal activity. Different behavior of thyme oil in two formulations (O/W – oil in water and W/O – water in oil) against Candida albicans was explained by the lower water bioavailability and higher pH in the W/O formulation, which might encourage a more potent fungicidal impact [2,15].

TABLE 19.4

Showing the Nano-Formulations of Essential Oil Used in Market

S. No.	Essential Oil	Polymer	Process	Food Based Applications
(1)	C. martini	Chitosan	Ionic gelation	Inhibit the growth of *Fusarium graminearum* and secretion of zearalenone and deoxynivalenol
(2)	Clove	Chitosan	-	Inhibit the growth of *Aspergillus niger* infestation
(3)	Cardamom	Chitosan	-	Effective against antibiotic-resistant pathogens
(4)	Tarragon	Chitosan, Gelatin	-	Encapsulated essential oil conserve pork slices
(5)	*Coriandrum sativum*	Chitosan	-	Encapsulation conserves rice during storage
(6)	Cinnamon	β-cyclodextrin	Inclusion complex	Food packaging agent in inhibiting *S. aureus* and *E. coli*
(7)	*Litsea cubeba*	β -cyclodextrin	-	Fungitoxic against *P. digitatum*, *P. italicum* and *Geotrichum citriaurantii* in
(8)	Black pepper essential oil	Hydroxypropyl-cyclodextrin	-	Antimicrobial potential against *E. coli* and *S. aureus*
(9)	*Melaleuca alternifolia*	β -cyclodextrin	-	Antimicrobial packaging to preserve the quality of beef
(10)	Orange essential oil	β -cyclodextrin and zein	-	Microbial and mould spoilage in cakes
(11)	*Pimenta dioca*	Chitosan/k-carageenan	Complex coacervation	Preservation against *B. subtilis*, *B. cereus* and Candida utilis in food
(12)	C. citratus	Polyvinylalcohol crosslinked glutaraldehyde	-	Food protection against *E. coli* ATCC25922
(13)	*Thymus vulgaris*	Sodium casein and maltodextrin	Spray drying	Preservative of hamburger-like meat products against *S. aureus*, *Listeria monocytogenes*
(14)	*Lavandula hybrida*	Soybean lecithin	-	Prevent *E. coli* contamination
(15)	Eugenol	Whey protein isolate, maltodextrin and lecithin	-	Suppress Listeria innocua and *E. coli*
(16)	Carvacrol	Sodium alginate and pectin	-	Inhibit *E. coli* K12
(17)	*Rosmarinus officinalis*	Maltodextrin and modified starch	-	Restrict the growth of Penicillium and Aspergillus spp.
(18)	*C.martini*	Polycaprolactone	Nanoprecipitation	Protection against *E. coli* and *S. aureus*
(19)	Thyme	Chitosan	-	Restrict food borne bacteria *B. cereus* and *S. aureus*
(20)	Cinnamon	Chitosan and polylactic acid	-	Inhibit the growth of *S. aureus* and *E. coli*
(21)	Oregano Lavender Cinnamon Citral Menthone	Starch	-	Same as above
(22)	*Cinnamon*	Chitosan and whey protein isolate	-	Chronic antibacterial activity against foodborne bacteria *S. putrifaciens*, *P. fragi*, *S. aureus* and *E. coli*
(23)	*Syzygium aromaticum*	Sodium alginate	Emulsification	In vivo bactericidal effect against foodborne pathogens *E. coli*, *S. typhimurium* and *S. aureus*
(24)	Cinnamaldehyde	Medium chain triglycerides	-	Antimicrobial activity against *S. aureus* and *E. coli*
(25)	Schinus terebinthifolius	Maltodextrin	-	Inhibition bacteria *B. subtilis*, *L. innocua*, and *L. monocytogenes*

The effectiveness of *Artemisia afra*, *Pteronia incana*, *Lavandula officinalis* and *Rosmarinus officinalis* essential oils in minimizing microbial contamination in aqueous cream formulation was established in research. The population of tested microorganisms was significantly restricted and under control (criteria A and B) during the challenge test up to the seventh day (criterion A and B).

At all concentrations, A. afra oil thujone (53%) was the most effective (0.5%, 1.0% and 1.5%). The scientists observed that the use of EDTA as a chelator affected the viability of bacteria and fungus. This encourages more research on a combination preservation strategy that contains not just EOs but also EDTA, synthetic preservatives, or surfactants.

The combination of *Calamintha officinalis* essential oil (1%) and 2mM EDTA in cetomacrogol cream resulted in good preservation activity and enabled the challenge test to be passed under all criteria. The inclusion of EDTA, which enabled oil penetration into bacterial cells, and a high quantity of carvon (64.3%), a major ingredient of essential oils with potent antibacterial effects, led to effective preservation. In the subsequent investigation, 1% and 2% *Calamintha officinalis* essential oil was used alone to cream and shampoo formulations. In the challenge test, the preservation effect was restricted to greater oil concentrations (2%) in cream preparations (meeting criteria A) and shampoos (meeting criterion B).

Maccioni investigated the synergistic impact of essential oils (*Laurus nobilis*, *Eucalyptus globulus* and *Salvia officinalis*) with and without the addition of a synthetic preservative (methyl-p-hydroxybenzoate – MPB) in cosmetic compositions. EOs (0.025% and 0.0125%) were administered alone, in combination, and in combination with MPB in three formulations: carbopol cream (O/W), carbopol hydrogel and hydrolyte. Low concentrations of the preservative were employed (0.01% and 0.001%, respectively) that were 20 and 200 times less than the typical quantity (0.2%). However, variations containing a mixture of MPB and two or three essential oils had a preservation effect, but not against all bacteria. Despite the presence of MPB, Candida albicans proved resistant to cream and hydrolate formulations. The synergistic impact of oils on grampositive bacteria was demonstrated, particularly in hydrogel formulations. This study shows that the use of blended essential oils can greatly minimize the use of synthetic preservatives in cosmetic preparations [15,16].

Complex preservative systems, such as EOs and synthetic preservatives or surfactants, have been thoroughly researched and enhanced. A comparative study of the efficacy of lavender, tea tree and lemon essential oils alone and in combination with synthetic preservatives (1,3-dimethylol-5,5-dimethyllhydantoin and 3-iodo-2-propyl butyl carbamate) in body milks (W/O) demonstrated the synergistic effect of the composed preservative system, allowing the reduction of synthetic preservative levels by up to 8.5%. In challenge tests, synthetic preservatives used without essential oils at a lower dose (0.1%) were unsuccessful against Pseudomonas aeruginosa and Candida sp. Within 14 days, the higher concentration (0.3%) sanitized the bodily milk. Tests conducted on a blend of lavender and tea tree oils (each at a concentration of 0.5%) and synthetic preservatives at a concentration of 0.1% yielded the same preservation efficiency.

A further investigation on the same essential oils and preservatives added to washing liquid and body balm (W/O) with a solubilizer verified the enhanced preservation effect. Only *S. aureus*, Candida sp. and A. niger met criteria A of the challenge test for washing solutions including essential oils (1%) without preservatives and solubilizer; however, in body balm, the combination had no preservative effect. The addition of 5% solubilizer (polysorbate 80) to tea tree essential oil boosted its bacteriostatic activity but not its fungiostatic activity. According to EP standards, a combination of tea tree oil (0.5%), solubilizer (5%) and synthetic preservatives (0.3%) in formulations for body balm satisfied criteria A of the challenge test [17].

A comparative study of the preservative efficacy of herbal extracts, essential oils (Lavandula officinalis, Melaleuca alternifolia and Cinnamonium zeylannicum) and methylparaben revealed that essential oils inhibit microbial growth in cosmetic emulsions more effectively than synthetic preservatives and extracts. The most powerful inhibitor of the development of *S. aureus*, *E. coli* and *C. albicans* was cinnamon oil (2.5%). Antimicrobial activity of essential oils (2.5%) was in certain cases 3.5 times greater than that of methylparaben, depending on tested microorganism strains (0.4%). This study proved that essential oil may be used at reasonably high doses to replace synthetic preservatives.

Recent research on antimicrobial emulsion formulations including collagen hydrolisate revealed that formulations containing 2% Thymus onites essential oil were efficient against tested bacteria and fungus when using the disc diffusion technique. The author observed that the formulation with antibacterial properties also had the lowest viscosity.

The majority of research suggest that EOs function better when combined with synthetic preservatives, chelators and solubilizers. The inclusion of EOs at a relatively low concentration or EOs combined with solubilizers facilitates the reduction of synthetic preservative content. When oil is applied on its own, its antibacterial action is typically inadequate and larger doses are necessary for a more potent impact. Such a high concentration of EOs causes several issues, including phase separation, unfavorable formulation viscosity, and a potent, unpleasant odor.

To produce a synergistic effect, each cosmetic formulation requires the development of a unique composition and proportion of the selected essential oils and synthetic additives, as well as a suitable concentration of all elements [18].

19.6.4 Antimicrobial Effects of Essential Oils

Selected essential oils appear to offer the benefit of limiting the growth of potential infections while having only a mild effect on the gut microbiota. Clostridium perfringens strains were shown to be sensitive to carvacrol, cinnamaldehyde, citral, limonene and thymol, especially at the highest concentration tested (500 mg/l), as well as to oregano oil, rosemary oil and thyme oil. In apples, clove oil, an essential oil isolated from the clove plant Syzygium aromaticum (L) Merr & Perry, has been shown to be bioactive, particularly its active component monoterpene eugenol, against B. cinerea, M. fructigena Honey, P. expansum Link and Phlyctema vagabunda Desm. In addition, the essential oils of basil (Ocimum basilicum L), fennel (Foeniculum sativum Mill), lavender (Lavandula officinalis Chaix), marjoram (Origanum majorana L), oregano (Origanum vulgare L), peppermint (Mentha piperita L), rosemary (Rosmarinus officinalis L), sage (Salvia officinalis L). The different efficacies of the various essential oils result from the varying antibacterial capabilities of each individual dynamic ingredient and their synergistic effect. In addition, although the antibacterial function of essential oils is frequently attributed to their primary elements, interactions between major and minor constituents may also play a crucial role in the antimicrobial activity of essential oils and should not be overlooked. Strawberry Colletotrichum acutum growth was inhibited by thyme, cinnamon bark and clove bud essential oils. The capacity of the hydrophobic molecules in the essential oils to break the microorganism's cell membrane, resulting in a change in cell shape, altered membrane permeability and electrolyte leakage, is responsible for the demise of these pathogens' cells. In addition, the addition of citrus film essential oils inhibited the growth of fungus and bacteria after 15 days of storage without altering quality metrics. The antimicrobial standards of some essential oils evaluated against food-borne pathogens and spoilage bacteria show a large potential for their application in the food sector, subject to rigorous tests to improve their efficacy. Recent research conducted over the past five years has shown that EOs and their bioactive compounds are highly effective against food-borne germs, moulds and oxidative degradation (Table 19.5) [19–21].

Table 19.5 summarizes some selected essential oils and their antibacterial activity against human pathogens.

TABLE 19.5

Plants, Major Constituents and Effective Against Microorganism

S. No.	Plants	Major Chemical Compound	Inhibited Microorganisms
(1)	*Artimisia cana* Aerial parts	Santolina triene, α pinene, camphen	*E. coli, S. aureus*, S. epidermidis
(2)	Achillea ligustica Aerial parts	Viridiflorol, terpin-4-ol	S. mutans
(3)	*Artimisia frigida* Aerial parts	1,8-cineole, methylchavicol, camphor	*E. coli, S. aureus*, S. epidermidi
(4)	Achillea clavennae Leaves and Flowers	Camphor, myrcene,1,8-cineole, βcaryophyllene,linalool, Gerenyl acetate	K. pneumonia, Streptococcus pneumonia, Haemophilus influenza, P. aeruoginosa
(5)	Cyperus longus Aerial parts	β-Himachalene, α-humulene, γ-himachalene	*S. aureus*, Listeria monocytogenes,, Enterococcus faecium, S. enterica, *E. coli*, Pseudomonas aeruginosa
(6)	*Cuminum cyminum* Leaves	γ-Terpin-7-al, γ-terpinene, β-pinene, Cuminaldehyde	Salmonella typhimurium, *E. coli*
(7)	Cymbopogon citrus Leaves	δ-cadinene, germacrene D, α-humulene, α-copaene, germacrene B, βcaryophyllene, β-bisabolene	*S. aureus*
(8)	*Dracocephalum foetidum* Leaves	limonene, n-menthal, 8-dien-10-al	E. hirae, *S. aureus*, Micrococcus luteus, *E. coli*, B. subtilis, S. mutans
(9)	*Eugenia caryophllata* Flower buds	Thymol, eugenol, carvacrol, cinnamaldehyde	S. epidermidis
(10)	*Eremanthus erythropapps* Leaves	viridiflorol, p-cymene germacrene D, γterpinene (Z)-caryophyllene	S. epidermidis
(11)	*Foeniculum vulgare* Leaves	limonene, methylchavicol, Trans-anthole	*E. coli*, Salmonella typhimurium
(12)	*Juniperus phoenicea* Arial part	α-terpinyl actate, α- pinene β-phellandrene	P. aeruginosa, *E. coli, S. aureus*, E. faecium, Salmonell Enteriditis
(13)	*Mentha piperita* Arial part	-	S. typhimurium, *S. aureus*, Vibrio parahaemolyticus
(14)	*Momordica Charantia* Seed	germacrene D, Trans nerolidol, cis-dihydrocarvacol	*S. aureus, E. coli*
(15)	*Laurus nobilis* Arial part	linalool, Eucalyptol (1,8-cineole)	Ecoli. Mycobacterium smegmatis
(16)	*Nigella sativa* Seeds	longifolene, Thymoquinone thymohydroquinone, α- thujene,p-cymen	P. aeruginosa, *S. aureus, E. coli*, Bacillus cereus
(17)	*Ocimum basilicum* Leaves, stems	methylchavicol, γ-terpinene	P. putida, Mariniluteicoccus flavus, Listeria innocua, *E. coli*, S. typhimurium, Brochothrix thermosphacta
(18)	P.amboinicus Leaves	viridiflorol, γ-terpinene, germacrene D,pcymene (Z)-caryophyllene	S. epidermidis

19.6.5 Current Trends on the Application of EOs in the Food Industry

There is an increasing interest in the manufacture of active packaging made from renewable and eco-friendly materials. It has been demonstrated that innovative active packaging extends the shelf life of food goods and inhibits the growth of certain germs.

Through the development of biocomposite systems composed of natural biopolymers or synthetic materials, these enhancements are possible. According to several sources, edible packaging has gained appeal among customers since an edible film or coating, in addition to extending the shelf life of the food product, may also be consumed with the meal. Other research, however, have shown that few examples of films can really be consumed (whereas others can just be destroyed more readily) and that customers will only accept edible films if they perceive them to be safe. These packaging techniques may include components designed to be immobilized on the film, discharged into the food, or capable of absorbing contaminants that cause deterioration. Packaging prevents food from dehydration and functions as a barrier against the surrounding atmosphere.

In addition to antimicrobials, antioxidants, texture enhancers, and essential nutrients, edible films can function as carriers for additional active chemicals. Moreover, these qualities can be improved by adding active chemicals, such as EOs, within the films.

EOs can be included in food and beverage packaging systems (edible films) as an extra component applied directly to edible films or encapsulated inside edible films. Various proportions of Origanum vulgare and Eugenia caryophyllata essential oils were integrated into cassava bagasse–polyvinyl alcohol-based trays, for example. EOs were integrated into the trays in two ways: either by adding them directly to the combination of components at proportions of 6.5%, 8.5% and 10% (w/w), or by coating the surface with EOs at concentrations of 2.5%, 5% and 7.5% (g oil/100g tray). The greatest results were obtained when trays were coated with the highest concentrations of EOs, particularly for O. vulgare, which exhibited 100% suppression of moulds, yeasts, and a few Gram-positive and Gram-negative bacteria, with a significant reduction in bacterial viability.

Consequently, EOs possess antibacterial and antioxidant capabilities and are compatible with several food kinds, thereby extending the shelf life of food items or boosting their organoleptic features. Several EOs that have been used in food packaging in the United States have been acknowledged as safe ingredients and included in the Food and Drug Administration's Generally Recognized as Safe (GRAS) category (FDA).

In addition, they are chemically unstable and readily oxidized, as well as sensitive to light, oxygen and temperature

TABLE 19.6

Recent Examples of Active Films Containing EOs as Main Constituents Showing Beneficial Properties for the Packaged Food Product

S. No.	Film	EOs	Main Results
(1)	Gelatin	Citrus sinensis	Antioxidant and antimicrobial activity Extension of shelf-life of nearly 10 days
(2)	Gelatin-	Oregano	Lowering of total volatile basic nitrogen, peroxide value, thiobarbituric acid and microbial growth
(3)	Gelatin-carboxymethyl cellulose-chitin nanofibers	*Trachyspermum ammi*	Growth inhibition of Pseudomonas spp., *S. aureus*, lactic acid bacteria (LAB), molds and yeasts. Stability of the chemical profile, color and sensory properties
(4)	Curdlan-PVA	Thyme	Antioxidant activity and extension of the shelf life
(5)	Chitosan-cassava starch	Myrcia ovata Cambessedes	Antimicrobial effect against Bacillus cereus, *B. subtilis* and *Serratia marcescens*
(6)	Chitosan	Cumin nanoemulsion	Growth inhibition of mesophilic and psychrophilic bacteria, *Enterobacteriaceae* and antioxidant effects
(7)	Chitosan	Clove	Microbial growth inhibition
(8)	Chitosan beads	Lavender or red thyme	Antifungal effect against *Botrytis cinerea*. Maintenance of appearance, color and firmness but odor, flavor and overall acceptability decrease
(9)	Chitosan and whey protein	Garlic nanoencapsulated	Retarded lipid oxidation and growth inhibition of main spoilage bacterial
(10)	Whey protein isolate	Oregano and garlic EOs	Antimicrobial effect against *Escherichia coli*, *Salmonella enteritidis*, *Listeria monocytogenes*
(11)	Whey protein isolate-cellulose nanofibers + TiO$_2$ nanoparticles	Rosemary EO	*Pseudomonas* spp., *Enterobacteriaceae*, *S. aureus*, *L. monocytogenes* and *E. coli*
(12)	Pectin	Oregano EO + resveratrol nanoemulsion	Lowering pH effect, color change, retarded lipid and protein oxidation, microbial growth
(13)	Sodium alginate	Citral and Eugenol EOs	Improvement of postharvest quality attributes during storage.
(14)	Sodium caseinate	Ginger EO nanoemulsion	Growth inhibition of total aerobic psychrophilic bacteria
(15)	Arabic gum- sodium caseinate	Cinnamon or lemongrass EO	Browning and decrease related enzymes

fluctuations. In this way, nanoencapsulation methods may be utilized to increase the chemical stability and physicochemical properties of essential oils, so enabling their application in food packaging. For instance, fish oil was encapsulated in polyvinyl alcohol nanofibers using emulsion electrospinning. The encapsulation efficiency for high omega-3 fatty acids resulted in an oil load capacity of 11.3% 0.3%, allowing for an equal distribution of oil throughout the fibres and fortification of the food matrix with less PUFAs. The kind and quantity of EOs included in the packing material will determine the biological activity.

Jasmine, rosemary, peppermint, cinnamon, oregano, thyme, cumin, eucalyptus, rosewood, clove, tea tree, palmarosa, geranium, lavender, lemongrass, mandarin, bergamot and lemon are some of the most often utilized EOs in packaging systems. On the food matrices where packaging solutions containing EOs have been utilized, a variety of food products, including fresh beef, butter, fresh octopus, ham and fish, may be found. Regarding their principal recognized components, they belong to the hydrocarbon monoterpenes group, which includes -pinene, -pinene, -selinene and p-cymene, or to the oxygenated monoterpenes group, which includes thymol, carvacrol, geraniol, borneol, eugenol, linalool, terpineol-4-ol, 1,8-cine.

Another research analyzed the varying quantities of main hydrocarbons and oxygenated monoterpenes found in EOs. At a concentration of 0.2 g/ml and a pH of 4.0, these compounds exhibited antibacterial activity against *E. coli* and *L.*

monocytogenes. This research demonstrated that oxygenated monoterpenes produce a more potent antibacterial action than their hydrocarbon counterparts. In fact, the effectiveness of these compounds against *E. coli* was examined in orange or apple juice in conjunction with heat treatments, revealing a synergistic deadly impact.

Thus, the objective of food packaging including biodegradable materials combined with EOs is to conduct antioxidant and antimicrobial tests to assess the final packaging system in contact with the food matrix in order to provide reliable findings. Table 19.6 below provides a summary of current tests utilizing films containing EOs for innovative food packaging solutions [22,23].

19.6.6 Role of Essential Oils in Perfumes

Essential oils and their separated constituents are frequently utilized in cosmetics due to their numerous advantages, including their pleasant scents. However, their attractive smells are the primary reason for their use in cosmetics. Frequently, the fatty acids, fatty oils and surfactants employed in the manufacturing of cosmetics have a disagreeable odor. The cosmetic and fragrance industries utilize essential oils and their terpenoids more than any other natural product. It has been discovered that essential oils, particularly terpenes and terpenoids, are widely used as active components in cosmetics and fragrance goods and procedures that have been patented (Table 19.7) [24,25].

TABLE 19.7

Aroma Profile of Essential Oils Used as Fragrances in Perfumes

S. No.	Plant Species	Part(s) Used	Principal Components	Appearance of Oils	Aroma Description
(1)	Amyris balsamifera L.	Wood	Elemol, 10,γ-epi-eudesmol, γ-eudesmol, valerianol, α-eudesmol, 7-α-epi-eudesmol, β-eudesmol, drimenol	Pale yellow to amber yellow	Characteristic
(2)	Angelica archangelica	Roots and rhizomes	Phellandrene, pinene, limonene, linalool and borneol; rich in coumarins	Colorless	Rich herbaceous-
(3)		Seeds	Including osthol, Angelicin, bergapten and imperatorin	Colorless	Fresher
(4)	Aniba rosaeodora	Wood/leaves	α-Pinene, β-pinene, limonene, 1,8-cineole, trans-linalool oxide (furanoid), cis-linalool oxide (furanoid)	Pale-yellow	Floral, sweet, woody and citric
(5)	Boswellia sacra Flueck.	Gum	α-Pinene (2.0%–64.7%), α-thujene (0.3%–52.4%), β-pinene (0.3%–13.1%), myrcene (1.1%–22.4%)	Clear or yellow	Balsamic, camphor-like
(6)	Cananga odorata Hook. F. and Thoms)	Flowers	Prenyl acetate, p-cresyl methyl ether, methyl benzoate, linalool, benzyl acetate, geraniol, geranyl acetate	Pale yellow to dark yellow	Characteristic, floral, recalling jasmine
(7)	Carum carvi L.	Ripe fruit	Myrcene, limonene, cis-dihyrocarvone, trans-carveol, cis-carveol, carvone	Colorless to pale yellow	Fresh, herbaceous
(8)	Cedrus atlantica (Endl.) Manetti ex Carrière	Wood/sawdust	Himachalene, α-himachalene, β-himachalene, cis-bisabolene, himachalol, allo-	Yellowish to orange-yellow	Camphoraceous–cresylic top note with a sweet, tenacious
(9)	Cinnamomum verum J. Presl	Bark	(E)-Cinnamaldehyde, eugenol, β-caryophyllene	Reddish-brown	Characteristically spicy burning
(10)	Citrus aurantiifolia (Christm.) Swingle	Peel	Limonene, β-pinene, γ-terpinene, citral	Colorless to greenish-yellow	Mild citrus, floral
(11)	Citrus aurantium var. amara L.	Flowers	Linalool, β-pinene, α-terpineol, limonene	Pale yellow to coffee brown	Sweet, fresh and floral odor
(12)	Citrus bergamia Risso	Rind of fruit	Linalyl acetate, limonene, α-terpinene, β-pinene, γ-terpinene, geraniol, linalool, lactones, bergaptene, β-bisabolene	Light green-yellow	Light, delicate citrus-like scent with slightly
(13)	Citrus limon (L.) Osbeck	Peel	Limonene, β-pinene, γ-terpinene, sabinene, α-pinene, geranial, neral, α-thujene, β-bisabolene, terpinolene, trans-α-bergamotene, α-terpineol, α-phellandrene and cis-sabinene-hydrate	Colorless to pale yellow	Fresh lemon peel
(14)	Citrus reticulata Blanco	Peel	Limonene, γ-terpinene, terpinolene, myrcene, α-pinene, p-cymene, α-thujene	Dark orange to reddish-orange or brownish-orange	Orange-like
(15)	Citrus sinensis (L.)	Peel	Limonene, α-pinene, β-pinene, sabinene, myrcene	Yellow to reddish-yellow	Characteristic, orange peel odor
(16)	Citrus paradisi	Peel of fruit	Limonene, citronellal, citral, sinensal, geranyl acetate, paradisol, geraniol, ketones	Yellow or greenish	Fresh, sharp, citrus aroma (top note)
(17)	Commiphora myrrha (Nees) Engl.	Gum	Furanoeudesma-1,3-diene, curzerene, β-elemene, lindestrene, furanodiene	Yellow to reddish-brown	Pleasant balsamic, camphor-like, musty, incense-like
(18)	Coriandrum sativum L.	Fruit	Borneol, 2E-decenal, camphor, dodecanal, n-decanal, neryl acetate, geraniol, linalool	Colorless to pale yellow	Floral, balsamic undertone and peppery-woody suave top-note

(Continued)

TABLE 19.7 (*Continued*)

Aroma Profile of Essential Oils Used as Fragrances in Perfumes

S. No.	Plant Species	Part(s) Used	Principal Components	Appearance of Oils	Aroma Description
(19)	Cupressus sempervivens L.	Leaves	α-Pinene, Δ-carene, limonene, sesquiterpene, α-terpinene, sabinene, carvone	Yellowish	Pleasingly smoky, woody smell and amber-like
(20)	Cymbopogon citratus (DC.)	Grass leaf	Citral, myrcene, dipentene, linalool, geraniol, nerol, citronellol and farnesol, sesquiterpenes, methyl heptenone, esters, acids and others	Yellow or pale sherry-colored	Intense sweet lemony, reminiscent of lemon drops (top note)
(21)	Cymbopogon martini	Leaves	Myrcene, linalool, β-caryophyllene, α-terpineol, geranyl acetate, geranyl isobutyrate, geraniol, (E,Z) farnesol, geranyl hexanoate	Yellow	Fresh rose-like
(22)	Daucus carota L.	Seeds	β-Bisabolene, elemicin, geranyl acetate and sabinene	Yellowish-brown color	Soft, sweet earthy grounding aroma
(23)	Elettaria cardamomum (L.) Maton	Fruits	α-Pinene, sabinene, myrcene, limonene, 1,8-cineole, linalool, linalyl acetate, terpinen-4-ol, α-terpineol, terpinyl acetate, trans-nerolidol	Almost colorless to pale yellow	Characteristic, spicy, cineolic
(24)	Eucalyptus globulus Labill.	Leaves	1,8-Cineole, α-pinene, limonene, aromadendrene, p-cymene, globulol	Yellow to red	Fresh balsamic camphor-like
(25)	Feniculum vulgare	Seeds	Phenols, trans-anethole, methyl chavicol, α-pinene, α-thujene, γ-terpinene, limonene, myrcene	Yellowish	Strong, anisey, camphoric
(26)	Helichrysum italicum	Aerial parts	Neryl acetate, g-curcumene, neryl propionate and ar-curcumene	Pale yellow to red	Strong, honey-like aroma
(27)	Hyssopus officinalis L.	Aerial parts	Isopinocamphone, pinocamphone, β-pinene	Light-yellow	Herbaceous, camphor-like odors with warm and spicy undertones
(28)	Illicium verum Hook.f.	Fruit	Trans-anethole (80%–90%)	Pale yellow	Warm, spicy, extremely sweet, licorice-like scent
(29)	Jasminum officinale L.	Flowers	Benzyl acetate, linalyl acetate, benzyl benzoate, methyl jasmonate, methyl anthranilate, linalool, nerol, geraniol, benzyl alcohol, farnesol, terpineol, phytols, eugenol, cis-jasmone; acids, aldehydes and others	Dark, orange-brown	Highly intense, rich, sweet, floral odor (base note)
(30)	Juniper communis L.	Ripe berries	α-Pinene (20%–50%), sabinene (<20%), β-pinene (1%–12%), β-myrcene 1%–35%, α-phellandrene (<1%), limonene 2%–12%, terpinen-4-ol (0.5%–10%)	Colorless or pale yellow-green	Fresh terebinth or turpentine-like/ conifer-like aroma
(31)	Laurus nobilis L.	Leaves	1,8-Cineole, sabinene, α-terpinyl acetate, linalool, eugenol, methyl eugenol, α-pinene	Yellow	Aromatic, spicy
(32)	Lavandula angustifolia Mill.	Flowering tops	Linalyl acetate (25%–47%), linalool (max. 45%), terpinen-4-ol (max. 8%), camphor (max. 1.5%), limonene (max. 1%) and 1,8-cineole (max. 3%)	Colorless to pale yellow	Sweet floral aroma
(33)	Litsea cubeba (Lour.) Pers.	Fruits	Citral (neral and geranial) (78.7%–87.4%), d-limonene (0.7%–5.3%)	Pale yellow	An intense lemon-like, spicy aroma
(34)	Matricaria chamomilla L.	Flowers and flower heads	(E)-β-Farnesene (4.9%–8.1%), terpene alcohol (farnesol), chamazulene (2.3%–10.9%), α-bisabolol (4.8%–11.3%) and α-bisabolol oxides A (25.5%–28.7%) and α-bisabolol oxides B (12.2%–30.9%)	Blue	Warm, fruity scent Sweet herbaceous odor (middle note)

(*Continued*)

TABLE 19.7 (*Continued*)

Aroma Profile of Essential Oils Used as Fragrances in Perfumes

S. No.	Plant Species	Part(s) Used	Principal Components	Appearance of Oils	Aroma Description
(35)	Melaleuca alternifolia (Maiden and Betche) Cheel	Leaves	Terpinen-4-ol (30%–48%), γ-terpinene (10%–28%), 1,8-cineole (traces-15%), α-terpinene (5%–13%), α-terpineol (1.5%–8%), p-cymene (0.5%–8%), α-pinene (1%–6%), terpinolene (1.5%–5%), sabinene (traces-3.5%), aromadendrene (traces-3%), δ-cadinene (traces-3%), viridiflorene (traces-3%), limonene (0.5%–1.5%), globulol (traces-1%), viridiflorol (traces-1%) [ISO 4730 (2004)]	Colorless to pale yellow	Intensive aromatic fresh camphoraceous odor
(36)	Melaleuca leucadendra (L.) L.	Fresh leaves and twigs	1,8-Cineole (14%–65%), terpenes (45%), alcohols (5%), esters	Pale yellow-green	Camphorous
(37)	Mentha x piperita L.	Aerial parts	Menthol, menthone, isomenthone, menthofuran, neomenthol, pulegone, menthyl acetate, 3-octanal, 1,8-cineole, limonene, trans-sabinene hydrate, β-caryophyllene	Almost colorless to pale greenish yellow	Characteristic
(38)	Myristica fragrans Houtt.	Seeds	Myristicin, α-pinene, sabinene, limonene, elemicin, eugenol, safrol and β-pinene	Pale yellow to nearly colorless	Spicy, sweet, woody
(39)	Ocimum basilicum L.	Flowering herb	Linalool, citronellol, geraniol, terpinen-4-ol, α-terpineol, methyl chavicol, eugenol, ethyl eugenol, limonene, camphene, α-pinene, β-pinene, methyl cinnamate, β-caryophyllene	Colorless or pale yellow	Light sweet, spicy odor
(40)	elargonium graveolens L'Hér.	Aerial parts	Citronellol, geraniol, linalool, citronellyl formate, geranyl formate, citral, guaiazulene	Pale or olive green	Sweet rose-like odor with a hint of mint or "greenness"
(41)	Pimenta dioica (L.) Merr.	Leaf	Mainly eugenol, less in the fruit (60%–80%) than in the leaves (up to 96%), also methyl eugenol, cineole	Yellowish-red or brownish liquid	Powerful sweet, spicy scent, similar to cloves
(42)	Pimpinella anisum L.	Seeds	Trans-anethole (75%–90%)	Colorless to pale yellow	Warm, spicy-sweet characteristic scent
(43)	Pinus massoniana Lamb.	Gum resin	α-Pinene, camphene, β-pinene, δ-3-carene, myrcene, limonene, p-cymene, longifolene	Colorless	Characteristic of gum turpentine
(44)	Piper nigrum L.	Unripe berries (peppercorns)	Sesquiterpenes (20%–30%) and monoterpenes (70%–80%) such as bisabolene, β-caryophyllene, thujene, pinene, camphene, sabinene	Light to pale olive	Dry, woody, warm, spicy, oriental (base note)
(45)	Pogostemon cablin (Blanco) Benth.	Leaves	α-Pinene (0.01%–0.3%), β-pinene (0.02%–1%), Limonene (0.01%–0.3%), δ-elemene (0.01%–1.9%), β-patchoulene (0.03%–12%), β-elemene (0.18%–1.9%), cycloseychellene (0.02–0.08%)	Yellow to reddish-brown	Earthy, woody and camphoraceous
(46)	Rosa damascena Mill.	Flowers	Citronellol, geraniol, nerol, nonadecane, heneicosane	Yellow to yellow-green	Sweet, floral, rosaceous
(47)	Rosmarinus officinalis L.	Leaves	Eucalyptol and α-pinene, camphor, bornyl acetate, camphene, β-pinene, β-myrcene, limonene and borneol	Colorless to pale yellow	Strong, warm, woody, balsamic aroma
(48)	Salvia sclarea L.	Inflorescences	Linalool, linalyl acetate	Yellow	Characteristic herbaceous odor
(49)	Santalum album L.	Heartwood	90% Santalol content (cis-α-santalol and cis-β-santalol) (ISO 3518:2002)	Yellow to light brown	Sweet woody
(50)	Syzygium aromaticum (L.) Merrill et. Perry	Bud	Eugenol (70%–95%), eugenol acetate (up to 20%) and β-caryophyllene (12%–17%)	Colorless or pale yellow	Clove aroma
(51)	Vetiveria zizanioides (Linn.) Nash	Roots	α-Vetivone (8.4%–13.3%), khusimol (0.6%–8.9%), β-vetivone (2.2%–3.7%), khusian-2-ol (1.8%–2.3%) and khusimone (1.2%–2.3%)	Pale-yellow to dark brown, olive or amber	Deep, smoky, earthy and woody with a sweet persistent undertone

19.7 Analysis of Essential Oils

Figure 19.9 depicts the complete analytical processes for the quality control analysis of essential oils. Table 19.8 depicts examples of plants along with phyto-constituents and essential oils [26].

19.8 Legislation of EO'S in Food

The legal framework governing the use of essential oils (EOs) in food includes two approaches: the use of EOs as a food component, which represents an additional nutrient to consume, and the use of EOs as part of the packaging. In the first case, EOs

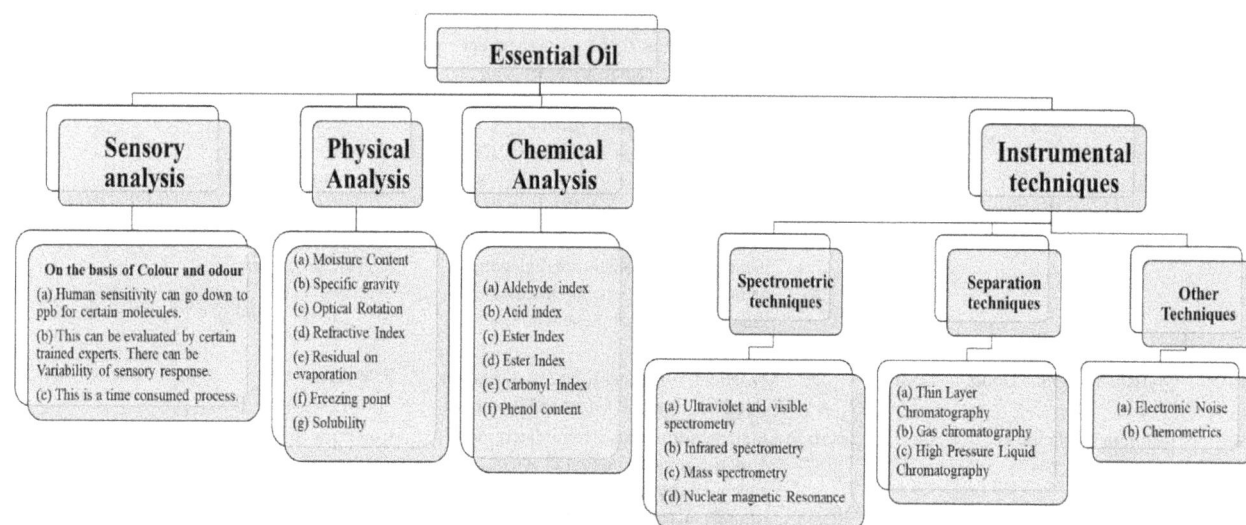

FIGURE 19.9 Analytical parameters of essential oil.

TABLE 19.8

List of Essential Oils in Cosmetics

S. No.	Application	Essential Oil	Plant	Phyto-constituents	Therapeutic Effect	Function
(1)	SKIN CARE	Chamomile	*M. chamomilla*	β-farnesene; bisabolol oxide; bisabolon oxide; chamazulene; 1,8-Cineole; α-Terpineol α-bisabolol;	In inflammatory wound healing	Anti-acne Anti-aging
(2)	SKIN CARE	Sandalwood	*S. spicatum*	β-santalol α-bisabolol; (*E*)-farnesol; nuciferol; α-santalol;	Antiseptic Antioxidant	Anti-aging
(3)		Evening primrose	*O. biennis*	β-amyrin; 1-hexacosanol	Same as above	Anti-acne Anti-wrinkles moisturizer
(4)		Camellia	*Camellia japonica*	β-amyrin; cycloartenol; lanosterol; lupeol; β-sitosterol; squalene	Same as above	Anti-aging
(5)		Rosemary	*Rosmarinus officinalis*	p-cymene, cineole; borneol; camphene; camphor; β-caryophyllene; 1,8-	Same as above	Anti-acne
(6)	HAIR CARE	Sweet orange	*Citrus sinensis*	limonene; myrcene; α-pinene; β-pinene	Same as above	Antidandruff
(7)		Lavender	*Lavandula officinalis*	borneol; caryophyllene; lavandulol; lavandulol acetate; linalool; linalyl acetate; α-terpineol; terpinene-4-ol	Same as above	Hair growth conditioning
(8)		Peppermint	*Mentha piperita*	carveone; 1,8-cineole; limonene; menthol; menthone; methyl acetate; neomenthol	Same as above	Hair growth conditioning
(9)		Thyme	*Thymus vulgaris*	α-cadinene; γ-cadinene; δ-cadinene; α-cadinol; δ-cadinol; β-caryophyllene; p-cymene; elemol; β-eudesmol; germacrene; limonene; γ-muurulene; myrcene; trans-β-ocimene; β-pinene; γ-terpinene; α-terpineol	Same as above	Antidandruff hair growth
(10)		Bergamot	*Citrus bergamia*	bergamottin; bergapten; citropten; limonene; linalool; linalyl acetate; α-pinene; β-pinene; γ-terpinene	Same as above	Antidandruff hair growth

must comply with regional food regulations. Depending on the purpose of their application, different laws will apply. EOs will be considered additives if they represent a "substance not normally consumed as a food in itself and not normally used as a characteristic ingredient of food, whether or not it has nutritive value, the intentional addition of which to food for a technological purpose in the manufacture, processing, preparation, treatment, packaging, transport, or storage of such food results, or may reasonably be expected to result, in it or its by-products becoming directly or indirectly toxic to humans or other animals." FDA has recognized linalool, thymol, eugenol, vanillin, carvacrol and limonene as flavoring agents in food [27].

Conclusion

This chapter explores the uses of terpenes in perfumery and oil industry. This chapter summarizes the information about volatile oils, classification, extraction, sources of natural essential oils, method of extraction such as hydro-distillation, hydro-diffusion, hydrolysis and decomposition by heat. It also includes traditional methods of producing attar through Hydro-distillation, direct steam distillation, essential oil extraction through expression, through enfleurage, de-fleurage. Different processes, namely Hot Maceration, Modern (Non-Traditional) Methods of Extraction, Headspace Trapping Techniques, Static Headspace Technique, Vacuum Headspace Technique, Dynamic Headspace Technique, Solid Phase Micro-Extraction, Supercritical Fluid Extraction, Simultaneous Distillation Extraction, Microwave Distillation, Controlled Instantaneous Decomposition, Thermo-Micro Distillation, Molecular Spinning Band Distillation have been explored. This chapter also emphasizes on Worldwide Current Trends in Essential Oils Market.

REFERENCES

1. Herman, R. A.; Ayepa, E.; Shittu, S.; Fometu, S. S.; Wang, J. Essential Oils and Their Applications -A Mini Review. *Advances in Nutrition & Food Science*, **2019**, 4, 4, 1 of 13.

2. Dreger, M.; Wielgus, K. Application of Essential Oils as Natural Cosmetic Preservatives. *Herba prolonica*, **2013**, 59(4), 142–156. doi: 10.2478/hepo-2013-0030.

3. Maurya, A.; Prasad, J.; Das, S.; Dwivedy, A. K. Essential Oils and Their Application in Food Safety. *Frontiers in Sustainable Food Systems*, **2021**, 5, 653420. doi: 10.3389/fsufs..2021.653420.

4. Teixeira, M. A.; Rodrigues, A. E. Coupled Extraction and Dynamic Headspace Techniques for the Characterization of Essential Oil and Aroma Fingerprint of Thymus Species. *Industrial & Engineering Chemistry Research*, **2014**, 53, 23, 9875–9882.

5. Jeleń, H. H.; Majcher, M.; Dziadas, M. Sample preparation for food flavor analysis (flavors/off-flavors). In: Pawliszyn J. (ed.) *Comprehensive Sampling and Sample Preparation*. Academic Press, Cambridge, Massachusetts, 2012, 119–145.

6. Gómez-Estaca, J.; López-de-Dicastillo, C.; Hernández-Muñoz, P.; Catalá, R.; Gavara, R. Advances in antioxidant active food packaging. *Trends in Food Science and Technology*, **2014**, 35, 42–51.

7. Werkhoff, P.; Güntert, M.; Krammer, G.; Sommer, H.; Kaulen, J. Vacuum Headspace Method in Aroma Research: Flavor Chemistry of Yellow Passion Fruits. *Journal of Agricultural and Food Chemistry*, **1998**, 46, 3, 1076–1093.

8. Jeleń, H.; Gracka, A.; Myśków, B. Static Headspace Extraction with Compounds Trapping for the Analysis of Volatile Lipid Oxidation Products. *Food Analytical Methods*, **2017**, 10, 2729–2734. doi: 10.1007/s12161-017-0838-x.

9. Dini, I. *Use of Essential Oils in Food Packaging*. Elsevier Inc., Amsterdam, the Netherlands, 2016.

10. Friedman, M.; Henika, P. R.; Mandrell, R. E. Bactericidal Activities of Plant Essential Oils and Some of Their Isolated Constituents Against Campylobacter jejuni, Escherichia coli, Listeria monocytogenes, and Salmonella enterica. *Journal of Food Protection*, **2002**, 65, 1545–1560.

11. Alizadeh Sani, M.; Ehsani, A.; Hashemi, M. Whey Protein Isolate/Cellulose Nanofibre/TiO$_2$ Nanoparticle/Rosemary Essential Oil Nanocomposite Film: Its Effect on Microbial and Sensory Quality of Lamb Meat and Growth of Common Foodborne Pathogenic Bacteria during Refrigeration. *International Journal of Food Microbiology*, **2017**, 251, 8–14.

12. Xiong, Y.; Li, S.; Warner, R. D.; Fang, Z. Effect of Oregano Essential Oil and Resveratrol Nanoemulsion Loaded Pectin Edible Coating on the Preservation of Pork Loin in Modified Atmosphere Packaging. *Food Control*, **2020**, 114, 107226.

13. Kumar, A.; Singh, P.; Gupta, V.; Prakash, B. *Application of Nanotechnology to Boost the Functional and Preservative Properties of Essential Oils*. Elsevier Inc.: Amsterdam, the Netherlands.

14. Herman, A.; Herman, A. P.; Domagalska, B. W.; Młynarczyk, A. Essential Oils and Herbal Extracts as Antimicrobial Agents in Cosmetic Emulsion. *Indian Journal of Microbiology*, **2013**, 53(2), 232–237. doi: 10.1007/s12088-012-0329-0.

15. Maccioni, A. M.; Anchisi, C.; Sanna, A.; Sardu, C.; Dessì, S. Preservative Systems Containing Essential Oils in Cosmetic Products. *International Journal of Cosmetic Science*, **2002**, 24, 53–59.

16. Cagliani, I. Consigli e suggerimenti, criteri di accettabilita. *Cosmetic News*, **1999**, 124, 25–27.

17. Camporese, A. Oli essenziali e terapia antimicrobica. *Leader for Chemist VIII*, **1997**, 1, 4–13.

18. Murmu, S. B.; Mishra, H. N. The Effect of Edible Coating Based on Arabic Gum, Sodium Caseinate and Essential Oil of Cinnamon and Lemon Grass on Guava. *Food Chemistry*. **2018**, 245, 820–828.

19. Noori, S.; Zeynali, F.; Almasi, H. Antimicrobial and Antioxidant Efficiency of Nanoemulsion-Based Edible Coating Containing Ginger (Zingiber officinale) Essential Oil and Its Effect on Safety and Quality Attributes of Chicken Breast Fillets. *Food Control*, **2018**, 84(18), 312–320.

20. Jafarzadeh, S.; Jafari, S. M.; Salehabadi, A.; Nafchi, A. M.; Uthaya Kumar, U. S.; Khalil, H. P. S. A. Biodegradable Green Packaging with Antimicrobial Functions Based on the Bioactive Compounds from Tropical Plants and Their By-Products. *Trends in Food Science and Technology*, **2020**, 100, 262–277.

21. Fernández-Pan, I.; Carrión-Granda, X.; Maté, J. I. Antimicrobial Efficiency of Edible Coatings on the Preservation of Chicken Breast Fillets. *Food Control*, **2014**, 36, 69–75.

22. Ni, Z. J.; Wang, X.; Shen, Y.; Thakur, K.; Han, J.; Zhang, J. G.; Hu, F.; Wei, Z. J. Recent Updates on the Chemistry, Bioactivities, Mode of Action, and Industrial Applications of Plant Essential Oils. *Trends in Food Science & Technology*, **2021**, 110, 78–89, doi: 10.1016/j.tifs.2021.01.070.

23. Santos, R. R.; Andrade, M.; de Melo, N. R.; Silva, A. S. Use of Essential Oils in Active Food Packaging: Recent Advances and Future TRENDS. *Trends in Food Science & Technology*, **2017**, 61, 132–140. doi: 10.1016/j.tifs.2016.11.021.

24. Irshad, M.; Subhani, M. A.; Ali, S.; Hussain, A. Biological Importance of Essential Oils. In: El-Shemy, H. A. (ed.) *Essential Oils - Oils of Nature*. IntechOpen, London, **2019**. Available from: https://www.intechopen.com/chapters/68027. doi: 10.5772/intechopen.87198.

25. Sharmeen, J. B.; Mahomoodally, F.M.; Zengin, G.; Maggi, F. Essential Oils as Natural Sources of Fragrance Compounds for Cosmetics and Cosmeceuticals. *Molecules*, **2021**, 26, 666. doi: 10.3390/molecules26030666.

26. Taherpour, A. A.; Khaef, S.; Yari, A. et al. Chemical composition analysis of the essential oil of *Mentha piperita* L. from Kermanshah, Iran by hydrodistillation and HS/SPME methods. *Journal of Analytical Science and Technology*, **2017**, 8, 11. doi: 10.1186/s40543-017-0122-0.

27. Barbieri, C.; Borsotto, P. Essential Oils: Market and Legislation. In: El-Shemy, H. A. (ed.) *Potential of Essential Oils*, IntechOpen, London, **2018**. https://www.intechopen.com/chapters/61798. doi: 10.5772/intechopen.77725.

Index

Note: **Bold** page numbers refer to tables; *italic* page numbers refer to figures.

abietanes 427, *427*
  antimalarial activity 108
  antimicrobial and anti-parasitic activity
    135
  immunomodulatory effects 133
abietic acid
  anti-inflammatory effects 113, **114**
  structure *114*
abscisic acid (ABA) 334
  structure *334*
  synthesis *335*
acetoxycrenulide 89, 135
4-acetoxydictyolactone 104, *104*
acetylation of alcohol 20
acetyl-CoA 182
3-O-acetyl-11-keto-β-boswellic acid
  anticancer activity 306
  structure *307*
aconicarmisulfonine A 277
aconitorientaline 276, *277*
acoradiene 157
acoranes 424, *425*, 502–503, *503*
actinidine 269, **269**, *269*, 271–272, *272*
acyclic monoterpenes 400
acyclic monoterpenoids 4, **4–5**
  skeletal structure 30, *30*
  structures *56*
  synthesis 31–35
acyclic sesquiterpenoid 332, 488
acyclic sesterterpenes 435
acyclic terpenes *400*
acyclic terpenoid 4, *23*
acyclovir 59
additives 599
adiananes 439, *439*
adsorption chromatography 19
adsorption, in purified fats 17
aframodial
  antimicrobial activity 121
  structure *122*
aframolins 121
ajmalicine 264
ajugarin-I 99, **100**
alcohol
  acetylation *20*
  esterification *20*
aliphatic imines 256, *256*
alisol W 299, *301*
alkaloids
  biological activitiy 278, **279**, *279*
  diterpenoid 275–278
  monoterpenoid 269–272
  sesquiterpenoid 272–275
  triterpenoid *268*, 278
allylketobenzoate 479, *480*
α-amyrin 386, *388*
α,β-epoxy-carvone (EC) 77
α-bergamotene 336–337

α-bisabolene 334, *335*
α-caryophyllene *see* humulene
α-curcumene 334–335, **335**, **336**
α-humulene *see* humulene
α-methylene-γ-lactone group (αMγL)
  526–527, *527*
α-pinene
  anti-inflammatory activity 60
  antimicrobial activity **54**
  cardiovascular effects **72**
  synthesis 568, *569*
α-terpineol 563–564, 578
  anticancer activity 203, *204*
  anticonvulsant activity 69
  antimicrobial activity **53**
  cardiovascular effects 71
  metabolism *578*
  synthesis 43–44, *44–45*, 564
α-thujone 579, *579*
alphitolic acid 377
alysine
  antidiabetic effects 118
  structure *119*
Alzheimer's disease (AD) 70, **71**
ammarane-type triterpene glycoside 239, *239*
amphilectane 172, *173*
amphotericin-B (AMB) 129, 137
ananosic acid B
  anticancer activity 303–304
  structure *304*
andrographolide (AND)
  anticancer activity 105, **106**, 216
  anti-inflammatory activity **368**
  anti-inflammatory effects 113
  antimalarial activity 109
  molecular structure *216*
  structure *113*
antcamphorol A 288, *290*
anthelmintic activity 132, 365–367
anthriscifolmine 275–276, *276*
anti-asthmatic activity 133
antibacterial activity **348**
anticancer activity
  ananosic acid B 303–304
  asiatic acid 306
  β-Escin 305
  betulin 301
  betulinic acid 301
  bivittoside D 302–303
  CDDO-Me 305
  celastrol 307
  cimicifoetisides 305–306
  cycloartane 299
  daedaleasides 304
  dehydrametenolic acid 301–302
  diterpenes 212–217
  ficutirucins 299, **300**
  ganoderic acid D 301

  glycyrrhizin 307
  impatienside A 302–303
  inonotsuoxides 304
  inotodiol 304
  limonoids 299
  lupeol 298–299
  monoterpenes 62–65, 199–206
  nimbolide 306
  3-O-acetyl-11-keto-β-boswellic acid 306
  protostane triterpene 299
  sesquiterpenes 206–212
  terpenes 379–382, *380–382*
  tetraterpenes 221–223
  tirucallane 299, **300**
  triterpenes 217–221
  triterpenoids 297–307, **298**
  ursolic acid 299–300
anticonvulsant activity 66–69
antidepressant activity 58
anti-diabetic activity
  alysine 118
  lupeol 308
  monoterpenoids 73–74
  oleanolic acid 308
  terpenes 383–385, *384–387*, **387**
  triterpenoids 307–310, **310**
  ursolic acid 310
  viburodorol A 310
antifilarial activity 109–112
antifungal activity **58**
anti-HIV drugs **369**, 369–376, *370–378*, 378
antihyperglycemic activity
  borapetoside-E 117
  clerodane 117
  rebaudioside-A 117
anti-inflammatory activity
  avicins 312
  escin 312–313
  ginsenosides 312
  glycyrrhizin 313
  lupeol 314
  monoterpenoids 59–62
  oleanolic acid 315
  platycodin D 315
  saikosaponins 315–316, **316**
  terpenes 367–369, **368**, *369*
  triterpenoids 312–316
  ursolic acid 315
antimalarial activity
  diterpenoids 108–109
  terpenes 363–365
antimicrobial activity
  of acyclic monoterpenes **56**
  β-amyrin 311–312, **312**
  betulinic acid 312, **312**
  of bicyclic monoterpenes **54**
  of *C. sempervirens* extract **138**
  diterpenoids 121–124

antimicrobial activity (*Cont.*)
 essential oils 602, **603**
 of monocyclic monoterpenes **53**
 monoterpenoids 55–58
 terpenes *378,* 378–379, **379**
 triterpenoids *311,* **311,** 311–312
antimicrobial resistance (AMR) 55
antioxidant activity
 fulgic acid 317, **317**
 madecassic acid 318
 monoterpenoids 62
 multiflorane triterpenoid 317–318
 terpenes 382
 triterpenoids 316–318
anti-proliferative activity
 kayadiol 103, **103**
 monoterpenoids 65
 sarcodonin-G (SG) 103
anti-tuberculosis 125
antiviral activity
 diterpenoids 129–132
 monoterpenoids 59, **59**
 triterpenoids 319–320, *320*
(+) apiosporamide 582, *582*
aplysinoplides A-C *349,* 349–350
apocarotenoids 440, *440*
apoptosis 65, 199
apotirucallanes *437,* 438
arglabin 542–543
argutine 269
Arndt-Eistert reaction *174,* 174–175
artemisinin 108, 340, 363, 449, 535, 540–541
 anticancer activity 206, *209, 210*
 biosynthesis *341*
 derivatives *536*
 structure *340, 363*
arterial hypertension 96
artesunate 207, 449
asbestinanes 431
ascaridole 448
 anticancer activity 204, *205*
 synthesis 49, *50*
asiatic acid 380
 anticancer activity 306
 molecular targets *307*
astaxanthin
 anticancer activity 221
 structure *222*
astragaloside II *372*
atisane 428, *430*
 diterpenoid alkaloids 277, *277*
atropurpuran 164, *165*
attar production, using hydro-distillation 594–595
aucubin
 antidiabetic effect **75**
 synthesis 49, *50*
aulacocarpinolide
 anticancer activity 104
 structure *104*
auraptene 448
automated Soxhlet extraction 12
avicins
 anticancer activity 219, *220*
 anti-inflammatory activity 312
 structure *313*
axamide-I 170, *170*

azoxystrobin 58

Baccatin III
 synthesis 472–473, *473*
 as taxol precursor 472
baccharanes 438
Baeyer–Villiger (BV) oxidation 162, 265, 458, *459,* 460
bakuchiol 364, *364*
bassic acid *386*
baueranes 438
β-carotene 468, *470*
Beckmann rearrangement 164, *166*
benzyl ethers 72
β-escin
 anticancer activity 305
 mechanism *305*
3β-acetoxy-copalic acid 125
β-amyrin 386
 antimicrobial activity 311–312, **312**
 structure *388*
β-carbolines 93, *94*
β-carotene
 anticancer activity 221–222
 structure *222*
β-caryophyllene (BCP) 337, **338**
β-caryophyllene oxide (BCPO) 337, **338**
7β-dihydroxy vouacapan-17β-oic acid
 antinociceptive effect 139
 structure *144*
β-elemene 449
 structures *339*
3β-hydroxy-copalic acid 125
β-pinene
 antimicrobial activity **54**
 cardiovascular effects **72**
 cationic polymerization of 262, *262*
 cytotoxicity 59, **59**
 for HSV-1 59, **60**
 L-menthol synthesis from 35, *37*
 myrcene synthesis from 31, *33*
β-thujone 579, *579*
betulin (BT)
 anticancer activity 301
 structure *303, 385*
betulinic acid (BA) 450
 anticancer activity 219, 301, *303*
 antimicrobial activity 312, **312**
 biosynthesis 294, *295*
 structure *219, 377*
beyeranes 428, *429*
bicyclic diterpenoid **7**
bicyclic monoterpenes 420, *420*
bicyclic monoterpenoids **7,** 256
 structures *55*
 synthesis 49–52
bicyclic sesquiterpenoid **7,** 336–339
bicyclic sesterterpenes 435, *436*
bicyclic sesterterpenoid **7**
bicyclic terpenoid 4, **4,** 7
bicyclophytanes 427
bifloranes *434,* 434–435
biogenesis
 carvone 185, *187*
 by engineered microbes 190–192
 geraniol 188, *188*
 iridoids 188, *188*
 limonene 185–188, *188*

 lycopene 188–190, *189*
 of monoterpenes via recombinant technology 192–194, *194*
 pimarane 188, *189*
 protein engineering strategies 192, *193,* **193**
 regulation 194
 terpene indole alkaloids 190, *191*
 via genome engineering 192, **194**
Birch reduction 41, 165, *167*
bisabolanes 421, *422,* 489
bisabolol 343–344, *344*
bischalcone *163*
bis-diterpenoid alkaloids 485
bivittoside D 302–303, *304*
blood pressure 98
bolivianine 354–355, *355*
bonducellpin-D 105
borapetoside-E 117
bornane 168, *169*
borneol
 anticancer activity 204, *205*
 antidiabetic effect **75**
 anti-inflammatory activity 61
 synthesis 50–51, *51*
bornyl diphosphate (BPP) 190
bornyl diphosphate synthase (BPPS) 190
boschnaloside 49
boswellic acids (BA) 320
 biological activity **322**
 molecular targets *322*
 structures *321*
brasilane 232, *233*
brevistylumsides 234, *237*
briaranes 431, *432*

C5 acetate 468, *470*
cacospongionolide-B 353, *353*
cacospongionolide-E 353, *353*
cacospongionolide-F 352, *352*
cadinanes 421–422, *423*
caesalminaxins 105
cafestol 468
(-) cafestol 472
(-)-cafestol retrosynthesis 468, *471*
Calamintha officinalis essential oil 601
C-alkyl group 19
camphanes 420, *421*
camphene 572, *572*
campherenanes 504, *505*
camphor 568, *569,* 572–573, *573*
 Beckmann rearrangement reaction 164, *166*
 bornane production from 168, *169*
 Meerwein-Ponndorf-Verley reduction 169, *169*
 substituted 1-norbornylamines from 256, *256*
 synthesis 50, *51,* 570, *570*
camptothecin 105
cananodine 274–275
 Craig-Henry synthesis of 275, *275*
cancer
 developmental stages 197, *198*
 diterpenoids in treatment 99–112
 hallmarks of 197, *198, 206*
 molecular targets of terpenoids in 199, *200*
 types 99–100, 197

cannabinoids 420
C20-anopterine 484, *484*
cantharidin (CTD) 206, *208,* **209**
capsianosides 234, *236*
caranes 420
cardioprotective activity
    betulinic acid 319
    lupeol 319
    triterpenoids 319–320
cardiovascular effects
    diterpenoids 96–99
    monoterpenoids 70–72, **72**
    rebaudioside-A 99, **99**
    stevioside **99**
    trans-dehydrocrotonin 99
carene 257, *258*
    antimicrobial activity **54**
carnosic acid **368**, *370*
carnosol
    anti-asthmatic activity 133
    anti-inflammatory activity **368**
carotenoids 4, 380, 440, *440*
    molecular mechanism *381*
carvacrol 449, 573, *574*
    antidiabetic effect **75**
    anti-inflammatory activity 61, **61**
    antimicrobial activity **53**
    anti-proliferative activity 202, *203*
    cardiovascular effects 70, **72**
    synthesis 40, *42*
carveol
    Oppenauer oxidation 160, *160*
    synthesis 45, *47*
carvone 257, 564, *565,* 573
    antidiabetic effect **75**
    antimicrobial activity **53**
    biogenesis 185, *187*
    catalytic hydrogenation 168
    (-)-cis-carveol transformation 259, *259*
    limonene production from 168, *168*
    oxidation and reduction 564, *566*
    synthesis 43, *43*
    terpene-based polymers production from
        262, *262*
    versatile transformation 162, *163*
carvone 7,8-epoxide, transformation of
        265, *268*
caryophyllanes 421, *422*
caryophyllene 580
casbane diterpene (CD) 123, **124,** *124*
casbanes 431, *431*
caseanigrescens-A
    anticancer activity 104
    structure *104*
cassanes 105, *105,* 427, *428*
catalpol
    antidiabetic effect **75**
    synthesis 45, *47*
cationic ring-opening polymerization
        (CROP). 265
C20-atisines diterpenoids 482, *482*
CDDO-Me
    anticancer activity 305
    structure *305*
C20-denudatine 482, *482*
C20-diterpenoid 481, *481*
ceanothic acid
    hepatoprotective activity 318–319

structure *318*
cedranes 425, *425,* 503, *503*
cedrene 510–511, *512, 513*
celastrol 320, 322
    anticancer activity 218, *219,* 307
    biological properties *323*
    biosynthesis 294, *296*
    cytotoxic effects *308*
cembranes 429, *430,* 431
cembrene-A 89, *92*
centrifugation 592
chamigranes *425,* 425–426, 503, *503*
chenopodium ambrosioides 204
C20-hetidine 482–483, *483*
Chikungunya virus (CHIKV) 129
chiral oxygen-containing heterocycles
        256, *256*
chiral terpenoids
    citral 560, 562, *562*
    geraniol 562–564
    metabolism reactions 572
    myrcene 560, *561*
    synthesis 582
chlorophyll 88
chromatographic methods 17, 19
chromen-4-ol 256, *257*
chromophores 22
CID *see* controlled instantaneous
        decomposition (CID)
cimicifoetisides
    anticancer activity 305–306
    structure *306*
cimicifugoside 237, *238*
cimifugine 238, *238*
cimigoside 237, *238*
cimiracemoside 238, *238*
1,4-cineole 573–574
1,8-cineole 574–575, *575*
    antimicrobial activity **53**
    cardiovascular effects **72**
cinmethylin 72
cinnamyl-alcohol dehydrogenase (CAD) 48
cisplatin 105
citral
    amine peroxides synthesis 159, *159*
    antimicrobial activity **56**
    citronellol synthesis from 33, *35*
    geometrical isomers *33*
    ¹H-NMR spectra *25*
    menthol synthesis from 36, *39*
    synthesis 26, *26,* 31, *33, 34*
    synthesis vis Claisen-Cope rearrangement
        173, *174*
citronellal 563, *563,* 575
    antimicrobial activity **56**
    menthol synthesis from 36, *39*
    synthesis 33, *35*
    Wolff-Kishner reduction 168, *168*
citronellol 258, 563, *563*
    anticonvulsant activity 66
    antidiabetic effect **75**
    antimicrobial activity **56**
    cardiovascular effects **72**
    iridoids preparation 272, *272*
    synthesis 33, *35*
Claisen rearrangement 33, 89, 163–164, *164*
Claisen-Schmidt condensation 162–163, *163*
cleisthanthanes 427, *428*

clerodane
    antihyperglycemic activity 117
    diterpene glycoside 236, *237*
    structure *118, 132, 139*
click reactions 154–155
2-C-methyl-D-erythritol 4-phosphate (MEP)
        pathway 286, *287,* 522
C20-napelline 483–484, *484*
CoA reductase (CCR) 48
codonolactone (CLT) 211, *212*
codonopsesquilosides 233, *235*
cohobation 595
cold fat, extraction with 595–596
columbin 139
compound action potential (CAP) 66
conjugated double bond 20
controlled instantaneous decomposition
        (CID) 598
copaiba oil
    anti-leishmanial activity 129, *130*
    molecular structure *130*
copalic acid 125
Cope rearrangement 33, 172–173, *173*
coprinol 171, *172*
corosolic acid *384*
costunolide *386,* 541
4-coumaric acid 47
C20-rearranged classes 484, *485*
crenulacetal 133, *137*
Crocetin 450
C₅ rule *see* isoprene rule
cryptone 3
cryptotanshinone
    for Alzheimer's disease 115
    structure *116*
cucurbitacins
    anticancer activity 220
    structure *221*
cucurbitanes 244, *244,* 438
curcumenone 209
curcumol *382*
curdione *381*
Curtius rearrangement 170, *170*
C20-vakognavine 483, *483*
C-15 Wittig Salt 468, *469*
cyanobacteria 191–192, **192**
cyathane 103, *104*
cyclic adenosine monophosphate (cAMP) 98
cyclic terpenes *400*
cyclin-dependent kinase 4 (CDK4) 202
cycloartanes 438
    anticancer activity 299
    biosynthesis 289, *292*
    structure *300*
cycloartane-type triterpene glycosides
        242, *242*
cyclocembranes 429, *430*
cyclofarnesanes 421, *422,* 489
cycloguaianes 422, *423*
cyclohexane monoterpenes 417, *419*
cyclohexenaldehyde 474, *475*
cyclopentane monoterpenes 417, *419*
cyclophosphamide (CP) 319
cyclosophoside A 240–241, *241*
cyclosporin A (CsA) 107
cymene 420, *420*
    antidiabetic effect **74, 75**
    cardiovascular effects **72**

p-cymene
 antimicrobial activity **53**
 synthesis 38, *40*
cynaropicrin 543
cytocromes P450 (CYPs) 87
cytotoxic activity
 abietane 103
 herical 108, ***108***
 labdane 103
 of royleanone 107
 sclareol 103

daedaleasides
 anticancer activity 304
 structure *304*
dammarane 436, *437*
 SAR study *320*
 synthesis 289, *291*
daphnanes 100, *101*, 431
daucanes 424, *424*, 501, *501*
decoction 11
defleurage 596
dehydroandrographolide succinic acid
 monoester (DASM) *370*
dehydroleucodine 541
dehydrolinalool 468, *469*
dehydrotrametenolic acid (DTA)
 anticancer activity 301–302
 structure *303*
delavayine 269
δ-cadinene 337–338
 mechanism for formation *338*
 structures *338*
dendrobine 272, *273*
14-deoxyandrographolide 96, *98*
14-deoxy-11,12-didehydroandrographolide 98
1-deoxy-D-xylulose 5-phosphate (DXP) 88
1-deoxy-D-xylulose 5-phosphate reductoisomerase (DXR) 88
1-deoxy-D-xylulose 5-phosphate synthase
 (GbDXS) 88
12-deoxyphorbol-13-(3E,5E-decadienoate) *371*
diabetes mellitus 117–120, 307
diacarnoxides-A and B 354, *354*
diacylglycerol acyltransferase (DGAT) 118
di-apocarotenoids 440, *440*
diastereomers 475, *476*, 558, *559*
diazoketoester 470, *471*
dichotomane diterpenes 129
dichrocephnoid-A 130
dictyolactone 104, *104*
dictyol-E 138, *140*
dictyone 104, *105*
dictyone acetate 104, *105*
dictyotalide-A and B 104, *104*
Dieckmann condensation 164, *165*
Diels-Alder reaction 2, 20, *21*, 95, 161, *162*
diethylcarbamazine (DEC) 109, **112**
digestion 10–11
dihydrotanshinone 135
dimerization 3, *3*
dimethylallyl diphosphate (DMADP) 182
dimethylallyl pyrophosphate (DMAPP) 35, 49,
 69, 88, 165
2,2-diphenym-1-picrylhydrazyl (DPPH)
 assay 76
direct steam distillation 595
diselenide 265

distillation 591
diterpene cyclases (DTCs) 93
diterpenes
 as anticancer agents 212–217, *214*, **214**
 on cell cycles 213, **215**, *215*
 chemical structure *215*
 classification 87, **87**
 mode of action **215**
 molecular structure 87, *88*
diterpene synthase 87
diterpenic acid A 170
diterpenoids 4, **4**, 426, 449–450, 481
 alkaloids 275–278
 in Alzheimer's disease management
 115–117
 in cancer treatment 99–112
 in cardiovascular disease treatment 96–99
 for diabetes 117–120
 glycosides 234–236
 harringtonolide 95, *95*
 for hyperlipidemia 117–120
 for inflammation 112–115
 labdane 92, *92*
 in leishmaniasis treatment 129–143
 for microbial infection 121–124
 mulinane 93–95, *94*
 neodolastane *95*, 95–96
 for pain and smooth muscle spasms
 128–129
 peucelinendiol 95, *95*
 sclarene 92, *92*
 structure *384*
 synthesis 88–96
 taxanes 92–93
 in tuberculosis treatment 124–128
D-limonene 448
DMAPP *see* dimethylallyl pyrophosphate
 (DMAPP)
dolabellanes 95, *96*, 431–432, *432*
dolastanes 432, *432*
drimanes 422, *423*
dynamic headspace technique 596

echinoside A 240, *241*
E-isomer 31
eleganolone 96, *98*
elemanolide 233
elemene 211, *211*
elemental analysis 26, *26*
elemol 173, *173*
elephantopin 341–342, *342*
eleutherobin 454
enantiomers (chirality) 558–559, *559*, *560*
enfleurage 595–596
ent-abietane 105
 cytotoxic activity **106**
 structure *106*
ent-abietane-type diterpenoids 100, *100*
ent-atisane 100, *101*
ent-kaurane 100, *102*
ent-kaur-16-en-19-oic acid 99
ent-rosane 100, *102*
enzyme-linked immunoabsorbent assay
 (ELISA) 232, *232*
EOs *see* essential oils (EOs)
epilepsy 66
epipomolic acid *376*
epoxidation 265, 455, *456*

epoxydictymene 161, *161*
equaling 595
eremophilanes 421, *423*
eremophilane-type glycosides 233, *233*, **234**
erogorgiaene 125, *127*
escin
 anti-inflammatory activity 312–313
 chemical structure *314*
E-selinene *509*, 509–510, *510*
essential oil of *Chenopodium ambrosioides*
 (EOCA) 76–77
essential oils (EOs) 590
 analysis 608, *608*
 antimicrobial effects 602, **603**
 in cosmetics **608**
 extraction methods *592*, 592–598
 food and food preservation 599, **599, 600**
 food industry, current trends on
 603–604, **604**
 heterogeneous chemical group in *591*
 hydrodistillation 592–595
 isolation from plant parts 17
 legislation in food 608–609
 nano-formulation 600
 as natural cosmetic preservatives 600–602
 in perfumes 604, **605–607**
 separation *18*
 chemical methods 17
 physical methods 17–19
 sources *590*, 591–592
 worldwide current trends in market
 598, *598*
esterification, of alcohol with acids *20*
ethambutol (EMB) 126
eucalyptol
 cardiovascular effects **72**
 synthesis 41, *43*
eudesmanes 421, *423*
eudesmanolide 233, 340, *341*
eugenol 78, **79**, *79*
 antibacterial activity 77
 antimicrobial activity **53**
 antioxidant activity 77
 minimum inhibitory concentration **80**
 synthesis 47–48, *48*
eunicellanes 431
euphanes 438
euphanes lanostanes *437*
euphorbesulins
 antimalarial activity 109, **111**
 structure 111
Euphorbia ebracteolata 125, *127*
evaporation 11
excavatolide-B 114
excisanin-A 388
 anticancer activity 217
 structure *217*, *390*
expression 17, 595
extraction *9*, *19*, 592
 advantages and disadvantages **16**
 automated Soxhlet extraction 12
 with cold fat 595–596
 controlled instantaneous decomposition
 598
 decoction 11
 digestion 10–11
 dynamic headspace technique 596
 efficiency 9

enfleurage and defleurage 596
by expression 596
headspace trapping technique 596
infusion 11
maceration 10, 596
microwave-assisted 12–13, *14,* 597
modern (non-traditional) methods 596
molecular spinning band distillation 598
percolation 11
pressurized hot water 12
principles **15**
simultaneous distillation extraction 597
solid phase micro-extraction 596–597
solvents for 15–16, **16**
Soxhlet extraction 12, *13*
stages 9
static headspace technique 596
supercritical fluid 12, 597
thermo-micro distillation 598
ultrasonic-assisted 13–15, *14*
vacuum headspace technique 596
volatile oil 590–591

farnesanes 421, 488, *488*
farnesene 332, 507, *507, 570*
farnesol 344, *344,* 507, *508,* 570, *571,*
    580–581, *581*
farnesyl diphosphate (FDP) 157, *158,* 182, 184
farnesyl phosphate (FPP) synthase 523
fast atom bombardment spectroscopy
    (FAB-MS) 24
Favorskii rearrangement 170, *171*
FDP *see* farnesyl diphosphate (FDP)
fenchanes 420, *421*
fenchone 575
fenchone synthesis 52, *52*
fernanes 439, *439*
ferruginol analogs, synthesis of *110*
feruloyl-CoA 48
ficutirucins
    anticancer activity 299, **300**
    structures *300*
filicanes 439, *439*
filifolinone
    anti-inflammatory activity **368**
fischerianoids 105
flexibilane 129
flexibilide
    antimicrobial activity 122
    structure *122*
fluconazole 56
food and food preservation, essential oils in
    599, **599, 600**
forskolin 96
    hypoglycemic effect 117
    structure *117*
fractional distillation 17
fractional inhibitory concentration (FIC) 56
fragrance *see* essential oil
free aglycones 286
friedelanes 438
friedelin 294, *296*
fulgic acid
    antioxidant activity 317, **317**
    structure *317*
furanocassanes 139
furanodiene 339
    structure *381*

structures *339*
furanodienone 386, *387*
furanoeremophilanes 421, *423*
furanoeudesmanes 421, *423*
furan ring transfer (FRT) reaction 550
furnolactone 139
furodysin 163, *164*
furodysinin 163, *164*
Furospinosulin-1 353, *354*
fuscosides 234, *236*
fusicoccanes 432, *433*
fusidanes 436, *437*

galbanic acid 207, *210*
gammaceranes *439,* 440
ganoderic acid D
    anticancer activity 301
    structure *303*
ganoderic acids 390, *391*
garsubellin A synthesis 167, *167*
gas-liquid chromatography (GLC) 19
gas-solid chromatography (GSC) 19
GDP *see* geranyl diphosphate (GDP)
genipin 74, **75**
geniposide 74, **75**
genome engineering 192, **194**
gentianine 271, *271*
gentiopicroside **74,** 271
geraniol 562–564, 575
    anticancer activity 205–206, **207**
    antidiabetic effect **75**
    antimicrobial activity **56**
    anti-nociceptive activity 76
    biogenesis 188, *188*
    cytotoxic and antitumor activity **76**
    geranyl acetate, transformation of
        258, *258*
    hydrolysis of glycosides 247, *247*
    ocimene synthesis from 35, *36*
    Oppenauer oxidation 159, *160*
    synthesis 33, *34*
    valerian alkaloid from 270–271, *271*
geranyl-*β*-D-glucopyranoside 247, *248*
geranyl diphosphate (GDP) 35, 182, 184
geranyl-geranyl diphosphate synthase
    (GGDPS) 88
geranyl geranyl-pyrophosphate (GGPP) 87,
    88, 184
geranyl pyrophosphate (GPP) 31, 49
germacranolide 233, 340, *342*
germacrene-A 338, *339*
gestational diabetes 73
GGPP *see* geranyl geranyl-pyrophosphate
    (GGPP)
gibberellanes 428–429, *430,* 481, *481*
ginkgolides 88, 115, 435
ginsenosides 238–239, *239*
    anti-inflammatory activity 312
    chemical structure *313*
gleditsia saponin-C *373*
glibenclamide 119
glutinanes 438
glyceraldehydes-3-phosphate 182
glycosides
    applications 249
    defined 229
    diterpenoid 234–236
    enzymatic glycosylation 230

functions **250**
hydrolysis 246–248
monoterpenoid 231–232
sesquiterpenoid 232–234
sesterpene 236–237
synthesis 229–231, *230*
tetraterpenoid 244
triterpenoid 237–244
glycosyltransferases 230
glycyrrhizin 320
    anticancer activity 307
    anti-inflammatory activity 313
    biosynthesis 296, *297*
    molecular targets *308, 314*
    pharmacological properties *321*
    structure *371*
glycyrrhizin analogs *372*
gnidimacrin
    antitumor activity 107, **108**
    structure *108*
GPP *see* geranyl pyrophosphate (GPP)
grayanane 143, **145**
grayanotoxanes 429, *430*
grayanotoxin-I 96, *98*
Grignard's reaction 160, *161*
GSC *see* gas-solid chromatography (GSC)
guaianes 422, *423*
guaianolides 342
    semi-synthesis 532–535, *533–535*
    synthesis 530, *530–532, 532*
guaiol 272
guaipyridine 272
    biogenesis *274*
    sesquiterpene alkaloids 274, *275*
    Van der Gen syntheses *274*
guanacastepenes 95, *96*
    antimicrobial activity 123, **124**
    structure *124*
guanine nucleotide pathway 117
gymnocladus saponin-G *374*

halimane 93, 137, *139*
halomin 380
haloselenuranes 265, *267*
harringtonolide 95, *95*
headspace trapping technique 596
head to tail fashion/1,4-linkage, in
    terpenoids 3, *3*
helenalin 541
heliosterpenoids 107
helminth fluorescent bioassay (HFB) 132
hemiterpenes 4, **4,** 400
hepatoprotective activity
    ceanothic acid 318–319
    terpenes 382–383, **384,** *384*
    triterpenoids 318–319
    triterpenoid saponins 318
    zizybrenalic acid 318–319
herbicidal activity 72–73
herical 108, *108*
    antibacterial spectrum **126**
    antifungal and antibacterial activity
        123–124
    structure *125*
herpes simplex virus (HSV) 59
heteronemin 352, *352*
high-performance liquid chromatography
    (HPLC) 19, 26, 181, 232, 249

high-performance thin-layer chromatography
(HPTLC) 19, 181, 287
himachalanes 422, *423,* 499–500, *500*
hinokiol 103, *103*
hippolide-A 350, *350*
hirsutanes 425–426, *426,* 505–506, *506*
HMG-CoA reductase 203
holothurins 241–242, *242*
hopanes 439, *439*
hopanoids 190, 291, *293*
horminone
    antimicrobial activity 122
    structure *123*
hot maceration 596
HPLC *see* high-performance liquid
        chromatography (HPLC)
HPTLC *see* high-performance thin-layer
        chromatography (HPTLC)
humulene 335–336, *336*
hydrazide motifs 62, **62**
hydrodiffusion 593
hydrodistillation (HD) 17, 592–595, *593*
    mechanism 593–595
    types 594–595
hydrolysis 246–248, 593–594
hydroxycopalic acid 129
7-hydroxyerogorgiaene 125, *127*
18-hydroxyferruginol 103, *103*
3-hydroxy-3-methylglutaryl-CoA
        (HMG-CoA) 49
hyperalgesia 59
hyperlipidemia 117–120
hypnotics 69
hyposantonin 157
hypotensive action **99**
hyppospongide-B 352, *352*
hyptadienic acid *384*

IDP *see* isopentenyl diphosphate (IDP)
ilexdunnoside 242, *242*
illudalanes 424, *424,* 501, *502*
illudanes 424, *424,* 501, *502*
illudinine 172, *173*
imbibition 11
iminoalcohols 256, *256*
immunity system 382
immunomodulatory activity 133, 382, **383**
impatienside A
    anticancer activity 302–303
    structure *304*
inflammation 312, 367
    acute 112
    chronic 112
    diterpenoids in treatment of 112–115
influenza type-A virus (IAV) 130
infrared spectroscopy (IR) 22, *23,* **24**
infusion 11
ingenane 100, *101*
inonotsuoxides
    anticancer activity 304
    structure *305*
inotodiol
    anticancer activity 304
    structure *305*
insulin-dependent diabetes mellitus
        (IDDM) 73
in vitro cytotoxicity
    diterpenoids **102**

terpenoid glycosides **245**
IPP *see* isopentenyl pyrophosphate (IPP)
irciformonin-I 354, *354*
iridoids 69, *69,* 164
    biogenesis 188, *188*
    synthesis 175, *176*
isoborneol 54
isocedranes 425, *425,* 503, *503*
isocomanes 506, *507*
isocomene 510, *511, 512*
isocopalanes 427, *428*
isodaucanes 424, *424,* 501, *501*
ISODIAL-linalool network, synthesis of *155*
isointermedeol *339*
isolactaranes 424, *424,* 501–502, *502*
isolation
    essential oils from plant parts 17
    separation from essential oils 17–19
isomadecassoside 239, *240*
isopentenyl diphosphate (IDP) 35, 88, 182
isopentenyl pyrophosphate (IPP) 49, 165, 287
isopimaranes 427
isoprene 2, *2,* 154, 400
    dimerization 3, *3*
    polymerization 3, *3*
isoprene rule 2–3
    special 3
    violations of 3
isoprene units 154
isoprenoid precursors 521
isoprenoids 182
isopulegol 256
    structure *76*
    synthesis 36–37
isopulegone synthesis 44, *45*
isosteviol 93, *93*
ivermectin 109, **112**

jatrophane 431, *431*
jolkinolide B 213

kalihinol
    antimalarial activity 109
    structure *112*
kappa opioid receptor (KOR) 139
kaurane diterpenoid *370*
kauranes 428, *429*
kayadiol
    anti-proliferative activity 103, **103**
    structure *103*
kempanes 433
ketal ester 479, *480*
ketonic terpenoid 3, *3*
kinetic resolution method 230, 460, *461,* 462,
        462
Knoevenagel reaction 170, *172*
Koenigs-Knorr reaction 229, *231*
kohamaic acid-A (KA-A) 350, *350,* **350**
kuraridine 118

labdanes 92, *92,* 427, *427*
    anti-inflammatory effects 113
    anti-leishmanial activity 137
    antispasmodic effects 128
    diterpene glycoside 234, *236*
    in-vivo antimalarial activity 109
    structural modification 125, *126*
    structure *114*

lactaranes 424, *424,* 501–502, *502*
    biosynthesis 547–548, *548, 549*
    cyclizations 553–554, *554, 555*
    isolation 548–549
    pharmacological applications 549–550
    ring enlargements 550–551, *550–553,* 553
    synthetic approaches 550
lactone ring biosynthesis 525–526, *527*
lactuside 341
lancilactone-C *378*
lanostanes 373, 438
lanosterol 288, *290,* 323
lathyranes 100, *102,* 431, *431*
laurene 514, *515*
lauric acid 157, *157*
lavandulol 4, 173, *173,* 563
leishmaniasis 129–143
leubethanol 125–126, *127,* **128**
leucosesterterpenone 354, *354*
leucosterlactone 354, *354*
leukotrienes (LTs) 133
levopimaradiene 88
levopimaradiene synthase (LPS) 88
liangshantine 276, *277*
licareol 33
limonene 566, *566, 567,* 576
    anticancer activity 202, *203*
    antidiabetic effect **74, 75**
    antimicrobial activity **53**
    anti-osteoarthritic effect 61
    biogenesis 185–188, *188*
    cardiovascular effects **72**
    cysteamine hydrochloride addition 155, *155*
    cytotoxicity 59, **59**
    for HSV-1 59, **60**
    isomerization 260
    polyamide synthesis 260, *260*
    polymerization 263, *263*
    synthesis 44
    terephthalic acid synthesis 260, *260*
limonene epoxide 159, *160*
limonene synthase (LimS) 184, 191
limonoids 263
    anticancer activity 299
    molecular structure *299*
linalool 563, *563,* 576, *576*
    anticancer activity 199, 202
    antimicrobial activity **56**
    biosynthesis 190
    cardiovascular effects **72**
    geranyl diselenide conversion 265, *267*
    hydrolysis of glycosides 247, *247, 248*
    as hypno-sedatives 69
    hypotensive activity 71
    structure *77*
    synthesis 33, *35, 36*
linalool-induced ERK phosphorylation 202
linalyl acetate **56,** 576
lindenane 233
linear triterpenes 435, *436*
lintenolides 354, *354*
liphagal 164
lipid oxidation 599
Lipinski's rule 132
liquid-liquid chromatography (LLC) 19
liquid-solid chromatography (LSC) 19
liver disorder 318
lobanes 434, *434*

longifolanes 422, *423,* 499–500, *500*
longifolene 581, *581*
longipinanes 422, *423,* 499–500, *500*
Lossen rearrangement 172
Luche reduction 170, *170*
luffariellolide 349, *349*
lupanes 438
lupeol 386, 450, 476
   anticancer activity 217, 298–299
   anti-diabetic activity 308
   anti-inflammatory activity 314, **368**
   biosynthesis 294, *295*
   molecular targets *309, 315*
   structure *217, 384, 388*
   synthesis 476, *477–480, 478–479*
lutein
   anticancer activity 221
   structure *222*
lycopene 244
   anticancer activity 221
   biogenesis 188–190, *189*
   structure *222*
lymphatic filariasis 109

maceration 12
   double 10
   process *10*
   simple 10
   steps involved in 10, *11,* 11–12
   triple 10
macrocyclic diterpenoid **8**
macrophages 113
macrorhynine 276, *276*
madecassic acid
   antioxidant activity 318
   structure *318*
malaria 363
Mannich reaction 158–159, *159*
manoalide 236, *237,* 349, *349*
maprouneacin 119, **121**
marasmanes 424, *424,* 501–502, *502*
marasmane sesquiterpenes
   biosynthesis 547–548, *548, 549*
   cyclizations 553–554, *554, 555*
   isolation 548–549
   pharmacological applications 549–550
   ring enlargements 550–551, *550–553,* 553
   synthetic approaches 550
marine terpenoids 453, *454,* **457**
marrubenol 96, *98*
marrulactone 129, *129*
marrulibacetal 128, *129*
maslinic acid *376*
mass spectrometry (MS) 24–25, *25*
matricarin 343, *343*
matrix metalloproteinase (MMP) 113
maximal electroshock-induced seizures
   (MES) 66
maytenfolone-A *375*
m-cresol, menthol synthesis from 35, *38*
Meerwein-Ponndorf-Verley reduction 51,
   169, *169*
megastigmanes *440,* 440–441
melazolide B synthesis 263, *264*
menstruum 9, 15
menthol 22, 447, *447,* 566, 576–577, *577*
   analgesic effect 60
   anticancer activity 203–204, *204*

antidiabetic effect **75**
antifungal activity 57
antimicrobial activity **53**
cardiovascular effects **72**
from citronellal 258, *259*
microbial transformation *457*
synthesis 35–36, *36–37,* 566, 566–567
terpene-based polymers production from
   262, *262*
L-menthol
   chemical synthesis and biosynthesis **38**
   synthesis from β-pinene 35, *37*
menthone 567, *567*
   Beckmann rearrangement reaction
     164, *166*
   Meerwein-Ponndorf-Verley reduction 169,
     169
   synthesis *566,* 566–567
menthyl glycosides 231, 232
merochlorin-A 348, *348*
metasequoic acid-A 112
metformin 117
methanesulfonic acid (MSA) 265
methyl copalate 129
2-C-methyl-D-erythritol 4-phosphate
     cytidyltransferase (MECT) 88
methylerythritol 4-phosphate (MEP) pathway
   88, *89,* 182
methylheptenone
   citral synthesis from 31, *34*
   linalool synthesis from 33, *35*
   myrcene synthesis from 31, *33*
methyl lambertianate 93
methyl-p-hydroxybenzoate (MPB) 602
methyl-substituted camphor ethylamine
   256, *256*
mevalonate-independent pathways 182
mevalonate (MVA) pathway 165, 182, 286,
   287, 521–522, *522*
mevalonic acid (MVA) pathway 88
Michael-Aldol annulation reaction 175, *175*
microbial transformation
   (−)-α-pinene *457*
   (4R)-carvone *459, 460*
   (4R)-(-)-carvone *459*
   (4S)-carvone *460*
   (R)-Limonene *458*
   menthol *457*
   (-)-menthone and (4S)-isopiperitenone *459*
microwave-assisted extraction (MAE) 12–13,
   14, 597
minimum inhibitory concentration (MIC) 125
mintlactone 161, *161*
mogrosides 288, *289, 291*
molecular rearrangements 155–157
molecular spinning band distillation 598
momordicoside *385*
monocrotaline (MCT) 71
monocyclic farnesane sesquiterpenes 421, 489
   bisabolanes 489, *489*
   cyclofarnesanes 489, *489*
   germacranes and elemanes 490, *490*
   humulanes 491, *491*
monocyclic monoterpenes *380,* 417, *418*
monocyclic monoterpenoids **6**
   structures *55*
   synthesis 35–49
monocyclic sesquiterpene **6**

monocyclic sesquiterpenoid 334–336
monocyclic sesterterpenes *434,* 435
monocyclic sesterterpenoid **6**
monocyclic terpenoids 4, **4,** 6, **6,** 6–7, *23,* 400
monoterpeglycosides *231*
monoterpene alcohol 4, **4,** *4*
monoterpenes 30, 380, 400, 418
   as anticancer agents 199–206
   biogenesis via recombinant technology
     192–194, *194*
   cytotoxicity 199, **200–202**
   mechanism action 199, *200*
monoterpennoids *400*
monoterpenoids 442, 447, *447*
   alkaloids 269–272
   biogenesis by engineered microbes
     190–192
   biosynthetic pathways in plant 184, *184*
   enzymes in production of 182, **182**
   factors affecting production of 31
   glycosides 231–232
   pharmacological activity 52–80, *54*
   polymerization of terpenes 261–263
   precursor pathways **33**
   production in microorganisms 191, **191**
   production pathways 32
   as starting compounds 255–261
   synthesis 31–52
   types 30, **30**
mooloolabenes (A-E) 352, *352*
mulinane 93–95, *94*
mulinolic acid
   antimicrobial activity 122
   structure *123*
multifloranes 438
   antioxidant activity 317–318
   structure *318*
myrcene 560, *561,* 577
   anticancer activity 204–205, *206*
   antimicrobial activity **56**
   polymerization 261, *261*
   synthesis 31, *33*
myrtenal 256
   antidiabetic effect **75**
   antifungal activity 57
myrtenalyl adamantane 257, *258*
myrtenic acid 57, 65
myrtenyl chloride 57
myxol glycosides 244, *244*

nano-formulation 600, **601**
natural cosmetic preservatives, essential oils as
   600–602
natural monoterpenes 361
Nε-diacylhydrazine 65
nemrutolone *388*
neodictyolactone 104, *104*
neodolastane *95,* 95–96, *97*
neohopanes 439, *439*
neopallavicinin 93, *94*
(+)- neosymbioimine *582–586, 583–584*
nepalactones 234, *235*
nepetalactone 259, *259*
nerolidol 344, *345, 346,* 509, *509*
   antimalarial drugs 363
   structure *363*
neurodegenerative disorders (NDDs) 69–70
neuroprotective activity 132–133

618   *Index*

nigranoic acid *378*
nimbolide
   anticancer activity 306
   molecular mechanism *306*
   synthesis 175, *176*
nitrosochloride (NOCl) 43
nomilin
   anticancer activity 219
   structure *220*
non-insulin-dependent diabetes mellitus
     (NIDDM) 73
nonpolar solvents 15
norcamphor 265
nordictyotalide 104, *104*
nor-diterpenes *see* terpenes
Nozaki reaction *260*
nuclear magnetic resonance spectroscopy
     (NMR) 22–24

OA *see* oleanolic acid (OA)
ocimene
   antimicrobial activity **56**
   synthesis 33, *36*
oleananes 438
   biosynthesis 291–293, *293*
   structures *293*
oleanolic acid (OA)
   anticancer activity 219
   anti-diabetic activity 308
   anti-inflammatory activity 315
   biosynthesis 296, *297*
   infrared spectrum 24
   mechanism of action *220*
   mechanism of actions *309*
   structure *377*
onitin synthesis 164, *167*
ophiobolins (Ophs) 346–347
   antibacterial activities **348**
   cytotoxic activities **348**
   nematocidal activity **348**
   sources **347**
Oppenauer oxidation 159–160, *160*
organoselenium compounds 265
oridonin
   for Alzheimer's disease 115
   anti-asthmatic activity 133
   anticancer activity 216
   molecular structure *216*
   structure *116*
oscillol diglycoside 244, *244*
otostegindiol
   antimalarial activity 109
   structure *111*
7-O-triethylsilylbaccatin-III 475–476, *477*
ovalbumin (OVA) 133
oxidation *20*
oxidative degradation 20, *21*, 21–22
oxidized derivative of hydroxy-cis terpenone
     (OHCT) 364
oxygen 19

PAB *see* pseudolaric acid B (PAB)
pachymic acid
   anticancer activity 220
   structure *221*
pachysananes 438
packing 11
paclitaxel; *see also* taxol

anticancer activity 100
   structure *100*
paeoniflorin 388–389
   antidiabetic effect **74**
   structure *391*
pain
   defined 128
   diterpenoids for 128–129
palauolol 350, *350*
palbinone *386*
pallavicinin 93, *94*
paludolactone 340, *342*
paralianones 114, *116*
parthenolide (PTL) 341, 541
   anticancer activity 212, *212*
   semi-synthesis 530, *530*
   structure *342*
   synthetic strategy 528–530, *529*
partition chromatography 19
patagonicoside A 241, *241*
patchouli alcohol 581–582, *582*
patchoulipyridine 275, *275*
patchoulol 345, *346*
Pauson-Khand reaction (PKR) 160–161
pentacyclic sesterterpenes *354, 435, 436*
pentacyclic terpenes 175, *175*
pentacyclic triterpenes *375, 437,* 438,
     439, *439*
pentacyclic triterpenoid 290–297, *292*
pentylenetetrazol (PTZ) 66
percolation 12
   modified 11
   reserved 11
   simple 11
perfumes, essential oils in 604, **605–607**
perillaldehyde (PAE) 392, *392*
perillic acid 448
perillyl alcohol (POH)
   anticancer activity 206, *207*
   antitumor activity 65, *65*
   cardiovascular effects **72**
   synthesis 44, *45*
periplanone B 511, *513,* 513–514, *514*
Perkin reaction 171–172, *172*
petrosaspongiolide-M 350–351, *351,* **351**
petrosaspongiolides 350, *350,* **350**
peucelinendiol 95, *95*
P-glycoprotein (P-gp) 107, 132
pharmacological activity 52–80, *54*
phellandrene synthesis 44, *46*
phomactin A 257–258, *258*
phosphphomevalonate decarboxylase 49
phytanes 88, 426–427, *427*
picrotoxanes 422, 500, *501*
picrotoxane sesquiterpene lactone
     glycosides 234
Pictet-Spengler reaction 93
piepunin 277, *277*
pimaranes 88, 188, *189,* 427
Pinacol-Pinacolone rearrangement 164–165,
     166–167
pinanes 420, *421*
pinene 50, *51,* 191, 567, 577–578
pinguisanes 506, *506*
piperitone
   synthesis 41–43, *43*
     Bergamanns synthesis 43, *43*
     via birch reduction 42, *43*

piscicidal 135, *137*
platanic acid *377*
platycodin D
   anti-inflammatory activity 315
   structure *315*
plumisclerin-A synthesis 95, *95*
P-menthane 7
POH *see* perillyl alcohol (POH)
polar solvents 15
polycyclic farnesane sesquiterpenes 421
   cadinanes 495, *496*
   caryophyllanes *491,* 491–492
   drimanes 497, *497*
   eremophilanes, furanoeremophilanes,
     valeranes 493–494, *493–495*
   eudesmanes and furanoeudesmanes *492,*
     492–493, *493*
   guaianes and cycloguaianes 497–499,
     497–499
polycyclic sesterterpenes 435
polymerization
   isoprene 3, *3*
   myrcene 261, *261*
   terpenes 261–263
polysciasoside A 239, *239*
polyterpenes **4**
poly-terpenes 441, *441*
pomolic acid *377*
prenylaromadendranes *434,* 435
prenyleudesmanes *434,* 434–435
prenylgermacranes *434, 434*
prenylguaianes *434,* 435
prenylquinones 441, *441, 442*
prenylsesquiterpenes 433, *433*
presilphiperfolianes 506, *507*
pressurized hot water extraction (PHWE) 12
pressurized liquid extraction (PLE) 12
prezizaanes 425, *426,* 504, *504*
pristimerin
   anticancer activity 217–218, **218**
   mode of action of 218, *218*
procurcumenol 209
propindilactone G synthesis 176, *177*
prostaglandins 380
prostratin
   antiviral activity 129–130
   structure *131, 371*
protein engineering 192, *193,* **193**
protein kinase (PKA) pathway 117
protoilludanes 424, *424,* 501, *502*
protostanes 436, *437*
protostane triterpene
   anticancer activity 299
   structure *301*
pseudoionone 163, *164*
pseudolaric acid B (PAB)
   anticancer activity 216
   molecular structure *217*
psiguadial B 169, *169*
ptaquiloside 233, *234*
pulegone 258, 578
   Favorskii rearrangement 170, *171*
   menthol synthesis from 36, *39*
   synthesis 44, *45*
   thymol synthesis from 40, *41*
   transformation 170, *171*
pyrolysis 35
pyruvic acid 182

quassinoids 323, *323*

rebaudioside-A (Reb-A)
    antihyperglycemic effect 117
    structure *118*
rebaudiosides 99
recombinant technology 192–194
Reformatsky reaction 31, 160, *161*
regioisomeric lactones 162
retinol acetate *see* vitamin A
rhamnofolanes 431
ring-closing metathesis (RCM) reaction 274
rosmanol **368**
rotundifolone 71, *71*, **72**
rotundine 275, *276*
Ruzicka, L. 2

sabinene synthesis 51–52
sacculatanes (prenyldrimanes) *434*, 435
S-adenosyl methionine (SAM) 48
Saegusa dehydrogenation reaction 93
Saegusa oxidation 161
safranal *384*
saikosaponins
    anti-inflammatory activity 315–316, **316**
    structure *316*
salaspermic acid *375*
Salkowski's test 2
salvinorin-A 129, *129*
Sanguisorba officinalis L., monoterpene
        glycosides from the roots of 232,
        232
santalanes 504, *505*
santalene 339
santamarine (STM) 340, *341*
santonin 535, *536*, 540
sarcodonin-G (SG) 164
    anti-proliferative activity 103
    synthesis 88–89, 164, *166*
scalaranes 351, *351*
scalarane sesterterpenoids 351, *352*
sclarene 92
sclareol 100, 103, 264
    cytotoxic activity 103
    structure *103*
sclareolide
    liphagal synthesis from 164, *166*
    structure *378*
secoemestrin D **348**, 349, *349*
secologanin 176, *176*
sedatives 69
selenolate 265
serrulatane 172, *173*
sesquiterpene 332
sesquiterpene lactones (STLs) 212, **212–213**,
        449, 518, *518*
    artemisinin 340
    biological activities 537, *538*
    biosynthesis 521–523
    characteristics 519, 521, *521*
    chemistry and medicinal uses of
        339–343, **340**
    classification 520, **520**
    diversity in structures of *520*
    enzymes 525–530
    eudesmanolide 340
    germacranolides 340–342
    guaianolides 342–343

hypothetical pathway *524*, 524–525
IIR signalling 540, *541*
JAK-STAT pathway 538, *540*
MAPK pathway 537
mechanism 540–543
NFκB pathway 537–538
pathway *523*, 523–524
phosphatydilinositol-3-kinase- Akt
    pathway 538
PPARγ and C/EBPα pathway 538
role 537, *539*
structure *519*, 520–521
structure–activity relationship 537, *537*
tenulin 343
sesquiterpenes 331, 420
    as anticancer agents 206–212, *209*
    biosynthesis 184, *187*
    synthesis 155
sesquiterpenoids 4, **4**, 331, *400*, 449
    alkaloids 272–275
    chemistry and medicinal uses of
        332–339, **333**
    classes **488**
    classification 332, **332**
    glycosides 232–234, *235*, 236–237
    as starting compounds 263–264
    structure *332*
sesquiterpenol 343–345, **344**
sesterpenes 332
    chemistry and medicinal uses of
        345–355, **347**
    medicinal uses of *355*
sesterstatins-4 and 5 351, *351*
sesterterpenes 435
sesterterpenoids 4, **4**
seven-keto-sempervirol
    anthelmintic activity 132
    structire 134
SFE *see* supercritical fluid extraction (SFE)
shionone 288, *290*
silphinanes 506, *507*
silphiperfolianes 506, *507*
Simmons-Smith cyclopropanation 89
simultaneous distillation extraction (SDE) 597
sinapyl-alcohol dehydrogenase (SAD) 48
size reduction 11
skytanthine 269, **269**, *270*
(S)-mecamylamine
    antidepressant activity 58
    chemical structure *59*
smooth muscle spasms 128–129
sobrerol 80, *80*
    cardiovascular effects **72**
    synthesis 48, *49*
solid phase micro-extraction (SPME)
    596–597, *597*
solvents
    for extraction 15–16, **16**
        factors consideration for selection 16
        properties 16
    nonpolar 15
    polar 15
    volatile 17
songorine 277–278, *278*
sonication *see* ultrasonic-assisted extraction
    (UAE)
Soxhlet extraction 12, *13*
sphenolobanes (prenyldaucanes) 435

spirocaesalmin
    antiviral activity 130
    chemical formulae *132*
SPME *see* solid phase micro-extraction
    (SPME)
spugna method 595
squalene 190, 265, *266*, 278, *279*
static headspace technique 596
steam distillation 595
steam distillation (SD) 17, *17*
stemarene 92, *92*
stemodene 92, *92*
stereodivergent asymmetric synthesis 45
stereoisomers 558
sterpuranes 424, *424*, 501–502, *502*
steviol 93, *93*
stevioside 96, *386*
STLs *see* sesquiterpene lactones (STLs)
streptozotocin (STZ) 117
striatal 96, *97*
striatin 96, *97*
strictosidine 266
strictosidine synthase 265
structure elucidation 19–26
suberosol *376*
substituted benzoate 476, *478*, 478–479, *479*
sulfonylhydrazone 474, *474*
supercritical fluid extraction (SFE) 12,
    14, 597
sweroside **74**
swertiamarin **74**
synthetic additives 599

talatisamine 157, *158*
tanshinones 135
    anticancer activity 216
    molecular structure *217*
taraxastanes 438
taraxeranes 438
taraxerol 294, *294*
taxadiene 92, *92*
taxanes 92–93, 432–433, *433*
Taxodium distichum 112
    compounds isolated from **113**
    in-vitro activity of **112**
taxol 380, *381*
taxoquinone
    antiviral activity 132
    structure *134*
tenulin 343, *343*
terephthalic acid 260, *260*
terpene indole alkaloids 190, *191*, 265, *265*
terpenes 2, 26, 361, *400*
    as anthelmintic drugs 365–367
    as anticancer drugs 379–382, *380–382*
    as anti-diabetic drugs 383–385, *384–387*,
        387
    as anti-HIV drugs **369**, 369–376, *370–378*,
        378
    as anti-inflammatory drugs 367–369,
        368, *369*
    as antimalarial drugs 363–365
    as antimicrobial drugs *378*, 378–379, **379**
    as antioxidants 382
    biosynthesis 182, *183*
        via cytosolic and plastidial pathways
        182, *184*
    chemical modifications 154

terpenes (*Cont.*)
 classification *417*
 with diverse therapeutic effects **362**
 as hepatoprotective 382–383, **384**, *384*
 immunomodulatory activity 382, **383**
 nomenclature **9**
 as useful drugs *363*
terpene synthases (TPS) 93
terpenoid glycosides 246, *246*, 286
terpenoids
 chemical properties 2
 as chemotherapeutics 441–442
 classification 442, **443–446**
 microbial synthesis 455
 occurrence and medicinal application **456**
 pharmacological properties 2
 physical properties 2
 phytochemical properties 2
 plant sources **401–416**
terpineol **72**
tetracyclic diterpenoids **8**
tetracyclic sesterterpenes 435, *436*
tetracyclic sesterterpenoids **8**, *354*
tetracyclic terpenoid 4, **4**, 7
tetracyclic triterpenes 436, *437*
tetracyclic triterpenoid 288–289
tetra-cyclophytanes 428, *429*
tetraterpenes **4**, 221–223, 440
tetraterpenoids 244
thapsigargicin 343
thapsigargin 342–343, *343*, 542
thermal decomposition 2, *3*
thermo-micro distillation 598
thin-layer chromatography (TLC) 286
thiol-ene photopolymerization 155, *156*
thiol-ene reaction *see* click reactions
thujanes 420
thujone **74**
thujopsanes 425, *426*, 505, *505*
thymol **368**, 579–580, *580*
 anticancer activity 204, *205*
 antidepressant activity 58
 antidiabetic effect **75**
 antimicrobial activity **53**
 cardiovascular effects **72**
 synthesis 40, *41–42*
  from *m*-cresol 40, *41*
  from menthol 40, *41*
  from pipertone 40, *41*
thymoquinone (TQ)
 anti-inflammatory effect **77**
 antitumor activity 63, *64*
 cardioprotective effects **78**
 synthesis 46–47, *48*
Tiffeneau-Demjanov rearrangement *260*
tigliane 100, *101*
 antiviral activity 129
 structure activity relationships *131*
tiglianes 431
Tilden's reagent 17, *18*
tirucallane

anticancer activity 299, **300**
 structures *300*
tirucallanes *437*, 438
TLC *see* thin-layer chromatography (TLC)
p-toluic acid 564, *565*
tolypodiol 190
tormentic acid *376*
totaranes 427, *427*
TQ *see* thymoquinone (TQ)
trans-dehydrocrotonin 99
tricyclic diterpenoid **8**
tricyclic sesquiterpenoids **8**, 339
tricyclic sesterterpenes 435, *436*
tricyclic sesterterpenoids **8**
tricyclic terpenoid 4, **4**, 7, *23*
tricyclophytanes 427
4,4,14-trimethyl-9,19-cyclo-5α *see*
  cycloartane
trinervitanes 433
Tripterygium wilfordii 113
triptolide 449–450
 anticancer activity 216
 anti-inflammatory effects 113
 structure *113*, *217*
triterpenes **4**, 217–221, 435, *436*, 476
triterpenoids 286, 450
 alkaloids 278
 as anticancer agent 297–307
 as anti-diabetic agents 307–310, **310**
 as anti-inflammatory agents 312–316
 as antimicrobial agents *311*, **311**, 311–312
 as antioxidant 316–318
 as antiviral agents 319–320, *320*
 biological activities *287*
 as cardioprotective agent 319
 characterization *287*
 classification *287*
 clinical trials **298**
 glycosides *237*, 237–244, *243*
 as hepatoprotective 318–319
 hepatoprotective activity 318–319
 with medicinal uses 297, **298**, *298*
 synthesis 287–297
triterpenoid saponins 318
tryptophan decarboxylase 190
tyrosine ammonia lyase (TAL) 47

ultrasonic-assisted extraction (UAE)
  13–15, *14*
ultraviolet spectrophotometry (UV) 22
unsaturation 19
ursanes 438
ursane triterpenoids
 biosynthesis 291–293, *293*
 structures *293*
ursolic acid (UA) 450
 anticancer activity 219, 299–300
 anti-diabetic activity 310
 anti-inflammatory activity 315
 molecular mechanism of action *302*
 molecular targets *302*, *316*

SAR study *310*
 structure *220*, *376*

vacuum headspace technique 596
valeranes 421, *423*
valerian 270–271, *271*
vancomycin 123
veatchine 277, *277*
verbenol 256
 antimicrobial activity **54**
 synthesis 52, *53*
vernodaline 207, *210*
vernolide-A 207, *210*
verticillanes 432–433, *433*
viburodorol A
 anti-diabetic activity 310
 structure *310*
villanovanes 428, *429*
vinblastine 264
vincristine 264
violaceuside 243, *243*
vitamin A 468
vitamin A acetate synthesis 158, *159*
volatile oil
 classification 590, **591**
 extraction 590–591
volatile solvents. extraction with 17
vouacapane 139
 antinociceptive activity 128
 anti-proliferative activity 143
 larvicidal activity 142, *144*
 structure *144*

Wagner-Meerwein migration 95
Wagner-Meerwein rearrangement 157,
  158, 265
 farnesyl diphosphate, vetispiradiene
  synthesis from 157, *158*
 geranyl diphosphate, camphene synthesis
  from 157, *158*
Wallach, O. 2
water 15
water distillation 594, *595*
Wieland-Miescher ketone 94
Wittig reaction 158, *159*
Wolff-Kishner reduction 168, *168–169*
Wolff rearrangement 169, *169*
Wurtz reaction 167

xeniaphyllanes 433–434, *434*
xenicanes 433–434, *434*
X-ray analysis 25

yohimbine 264
yunnanenseine A 277, *277*

zingiberene 336, *337*
Z-isomer 31
zizaanes 425, *426*, 504, *504*
zizybrenalic acid 318–319
zone of inhibition (ZOI) 77